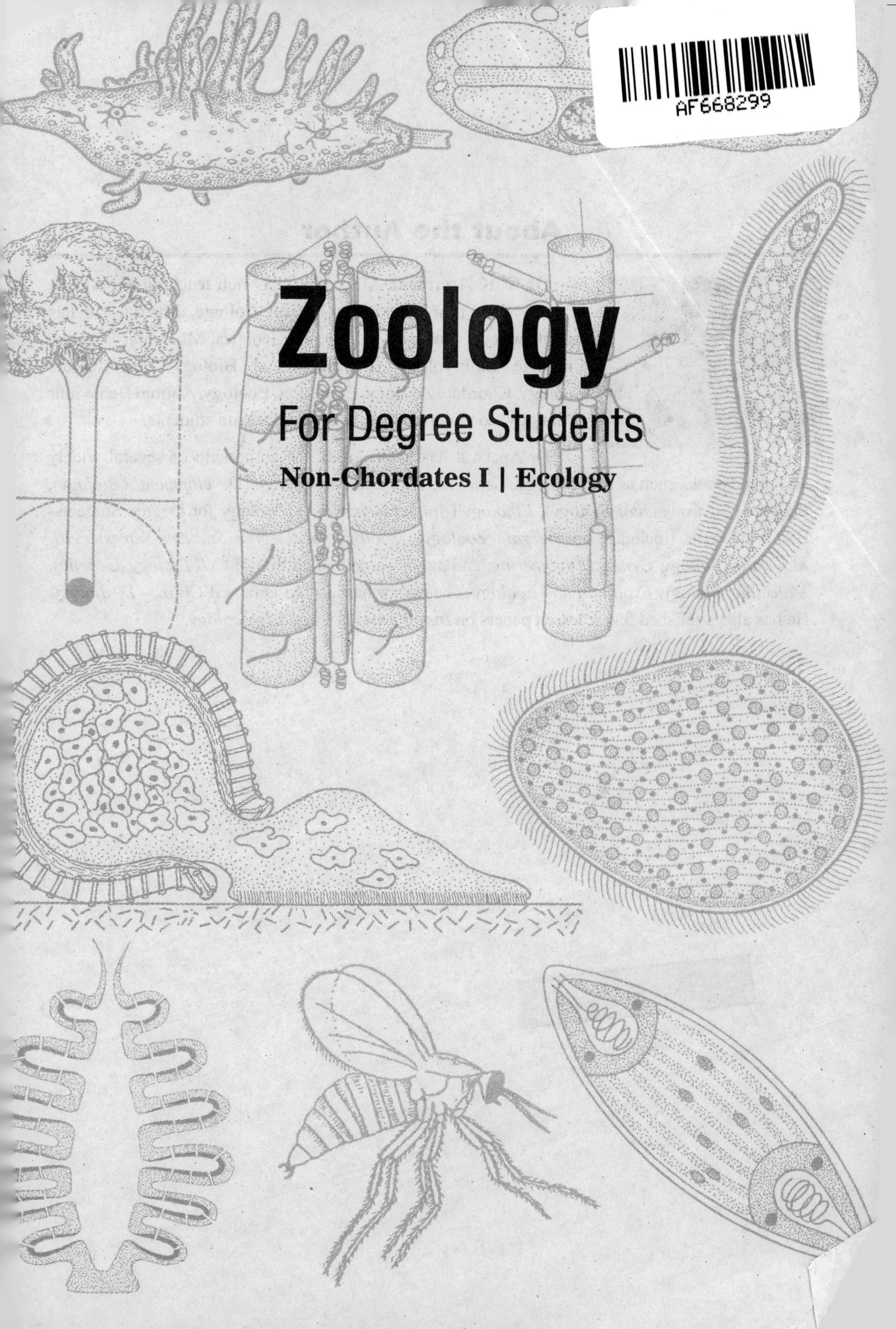

Zoology

For Degree Students

Non-Chordates I | Ecology

About the Author

Dr V K Agarwal, MSc, PhD, has rich teaching experience of more than four decades at Meerut College, Meerut. He has taught subjects such as Cell Biology, Genetics, Molecular Biology, Chordate Embryology (Developmental Biology), Invertebrate Zoology, Chordate Zoology, Evolution, Ecology, Animal Behaviour (Ethology) to undergraduate and postgraduate students.

Dr Agarwal has to his credit the authorship of several widely acclaimed books such as *Cell Biology*, *Genetics*, *Chordate Embryology* (*Developmental Biology*), *Ecology* (*Environmental Biology*), *Ethology* (*Animal Behaviour*), Zoology for Degree Students-BSc First Year (includes *Invertebrate Zoology*), *Zoology for Degree Students-Semester III*, *Molecular Biology, Genetic Engineering,* and a multi-coloured edition of *Cell Biology, Genetics, Molecular Biology, Evolution and Ecology.* The author has also co-authored *CBSE – IX Biology.* He has also published few research papers on *Insect Physiology* and *Toxicology*.

Zoology

For Degree Students

Non-Chordates I | Ecology

BSc (Hons) Semester I

As per UGC Choice Based Credit System (CBCS)

V K Agarwal
MSc, PhD,
Former Head of the Zoology Department
Meerut College, Meerut, Uttar Pradesh

S. CHAND
PUBLISHING

S Chand And Company Limited
(ISO 9001 Certified Company)

S Chand And Company Limited

(ISO 9001 Certified Company)

Head Office: D-92, Sector–2, Noida – 201301, U.P. (India), Ph. 91-120-4682700

Registered Office: A-27, 2nd Floor, Mohan Co-operative Industrial Estate, New Delhi – 110 044, Phone: 011-49731800

www.**schandpublishing.com**; e-mail: **info@schandpublishing.com**

Marketing Offices:

Chennai	:	Ph: 23632120; chennai@schandpublishing.com
Guwahati	:	Ph: 2738811, 2735640; guwahati@schandpublishing.com
Hyderabad	:	Ph: 40186018; hyderabad@schandpublishing.com
Jalandhar	:	Ph: 4645630; jalandhar@schandpublishing.com
Kolkata	:	Ph: 23357458, 23353914; kolkata@schandpublishing.com
Lucknow	:	Ph: 4003633; lucknow@schandpublishing.com
Mumbai	:	Ph: 25000297; mumbai@schandpublishing.com
Patna	:	Ph: 2260011; patna@schandpublishing.com

First Edition 2017
Reprints 2018 (Thrice), 2019 (Thrice), 2020 (Twice), 2021, 2022 (Twice)

Reprint 2023

ISBN: 978-93-525-3438-8 **Product Code:** H6ZOO68ZOOL10ENAA17R

PRINTED IN INDIA

By Vikas Publishing House Private Limited, Plot 20/4, Site-IV, Industrial Area Sahibabad, Ghaziabad – 201 010 and Published by S Chand And Company Limited, A-27, 2nd Floor, Mohan Co-operative Industrial Estate, New Delhi – 110 044.

Preface

It gives me great pleasure and satisfaction to present this textbook, **Zoology for Degree Students – Semester I**, covering subjects **Non-Chordates: Protista to Pseudocoelomata** and **Ecology**. This comprehensively written textbook provides up-to-date information on the subject in an informative and systematic manner.

This book has been designed as per the UGC Choice Based Credit System (CBCS) syllabus of Zoology, BSc (Honours) - Semester I. It includes the following two core courses:

1. **CC–I Non-Chordates: Protista to Pseudocoelomata**
2. **CC–II Ecology**

"**Non-Chordates or Invertebrate Zoology** is for those who love animated life in all its myriad forms. It includes most animals and even animals such as protists (Protozoa). It is an opportunity to revel in the sheer joy of an astounding diversity, which ranges from the infinitesimal to the gigantic, and from sluggish to the swift. Invertebrate Zoology not only delights the eye with variety, it also challenges the mind to seek unity in diversity" (**Ruppert**, **Fox** and **Barne**, 2003).

"**Ecology** is a discipline that addresses the highest and most complex levels of biological organization. It is the study of holism and emergence of properties of life taken from the top down. Even hardnosed laboratory scientists, who were focused on less complex levels of molecules and cells, knew in their hearts that in time, biologists must eventually arrive at this discipline. To understand ecology thoroughly would be to understand all of biology, and to be a complete biologist is to be an ecologist" (**E.O. Wilson** 2005).

Ecology is now seen as not just a biological but a human science. The future of our species depends on how well we understand that extension and employ it in the wise management of our natural resources. Ecologists are increasingly being asked to apply their knowledge towards efforts to solve current environmental problems. So students of ecology need to master a heady mix of abstract concepts, experimental reasoning, mathematical equations, and details about particular organisms and their habitats.

Some of the special features of this textbook are:

1. It includes Invertebrate Zoology up to pseudocoelomates and minor phyla
2. Ecology contains certain topics of human made ecosystems, environmental pollution and wildlife conservation
3. Ecology also includes certain modern/current topics such as landscape ecology, artificial ecosystems and ecological footprints
4. Certain current environmental problems such as eutrophication, algal blooms, greenhouse gases, global warming, peeling of ozone umbrella, toxicity due to xenobiotic substances such as DDT and plastic bags have also been discussed
5. Alarming issues such as soil erosion, floods, sewage treatment, sanitary landfills, management of soild waste to make a clean and green India have also been covered

Undergraduate students will find this book immensely useful as it lucidly explains the concepts with the help of well-illustrated diagrams.

To friends and family, I remain grateful and thank them for their support during development of this textbook. Special thanks to my wife Usha Gupta for unending inspiration during crafting of this textbook. I wish to thank Trilok Jain who has illustrated the figures of this book.

Many thanks to the editorial team at S. Chand, especially, Surabhi Khare, Shalini Negi, Rupesh Gupta, Priyanka Sharan, Sanjeev Arora, Shagun Mishra and Rajesh Singh for timely release of the book. Their zeal to excel is really heartwarming.

Suggestions and feedback from the readers will be gratefully acknowledged.

V K Agarwal
Villa–20
Green Wood City
Haridwar Bypass Road
Meerut - 250 005
Uttar Pradesh (India)

Syllabus

BSc (Hons) Zoology – Semester I

CORE COURSE I

NON-CHORDATES I: PROTISTS TO PSEUDOCOELOMATES

Unit 1: Protista, Parazoa and Metazoa **19**

General characteristics and Classification up to classes
Study of *Euglena, Amoeba* and *Paramecium*
Life cycle and pathogenicity of *Plasmodium vivax* and *Entamoeba histolytica*
Locomotion and Reproduction in Protista
Evolution of symmetry and segmentation of Metazoa

Unit 2: Porifera **7**

General characteristics and Classification up to classes
Canal system and spicules in sponges

Unit 3: Cnidaria **12**

General characteristics and Classification up to classes
Metagenesis in *Obelia*
Polymorphism in Cnidaria
Corals and coral reefs

Unit 4: Ctenophora **4**

General characteristics and Evolutionary significance

Unit 5: Platyhelminthes **10**

General characteristics and Classification up to classes
Life cycle and pathogenicity of *Fasciola hepatica* and *Taenia solium*

Unit 6: Nemathelminthes **8**

General characteristics and Classification up to classes
Life cycle, and pathogenicity of *Ascaris lumbricoides* and *Wuchereria bancrofti*
Parasitic adaptations in helminthes

CORE COURSE II

PRINCIPLES OF ECOLOGY

Unit 1: Introduction to Ecology **6**

History of ecology, Autecology and synecology, Levels of organization, Laws of limiting factors, Study of physical factors

Unit 2: Population **24**

Unitary and Modular populations

Unique and group attributes of population: Density, natality, mortality, life tables, fecundity tables, survivorship curves, age ratio, sex ratio, dispersal and dispersion

Exponential and logistic growth, equation and patterns, r and K strategies

Population regulation - density-dependent and independent factors

Population interactions, Gause's Principle with laboratory and field examples, Lotka-Volterra equation for competition and Predation, functional and numerical responses

Unit 3: Community **12**

Community characteristics: species richness, dominance, diversity, abundance, vertical stratification, Ecotone and edge effect; Ecological succession with one example

Theories pertaining to climax community

Unit 4: Ecosystem **14**

Types of ecosystems with one example in detail, Food chain: Detritus and grazing food chains, Linear and Y-shaped food chains, Food web, Energy flow through the ecosystem, Ecological pyramids and Ecological efficiencies

Nutrient and biogeochemical cycle with one example of Nitrogen cycle

Human modified ecosystem

Unit 5: Applied Ecology **4**

Ecology in Wildlife Conservation and Management

Contents

NON-CHORDATES I: PROTISTA TO PSEUDOCOELOMATES (CC–I)

PRINCIPLES OF ECOLOGY (CC–II)

NON-CHORDATES I: PROTISTA TO PSEUDOCOELOMATES

(CC–I)

1

Introduction

A living organism is primarily physico-chemical material that shows a high degree of complexity, is capable of self-regulation, possesses a metabolism and perpetuates itself through time. Many biologists believe that life is an arbitrary stage in the growing complexity of matter, with no sharp dividing line between the living and non-living worlds.

Living substance is composed of a highly structured array of macromolecules, such as proteins, lipids, nucleic acids and polysaccharides, as well as, smaller organic and inorganic molecules. A living organism has built-in regulatory mechanisms and interacts with the environment to sustain its structural and functional integrity. All reactions occurring within an individual living unit are called **metabolism**. Specific molecules (DNA or RNA) containing information in their structure are utilized both in the regulation of internal reactions and in the production of new living units.

Biology (Gr., *bios* = life + *logos* = study) is the study of life. It ranges from the study of chemicals in microscopic cells to the study of communities of organisms interacting with their physical environment. The biologist explains that how organisms and their parts look, work and behave. But this is not all. The biologist seeks to explain the mechanisms by which the present day world developed and attempts to predict where it is headed—in other words, the historical element, the evolution of the system, must be understood, for it sets the limits for life itself and shows us glimpses of our future.

Some of the important branches of biology are the following:

1. Zoology (Gr., *zoon* = animal + *logos* = study) which deals with the study of animals.

2. Botany (Gr., *botane* = herb) which deals with the study of plants.

3. Microbiology which deals with the study of microorganisms such as viruses, bacteria, cyanobacteria (blue green algae), archaebacteria, protozoa, diatoms, unicellular algae and yeasts and other microscopic fungi.

1.1 NATURAL AND UNNATURAL GROUPS

One of the first evolutionary trees of life conceived from Darwinian (genealogical) perspective was published by **Ernst Haeckel** in 1866. Haeckel's famous **tree of life** began a tradition of depicting phylogenetic hypotheses as branching diagrams, or trees, a tradition that has persisted since that time. Since Haeckel's day, many names have been coined for the larger branches that sprout from these trees. Some of these names refer to groups of organisms that are probably **natural phylogenetic groups** (*i.e.*, groups that include an ancestor and all of its descendants), such as **metazoa** (the Animal Kingdom). Other names refer to **unnatural**, or **composite, groupings of organisms**, such as "**microbes**" (*i.e.*, any organism that is microscopic in size, such as bacteria, most protists and unicellular fungi) and "**protozoa**" (a loose assemblage of primarily unicellular heterotrophic eukaryotes).

The discovery that organisms with a cell nucleus constitute a natural group divided the living world neatly into two categories, the **prokaryotes** (those organisms lacking membrane-enclosed organelles and a nucleus, and without linear chromosomes) and the **eukaryotes** (those organisms that do possess membrane-bound organelles and a nucleus and linear chromosomes). Investigations by **Carl Woese** and others, beginning in the 1970s, led to the discovery that the prokaryotes actually comprise two distinct groups, called **Eubacteria** and **Archaea** (= Archaebacteria), both quite distinct from eukaryotes (Table 1.1).

Table 1.1 The six kingdoms of life (Source. Brusca and Brusca, 2003).

A. The Prokaryotes (the "domains" Eubacteria and Archaea)*

1. **Kingdom Eubacteria (Bacteria).** The "true" bacteria, including cyanobacteria (or blue-green algae) and spirochetes. Never with membrane-enclosed organelles or nuclei, or a cytoskeleton; none are methanogens; some use chlorophyll-based photosynthesis with peptidoglycan in cell wall; with a single known RNA polymerase.
2. **Kingdom Archaea (Archaebacteria).** Anaerobic or aerobic, largely methane-producing micro-organisms. Never with membrane enclosed organelles or nuclei, or a cytoskeleton; none use chlorophyll-based photosynthesis, without peptidoglycan in cell wall; with several RNA polymerases.

B. The Eukaryotes (the "domain" Eukaryotes, or Eukarya). Each with a variety of membrane-enclosed organelles (*e.g.*, mitochondria, lysosomes, peroxisomes) and a membrane-enclosed nucleus. Cells gain structural support from an internal network of fibrous proteins called a **cytoskeleton.**

1. **Kingdom Fungi.** The Fungi probably monophyletic group** that includes molds, mushrooms, yeasts and others. Saprobic, heterotrophic multicellular organisms. The 72,000 described species are thought to represent only 5 – 10 percent of the actual diversity.
2. **Kingdom Plantae (= Metaphyta).** The multicellular plants. Photosynthetic autotrophic, multicellular organisms that develop through embryonic tissue layering. Includes some groups of algae, the bryophytes and their kin, and the vascular plants (about 240,000 of which are thought to represent about half of Earth's actual plant diversity).
3. **Kingdom Protista.** Eukaryotic single-celled micro-organisms and certain algae. A polyphyletic grouping** of perhaps 18 phyla, including euglenids, green algae, diatoms and some other brown algae, ciliates, dinoflagellates, foraminiferans, amoebae and others. The 80,000 described species probably represent about 10 percent of the actual protist diversity on earth.
4. **Kingdom Animalia.** The multicellular animals. A monophyletic taxon, containing 34 phyla of ingestive, heterotrophic, multicellular organisms. About 1.3 million living species have been described, estimates of the number of undescribed species range from lows 10 – 30 million to high of 100 – 200 million.

* Portion of old "Kingdom Monera" are now included in the Eubacteria and the Archaea. Viruses (about 5000 described "species") and subviral organisms (Viroids, and Prions) are not included in this classification.

** See Box 1.1.

Box 1.1

1. **Monophyletic group** is a group of species that includes an ancestral species and all of its descendants-that is, a **natural group.**
2. **Paraphyletic group** is a group whose member species are descendants of a common ancestor, but that does not contain all the species descended from that ancestor.
3. **Polyphyletic group** is a group comprising species that arose from two or more different immediate ancestors. Such composite taxa have been established primarily because of insufficient knowledge concerning the species in question, *e.g.*, protozoa, radiata, etc.

It has been recently been suggested that the three main divisions of life (Eubacteria, Archaea and Eukaryota) should be recognized at a new taxonomic level, called **domains**. However, fundamental questions remain about these three "domains", including how many natural groups (kingdoms) exist in each domain, whether the domains themselves represent natural (= monophyletic) groups, and what the phylogenetic relationships are among these domains and the kingdoms, they contain. Current evidence suggests that Eukaryotes are a natural group, defined by the unique trait of a nucleus and linear chromosomes, whereas Eubacteria and Archaea may not be natural groups.

Courses and textbooks on Invertebrate Zoology often include discussion of two eukaryotic kingdoms, the **Anemalia** (= Metazoa) and certain "animal-like" (*i.e.*, heterotrophic) protist phyla loosely referred to as "**Protozoa**". Following this tradition, we intend to treat 13 out of 33 phyla of metazoa and 18 phyla of protists (many of which have traditionally been viewed as "protozoa") in this text. The vast majority of kinds (species) of living organisms that have been described are animals. The **kingdom Animalia** or **Metazoa**, is usually defined as the multicellular, integestive, heterotrophic (*Heterotrophic organisms* are those that consume other organisms or organic material as food) eukaryotes. However, its members possess other unique features as well, such as an acetylcholine cholinsterase-based nervous system, special types of cell-cell junctions, and a unique family of connective tissue protein called **collagens**. Among the metazoa are some species that possess a backbone (or vertebral column), but most do not. Those that possess a backbone, constitute the subphylum Vertebrata of the phylum **Chordata** (= animals with notochord). Those that do not possess a backbone (the remainder of the phylum chordata, plus 33 additional animal phyla) constitute **invertebrates** (see Table 1.2). Thus we can see that the division of animals (= Metazoa) into invertebrates and vertebrates is based more on tradition and convenience, reflecting a dichotomy of zoologists' interests, than it is on the recognition of natural biological groupings (see **Brusca** and **Brusca**, 2003).

Table 1.2 Approximate numbers of known extant species in various groups (Source. Brusca and Brusca 2003).

Of the 1,335,188 estimated described species of Animalia or Metazoa (excluding protists), 1,228,518 (96 per cent) are invertebrates.

A. Kingdom Protista (80,000)

B. Kingdom Animalia (Metazoa)

(*a*) *Invertebrata*

1. Porifera (5,500)	2. Cnidaria (10,000)
3. Ctenophora (100)	4. Placozoa (1)
5. Monoblastozoa (1)	6. Rhombozoa (70)
7. Orthonectids (20)	8. Platyhelminthes (20,000)
9. Nemertea (900)	10. Gnathostomulida (80)
11. Rotifera (1,800)	12. Gastrotricha (450)
13. Kinorhyncha (150)	14. Nematoda (25,000)
15. Nematomorpha (230)	16. Priapula (16)
17. Acanthocephala (700)	18. Cycliophora (1)
19. Entoprocta (150)	20. Loricifera (10)
21. Annelida (16,500)	22. Echiura (135)
23. Sipuncula (320)	24. Tardigrada (600)
25. Onychophora (110)	26. Arthropoda (1,097,631)
27. Mollusca (93,195)	28. Brachiopoda (335)
29. Ectoprocta (4,500)	30. Phornida (20)
31. Chaetognatha (100)	32. Echinodermata (7000)
33. Hemichordata (85)	

(*b*) *Chordata* (49,693)

1. Urochordata (3,000)	2. Cephalochordata (23)
3. Vertebrata (46,670)	

Major and Minor Invertebrate Phyla

The invertebrate animals have been classified into about 33 phyla which are often categorized as **major invertebrate phyla** and **minor invertebrate phyla**. Such a classification is based on two criteria:

(1) **number of animals and species present in** a phylum and (2) **participation in biotic community**, *i.e.*, representation of the phyla in great majority of biotic communities.

Major Invertebrate Phyla

These include the following ten phyla: (numerical strength of each phylum has been indicated against it (Table 1.2):

1. Porifera; 2. Cnidaria; 3. Platyhelminthes; 4. Rotifera; 5. Nematoda; 6. Annelida; 7. Arthropoda; 8. Mollusca; 9. Ectoprocta; 10. Echinodermata.

Minor Invertebrate Phyla

These are 22 in number (Table 1.2). These are 1. Ctenophora; 2. Placozoa; 3. Monoblastozoa; 4. Rhombozoa; 5. Orthonectids; 6. Nemertea; 7. Gnathostomulida; 8. Gastrotricha; 9. Kinorhyncha; 10. Nematomorpha; 11. Priapula; 12. Acanthocephala; 13. Cycliophora; 14. Entoprocta 15. Loricifera; 16. Echiura; 17. Sipuncula; 18. Tardigrada; 19. Onychophora; 20. Brachiopoda; 21. Phornida; 22. Chaetognatha.

On the basis of community-participation, two phyla—the Rotifera and the Ectoprocta are not fit to be considered as the major invertebrate phyla, because they are limited in the participation in animal communities. Thus, by substracting these two phyla from the category of 10 major phyla, the total number of major phyla falls to only 8.

Differences in Invertebrates and Vertebrates

In vertebrates, a characteristic structure the **notochord**, develops during their embryonic period and later on the same is replaced by a series of **vertebrae**, forming the vertebral column. In the invertebrate animals since the notochord is totally absent, they are also known as **non-chordates**.

The invertebrate animals differ from the vertebrate animals in many respects (see Table 1.3).

Table 1.3 Differences between invertebrates and vertebrates.

	Invertebrates		Vertebrates
1.	Body symmetry may be asymmetrical, radial or bilateral.	1.	Body is bilaterally symmetrical.
2.	Appendages if present are more than two pairs.	2.	There are never more than two pairs of appendages.
3.	May have non-living exoskeleton which is formed as a secretion of the ectoderm. Endoskeleton absent.	3.	There is always an endoskeleton which is made of living tissues such as cartilage and bones.
4.	Skin consists of only one layer.	4.	Skin consists of two distinct layers, an outer epidermis overlying a thick dermis.
5.	Alimentary canal is present or absent.	5.	Alimentary canal is always present.
6.	True tail is absent. Anus is terminal and post-anal opens at the posterior end of body.	6.	A true tail, which may be defined as the continuation of the body axis, may be present.
7.	The breathing or respiratory organs normally develop from the ectoderm. No gill-slits are present in pharynx.	7.	The breathing or respiratory organs develop in relation to the wall of pharynx (an endodermal organ) and consist of gills or lungs. Pharyngeal gill-slits are present in the embryos of all vertebrates.

	Invertebrates		Vertebrates
8.	The circulatory system is mostly open type.	8.	The circulatory system is closed type containing arteries, veins and capallaries.
9.	The heart is always dorsal in position.	9.	The heart is always ventral in position.
10.	The respiratory pigment such as haemoglobin, if present, is dissolved in the plasma.	10.	The haemoglobin is contained in the red blood corpuscles.
11.	The central nervous system consists of solid ganglia connected by nerve fibres to form a chain.	11.	The central nervous system is in the form hollow thick-walled tube.
12.	The nerve cord lies ventrally to the digestive tract.	12.	The neural tube (brain + spinal cord) lies dorsally to the digestive tract.
13.	Eyes develop from the skin.	13.	Eyes develop as outgrowths of the brain.
14.	Reproduction, sexual, asexual or both.	14.	Reproduction is only sexual.

1.2 ANIMAL HABITATS

Habitat is a place where a community of organisms is found. Rocky seashores, the surface of soil and human skin are all habitats for particular communities of living organisms (**Indge**, 1997). Animals are found in all habitats. Many animals live in water. They are called **aquatic animals**. Other animals live on land. They are known as **terrestrial animals**.

1. Aquatic animals. Some aquatic animals live in the sea. They are known as **marine animals**. Other animals inhabit pools, ponds, lakes, streams and rivers. They are called **freshwater animals**. Many aquatic animals are adapted to live in **estuaries**, the regions where rivers enter the sea. The animals are classified ecologically into three categories: plankton, nekton and benthos.

***(i)* Plankton.** These are small organisms that float and drifted passively in surface waters (pelagic zone; Fig. 1.1) by waves, currents and wind. They have very weak or no locomotory organs. They are confined to the photic (well lighted) zone only. Both plants (**phytoplankton**) and animals (**zooplankton**) are included among the plankton. The zooplankton includes jellyfishes, comb jellies, arrow worms, many small crustaceans, larvae and eggs.

***(ii)* Nekton.** These are actively swimming animals. They have well-developed locomotory organs. They occur in surface (pelagic zones) as well as deep (benthic zone) waters and in both photic and aphotic zones. They include fishes and squids, cuttlefishes, etc.

***(iii)* Benthos.** These are the bottom dwelling animals that crawl over, burrow into or attached to the bottom. Benthic animals such as most sea anemones, sponges, many snails and barnacles may live on surface of substratum, they are called **epifauna** or **epibenthic** forms. Clams, worms and sea cucumbers burrow within soft substrata. These burrowing benthos form **infauna**.

Benthic animals may also be categorized by their locomotory capabilities. Animals that are generally quite motile and are active are described as being **errant** (*e.g.*, crabs, many worms), whereas those that are firmly attached to the substratum are **sessile** (*e.g.*, sponges, corals, barnacles, etc.). Others are unattached or weakly attached, but generally do not move around much (*e.g.*, crinoids, solitary anemones, most clams); these animals are said to be **sedentary**.

2. Terrestrial animals. These animals include arthropods (isopods, insects, spiders, mites, scorpions) and some animals which are restricted to relatively moist areas, *e.g.*, snails, nematoda, and earthworms. Some terrestrial animals live underground and are called **fossorial animals**. These include earthworms, termites and other ants. Others climb walls and rocks and are termed **scansorial animals**. Some animals live on trees and are known as **arboreal animals**. Certain terrestrial animals can take to flight. These are called **volant, aerial** or **flying** animals, *e.g.*, insects. The land animals that run fast are called **cursorial animals**, *e.g.*, cockroaches. The terrestrial forms may occur in forests, grasslands, deserts and tundras.

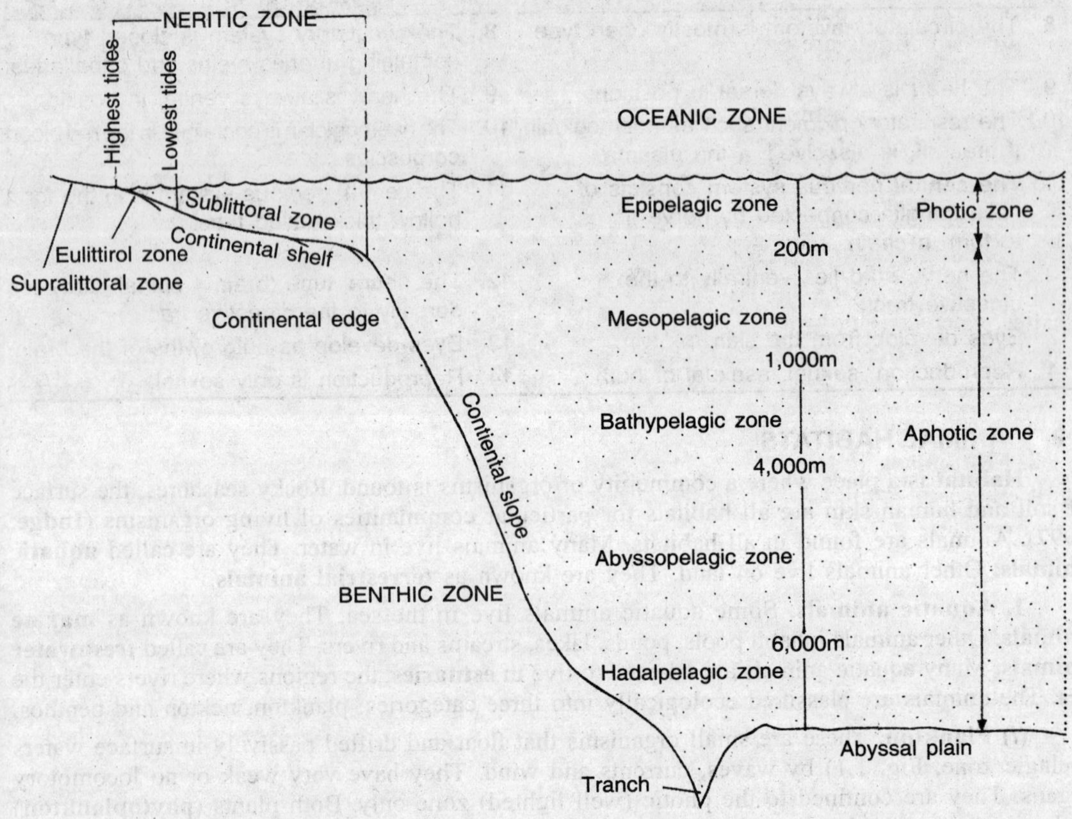

Fig. 1.1. A schematic cross section of the major habitat regions of the ocean (after Brusca and Brusca, 2003).

Box 1.2

Terrestrial animals have to pay prize for leaving the water habitat

Water provides a medium for support, for dispersing gametes, larvae and adults, and for diluting waste products, and is a source of dissolved materials needed by animals. Animals living in terrestrial environment do not enjoy these benefits of water, and must pay the prize (see Brusca and **Brusca**, 2003).

Mode of Existence

Majority of animals live singly and are called **solitary animals**. Some live in groups but without helping one another, *e.g.*, sea squirts. They are known as **gregarious animals**. Certain animals live in groups and help each other. They are called **colonial animals**. They include termites, ants, bees and wasps. Many animals lead an independent life and are termed **free-living animals**. Some, however, live in association with other species. An association of two different species is called **symbiosis** (Gr. *syn* = together; *bios* = life).

A Special Type of Environment: Symbiosis

As indicated earlier, many invertebrates live in an intimate association with other animals or plants. This kind of association is termed as symbiotic relationship, or simply **symbiosis**. Symbiosis was first defined in 1879 by German mycologist **H.A. DeBary** as "*unlike organisms* *living together*". In most symbiotic relationships, a larger organism (called the **host**) provides an environment (its body, burrow, nests, etc.) on or within which a smaller organism (the **symbiont**) lives. Some symbionts are opportunistic (**facultative**), whereas others cannot survive without their host (**obligatory**).

Symbiotic relationships can be sub-divided into following categories based on the nature of interaction between the symbiont and its host: parasitism, mutualism and commensalism.

1. Parasitism. In this form of symbiosis, the symbiont (a parasite) receives benefits at the expense of the host. Parasite may be external (**ectoparasite**) such as lice, ticks and leeches; or internal (**endoparasites**), such as liver flukes, some roundworms and tapeworms. Some parasites (*e.g.*, lice, tapeworm) live their entire adult life in association with their hosts are **permanent parasites**, whereas **temporary**, or **intermittent parasites**, such as bedbugs only feed on the host and then leave it. A **definitive host** is one in which the parasite reaches reproductive maturity. An **intermediate host** is one that is required for parasite's development, but in which the parasite does not reach reproductive maturity.

2. Mutualism. In this type of symbiosis both host and symbiont benefit. Such relationship may be extremely intimate and important for the survival of both parties; for example, the bacteria in our own large intestine are important in the production of certain vitamins and in processing material in gut. Another example is the relationship between termites and certain protists (*e.g.*, *Trichonympha*) that inhabit their digestive tracts and are responsible for the breakdown of cellulose into compounds that can be assimilated by their insect hosts (see *Animal Behaviour*, **Agarwal, V.K.**, 2009). Other mutualistic relationships may be less binding on the organism involved. Cleaner shrimps, for example, inhabit coral reef environments where they establish "cleaning stations", that are visited regularly by reef-dwelling fishes that present themselves to the shrimps for the removal of parasites. Obviously, even this rather loose association results in benefits for the shrimps (a meal) as well as for the fishes (removal of parasites). The mutualistic relationships between plants and their pollinators are essential to the survival of most flowering plants and their insect partners (and in some cases, their bird or nectar-feeding bat partners).

3. Commensalism. This type of symbiosis is usually described as an association that is advantageous to one party (the symbiont) but leaves the other (the host) unaffected. For instance, among invertebrates there are numerous examples of one species inhabiting the tubes or burrow of another (**inquilism**); the former ('symbiont–I') obtains protection, food; or both with little or no apparent effect on the latter ('symbiont–II'). A special type of commensalism is **phoresis**, where in the two symbionts "travel together", but there is no physiological or biochemical dependency on the part of either participant. Usually one **phoront** is smaller than the other and is mechanically carried about by its larger companion.

Evolutionary Role of Symbiosis—The Symbiogenesis

The Russian biologist **Konstantin Mereschkovsky** (1855 – 1921) developed the **"two plasm"** (cell within a cell) **theory**, claiming that chloroplasts originated from blue-green algae (cyanobacteria). For this process, he invented the term **symbiogenesis** (which is a macroevolutionary event). According to **serial endosymbiotic theory** (**SET**) of symbiogenesis, three classes of organelles of eukaryotic cells, namely mitochondria, cilia/flagella and plastids, are hypothesized to have once been free-living prokaryotes that were acquired endosymbiotically and in certain sequence by another (host) prokaryote (Fig. 1.2). Although symbiogenesis is an old idea, it was **Lynn Margulis** (1989) who most vigorously championed it in the twentieth century.

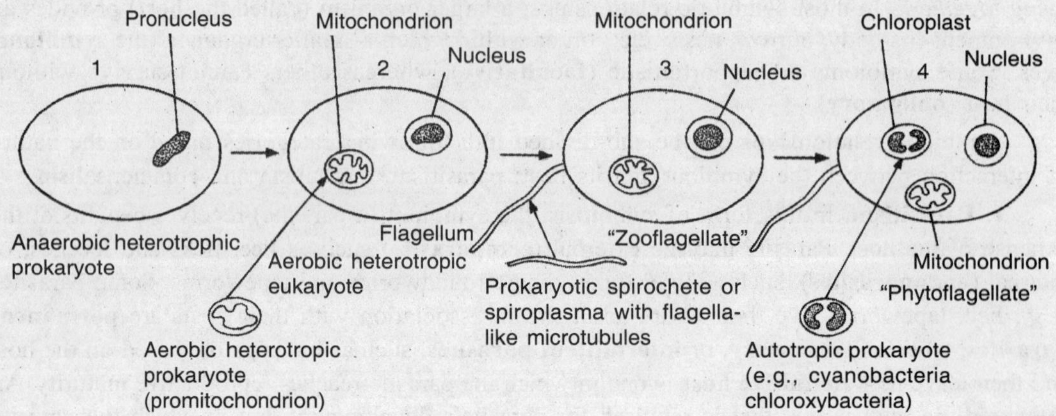

Fig. 1.2. A simple model of the origin of eukaryotic cells by symbiosis (the serial endosymbiotic theory). The three major events depicted are: acquisition of an aerobic heterotrophic prokaryote (origin of mitochondrion); acquisition of a spirochaete or spiroplasma-like prokaryote (origin of flagellum) and acquisition of an autotrophic prokaryote (origin of chloroplast) (after Brusca and Brusca, 2003).

Beyond the origin of the eukaryotic cell, symbiogenesis may be at work in many other systems. Many invertebrates are involved in such relationships, including corals and other cnidarians that serve as hosts for symbiotic dinoflagellates (called **zooxanthallae**) that live within their tissues. Various animals that harbor (and exploit) tetrodotoxin-secreting bacteria (many chaetognaths, the blue-ringed octopus, a sea star and horseshoe crab, and certain tetraodontid fishes; Box 1.3), and squids with luminous bacteria, and lichens (an intimate association between fungi and cyanobacteria or green algae) are other examples.

Box 1.3

Tetrodotoxin is a neurotoxin $C_{11}H_{17}N_3O_8$ that is found in pufferfishes and that blocks nerve conduction by suppressing permeability of nerve fiber to sodium ions.

1.3 CONCEPT OF BAUPLAN AND RELATED CONCEPTS

Stability of organismal morphology is a deep-seated notion that dates at least from the eighteenth century; the idea of limited number of plants, types or archetypes of animal forms is thus an old one. **Richard Owen** introduced the term **archetype** in 1848 to represent a model organism, or the sum of features shared by a group of related organisms. The concept of the embryological archetype, and the fact that adults are nothing more than the accumulation of features added during their development, was formalized by **Karl von Baer** and **Ernst Haeckel** in the second half of the nineteenth century. Today this concept has grown into the notion of **conserved body plans** or **bauplane** (or **Bauplan**). This concept speaks to *a stability in form that maintains itself through evolutionary time and phylogenetic divergence.*

The term **bauplan** (German for "ground plan" or "blueprint") was introduced as a technical term in Zoology in 1945 by the embryologist-turned-philosopher **Joseph Henry Woodger**. More recently **Niles Eldredge** (1989) discussed the bauplan as the common structural plan within a monophyletic taxon; **Valentine** (1986) distinguished bauplan as assemblages of homologous architectural and structural features distinguished phyla and classes; and **Gould** (1977, 1980, 1992) spoke of structural constraints leading to fundamental ground plan of anatomy. In his review of the reunion of developmental and evolutionary biology, **Atkinson** (1992) claimed that "*the single most critical concept of the reunions is that of the bauplan*".

The concept of the bauplan expresses both a notion of morphological stability and the fact that some aspects of embryonic and/or adult morphology are more free to vary than are others. That is, some stages of development are more constrained than others. The most striking evidence of developmental constraints is simple fact that despite the great variety of animals, there are relatively few basic types of animal body plans (**Brusca** and **Brusca**, 2003).

Development canalization. It is a form of constraint that channels ontogeny into restricted sets of pathways that lead to a standard phenotype in spite to genetic or environmental disturbances. The preservation of Hox gene function across phyla is a good example of developmental canalization (Box 1.4).

Box 1.4 Homeobox (Hox) Genes

Homeobox or Hox genes have played a macroevolutionary role in the origin of invertebrates. Hox genes are master regulatory genes and doing so, "select" the developmental pathways that are followed by dividing cells. Hox genes have two functions: 1. they encode short regulatory proteins that bind to particular sequences of bases of DNA and either enhance or repress gene expression, and 2. They encode proteins that are expressed in complex patterns that determine the basic geometry of the organisms. The term Hox genes refers specifically to those genes that *are clustered in an array on the chromosome and function primarily in establishing regional or segmental identities*. In all metazoa that have been examined, regional or segmental specialization is controlled by the spatially localized expression of these genes, which play crucial roles in determining body patterns. They underlie such fundamental attributes as anterior-posterior differentiation (in both invertebrates and vertebrates) and the positioning of body wall.

Hox genes have been conserved to a remarkable degree throughout the animal kingdom, and they are now known from all animal phyla that have been examined. There is a striking corelation between the order of these genes on their chromosomes and the position of their expression in the developing animal along the main body axis. Hox proteins regulate the genes that control the cellular processes involved in morphogenesis (*i.e.*, development of the architectural features of organisms; formation and differentiation of tissues and organs). In doing so, they demarcate relative positions in animals—they do not specify the precise nature of particular structure. For example, in arthropods, Hox genes regulate where body appendages form, and they can either suppress limb development or modify it (in concert with other regulatory genes) to create unique appendage morphologies. Mutations in Hox genes, and other developmental genes, can create gross mutations (homeotic mutations or homeosis).

The characteristics of an organism's bauplan are not the same thing as its phylogenetically unique features. Instead, bauplans must be viewed as nested sets of conserved body plans, as would be predicted within an ancestor – descendant hiearchial system such as animal phylogenesis. For example, snakes possess a bauplan that differs from the bauplan of lizards, turtles or crocodiles — yet each share the reptilian bauplan. Reptiles, birds and mammals each have individual bauplan but share the vertebrate bauplan. Thus, bauplans consist of a mix of ancestral and derived characters. To understand their origin requires a knowledge of adult, larval and embryonic phases of the life cycle.

Woodger clearly asserted, as had **von Baer**, that the most basic structures defining bauplan develop early in embryonic life. Consequently, deviations early in development would have much more drastic consequences for morphology than deviations later in development. Mechanisms that establish bauplans buffer development against environmental and genetic disorders. They constrain development. **Ernst Mayr** repeatedly drew attention to the importance of such constraints, specifically in relation to bauplans, conserved morphological features and the taxonomic features used in classification.

Due to field of molecular evolutionary developmental biology, we now know that much of an animal's initial embryogeny is under maternal cytoplasmic control rather genomic control. However, at some point early in embryogenesis (*i.e.*, the origin and development of the embryo), the zygote's parental (nuclear) genome takes primary control of development. Recent work suggests that this may be one of several **pivotal points** in the control of animal ontogeny—*occasions that demarcate the fixation of bauplan*. The point at which the zygotic genome takes over control of embryogenesis has been referred to as **zootypic stage**. It may be here that *Hox* genes establish the most basic, or primary, animal body patterning (*e.g.*, anterior-posterior axis and dorsal-ventral surface).

At a later stage of embryogenesis another critical point is reached, which has been called the **phylotypic (phyletic) stage**. The *phylotype* theoretically represents the stage when the gene responsible for secondary patterning of a body plan are first fully expressed and the adult *morphogenetic fields* are positioned. This juncture is not well understood. **Anderson** (1973) identified the blastula as the phylotypic stage for the annelids and arthropods, whereas **Sander** (1976, 1983) identified the germ band stage (a 20-segmented larval stage with head, thorax and abdomen already outlined and segmented) as the phylotypic stage of insects. **Cohen** (1977, 1979), on the other hand, distinguished **phylotypic larvae** (the trochophore of annelids, for instance) from **adaptive larvae**. Phylotypic stage larvae have a simple morphology determined more by developmental (genetic) programs than by physiological requirements.

The phylotypic stage is usually thought of as the stage at which embryos within a phylum shows the greatest level of morphological similarity. Beyond this stage, the zygotic genome begins moving embryos down the individual tracks of the various lineages. In other words, early developmental stages of closely related taxa converge on a phylotype in the course of their ontogeny, only to diverge again as the adult form unfolds.

Thus, it seems likely that there are several fundamental levels of body patterning during ontogeny, and closely related taxa share critical junctures of this process in ways that recall Haeckel's "law of recapitulation".

It is already becoming clear that such developmental stages (*e.g.*, zootypic stage and the phototypic stage) canalize ontogenetic events to produce the adult bauplan. This canalization comes about in part, due to following constraints that work to maintain overall body plan:

1. **Structural constraints** (*e.g.*, constraints imposed by the limitations of patterns in early developmental stages).
2. **Genetic constraints** (*e.g.*, rate of mutation and recombination of individual alleles).
3. **Direct developmental constraints** (*e.g.*, obligatory tissue interactions).
4. **Cellular constraints** (*e.g.*, limits to rate and number of cell divisions, secretions of cell products, cell migration).
5. **Metabolic constraints** (*e.g.*, dependence on particular metabolic pathways).
6. **Functional constraints** (*e.g.*, the interconnectedness of parts of different organ systems involved in critical functions).

1.4 BODY SYMMETRY

A fundamental aspect of an animal's bauplan is its overall shape or geometry. In order to discuss invertebrate architecture and function, we must first familiarise ourselves with a basic aspect of body form: symmetry.

Symmetry. It refers to the regular arrangement of body structures relative to axis of the body. Animals that can be bisected or split along at least one plane, so that the resulting halves are similar to one another, are said to be **symmetrical**. For example, a shrimp can be bisected vertically through its midline, head to tail, to produce right and left halves that are mirror images of one another.

A few animals have no body axis and no plane of symmetry. They are said to be **asymmetrical**. Many sponges, for example, have irregular growth form and lack any clear plane of symmetry. Similarly, many protists, particularly the amoeboid forms, are asymmetrical.

Spherical symmetry. It is seen in creatures whose bodies lack an axis and have the form of a sphere, with the body parts arranged concentrically around, or radiating from, a central point (Fig. 1.3A). A sphere has an infinite number of planes of symmetry that pass through its center to divide it into like halves. Spherical symmetry is rare in nature; in the strictest sense, it is found in certain protists. Organisms with spherical symmetry share an important functional attribute with asymmetrical organisms, in that both groups lack **polarity**. That is, there exists no clear differentiation along an axis. In all other forms of symmetry, some level of polarity has been achieved and with polarity comes specialization of body regions and structures.

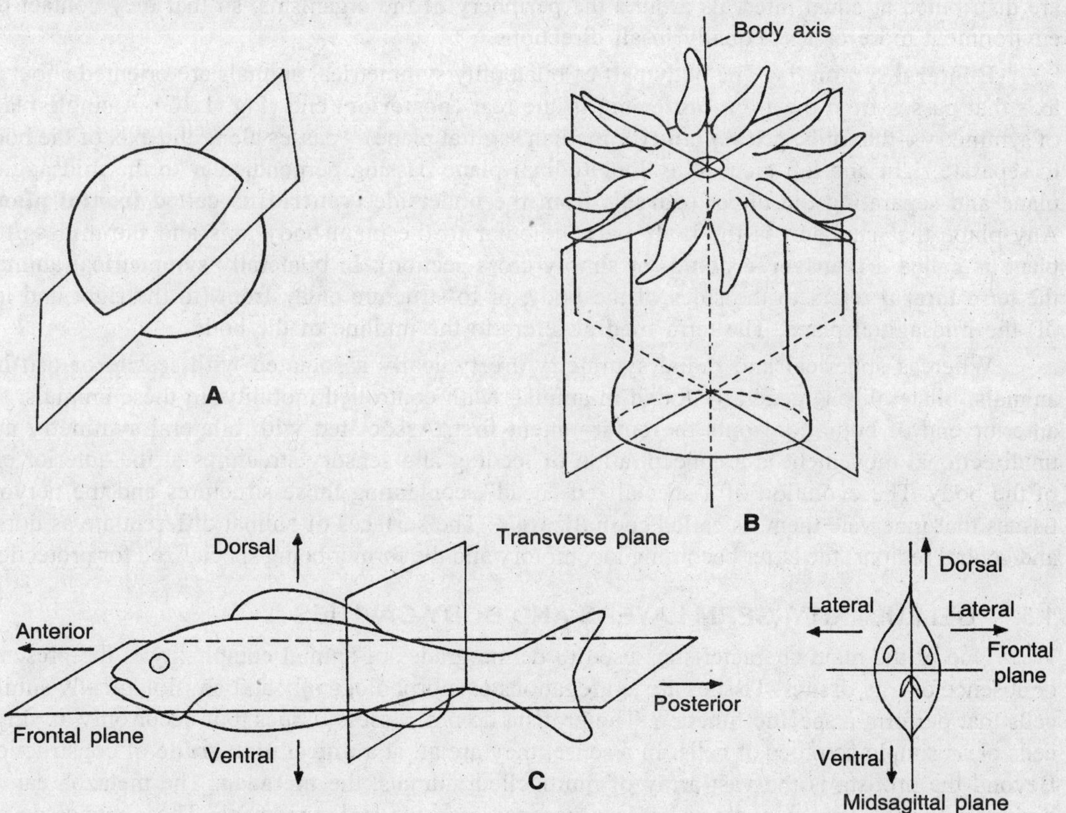

Fig. 1.3. A—Spherical symmetry; B—Radial symmetry; C—Bilateral symmetry.

Radial symmetry. A body displaying this type of symmetry has the general form of a cylinder, with one main axis around which the various parts are arranged (Fig. 1.3B). In a body displaying perfect radial symmetry, the body parts are arranged equally around the axis, and any plane of sectioning that passes along that axis results is similar halves (rather like a cake being divided and sub-divided into equal halves and quarters). Nearly perfect radial symmetry occurs in some sponges (*Xetospongia*) and in many cnidarian polyps. Perfect radial symmetry is relatively rare, however, and most radially symmetrical animals have evolved modifications on this theme. **Biradial symmetry**, for example, occurs where portions of the body are specialized and only two planes of sectioning can divide the animal into perfectly similar halves. Common examples of biradial organisms are ctenophores and many sea anemones (*Epiactus*).

Further specializations of the basic radial body plan can produce nearly any combination of multiradiality. For example, many jelly fishes (*i.e.*, hydromedusa of *Scrippsia*) possess **quadriradial symmetry**. Most echinoderms (*e.g.*, sea star *Patiria* and sea bisquit *Clypeaster*) are said to display **pentaradial symmetry**.

A radially symmetrical animal has no front or back end; rather it is organized about an axis that passes through the center of its body, like an axle through a wheel. When a gut is present, this axis passed through the mouth-bearing (**oral**) surface to the opposite (**aboral**) surface. Radial symmetry is most common in sessile and sedentary animals (*e.g.*, sponges, sea stars and sea anemones) and drifting pelagic species (*e.g.*, jellyfishes and ctenophores). Given these life styles, it is clearly advantageous to be able to confront the environment equally from a variety of direction. In such creatures the feeding structures (tentacles) and sensory receptors are distributed at equal intervals around the periphery of the organisms, so that they contact the environment more or less equally in all directions.

Bilateral symmetry. The body part of bilaterally symmetrical animals are oriented about an axis that passes from the front (**anterior**) to the rear (**posterior**) end (Fig. 1.3C). A single plane of symmetry—the **midsagittal plane** (or median sagittal plane)—passes along the axis of the body to separate right and left sides. Any longitudinal plane passing perpendicular to the midsaggital plane and separating the upper (**dorsal**) from the underside (**ventral**) is called **frontal plane**. Any plane that cuts across the body perpendicular to the main body axis and the midsagittal plane is called a **transverse plane** (or simply cross section). In bilaterally symmetrical animals the term **lateral** refers to the sides of the body, or to structure away from (to the right and left of) the midsagittal plane. The term **medial** refers to the midline of the body.

Whereas spherical and radial symmetry are typically associated with sessile or drifting animals, bilaterality is generally found in animals with controlled mobility. In these animals, the anterior end of body confronts the environment first. Associated with bilateral symmetry and unidirectional movement is a concentration of feeding and sensory structures at the anterior end of the body. The evolution of a specialized "head", containing those structures and the nervous tissues that innervate them, is called **cephalization**. The surfaces of animal differentiate as dorsal and ventral regions, the latter becoming locomotory and the former being specialized for protection.

1.5 CELLULARITY, GERM LAYERS AND BODY CAVITIES

One of the main characteristics used to define grades of animal complexity is the presence or absence of true **tissues**. Tissues are aggregations of morphologically and physiologically similar cells that perform a specific function. The protista do not possess tissues, but occur only as single cells or as simple colonies of cells. In a sense, they are all at a **unicellular grade** of construction. Beyond the protista is the vast array of multicelled animals, the **metazoa**. The metazoa can be divided into three major levels or grades: mesozoa, parazoa and eumetazoa. These names do not represent formal taxa, but may be used to group the metazoa by their level of overall complexity. The mesozoa and parazoa are not generally considered to possess true tissues, and for this reason they are separated from the rest of the metazoa. The eumetazoa pass through distinct embryonic stages during which tissue layers form (Table 1.4).

The embryonic tissue layers of eumetazoa are called **germ layers** (from the Latin *german* = a sprout, bud, or embryonic primordium) and it is from these germ layers that all adult structures develop. The germ layers initially form as outer and inner sheets or masses of embryonic tissues, termed **ectoderm** and **endoderm** (or entoderm), respectively. In the embryogeny of the radiata phyla Cnidaria and Ctenophora, only these two germ (or germinal) layers develop (or if a middle layer does develop, it is produced by ectoderm, is largely non-cellular, and is not considered a true germ layer). These animals are regarded as **diploblastic**. In the embryogeny of most animals, however, a third cellular layer, the **mesoderm**, arises between the ectoderm and the endoderm; these metazoan groups are said to be **triploblastic.**

Table 1.4 Organization of the Protista and Animal phyla on the basis of their body construction (Brusca and Brusca, 2003).

1. **Unicellular organisms (Protista).** Phyla Euglenoida, Kinetoplastida, Ciliophora, Apicomplexa, Dinoflagellata, Stramenopiles, Rhizopoda, Actinopoda, Granuloreticulosa, Diplomonadida, Parabasilida, Chlorophyta, Cryptomonada, Microspora, Ascetospora, Myxozoa, Opalinida, Choanoflagellata.
2. **Multicellular organisms.** The **Metazoa** or **Animalia**
   A. Without tissue
   1. **The Mesozoa.** Phyla Orthonectida, Rhombozoa, Placozoa and Monoblastozoa (*Salinella*).
   2. **The Parazoa.** Phylum Porifera (sponges)

   B. With true tissues. Eumetazoa
   1. **The diploblastic eumetazoa** (without true mesoderm). The **"Radiata"**
   2. **The triploblastic eumetazoa** (with true mesoderm). The "**Bilateria**"
      (*i*) **Acoelomate.** Without a body space other than digestive tract; mesenchyme muscle fill region between gut and epidermis.
      Phyla Platyhelminthes, Gastrotricha, Entoprocta, Gnathostomulida.
      (*ii*) **Blastocoelomate** (formerly called **Pseudocoelomata**). With a persistent blastocoel between gut and body wall.
      Phyla Acanthocephala, Kinorhyncha, Loricifera, Nematoda, Nematomorpha, Rotifera
      (*iii*) **Coelomates (or eucoelomates).** With a true coelom (*i.e.*, mesodermal cavity)
      Phyla Nemertea, Phoronida, Ectoprocta, Brachiopoda, Sipuncula, Echiura, Mollusca, Priapula, Onychophora, Tardigrada, Annelida, Arthropoda, Echinodermata, Chaetognatha, Hemichordata, Chordata*

* Some of these groups (*e.g.*, Arthropoda, Mollusca) have greatly reduced coelomic spaces; often the main body cavity is a blood filled space called a **hemocoel** and is associated with an open circulatory system.

The evolution of mesoderm greatly expanded the evolutionary potential for animal complexity. Three major grades of construction are recognizable among the triploblastic metazoa: **acoelomate, blastocoelomate (pseudocoelomate)** and **eucoelomate**. A acoelomate grade occurs in several triploblastic phyla: Platyhelminthes, Entroprocta, Gnathostomulida and Gastrotricha. In these animals, the mesoderm form a more or less solid mass of tissue, sometimes with small open spaces (lacunae), between the gut and body wall (Fig. 1.4 A). In nearly all other triploblastic animals, an actual space develops as a fluid filled cavity between the body wall and the gut. In many phyla (*e.g.*, annelids and echinoderms), this cavity arises within the mesoderm itself and is completely enclosed within a thin lining called the **peritoneum**, which is derived from the mesoderm. Such a cavity is called a true **coelom** (**eucoelom**; Fig. 1.4C).

Several groups of triploblastic metazoa (*e.g.*, rotifers, roundworms and others) possess small or large body cavities that are neither from the mesoderm nor fully lined by peritoneum or any other form of mesodermally derived tissue. Such a cavity is used to be called a **pseudocoelom** (= **blastocoelom;** Fig. 1.4B). The organs of these animals actually lie free within the body cavity and bathed directly in its fluid. In most cases the space represents persistent remnants of the embryonic *blastocoel* and since there is nothing "false" about it, **Bursca** and **Bursca** (2003) used the more descriptive term **blastocoelom** instead of pseudocoelom.

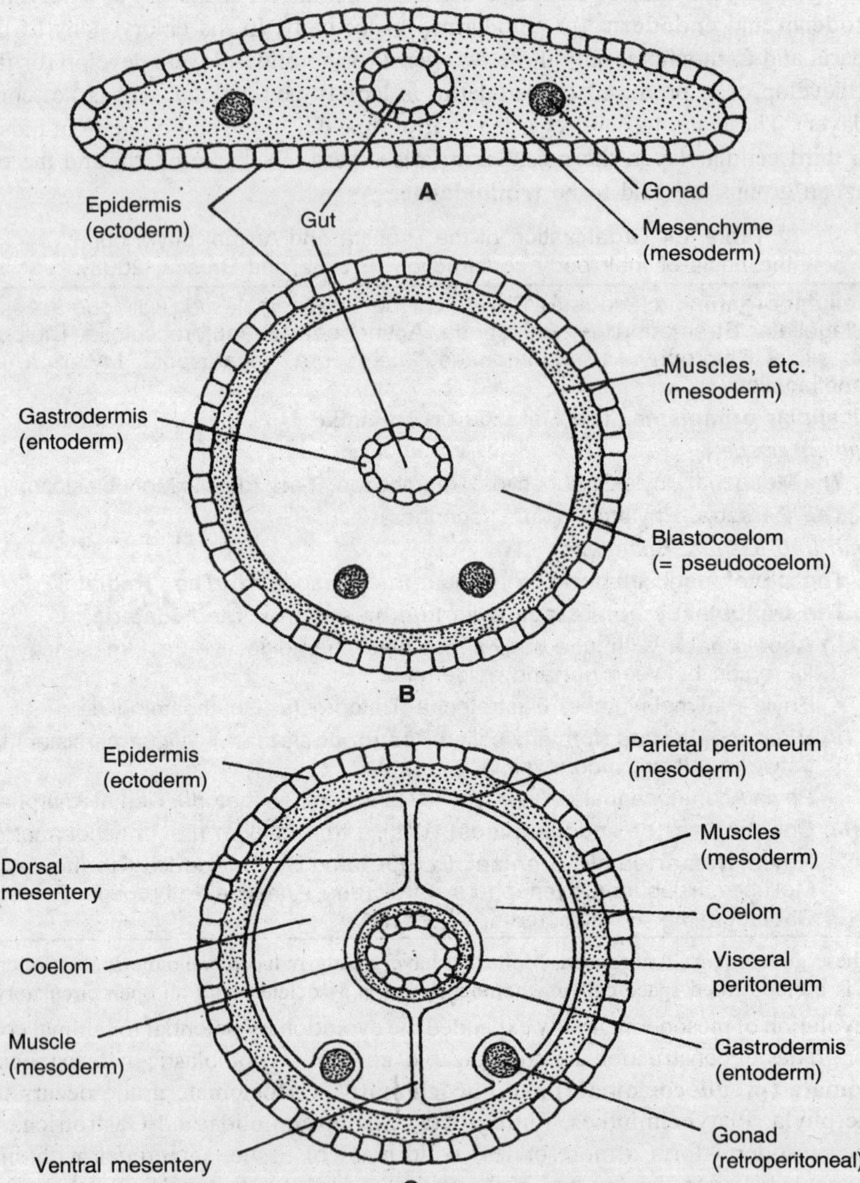

Fig. 1.4. Principal body plans of triploblastic metazoa (diagrammatic cross section). A—Acoelomate body plan; B—Blastocoelomate (= pseudocoelomate) body plan and C-Eucoelomate body plan (after Brusca and Brusca, 2003).

The evolution of this space (blastocoelom and eucoelom) created a radically new architecture, a tube-within-a tube design in which the inner tube (the gut and its associated organs) was freed from the constraint of being attached to the outer tube (the body wall), except at the very ends. The fluid-filled cavity not only served as a mechanical buffer between these two largely independent tubes, but also allowed for the development and expansion of new structures within the body, served as a storage chamber for vary body products (*e.g.*, gametes), provided a medium for circulation and was in itself an incipient hydrostatic skeleton.

1.6 GENERAL CHARACTERS OF SOME PHYLA OF INVERTEBRATA

1. Phylum Protozoa. It includes one celled, free-living, mostly aquatic animals. They are small in size, usually microscopic with distinct nucleus. Each individual consists of only one cell which has to carry on all the vital activities. They are abundantly found in water containing decaying organic matter but others live in damp soil, in fresh water or in the sea. Some are commensals, mutualists or parasites and pathogens. The examples are *Amoeba, Euglena, Paramoecium, Vorticella, Polystomella, Entamoeba histolytica* (causes amoebic dysentry in humans), *Trypanosoma gambiense* (causing sleeping sickness in humans), *Plasmodium vivax* (causing malaria in humans), *Monocystis* (a parasite of earthworms), *etc*.

2. Phylum Porifera. It includes the sponges which are found both in ocean and fresh water. These are multicellular animals having two germ layers (*i.e.*, external **pinacoderm** and internal **choanoderm**). Both layers have cells which are arranged loosely aggregated and a mostly gelatinous intermediary layer (called **mesohyl**) which carries **archaeocytes** (large amoeboid **totipotent cells** that can differentiate into all various other cell types). The body is perforated with many pores and canals through which water flows. Within sponges, water currents are generated by collared, flagellated cells called **choanocytes**, whose flagella move water out through an exhalant body opening (osculum) thereby drawing water in through external pores (ostia). This is done for filtering water for food and oxygen. The body wall of sponge is supported by a framework which is composed of lime or of silica (**spicules**) or of an organic substance known as **spongin** (collagen). The examples are *Sycon, Hyalonaema, Euplectella*, etc.

3. Phylum Cnidaria. These are lowest metazoan animals with radial symmetry of body and two definite body layers, ectoderm (epidermis) and endoderm (gastrodermis) with mesoglea in between. They are both marine and fresh water and carnivorous in habit. Their tentacular epidermis is armed with specialised cells, called **cnidocytes** that contain miniature stinging, harpoon-like organelles called **nematocysts**. Nematocysts immobilize prey and let the animal bring them to the gastric cavity (coelenteron). The examples are *Hydra, Obelia, Aurelia*, sea anemone, corals, etc.

4. Phylum Ctenophora. Ctenophora is a small group. They are all marine and planktonic. Their biradially symmetrical bodies contain eight meridionally placed ciliated plates. They resemble the cnidarians on many counts but differ from them in not having the nematocysts. Their mesoglea contain mesenchymal muscle cells. They possess a specialised aboral sense organ and the tentacles bear adhesive cells. The examples are *Beroe, Pleurobrachia, Hormiphora*, etc.

5. Phylum Platyhelminthes. It includes the free-living flatworms (Turbellaria) and parasitic flatworms (flukes and tapeworms). They have dorso-ventrally flattened body. These animals possess ectoderm and endoderm and another cell layer called mesoderm in between. For this reason they are called **triploblastic animals**. They lack a body cavity, so are called **acoelomates**. Digestive canal is incomplete with only one opening known as **mouth** and there is no anus. Excretion of waste products is done by peculiar flame cells (protonephridia). They possess extensive hermaphroditic reproductive system. The examples are leaf-like *Dugesia tigrina* (free living), leaf-like fluke (*Fasciola hepatica*), ribbon-like tape worm (*Taenia solium*), etc.

6. Phylum Nematoda. It includes round worms or thread worms having soft, tubular shape and unsegmented bodies. They contain a distinctive, fluid-filled body cavity called **pseudocoel**. They have epidermal cuticle, adhesive organs, constant cell numbers (eutely) and a digestive tract with mouth, anus and a muscular pharynx that pumps food into flaccid gut cavity. They are mostly aquatic in habit, some are endoparasitic. The examples are *Ascaris lumbricoides, Wuchereria bancrofti* (filaria), *Ancylostoma duodenale* (hook worm), *Meloidogyne* (root-knot nematode), etc.

7. Phylum Mollusca. These are unsegmented coelomates (have true body cavity). The soft parts of body are enclosed in a hard calcareous shell, as in snails and oysters. A fleshy muscular foot for locomotion is often present (they have a body plan based on creeping over shallow marine substrates). They are earliest known herbivorous mostly having a **radula**. Radula is a rasplike organ bearing chitinous teeth that unrolls from the mouth and scrapes algae from rocks. Molluscs are found on land, specially in moist places and in fresh water and salt water. The examples are freshwater mussel (*Lamellidens*), Apple snail (*Pila*), Pearl oyster (*Pinctada*), Cuttlefish (*Sepia*), etc.

8. Phylum Annelida. The soft, elongated, worm-like bilaterally symmetrical coelomate body is metamerically segmented. They can localise hydrostatic pressures to specific segments to do active sustained burrowing. Annelids include the earthworms, leeches and sandworms that live on soil, fresh water and sea water respectively.

The examples are terrestrial clitellum-bearing hermaphrodite earthworm (*Pheretima posthuma*) and fresh water leech (*Hirudinaria granulosa*) and marine *Aphrodite, Arenicola, Chaetopterus, Nereis*, etc.

9. Phylum Arthropoda. The body is bilaterally symmetrical, segmented, with definite head, thorax and abdomen. Hardened, chitinous exoskeletons with their inner projections (apodemes or endophragma) for muscle attachments, jointed appendages and development of new kinds of cephalic structures have made this phylum largest of the animal phyla (80 per cent of all known animal species today are arthropods). The examples are horseshoe crab, prawn, crab, centipede, millipede, scorpion, spider, ticks, mites, cockroach, grasshopper, locust, silkmoth, honeybee, termite, wasp, ant, house fly, mosquito, body louse, lac-insect, rice-bug, beetle, etc.

10. Phylum Echinodermata. These are spiny-skinned, marine triploblastic, unsegmented, coelomate animals. They have a mesodermal skeleton of small calcite plates or spicules lying just below the outer epidermis. They have bilateral symmetry in larvae, followed by a five-rayed (pentameral) symmetry in adults. They contain a **water vascular system** derived from coelom in which small surface **tube feet** used for locomotion and feeding connect to internal canals of circulating sea water. The examples are star fish (*Asterias*), sea-lily, sea-cucumber, sea-urchin, etc.

11. Phylum Hemichordata. They are bilateral deuterostomes, body vermiform or saccate and fundamentally trimeric, with prosome, mesosome and metasome, each with coelomic compartments. Solitary or colonial. With pharyngeal gill slits/pores. Well developed open circulatory system. Excretion by glomerulus. Gonads extra coelic. Complete gut with preoral (buccal) gut diverticulum. Deposit or suspension feeders (Box 1.5). Without a notochord. Have short, dorsal, mesosomal, occasionally hollow nerve cord. Dioecious, with external fertilization and indirect development producing tornaria larva. Strictly marine, except for *Planctosphaera pelagica*. The examples are acorn worms such as *Saccoglossus, Balanoglossus, Cephalodiscus, Rhabdopleura, Atubaria, Planctosphaera*, etc.

Box 1.5

Suspension feeding. It is the removal of suspended food particles from the surrounding medium by some sort of capture, trapping of filteration mechanism.

Deposit feeders. They are microphagous and they obtain nutrients from the sediments of soft-bottom habitats (muds and sands).

QUESTIONS

Long Answer Questions

1. Describe characteristic features of invertebrates.
2. What is the concept of major and minor phyla? Enumerate major invertebrate phyla and minor invertebrate phyla.
3. Classify invertebrate animals upto phyla. Mention main characters and one or two example of each phylum.
4. Write an essay on 'Bauplan concept'.

Short Answer Questions

1. What is a living organism?
2. Differentiate between invertebrates and vertebrates.
3. Differentiate between bilateral and radial symmetry.
4. Write a note on habitat.
5. What is symbiosis? How does it serve as a habitat of some animals?
6. Give an account of symmetry in animals.
7. What is body cavity? Describe its various types.
8. "The group Invertebrata is unnatural". Comment on this statement.
9. Write down the six constraints which maintain body plan of the animals.
10. Write short notes on the following:
    (*i*) Bauplan concept (*ii*) Hox gene
    (*iii*) Cephalization.

Very Short Answer Questions

1. Define the term biology and zoology.
2. What do you mean by zoology?
3. Define Invertebrata.
4. Give examples of sessile animals.
5. Give examples of sedentary animals.
6. What is sagittal section.
7. Name the Invertebrate phyla which show metamerism.
8. What are triploblastic animals?
9. Name three types of body cavity found in animals.
10. Give some examples of zooplankton.
11. Give some examples of nekton.

Fill in the Blanks

1. Marine animals live in the ....
2. Small organisms that passively float in surface water are referred to as ....
3. Animals which actively swim about in water are referred to as ....
4. Symmetry in *Hydra* is ....
5. Animal that runs fast on land is called ....
6. Fossorial animals live in ....
7. Protozoa form a .... group.
8. Metazoa form a .... group.
9. Kingdom Protista is a .... group.
10. Kingdom Animalia is a .... group.
11. Tardigrade is a minor ....
12. Term bauplan was coined by .... in 1945.

Multiple Choice Questions

1. Which phylum has highest number of species?
   (*a*) Protozoa (*b*) Porifera
   (*c*) Arthropoda (*d*) Mollusca

2. Platyhelminthes are commonly called
   (*a*) round worm (*b*) flat worm
   (*c*) blind worm (*d*) tubiculous worm
3. Which group includes all marine animals?
   (*a*) Mollusca (*b*) Porifera
   (*c*) Echinodermata (*d*) Cnidaria
4. A mesodermal endoskeleton among invertebrata is found in
   (*a*) Porifera (*b*) Mollusca
   (*c*) Echinodermata (*d*) Cnidaria
5. Animals living at the bottom of sea are called
   (*a*) plankton (*b*) nekton
   (*c*) benthos (*d*) none of these
6. Protozoans are now regarded as
   (*a*) an independent phylum (*b*) a group of monera
   (*c*) a group of protista (*d*) a group of animals

ANSWERS

Very Short Answer Questions

1. Biology is study of life. Biology investigates the origin, structure, functions and distribution of life. Biology includes *Botany* (study of plant life) and *Zoology* (study of animal life).
   Zoology is the scientific study of animal life.
2. Zoology is the branch of biology or life sciences that deals with the animal organisms.
3. Animals which do not have a notochord and vertebral column (backbone) are called **invertebrates.** (**Note.** The division of animals into invertebrates and vertebrates is based more on tradition and convenience, reflecting a dichotomy of zoologist's interest, than it is on the recognition of natural biological groupings; see **Brusca** and **Brusca**, 2003).
4. Sponges, corals and barnacles.
5. Crinoids, solitary anemones and most clams.
6. Median longitudinal vertical section.
7. Annelida and Arthropoda.
8. Triploblastic animals are those which develop from three germ layers namely ectoderm, mesoderm and endoderm.
9. True coelom (Eucoelom), psuedocoelom and haemocoel.
10. Jelly fishes, comb jellies, arrow worms and small crustaceans and larvae.
11. Squids, cuttlefishes and fishes.

Fill in the Blanks

1. sea	**2.** plankton	**3.** nekton	**4.** radial
5. cursorial	**6.** burrow	**7.** unnatural	**8.** natural
9. polyphyletic	**10.** monophyletic	**11.** phylum	**12.** J.H. Woodger.

Multiple Choice Questions

1. (*c*)	**2.** (*b*)	**3.** (*c*)	**4.** (*c*)
5. (*c*)	**6.** (*c*).		

2

Protozoa: Characters, Classification and Types

The **protozoa** (Gr., *protos* = first or primitive; *zoon* = animal) are microscopic, eukaryotic, unicellular, solitary or colonial, commonly motile and phagocytic heterotrophic animalcules. The term protozoa was coined by **Goldfuss** (1818) and unicellular nature of protozoa was established by **von Siebold** (1845). There are about 50,000 species of protozoa (*i.e.*, 30,000 living and 20,000 extinct) which are found in all moist places such as sea, freshwater and soil. Each protozoan functions as an independent unit and is able to perform effectively all the activities necessary for life.

2.1 GENERAL CHARACTERS

1. Protozoa are small, generally microscopic animalcules. Individual independent protozoa range in size from a micrometer (*e.g., Micromonas*) to a few millimeters in some dinoflagellates, amoebas and ciliates.
2. They are simplest and most primitive animals having **cellular level of organisation.**
3. **Unicellular** or **single-celled** animals having no tissues or organs. Specialised parts of living material (*i.e.*, protoplasm) form membrane-bound organelles or organoids.
4. The single cell performs all necessary functions. There is no **physiological division of labour**.
5. The body is bounded by the plasma membrane. The rigidity or flexibility of the body and its shape are largely dependent on the nature of the **cytoskeleton** (*e.g.*, pellicle;) and **exoskeleton** (*e.g.*, test).
6. Animalcules may be **free-living, commensal, mutualistic** or **parasitic** (both ecto-parasitic and endoparasitic). The free-living protozoans are aquatic, inhabiting fresh and salt waters. These may be **solitary** or **colonial**; in colonial forms the individuals are alike and independent.
7. Body shape variable; it may be spherical, oval, elongated or flattened.
8. Cytoplasm is differentiated into an outer **ectoplasm** (or cell cortex) and inner **endoplasm.**
9. Protozoans may have one or more **nuclei** which may be **monomorphic** or **dimorphic**. In case of dimorphic nuclei, two basic types are recognized: 1. vesicular, containing considerable nucleoplasm and 2. compact, containing little nucleoplasm.
10. Three major types of locomotor organelles namely flagella, pseudopodia and cilia occur among protozoa. The classification of phylum Protozoa mainly depends on the type and distribution of these locomotory parts. **Pseudopodia** are semipermanent to transitory extensions of protoplasm of body surface and used by sarcodina. **Flagella** are permanent

or semipermanent locomotory organelles of mastigophora. **Cilia** are many and occur in ciliata. Some protozoans are **sessile** (attached or fixed) and do not move.

11. All types of **nutrition** occur in protozoa. Thus, nutrition may be **holozoic** (animal-like), **holophytic** (plant-like, photoautotrophy), **saprozoic** (absorb soluble organic nutrients from their environment) or **parasitic**.
12. Holozoic protozoans have two general kinds of feeding: **hunting**, which involves seeking and capturing prey; and **trapping** or **suspension feeding**, which is often used by sessile protozoans.
13. Digestion occurs intracellularly within a **food vacuole** by the help of enzymes of lysosomes, and food reaches the vacuole through a cell mouth (**cytostome**) or by engulfment by pseudopodia (endocytosis and phagocytosis).
14. Respiration, excretion and egestion of waste products occur through body surface. Protozoa are mostly **ammonotelic**, *i.e.*, they excrete their nitrogenous wastes in the form of ammonia.
15. Many protozoa osmoregulate to remove excess water (called **water volume regulation**) and to adjust the concentration and proportion of their internal ions (**ionic regulation**). Osmoregulation is accomplished by **contractile vacuoles**. There is no evidence that the contractile vacuoles have any specific excretory function.
16. **Cyclosis** or cytoplasmic streaming distributes materials throughout the continuous protoplasmic mass of the organism. For example, in ciliates cyclosis occurs due to movements of food vacuoles along microtubules extending from the cytostome.
17. Protozoa are sensitive to many kinds of stimuli including touch (**thigmotaxis**), temperature changes (**thermotaxis**), light (**phototaxis**), and many chemicals (**chemotaxis**).
18. Intercellular chemical signalling (**pheromones**) in protozoans involves signal molecules such as serotonin, β-endorphin, acetylcholine and cyclic-AMP (cAMP).
19. Reproduction is asexual or sexual; **asexual reproduction** by **binary fission, multiple fission** and **budding** or **sporulation** and **sexual reproduction** by gamete formation or conjugation of adults.
20. Resistant coverings are sometimes formed for brief parts of a life cycle. This is especially true of parasites, which usually pass from one host to another as cysts or spores, covered by a resistant membrane that protects them while out of the host. Many free-living protozoa also encyst to survive unfavourable environmental conditions and be dispersed by air or water currents in a protected stage.
21. In some protozoans, the life cycles are complicated and show **alternation of generation**, *i.e.*, life cycle includes asexual and sexual phases.

2.2 CLASSIFICATION

Outline Classification of Phylum Protozoa

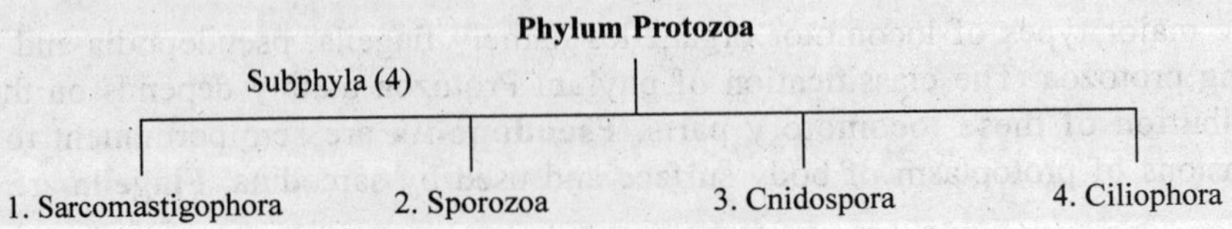

Subphylum
1. Sarcomastigophora

Superclasses (3)

1. Mastigophora

2. Opalinata
Example. *Opalina*

3. Sarcodina

Classes (2)

1. Phytomastigophorea

Orders (7)

1. Chrysomonadida
Examples. *Chromulina, Chrysamoeba*
2. Cryptomonadida
Examples. *Chilomonas, Cryptomonas*
3. Dinoflagellida
Examples. *Ceratium, Noctiluca*
4. Euglenida
Examples. *Euglena, Paranema*
5. Heterochlorida
Example. *Heterochloris*
6. Chloromonadida
Example. *Coelomonas*
7. Volvocida
Examples. *Volvox, Chlamydomonas*

2. Zoomastigophorea

Orders (8)

1. Choanoflagellida
Example. *Protospongia*
2. Rhizomastigida
Example. *Mastigoamoeba*
3. Kinetoplastida
Example. *Bodo, Trypanosoma, Leishmania*
4. Retortamonadida
Example. *Chilomastix*
5. Diplomonadida
Example. *Giardia*
6. Oxymonadida
Example. *Oxymonas*
7. Trichomonadida
Example. *Trichomonas*
8. Hypermastigida
Example. *Trichonympha*

Superclass 3. Sarcodina

Classes (3)

1. Rhizopodea

2. Piroplasmea
Example. *Babesia*

3. Actinopodea

Subclasses (4)

1. Lobosia
Orders (2)

1. Amoebida
Examples. *Amoeba, Entamoeba*
2. Arcellinida
Examples. *Arcella, Difflugia, Euglypha*

2. Filosia
Examples. *Allogromia, Penardia*

3. Granulo-reticulosia
Order (1)
1. Foraminiferida
Example. *Eliphidium* (= *Polystomella*)

4. Mycetozoia
Example. *Plasmodiophora*

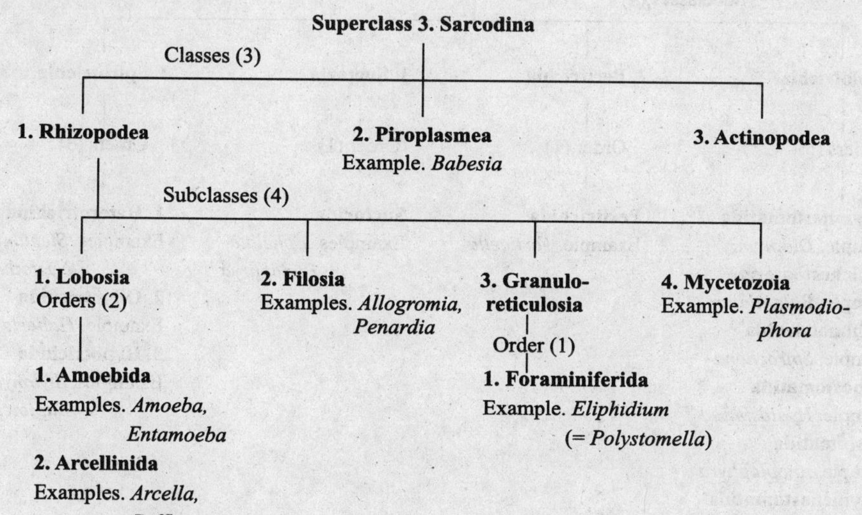

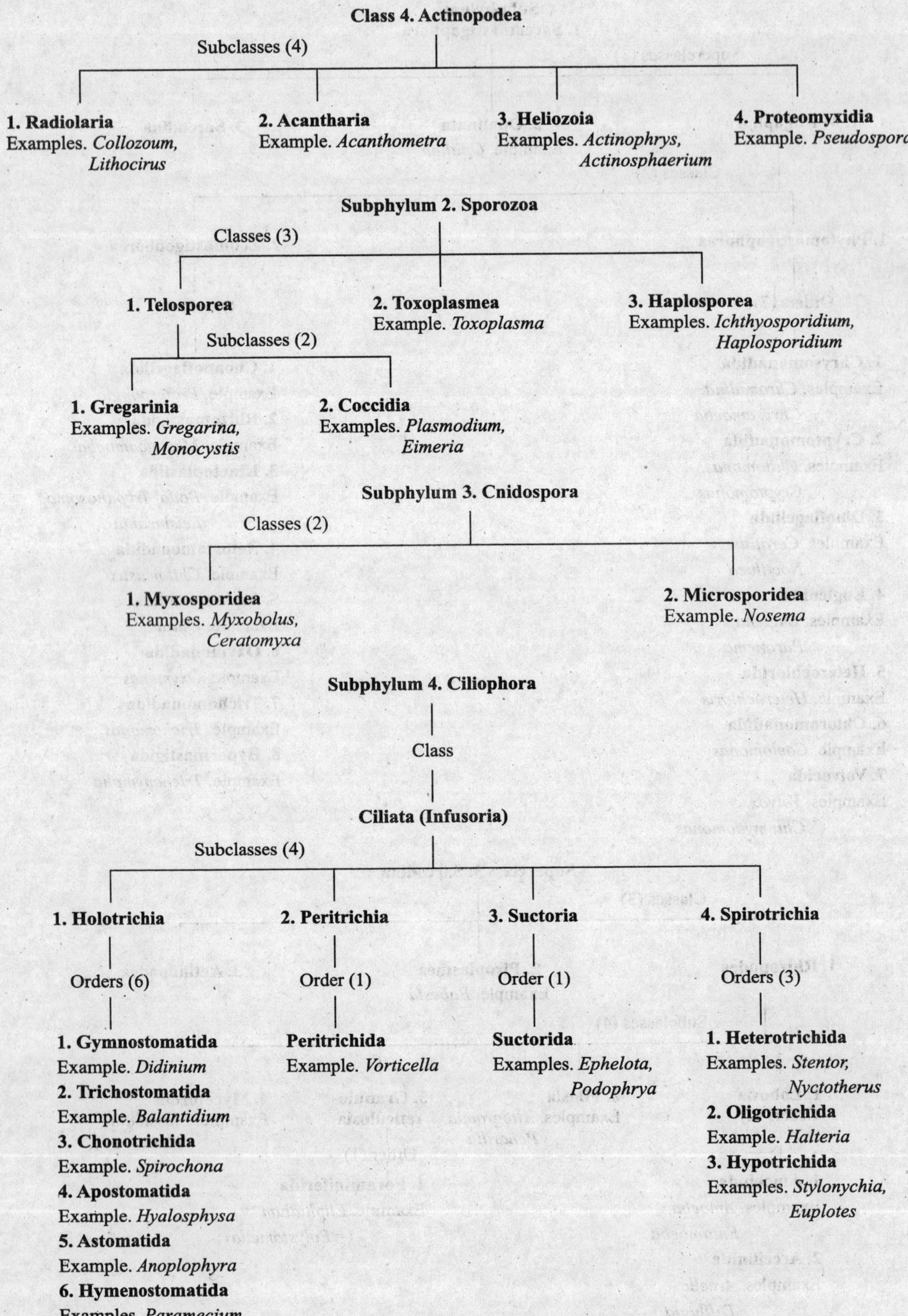

Class 4. Actinopodea
Subclasses (4)
1. Radiolaria
Examples. Collozoum, Lithocirus
2. Acantharia
Example. Acanthometra
3. Heliozoia
Examples. Actinophrys, Actinosphaerium
4. Proteomyxidia
Example. Pseudospora
Subphylum 2. Sporozoa
Classes (3)
1. Telosporea
Subclasses (2)
1. Gregarinia
Examples. Gregarina, Monocystis
2. Coccidia
Examples. Plasmodium, Eimeria
2. Toxoplasmea
Example. Toxoplasma
3. Haplosporea
Examples. Ichthyosporidium, Haplosporidium
Subphylum 3. Cnidospora
Classes (2)
1. Myxosporidea
Examples. Myxobolus, Ceratomyxa
2. Microsporidea
Example. Nosema
Subphylum 4. Ciliophora
Class
Ciliata (Infusoria)
Subclasses (4)
1. Holotrichia
Orders (6)
1. Gymnostomatida
Example. Didinium
2. Trichostomatida
Example. Balantidium
3. Chonotrichida
Example. Spirochona
4. Apostomatida
Example. Hyalosphysa
5. Astomatida
Example. Anoplophyra
6. Hymenostomatida
Examples. Paramecium, Tetrahymena
2. Peritrichia
Order (1)
Peritrichida
Example. Vorticella
3. Suctoria
Order (1)
Suctorida
Examples. Ephelota, Podophrya
4. Spirotrichia
Orders (3)
1. Heterotrichida
Examples. Stentor, Nyctotherus
2. Oligotrichida
Example. Halteria
3. Hypotrichida
Examples. Stylonychia, Euplotes

Detailed Classification

According to **Honigberg** *et al.*, (1964) phylum Protozoa has been classified into **four** subphyla: Sarcomastigophora, Sporozoa, Cnidospora and Ciliophora.

Subphylum 1. Sarcomastigophora

1. The locomotory organs are either flagella or pseudopodia or both.
2. Single type of nucleus (monomorphic).
3. Asexual reproduction by binary fission and multiple fission.
4. Sexual reproduction by syngamy.

Subphylum 2. Sporozoa

1. All are endoparasites.
2. Body covered with thick pellicle.
3. Locomotory organs are absent in adult. Flagella present only in microgametes of some group.
4. Nucleus is of the single type.
5. Saprozoic nutrition.
6. Asexual reproduction by multiple fission.
7. Sexual reproduction by syngamy followed by spore formation.
8. Spores contain one to many sporozoites which represent the infective phase.

Subphylum 3. Cnidospora

1. All are parasitic.
2. Spore formation occurs throughout life.
3. Spores have one to many cells, each with one or more **polar filaments** (which are coiled threads and can be shot out for attaching to the host) and one or more **sporoplasms** (analogous structures to sporozoites).
4. Without sporogony, zygote gives rise to one or more trophozoites, each with many nuclei.

Subphylum 4. Ciliophora

1. Most species are free-living, some are commensal or parasitic.
2. Body organization complex.
3. Presence of simple **cilia** or compound ciliary structures as locomotory or feeding organelles at some stage in the life cycle. Subpellicular cilia present even when cilia absent.
4. Nuclear dimorphism occurs, *i.e.*, they have two nuclei, a large vegetative or trophic **macronucleus** and a small generative or reproductive **micronucleus**.
5. Nutrition heterotrophic.
6. Contractile vacuole typically present.
7. Asexual reproduction by transverse binary fission; or by budding or multiple fission.
8. Sexual reproduction by conjugation.
9. Gametes are always absent.

Subphylum 1. Sarcomastigophora

It includes **three** superclasses:

Superclass 1. Mastigophora

1. They are commonly called **flagellates.**
2. Body is covered by pellicle.

3. Locomotary organs of adults are flagella.
4. Asexual reproduction by longitudinal binary fission.
5. Free living or parasitic.

Superclass 2. Opalinata

1. Entire body surface is covered by longitudinal, oblique rows of cilia. Infraciliature characteristic of the true ciliates is lacking.
2. Cytostome absent; saprozoic nutrition.
3. Two or many monomorphic nuclei.
4. Asexual reproduction by longitudinal binary fission.
5. Sexual reproduction involves syngamy with flagellated gametes.
6. Gut (*i.e.*, rectum) parasites/commensals of anurans (frogs, toads, tadpoles) and also of fishes, salamanders and reptiles.

Examples. *Opalina*, *Cepedea*, *Zelleriella*.

(**Note.** Superclass Opalinata includes single class **Opalinatea** and single order **Opalinida**, both having characters of superclass).

Superclass 3. Sarcodina (or Rhizopoda)

1. The amoeboid form is predominant.
2. Locomotion and feeding by pseudopodia, or by locomotive protoplasmic flow without distinct pseudopodia. Flagella when present, usually restricted to developmental stages of life cycle.
3. Pseudopodia may be lobopodia, filopodia or reticulopodia.
4. Body naked or with external or internal test or skeleton.
5. Nutrition holozoic and saprozoic.
6. Asexual reproduction by fission.
7. Sexual reproduction by flagellate (rarely amoeboid) gametes.
8. Most species are solitary and free-living.

Superclass 1. Mastigophora

It is divided into **two** classes:

Class I. Phytomastigophorea

1. Chlorophyll containing chromatophores (or chloroplasts) are present.
2. There are one or two flagella.
3. Nutrition mostly holophytic (by photosynthesis).
4. Reserve food is starch or paramylon.

Class II. Zoomastigophorea

1. Chloroplasts (chromatophores) are absent.
2. There are one to many flagella; often undulating membrane is present.
3. Nutrition holozoic or saprozoic.
4. Reserve food is glycogen.
5. Parasitic, commensal or symbiotic.

Class I. Phytomastigophorea

It is divided into following **seven** orders:

Order 1. Chrysomonadida

1. Two unequal flagella; one hairy and directed anteriorly, other trailing and smooth.
2. Chloroplasts golden brown containing chlorophyll a and c. Some species lack chloroplasts and feed holozoically.
3. Starch absent. Storage products are **chrysolaminarin** and fat.
4. Cells are either naked or with richly patterned silicified scales (cysts) or with plate-like lorica.
5. Sexual reproduction present.
6. Marine or freshwater animals.

Examples. *Chromulina*, *Chrysamoeba*, *Ochromonas*.

Order 2. Cryptomonadida

1. Cells flattened, naked, without wall or pellicle.
2. They have a gullet that reaches upto the middle of the body.
3. They have two unequal flagella.
4. Chloroplasts brown, red, olive-green, blue or yellow.
5. Reserve food includes starch and fat.
6. Sexual reproduction unknown.
7. Stigma present.
8. Marine or freshwater.

Examples. *Chilomonas*, *Cryptomonas*.

Order 3. Dinoflagellida

1. Cells flattened or of complex symmetry with transverse and ventral grooves and with armour of cellulose plates. Cytoplasm is vacuolated.
2. Two heterodynamic flagella, inserted apically or laterally, one ribbon-shaped with paraxial rod and single row of fine hairs, other smooth or with two rows of stiffer hairs.
3. Chloroplasts typically golden brown (xanthophyll) and green (chlorophyll a and c present).
4. Reserve food includes starch and fat.
5. Nucleus unique among eukaryotes in having permanently condensed chromosomes having less amount of histone proteins.
6. Mitosis intranuclear; asexual reproduction by fission.
7. Sexual reproduction present.
8. Stigma is present in many species.
9. Free living, pelagic, marine, freshwater; some parasitic, some mutualistic (*e.g.*, *Symbiodinium microadriaticum* is symbiont of corals).
10. Nutrition holophytic in most species, holozoic in some.
11. Some are bioluminscent (*e.g.*, *Noctiluca*).

Examples. *Ceratium, Noctiluca, Glenodinium, Gymnodinium*.

Order 4. Euglenida

1. Cells with helical symmetry, naked but with complex and thick pellicle of interlocking proteinaceous strips.

2. Two (rarely more) unequal flagella, one or both arising from an anterior invagination (**reservoir**). The reservoir contains the roots of flagella and receives the contents of the contractile vacuole. It leads to so-called gullet or cytopharynx which opens to outside as mouth (cytostome). Mouth is not used in ingestion of food.
3. One flagellum bears mastigonemes distally.
4. Stigma is present.
5. Chloroplasts grass-green, but absent in many genera.
6. Reserve food is paramylon (secreted by pyrenoids of chloroplasts) in *Euglena*, fat and cyclic metaphosphates.
7. Non-spindle intranuclear mitosis.
8. Solitary or colonial, most freshwater.

Examples. *Euglena, Peranema, Phacus.*

Order 5. Heterochlorida

1. Two unequal flagella.
2. Chloroplasts yellow-green.
3. Reserve food is oil and possibly chrysolaminarin.
4. Supposedly related to xanthophyceae algae.
5. Siliceous cysts.

Examples. *Heterochloris, Myxochloris.*

Order 6. Chloromonadida

1. Dorso-ventrally flattened forms with delicate pellicle.
2. Gullet present, stigma absent.
3. Two flagella.
4. Chloroplasts are green and numerous.
5. Characteristic ring of Golgi apparatuses at anterior end.
6. Sexual reproduction by fusion of two individuals.

Examples. *Coelomonas, Vacularia.*

Order 7. Volvocida

1. Small sized cells with rigid cellulose covering.
2. Mouth and gullet absent.
3. Two or four equal, smooth, apical flagella.
4. Chloroplasts (= chromatophores) are grass-green.
5. Reserve food is starch and fat.
6. Stigma present.
7. Sexual reproduction by syngamy.
8. Mostly freshwater; some form colonies.

Examples. *Volvox, Chlamydomonas, Eudorina.*

Class II. Zoomastigophorea

It is divided into following **eight** orders:

Order 1. Choanoflagellida

1. A collar (or ring of microvilli) at the base of a single apical flagellum.
2. Cell with membraneous sheath or basket-like siliceous lorica.
3. Nutrition holozoic.

4. They are solitary or colonial and are free-living or stalked.

Example. *Proterospongia.*

Order 2. Rhizomastigida

1. Small and amoeboid.
2. Flagella one to four.
3. Locomotion by flagella and pseudopodia.
4. Freshwater and free-living.

Examples. *Mastigamoeba, Dimorpha.*

Order 3. Kinetoplastida

1. One or two flagella arising from depression.
2. Single mitochondrion extending entire length of body as single tube, hoop or network of branching tubes.
3. Gullet is absent.
4. They contain a single large **kinetoplast** which is genetically autonomous, DNA-containing, membrane-bound organelle (mitochondria) associated with the basal body at the base of flagella.
5. Most species are parasitic; some are free-living; freshwater forms are sessile.
6. Solitary or colonial.

Examples. *Oikomonas, Bodo, Trypanosoma, Leishmania.*

Order 4. Retortamonadida

1. Two or four flagella, one turned posteriorly and associated with ventrally located cytostome.
2. Mitochondria and Golgi apparatus absent.
3. Cysts present.
4. Gut parasites of insects and vertebrates.

Example. *Chilomastix.*

Order 5. Diplomonadida

1. Flagella 3 to 8, one of them directed backwards over the body and associated with cytostome or with organelles forming cell axis.
2. Mitochondria and Golgi apparatus absent.
3. One or two karyomastigont (nucleus + mastigont) are present. **Mastigont system** is a complex system formed by groups of flagella which are associated with several microtubular and fibrillar organelles.
4. Cysts present.
5. Free-living or parasitic.

Examples. *Giardia*, *Hexamita*.

Order 6. Oxymonadida

1. Uninucleate or multinucleate forms.
2. One or more karyomastigonts are present, each containing four flagella typically arranged in two pairs in motile stage.
3. Mitochondria and Golgi apparatus absent.
4. Cysts in some species.
5. Sexual reproduction in some species.
6. Parasitic.

Examples. *Foaina, Oxymonas.*

Order 7. Trichomonadida

1. Typically karyomastigonts with four to six flagella, but one genus only with one flagellum and another with no flagella at all.
2. One flagellum recurrent, free, or with proximal or entire length adherent to body surface as undulating membrane.
3. They have a longitudinal supporting structure known as an **axostyle.**
4. No cysts are formed.
5. Parasitic in humans and other animals.

Examples. *Trichomonas tenax* (mouth); *T. hominis* (intestine), *T. vaginalis* (vagina); *Chilomastix mesnili* (large intestine).

Order 8. Hypermastigida

1. Complex forms with numerous flagella.
2. Flagella bearing kinetosomes distributed in circle, in plates or in longitudinal or spiral rows meeting in a centralized structure.
3. One nucleus per cell.
4. Cysts in some forms.
5. Some species have sexual reproduction.
6. Gut parasites of termites and cockroaches.

Examples. *Trichonympha campanula* (termite gut), *Lophomonas* (cockroach intestine).

Superclass 3. Sarcodina

It includes **three** classes:

Class 1. Rhizopodea

1. Locomotory organelles are lobopodia or filopodia but never axopodia.
2. They are creeping forms.

Class 2. Piroplasmea

1. Small, round rod-shaped or amoeboid organisms.
2. Parasites in red blood cells of vertebrates.
3. Spores are not formed.

Example. *Babesia.*

(**Note.** Class Piroplasmea is not divided into orders).

Class 3. Actinopodea

1. Pseudopodia are ray-like axopodia. Each of the pseudopodia is supported by a central axial rod comprising a bundle of microtubules. Axopods are commonly used in food capturing.
2. Primarily sessile or floating (planktonic) spherical forms.
3. Test is present or absent. Test is composed of organic matter and/or silica or strontium sulphate.
4. Both asexual and sexual reproduction present.

Class 1. Rhizopodea

This class is divided into **four** subclasses:

Sub-class I. Lobosia

1. Locomotory organelles are lobopodia which are broad and blunt pseudopodia containing both ectoplasm and endoplasm.

2. Usually uninucleate forms.
3. Sporangia absent.

This sub-class is divided into **two** orders:

Order 1. Amoebida

1. Typically uninucleate amoeboid forms.
2. Test (skeleton) absent.
3. Ectoplasm and endoplasm distinct.
4. Mitochondria typically present.
5. No flagellate stage in life cycle.
6. Largely freshwater, some parasitic.

Examples. *Amoeba*, *Entamoeba*.

Order 2. Arcellinida

1. The body is encased in an one-chambered test having a single aperture through which animal extends or retracts its lobopodia.
2. Test is composed of either organic material (*e.g.*, *Arcella*) or inorganic foreign particles (*e.g.*, *Difflugia*) or both.
3. Free-living and mostly freshwater.

Examples. *Arcella*, *Difflugia*, *Euglypha*.

Sub-class II. Filosia

1. Pseudopodia are filopodia which are tapering and branching.
2. Body naked or with a shell with single aperture.
3. Ectoplasm is not distinct.
4. No flagellate or spores known.
5. Marine or freshwater.

Examples. *Allogromia*, *Penardia*, *Gromia*.

Sub-class III. Granuloreticulosia

1. Pseudopodia are reticulopodia which are delicate and finely granular.

This sub-class includes the following significant order:

Order Foraminiferida

1. Large sized protozoans with uni-or multichambered perforated calcareous test or shell.
2. Reticulopodia protruding from test's aperture, wall perforation or both.
3. Reproduction with alternation of sexual and asexual generations.
4. Gametes usually flagellate, rarely amoeboid.
5. Mostly marine.

Examples. *Globigerina*, *Elphidium* (= *Polystomella*), *Pavonina*.

Sub-class IV. Mycetozoia

1. Body large, amoeboid and multinucleate.
2. Pseudopodia numerous and blunt.
3. Form sporangia which liberate spores.
4. Nutrition phagocytic.

Examples. *Plasmodiophora* and other slime molds.

Class 3. Actinopodea

It is divided into **four** subclasses:

Subclass I. Radiolaria

1. A perforated central spherical chitinoid capsule separates the protoplasm into inner endoplasm and outer vacuolated ectoplasm.
2. Skeleton of siliceous spicules.
3. Pseudopodia are axopodia.
4. Contractile vacuole absent.
5. Marine.

Examples. *Collozoum*, *Thalassicola*, *Lithocircus*.

Subclass II. Acantharia

1. Strontium sulphate skeleton, usually composed of 20 radial or 10 dimetrical spines.
2. Spines more or less joined in cell centre.
3. Extracellular outer (cortex) and inner envelopes usually present. Inner envelope is called **capsular membrane** closely lining cell mass.
4. Locomotion by axopodia.
5. Marine, mostly planktonic.

Example. *Acanthometra*.

Subclass III. Helizoia

1. Spherical protozoans without central capsule.
2. Skeletal structures, if present, siliceous and organic.
3. Body of most species is naked and differentiated into outer vacuolated ectoplasm and inner dense endoplasm.
4. Axopodia radiate on all sides.
5. There may be more than one nucleus.
6. Most species freshwater, some marine.

Examples. *Actinophrys* (uninucleate), *Actinosphaerium* (multinucleate), *Clathrulina* (with perforated test).

Subclass IV. Proteomyxidia

1. Pseudopodia are filopodia which are slender with pointed tips and are composed of ectoplasm only.
2. Reproduction by binary or multiple fission.
3. Mostly parasite of algae and higher plants.
4. Marine or freshwater.

Examples. *Pseudospora*, *Vampyrella*.

Subphylum 2. Sporozoa

The subphylum sporozoa is divided into **three** classes:

Class I. Telosporea

1. Spores without polar capsules or filaments.
2. Sporozoites elongated, microgametes flagellated.
3. Trophozoites with one nucleus only.
4. Pseudopodia are absent and mature organism do locomotion by body flexion or gliding.
5. Reproduction is both asexual and sexual.

Class II. Toxoplasmea

1. Spores are not formed.
2. There are no flagella or pseudopodia at any stage.
3. Only asexual reproduction, *i.e.*, binary fission.
4. Cysts containing many naked sporozoites.
5. They infect birds and mammals causing mental retardation in humans.

Examples. *Toxoplasma*, *Sarcocystis*.

Class III. Haplosporea

1. Spores are present and have spore cases.
2. Pseudopodia may be present but flagella are absent.
3. Reproduce only asexually by schizogony (*i.e.*, multiple fission of a trophozoite).
4. Parasites of fish and invertebrates.

Examples. *Ichthyosporidium*, *Haplosporidium*, *Caelosporidium*.

Class I. Telosporea

It is subdivided into **two** subclasses:

Subclass 1. Gregarinia

1. Mature trophozoites (gamonts) are large and extracellular living as parasites in the gut and body cavity of invertebrates or lower chordates such as annelids, sipunculids, arthropods, hemichordates or ascidians.
2. Reproduction is entirely sexual with sporogony. Life cycle consisting of gametogony and sporogony.
3. Gametes similar (isogamous).
4. Gametocyte show syzygy (end-to-end union of sporonts).
5. Zygote non-motile forming oocysts within gametocytes.
6. Spores contain eight sporozoites.

Examples. *Gregarina*, *Monocystis*, *Nematocystis*, *Lankestrella*.

Subclass 2. Coccidia

1. Mature trophozoites (gamonts) are small and intracellular.
2. Gametocytes are dimorphic.
3. Sporozoites multiply by schizogony in tissue cells.
4. Life cycle characteristically consisting of merogony, gametogony and sporogony.
5. Blood or gut parasites of vertebrates.

Examples. *Eimeria*, *Isospora* and *Plasmodium*.

Subphylum 3. Cnidospora

It is divided into **two** classes:

Class 1. Myxosporidea

1. Spores of multicellular origin.
2. Spores with two or more valves in their membranes.
3. Trophozoite stage well developed and is the main site of proliferation.
4. They are parasites of cold-blooded vertebrates such as fish.

Examples. *Myxobolus*, *Ceratomyxa*.

Class 2. Microsporidea

1. Spores of unicellular origin.
2. Spores with imperforate wall, having only valve.
3. There is one long tubular polar filament through which the sporoplasm emerges.
4. Mitochondria absent.
5. They are intracellular (**cytozoic**) parasites in arthropods (*e.g.*, honey bees, silk worms) and vertebrates (*e.g.*, fish).

Example. *Nosema.*

Subphylum 4. Ciliophora

The subphylum has the following class:

Class Ciliata

1. Most species are free-living and abundant in both marine and freshwater habitats.
2. Ciliate protozoans have a definite body form and size.
3. Body bounded externally by a firm and elastic pellicle.
4. Locomotory and feeding organelles are cilia.
5. Definite mouth (cytostome) and gullet present. Anal aperture (cytopyge) permanent.
6. Contractile vacuoles are always present at fixed positions.
7. Have two types of nuclei: larger vegetative **macronucleus** and small reproductive **micronucleus**.
8. Binary fission transverse.
9. Sexual reproduction never involves the formation of gametes.

The class Ciliata (or Infusoria) is divided into **four** subclasses:

Subclass 1. Holotrichia

1. Body cilia uniform (*i.e.*, no distinction between oral cilia and somatic cilia).
2. Cytostome often apical (or subapical) or midventral, on surface of body or at bottom of atrium or vestibulum.

Subclass 2. Peritrichia

1. Oral ciliary field prominent; body or somatic cilia absent or reduced to temporary posterior circlet of locomotor cilia.
2. Widely distributed species; many are stalked and sedentary (sessile), others are mobile.
3. Dispersal by migratory larval form called **telotroch**.
4. Conjugation total, involving fusion of microconjugants and macroconjugants.

Subclass 3. Suctoria

1. Sessile (sedentary), stalked adult ciliates with the distal end bearing few to many suctorial tentacles (each ending into a sucker-like knob).
2. Adult without cilia; migratory young larva with cilia which originate by budding from parent suctorian.

Subclass 4. Spirotrichia

1. Larged-sized free-swimming ciliates.
2. Body (somatic) cilia reduced.
3. Oral or buccal cilia well marked.
4. Cysts common in some groups.

Subclass 1. Holotrichia

This subclass is divided into following **six** orders:

Order 1. Gymnostomatida

1. Buccal (or oral) cilia absent; body or somatic ciliation uniform.
2. Cytostome opens directly to outside, vestibule (buccal cavity) absent.
3. Toxicysts (a type of trichocyst which induce paralysis or lysis of the prey) common.

Examples. *Didinium*, *Coleps*, *Nassula*.

Order 2. Trichostomatida

1. Cytostome at the bottom of a ciliated vestibule.
2. No buccal cilia; spiral rows of cilia present in the vestibule.
3. Endocommensal of vertebrate hosts such as cattle, pig, monkeys and human beings.

Examples. *Balantidium*, *Colpoda*, *Isotricha*.

Order 3. Chonotrichida

1. Vase-shaped ciliates lacking in body cilia.
2. Sessile (sedentary) forms.
3. Vestibular cilia form a spirally coiled apical funnel.
4. Marine and ectocommensal on crustaceans.

Examples. *Spirochona*, *Lobochona*, *Chilodochona*.

Order 4. Apostomatida

1. Body with spirally arranged cilia.
2. Cytostome mid-ventral, rosettle-shaped in *Foettingeria actinarium*.
3. Marine and parasites or commensals.
4. Life cycle complex, sometimes involving alternation of hosts (unique in phylum), one host is commonly a crustacean.

Examples. *Hyalosphysa*, *Polyspira*, *Foettingeria*.

Order 5. Astomatida

1. Body large or long.
2. Mouth absent.
3. Body ciliation uniform.
4. Complex infraciliary endoskeleton and often with elaborate holdfast, organelles (hooks, spines or sucker) at anterior end.
5. Intestinal endoparasites of oligochaetes (terrestrial, freshwater or marine).
6. Asexual reproduction by fission (budding).

Examples. *Anoplophrya*, *Hoplitophrya*.

Order 6. Hymenostomatida

1. Body small and with uniform somatic cilia.
2. Buccal cavity well defined; buccal ciliature consists of an undulating membrane and an adoral zone of membranelles.
3. Oral area on ventral surface, usually in anterior half of body.

Examples. *Tetrahymena*, *Paramecium*, *Colpidium*.

Subclass 2. Peritrichia

Order Peritrichida

1. Mostly attached stalked bell-shaped ciliates; the stalk is secreted by a region at the aboral end known as the **scopula**.
2. Feeding apparatus is highly specialised.

Examples. *Vorticella*, *Trichodina*.

Subclass 3. Suctoria

1. Body sessile and stalked.
2. Young with cilia, adult with tentacles.

Order Suctorida

Characters as of subclass suctoria.

Examples. *Ephelota*, *Podophrya*, *Acineta*.

Subclass 4. Spirotrichia

It is divided into **three** orders:

Order 1. Heterotrichida

1. Generally large to very large forms, often 'highly contractile, sometimes pigmented (*e.g.*, bluish *Stentor*).
2. Body encased in lorica.
3. Body cilia usually absent.
4. Macronucleus oval or often beaded.
5. Parasitic and free-living species.

Example. *Stentor*, *Bursaria*, *Blepharisma*, *Nyctotherus*.

Order 2. Oligotrichida

1. Body ovoid to elongate, sometimes with tail.
2. Pellicle thickened; body (=somatic) cilia reduced.
3. Oral (=buccal) membranelles are limited to front end.
4. Free-swimming, macrophagous, mainly pelagic.

Example. *Halteria*.

Order 3. Hypotrichida

1. Dorsoventrally flattened, highly mobile, with unique cursorial (or running) type of locomotion.
2. Body dominated by compound ciliary structures such as **cirri** which are located on the ventral side of body and used as "legs" for creeping over the substratum.

Examples. *Stylonychia*, *Euplotes*, *Oxytricha*.

2.3 RECENT CLASSIFICATION SCHEME OF PROTISTA

Brusca and **Brusca** (2003) noted in his book Invertebrates that "Ideas about protist phylogeny and classification are in such a state of flux that any scheme used at the time of this writing would be out-of-date by the time this text reached the book shelves". Thus, on the basis of modes of nutrition and locomotion, protists were divided into the *Mastigophora* (locomotion with flagella), *Ciliophora* (locomotion with cilia), *Sarcodina* (locomotion with pseudopodia), and *Sporozoa* (parasites with no obvious locomotory structures). While these divisions might accurately describe protist's roles in ecosystems, *they do not accurately reflect evolutionary relatedness*. Pseudopodia and flagella are present in many different kinds of cells (including plant and animal cells) and their presence does not indicate unique relatedness (*i.e.*, they are shared primitive features, or

symplesiomorphies). Photosynthetic protists contain many different types of chlorophylls and have differently constructed chloroplasts, indicating that they are not closely related. Accordingly most recent protist classification has included following seventeen phyla (see **Brusca** and **Brusca** 2003). [**Note.** Former protist group Myxozoa or Myxosporidea (*e.g.*, *Myxobolus cerebralis*) is found to include highly modified cnidarians, parasitic in certain invertebrates and vertebrates. This revelation was made possible by discovery of certain metazoans and cnidarian features (*e.g.*, collagen, nematocysts) in these animals as well as ribosomal DNA data (Box 2.1)].

Box 2.1

Myxozoan cnidarian (*e.g., Myxobolus cerebralis*, Fig. 2.1) is a parasite of fish (primary host) and of *Tubifix* worm (Intermediate host). Previously it was considered to be protist. *M. cerebralis* and similar other cnidarians are now viewed as highly specialized parasitic cnidarians (Brusca and Brusca, 2003).

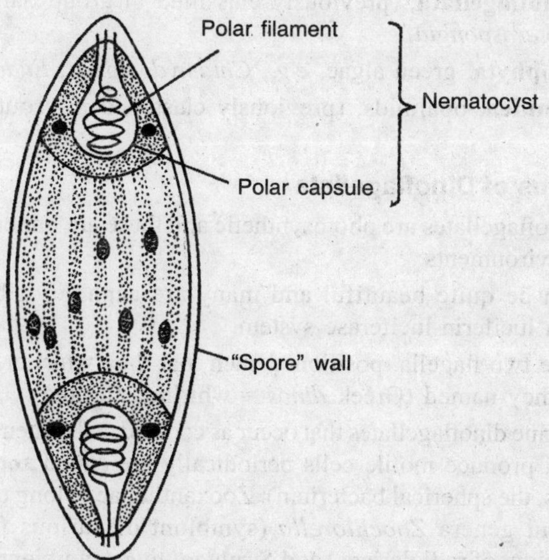

Fig. 2.1. Previously considered to be protists, myxozoans are now viewed as highly specialized, parasitic cnidarians (after Bursca and Bursca, 2003).

Current Phyla of Kingdom Protista

1. Phylum Euglenida, (previously classified in group Sarcomastigophora), *e.g.*, *Ascoglena*, *Colacium*, *Euglena*.

2. Phylum Kinetoplastida, trypanosomes (previously classified in group Sarcomastigophora) *e.g.*, *Bodo*, *Leishmania*, *Trypanosoma*.

3. Phylum Ciliophora, ciliates, *e.g.*, *Balantidium*, *Euplotes*, *Paramecium*, *Stentor*, *Vorticella*.

4. Phylum Apicomplexa (old name Sporozoa), *e.g.*, *Eimeria*, *Gregarina*, *Plasmodium*, *Toxoplasma*.

5. Phylum Dinoflagellata (previously classified in group Sarcomastigophora), *e.g.*, *Ceratium*, *Noctiluca*, *Zooxanthella*.

6. Phylum Stramenopila, *e.g.*, diatoms, brown algae, golden algae, oomycetes.

7. Phylum Rhizopoda, amoebas (previously classified in group Sarcomastigophora), *e.g.*, *Amoeba*, *Arcella*, *Entamoeba*.

8. Phylum Actinopoda, radiolarians, heliozoans, (previously classified in group Sarcomastigophora), *e.g.*, *Actinophrys*, *Actinosphaerium*, *Litocolla.*

9. Phylum Granuloreticulosa, foraminiferans, (previously classified in group Sarcomastigophora), *e.g.*, *Elphidium*, *Gromia*, *Iridia.*

10. Phylum Diplomonadida, (previously classified in group Sarcomastigophora), *e.g.*, *Giardia*, *Hexamida.*

11. Phylum Parabasilida, (previously classified in the group Sarcomastigophora), *e.g.*, *Dientamoeba*, *Trichomonas*, *Trichonympha.*

12. Phylum Cryptomonada, (previously classified in group Sarcomastigophora), *e.g.*, *Chilomonas.*

13. Phylum Microspora, microsporans, *e.g.*, *Nosema.*

14. Phylum Ascetospora, ascetosporans, *e.g.*, *Haplosporidium.*

15. Phylum Choanoflagellata, (previously classified in group Sarcomastigophora), *e.g.*, *Codosiga*, *Monosiga*, *Proterospongia.*

16. Phylum Chlorophyta, green algae, *e.g.*, *Chlamydomonas*, *Eudorina*, *Volvox.*

17. Phylum Opalinidea, opalinids, (previously classified in group Sarcomastigophora), *e.g.*, *Opalina.*

Certain Unique Features of Dinoflagellata

1. Many species of dinoflagellates are photosynthetic and these are important primary producers in many aquatic environments.
2. Dinoflagellates can be quite beautiful and many are capable of bioluminescence (*e.g.*, *Gonyaulax*) using a luciferin-luciferase system.
3. Dinoflagellates have two flagella, positioned such that they whirl or spin as they swim, the attribute to which they named (Greek *dinos* = whirling, turning).
4. **Endosymbiotic** marine dinoflagellates that occur as coccoid cells when inside their invertebrate or protist hosts, but produce motile cells periodically are called **zooxanthellae** (*coccoid* = resembling a coccus, the spherical bacterium). Zooxanthellae belong to the poorly understood but vastly important genera *Zoochlorella* (symbiont of various freshwater organisms), *Zooxanthella* (symbiont of radiolarians) and *Symbiodinium* (symbiont of cnidarians and some other metazoans). Species of *Symbiodinium* (= *Zooxanthella*) are best known as critically important mutualistic symbionts of hermatypic corals (reef-builder corals of Medreporaria and Scleractina). There are several species of *Symbiodinium* in corals. All are photosynthetic and provide nutrients to the corals and help create the internal chemical environment necessary for coral to secrete its calcium carbonate skeleton. Zooxanthellae also occur in many cnidarians other than scleractinian corals, such as milleporinids, chondrophorans, sea anemones and various medusae.
5. Some planktonic dinoflagellates occasionally undergo periodic bursts of population growth and are responsible for a phenomenon known as **red tide**. Red tides have nothing to do with actual tides, and they are only rarely red. A red tide is simply a streak or patch of ocean water discoloured (generally a pinkish orange) by the presence of billions of dinoflagellates. During a red tide, densities of these dinoflagellates may be as high as 10 to 100 million cell per liter of sea water. Exactly why the population explosions of these specific organisms occur is not entirely clear, but **organic pollutants** from terrestrial runoff may be a key culprit.

   Toxic effects of red tides. Many red tide organisms manufacture highly toxic substances. One group of toxins produced by dinoflagellate species such as *Alexandrium* spp., *Gymnodinium* *catenatum* and *Pyrodinium behamense* is called **saxitoxins**. Saxitoxins block the sodium potassium pump of nerve cells and prevent normal impulse transmission. When suspension feeders such as mussels and clams eat these protists, they store the toxins in their bodies. Extremely high concentrations of toxic dinoflagellates will even kill suspension feeders and occasionally also fish caught in the thick of the bloom. The shell fish feeding on the dinoflagellates become toxic to animals that eat them. In humans, the result is a disease known as **paralytic shellfish poisoning (PSP)**. Extreme cases of PSP result in muscular paralysis and respiratory failure. Over 300 human deaths worldwide have been documented (see **Brusca** and **Brusca**, 2003).

Gymnodinium breve releases a family of toxins called **brevetoxins** that result in **neurotoxic shellfish poisoning (NSP).** Red tides formed by *G. breve* have produced massive fish kills in Florida and along Gulf of Mexico in the recent years. Humans consuming animals that have brevetoxins accumulated in their tissues experience uncomfortable gastrointestinal side effects such as diarrhoea, vomiting and abdominal pain and also neurological problems including dizziness and reversal of temperature sensation.

A newly discovered species of dinoflagellate, *Pfiesteria piscicida* have caused in 1997 massive fish kills on the eastern shore of Maryland. The toxin in *Pfiesteria* is 1,000 times more powerful than cyanide. Outbreaks of *Pfiesteria* are thought to be linked to faecal material and urine of hogs of hog-farms which are dumped in the sea in North Carolina sea coasts (see **Brusca** and **Brusca**, 2003).

6. The shape of dinoflagellates is maintained by alveoli beneath cell's surface, and a layer of supporting microtubules. In some, the alveoli are filled with polysaccharides, typically cellulose, and these dinoflagellates are said to be **thecate** or **armored** (*e.g.*, *Ceratium*). Dinoflagellates that have empty alveoli are said to be **athecate** or **naked** (*e.g.*, *Noctiluca*).
7. In dinoflagellates, osmoregulation is performed by a unique system of double-membrane-bound tubules, called **pusules**, which open to the outside via a canal.
8. The chloroplasts of photosynthetic dinoflagellates are surrounded by three membranes, and thylakoids are arranged in stacks of three. Photosynthetic pigments include chlorophylls a and c_2, phycobilins, carotenoids (*e.g.*, β carotene) and also the xsnthophylls (*e.g.*, peridinin, found only in dinoflagellates), neoperidinin, dinoxanthin, and neodinoxanthin. These xanthophylls mask the chlorophyll pigments and account for the golden or brown colour that is commonly seen in dinoflagellates.
9. A few dinoflagellates (*e.g.*, *Noctiluca*) have a permanent cell mouth or **cytostome** supported by sheets of microtubules. The cytosome is often surrounded by **extrusomes** which are of three types: trichocysts, mucocysts and nematocysts. The most common are **trichocysts** (similar to those found in ciliates), believed to be fired in defense or to capture and bind prey. Sac-like **mucocysts** secrete mucous material on to the surface of cell. This may help in attachment to substratum (*e.g.*, *Amphidinium*) or may help to capture prey (*e.g.*, *Noctiluca*). Other dinoflagellates (*e.g.*, *Nematodinium*, *Nematopsides*, *Polykrikos*) have nematocysts that resemble, but not homologous to the stinging organelles of cnidarians of the same name.
10. The nuclei of dinoflagellates have three unusual features: (*i*) they contain five to ten times the amount of DNA that is found in most eukaryotic cells; (*ii*) the five histone proteins that are typically associated with the DNA of other eukaryotic cells are absent; and (*iii*) the chromosomes of dinoflagellates remain condensed and (*iv*) the nucleolus remains intact during interphase and mitosis. Most dinoflagellates (except *Noctiluca*) spend much of their lives as haploid cells (called **vegetative cells** to distinguish them from haploid gametes).

11. Nuclear division is by closed extranuclear pleuromitosis. No centrioles are present, and the organization center for the mitotic spindle is not obvious. Meiosis is of unusual type. Both asexual and sexual reproduction occur in the dinoflagellates.

Thus, dinoflagellates are unlike all other eukaryotes in many respects.

2.4 TYPE STUDIES

1. *Trypanosoma*

Classification

Phylum	–	**Protozoa**
Subphylum	–	**Sarcomastigophora**
Superclass	–	**Mastigophora (Flagellata)**
Class	–	**Zoomastigophora**
Order	–	**Kinetoplastida**
Genus	–	***Trypanosoma***

African sleeping sickness is a fatal human disease that is caused by protozoan parasite *Trypanosoma gambiense* (Fig. 2.2; Box. 2.2).

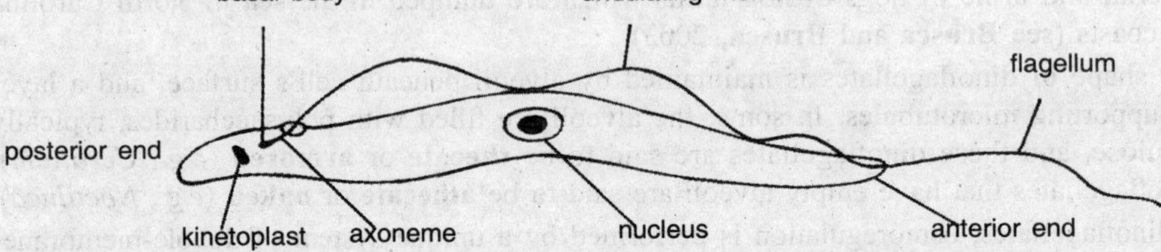

Fig. 2.2. *Trypanosoma*. Morphology of trypomastigote form.

T. gambiense is found in central Africa, West Africa (from Gambia to the Congo) and Uganda. Commonly low marshy lands near the rivers and lakes have the greatest incidence of infection because of the blood sucking insect vector, the tsetse fly (*Glossina palpalis*) inhabit these areas.

Box 2.2

The genus *Trypansoma* is parasitic in the blood of most of the vertebrates such as fishes, amphibians, reptiles, birds and mammals. Many species of *Trypanosoma* are **pathogenic** causing human disease called **trypanosomiasis**. Only three species of it are pathogenic in human beings, *viz.*, *T. gambiense, T. rhodesiense* and *T. cruzi. T. gambiense* and *T. rhodesiense* cause a fatal disease in human beings (hosts) called **sleeping sickness** in Africa, while *T. cruzi* causes **chagas disease** in children of South America. Their transmission from one vertebrate host to other takes place by invertebrate blood-sucking animals such as insects and leeches. These animals are called **vectors**.

Surra disease of horses and cattles is caused by *Trypanosoma evansi* and is transmitted by blood-sucking flies, celled *Tabanus* and *Stomoxys*.

Habit and Habitat

Trypanosoma gambiense lives as a digenetic endoparasite in the blood, lymph, lymph nodes, spleen or cerebrospinal fluid of human beings and in the intestine of blood-sucking insect, *Glossina palpalis*.

Morphology

T. gambiense is a unicellular, microscopic, slender, elongated, colourless, leaf-like protozoan (Fig. 2.3). Its body measures about 10 μm to 40μm in length and 1μm to 5 μm in breadth. The anterior end of body is pointed and the posterior end is blunt.

The body is covered by tough but elastic **pellicle** which is composed of numerous microtubules. Pellicle is followed by cytoplasm which is differentiated into an outer clear layer called **ectoplasm** and an inner granular **endoplasm**. The endoplasm contains single nucleus, single mitochondrion, single Golgi apparatus, endoplasmic reticulum and ribosomes. It lacks contractile vacuole and food vacuoles, however, contains many scattered greenish refractile bodies, called **volutin granules**. These granules contain glycogen and phosphates and represent reserve food material of *Trypanosoma*.

Nucleus is single, oval, vesicular and is located in the middle of its body. It contains a large **endosome, karyosome** or **nucleolus**. As nucleus is concerned with the nutritive functions of the parasite, it is often called **trophonucleus. Mitochondrion** is single, elongated, giant and extends from anterior to the posterior ends of the body. It is called **chondriome**. A conspicuous mass of DNA, called **kinetoplast**, is embedded in the mitochondrion just near the gullet.

A small pocket-like depression on one side near the posterior end represents gullet and reservoir. *Trypanosoma* being **monoflagellate** contains single long flagellum which originates from a **basal body** which is located close to the gullet near posterior end of body. It consists of an **axoneme** comprising of microtubules in 9 + 2 arrangement and bounded

Fig. 2.3. *Trypanosoma gambiense*. Details of cellular structure.

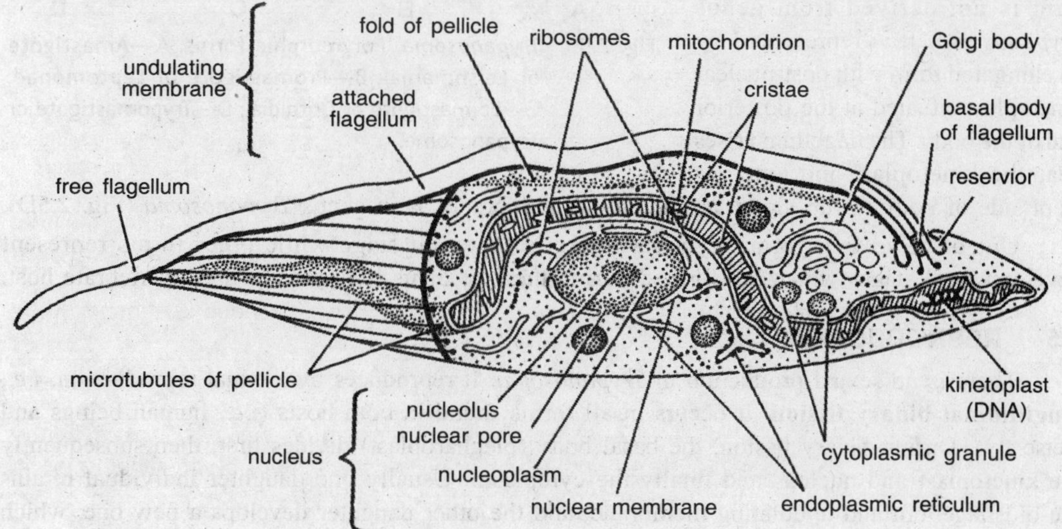

Fig. 2.4. *Trypanosoma*. An electron microscopic structure.

by the plasma membrane. Flagellum extends forward along the body to project beyond its front end, in the form of free flagellum. All along its length over the body, it is connected to the body by a fin-like **undulating membrane** formed of cytoplasm and folded pellicle. Undulating membrane is supposed to be an adaptive structure for locomotion in a viscous environment (blood, lymph) where it lives.

Polymorphism. *Trypanosoma* is a **polymorphic form**; it occurs in following four morphological forms:

1. Amastigote form (old name = Leishmanial stage). It is represented by rounded form which is without any external flagellum as in the genus *Leishmania*. It is a rounded or oval form in which reduced flagellum remains in the cytoplasm; basal body and kinetoplast are situated in the anterior part of body (Fig. 2.5A).

2. Promastigote form (old name = Leptomonad stage). It is represented by an elongated form with the kinetoplast lying to the anterior to the nucleus (antenuclear kinetoplast). The flagellum arises near kinetoplast and emerges from the anterior end of body, as in the genus *Leptomonas* and others. In it, the flagellum is short, unattached and free (Fig. 2.5B).

3. Epimastigote form (old name = Crithidial stage). It is represented by an elongated form with the kinetoplast lying anterior and close to the nucleus (*i.e.*, juxtanuclear kinetoplast); the flagellum arising near it emerges from the side of the body to run along a short undulating membrane (Fig. 2.5C).

4. Trypomastigote form (old name Trypanosome stage). This term is not derived from genus *Trypanosoma*. It is represented by an elongated form with post-nuclear kinetoplast situated at the posterior end of the body. The flagellum arises near the kinetoplast and emerges from side of body to run along an undulating membrane, as in genus *Trypanosoma* (Fig. 2.5D).

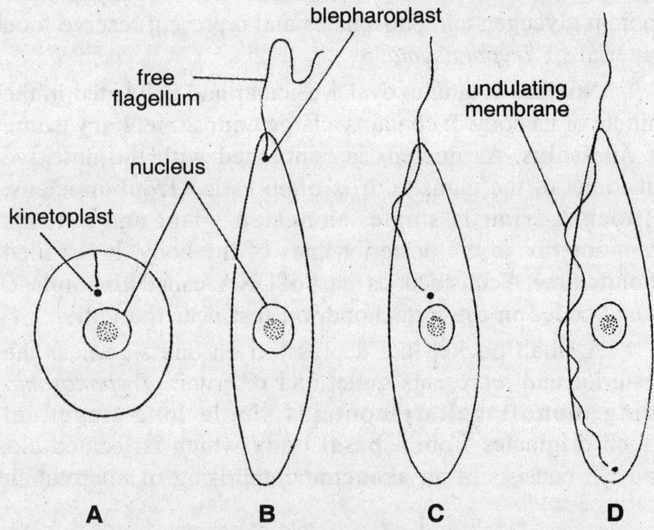

Fig. 2.5. *Trypanosoma*. Polymorphic forms. A—Amastigote or Leishmanial; B—Promastigote or Leptomonad; C—Epimastigote or Crithidial; D—Trypomastigote or Trypanosome.

Out of these four forms trypomastigote is the adult stage while other forms represent developmental stages which are formed during the part of its life cycle in the invertebrate host.

2.5 REPRODUCTION

There is no sexual production in *Trypanosoma*. It reproduces by asexual reproduction, *i.e.*, **longitudinal binary fission**. It occurs in all forms found in both hosts (*i.e.*, human beings and tsetse fly). During binary fission, the basal body (blepharoplast) divides first, then subsequently the kinetoplast and nucleus and finally the cytoplasm. Usually one daughter individual retains the old flagellum and undulating membrane, and the other daughter develops a new one, which grows out before the complete separation of the daughters.

Life Cycle

Trypanosoma gambiense is a digenetic pathogen; it passes its life cycle in two different types of hosts: two vertebrate hosts and one invertebrate host. These three hosts fall into following types:

1. Primary host or definitive host in which parasite lives initially in blood outside the red blood cells (*i.e.*, extracellularly). Later on parasite lives in cerebrospinal fluid of the host, *e.g.*, human beings.

2. Reservoir host in which parasite lives in the blood but causes no harm. *e.g.*, domestic and wild animals such as pigs, antelopes, buffaloes and reed bucks.

3. Secondary host or intermediate host in which polymorphic forms of the parasite are developed. *e.g.*, blood sucking insect, tsetse fly, *Glossina palpalis* and *Glossina tachinoides*. Since tsetse fly carries the trypanosomes from person to person, it is known as **vector** or **carrier**. Both male and female tsetse fly bite by daylight, usually in the early morning and evening.

A. Life Cycle in Human Beings

1. Infection. Infective stage of *Trypanosoma* exists in the salivary glands of tsetse fly and is called **metacyclic form** (trypomastigote). When such infected fly bites a human being to suck his/her blood, it punctures the skin with its proboscis (which is a needlelike mouth part). It injects its saliva in the wound of human being to check blood clotting (the saliva of tsetse fly contains an anticoagulant). The infective stage of *Trypanosoma* enters human blood along with saliva.

2. Multiplication. On reaching the blood of human, the metacyclic forms change into long slender flagellated forms. These swim actively by beating their flagellum and multiply asexually by longitudinal binary fission. During this phase of life cycle, parasites aborb glucose from blood of the host and get energy by anaerobic oxidation of food (glycolysis). Upto this stage, the mitochondrion of each parasite has not developed cristae.

3. Pleomorphism. Gradually blood of human beings develop antibodies against the multiplying slender forms of *Trypanosoma*. Due to antibodies, slender forms stop absorbing glucose and their glycolysis too stops. The slender forms also stop their multiplication and 'change into short stumpy forms, each of which lacks free flagellum (Fig. 5.4). During this transformation certain **intermediate forms** also appear; they have shortened body and reduced free flagellum. Production of individuals with slight variations in body form (*e.g.*, slender, intermediate and stumpy form) is called **pleomorphism.**

The stumpy forms get their energy from aerobic oxidation of pyruvic acid which is the end product of glycolysis. Their mitochondria develops cristae and produces enzymes of Krebs cycle. The stumpy (ovoid leishmania forms) are sucked by the tsetse fly along with blood meal.

4. Relapse of infection. Some of the slender forms of *Trypanosoma* are not metamorphosed. They escape from the action of host's antibodies by changing the antigens which are present on their surface. Such slender forms survive and continue to multiply and cause relapse of infection.

B. Life Cycle in Tsetse Fly

1. Transfer to the tsetse fly. When the blood of an infected person or reservoir host is sucked up by a male or female tsetse fly, short and stumpy forms of *Trypanosoma* (*i.e.*, trypomastigote) enter its alimentary canal along with the blood. These exist in an unchanged form in the midgut for few hours.

2. Development in midgut. After tenth to fifteenth day short-stumpy forms develop into long slender forms. These forms pass into extra-peritrophic space (a space between the peritrophic membrane and the epithelial cells). The slender forms multiply by longitudinal binary fission. By fifteenth day they enter the lumen of the proventriculus.

3. Development in salivary glands. The long slender forms then migrate forward to the buccal cavity. They pass on the hypopharynx and eventually reach the salivary glands through the opening of salivary ducts. Here they multiply and change their morphology, first into **epimastigote** (or crithidial stage) and then into **metacyclic stage** (short stumpy form of trypomastigote) which are infective to human beings.

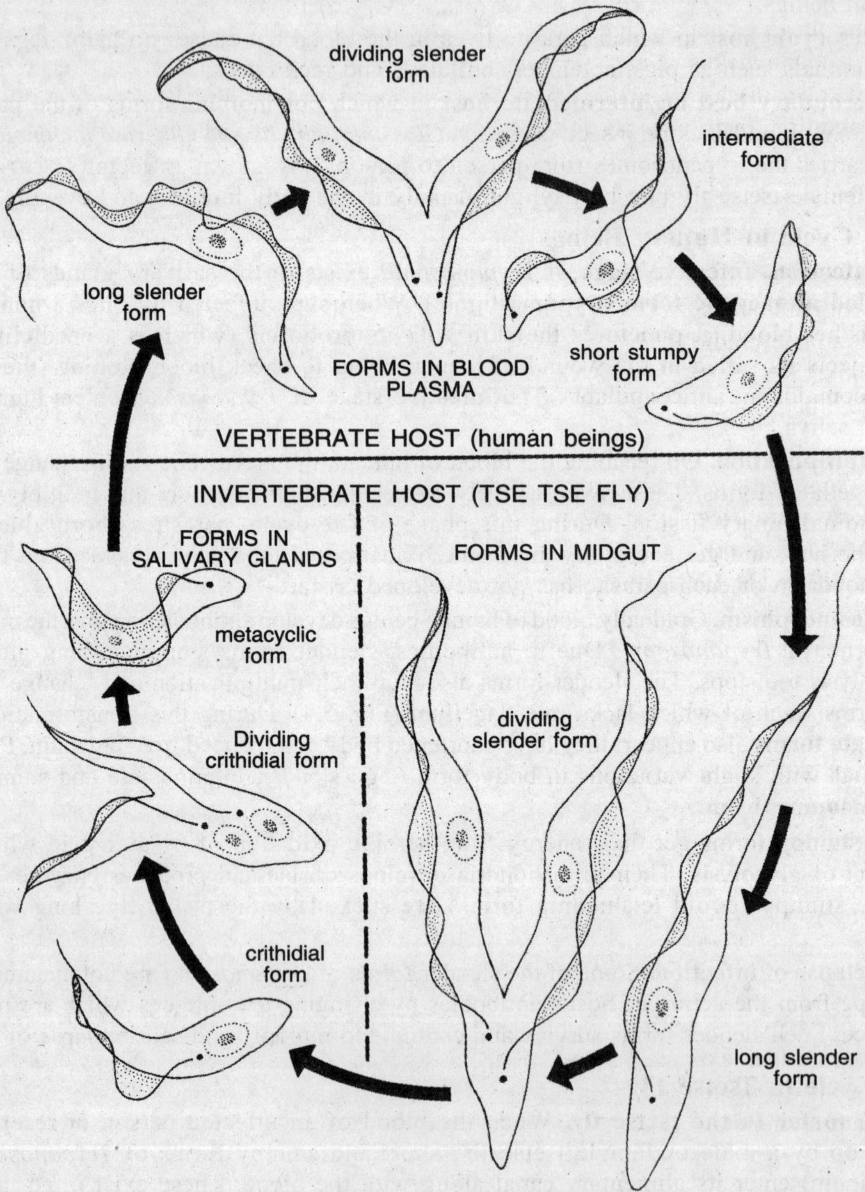

Fig. 2.6. *Trypanosoma gambiense*. Life cycle.

Economic Importance

Presence of trypanosomes in human blood causes **gambian fever** which persists only for few days and characterized by fever and severe headache, both of which recur at regular intervals accompanied by loss of nocturnal sleep, increasing weakness, loss of weight and anaemia. Usually, the pathogenic parasite succeed in penetrating the lymphatic glands. Because of infection of trypanosome the lymphatic glands swell; enlargement of spleen and liver too becomes evident. Ultimately parasite makes its entry into the cerebrospinal fluid of brain and spinal cord and causes **sleeping sickness** disease. This disease is characterized by following symptoms: development of lethargic condition, recurrence of fever, loss of consciousness first at regular intervals and finally patient enters into a state of coma ending in death.

Prevention. Direct attacks on the vector of sleeping sickness, tsetse fly, help in checking the spread of sleeping sickness. These include:

1. Elimination of breeding grounds of tsetse fly.
2. The endemic areas should be kept clean and regularly sprayed by insecticides (*e.g.*, DDT) for complete eradication of vector (tsetse fly).
3. Reservoir hosts (domestic animals) should be kept free from its infection.

Treatment (Therapy). The sleeping sickness can be treated only in the early stages of infection, since once the parasite has entered the cerebrospinal fluid it is difficult to cure sleeping sickness. Compounds of **arsenic (tryparsamide, melarson oxide, melarsoprol** and **trimelarsen)** and **antimony** were used until recently to cure sleeping sickness (now these druge are used for final stages of infection). Two drugs, **Bayer 205** (also called **Antrypol, Germanin** or **Suramin)** and **Pentamidine** (or **Lomidine**) are now widely used for both treatment and prophylaxis of human infections of *Trypanosoma*. Periodic injections of **Naphuride** can save a person from catching infection.

2. *Monocystis*

Classification

Phylum	–	**Protozoa**
Subphylum	–	**Sporozoa (Apicomplexa)**
Class	–	**Telosporea**
Subclass	–	**Gregarinia**
Genus	–	***Monocystis***

Habit and Habitat

Monocystis is an **extracellular endoparasite** in the seminal vesicle of earthworms (*Pheretima, Eutyphoeus* and *Lumbricus*). *Monocystis pheretimi* is parasite of *Pheritima posthuma*. Its adult feeding and growing stage is called **trophozoite** and represents vegetative phase of parasite. The stages of *Monocystis* that are usually present in the seminal vesicles are the trophozoite, cysts containing two individuals, gametes and spores in various stages of development and isolated spores. Trophozoite develops within the **sperm morula** (*i.e.*, group of developing spermatozoa) of earthworm. Developing trophozoite tends to live at the expense of the cells among which it lies. The spermatozoa of the earthworm which are deprived of nourishment by the parasite, slowly shrivel up finally becoming tiny filaments on the surface of trophozoite. As a result, young trophozoite appears ciliated because of attached tails of the dead sperms.

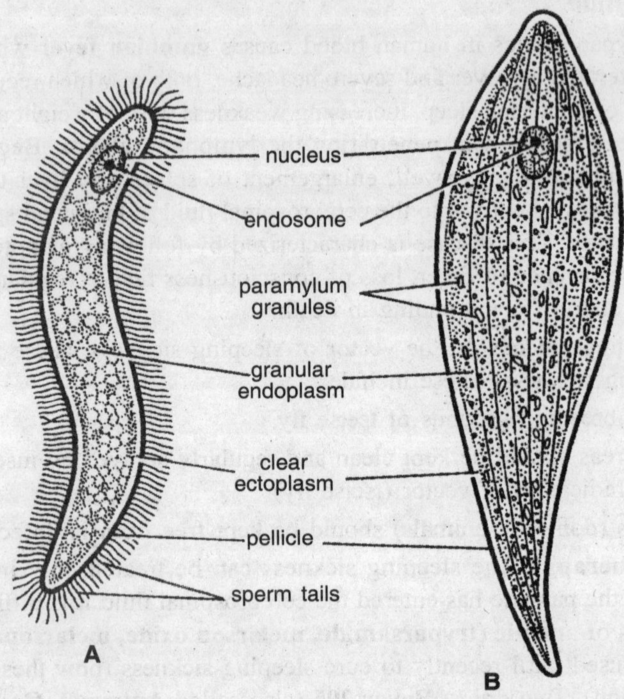

Fig. 2.7. *Monocystis*. A—Young trophozoite; B—Mature trophozoite.

Morphology

A full grown or adult trophozoite of *Monocystis* is known as **sporadin**. The mature trophozoite is an elongated spindle or club-shaped, dorso-ventrally flattened worm-like organism (Fig. 2.7). It is pointed at the both ends of body. The body of *Monocystis* is bounded externally by a thick, smooth, elastic **pellicle** containing longitudinally arranged contractile microtubules. The body cytoplasm is distinctly differentiated into following two layers:

1. Cortex or ectoplasm. This is outer or peripheral, clear and firm layer of cytoplasm. It is further subdivided into three layers—the outer **epicyte**, the middle **sarcocyte** and inner **myocyte**. Myocyte consists of longitudinal and transverse contractile fibres, called **myonemes**. These cause metabolic contraction of body and characteristic wriggling or gliding movements called **gregarine movement**.

2. Medulla or endoplasm. It is inner, more fluid-like, granular part of cytoplasm. Endoplasm is rich in granules of **paraglycogen** (a special form of glycogen; reserve food of the animal), fat globules and sometimes **volutin** (a protein rich in phosphorus and nucleic acid).

Nucleus. The nucleus is single, large, vesicular and ovoid occurring in the endoplasm. It lies toward the anterior end of trophozoite's body and is bounded by a distinct **nuclear membrane** and contains a large distinctive **nucleolus**. It is haploid containing four chromosomes.

Due to its parasitic mode of life, the trophozoite of monocystis lacks parapodia, cilia, flagella, mouth, gullet and the food vacuoles and contractile vacuoles.

3. Physiology

1. Locomotion. It moves very slowly by rhythmic contraction and relaxation of myonemes, causing **gregarine movement.**

2. Nutrition. Nutrition in *Monocystis* is saprozoic; it absorbs nourishment from the fluid of the seminal vesicle by osmotrophy through the general body surface.

3. Respiration and excretion. Respiration and excretion occur by diffusion through general body surface of *Monocystis*.

4. Reproduction. *Monocystis* reproduces sexually and the process is always followed by **asexual reproduction.**

Life Cycle

Monocystis is **monogenetic**, *i.e.*, its life cycle is completed in a single host, *i.e.*, earthworm. Life cycle of *Monocystis* involves **gametogony, syngamy** and **sporogony** (Fig. 2.8).

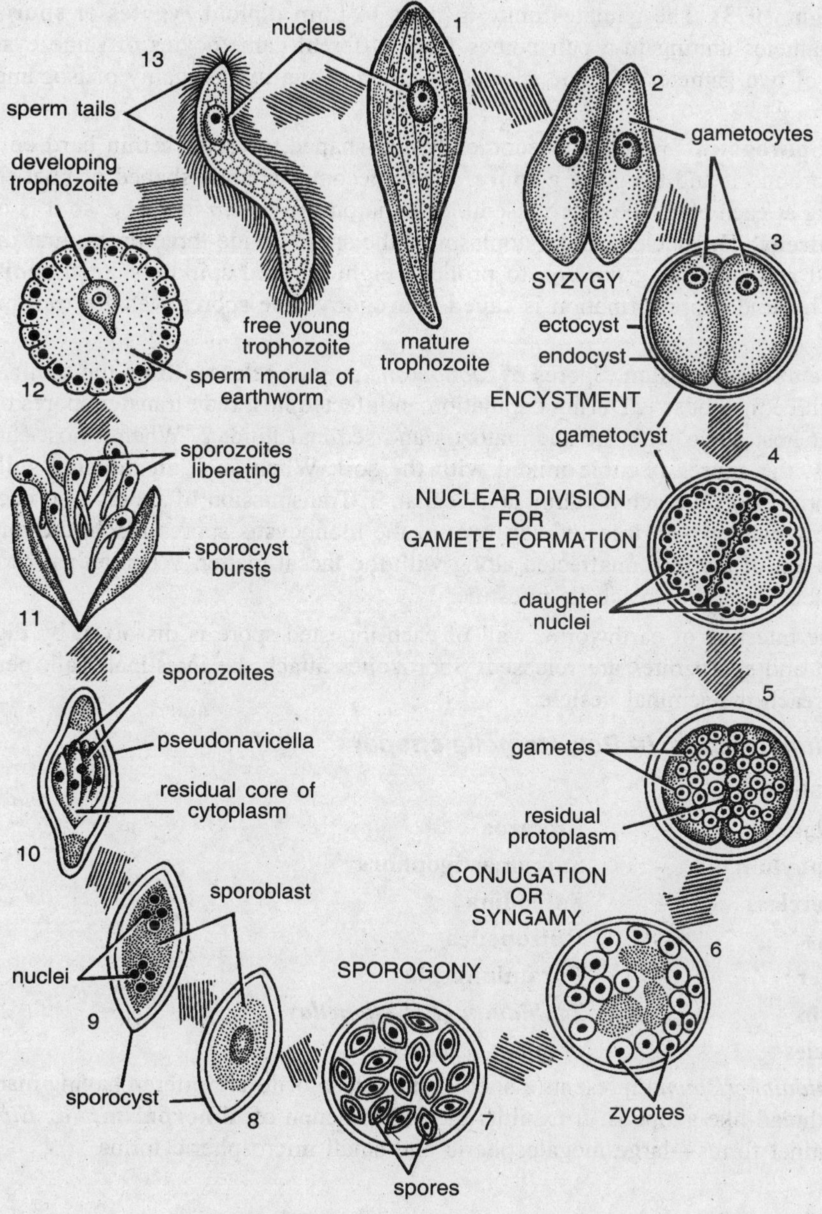

Fig. 2.8. *Monocystis*. Life cycle.

(*i*) **Gametogony (or gamontogony).** It is a method of sexual reproduction in which two trophozoites come together, become short and rounded and are called **gamonts** or **gametocytes.** The pair secretes a common protective two-layered cyst wall and becomes **gametocyst**. The outer layer of cyst is thick and hard and is called **ectocyst** and inner layer is thin and soft and is called **endocyst**.

The two trophozoites in the gametocyst never fuse or conjugate but remain together in pair. This stage is called **syzygy** (Fig. 2.8). Ultimately, each gametocyte of gametocyst divides by repeated mitotic cell divisions, producing a number of small cells called **gametes**.

(*ii*) **Syngamy.** Gametes of *Monocystis* are identical morphologically, so are called **isogametes** (**M.A. Sleigh**, 1973). The gametes unite in pairs to form diploid **zygotes** or **sporoblasts**. As a rule, two gametes uniting in a pair comes from different gametocytes of same cyst. The union and fusion of two gametes of same species is called **syngamy;** syngamy of isogametes is called **isogamy.**

(*iii*) **Sporogony.** Sporoblast becomes lemon-shaped, secretes a thin hard covering called **sporocyst** around it and becomes a **spore**. Spore becomes spindle-shaped or boat-shaped with a mucoid plug at each end. The spore resembles a diatom *Navicella* in shape, so it is often called a **pseudonavicella**. The nucleus and cytoplasm of the spore divide three times, first division being *meiotic* and rest two being *mitotic*, to produce eight haploid spindle-shaped **sporozoites**. The process of haploid spore formation is called **sporogony**. The spores wait for transmission to the new host.

Mode of transmission. Spores of *Monocystis* may be released from the seminal vesicles by following three methods: 1. During copulation an infected host may transfer spores of monocystis to the other host along with the spermatozoa and seminal fluid. 2. When a host earthworm dies and decays, the spores become mixed with the soil. When such an infected soil is eaten by another worm, spore's reach its alimentary canal. 3. Transmission of spores also occurs by birds. When a bird swallows an infected earthworm, the monocystis spores reach, the gut of the bird. These pass out of the bird unaffected along with the faecal matter. When an earthworm eats this faecal matter, the spores reach its intestine.

In the intestine of earthworm, wall of each ingested spore is dissolved by digestive juices of the host and sporozoites are released. Sporozoites attack the intestinal wall, penetrate it and somehow reach the seminal vesicle.

3. *Elphidium crispum* (= *Polystomella crispa*)

Classification

Phylum	–	**Protozoa**
Subphylum	–	**Sarcomastigophora**
Superclass	–	**Sarcodina**
Class	–	**Rhizopodea**
Order	–	**Foraminiferida**
Genus	–	***Elphidium* (*Polystomella*)**
Species	–	***crispum* (*crispa*)**

Elphiduim crispum represents a shelled-protozoan or foraminiferan having many chambered shell and thread-like axopods. It exhibits the phenomenon of **dimorphism**, *i.e., Elphidium* exists in two distinct forms—large megalospheric and small microspheric forms.

1. Habit and habitat. *Elphidium* is a marine and free-living protozoan. It is found on the bottom of sea at about the depth of 300 fathoms (one fathom = 1.8 metres). It creeps among the sea weeds.

2. Morphology

(i) **Shell.** *Elphidium* is shelled, unicellular and microscopic protozoan (having 3 to 4 mm diameter). Its shell is hard, transluscent, pale yellow in colour, and chemically is made up of calcium carbonate and small amounts of silica and magnesium sulphate. The shell is biconvex, coiled, multichambered or multilocular (= polythalamus) and perforated. Shell has a central rounded part, called **umbo** and a peripheral rigid continuous rim, called the **keel.**

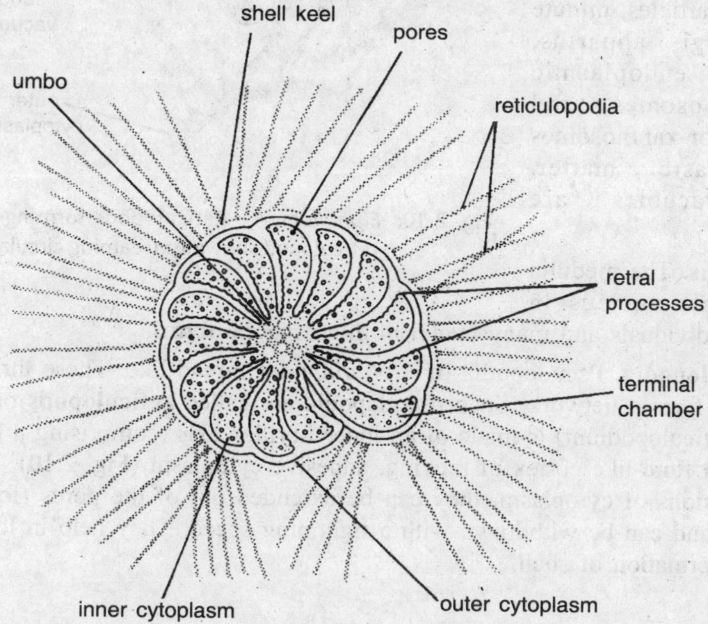

Fig. 2.9. *Elphidium crispum.*

The shell is coiled in a **planospiral** manner, *i.e.*, all the V-shaped chambers or coils lie in the same plane or a flat spiral. All the chambers are filled with cytoplasm. Walls of each chamber bear minute pores through which the cytoplasm remains continuous from one chamber to another, and also extends to outside, to form pseudopodia. Each chamber of shell has a convex anterior surface and a concave posterior surface bearing **retral processes** filled with cytoplasm. The formation of shell begins with an initial single chamber, called **proloculum**. As the animal grows in size, successive chambers are laid down in a spiral manner, forming whorls. Each new chamber is larger than the preceding one and each new whorl partially overlaps and conceals the older one. Thus, only the last whorl is visible from outside. Overlapped portions of shell-chambers are called **alar processes.**

Shell and polymorphism. The small or large size of proloculum results in **dimorphism (polymorphism)** in *Elphidium*. There occur two forms: the large **megalospheric form (gamont)** which has a large central chamber (proloculum), a large single nucleus and sexual mode of reproduction.

The small **microspheric form** (**agamont**) has a small central chamber, proloculum, many small nuclei and an asexual mode of reproduction.

(*ii*) **Cytoplasm.** The shell's chambers are filled with inner cytoplasm or **medulla**. A thin layer of **outer cytoplasm** or **cortex** covers the shell from outside. The inner cytoplasm contains nucleus or nuclei, food particles, minute vacuoles, Golgi appartus, mitochondria, endoplasmic reticulum, ribosomes and brown granules or xanthosomes containing waste matter. Contractile vacuoles are absent.

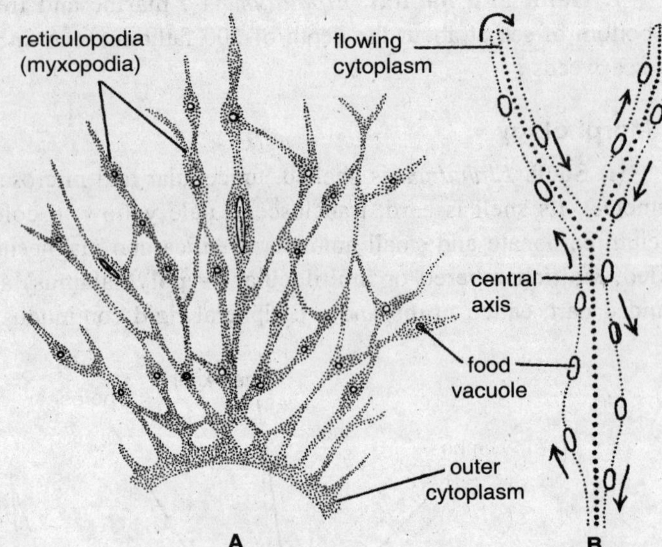

Fig. 2.10. *Elphidium*. A—Reticulopods forming a network. B—A reticulopod showing streaming circulation in the cortex.

(*iii*) **Nucleus.** The medulla contains a single nucleus in megalospheric individuals and many nuclei in microspheric forms.

(*iv*) **Reticulopodia.** Pseudopodia of *Elphidium* are thread-like. These threads branch and branches fuse to form a network. Such pseudopodia are called **reticulopods** or **axopods**. Each reticulopod (= reticulopodium) is made of an inner fibrillar **axis** (comprising a bundle of microtubules) and outer fluid-like cortex of moving, adhesive cytoplasm (Fig. 2.10). Reticulopods are temporary extensions of cytoplasm, they can be extended out of the pores (foramina) of shell like tiny rockets and can be withdrawn with a lightining speed. They help in locomotion, nutrition and in the formation of shell.

3. Physiology

(*i*) **Locomotion.** *Elphidium* creeps slowly with the help of its reticulopods on sea weeds at the bottom of the ocean. Locomotion of animal is performed by contraction and expansion (extension) of the reticulopods.

(*ii*) **Nutrition.** Nutrition of *Elphidium* is **holozoic**. It feeds upon minute organisms such as diatoms, algae, protozoans and microcrustaceans. The network of reticulopods secrete a layer of adhesive cytoplasm to trap the food organisms. This adhesive cytoplasm contains discharged lysosomes and their proteolytic secretions which paralyze the prey and initiate digestion even during capture (**Ruppert** and **Barnes**, 1994). The food is enclosed in food vacuoles existing in the cortex of reticulopods. Digestion of food is normally completed outside the shell and the products of digestion pass into the medulla.

4. Life cycle. The life cycle of *Elphidium* involves alternation of asexual generation (microspheric forms) with sexual generation (macrospheric or megalospheric forms). The macrospheric forms or gamonts produce numerous identical biflagellate gametes called **isogametes**. The gametes are released into sea water through the pores of shell. They swim for sometime and then gametes fuse in pairs (**syngamy**) to form **zygotes**. Each zygote secretes a small initial chamber (proloculum) which develops into a rnicrospheric form (Fig. 2.11).

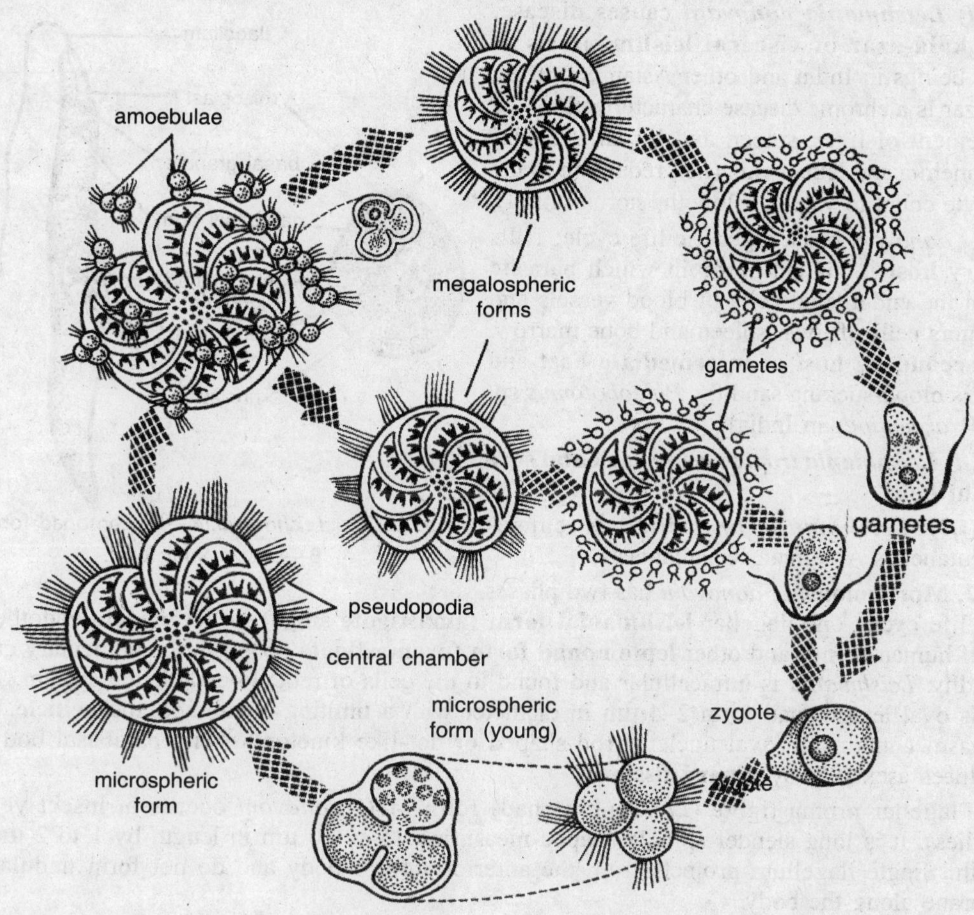

Fig. 2.11. *Elphidium crispa*. Life cycle.

5. *Leishmania*

Classification

Phylum	–	**Protozoa**
Subphylum	–	**Sarcomastigophora (Plasmodroma)**
Superclass	–	**Mastigophora**
Class	–	**Zoomastigophorea**
Order	–	**Kinetoplastida (Protomonadina)**
Genus	–	***Leishmania***

1. Habit and habitat. *Leishmania* is a pathogenic intracellular endoparasite of humans and other mammals, causing the following three types of diseases in humans:

Box 2.3

1. The genus *Leishmania* was created by **Ross** in 1903 to include parasite of Indian kala-azar, *Leishmania donovoni*.
2. The word kala-azar has been derived from two Indian words—*kala* and *azar*, meaning "black sickness", an illness in which the colour (pigmentation) of the skin turns black. The *kala* also means "deadly" thereby signifying a fatal illness.

(*i*) ***Leishmania donovani*** causes disease, called **kala-azar** or **visceral leishmaniasis** in human beings in India and other Asian countries. Kala-azar is a chronic disease characterized by the enlargement of liver, spleen and by an irregular fever, anemia and leucopenia (*i.e.*, reduction in the leucocyte count to values below the normal limit).

L. donovoni has a digenetic life cycle: 1. Its **primary host** is human being in which parasite lives in the endothelial cells of blood vessels and also enters cells of liver, spleen and bone marrow. 2. Its **secondary host** or **intermediate host** and vector is blood-sucking sand fly, *Phleobotomus* sp. (*e.g.*, *P. argentipes* in India).

(*ii*) ***Leishmania tropica*** causes **oriental sore** or **Delhi boils**.

(*iii*) ***Leishmania brassiliensis*** causes mucocutaneous American leishmaniasis.

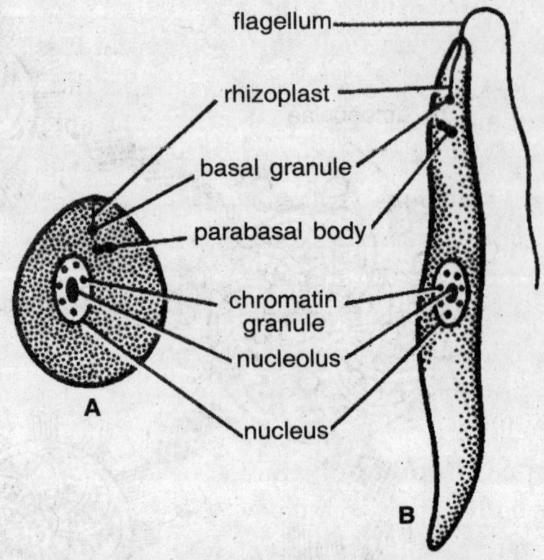

Fig. 2.12 *Leishmania*. A—Leptomonad form; B—Leishmanial form.

2. Morphology. *L. donovoni* has two phases in the life cycle, one flagellar **leishmanial form (amastigote stage)** in the reticulo-endothetial cells of human beings and other **leptomonad form (promastigote stage)** in the alimentary canal of sandfly. *Leishmania* is intracellular and found in the cells of reticulo-endothelial system. This form is oval leptomonad form 2–4 μm in diameter with a limiting membrane, the **pellicle**. The cytoplasm contains an oval nucleus, rod-shaped or dot-like kinetoplast and parabasal body. It reproduces asexually by **binary fission**.

Flagellar promastigote (*i.e.*, leptomonad) form of *L. donovoni* occurs in insect vector (sandflies). It is long slender spindle shaped measuring 15 to 20 μm in length by 1 to 2 μm in breadth. Single flagellum projects from the anterior side of body and do not form undulating membrane along the body.

3. Transmission. Amastigote forms of the pathogen are liberated in human blood stream as a result of the rupture of reticulo-endothelial cells. In the blood stream, some of free amastigote are phagocytosed by the neutrophilic granulocytes and monocytes (macrophages). In India, the vector, the blood sucking sandfly (*Phlebotomus argentipes*) draws these free amastigote forms during its blood-meal. Inside the midgut of sandfly these amastigote forms develop into **promastigote forms** which again multiply by binary fission producing numerous flagellates. Flagellates tend to spread forwards to anterior part of alimentary canal (pharynx and buccal cavity; salivary glands are not infected). The transmission of pathogen is thereby affected through bite of infected sand fly from human being to human being.

6. *Ceratium*

Classification

Phylum	–	**Protozoa**
Subphylum	–	**Sarcomastigophora (Plasmodroma)**
Class	–	**Mastigophora**
Order	–	**Dinoflagellida**
Genus	–	***Ceratium***

1. Habit and habitat. *Ceratium* (*e.g., Ceratium hirudinella*) is found in freshwater plankton. The nutrition in pigmented species is mainly holophytic (autotrophic), sometimes holozoic. Many species of *Ceratium* are known to be **luminiscent**.

2. Morphology. The body of *Ceratium* is enclosed within a thick and rigid pellicle of cellulose called **shell, theca** or **lorica**. It is made of many closely fitting platelets. In the lorica, there are three (*i.e.*, one anterior or apical and two posterior or lateral) long projections or wing-like extensions (spines) which protrude from the body, creating a bizarre shape (Fig. 2.13). The body of *Ceratium* bears two flagella. One is attached a short distance behind the middle of the body, is directed posteriorly and lies in a longitudinal groove (called **sulcus**). The other flagellum is transverse and located in a groove (called **annulus** or **cingulum**) that encircles the body. Thus, cingulum divide the body into an anterior region the **epicone** and a posterior **hypocone**. The transverse flagellum causes both rotation and forward movement; the longitudinal one drives water posteriorly and contributes to forward motion.

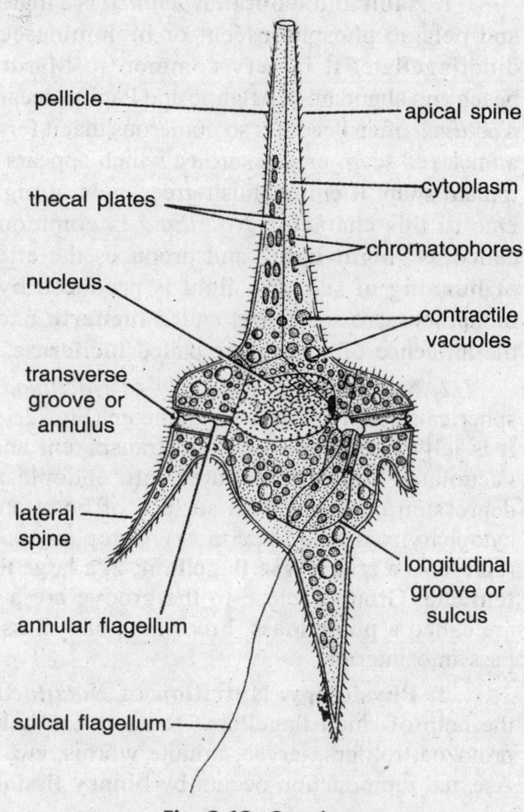

Fig. 2.13. *Ceratium.*

Cytoplasm contains a single large and central **nucleus** containing prominent **endosome** (or nucleolus) and many rod-shaped **chromatophores** which are divided into five distinct groups. Chromatophores are green (having chlorophyll) in freshwater forms, but they are yellowish-brown in marine forms. Reserve food material is starch, glycogen and fat droplets. Cytoplasm also contains non-contractile vacuole, called **pusule**, a stigma and foreign bodies such as bacteria, flagellates and diatom.

3. Physiology. Nutrition of *Ceratium* is both **holophytic** and **holozoic**. Feeding for holozoic nutrition is done by extruding large parts of cytoplasm through the pores of lorica to form an anastomosing net on the body (pseudopodia called **reticulopodia**) which entangles the prey. The partly digested food is withdrawn along with the cytoplasm.

4. Life Cycle. *Ceratium* reproduces asexually by **binary fission** and **cyst** formation.

7. *Noctiluca*

Classification

Phylum	–	**Protozoa**
Subphylum	–	**Sarcomastigophora (Plasmodroma)**
Class	–	**Phytomastigophorea**
Order	–	**Dinoflagellida**
Genus	–	***Noctiluca***

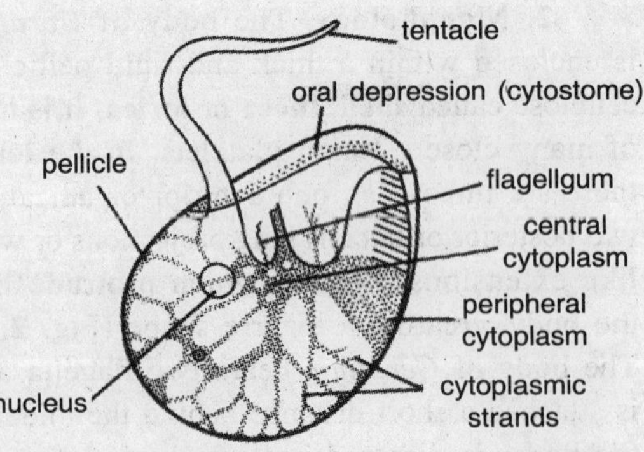

Fig. 2.14. *Noctiluca*.

1. Habit and habitat. *Noctiluca* is a marine and pelagic phosphorescent or bioluminescent dinoflagellate. It is very common in Mardras beach and abundant in Atlantic and Pacific oceans. *Noctiluca* often becomes so numerous that it forms a thick red scum on sea surface which appcars as tomato soup. It emits bluish-green light at night. Due to this character, *Noctiluca* is commonly called as "**night light**" and produces the effect of **burning of sea**. The light is produced by a bioluminescent substance, called **luciferin** under the influence of an enzyme called **luciferase**.

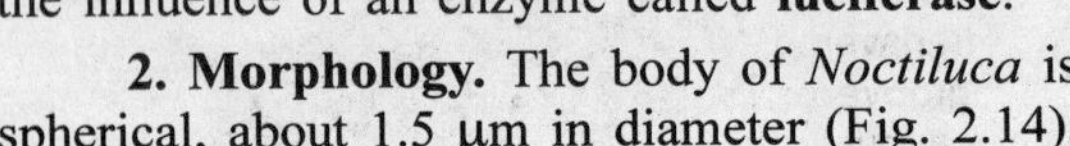

2. Morphology. The body of *Noctiluca* is spherical, about 1.5 μm in diameter (Fig. 2.14). It is jelly-like, pink in colour, transparent and bounded by thick **pellicle**. Cytoplasm is highly vacuolated and differentiated into endoplasm and ectoplasm. **Cytopharynx** is a pouch-like depression on the ventral surface of body and leads into an elongated mouth (during floating cytopharynx is held uppermost). Cytopharynx contains two flagella: 1. a soft flap or **tooth** which represents a **transverse flagellum**; 2. a large thick flagellum which is modified into stout, striated **tentacle**. Grouped close to the groove are a nucleus, flagella and mouth, all of them together are called a **polar mass**. From the polar mass branching and anastomosing strands of cytoplasm pass into interior.

3. Physiology. Nutrition of *Noctiluca* is **holozoic**. It being predator collects food by the help of small flagellum. Tentacle too helps in food collection. Food of *Noctiluca* includes protozoa, rotifers, larvae, minute worms, etc. *Noctiluca* reproduces both asexually and sexually. Asexual reproduction occurs by **binary fission** and sexually by **sporulation** or spore formation.

4. Economic importance. Blooms of *Noctiluca* are toxic to fish and other aquatic life.

8. *Nyctotherus*

Classification

Phylum	–	**Protozoa**
Subphylum	–	**Ciliophora**
Class	–	**Ciliata**
Order	–	**Heterotrichida (Spirotrichia)**
Genus	–	***Nyctotherus***

1. Habit and habitat. *Nyctotherus* is a common intestinal endo-commensal of invertebrates, *e.g.*, apple snail, cockroach, frog, human beings, etc. Its significant species include *N. cordiformis* (found in the rectum of frog) and *N. ovalis* (found in the intestine of cockroaches). In human beings, its infection causes dysentery.

2. Morphology. The body of *Nyctotherus* is kidney-shaped, dorso-ventrally compressed and ranges from 60 μm to 120 μm in length. Body is covered by longitudinal and slightly spiral rows of equal sized cilia. The **oral groove** or **peristome** arises from the anterior end and extend along the right side of body upto the middle of body. It is lined by a row of specially large cilia which form **adoral zone of the membranelle**. The cytopharynx is long and spirally coiled like the watch-spring (Fig. 2.15). It is lined with undulating membrane.

The cytoplasm is differentiated into a thin clear peripheral layer of cytoplasm, called **ectoplasm** and a central mass of granular cytoplasm, the **endoplasm**. The endoplasm contains a large bean-shaped **macronucleus** and a small round **micronucleus** (both in the anterior half of body), a contractile vacuole (near the posterior end) and food vacuoles. The cell anus, cytopyge or **cytoproct** is a permanent aperture at the posterior end of body. Glycogen granules of yellowish-brown colour are found scattered in the endoplasm. *Nyctotherus* feeds upon the intestinal debris (saprozoic).

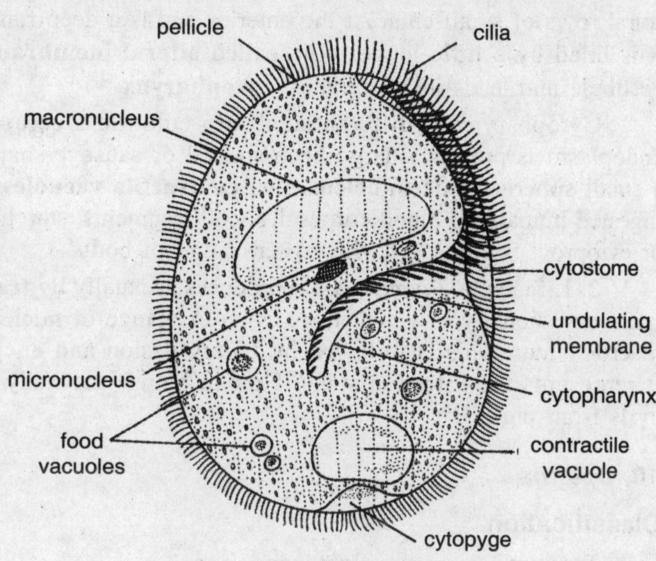

Fig. 2.15. *Nyctotherus cordiformis*.

3. Life cycle. *Nyctotherus* reproduces asexually by **binary fission**. In the rectum of tadpole these produce **exconjugant forms** which **conjugate** (to exchange the nuclear material) during metamorphosis of tadpole. In the newly metamorphosed frog, the exconjugants undergo binary fission to produce normal individuals which undergo **encystment**. The cysts pass out of the host along with its faecal matter. These cysts may be eaten up by the tadpoles.

9. *Balantidium*

Classification

Phylum	–	**Protozoa**
Subphylum	–	**Ciliophora**
Class	–	**Ciliata**
Order	–	**Holotrichia**
Genus	–	***Balantidium***

1. Habit and habitat. *Balantidium* is an intestinal endoparasite of various animals, mainly anuran amphibians and mammals. Several species are known to occur in frogs. *Balantidium coli* occurs in the colon of human beings, monkeys and pigs. It causes **balantidial dysentery** in human beings and often results in ulcers in human intestine.

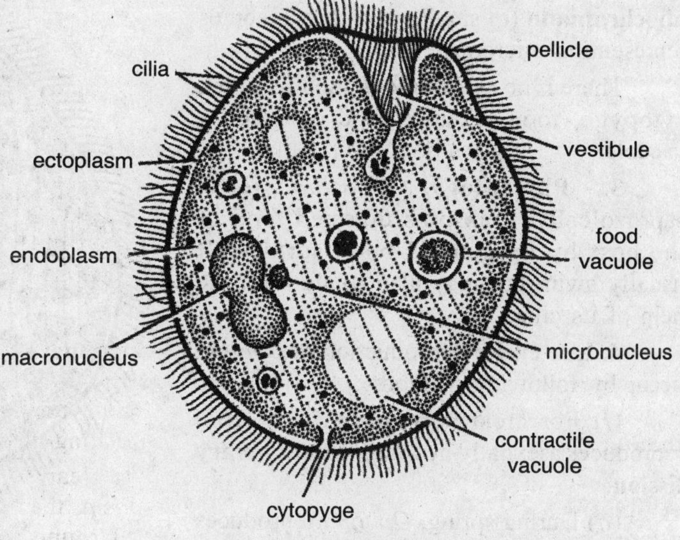

Fig. 2.16. *Balantidium coli*.

2. Morphology. *Balantidium* is an active motile ciliate. It is oval in shape and measures 60-70 μm in length and 40-60 μm in width. The entire outer surface of parasite is covered by longitudinal, slightly spiral rows of small cilia. At the anterior end is a deep funnel shaped depression, the **vestibule.** It is lined by a little longer cilia called **adoral membranelle**. Cytostome is in the bottom of vestibule and leads into a cilialess **cytopharynx**.

Cytopharynx is differentiated into colourless **ectoplasm** and dense **endoplasm**. Inside endoplasm is present a large bean-shaped or sausage-shaped obliquely placed **macronucleus,** a small spherical **micronucleus**, two **contractile vacuoles** and many **food vacuoles** containing ingested human red blood corpuscles, cell fragments, starch and yeast. A permanent anal opening or cytoproct is situated at the posterior end of body.

3. Life cycle. *Balantidium* reproduces asexually by **transverse binary fission**. **Conjugation** takes occasionally in which there is an exchange of nuclear material and reorganization of the macronucleus. This is followed by binary fission and **encystment**. The cyst is double-layered. It represents the infective stage. Human beings probably become infected by swallowing the cysts from pigs.

10. *Opalina*

Classification

Phylum	–	**Protozoa**
Subphylum	–	**Sarcomastigophora**
Superclass	–	**Opalinata**
Genus	–	***Opalina***

1. Habit and habitat. *Opalina* lives as a **commensal** in the large intestine of frogs and toads. The most common species is *Opalina ranarum* which is found in the rectum of frogs or toads.

2. Morphology. Body of *Opalina* is flat, leaf-like and oval, having body length upto 1 millimeter. Body is covered with parallel longitudinal oblique rows of many equal-sized motile fibrils which are called **cilia** or shortened **flagella**. The cytoplasm is distinguished into a clear outer **ectoplasm** and granular inner **endoplasm**. *Opalina* is **multinucleate**, *i.e.*, body contains a large number of monomorphic spherical nuclei evenly distributed throughout the cytoplasm. Each nucleus has both **trophochromatin** (of large sized chromosomes representing macronucleus) and **idiochromatin** (of small sized chromosomes representing micronucleus).

There is no cytostome or mouth, gullet, cytopyge, food vacuoles and contractile vacuole.

3. Physiology. *Opalina* feeds saprozoically, it absorbs digested food of the host by its general body surface. It usually swim but may also rotate with the help of its cilia.

4. Life cycle. Reproduction of *Opalina* occur by following methods:

(*i*) For most part of year, *Opalina* reproduces asexually by **longitudinal binary fission**.

(*ii*) During spring, *Opalina* reproduces sexually by **plasmotomy**, in which cell

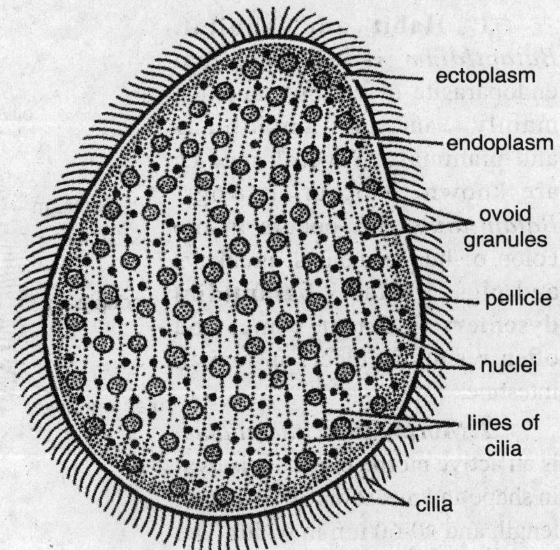

Fig. 2.17. *Opalina ranarum*.

division involves repeated **cytokinesis** (cytoplasmic division) but no **karyokinesis** (nuclear division). As a result many daughter cells are produced, each with a few nuclei, generally 3 to 6. These daughter cells become encysted and pass out of the host into water, from where they are swallowed by tadpoles. The cysts dissolve in the intestine of tadpole and the cells (called **gametocyte**) divide to form uninucleate **microgametes** (male gamete) or **macrogametes** (female gamete). These gametes are unequal in size and are called **anisogametes**. The male and female gametes fuse to form diploid **zygotes** which become multinucleated by the division of nucleus, and it increases in size to form an adult form.

Previously *Opalina* was placed in **Ciliophora**, then it was placed in **Flagellata**, but now visualizing its several peculiar features, it is placed in a separate superclass **Opalinata.**

11. *Vorticella*

Classification

Phylum	–	**Protozoa**
Subphylum	–	**Ciliophora**
Class	–	**Ciliata**
Subclass	–	**Peritrichia**
Order	–	**Peritrichida**
Genus	–	***Vorticella***

1. Habit and habitat. *Vorticella* is a freshwater sedentary (fixed) ciliate found attached to submerged aquatic plants, animals (**epizoic**) and stones of ponds, tanks, lakes, rivers and streams. It is a solitary animal, *i.e.*, free and independent even when occurring in gregarious form. *Vorticella* is worldwide in distribution. Its most common species is *Vorticella campanula.*

2. Morphology. Body of *Vorticella* is inverted bell-shaped with a long, contractile stalk. Stalk is secreted by a region at the aboral end of bell and is called **scopula**. The colour of the animalcule is yellowish, greenish or bluish. The rim or margin of the broad free end of the bell is thickened, called **peristomial lip** or **collar** (Fig. 2.18). Inside the peristomial collar is a narrow, shallow, circular and marginal depression called the **peristome** or **oral groove**. The peristome surrounds a broad, slightly convex circular central disc, the **peristomial disc** or **oral disc** that seems to close the opening of the bell. At one side, the peristomial disc is fused with the collar.

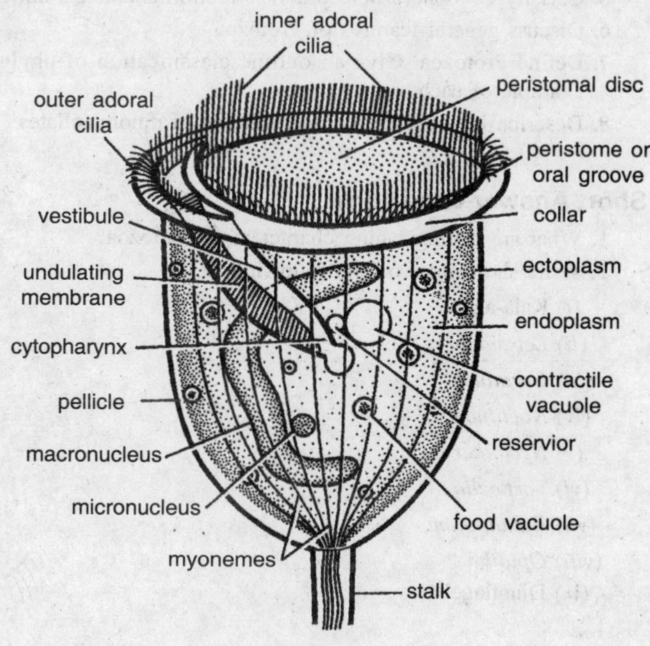

Fig. 2.18. *Vorticella campanula*.

Between the peristome and the peristomial disc is a permanently open sunken space on the left side, it is called **vestibule** or **infundibulum**. From the vestibule a narrow **cytopharynx** leads inwards endoplasm and opens into the endoplasm. The cytopharynx lacks cilia. Between the vestibule and the cytopharynx is a **cytostome** which can open or close. **Cytoproct** is a temporary opening for expelling out the undigested food.

Cilia are present only in the peristomial region. The rest of body is without cilia. The cilia of peristomial groove are arranged in three concentric rows (circlets of cilia) and form 1½ spiral around the peristomial disc. Cilia of vestibule are fused in to an **undulating membrane**.

The entire body of *Vorticella* is covered by a **pellicle** which is very thick at the base of the bell. Cytoplasm is divided into an outer layer of clear, firm, **ectoplasm** or **cortex** containing **myoneme system** and an inner fluid or granular **endoplasm** or **medulla**. The endoplasm contains a large horse-shoe-shaped **macronucleus**, small rounded **micronucleus**, a few **food vacuoles** and one **contractile vacuole**. From the bell, myonemes extend into the stalk, forming a loose spiral running through the centre of the stalk. It is called **axial filament** or **spasmoneme**.

3. Life cycle. *Vorticella* reproduces asexually by **longitudinal binary fission** and sexually by conjugation. Sometimes during unfavourable conditions, *Vorticella* develops a posterior ring of cilia to form free swimming form, called **telotroch**. It becomes detached from the stalk to swim free until a new place of attachment is found. The telotroch helps in its dispersal.

QUESTIONS

Long Answer Questions

1. Classify the phylum Protozoa upto orders giving diagnostic characters and examples of each group. (*C.C.S.U. Meerut 2005*)

2. Mention the distinguishing features of phylum Protozoa and give an outline classification of the same with characters and examples of its subclasses.

3. Classify Protozoa upto classes, giving characters and examples of each class.

4. Give main characteristics of phylum Protozoa and classify it upto orders giving suitable examples.

5. Classify Ciliophora upto orders. Mention characters and suitable examples of each order.

6. Discuss general features of Protozoa.

7. Define Protozoa. Give an outline classification of phylum Protozoa upto orders with at least one example of each.

8. Describe in detail the unique features of dinoflagellates.

Short Answer Questions

1. What are distinguishing characters of Protozoa?

2. Write short notes on the following:

(*i*) Kala-azar
(*ii*) *Leismania*
(*iii*) *Ceratium*
(*iv*) *Noctiluca*
(*v*) *Nyctotherus*
(*vi*) *Vorticella*
(*vii*) *Balantidium*
(*viii*) *Opalina*
(*ix*) Dinoflagellates

Very Short Answer Questions

1. Name a sedentary protozoan.
2. Which disease is caused by *Leishmania donovani*?
3. Which group of protozoans has trichocysts?
4. Mention one colonial protozoan.
5. Name the group of Protozoa in which locomotory organelles are lacking.
6. Protozoa are immortal. Is it correct?
7. What is the basis of classification of Protozoa?
8. Whether *Myxobolus cerebralis* is a protozoan or cnidarian.

Fill in the Blanks

1. *Giardia* produces the disease ....
2. Trichocyst is present in the body of ....
3. Flagellum is the main locomotory organ of the class ....
4. The chief mode of reproduction in Sporozoa is ....
5. The insect vector required for the transmission of African sleeping sickness is the ....
6. *Euglena* belongs to class ....
7. *Trypanosoma gambiense* lives in human body in ....
8. *Monocystis* belongs to order ....
9. The trophozoites of *Monocystis* develop inside the ....
10. The pairing of gametocytes in *Monocystis* is called ....
11. The spores of *Monocystis* are called ....
12. Zygote formed by the fertilization of megagamete and microgamete in *Monocystis* is called ....
13. The causative species of sleeping sickness is ....
14. Nutrition in *Monocystis* is ....
15. Locomotory organ of *Paramecium* is ....

Multiple Choice Questions

1. Classification of protozoa is based on
   (*a*) food vacuole (*b*) cyst structure
   (*c*) types of locomotory organelles (*d*) habit and habitat
2. *Amoeba* belongs to class
   (*a*) rhizopoda (*b*) zoomastigophorea
   (*c*) myxosporida (*d*) ciliata
3. Which of the following belongs to the class Mastigophora?
   (*a*) *Amoeba* (*b*) *Monocystis*
   (*c*) *Noctiluca* (*d*) *Paramecium*
4. Which of the following statement is true of amoeba?
   (*a*) they form pseudopodia
   (*b*) none has an outer shell
   (*c*) they have a flagellated phase in their life cycle
   (*d*) they do not have food vacuoles
5. Tongue-like projection of body of *Amoeba* is known as
   (*a*) rhizopodium (*b*) axopodium
   (*c*) lobopodium (*d*) contractile vacuole
6. The process of cell drinking of *Amoeba* is called
   (*a*) ingestion (*b*) osmoregulation
   (*c*) pinocytosis (*d*) none of the above

7. Which of the following animals is able to manufacture its own food?
(*a*) *Amoeba* (*b*) *Paramecium*
(*c*) *Euglena* (*d*) *Plasmodium*

8. *Trypanosoma gambiense* lives in human body in
(*a*) lymph (*b*) blood
(*c*) cerebrospinal fluid (*d*) blood and cerebrospinal fluid

9. Trypanosomiasis is transmitted by
(*a*) louse (*b*) may fly
(*c*) tsetse fly (*d*) fruit fly

10. *Trypanosoma gambiense* is transmitted by
(*a*) house fly (*b*) may fly
(*c*) sand fly (*d*) tsetse fly

11. Intermediate host of sleeping sickness is
(*a*) sand fly (*b*) tsetse fly
(*c*) male *Anopheles* (*d*) none of the above

12. Sleeping sickness is caused by
(*a*) *Trypanosoma* (*b*) *Entamoeba*
(*c*) *Plasmodium* (*d*) *Leishmania*

13. Polymorphism is shown by
(*a*) *Amoeba* (*b*) *Paramecium*
(*c*) *Trypanosoma* (*d*) *Sycon*

14. Which of the following protozoans lives in the gut of termites and help them digest cellulose?
(*a*) *Plasmodium* (*b*) *Amoeba proteus*
(*c*) *Trichonympha* (*d*) *Trypanosoma*

15. *Glossina palpalis* transmits the infective stages of
(*a*) *Leishmania* (*b*) *Trypanosoma*
(*c*) *Plasmodium* (*d*) *Wuchereria*

16. Chagas' disease is caused by
(*a*) *Trypanosoma cruzi* (*b*) *Trypanosoma gambiense*
(*c*) *Trypanosoma rhodesiensi* (*d*) *Trypanosoma brucei*

17. The disease kala-azar is caused by
(*a*) *Trypanosoma* (*b*) *Giardia*
(*c*) *Trichomonas* (*d*) *Leishmania*

18. The adult feeding stage in the life cycle of *Monocystis* is.
(*a*) sporocyst (*b*) gametocyst
(*c*) trophozoite (*d*) sporozoite

19. *Amoeba* has (*Rohilkhand 2000*)
(*a*) lobopodia (*b*) filopodia
(*c*) reticulopodia (*d*) axopodia

20. Sleeping sickness is caused by (*Rohilkhand 2000*)
(*a*) tse-tse fly (*b*) house fly
(*c*) dragon fly (*d*) may fly

21. African sleeping sickness is caused by (*Kanpur 2001*)
(*a*) *Trypanosoma* (*b*) *Entamoeba coli*
(*c*) *Entamoeba histolytica* (*d*) *Monocystis*

ANSWERS

Very Short Answer Questions

1. *Vorticella.*
2. Kala-azar (Black fever).
3. Ciliophora and *Noctiluca* (Dinoflagellata).
4. *Volvox.*
5. Sporozoa.
6. Yes.
7. Locomotory organelles.
8. It is a cnidarian.

Fill in the Blanks

1. Diarrhoea
2. *Paramecium*
3. Mastigophora
4. Multiple fission
5. Tsetse fly (*Glossina sp.*)
6. Mastigophora
7. Cerebrospinal fluid and blood
8. Eugregarinida
9. Seminal vesicles (of earthworm)
10. Syzygy
11. Pseudonavicella
12. Sporoblast
13. *Trypanosoma*
14. Saprozoic
15. Cilium

Multiple Choice Questions

1. (*c*) **2.** (*a*) **3.** (*c*) **4.** (*a*)
5. (*c*) **6.** (*c*) **7.** (*c*) **8.** (*d*)
9. (*c*) **10.** (*d*) **11.** (*b*) **12.** (*a*)
13. (*c*) **14.** (*c*) **15.** (*b*) **16.** (*a*)
17. (*d*) **18.** (*c*) **19.** (*a*) **20.** (*a*)
21. (*a*).

3

Euglena (Phylum Protozoa)

3.1 CLASSIFICATION

Phylum	–	Protozoa
Subphylum	–	Sarcomastigophora
Superclass	–	Mastigophora
Class	–	Phytomastigophora
Order	–	Euglenida
Genus	–	*Euglena*
Species	–	*viridis*

3.2 HABIT AND HABITAT

Euglena viridis is a solitary, free-living (free-swimming) and freshwater protozoan. It is commonly found in ponds, ditches, pools, lakes, etc., especially in any other water that is rich in soluble organic matter (*i.e.*, stagnant waters containing decaying nitrogenous wastes such as dead bodies of plants and animals, and urine and faeces of animals). Under favourable environmental conditions it multiplies rapidly and forms green scum on the water surface. Other common Indian species of *Euglena* are *E. agilis*, *E. orientalis*, *E. gracilis*, *E. spirogyra* and *E. fusiformis*.

Euglena needs sunlight for photosynthesis, exhibits reactivity and rapidly multiplies by longitudinal binary fission.

3.3 MORPHOLOGY

1. Shape and size. *Euglena viridis* is green, slender, elongated and spindle-shaped in appearance. Its anterior end is bluntly rounded, the middle part is wider and posterior end is pointed.

Euglena viridis is about 4–60 μm in length and 14–20 μm in breadth at the thickest part of the body.

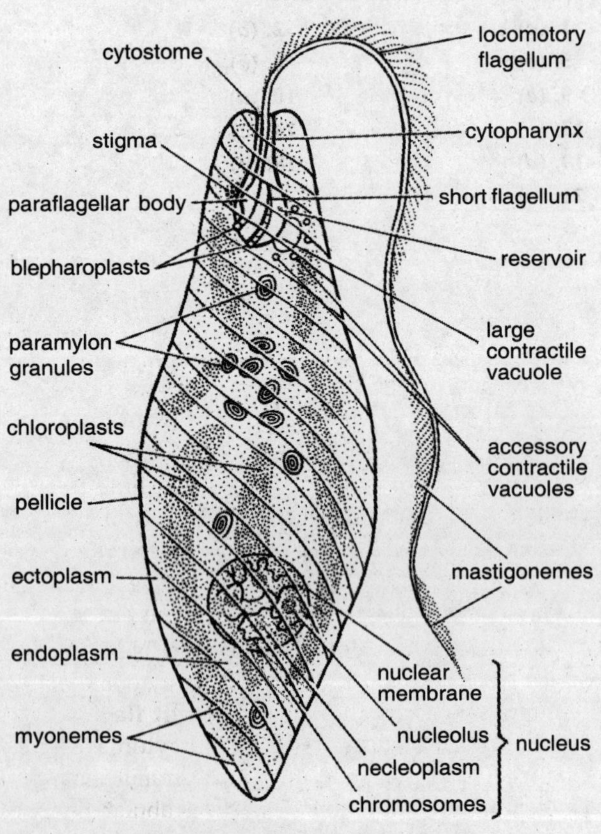

Fig. 3.1. *Euglena viridis*. Structure.

2. Pellicle. *Euglena* lacks a cell wall. The outer layer of the cell is the plasma membrane, immediately below which is the thin, flexible, strong, proteinaceous **pellicle**. Pellicle consists of a number of thickened longitudinal strips and microfibrils (microtubules) articulating with each other. It consists of parallel helical striations or sculpturing which can be observed by electron microscopy (Fig. 3.3). These striations extend counter-clockwise from anterior to posterior ends of body and involve longitudinal rigid strips which are interlocked with each other along their margins. When minute fibrils, called **myonemes**, occurring just beneath the pellicle in the cytoplasm contract, they cause the strips of pellicle to slide over one another and affect a change in body shape of *Euglena*.

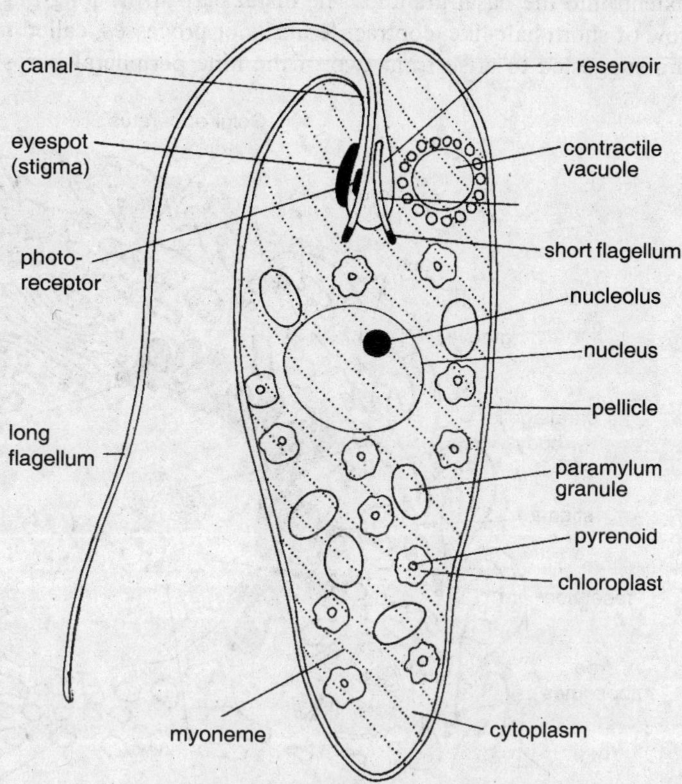

Fig. 3.2. *Euglena gracilis*. Structure.

Pellicle protects the body (acts like **exoskeleton**), maintains its definite shape and helps in movements. Pellicle of *Euglena* is flexible enough to permit temporary changes in the body shape; these changes of body shape are called **metaboly** or **euglenoid movement**.

Box 3.1 Myonemes

Contraction of *Euglena* body is brought about by bundles of contractile filaments, or **myonemes** that lie in the pellicle. The myoneme fibre is not composed of proteins actin and myosin, as in animal muscles, but rather another protein called **spasmin**, which requires ATP for extension (see **Ruppert** and **Barnes**, 1994).

3. Cytostome and cytopharynx. A little to one side of the centre of the anterior blunt end of the body is a funnel-shaped depression or invagination, called **cytostome** (cell mouth; Fig. 3.1). Cytostome leads into a short tubular **cytopharynx** or **gullet** which in turn opens into a insert; instead of, spherical vesicle, called **reservoir** or **flagellar sac**. Quite curiously, the mouth or cytostome and gullet are not used for the ingestion of food but as a canal for the escape of fluid from the reservoir.

4. Flagella. *Euglena* contains two flagella. A long and thin and whip-like flagellum which emerges out of the cytostome through cytopharynx. Another much shorter flagellum which lies within the reservoir (called **nonemergent flagellum**). Each flagellum arises from a minute **basal granule** or **kinetosome** lying in the cytoplasm just inside the border (plasma membrane of reservoir). The flagella and their basal granules are fibrillar structures; each flagellum contains 11 (eleven) longitudinally arranged microtubular fibres which are arranged in a specific pattern of two central singlet fibres and nine peripheral doublet fibres (9 + 2). Only the 9 peripheral fibres extend into the basal granule. The distal part of the long flagellum bears along its one margin a row of short, hair-like, contractile fibrils or processes, called **mastigonemes**. These mastigonemes are suggested to arise from two of the nine peripheral fibres (**Manton**, 1959).

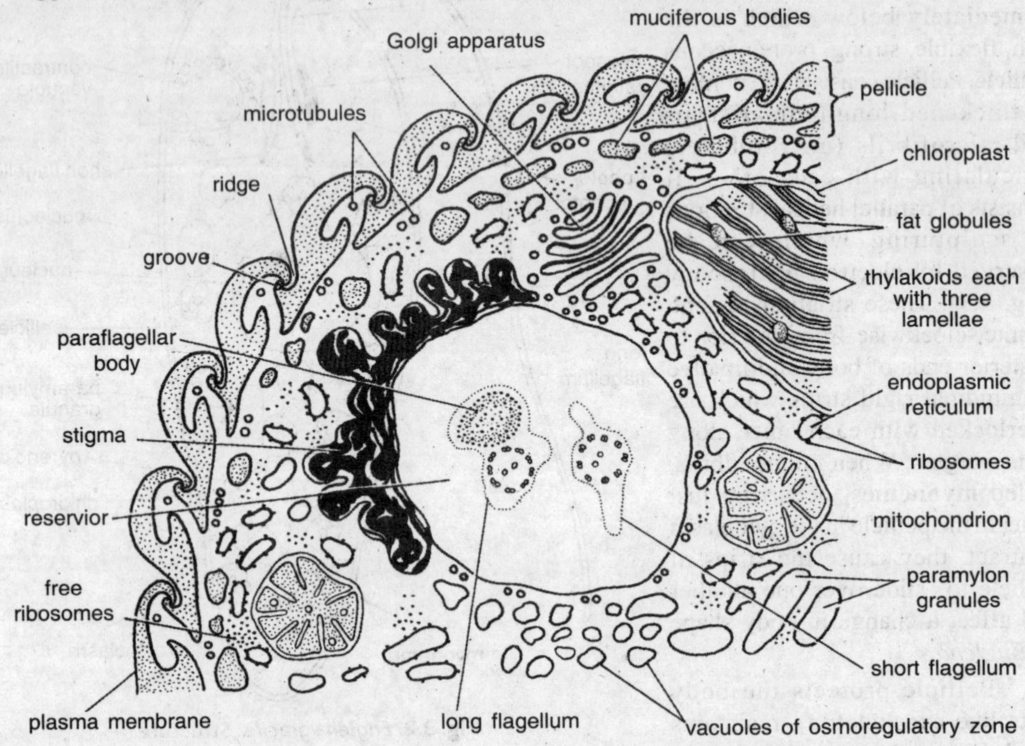

Fig. 3.3. *Euglena*. Electron microscopic structure of a part of T.S. of body passing through reservoir.

5. Cytoplasm and organelles. The cytoplasm of *Euglena* is differentiated into a thin, clear and dense outer layer of **ectoplasm** and a more fluid and granular inner (central) layer of **endoplasm**. Endoplasm contains a variety of cellular organelles such as stigma, chloroplasts (chromatophores), mitochondria, nucleus, contractile vacuole and also usual cellular organelles such as endoplasmic reticulum, Golgi apparatus, ribosomes, etc.

***(i)* Nucleus.** A single, large, spherical, vesicular nucleus lies in the middle or toward posterior end of body. It contains only one chromosome (in the form of chromatin) and a few large **endosomes** or **nucleoli**. Nuclear membrane is not continuous with endoplasmic reticulum (unusual among protists; see **Brusca** and **Brusca**, 2003).

***(ii)* Contractile vacuole.** A single large contractile vacuole lies near the reservoir on one side (opposite to stigma). It is surrounded by several minute **accessory** contractile vacuoles which fuse together to form the larger vacuole. The contractile vacuole discharges its watery contents into the reservoir; from there fluid escapes through cytopharynx and cytostome.

***(iii)* Photoreceptor apparatus.** It includes stigma and paraflagellar body.

1. Stigma or eyespot is a small discoid red or orange-red body which is located close to the inner end of the cytopharynx adjacent to reservoir. It consists of a plate of **lipid droplet** and a granule of red **carotenoid pigment**. Stigma is a cup-shaped structure having colourless mass of oily droplets in its concavity which function as a **lens**. Stigma has a one-sided light filtering action and thus, helps in phototactic response.

2. Paraflagellar body (photoreceptor). It is a small swelling at the root of the long flagellum and is sensitive to light. The paraflagellar body acts as a photoreceptor (*i.e.*, detects light) and enables organisms to swim towards light of optimum intensity, called **phototaxis**. It also gives a directional light response in connection with stigma.

***(iv)* Chloroplasts or chromatophores.** *Euglena* contains many, slender, band-like elongated chloroplasts which are arranged in a **stellar fashion** or like the rays of a star. Chloroplasts are bounded by double unit membranes and contain membrane-bounded chlorophyll containing **thylakoids** in the **matrix** or **stroma**. Stroma contains DNA molecules, small-sized ribosomes and fat globules. Chloroplasts contain three types of photosynthetic pigments such as chlorophyll-a, chlorophyll-b and β-carotene. Located in the centre of each chloroplasts is clear area, called **pyrenoid** which is the centre for **paramylum** - biosynthesis and storage. Paramylum (starch paramylon) is a polysaccharide (β-1, 3 glycan) starch which is the reserve food of *Euglena* and is formed by the product of photosynthesis. Paramylum bodies of various shapes and sizes are found scattered throughout the endoplasm.

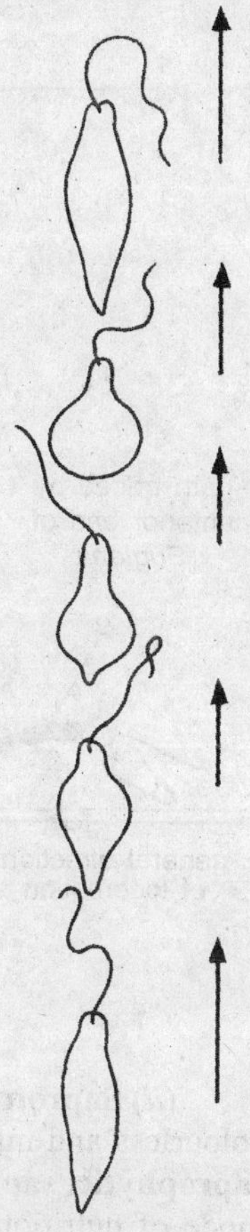

Fig. 3.4. Euglenoid movement or metaboly. It is thought to be accomplished by the sliding of microtubules against the protein strips (after Brusca and Brusca, 2003).

3.4 PHYSIOLOGY

1. Locomotion. *Euglena* exhibits following two types of movements:

***(i)* Euglenoid movement.** This is a characteristic slow and limited movement of *Euglena*, called **metaboly** or **euglenoid movement**. Upon a solid substratum, *Euglena* slowly wriggles like a worm by means of **peristalsis**, *i.e.*, waves of contraction and expansion sweeping over the body from anterior to posterior ends and animal moves forward. Peristalsis is performed by the help of myonemes which are located just beneath the pellicle.

***(ii)* Flagellar movement.** The locomotory flagellum is at the anterior end of the *Euglena* body and pulls the organism forward. This sort of flagellum is called **tinsel flagellum**, since it possesses minute lateral projections, called **mastigonemes** on its surface which aid this type of locomotion. The flagellum propagates an **undulatory wave** from the cell to the flagellar tip. As such waves of activity pass through the flagellum in spiral fashion from its base to its tip, they increase in amplitude and velocity. Such an activity of the flagellum causes the body of *Euglena* to rotate about its axis, at about one complete body turn per second, as well as make a corkscrew pathway through water (Fig. 3.6). The organism can travel forward at a rate of 0.5 mm S^{-1} which is approximately four times its body length. *Euglena* is able to change its direction by the action of contractile myonemes.

2. Nutrition. The mode of nutrition in *Euglena* is **mixotrophic**, *i.e.*, its mode of nutrition is either holophytic or saprophytic.

***(i)* Holophytic nutrition.** *Euglena viridis* being a chlorophyll containing green organism is autotrophic. It synthesizes its own food from carbon dioxide, water, and mineral salts by the help of chlorophyll and sunlight. However, *Euglena* is dependent on an external supply of vitamins B_1 and B_{12}, which like animals, it cannot synthesize for itself. Its reserve food is **paramylum** which is stored in the form of refractile paramylum bodies.

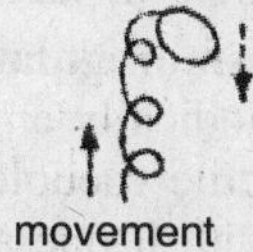

Fig. 3.5. *Euglena*. Flagellar movement by helical beat.

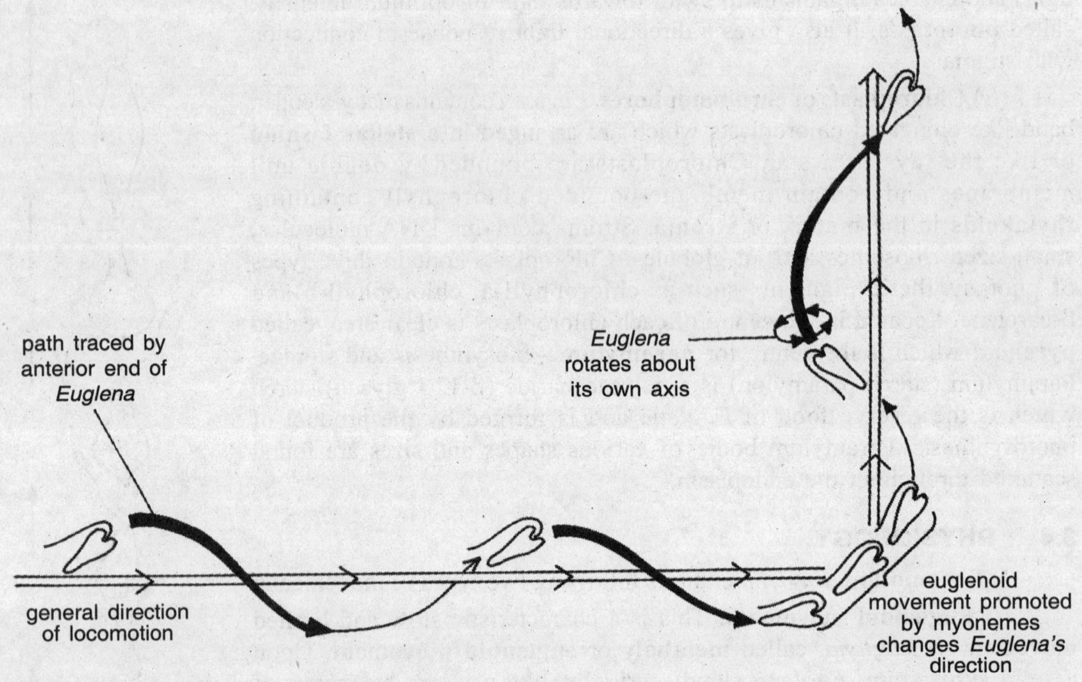

Fig. 3.6. *Euglena*. Path of flagellar movement.

(*ii*) Saprotrophic nutrition. In the absence of sunlight *Euglena* loses its chloroplasts, become colourless and manages its food by **saprotrophic** mode of nutrition. Certain other terms such as **saprophytic, saprozoic** or **osmotrophic** nutrition are also used for saprotrophic nutrition. In this mode of nutrition, animal absorbs through the general body surface some organic substances in solution from decaying matter. However, *Euglena* requires ammonium salts, instead of nitrates for their source of nitrogen. Chloroplasts return when the organisms are returned to light.

Box 3.2 Holozoic *Euglena*

Some species of *Euglena* (*e.g.*, *Euglena gracilis*) are colourless and non-photosynthetic; they are heterotrophic and have following two types of nutrition:

1. Holozoic nutrition. (Involving intracellular digestion). They ingest small food particles at the base of the gullet by pinocytosis or phagocytosis where the pellicle is absent. Food may be driven into the gullet by the action of the flagella.

2. Saprotrophic nutrition. They carry extracellular digestion and grow most abundantly in water where putrefaction is taking place (since decaying material is rich in organic compounds).

3. Respiration. *Euglena* respires through its general body surface. Oxygen dissolved in the water diffuses in through the pellicle and carbon dioxide liberates to outside.

4. Osmoregulation. The water contents of body are regulated by the **contractile vacuole** which periodically empties into the reservoir. This is called **osmoregulation**. During its **diastole**, the contractile vacuole grows in spurts because smaller accessory vacuoles, formed in its vicinity, coalesce with it one after another. After its maximum growth, the contractile vacuole undergoes **systole** by fusing with the lining of the reservoir and bursting into it. To recall, water tends to enter the animal body by **endosmosis**.

5. Excretion. *Euglena* is an ammonotelic animal; its main excretory waste is ammonia. Excretion occurs through the general body surface and contractile vacuole both.

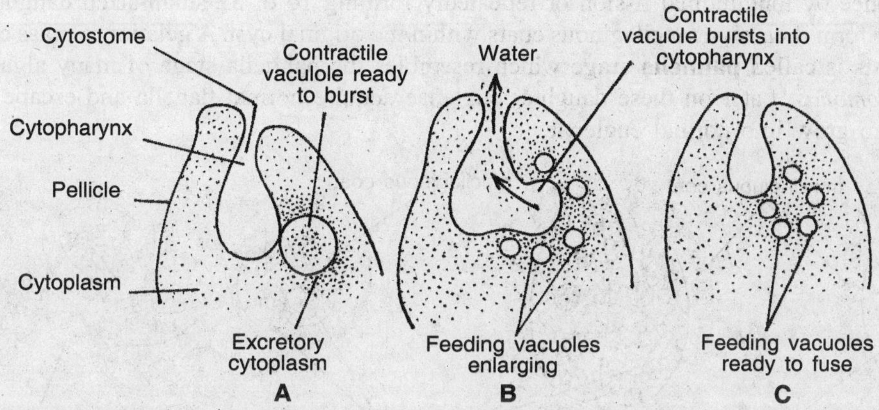

Fig. 3.7. Working of contractile vacuole in *Euglena*.

3.5 REPRODUCTION

No sexual reproduction occurs in *Euglena*. Asexual reproduction of *Euglena* takes place by the following methods:

1. Longitudinal binary fission. Under favourable ecological conditions such as food availability, optimum light, temperature and water, an active *Euglena* reproduces by **longitudinal binary fission (symmetrogenic)**. It is initiated by division of basal bodies. Nucleus is divided by peculiar **mitosis**, in which, the mitotic spindle is formed inside the intact nucleus; *i.e.*, nuclear envelope and endosome or nucleolus are not disintegrated during mitosis. Karyokinesis (nuclear division) is followed by cytokinesis (cytoplasmic division) which involves duplication of flagella, stigma, contractile vacuole, chloroplasts, mitochondria, etc. During later stages reservoir, cytopharynx and cytostome are duplicated by a longitudinal furrow. During final step of cytokinesis, a longitudinal furrow appears at anterior end of body and proceeds backward, dividing the parent body into two daughter euglenae.

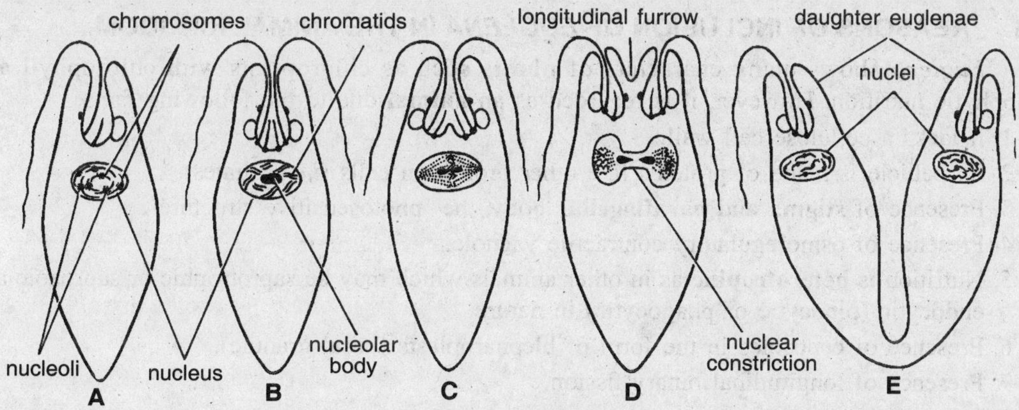

Fig. 3.8. *Euglena*. Longitudinal binary fission.

2. Multiple fission and palmella stage. Under unfavourable environmental conditions such as lack of food and oxygen, excessive heat or cold, draught, etc., *Euglena* becomes inactive and throws off its long flagellum and forms a protective thick (two layered yellowish – brown mucilaginous) **cyst** around it. A number of such encysted euglenae come together and become embedded in a common gelationous mass on the water surface. Within the cyst, each *Euglena* divides once by longitudinal fission or repeatedly forming 16 or 32 small-sized daughter forms. The latter form their own mucilaginous coats within the original cyst. A gelatinous mass containing many cysts is called **palmella stage** which resembles the palmella stage of many algae such as *Chlamydomonas*. Later on these daughter euglenae acquire normal flagella and escape out from the cyst to grow into normal euglenae.

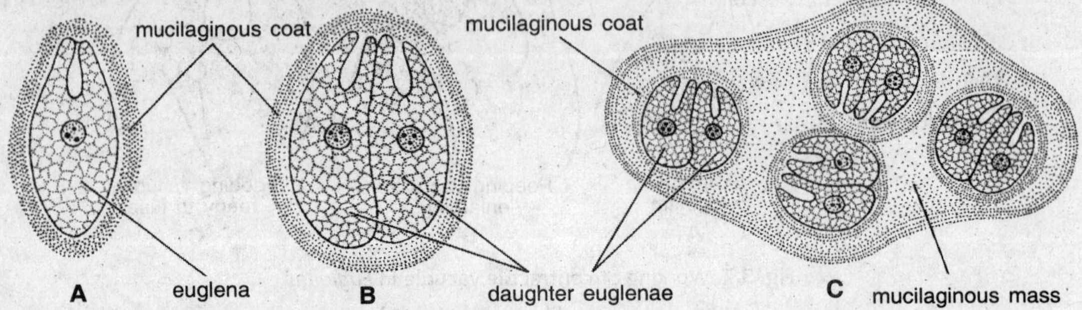

Fig. 3.9. *Euglena*. Multiple fission. A—Encysted individual; B—Fission in encysted individual; C—Palmella stage.

3. Encystment. During unfavourable conditions, *Euglena* undergoes **encystment**. For encystment, first *Euglena* becomes inactive, loses its flagella and secretes a thick rounded, red-coloured (red colour is due to **haematochrome** pigment) protective **cyst**. The cyst is secreted by the **muciferous bodies** lying below the pellicle of *Euglena*. Encysted euglenae stand a fair chance of wide dispersal. On return of favourable conditions, encysted euglenae emerge out and resume normal active life.

Economic importance. *Euglena* forms an important trophic level of an aquatic food chain. Being producer, it is consumed by herbivorous and plankton-eating fishes (primary consumers). Some species of *Euglena* have been used in experiments for waste water treatment and have been reported to extract heavy metals such as magnesium, iron and zinc from sludge (see **Brusca** and **Brusca**, 2003).

3.6 REASONS OF INCLUSION OF *EUGLENA* IN THE ANIMAL KINGDOM

Euglena shows many characters of **plants** such as chloroplasts with chlorophyll and holophytic nutrition. However, it is regarded as an **animal** due to the following facts:

1. It lacks a cellulose cell wall.
2. Its pellicle is made of proteins like other protozoan cells (*i.e.*, ciliates).
3. Presence of **stigma** and **paraflagellar** body, the 'photosensitive structure'.
4. Presence of osmoregulatory contractile vacuole.
5. Nutrition is **heterotrophic** as in other animals which may be saprotrophic or saprozoic and endocytic (pinocytic or phagocytic) in nature.
6. Presence of centrioles in the form of blepharoplasts (basal granule).
7. Presence of longitudinal binary fission.

QUESTIONS

Long Answer Questions

1. Write about the different habits of *Euglena*. Describe its morphological and anatomical features.
2. Give an account of structure, nutrition and reproduction in *Euglena*.

3. Describe the structure of *Euglena* and explain its movements.

4. Draw a well labelled diagram of a transverse section of flagellum and write functions of following in *Euglena*: 1. Stigma; 2. Paramylum; 3. Contractile vacuole; and 4. Cyst.

5. Give an illustrated account of the structure and life history of *Euglena viridis*.

6. Describe structure and reproduction of *Euglena*.

7. Give the structure of *Euglena* and point out the reasons for including it in Animal Kingdom.

8. Describe functional anatomy of *Euglena* giving reasons of its inclusion in Animal Kingdom.

9. Describe the structure, locomotion and mode of nutrition in *Euglena*. (*Bundelkhand 2000*)

Short Answer Questions

1. What do you mean by mixotrophic nutrition?

2. Discuss in brief chromatophores of *Euglena*.

3. Draw a well labelled diagram of *Euglena viridis*.

4. Write a detailed account of binary fission in *Euglena*.

5. Write short notes on the following:

(*i*) Locomotion in *Euglena*,

(*ii*) Nutrition in *Euglena*

(*iii*) Functions of flagellum in *Euglena*

(*iv*) Euglenoid movement

(*v*) Palmella stage

(*vi*) Flagellar movement

Very Short Answer Questions

1. Name the parts of oral apparatus of *Euglena*.

2. Pellicle of *Euglena* consists of two layers. Give their names.

3. Mention the alternative terms for basal bodies.

4. Where is stigma is located in *Euglena*?

5. What is endosome?

6. Name the species of *Euglena* that has permanently lost chlorophyll.

7. Where does the contractile vacuole discharge its contents in *Euglena*?

8. In *Euglena*, the binary fission is symmetrogenic. Is it so?

9. What is palmella stage?

Fill in the Blanks

1. Paramylum bodies in *Euglena* store .... as food reserve.

2. Paramylum bodies are the bioproduct of .... in *Euglena*. (*CCSU Meerut 2001*)

3. The nucleoli of *Euglena* are also known as ....

4. Sporulation in *Euglena* is found during ....

5. Gliding movement shown by *Euglena* is called ....

6. The mode of nutrition in *Euglena* is ....

7. A basal body is derived from ....

8. Paramylum is stored strach in ....

Multiple Choice Questions

1. In *Euglena*, food is stored in the form of (*Kanpur 2001*)

(*a*) paramylum bodies (*b*) chromatophores

(*c*) mitochondria (*d*) nucleus

2. *Euglena* moves with the help of

(*a*) cilia and flagella (*b*) undulating membrane

(*c*) pseudopodia (*d*) flagellum

3. Osmoregulation in *Euglena* is done by
(*a*) flagella (*b*) nucleus
(*c*) contractile vacuole (*d*) general body surface

4. Reserve food in *Euglena* is in the form of
(*a*) starch (*b*) glycogen
(*c*) fatty acid (*d*) paramylum

5. *Euglena* lacks
(*a*) cytostome (*b*) cytopharynx
(*c*) cytopyge (*d*) contractile vacuole

6. In *Euglena* metaboly is performed by
(*a*) flagellum
(*b*) pseudopodium
(*c*) sliding of microtubules against protein strips
(*d*) none of these

7. Nutrition in *Euglena* is
(*a*) autotrophic only (*b*) saprotrophic only
(*c*) holotrophic only (*d*) mixotrophic

8. *Euglena* contains chlorophyll
(*a*) *a* and *c* (*b*) *a* and *c*
(*c*) *b* and *c* (*d*) *a*, *b* and *c*

9. In *Euglena*, the binary fission is
(*a*) longitudinal (*b*) transverse
(*c*) irregular (*d*) lacking

10. Pyrenoid of chloroplast acts as a center for
(*a*) enzyme secretion (*b*) paramylum formation
(*c*) oxygen release (*d*) none of these

11. Locomotion of *Euglena* takes by
(*a*) pseudopodia (*b*) flagellum
(*c*) cilia (*d*) by gliding

12. *Euglena* reproduces
(*a*) both asexually and sexually (*b*) only asexually
(*c*) only sexually (*d*) conjugation

ANSWERS

Very Short Answer Questions

1. Cytosome, cytopharynx, reservoir.
2. Plasma membrane and epiplasm (or protein plates plus microtubules).
3. Blepharoplast, Kinetoplast.
4. On the flagellum in the cytopharyx.
5. Nucleolus in the nucleus of *Euglena*.
6. *Euglena gracilis*.
7. Into cytopharynx.
8. Yes.
9. Multiplication of *Euglena* within gelatinous cyst.

Fill in the Blanks

1. Starch
2. Photosynthesis
3. Endosomes
4. Unfavourable conditions.
5. Metaboly or Euglenoid movement
6. Mixotrophic
7. Centriole
8. *Euglena*

Multiple Choice Questions

1. (*a*) **2.** (*d*) **3.** (*c*) **4.** (*d*)
5. (*c*) **6.** (*c*) **7.** (*d*) **8.** (*a*)
9. (*a*) **10.** (*b*) **11.** (*b*) **12.** (*b*).

4

Amoeba (Phylum Protozoa)

4.1 CLASSIFICATION

Phylum	—	**Protozoa**
Subphylum	—	**Sarcomastigophora**
Superclass	—	**Sarcodina**
Class	—	**Rhizopodea**
Subclass	—	**Lobosia**
Order	—	**Amoebida**
Genus	—	***Amoeba***
Species	—	***proteus***

The Greek word *'amoibe'* means *change*. *Amoeba proteus* was discovered by **Russel von Rosenhoff** in 1755. He called it *"the little proteus* after the mythical sea-god **Proteus** who is believed to be capable of changing shape. **Leidy** was the first to describe *Amoeba proteus* in 1878.

4.2 HABIT AND HABITAT

1. Shape and Size

Amoeba proteus is an one-celled microscopic animal. It is about 0.6 mm across the body and about 0.25 mm in diameter. Under the compound microscope, living *A. proteus* appears as an irregular, jelly-like tiny mass of translucent protoplasm that is constantly changing its shape by sending out and withdrawing finger-like outgrowths, called **pseudopodia** or **"false feet".**

2. Pseudopodia

Amoeba constantly changes shape of its body. This is brought about by temporary outpushings of the cytoplam, called **pseudopodia**, which are regularly formed at any part of animals surface. Pseudopodia of *Amoeba* are broad, cylindrical with blunt rounded tips and composed of both ectoplasm and endoplasm. Such type of pseudopodia are called **lobopodia**. These lobopodia are used for locomotion and feeding. In *Amoeba proteus*, since many pseudopodia are formed simultaneously, it is called a polypodial species.

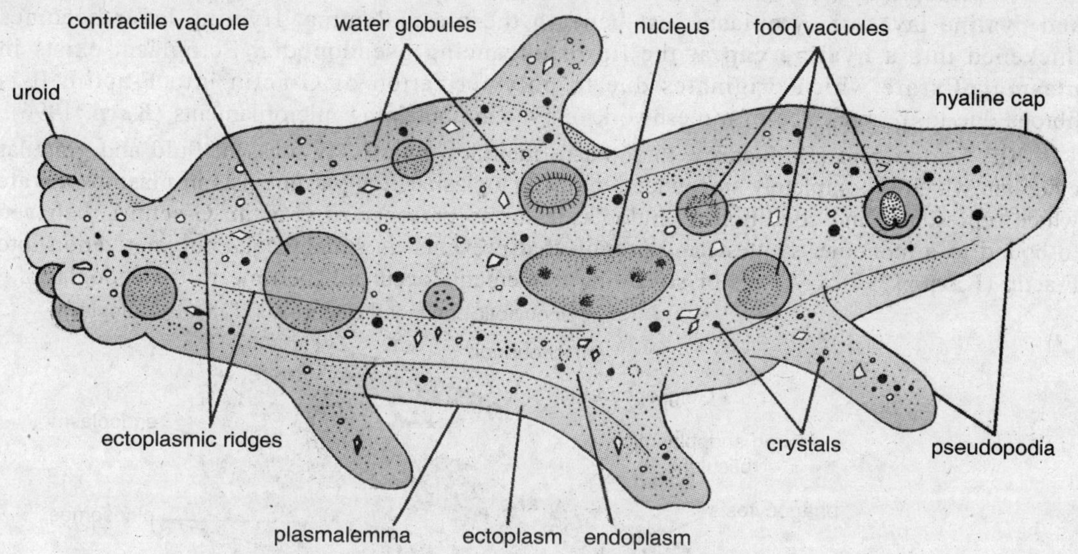

Fig. 4.1. *Amoba proteus.* Structure as seen under compound light microscope.

4.3 MORPHOLOGY

Amoeba is a free-living microscopic animal found living on the mud at the bottom of shallow freshwater ponds and streams, where there is some movement of water and plenty of decaying organic matter. It is **omnivorous** feeding on a varied diet of algae, ciliates, flagellates and bacteria. Normally, *Amoeba* creeps upon solid objects (*e.g.,* underside of aquatic vegetation), feeds upon a variety of food, responds to various stimuli and multiplies by mitosis.

Cellular Structure

The unicellular body of *Amoeba* has the following three main parts: 1. Plasma membrane; 2. Cytoplasm and 3. Nucleus.

1. Plasma membrane. *Amoeba* lacks a **cell wall**. Being enveloped merely by a delicate, 1 μm to 2μm thick, **plasma membrane** or **plasmalemma**, amoeba is said to be 'naked' like an ordinary animal cell. Plasma membrane is an unit membrane of fluid mosaic architecture. It consists of a lipid bilayer and many peripheral (extrinsic) and integral (intrinsic) **protein molecules** which are embedded in lipid bilayer. Due to presence of a lipid bilayer on its surface, amoeba does not get wet by water. Molecules of oligosaccharides (carbohydrates) often remain attached either to lipid or protein molecules of outer surface of plasma membrance. They form a adhesive layer, called **"glycocalyx"**.

Plasmalemma of *Amoeba* is a delicate but firm, semipermeable membrane, having remarkable power of regeneration, since, if injured it regenerates fast enough to prevent any loss of protoplasm. It is capable of evagination (for pseudopodia formation) and of invagination for **endocytosis,** which includes **pinocytosis** (intake of liquid food) and **phagocytosis** (intake of solid food).

2. Cytoplasm. The cytoplasm of *Amoeba* consists of various formed structures suspended in a colloidal **matrix, hyaloplasm or cytosol**. The cytosol is differentiated into outer ectoplasm and inner endoplasm.

(i) Ectoplasm (Ectosarc). The ectoplasm forms the outer, thin, clear (non-granular) and hyaline layer of cytoplasm just beneath the plasmalemma. Hyaline layer becomes thickened into a **hyaline cap** at the tip of advancing pseudopodia. Ectoplasm exists in **plasmagel state** which originates due to polymerization of G-actin into F-actin. It is fibrous due to the presence of a meshwork of F-actin containing microfilaments (**Karp**, 1996).

(ii) Endoplasm (Endosarc). Endoplasm includes main body mass of fluid and granular cytoplam which is completely surrounded by the ectoplasm. Endoplasm exists in **plasmosol state** which represents fluid condition of endoplasm due to presence of G-actin. G-actin is believed to bound to a monomer-sequestering protein that prevents its spontaneous polymerization into F-actin (**Karp**, 1996). Cyclosis or streaming movements occur all the time in the endoplasm.

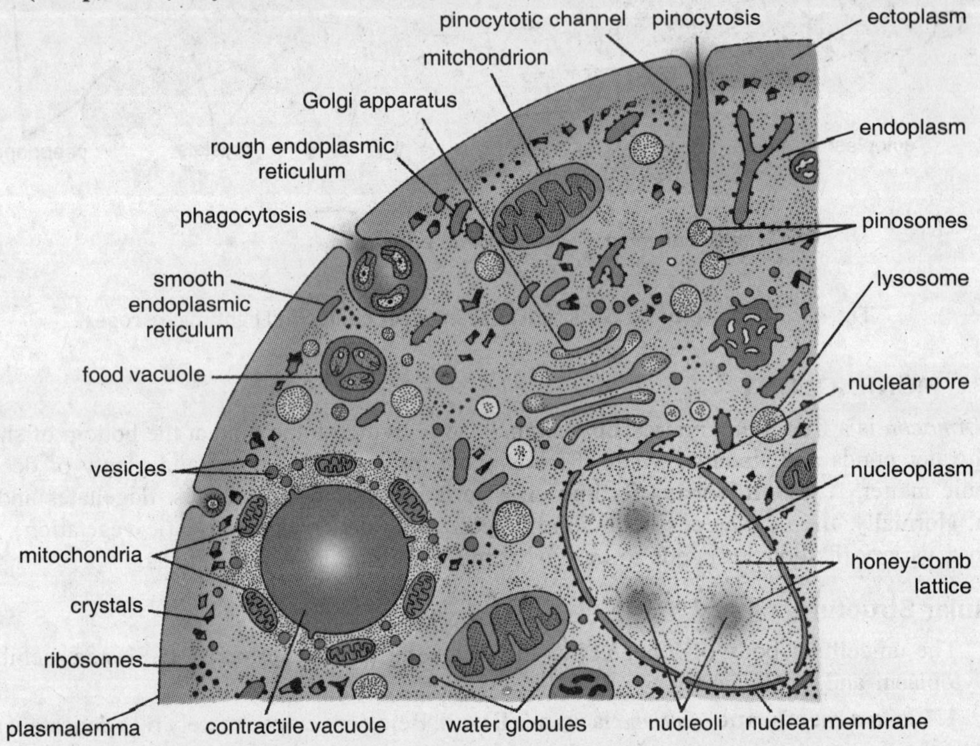

Fig. 4.2. *Amoeba.* Ultrastructure as revealed by electron microscope.

Endoplasmic organelles and inclusions. Endoplasm contains a nucleus, cytoplasmic organelles, inclusions and three types of vacuoles. **Cytoplasmic organelles** are membrane bound living structures performing various metabolic activities, *e.g.,* endoplasmic reticulum, mitochondria, two or three sets of Golgi apparatuses, lysosomes, ribosomes, etc.; the presence of all of these organelles in *Amoeba* is revealed only by electron microscope (Fig.4.2.). **Inclusions** of *Amoeba* are non-living storage organic or inorganic molecules such as glycogen granules, fat droplets and minute crystals of carbonyl diurea, etc.

Contractile vacuole. A single contractile vacuole, exists in the outer part of the endoplasm near the posterior end of body. It appears like a bubble of clear, watery fluid, enclosed within a delicate and elastic unit membrane (sometimes called **condensation membrane**). Numerous mitochondria exist around the contractile vacuole; they provide energy for the working of contractile vacuole. Many tiny vacuoles of water, called **accessory vacuoles**, appear in the vicinity of main vacuole which coalesce to form the main vacuole. The function of contractile vacuole is **osmoregulation**: excess quantity of water regularly collects into it from cytoplasm and is periodically discharged out by it.

Food vacuoles. Numerous small and noncontractile food vacuoles, concerned with **nutrition** exist in endoplasm. They move about with cyclosis (*i.e.*, rotatory streaming movement of endoplasm). Food vacuoles are not permanent structures; each is formed when *Amoeba* ingests some food by endocytosis.

Food vacuoles vary in shape and size according to the ingested food particles. Food is digested within these food vacuoles by the help of lysosomal enzymes. When contained food is fully digested, the concerned vacuole reaches body surface and bursts to discharge out its contents including indigestible remains of food (exocytosis).

Water vacuoles. These are several small, spherical, non-contractile vacuoles filled with water.

3. Nucleus. *Amoeba proteus* contains a single, circular, disc-like or concavo-convex **nucleus**. Nucleus is embedded in the endoplasm but occupies no fixed position. Its delicate, firm and porous **nuclear envelope** (or **karyotheca**) consists of two unit membranes. Its **nucleoplasm** which has honey-comb lattice, contains many **nucleoli** and about 500 scattered chromatin granules, called **chromidia**. Nucleus is concerned with the organization and integration of the animal's metabolic and reproductive processes.

4.4 PHYSIOLOGY

In spite of its simplicity, *Amoeba* performs all the vital activities of life such as locomotion, osmoregulation, nutrition, digestion, assimilation, egestion, metabolism, respiration, excretion, response to stimuli, reproduction and growth.

1. Locomotion (= Amoeboid Movement)

Amoeba exhibits characteristic **amoeboid movement** by the formation of temporary finger-like projections of body, the pseudopodia. Since pseudopodia of amoeba are thick blunt, finger-like in appearance having rounded tips, they are called **lobopodia**. Amoeboid movement is a slow type of movement and also occurs in many other protozoans, in amoebocytes of sponges and in leucocytes of vertebrates.

Amoeboid movement in amoebae depends upon following three basic factors:

1. Attachment to substratum. Since amoeboid movement involves crawling, it needs contact of body with a solid substratum. **Adhesion** with substratum is facilitated by glycocalyx and also by traces of inorganic ions such as Ca^{++}, Mg^{++} and K^{+}, that are present in the surrounding water.

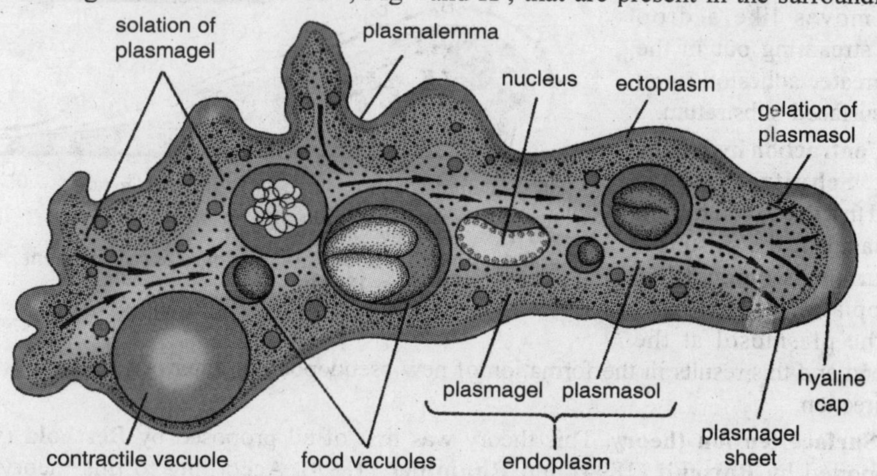

Fig. 4.3. *Amoeba proteus*. Solation and gelation of cytoplasm during amoeboid movement.

2. Continuous viscosity conversions. Conversion of **plasmagel** (ectoplasm) into sol at the trailing end and of **plasmosol** (endoplasm) into gel at the advancing end (Fig.4.3.) must constantly occur. This process involves depolymerization of F-actin (gel) into G-actin (sol) at the posterior end of body and polymerization of G-actin (sol) into F-actin (gel) at the anterior end of body.

3. Force and flow. There must be some force to cause forward flow of plasmosol (endoplasm) from behind into a pseudopodium.

Mechanism of amoeboid movement. In amoeba, locomotion may involve a single large, tubular **lobopodium** or several small one with caps of hyaline protoplasm at the tips. At various points, a pseudopodium is anchored to the substratum.

The theories of amoeboid movement accepted by most zoologists at present assume that cytoplasmic flow is related to the changes between sol and gel states of the peripheral cytoplasm. The pseudopodial tip controls the change. As a result of some initial stimulus, the outer gelled ectoplasm becomes fluid (sol) at the site where the pseudopodium will form and internal pressure causes the inner fluid endoplasm (= plasmasol) to flow out at this point, forming a pseudopodium. In the interor of the pseudopodium, the endoplasm flows forward along the line of progression. Around the periphery, endoplasm is converted to ectoplasm (= gel), thus, building up and extending the sides of pseudopodium like a well starched sleeve.

In the conversion of endoplasm to ectoplasm **actin** subunits (called **G-actin**; G = globular) polymerize, *i.e.*, become longer **F-actin** (F= filamentous; *e.g.*, microfilament) and bind to each other at more or less right **angles**, creating a **mesh** of microfilaments. It is this mesh that accounts for the rigid gelatinous state of ectoplasm. The small mesh size excludes organelles and thus, accounts for the hyaline appearance. At the posterior end of body, ectoplasm is converted to endoplasm by depolymerization (**Karp**, 1996; **Ruppert** and **Barnes,** 1994).

The protein **filamin** and **α-actinin** link the actin microfilaments together into a network. This converts the sol into gel. The protein **gelsolin** and **villin** break up actin network. This changes the gel to sol. Ca^{2+} (ions) concentration and pH regulate the sol-gel transitions.

Theories regarding amoeboid locomotion. How pseudopodia are formed and bring about locomotion is explained by the following theories:

1. Adhesion theory. According to this theory, amoeba moves like a drop of water streaming out in the path of greater adhesion upon a rough-surfaced substratum.

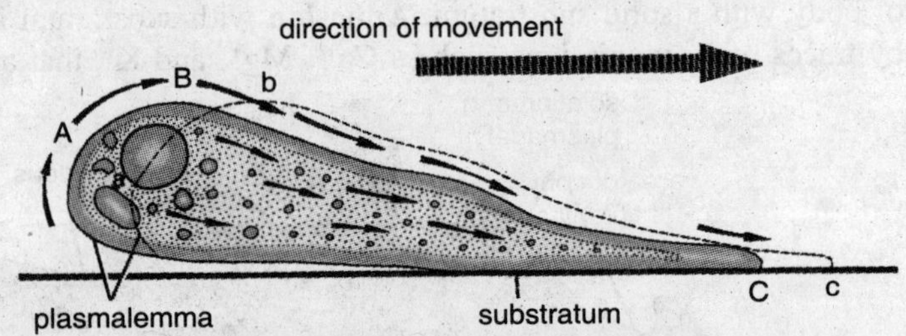

Fig. 4.4. *Amoeba verrucosa* showing rolling movement (**Jennings,** 1904).

2. Contraction hydraulic theory. Schultze (1875) postulated that due to contraction of plasmagel at the posterior end, a current is produced in the protoplasm. This current pushes the plasmosol at the anterior end and this results in the formation of new pseudopodium anteriorly and the body moves in that direction.

3. Surface tension theory. This theory was first of all proposed by **Berthold** (1886) and later supported by **Butschii** (1894) and **Rhumbler** (1898). According to this theory, constant internal tension forces the cytoplasm to shoot out from surface in the form of pseudopodium wherever surface tension suddenly lowers due to local effects.

4. Rolling movement theory. By placing a particle of lamp-black (carbon) on the surface of *Amoeba verrucosa* (which is a unipodal species forming a single pseudopodium at a time), **Jennings** (1904) noted that amoeba simply rolls like a ball during locomotion (Fig. 4.4).

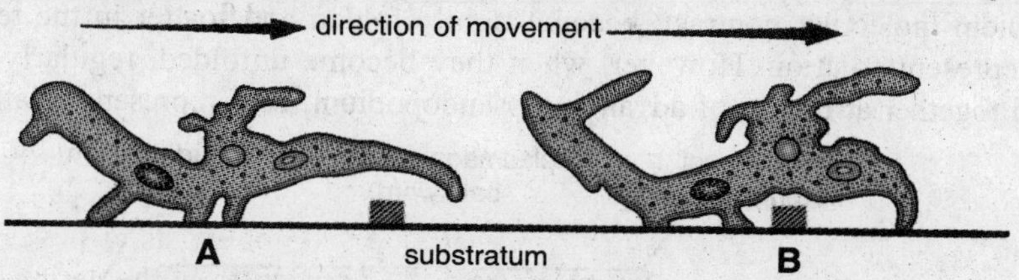

Fig. 4.5. *Amoeba proteus*. Walking movement due to contraction in the protoplasm.

5. Walking movement theories. Dellinger (1906) proposed that *Amoeba proteus* virtually *"walks on pseudopodia"* just as higher animals move on their legs. Thus, as the pseudopodium strikes an object, it sticks to it. This causes contraction of an edoplasmic fibrous network of contractile elements, pulling the whole body mass close to the pseudopodium (Fig. 4.5).

6. Sol-gel theory. This theory is the most accepted theory of pseudopodium formation. It was proposed only when cytoplasm was recognized as a **colloidal** system. Sol-gel theroy was strongly advocated by **Hyman** (1917) and also adopted by **Pantin** (1923-1926; see Box 4.1) and **Mast** (1925-1931). This theory postulates that pseudopodia are formed and withdrawn due to **cyclic viscosity change** in the collodial cytoplasm from sol to gel and gel to sol states. Acording to **Mast**, amoeboid movement is brought about by following four processes:

1. Contact of plasma membrane of advancing side of body with a substratum and resultant weakening of plasmagel (ectoplasm) layer here due to its partial solation.

Box 4.1 Pantin's Theory

Pantin (1923) studied the locomotion of the marine *Amoeba limax*. It forms a pseudopodium by swelling of protoplasm due to secretion of acid and absorption of water at that place. As the pseudopodium forms and extends in front, a gelatinous ectoplasmic tube is formed. At the posterior end this ectoplasmic tube changes into endoplasm. This ectoplasmic tube contracts and forces the endoplasm to stream to the front, this brings about locomotion.

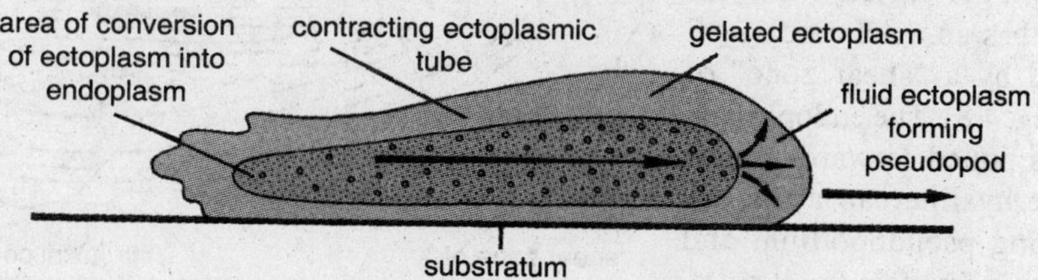

Fig. 4.6. *Amoeba limax*. Ectoplasmic tube forces endoplasm forward and causes locomotion.

2. Sudden forward streaming of plasmasol (endoplasm) under tension, which breaks the weakened plasmagel sheet and is converted into gel after interacting with the hyaline cap.
3. Conversion of plasmagel into plasmasol in trailing part of body.
4. Simultaneous contraction in liquifying plasmagel at the rear end.

According to **Pantin**, fluid plasmasol pushes against hyaline cap and fans out to all sides in a fountain flow and gelates.

Sol-gel theory of **Mast** received wide support and become a classic textbook explanation for amoeboid locomotion.

7. Molecular folding-unfolding theory. Goldacre and **Lorch** (1950,52) provided a strong support to sol-gel theory by explaining the biophysical and biochemical processes involved in reversible 'sol-gel' changes and contraction of plasmagel. According to this theory, when the long-chain protein molecules contract, become tightly folded and scatter in the temporary hind region, they represent solation. However, when they become unfolded, regularly arranged and tightly bonded together at the tip of advancing pseudopodium, they represent gelation (Fig. 4.7).

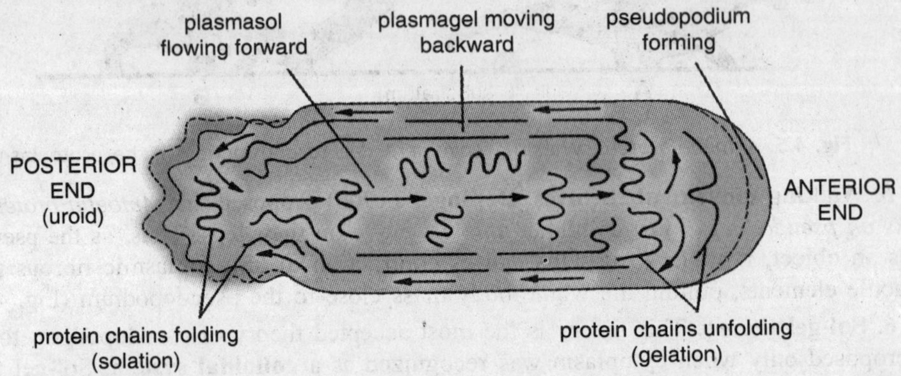

Fig. 4.7. Molecular folding-unfolding during solation and gelation of cytoplasm for amoeboid movement.

Quite recently, role of polymerization of actin subunits (G-actin) into actin microfilaments (F-actin) and vice-versa have been established for folding-unfolding or sol-gel theroy (**Karp,** 1996).

8. Sliding theroy. According to **Huxley** (1954), the proteins participating in solation and gelation are similar to **actin** and **myosin** of muscle fibres. Their folding-unfolding occurs due to sliding of actin and myosin filaments and expends great amount of ATP molecules.

9. Fountain zone contraction theory (or Front contraction theory or Pull hypothesis). Allen (1962) postulated that ectoplasm of *Amoeba* is distinguished into an outer fluid hyaline layer and an inner viscid "gel" layer, while the endoplasm has an axial core of gel surrounded by a "shear zone" of fluid sol (Fig. 4.8). The endoplasmic axial gel is pulled forward, strikes against the hyaline cap at the tip of advancing pseudopodium and fans out like a fountain to all sides, becoming continuous with the gel layer of ectoplasm. In the "fountain zone", the **gel contracts** by releasing some water into the hyaline cap, thus, pulling the axial core of endoplasm farther forward.

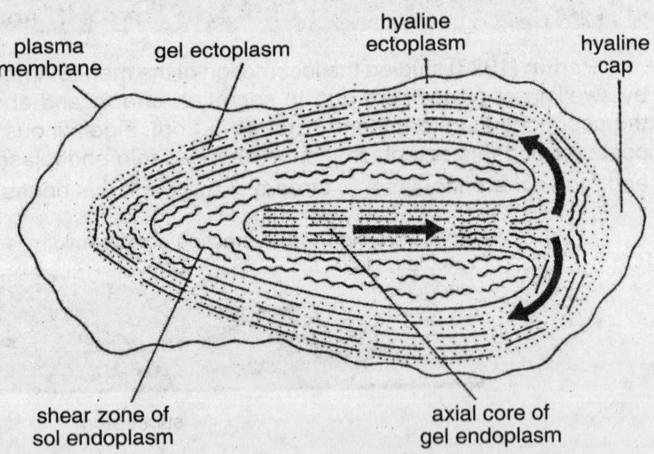

Fig. 4.8. Amoeboid movement on the basis of Allen's fountain zone contraction theory.

10. Ectoplasm contraction theory (or push hypothesis). According to **Goldacre**, the gel portion (ectoplasm) contracts at the rear end of the body and squeezes the sol portion (endoplasm) forward, *i.e.*, force is created from behind, like the squeezing of the back of a tooth paste tube. On reaching the anterior end the sol is converted into gel.

11. Contraction-hydraulic theory. This theory was postulated by **Rinaldi** and **Jahn** (1963). It discarded Allen's fountain-zone contraction theory and supported Mast's sol-gel theoy. According to it, contraction of plasmagel layer at trailing end exerts a **hydraulic pressure** upon the plasmasol. Resistance to this pressure is minimum at the front end, highest in the middle and medium at the rear. Due to this reason, more and more of plasmasol tends to flow forward into the advancing pseudopodium.

12. Contractile jacket theory. This is modern version of Huxley's sliding theory. According to it, bonding of **myosin** with the **actin mesh** could convert the mesh into a **contractile jacket**, forcing the interior fluid endoplasm forward. **Lee** and **Jacobson** (1993) have demonstrated that **myosin I** (bipolar filaments) act as **primary motor** which drives cell locomotion. Myosin motor acts on actin filaments (microfilaments) and causes them to slide toward the leading edge of the cell.

13. Osmotic theory. According to **Ruppert** and **Barnes** (1994), the initial depolymerization of the actin mesh (*i.e.*, plasmasol formation) at the pseudopodium tip would increase the number

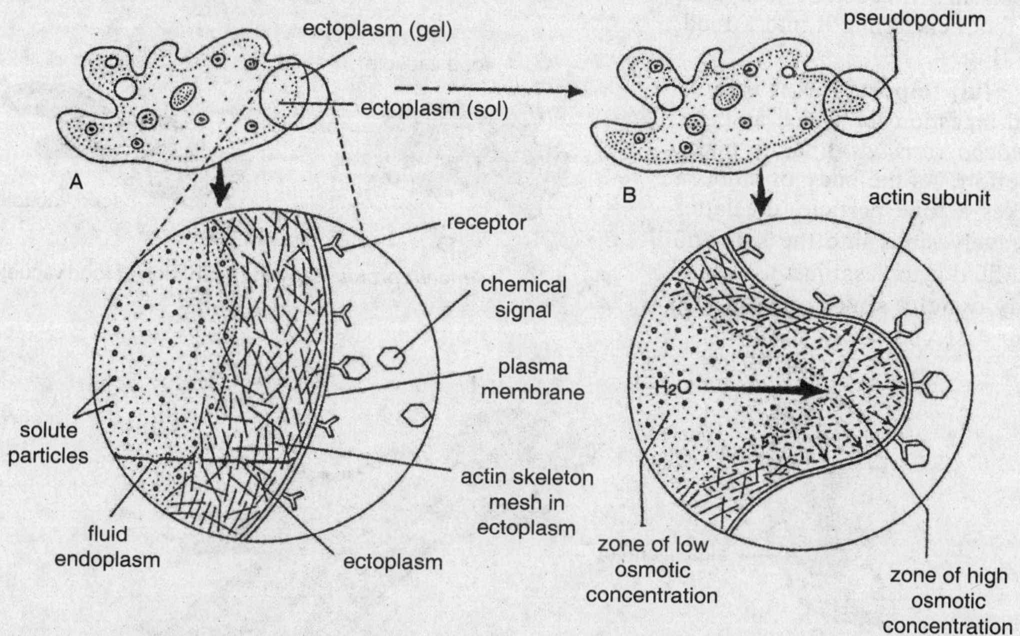

Fig. 4.9. *Amoeba*. Osmotic theory of flow of endoplasm during amoeboid locomotion. Actin mesh creates gel state of ectoplasm. A—Chemical signals bind to membrane receptors and initiate depolymerization of the actin mesh. B—Depolymerized actin subunits raise osmotic concentration of this region of ectoplasm and water flows in. The influx of water causes the extension of the cell as pseudopodium. Peripheral to the pseudopodial tip, actin subunits repolymerize as skeletal mesh, extending the ectoplasmic gel state forward like a progressive sleeve.

of particles (actin subunits). Particle increase would raise the **osmotic concentration** and water would flow from the endoplasm out into the tip. The influx of water causes the extension of the cell as a pseudopodium.

2. Nutrition

Amoeba is **holozoic** (= ingest solid organic particles) and **omnivorous**. It ingests all kinds of microorganisms such as algal cells and filaments, bacteria, other protozoans (smaller flagellates and ciliates) and small metazoans such as rotifers and nematodes. Nutrition of *Amoeba* involves the following activities.

1. Ingestion. *Amoeba* lacks any mouth and food is ingested at any point of body surface, but it is usually at the advancing anterior end. In *Amoeba*, food is captured by pseudopodia. Ingested food along with some water occupies a vesicle, called **food vacuole** (Fig. 4.10). According to the nature of food *Amoeba* employs the following five methods of ingestion.

(i) Circumvallation. When an amoeba comes near an actively moving prey, the body part immediately in line with it stops moving and pseudopodia are formed above, below and on the sides of the food to form a **food cup**. Soon edges of food cup fuse around the food to form a non-contractile membrane-bound food vacuole (Fig. 4.11C).

(ii) Circumfluence. This method of ingestion is used for immobile or less active food particles such as bacteria. For ingesting such food, the whole cytoplasm of *Amoeba* flows around the food, encircling it into a food cup (Fig. 4.11B).

(iii) Import. This is passive food-ingestion for which amoeba (*Amoeba verrucosa*) has to make no effort. As the body of amoeba strikes a food particle, the latter passively sinks into the soft and semifluid cytoplasm just as a solid body would sink in a swamp (Fig. 4.11A).

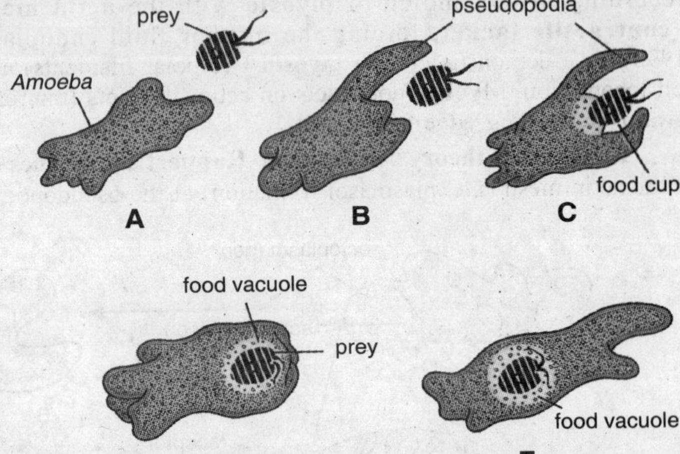

Fig. 4.10. *Amoeba proteus* showing formation of food vacuole.

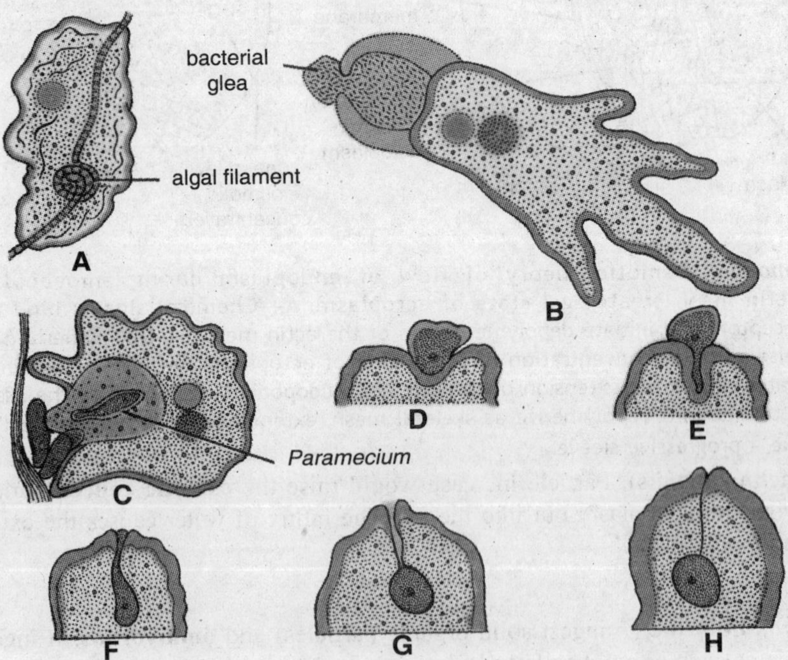

Fig. 4.11. *Amoeba*. A—Ingestion by import; B—Ingestion by circumfluence; C—Ingestion by circumvallation; D,E,F,G and H—Ingestion by invagination.

(iv) Invagination. Firstly, the prey is killed and held near the body surface by adhesive secretion of pseudopodia of amoeba (*Amoeba verrucosa*) and plasma membrane invaginates at the point of contact and encloses the prey into a food vacuole.

(v) Pinocytosis. Pinocytosis (cell drinking) is the process of ingestion of fluid material in bulk by the plasma membrane. It was first of all observed by **Mast** and **Doyle** (1934) in *Amoeba proteus*. By pinocytosis, *Amoeba* can ingest colloids from surrounding medium. **Pinocytic vesicles** or **pinosomes** form by invagination of plasma membrane particularly at the tips of pseudopodia (Fig. 4.12). Pinosomes containing engulfed food material pinch off from internal ends of pinocytic channels and sink into the endoplasm.

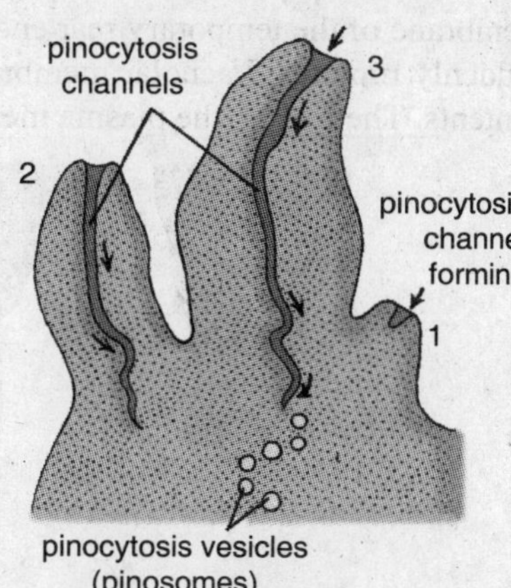

Fig. 4.12. *Amoeba. Ingestion by pinocytosis*

2. Digestion. Digestion occurs within the food-vacuoles by the help of lysosomal enzymes. These food vacuoles are analogous to the alimentary canal of higher animals such as earthworm and frog, except that digestion in food vacuoles is **intracellular** and that in alimentary canal is **extracellular**.

The contents of food vacuoles (called **primary food vacuoles**) become acidic due to HCl but later they become alkaline (current reports do not agree with this fact; **Karp**, 1996). Digestion of food in food vacuoles occurs by means of lysosomal enzymes which work in an acidic medium (see Box 4.2). Lysosomes fuse and merge with food vacuoles to release their enzymes into them. The presence of some enzymes such as **proteases** (protein-splitting enzymes), **amylase** (starch-digesting enzymes) and lipases (fat-splitting enzymes) have been demonstrated in *Amoeba*. Thus, due to digestion proteins are hydrolysed into amino acids, starch into glucose and fats into fatty acids and glycerol. When the digested food is reduced to molecular form, the food vacuole buds off into smaller and smaller **secondary food vacuoles** which carry away the digested food.

Box 4.2 Lysosomes

Lysosomes act as cell's digestive organelles. Included within the boundaries of a typical lysosome is a collection of about 50 different hydrolytic enzymes. These enzymes are capable of hydrolyzing virtually every type of biological macromolecules. The enzymes of a lysosome share an important property: **all have their optimal activity at an acid PH**—they are **acid hydrolases**. The pH optimum of these enzymes is correlated with the low pH of lysosomal compartment, which is approximately 4.6. The high internal proton (H^+ions) concentration is maintained by a proton transporter (an H^+-ATPase) present in lysosomes boundary membrane. Lysosomal membranes contain two groups of acidic, highly glycosylated integral proteins, called **Igp-A** and **Igp-B**, whose function may be to protect the lysosomal membrane from attack by the enclosed enzymes (**Karp**, 1996).

3. Absorption and assimilation. Each food vacuole exists in endoplasm for about 15 to 30 hours; it is the time taken for complete digestion of food in it. During this period, the food vacuole continuously circulates in the body with cyclosis of plasmosol and directly supply nourishment to all parts of body. The digested food, water and minerals are absorbed by the surrounding endoplasm by simple process of **diffusion** and immediately get **assimilated** to build new protoplasm. Amino acids are used in making new types of proteins; excess amino acids are changed into sugars and stored as reserve food in the form of **glycogen**. Fats are also stored.

4. Egestion. *Amoeba* lacks an anus (vent) or fixed place for the act of egestion. After its food contents are completely digested, a vacuole falls out of cyclosis and becomes stationary.

Thus, it is simply left behind as the advancing body flows forwards (Fig. 4.13). As the plasma membrane of the temporary rear end touches such a vacuole containing undigested food residue, it suddenly ruptures. Vacuolar membrane also simultaneously ruptures to allow exit of the vacuolar contents. There after, the plasma membrane immediately heals up to prevent any loss of cytoplasm.

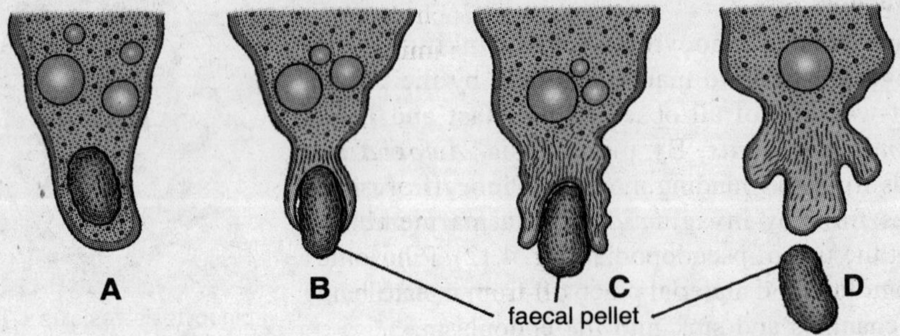

Fig. 4.13. *Amoeba verrucosa*. Various stages of egestion.

3. Osmoregulation

The process by which the water contents of the protoplasm are controlled is called **osmoregulation**. In *Amoeba proteus* the process of osmoregulation is performed by the single contractile vacuole.

Since, freshwater is **hypotonic** to the cytoplasm of amoeba, there occurs a continuous **endosmosis** or inflow of water into the body through the semipermeable plasma membrane (= body surface). Also some water is added to cytoplasm from food vacuoles and from catabolic reactions. If all this water is allowed to accumulate, the body of amoeba will soon swell and burst. Hence, amoeba possesses a contractile vacuole which regularly pumps out excess water and it needs a lot of energy for this pumping (This is evident by the presence of many mitochondria in the vicinity of contractile vacuole). Contractile vacuole operates into two phases—**diastole** and **systole**.

In the phase of diastole, the contractile vacuole grows in spurts to its optimum size by coalescing with small **accessory vacuoles** which collect water from the cytoplasm and appear in its close vicinity. Then, the contractile vacuole becomes stationary and reaches the temporary hind end of body, as the moving body of amoeba flows forward. As the vacuole touches the rear plasma membrane, it bursts out and collapses. This is **systole phase**. Soon contractile vacuole reappears at its usual place as a minute vacuole and starts growing in its **diastole phase** as before.

4. Respiration

Amoeba is an aerobic animal, it takes in oxygen and gives off carbon dioxide. Oxygen is required for metabolism (*i.e.*, oxidation of food and for the production of energy-rich compound ATP) and is obtained from surrounding water. Similarly, carbon dioxide formed in body as a by product of oxidation, also leaves the body in solution. Such a **gaseous exchange** occurs through the general body surface (plasma membrane) of *Amoeba* by **diffusion** in accordance with diffusion gradients. Thus, due to active metabolism (=catabolism) the concentration of O_2 is always lower and that of CO_2 always higher in the cytoplasm than in surrounding water. So, O_2 continuously enters into the body and CO_2 leaves it in exchange.

5. Excretion

Elimination of nitrogenous byproducts (toxic metabolic wastes) from the body is called **excretion**. *Amoeba* is an **ammonotelic** animal, since its principal excretory waste is **ammonia**.

Amoeba lacks speical excretory organelles to excrete ammonia. Like CO_2, ammoina also continuously diffuses out, in solution, through plasma membrane, because its concentration is always higher in the body than in surrounding water.

Sometime, nitrogenous waste of *Amoeba* also includes **urea**, which is transformed into storable minute crystals of **carbonyl diurea**. Such crystals are found in the endoplasm and are called **biurets** (bipyramidal) and **triurets** (tripyramidal).

4.5 BEHAVIOUR

Amoeba does not possess any specific sensory organelles, however, it is able to respond to a variety of stimuli. The responses to different forms of stimuli are called **taxes** (**singular taxis**). A taxis may be either **positive**, in which animal moves towards the stimulus, or **negative** in which the animal moves away from the stimulus. *Amoeba proteus* exhibits both types of taxes, positive as well as negative. For example, it can discriminate between different types of food (positive chemotaxis for food and negative chemotaxis for sand), it moves away from bright light (negative phototaxis), strong chemical solutions (negative chemotaxis) and peristent mechanical irritation (negative thigmotaxis). Violent disturbances causes it to withdraw all pseudopodia and remain stationary for some time. Response of *Amoeba* to gravity is positive. This is called **geotaxis**.

4.6 REPRODUCTION

Amoeba proteus reproduces mainly by asexual reproduction which takes place only by **binary fission** (**Green** *et al.*, 1990). Other species of *Amoeba*, however, reproduce asexually by **multiple fission, sporulation** and **cyst formation** in addition to the common binary fission. Amoeba does not reproduce **sexually** by mating, *i.e.*, by the fusion of gametes.

1. Binary Fission

Amoeba undergoes binary fission during favourable conditions of food and temperature. Binary fission includes **mitotic cell division**, in which there is a correlation between nuclear division and changes in external (*i.e.*, cytoplasmic) structures. Mitosis of binary fission is triggered off in response to limits imposed either by the surface area to volume ratio, and/or the ratio between cytoplasmic volume and nuclear volume. Thus, when *Amoeba* attains a maximum size, *i.e.*, 0.25 mm, it starts to reproduce. Just prior to occurrence of binary fission, amoeba becomes sluggish and spherical with its surface covered with small radially arranged pseudopodia. During occurrence of binary fission, contractile vacuole ceases to function and disappears in prophase. In prophase, nucleus undergoes following changes: nuclear honey-comb lattice and nucleoli disintegrate, nucleus elongates to form an **intranuclear spindle** and 500 chromidia becomes duplicated. In **metaphase a multipolar spindle** is formed and chromosomes arrange at the equator of the spindle to form **metaphase plate**. In **anaphase**, the chromatids (= daughter chromosomes) separate and move towards opposite poles of spindle which becomes **tripolar** during mid anaphase and finally becomes **unipolar** during late anaphase. During late anaphase, the pseudopodia become larger and irregular in shape; and the nucleus first becomes dumb-bell shaped by a middle constriction and finally divides into two daughter nuclei. In **telophase**, pseudopodia assume normal shape, the body first elongates, then constricts in the middle, and finally divides into daughter amoebae, each having a daughter nucleus. Under the ideal conditions (*i.e.*, about 24°C) the whole process of binary fission can be completed within 30 minutes. Each daughter *Amoeba* then proceeds to feed and grow to maximum size.

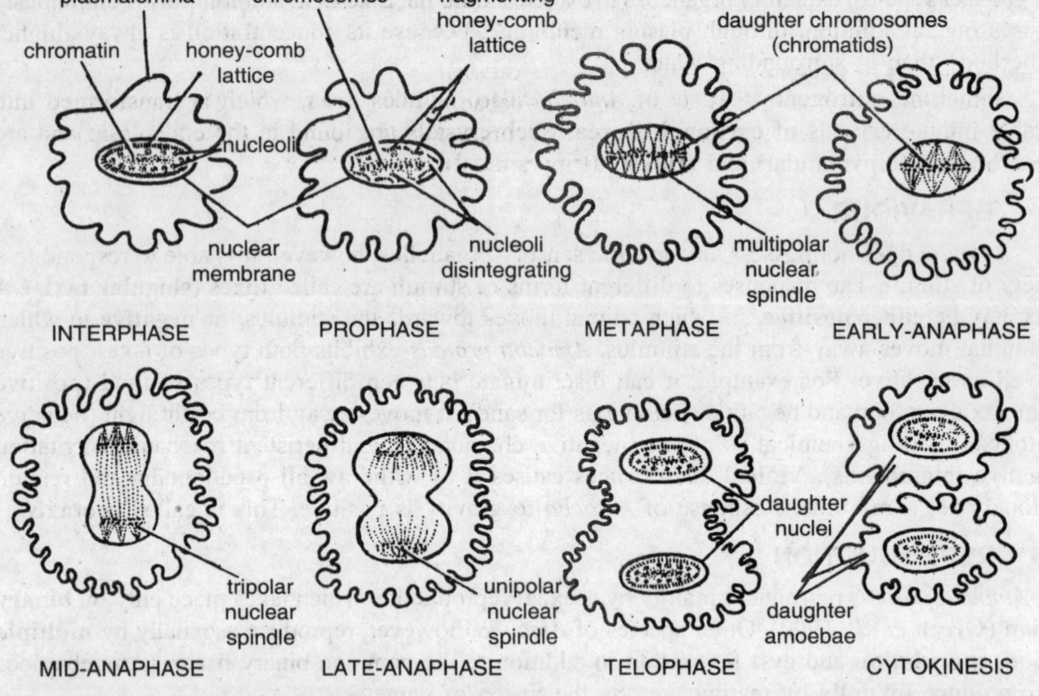

Fig. 4.14. Amoeba. Stages of binary fission.

Immortality of Amoeba

According to **Weismann's** classical **germ plasm theory**, the body of a multicellular animal (metazoa) is divisible into two parts: somatic or germinal. The **somatic part** or **soma** is the vegetal part, concerned merely with the maintenance of body itself. Due to continuous wear and tear, soma stops functioning after a certain age. Thus, soma is **mortal** and lasts only upto death of the organism.

The **germinal part** or **germ** is, on the other hand, concerned with reproduction; at least some part of it is transferred in the form of gametes to the progeny before death of the organism. Germ is, therefore, **immortal**.

The body of *Amoeba*, however, is not divisible into soma and germ. So its whole body gets divided into two daughter amoebae in binary fission. Although the parent amoeba loses its identity in binary fission, *i.e.*, does not exist thereafter, but no part of its body has been wasted or destroyed (parent becomes completely merged in the offspring). Thus, there is a continuity of life; *Amoeba* is potentially **immortal** and it never dies. Death of *Amoeba* can, however, take place due to starvation, accident or some other gross ecological hazards.

2. Multiple Fission

This mode of asexual reproduction occurs in *Amoeba* when there exist unfavourable ecological conditions such as scarcity of food, rise and fall in temperature, etc. Pseudopodia are withdrawn, the animal becomes rounded, streaming movements (cyclosis) of endoplasm ceases, protoplasm becomes minutely granular, and distinction between ectoplasm and endoplasm is lost. The animal secretes a highly resistant 3-layered **cyst** around itself. Now, its nucleus undergoes repeated mitotic divisions, forming about 500 to 600 daughter nuclei. Each daughter nucleus becomes surrounded by a small segregated mass of cytoplasm and forms **pseudopodiospores** or **amoebulae**. On the return of favourable conditions, the cyst absorbs water and ruptures to release the amoebulae which grow into adult amoebae. According to current reports, no multiple fission occurs in the cyst, but cyst formation occurs.

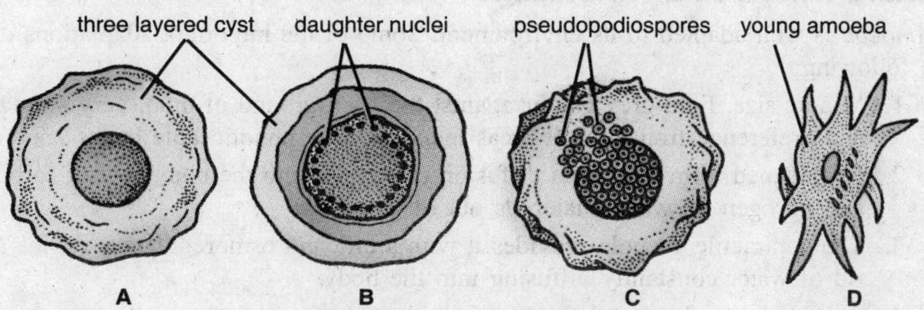

Fig. 4.15. *Amoeba*. Encystment and multiple fission.

3. Encystment

In *Amoeba*, encystment occurs regularly for **survival** under unfavourable conditions, *i.e.*, to tide over draught and extreme temperature, etc. Encysted amoeba lives under "suspended animation" with extremely slow metabolism. Encystment is also a mean of **dispersal**; cysts are blown far and wide by winds. On return of favourable conditions, amoeba hatches out from its cyst **(excystment)** and resumes active life.

4. Sporulation

Some species of *Amoeba* multiplies by a process of **sporulation** without encystment during unfavourable conditions. Sporulation begins with break down of nuclear membrane and release of **chromatin blocks** into the cytoplasm. Each chromatin block acquires a nuclear membrane to form a nucleus. Now the cytoplasm of parent body segregates into small masses, each around a daughter nucleus (Fig. 4.16). Thus, about 200 minute spore-like daughter amoebulae are formed in the body of parent amoeba. The peripheral cytoplasmic layer of each amoebulae changes into a tough and resistant spore-membrane. Finally, the parent body disintegrates and spores are set free. Spores sink down at the bottom of the pond. On return of favourable conditions, their spore-membranes rupture. Thus, a minute amoeba emerges out from each spore and repidly grows into an adult amoeba.

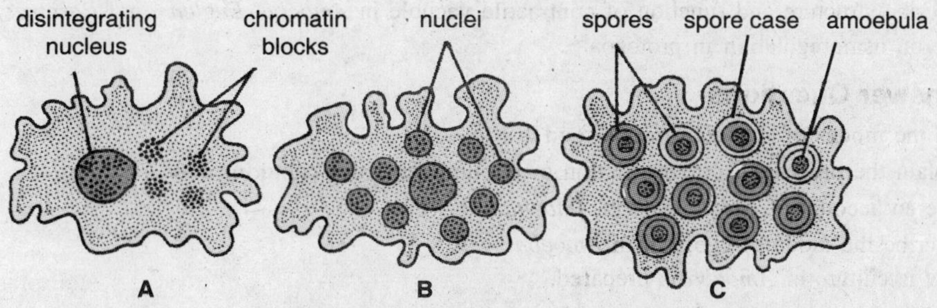

Fig. 4.16. *Amoeba*. Sporulation.

5. Regeneration

Amoeba has a tremendous power of regeneration. If it is cut into small pieces, each piece containing a part of the nucleus, rapidly regenerates into a complete amoeba. A piece without a nuclear fragment does not regenerate and soon perishes.

4.7 ADAPTATION TO ENVIRONMENT

Amoeba is well adapted to its environment. Some of the important adaptations of Amoeba are the following:

1. Minute size, light-grey colour against the back-ground of mud, very slow movement and preference from dim-lit areas make amoeba unnoticeable by its enemies.
2. Its flattened form facilitates diffusion of oxygen into the body, and of carbon dioxide and nitrogenous waste materials out of it.
3. The contractile vacuole provides it with a efficient osmoregulatory device for getting rid of water constantly diffusing into the body.
4. Amoeba's omnivorous habit ensures adequate food everywhere.
5. Good degree of irritability ensures its survival.
6. High rate of binary fission maintains its population during favourable conditions.
7. Cyst formation affords a good means of tiding over unfavourable conditions and dispersal to new localities.

Amoeba is a modern organism (animal), well adapted to its environment, but with little morphological differentiation. Its simplicity is confined to structure only. Physiologically it is as complex as any metazoan, because it carries on all the vital functions of life.

QUESTIONS

Long Answer Questions

1. Write about the habitat, habits and adaptations of *Amoeba*.
2. Describe the structure of *Amoeba*.
3. Give the process of locomotion and nutrition in *Amoeba proteus*.
4. State the process of digestion and the function of the contractile vacuole in *Amoeba*.
5. Describe different theories of locomotion in *Amoeba*.
6. Justify the statement that *Amoeba* is immortal.
7. Describe locomotion in *Amoeba* giving various theories connected with amoeboid movements.
8. Explan how respiration, excretion and osmoregulation occur in *Amoeba*.
9. Describe the formation of pseudopodia.
10. Discuss the mode of reproduction in *Amoeba*.
11. Describe structure and function of contractile vacuole in *Amoeba*, *Euglena* and *Paramecium*. Add a note on osmoregulation in protozoa.

Short Answer Questions

1. List the modes of *Amoeba* ingests food?
2. Explain the process of circumvalation or circumfluence for capturing food particle.
3. Give an account of endocytosis in *Amoeba*.
4. Describe the mode of egestion in *Amoeba*.
5. How is culture of *Amoeba* is prepared.

Very Short Answer Questions

1. Name the proteins that help in changing sol to gel.

2. Which proteins play a role in converting gel to sol?
3. Mention the modes of reproduction in *Amoeba*.
4. What is geotaxix?

Fill in the Blanks

1. *Amoeba* was first recorded by ....
2. Ectoplasm of *Amoeba* contains a mesh of ....
3. Nucleus of *Amoeba* has the form of a ....
4. Most zoologists at present accept .... theory of pseudopodia formation.
5. Amoeboid locomotion is shown by .... and .... also in vertebrates.
6. Nutrition in *Amoeba* is ....
7. The main nitrogenous waste in *Amoeba* is ....
8. Cyst formation in *Amoeba* serves two functions: .... and dispersal.
9. *Amoeba* reproduces by .... and multiple fission.
10. Mitosis in *Amoeba* is ....

Multiple Choice Questions

1. *Amoeba* has
(*a*) filopodia (*b*) lobopodia
(*c*) axopodia (*d*) reticulopodia

2. Ingestion of food in *Amoeba* occurs by
(*a*) circumvalation (*b*) invagination
(*c*) import (*d*) all the above

3. Pseudopodia of *Amoeba* are important for
(*a*) feeding only (*b*) feeding and locomotion
(*c*) locomotion only (*d*) offence

4. Osmoregulation *Amoeba* takes place by
(*a*) food vacuole (*b*) contractile vacuole
(*c*) plasmalemma (*d*) plasmagel

5. Of all the vital activities that keep an amoeba in live state, the most important one is
(*a*) food ingestion (*b*) metabolism
(*c*) excretion (*d*) osmoregulation

6. *Amoeba* excretes by
(*a*) contractile vacuole (*b*) hyaline cap
(*c*) plasmalemma (*d*) pseudopodia

7. In *Amoeba*, contractile vacuole remains
(*a*) near advancing end (*b*) in middle of body
(*c*) near trailing end (*d*) anywhere in the body

8. In *Amoeba*, conversion of plasmagel into plasmasol at trailing end and of plasmasol into plasmagel at advancing end explains
(*a*) contraction theory (*b*) attachment to substratum
(*c*) theory of rolling movement (*d*) continuous viscosity change

9. Idea of conversion of sol into gel and vice-versa in *Amoeba*, was first given by
(*a*) Mast (*b*) Best
(*c*) Pantin (*d*) Hyman

10. *Amoeba* is called " immortal" due to regular
    (*a*) sporulation (*b*) conjugation
    (*c*) encystment (*d*) regeneration

11. Reproduction in amoeba during unfavourable conditions occurs in three-layered cysts. This Phenomenon is called
    (*a*) sporulation (*b*) conjugation
    (*c*) encystment (*d*) regeneration

ANSWERS

Very Short Answer Questions

1. Filamin and α-actinin. 2. Gelsolin and villin.
3. Binary fission and multiple fission. 4. Response to gravity.

Fill in the Blanks

1. Roesel von Rosenhof 2. Microfilaments 3. Biconvex disc
4. Sol-gel (change in viscosity). 5. Leucocytes, Macrophages. 6. Holotrophic.
7. Ammonia (NH_3). 8. Perennation. 9. Binary fission.
10. Intranuclear (closed).

Multiple Choice Questions

1. (*b*) 2. (*d*) 3. (*b*) 4. (*b*)
5. (*b*) 6. (*c*) 7. (*b*) 8. (*d*)
9. (*d*) 10. (*c*) 11. (*c*)

5

Paramecium (Phylum Protozoa)

5.1 CLASSIFICATION

Phylum	–	**Protozoa**
Subphylum	–	**Ciliophora**
Class	–	**Ciliata**
Subclass	–	**Holotrichia**
Genus	–	***Paramecium***
Species	–	***caudatum***

Paramecium is well-known ciliate protozoan. It demonstrates a very high degree of **cellular differentiation**, exhibiting many complex organelles which have been designed to perform specific functions for the organism. Not only its structure show specialization, but its reproductive activity is also complex. There are about 10 species of *Paramecium*. Its common species are *P. caudatum* and *P. aurelia*.

5.2 HABIT AND HABITAT

Paramecium caudatum (Gr., *paramekes* = oblong; L. *caudata* = tail) is a free-living organism having cosmopolitan (world-wide) distribution. It lives in stagnant water of freshwater, ponds, pools, ditches, and slow flowing streams rich in decaying organic matter.

5.3 STRUCTURE (MORPHOLOGY)

Size and shape. *Paramecium caudatum* is a unicellular and microscopic protozoan. It measures 170 to 290 mm. It is visible to the naked eye. Its body has a constant elongated, slipper-like shape, so the animal is commonly called **slipper animalcule**. The anterior end of animal's body is blunt and posterior end is thick and pointed. Body is **asymmetrical** exhibiting flat **ventral** or **oral surface** and convex **dorsal** or **aboral surface**. Cellular body of *Paramecium* has the following structures:

1. Pellicle. The whole body of *Paramecium* is covered by a thin, firm, flexible membrane, the **pellicle**. It maintains a definite shape of the animal and also serves as a skeletal structure. The pellicle has a uniform appearance, being composed of a lattice of hexagonally shaped pits, each centrally perforated by a single cilium. The margins of hexagonal depressions are slightly raised up as ridges. Trichocysts open in the ridges (Fig. 5.2).

Ultrastructure of pellicle. The electron microscopic study of pellicle (Fig. 5.3) by **Ehrat** and **Powers** (1957) has revealed that the hexagonal depressions correspond to regular series of cavities, the **alveoli**. All alveoli collectively form a continuous alveolar layer, which is delimited by an **outer alveolar membrane**. The outer alveolar layer is covered by **plasma membrane** which also surrounds the cilia. Thus, pellicle comprises of three membranes—1. Outer plasma membrane (outermost). 2. Outer alveolar membrane (middle membrane), and 3. Inner alveolar membrane (innermost).

2. Cilia. Entire body surface of *Paramecium* is covered with numerous (10,000 to 14,000), tiny (10–12 μm in length and 0.27 μm in diameter), hair-like projections of cytoplasm, called **cilia**.

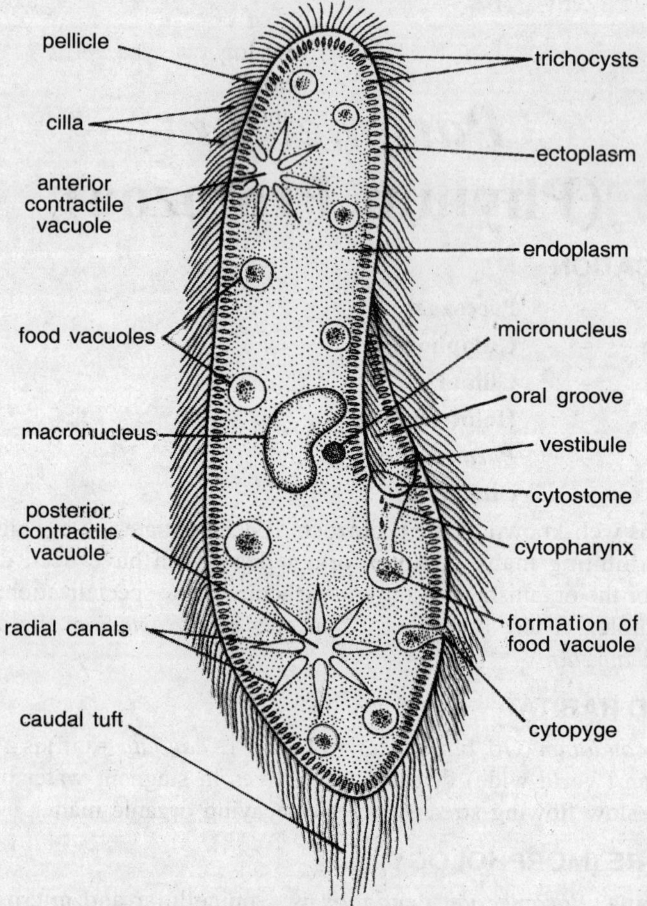

Fig 5.1. *Paramecium caudatum*. Structure.

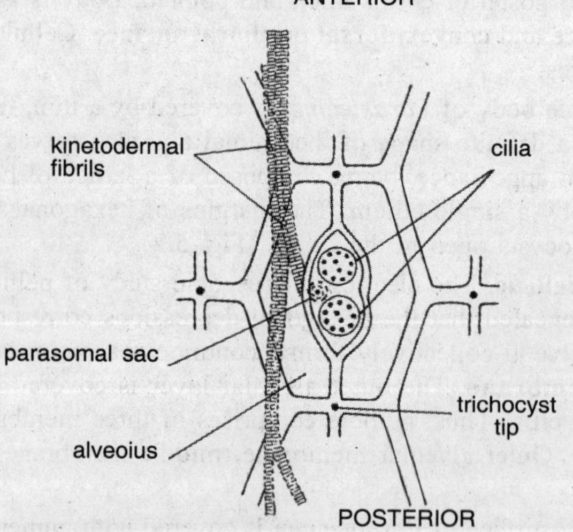

Fig. 5.2. *Paramecium aurelia*. Hexagonal area or ciliary field.

They are arranged in longitudinal rows, which are almost parallel on the dorsal surface and slightly oblique on the ventral surface. The length of cilia is unifrom throughout the body (a condition called **holotrichous**), but there are a few longer cilia at the posterior end of body. These form **caudal tuft** of cilia (hence the name **caudatum**). According to **Green** *et al.*, 1990, certain caudal cilia are **tactile cilia**, they are stiff and non-locomotory ones and are used in the detection of external stimulus.

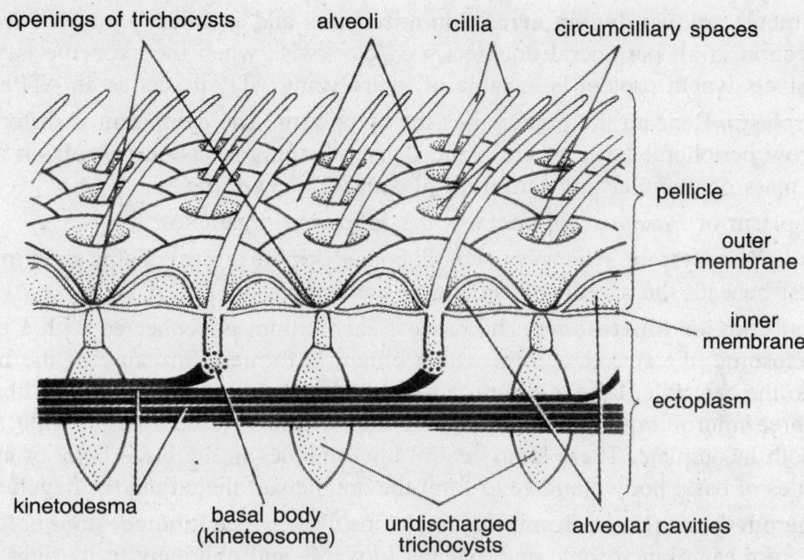

Fig. 5.3. *Paramecium*. Electron microscopic structure of pellicle, infraciliary system and associated structure.

Ultrastructure of a cilium. The electron micrograph (Fig. 5.4) of cross section of a cilium (or flagellum) exhibits three major parts: a central axoneme, the surrounding **plasma membrane** and the interposed cytoplasmic **matrix** (fluid). The entire ciliary projection is covered by a membrane that is continuous with the outer plasma membrane of pellicle. The core or shaft of the cilium, called the **axoneme**, contains an array of microtubules that run longitudinally through the entire organelle. Axoneme consists of nine peripheral doublet microtubules surrounding a central pair of single microtubules (*i.e.*, typical 9 + 2 arrangement of microtubules). All microtubules of the axoneme have the same polarity: their **plus ends** are the tip of the projection and their

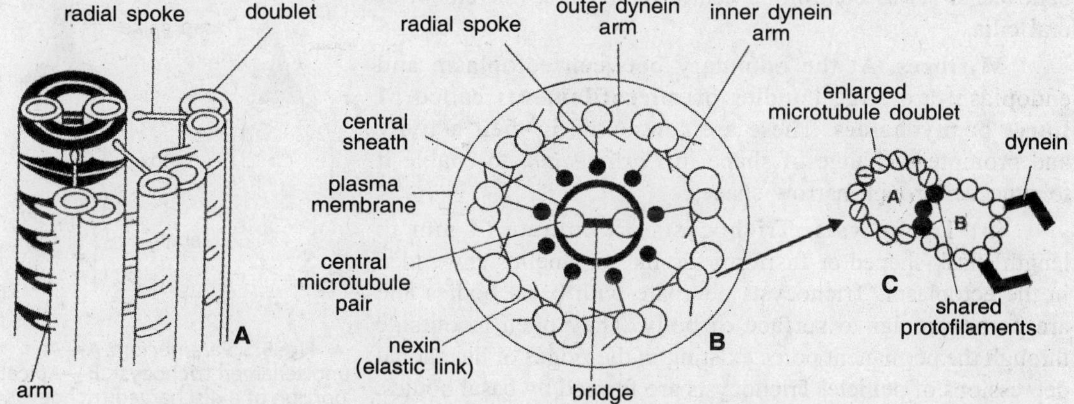

Fig. 5.4. Internal organization of cilia. A—Mode of arrangement of tubules and associated material as seen from outside. B—Cross section of cilia. C—An enlarged microtubule doublet.

minus ends at the base. Each peripheral doublet consists of one complete microtubule, the **A tubule**, and one incomplete microtubule, the **B tubule**, which contains 10 or 11 subunits (protofilaments of a and b tubulin proteins) rather than usual 13.

The central tubules are enclosed by projections that form **central sheath**, which is connected to A tubules of the peripheral doublets by a set of nine **radial spokes**. The doublets are connected to one another by an **interdoublet bridge** composed of an elastic protein, called **nexin**. Extending from each A tubule are two **dynein arms**—an **outer arm** and an **inner arm**, that are oriented in the same direction in all peripheral doublets, *i.e.*, clockwise when the axoneme is viewed from base to tip. Since dynein protein is capable of hydrolyzing ATP, it acts as an ATPase enzyme.

3. Cytoplasm. Beneath the pellicle lies the cytoplasm. The cytoplasm is differentiated into an outer narrow peripheral layer of clear and dense **plasmagel**, called **ectoplasm** and an inner large central mass of granular and semifluid **plasmasol** or **endoplasm**.

A. Ectoplasm of *Paramecium* contains the following organelles:

(*a*) Infraciliary system. It consists of basal bodies (**kinetosomes**) of cilia and **kinetodesmata**, all located just beneath the alveoli of pellicle in the ectoplasm (Fig. 5.3).

(*i*) Basal body or kinetosome. The base of each cilium is connected with a tubular **basal body** or **kinetosome**, the structure from which cilium is formed. Structure of the basal body is comparable to the centriole. Like a cilium, a basal body contains nine peripheral fibres, but each consists of three microtubules rather than two. The A tubule is complete, while the B and C tubules are both incomplete. There is no central microtubules in the basal body or centriole. The A and B tubules of basal body elongate to form the doublets of the cilium (or flagellum, Fig. 5.4).

(*ii*) Kinetodesmata (kinetodesma). A single fine fibril, called **kinetodesmos** or **kinetodesmal fibril**, arises from each kinetosome and extends forwards and obliquely to its right (Fig. 5.3). It joins other fibrils from adjacent kinetosomes to form a longitudinal bundle of striated fibrils, called a **kinetodesma**. The individual fibrils do not run anteriorly farther than the five kinetosomes. Thus, the number of fibrils in each kinetodesma remains five. The kinetosomes and fibrils of a particular row are collectively known as a **kinety**. All the kinety systems together form **infraciliary system** which coordinate beating of cilia and also plays an important role in the morphogenesis of *Paramecium* and other ciliates.

Motorium. There is also a dense bilobed mass or network of fibrils situated in the endoplasm near the cytostome. This is called **neuromotorium (Lund,** 1933) or **motorium**. It has branches which interconnect with the fibrils of the ectoplasm and the whole fibrillar system is believed to act as controlling centre for feeding movement of oral cilia.

M fibres. At the boundary between ectoplasm and endoplasm are large bundles of **microfilaments** called **M fibres** or **myonemes**. These are contractile in their activity and promote a change of shape in *Paramecium* to enable it to squeeze through narrow spaces.

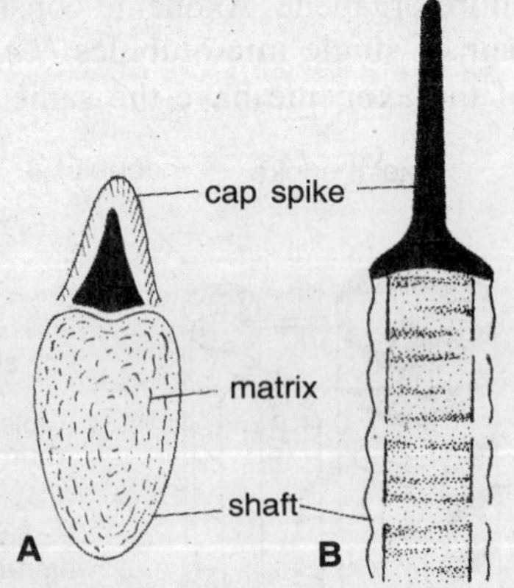

Fig. 5.5. *Paramecium*. A—An undischarged trichocyst. B—Apical portion of a discharged trichocyst.

(*b*) Trichocysts. Trichocysts are minute (4 mm in length) flask-shaped or fusiform sac-like organelles embedded in the ectoplasm. Trichocysts alternate with basal bodies and are perpendicular to surface of body. They open to outside through the permanent pores existing in the ridges of hexagonal depressions of pellicle. Trichocysts are formed by basal bodies. Each trichocyst consists of an elongated **shaft** and a terminal conical **barb** or **spike** covered with a **cap** (Fig. 5.5). The shaft is filled with a dense fluid having swelling substance, an unfolded fibrous protein, called **trichinin.**

Upon stimulation, the ectoplasm suddenly contracts. This separates the cap of the trichocyst to let the water enter the sac. This results in unfolding of trichinin protein of shaft and formation of a fine 40 mm long striated filament having pointed sticky spike (Fig. 5.5). Once discharged trichocysts cannot be withdrawn, they are replaced by new ones. Trichocysts serve as a means of anchorage during feeding activity. Previously they were considered as organelle of defence, *i.e.*, the long filaments from exploded trichocysts form a sort of screen that gives some degree of safety to the organism.

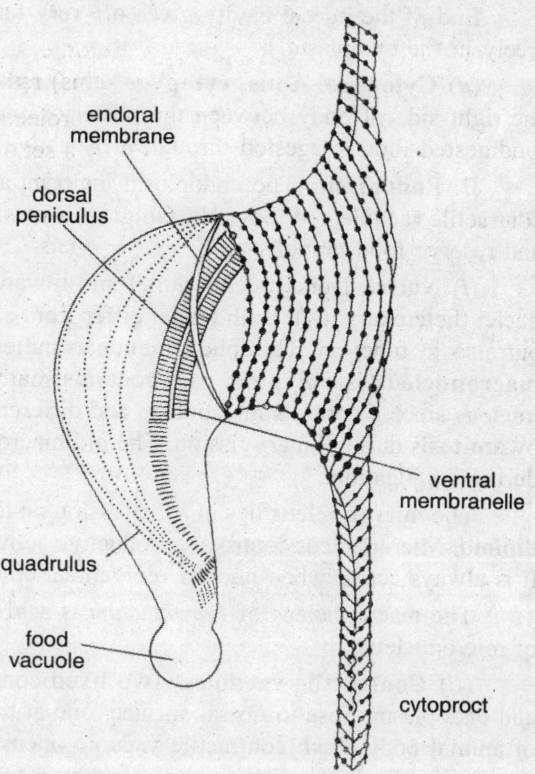

Fig. 5.6. *Paramecium*. Buccal ciliature.

(*c*) Oral apparatus. Near the anterior end of the animal and on its vental surface is a permanent ciliated broad and shallow longitudinal depression called the **oral groove** or **peristome.** It starts from the left side of the anterior end of the body and extends obliquely backwards towards the right side upto the middle of the body. There, it leads into conical funnel-like depression, called **vestibule.** The vestibule leads into a wide tubular passage, the **buccal cavity** or **gullet**. Buccal cavity extends backwards and ends in a small oval aperture, the **mouth** or **cytostome.** The mouth leads into a short tubular **oesophagus** or **cytopharynx** that opens into the endoplasm. The cytopharynx forms a food vacuole at its proximal end.

Buccal ciliature. The ciliature of buccal cavity is different and specialized. It includes the following three types of structures:

(*i*) Endoral membrane. At the junction of the vestibule and the buccal cavity is a crescentic **undulating membrane** or **endoral membrane** which runs transversely along the right wall. This membrane is formed due to sticking together of the cilia of a single row.

(*ii*) Peniculi and quadrulus. There are three membranelles (see Box 5.1) in the left wall of the buccal cavity. Each membranelle is formed by adhereing of four rows of cilia. These membranelles are of three types: a **dorsal peniculus**, a **ventral peniculus** and **quadrulus.** Ventral peniculus is short and ends at the cytostome. The dorsal peniculus is longer and crosses over to right wall at the cytostome and ends on the right wall of the cytopharynx. Cilia of two peniculi beat longitudinally in opposite direction.

In quadrulus four rows of cilia are separated and united at the ends. Quadrulus extends along the dorsal wall (roof) of the buccal cavity, crosses over to the right wall near the cytostome.

Box 5.1 Definitions

Membranelle. It is formed from two, three or four rows of cilia, all of which adhere together; actual fusion of cilia do not occur. This adherence causes all cilia of a membranelle to beat together.

Undulating membrane. It is formed by adherence of cilia of a single row.

End of the buccal cavity gives off very long **dorsal cilia** or **post buccal cilia** that project freely in the cytoplasm.

(*d*) Cytopyge. Anus, cytopyge or **cytoproct** exists as a minute pore at a fixed point on the right side of body between the posterior end of oral groove and posterior end of the body. Undigested food is egested through it by active vacuolar activity (called **exocytosis**).

B. Endoplasm. The endoplasm includes a variety of important organelles such as nuclei, contractile vacuole, food vacuole, Golgi apparatus, mitochondria, endoplasmic reticulum, ribosomes and reserve food granules.

(*i*) Nuclei. Dorsal to the gullet and towards the centre of body, lie two different types of nuclei (heterokaryotic). Both nuclei differ from each other not only in appearance (size and shape) but also in function. This phenomenon is called **nuclear dimorphism.** The large bean-shaped **macronucleus** is **polyploid** and contains many nucleoli. It is termed **somatic** or **vegetative nucleus** since it controls metabolism and differentiation within the animals. Macronucleus divides by **amitosis** during binary fission. The old macronucleus disappears and a new one is reorganized during conjugation.

The **micronucleus** lies in a depression on the surface of the macronucleus. It is spherical and **diploid.** Micronucleus controls reproductive activities of animal and gives rise to new macronuclei. It is always active when nuclear reorganization takes place during the life history of the animal.

The macronucleus of *Paramecium* is said to contain about 500 times more DNA than that of micronucleus.

(*ii*) Contractile vacuoles. Two fixed contractile vacuoles are present between ectoplasm and endoplasm close to dorsal surface, one at the anterior end and the other at the posterior end of animal body. Each contractile vacuole opens through a discharge canal in the pellicle on the dorsal side. Each vacuole is surrounded by 6 to 10 long and narrow **radiating** or **feeder canals.** Each feeder canal contains the following three parts:

1. Injector canal. It is the innermost part of feeder canal which opens into the vacuole.

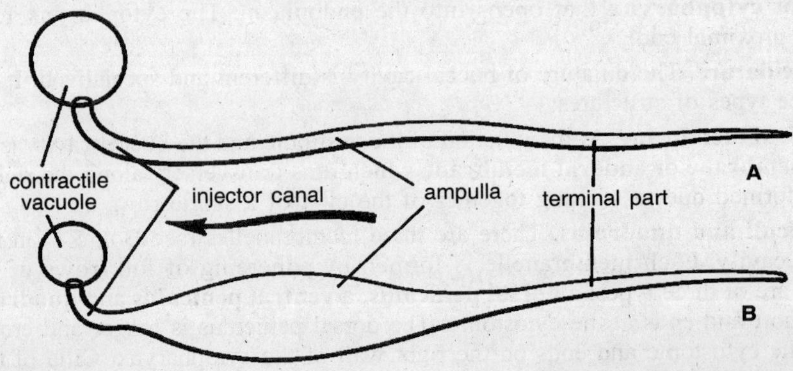

Fig. 5.7. *Paramecium*. Radial canal. A—Empty (systole); B—Full (diastole).

2. Ampulla or accessory vacuole. It is middle part of the feeder canal. It swells like a baloon when filled with water, *i.e.*, when it is in **diastole.** It collapses when it is empty, *i.e.*, when it is in **systole** (Fig. 5.7).

3. Terminal part. It is outer part of the radial or feeder canal which extends into the cytoplasm. It is also called **nephridial canal** or **collecting tubule.** The distal end of collecting tubule is surrounded by a network of fine coiled **nephridial tubules** or **spongiome tubules** (**Ruppert** and **Barnes,** 1994). These tubules form a sponge around the collecting tubules.

Spongiome or nephridial tubules collect water from the surrounding cytoplasm and transfer it to collecting tubule. Endoplasmic reticulum and mitochondria are also found around the spongiome tubules.

Physiology of contractile vacuole. Contractile vacuoles maintain water balance (called **osmoregulation**) of the body. It pumps excess of body water out of the organism. The wall of contractile vacuoles and radiating canals contain bundles of microtubules and microfilaments (actin), called **myofibrils** and are contractile. Radiating canals become swollen on receiving water (**diastole**) and contract (**systole**) to ultimately empty into the contractile vacuole. Vacuole is reconstituted and grows in size (diastole). When the contractile vacuole reaches its maximum size it contracts suddenly (systole) and discharges its contents through the permanent pore in the pellicle. This cycle is repeated again and again to do osmoregulation.

It is reported that posterior contractile vacuole empties and fills at a faster rate than the anterior one, because of greater endosmosis in the region of gullet. *P. caudatum* can complete a cycle of vacuole filling and discharge as rapidly as every 6 seconds and expel a volume equivalent to its entire body every 15 minutes.

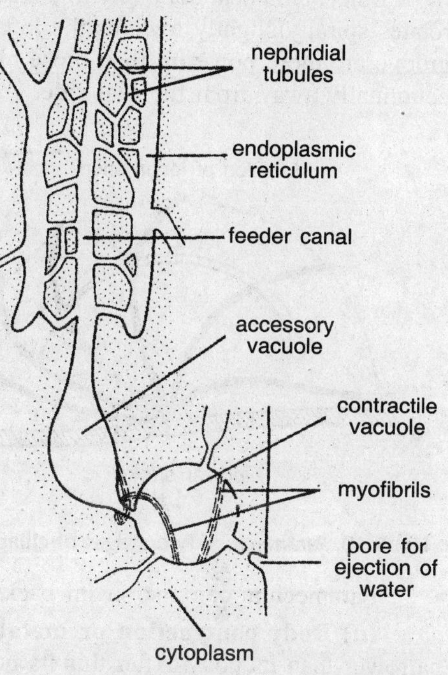

Fig. 5.8. *Paramecium*. Contractile apparatus.

***(iii)* Food vacuoles (or gastrioles).** Endoplasm contains a variety of food vacuoles containing food in various stages of digestion. There may be as many as 50 food vacuoles at one time in a single individual. Food vacuoles are spherical in form, contain ingested food particles and a small amount of fluid. They are bounded by a unit membrane of lipoproteins. Food vacuoles move (circulate) in the endoplasm by rotatory streaming movement, called **cyclosis**, of endoplasm.

5.4 PHYSIOLOGY

1. Locomotion. *Paramecium* is a free living aquatic animal. It does locomotion by the following methods:

Fig. 5.9. A longitudinal row of cilia showing metachronous movement.

***(i)* Ciliary movement.** Cilia are main locomotory organelles in *Paramecium*. It swims in water by rhythmic beating of cilia. Each cilium beats a little in advance of the one immediately behind so that waves of ciliary activity pass over the animal (called **metachronous rhythm**). A cilium can be compared with an oar as it moves the cell in a direction perpendicular to the cilium itself. In its power stroke or effective stroke, the cilium is maintained in a rigid state (Fig. 5.10) as it pushes against the surrounding medium. In its **recovery stroke**, the cilium becomes flexible, offering relatively little resistance to the medium. The movement of cilia is controlled by the neuromotor system. Quite recently certain chemical messengers such as **Ca^{2+}ions** and **cyclic AMP (cAMP)** are reported to influence the rate and form of ciliary beat. *Paramecium* swims at the rate of about 1mm per second.

Since, cilia beat obliquely to the right, so the body of *Paramecium* while swimming forward rotates spirally slightly towards the left (Fig. 5.11). Further, effective strokes of the cilia of oral groove are more powerful than those of body cilia. This makes the anterior end of body to turn continually away from the oral side.

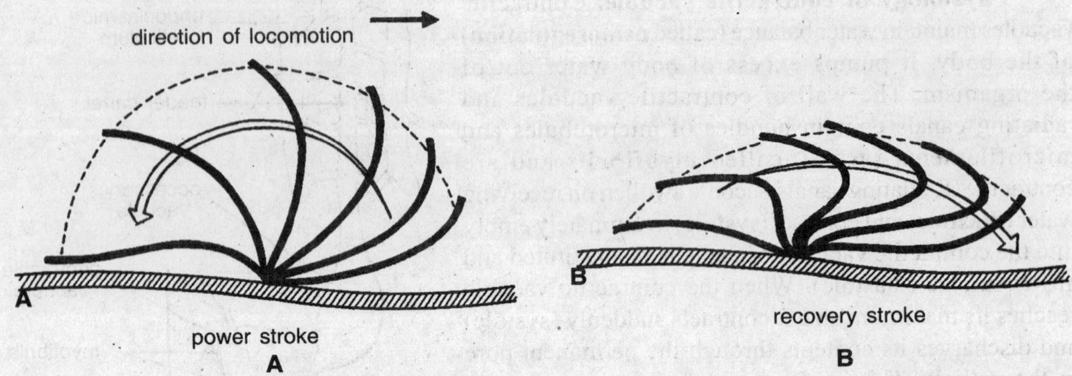

Fig. 5.10. *Paramecium*. Two stages of ciliary movement. A—Effective or power stroke; B—Recovery stroke.

Paramecium can also swim backwards, in which case its cilia beat forwards.

***(ii)* Body contraction or metaboly.** *Paramecium* can squeeze itself through a passage narrower than its body. After this its body assumes normal shape. Such a temporary change in body shape is called **metaboly**; it is brought about by **M fibres** or **myonemes**.

2. Nutrition. *Paramecium* is **holozoic** in nutrition. The process of nutrition includes the following steps:

***(i)* Food.** *Paramecium* is an **omnivorous** and **microbivorous** animal. It feeds on a variety or microscopic biological materials such as bacteria, algae, diatoms, yeasts, small protozoans and tiny fragments of large-sized animals and plants.

***(ii)* Ingestion.** Food is ingested through **cytostome**. A current of water in the form of a vortex (whirl pool) is produced by the constant lashing movement of cilia of oral groove. Due to this the food particles are swept towards the cytostome and are carried down into the cytopharynx. These food particles are aggregated into a rounded mass (ball) by the movement of penniculi and quadrulus. Food mass along with a drop of water is enclosed in a **food vacuole** (by endocytosis) and ingested into the endoplasm.

***(iii)* Digestion and assimilation.** After their separation from the cytopharynx, the food vacuoles circulate through the endoplasm along a definite circular path (often resembling with the figure of 8). Their circulation is brought about by the rotatory streaming movements, called **cyclosis**, of endoplasm. The food vacuoles starting from cytopharynx are carried firstly behind, then forward to the dorsal surface and finally backward down to the oral surface. During its journey, the food is digested inside the food vacuole by a variety of lysosomal enzymes such as **amylase, proteinase, dipeptidase, lipase,** etc. The food vacuole is first acidic but becomes basic later on. Digested food is absorbed and assimilated by the endoplasm during cyclosis.

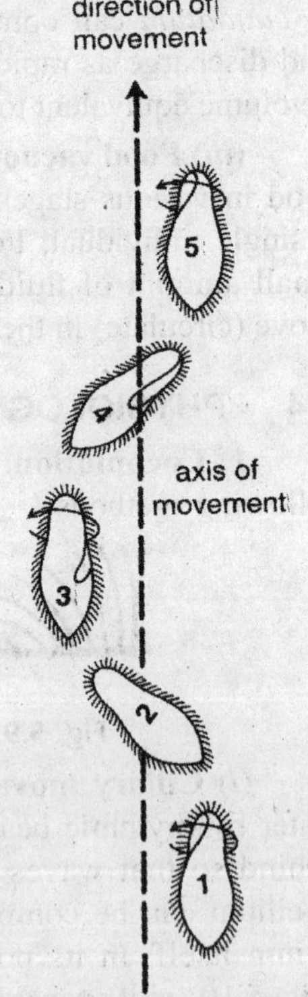

Fig. 5.11. *Paramecium*. Spiral path of movement.

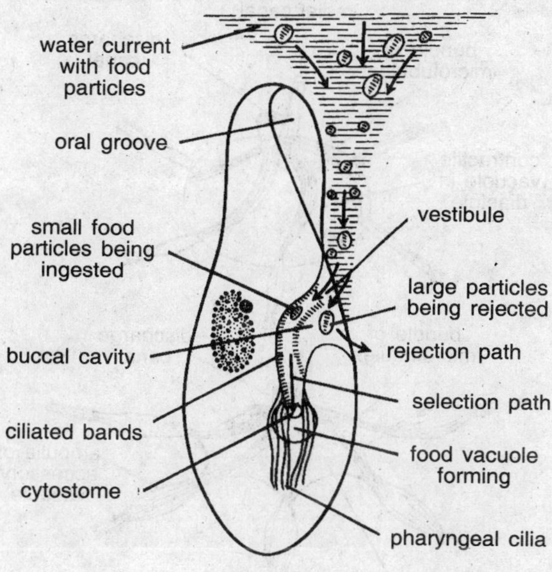

Fig. 5.12. *Paramecium*. Ciliary action creating vortex (whirl pool) and drawing food particles into the oral groove with the water current.

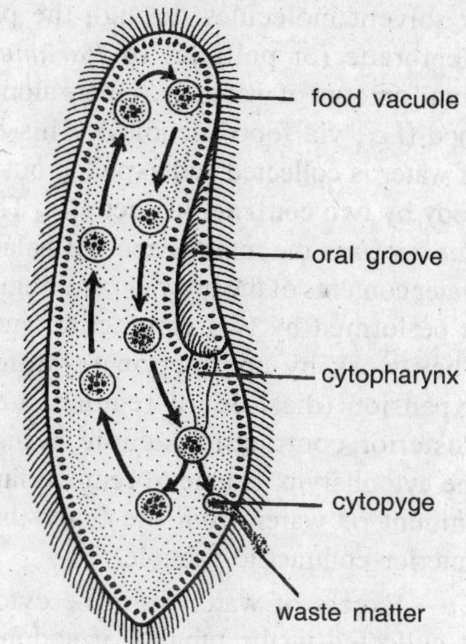

Fig. 5.13. *Paramecium*. Movement of food vacuoles due to cyclosis in the endoplasm.

(*iv*) Egestion. The undigested food or faecal matter is discharged outside through a definite anal spot or cytopyge by active vacuolar activity called **exocytosis**.

3. Respiration. In *Paramecium*, respiration occurs through the surface of body. The oxygen dissolved in surrounding water diffuses in through the general body surface and carbon dioxide is liberated in a similar way.

4. Excretion. *Paramecium* is **ammonotelic**; its excretory waste is **ammonia.** *Paramecium aurelia* has been shown to liberate ammonia, presumably as a result of amino acid **deamination** (catabolism), but not from purine metabolism. Purines are probably excreted as **hypoxanthine**, with some **guanine** and **adenine** (**Meglitsch** and **Schram**, 1991). Nitrogenous wastes are removed by diffusion through entire body surface.

5. Osmoregulation. The concentration of solutes of body cytoplasm of *Paramecium* is higher than that of the surrounding water (*i.e.*, cytoplasm is a **hypertonic solution** and water is a **hypotonic solution**). This results in **endosmosis** (*i.e.*, inward movement of water

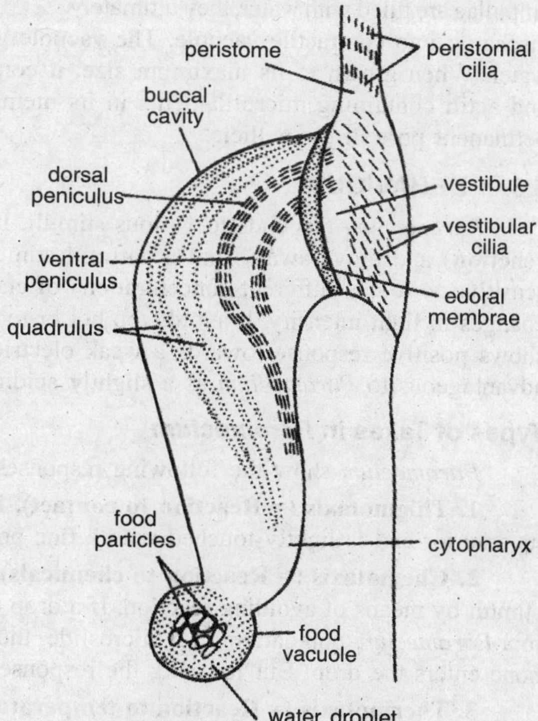

Fig. 5.14. *Paramecium*. Oral groove with food vacuole under formation.

or solvent molecules through the plasma membrane (or pellicle) of *Paramecium*. Small amount of water also enters along with food (*i.e.*, via food vacuoles). This excess of water is collected and expelled out of the body by two **contractile vacuoles**. Thus, in *Paramecium* the maintenance of balance of water contents of the body or osmoregulation is performed by two contractile vacuoles. They work by contraction (systole) and expansion (diastole) at regular intervals. Posterior contractile vacuole being near the cytopharynx is able to remove far more amount of water from the body than the anterior contractile vacuole.

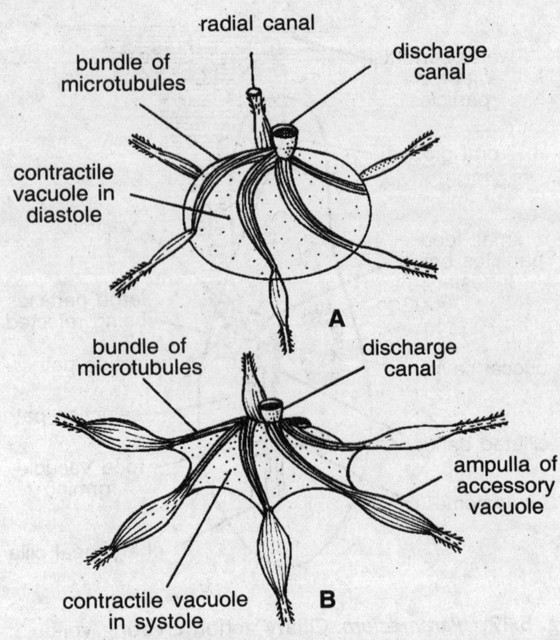

Fig. 5.15. *Paramecium*. Working of contractile vacuole. A—In diastole; B—In systole.

Excess of water from the cytoplasm is collected in the tubules of endoplasmic reticulum. From them water is passed into nephridial tubules or spongiome tubules. Collecting tubules (= nephridial canals) of radiating canals collect water from these tubules and convey it to the ampullae. When ampullae are filled with water, they ultimately discharge into contractile vacuole. The vacuole gradually expands by collecting more and more water. When grown to its maximum size, it contracts (by the help of bundles of microtubules and actin containing microfilaments in its membranes) and expels water to outside through a permanent pore in the pellicle.

5.5 BEHAVIOUR

Paramecium responds to various stimuli. It moves towards a beneficial stimulus (**positive reaction**) and move away from a harmful stimulus (**negative reaction**). Thus, *Paramecium* is sensitive to touch, different concentrations of chemicals, oxygen and carbon dioxide levels, and changes in light intensity. It avoids too hot or too cold waters and also strong light. *Paramecium* shows positive response towards a weak electric charge and weak acidic environment (This is advantageous to *Paramecium* as a slightly acidic medium contains more bacteria).

Types of Taxes in *Paramecium*

Paramecium show the following responses to different stimuli (Fig. 5.16):

1. Thigmotaxis (= Reaction to contact). Response to contact is varied in *Paramecium*. If its anterior end is lightly touched with a fine point, a strong avoiding reaction occurs.

2. Chemotaxis (= Reaction to chemicals). Generally *Paramecium* responds to a chemical stimuli by means of avoiding reaction. If a drop of weak salt solution (0.5 per cent) is introduced in a *Paramecium* population on a microslide, the animals respond with the avoiding reaction and none enters the drop. But for acid, the response of paramecia is positive.

3. Thermotaxis (= Reaction to temperature). *Paramecium* seeks an optimum temperature of 24 to 28°C. Greater heat stimulates rapid movement and avoiding reaction.

4. Phototaxis (= Reaction of light). Except green *Paramecium bursaria* which is positively phototactic, paramecia exhibit a negative phototactic reaction to bright light and to ultraviolet rays.

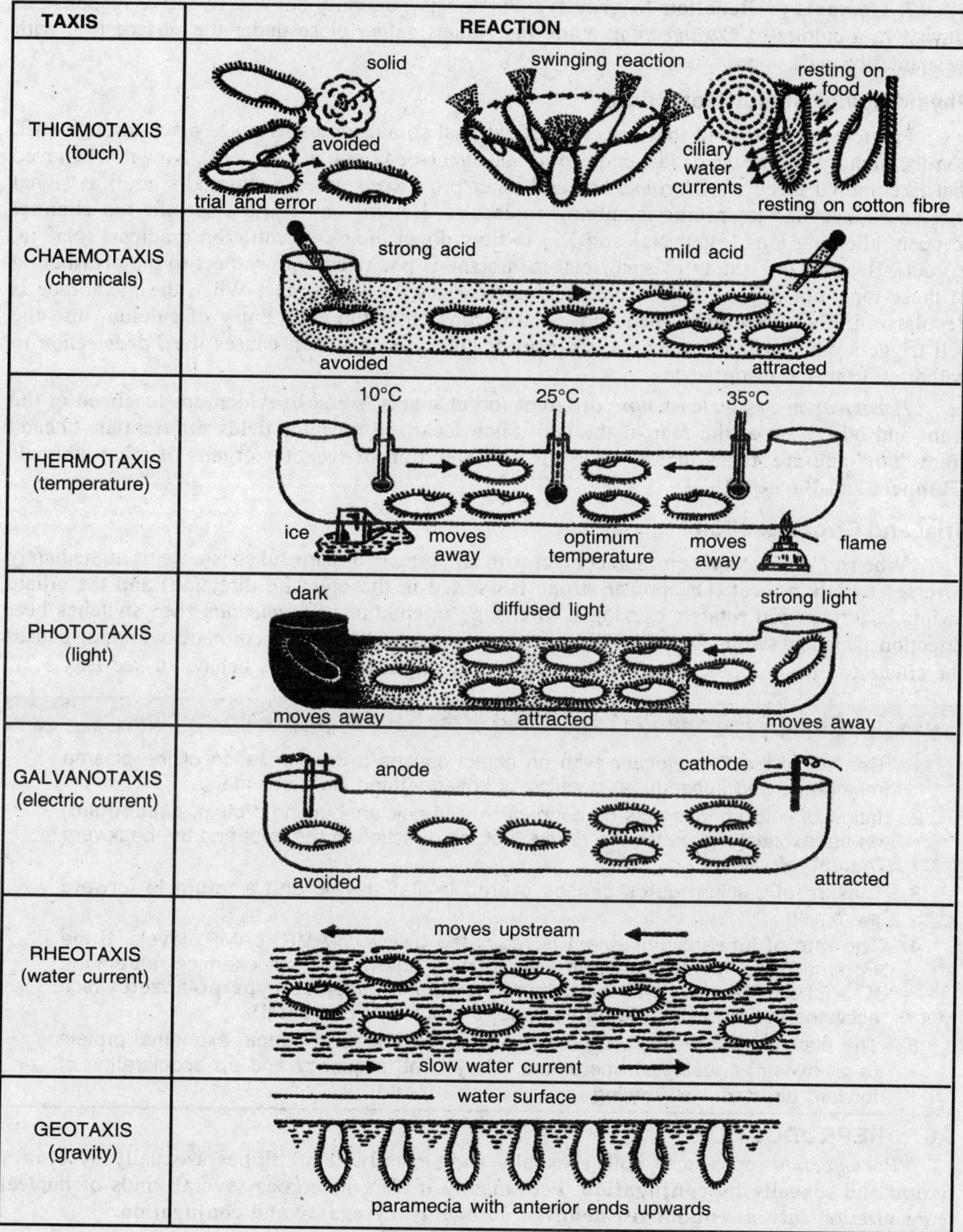

Fig. 5.16. Various taxes of *Paramecium*.

5. Galvanotaxis (= Reaction to electric current). Paramecia show positive response to negative electric pole, *i.e.*, they swim toward and concentrate at cathode (negative electrode).

6. Rheotaxis (= Reaction to water current). Paramecia show positive rheotaxis.

7. Geotaxis (= Reaction to gravity). Paramecia generally show a negative response to gravity. In a culture of *Paramecium*, many individuals gather close under the surface film with their anterior ends upper most.

Physiology of Stimuli Detection

Previously it was held that detection of external stimuli probably takes place through stiff, non-locomotory cilia at the caudal end of *Paramecium* (see **Green** *et al.*, 1990). Now it is believed that like animal receptors, *Paramecium* and other protozoans receive external stimuli as signal substances that bind to specific membrane molecules. Binding can cause a specific **ion channel** to open, allowing ions (often Na^+ and K^+) to flow down their concentration gradients (Na^+ in, K^+ out). Because the resting plasma / cell membrane is polarized with respect to the distribution of these ions, opening of the ion channels depolarizes the membrane. When the membrane is depolarized, **Ca^{++} channels** open and calcium ions enter the cell. Entry of calcium into the cell triggers other changes, such as, reversal of ciliary beat, which causes the *Paramecium* to withdraw from the disturbance.

Paramecium has at least *nine* different ion channels, some of which are localized at the front and others are at the rear of the cell. Such localized receptor fields differentiate "head" from "tail" and are, thus, analogous to the concentration of receptor organs of many animals (**Ruppert** and **Barnes**, 1994).

Trial and Error Reaction

When a *Paramecium* comes in contact with an obstacle or harmful substance, it immediately reverses its ciliary beat (*i.e.*, power stroke is exerted in the opposite direction) and the ciliate swims backward and rotates, causing it to change orientation. *Paramecium* then switches beat direction again and swims forward. This process continues until the orientation of swimming causes the ciliate to miss the object and swim pass it (For physiology of this behaviour see Box 5.2).

Box 5.2 Physiology of trial and error behaviour of *Paramecium*

1. Contact of the *Paramecium* with an object causes a depolarization of the plasma membrane and subsequent opening of voltage-gated Ca^{2+} channels.
2. Inflow of calcium ions acts (in connection with calcium binding protein, **calmodulin**) on the axoneme to cause the reversal of the direction of the beat and the backward "retreat" of ciliate.
3. Closure of calcium gates causes a drop in Ca^{2+}-levels and a return to forward swimming.
4. The rate of forward movement is regulated by cyclic AMP (cAMP) levels. If the *Paramecium* should encounter a stimulatory agent, such as chemical attractant, **K^+ – channels** open and the plasma membrane becomes **hyperpolarized** which activates the enzyme, **adenyl cyclase**, that synthesizes **cAMP**.
5. The elevated cAMP level triggers the phosphorylation of certain axonemal protein (*e.g.*, dynein) causing an increased ciliary beat frequency and an acceleration of forward directed – swimming.

5.6 REPRODUCTION

Paramecium reproduces both asexually and sexually. It multiplies asexually by **binary fission** and sexually by **conjugation**. Periodically it also undergoes several kinds of nuclear reorganization such as **endomixis, hemixis, autogamy, cytogamy** and **conjugation**.

I. Binary Fission

During favourable conditions *Paramecium* multiplies by **transverse binary fission**. It is most common method of reproduction in *Paramecium*. In this process, a full grown individual divides into two daughter individuals without leaving a parental corpse behind. Before the beginning of fission, *Paramecium* stops feeding, developing two minute pores near the middle of body for the contractile vacuoles that are to appear later for the daughter individuals. Both nuclei increase in size, elongate and pull themselves apart. The macronucleus divide **amitotically**, randomly distributing its chromosomes between the two newly formed macronuclei. The micronucleus undergoes **mitosis**. A spindle forms within its nuclear envelope and chromosomes are shared between the two micronuclei. A transverse cytoplasmic constriction furrow forms (by the help of microtubules and microfilaments) which deepens gradually dividing it into two equal sized daughter paramecia. Of the two paramecia produced, the anterior one is called **proter** and posterior one is called **opisthe**. Each half of the animal contains only those cell organelles which were originally present in it. For example, original feeding apparatus (buccal cavity, cytopharynx, etc.) remains with the anterior half while new feeding apparatus develops in the posterior half. Two new contractile vacuoles also appear, one near anterior end of opsithe and another near posterior end of proter. By the time the separation is completed, both daughters have a full complement of cell organelles. Ultimately both paramecia separate from each other and start their free-living existence. The entire process of binary fission is completed within ½ to 2 hours. All daughter paramecia produced asexually from one parent paramecium form a **clone**. All the members of a clone are hereditarily alike.

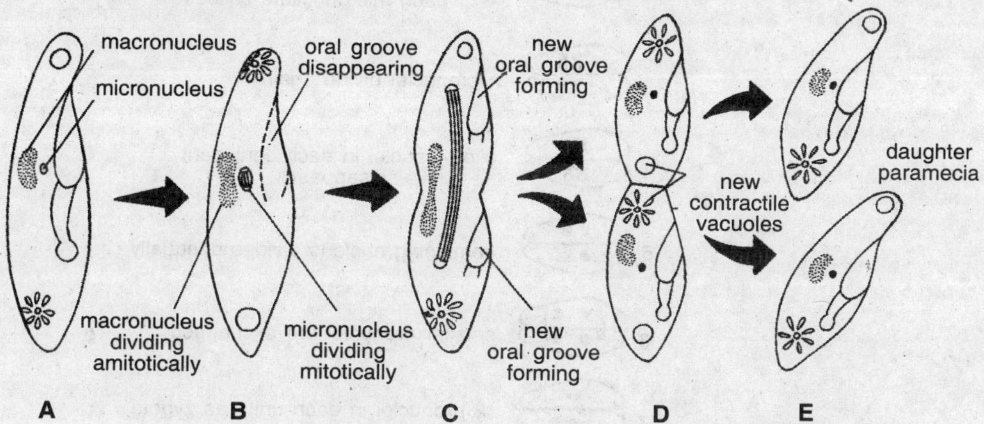

Fig. 5.17. *Paramecium*. Asexual reproduction by transverse binary fission.

II. Conjugation

Conjugation is a form of sexual reproduction. It is a temporary union of two individuals of a species for the purpose of exchange of their micronuclear material. Basically, conjugation involves **meiosis** and **nuclear exchange** and results in production of offspring with a wide variety of genotypes. Conjugation corresponds to cross-fertilization of higher animals and is essential for rejuvenation and continuity of race. It was described in *Paramecium caudatum* in 1889 by **Hertwig** and **Maupas**.

1. Mating types. According to **Sonneborn** (1947), *Paramecium caudatum* has 13 varieties (called **syngens**) which differ only in minor physiological features such as optimum temperature for growth, ability to conjugate and mating time during a 24 hour period. Each variety contains two **mating types** which differ in chemical characteristics (probably proteins) of their surface membranes. The mating types are indicated by Roman numerals, *e.g.*, variety 1 has two mating types I and II, the variety 2 has mating types III and IV and so on. Conjugation can occur between the individuals of two mating types of same variety.

2. Factors Inducing Conjugation

(*i*) Conjugation occurs usually under unfavourable living conditions such as starvation or shortage of food (*e.g.*, bacteria).

(*ii*) It occurs in individuals which have passed through desirable number (normally 300) of asexual generations. It is believed that during asexual reproduction paramecia exist in a **period of immaturity** and due to repeated binary fissions, they become **sexually mature** to conjugate.

(*iii*) The individuals entering into conjugation are generally of the same size. Both are also small sized than individuals reproducing asexually.

(*iv*) Conjugation occurs when conjugants are sticky, free of food vacuoles and slow in movement.

(*v*) Conjugation does not take place during night or darkness; it starts in early morning (dawn) and continues till afternoon.

(*vi*) A proteinous substance in the cilia of mating type individuals is said to induce conjugation.

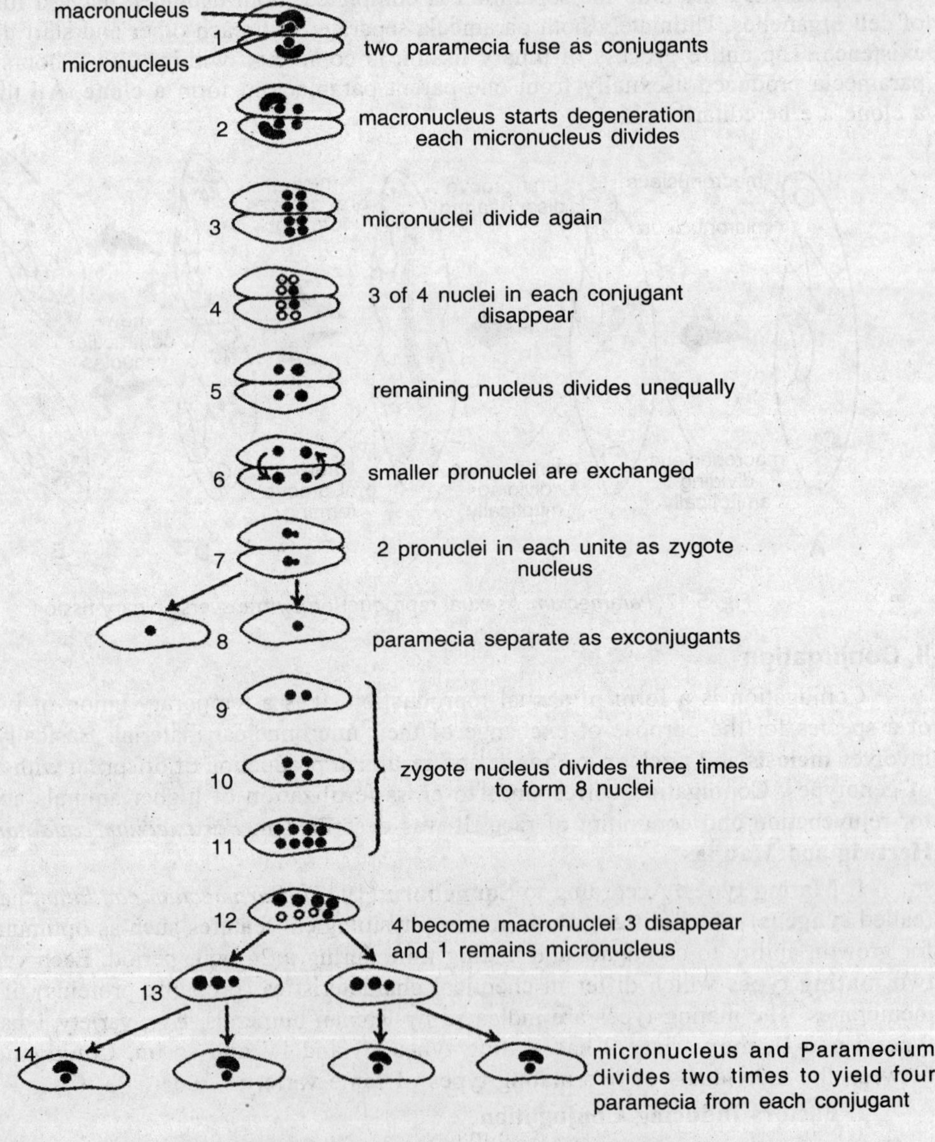

Fig. 5.18. *Paramecium caudatum*. Stages of conjugation.

3. Process of conjugation. Two different compatible mating types (**conjugants**) adhere to each other at their oral grooves. Their cilia produce a substance on the surface of body which causes adhesion of the two conjugating paramecia. At the region of contact a cytoplasmic continuity, called **cytoplasmic bridge** or **conjugation bridge** is established by the local destruction of cilia, pellicle, trichocyst, expect the basal bodies. At nuclear level, in each conjugant, following changes occur:

(A) Macronuclear changes. The macronucleus breaks up into fragments and is absorbed in the cytoplasm during the process of conjugation.

(B) Micronuclear changes.

(*i*) The diploid micronucleus divides by **meiotic cell division** and produces four haploid micronuclei.

(*ii*) Three of the four daughter micronuclei degenerate.

(*iii*) The remaining micronucleus of each conjugant divides **mitotically** into two unequal pronuclei, called **gametic nuclei:** 1. a larger stationary **female pronucleus;** 2. a smaller, active and migratory **male pronucleus.**

(*iv*) The male pronuclei of two conjugants are exchanged so that the male pronucleus of one individual passes through cytoplasmic bridge into the other individual and fuses with the female pronucleus forming the **zygotic nucleus** or **synkaryon.**

This event (called **amphimixis**) restores diploidy of the conjugants and initiates genetic variability in both zygotic nuclei.

(*v*) The conjugants now (*i.e.*, after about 12–48 hours) separate and are called **exconjugants.**

(*vi*) The zygotic nucleus of each exconjugant divides mitotically to form eight daughter nuclei.

(*vii*) Four of the eight nuclei grow in size to become **macronuclei**. Remaining four nuclei are called **micronuclei**.

(*viii*) Three of the four micronuclei disintegrate.

(*ix*) Transverse binary fission of each exconjugants takes place (involving mitosis of the micronucleus). Two macronuclei and one micronucleus enter in each daughter cell.

(*x*) Further binary fission of these daughter cells results in separation of macronuclei and another mitotic division of micronucleus.

(*xi*) The end result of conjugation is that four daughter paramecia are formed from each exconjugant. Each of them contains one macronucleus and one micronucleus.

(*xii*) Thus, process of conjugation produces eight small sized daughter paramecia, each of which grow and reform lacking body organelles.

Significance of Conjugation

Conjugation resembles the sexual reproduction of higher animals in many respects. It is an important process of nuclear reorganization and nuclear exchange which recur in between the binary fissions (asexual reproduction). The end products of this process are eight daughter paramecia which are formed after the asexual multiplication of both exconjugants. The sexual process in *Paramecium* ends with the fusion of two gametic nuclei. Thus conjugation in *Paramecium* has following two-folds significance:

1. Genetic significance

(*i*) Nuclear reorganization. The macronucleus is formed of trophocromatin which controls all the physiological activities of the individual. Continued amitotic division of the macronucleus during binary fission leads to unequal distribution of its chromosomes. This upsets the normal coordinated activity of *Paramecium* and leads to **depression**. The worn out and degenerated macronucleus is replaced by a new one during conjugation. This results in restoration of vigour and vitality of the old individuals (exconjugants).

***(ii)* Hereditary variations.** Conjugation involves meiosis and fusion of haploid pronuclei of different individuals and this results in the formation of new combinations of genes in the descendants. This ensures their better adaptability to the new conditions of life.

2. Physiological significance. It involves **rejuvenation**, *i.e.*, restoration of vitality, of *Paramecium*. It has been found that individuals cannot continue to multiply indefinitely by asexual methods. After a definite number of asexual generations (binary fissions), individuals start losing their vigour and physiological efficiency. They appear unhealthy and ultimately die. The conjugation is the only process which infuses former vigour, restores its healthy organization and thus, saves the race from senile decay.

III. Endomixis

Endomixis is a method of nuclear reorganization followed by the division of body within single individual, without any nuclear exchange or fusion. It occurs only in the absence of conjugation and has been described by **Woodruff** and **Erdmann** (1914) in *Paramecium aurelia*, a binucleate species. Endomixis involves the following steps:

1. The macronucleus disintegrates and is absorbed in the surrounding cytoplasm.
2. The two micronuclei divide twice forming eight daughter micronuclei.
3. Six of these micronuclei disintegrate and get absorbed, leaving two micronuclei.
4. The animal with two micronuclei divides by transverse binary fission and two daughter individuals, each with one micronucleus, are formed.
5. In each daughter *Paramecium*, the micronucleus divides twice, forming four nuclei.
6. Of these four nuclei, two enlarge to become macronuclei and two remain small micronuclei.
7. Both micronuclei in each daughter individual again divide. This nuclear division is accompanied by the division of body.

Ultimately four daughter paramecia are formed from a single parent and each possesses one macronucleus and two micronuclei.

Many protozoologists (*viz.* **Diller**, 1936; **Sonneborn**, 1947; **Wichterman**, 1953; etc.) do not believe in the validity of endomixis; they consider it as a special case of autogamy or a mixture of hemixis and autogamy.

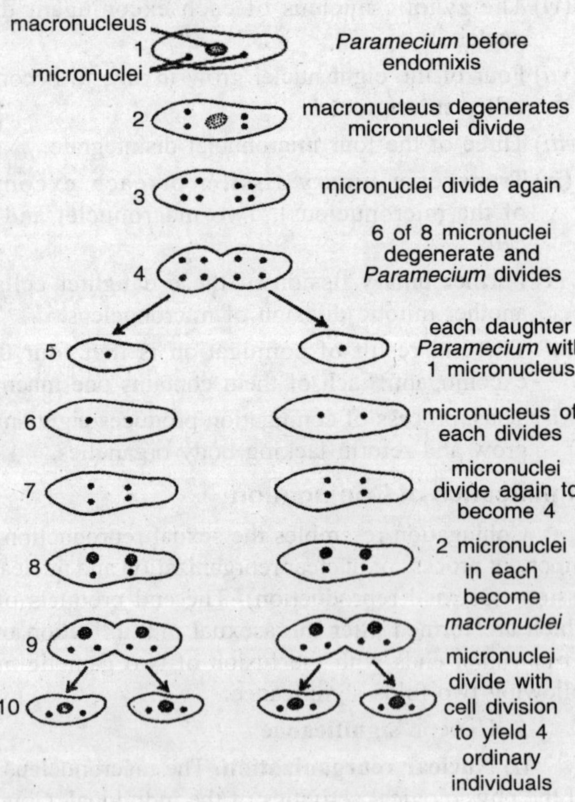

Fig. 5.19. *Paramecium aurelia*. Various stages of endomixis.

Significance of Endomixis

Like conjugation, endomixis is a process of multiplication and rejuvenation and occurs as a substitute of conjugation.

IV. Hemixis

Hemixis was described by **Diller** (1936) in *Paramecium aurelia*. It is a simple phenomenon of nuclear reorganization. In this process the micronuclei do not participate. The macronucleus gives off small fragments of chromatin, called **chromatin balls** which are absorbed in the surrounding and micronuclei behave normally.

Significance of Hemixis

The Hemixis is regarded as a type of nuclear reorganization in which old macronuclei are purified. The discarded chromatin balls probably contain accumulated waste substances.

V. Autogamy

Autogamy occurs within a single individual and involves fusion of two micronuclei of same individual into a synkaryon. It was described in *Paramecium aurelia* by **W.E Diller** (1936) and he called autogamy as **self-fertilization**. Similar nuclear changes occur in autogamy and conjugation, but there is no exchange of nuclei in autogamy. Autogamy involves the following steps (Fig. 5.20):

1. Macronucleus grows into a hapazard mass and break up into fragments which are absorbed by endoplasm.
2. The two micronuclei (which are diploid) divide meiotically and produce eight haploid daughter nuclei. Seven of these daughter micronuclei degenerate.
3. The remaining micronucleus divides by mitosis and produce two nuclei, now called **gamete nuclei**.
4. A temporary **cytoplasmic cone** develops near the cytostome. The two gamete nuclei enter this cone and fuse together to produce a diploid **zygote nucleus** or **synkaryon**.
5. The zygote nucleus divides twice by mitosis and produce four nuclei.
6. Two of these nuclei grow to become two **macronuclei**, and remaining two become **micronuclei**.

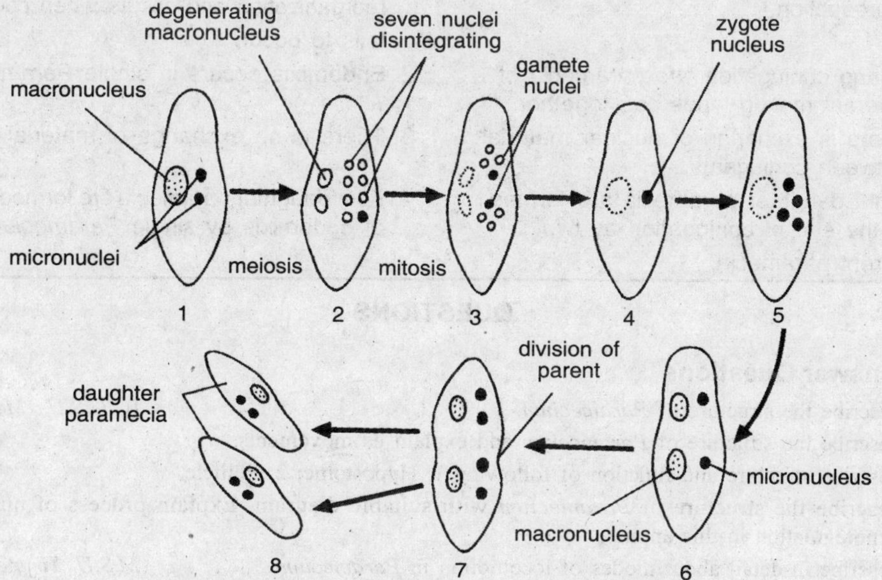

Fig. 5.20. *Paramecium aurelia*. Various steps of autogamy.

7. Each of the micronuclei and the body of *Paramecium* divide simultaneously to produce two daughter paramecia, each with a new macronucleus and two micronuclei.

Significance of Autogamy

1. Autogamy results in the production of a new macronucleus which rejuvenates *Paramecium* by increasing its vitality.
2. It is a type of sexual reproduction (by self fertilization), involving production and fusion of two haploid gamete nuclei. This may result in homozygosity of certain genes.

VI. Cytogamy

Cytogamy type reproduction was reported by **R. Witcherman** (1940) in *Paramecium caudatum*. In cytogamy, two paramecia (small, 200 mm long) form a pair and become temporarily attached at their **oral surfaces**, as in conjugation. Both of these individuals are called **cytogamonts**. Their nuclei undergo similar changes as occurring during conjugation, but there is no nuclear exchange between the cytogamonts. The two haploid gamete nuclei in each cytogamont are fused to form a synkaryon as in **autogamy**. The individuals now separate and divide as in conjugation. A new macronucleus is formed in each individual.

Significance of Cytogamy

1. During cytogamy, since a new macronucleus is formed, it rejuvenates *Paramecium*.
2. Self-fertilization also occurs during cytogamy due to which homozygosity of certain genes may occur.

Paramecium is believed never to **encyst**.

Distinction Between Conjugation and Endomixis

Conjugation and endomixis are two methods or reproduction and nuclear reorganization in *Paramecium*. Both have the following differences (see Table 5.1).

Table 5.1. Differences between conjugation and endomixis.

Conjugation	Endomixis
1. Conjugation is a process of sexual reproduction.	1. Endomixis is a process of nuclear reorganization and occurs when conjugation fails to occur.
2. During conjugation two paramecia of different mating types pair together.	2. Endomixis occurs in single *Paramecium*.
3. There is exchange of nuclear material between conjugants.	3. There is no exchange of material.
4. Eight daughter paramecia are formed at the end of conjugation by two parent paramecia.	4. Four daughter paramecia are formed at the end of endomixis by single *Paramecium*.

QUESTIONS

Long Answer Questions

1. Describe the structure of *Paramecium*. (*C.C.S.U., Meerut 2008*)
2. Describe the structure of *Paramecium* and explain its movements.
3. Describe structure and function of following 1. Hypostome; 2. Pellicle.
4. Describe the structure of *Paramecium* with suitable diagram. Explain process of nutrition and osmoregulation in this animal.
5. Describe in detail about modes of locomotion in *Paramecium*. (*M.S.U. Tirunelveli 2006*)
6. Describe the food passage in *Paramecium* showing arrangement of cilia. (*C.C.S.U. Meerut 2003*)

7. Describe the process of conjugation and its significance in *Paramecium*. (*C.C.S.U. Meerut 2006, 2009*)

8. Give an account of asexual and sexual reproduction in *Paramecium*.

9. Describe the behaviour of *Paramecium* under the following heads: 1. Thigmotaxis; 2. Thermotaxis; 3. Geotaxis; 4. Galvanotaxis; 5. Chemotaxis.

10. What is cyclosis? Describe feeding, digestion and excretion of *Paramecium*.

11. Describe sexual reproduction in *Paramecium*.

12. Describe conjugation, autogamy and endomixis in *Paramecium*.

Short Answer Questions

1. Describe the process of nutrition of *Paramecium*. (*C.C.S.U. Meerut 2001; M.S.U. Tirunelveli 2006*)

2. Describe locomotion in *Paramecium*.

3. Describe the digestion in *Paramecium*. (*C.C.S.U. Meerut 2003*)

4. Give an account of binary fission in *Paramecium*.

5. Give diagrams of process of conjugation in *Paramecium*. (*C.C.S.U. Meerut 2005*)

6. Define conjugation. Describe the process and give its significance. (*Rohilkhand 2003*)

7. How is ciliary locomotion performed in *Paramecium*? (*Purvanchal 2001*)

8. Distinguish between conjugation and autogamy in *Paramecium*. (*Avadh 2001*)

9. Write Brief/Short notes on the following:

(*i*) Structure of *Paramecium*.
(*ii*) Infra-ciliary system.
(*iii*) Trichocyst
(*iv*) Contractile vacuole
(*v*) Osmoregulation
(*vi*) Cyclosis (*Kanpur 2001*)
(*vii*) Ciliary movement
(*viii*) Filter feeding in *Paramecium*
(*ix*) Autogamy in *Paramecium*.
(*x*) Binary fission
(*xi*) Conjugation in *Paramecium*
(*xii*) Endomixis

Very Short Answer Questions

1. Write down zoological name of following animal – slipper animalcule.

2. What is axoneme?

3. What are trichocysts? (*C.C.S.U. Meerut 2004, 2006*)

4. Name the material found in a trichocyst.

5. What is cyclosis?

6. Give the location of cytostome in *Paramecium*.

7. Where is the endoral membrane located in *Paramecium*?

8. What are Kappa particles? Describe. (*C.C.S.U., Meerut 2007*)

9. Which nucleus controls sexual reproduction in *Paramecium*?

10. Is binary fission in *Paramecium* homothetogenic?

11. What do the term proter and opisthe indicate?

12. Who discovered conjugation in *Paramecium*?

13. Define synkaryon or zygote. (*C.C.S.U., Meerut 2008, 2009*)

Fill in the Blanks

1. *Paramecium* is commonly called .... animalcule/organism.
2. Pellicle of *Paramecium* is .... layered.
3. A layer of alveoli lies under the .... in *Paramecium.*
4. Pellicle of *Paramecium* has pores for .... and .... to emerge.
5. Infraciliature of *Paramecium* consists of .... and kinetodesma.
6. In *Paramecium,* the kinety comprises .... , kinetosome and ....
7. In *Paramecium,* the cytopharynx is separated from the cytoplasm by a ....
8. In *Paramecium,* the cytopharynx is lined with microtubules called ....
9. Cilia in *Paramecium* beat in a .... rhythm.
10. Macronucleus of *Paramecium* contain .... nucleoli.
11. Autogamy occurs only in *Paramecium* ....
12. In *Paramecium* the .... function is controlled by macronucleus.
13. In *Paramecium,* reproduction is controlled by ....
14. The process of self-fertilization is called .... in *Paramecium.*
15. In *Paramecium* rotatory endoplasm carries food vacuole along a definite path, this process is called .... (*C.C.S.U., Meerut 2000*)
16. The two types of nuclei in *Paramecium* are the .... which controls normal cellular metabolism, and the .... which involves in reproduction.
17. Intracellular digestion takes place inside ....
18. Contractile vacuole helps in ....
19. Quadrulus are specialized long cilia in the .... of *Paramecium.*
20. Response to light is called ....
21. Conjugation in *Paramecium* is a type of .... reproduction.
22. An exchange of genetic information occurs in ciliates when cells .... and exchange haploid ....
23. The unicellular algae present in the endoplasm of *Paramecium bursaria* are called ....

Multiple Choice Questions

1. Feeding in *Paramecium* takes place through
   (*a*) undulating membrane (*b*) general body surface
   (*c*) oral groove (*d*) none of the above
2. Which of the following one is filter feeder?
   (*a*) *Amoeba* (*b*) *Trypanosoma*
   (*c*) *Monocystis* (*d*) *Paramecium*
3. Method of food intake in *Paramecium* is
   (*a*) autozoic (*b*) saprozoic
   (*c*) holozoic (*d*) holophytic
4. Movement of food vacuole in *Paramecium* along a definite path is known as (*M.S.U. Tirunelveli 2006*)
   (*a*) cytokinesis (*b*) cyclosis
   (*c*) endomixis (*d*) metagenesis
5. Cytopyge is found in
   (*a*) *Paramecium* (*b*) *Amoeba*
   (*c*) *Trypanosoma* (*d*) *Entamoeba*
6. *Paramecium* moves about by
   (*a*) floating (*b*) looping
   (*c*) crawling (*d*) beating of cilia

7. Which of the following helps in anchorage and defence of *Paramecium*?
   (*a*) nematocyst (*b*) mucocyst
   (*c*) trichocyst (*d*) statocyst
8. The number of nucleus in *Paramecium* is
   (*a*) one (*b*) two
   (*c*) three (*d*) four
9. In *Paramecium*, the essential genetic information is carried by
   (*a*) micronucleus (*b*) macronucleus
   (*c*) mitochondria (*d*) both (*a*) and (*b*)
10. The main nitrogenous waste in *Paramecium* is
    (*a*) urea (*b*) uric acid
    (*c*) ammonia (*d*) amino acid
11. The main function of the contractile vacuole is (*M.S.U. Tirunelveli 2006*)
    (*a*) pumping out excess water (*b*) excretion
    (*c*) osmoregulation (*d*) respiration
12. Mention the function of the neuromotor system
    (*a*) coordination of ciliary beat (*b*) coordination of various stimuli
    (*c*) control of digestion (*d*) control of respiration movement
13. Binary fission in *Paramecium* is
    (*a*) horizontal (*b*) longitudinal
    (*c*) transverse (*d*) irregular
14. Rejuvenation of weakened *Paramecium* occurs by
    (*a*) conjugation (*b*) autogamy
    (*c*) hemixis (*d*) all of these
15. How many paramecia are produced after the conjugation?
    (*a*) two (*b*) four
    (*c*) eight (*d*) sixteen
16. Cyclosis in *Paramecium* is associated with (*Kanpur 2001*)
    (*a*) respiration (*b*) excretion
    (*c*) digestion (*d*) locomotion
17. Slipper animalcule is the popular name of
    (*a*) *Euglena* (*b*) *Amoeba*
    (*c*) *Paramecium* (*d*) *Trypanosoma*
18. *Paramecium caudatum* has
    (*a*) single nucleus (*b*) two micronuclei of the same size
    (*c*) one micronucleus and one macronucleus (*d*) only one meganucleus
19. Paramecium and other ciliates differ from all other protozoans in
    (*a*) having two types of nuclei
    (*b*) capturing prey with pseudopodia
    (*c*) having contractile vacuoles
    (*d*) not using flagella for locomotion
20. Which of the following statement is not true of *Paramecium*?
    (*a*) they are predatory ciliates
    (*b*) they use pseudopodia for capturing prey
    (*c*) they are preyed upon by *Didinium*
    (*d*) they use contractile vacuoles for removing excess water

21. Animal-like nutrition is known as
(*a*) holophytic (*b*) holozoic
(*c*) saprophytic (*d*) parasitic

22. How *Paramecium* takes its food?
(*a*) by oral groove (*b*) by undulating membrane
(*c*) by pseudopodia (*d*) through body surface

23. Cyclosis is present in
(*a*) *Amoeba* (*b*) *Euglena*
(*c*) *Paramecium* (*d*) *Plasmodium*

24. The food vacuole of *Paramecium* is
(*a*) first alkaline then acidic
(*b*) only alkaline
(*c*) only acidic
(*d*) first acidic and then alkaline

25. Intracellular digestion occurs inside
(*a*) stomach (*b*) food vacuole
(*c*) contractile vacuole (*d*) alimentary canal

26. What is main function of contractile vacuole?
(*a*) food regulation (*b*) excretion
(*c*) osmoregulation (*d*) reproduction

27. Asexual reproduction of *Paramecium* is brought about by
(*a*) simple binary fission (*b*) multiple fission
(*c*) sporulation (*d*) transverse binary fission

(*Rohilkhand 2000*)

28. In *Paramecium* nutrition is
(*a*) holophytic (*b*) holozoic
(*c*) mixotrophic (*d*) parasitic

(*Rohilkhand 2000*)

29. *Paramecium caudatum* has
(*a*) one macro-and two micronuclei (*b*) one macro-and one micronuclei
(*c*) two macro-and one micronuclei (*d*) two macro-and two micronuclei

ANSWERS

Very Short Answer Questions

1. *Paramecium caudatum*.
2. Axial filament of a cilium or flagellum. It has 9 + 2 arrangement of microtubules.
3. Trichocysts are oval, rod or spindle-shaped minute bodies situated to right angle to the body surface of *Paramecium* in ectoplasm. They serve the purpose of defence and keeping the animal attached to substratum while it feeds.
4. Swelling substance.
5. Rotary streaming movement of food vacuoles in the endoplasm (cytoplasm) is called **cyclosis.** The food vacuoles are regularly circulated by streaming endoplasm.
6. Between the buccal cavity and cytopharynx.
7. Along the right border of the junction between the vestibule and the buccal cavity.
8. Kappa particles are cytoplasmic factors in certain paramecia (called **killers**) that mediate production of **paramecin** (a toxin) and thereby makes the medium in which such microorganisms are grown toxic to members of strains not possessing the factors (sensitives).
9. Micronucleus.
10. Yes.

11. Anterior and posterior daughter paramecia formed by transverse binary fission.
12. **Hertwig** and **Maupas**.
13. Synkaryon is a cell nucleus formed by the fusion of two pre-existing nuclei. Zygote is a cell formed by the union of two gametes.

Fill in the Blanks

1. Slipper
2. Three
3. Plasma membrane
4. Cilia, Trichocysts
5. Kinetosome
6. Cilia, Kinetodesma
7. Membrane
8. Nematodesmata
9. Metachronous
10. Several
11. *aurelia*
12. Vegatative
13. Micronucleus
14. Autogamy
15. Cyclosis
16. Macronucleus, micronucleus
17. Food vacuole
18. Osmoregulation
19. Dorsal wall of vestibule
20. Phototaxis
21. Sexual
22. Conjugate micronuclei
23. *Zoochlorella*.

Multiple Choice Questions

1. (*c*)	**2.** (*d*)	**3.** (*c*)	**4.** (*b*)	**5.** (*a*)	**6.** (*d*)
7. (*c*)	**8.** (*b*)	**9.** (*a*)	**10.** (*c*)	**11.** (*c*)	**12.** (*a*)
13. (*c*)	**14.** (*d*)	**15.** (*b*)	**16.** (*c*)	**17.** (*c*)	**18.** (*c*)
19. (*a*)	**20.** (*b*)	**21.** (*b*)	**22.** (*a*)	**23.** (*c*)	**24.** (*d*)
25. (*b*)	**26.** (*c*)	**27.** (*d*)	**28.** (*b*)	**29.** (*b*).	

6

Plasmodium (Malaria Parasite) (Phylum Protozoa)

6.1 CLASSIFICATION

Phylum	–	**Protozoa**
Subphylum	–	**Sporozoa (Apicomplexa*)**
Class	–	**Telosporea**
Subclass	–	**Coccidia**
Order	–	**Eucoccidia (Haemosporidida)**
Genus	–	***Plasmodium***

[*The term Apicomplexa is now used for Sporozoa; it is so named because of a complex of ring-like, tubular filamentous organelles at the apical end; see **Ruppert** and **Barnes**, 1994] (Fig. 6.1).

6.2 GEOGRAPHICAL DISTRIBUTION

The tropical zone of earth is the endemic home of all malaria parasites (*Plasmodium*).

6.3 HABIT AND HABITAT

Plasmodium or malaria parasite is a pathogenic **intracellular blood parasite** (= cytozoic endoparasite) of humans and other vertebrates. It lives in hepatocytes (liver cells) and red blood corpuscles (or erythrocytes). Its life cycle is **digenetic**, *i.e.*, the life cycle of the parasite is completed on the following two hosts:

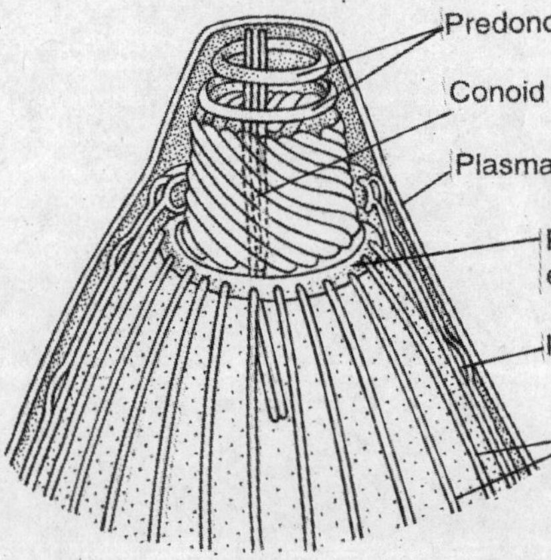

Fig. 6.1. Microstructure of the apical complex of sporozoa (after Brusca and Brusca, 2003).

(*i*) Humans. They are **primary hosts** or definitive hosts where it lives in erythrocytes (RBC) and completes its **endogenous** or **asexual cycle** (*e.g.*, schizogony).

(*ii*) Female *Anopheles* mosquito. It is the **secondary** or **intermediate host** where *Plasmodium* completes its **exogenous** or **sexual cycle** (*e.g.*, gametogony, syngamy and sporogony) inside the alimentary canal.

At present, mosquito is considered as primary host (due to presence of sexual phase of life cycle) and human being is condisered as secondary host (due to presence of asexual phase of life cycle of the parasite).

6.4 MORPHOLOGY

Plasmodium exhibits typical parasitic adaptations of sporozoa, *e.g.*, feeding trophozoite of human RBC is characterized by the absence of locomotory organs, a mouth, anal opening, excretory pore and vacuoles. However, in it reproductive capacity becomes more prominent. The mode of nutrition is **saprozoic** which occurs by **osmotrophy**. Respiration is **anaerobic**.

While plasmodia are the actual infectious agents of disease (malaria), the female mosquito is the **vector** that passes the portozoan (**parasite**) from one host to another. Following four species of *Plasmodium* are pathogenic to human beings causing different types of malaria:

1. *Plasmodium vivax* — Causes benign tertian malaria.
2. *Plasmodium falciparum* — Causes malignant tertian malaria, pernicious malaria, falciparum malaria or estivoautmnal malaria.
3. *Plasmodium malariae* — Causes quartan malaria.
4. *Plasmodium ovale* — Causes ovale tertian malaria.

The most virulent in human populations is *Plasmodium falciparum*. Its infection produces severe anemia, cerebral symptoms, renal failure, pulmonary edema and death.

6.5 LIFE CYCLE

It involves the following two cycles which alternate with each other:

A. Asexual cycle in humans. It is completed in the following steps:

1. Pre-erythrocytic cycle (primary exo-erythrocytic cycle or tissue phase). Parasite (malaria sporozoite; Fig. 6.2) finds its way into the blood of healthy person along with the salivary secretion (containing anticoagulant) of infected female *Anopheles* mosquito which bites for getting a blood meal. The parasite enters liver cells (hepatocytes) and asexually multiplies there by **liver schizogony** producing numerous haploid asexual blood forms called **cryptomerozoites** or **metacryptomerozoites**. For example, in case of *P. falciparum* upto 30,000 cryptomerozoites are released when a hepatocyte ruptures (**Robbin**, 1994).

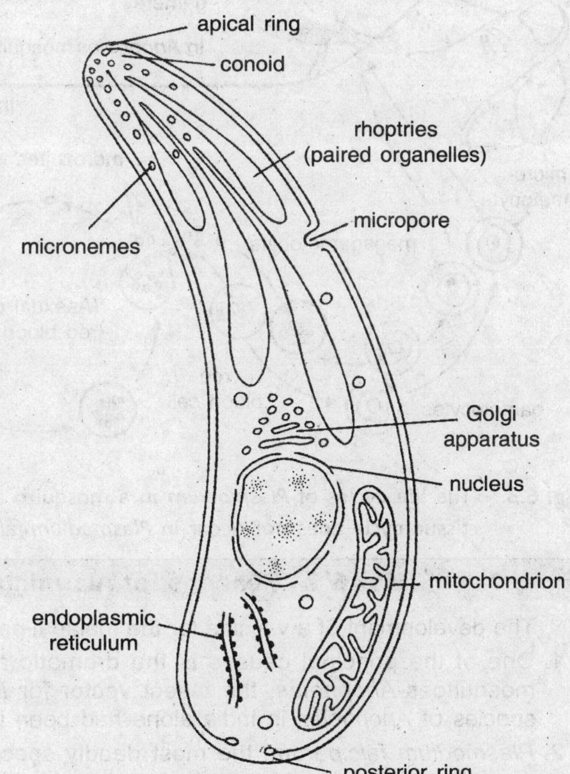

Fig. 6.2. Lateral view of a generalized sporozoan (apicomplexan); the sporozoite showing apical complex.

2. Erythrocytic cycle (erythrocytic schizogony and gametogony). Some cryptomerozoites are transported to red blood corpuscles (or erythrocytes). Here, they become feeding **trophozoites** which metabolise haemoglobin, change morphology from ring (signet-ring stage) to schizont (full-grown trophozoite), produce **malaria pigment** or **hemozoin**

and multiply asexually by **schizogony** to produce **merozoites** (Fig. 6.3). Merozoites are liberated in the blood by the rupture of RBCs, causing chills and fever. Merozoites enter fresh RBCs to repeat schizogony.

B. Sexual cycle in female *Anopheles*. During erythrocytic cycle in human blood, some merozoites develop into **gametocytes**. The male gametocytes are called **microgametocytes** and female as **macrogametocytes**. When the mosquito (female *Anopheles*) bites an infected human, it sucks up gametocytes of *Plasmodium* along with its blood meal. In the digestive tract of the mosquito the gametocytes develop into haploid male and female **gametes.** Fertilization occurs within stomach and the **zygote** (called **ookinete**) burrows into the stomach wall (Fig. 6.3). Following zygotic meiosis a series of mitosis occurs, resulting in the formation of many **sporozoites,** which migrate to the insect's salivary gland, from where they are introduced into the new host.

Thus, in the life cycle of *Plasmodium* there occurs an alternation of sexually reproducing and asexually reproducing generations or **metagenesis**.

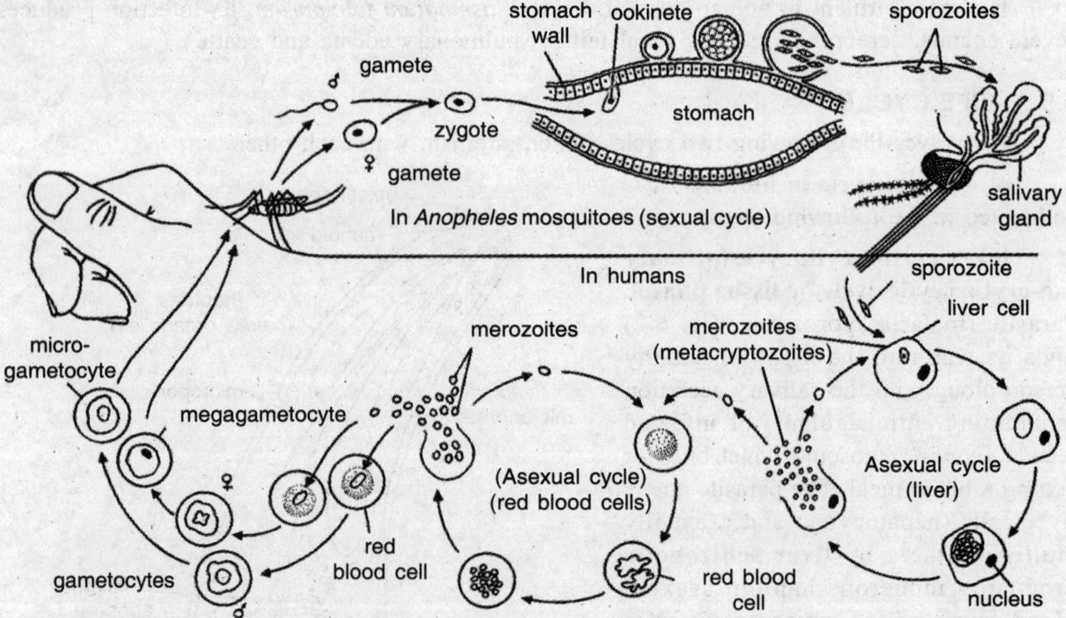

Fig. 6.3. The life cycles of *Plasmodium* in a mosquito and in a human. Reinvasion of liver cells in the tissue cycle does not occur in *Plasmodium falciparum*.

Box 6.1 Reasons for Alarming Comeback of Malaria

The development of a vaccine for the malarial parasite continues to elude researchers.

1. One of the principal causes is the dramatic rise in number of pesticide-resistant mosquitoes-*Anopheles*, the insect vector for *Plasmodium*. By 1968, 38 strains or species of *Anopheles* in India alone had been identified as largely DDT-resistant.
2. *Plasmodium falciparum*, the most deadly species of malarial parasite has become recently resistant to **chloroquine**, once a backbone of malarial drug therapy.
3. Researchers have recently discovered that *Plasmodium* species, like *Trypanosoma*, have the ability to avoid detection by the human immune system by switching among as many as

150 genes that code for different versions of the protein that coats the surface of their cells (it is this protein coat that the human immune system relies on as a recognition factor). Furthermore, because the parasite sequesters itself inside red blood cells, it is largely protected from most malarial drugs.

4. Recent evidence also suggested that the parasite may affect its mosquito host by inducing it to bite more frequently than uninfected mosquitoes (see **Brusca** and **Brusca**, 2003).

QUESTIONS

Long Answer Questions

1. Name the parasite that causes malaria in human beings. How is this fever transmitted from an infected to healthy person? Suggest measures to control malaria.
2. Give a brief account of the life history of *Plasmodium vivax*.
3. Decribe in detail part of life cycle of *Plasmodium vivax* in mosquito.
4. Give an account of asexual cycle of *Plasmodium vivax* in human beings.
5. Explain the following terms.

(*i*) Parasite; (*ii*) *Ookinete;*
(*iii*) Intermediate host; (*iv*) Incubation period;
(*v*) Alternation of generations; (*vi*) Malaria.

Short Answer Questions

1. What is a primary host?
2. What is a vector?
3. Where do schizogony, gamogony and sporogony of *Plasmodium* take place?
4. List the stages of malarial parasite found in the mosqnito's stomach.

Very Short Answer Questions

1. Who discovered malaria parasite?
2. Name the four species of *Plasmodium.*
3. Name the primary and secondary host of *Plasmodium vivax*.
4. Define incubation period in relation with *Plasmodium*.
5. How is female *Anopheles* affected by *Plasmodium* infection?
6. What is infective stage of malarial parasite called?
7. What name is given to malarial parasite when it lives the liver cells?
8. Where does fertilization occus in *Plasmodium*?
9. What is mosquito day?

Fill in the Blanks

1. *Plasmodium* has two hosts: humans and female .... mosquito.
2. Malarial parasite completes .... cycle in human beings.
3. *Plasmodium* completes its sexual cycle in ....
4. Cryptozoites live in the .... cells.
5. Motile zygote of *Plasmodium* is called ....
6. Malaria is caused by .... species of ....

Multiple Choice Questions

1. An intracellular protozoan parasite is
   (*a*) Entamoeba (*b*) Ascaris
   (*c*) Trypanosoma (*d*) *Plasmodium*
2. Life cycle of malarial parasite is
   (*a*) monogenetic (*b*) digenetic
   (*c*) not in *Anopheles* (*d*) none of these
3. Vector of *Plasmodium* is
   (*a*) *Culex* (*b*) *Aedes*
   (*c*) *Anopheles* (*d*) *Musca*
4. The reserve stock of malarial parasite is found in the human
   (*a*) blood cells (*b*) blood plasma
   (*c*) sinusoids (*d*) liver cells
5. The most dangerous malaria is caused by
   (*a*) *Plasmodium vivax* (*b*) *Plasmodium malariae*
   (*c*) *Plasmodium ovale* (*d*) *Plasmodium falciparum*
6. The adult trophozoite in a red blood corpuscle (erythrocyte) is called
   (*a*) sporont (*b*) cryptoschizont
   (*c*) metacryptoschizont (*d*) erythrocytic schizont
7. The stage of *Plasmodium* that appears in human red blood cell during night only
   (*a*) schizont (*b*) sporozite
   (*c*) gamont (*d*) microgametocyte
8. Toxic material formed after proteolysis of haemoglobin is
   (*a*) haematin (*b*) haemozoin
   (*c*) haem (*d*) Schuffner's dot
9. Which of the following fish is helpful in biological control of *Plasmodium*?
   (*a*) *Catla* (*b*) dog fish
   (*c*) *Gambusia* (*d*) silver fish
10. To which of the following classes will you place *Plasmodium*?
    (*a*) sporozoa (*b*) mastigophora
    (*c*) rhizopoda (*d*) ciliata
11. An insect (*e.g.*, mosquito) which transmits a disease is known as
    (*a*) intermediate host (*b*) parasite
    (*c*) vector (*d*) prey
12. Which one is a communicable disease?
    (*a*) malaria (*b*) diabetes
    (*c*) hypertension (*d*) kwashiorkor
13. Malaria fever is caused by
    (*a*) female *Culex mosquito* (*b*) male *Culex* mosquito
    (*c*) female *Anopheles* mosquito (*d*) male *Anopheles* mosquito
14. Causative organism of malaria is
    (*a*) *Trypanosoma gambiense* (*b*) *Entamoeba histolytica*
    (*c*) *Leishmania donovani* (*d*) *Plasmodium vivax*
15. Syzygy is
    (*a*) union of gametes in Protozoa (*b*) pairing of gametocytes
    (*c*) pairing of homologous chromosomes (*d*) formation of spores

16. The protozoans that cause malaria in humans are

(*a*) radiolarians (*b*) dinoflagellates

(*c*) sporozoans (*d*) ciliates

17. Schizogony and schizont stage of *Plasmodium* occur in (*M.S.U. Tiruneleveli 2006*)

(*a*) erythrocytes and liver cells of human beings

(*b*) blood of human beings

(*c*) stomach of human beings

(*d*) only liver in human beings

18. Exflagellation in malarial parasite takes place during the formation of

(*a*) ova (*b*) sperms

(*c*) gametocytes (*d*) sporozoites

19. Sporogony of malarial parasite occurs in

(*a*) liver of human beings (*b*) erythrocytes of human beings

(*c*) salivary glands of mosquito (*d*) stomach of mosquito

ANSWERS

Very Short Answer Questions

1. Laveran in 1880.

2. *Plasmodium vivax, P. ovale, P. malariae and P. falciparum.*

3. Human beings are the primary host and female *Anopheles* mosquito is the **secondary *or* intermediate** host of *Plasmodium vivax*.

4. The period between inoculation of sporozoites into human blood and first appearance of symptoms of malaria is called incubation period. On an average it is 14 days in *Plasmodium vivax* and *P. ovale*, 12 days in *P. falciparum* and 28 in *P. malariae*.

5. *Plasmodium* is non-pathogenic in mosquito host.

6. Sporozoite.

7. Cryptozoite and metacryptozoite.

8. Lumen of stomach of female *Anopheles* mosquito.

9. August 10, 1897, when oocysts of *Plasmodium* were found on the mosquito's stomach by **sir Ronald Ross** (1857-1932). He established that malarian parasites are sucked up by female *Anopheles* mosquito and later on injected in the human blood. For this discovery he earned Nobel price in Physiology and Medicine, 1902.

Fill in the Blanks

1. *Anopheles* **2.** Asexual

3. Female *Anopheles* mosquito **4.** Liver

5. Ookinete **6.** Four, *Plasmodium*

Multiple Choice Questions

1. (*d*)	**2.** (*b*)	**3.** (*c*)	**4.** (*d*)
5. (*d*)	**6.** (*d*)	**7.** (*a*)	**8.** (*b*)
9. (*c*)	**10.** (*a*)	**11.** (*c*)	**12.** (*a*)
13. (*c*)	**14.** (*d*)	**15.** (*b*)	**16.** (*c*)
17. (*a*)	**18.** (*b*)	**19.** (*d*).	

7

Entamoeba histolytica (Phylum Protozoa)

7.1 CLASSIFICATION

Phylum	–	**Protozoa**
Subphylum	–	**Sarcomastigophora**
Superclass	–	**Sarcodina**
Class	–	**Rhizopoda**
Order	–	**Amoebida**
Genus	–	***Entamoeba***
Species	–	***histolytica***

Entamoeba histolytica is an amoeba-like microscopic disease-producing (pathogenic) endoparasite of human beings. It inhabits colon of human beings and produces the disease called **amoebiasis** or **amoebic dysentery**. **Lamble** (1859) discovered this species. Russian Zoologist **Losch** (1875) discovered its pathogenic nature.

7.2 DISTRIBUTION

Entamoeba histolytica is cosmopolitan (world-wide or universal) in distribution but is more common in tropical countries such as India, China, Mexico, etc., due to poor sanitary conditions.

7.3 HABIT AND HABITAT

Entamoeba histolytica lives in the lumen of the lower portion of small intestine and the entire large intestine (colon) of humans. It may invade the lining (*i.e.*, mucosa and submucosa) of the host's large intestine. There it feeds on tissues and red blood corpuscles of intestinal wall and produces ulcers and dysentery which involves inflammation of intestine, loose stools or diarrhoea and 6 to 10 blood-stained painful mucus motions per day. In cronic cases, it may enter the blood circulation and may infect liver, lungs, kidneys, testes and even the barin. It proves fatal when it produces large abscesses in liver or perforates the intestinal wall of the host.

7.4 MORPHOLOGY

The adult *E. histolytica* occurs in following two forms:

A. Trophozoite or magna form. It is tissue dwelling large-sized (having 20–30 μm diameter), trophic or feeding form. It lives in the mucosa of intestinal wall and feeds on fragments of tissue cells, erythrocytes (RBC), leucocytes (WBC) and bacteria. By the help of secretion of a proteolytic enzyme called **cytolysin** or **histolysin**, it destroys mucosa and produces disease. Thus, it is a **pathogenic form**.

B. Precystic or minuta form. It is lumen-dwelling, small sized (having diameter of 7–10 μm or microns), non-feeding and **non-pathogenic form**.

A. Trophozoite. Trophozoite (Fig. 7.1) represents mature parasite which is active or motile adult resembling *Amoeba* in some structural details. It is irregular or asymmetrical in form and is covered by a thin and flexible or elastic **plasmalemma** or **plasma membrane**. The cytoplasm of trophozoite is differentiated into two distinct regions: ectoplasm and endoplasm. **Ectoplasm** forms a thin, clear (hyaline), non-granular and relatively firm outer layer. It contains no organelles. The **endoplasm** is the inner, granular, fluid-like and constitutes the bulk of the body. Endoplasm contains two types of organelles: nucleus and food vacuoles. The contractile vacuole is absent in *E. histolytica* since it lives in an isotonic environment and does not face the problem of osmoregulation.

Nucleus of trophozoite is large (4–6 μm in diameter), rounded and vesicular containing six chromosomes (**Kofoid** and **Swezy**). It is bounded by a delicate **nuclear envelope** having **chromatin granules** on its inner surface. Nucleus has a central dot-like **nucleolus** (**endosome** or **karyosome**) surrounded by a clear area called **halo**. Nucleoplasm contains distinct spoke-like striations existing between nucleolus and peripheral chromatin.

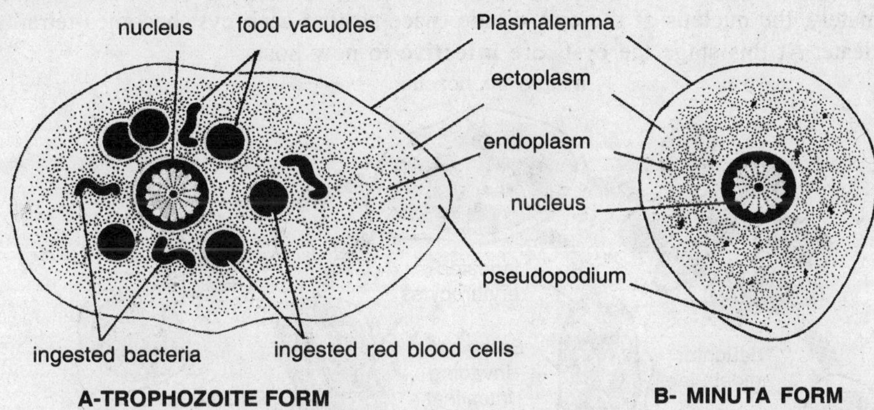

Fig. 7.1. *Entamoeba histolytica*. A—Trophozoite or magna form; B—Minuta form.

Food-vacuoles vary in number and size. They contain red blood corpuscles (a characteristic feature), white blood cells or bacteria in various stages of digestion.

7.5 PHYSIOLOGY

***(i)* Locomotion.** *E. histolytica* performs locomotion by the help of a single, large, thick and blunt pseudopodium called **lobopodium**. It is called **monopodial locomotion**. Lobopodium is formed largely of ectoplasm and protrudes at the advancing end of body.

***(ii)* Nutrition.** Trophozoite has **holozoic nutrition**. It dissolves the intestinal mucous membrane of the host and phagocytoses the debris, *i.e.*, blood and epithelial cells, at the level of its plasma membrane. Food is digested intracellularly inside the food vacuoles by the help of lysosomal enzymes.

***(iii)* Respiration.** Trophozoite is a **facultative anaerobe**, *i.e.*, it requires very little oxygen for the breathing and can survive without oxygen. It is an adaptation for its endoparasitic mode of existence inside the colon.

***(iv)* Excretion.** Trophozoite excretes its nitrogenous metabolic wastes in the form of **ammonia**. The ammonia is got rid off by diffusion through general body surface into the surrounding medium.

B. Minuta form. It represents precystic form which is small, spherical, non-motile, non-feeding and non-pathogenic. Minuta form lacks pseudopodium and food vacuoles. It contains many glycogen granules (reserve food) and black, rod-like **chromatoid bodies** (probably ribonucleoproteins). Chromatoid bodies occur singly or in the multiple of two or more but disappear in the old cysts. Minuta form undergoes encystation and helps in the transmission of parasite.

7.6 REPRODUCTION

Entamoeba histolytica shows only asexual reproduction which involves mitosis having intranuclear karyokinesis and the cytokinesis. Asexual reproduction takes place by **binary fission**.

7.7 LIFE CYCLE

E. histolytica is a **monogenetic parasite**, *i.e.*, its life cycle involves only one host; the host being the human beings. Its life cycle includes the following steps:

(*i*) Encystment. Some of the precystic minuta forms exist in intestinal lumen and undergo **encystment** or **encystation**. However, before the encystment, they become round, eliminate food vacuoles and accumulate considerable amount of reserve food materials in the form of glycogen granules and chromatoids. Soon each parasite secretes a thin, rounded, resistant, colourless and transparent cyst wall around it. The cysts of *Entamoeba histolytica* have average size of 12 mm (or 12 microns). Each cyst has a clear cytoplasm and single nucleus and is called **mononucleate cyst**. Ultimately, the nucleus of the cyst divide twice so that each cyst becomes tetranucleate or quadranucleate. At this stage the cysts are **infective** to new hosts.

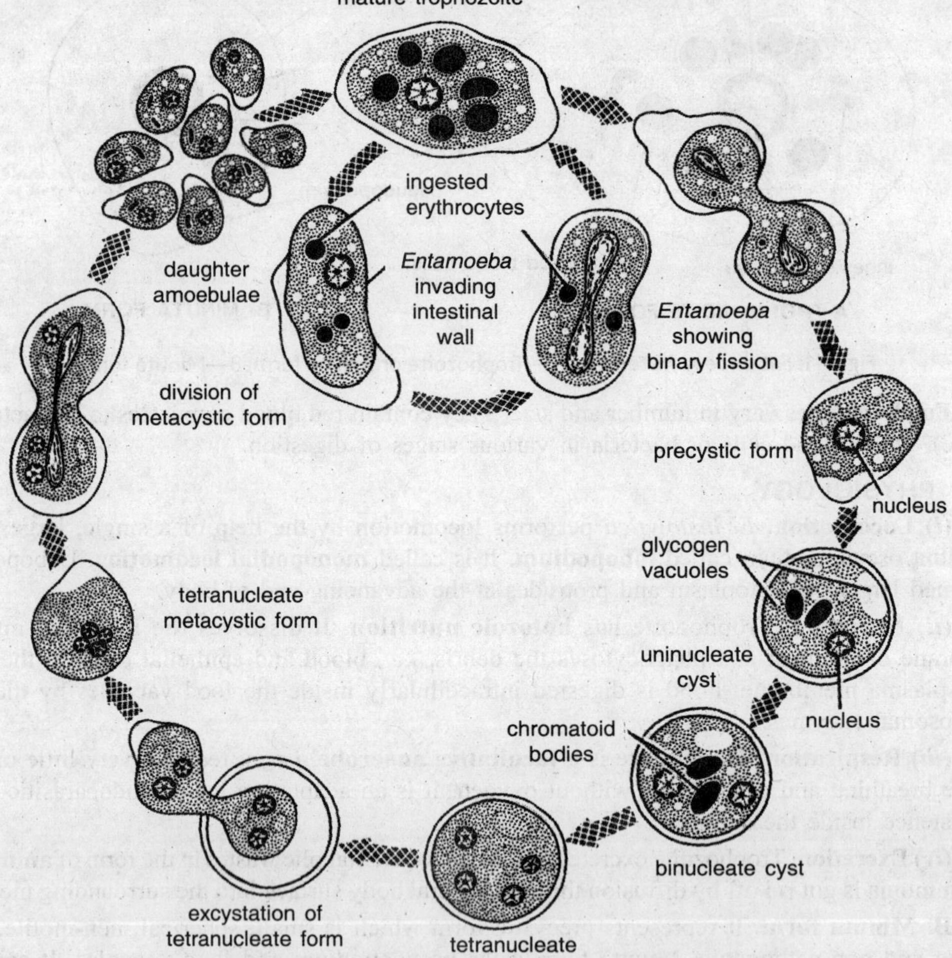

Fig. 7.2. *Entamoeba histolytica*. Life cycle.

(*ii*) Transfer to new host. The tetranucleate cysts pass out with the faecal matter of the host (These are used for diagnosis of the infection). Cysts remain alive for 10 days in the moist stool. They die if dried or if the atmospheric temperature rises to about 50°C. Cysts are also resistant to chemicals. Infection of fresh human host takes place by swallowing the infective cysts with contaminated food and drinks. Contamination of food and drinks is brought about by houseflies, cockroaches and dirty fingers of food handlers.

Through contaminated food, infective cysts pass into the lower portion of the small intestine of the host.

***(iii)* Excystment.** Excystation of the parasite occurs in host's intestine according to any of two methods: 1. The cyst wall is digested by alkaline juice (trypsin enzyme) of host's small intestine. 2. In culture, a small portion of cyst is dissolved by some enzymes secreted by the pseudopodium of the parasite. Anyhow, a tetranucleate amoebae is hatched out and is called **metacystic form**. Metacystic entamoeba produces a new generation of trophozoite by a diverse series of nuclear and cytoplasmic divisions which result in the production of eight uninucleate daughter amoebulae which represent young trophozoites. Young trophozoites feed on the contents (bacteria) of the lumen of intestine and grow in size to form the normal parasitic trophozoite.

7. Pathogenecity. *Entamoeba histolytica* causes **amoebic dysentry**. It causes ulcers and abscesses (*i.e.*, a localised collection of pus surrounded by inflammed tissue) in the intestinal wall by eroding submucosa. As the ulcers burst, the blood and mucus are passed out.

8. Symptoms. The common symptoms of infection of amoebic dysentery are abdominal pain, nausea, flatulance and bowel irregularity with fatigue and headache.

9. Treatment (therapy) and prevention (prophylaxis). Prevention from infection involves personal health and hygiene. One must keep his food-stuff, drinking water and utensils clean and well-covered, preventing their contamination by houseflies, ants, mice, cockroaches, etc. Also, one must keep his nails trimmed and wash hands with soap and water before meals.

Metronidazole, Emetin, iodine compounds (**Yatren, Diodoquin**) and arsenic compounds (**Carbarsone, Thiocarbarsone**) are used in amoebic dysentry. **Bismuth subritrate** is found beneficial in controlling amoebic dysentery. Certain **antibiotics**, such as **Fumagillin, Erythromycin, Terramycin** and **Auromycin** have proved to be more effective in case of severe infection. In case of secondary infection, *i.e.*, infection of liver, lungs, etc., **Chloroquine** is given.

7.8 *ENTAMOEBA GINGIVALIS*

Classification

Phylum	–	**Protozoa**
Subphylum	–	**Sarcomastigophora**
Superclass	–	**Sarcodina**
Class	–	**Rhizopodea**
Order	–	**Amoebida**
Genus	–	***Entamoeba***
Species	–	***gingivalis***

Entamoeba gingivalis is a nonpathogenic human parasite. It was first parasitic amoeba reported from human beings and was discovered by **Gross** in 1849.

1. Habit and habitat. *E. gingivalis* is commonly known as **mouth amoeba**. It lives in the cavities of teeth; in the tartar and plaque deposited around the base of teeth. It may also live in the abscesses of gums and pus-pockets of tonsils.

2. Morphology. The subspherical **trophozoite** of *E. gingivalis* measure 12 to 20 mm in diameter. It bears 2 or 3 small and blunt (*i.e.*, broad and rounded) pseudopodia, called **lobopodia**. The cytoplasm is divisible into a clear homogeneous peripheral **ectoplasm** and central, granular, highly vacuolated **endoplasm**. Endoplasm contains a vesicular **nucleus** with central **endosome** and several **food vacuoles**.

3. Physiology. Trophozoites has pseudopodial locomotion. It feeds mainly on bacteria present in the cellular debris around the roots of teeth and also on white blood cells (WBC) by phagocytosis.

4. Life cycle. Trophozoite reproduces only by asexual means, *i.e.*, by binary **fission**. It does not form cysts and is transmitted from mouth to mouth of persons either by contact during kissing or while eating or drinking by same utensils.

5. Pathogenicity. *E. gingivalis* is found in nearly 70 per cent of human population. Its occurrence is more common in the mouth of persons suffering from pyorrhoea. It was believed in the past that pyorrhoea was caused by *E. gingivalis*. However, it is now certain that pyorrhoea is caused by infection of bacteria. *E. gingivalis* is known to aggravate pyorrhoea disease by destroying the gum tissues.

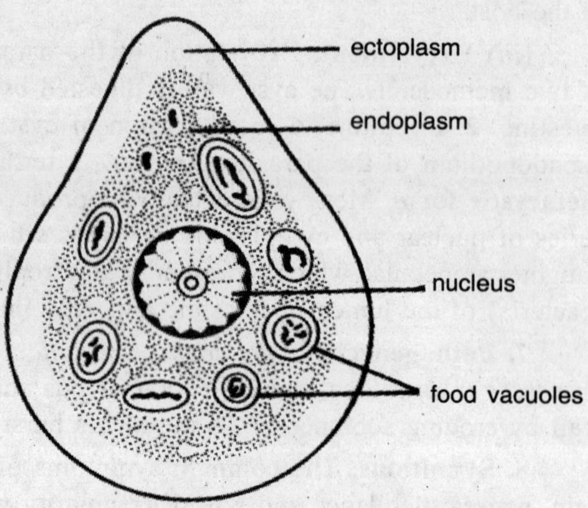

Fig. 7.3. *Entamoeba gingivalis*. Trophozoite.

QUESTIONS

Long Answer Questions

1. Describe in detail about life cycle, pathogeny and control measures of *Entamoeba*. (*M.S.U. Tirunelveli 2006*)
2. Give an account of habit, habitat, classification, characters and economic importance of *Entamoeba histolytica*. (*C.C.S.U. Meerut 2000*)

Short Answer Questions

1. Describe the pathogenicity, treatment and prevention of *Entamoeba* species.
2. Describe the structure of *Entamoeba histolytica*.
3. Give an account of the life-cycle of *Entamoeba*.
4. Write short notes on:
   (*i*) *Entamoeba histolytica* (*ii*) Amoebiasis
   (*iii*) *Entamoeba gingivalis*

Very Short Answer Questions

1. Name the enzyme secreted by *Entamoeba histolytica* to erode mucous membrane of host's intestine.

Fill in the Blanks

1. *Entamoeba histolytica* produces the disease ....
2. Amoebic dysentery is caused by ....
3. Trophozoite stage of *Entamoeba histolytica* produces .... to destroy tissue. (*C.C.S.U. Meerut 2001*)
4. The feeding process of *Entamoeba histolytica* is called as ....

Multiple Choice Questions

1. *Entamoeba histolytica* is found in human beings in
   (*a*) colon (*b*) oral cavity
   (*c*) stomach (*d*) intestine

2. Which one causes amoebic dysentery in human beings?
(*a*) *Amoeba proteus* (*b*) *Taenia solium*
(*c*) *Entamoeba histolytica* (*d*) *Ascaris*

3. *Entamoeba histolytica* lives in human body in
(*a*) stomach (*b*) duodenum
(*c*) blood (*d*) colon

4. Which of the following animals has only one host?
(*a*) *Plasmodium vivax* (*b*) *Entamoeba histolytica*
(*c*) *Trypanosoma gambiense* (*d*) none of the above

5. Trophozoite is
(*a*) the sexual stage of parasite that produces spores
(*b*) the sexual stage of parasite that produces gametes
(*c*) the adult feeding stage of protozoan parasite
(*d*) none of the above

6. Infective stage of *Entamoeba histolytica* is
(*a*) sporozoite (*b*) quadrinucleate cyst
(*c*) trophozoite (*d*) spore

7. How does locomotion takes place in *Entamoeba*?
(*a*) by lobopodia (*b*) by filopodia
(*c*) by rhizopodia (*d*) by axopodia

8. It causes amoebiasis
(*a*) mosquito (*b*) *Trypanosoma*
(*c*) *Paramecium* (*d*) *Entamoeba*

ANSWERS

Very Short Answer Questions

1. Cytolysin.

Fill in the Blanks

1. Amoebic dysentery **2.** *Entamoeba histolytiea* **3.** Cytolysin
4. Holozoie

Multiple Choice Questions

1. (*a*) **2.** (*c*) **3.** (*d*) **4.** (*b*) **5.** (*c*) **6.** (*b*)
7. (*a*) **8.** (*d*).

8 Locomotion in Protozoa

8.1 LOCOMOTORY ORGANELLES IN PROTOZOA

Four major types of locomotor organelles occur among protozoa; and usually each type of them is characterized by a class:

1. **Pseudopodia** are characteristics of **Sarcodina**.
2. **Flagella** are characteristics of **Flagellata** (**Mastigophora**).
3. **Cilia** are characteristics of **Ciliata**.
4. **Myonemes** are characteristics of **Sporozoa**.

I. Pseudopodia

Pseudopodia are temporary extensions of cytoplasm from any part of the body. They occur in those protozoa which are "naked" or have a very thin pellicle. Pseudopodia are formed from ectoplasm but also have a core of endoplasm. Among Sarcodina, the following kinds of pseudopodia are found.

1. Lobopodia. These are short finger-shaped, round-tipped pseudopodia containing both ectoplasm and endoplasm. *e.g.*, *Amoeba*, *Entamoeba*, *Arcella* and *Difflugia*.

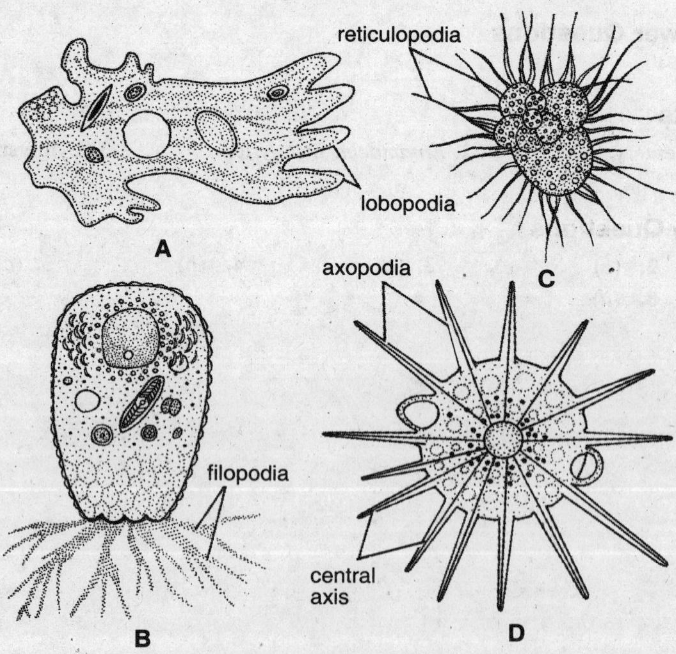

Fig. 8.1. Types of pseudopodia. A—Lobopodia of *Amoeba*; B—Filopodia of *Euglypha*; C—Reticulopodia of *Globigerina*; D—Axopodia of *Actinophrys*.

2. Filopodia. These are slender, filamentous hyaline projections, formed exclusively of ectoplasm. Filopodia have pointed tips and a tendency to branch and radiate in all directions. They never anastomose to form network, *e.g.*, *Euglypha*, *Chlamydophorus*.

3. Reticulopodia (= Rhizopodia). These are thin, long filamentous structures of endoplasm which branch extensively and fuse together to form meshworks that act as food traps, *e.g.*, *Elphidium*, *Globigerina* and *Allogromia*.

In case of *Microgromia*, the reticulopodia of several individuals unite to form a multicellular reticulum (Fig. 8.2).

4. Axopodia. These are long, slender, stiff extensions of ectoplasm which radiate from circular body in all directions. Axopodia contain a slender axial filament, composed of parallel microtubules rolled into a spiral when seen in the electron microscope. The fibrils are like flagellar fibrils and cause axopodial movement, *e.g.*, *Actinophrys*, *Actinospherium*.

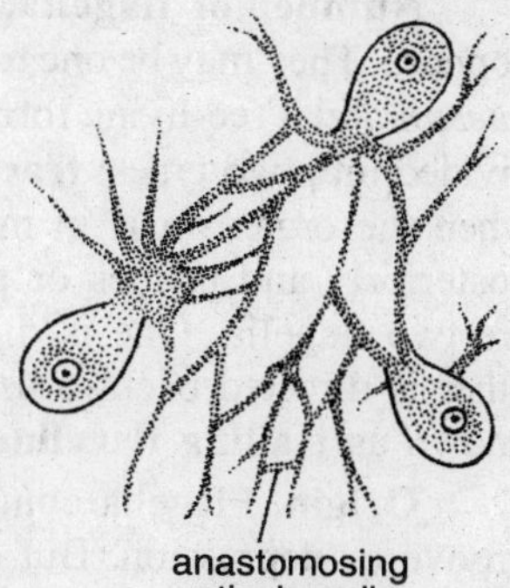

Fig. 8.2. Anastomosing reticulopodia of *Microgromia socialis*.

II. Flagella

Flagella are long, delicate, whip-like structures. They are found in flagellates (class Mastigophora) and the flagellate stages of sarcodine protozoans. Each flagellum arises from a **basal granule/body** or **kinetosome**.

Ultrastructure. The outer covering of a flagellum is continuous with the plasma membrane of the body surface. It contains a **matrix** in which a ring of nine doublet microtubules surrounds two central microtubules; a 9 + 2 arrangement also seen in cilia. Both central fibrils are connected by a bridge and are enclosed in a common central sheath. They confer bilaterality and the plane of flagellar beat may be associated with the orientation of central plane.

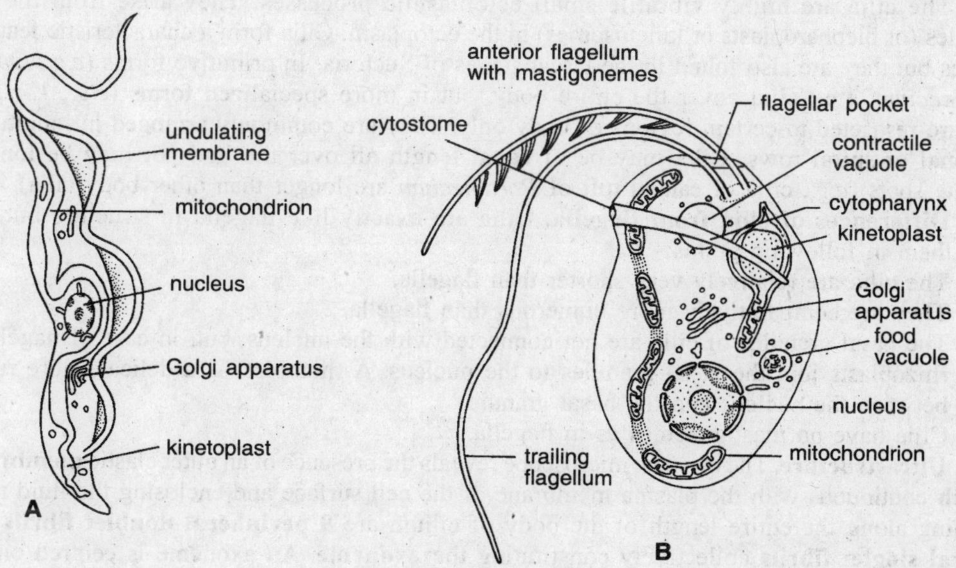

Fig. 8.3. A—*Trypanosoma* showing undulating membrane; B—*Bodo saltans* showing tractellum and trailing flagella.

At the tip of the main flagellum may be a very fine end piece or the main axis of the flagellum. Flagellum may bear fine, flexible lateral processes or **mastigonemes** on one side (Fig. 8.3B) or on both sides. Mastigoneme-bearing flagellum called **flimmer** or **ciliary flagellum.**

Number of flagella. The number and arrangement of flagella varies greatly in Mastigophora. They may be one to eight in number; several flagella occur in parasitic forms (*e.g.*, *Giardia intestinalis*). Free-living forms (*e.g.*, *Euglena*) are usually with one to two flagella. Flagella are often divided into two types: **tractella** and **pulsella**. The tractellum is situated at the end that is anterior when the organism is in motion and drags the body along; the pulsellum is generally situated posteriorly and pushes or propels the body forward. In *Bodo* (Fig. 8.3B) and *Aesonema* there are two flagella. One is directed forward, drags the body and is called **tractellum**. while the other is directed backwards along the side of the body and acts as an organ of propulsion, it is known as **trailing flagellum**.

Origin. Flagella commonly arise from the anterior end of body either directly or from a groove or depression. But in *Trypanosoma* (Fig. 8.3A), flagellum originates from the posterior end. Trypanosomes have two kinetosomes, but only one has a flagellum. In many forms this flagellum lies against the side of the cell body's membrane. When the flagellum beats, the membrane of the cell is pulled up into a fold and looks like a waving or undulating membrane. This arrangement appears to be relatively efficient in moving the cell through viscous media (such as blood). Although trypanosomes can change the direction of flagellar beat in response to chemical tip of the flagellum and proceeds towards the kinetosome. This reverse of the way flagella usually beat (from base to tip).

The kinetosomes are centriole like structures and often lie at the bottom of flagellar pockets or reservoirs. Kinetoplast (Fig. 8.3A) is the dark-staining concentration of mitochondrial DNA (mDNA). It generally is found in the part of mitochondrion lying close to the kinetosomes, although there is no known relationship between these two structures. The size, shape and position of the kinetoplast is important in taxonomy of trypanosomes and bodonids (see **Brusca** and **Brusca** 2003).

III. Cilia

The cilia are highly vibratile small ectoplasmic processes. They arise from the basal granules (or blepharoplasts or kinetosomes) in the ectoplasm. Cilia form a characteristic feature of ciliates but they are also found in the larval stages of Suctoria. In primitive forms (*e.g.*, *Opalina*, *Paramecium*, etc.) cilia cover the entire body, but in more specialized forms (*e.g.*, *Vorticella*) cilia are restricted to certain regions of body only. They are commonly arranged in longitudinal, diagonal or spiral rows. Cilia may be of equal length all over the body or may be longer at definite spots (*e.g.*, cilia of caudal tuft of *Paramecium* are longer than other body cilia).

Differences of cilia from flagella. Cilia are exactly like flagella in structure but differ from them in following points:

1. The cilia are relatively very shorter than flagella.
2. They are comparatively more numerous than flagella.
3. The basal granules of cilia are not connected with the nucleus, but, in case of flagella, the rhizoplasts join the basal granules to the nucleus. A rhizoplast is a delicate fibre running between the nucleus and the basal granule.
4. Cilia have no mastigonemes as in flagella.

Ultrastructure. The electron microscope reveals the presence of an outer elastic **membranous sheath** continuous with the plasma membrane of the cell surface and enclosing the fluid matrix. Running along the entire length of the body of cilium are **9 peripheral doublet fibrils** and **2 central singlet fibrils** collectively constituting the **axoneme**. An axoneme is centred on basal granule (= blepharoplast) which exhibits 9 peripheral subfibre **triplets**, each disposed in a twist-like fashion (Fig. 8.4).

Specialized organelles. The cilia may form the following composite motile organelles which are found in several protozoan groups. These are **membranelles** of Ciliata (*Paramecium*), **undulating membrane** of Spirotricha and Peritricha and **cirri** of Hypotricha. **Cirri** are tuft-like brushes of functionally fused somatic cilia. Their kinetosomes form a basal fibre plate. **Undulating membranes** are formed of a row of oral cilia. **Membranelle** is flat plate of oral cilia with a basal fibre plate. They are arranged in a longitudinal series and beat in a coordinated fashion.

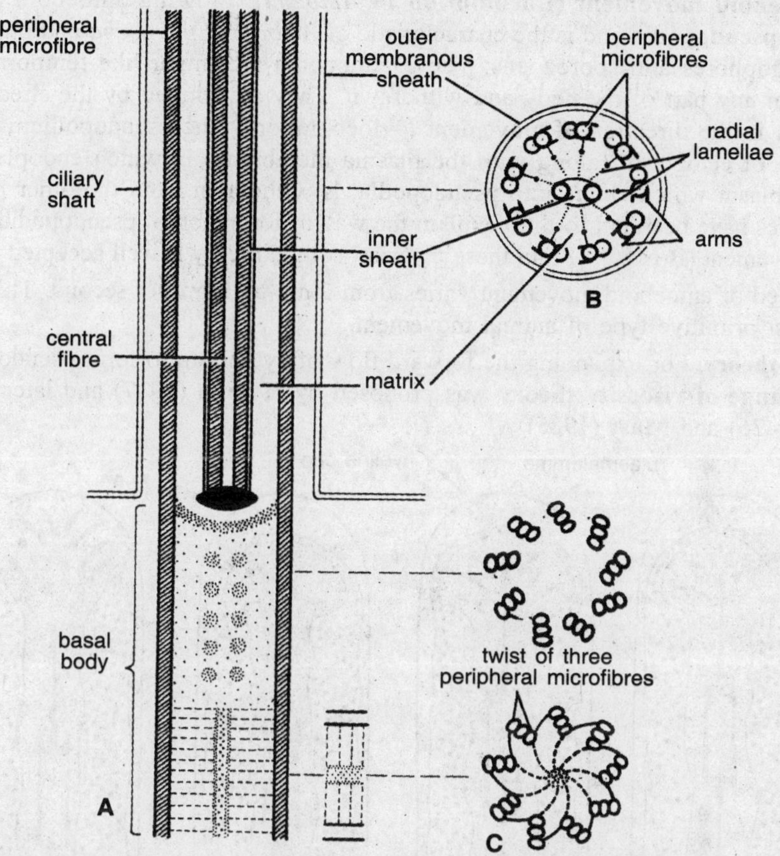

Fig. 8.4. Structure of cilium. A—L.S. of cilium; B—T.S. cilium; C—T.S. of basal granule or kinetosome.

IV. Myonemes

The myonemes are very fine contractile fibrils in the pellicle of flagellates, ciliates, sporozoans and infusorians. In more complex ciliates, myonemes may extend into the endoplasm. They may be arranged longitudinally, transversely or spirally. Myonemes are primarily organelles for the **metaboly**, *e.g.*, *Euglena*, *Paramecium* and secondarily for **locomotion** by muscle-like contraction in *Monocystis* and sporozoan such as *Plasmodium*.

8.2 MODES OF LOCOMOTION

Following four different modes of locomotion occur in protozoa:

1. Amoeboid movement

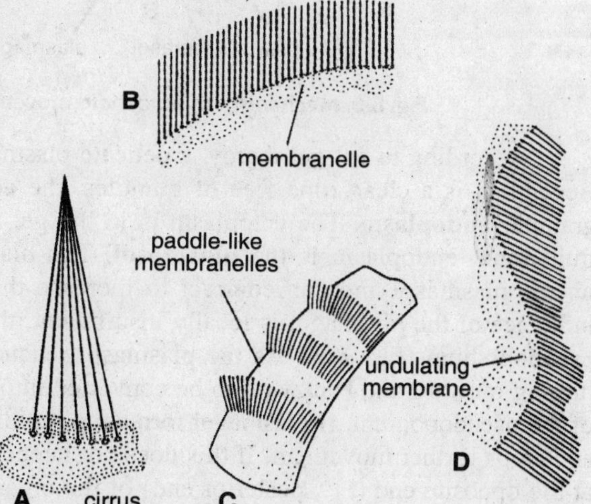

Fig. 8.5. Specialized organelles of cilia in protozoa.

2. Flagellar movement
3. Ciliary movement
4. Metabolic movement

1. Amoeboid movement (Locomotion in *Amoeba*). Flowing amoeboid movement is performed by **pseudopodia** and is the characteristic of *Amoeba*, *Entamoeba* and other Sarcodina, certain mastigophores and sporozoans. The pseudopodia are finger-like temporary processes given out from any part of the body and withdrawn. They are formed by the streaming flow of the cytoplasm in the direction of movement (= locomotion). The pseudopodium is formed by the projection of ectoplasm (along with the plasma membrane) in which endoplasm flows. In fact, the cytoplasm which enters into pseudopodia, is withdrawn from the other parts of body. Various theories have been put forth to explain the way of formation of pseudopodia and resultant amoeboid movement (Box 8.1). From these theories, sol-gel theory is well accepted by zoologists.

The speed of amoeboid movement varies from 2mm to 3mm per second. This is supposed to be the most primitive type of animal movement.

Sol-gel theory. For explaining the forward flow of cytoplasm to form pseudopodia, **sol-gel theory** or **change of viscosity theory** was proposed by **Hyman** (1917) and later supported by **Pantin** (1923–26) and **Mast** (1925).

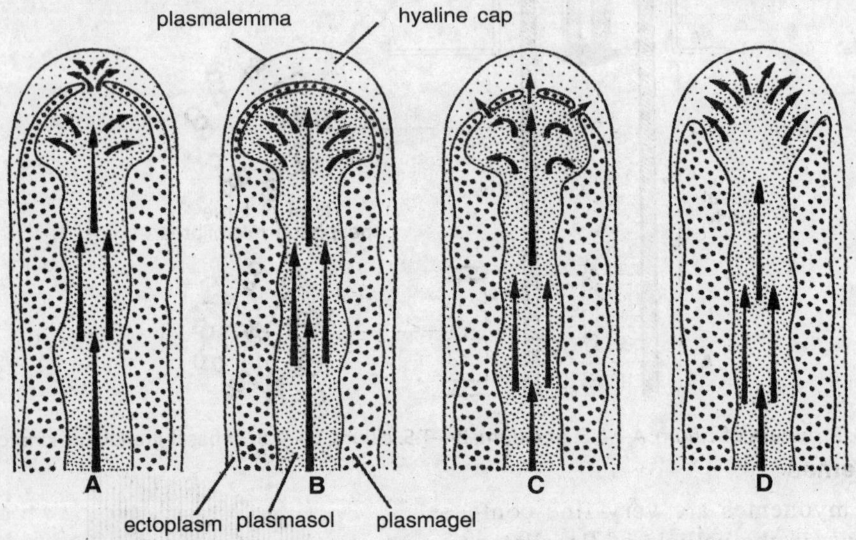

Fig. 8.6. Mechanism of amoeboid movement as explained by sol-gel theory.

According to sol-gel theroy, a delicate plasma membrane covers the surface of an amoeba. Beneath it is a clear zone free of granules, the **ectoplasm**, differentiated from the inner most granular **endoplasm**. The ectoplasm is in the gel state and makes up the **plasmagel**. The more fluid inner endoplasm is the **plasmasol**. The plasmagel exercises a constant, slight pressure on the plasmasol and can contract to increase this pressure. If for any reason the restraining influence of the plasmagel is locally insufficient, the plasmasol erupts, flowing into an extending pseudopodium (Fig. 8.6). As the plasmasol reaches the advancing tip, it spreads fountain-like, turning outward and backward to be converted into plasmagel and thus extend the plasmagel tube of the pseudopodium. If plasmagel forms too rapidly at the tip, a continuous retaining layer forms and stops further movement. If this does not happen and if new plasmasol is continuously formed at the opposite end (*i.e.*, posterior end) of the organism, movement can continue indefinitely. The change of plasmasol to plasmagel involves a loss of volume; therefore, the protoplasm may stream forward to occupy the space made available in this way. Thus, for the amoeboid movement, the following four types of changes occur in *Amoeba* (Fig. 8.12):

1. The plasma membrane of body of *Amoeba* adheres to the substratum.
2. Conversion of plasmasol to plasmagel (*i.e.*, gelation) at the "anterior" advancing tip.
3. Conversion of plasmagel into plasmasol (*i.e.*, solation) at the "posterior" end of body.
4. Contraction of the gelated protoplasm, and
5. Forward streaming of the plasmasol.

Box 8.1 Theories Regarding Amoeboid Movement

1. Adhesion theory. According to this theory, amoeba moves like a drop of water streaming out in the path of greater adhesion upon a rough-surfaced substratum.

2. Contraction hydraulic theory. Schultze (1875) postulated that due to contraction of plasmagel at the posterior end, a current is produced in the protoplasm. This current pushes the plasmosol at the anterior end and this results in the formation of new pseudopodium anteriorly and the body moves in that direction.

3. Surface tension theory. This theory was first of all proposed by **Berthold** (1886) and later supported by **Butschii** (1894) and **Rhumbler** (1898). According to this theory, constant internal tension forces the cytoplasm to shoot out from surface in the form of pseudopodium wherever surface tension suddenly lowers due to local effects.

4. Rolling movement theory. By placing a particle of lamp-black (carbon) on the surface of *Amoeba verrucosa* (which is a unipodal species forming a single pseudopodium at a time), **Jennings** (1904) noted that amoeba simply rolls like a ball during locomotion (Fig. 8.7).

direction of movement

plasmalemma substratum C c

Fig. 8.7. *Amoeba verrucosa* showing rolling movement (Jennings, 1904).

5. Walking movement theories. Dellinger (1906) proposed that *Amoeba proteus* virtually "*walks on pseudopodia*" just as higher animals move on their legs. Thus, as the pseudopodium strikes an object, it sticks to it. This causes contraction of an endoplasmic fibrous network of contractile elements, pulling the whole body mass close to the pseudopodium (Fig. 8.8).

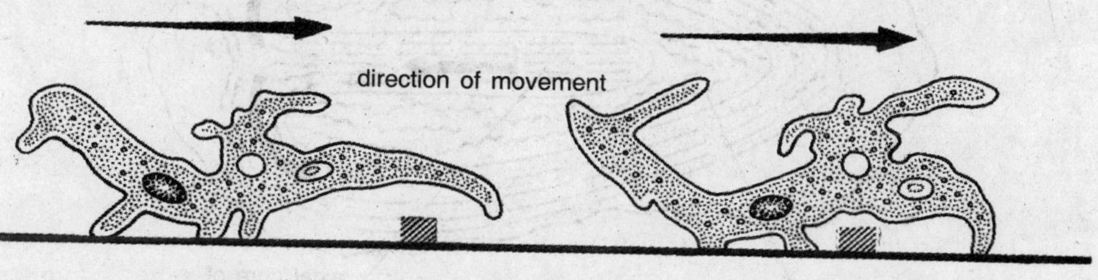

Fig. 8.8. *Amoeba proteus*. Walking movement due to contraction in the protoplasm.

6. Sol-gel theory. This theory is the most accepted theory of pseudopodium formation. It was proposed only when cytoplasm was recognized as a **colloidal** system.

7. Molecular folding-unfolding theory. Goldacre and **Lorch** (1950,52) provided a strong support to sol-gel theory by explaining the biophysical and biochemical processes involved in reversible 'sol-gel' changes and contraction of plasmagel. According to this theory, when the long-chain protein molecules contract, become tightly folded and scatter in the temporary hind region, they represent solation. However, when they become unfolded, regularly arranged and tightly bonded together at the tip of advancing pseudopodium, they represent gelatin (Fig. 8.9).

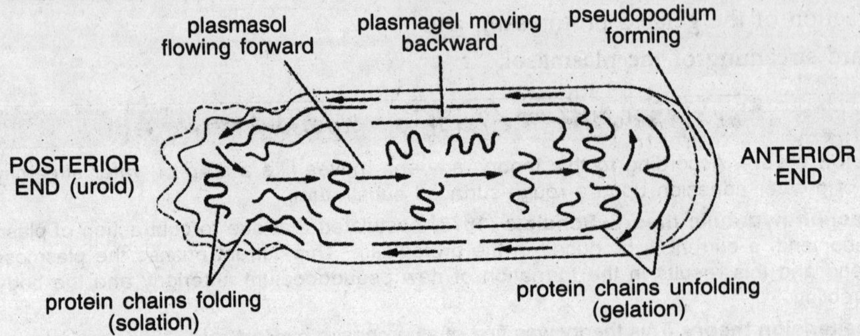

Fig. 8.9. Molecular folding-unfolding during solation and gelation of cytoplasm for amoeboid movement.

Quite recently, role of polymerization of actin subunits (G-actin) into actin microfilaments (F-actin) and vice-versa have been established for folding-unfolding or sol-gel theroy (**Karp**, 1996).

8. Sliding theroy. According to **Huxley** (1954), the proteins participating in solation and gelation are similar to **actin** and **myosin** of muscle fibres. Their folding-unfolding occurs due to sliding of actin and myosin filaments and expends great amount of ATP molecules.

9. Fountain zone contraction theory (or Front contraction theory or Pull hypothesis). Allen (1962) postulated that ectoplasm of *Amoeba* is distinguished into an outer fluid hyaline layer and an inner viscid "gel" layer, while the endoplasm has an axial core of gel surrounded by a "shear zone" of fluid sol (Fig. 8.10). The endoplasmic axial gel is pulled forward, strikes against the hyaline cap at the tip of advancing pseudopodium and fans out like a fountain to all sides, becoming continuous with the gel layer of ectoplasm. In the "fountain zone", the **gel contracts** by releasing some water into the hyaline cap, thus, pulling the axial core of endoplasm farther forward.

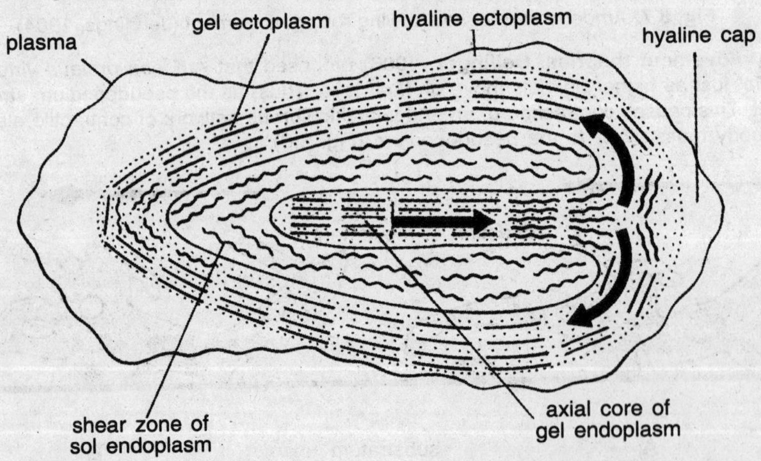

Fig. 8.10. Amoeboid movement on the basis of Allen's fountain zone contraction theory.

10. Ectoplasm contraction theory (or Push hypothesis). According to **Goldacre**, the gel portion (ectoplasm) contracts at the rear end of the body and squeezes the sol portion (endoplasm) forward, *i.e.*, force is created from behind, like the squeezing of the back of a tooth paste tube. On reaching the anterior end the sol is converted into gel.

11. Contraction-hydraulic theory. This theory was postulated by **Rinaldi** and **Jahn** (1963). It discarded Allen's fountain-zone contraction theory and supported Mast's sol-gel theoy. According to it, contraction of plasmagel layer at trailing end exerts a **hydraulic pressure** upon the plasmasol. Resistance to this pressure is minimum at the front end, highest in the middle and medium at the rear. Due to this reason, more and more of plasmasol tends to flow forward into the advancing pseudopodium.

12. Contractile jacket theory. This is modern version of Huxley's sliding theory. According to it, bonding of **myosin** with the **actin mesh** could convert the mesh into a **contractile jacket**, forcing the interior fluid endoplasm forward. **Lee** and **Jacobson** (1993) have demonstrated that **myosin I** (bipolar filaments) act as **primary motor** which drives cell locomotion. Myosin motor acts on actin filaments (microfilaments) and causes them to slide toward the leading edge of the cell.

13. Osmotic theory. According to **Ruppert** and **Barnes** (1994), the initial depolymerization of the actin mesh (*i.e.*, plasmasol formation) at the pseudopodium tip would increase the number of particles (actin subunits). Particle increase would raise the **osmotic concentration** and water would flow from the endoplasm out into the tip. The influx of water causes the extension of the cell as a pseudopodium.

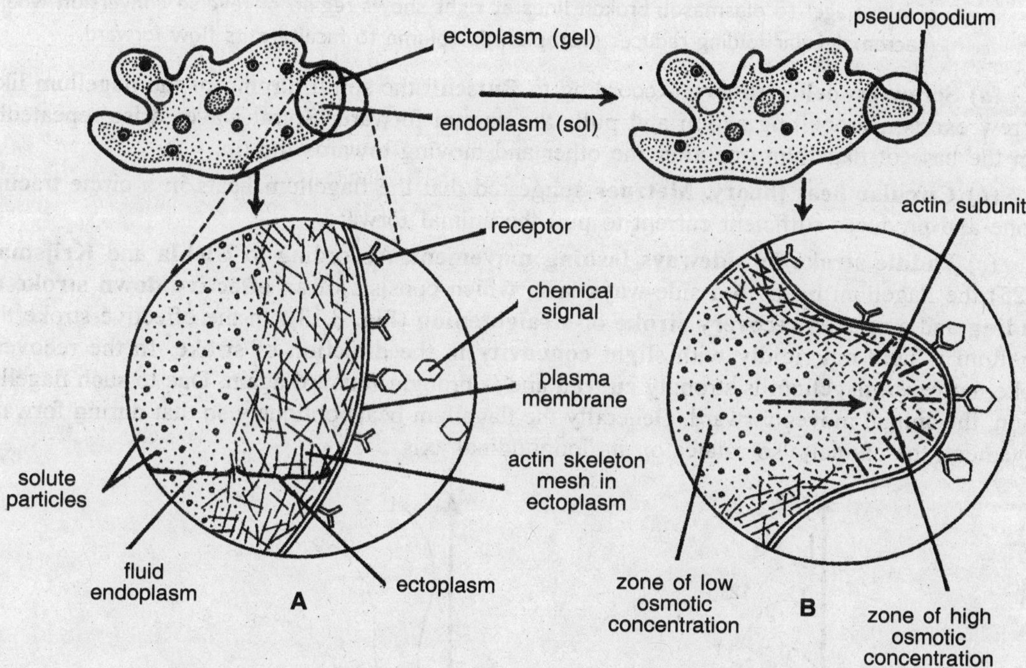

Fig. 8.11. *Amoeba*. Osmotic theory of flow of endoplasm during amoeboid locomotion. Actin mesh creates gel state of ectoplasm. A—Chemical signals bind to membrane receptors and initiate depolymerization of the actin mesh. B—Depolymerized actin subunits raise osmotic concentration of this region of ectoplasm and water flows in. The influx of water causes the extension of the cell as pseudopodium. Peripheral to the pseudopodial tip, actin subunits repolymerize as skeletal mesh, extending the ectoplasmic gel state forward like a progressive sleeve.

2. Flagellar movement (Locomotion in *Euglena*). This sort of movement is the characteristic of Mastigophora or flagellated protozoans, *e.g.*, *Euglena*. Flagellar movement is produced by continuous beating (*i.e.*, lashing movement) of long, elastic flagella. Following four types of flagellar movements have been recognized:

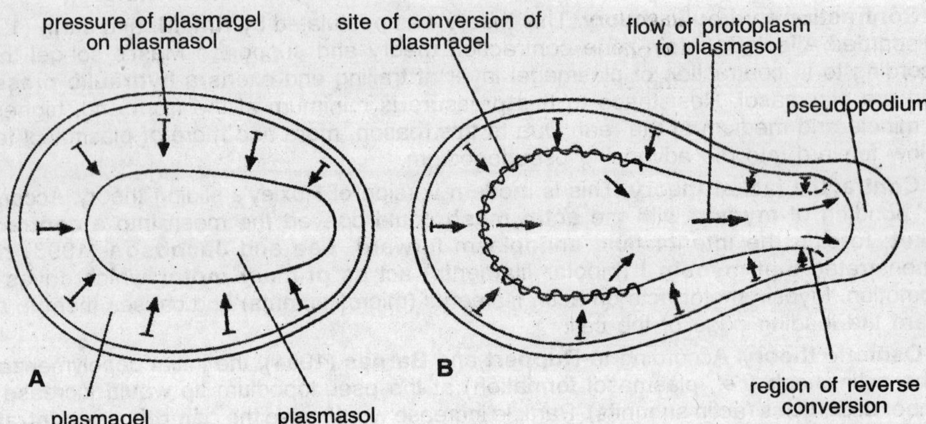

Fig. 8.12. Mode of formation of pseudopodium. Arrows with square tails show pressure of plasmagel on plasmasol; straight arrows show flow of protoplasm. Wavy line in B indicates site of conversion of plasmagel to plasmasol; broken lines at right shows region of reverse conversion where macromolecular folding reduces protoplasmic volume to facilitate its flow forward.

***(a)* Screw propeller theory.** According to **Butschli** the spiral turning of the flagellum like a screw exerts a propelling action and pulls the animal forward. Spiral waves arise repeatedly from the base of flagellum one after the other and moving towards the tip.

***(b)* Circular beat theory. Metzner** suggested that the flagellum beats in a circle tracing a cone and produces sufficient current to pull the animal forward.

***(c)* Paddle stroke or sideways lashing movement.** According to **Ulehla** and **Krijsman** (1925) the flagellum beats in a side-ways lash, which consists of an **effective down stroke** or **bending** and a relaxed **recovery stroke** or **straightening** (Fig. 8.13). In the effective stroke the flagellum is held out rigidly with slight **concavity** in the direction of stroke. In the recovery stroke, however, flagellum is strongly **curved** and is brought forward again. Due to such flagellar action, the animal moves forward. Generally the flagellum beats obliquely so that during forward movement, the animal also rotates on its longitudinal axis.

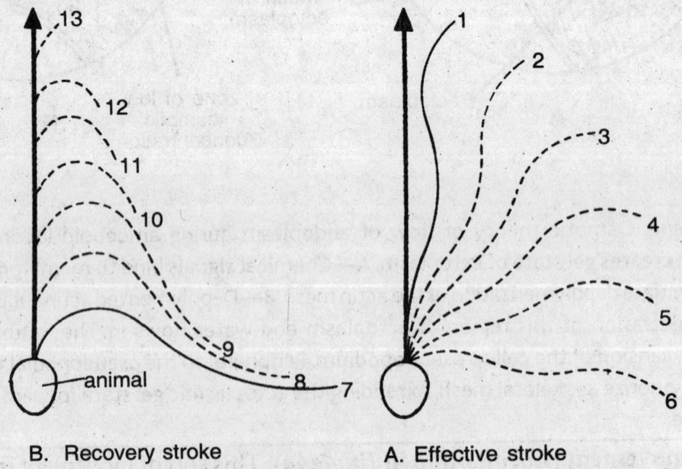

Fig. 8.13. Two stages in the lashing movement of flagellum.
A—Effective stroke; B—Recovery stroke.

(*d*) **Undulating movement.** Flagellar propulsion follows essentially the same principle as that of a propellar. The flagellum undergoes wave-like **undulations** that either push or pull. The undulatory waves pass from base to tip and drive the organism in the opposite direction (Fig. 8.14), or more rarely, the undulations pass from tip to base and pull the organism.

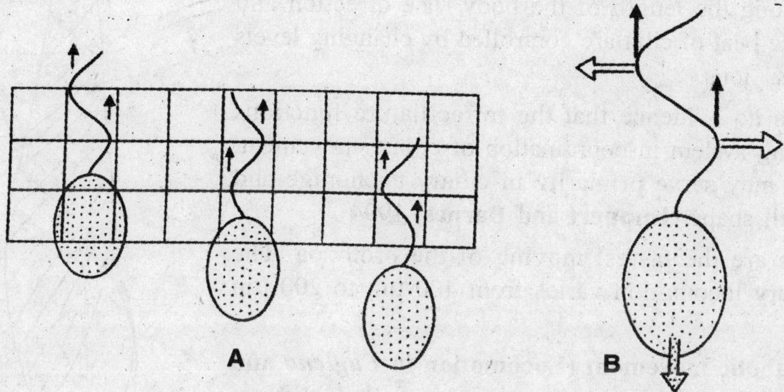

Fig. 8.14. Flagellar propulsion. A—Base to tip wave propagation. B—Forces generated by base - to - tip wave propagation. Lateral forces (outlined arrows) cancel each other. Longitudinal forces (solid arrows) combine to produce forward thrust in the opposite direction.

Many flagellates have flagella with fine lateral branches, called **mastigonemes**. The hydrodynamic effect of the branches is to cause the flagellum to **pull** rather than push even though the flagellar waves are passing from base to tip.

3. Ciliary movement (Locomotion in *Paramecium*). Ciliary movement is performed by extremely fine and very small thread-like processes of ectoplasm in ciliates (*e.g.*, *Paramecium*) and larval suctorians. The ciliary movement is very much similar to that of flagellar movement, consisting of the same effective stroke and recovery stroke, but the cilia bend throughout their length (Fig. 8.15). The water moves in the direction of beat, while the animal moves in the opposite direction.

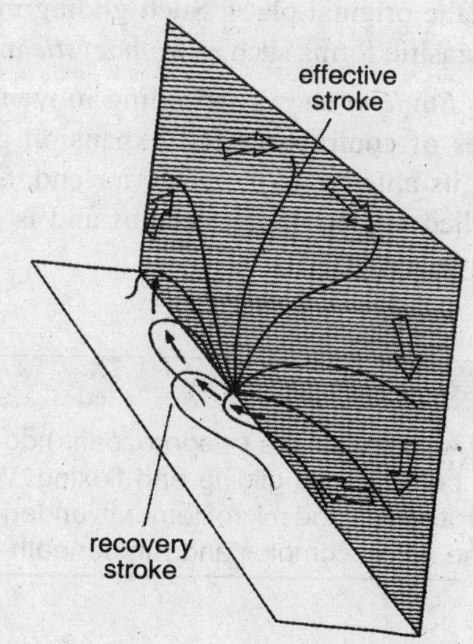

Fig. 8.15. The effective (outlined arrows) and recovery (solid arrows) stokes in the beat cycle of a single cilium.

Every cilium of body moves independently and slightly in advance of the cilium behind it. Thus, the cilia of a longitudinal row do not move simultaneously but **metachronously**, *i.e.*, one after another, so that a wave of propulsion passes from one end of body to the other, whereas cilia of a transverse row beat synchronously (all together). Since in *Paramecium* the direction of the effective ciliary stroke is oblique to the long axis of the body (Fig. 8.16), this causes the ciliate to swim in a spiral course and at the same time to rotate on its longitudinal axis.

Fig. 8.16. A single longitudinal row of cilia showing metachronous movement.

In fact, the movement of adjacent cilia are coupled as a result of **interference effects** of the surrounding water layers. Thus, hydrodynamic forces impose a coordination of the cilia. The beat of individual cilia is part of the metachronal waves that sweep along the length of the body. The direction and intensity of the beat of cilia are controlled by changing levels of Ca^{2+} and K^{+} ions.

There is no evidence that the infraciliature functions as a conducting system in coordination of ciliary movement. Infraciliature may serve primarily in ciliary anchorage and controlling cell shape (**Ruppert** and **Barnes**, 1994).

Ciliates are the fastest moving of the protozoa. The speed of ciliary locomotion varies from 400 μm to 200 μm per second.

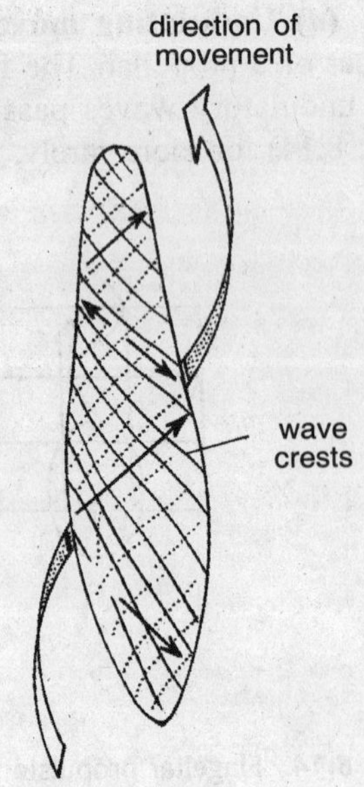

Fig. 8.17. Metachronal waves in *Paramecium* during forward swimming. Wave crests are shown by lines and their direction by the small solid arrows. Dotted lines indicate direction of effective stroke of cilia. Movement of ciliate is indicated by large arrow.

4. Metabolic movement (Locomotion in *Euglena* and *Monocystis*). Metabolic movement is also called **gliding, wriggling, peristaltic** or **gregarine movement**. It occurs in sporozoans (*Monocystis*) and certain flagellates (*Euglena*). The contractile **myonemes** of body contract and expand to bring about a change in shape of the body and relative displacement from the original place. Such gliding movement is performed by parasitic forms such as *Monocystis* in the body fluid of host.

Euglena shows wriggling movement by the alternative waves of contraction and expansion passing over the body from its anterior to the posterior end. Such sort of movement is called **euglenoid movement** and is performed by the help of myoneme fibrils.

Box 8.2

Apicomplexans (= sporozoans) do not have cilia, flagella or pseudopodia. Nevertheless, they can be observed gliding and flexing. What causes these movements is not well understood, but microtubules and microfilaments underneeth alveoli may play a role. In them, microtubules originate at the apical complex and run beneath the alveoli (**Brusca** and **Brusca**, 2003).

QUESTIONS

Long Answer Questions

1. Write an essay on locomotion in protozoa.
2. Describe structure and function of locomotory organs of protozoa.

Short Answer Questions

1. Explain the mechanism of ciliary movement.
2. Explain the flagellar locomotion.
3. Compare the movements of *Amoeba* and *Paramecium*.
4. Write a short note on amoeboid locomotion.

Very Short Answer Questions

1. What are pseudopodia?

Fill in the Blanks

1. Locomotory organelles in Protozoa are .... , pseudopodia and ....

2. Lobopodia are found in .... and ....

3. Organelles of locomotion are .... in radiolarians.

4. Locomotory organ of *Paramecium* is ....

5. Gliding movement shown by *Euglena* is called ....

Multiple Choice Questions

1. Axopodia are locomotory organelles in
(*a*) amoebida (*b*) foraminifera
(*c*) helizoea (*d*) radiolarians

2. Lobopodia occurs in
(*a*) *Amoeba* (*b*) *Entamoeba*
(*c*) *Arcella* (*d*) all of these

3. Protozoans are mainly classified on the basis of
(*a*) locomotory organ (*b*) types of nuclei
(*c*) contractile vacuole (*d*) mode of nutrition

4. Protozoans lack which of these as locomotory organs
(*a*) pseudopodia (*b*) flagella
(*c*) cilia (*d*) parapodia

5. In *Amoeba* pseudopodia are formed due to
(*a*) contact with food
(*b*) sol-gel change
(*c*) movement towards area of higher temperature
(*d*) all the above

ANSWERS

Very Short Answer Questions

1. Broad, rounded protrusions formed during amoeboid movement as portions of cell surface are pushed outward by a column of cytoplasm that flows through the interior of the cell toward the periphery.

Fill in the Blanks

1. Flagella, Cilia **2.** *Amoeba, Arcella* **3.** Filopodia **4.** Cilium
5. Metaboly.

Multiple Choice Questions

1. (*c*) **2.** (*d*) **3.** (*a*) **4.** (*d*)
5. (*b*).

9

Reproduction in Protozoa

The mode of reproduction in protozoa is variable among different groups, however, it is basically a cell division. In protozoa, reproduction takes place by **asexual** (involving **mitosis**) and **sexual** (involving **meiosis**); however, some reproduce asexually only.

9.1 ASEXUAL REPRODUCTION

Asexual reproduction involves the mitotic cell division of parent body into two or more young individuals that develop into mature organisms. It always involves a single parent and neither meiosis nor fertilization occurs. Protozoa reproduce asexually by following four different methods: 1. Binary fission; 2. Multiple fission; 3. Budding; 4. Plasmotomy.

I. Binary fission. Division of the organism into two or more progeny is called **fission**. Fission typically involves **karyokinesis** or nuclear division and **cytokinesis** or cytoplasmic division. After or during later part of karyokinesis, cytoplasmic constriction begins. The two daughters then enter a phase of **growth** and **differentiation** following cytokinesis.

When fission results in two similar and equal daughter cells, it is called **binary fission**. Where there is a definite body axis, binary fission may be of following types:

1. Simple binary fission. As in Sarcodina (*Amoeba* and *Entamoeba*) the plane of division cannot be recognized.

2. Longitudinal binary fission. In flagellates (*Euglena*) the plane of division is along the long axis of the animal's body. In this case, division is said to be **symmetrogenic**, that is, longitudinal division cuts between the rows of basal bodies and producing mirror-image of daughter cells.

3. Oblique binary fission. In dinoflagellates (*Ceratium*) the division plane is oblique. In armoured dinoflagellates, the two fission products regenerate the missing plate or form a completely new armour, the old one having been shed before division.

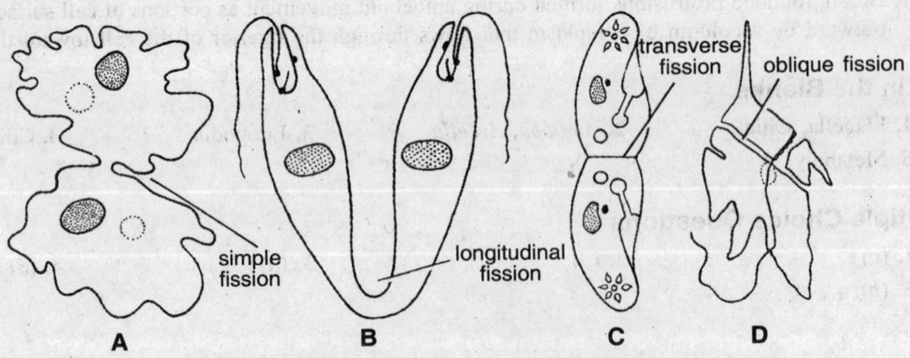

Fig. 9.1. Binary fission in protozoa. A—*Amoeba* (simple); B—*Euglena* (longitudinal); C—*Paramecium* (transverse) D—*Ceratium* (oblique).

4. Transverse binary fission. In ciliates (*Paramecium*) the division plane cuts across the kinetics *i.e.*, longitudinal rows of cilia and basal bodies. This type of division is called **homothetogenic.** Mitotic spindles are formed only in the division of the micronuclei. Division of macronucleus is usually accomplished by constriction (by amitosis).

In *Arcella* and *Euglypha*, a mass of cytoplasm extrudes from the opening of shell prior to division. This extruded mass secretes a new shell. The double-shelled animal now divides (Fig. 9.2).

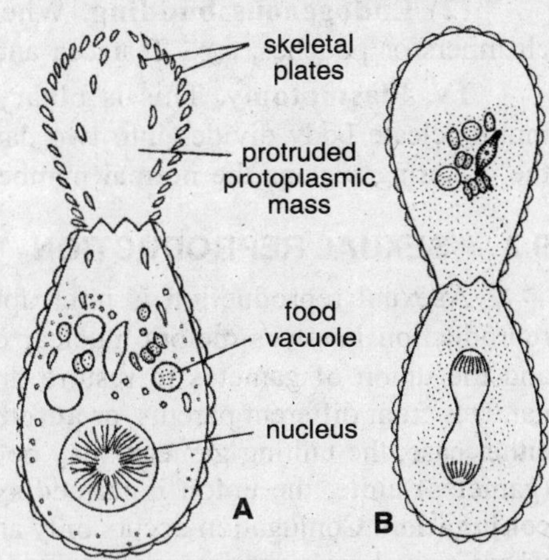

Fig. 9.2. Transverse binary fission in *Euglypha*, a shelled amoeba. A—Formation of skeletal plates on the protoplasmic mass protruding from aperture. B—Division of nucleus.

II. Multiple fission. Multiple fission is common in sporozoan parasites (*Monocystis*, *Plasmodium*), foraminiferans and radiolarians. Preparation for division involves repeated nuclear divisions (mitosis) to form a multinucleate organism with no cytokinesis. The cytoplasm then cleaves to form many tiny uninucleated individuals. These are set free by rupture of parent body. It is common for multiple fission to leave a mass of residual protoplasm that degenerates.

The multiple fission occurs at different phases in the life cycle of different individuals and the products are named accordingly.

(*i*) Schizogony. In schizogony, multiple fission occurs during sexual cycle in trophozoites or schizonts existing inside host cells (liver and erythrocytes) to produce primarily merozoites and later on gametocytes, *e.g.*, *Plasmodium*.

(*ii*) Gamogony. When multiple fission produces sex cells or gametes, it is termed as gamogony, *e.g.*, *Monocystis*.

(*iii*) Sporogony. The multiple fission in the zygote is known as sporogony and the offsprings as **spores**, *e.g.*, *Plasmodium*, *Eimeria*, *Monocystis*, etc.

III. Budding. Budding is a type of asexual reproduction in which one or more small daughters are produced by the parent. It may be considered a form of unequal fission. Like the fission, it involves preparatory phases before budding and a later period of growth and differentiation. Budding may be of following two types:

(1) Exogenous budding. When budding occurs at the body surface, *e.g.*, *Acanthocystis*, *Noctiluca*, *Ephelota* (Fig. 9.3) and *Entamoeba*.

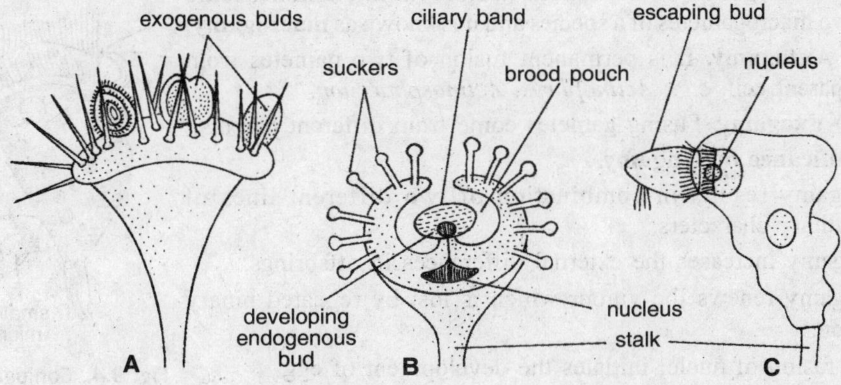

Fig. 9.3. Asexual reproduction by budding A—Formation of exogenous bud in *Ephelota*. B—Formation of endogenous bud in a Suctoria; C—Expulsion of bud from brood pouch in *Tochophyra*.

(2) Endogenous budding. When budding occurs in an internal cavity such as brood chambers or pouches, *e.g.*, Testacea and Suctoria.

IV. Plasmotomy. This is binary fission in multinucleate form such as *Opalina*. The multinucleate body divides into two daughter individuals without any nuclear division. Each of the offspring regains the normal number of nuclei by the division of parental nuclei.

9.2 SEXUAL REPRODUCTION

Asexual reproduction is interrupted at regular intervals by sexual reproduction. Sexual reproduction involves meiotic nuclear division that results in a change from diploid to haploid and the union of gametes to restore diploidy. It may be **amphimictic**, involving the union of gametes from different parents, or **automictic**, in which the gametes arise from the same parent. In either case, the uniting gametes may be whole organisms or nuclei only. Where whole organisms (gametes) unite, the union is termed **syngamy**. Where only nuclei unite, the process is called **conjugation**. Conjugation occurs only among ciliates, whereas syngamy occurs in all other groups where sexual reproduction occurs. Among protozoans the following types of sexual reproduction have been recognized: 1. Syngamy 2. Conjugation; and 3. Automixis.

1. Syngamy. It is the complete fusion of the **gametes** to form a **zygote** which gives rise to adult. The fused diploid nucleus is called **synkaryon**. Depending upon the nature of fusing gametes, a syngamy may be of the following types:

(*i*) Hologamy. It is the fusion of two mature individuals which do not produce gametes, *e.g.*, rhizopods and flagellates.

(*ii*) Paedogamy. It is the fusion of two young individuals.

(*iii*) Isogamy. The fusing gametes are similar in form and size and are known as **isogametes**. Their fusion is called **isogamy**. It is common in Foraminifera, Phytomonadida and Gregarinida.

In some isogamous species, the gametes are chemically differentiated, *i.e.*, a gamete belongs to a mating type and can unite only with a gamete from a different mating type, thus, ensuring **outbreeding**.

(*iv*) Anisogamy. The fusing gametes, dissimilar in form and size, are known as **anisogametes** and their union is called **anisogamy**. Usually the male gamete is small, active and may be flagellated and is known as **microgamete**. The female gamete is large, immotile and food-laden and is known as **macrogamete**. Anisogamy occurs in Sporozoa (*Plasmodium*) and Volvocida (= Phytomonadida; *e.g.*, *Volvox*).

(*v*) Microgamy. Certain foraminiferans and *Arcella* produce two types of **microgametes** which unite together to form synkaryon. This union is known as **microgamy**.

(*vi*) Macrogamy. In Heliozoia (*Actinophrys*), union occurs between two macrogametes of a species and it is known as **macrogamy**.

(*vii*) Autogamy. It is permanent fusion of two gametes from the same parent cell, *e.g.*, *Actinophrys*, *Actinosphaerium*.

(*viii*) Exogamy. Fusing gametes come from different parents.

Significance of syngamy.

1. Syngamy results in combination of two different lines of hereditary characters.
2. Syngamy increases the external differences in offsprings.
3. Syngamy renews the vigour which is lost by repeated binary fission.
4. The fusion of nuclei initiates the development of egg.

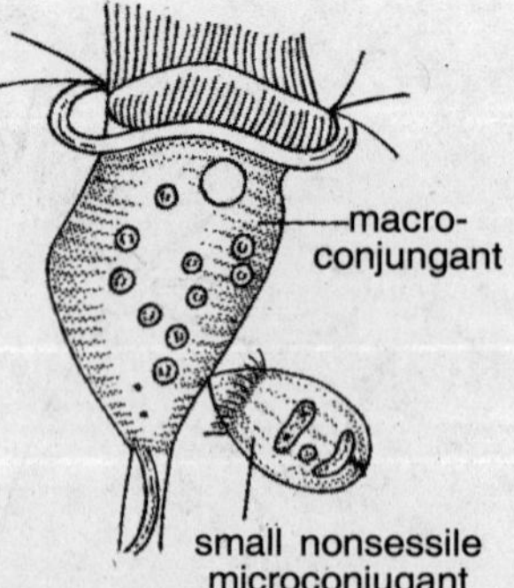

Fig. 9.4. Conjugation in *Vorticella*.

2. Conjugation. Conjugation is the temporary union of two individuals of different mating types of a species to facilitate exchange of the nuclear material. These individuals separate out after nuclear exchange. The pairing gametes (individuals) are called **conjugants** and may be either isogamous (equal) as in *Paramecium* or **anisogamous** (unequal) as in *Vorticella*. In case of *Vorticella*, the two members of a conjugating pair are of striking different sizes (Fig. 9.4). Such "dioecious" macro-and micro-conjugants represent an adaptation for conjugation in sessile species. The macroconjugant remains attached while the small bell of the microconjugant breaks free from its stalk and swims about. A synkaryon forms only in the macroconjugant from one gametic nucleus contributed by each conjugant. However, there is no separation after conjugation, and little migratory conjugant degenerates.

In the Suctoria, conjugation takes place between two attached individuals that happen to be located side by side.

Significance of conjugation. In fact, conjugation is not a mode of multiplication but instead is an important episode of reproduction which leads to:

1. New nuclear combinations and new hereditary combinations.
2. Rejuvenation and revival of the lost vigour and overcoming the period of depression resulting due to repeated binary fission. For example, it has been shown that some species of *Paramecium* can pass through only approximately 350 continuous asexual generations. If nuclear organization does not occur, the asexual line (clone) dies out, apparently because of decline in function of the macromolecules.
3. Reorganisation of a fresh macronucleus to accelerate the metabolic activities.

3. Automixis. It is a process of self-fertilization or **autogamy** that results in **homozygosis**. Automixis is the fusion of two gametic nuclei originating by the division of single nucleus of an individual. It is conjugation-like method of nuclear reorganization. The macronucleus degenerates and micronucleus divides a number of times to form eight or more nuclei. Two of these nuclei fuse to form a synkaryon; the others degenerate and disappear. The synkaryon then divides to form a new micronucleus and macronuclei as occurs in conjugation.

4. Endomixis. It is a type of nuclear reorganization which does not involve meiosis and fusion of nuclei (fertilization). Endomixis is a regular feature in the life history of *Paramecium aurelia*. It occurs in a single individual. As a result of nuclear changes a new macronucleus is formed from the micronucleus which accelerates metabolic activities and brings renewed vigour and vitality.

5. Hemixis. It is also a nuclear purification method of *Paramecium*. In hemixis, the macronucleus throws out fragments of different sizes which are absorbed in the cytoplasm. The remaining part of macronucleus functions as normal macronucleus.

9.3 REGENERATION

The capacity of regeneration (*i.e.*, replacement of lost parts) is wide-spread among free-living protozoa (*e.g.*, *Amoeba*), but is slight in parasitic protozoa. A proper part of cytoplasm and nucleus can regenerate the entire individual.

9.4 PARTHENOGENESIS

The production of young from unfertilized eggs is called **parthenogenesis.** In certain protozoa such as *Actinophrys* the gametes which fail to fertilize start their development parthenogenetically.

9.5 PLASMOGAMY

In certain rhizopods, two individuals unite partially and their nuclei do not fuse. As a result a multinucleated mass of protoplast is produced.

QUESTIONS

Long Answer Questions

1. Describe the various modes of reproduction found in protozoa studied by you.

2. Describe the various modes of reproduction found in *Euglena*, *Monocystis* and *Paramecium*.

3. Describe the different types of asexual reproduction in protozoa.

Short Answer Questions

1. Write short notes on:

(*i*) budding (*ii*) binary fission

(*iii*) syngamy (*iv*) automixis

(*v*) conjugation

2. Distinguish between symmetrogenic and homothetogenic binary fission.

Very Short Answer Questions

1. Name the processes in *Paramecium* that correspond to cross-fertilization and self-fertilization in higher animals.

2. Define syngamy.

Fill in the Blanks

1. *Amoeba* reproduces by binary fission and ....

2. Homothetogenic division occurs in .... binary fission.

3. Symmetrogenic division occurs in .... binary fission.

Multiple Choice Questions

1. If two fusing nuclei are present in two different cells, then the process is known as

(*a*) autogamy (*b*) paedogamy

(*c*) hologamy (*d*) none of these

2. Asexual reproduction is of common occurrence in protozoa because

(*a*) sexes have not been differentiated (*b*) it helps to multiply rapidly

(*c*) majority of protozoans are parasites (*d*) large number can localise source of food efficiently

3. Which of the following is lacking in *Amoeba*?

(*a*) sexual reproduction (*b*) multiple fission

(*c*) binary fission (*d*) cyst formation

ANSWERS

Very Short Answer Questions

1. Conjugation and autogamy.

2. It is fusion of two sex cells or gametes with the formation of zygote; the zygote nucleus is called **synkaryon**. Syngamy may include hologamy, isogamy, anisogamy and autogamy depending upon the fusing gametes.

Fill in the Blanks

1. Multiple fission **2.** Transverse **3.** Longitudinal.

Multiple Choice Questions

1. (*b*) **2.** (*b*) **3.** (*a*).

10

Protozoa and Disease

A large number of protozoa are parasitic to human beings and animals. Most parasites, both ectoparasites and endoparasites, are **pathogenic**, causing various types of diseases in human beings and animals. Most of these diseases are more common in the tropical and substropical regions of world.

The diseases occur as a result of the interaction between parasitic protozoans and the hosts (human beings and animals) under specific environmental conditions. This interaction is called **infection** which is always expressed in the form of **infectious disease**. All pathogenic parasites have the following common characteristics: 1. All are **obligatory parasites,** *i.e.*, they depend for their existence upon their hosts. 2. All show **host specificity**, *i.e.*, they harbour selective hosts. 3. They complete their life cycle in one or more than one host. 4. They secrete toxic substances which cause the particular disease. Parasitic species are found in all classes of protozoa and cause the following diseases in human beings and animals.

10.1 PROTOZOA AND HUMAN DISEASES

About 15 species of protozoa are pathogenic parasites of human beings causing some of the fatal diseases. They spend all or at least part of their life cycles in the body tissues, alimentary canal or blood of the host (human beings). Certain significant protozoan diseases of human beings are as follows:

1. Amoebiasis (Amoebic dysentery). It is caused by *Entamoeba histolytica* which is an endoparasite and has world-wide geographical distribution but is common in tropics and subtropics. Infection of *E. histolytica* generally takes place through drinking water. The trophozoite (= tissue-invading form) of the pathogen penetrates the walls of large intestine or colon, secretes histolytic enzymes and feeds upon its cells causing ulcers. These ulcers rupture and discharge mucus and blood along with numerous trophozoites which pass along with stools causing acute dysentery (ulcerative colitis). If the infection is allowed to continue the parasite may reach the liver, heart, lungs and brain where it causes abscesses which prove fatal.

There is no intermediate host in the life cycle of *E. histolytica.* Transmission of the parasite from person to person takes place through the tetranucleate cysts with contaminated food and drinks. These cysts are also transmitted by houseflies and cockroaches. Amoebiasis can be treated with drugs such as Emetine, Fumagillin, Metronidazole, etc.

2. Pyorrhoea (or Periodontitis). This bacterial disease is aggravated by *Entamoeba gingivalis* which lives in the tartar of teeth and abcesses of gum. Infection is caused directly by kissing or feeding in the same bowl.

Entamoeba coli is another species of *Entamoeba* which lives as a commensal in the large intestine of human beings and is non-pathogenic. It feeds on bacteria.

3. Diarrhoea. It is caused by a flagellate endoparasite, *Giardia intestinalis*. The parasite has world-wide geographical distribution and lives in small intestine (*i.e.*, duodenum and upper part of jejunum) of human beings. When conditions in the duodenum are unfavourable, encystment occurs and cysts are voided with faeces.

The parasite attaches itself with the help of its sucking disc on the convex surfaces of the epithelial cells in the intestine and may cause a disturbance of intestinal function. This leads to malabsorption of fat, persistent looseness of bowels and mild steatorrhoea (*i.e.*, passage of yellowish and greasy stools in which there is axcess of fat). The patient also suffers epigastric pain, abdominal discomfort, loss of appetite, headache and toxic effects (allergy). Transmission of *Giardia* to new host occurs via contaminated water and food. For the treatment of infection of *Giardia* the following drugs are effectively used: Atebrin, Chloroquine and Acranil.

4. Trypanosomiasis (African sleeping sickness). It is a disease of central nervous system. It is caused by the species of *Trypanosoma* (such as *T. gambiense*, *T. rhodesiense*, *T. cruzi*) which are flagellate parasites of blood plasma (in humans, the **definitive host**) and gut (in invertebrate, the **intermediate host**). The pathogen of this dreaded disease is transmitted among humans either by bite of blood-sucking insects or by bug's faeces. Following three species of *Trypanosoma* cause sleeping sickness in human beings:

(i) Trypanosoma gambiense is transmitted by tsetse fly, *Glossina palpalis*. It causes Gambian or Central African sleeping sickness.

(ii) Trypanosoma rhodesiense is transmitted by tsetse fly, *Glossina morsitans*. It causes Rhodesian or East African sleeping sickness. Infection of both of these pathogens are confined to those parts of Africa where tsetse flies (the vectors) are found. On infection by the *Trypanosoma*, initially fever is caused during which the parasite resides freely in the blood of host, where it consumes an enormous amount of glucose; then the parasite collects in the lymph glands, spleen and liver causing their enlargement; and finally they enter the cerebrospinal fluid causing **sleeping sickness**. Sleeping sickness results in coma and eventually death. In case of early and acute infection, sleeping sickness can be treated by drugs such as Suramin and Pentamidine. However, when infection of *Trypanosoma* reaches the central nervous system, arsenic containing drugs such as Tryparsamide, Melarsen, Melarsoprol (Mel B) and Trimelarsen are effectively used.

(iii) Trypanosoma cruzi is transmitted by faeces of a bug, called *Triatoma megista*. It causes **Chagas' disease** (a sleeping-sickness-like disease) or American trypanosomiasis in South and Central America. Symptoms of Chagas' disease include continuous fever, swollen lymph glands, spleen and liver. This disease can be cured by the drug Nitrofurazone.

5. Leishmaniasis. It is caused by the species of *Leishmania*, the flagellate intracellular endoparasite in the reticulo-endothelial cells of vertebrate host (*e.g.*, human beings) and the gut of an invertebrate host, the blood-sucking sandfly, *Phleobotomus*. The following three species of *Leishmania* cause different types of diseases in human beings:

(i) Leishmania donovani. It causes **kalaazar** or visceral leishmaniasis which is widespread and endemic in many places in India, China, Africa, Southern Europe, South America and Russia. In India kalaazar is common is Assam and Bengal along the coasts of the rivers Ganges and Brahmputra. In India, it is also endemic in Bihar, Orissa, Madras, and eastern U.P. In India, vector of *L. donovani* is the sandfly *Phlebotomus argentipes*.

Pathogen of kalaazar attacks the endothelial cells, bone marrow, liver, lymph glands and blood vessels of the spleen. These organs become enlarged. Symptoms of anaemia, high fever and pigmented dry and rough skin appear in the patient. Death of untreated patient occur due to profound immuno-suppressive effect inviting secondary bacterial infections. This disease has been effectively treated by following drugs: Urea stibamine, Aminostiburea, Neostibosan, etc.

(ii) Leishmania tropica. It causes **cutaneous leishmaniasis, Oriental sore (Tropical sore),** or **Delhi boil**. Its vector is sandfly, *Phlebotomus sergenti*, in India, Iraq and Persia. The pathogen is found along the shores of Mediterranean through Syria, Arabia, Mesopotamia, Iran to Central Asia and drier parts of Central and Western India. The infection is limited to a local lesion of the skin and subcutaneous tissues which turn into ulcerating wounds. Treatment of Delhi boil includes regular cleaning and dressing of the boils by pentavalent preparation of antimony.

(iii) Leishmania brasiliensis. It causes **espundia** or **muco-cutaneous leishmaniasis**. The pathogen is confined in Central and South America. It is intracellular parasite of macrophage cells of skin and mucous membrane of nose and buccal cavity. It is transmitted by the vector, the anthrophilic sandflies. Symptoms of espundia include multiple sores over large areas of the skin and oro-nasal mucosa. Ulceration in nasal cavities, mouth and pharynx is quite frequent. Treatment includes pentavalent preparation.

6. Trichomoniasis. It is caused by the species of flagellate *Trichomonas* which is pathogenic parasite of vertebrates and many invertebrates. Its life cycle includes only tropozoite phase; there is no cystic phase. Following three species of *Trichomonas* are pathogenic in human beings:

(i) Trichomonas hominis. It inhabits the ileo-caecal region of human beings and causes diarrhoea and dysentry.

(ii) Trichomonas tenax. It lives in the oral cavity, found around the tartar of the teeth and in pyorrhoeic sockets.

(iii) Trichomonas vaginalis. It lives in the urino-genital system of male and female human beings. In women, it inhabits the vagina causing **vaginitis**. The disease is characterized by inflammation of vaginal mucosa, burning sensation, annoying itch and abnormal greenish yellow discharges (called **leucorrhoea**). Transmission of pathogen is direct, taking place during coitus. Drugs containing arsenic and iodine and antibiotics such as Terramycin and Auromycin have proved effective in the treatment of this disease.

7. Malaria. This notorious and fatal disease of human beings is caused by the species of a sporozoan parasite, *Plasmodium*. It is transmitted through the bite of female *Anopheles* mosquito which is vector and secondary host. In human beings, which are the primary host, the parasite attacks and multiply asexually in liver cells and red blood cells. A toxic substance, called **hemozoin**, is released by the parasite during its asexual cycle in human beings and causes the symptoms and signs of malaria. Following four species of *Plasmodium* cause malaria in human beings:

(i) Plasmodium vivax causes benign malaria in which fever comes in every 48 hours.

(ii) Plasmodium malariae causes quartan malaria in which fever comes every 72 hours.

(iii) Plasmodium ovale causes mild tertian malaria in wich fever occurs on every 48 hours.

(iv) Plasmodium falciparum causes malignant subtertian malaria in which fever is more or less continuous (24-48 hours).

In malaria, the spleen is enlarged, erythrocytes decrease in number, pigment granules collect in the spleen and there is high temperature accompanied by chills and shivering. This fatal disease is treated by the following drugs: Quinine, Camoquine, Chloroquine, Plasmoquine, Resochin, etc.

8. Toxoplasmosis. It is caused by a sporozoan parasite, *Toxoplasma gondii* and common in Europe and United States. The pathogen is intracellular parasite of various organs and causes hydrocephalus and chorioretinitis (= inflammation of the choriod and retina of the eye).

9. Balantidial dysentery. It is caused by a ciliate parasite, *Balantidium coli*. The pathogen lives in the large intestine of human beings causing dysentery, diarrhoea and ulceration of colon. Infection is by cysts through contaminated water and food. This disease is effectively treated by drugs/ antibiotics such as Carbarsone, Diiodoquin and Oxytetracycline.

10.2 PROTOZOA AND ANIMAL DISEASES

Many parasitic protozoa of domestic and wild animals are pathogenic, causing a variety of diseases in them. Some of the pathogenic protozoa of animals are as follows:

1. *Entamoeba gingivalis*. It lives as parasite in the tartar around the teeth of dogs and cats.
2. *Trichomonas hominis*. It lives in the ileo-caecal region of monkeys, cats and dogs. It causes diarrhoea and dysentery. Infection occurs during encysted stage through contaminated food and water.
3. *Trichomonas gallinae*. It lives in the cells of epithelial lining of the oesophagus and crop of birds such as pigeons, turkeys, fowls, etc. It causes necrotic nodules in these organs.
4. *Trichomonas foetus*. It occurs in the urino-genital system of cattle, horse and sheep. It causes abortions, delayed conception and inflammation of perputial sacs. Transmission is direct during sexual intercourse.
5. *Histomonas maleagridis*. It lives in the caecum and liver of birds such as a turkeys and fowls. It causes black head disease in them.
6. *Trypanosoma*. Its following five species cause various diseases in domestic animals:
   (*i*) *Trypanosoma brucei*. It cause **nagana** disease in domestic animals. Its vector is *Glossina morsitans*.
   (*ii*) *Trypanosoma evansi*. It causes **surra** disease in horses, camels, dogs, mules and cattles. Its vectors are flies *Tabanus* and *Stomoxys*.
   (*iii*) *Trypanosoma equiperdum*. It causes **dourine** which is a disease of sex organs of horses. Disease is transmitted directly during copulation (*i.e.*, there is no vector).
   (*iv*) *Trypanosoma equinum*. It cause the disease, called **Mal de Caderas** in horses in South America. Its vector is a tabanid fly.
   (*v*) *Trypanosoma hippicum*. It causes the disease **Murina** or **Deren gadera** in horses and mules in Panama. Its vector is fly.
7. *Isospora hominis*. It lives as an endoparasite of epithelial cells of gut of dogs and cats. It causes **diarrhoea** and other abdominal discomforts.
8. *Eimeria* species. This sporozoan lives in the digestive tract of birds, sheep and cattle. It causes **dysentery** and **diarrhoea** in them. *Eimeria truncata* infests the kidneys of goose (bird) and often causes death of its host.
9. *Babesia bigemina*. It is a sporozoan parasite in erythrocytes (RBC) of cattle. It is a widespread fatal diseases of cattle in America and Australia and causes **texas fever** or **red water fever** destroying red blood cells. Its infection occurs through the bite of vector female tick, *Boophilus annulatus*.
10. *Babesia canis*. It is endoparasite of red blood corpuscles of dogs. It causes anaemia, jaundice and fever in dogs. Its infection occurs due to bite of ticks.
11. *Babesia equi*. It is an endoparasite of red blood corpuscles of horses. It causes anaemia, jaundice and paralysis of hindlimbs. Its infection occurs through the bites of ticks.
12. *Sarcocystis linedamanni*. It is a sporozoan inhabiting the striped muscle fibres of sheep's oesophagus. It causes degeneration of these muscles.
13. *Haenoproteus*. It lives in the RBC and the endothelial cells of blood vessels of birds.
14. *Plasmodium* species. Its many species live as parasite of RBC of lizards, birds and mammals and causes malaria in them.
15. *Nosema apis*. It is a sporozoan parasite which lives in the intestinal epithelium and Malpighian tubules of honey bees. It causes muscular disease in them.

QUESTIONS

Long Answer Questions

1. Write a short essay on protozoa and human diseases.
2. Describe briefly the pathogenic protozoa of human beings.
3. Write an essay on "Protozoa and disease."
4. Discuss the diseases produced in human beings by flagellate protozoans.
5. Name the diseases caused by protozoans in human beings. Give an account of any two of these.

Short Answer Questions

1. Write short notes on:
   (*i*) Sleeping sickness
   (*ii*) Giardiasis
   (*iii*) Amoebic dysentery
   (*iv*) Malaria

Very Short Answer Question

1. Mention four essential characters of a pathogen.
2. Where is the location of *Giardia intestinalis* in human host?
3. Give the name of the vector of East African sleeping sickness.
4. Give generic name of sandfly, the vector of kalaazar.
5. What is the name of vector of pathogen of oriental sore?
6. Name the protozoan pathogen of honey bee.
7. What is the primary host of malarial parasite?
8. The drug Resochin is used for treatment of which fatal human disease?
9. Which protozoan causes leucorrhea of human females?

Fill in the Blanks

1. *Giardia* produces the disease ....
2. *Entamoeba histolytica* produces the disease ....
3. *Entamoeba gingivalis* may aggrevate ....
4. Amoebic dysentery is caused by ....
5. *Plasmodium* is an intracellular parasite in a .... host.
6. Vector of pathogen of sleeping sickness is ....
7. Parasitic protozoans lack .... and .... organs.
8. Two species of *Trypanosoma* cause sleeping sickness: *T.* .... and *T.* ....
9. Kalaazar is caused by ....
10. Transmission of protozoan .... occurs during coitus.

Multiple Choice Questions

1. Female *Anopheles* mosquito is a vector of a pathogen that causes
   (*a*) yellow fever
   (*b*) filariasis
   (*c*) malaria
   (*d*) dengue
2. Kalaazar is caused by
   (*a*) *Plasmodium vivax*
   (*b*) *Leishmania donovani*
   (*c*) *Trypanosoma gambiense*
   (*d*) *Wuchereria bancrofti*.
3. Kalaazar, a severe liver infection is transmitted mainly by the biting of
   (*a*) sand fly
   (*b*) mosquito
   (*c*) tse-tse fly
   (*d*) dog
4. *Leishmania tropica* causes
   (*a*) kalaazar
   (*b*) oriental sore
   (*c*) sleeping sickness
   (*d*) none of these
5. *Trypanosoma gambiensis* causes sleeping sickness in human beings. It infects the
   (*a*) blood
   (*b*) liver
   (*c*) brain
   (*d*) cerebro-spinal fluid
6. Amoebiasis is caused by
   (*a*) biting of mosquito
   (*b*) contaminated food and water
   (*c*) by virus
   (*d*) sexual transmission

7. Infective stage of *Entamoeba histolytica* for human beings is
   (*a*) uninucleate cyst (*b*) binucleate cyst
   (*c*) quadrinucleate cyst (*d*) trophozoite cyst
8. Motile elongated zygote of *Plasmodium* is found in
   (*a*) human RBCs (*b*) human liver
   (*c*) gut of mosquito. (*d*) salivary gland of mosquito
9. Maurer's dots are observed in the erythrocytes of human beings if these are infected with
   (*a*) *Plasmodium vivax* (*b*) *P. ovale*
   (*c*) *P. falciparum* (*d*) *P. malariae*
10. Mepacrine and paludrine are effective drugs in treatment of
   (*a*) amoebic dysentery (*b*) malaria
   (*c*) diarrhoea (*d*) sleeping sickness
11. Bouts of malaria is due to
   (*a*) release of merozoites and their metabolites hemozoin
   (*b*) cryptozoites
   (*c*) sporozoites
   (*d*) micro-cryptomerozoites
12. Who for the first time discovered *Plasmodium vivax* in erythrocytes of human beings?
   (*a*) Charles Laveran (*b*) Ronald Ross
   (*c*) Grass (*d*) Stephans
13. Select the matching pair
   (*a*) *Plasmodium* – Pyorrhoea (*b*) *Leishmania* – Dysentery
   (*c*) *Giardia* – Diarrhoea (*d*) *Trypanosoma* – Kalaazar
14. *Histomonas* causes this disease in poultry
   (*a*) coryza (*b*) coccidiosis
   (*c*) black head (*d*) chronic ashthma
15. Which of the following disease continues to be a major health problem in India?
   (*a*) malaria (*b*) polio
   (*c*) plague (*d*) tuberculosis

ANSWERS

Very Short Answer Question

1. 1. Obligate parasitism; 2. Host specificity; 3. Completion of life cycle of the parasite in one or two hosts; and 4. Secretion of toxin by the pathogen which results in the symptoms of the disease.
2. Small intestine (*i.e.*, duodenum and jejunum).
3. Tse-tse fly (*Glossina marsitans*).
4. *Phleobotomus argentipes*.
5. Sandfly, *Phleobotomus sergenti*.
6. *Nosema apis*.
7. Female *Anopheles* mosquito.
8. Malaria.
9. Protozoan, *Trichomonas vaginalis*.

Fill in the Blanks

1. Diarrhoea
2. Amoebic dysentery
3. Periodontitis
4. *Entamoeba histolytica*
5. Vertebrate
6. *Glossina palpalis*
7. Locomotory, osmoregulatory
8. *gambiense, rhodesiense*
9. *Leishmania donovani*
10. *Trichomonas vaginalis*.

Multiple Choice Questions

1. (*c*) 2. (*b*) 3. (*a*) 4. (*b*) 5. (*d*) 6. (*b*)
7. (*c*) 8. (*c*) 9. (*c*) 10. (*b*) 11. (*a*) 12. (*a*)
13. (*c*) 14. (*c*) 15. (*a*).

11

Phylum Porifera: Characters, Classification and Types

Phylum Porifera includes animals having pores in the body wall. The name porifera (L., *porous* = a pore; *ferre* = to bear) was coined by **Robert Grant** (1836) for the plant-like sessile (permanently attached, hence nonmotile) multicellular creatures, called **sponges. Aristotle, Pliny** and other ancient naturalist considered sponges to be plants. In fact, it was not until 1765, when internal water currents and movements of the oscula were first observed by **Ellis**, that the animal nature of sponges became clearly established. The morphology and physiology of sponges were first worked out by **Grant** (1875). **Huxley** (1875) and **Solas** (1884) suggested their complete separation from other Metazoa on the basis of numerous peculiar features of sponges. **Solas** created an isolated branch of Metozoa called **Parazoa** for the sponges. In evolutionary terms, phylum Porifera is considered as **"dead-end" phylum** since it has not given rise to any group of organisms.

11.1 GENERAL CHARACTERS

Poriferans are sessile, suspension-feeding, multicellular animals that utilise flagellated cells called **choanocytes** to circulate water through a unique system of water canals. (**Note.** Suspension feeding is the removal of suspended food particles – bacteria, phytoplankton, zooplankton and some detritus – from the surrounding medium by some sort of capture, trapping or filteration mechanism). They are metazoa lacking true embryological germ layering. Not only true tissues are absent, but most of the body cells are **totipotent** – they are capable of changing form and function. Despite being multicellular, poriferans function like organisms at the unicellular grade of complexity, *i.e.*, their nutrition, cellular organization, gas exchange, and response to environmental stimuli are all protist like.

Box 11.1

Two unique organizational attributes *define* sponges and have played major roles in poriferan success: the **water current channels** or **aquiferous system** (and its **choanocytes**), and the highly **totipotent** nature of sponge cells (**Brusca** and **Brusca**, 2003). A totipotent cell is capable of developing into a complete organism or differentiation into any of its cells or tissues.

General Characters

1. Multicellular animals having **cellular grade of construction**. The cells are arranged in loose aggregations, hardly forming any tissue, organs or systems.
2. There occurs some cellular differentiation or **division of labour**, but no tissue organization. Unlike epithelium of most other animals, a basal lamina and intracellular junctions are absent.

3. All are aquatic, mostly marine, some are freshwater animals.
4. Solitary or colonial, all sessile—attached to various objects in water. (**Note.** According to **Bond** (1997, 1998) some sponges can move from one place to another).
5. Shape and size of body of sponges are variable. Size ranges between 1cm to 1m. Shape may be cylindrical, vase-shaped, cup-shaped, spherical or disc-shaped.
6. Body is either asymmetrical or superficially radially symmetrical.
7. Sponge's body is **diploblastic** with an outer ectodermal layer of flattened pinacocyte cells (called **pinacoderm**) and an inner endodermal layer of collared flagellate cells or choanocytes (called **choanoderm**). In between is a mesoglea, mesenchyme or **mesohyl** which is gelatinous and proteinaceous matrix containing collagen fibres, skeletal material and amoeboid cells (which are migratory cells of both layers). (For contradiction of these facts see Box 11.2).

Box 11.2 Sponges are Monoblastic—They Lack an Endoderm

According to current reports body of sponges is **monoblastic**, consisting of a layer of cells which lines a non-living matrix (the mesohyl). According to them, the spongocoel and choanocyte-lining are not homologous to the gut of other metazoan animals. The choanocytes are not endodermal derivatives. **Indeed there is no endoderm in sponges**. Sponges are gutless animals and the gutless condition is primary (**Ruppert** and **Barnes,** 1994).

8. Sponge's architecture is unique, being constructed around a system of water canals, called **canal system** or **aquiferous system**, an arrangement that is correlated with sponge's sessility.
9. Body surface of sponge is perforated by numerous pores, called **ostia**, serving for the inflow of water. By collective lashing, the flagellated collar cells draw water through these pores. The water currents pass through ostia into the chambers (some of which are lined by choanocytes) and finally in the central cavity called **atrium** or **spongocoel**. From spongocoel water currents escape out through a single large exhalent opening, the **osculum**. Sponge cells obtain food and oxygen from these water current, and release carbon dioxide, excretory substance (ammonia) and reproductive products (sperm, ova) into it.

Box 11.3

The one-cell thick choanoderm may remain simple and continuous (the **asconoid** condition), or it may become folded (the **syconoid** condition), or it may become greatly subdivided into separate flagellated chambers (the **leuconoid** condition) (Fig. 11.1).

10. Sponges contain a single body cavity (*i.e.*, spongocoel). They lack gut (*i.e.*, digestive cavity) and coelom.
11. The body is strengthened by an internal skeleton of calcareous or siliceous **spicules** or horny **spongin** fibres.
12. Digestion is **intracellular** in food vacuoles as in protozoa.
13. Sponges are **ammonotelic**; their chief excretory waste is **ammonia**.
14. The sensory cells and nerve cells are absent, but each cell is directly stimulated and also transmits sensations to other cells.
15. Amoebocytes are the workhorses of sponge economy. They function in internal transport of food and for a variety of other metabolic activities.
16. Asexual (or **clonal**) reproduction occurs by **fragmentation** (regeneration), **budding**, **gemmules** and **reduction bodies**.

17. All are **hermaphrodite**, most are **protandrous**. Sexual reproduction is carried by the formation of sperm and ova. **Fertilization** is internal but cross-fertilization occurs as a rule.
18. Cleavage is **holoblastic**, development indirect through a free-swimming ciliated larva called **diploblastula, amphiblastula** or **parenchymula larva**. The larva is usually at the blastula stage of development.

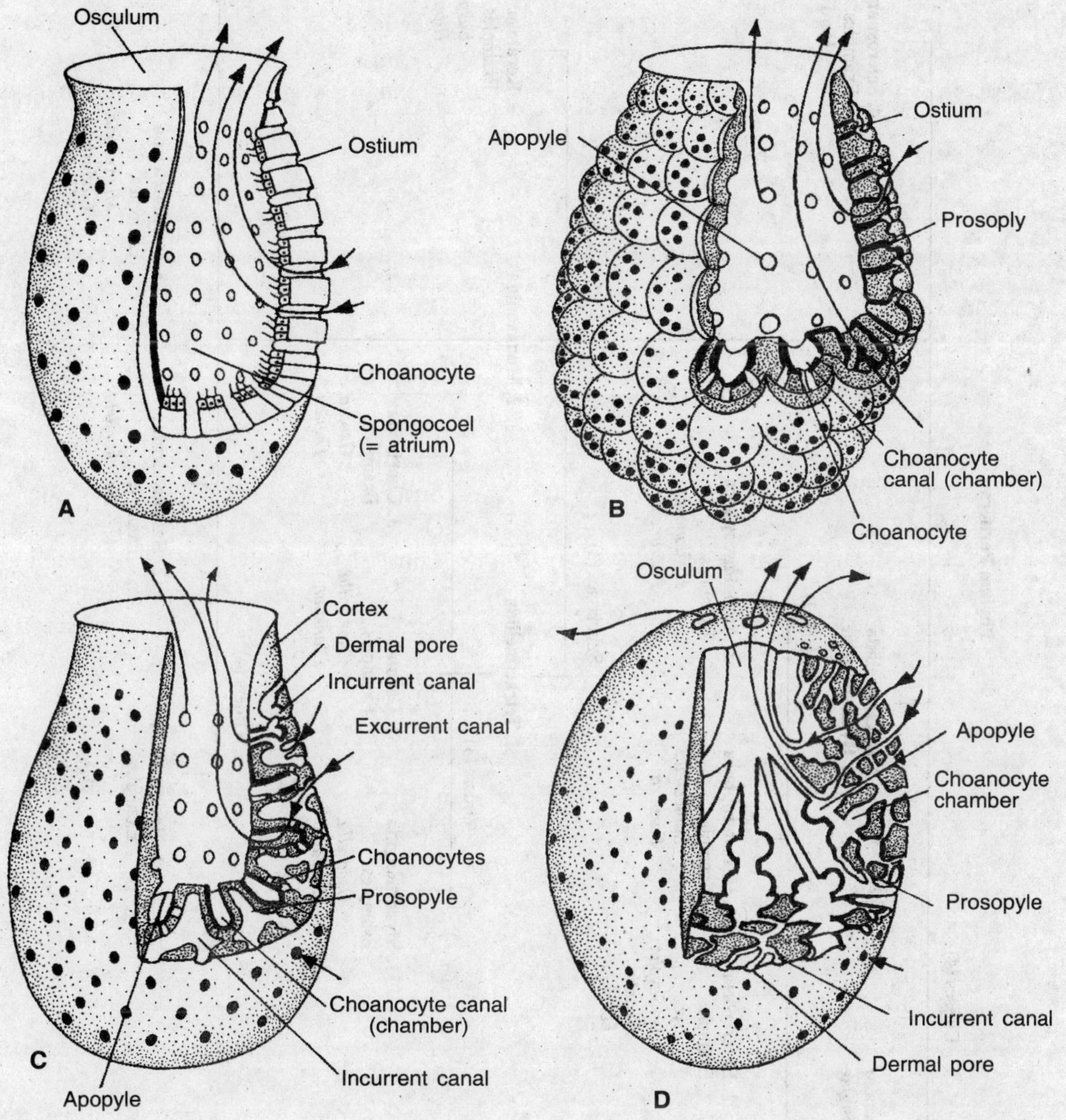

Fig. 11.1. Body complexity in sponges (arrows indicate flow of water). A—The asconoid condition; B—A simple syconoid condition; C—A complex syconoid condition with cortical growth; D—A leuconoid condition (after Brusca and Brusca, 2003).

11.2 OUTLINE CLASSIFICATION OF PHYLUM PORIFERA

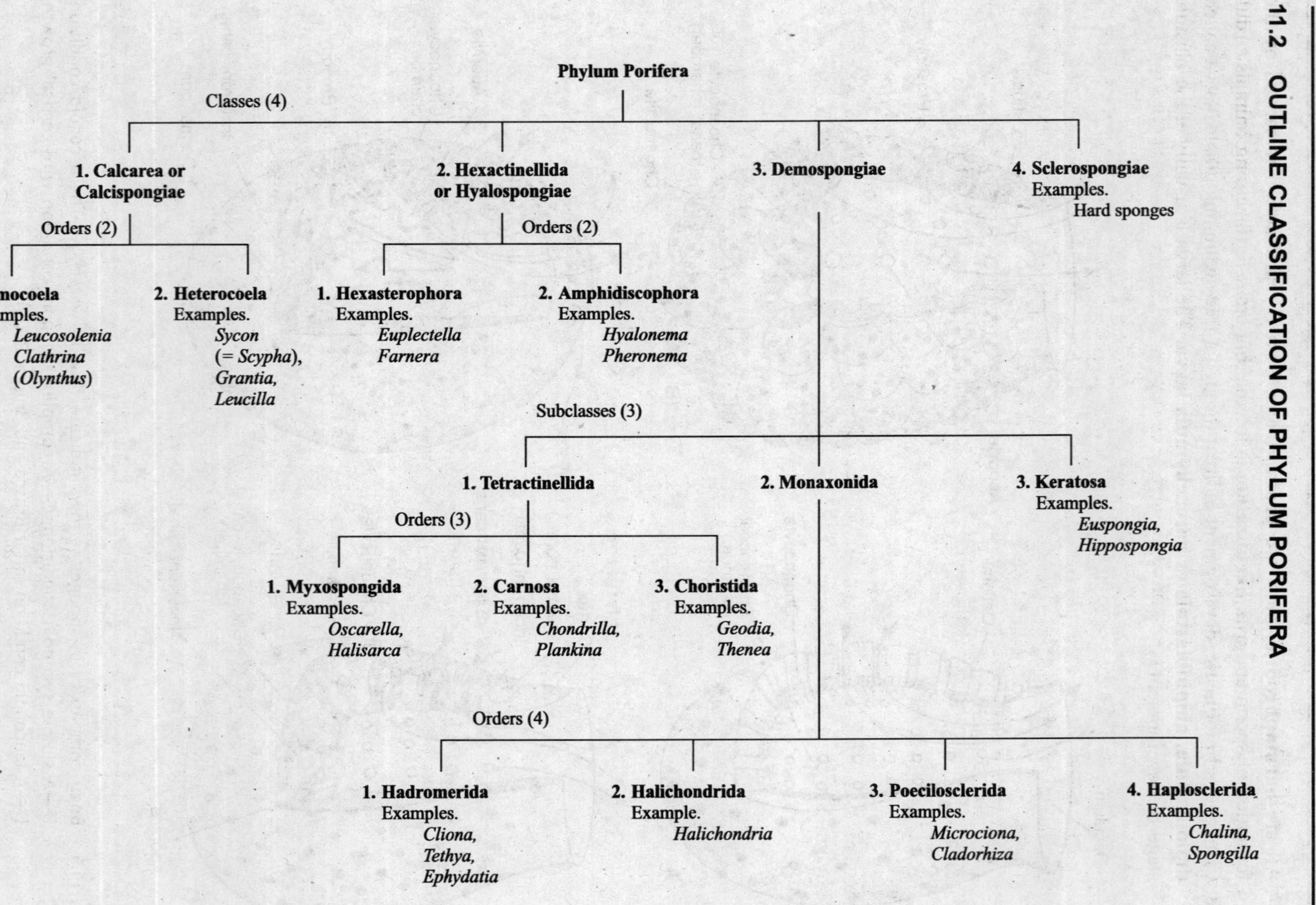

11.3 CLASSIFICATION

Sponges are the most primitive multicellular, heterotrophic, pore-bearing animals. Adult sponges are always attached (sessile) and motionless. Except for about 150 freshwater species, about 5000 described species of sponges are marine animals. They abound in all seas, whereever rocks, shells, submerged timbers, or corals provide a suitable substratum. Some species even live on soft sand or mud bottoms. Most sponges prefer relatively shallow water, but some groups, including most glass sponges, live in deep water.

On the basis of different types of spicules, sponges are classified into four classes (**Berquist**, 1978): Calcarea, Hexactinellida, Desmospongiae and Sclerospongiae. Many workers consider this classification plan as artificial one; in consequence, according to **Meglitsch** and **Schram** (1991), phylum Porifera includes two subphyla: 1. **Cellularia** (cellular organization; epidermis with pinacocytes; mesoglea of collagenous matrix, amoeboid cells and skeletal elements, various choanocyte, distinct and nucleated; myocytes present). 2. **Symplasma** (symplastic or syncytial organization; epidermis lacking pinacocytes, mesoglea as thin mesolamella, collar elements enucleate connected in reticular networks to nuclear choanoblasts, has secondary reticulum about choanoderm). Subphylum Cellularia includes 3 classes: Calcarea, Demospongie and Sclerospongiale. Subphylum Symplasma includes single class Hexactinellida.

Class I. Calcarea or Calcispongiae

1. Calcareous sponges.
2. Small sized sponges, less than 15 cm in height.
3. Skeleton composed of discrete calcareous spicules of calcium carbonate. Spongin fibre absent.
4. Spicules are **monaxon** or **tetraxon**. Tetraxon spicules lose one ray to become triradiate.
5. Spicules are differentiated into microscleres and megascleres.
6. Body organization is **asconoid**, **syconoid** and **leuconoid** type (Fig. 11.1).
7. Body is vase-like cylindrical with bristly outer surface.
8. Choanocytes are distinct, large and nucleated. They line the spongocoel in primitive forms and restricted to flagellated chambers in others.
9. They are solitary or colonial, marine occurring in shallow waters in all oceans.

Class II. Hexactinellida or Hyalospongiae

1. They are commonly called **glass sponges**.
2. Their skeleton is composed of six-rayed (hexactine), siliceous spicules or triaxon spicules. In some hexactinellids the spicules are fused to form a lattice-like skeleton.
3. Epidermis lacks pinacocytes.
4. Mesoglea exists as thin mesolamella having collagen and spongin.
5. Spongocoel is well developed and opens to exterior by a single **osculum**. Osculum is sometimes covered by a **sieve plate**–a grate-like covering formed from spicules.
6. Glass sponges have unique structural plan. They have symplastic or **syncytial organization**. Thus, all surfaces exposed to water are covered not by pinacoderm but by a framework of syncytial strands (called **trabecular syncytium**) through which long spicules may project. Another syncytium (called **choanocyncytium**), containing flagella with collars, lines the flagellated chambers. These collar bodies lose their nuclei after being formed (Fig. 11.2).
7. Glass sponges are well-formed, vase-to-funnel-shaped individuals, attached to the substrate by a tuft of root spicules.

8. They are found in deep tropical seas (between depths of 200 to 1000 m).
9. Glass sponge may grow upto one metre.
10. Recent histological insight in glass sponges has been gained by work on *Rhabdocalyptus dawsoni* which lives in shallower waters (**Mackie, G.O.**, and **C.L. Singla**, 1983).

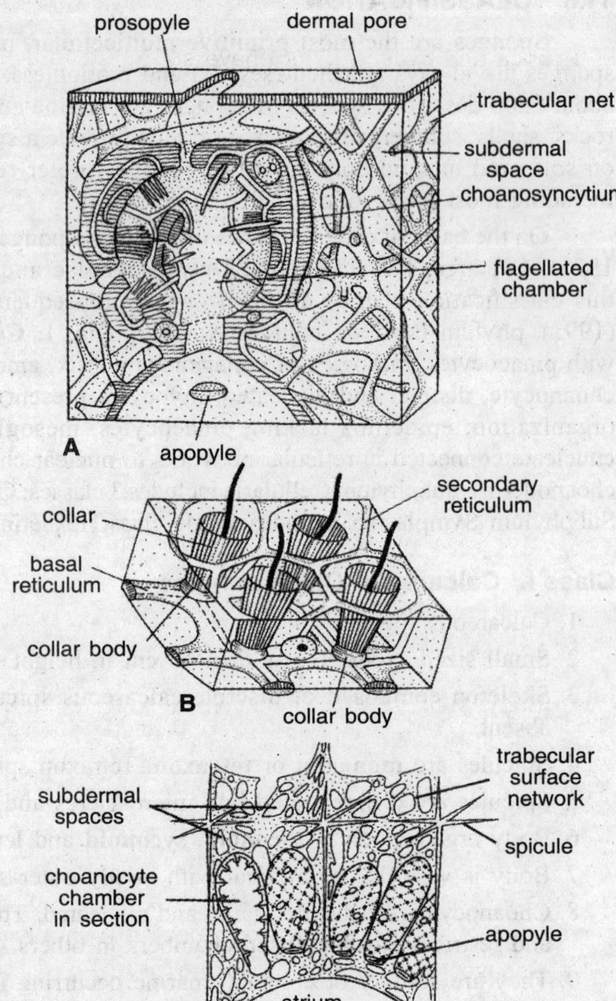

Fig. 11.2. Internal structure of glass sponge. A—Body plan of idealized hexactinellid. Water enters dermal pore, flows around trabecular net, goes through prosopyl in flagellated chambers, exits through apopyle into excurrent lacunae. B—Details of wall of flagellated chamber having secondary reticulum (internal part of trabecular system) around collars. C—Choanocyte chambers within the body wall of the *Euplectella*.

Class III. Demospongiae

1. This large-sized class includes about 90 per cent of sponge species.
2. Generally marine, a few freshwater forms. Marine forms range in distribution from shallow water to great depth. Body shape is irregular.
3. Colouration is brilliant because of pigment granules located in the amoebocytes.
4. Skeleton may consist of siliceous spicules or spongin fibres or a combination of both. The genus, *Oscarella* is unique in lacking both spongin and spicule.
5. Siliceous skeleton of Demospongiae differ from the Hexactinellida in that their larger spicules are monaxon or tetraxon, never triaxons (hexactines).
6. Generally with small, microscleres, **tissue spicules** and large, megascleres, **skeletal spicules**.
7. Body organization is **leuconoid** type.

Class IV. Sclerospongiae

1. Demospongiae-like ancient sponges, called **coralline sponges**.
2. Skeleton composed of siliceous and aragonitic (calcite) spicules, and spongin fibres.
3. Leuconoid sponges.
4. Found in caves, tunnels, crevices and deep waters overhanging on coral reefs.

Examples. Hard sponges.

Box 11.4

Brusca and **Brusca** (2003) have suggested following classification of sponges:

Phylum Porifera

1. Class Calcarea

(*i*) **Subclass Calcinea**, *e.g.*, *Clathrina*, *Dendya*, *Soleniscus*.

(*ii*) **Subclass Calcaronea**, *e.g.*, *Grantia*, *Leucilla*, *Leucosolenia*, *Scypha* (*Sycon*).

2. Class Hexactinellida

(*i*) **Subclass Amphidiscophora**, *e.g.*, *Hyalonema*, *Monorhphis*, *Pheronema*.

(*ii*) **Subclass Hexasterophora**, *e.g.*, *Euplectella*, *Hexactinella*, *Rosella*, *Sympagella*

3. Class Demospongiae

(*i*) **Subclass Homoscleromorpha**, *e.g.*, *Corticium*, *Oscarella*, *Plakia*.

(*ii*) **Tetractinomorpha**, *e.g.*, *Chondrilla*, *Cliona*, *Tethya*, *Tetilla*, *Merila*

(*iii*) **Ceractinomorpha**, *e.g.*, *Clathria*, *Spongia*, *Spongilla*, *Tedania*.

The subclass ceractinomorpha now contain some sponges previously assigned to the class sclerospongiae.

Class I. Calcarea or Calcispongiae

It is divided into following **two** orders:

Order 1. Homocoela

1. Sponges with radially symmetrical cylindrical bodies.
2. Body wall is thin and not folded. The spongocoel is lined with choanocytes.
3. Ascon type canal system.

Examples. *Leucosolenia*, *Clathrina*.

Order 2. Heterocoela

1. Sponges with vase-shaped body.
2. Body wall thick and folded.
3. Choanocytes are restricted to flagellated chambers or radial canals.
4. Spongocoel is lined by pinacocytes (pinacoderm).

Examples. *Scypha* (*Sycon*), *Grantia*, *Leucilla*.

Class II. Hexactinellida or Hyalospongiae

It is divided into following **two** orders:

Order 1. Hexasterophora

1. Spicules are six-rayed and star-shaped (hexasters).
2. Amphidiscs are absent.

Examples. *Euplectella* (Venus's flower basket).

Order 2. Amphidiscophora

1. Spicules in the form of "anchor-like" amphidiscs.
2. Hexasters absent.

Examples. *Hyalonema* (glassrope sponge), *Pheronema* (bowl sponge).

Class III. Demospongiae

This large-sized class is divided into following **three** subclasses:

Subclass 1. Tetractinellida

1. Skeleton of siliceous tetraxon (four-rayed) spicules. (These spicules are absent in order Myxospongida).
2. Spongin is absent.

Subclass 2. Monaxonida

1. Skeleton of siliceous monaxon spicules.
2. Spongin fibre may be present or absent.
3. Micro-and megascleres may be present or absent.

Subclass 3. Keratosa

1. Spicules absent.
2. Skeleton contains horny spongin fibres.
3. Includes horny sponges.

Examples. *Euspongia* (bath sponge), *Hippospongia* (horse sponge), *Hircinia* (horny sponge).

Subclass 1. Tetractinellida

It is divided into following **three** orders.

Order 1. Myxospongida

1. Simple structure.
2. Spicules absent.

Examples. *Oscarella*, *Halisarca.*

Order 2. Carnosa (or Homosclerophora)

1. Simple structure.
2. Microscleres (tissue spicules) and megascleres (skeletal spicule) are similar in shape and size.

Examples. *Alina*, *Plankina*, *Chondrilla.*

Order 3. Choristida

1. Microscleres and megascleres differ in size and shape.

Examples. *Geodia*, *Thenea*, *Stelletta.*

Subclass 2. Monaxonida

It is divided into following **four** orders.

Order 1. Hadromerida

1. Microscleres absent or aster-shaped.
2. Monaxon megascleres are in the form of tylostyles (in which both ends of the spicule are knob-like).
3. Spongin absent.

Examples. *Cliona* (boring sponge), *Tethya*, *Poterion* (Neptune's goblet), *Ephydatia* (a freshwater sponge).

Order 2. Halichondrida

1. Microscleres absent.
2. Very little spongin fibre.
3. Monaxon megascleres are often of two types, *i.e.*, **monactines** and **diactines.**

Example. *Halichondria* (crumb of bread sponge).

Order 3. Poecilosclerida

1. Microscleres include **sigmas** (C or S-shaped), **toxas** (bowl-shaped) and **chelas** (with recurved hooks or plates).
2. Monaxon megascleres are of two types, one type is found in the ectoderm and another type in the choanocyte layer.

   Example. *Cladorhiza.*

Order 4. Haplosclerida

1. Monaxon megascleres of only one kind (diactinal or two-rayed).
2. Microscleres absent.
3. Spongin fibres present.

   Examples. *Chalina* (mermaid's gloves or deadman's finger), *Spongilla.*

11.4 TYPE STUDIES

1. *Euplectella*

Classification

Phylum	–	**Porifera**
Class	–	**Hexactinellida**
Order	–	**Hexasterophora**
Genus	–	***Euplectella***
Species	–	***aspergillum***

Euplectella aspergillum (Fig. 11.3) is commonly known as **Venus's flower basket**. It is a beautiful white marine glass sponge which is found attached by its siliceous **root spicules** to the bottom of deep seas around Philippines and Japan. *Euplectella* is a solitary animal.

The size of individuals varies from 15–30 cm in length. The body is cylindrical and slightly curved (Its curved structure is an adaptation to the slow, constant water current at great depth of sea). The entire body is supported by a skeleton of six-rayed siliceous spicules which are bound together by siliceous cement so that they form a filmy lattice-like skeleton. The canal system is of syconoid type. The **parietal gaps** situated in between the lattice lead into the paragastric cavity (spongocoel). Osculum is covered with an **oscular sieve** formed of fused spicules. There is no epidermis of pinacocytes. Flagellated chambers lie radially in the sponge wall (Fig. 11.2C).

Venus's flower basket displays an interesting commensal relationship with certain species of shrimp (*Spongicola venusta*). A young male and a young female shrimp enter the spongocoel along with water current where they grow and cannot escape through the sieve plate covering the osculum. Thus, their entire life is spent within the sponge, where they feed on plankton brought in by the sponge's water currents. The partners remain unseparated till one of

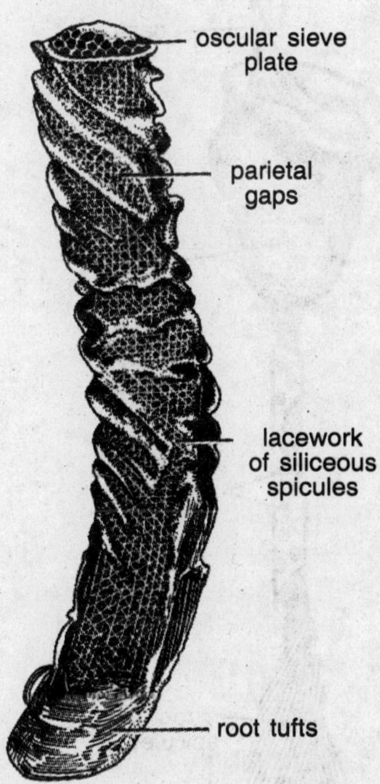

Fig. 11.3. *Euplectella.*

them dies. Such Venus's flower baskets were once used by Japanese as a wedding gift since each carriers felicitation of decapod couple bound in permanent marital union. The Japanese call it "*together unto old age unto same grave.*"

2. *Hyalonema*

Classification

Phylum	–	**Porifera**
Class	–	**Hexactinellida**
Order	–	**Amphidiscophora**
Genus	–	***Hyalonema***

It is one of the most common hexactinellid sponge, with many species. *Hyalonema* (Fig. 11.4) is commonly called **glass rope sponge**. It is marine and found in the sea (Atlantic and Pacific near New England) at a depth of 10 to 15 metre.

Hyalonema has a round or oval body with a simple spirally-twisted rope-like root-tuft. The root-tuft is long, stalk-like and is made up of a bundle of very long glassy silicious monaxon spicules. The sponge remains attached to the substratum (sea-bottom) by the root-tuft. The root-tuft extends into the body as an axis or **columella** and often project out of the body in the form of **gastral cone**.

The middle part of the columella commonly has attached symbiotic sea anemones (of genus *Epizoanthus* or *Polythoe*). The skeleton of *Hyalonema* contains small branching five-rayed or six-rayed spicules and large and small amphidiscs. Excurrent canal opens on the gastral cone surface, spongocoel is absent.

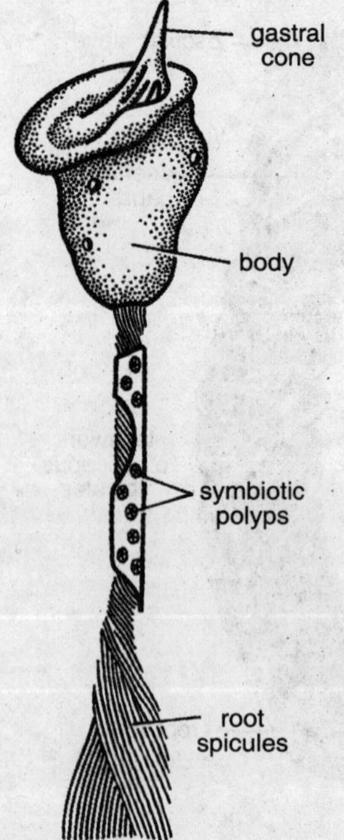

Fig. 11.4. *Hyalonema.*

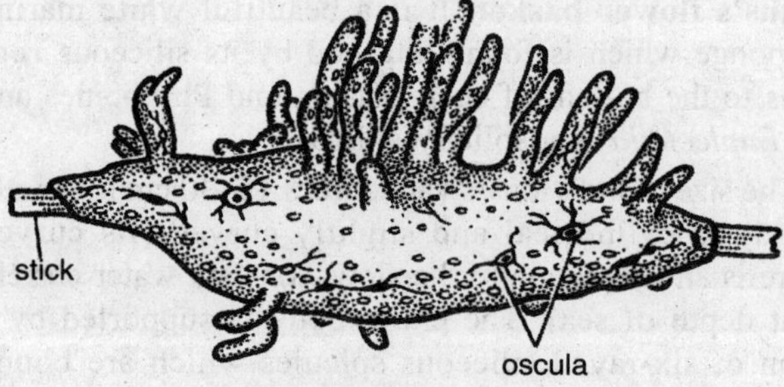

Fig. 11.5. *Spongilla.*

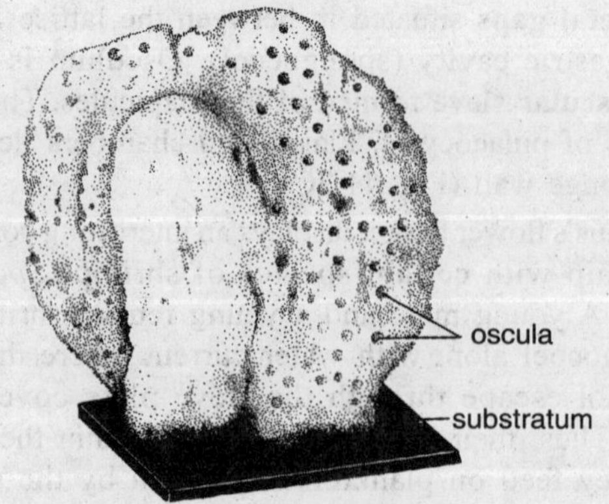

Fig. 11.6. *Euspongia.*

3. *Spongilla*

Classification

Phylum	–	**Porifera**
Class	–	**Demospongiae**
Subclass	–	**Monaxonida**
Order	–	**Haplosclerina**
Genus	–	***Spongilla***

It is a well known freshwater sponge which is colonial and is abundantly found in ponds, lakes, slow streams where the water is not turbid. Profusely branched colony of *Spongilla* (Fig. 11.5) grows on submerged plants, sticks and stones; it shows an encrusting growth pattern. Colony of *Spongilla* is green in colour because of the presence of symbiotic *Zoochlorellae* (a green algae) in various cells of sponge such as pinacocytes, choanocytes, amoebocytes, etc.

The body wall of spongilla consists of a very thin **dermal membrane** scattered with numerous **dermal pores** and **oscula**. The canal system is **rhagon type**. The skeleton contains siliceous spicules that form a network of monaxon spicules, microscleres and megascleres embedded in spongin. Asexual reproduction takes place by gemmules. Sexual reproduction includes ciliated free-swimming larva, called **parenchymella or parenchymula.**

4. *Euspongia*

Classification

Phylum	–	**Porifera**
Class	–	**Demospongiae**
Subclass	–	**Keratosa**
Genus	–	***Euspongia***

Euspongia (Fig. 11.6) is the common **bath sponge**. It is a horny sponge which is found attached with rocky bottom in warm shallow sea waters. It is found in plenty in the Mediterranean sea, the Gulf of Mexico and West Indies. Body is rounded and large and light brown in colour. Body surface contain many oscula.

Inside the bath sponge are numerous inhalent canals, flagellated chambers, exhalant canals, and spaces; Skeleton is made of spongin fibres only. Spongin fibres form a network, the thickest fibres run radially and each ends in small projection, called **conulus**. Body contains many conuli.

QUESTIONS

Long Answer Questions

1. Classify phylum Porifera and give diagnostic or general characters of the phylum, and examples of its various classes.
2. Enumerate the distinguishing features of phylum Porifera. Classify it giving chief characters and examples of each order. (*Rohilkhand 2001*)
3. Give an outline classification of phylum Porifera.

Short Answer Questions

1. Classify and comment on the following:

   (*i*) *Euplectella* (*ii*) *Hyalonema*

   (*iii*) *Spongilla* (*iv*) *Euspongia*

Very Short Answer Questions

1. Mention the scientific name of Venus's flower basket.
2. Give the name of a sponge having commercial value.
3. On which main characters is the classification of sponges based.
4. What is canal system?
5. How do fresh water sponges manage to survive desiccation in summer?
6. Give the order and class of *Hyalonema*.
7. Cite two common examples of class Demospongiae.
8. Name the cell which secretes spongin fibres.
9. What is choanocyte?
10. Scleroblasts are of two types. Name them.
11. Name any freshwater sponge. Also give its order and class.
12. What are gemmules?
13. Name the two pores connecting the canals of sponge internally.

Fill in the Blanks

1. *Spongilla* is an example of .... sponge.
2. .... is the costliest Turkish bath sponge.
3. Sponges have .... level of organisation.
4. Asexual reproductive bodies of sponges are ....
5. Life cycle of sponges includes a free-swimming larva named .... or ....
6. Phylum Porifera is divided into four classes Calcarea, Hexactinellida, ... and ....
7. *Leucosolenia* belongs to order .... of class ....
8. A common heterocoel sponge is ....
9. The amoeboid cells that secrete silicious spicules are termed ....
10. .... secrete spongin fibres.
11. .... in sponges form the basis of classification in Porifera.

Multiple Choice Questions

1. Body organization in sponges is of
   (*a*) protoplasmic grade (*b*) tissue grade
   (c) cellular grade (*d*) organ grade
2. In sponges food is ingested by
   (*a*) scleroblasts (*b*) choanocytes
   (*c*) porocytes (*d*) pinacocytes
3. The zoological name of common bath sponge is
   (*a*) *Leucosolenia* (*b*) *Sycon*
   (*c*) *Spongia* (*d*) *Euplectella*
4. Which one of the following cells maintains current of water in the canal system of sponge?
   (*a*) porocyte (*b*) myocyte
   (*c*) chromocyte (*d*) choanocyte
5. Incurrent canals communicate with the radial canals by
   (*a*) apopyles (*b*) prosopyles
   (*c*) dermal ostia (*d*) gastral ostia
6. Glass sponges belong to the class
   (*a*) demospongiae (*b*) calcarea
   (*c*) tetractinellids (*d*) hexactinellida
7. Which of the following belongs to class Hexactinellida?
   (*a*) *Sycon* (*b*) *Euplectella*
   (*c*) *Spongilla* (*d*) *Leucosolenia*

8. Ability of sponges to replace the lost body part is
(*a*) growth (*b*) autotomy
(*c*) reproduction (*d*) regeneration

9. Glass rope sponge is the common name of
(*a*) *Hyalonema* (*b*) *Sycon*
(*c*) *Euspongia* (*d*) *Leucoslenia*

10. Gemmules play a role in
(*a*) perennation (*b*) dispersal
(*c*) reproduction (*d*) all of these

11. Who described the animal nature of Porifera?
(*a*) Ellias (*b*) Ross
(*c*) Robert Grant (*d*) Huxley

12. Parazoology is the study of
(*a*) Protozoa (*b*) Porifera
(*c*) Cnidaria (*d*) parasite

ANSWERS

Very Short Answer Questions

1. *Euplectella.*
2. *Spongia* (= *Euspongia*).
3. Skeleton and canal system.
4. It is a system of intercommunicating cavities found in a sponge body.
5. By the formation of resistant gemmules.
6. Amphidiscophora (order), Hexactinillida (class).
7. *Euspongia*, *Spongilla.*
8. Spongioblasts.
9. Any of the collar cells or choanate, flagellate cells lining the cavities of sponge. Choanocytes drive water through canals and chambers constituting the canal system or **aquiferous system** (**Brusca** and **Brusca**, 2003).
10. Calcoblasts, silicoblasts.
11. *Spongilla*, *Haplosclerida* (order), *Demospongiae* (class).
12. Asexual reproductive bodies.
13. 1. **Prosopyle** which connects the incurrent canal to radial canal.
    2. **Apopyle** which connects the radial canal to excurrent canal or spongocoel.

Fill in the Blanks

1. Fresh water
2. *Spongia officianalis*
3. Cellular
4. Gemmules
5. Amphiblastula, parenchymula
6. Demospongiae, Sclerospongiae
7. Homocoela, Calcarea
8. *Sycon*
9. Silicoblasts
10. Spongioblasts
11. Skeleton

Multiple Choice Questions

1. (*c*) **2.** (*b*) **3.** (*c*) **4.** (*d*) **5.** (*b*) **6.** (*d*)
7. (*b*) **8.** (*d*) **9.** (*a*) **10.** (*d*) **11.** (*a*) **12.** (*b*).

12

Scypha (*Sycon*)
(Phylum Porifera)

12.1 CLASSIFICATION

Phylum	–	**Porifera**
Class	–	**Calcaria**
Order	–	**Heterocoela**
Type	–	***Scypha* (*Sycon*)**

Scypha or *Sycon* is commonly called **urn sponge** or **crown sponge**. It is a marine calcareous sponge. It is worldwide in distribution and found in abundance near North Atlantic shores. *Scypha* is found attached to submerged rocks, shells of molluscs and corals in shallow waters (upto the depth of 50 fathoms; 1 fathom = 6 feet) near the coast. It prefers places where waves provide the animal with plenty of food and well-oxygenated water. *Scypha* often occurs in the form of branching colonies.

The name of genus *Sycon* is replaced by *Scypha* by **de Laubenfels** (1936). The different species of *Scypha* are *S. ciliatum*, *S. elegans*, *S. coronata*, etc.

12.2 EXTERNAL FEATURES

1. Shape and size. *Scypha* is a small vase-shaped sponge. It measures 2.5 to 8 cm in length. Its colony has two to seven cylinders which are connected together at the base (Fig. 12.1). The base of *Scypha* is attached by a sticky secretion to some submerged solid substrata.

2. Colour. It is grey or light brown in colour.

3. Osculum. The distal free end of each cylindrical branch of *Scypha* has a single large opening, called **exhalent pore, excurrent pore** or **osculum**. The paragastric cavity or spongocoel opens to exterior through osculum. Water currents exit the sponge through this opening.

Osculum is surrounded by numerous straight needle-like calcareous monaxon

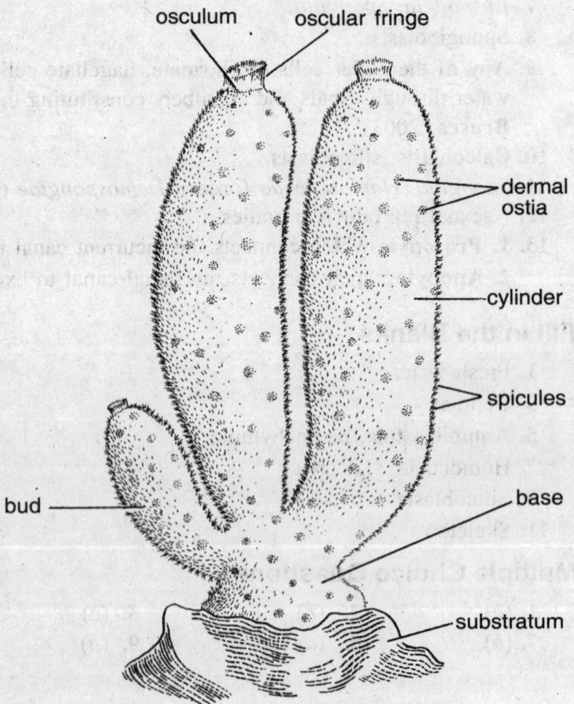

Fig. 12.1. *Scypha*. Colony

spicules arranged in a circlet or crown. This is called **oscular fringe** and it prevents the entry of other animals into the sponge.

4. Ostia. The body surface of *Scypha* has numerous regularly arranged microscopic polygonal elevations. Every elevations are separated from the others by a system of depressed lines of furrows. In these furrows are located numerous microscopic apertures, the **ostia** which are **inhalent** or **incurrent pores**. Water enters the cavity of sponge through these pores. The entire body surface of *Scypha* appears bristly as numerous monaxon spicules (called **oxeote**) protrude out from it.

5. Spongocoel. Inside each cylindrical branch exists a central cavity which leads to osculum; it is called **spongocoel**, **atrium** or **paragastric cavity** (Fig. 12.2). It is not a digestive cavity and is not lined by endoderm. The wall of cylinder is thick due to increase in the amount of **mesogloea mesenchyme** or **mesohyl**. The wall of *Scypha* has folded in such a way as to form two types of canals, the **incurrent canals** and **radial canals**, both lie alternately and radially around the spongocoel. No ostium opens directly to spongocoel (so ostia are absent from the collar and basal regions of *Scypha*).

12.3 MICROSCOPIC ORGANIZATION (HISTOLOGY OR DIFFERENT TYPES OF CELLS IN *SCYPHA*)

Scypha exhibits **cellular grade of organization**, *i.e.*, its cells form loose aggregation and are not organized into tissues. It is a **monoblastic** animal having only one layer of cells lining a non-living matrix, **mesohyl** from both sides (**Note.** Previously *Scypha*, like other Porifera, was considered as a diploblastic animal having two embryonic layers, the ectoderm (dermal layer) and endoderm (gastral layer) but recent embryological data have proved that sponges lack endoderm (see **Ruppert** and **Barnes,** 1994).

According to the nature of cells the single embryonic layer of sponges forms two types of layers: **pinacocyte layer (pinacoderm)** and **choanocyte layer (choanoderm)**. Both of these layers secrete and cover the mesohyl from all sides. Pinacocyte layer and choanocyte layer lack basal laminae, inter-cellular junctions and most other ultrastructural connections that make epithelial layers of other animals (eumetazoa). Mesohyl consists of a gelatinous proteinaceous matrix containing **collagen fibres** (or spongin A), skeletal material (spicules) and amoebocytes. It is equivalent to connective tissue of other metazoans.

The pinacoderm includes **exopinacoderm** covering the body surface except ostia and osculum and **endopinacoderm** lining the incurrent canals and spongocoel. Choanoderm lines the radial canals and consists of a single layer of choanocytes. Structure and function of various types of cells present in the sponges can be considered under the following headings:

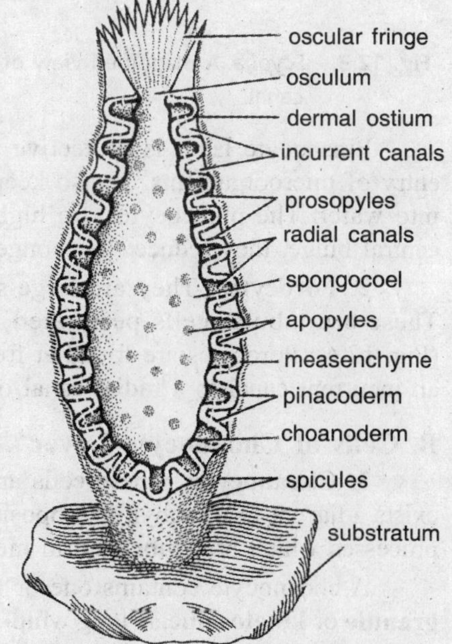

Fig. 12.2. *Scypha*. A diagrammatic L.S. of a cylinder showing gross internal structure.

A. Cells of Pinacocyte Layer

1. Pinacocytes. Pinacocytes are large thin polygonal scale-like flat cells. They are cemented together by their edges like the tiles in a floor. Each pinacocyte contains a nucleus at the centre which is much thicker than the periphery. Pinacocytes cover entire outer surface of the body and also lines the incurrent canal, excurrent canal and spongocoel (Fig. 12.3).

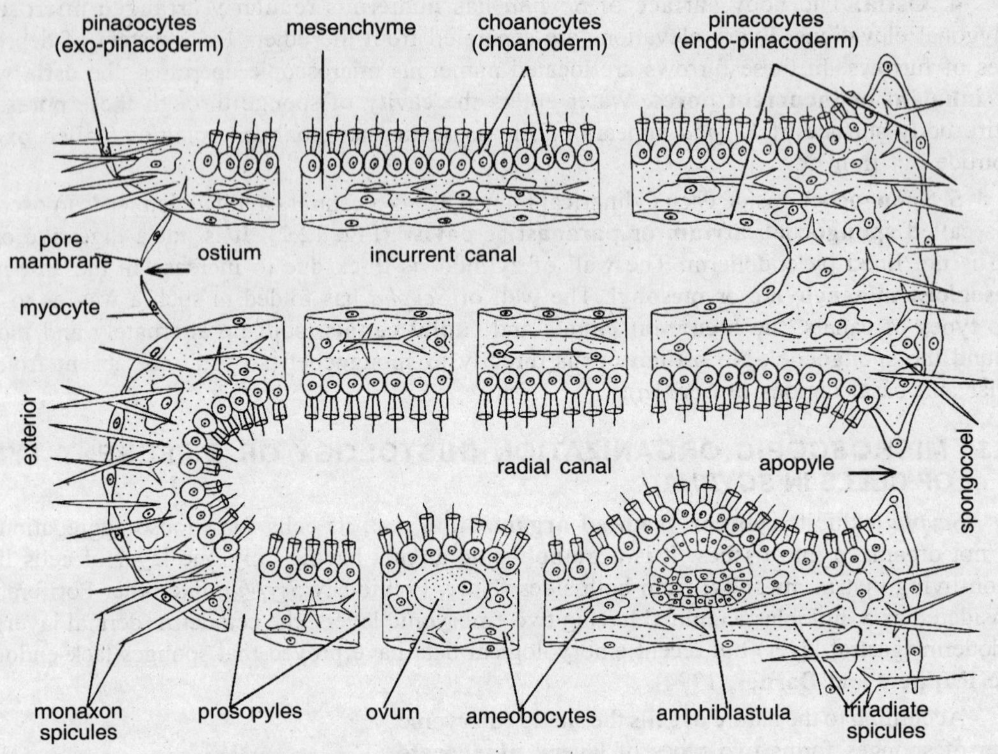

Fig. 12.3. *Scypha*. A magnified view of L.S. of a part of body wall showing one incurrent and one radial canal.

Pinacocyte layer is protective in function. It prevents mechanical injury and checks the entry of microorganisms. It also keeps semifluid mesohyl in position, preventing its dispersion into water. The pinacocytes are highly contractile and by withdrawing their margins into the central bulge, they reduce the sponge surface to an appreciable extent.

2. Porocytes. They are large-sized contractile pore cells and are modified pinacocytes. These are tubular cells perforated along their long axis by a central canal, the **prosopyle** (Fig. 12.4). Porocytes are found at frequent intervals among the pinacocytes. Prosopyle connects an incurrent canal to a radial canal or flagellated chamber.

B. Cells of Choanocyte Layer

3. Choanocytes. These cells are large, ovoid and nucleated. One end of each choanocyte exists adjacent to mesohyl, its opposite end projects into radial canal. A choanocyte has root-like processes which are embedded in mesohyl (Fig. 12.5A).

A choanocyte contains one or two contractile vacuoles, food vacuoles and a single **basal granule** or **kinetosome**. A long whip-like flagellum arises from basal granule. The basal granule is also joined to centriole and the two together forming typical **centroblepharoplast**. The flagellum contains typical 9 + 2 microtubular arrangement. It is surrounded at its base by a transparent contractile collar which is made up of a ring of 20 to 30 erect and closely set microvilli that extend from the cytoplasm (Fig. 12.5B). The microvilli or villi have microfilament cores and

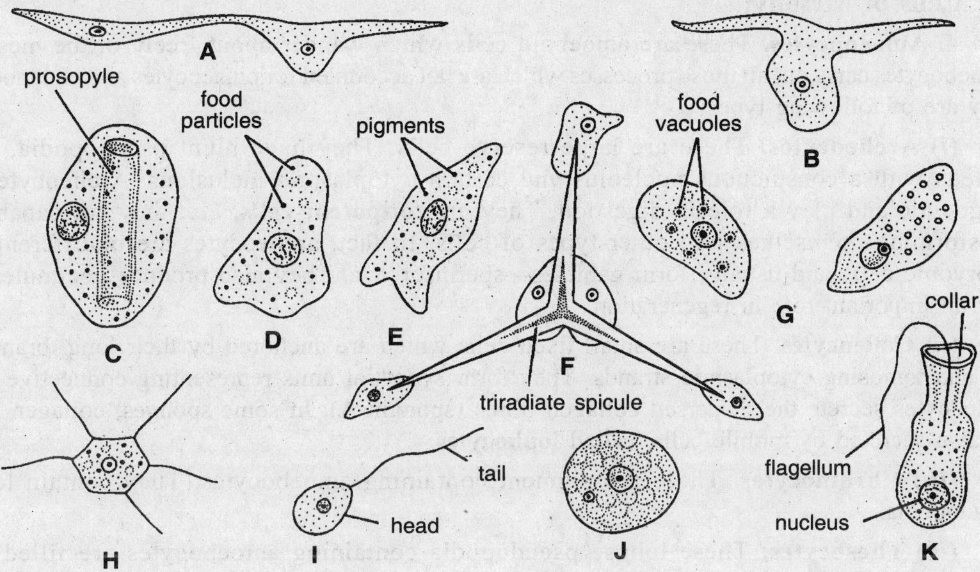

Fig. 12.4. *Scypha*. Cellular elements. A—Pinacocytes; B—Myocyte; C—Porocyte; D—Thesocyte; E—Chromocyte; F—Scleroblast; G—Archeocytes; H—Collencyte; I—Sperm; J—Ovum; K—Choanocyte.

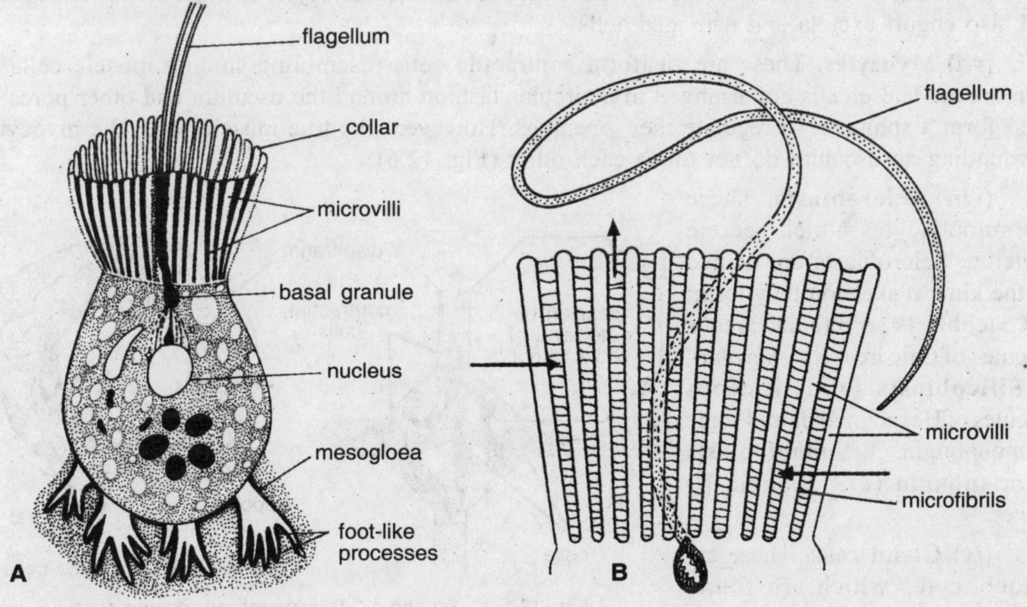

Fig. 12.5. *Scypha*. A—Ultrastructure of a choanocyte; B—Details of collar of choanocyte. Microvilli are connected by delicate microfibrils. Arrow indicates assumed flow of water through collar-sieve.

are connected to one another by anastomosing mucous strands (a mucous reticulum). Choanocytes are held in place by inter-digitation of adjacent basal surfaces. The choanocytes are responsible for moving water through the sponge and for obtaining food. In addition, the choanocytes play an important role in digestion, excretion and reproduction (*i.e.*, fertilization).

C. Cells of Mesohyl

4. Amoebocytes. These are amoeboid cells which wander about freely in the mesohyl. Amoebocytes carry out all those processes which are not accounted for pinacocytes and choanocytes. They are of following types:

(*i*) **Archeocytes.** These are large reserve cells. They have blunt pseudopodia, large nucleus with a conspicuous nucleolus and certain cytoplasmic inclusions. Archeocytes are phagocytic and play a role in digestion. They are **totipotent cells**, *i.e.*, they are capable of transforming themselves into other types of cells. In fact, archeocytes are undifferentiated embryonic cells and usually form gametes—sperm or ova. They also produce gemmules and play an important role in regeneration.

(*ii*) **Collencytes.** These are small fixed cells which are anchored by their long, branching and anastomosing cytoplasmic strands. They form syncytial units representing connective cells. Collencytes secrete the dispersed collagen fibres (spongin A). In some sponges, collagen fibres are also secreted by mobile cells, called **lophocytes**.

(*iii*) **Chromocytes.** These are pigment containing amoebocytes. They contain lobose pseudopodia.

(*iv*) **Thesocytes.** These lobose pseudopodia containing amoebocytes are filled with reserve food. Thesocytes store food in the form of glycogen, fats and proteins (glycoproteins and lipoproteins).

(*v*) **Trophocytes.** These amoebocytes are found loaded with digested food and serve to transfer it from one place (or cell) to other.

(*vi*) **Phagocytes.** These amoebocytes collect food from choanocytes through their pseudopodia and also engulf excreta and damaged cells.

(*vii*) **Myocytes.** These are fusiform contractile cells resembling smooth muscle cells of vertebrates. These cells are arranged in a circular fashion around the osculum and other pores so as to form a sphincter to regulate their openings. However, like true muscle cells, the myocytes surrounding an osculum do not touch each other (Fig. 12.6).

(*viii*) **Scleroblasts.** These are amoebocytes which secrete skeleton. Scleroblasts are named by the kind of skeleton they form: 1. **Calcoblasts** secrete calcareous spicules of Calcarea (*e.g., Scypha*). 2. **Silicoblasts** form siliceous spicules of Hexactinellida and some Demospongia. 3. **Spongioblasts** form **spongin** (*i.e.*, spongin B) **fibres.**

(*ix*) **Gland cells.** These are amoebocytes which are found attached to body surface by means of long strands. They secrete slime (mucus).

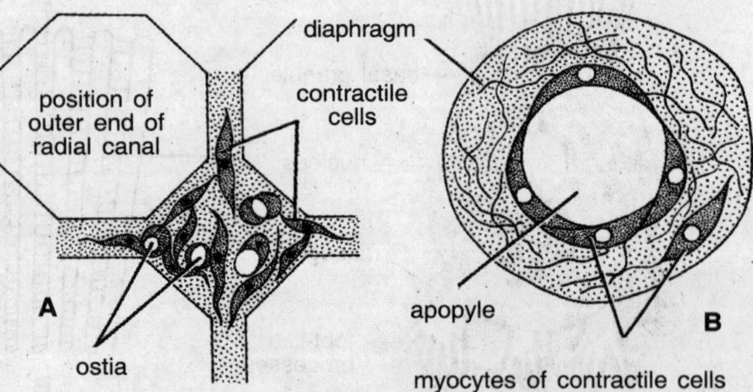

Fig. 12.6. *Scypha*. A—Pore membrane showing ostia; B—An apopyle surrounded by its diaphragm.

Box 12.1

In almost all sponges, the entire supportive matrix is built on a framework of fibrillar collagen. The cells that secrete this material are called **collencytes, lophocytes** and **spongocytes**. Collencytes are morphologically nearly indistinguishable from pinacocytes, whereas lophocytes are large highly motile cells that can be recognised by a collagen they typically trail behind them (Fig. 12.7). The primary function of both cell types is to secrete the dispersed fibrillar **collagen** found intercellularly in virtually all sponges. Spongocytes produce the fibrous supportive collagen referred to as spongin. Spongocytes operate in groups and are always found wrapped around a spicule or spongin fibre (see **Brusca** and **Brusca**, 2003).

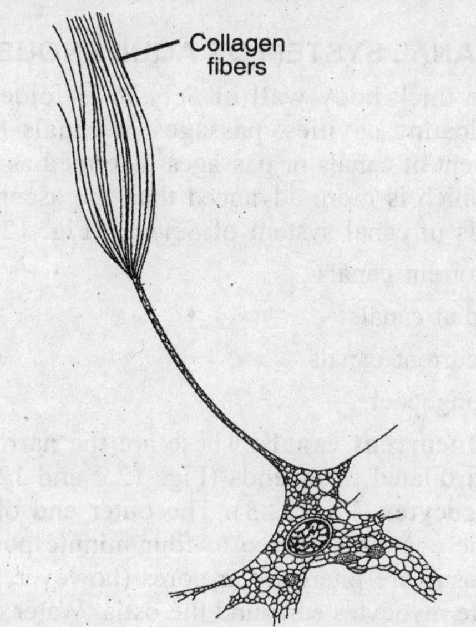

Fig. 12.7. A lophocyte with its tail of collagen fibres.

Skeleton of *Scypha*

In *Scypha* the skeleton consists of minute, crystalline structures called **sclerites** or **spicules** (Fig. 12.8). These spicules are **calcareous**, *i.e.*, they are made of calcium carbonate. Spicules of *Scypha* are composed of calcium carbonate (about 87%), magnesium carbonate and other minerals (10%), water (3%) and organic materials (in traces). In *Scypha* spicules are megascleres (large) and remain separate. Depending upon the shape and size, the spicules of *Scypha* are of following types:

1. One-rayed or monaxon spicules are elongated straight needle-like spicules having only one axis. They are of following two types:

(*a*) Large one-rayed monaxon spicules are found around the osculum and are arranged in a circlet, called **oscular fringe**.

(*b*) Spear-like or club-shaped oxeote spicules occur in groups and project from the surface of *Scypha*. They give a bristly appearance to the sponge and partly conceal and protect the ostia.

2. Three-rayed or triaxon spicules consist of three axes. They occur along the radial canals with one ray pointed towards the distal end. Triaxon spicules form body supporting network of *Scypha*.

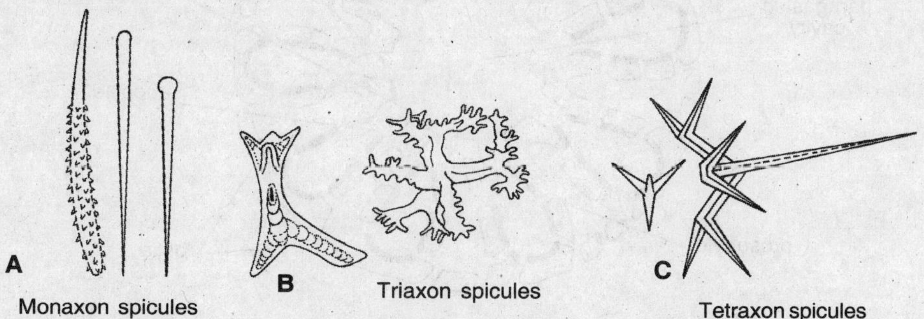

Fig. 12.8. *Scypha*. A—Monaxon spicules; B—Triaxon spicules; C—Tetraxon spicules.

3. Four-rayed or tetraxon spicules consist of four axes. They are embedded in the thick gastral cortex surrounding the spongocoel.

12.4 CANAL SYSTEM OR AQUIFEROUS SYSTEM

The thick body wall of *Scypha* is folded to produce an arrangement of complex intercommunicating cavities, passages or canals for the entrance and exit of water current. This arrangement of canals or passages is termed as **canal system**. *Scypha* has **syconoid** type of canal system which is more advanced than the asconoid canal system of *Leucosolenia*. Various pores and canals of canal system of *Scypha* (Fig. 12.3) are of following types:

1. Incurrent canals
2. Radial canals
3. Excurrent canals
4. Spongocoel

1. Incurrent canals. These are the narrow spaces between the radial canals with narrow inner and dilated outer ends (Fig. 12.2 and 12.3). Incurrent canals are lined by flattened cells, the **pinacocytes** (Fig. 12.3). The outer end of each incurrent canal is covered by a thin pore membrane containing three to four minute pores, called **dermal pores** or **ostia** (Fig. 12.6). In *Scypha*, ostia are intercellular pores (however, in *Leucosolenia* they are intracellular pores). The contractile myocytes surround the ostia. Water current enters the incurrent canal through the ostia and myocytes regulate the amount of water entering the canals by increasing or decreasing the diameter of the ostia.

The inner end of each incurrent canal is closed. However, the wall separating an incurrent canal and radial canal is pierced by many minute pores, called **prosopyles**. The prosopyles are perforations in **porocytes**. Incurrent canals open into the radial canals by prosopyles.

2. Radial canals or flagellated chambers. These are radially arranged, straight, elongated and octagonal canals of *Scypha* wall. They alternate with incurrent canals. Radial canals are lined by flagellated collar cells or **choanocytes**. They end blindly towards the outer surface but opens in the **excurrent canal** at their inner ends through the **apopyles** (Fig. 12.9). The radial canals receive water current from incurrent canals through the prosopyles and convey it to spongocoel via the excurrent canals. The flagella of choanocytes beat continuously and direct the path of water current.

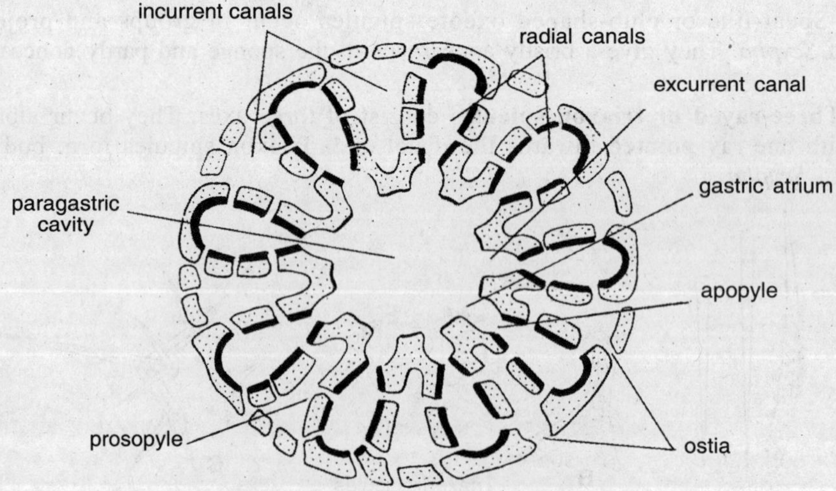

Fig. 12.9. *Scypha*. T.S. body to show canal system.

3. Excurrent canals. These are short but wide passages between the radial canals and spongocoel. Each of them is lined by flattened **pinacocytes**. The excurrent canal is separated from the radial canal by a thin elastic **diaphragm**. The diaphragm is perforated by a large central aperture, called **apopyle**. The apopyle is surrounded by myocytes and is capable of contraction and expansion, thus, regulating the diameter of apopyle. The excurrent canal opens into the spongocoel through a wide opening called **gastric ostium**.

4. Spongocoel. It is the central cavity of a cylindrical branch of colony of *Scypha*. The spongocoel of all branches remains continuous at the base of sponge. Spongocoel is lined by flattened cells or **pinacocytes**. Into it opens the excurrent canals. The spongocoel itself opens to outside through a terminal opening, the **osculum**. A group of contractile cells, called **myocytes** surround the osculum. These form an **oscular sphincter**. This sphincter regulates the diameter of the oscular opening.

Mechanism of Water Circulation

The physiology of *Scypha* and other sponges largely depends on the current of water flowing through the body. The water brings in oxygen and food and removes waste. Even sperms are moved in and out and larvae released by the water current.

The water current is produced and water is pumped into the body through ostia by the asynchronous beating of flagella of choanocytes which lines the radial canals and are oriented towards apopyle. A wave of spiral undulations passes from the base to the tip of each flagellum independently driving water from the radial canal. As a result, water is sucked into the radial canal from the incurrent canals through the prosopyles located between the bases of the choanocytes. It is then driven to the centre of the radial canal and out of the larger apopyle into an excurrent canal. Finally, the water leads into the spongocoel and is discharged to the exterior through osculum. The path of water current can be represented as follows:

Water from outside —through ostia→ **incurrent canal** —through prosopyles→ **radial canals** —through apopyle→ **excurrent canal** —through ostium→ **spongocoel** —through osculum→ **outside**

Physiological importance of canal system. The water current is of great physiological importance for the sponges because it serves the following functions:

1. Nutrition. The flowing water provides food particles to nearly all the cells of *Scypha*. Silica, calcium salts and other substances are absorbed as the water passes through the sponge canal.

2. Respiration. The water entering the body of *Scypha* is rich in oxygen. Gaseous exchange occurs by simple diffusion between the flowing water and all cells of *Scypha*. Carbon dioxide is removed from sponge cells by outgoing water currents.

3. Excretion. While leaving the body, the water current removes egested waste and nitrogenous excretory products (mainly ammonia).

4. Reproduction. The spermatozoa enter other sponges along with the water current.

12.5 NUTRITION

Scypha is a **filter feeder** having **holozoic nutrition**. It depends on the stream of water passing through the body as a source of food.

Food. *Scypha* feeds on both living and dead organic food. It feeds on extremely fine particulate material which forms about 80 per cent of organic matter (organic detritus) consumed by the sponge. The remaining 20 per cent consists of bacteria and protists such as protozoa (dinoflagellates, etc.) and other fine plankton (diatoms). There is also evidence that *Scypha* and other sponges can utilize dissolved nutrients.

Feeding. Food enters the body of *Scypha* along with water currents through ostia. It is picked up by various sponge cells by following two methods:

1. Screening of food
2. Phagocytosis

1. Screening of food. Food particles are apparently selected largely on the basis of size and are screened in the course of their passage into flagellated chamber (radial canal). Only particle smaller than a certain size can enter the ostia or pass through the prosopyles. The finest particles are removed by the choanocytes, perhaps by filteration across the microvilli of collar of the choanocyte (the microvilli are 0.03 to 0.10 mm apart).

Large particles (5–50 mm) are phagocytized by pinacocytes and porocytes lining the inhalent pathways of canal system or by archeocytes that move out of the mesohyl. Indeed, populations of archeocytes are found to clean the outer pinacoderm surface. Particles of bacterial size and below (1 mm) are removed and engulfed (phagocytized) by choanocytes on the surface of the cell and not on the microvillar collars.

2. Phagocytosis. All sponge cells can phagocytize food particles like the protozoans, *i.e.*, food particles are engulfed by the help of pseudopodia, food vacuoles are formed and **intracellular digestion** of food occurs inside food vacuoles by the help of lysosomal enzymes. Beating of flagellum of a choanocyte draws water with food particles towards the collar. Food particles adhere to the outer sticky surface of collar while water escapes through sieve-like microvilli of collar (Fig. 12.5B). These particles are then engulfed by pseudopodia arising from choanocytes (Fig. 12.10). Both archeocytes and choanocytes can transfer their engulfed particles to other cells (amoebocytes).

Digestion. Digestion of food is intracellular. It takes place inside the food vacuoles of archeocytes and digested food is absorbed across the vacuolar membrane. In *Scypha*, food digestion occurs in large sized choanocytes. The food vacuoles are first acidic and then become alkaline. The enzymes are supplied by lysosomes. Three main types of enzymes, namely proteases, carbohydrases and lipases have been detected in sponge extracts.

Assimilation. The digested food is distributed from cell to cell like the postman distributing mail by the wandering amoebocytes found in the mesohyl. The digested food is stored as glycogen, fat and glycoproteins in the storage cells called **archeocytes.** Food is also stored in **nurse cells,** called **trophocytes** that build up food reserves for use by the young oocyte or by amoebocytes preparing to form gemmules.

Egestion. The undigested food residue is egested out from the amoebocytes into the outgoing water current and ultimately expelled out of body.

12.6 REPRODUCTION AND DEVELOPMENT IN *SCYPHA*

Scypha reproduces both asexually and sexually.

1. Asexual reproduction. In *Scypha* asexual reproduction occurs by

(*i*) Budding

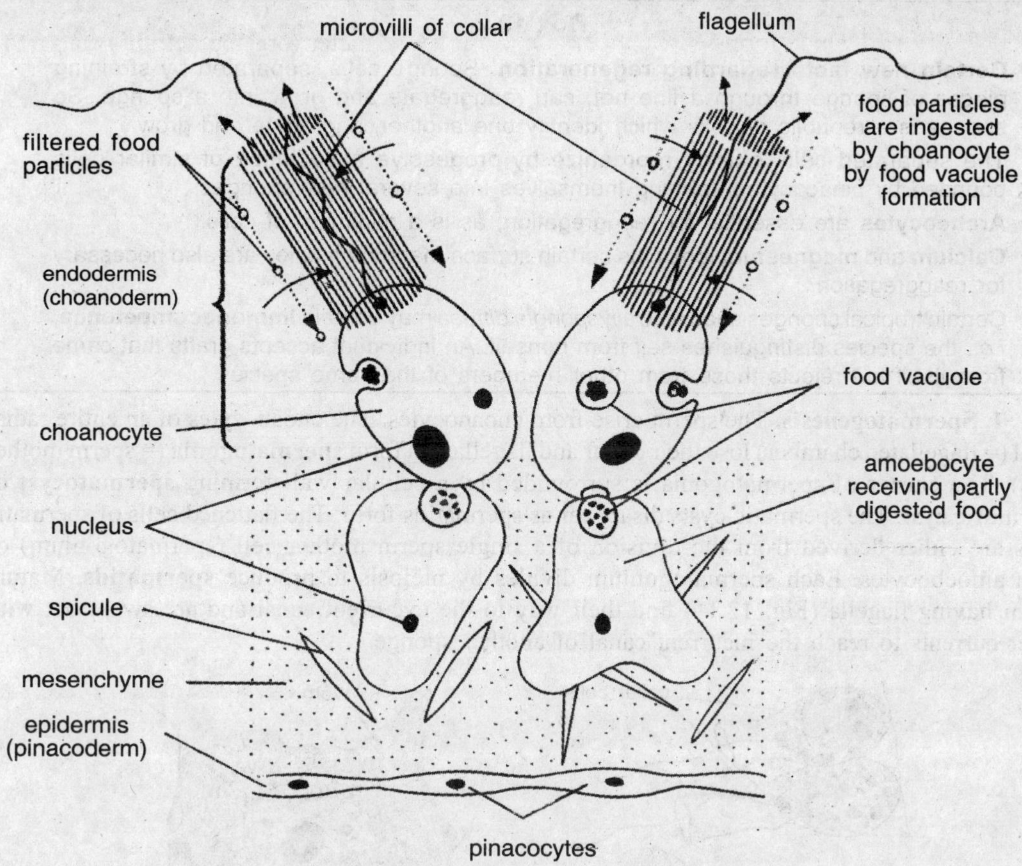

Fig. 12.10. Nutrition by choanocyte in sponge.

(*ii*) Regeneration

Gemmules are not formed in *Scypha*.

(*i*) Budding. During favourable conditions *Scypha* reproduces by **budding**. A small bud appears at the base of the adult *Scypha* and grows into full size. An osculum develops at upper end of this bud. It may remain attached to the parent helping in the formation of colony or it may break free to lead an independent existence of an individual.

(*ii*) Regeneration. All sponges including *Scypha* have great power of regeneration. These sponges not only replace the parts lost during injury, but any piece of the body is capable of growing into a complete sponge. However, the process of regeneration is very slow and is completed in months or years (see Box 12.2).

2. Sexual reproduction. *Scypha* is **hermaphrodite (monoecious)**, *i.e.*, both male and female gametes are formed inside the body of same animal. It lacks definite sex organs or gonads. Both sperms and ova are produced from the **archeocytes** or **choanocytes**. Gamete production appear to be initiated by changes in water temperature, photoperiod or cellular regression depending on the species of sponge.

Box 12.2

Certain new facts regarding regeneration. Sponge cells, separated by straining pieces of sponge through a fine net, can reaggregate and grow into a sponge. So sponge is a republic of cells which identify one another, aggregate and grow.

The separated cells quickly reorganize by progessive association of similar cells bounded by pinacocytes, forming themselves into several new sponges.

Archeocytes are essential for reaggregation, as is a minimum of cells.

Calcium and **magnesium** ions plus certain surface macromolecules are also necessary for reaggregation.

Certain tropical sponges such as *Callyspongia diffusa* may contain **immunocompetence**, *i.e.*, the species distinguishes self from nonself. An individual accepts grafts that came from itself but rejects those from other members of the same species.

1. Spermatogenesis. The sperm arise from choanocytes. The choanocytes of an entire radial canal (= flagellated chamber) lose their collar and flagella and form **spermatogonia** (= sperm mother cells). The cluster of spermatogonia is surrounded by a cellular wall forming **spermatocyst** or **spermatic cyst**. The spermatic cysts disappear as spermatids form. The flattened cells of spermatic cysts are either derived from the division of a single sperm mother cell (spermatogonium) or from amoebocytes. Each spermatogonium divides by meiosis to produce **spermatids.** Mature sperm having flagella (Fig. 12.11) find their way to the excurrent canal and are swept out with water currents to reach the incurrent canal of another sponge.

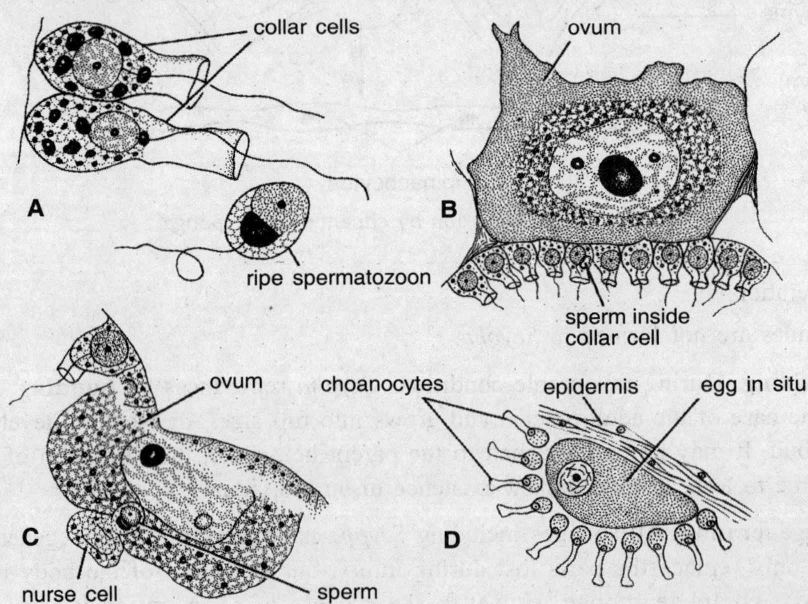

Fig. 12.11. Fertilization in *Grantia*. A—A ripe spermatozoon near two choanocytes; B—Entry of sperm in choanocyte; C—Carrier or nurse cells giving up sperm nucleus into ovum; D—Zygote.

2. Oogenesis. The egg mother cell or **ovocyte** is derived from an enlarged amoebocyte (*i.e.*, archeocyte) having a large nucleus with prominent nucleolus. It grows and acquires food for storage either by engulfing/fusing with similar amoebocytes or by engulfing nearby **nurse cells** or **trophocytes**. Meiosis occurs at the end of the growth period to produce an **ovum**. Ovum lies in the wall of radial canal (in mesohyl) ready to be fertilized by a sperm from another sponge.

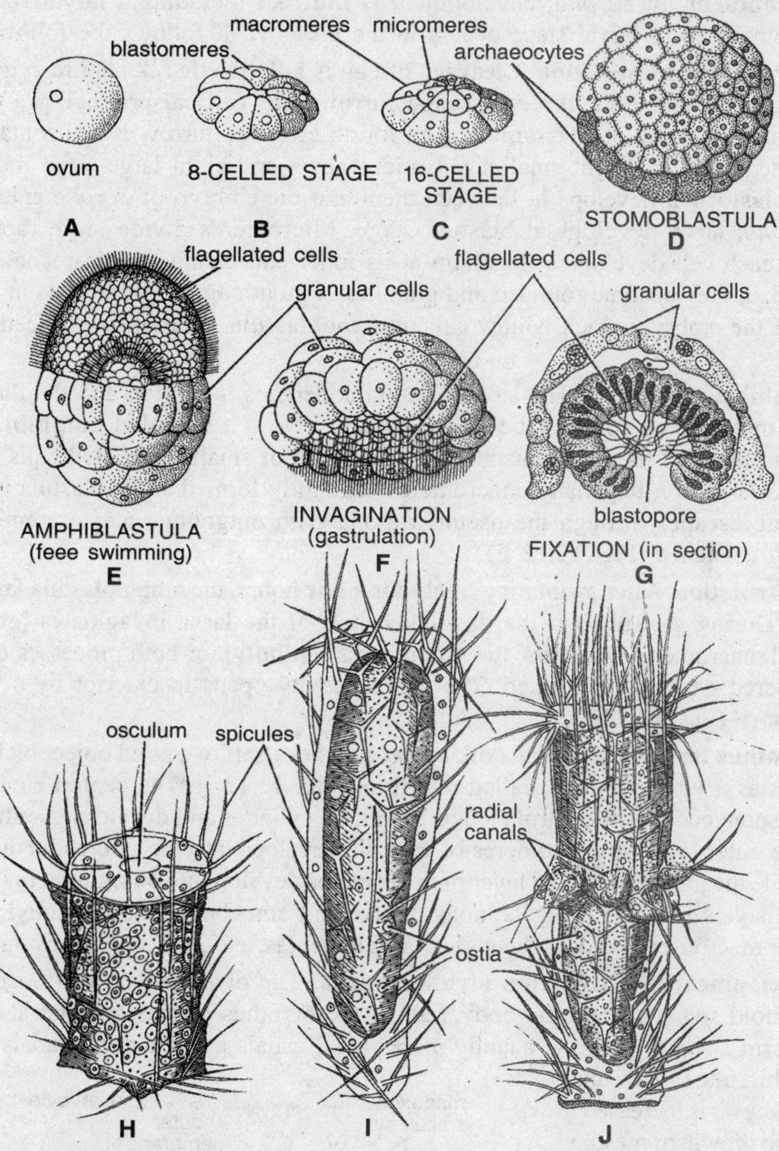

Fig. 12.12. *Scypha*. Development. A—Zygote; B—8-celled stage; C—16-cell stage; D—Stomoblastula; E—Amphiblastula; F—Gastrulation; G—Fixation of gastrula to substratum; H—Young sponge; I—Olynthus stage; J—Syconoid stage.

3. Fertilization. In *Scypha*, like the most sponges, the fertilization is **internal** and is of **cross fertilization** type. After a sperm has reached a radial canal (of another sponge), it is engulfed by a choanocyte occurring adjacent to the ripe ovum. This choanocyte has to transport the sperm to egg, it is called **nurse cell** or **carrier**. Sperm loses its flagellum. Carrier choanocyte too loses its collar and flagellum, becomes amoeboid and moves towards the ovum. After the carrier with its sperm has reached an egg, it either transfers the sperm nucleus to ovum or the carrier and sperm nuclei are engulfed by the egg. The nuclei of ovum and sperm fuse to form **zygote**.

4. Development. In *Scypha*, development is **indirect** including a larval stage in its life cycle. Development up to larval stage occurs in the mesohyl. It follows the following steps:

***(i)* Cleavage and blastulation.** Cleavage of egg is **holoblastic** (*i.e.*, entire zygote is divided by the cleavage furrow). First three cleavage furrows are vertical producing a disc of eight equal sized pyramidal cells or **blastomers**. The fourth cleavage furrow is horizontal and divides blastomeres unequally into eight small-sized **micromeres** and eight large-sized **macromeres**. A cavity called **blastocoel**, develops in between them and the embryo of *Scypha* enters a **blastula stage**. In *Scypha*, larva develops at blastula stage. Micromeres divide more rapidly, become elongated and each cell develops a flagellum at its inner end facing the blastocoel. Macromeres remain undivided, they become rounded and granular. A mouth opening develops in the middle of macromeres of the embryo which is now called **stomoblastula**. Stomoblastula feeds on adjacent maternal cells.

***(ii)* Amphiblastula larva.** The stomoblastula undergoes **inversion** and the flagellated ends of flagellated micromeres come to lie to the exterior. Now it is called **amphiblastula larva**. This larva is hollow and its one hemisphere is composed of small flagellated cells and the other hemisphere has large non-flagellated macromeres. The fully formed amphiblastula is set free into the radial canal, escapes through the osculum along with outgoing water currents and leads a free swimming existence (Fig. 12.12 E).

***(iii)* Gastrulation.** After swimming freely for a few hours, the amphiblastula larva undergoes **gastrulation**. During gastrulation, the flagellated half of the larva invaginates (**emboly**),or the non-flagellated macromeres overgrow the micromeres (**epiboly**), or both processes occur together and a two layered **gastrula** is formed. The central cavity opens to exterior by a wide opening called **blastopore** (Fig. 12.12 G).

***(iv)* Olynthus larva.** The gastrula of *Scypha* attaches itself to a solid object by its blastoporal end and grows as a young sponge, called **olynthus** (Fig. 12.12 I). The central cavity of gastrula develops into spongocoel. The gastrula lengthens into a cylinder and develops **osculum** at the free distal end. The outer layer of macromeres of gastrula develops into the **pinacoderm** (pinacocytes, porocytes) while the inner flagellated layer of micromeres develop into **choanoderm** (choanocytes). Both of these layers produce the gelatinous matrix and amoebocytes of mesohyl. **Scleroblasts** are formed by macromeres and **archeocytes, amoebocytes,** etc., are formed by micromeres.

***(v)* Development of adult from olynthus larva.** The olynthus stage of *Scypha* resembles a simple asconoid sponge. Soon the body wall of the olynthus pushes out first at the middle of the body to form a radial canal. Gradually more radial canals and incurrent canals develop. The choanocytes shift into radial canal. Body wall of *Scypha* is increased in thickness due to growth of mesohyl and formation of spicules. Thus, an adult *Scypha* is formed.

12.7 GEMMULES

The gemmules are asexual reproductive bodies of sponges. They are regarded as the **internal buds**. Gemmules are resistant packets of essential cells and are formed by all freshwater sponges (*Spongilla*, *Ephydatia*, etc.) and a few marine sponges (*Ficulina*, *Tethya*) to tide over winter and drought (*i.e.*, adverse ecological conditions).

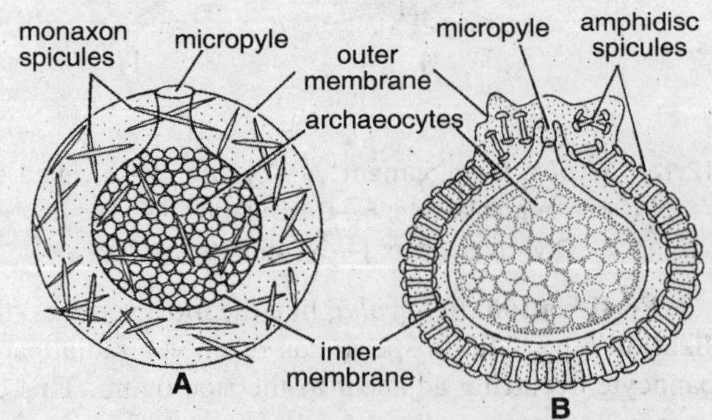

Fig. 12.13. Gemmules of freshwater sponges. A—Gemmule of *Spongilla*; B—Gemmule of *Ephydatia*.

Formation of Gemmules

Freshwater gemmules begin as masses of **archeocytes** in (called **thesocyte**)mesohyl (mesogloea). They are filled with food from special nurse cells, called **trophocytes.** Other amoebocytes called **spongocytes** form a covering around the central mass and differentiate into a layer of columnar cells that secrete a hard two-layered **shell** containing spongin-like material. **Scleroblasts** from the parent mesohyl form special gemmule spicule, depositing them into double-layered shell. The thick resistant double-layered shell of gemmule is broken at the **micropyle,** a pore through which the cells emerge during germination of gemmule for the formation of a new sponge.

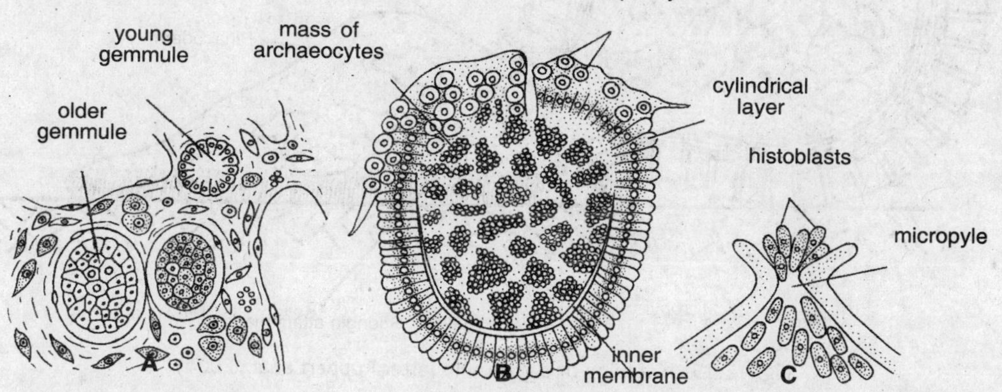

Fig. 12.14. Gemmules of marine sponges. A—Early stages in the formation of gemmule in marine sponge; B—A gemmule of marine sponge; C—Germination of gemmule.

Structure of gemmule of freshwater sponge. A gemmule is a rounded or flask-shaped structure (Fig. 12.13). It is a rounded mass of food-laden archeocytes, enclosed in a double membrane capsule or shell. The inner membrane is thin and outer thick. Between the two membranes are radially arranged **spicules**. These are of monaxon type in freshwater sponges (*Spongilla*) but of amphidisc type in the marine sponges. The amphidisc spicule (of gemmule) have spiny arms and have hooks on their both ends.

Formation of marine gemmules. Marine gemmules also begin as a mass of amoebocytes (archeocytes), surrounded by other cells. The outer layer of cells develops into a flagellated surface epithelium. At maturity, gemmules have a flagellated and a non-flagellated pole. After a free-swimming period, they settle and attach at the non-flagellated end to develop into young sponges (see **Meglitsch** and **Schram**, 1991).

Germination of Gemmules

Gemmules are formed in unfavourable climatic conditions especially in winter. After the formation of gemmules, the sponge body disintegrates and its remains along with gemmules are deposited at the bottom. During the spring gemmule "hatch", the peripheral thesocytes differentiate into a pinacoderm that balloons out, like a bubblegum bubble, through the micropyle (Fig. 12.15). This pinacoderm bubble makes contact with and attaches to the substratum. Next, the deeper thesocytes issue from the micropyle into the bubble and establish, after differentiation, the interior of juvenile sponge.

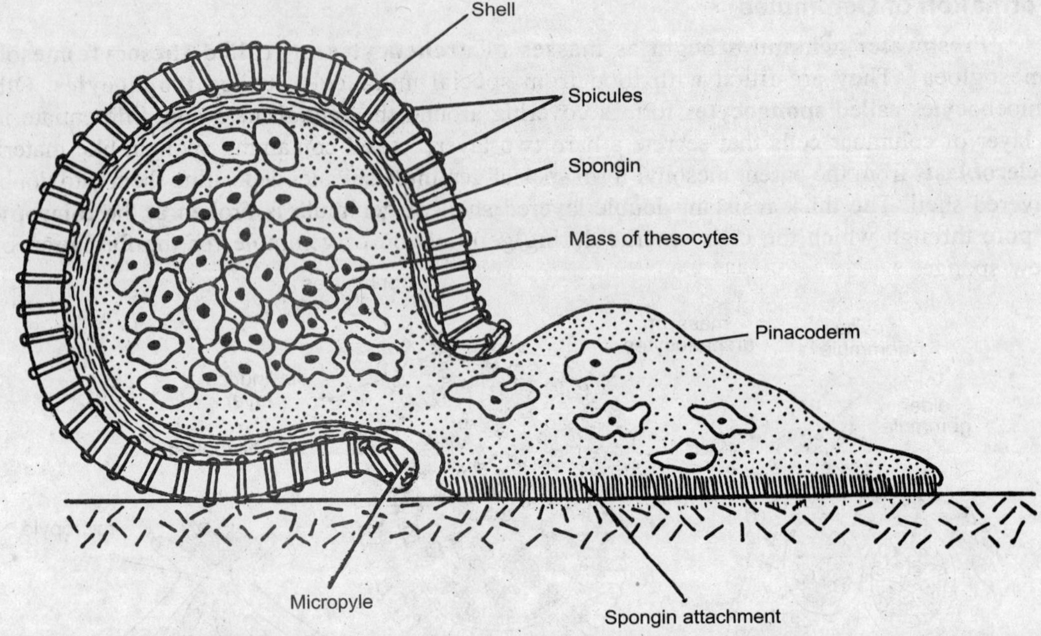

Fig. 12.15. Hatching of a gemmule (after Ruppert *et al.*, 2004).

Significance. Formation of gemmules is a life-saving mechanism. These are highly resistant and are not harmed by freezing temperature and drying while the parent sponge dies. They help in multiplication and dispersion of races. Thus, gemmules form a significant means for the continuation and multiplication of race.

Box 12.3

Meaning of some technical terms of Porifera (Source **Ruppert** *et al.*, **2004**)

Apopyle = back gate; Archeocyte = progenitor cell; Choanocytes = collar cells; Lophocytes = crest cells; Mesohyl = middle wood; Myocytes = muscle cells; Pinacocytes = platter cells; Pinacoderm = platter skin; Prosopyles = front gates; Sclerocytes = hard cells.

QUESTIONS

Long Answer Questions

1. Describe the main histological elements found in *Sycon* and indicate their functions.
2. Draw neat and labelled diagram of L.S. of *Sycon*. Describe the structure and functions of different types of cells in it.
3. Describe different kinds of cells found in the body wall of *Sycon*/Porifera and mention their functions.
4. With the help of suitable diagrams describe the canal system in *Sycon*.
5. Which system distinguishes sponges from coelenterates? Describe the canal system with appropriate diagrams.
6. Describe the mode of feeding and digestion in *Sycon*.
7. How do sponges take nourishment? Give a detailed account of feeding in *Sycon*.
8. Give an account of reproduction and development in *Scypha* (*Sycon*).
9. Describe how reproduction takes place in *Sycon*.

10. Give an account of reproduction in the sponge type that has been studied by you.

11. Give an illustrated account of sexual reproduction in *Sycon.*

12. How does sexual and asexual reproduction takes place in *Sycon*?

13. Describe the sexual reproduction and development of *Sycon.* (*C.C.S.U. Meerut 2008*)

Short Answer Questions

1. Describe habits, habitat and external structure of *Scypha.*

2. Give the natural history and morphology of *Sycon.*

3. Give the systematic position of the *Sycon.*

4. Describe the structure of *Sycon.*

5. Describe the geographical distribution and ecology of *Sycon.*

6. Write short note on the importance of canal system of *Sycon.*

7. Give an account of canal system of *Sycon* and explain the mechanism and importance of circulation of water through it.

8. Explain the physiological importance of canal system in sponges. Describe the canal system of *Sycon.*

9. Write short note on Gemmule (*Purvanchal 2001; C.C.S.U. Meerut 2005*)

10. Draw a neat and labelled diagram of T.S. of *Sycon* (No description is needed). (*C.C.S.U. Meerut 2007*)

Very Short Answer Questions

1. What is the common name of *Scypha*?

2. Name the order of *Scypha.*

3. What is the class of *Scypha*?

4. Which cells maintain a current of water in *Scypha*?

5. Where are amoebocytes located in *Scypha*?

6. Name the cell that carries a sperm to an ovum during fertilization of *Scypha.*

7. Which embryonic stage undergoes inversion in *Scypha*?

8. Name the phenomenon that prevents self fertilization in *Scypha.*

9. Which is the main excretory product of *Scypha*?

10. Name the spicules which form the oscular fringe.

11. How many scleroblasts produce a triradiate spicule?

12. Where are oxeotes located in *Scypha*?

13. What is the name of free-swimming larval form of *Scypha*?

14. What is the function of choanocytes? (*C.C.S.U., Meerut 2004*)

Fill in the Blanks

1. Water current enters the *Scypha* (*Sycon*) through .... and leaves via ....

2. In *Scypha*, choanocytes line the .... canals.

3. In *Scypha*, the pinacocytes line the .... , excurrent canals and ....

4. Surface of the *Scypha* is covered by cells called ....

5. In *Scypha*, the incurrent canals communicate with the radial canals by ....

6. In *Scypha*, gastric ostia / atria lead from .... into the ....

7. In *Scypha*, the apertures between radial canal and excurrent canals are called ....

8. Spicule-secreting cells of *Scypha* are called ....

9. The food storing cells of *Scypha* are known as ....

10. The spicules of *Scypha* include monaxon, ... and tetraradiate.

Multiple Choice Questions

1. The common name of *Scypha* or *Sycon* is
(*a*) bath sponge (*b*) crown sponge
(*c*) venus flower basket (*d*) glass rope sponge

2. Canal system of *Scypha* is
(*a*) ascon type (*b*) syconoid type
(*c*) leucon type (*d*) rhagon type

3. The mesohyl is largely secreted by
(*a*) collencytes (*b*) trophocytes
(*c*) choanocytes (*d*) chromocytes

4. In *Scypha*, food is firstly ingested by choanocytes and digestion is completed in
(*a*) amoebocytes (*b*) chromocytes
(*c*) thesocytes (*d*) archaeocytes

5. The exit pore of *Scypha* is
(*a*) ostia (*b*) osculum
(*c*) spongocoel (*d*) prosopyle

6. In *Scypha*, water currents are produced by
(*a*) choanocytes (*b*) pinacocytes
(*c*) archaeocytes (*d*) thesocytes

7. What is the name of hollow blastula of *Scypha*?
(*a*) streogastrula (*b*) coeloblastula
(*c*) stomoblastula (*d*) amphiblastula

8. Name of the larva of *Scypha* is
(*a*) amphiblastula (*b*) parenchymula
(c) stomoblastula (*d*) ephyra

9. Young stage of syconoid sponge is
(*a*) coeloblastula (*b*) parenchymula
(*c*) olynthus (*d*) stomoblastula

ANSWERS

Very Short Answer Questions

1. Crown sponge.

2. Heterocoela.

3. Calcarea.

4. Choanocytes.

5. Mesohyl (Mesenchyme).

6. Carrier choanocyte.

7. Stomoblastula.

8. Protogyny.

9. Ammonia.

10. Long monaxons.

11. Three.

12. Oxeotes are monaxon spicules which project out from dermal cortex of mesohyl.

13. Amphiblastula larva.

14. (*i*) Produce water current by beating of flagella and direct it into the canal system.

(*ii*) Draw food from water current toward the collar where it is captured by phagocytosis and transferred to amoebocytes for digestion.

(*iii*) Take part in reproduction by modifying into spermatogonia.

(*iv*) Help in internal fertilization by transferring sperm to oocyte.

Fill in the Blanks

1. Dermal ostia, osculum
2. Radial
3. Incurrent canal, spongocoel
4. Pinacocytes
5. Prosopyles
6. Excurrent canals, spongocoel
7. Apopyles
8. Calcoblasts
9. Thesocytes
10. Triradiate.

Multiple Choice Questions

1. (*b*) **2.** (*b*) **3.** (*c*) **4.** (*a*)
5. (*b*) **6.** (*a*) **7.** (*c*) **8.** (*a*)
9. (*c*).

13

Canal System in Sponges

The passages through which water constantly flows from outside the body to the interior of the body and then outside again, form the **canal system** or **aquiferous system**; see **Ruppert** *et al.*, 2004) in sponges. Thus, canal system or aquiferous system is the circulatory system of choanoderm, pores and chambers. In majority of sponges the body wall is folded and is traversed by a complex system of canals (*e.g.*, incurrent canals, radial canals and excurrent canals) through which water current travels to reach the spongocoel and ultimately exit through osculum. Sponges contain the following three types of canal system:

1. Asconoid or ascon type canal system
2. Syconoid or sycon type canal system
3. Leuconoid or leucon type canal system.

13.1 ASCONOID TYPE CANAL SYSTEM

It is simplest type of canal system which is found in *Leucosolenia* and olynthus larval stage of *Scypha*. In this type of canal system, there is a radially symmetrical vase-like body consisting of a thin body wall enclosing a large central cavity, the **spongocoel** (Fig. 13.1). The spongocoel opens at the free end by a narrow circular aperture, the **osculum**. An epidermis of close-fitting pinacocytes (pinacoderm) covers the surface of the ascon wall. Scattered among the pinacocytes are tubular **porocytes** with the **ostia**. Collared choanocytes (forming choanoderm) line the spongocoel. Their flagella force water toward the osculum and draw it into spongocoel through the ostia. A jelly-like **mesogloea** or **mesohyl** fills the space between the pinacoderm and choanoderm and contains wandering amoebocytes, calcareous monaxon and triaxon spicules, etc.

Thus, in asconoid type canal system, the water current enters through ostia, reaches the spongocoel and finally leaves the body through osculum.

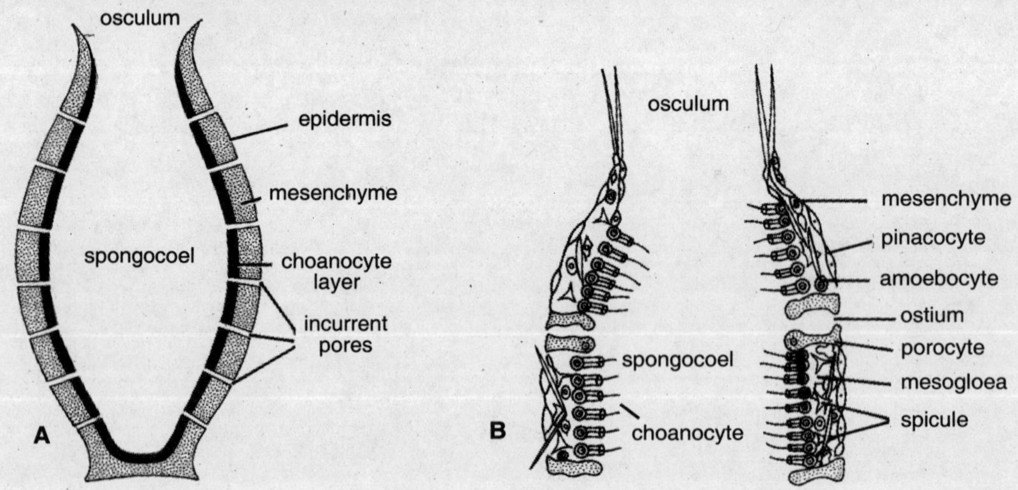

Fig. 13.1. Asconoid canal system A—Diagrammatic view; B—Schematic cellular structure of upper part of asconoid canal system.

13.2 SYCONOID TYPE CANAL SYSTEM

It is a step forward in the evolution of canal system. Syconoid canal system is formed when asconoid body wall gives out finger-like processes at regular intervals. These finger-like outgrowths are lined by choanocytes and they are called **flagellated** or **choanocyte chambers** or **radial canals**. They open to spongocoel via apopyle and excurrent canals. As choanocytes come to line the radial canals, the spongocoel is lined with flattened pinacocytes (called **endopinacocytes**). The wide opening between two adjacent radial canals forms the **ostium** and space between them is called **incurrent canal**. Incurrent canals communicate with radial canal through many **prosopyles** (Fig. 13.2).

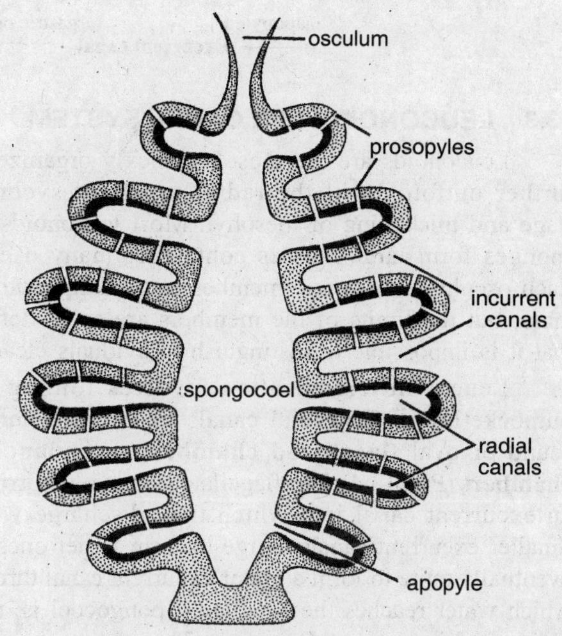

Fig. 13.2. Syconoid I type canal system.

Syconoid type canal system is of two types: **syconoid I** and **syconoid II**. Syconoid I type canal system is simple and found in *Sycetta* (*i.e.*, heterocoelous sponges). In *Sycetta*, each radial canal is a simple out-pocketing of the spongocoel and is exposed directly to the surrounding water (Fig. 13.2).

In more complex forms (**syconoid II type canal system**), the outer margins of all radial canals are attached, *i.e.*, the pinacoderm and mesohyl spread over the outer surface forming **cortex** (Fig. 13.3). Cortex may be thin in simple forms (*e.g.*, *Scypha*) and becomes thick in still more complex forms (*e.g.*, *Grantia*). Due to the development of cortex, the wide spaces in between adjacent radial canals acquire tubular appearance and they form **incurrent canals**. These canals are lined by pinacocytes. The epidermis (= pinacoderm) at free distal ends of incurrent canals become pierced by minute pores which are known as **incurrent pores** or **ostia**. Through these ostia, the water current enters the incurrent canals. The passage of water current in syconoid II type canal system occurs by the following route:

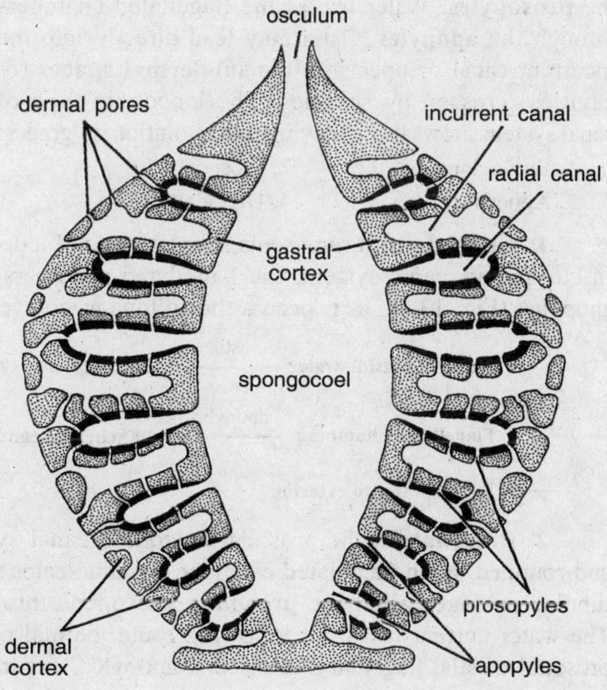

Fig. 13.3. Syconoid II type canal system having cortex.

Entry of outside water —ostia→ **Incurrent canal** —prosopyles→ **Radial canal** →

—apopyle→ **Excurrent canal** —gastric ostia→ **Spongocoel** —osculum→ **Exit of water to outside**

13.3 LEUCONOID TYPE CANAL SYSTEM

Leuconoids are the most complexly organized sponges. Leuconoid structure develops by further outfolding of the radial canals of syconoid stage and thickening of mesohyl. Most leuconoid-type sponges form large masses containing many **oscula**. Each osculum marks one member of the large colonial mass, but the limits of the members are so ill-defined that it is impossible to distinguish individuals clearly.

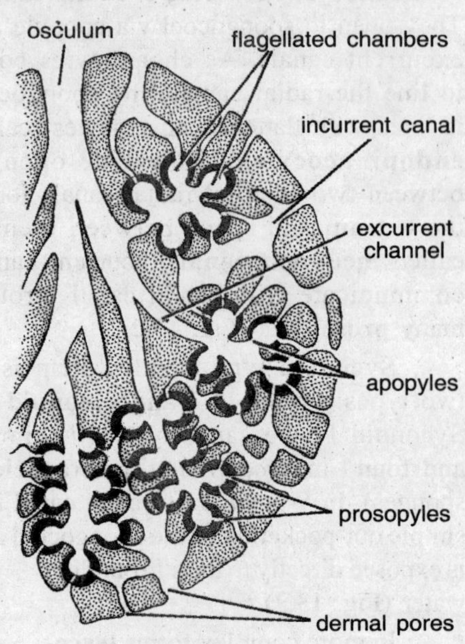

Fig. 13.4. Leuconoid type canal system showing eurypylous chambers.

Leuconoid organization involves folding and outpocketing of the radial canals to form clusters of round or oval **flagellated chambers** or **choanocyte chambers** (Fig. 13.4). The flagellated chambers surround an excurrent canal into which they discharge water. Smaller excurrent canals merge to form larger ones; all eventually unite to form a major excurrent canal through which water reaches the osculum. Spongocoel is, thus, obliterated in leuconoid sponges. The cortex of sponge contains a system of branching incurrent canals, delivering water in the flagellated chambers by way of small pores, the **prosopyles**. Water leaves the flagellated chambers through the **apopyles**. Ostia may lead directly into the incurrent canal or open into the **subdermal spaces** (or cavities) crossed by spicules. The leuconoid type of canal system shows the following four evolutionary grades:

1. Eurypylous
2. Prosodal
3. Aphodal
4. Diplodal

1. Eurypylous. It represents simplest type of leuconoid canal system and occurs in *Leucilla*. In eurypylous canal system, the flagellated chambers lead into excurrent canals through wide apopyles (Fig. 13.4). In it occurs the following course of water currents:

Entry of outside water —ostia— **Subdermal cavity and incurrent canals** —prosopyles→

Flagellated chambers —apopyles→ **Excurrent canals** → **Large channels** —osculum→

→ **Exit of water to exterior.**

2. Prosodal. In the prosodal leuconoid canal system, the flagellated chambers are small and rounded. Each flagellated chamber communicates with an incurrent canal by a long, narrow, tubular passage called the **prosodus** and opens into the excurrent canal by a wide apopyle. The water current takes the following route: dermal ostia, hypodermal spaces, incurrent canals, prosodal canals, flagellated chambers, apopyles, excurrent canals, excurrent spaces and oscula.

3. Aphodal. In certain leuconoid sponges such as *Geodia* and *Stelleta*, the apopyles instead of being wide openings are drawn into narrow canals. These canals are called **aphodus** and open into excurrent canals (Fig. 13.5).

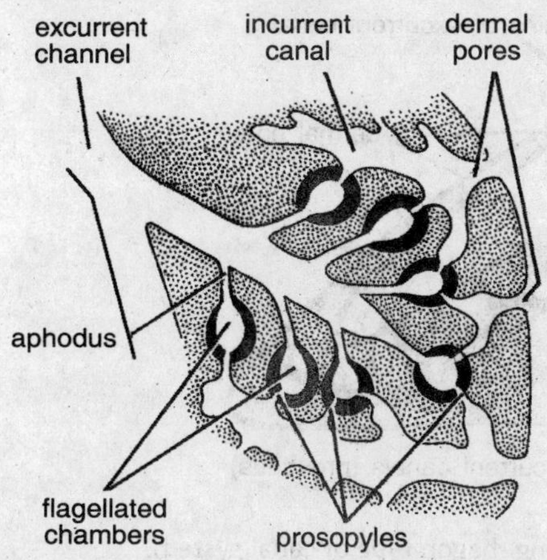

Fig. 13.5. Leuconoid type canal system with aphodal chambers.

Fig. 13.6. Leuconoid type canal system with diplodal chambers.

4. Diplodal. This type of leuconoid canal system is found in *Spongilla* and *Oscarella*. In this case, both apopyles and prosopyles are drawn out into narrow canals called **aphodus** and **prosodus**, respectively (Fig. 13.6).

Rhagon Type Canal System

In Demospongiae (*e.g.*, *Spongilla*), the existing leuconoid type canal system is not derived by way of asconoid or syconoid stages as are evident in calcareous sponges. Instead it is derived from a larval stage, called **rhagon** whose canal system is **rhagon type** (Fig. 13.7). The larval canal system includes flattened pyramid-like form with broad base, called **hypophare**, a spacious **spongocoel** and lateral walls, called **spongophares**. Flagellated chambers occurs in each spongophare. The dermal pores leading to flagellated chambers are known as **prosopyles** and their openings into spongocoel are termed as **apopyles**. The spongocoel opens to the exterior by **osculum**.

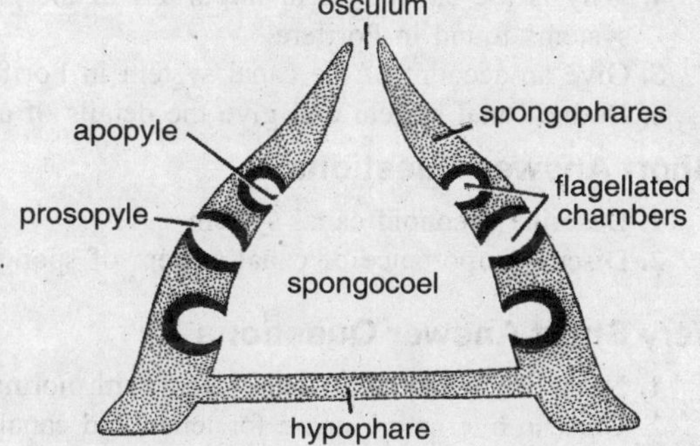

Fig. 13.7. *Rhagon* larva. V.S. of body showing rhagon type canal system.

Thus, in *Spongilla* (Fig. 13.8), between the flagellated chambers and the pinacoderm (or epidermis) lies a considerable thickness of mesohyl which is traversed by subdermal spaces and incurrent canals. In this sponge, the course of current of water is as follows:

Entry of outside water —ostia→ **Subdermal spaces** → **Incurrent channels** —prosopyles→ **Flagellated chambers** —apopyles→ **Excurrent canals** → **Spongocoel** —osculum→ **Exit of water to exterior.**

Lastly, the volume of water pumped by a sponge is remarkable. For example, a specimen of *Leuconia*, a leuconoid sponge 10 cm in height and 1 cm in diameter has roughly 2,250,000 radial canals and pumps 22.5 litre of water per day through its body. The flow velocity is greatest through the osculum and slowest through the radial canals, because these two regions have, respectively, the smallest and the largest total cross sectional areas of the various passageways.

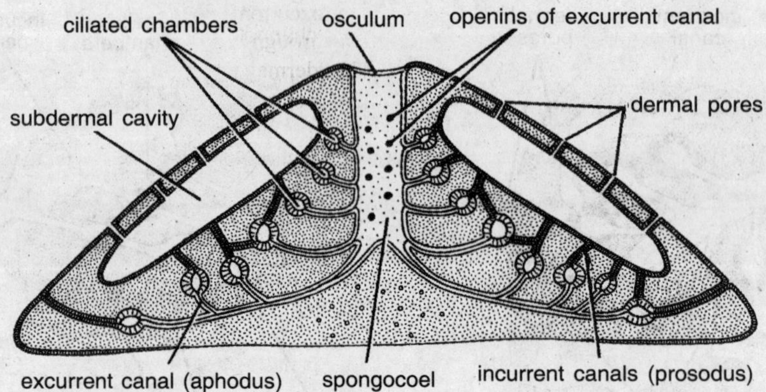

Fig. 13.8. *Spongilla*. Diagrammatic V.S. of body showing rhagon type of canal system.

QUESTIONS

Long Answer Questions

1. Describe the various types of canal systems found in sponges.
2. Give an account of rhagon canal system.
3. Define canal system. Give an account of canal system of Porifera.
4. Why is the canal system important in the phylum Porifera? Describe the different types of canal systems found in Porifera.
5. Give an account of the canal system in Porifera and discuss its importance.
6. Define canal system and give the details of a leuconoid type of canal system.

Short Answer Questions

1. Describe syconoid canal system.
2. Discuss importance of canal system of sponges.

Very Short Answer Questions

1. Name the leuconoid canal system with normal prosopyles and apopyles.
2. Give an alternative name for leuconoid canal system in Demospongiae.

Fill in the Blanks

1. Sponges acquire food by filtering water. The food-containing water current enters a sponge through .... and is then moved through the internal cavity to the .... by the flagella of .... As the water moves, food is removed by the ....
2. In sponges a constant current of water passes through ostia into spongocoel and comes out through oscula. This is known as .... (water vascular system/canal system)
3. The simplest type of canal system in sponges is ....
4. Canal system found in Demospongia is ....
5. In the leucon type of canal system when the flagellated chambers open directly by wide apopyles into the excurrent canal, it is called ....
6. The cells of a sponge that regulate the entry of water by closing pores in response to irritating stimuli are the .... cells.
7. In the leucon type of canal system when another narrow tube, called prosodus is present between the incurrent canal and the flagellated chamber, it is called .... (rhagon/diplodal/apodal)
8. Rhagon type of canal system is similar to .... (leucon type/ascon type)
9. There are four grades of leuconoid canal system: eurypylous, ... , ... , and diplodal.

Multiple Choice Questions

1. Incurrent canal communicates with the radial canal through
   (*a*) apopyle (*b*) dermal pores
   (*c*) prosopyle (*d*) none of the above
2. In *Sycon*, the choanocytes line the
   (*a*) excurrent canal (*b*) incurrent canal
   (*c*) radial canal (*d*) spongocoel or paragastric cavity
3. Which type of canal system is found in *Leucosolenia*?
   (*a*) asconoid (*b*) syconoid
   (*c*) leuconoid (*d*) rhagon
4. In sponges, water current passes through ostia into spongocoel and comes out through oscula. This is called
   (*a*) water vascular system (*b*) canal/aquiferous system
   (*c*) feeding current (*d*) circulating current
5. The canal system in the sponge *Olynthus* is of
   (*a*) asconoid type (*b*) syconoid type
   (*c*) rhagon type (*d*) leuconoid type
6. Diplodal canal system is identified by the presence of
   (*a*) aphodus only (*b*) prosodus only
   (*c*) both aphodus and prosodus (*d*) none of the above
7. Which of the following is the exhalant pore?
   (*a*) ostium (*b*) prosopyle
   (*c*) apopyle (*d*) osculum
8. Flagellated cells of sponges are
   (*a*) pinacocytes (*b*) choanocytes
   (*c*) monocytes (*d*) thesocytes
9. Which one of the following cells maintains water currents in sponges?
   (*a*) porocyte (*b*) choanocyte
   (*c*) myocyte (*d*) chromocyte
10. The system through which water travels through body of a sponge is called
    (*a*) blood vascular system (*b*) canal or aquiferous system
    (*c*) excretory system (*d*) water vascular system
11. Spongocoel is lined by choanocytes in the canal system of
    (*a*) asconoid type (*b*) syconoid type
    (*c*) rhagon type (*d*) syconoid and leuconoid type
12. The canal system of *Euplectella* is
    (*a*) asconoid (*b*) syconoid
    (*c*) eurypylous (*d*) leuconoid

ANSWERS

Very Short Answer Questions

1. Eurypylous.
2. Rhagon.

Fill in the Blanks

1. Ostia, osculum, choanocytes, choanocytes
2. Canal system
3. Ascon type
4. Leucon type
5. Euryplous
6. Epidermal
7. Diplodal
8. Leucon type
9. Prosodal, aphodal

Multiple Choice Questions

1. (*c*)	**2.** (*c*)	**3.** (*a*)	**4.** (*b*)	**5.** (*a*)	**6.** (*c*)
7. (*d*)	**8.** (*b*)	**9.** (*b*)	**10.** (*b*)	**11.** (*a*)	**12.** (*c*).

14

Skeleton of Sponges

The skeleton of sponges is relatively complex and provides a supporting framework for the living cells of the animal. It is formed of spicules or organic fibre and is primarily located in the mesohyl, but spicules frequently project through the pinacoderm. The skeleton may be composed of **calcareous spicules** (of calcium carbonate), **siliceous spicules** (of silicon dioxide) or fibres of **spongin B** (a scleroprotein), or a combination of the latter two. Spicules have an axis of organic material around which is deposited the inorganic substance. The fully formed calcareous spicule is surrounded by a sheath but lacks any axial filament of an organic nature. Siliceous spicule, on the other hand, has a core in the form of an axial filament. (Box 14.1).

Box 14.1

Skeletal elements of sponges are of two types, **organic** or **inorganic**. The organic element is always **collagenous** and inorganic element is either **siliceous** (hydrated silicon dioxide) or **calcareous** (calcium carbonate in the form of calcite or aragonite). Sponges are the only animals that use hydrated silica as a skeletal material (**Bursca** and **Bursca**, 2003).

14.1 TYPES OF SPICULES

The skeletal elements of sponges exist in a variety of form and are important in the identification and classification of species. Spicules are of two general kinds—megascleres and microscleres. The microscleres are small to minute reinforcing (or packing) spicules; the megascleres are large structural spicules. The demosponges and hexactinellids have both types but calcareous sponges often have only megascleres. The suffix—axon refers to the number of axes a spicule has, while—actine refers to the number of rays or points.

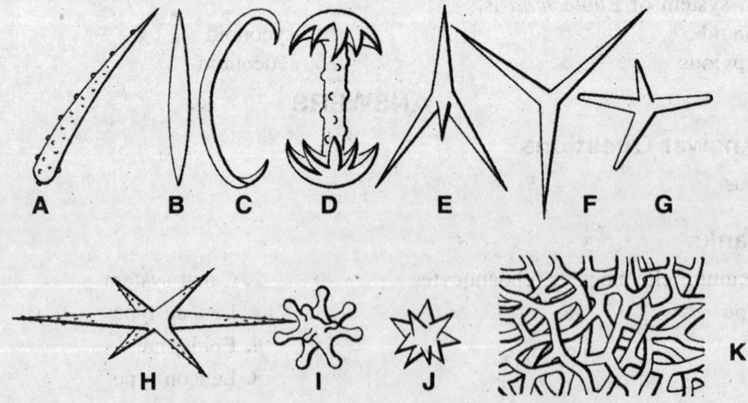

Fig. 14.1. Skeleton of sponges. A—Monactinal monaxon (style); B—Diactinal monaxon (oxea); C—Curved monaxon (sigma); D—Monaxon with hooked ends; E—Tetraxon; F—Triradiate; G—Calthrops; H—Hexactinal triaxon; I and J—Polyaxon; K—Spongin fibres.

A. Megascleres. The megascleres are the larger 'skeletal spicules' that constitute the chief supporting framework of the sponge. These are of following six types:

(1) Monaxons. These spicules have single axis and are shaped like needles or rods or may be curved or straight. They may contain pointed, knobbed or hooked ends. When growth of spicule has occurred in one direction only, the spicule is called **monactinal monaxon** (Fig. 14.2) or **style**. Styles are typically rounded (called **strongylote**) at one end and pointed (called **oxeote**) at the other end. Styles in which broad end is knobbed are called **tylostyles;** those curved with thorny processes are called **acanthostyles.**

Monaxons that develop by growth in both directions from a central point are named **diactinal monaxons, diactines** or **rhabds**. Rhabds may be of following four types:

(*i*) **Oxeas** which are pointed at both ends. (*ii*) **Tornotes** which are lance-headed on both ends. (*iii*) **Strongyles** which are rounded at both ends. (*iv*) **Tylotes** which are knobed at each end, like a pin head.

2. Tetraxons. These four-rayed spicules are also called **tetractines** or **quadriradiates** spicules. Four rays of the tetraxon spicule radiate from a common point and may be equal in size; such a spicule is called **calthrops**. In **triaenes** spicule, one ray is elongated, called **rhabdome** which bears a crown of three smaller rays. When triaenes loses one smaller ray, it is called **diaene**. If the elongated ray bears a disc at both ends, it is called **amphidisc**. The **triod, triradiate** or **triactinal** spicule which is common in calcareous sponges, lacks an elongated ray.

3. Triaxons. The triaxon (three-rayed) or **hexactinal** (six-rayed) spicule consists of three axes crossing at right angles, producing six rays extending at right angles from a central point. These spicules too become modified secondarily to produce many varieties. Triaxon spicules are characteristics of class Hexactinellida.

4. Polyaxons. In polyaxon spicules several equal rays radiate from a central point.

5. Spheres. These are rounded bodies in which growth is concentric around a centre.

6. Desma. A desma consists of an ordinary minute monaxon, triradiate or tetraxon spicule called **crepis**, on which layers of silica have been deposited irregularly. Based on different shapes of crepis, desmas may be called **monocrepids, tricrepids** and **tetracrepids**. They become united into a network to form a reticulated skeleton, called **lithistid.**

B. Microscleres. The microscleres are the smaller 'flesh spicules' that occur strewn throughout the mesohyl. They support the pinacoderm lining the water canals. Microscleres are of following types:

1. Spires. Spires are curved in one plane or spirally twisted. They show many shapes:

(*i*) **Sigmas** which are C-shaped spicules. Spirally twisted sigmas are called **sigmaspires**. (*ii*) **Toxas** which are bow-shaped spicules. (*iii*) **Chelas** which contain recurved hooks, plates or flukes at each end. If two ends are alike, chelas are called **isochelas**, when unlike, **anisochelas**.

2. Asters. Aster contains either a small centre and long rays or a large centre and small rays. Among the small-centred asters are **oxyasters** with pointed rays, **strongylaster** with rounded ends and **sterrasters** with rays reduced to small projections from the spherical surface. Short spiny microscleric monaxons are known as **streptasters.**

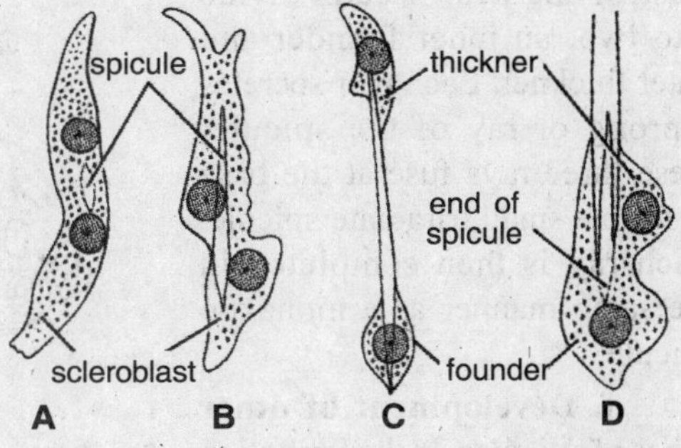

Fig. 14.2. Formation of a monaxon spicule.

14.2 DEVELOPMENT OF SPICULES

Spicules are secreted by **scleroblasts** or **sclerocytes** which are amoebocytes of mesohyl. Sclerocytes are active cells that possess abundant mitochondria, cytoplasmic microfilaments and small vacuoles. The process of development of spicules is best known only for calcareous spicules.

1. Development of monaxon spicules. In calcareous sponges monaxon spicule is secreted within a binucleate **calcoblast** which is formed due to incomplete division of an ordinary calcoblast (*i.e.*, calcareous spicule forming scleroblast). The calcium carbonate is deposited around an organic axial thread contained in the cytoplasm between two nuclei. As the spicule lengthens, the two nuclei draw apart until calcoblast divides into two cells. One cell, the **founder** is located at the inner end, the other, the **thickner** at the outer end of the spicule. The spicule is laid down chiefly by the founder cell which establishes the shape and length of spicule. The thickner cell deposits additional layers of calcium carbonate. When the formation of spicule is completed, both cells (founder and thickner) leave from the inner end of the spicule and return to mesohyl (Fig. 14.2).

The development of siliceous monaxon spicule takes place in one scleroblast called **silicoblast**.

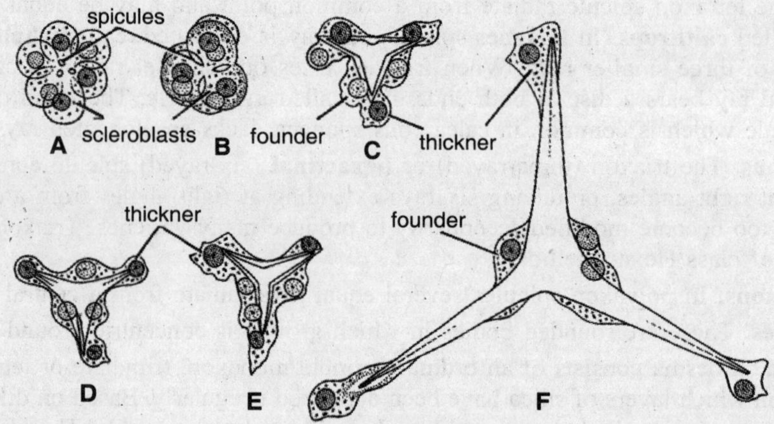

Fig. 14.3. Formation of a triaxon spicule.

2. Development of triaxon spicules. Triaxon calcareous spicules are secreted by three calcoblasts which come together in a triangle (Fig. 14.3). Each calcoblast of the trio of cells divide into two, an inner founder and outer thickner. Each pair secretes a prong or ray of the spicule; these three rays fuse at the base to form a small triradiate spicule. Each ray is then completed in the same manner as a monaxon spicule.

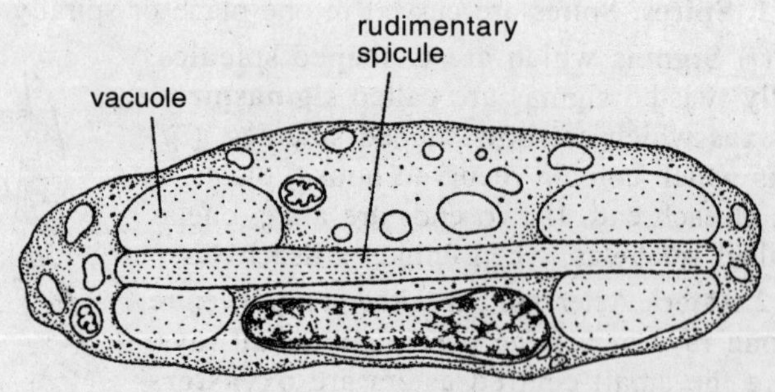

Fig. 14.4. A sclerocyte of *Mycale* (Demospongiae) with rudimentary siliceous spicule extending between two vacuoles.

3. Development of other types of spicules. In the formation of tetraxon or quadriradiate spicules, the fourth prong or ray is added to developing triradiate spicule by an additional scleroblast. The hexactinal (siliceous) spicule arises in the centre of a multinucleate syncytial mass which is probably formed by repeated nuclear divisions of an original silicoblast.

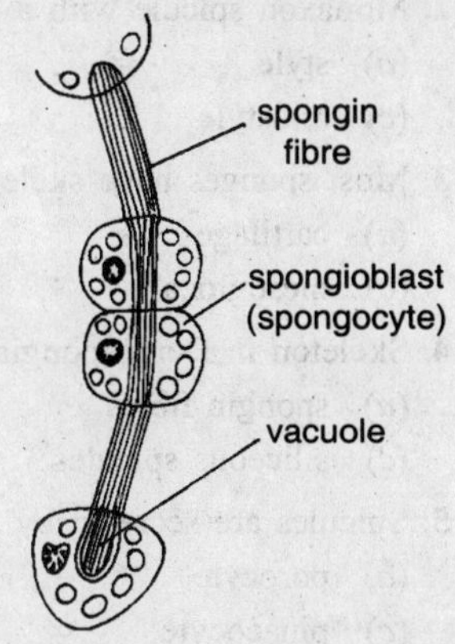

Fig. 14.5. Formation of spongin fibre by a group of spongioblasts.

Spongin: Structure and Secretion

Skeleton of sponges of class Demospongiae is composed of fibres of halogenated scleroprotein termed **spongin B**. Spongin exists in two forms: 1. As a cuticular secretion of a tenacious but elastic cementing substance which glues siliceous spicules together into a definite system of skeletal fibres. 2. In the form of elastic fibrillae secreted in the cells. In most sponges, spongin fibres incorporate some siliceous spicules or detritus such as sand grains, radiolarian and foraminiferan skeletons. Some sponges contain so much spongin that they are tough and rubbery.

The spongin is secreted by **spongioblasts** or **spongocytes**. Each spongioblast develops a vacuole, within which spongin material collects. Spongioblasts line up so that the short sections of spongin are united into a single long fibre. Spongin is made up of **collagen** fibrils.

QUESTIONS

Long Answer Questions

1. Describe in brief the spicules.
2. Give an account of skeleton of sponges.
3. What is the chemical nature of skeletal elements found in sponges? How does the nature and form of the skeleton act as basis for the classification of the phylum Porifera? Discuss.

Short Answer Questions

1. Write a short note on spicule.

Very Short Answer Questions

1. Name any two of the six major types of spicules regarding the shape.
2. What type of protein in found in spongin.

Fill in the Blanks

1. .... in sponges form the basis of classification.
2. The soft body of sponges is supported by .... (setae/spicules)
3. A skeleton of *Euspongia* is formed of .... (spongin fibre/calcareous spicules)
4. The spicules of a sponge body are secreted by .... (scleroblasts/archeocytes)
5. Spicules are of two types regarding size ....
6. Spongin fibres occur in the sponges of class ....
7. Siliceous spicules are secreted by .... (calicoblast/silicoblasts)

Multiple Choice Questions

1. Which type of spicules are found in gemmule of *Spongilla*?

(*a*) amphidisc (*b*) monaxon
(*c*) triaxon (*d*) tetraxon

2. Monaxon spicule with a knob at one end is called
   (*a*) style (*b*) diactyle
   (*c*) telostyle (*d*) acanthostyle
3. Most sponges have skeleton made of
   (*a*) cartilage (*b*) spicules or spongin fibres
   (*c*) mesogloea (*d*) pinacocytes and choanocytes
4. Skeleton in Demospongia is
   (*a*) spongin fibres (*b*) calcareous spicules
   (*c*) siliceous spicules (*d*) both spongin fibres and siliceous spicules
5. Spicules are secreted by
   (*a*) porocyte (*b*) choanocyte
   (*c*) pinacocyte (*d*) scleroblasts
6. Hexactine spicules occur only in
   (*a*) calcarea (*b*) hexactinellida
   (*c*) demospongiae (*d*) calcarea and demospongiae

ANSWERS

Very Short Answer Questions

1. Monaxons, triaxons.
2. Collagen.

Fill in the Blanks

1. Spicules
2. Spicules
3. Spongin fibre
4. Scleroblasts
5. Microscleres, macroscleres
6. Demospongiae
7. Silicoblasts

Multiple Choice Questions

1. (*a*) **2.** (*a*) **3.** (*b*) **4.** (*a*)
5. (*d*) **6.** (*b*).

15

Affinities of Sponges (Porifera) and Origin of Metazoa

15.1 AFFINITIES OF PORIFERA

The evolutionary origin of sponges poses a number of interesting problems. The absence of organs and the low level of cellular differentiation (= division of labour) and interdependence in sponges certainly seem to be primitive characteristics. But a specialized body structure built around a canal system and lacking distinct anterior and posterior ends is found in no other groups of animals. Moreover, the cellular differentiation of sponges is unlike that of other metazoans. All these features suggest that sponges are phylogenetically remote from other metazoans (called eumetazoans).

Sponges show affinities with Protozoa as well as with Metazoa:

A. Affinities of Porifera with Protozoa

Similarities. Sponges show following similarities with *Proterospongia* (Choanoflagellida, Mastigophora; this colonial protozoa is regarded as **ancestor of Metazoa** in general since it contains a collar-ring of microvilli around a flagellum):

1. Intracellular digestion.
2. Presence of collar cells and amoeboid cells as found in *Proterospongia.*
3. Formation of skeletal spicules by independent cells.
4. The cells of sponge body are independent in their function.
5. A process known as **inversion** occurs in amphiblastula larva like those of *Volvox* (a colonial protozoan).

Differences. Sponges differ from colonial and other protozoa in following aspects:

1. Presence of canal system.
2. Presence of a characteristic skeleton.
3. The development of the multicellular body from a fertilized egg by cleavage. Due to this feature sponges become more complex than a colonial protozoan and show resemblance with Metazoa. Thus, this lone feature negates any affinity of sponges with the protozoa.

B. Affinities of Porifera with Metazoa

Similarities. Among Metazoa, coelenterates (cnidarians) show following similarities with sponges.

1. Sedentary in habit.
2. Diploblastic acoelomates, absence of cellular mesoderm.

3. Presence of non-cellular mesogloea between outer and inner layers.
4. The single body cavity, the spongocoel of sponges which opens out through a single aperture, the osculum, can be compared with the gastrovascular cavity, opening to the exterior by terminal hypostome (single opening) of coelenterates.
5. Asexual reproduction by budding to form colony.
6. The parenchymula larva of sponges is comparable with planula of Coelenterata.

Differences. The sponges differ from Metazoa in having the following features:

1. In sponges the cells are less specialized and interdependent than the cells of Metazoa.
2. Tissue formation is restricted to the formation of epithelial lining on the surface. But this epithelium lacks basal lamina and intercellular junction.
3. Sponges are diploblastic but most Metazoa are triploblastic (having three germinal layers in their body namely ectoderm, mesoderm and endoderm).
4. Osculum acts only as an exhalent aperture.
5. The sponges do not possess an anterior end or head like those of metazoa.
6. Stinging cells (cnidoblasts containing nematocysts) are not found in sponges.
7. Body surface of sponges is perforated by numerous minute pores and they possess a unique canal system in their body. All these characters are absent in Metazoa.
8. Sponges lack nervous tissue to coordinate the body activities.
9. In the development of sponges, the flagellated upper cells of the embryo situated at the animal pole become the choanoderm (= endoderm) and the granular cells of vegetal pole becomes the pinacoderm (= ectoderm). Such sort of inversion is unique and does not occur in other Metazoa. This character indicates that sponges and coelenterates have separate origin from Protozoa.

Conclusion. Visualizing the above affinities of Porifera, there can be little doubt that sponges diverged early from main line of metazoan evolution and have given rise to no other members of the Animal Kingdom. Because of their isolated phylogenetic position, the sponges often been placed in a separate subkingdom, the **Parazoa**, distinct from the other multicellular animals, the **Eumetazoa**.

15.2 ORIGIN OF METAZOA

Unraveling the origin of the multicellular animals (metazoans) has presented many problems. There is no direct proof about the ancestors of Metazoa. But all zoologists agree that the metazoans have evolved from some unicellular or acellular organisms, such as Protozoa. This is strongly evident by the facts of the embryonic development in which each metazoan passes from an unicellular (zygote) to a cellular condition (morula, blastula, gastrula, etc.) There is much disagreement over which group of unicellular organisms gave rise to them and how. The following three prevalent hypotheses (or theories) in current use are

1. That the metazoans arose from a syncytial (multinucleate) ciliated form from which cell boundaries later evolved. This is known as *syncytial ciliate hypothesis* or *syncytial theory*.
2. That the origin of metazoans was polyphyletic, or derived from more than one group of unicellular organisms. This is known as *polyphyletic origin hypothesis*.
3. That they arose from a colonial flagellated form in which the cells gradually became more specialized and interdependent. This is known as *colonial flagellate hypothesis*.

1. Syncytial Ciliate Hypothesis

Resemblance between multinucleate ciliates and acoelomate flatworms led **E.D. Hanson** (1958) and **Jovan Hadzi** (1963) to propose a ciliate origin for metazoans. Proponents of syncytial theory believed that metazoans arose from an ancestor shared with the single-celled ciliates. The common ancestor of metazoans acquired multiple nuclei with a single cell membrane and later became compartmentalise into the multicellular condition. It is assumed that the body form of the ancestor resembled that of modern ciliates and thus tended toward bilateral symmetry. Therefore the earliest metazoans would have been bilateral and similar to the present primitive flatworms.

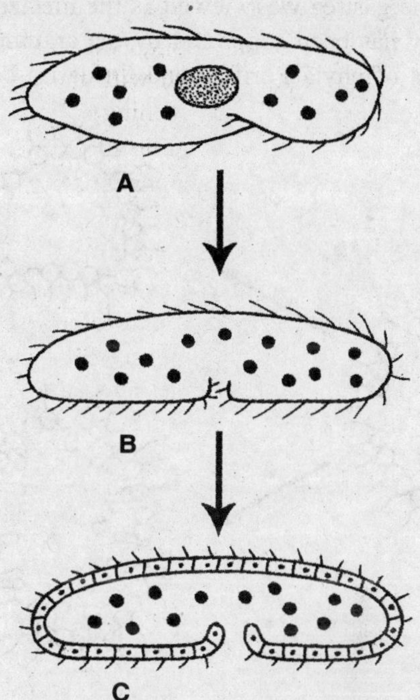

Fig. 15.1. Syncytial theory. A—Schematic representation of the hypothetical transition from a multinucleate, ciliated protist. B—The hypothetical metazoan precursor (sagittal section) as it assumed a benthic, crawling lifestyle and developed a ventral mouth and simple pharynx. C—The hypothetical metazoan precursor (sagittal section) after it achieved the acoel grade via cellularization of the epidermis, which surrounded a syncytial endodermis or entodermis (after Brusca and Brusca, 2003).

There are several objections to this hypothesis. It ignores the embryology of the flatworms in which nothing similar to cellularization occurs, it does not explain the presence of flagellated sperm in the metazoans, and perhaps more important, it implies that the radial symmetry of cnidarians is derived from a primary bilateral symmetry, for which there is no evidence.

2. Polyphyletic Origin Hypothesis

This hypothesis was proposed by **Greenberg** (1959) and **Preston** (1967); they held that probably all Metazoa have not originated from any single ancestor. So some zoologists prefer the idea that the metazoans has a **polyphyletic origin** and suggest that the sponges, cnidarians, ctenophores and remaining eumetazoans evolved independently. Thus no single scheme might account for them all.

3. Colonial Flagellate Theory

(*i*) Haeckel's gastrea theory. The colonial theory is based upon ideas first expressed by **Ernst Haeckel** (1874), who suggested that a colonial flagellated protist gave rise to a planuloid metazoan ancestor (the planula is the basic larval type of cnidarians). The ancestral protist in this theory was a hollow sphere of flagellated cells that developed some degree of anterior-posterior locomotor orientation, and also evolved some level of specialization of cells into separate somatic and reproductive (or germinal) function. Such conditions are common in living colonial protists (*e.g.*, *Volvox*). Haeckel called this hypothetical protometazoan ancestor the **blastea** and supported its cogency by noting the widespread occurrence of **coeloblastula** among modern animals. In this projection, the first metazoan arose by invagination of the blastea; the resulting animals had a double-layered gastrula-like body (Haeckel's **gastrea**), with a blastopore-like opening to outside (Fig. 15.2 B). Similar to the gastrulae of many modern animals. Haeckel believed that these ancestral creatures (the blastea and gastrea) were recapitulated in the ontogeny of modern animals and the gastrea was viewed as the metazoan precursor to the cnidarians. In addition, the colonial theory has been supported by the argument that the body walls of many lower animals (*e.g.*, members of phyla Porifera and Cnidaria) bear monoflagellated or monociliated cell.

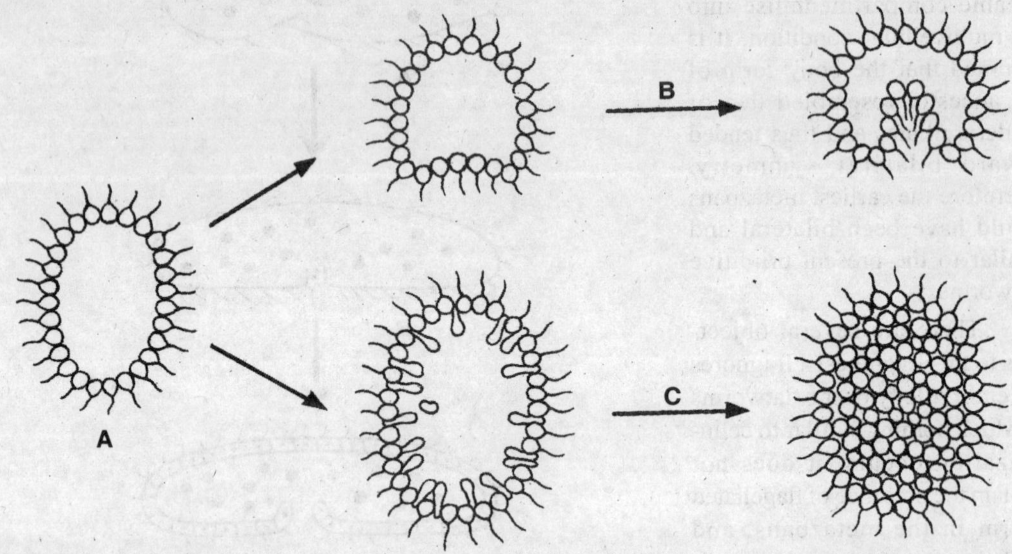

Fig. 15.2. Two versions of the colonial flagellate theory of the origin of the Metazoa. A—The hypothetical colonial flagellate ancestor, Haeckel's blastea (section). B—According to Haeckel, the transition to a multicellular condition occurred by invagination, a developmental process that resulted in hollow gastrea. C—According to Metschnikoff, the formation of a solid gastrea occurred by ingression (after Brusca and Brusca, 2003).

Modification of Haeckel's gastrea theory. Haeckel's original ideas have been modified over the years by various authors investigating a colonial protist ancestry to the metazoan condition (**Metschnikoff**, 1883; **Hyman, 1940**). Some have argued that the transition to a layered construction occurred by ingression rather than by invagination, and that the original Metazoa were solid not hollow (Fig. 15.2 C). This idea is based on the fact that ingression is the primitive form of gastrulation among cnidarians.

(*ii*) Butschli's plakula hypothesis. An interesting offshoot of the colonial theory was advanced by **Otto Butschli** in 1883. He suggested that the primitive metazoan was a bilaterally symmetrical flattened creature of two layers of cells; he called this hypothetical animal a ***plakula.*** According to Butschli, the plakula crawled about ingesting food through its "ventral" cell layer. Eventually this animal becomes somewhat hollow by the separation of its dorsal and ventral cell layers; this development allowed invagination of the nutrititive cells (Fig. 15.3). This formation of a "gut" chamber increased the digestive surface area and at the same time produced inner and outer cell layers, an arrangement approaching the metazoan grade of complexity.

Most evidence today points to the protist phylum protozoa's choanoflagellata as the likely ancestal group from which the Metazoa arose. Choanoflagellates possess collar cells essentially identical to those found in sponges. Genera such as *Proterospongia*, *Sphaeroeca* and other colonial choanoflagellates (Fig. 15.4 C, D), and are commonly cited as typifying a potential metazoan precursor (see **Brusca** and **Brusca**, 2003).

Recently, **Wainright** and coworkers (1993) have advanced molecular biological evidence in support of colonial flagellate theory. Their evidence is based on small subunit RNA sequences and on similarities in complex biochemical pathways. Thus, according to them, the metazoans represent a monophyletic assemblage including choanoflagellates ("collared" flagellates such as *Codosiga*). The sister group of metazoans appears to be the fungi (see **Hickman** *et al.*, 1997).

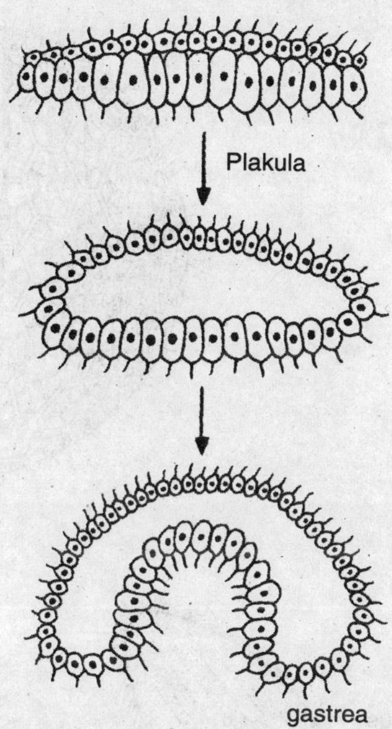

Fig. 15.3. Butschli's hypothetical plakula and its transformation to a metazoan "gastrea" by invagination of a digestive chamber (after Brusca and Brusca, 2003).

Search for the Living Organisms Serving as Protists-Metazoan Intermediates

Among all of the ideas concerning the evolutionary origin of the metazoan condition, there exists a common problem: the search for intermediates between the Protista and Metazoa. Search is on of a modern day metazoan creature which combine protist and metazoan traits. These organisms include multinucleate ciliates (*Paramecium*), the colonial *Volvox* and *Trichoplex*, a mesozoan.

About a century ago a tiny, flagellated, but multicellular creature was discovered in marine aquarium. This animal named *Trichoplax adhaerens*, was placed in its own phylum, the Placozoa (**K.Grell**, 1972, 1982). For many years, *Trichoplax* was believed to be a larval stage of some invertebrate, but workers are now convinced that it is an adult of uncertain affinity at the mesozoan grade of organisation. *Trichoplax* has an outer, partly flagellated epithelium surrounding an inner mesenchymal cell mass. Its body margins are irregular, and animal changes shape like an amoeba. There exists some division of labour among various cells and areas of the body. When feeding, *Trichoplax* "*hunches up*" to form a temporary digestive chamber on its underside (Fig. 15.4 E) – producing a form strikingly similar to Butschli's hypothetical plakula. *Trichoplax* may represent the most primitive of all living Metazoa and perhaps even be a conservative descendant line of the ancestral metazoan type.

J. Frenzel (1892) described a tiny organism reportedly collected from salt beds in Argentina. He named this mesozoan animal *Salinella* (Fig. 15.4 F). Although *Salinella* does not possess the layered organisation of the Metazoa, it appears to display a higher level of functional organization than colonial protists. In *Salinella*, a single layer of cells forms the entire body wall and separates the digestive cavity from the outside. The digestive cavity was described as open at both ends as a mouth and anus and the animal fed on organic detritus. The phylum name **Monoblastozoa** has been proposed for this curious animal. Unfortunately, *Salinella* has not been seen since Frenzel's original report, and many zoologists now suspects that Frenzel seriously misinterpreted whatever creature he observed (see **Brusca** and **Bursca**, 2003).

Finally, animals of two mesozoan phyla, Rhombozoa and Orthonectida (Fig. 15.4 G , H), are structurally simple but display complicated life cycles; they are all endoparasites of invertebrates. Some workers believed that they are derived from another group (or groups) of Metozoa, perhaps

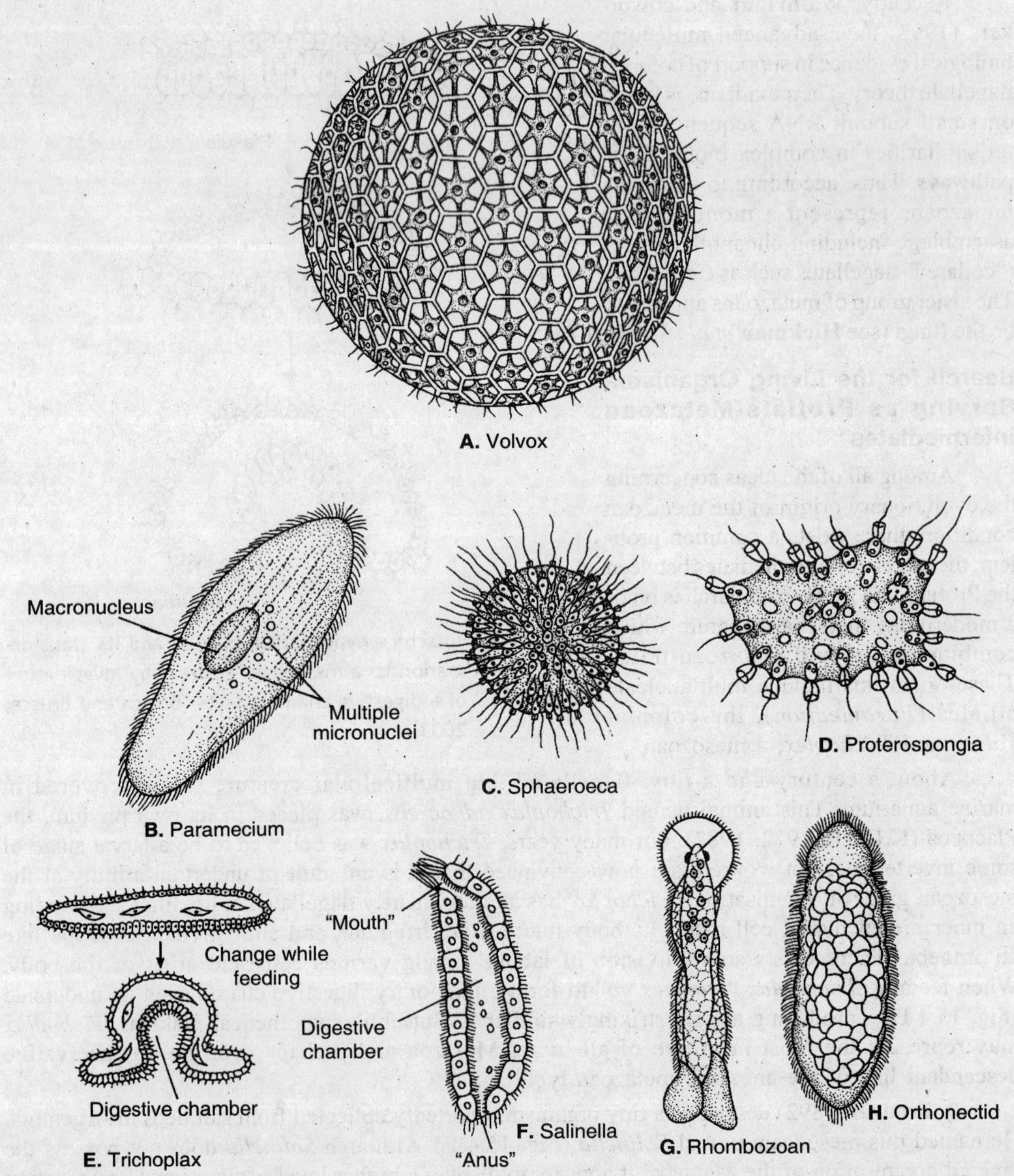

Fig. 15.4. Living organisms that have been considered as protist-metazoan intermediates or that play major roles in various hypotheses concerning the origin of the metazoan condition. A—*Volvox*, a colonial flagellate; B—A multinucleate ciliate (*Paramecium*); C, D—*Sphaeroeca volvox* and *Proterospongia*, two choanoflagellates; E—*Trichoplax*; F—*Salinella*; G—A decymid rhombozoan; H—An orthonectid (after Brusca and Brusca, 2003).

from parasitic trematodes (flukes) which undergo similar complex life cycles. Other workers have suggested that these mesozoan animals are primitively simple and thus may have been derived from early metazoan or premetazoan stock.

Origin of Bilateral Symmetry

Bilateral symmetry could have evolved later when some of the planula-like ancestors became adapted for a creeping form of locomotion on the ocean floor. Dorsal and ventral surfaces would have differentiated, a ventral mouth would have appeared and a beginning would have been made toward cephalisation (a concentration to neurons and sensory structures at the anterior). This would have led to primitive bilateral symmetry, resembling that of the flatworms (see **Hickman** *et al.*, 1997).

QUESTIONS

Long Answer Questions

1. Give reasons for including sponges in the animal kingdom.
2. Explain "Phylum Porifera is blind offshoot of Metazoa".
3. Discuss affinities of Parazoa.
4. Describe the affinities of Porifera.
5. Discuss the origin of Metazoa. (*M.S.U. Tirunelveli 2006*)

Short Answer Questions

1. Describe Butschli's plakula hypothesis.
2. Describe the search for living Protist-Metazoan intermediate.
3. Write short notes on:
   (*i*) Blastea
   (*ii*) Gastrea
   (*iii*) Metschnikoff's theory of origin of Metazoa
   (*iv*) Affinities of Porifera.

Very Short Answer Questions

1. What is the name of theory proposed by Haeckel.
2. What is meant by the word syncytial?
3. Who modified the Haeckel's theory?

Fill in the Blanks

1. The term blastea was given by ....
2. Haeckel's .... is basically a colonial theory.
3. *Trichoplax adaerens* belongs to Phylum ....
4. *Salinella* was reported by ....
5. Butseli's hypothetical animal is ....

Multiple Choice Questions

1. Syncytial theory explains
   (*a*) formation of coelom (*b*) symmetry of animal
   (*c*) origin of life (*d*) origin of metazoa
2. Polyphyletic theory is given by
   (*a*) Greenberg and Preston (*b*) Lang and McBridge
   (*c*) Hanson and Hadzi (*d*) Lankester
3. Colonial theory was proposed by
   (*a*) Butschli and Haeckel (*b*) Hanson and Hadzi
   (*c*) L.H. Hymen (*d*) Metschnikoff

4. Gastrea theory of metazoan ancestory was presented by

(*a*) E.H. Haeckel (*b*) L.H. Hymen

(*c*) J. Hadzi (*d*) Lankester

5. Syncytial theory was proposed by

(*a*) Hanson and Hadzi (*b*) L.H. Hymen

(*c*) Lankester (*d*) E.H. Haeckel

6. Which group of protista (protozoa) support and give best explanation of colonial theory of origin of Metazoa?

(*a*) ciliates (*b*) sarcodina

(*c*) ciliophora (*d*) flagellate

ANSWERS

Very Short Answer Questions

1. Gastrea theory.

2. Syncytial means something constituting syncytium. Syncytium is a multinucleate cell.

3. Metschnikoff.

Fill in the Blanks

1. Haeckel **2.** Gastrea theory **3.** Placozoa **4.** Frenzel

5. Plakula

Multiple Choice Questions

1. (*d*) 2. (*a*) 3. (*a*) 4. (*a*)

5. (*a*) 6. (*d*).

16

Phylum Cnidaria (Coelenterata): Characters, Classification and Types

Cnidaria or **Coelenterata** (see Box 16.1) may be defined as '*diploblastic tentaculate metozoa with tissue grade of construction having nematocysts, a single body cavity, the coelenteron or gastrovascular cavity and a single mouth opening to the gut (= coelenteron) derived from the blastopore.*' Cnidaria includes jelly fish, sea anemones, corals and various *Hydra*-like polyps. It is a very successful phylum with about 10,000 living species and almost as many fossils.

Box 16.1

In certain old text books, the name Coelenterata was used as a phylum designation to include both the **Cnidaria** and **Ctenophora**. Later on, it was found better to separate the groups as discrete phyletic entities. The Cnidaria closely approximate an idealised diploblastic condition and possess highly diagnostic intracellular structures, the nematocysts, while the Ctenophora are triploblastic and do not possess nematocysts (**Marshall** and **Williams**, 1995).

16.1 GENERAL CHARACTERS

1. Primitive metazoans having **tissue grade of organization**. For example, epitheliomuscular tissue has **basal lamina** and intercellular junctions such as **belt junctions, septate sealing junctions** and **gap junctions** (Gap junctions are found only in class Hydrozoa).
2. They are aquatic, mostly marine except few freshwater forms such as *Hydra* and *Craspedacusta*.
3. They are sessile or free swimming and solitary or colonial.
4. Exhibit primary radial symmetry of body, so are placed in division **Radiata** of Animal Kingdom.
5. Cnidarians are **diploblastic**, that is, their bodies are constructed from only two germ layers, **ectoderm** and **endoderm**, from which two adult epithelia arise.
6. Their body wall consists of two cellular layers and one extracellular layer, *i.e.*, an outer epithelium, the **epidermis**, and an inner epithelium, the **gastrodermis**, lining the gastrovascular cavity and between these occur an extracellular gelatinous **mesogloea**.
7. Mesogloea ranges from a thin, noncellular basal lamina (called **mesolamella**) as in hydras (Hydrozoa) to thick, fibrous, jelly-like, connective tissue with or without mesenchymal cells (which have migrated from both germ layers).
8. The body wall encloses a single, central, blind sac-like body cavity called **coelenteron** or **gastrovascular cavity**. This cavity serves as gut (for digestion of food) and as circulatory system (for circulation of digested body).

9. Gastrovascular cavity lies along the polar axis of the animal and opens to outside at one end of body to form a **mouth**. Mouth serves both in ingestion and egestion. **Anus** is absent.
10. Acoelomate animals because they do not possess a second body cavity, the coelom.
11. A circle of short and slender **tentacles** surround the mouth to aid in the capture and ingestion of food. Tentacles represent evaginations of the body wall.
12. Most cnidarians show the phenomenon of **polymorphism** (see Box 16.2) but most individuals (zooids) are reducible either to hydroid (= polypoid) or medusoid types. The **medusoid** or jelly-fish form represents the sexual phase and is adapted for free-swimming mode of life. The **polypoid** or **hydroid** form represents asexual phase and is adapted for sessile mode of life. Polypoid forms may be solitary or colonial but medusoid forms are always solitary.

Box 16.2 Polymorphism

The occurrence of structurally and functionally different types of individuals, called **zooids**, within the same organism during its life history is called **polymorphism**. Cnidaria contains nine types of zooids such as **trophozooid** or **gastrozooid** (feeding), **gonophore** (sessile medusae), **gonozooid** (gonophore production), **dactylozooid** (protection, food capture), **tentaculozooid** (same as dactylozooid), **nematophore** (food capture), **pneumatophore** (float), **nectophore** (swimming) and **phyllozooid** (protective cover).

13. All cnidarians have unique stinging organs, called **nematocysts** which are a type of **cnide** (see Box 16.3). A cnide is an organ of **cnidocyte** or **cnidoblast** which is capable of eversion from cnidocyte. Cnidocytes are located in ectoderm (= epidermis) especially in tentacles. Nematocysts function in prey capture and may inject a protein toxin into the prey. About 18 types of nematocysts are found in cnidaria to serve different functions. They stun, paralyze or kill prey, entangle food and attach it to tentacles and perform other functions.

Box 16.3

Cnidarian contain following three types of **cnide** which are cellular organelles of cnidocytes and are capable of eversion from the cell:

1. Nematocysts
2. Spirocysts
3. Ptychocysts

Nematocysts are found in all cnidarians. Spirocysts and ptychocysts, however, are found only in anthozoa.

14. Cnidarians are **carnivorous**; digestion is extracellular and intracellular.
15. Skeleton either calcareous or horny and external (exoskeleton) or internal (endoskeleton).
16. Locomotion is brought about by smooth muscle fibres.
17. Respiratory, circulatory and excretory systems are absent.
18. Nervous system is a collection of multipolar and bipolar nerve cells forming an irregular net or plexus.
19. Cnidarians contain very simple sensory organs such as **dermoptic light sensitive organs, ocelli** and **statocysts**.
20. Reproduction is both by **asexual** and **sexual** methods. Asexual reproduction by **budding** or **strobilation**. Sexual reproduction produces characteristic ciliated **planula** larva.
21. Their life history shows **alternation** of a sessile and asexual **polypoid generation** and a free swimming and sexual **medusoid generation**, both being diploid phases. It is called **metagenesis** (see Box 16.4).

Box 16.4 What is metagenesis?

True type of alternation of generation occurs in the plants where a regular alternation between a **haploid** gametophyte and **diploid** sporophyte occurs. Such sort of alternation is not found in the cnidarians. In them occurs a somewhat modified type of alternation of generation, called **metagenesis**. Thus, in Cnidaria, alternation of two **diploid generations** occur, *i.e.*, diploid sessile asexual polypoid generation alternates with diploid free-swimming sexual generation. In them, the only haploid cells in the life cycle are the gametes. In fact, metagenesis implies postponement of sexual reproductive phases.

16.2 CLASSIFICATION

The classification of phylum Cnidaria is based on the duration of polypoid and medusoid generations. An outline classification of phylum Cnidaria upto orders and suborders including examples have been given on page 200.

Detailed Classification

Phylum Cnidaria is divided into following **four** classes:

Class 1. Hydrozoa

1. Hydrozoa (Gr., *hydra* = water; *zoios* = animal) include a polypoid, medusoid or both forms in the life cycle. However, polypoid phase is usually predominant in the life cycle.
2. Polyps are solitary or colonial.
3. Aquatic, mostly marine, a few freshwater.
4. Mesogloea is acellular or noncellular and jelly like.
5. A calcareous or cuticular skeleton, the **perisarc**, covers the body of most of the hydrozoans.
6. Gastrovascular cavity without gastric tentacles or septa (or mesentries); without gullet or stomodeum.
7. Nematocyst-containing cnidocytes are found only in epidermis.
8. Gonads develop only in epidermis. Sperm and ova are shed directly to the outside.
9. Medusae typically with **craspedon** or **velum** (called craspedote) which improves swimming efficiency of medusa.
10. Life history of some includes alternation of asexual polypoid and sexual medusoid generations.
11. Phenomenon of polymorphism is common.
12. Class Hydrozoa includes about 2700 species of most primitive cnidarians.

Box 16.5 Taxonomy of Cubozoa

Cubozoa forms a new class which was created by **Werner** (1975) and **Calder** and **Peters** (1975). They were separated from Scyphozoa on the following affinities. Their hydrozoan features include. 1. The presence of velum; 2. The possession of a unique type of nematocyst (stenotele); 3. Lack of a notched bell margin and rhopalia.

Chironex (sea wasp) contain deadly poisonous nematocysts. Its sting can cause death of a human being within 3 to 20 minutes.

Class 2. Scyphozoa

1. In Scyphozoa (Gr., *skyphos* = cup + *zoios* = animal) medusa stage is predominant. Animal body is umbrella or bell-shaped.
2. All are marine, solitary, free-swimming or attached by aboral stalk. They float in cooler waters.
3. They lack velum (acraspedote).
4. Mesogloea is much thickened and cellular.

Outline Classification

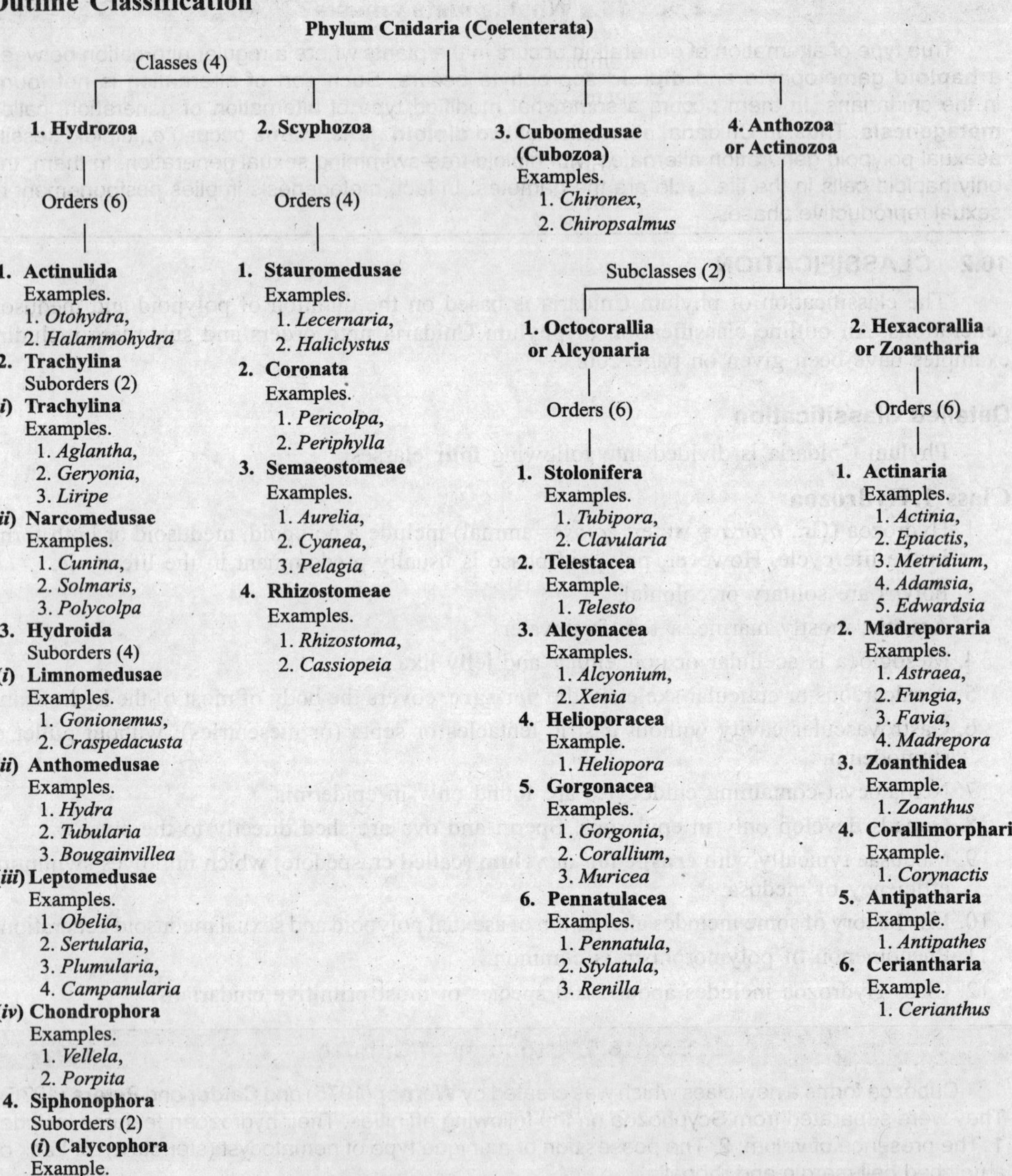

5. Nematocyst-containing cnidocytes are found both in the epidermis and gastrodermis.
6. The gastrovascular cavity is without stomodaeum but with gastric tentacles and may or may not be divided into four compartments (inter-radial pockets) by four septa or ridges.
7. Medusae with marginal sense organs, called **tentaculocysts** which have endodermal statoliths.
8. Perisarc absent.
9. The gonads are endodermal in origin. Gametes are shed in the gastrovascular cavity.
10. Polypoid stage missing or sometimes present, developing directly in medusae by transverse fission or **strobilation.**
11. Polyp (called **scyphistoma**) is hydrotuba; mesentries are present only in young polyp.
12. Class Scyphozoa includes about 200 species of quite large and brilliantly coloured jellyfishes or true medusae. The largest jellyfish, *Cyanea arctica*, weighs nearly a ton and may possess tentacles upto a hundred feet long.

Class 3. Cubozoa or Cubomedusae

1. All marine, found in tropical and subtropical oceans (Box 16.5).
2. An attached polypoid stage of short duration arises from the planula.
3. The medusa phase is long and dominant.
4. The medusae are squarish in cross section.
5. Each corner of medusae bears one or a group of tentacles.
6. Contain a peculiar and complicated eye which can detect very accurately the direction of incoming light.
7. Medusae contain flaplike **velarium** which functions like the velum.
8. Radial canals are replaced by four radial pouches.

Examples. *Charybdaea*, *Chiropsalmus*, *Chironex*.

Class 4. Anthozoa

1. All are marine.
2. Solitary or colonial. Mostly colonial.
3. Anthozoa (Gr., *anthos* = flower; *zoios* = animal) or Actinozoa are exclusively polypoid, medusa phase is absent.
4. Their radial symmetry is modified. Body is cylindrical with hexamerous or octamerous or biradial or radiobilateral symmetry.
5. Mesogloea is thin but is richly cellular and fibrous.
6. Gastrovascular cavity is divided into compartments by complete or incomplete longitudinal or raidal septa or mesentries (which are endodermal structures).
7. Mouth is not rounded. It is slit-like and opens into a tubular pharynx lined by ectoderm. Pharynx opens into the gastrovascular cavity.
8. The mesentries bear nematocysts and gonads.
9. Skeleton either external or internal.
10. Exoskeleton is formed from calcium carbonate. It often forms massive corals.
11. Gametes are shed in coelenteron; fertilization is external.
12. Class Anthozoa is the largest class of phylum Cnidaria. It contains over 600 species of sea anemones, corals, sea fans and sea pansies.

CLASS 1. HYDROZOA

Class Hydrozoa is divided into the following **six** orders.

Order 1. Actinulida

1. Tiny, solitary hydrozoans.
2. Body form resembles actinula larva.
3. Medusae phase absent.
4. Unlike other mature polypoid and medusoid hydrozoans, the epidermal cells are ciliated.
5. They are regarded as **neotenic** forms because they persist in the larval form (*i.e.*, actinula) and reproduce (*i.e.*, actinula does not metamorphose into medusa and becomes sexually mature).

Examples. *Otohydra*, *Halammohydra*.

Order 2. Trachylina

1. Hydroid or polypoid phase is reduced or absent.
2. Medusoid phase dominant.
3. Medusa develops from the fertilized egg. Actinula larva develops from the planula larva and the actinula develops into medusa.
4. Medusae are large, provided with marginal sense organs such as tentaculocysts and statocysts. Statoliths are endodermal.
5. This order includes most primitive members of the class Hydrozoa.

Order Trachylina is divided into the following **two** suborders.

Suborder 1. Trachymedusae

1. Margin of bell or umbrella is smooth (not scalloped).
2. With four, six, or eight radial canals.
3. Manubrium is long.
4. Gonads borne on the radial canals.

Examples. *Aglantha*, *Gonionemus*, *Geryonia*, *Liriope*.

Suborder 2. Narcomedusae

1. Margin of bell or umbrella is scalloped, *i.e.*, clefted by tentacle bases.
2. Radial canals are reduced; redial pouches are present.
3. Manubrium is absent.
4. Gonads borne in the floor of gastric cavity or stomach.
5. Tentaculocysts naked.

Examples. *Cunina*, *Solmaris*, *Polycolpa*.

Order 3. Hydroida

1. Polypoid generation is well developed. Medusa phase is present or absent.
2. Polyps are solitary or colonial.
3. Medusa phase is short lived, if present. Medusae are produced by polyp by budding.
4. Sense organs of medusae are ocelli and statocysts, all are ectodermal in origin.

Order Hydroida is divided into following **four** suborders:

Suborder 1. Limnomedusae

1. Mostly fresh water forms, some are marine.
2. Polyps are small in size and solitary.

3. Polyps lack perisarcal skeleton.
4. Some are with medusae.

Examples. Marine *Gonionemus* and freshwater *Craspedacusta* (freshwater jellyfish).

Suborder 2. Anthomedusae (or Athecata or Gymnoblastea)

1. Stalk typically covered with perisarc.
2. Perisarc does not form thecal cups for zooids, *i.e.*, hydrotheca and gonotheca are absent.
3. Tall and bell-shaped free swimming medusae.
4. Medusae bear eye spots or ocelli at the bases of tentacles.

Examples. *Hydra*, *Tubularia*, *Bougainvillea*, *Pennaria*, *Eudendrium*, *Sarsia*.

Suborder 3. Leptomedusae (Thecata or Calyptoblastea)

1. Colonial Hydrozoa.
2. Zooids are contained in skeletal cups, *i.e.*, polyps are enclosed in **hydrotheca** and blastostyles with developing medusae are enclosed in **gonotheca**.
3. Medusae, if present, are typically flattened, bowl or saucer-shaped.
4. Medusae bear gonads on the radial canals.
5. Medusae are with statocysts but without eyespots or oceli.

Examples. *Obelia*, *Sertularia*, *Plumularia*, *Campanularia*.

Suborder 4. Chondrophora

1. Free-floating polypoid colonies with high degree of polymorphism.
2. These cnidarians can also be interpreted as large, single, inverted polyps.

Examples. *Velella* (by-the-wind sailor or purple sailor), *Porpita* (blue button).

Order 4. Siphonophora

1. They form free-swimming or floating colonies of polypoid and medusoid individuals.
2. Colonies with gas-filled floats (pneumatophores) or large swimming bells which help in floating of the colony.
3. Siphonophora exhibits polymorphism of high degree. Thus, colony is composed of different types of modified polyps such as gastrozooids, gonozooids, dactylozooids and medusae. These zooids are either attached to stem or disc.
4. Polyps are without tentacles.
5. Medusae are incomplete and rarely freed.
6. Perisarc absent.
7. They are found largely in warm seas.

This order includes the following **two** suborders:

Suborder 1. Calycophora

1. Apical float or pneumatophore absent.
2. One or more swimming bells or **nectocalyces** (or nectophores) present at the upper end of the colony.

Examples. *Diphyes*, *Abyla*.

Suborder 2. Physophorida

1. Upper end of the colony bears a float or pneumatophore.

Examples. *Physalia* (Portuguese man-of-war), *Muggiae*, *Halistemma*.

Order 5. Milleporina

1. Colonial coral-like Hydrozoa without perisarc.
2. Their massive porous calcareous skeleton is secreted by ectoderm. Polyps protrude out through the pores of skeleton.
3. Colony has two kinds of zooids, the gastrozooids and the dactylozooids.
4. Gastrozooids (nutritive zooids) are short and provided with mouth and tentacles.
5. Dactylozooids are elongated, hollow, slender tactile zooids. They contain tentacles but lack mouth.
6. Medusae develop in small chambers, becoming free. They lack mouth radial canals and tentacles (*i.e.*, they become degenerated).
7. Milleporina are commonly called **stinging** or **fire corals** due to the presence of powerful nematocysts in them.

   Example. *Millepora* is the only genus.

Order 6. Stylasterina

1. Colonial coral-like hydrozoa; commonly called **hydrocorals**.
2. Colony has two types of zooids, the feeding polyps or **gastrozooids** and defensive polyps or **dactylozooids**. Both types of polyps are located within star-shaped openings of the skeleton.
3. Dactylozooids are small, solid without tentacles.
4. Gastrozooids have a cup with pointed spine.
5. Medusae are permanently sessile and degenerated.

   Examples. *Stylaster*, *Allopora*.

CLASS 2. SCYPHOZOA

Class Scyphozoa is divided into following **four** orders.

Order 1. Stauromedusae

1. Adult developing directly from scyphistoma (see Box 16.6) remaining attached to seaweeds by aboral stalk.
2. Body globet or trumpet shaped.
3. Mouth cruciform (four-cornered) with small oral lobes and a short quadrangular manubrium.
4. Gastrovascular system is divided into central stomach and four preradial pouches by the four inter-radial septa.

Box 16.6

Scyphistoma. A sessile, polypoid or hydroid larva of many scyphozoa which may produce either more scyphistomae or free-swimming medusae.

5. Gonads are elongated band-like borne on the faces of septa.
6. Marginal sense organs absent or modified as tentacles.
7. Fertilization is external.
8. Larva is **planula** but lacks cilia.
9. They inhibit coastal water of colder regions. All are marine.

   Examples. *Haliclystus*, *Lucernaria.*

Order 2. Coronata

1. Free swimming Scyphozoa, found inhabiting the deep waters of ocean.
2. Medusoid body is conical, dome-shaped or flattened and grooved.

3. The umbrella is divided by a horizontal furrow into an upper cone and a lower cone.
4. The crown consists of flattened blades (pedal lobes) called **pedalia** which bear solid tentacles.
5. The bell margins are scalloped (*i.e.*, cut into curves) into **lappets** which alternate with pedalia.
6. Gastrovascular cavity is divided by septa; mouth is four cornered.
7. Tentaculocysts are four to sixteen.

Examples. *Pericolpa, Periphylla, Nausithoe, Atolla.*

Order 3. Semaestomeae

1. Most common free-swimming medusae found inhabiting the coastal waters of all oceans.
2. Umbrella is flat, saucer or bowl-shaped.
3. Mouth is square.
4. The corners of the mouth are prolonged into frilly **oral tentacles** or **oral arms.**
5. Without furrow on pedalia (exumbrella)
6. The margin of umbrella is fringed with hollow tentacles.
7. Eight to sixteen **rhopalia** (= little club-shaped compound sensory organs, each containing a statocyst, an ocellus and two sensory pits) are present.
8. Margin of umbrella is scalloped.
9. Gastric pouches and filaments are absent.

Examples. *Aurelia, Cyanea, Pelagia.*

Order 4. Rhizostomae

1. Free-swimming Scyphozoa found in shallow waters of tropical and subtropical oceans.
2. The umbrella is saucer or bowl-shaped or flattened. It may be concave on the top.
3. Medusae with fused oral tentacles, containing many canals and small mouths (secondary mouths). Original or primary mouth is lost due to fusion of oral arms except in *Stomolophus.*
4. Gastrovascular cavity is not divided by septa.
5. Marginal tentacles are absent but eight or more rhophalia are present.
6. Margin of umbrella is scalloped.

Examples. *Rhizostoma, Cassiopeia, Stomolophus.*

CLASS 4. ANTHOZOA

Class Anthozoa is divided into following **two** subclasses.

Subclass 1. Octocorallia (or Alcyonaria)

1. Colonial marine forms; polyps usually connected by coenenchyme (see Box 16.7).

Box 16.7 Coenenchyme

It is a tissue comprising mesogloea surrounding and uniting the polyps in compound anthozoans. This tissue consists of a thick mass of mesogloea perforated by gastrodermal tubes that are continuous with the gastrovascular cavity of the polyps. The surface of entire flesh mass is covered by a layer of epidermis, which joins the epidermis of the polyp column. Only the upper portion of the polyp projects above the coenenchyme.

2. Each polyp is long or short cylinder terminating orally into a flat circular oral disc having oval or elongated mouth in the centre.
3. Each polyp bears eight feathery or pinnate, hollow tentacles.
4. Gastrovascular cavity is divided by eight single, complete septa (mesentries).
5. Single ventral **siphonoglyph** (see Box 16.8) is present.

Box 16.8 Siphonoglyph

It is a ciliated groove leading from mouth to gullet in certain anthozoans. It provides for the entry of water into gastrovascular cavity. This current of water functions to maintain an internal fluid, or hydrostatic skeleton against which the muscular system can act. Water also helps in exchange of gases through the gastrodermal surface.

6. Endoskeleton (of spicules) is the product of mesogleal cells.
7. Spicules may be calcareous or horny in nature.

Subclass 2. Hexacorallia (or Zoantharia)

1. Solitary or colonial marine forms.
2. Polyps with more than eight **tentacles** or **septa** (mesentries), typically in the cycle of 12.
3. Tentacles simple, rarely branched, hollow and cone-shaped.
4. Mesentries may be complete or incomplete.
5. Two siphonoglyphs are commonly present.
6. Endoskeleton when present is calcareous, derived from ectoderm.

Subclass 1. Octocorallia (or Alcyonaria)

Subclass Octocorallia is divided into following **six** orders.

Order 1. Stolonifera

1. They are inhabitants of shallow waters in the temperate and tropical oceans.
2. Coenenchymal mass is absent; simple polyps arise singly from creeping mats or **stolons**.
3. Skeleton of calcareous **tubes** of separate or fused calcareous spicules or horny external cuticle.
4. Spicules are formed by mesogleal cells.
5. Tubes are connected by horizontal platforms.

Examples. *Clavularia*, *Tubipora* (organ-pipe coral).

Order 2. Telestacea

1. Colony consists of simple or branched erect stems arising from a creeping base (stolons).
2. Each stem is a very elongated polyp bearing lateral polyps.
3. They contain a skeleton of calcareous spicules.

Example. *Telesto*.

Order 3. Alcyonacea

1. **Soft corals** which are mostly tropical.
2. Coenenchyme forming a rubbery or fleshy mass.
3. Colony may have a massive mushroom shape or an encrusting growth form.
4. Lower ends of their polyps fused with coenenchyme and only oral ends are protruding.
5. In some forms the polyps are dimorphic, *i.e.*, autozooids and siphonozooids.
6. Skeleton consists of separate mesogleal spicules, often consolidated, without axial core.

Examples. *Alcyonium*, *Xenia*, *Gersemia*, *Sarcophyton*.

Order 4. Helioporacea

1. They include the Indo-pacific **blue coral**.

2. They contain a massive, blue calcareous skeleton.

Example. *Heliopora* is the only genus.

Order 5. Gorgonacea

1. **Horny corals** which are found in tropical and subtropical sea shores.
2. The colony is usually upright plant-like, consists of a main stem arising from a basal plate or tuft of stolons and number of branches bearing polyps.
3. Skeleton of mesogleal spicules and axial core of horny material called **gorgonin** or calcareous material.
4. Polyps short (Fig. 16.1)

Examples. *Gorgonia* (sea fan), *Corallium* (precious red coral), *Muricea* (sea rod), *Leptogorgia* (sea whip).

Order 6. Pennatulacea

1. **Sea pens.**
2. Colony consists of a fleshy, flattened or elongated body or **rachis** (called **axial** or **primary polyp**). Secondary polyps are many small lateral dimorphic polyps attached at sides of rachis.

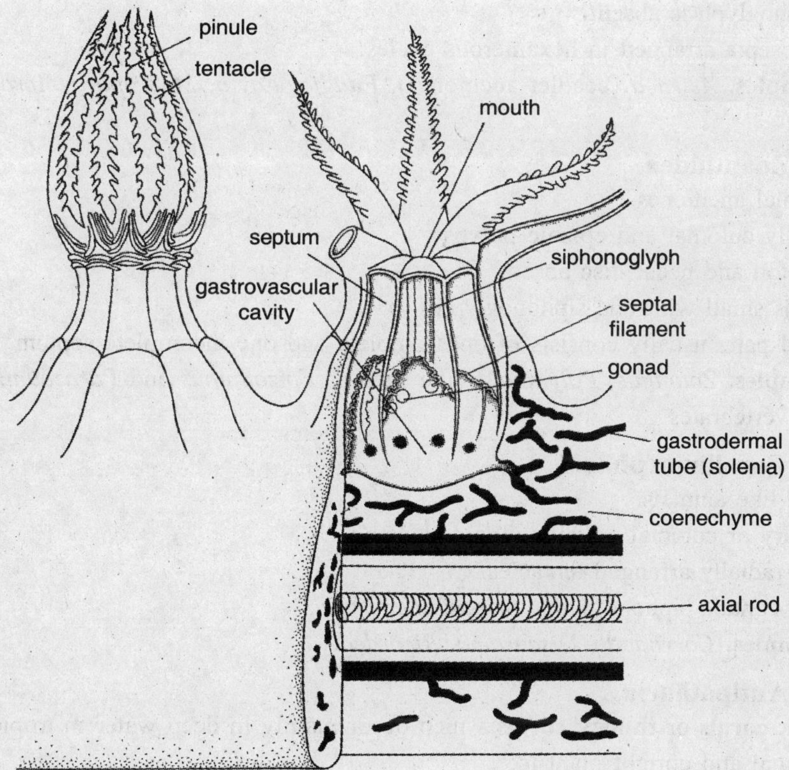

Fig. 16.1. Structure of a gorgonian coral of the anthozoan subclass Octocorallia. Note the position and structure of siphonoglyph and coenenchyme.

3. Rachis is supported by skeleton of separate calcareous spicules and axial rod.

Examples. *Pennatula* (sea pen or sea feather), *Stylatula* (sea pen); *Renilla* (sea pansy).

Subclass 2. Hexacorallia (or Zoantharia)

Subclass Hexacorallia is divided into **six** orders.

Order 1. Actiniaria

1. **Sea anemones.**
2. Solitary anthozoans with no skeleton.
3. Cylindrical polypoid body is divided into oral disc, column and base.
4. The aboral end is provided with pedal disc.
5. Tentacles and mesentries are numerous and often arranged in the multiple of six.
6. Usually with two siphonoglyphs.
7. Septa (mesentries paired, complete and incomplete).

Examples. *Actinia*, *Metridium*, *Adamsia*, *Edwardsia*, *Epiactis*.

Order 2. Madreporaria (or Scleractinia)

1. **Stony corals.**
2. Colonial forms.
3. Exoskeleton compact, hard, heavy calcareous and secreted by ectoderm (epidermis).
4. Polyps small and enclosed in cup-like depressions of the exoskeleton.
5. Siphonoglyph is absent.
6. Sclerosepta arranged in hexamerous cycles.

Examples. *Astraea* (beadlet anemones), *Fungia*, *Favia*, *Madrepora*, *Acropora* (staghorn coral).

Order 3. Zoanthidea

1. Colonial anemones.
2. Usually colonial and epizoic polyps.
3. Skeleton and pedal disc absent.
4. Polyps small with one siphonoglyph.
5. Septal pairs usually consists of one complete and one incomplete septum.

Examples. *Zoanthus*, *Palythoa*. Some such as *Epizoanthus* and *Parazoanthus* are epizoic on other invertebrates.

Order 4. Corallimorpharia

1. Coral-like animals.
2. Solitary or colonial polyps without skeleton.
3. With radially arranged tentacles.
4. They contain siphonoglyphs or basilar muscles.

Examples. *Corynactis*, *Discosoma*, *Ricordea*.

Order 5. Antipatharia

1. **Black corals** or **thorny corals** which occur mainly in deep water in tropics.
2. Colonial and upright plant-like.
3. Skeleton in the form of branched and thorny chitinous axis.
4. Polyps with two siphonoglyphs.
5. Polyps with six, ten or twelve single, complete septa and six simple or eight branched tentacles.

Examples. *Antipathes*.

Order 6. Ceriantharia

1. Long, solitary polyps adapted for burrowing. Live in self-secreted tubes which are buried in sand or mud.
2. Polyps lack skeleton and pedal disc.
3. On the oral disc, tentacles are arranged in two whorls — **marginal whorl** and **oral whorl**.
4. Only single siphonoglyph present.
5. Many single, complete septa are present.

Example. *Cerianthus.*

16.3 TYPE STUDIES

1. *Physalia*

Classification

Phylum	–	**Cnidaria (Coelenterata)**
Class	–	**Hydrozoa**
Order	–	**Siphonophora**
Suborder	–	**Physophorida**
Genus	–	***Physalia***

Physalia is a marine, polymorphic, pelagic floating colony of various modified polypoid and medusoid zooids. It is one of the most famous cnidarian and is commonly known as **Portuguese man-of-war** (see Box 16.9). *Physalia* is found in tropical and sub-tropical oceans; it is seen abundantly on North Atlantic shores, both in Europe and North America. *Physalia* contains a large (may be upto 30 cm in length) bright-blue gas-filled **pneumatophore** or **float** which is formed by several **medusoids**. The float is chitinous, muscular and elongated bladder-like and is filled by a variety of gases such as carbon monoxide (CO), nitrogen (N_2), oxygen (O_2), argon (Ar), etc. Both ends of float are conical, at one end is an **apical pore**; the upper edge is produced into a median **crest** or **sail**. Float contains **gas glands** on the inner surface which secrete gases that inflates the float. The float lacks the swimming bells or nectocalyces, so is drifted to and fro by the wind. During rough weathers gas may be expelled out through the pore causing the entire colony to sink; when fresh gas is secreted, colony rises again. According to **Hyman** (1959) *Physalia* cannot go below the surface.

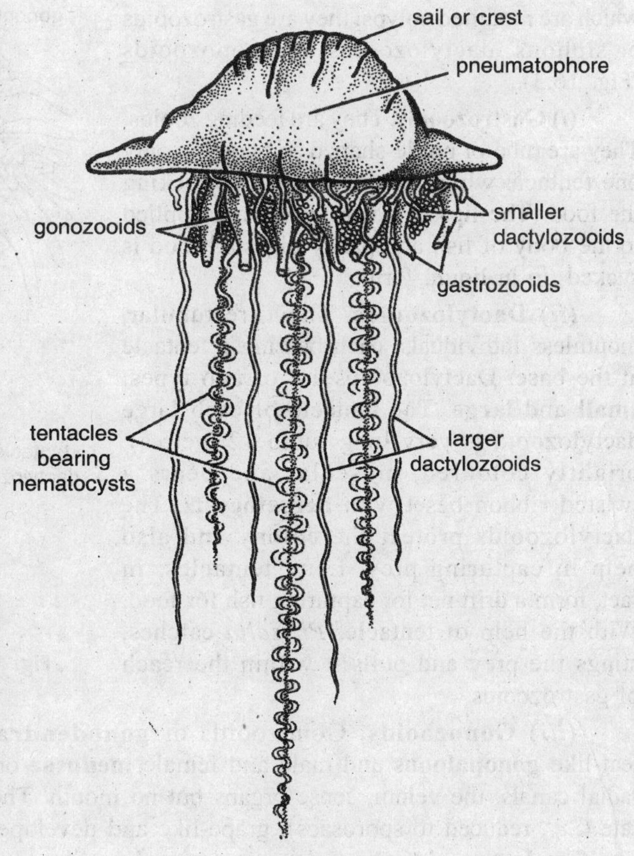

Fig. 16.2. *Physalia.*

Thus, *Physalia* is *pleustonic*, *i.e.*, adapted entirely to surface existence, partly in water and partly outside.

Box 16.9 Star attractions of *Physalia*

Physalia's common name portuguese man-of-war may be coined because of the following reasons: (*i*) its sudden appearance and disappearance like the portuguese warships during war; (*ii*) it appears like upturned warships during war; (*iii*) brilliant peacock blue green colour of its crest resembles the cap of portuguese soldiers/or of Napoleon Bonapart.

Physalia is famous for its following three salient features:

(*i*) It forms a very good example of the phenomenon of polymorphism. It exhibits many types of zooids.

(*ii*) It has a large gas-filled, horizontal float for the colony. This float may have a brilliant peacock-blue colour or sometimes orange colour.

(*iii*) An accidental encounter with its tentacles (having batteries of nematocysts) can be a painful and even a fatal experience for the swimmers. Its nematocysts contain a proteinous **neurotoxin** which affects nervous system of victim causing severe burning pain, skin lesions, eruptions, fever, prostration and respiratory interference.

In *Physalia*, the coenosarc is not in the form of a stem or stolon, but it is flattened disc-like occurring at the under surface of the float. From the coenosarc hangs numerous **cormidia.** A cormidium is a group of polymorphic individuals which are modified polyps; they are gastrozooids or siphons, dactylozooids and gonozooids (Fig. 16.3).

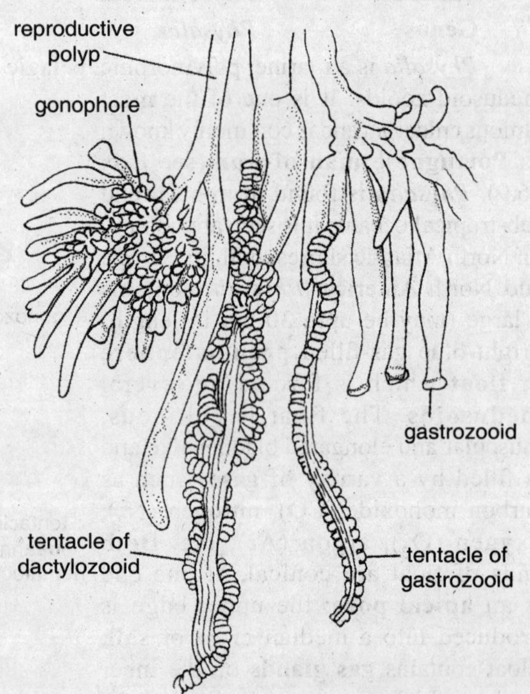

Fig. 16.3. *Physalia*. Various types of zooids.

(*i*) Gastrozooids. They are feeding zooids. They are tube-or bottle-shaped and contain only one tentacle which is also meant for sucking the food. The lips of gastrozooids are applied to the body of fish and partly digested food is sucked up in liquid form.

(*ii*) Dactylozooids. They are tubular, mouthless individuals, each bearing a tentacle at the base. Dactylozooids are of two types: **small** and **large**. The tentacle of each large dactylozooid is very long (upto 12 metres), brightly coloured, muscular and bears a twisted ribbon beset with **nematocysts**. The dactylozooids protect the colony and also help in capturing prey. Long tentacles, in fact, form a drift net for capturing fish for food. With the help of tentacle, *Physalia* catches, stings the prey and pulls it within the reach of gastrozooids.

(*iii*) Gonozooids. Gonozoóids or **gonodendra** are branching blastostyles having leaf-like **gonopalpons** and male and female **medusae** or **gonophores**. Each medusa bears four radial canals, the velum, sense organs but no mouth. The male gonophores are sessile, degenerate (*i.e.*, reduced to sporosacs), grape-like and developed on the proximal part of gonodendra. The female gonophores are degenerate, medusa-like, develop on the distal part of gonodendra and are set free.

The gastrozooids, dactylozooids, blastostyles and gonopalpons are modified polyps, while the gonophores and the float (pneumatophore) are modified medusae.

2. *Velella*

Classification

Phylum – **Cnidaria (Coelenterata)**
Class – **Hydrozoa**
Order – **Hydroida**
Suborder – **Chondrophora**
Genus – ***Velella***

Velella is commonly called **Purple Sail or By-The-Wind Sailor**. It is a polymorphic, marine and colonial form; it looks like a single medusae (Fig. 16.4). It is found in warm seas (*i.e.*, Southern Atlantic coast). *Velella* is a pelagic hydroid which floats near the surface of water.

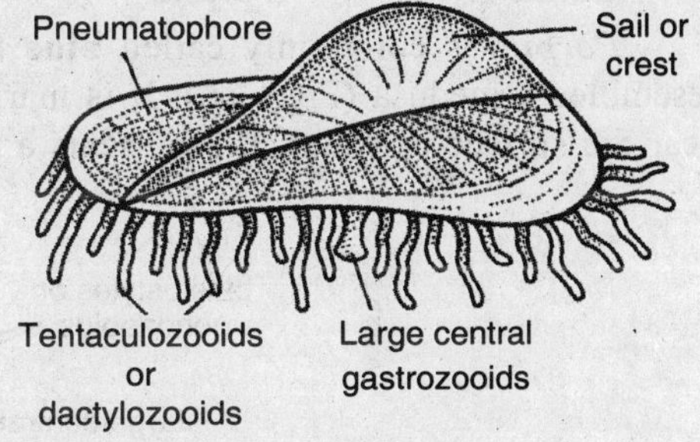

Fig. 16.4. *Velella*.

Velella has a rhomboidal body, on the upper surface of which is an oblique contractile **sail**. Body dorsally contains a **float** or **pneumatophore** which is flattened (5–8 cm long), chitinous and internally divided into a number of concentric air-filled **chambers** (Fig. 16.5). These chambers keep the colony permanently afloat and open above ('aborally') by pores and below ('orally') by canals, called **tracheae**. Air is pumped into float and sail through the tracheal system. The coenosarc forms a flat disc beneath the float and buds off the following zooids which hang downwards:

1. Gastrozooids. From the middle of ventral side of body hangs a single large polyp, called **gastrozooid**. It possesses mouth but lacks tentacles. It feeds the entire colony.

2. Gonozooids. All around the gastrozooid are numerous gonozooids or blastostyles, each having a mouth. They bear clusters of **medusae** or **gonophores** which bear gonads on the manubrium. These medusae are set free, they shed gametes and soon die as they cannot feed.

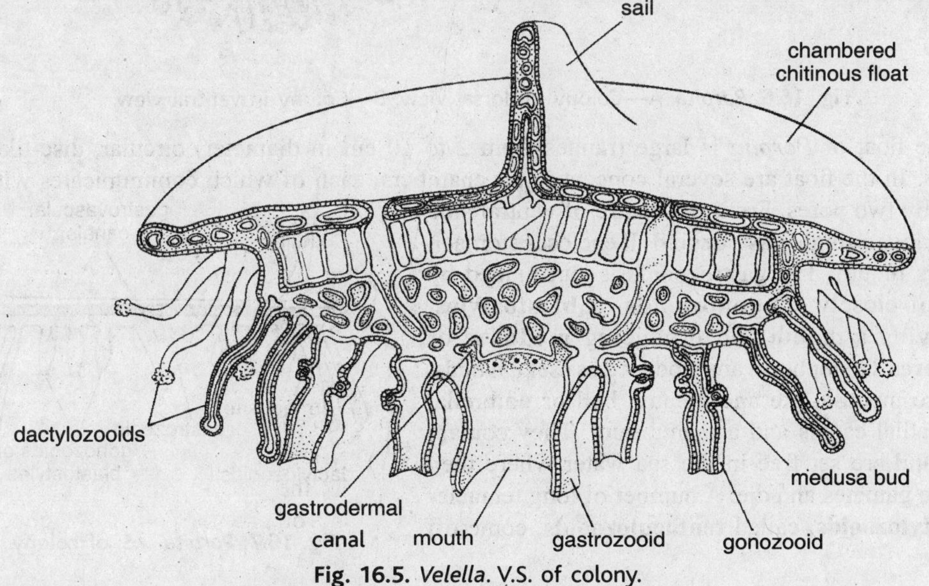

Fig. 16.5. *Velella*. V.S. of colony.

3. Dactylozooids. On the margin of colony is a circle of dactylozooids. They are long, hollow and tentacle-like and have nematocysts.

The pneumatophore (float) and its sail are a modified medusoid, while the gastrozooids, gonozooids and dactylozooids are modified polyps.

3. *Porpita*

Classification

Phylum	–	**Cnidaria (Coelenterata)**
Class	–	**Hydrozoa**
Order	–	**Hydroida**
Suborder	–	**Chondrophora**
Genus	–	***Porpita***

Porpita is commonly called **blue button**. It has an organization similar to *Velella* and resembles a medusa (Fig.16.6). It is marine, found in all seas. *Porpita* is pelagic, floats on or near the surface of water. It represents a polymorphic floating colony.

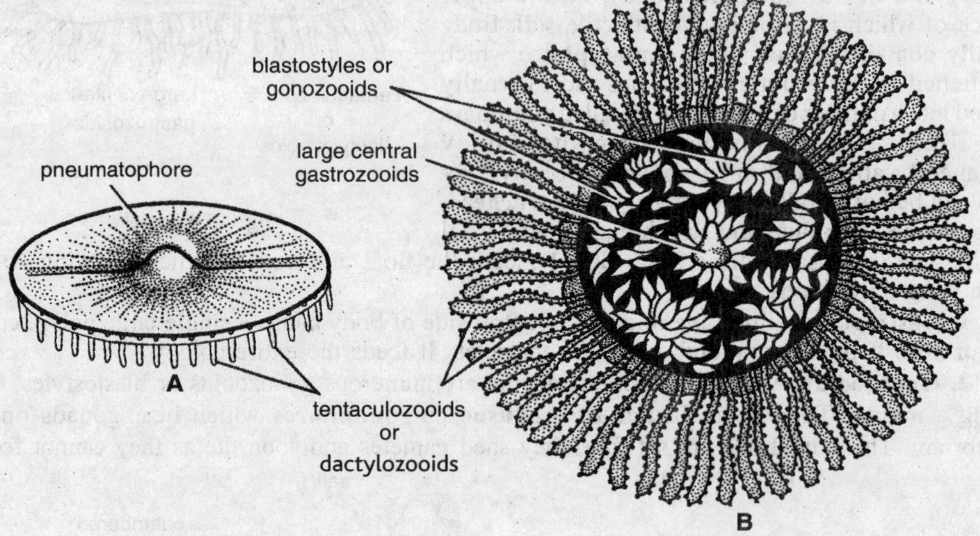

Fig. 16.6. *Porpita*. A—Colony in dorsal view; B—Colony in ventral view.

The float of *Porpita* is large (ranges from 2 to 10 cm in diameter) circular, disc-like and chitinous. In the float are several concentric air chambers, each of which communicates with the exterior by two pores. From the centre of ventral side of float hangs a single **gastrozooid**. It encloses stomach and bears mouth. The gastrozooids is surrounded by groups of closely set **gonozooids** or **blastostyles**, each having a mouth and producing **medusae** or **gonophores**. Gonophores are produced as asexual buds. These are medusa-like and contain bell or umbrella, velum, radial canals and a manubrium. They contain gonads and are set free in the sea water where they discharge gametes and die. A number of long tentacle-like **dactylozooids**, called **tentaculozooids**, come off

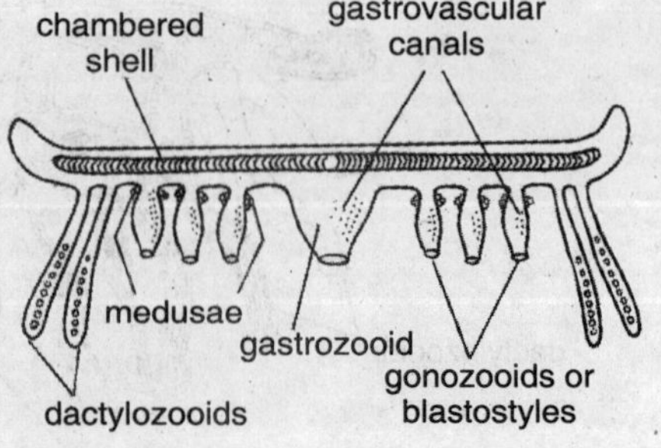

Fig. 16.7. *Porpita*. V.S. of colony.

from the edge of disc; they are loaded with **nematocysts**. The thick fleshy coenosarc below the float is traversed by endodermal tubes communicating with the cavities of the gastrozooid and gonozooids. Ectodermal tubes, called **tracheae**, communicate the air chambers with the outside. Each air chamber dorsally opens to outside by several pores. Contractions of the zooids are believed to circulate air in the tracheae and the air chambers, thus, helping respiration.

4. *Corallium* (Red Coral)

Classification

Phylum	–	**Cnidaria (Coelenterata)**
Class	–	**Anthozoa**
Subclass	–	**Octocorallia**
Order	–	**Gorgonacea**
Genus	–	***Corallium***
Species	–	***rubrum***

Corallium rubrum (Fig. 16.8) is commonly called **red coral** or **red moonga**. It is a marine colonial antho-zoan which inhabits rocky bottom chiefly in the Mediterranean and off coast of Japan at the depth of 30–100 metres. Colony of red coral is branched upright vertically and reaches upto 30 cm in height.

The body of red coral contains a central **axial rod** which is composed of an organic substance called **gorgonin** (which is a tanned collagen). This axial rod is impregnated with calcium carbonate. Around the axis occurs **coenenchyme**, in which are embedded red coloured calcareous **spicules**. Colony of red coral is **dimorphic**, containing following two types of polyps or anthozooids which are white in colour, retractile and embedded in the coenosarc. The free oral end of an anthozoan polyp is called **anthocodium**.

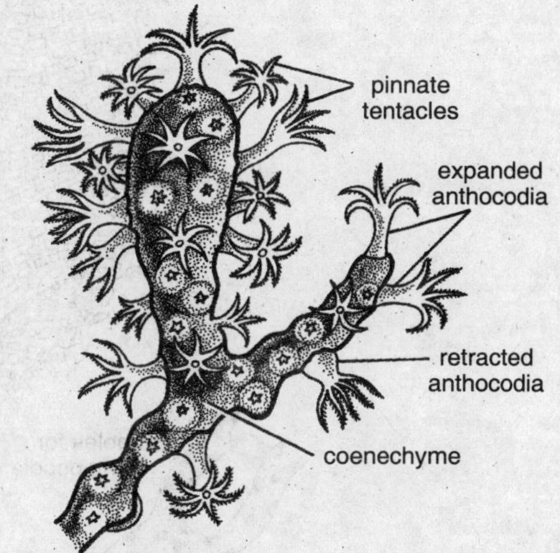

Fig. 16.8. *Corallium rubrum*. A colony.

1. Autozooids. They are nutritive zooids. They bear eight pinnate tentacles, have mesentries and gonads.

2. Siphonozooids. These small sized zooids have no tentacles, contain reduced mesentries and pump water into the canals of the colony.

Corallium rubrum has great economic importance. It provides precious red coral which is used extensively in jewellery.

5. *Gorgonia*

Classification

Phylum	–	**Cnidaria (Coelenterata)**
Class	–	**Anthozoa**
Subclass	–	**Octocorallia**
Order	–	**Gorgonacea**
Genus	–	***Gorgonia***

Gorgonia is commonly called **sea fan**. It is marine, colonial and found in shallow waters on the rocks of tropical and sub-tropical shores (*i.e.*, Indo-Pacific Ocean especially in Malay Archipelago).

Gorgonia produces a large (50 cm in height) upright, fan-shaped colony of yellow or red colour. In this colony, the branches unite to form a network in one plane. The colony contains a short central axial rod (the **stalk**) composed of an organic substance called gorgonin (a collagen-like protein). The stalk is attached with the substratum by a **hold fast** or **pedal/basal plate**. The colony produces branches in one plane only (like the feathers of a bird). These branches are united by numerous cross sections. This skeleton is flexible and allows bending.

Numerous small polyps, called **anthocodia** (= retractile polyps) are found in rows on two sides of branches (Fig. 16.9B). The skeleton is covered with a layer of **coenenchyme** which contains gastrovascular cavities of the polyps and loose calcareous spicules. Polyps feed on zooplankton and other smaller particles. Sexes are separate.

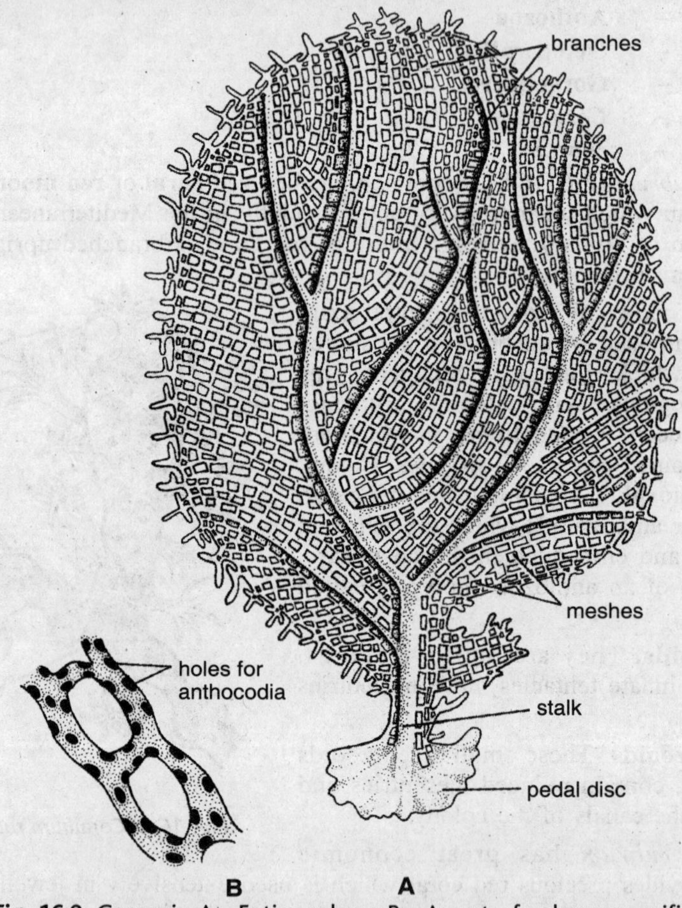

Fig. 16.9. *Gorgonia*. A—Entire colony; B—A part of colony magnified.

6. *Tubipora*

Classification

Phylum	–	**Cnid**
Phylum	–	**Cnidaria (Coelenterata)**
Class	–	**Anthozoa**
Subclass	–	**Octocorallia**
Order	–	**Stolonifera**
Genus	–	***Tubipora***
Species	–	***musica***

Tubipora is commonly called an **organ pipe coral**. It is widely distributed on coral reefs in warm waters of Atlantic, Indian and Pacific oceans. It is a colonial polypoid anthozoan. The colony consists of long parallel upright polyps enclosed in vertical (longitudinal) skeleton tubes arising from a basal plate having creeping mat of **stolons**. The mesogloeal **spicules** fuse to form a continuous vertical tube of each polyp. All the vertical tubes of different polyps are united at definite intervals by transverse calcareous **platforms** (Fig. 16.10). In the living condition this skeleton is covered by ectoderm. The polyps are bright green in colour and skeleton is dull red in colour due to the presence of iron salts. *Tubipora* reproduces asexually by budding.

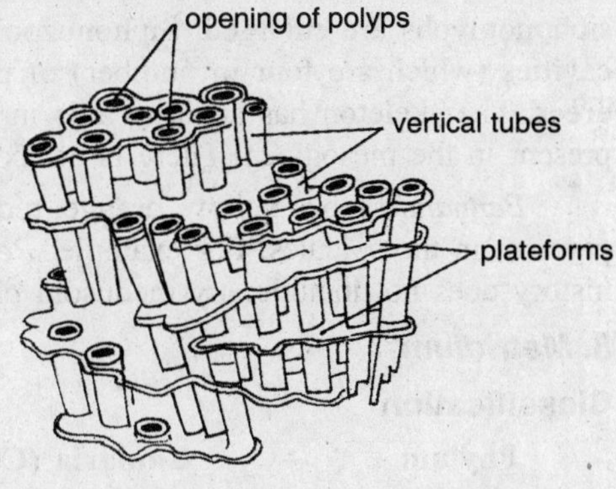

Fig. 16.10. *Tubipora*. Skeleton of organ-pipe coral.

7. *Pennatula*

Classification

Phylum	–	**Cnidaria (Coelenterata)**
Class	–	**Anthozoa**
Subclass	–	**Octocorallia**
Order	–	**Pennatulacea**
Genus	–	***Pennatula***

Habit and Habitat

Pennatula is commonly called **sea pen** or **sea feather**, because of its pen or feather-like appearance. It is a colonial, marine animal found on the soft bottom of warmer seas (Atlantic and Pacific oceans).

Structure

Pennatula contains an elongated, fleshy **primary polyp** (Fig. 16.11). This polyp is modified to form a stem-like basal stalk or **peduncle** and a distal **rachis**. The end of the peduncle is enlarged to form the **end bulb**. The bulb and peduncle are buried in sea mud. The peduncle has special musculature due to which peristaltic contractions occur in the peduncle and sea pen can burrow in the mud or even change its position. The rachis bears two rows of long, fleshy, leaf-like, lateral branches, called **pinnules**. The colony of *Pennatula* is **dimorphic**, containing following two types of **secondary polyps:**

1. Autozooids or anthocodia. They are nutritive zooids and occur in a single row along the upper margin of pinnule. Autozooids have tentacles, mesentries and gonads.

2. Siphonozooids. These are highly modified or reduced types of zooids. They occur on the dorsal side of rachis and between the bases of the pinnules. These are without tentacles and gonads. Their mesentries are reduced, but their

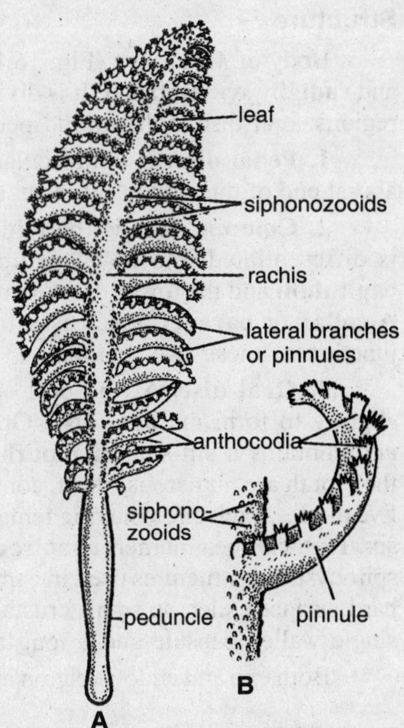

Fig. 16.11. *Pennatula*. A—Entire colony; B—A portion of colony magnified.

siphonoglyphs are enlarged. Siphonozooids pump water into the interconnected gastrovascular cavities (which are four in number) of pirmary polyp, thereby keeping the colony **turgid** and **erect**. The skeleton has a horny axis in the stem (primary polyp) and calcareous spicules are present in the mesogloea. There is no axis in the pinnules.

Pennatula looks yellow, orange, red or purple in colour due to the presence of **lipochrome** pigment in the spicules. For example, *Pennatula sulcata* has a bright orange-red colour. Its life history does not include any medusoid phase.

8. *Metridium*

Classification

Phylum	–	**Cnidaria (Coelenterata)**
Class	–	**Anthozoa**
Subclass	–	**Hexacorallia**
Order	–	**Actiniaria**
Genus	–	***Metridium***

Habit and Habitat

Metridium is commonly known as **sea anemone**. It is a large sessile, brightly coloured, solitary, flower-like animal. *Metridium* is a marine animal, found attached to the rocks, piles of wharves (= landing-stages building especially along the shore for loading and unloading vessels) and solid objects from tide pools to a depth of 90 fathoms. It is found on the Atlantic Coast, Pacific Coast and Europe.

Structure

Body of *Metridium* (Fig. 16.12) is short, cylindrical and **radially symmetrical**. It is divisible into three distinct regions: oral disc, column and pedal disc.

1. Pedal disc. It is muscular, flattened base at the aboral end of the column. It is attached to the substratum.

2. Column. This is the middle cylindrical part. It is differentiated into distal or upper thin-walled short **capitulum** and the lower or proximal thickwalled **scapus**. A **collar** or **parapet** arises as a fold from scapus at the junction of these two regions.

3. Oral disc. At the oral end, the column flares slightly to form the **oral disc**. Oral disc is flat, circular and contains a slit-shaped **mouth** in its centre. Around the mouth are numerous, short, conide hollow **tentacles** in five circles, each circle having tentacles in the multiples of six. Tentacles bear numerous **spirocysts** and **nematocysts**, spirocysts are structures used in capturing of prey that have hard surfaces such as small crustaceans; each contains a single-walled capsule and a long adhesive thread.

Both exo and endo-skeleton are absent in *Metridium*.

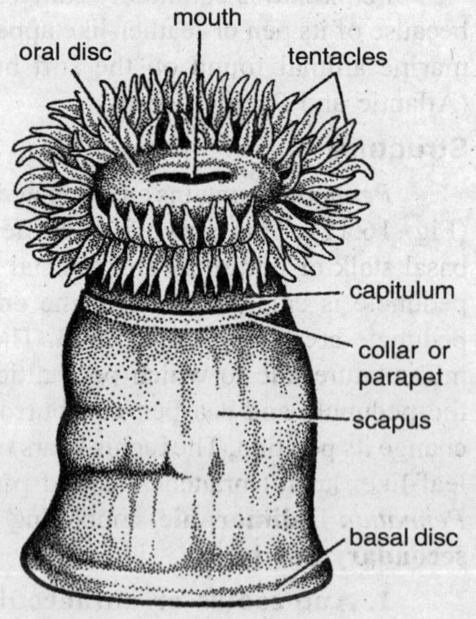

Fig. 16.12. *Metridium*.

Internal Structure

1. Siphonoglyphs. The mouth leads downwards into a long, stout tube called **pharynx** which is lined by invaginated ciliated epidermis and hangs in the coelenteron (Fig. 16.13). In the pharynx are two longitudinal ciliated grooves called **siphonoglyphs** which are meant for the entry of water into the gastrovascular cavity. This current of water functions to maintain an internal fluid skeleton or **hydrostatic skeleton** against which the muscular system can act and also provides for the exchange of gases through the gastrodermal surface.

2. Mesentries. The gastrovascular cavity of the sea anemone is partitioned by longitudinal, radiating septa or **mesentries** which increase the internal surface area. In the sea anemone, there are two types of septa—complete and incomplete. **Complete septa** are connected to the

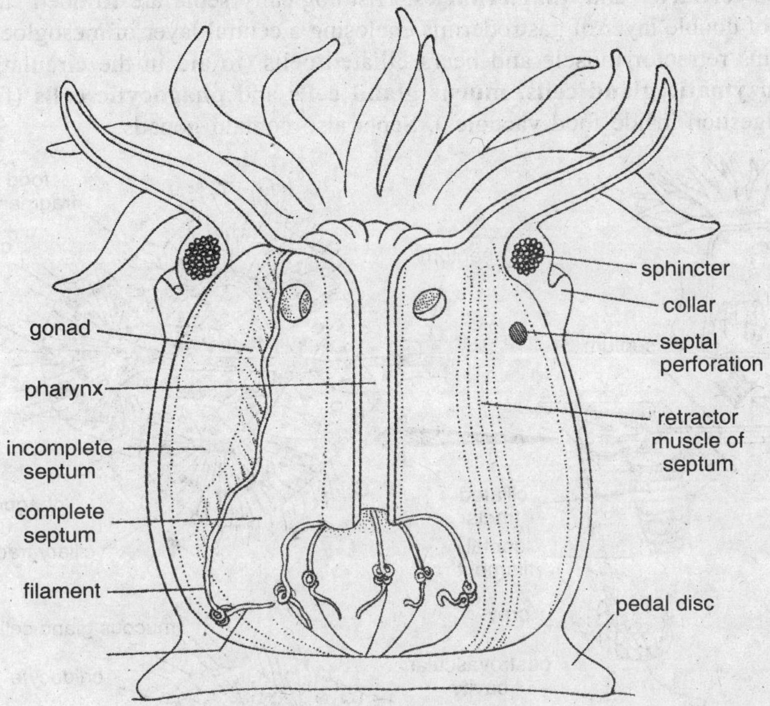

Fig. 16.13. L.S. of a sea anemone.

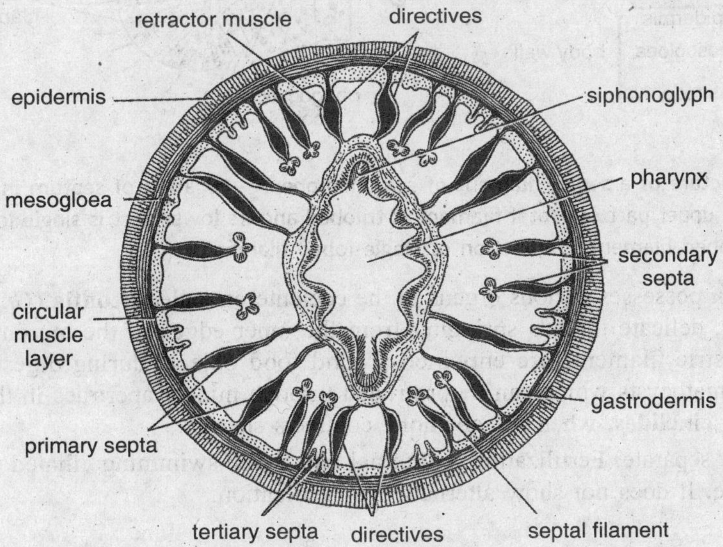

Fig. 16.14. *Metridium*. Cross section of body through the pharynx.

body wall on one side and to the wall of the pharynx on the other (Fig. 16.14). **Incomplete septa** are connected only to the body wall and extend only partially in the gastrovascular cavity. Both complete and incomplete septa are arranged in adjacent pairs. In the *Metridium*, there are six pair of complete or **primary mesentries**. Two pairs of primary mesentries are joined to the siphonoglyphs and are called **directives** (Fig. 16.14). Between adjacent primary mesentry pairs are **secondary mesentries** or incomplete septa; they are six pairs. Further smaller types of incomplete septa are called **tertiaries** and **quarternaries**. Histologically septa are trilobed structures. They are composed of double layer of gastrodermis enclosing a central layer of mesogloea (Fig. 16.15). A septa contains retractor muscle and bears ciliated cells (to aid in the circulation of water), **cnidocytes, enzymatic gland cells, mucus gland cells** and **phagocytic cells** (for performing intracellular digestion inside food vacuoles). Septa also contain gonads.

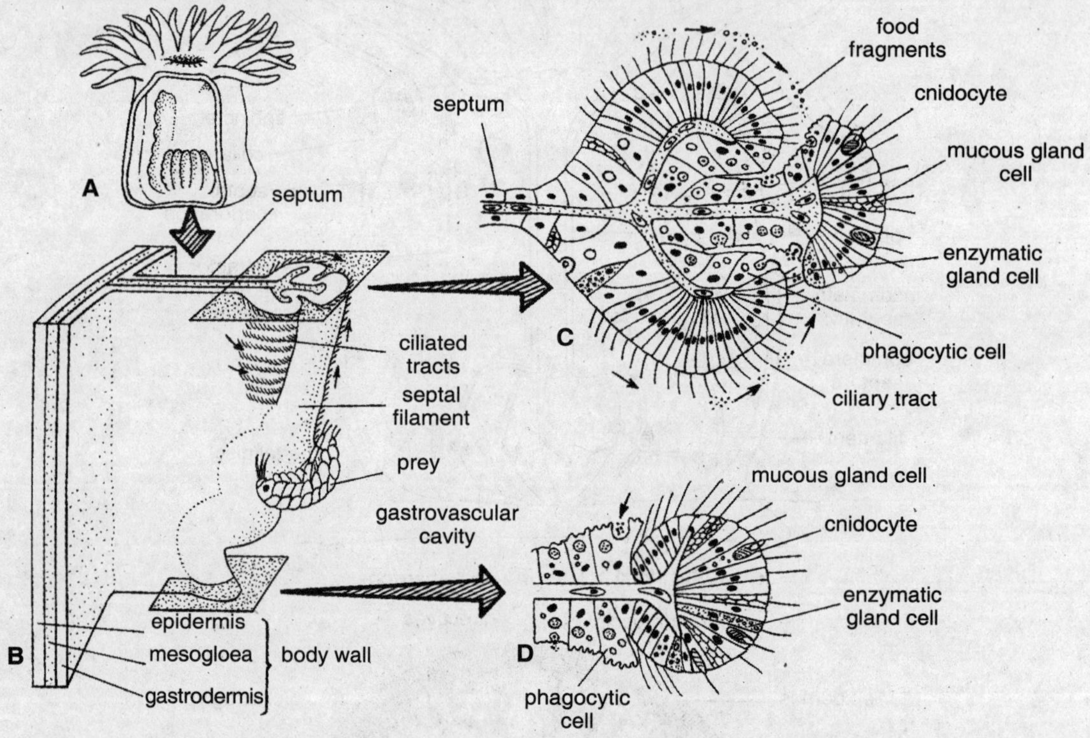

Fig. 16.15. Structure of a septal filament of sea anemone. A—Location of septum in sea anemone; B—The upper part of septal filament is trilobed and its lower part is single lobed; C—Section of trilobed filament; D—Section of single-lobed filament.

Metridium possesses curious organs in the coelenteron called **acontia** (*Gr., akontion* =dart). These are long, delicate threads springing from the inner edges of the mesentries and like the mesenterial gastric filaments are entwined around food objects during digestion. Acontia are loaded with nematocysts which can be protruded through minute apertrues in the column, called **'port holes'** or **cinclides**, when the anemone contracts sharply.

Sexes are separate. Fertilization is internal and a free-swimming ciliated **planula** originate from the zygote. It does not show alternation of generation.

9. *Adamsia*

Classification

Phylum	–	**Cnidaria (Coelenterata)**
Class	–	**Anthozoa**
Subclass	–	**Hexacorallia**
Order	–	**Actiniaria**
Genus	–	***Adamsia***

Adamsia is commonly called **sea anemone.** *Adamsia palliata* is a marine **commensal** (Fig. 16.16), found attached to the empty shells of gastropods inhabited by hermit crab, *Eupagurus* (Box 16.10). It is found in Indian ocean, USA (North Carolina to Florida) in shallow waters.

Box 16.10 Definitions

1. **Commensalism.** It is a relationship between two organisms of different species in which one organism—the **commensal**—benefits from the partnership while the other neither gains nor loses. The commensal often gets shelter, transport or food from its host.
2. **Cinclides.** Pores in the body wall of some sea anemones for the release of water and stinging cells.

The body of *Adamsia* is large and cylindrical and divisible into three distinct regions: the pedal disc, column and oral disc. **Pedal disc** is flat, bilobed and sucker-like. It is attached to the molluscan shell. **Column** is cylindrical and bears a band of cinclidal tubercles at its base. Two siphonoglyphs and six pairs of mesentries are present. **Oral disc** bears an elongated slit-like **mouth** in its centre. Mouth is surrounded by numerous, short, conical tentacles bearing nematocysts. Tentacles are arranged into five whorls.

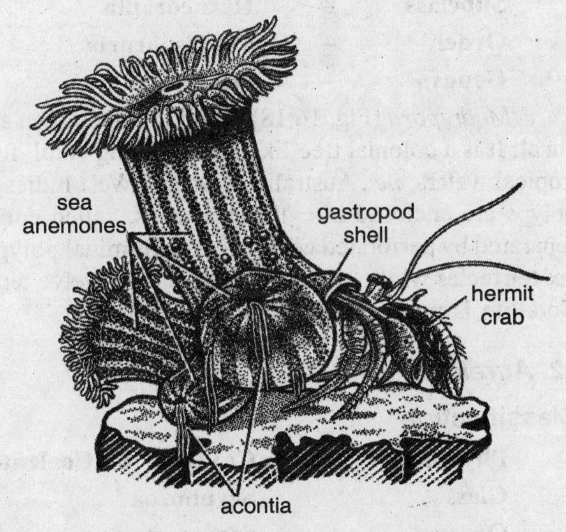

Fig. 16.16. *Adamsia*. Commensalism.

Adamsia provides a good example of the phenomenon of **commensalism**. *Adamsia* protects the hermit crab from the attack of its enemies by stinging with help of nematocysts. In return, *Adamsia* gets the advantage of transportation from one place to another by the hermit crab. Due to this action, sea anemone gets a variety of food supply.

10. *Fungia*

Classification

Phylum	–	**Cnidaria (Coelenterata)**
Class	–	**Anthozoa**
Subclass	–	**Hexacorallia**
Order	–	**Madreporaria**
Genus	–	***Fungia***

Fungia is commonly called **mushroom coral** (Fig. 16.17). It is solitary marine stony or scleractinian coral of coral reefs. It occurs in warm sea water of Indo-Pacific region and Gulf of California. Its skeleton is called **corallite** and it is discoidal and mushroom-like with upper surface convex and lower concave. A large number of septa (called **sclerosepta**) radiate out from the centre and are connected by calcareous rods, called **synapticulae**. Adult animal bears a single large polyp (25 cm in diameter) bearing many tentacles. Siphonoglyph is absent. It reproduces asexually (by transverse fission) and sexually (by producing a planula larva).

septa

Fig. 16.17. *Fungia*.

11. *Madrepora*

Classification

Phylum	–	**Cnidaria (Coelenterata)**
Class	–	**Anthozoa**
Subclass	–	**Hexacorallia**
Order	–	**Madreporaria**
Genus	–	***Madrepora***

Madrepora (Fig. 16.18) is commonly known as **horn coral**. It is a colonial tree-like reef-building coral, found in tropical waters, *i.e.*, Australian seas and West Indies. Small polyps are enclosed in cylindrical cups, called **corallites,** separated by perforated **coenosteum**. Terminal polyps have six **tentacles** while lateral polyps contain twelve tentacles. Corallite is made up of calcium carbonate.

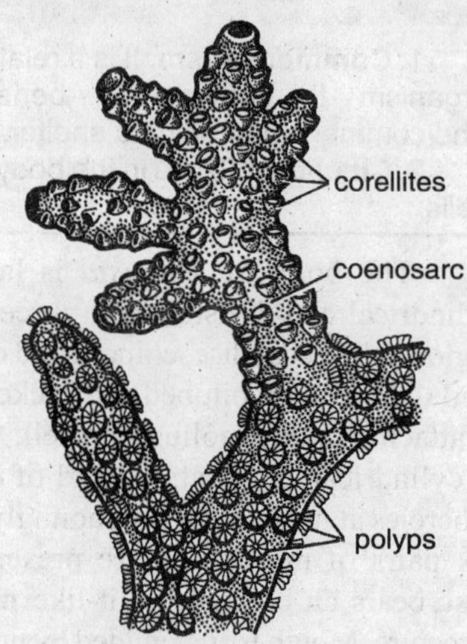

Fig. 16.18. *Madrepora*.

12. *Aurelia*

Classification

Phylum	–	**Cnidaria (or Coelenterata)**
Class	–	**Scyphozoa**
Order	–	**Semaeostomeae**
Genus	–	***Aurelia***
Species	–	***aurita***

Aurelia is a **jellyfish**; its body is composed of jelly-like material. However, it is not a fish.

Habit and Habitat

Aurelia aurita is commonly known as **moon-jelly**. It occurs in coastal waters of tropical and temperate oceans of the world. It is **pelagic**, floating or swimming freely close to the surface of water. It lives either singly or in large groups, called **shoals**. It is carnivorous and reproduces both asexually and sexually. Life cycle includes both polypoid and medusoid phases; but medusoid phase is dominant one.

Structure

The medusoid body form of *Aurelia* measures 7 to 30 cm in diameter. It is flattened, saucer-shaped structure, called **bell** or **umbrella**. It exhibits **tetramerous radial symmetry**, *i.e.*, body parts are symmetrically arranged around the oral-aboral axis in fours or some multiple of four. It is almost transparent having bluish-white body with pinkish gonads.

Umbrella-shaped body is distinctly divided into a slightly convex upper surface, called **exumbrella** and the concave lower surface, called **subumbrella**. The margin of umbrella is circular which is **scalloped**, *i.e.*, broken by eight notches into eight lobes. Each notch contains two leaf-like processes, called **marginal lappets**. Between the lappets lies a balancing sensory organ, the **statocyst** or **tentaculocyst**. The statocyst and lappets together form a **rhopalium**. The margin of umbrella in between notches is provided with numerous, small hollow marginal **tentacles**. *Aurelia* is a **acraspedote cnidarian** (Box 16.11), it lacks a true velum, however, contains a **pseudovelum** or **velarium**.

Box 16.11 Definitions

1. **Craspedote.** Cnidarians with velum or **craspedon** in their medusae, *e.g.*, hydromedusae (or medusae) of *Obelia* and other hydrozoans.
2. **Acraspedote.** Cnidarians without any velum or craspedon in their medusae, *e.g.*, *Aurelia* and other scyphozoans.

The subumbrellar surface of *Aurelia* bears in its centre a short **manubrium**. At its free end the manubrium contains a square mouth (Fig. 16.19B). All the four corners of the mouth are drawn out into four, long, mobile, delicate, frilled and tapering processes called **oral arms**. Each oral arm has a ventral ciliated groove and is provided with numerous nematocysts. Mouth leads into a short **gullet** which opens into **stomach**. The stomach gives rise to four inter-radial **gastric pouches**.

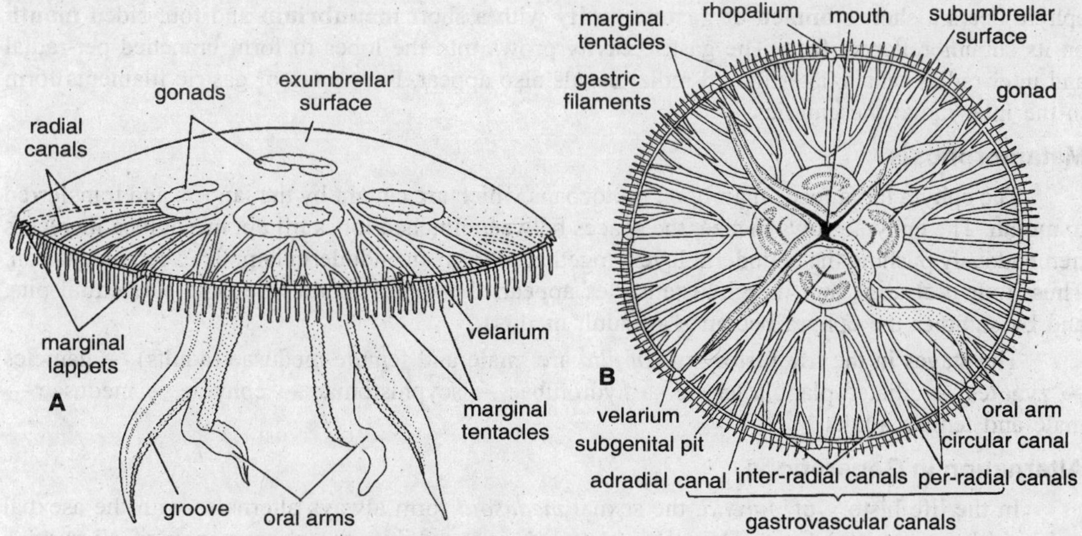

Fig. 16.19. *Aurelia aurita*. A—Lateral view; B— Ventral view.

Gonads and Subgenital Pits

Four horse-shoe-shaped pinkish gonads which later become circular are situated between oral arms in the center of the subumbrellar surface. Gonads are, in fact, internal lying in gastric pouches but they are visible externally. Just below each gonad in the subumbrellar surface, is a circular aperture called **subgenital pit** of unknown function. The gonads, however, have no connection with subgenital pits.

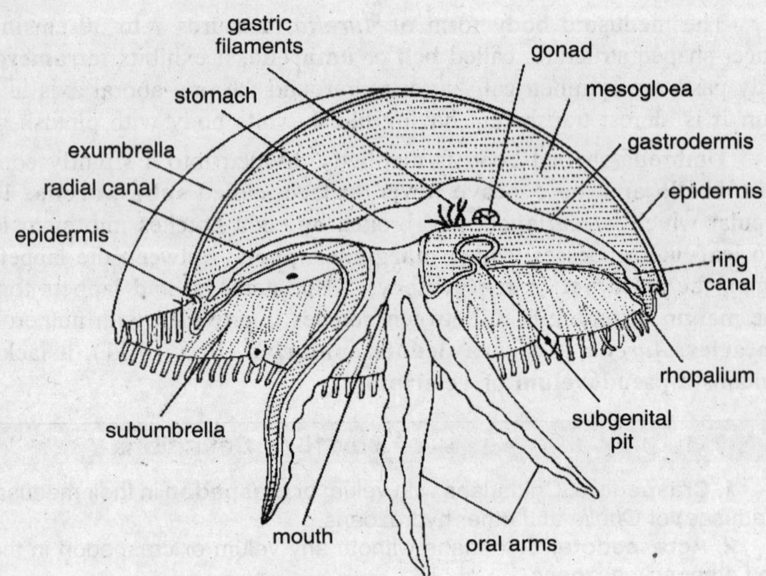

Fig. 16.20. *Aurelia*. Gastrovascular system.

Ephyra Larva

The ephyra is a young medusoid form of *Aurelia* (Fig. 16.21). It has eight **notched lobes** or **arms** which are per-radial and inter-radial. Each lobe has two over grown **marginal lappets** with a deep groove between them having a short tentacle which becomes a **tentaculocyst**. An ephyra bears a short **stomach** or **gastric cavity** with a short **manubrium** and four-sided **mouth** on its subumbrellar surface. The gastric cavity grows into the lobes to form branched per-radial and inter-radial canals, later the ad-radial canals also appear. Four pairs of gastric filaments form in the inter-radial mesenteries.

Metamorphosis

The ephyra larva feeds mainly on protozoans which are caught by the lappets and transferred to mouth. The ephyra grows in size, the spaces between notched lobes fill up, mesogloea increases tremendously pushing the endoderm layers together to form an endoderm lamella except in enteron. Thus, four oral arms and marginal tentacles appear, the septal funnels become subgenital pits, and by summer the ephyra becomes an adult medusa.

The stages in the life history of *Aurelia* are: male and female medusae (adults) → gametes → zygotes → ciliated planula larvae → hydrotubae → scyphistomae → ephyrae → medusae → male and female adults.

Alternation in Generations

In the life history of *Aurelia*, the sexual *medusoid* form always alternates with the asexual *polypoid* (scyphistoma) form. Thus, in a sense, *Aurelia* exhibits the phenomenon of *alternation of generations* in its life history. The free swimming adult *Aurelia*, representing the medusoid form, is the **sexual generation** which reproduces by the formation of gametes. The zygote develops into fixed polypoid form called scyphistoma through a free swimming planula larval stage. The polypoid form scyphistoma represent **asexual generation**, which reproduces asexually by budding or fission.

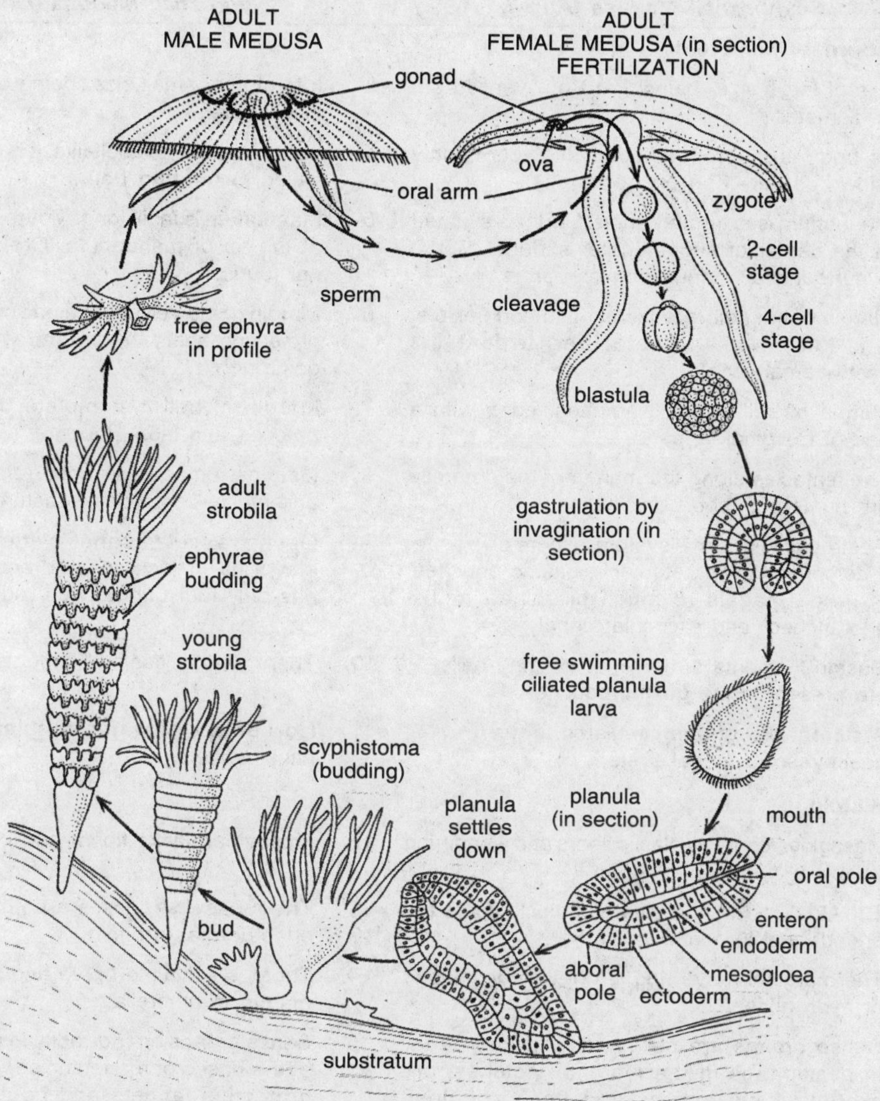

Fig. 16.21. *Aurelia aurita*. Stages in development and life history.

DIFFERENCES BETWEEN MEDUSAE OF *AURELIA* AND *OBELIA*

Table 16.1 Differences between scyphozoan medusa (*e.g.*, *Aurelia*) and hydrozoan medusa (*e.g.*, *Obelia*).

	Scyphozoan Medusa (Aurelia)		***Hydrozoan Medusa (Obelia)***
(a)	***Status***		
1.	Adult *Aurelia* represents large medusoid phase of life history; it is dominant stage.	1.	The medusa represents one of the zooids in the life history of polymorphic colony of *Obelia*.
2.	It is formed by metamorphosis of an ephyra larva.	2.	It is formed from a bud arising in the colony.

	Scyphozoan Medusa (*Aurelia*)		*Hydrozoan Medusa* (*Obelia*)
(*b*)	**External Characters**		
3.	It is of large size, being 7–10 or even 30 cm in diameter.	3.	It is of very small size, being only 6 mm in diameter.
4.	Its body is umbrella-shaped. It is commonly known as **jelly fish** or **moon-jelly**	4.	Its body is inverted bell-like. It is commonly called **swimming bell.**
5.	Manubrium is extremely short. Mouth is situated at the center of subumbrellar surface. Mouth is surrounded by four long oral arms.	5.	Manubrium is fairly long. Mouth is situated at the tip of manubrium. Oral arms are not found.
6.	Margin of the umbrella is very thin and fiexible. It is known as the **velarium**. It contains gastrodemal canals.	6.	Margin of the bell bears a narrow inwardly projecting shelf, the **velum**. which lacks gastrodemis.
7.	Margin of bell bears 8 notches, each with a pair of marginal lappets.	7.	Margin of bell is complete, having no notches and lappets.
8.	The tentacles along the margin of the umbrella are numerous, short and hollow.	8.	Marginal tentacles are solid, relatively long and highly contractile structures.
9.	Gastrovascular canal system is more elaborate. It comprises gullet, stomach, gastric pouches, 16 raidal canals (8 much branched and 8 unbranched) and a circular canal.	9.	Gastrovascular canal system is less elaborate. It lacks gastric pouches and has only four unbranched radial canals.
10.	Gastric filaments armed with stinging cells are present in the gastric pouches.	10.	There are no gastric filaments.
11.	A current of water is circulated in the gastrovascular canal system.	11.	No current of water circulates in the canal system.
(*c*)	**Histology**		
12.	Mesogloea contains elastic fibers and wandering cells.	12.	Mesogloea lacks fibers and cells.
13.	The striated and non-striated muscle fibers are endodermal in origin.	13.	The muscle strands are both ecto- and endo-dermal in origin.
14.	The marginal nerve ring is not well developed or even absent.	14.	There is a double nerve ring around the margin of umbrella.
15.	Sense organs are per-radial and inter-radial in position, lie in the marginal notches, are protected by lappets and hood, enclose gastrodermal canal and are called **rhopalia**. They maintain equilibrium and also have spots sensitive to light and chemicals.	15.	Sense organs are ad-radial in position, lie at the bases of the tentacles, are without protective lappets and hood, and are purely ectodermal sacs and are called statocysts. They are concerned only with equilibrium.
16.	The nematocysts are found all over umbrella, tentacles and oral arms. These are of three types.	16.	The nematocysts are only found on manubrium and tentacles. These are of only one type.
(*d*)	**Gonads**		
17.	Gonads are situated on the floor of the gastric pouches and are formed of gastrodermis.	17.	Gonads are on the radial canals and are formed of epidermis.
18.	Gamets are shed in the gastric pouches. whence they escape into the sea by water current.	18.	Gametes are shed directly in sea water.

QUESTIONS

Long Answer Questions

1. Mention the distinguishing features of phylum Coelenterata. Give its general characters, classification and examples of each class.

2. What is the basis of classification of phylum Coelenterata. Classify it upto orders giving examples.

3. Give classification of class Hydrozoa upto orders. Mention the characteristics of each order.

Short Answer Questions

1. Give an outline classification of phylum Cnidaria or Coelenterata.

2. Classify and comment on the following:

(*i*) *Physalia*

(*ii*) *Velella*

(*iii*) *Porpita*

(*iv*) *Aurelia*

(*v*) *Corallium* (Red coral)

(*vi*) *Gorgonia*

(*vii*) *Tubipora*

(*viii*) *Pennatula*

(*ix*) *Metridium*

(*x*) *Adamsia*

(*xi*) *Fungia*

(*xii*) *Madrepora*

3. Draw a well labelled diagram of subumbrella view of *Aurelia*. (*Purvanchal, 2001*)

4. Draw a labelled diagram of medusa of *Aurelia* (No description is required). (*Rohilkhand, 2003*)

Very Short Answer Questions

1. What is coral?

2. Name some coral building cnidarians.

3. Name the famous reef near Australia.

4. What are coral reefs?

5. Name three kinds of coral reefs.

6. Suggest the term for the occurrence of two or more types of individuals in an animal.

7. What is polymorphism? (*C.C.S.U. Meerut 2008*)

8. What is the zoological name of Portuguese man of war?

9. Name the polymorphic hydrozoan in which the penumatophore lacks sail.

10. What is the function of dactylozooids?

11. Name the phylum in which nervous system appears for the first time.

12. In which class of the phylum Cnidaria is stomodaeum found?

13. Name a freshwater cnidarian.

14. Whether gonads are endodermal in *Aurelia*.

15. What are mesentries?

16. Define the siphonoglyph.

17. Give zoological name of sea anemone studied by you.

18. What is corallite?

Fill in the Blanks

1. Cnidarians have .... symmetry and .... level of organisation.

2. Cnidarians have a single cavity called ....

3. Cells characteristic of cnidarians are .... and ....

4. The free-swimming larva of cnidarians is called ....

5. Cnidarians are first metazoans to have a digestive tract and ....

6. The phylum Cnidaria is divided into 4 classes .... , ... , Cubomodusae and Anthozoa.

7. The common name of *Tubipora* is ....

8. Symmetry in Anthozoa is ....

9. The true corals belong to the order ....

10. The zoological name of the red coral of commerce is ....

11. The sense organs of *Aurelia* are called ....

12. Hydrozoa have both polyp and .... form.

13. The most highly polymorphic hydrozoans form the order ....

14. Predominant body form is Scyphozoa is ....

15. .... exist in polyp form only.

16. *Vellela* forms a .... colony.

17. Zooids of *Porpita* include pneumatophore, ... , gonozooids and ....

18. Function of dactylozooids is ....

19. Members of the subclass *Octocorallia* have ....tentacles and a single .... in the stomodaeum.

20. Hexacorallina have .... tentacles and .... siphonoglyphs.

Multiple Choice Questions

1. Sea anemone is a

(*a*) anthozoan (*b*) scyphozoan

(*c*) hydrozoan (*d*) none of these

2. Sea pen is the common name of

(*a*) *Corallium* (*b*) *Pennatula*

(*c*) *Gorgonia* (*d*) *Fungia*

3. Jellyfish is a

(*a*) hydrozoan (*b*) scyphozoan

(*c*) anthozoa (*d*) fish

4. An example of soft coral is

(*a*) *Gorgonia* (*b*) *Tubipora*

(*c*) *Heliopora* (*d*) *Alcyonium*

5. The fire coral is the
(*a*) *Tubularia* (*b*) *Pennaria*
(*c*) *Millepora* (*d*) *Physalia*

6. *Gorgonia* is generally known as
(*a*) sea pen (*b*) sea fish
(*c*) sea fan (*d*) sea disc

7. Dead man's finger is
(*a*) *Heliopora* (*b*) *Fungia*
(*c*) *Alcyonium* (*d*) *Corallium*

8. Which is known as mushroom coral?
(*a*) *Fungia* (*b*) *Aurelia*
(*c*) *Adamsia* (*d*) *Tubipora*

9. Function of the polyp is
(*a*) reproduction (*b*) excretion
(*c*) respiration (*d*) nutrition

10. *Pennatula* is
(*a*) dimorphic (*b*) trimorphic
(*c*) tetramorphic (*d*) polymorphic

11. A tetramorphic colony is formed by
(*a*) *Gorgonia* (*b*) *Obelia*
(*c*) *Velella* (*d*) *Physalia*

12. Scyphistoma of *Aurelia* multiplies by
(*a*) budding (*b*) strobilation
(*c*) binary fission (*d*) none of these

13. Metagenesis is exhibited by
(*a*) *Aurelia* (*b*) *Metridium*
(*c*) *Obelia* (*d*) *Porpita*

14. The function of a dactylozooid in polymorphic colony is
(*a*) reproduction (*b*) offence and defence
(*c*) nutrition (*d*) locomotion

15. The larva of *Physalia* is
(*a*) planula (*b*) ephyra
(*c*) scyphistoma (*d*) cydippid larva

16. Polymorphism is found in
(*a*) *Halistemma* (*b*) *Leucernaria*
(*c*) *Zoanthus* (*d*) *Alcyonium*

17. Horn coral is
(*a*) *Madrepora* (*b*) *Fungia*
(*c*) *Astraea* (*d*) *Adamsia*

18. Hypnotoxin is a piosonous fluid produced by
(*a*) parasitic protozoa (*b*) nematocysts
(*c*) *Ascaris* (*d*) sponges

ANSWERS

Very Short Answer Questions

1. The skeleton of certain solitary and colonial anthozoan coelenterates: composed chiefly of calcium carbonate.

2. *Millepora, Tubipora, Gorgonia.*

3. The Great Barrier Reef.
4. A ridge or mass of limestone built up of detritus material deposited around a framework of skeletal remains of molluscs, colonial corals, and massive calcareous algae.
5. Fringing reef, barrier reef and atoll.
6. Polymorphism.
7. Polymorphism is a phenomenon of occurrence in an animal of more than one type of individuals, other than male and female, all with different form and function. The individuals of polymorphic cnidarians are known as **zooids.** Some cnidarians have only two types of zooids-polyp and medusae, while other colonial forms may contain many different types of modified polyps or zooids such as gastrozooids, gonozooids and dactylozooid.
8. *Physalia.*
9. *Porpita.*
10. Protection (They are defensive polyps).
11. Cnidaria (= Coelenterata).
12. Anthozoa.
13. *Hydra.*
14. Yes.
15. The gastrovascular cavity of sea anemone is partitioned by longitudinal, radiating septa, called **mesentries**. They increase the internal surface area.
16. It is a ciliated groove leading from the mouth to the gullet in certain anthozoans.
17. *Adamsia palliata* (or *Metridium*).
18. Exoskeleton of individual zooid of Cnidaria.

Fill in the Blanks

1. Raidal, tissue
2. Coelenteron
3. Cnidoblast, epitheliomuscular cells
4. Planula
5. Nervous system
6. Hydrozoa, Scyphozoa
7. Organ-pipe coral
8. Biradial
9. Madreporaria
10. *Corallium rubrum*
11. Rhopalia
12. Medusa
13. Siphonophora
14. Medusa
15. Anthozoa
16. Tetramorphic
17. Gastrozooid, dactylozooids
18. Protection
19. Eight, siphonoglyph
20. Numerous, two.

Multiple Choice Questions

1. (*a*)	2. (*b*)	3. (*b*)	4. (*d*)	5. (*c*)	6. (*c*)
7. (*c*)	8. (*a*)	9. (*d*)	10. (*a*)	11. (*c*)	12. (*b*)
13. (*c*)	14. (*b*)	15. (*a*)	16. (*a*)	17. (*a*)	18. (*b*).

17

Obelia (Phylum Cnidaria)

17.1 CLASSIFICATION

Phylum	–	**Cnidaria (or Coelenterata)**
Class	–	**Hydrozoa**
Order	–	**Hydroida**
Suborder	–	**Leptomedusae**
Genus	–	***Obelia***
Species	–	***geniculata***

17.2 HABIT AND HABITAT

Obelia is commonly referred to as the **sea fur**. It is a marine, sedentary, colonial, delicate and filamentous organism. Its small (2 to 3 cm in height) branched colonies are attached to the rocks, stones, molluscan shells, wooden piles and seaweeds in shallow sea water. *Obelia* looks like whitish or light brown fur on the substratum. It exhibits phenomena of **polymorphism, alternation of generations** and **metagenesis**.

Geographical Distribution

Obelia has a cosmopolitan distribution. It is abundant both in the Atlantic and Pacific coastal waters. It is usually found upto the depth of 80 metres.

17.3 STRUCTURE

Obelia exists in following two distinctly different forms during its life history: 1. **Polypoid, hydroid** or **hydranth phase;** 2. **Medusoid phase.**

1. Hydroid Colony

A hydroid colony of *Obelia* contains different types of individuals, called **zooids**. Each colony of *Obelia* consists of a horizontal thread-like branched root, called **hydrorhiza** which is attached to the substratum. The hydrorhiza gives off several vertical stems called **hydrocauli.** Both hydrorhiza and hydrocauli are hollow tubes. Each hydrocaulus gives off short lateral branches in an alternate manner. Each lateral branch bears a zooid.

Microscopic Structure of *Obelia* Colony

1. Coenosarc. All branches and zooids of *Obelia's* colony contain an inner tubular living portion, called **coenosarc**. Coenosarc encloses a cavity or canal, called **enteron** or **gastrovascular cavity.** Through this canal digested food is distributed throughout the hydroid colony. The wall of gastrovascular cavity is formed of outer **epidermis** (ectodermal in origin) and an inner

gastrodermis (endodermal in origin). A gelatinous and acellular **mesogloea** lies in between these two layers of the diploblastic animal.

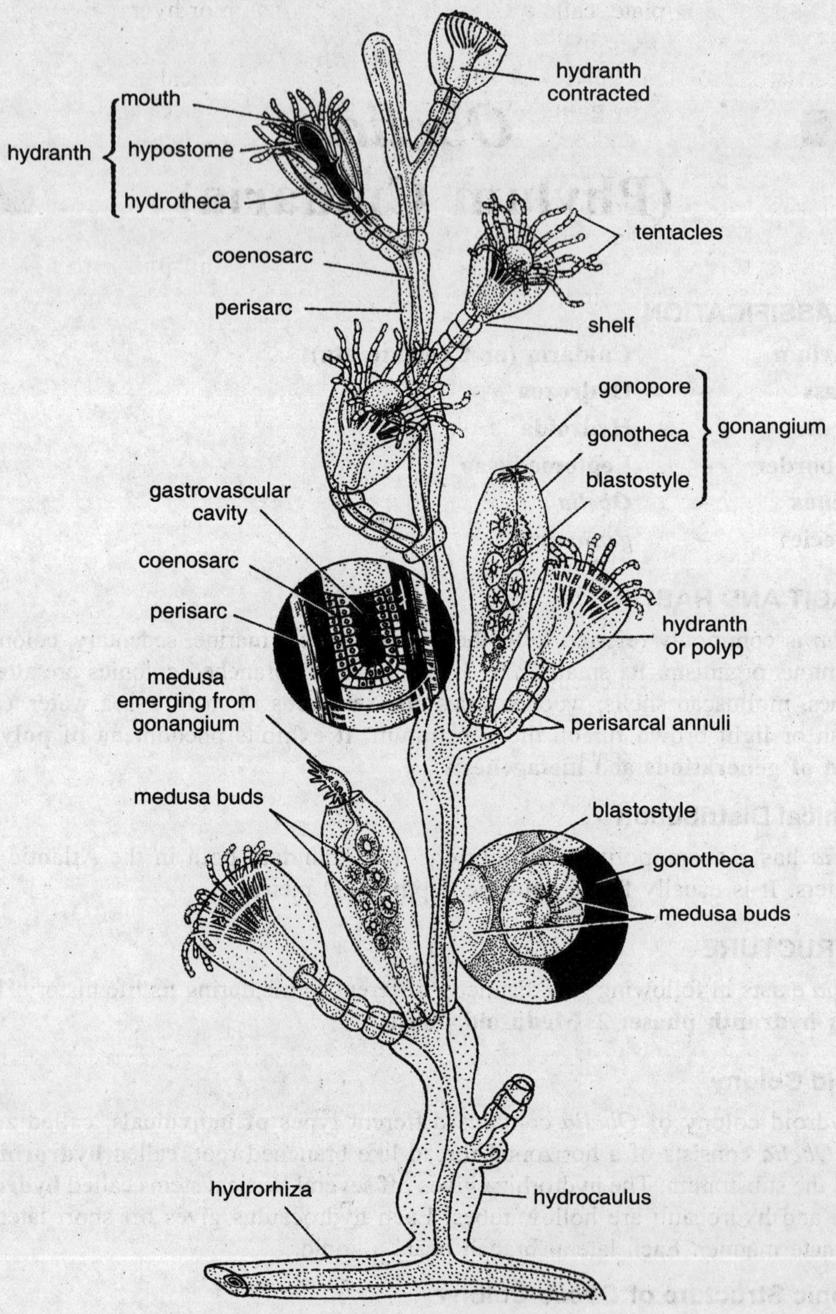

Fig. 17.1. *Obelia*. A portion of colony.

2. Perisarc. The coenosarc of entire colony is covered by a tough, yellow or brown transparent non-living protein (=chitin) layer, called **perisarc**. It is secreted by epidermis over the coenosarc and forms the **exoskeleton** of the colony. Perisarc covers hydrorhiza, hydrocauli, their branches and zooids. Around each polyp, perisarc expands to form a clear widely open wine glass-shaped structure, called **hydrotheca**. Inside the hydrotheca is present a circular, centrally perforated disc-like basal plate, called **shelf**, to support the polyp or hydranth. The hydranth can contract and withdraw into the hydrotheca.

The perisarc around the blastostyle forms a transparent cylindrical capsule called **gonotheca** which is open at its free end by **gonopore**. The blastostyle and gonotheca are called a **gonangium**. At the base of each branch and each zooid, perisarc has annular (ring-like) constrictions which permit some swaying of the branches and zooids caused due to water currents.

3. Zooids and polymorphism. The colony of *Obelia* is **polymorphic** showing the phenomenon of polymorphism. When two or more members of a species or zooids of a colony are structurally modified for different functions, it is called **polymorphism**. A young colony of *Obelia* is termed **dimorphic**, since it contains only two types of zooids, namely **gastrozooids** (*i.e.*, polyps or hydranths) and **gonozoids** (*i.e.*, blastostyles). Mature blastostyles produce sexual zooids, the **medusae** by asexual budding. So a mature *Obelia* colony is termed **trimorphic**, since it contains three types of zooids: gastrozooids, gonozooids and medusae (Fig. 17.1).

I. Gastrozooids

They are also called **polyps** (Gr., *polypus*= many-footed), **hydranths** (Gr., *hydra* or *hydor* = water + *anthos* = flower) or **trophozooids**. All polyps are feeding structures and their collective activity supplies the whole colony with an adequate means of nourishment. Polyps are situated singly at apices of each main stem (hydrocaulus) and all branches. Each polyp is cylindrical vase-like structure attached to the axis of the hydrocaulus by its proximal end and remains free at its distal end (Fig. 17.2). The free distal end of the polyp is produced into a conical elevation called **manubrium** or **hypostome** which is about one-third of the total length of hydranth. The apex of the hypostome bears a circular terminal aperture, called **oral aperture** or **mouth**. A circlet of numerous (about 24) slender and solid **tentacles** occur around the base of the hypostome.

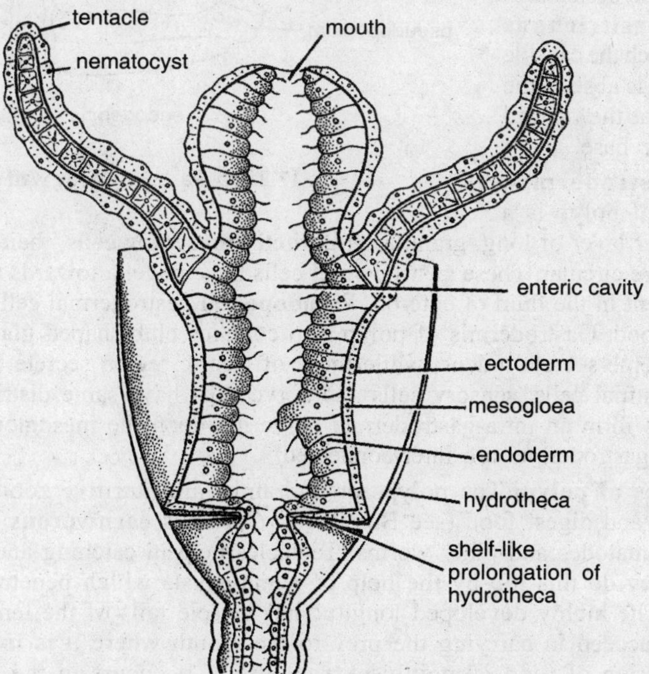

Fig. 17.2. *Obelia*. V.S. of a polyp or hydranth.

Mouth is capable of great dilation and contraction. Below the hypostome is the **stomach** or **gastric region** of polyp which forms about two-third of the length of hydranth.

Mouth opens into a spacious gastrovascular cavity or **enteron** of the polypoid body, *i.e.*, cavity of hypostomal and gastric part of polyp. At the base of the polyp, the cavity is continuous with coenosarcal cavity of concerned branch (hydrocaulus). The polyp is covered by cup-shaped perisarcal hydrotheca.

Histology of polyp. The wall of hydranth is composed of an outer ectodermal **epidermis** and an inner endodermal layer of **gastrodermis**. There is an acellular, thin, transparent and gelatinous **mesogloea** in between the two layers (Fig. 17.2). The tentacles of polyp are solid with no enteron. Each tentacle consists of an outer layer of epidermal cells and a single row of thick walled cubical vacuolated gastrodermal cells.

1. Epidermis. Epidermis of polyp is a continuous layer of **epithelio-muscular cells** which are long, conical and columnar in shape (Fig. 17.3). The inner ends of these cells are produced into muscular processes which run longitudinally. Epidermis of hypostome and tentacles also consists of **sensory** and **epithelio-glandular cells** and 'batteries' of nematocyst containing **cnidocytes**. Epidermis of polyp contains very few **interstitial cells** (*i.e.*, undifferentiated cells). Branched **nerve cells** occur in deeper zones of epidermis. *Obelia* has only one kind of nematocyst called **basitrichous isorhizas** in which the capsule is oval, the butt is absent, the thread is open at the tip and has spines at its base.

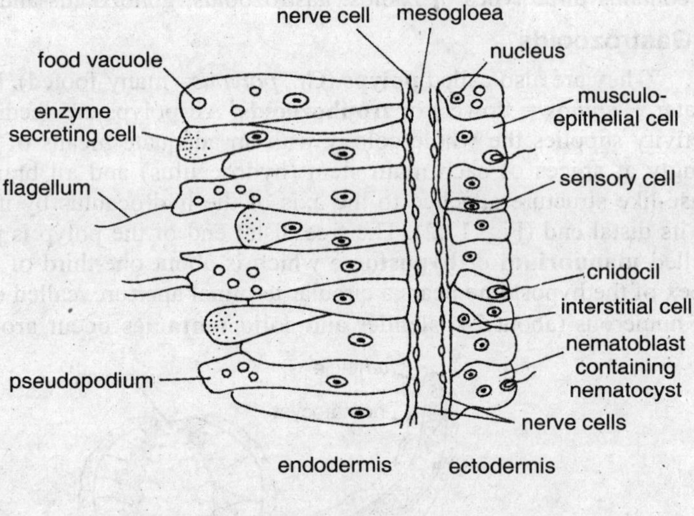

Fig. 17.3. *Obelia*. L.S. of body wall of hydranth.

2. Gastrodermis. Gastrodermis of polyp is a continuous inner layer of long, granular **endothelio-muscular cells**, their muscle processes point outwards and are circular. These gastrodermal cells have flagella towards their inner ends; flagella produce a current in the fluid of enteron. **Pseudopodial** gastrodermal cells can form pseudopodia for engulfing food. Gastrodermis of polyp also contains club-shaped **glandular cells**. Glandular cells of hypostome secrete mucus, while those of gastric region secrete the proteolytic digestive enzymes. Interstitial cells, sensory cells and nerve cells have same distribution as in epidermis. The nerve cells form an intra-gastrodermal nerve net upon the mesogloea. The two nerve nets: epidermal and gastrodermal are inter-connected.

Functions of polyp. The polyps or hydranths are nutritive zooids of the colony. They capture, ingest and digest food (see Box 17.1). *Obelia* is **carnivorous** and feeds upon aquatic crustaceans, nematodes and other worms. Tentacles help in catching and conveying the prey to the mouth. They do this job by the help of nematocysts which penetrate, hold and generally kill the prey. The highly developed longitudinal muscle tails of the tentacles holding the prey contract and succeed in carrying the prey to the mouth where it is **ingested** and passed into enteron. Digestion of food is mainly **extracellular**. It occurs in the enteron by the help of enzymes secreted by glandular gastrodermal cells. Pseudopodial gastrodermal cells ingest partly digested food (food particles) by **phagocytosis**. **Intracellular digestion** within the food vacuoles completes the digestive process. After digestion has been completed in the gastrodermis, soluble food diffuses to the epidermis through the mesogloea. Any unwanted/undigested food remains are **egested** via the oral aperture.

Box 17.1 Radial symmetry helps *Obelia* in feeding

Being sessile, *Obelia* has to rely for its food source on water currents carrying prey or prey swimming towards it. The radial symmetry of the hydranths and the branched nature of the whole colony enable the tentacles to sweep through large quantities of water and to encounter food from all directions.

Beating of flagella of gastrodermal cells and rhythmic contraction (= peristaltic movement) of hydranth circulates the digested food (along with food particles) throughout the coenosarc of the colony.

II. Gonozooid

When *Obelia* colony has reached its full development, it produces special non-feeding zooids called **gonangia** (Fig. 17.4). Gonangia occur in the axils of older hydranths, towards the proximal end of hydrocauli. Each gonangium comprises of medusae-bearing blastostyle and gonotheca. A **blastostyle** is a club-shaped cylindrical and hollow extension of the coenosarc. It has no mouth or tentacles and is surrounded by a portion of perisarc, called the **gonotheca**. The gonotheca opens at its distal end by a **gonopore**. Blastostyle bears many saucer-shaped, hollow buds, called **medusae** or **gonophores**. Medusae are budded-off from the blastostyle and eventually expelled via the gonopore into the sea.

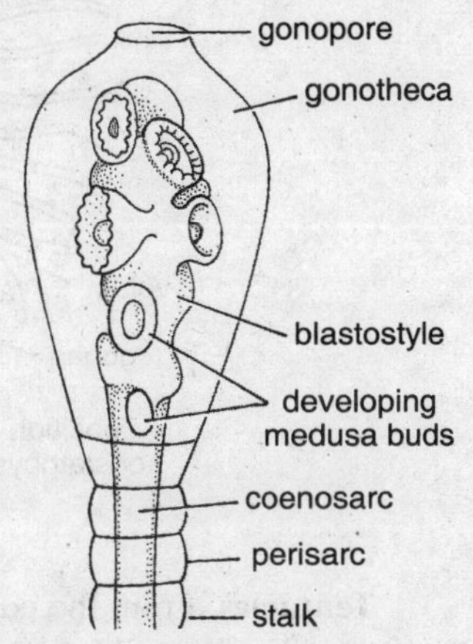

Fig. 17.4. *Obelia*. A gonangium.

Histology of blastostyle. A blastostyle has two layered body wall containing outer epidermis and inner gastrodermis, both layers being separated by mesogloea. Its cavity, the gastrovascular cavity, is continuous with the coenosarcal canal of the hydrocaulus and hydrorhiza. Blastostyle is rudimentary and does not open to exterior.

Functions of gonozooid. The main function of gonozooids is to take part in reproduction, hence are called **reproductive zooids**. Each blastostyle produces numerous lateral buds (by **asexual budding**) which develop into **medusae**. Each saucer-shaped medusa is attached by middle of its convex surface to the blastostyle, when fully formed it breaks free and escapes out in water through the gonopore. The free-swimming medusae eventually develop gonads and represents the sexual phase of the life cycle of *Obelia*.

III. Medusa

Medusae are the third type of zooids of *Obelia* colony. These are modified zooids. Medusae are produced as hollow buds from the coenosarc of blastostyle by asexual budding during spring and summer seasons. They are solitary and **pelagic**, *i.e.*, they swim freely on the surface of water. Mature madusae bear gonads and reproduce sexually. They are **unisexual (dioecious),** *i.e.*, male gonads (testes) and female gonads (ovaries) are borne by separate individuals.

A. Structure of medusa. A medusa is radially symmetrical. It is tiny (microscopic) and circular having umbrella-like shape. The convex outer surface is known as the **exumbrella** (Fig. 18.7) and the concave inner surface is the **subumbrella**. Hanging down from the centre of the subumbrellar surface is a short, hollow **manubrium** (L.*manus* = handle) at the end of which is a four-lobed oral aperture or **mouth**. Mouth opens into a part of gastrovascular cavity, called **stomach** or **gastric cavity** situated inside the manubrium. From the stomach arise four **radial**

canals which are delicate flagellated tubes, they pass outwards to the edge of umbrella and join a flagellated **circular canal** running near the margin of umbrella. The free edge of umbrella becomes folded inward towards manubrium to form a rudimentary **velum**. The medusae with a velum are called **craspedote** (*e.g.*, *Obelia* and other hydrozoans) and those without the velum are called **acraspedote** (*e.g.*, Scyphozoa).

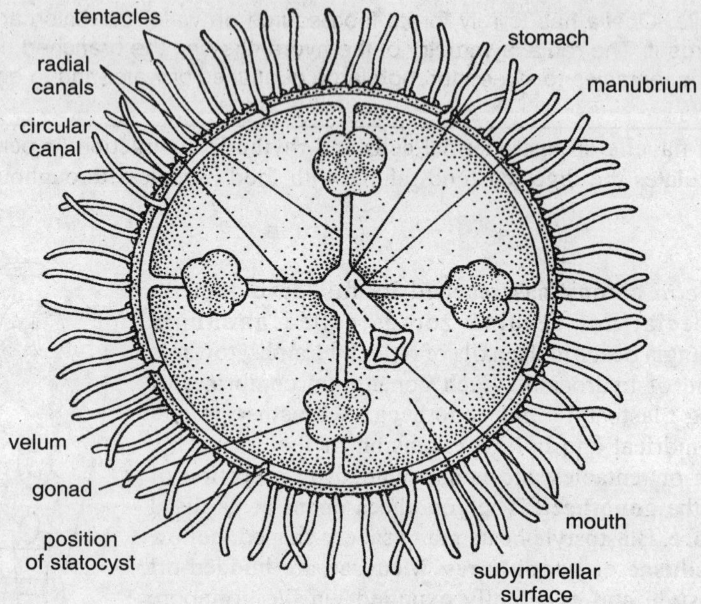

Fig. 17.5. *Obelia*. Medusa in oral view.

Tentacles. From the edge of the umbrella arise numerous small, solid and highly contractile **tentacles**. There are 16 tentacles in young medusa but this number increases in the adult stage. According to their position, 16 tentacles are classified as follows (Fig. 17.6):

1. Per-radial tentacles. These are four in number and they are situated against the four radial canals.

2. Inter-radial tentacles. These are four in numbers and are located on radii, bisecting the angle between two per-radial tentacles.

3. Ad-radial tentacles. These are eight in number and found in between the per-radial and inter-radial tentacles.

4. Sub-radial tentacles. These tentacles are found in between the ad-radial and per-or inter-radial tentacles.

B. Histology of medusa. Like the hydranth, medusa is **diploblastic** having outer ectodermal **epidermis** and inner endodermal **gastrodermis**. A gelatinous **mesogloea** exists between the two epithelia; it becomes thickened towards exumbrella forming the main bulk of the body.

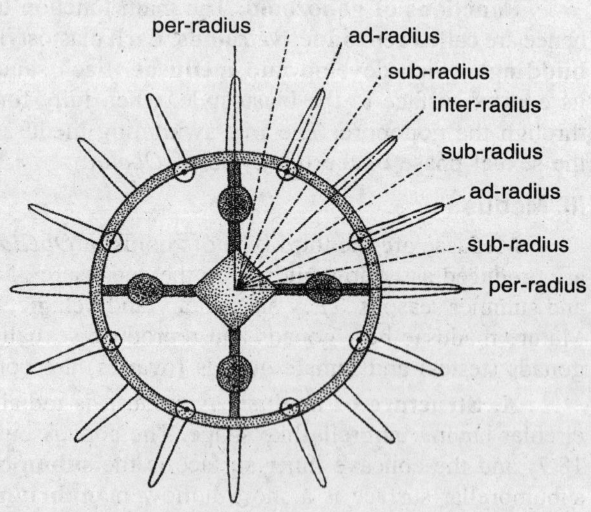

Fig. 17.6. *Obelia*. Radial symmetry and types of tentacles.

1. Epidermis. The epidermis covers manubrium, subumbrella surface, the margins of the umbrella and entire exumbrella surface of medusa (Fig. 17.7). It consists of **epitheliomuscular cells** which are produced into muscle processes. Muscle processes run longitudinally in the manubrium and tentacles. They form striated circular muscles and some radial muscles of subumbrella and bring about locomotory movements. The epidermis of exumbrella is devoid of musculature. Velum is the double fold of epidermis. It contains no endoderm but has thick mesogloea in the middle.

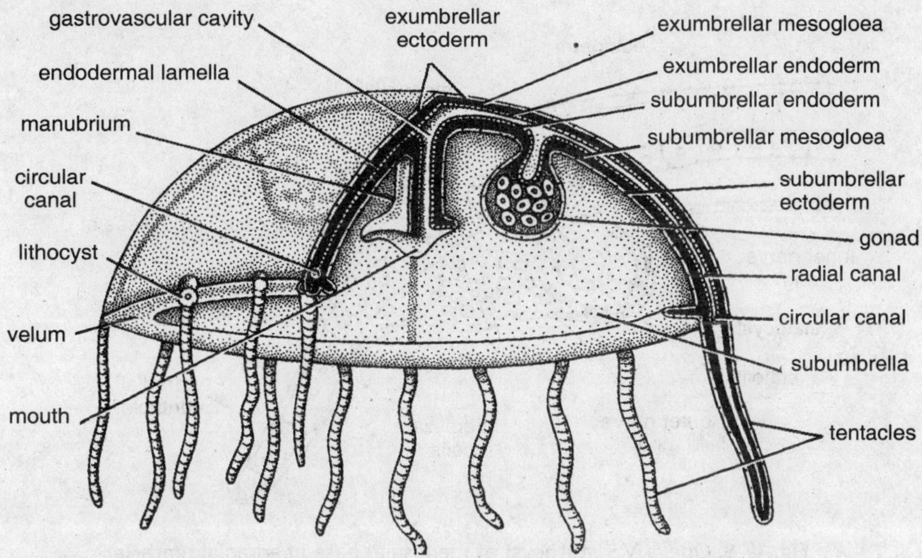

Fig. 17.7. *Obelia*. Medusa with a part cut away to show its histology and structure.

The structure of tentacles is essentially the same as in the polyp, but in addition they each possess a swelling at their base where interstitial cells have accumulated. The interstitial cells are used to replace lost or damaged tentacular **nematoblasts.**

2. Gastrodermis. The endodermis or gastrodermis lines the gastrovascular cavity of manubrium, stomach, radial canals and circular canals. The epithelial cells of endodermis have no muscle processes, many of them are flagellated cells, and some are glandular cells secreting digestive enzymes.

Certain Organ Systems of Medusa

1. Nervous system. In order to coordinate the over all activity of the medusa the nervous system has necessarily to be more complex than that of the polyp. In the medusa the epidermal nerve cells and fibres form two **circular nerve rings**, one external and one internal to circular canal. Primarily the inner nerve ring controls the subumbrellar muscle fibres, whilst the outer nerve ring receives impulses from the statocyst. However, they are also interconnected with each other, the sensory cells and with processes innervating the tentacles.

2. Sense organs (statocysts). Medusa contains eight receptor organs, called **statocysts.** Statocysts are located inside the bases of eight adradial tentacles. Each contains a fluid-filled sac lined with ectodermal cells. Each sac possesses movable particle of calcium carbonate, called **statolith** or **otolith**. Alongside the statolith is a series of sensory protoplasmic processes. The whole apparatus of the statocyst is an organ of equilibrium and helps to bring about coordination in the movement of medusa (see Box 17.2).

Box 17.2

The medusa of *Obelia* normally swims through the water with the margins of its bell horizontally aligned. If the bell is tilted, angular displacement of the statolith will cause impulses to be produced by the sensory processes of statocyst which are passed to the outer nerve ring. These impulses are then relayed to the inner nerve ring and subumbrellar musculature will be stimulated to take the appropriate **reflex corrective action** to restore equilibrium.

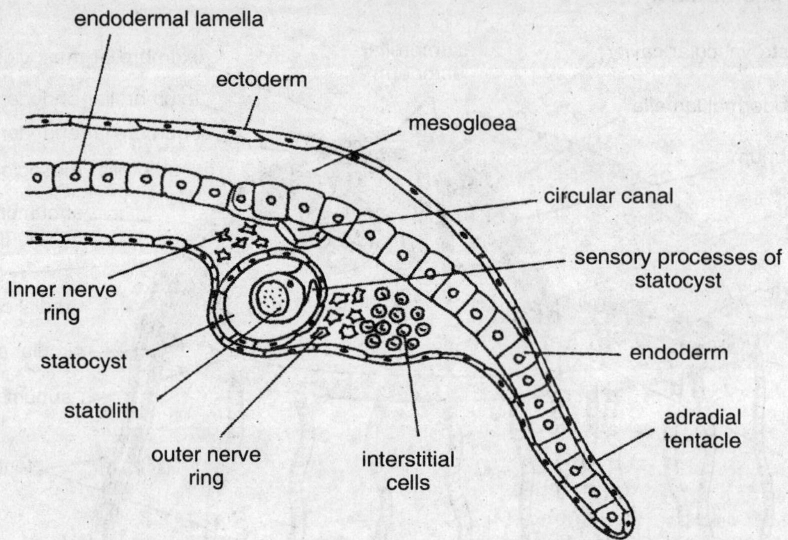

Fig. 17.8. *Obelia*. V.S. statocyst of medusa at base of adradial tentacle.

3. Reproductive organs. Each medusa of *Obelia* contains only four **gonads** which may be either testes or ovaries depending on the sex of the medusa. Halfway along each radial canal and protruding from the subumbrellar surface are the gonads. They consist of an outer layer of **ectodermal cells** and a core of **endodermal cells** derived from the radial canal itself. Germ cells originate in the ectoderm of manubrium and migrate to gonads. It is here that **meiosis** occurs. Each medusa produces either flagellate **sperms** or spherical **ova**. Liberation of the gametes occurs when the ripe gonads burst. After the release of gametes the medusa dies.

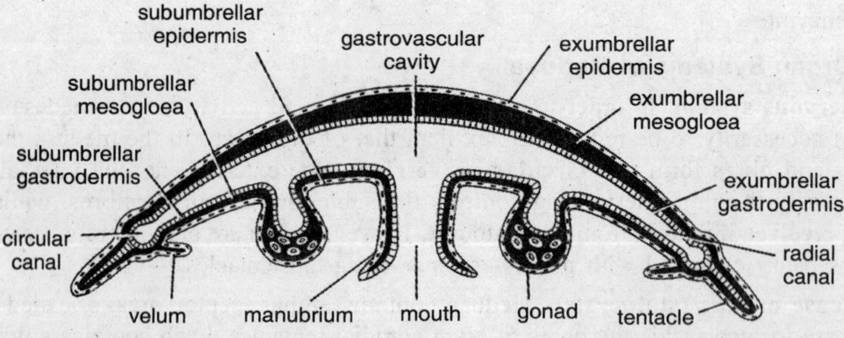

Fig. 17.9. *Obelia*. V.S. of medusa showing gonads.

Functions of medusa. Medusa is the sexual zooid. It being dioecious produces one type of gamete either sperms or ova.

Division of Labour in *Obelia*

In *Obelia* different functions are performed by different zooids:

1. The polyp or hydranth is nutritive in function. It catches and digests foods.
2. Blastostyles are reproductive zooids. They produce medusae by asexual budding.
3. Medusae are sexual zooids. They contain gonads and produce gametes. They being free swimming shed off their gametes over a wider area, ensuring a wide dispersal.

COMPARISON OF POLYP AND MEDUSA OF *OBELIA*

The blastostyles produce medusae by asexual budding in spring and summer seasons. During these seasons a number of medusae may be seen in various stages of development in a blastostyle of *Obelia*. The wall of blastostyle (coenosarc) is pushed out in the form of a small **protuberance** or **bud** (Fig. 17.10). It contains the epidermis, mesogloea, gastrodermis and encloses a part of the gastrovascular cavity or enteron. The bud grows larger and its coenosarc becomes like a **vesicle** which is attached to the blastostyle by a narrow **stalk**. The cavity of the **vesicle** is continuous with the enteron of blastostyle.

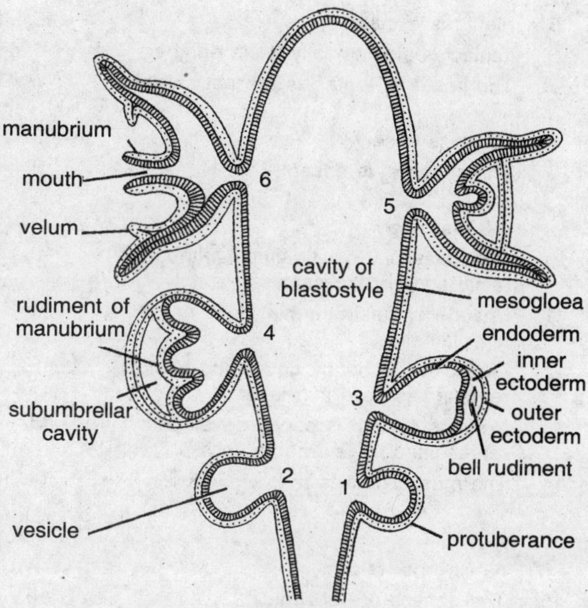

Fig. 17.10. *Obelia*. Stages of development of medusa from a blastostyle.

The epidermis at the top of the vesicle now splits into two layers. The inner of the two again splits and a small cavity, called **bell rudiment**, appears between the newly splitted epidermal layers. The innermost epidermis and the gastrodermis, now, evaginate to form **rudiment** of **manubrium**. This rudiment elongates and breaks the outer epidermis. The broken epidermis (double layered) becomes **velum** (or **craspedon**). Later the mouth opens at the free end of manubrium.

Tentacles arise as finger-like outgrowths from the margin of the umbrella. Finally the stalk connecting the developing medusa with the blastostyle breaks and its aperture closes up. Thus, a medusa is formed which is set free. Its upper convex surface becomes **exumbrella,** while its inner concave surface becomes **subumbrella**. The fully formed medusae come out of the aperture of gonotheca. It swims freely in the sea water.

3. Differences between Polyp and Medusa. The polyp and medusa are two different types of zooids of *Obelia*. Their comparison can be made under the following steps (see Table 17.1).

Table 17.1 Differences between polyp and medusa of *Obelia*.

	Polyp		*Medusa*
	I. Habit		
1.	It is fixed or sessile.	1.	It is always free-swimming and pelagic.
	II. Position		
2.	It is attached to the vertical branches of the colony.	2.	It develops as a bud on the axial cylinder, the blastostyle, and detaches from it to live freely.

	Polyp		*Medusa*
	III. Morphology		
3.	Body cylindrical, elongated and vase-like.	3.	Body saucer-shaped or umbrella-like.
4.	The hypostome or manubrium is conical in form and constitutes one third of the whole body. It is directed upward.	4.	Manubrium is comparatively small and quadrangular and hangs downwards.
5.	Mouth is circular.	5.	Mouth is square.
6.	Tentacles are normally 24 in number.	6.	Tentacles are 16 in young but numerous in adult.
7.	The bases of tentacles are not swollen.	7.	The bases of tentacles are swollen and contain reserve interstitial cells.
8.	Velum is absent.	8.	Velum is present around the margin of umbrella.
9.	The statocyst is absent.	9.	The statocyst is present in the base of eight adradial tentacles.
	IV. Histology		
10.	Gastrovascular cavity simple without radial and circular canals.	10.	Gastrovascular cavity is represented by stomach, four radial canals and one circular canal.
11.	Endoderm (gastrodermis) does not form lamella.	11.	Endoderm of the oral and aboral region constitutes endodermal lamellae.
12.	Mesogloea is poorly developed being represented as a thin layer.	12.	Mesogloea is enormously developed.
13.	Nervous system is poorly developed; represented by few nerve cells.	13.	Nervous system is well developed with a double ring of irregular nerve cells.
14.	The musculature is not developed.	14.	The muscle processes of epithelio-muscular cells are well developed and constitute a well developed musculature.
	V. Reproduction		
15.	The gonads are absent.	15.	The gonads are present on the radial canals in four groups.
16.	The polyp is a **nutritive zooid** and represent an **asexual** generation.	16.	The medusa is a **reproductive zooid** which represents the **sexual** generation.
17.	The polyp reproduces asexually by budding.	17.	The medusa reproduce sexually by the formation of gametes.
18.	It does not cause wide dispersal of species.	18.	It causes wide dispersal of species.

4. Similarities between Polyp and Medusa. Despite their remarkable differences in structure and function, polyp and medusa are basically similar structures; they are **homologous structures** as shown by the following similarities between the two.

1. Both are **radially** symmetrical.
2. Both are **diploblastic** having outer epidermis (ectodermal) and inner gastrodermis (endodermal).
3. **Mouth** and **hypostome** (manubrium) are homologous structures in both.
4. The stomach, radial canals and circular canals of medusae are homologous to the gastrovascular cavity of polyp.
5. Both are **carnivorous**; in both the food is captured and ingested with the help of tentacles.
6. Digestion is **extracellular** as well as **intracellular** and digested food diffuses throughout the body without a circulatory system.
7. The convex outer surface of umbrella of medusa is homologous with the base of the polyp.

Advancement of Medusa over Polyp

Medusa shows many advanced features than the polyp. Few of them are as follows:

1. The epidermis of medusa resembles the epithelium of higher Metazoa. It forms a thin, protective and sensitive layer of small cells.
2. The excessive development of **mesogloea** reduces the gastrovascular cavity to a system of canals and also provides lightness which helps in buoyancy.
3. The nervous system shows differentiation into two nerve rings, which form the central nervous system, and nerve nets which form the peripheral nervous system.
4. The marginal sense organs (statocysts) present at the bases of eight tentacles are of great advantage to the free-swimming habit of medusa.
5. The mode of sexual reproduction increases the genetic variations of the species and the free swimming habit of medusa provides wide dispersal of the species.

Derivation of Medusa from Polyp or Evolutionary Significance

Despite the medusa showing slight advancement over the polyp in structural details, a medusa can be said to be a modified polyp. A medusa can be derived from polyp by assuming that the polyp is inverted and its tentacular region is pulled out. The result will be that the manubrium is now directed downwards and a disc-shaped appearance is assumed (Fig. 17.11). A saucer shaped form is developed by basal region. It, therefore, consists of two surfaces, the outer convex surface (**exumbrellar surface**) and the inner concave surface (**subumbrellar surface**). The gastrovascular cavity of the polyp can be said to form the exumbrellar surface.

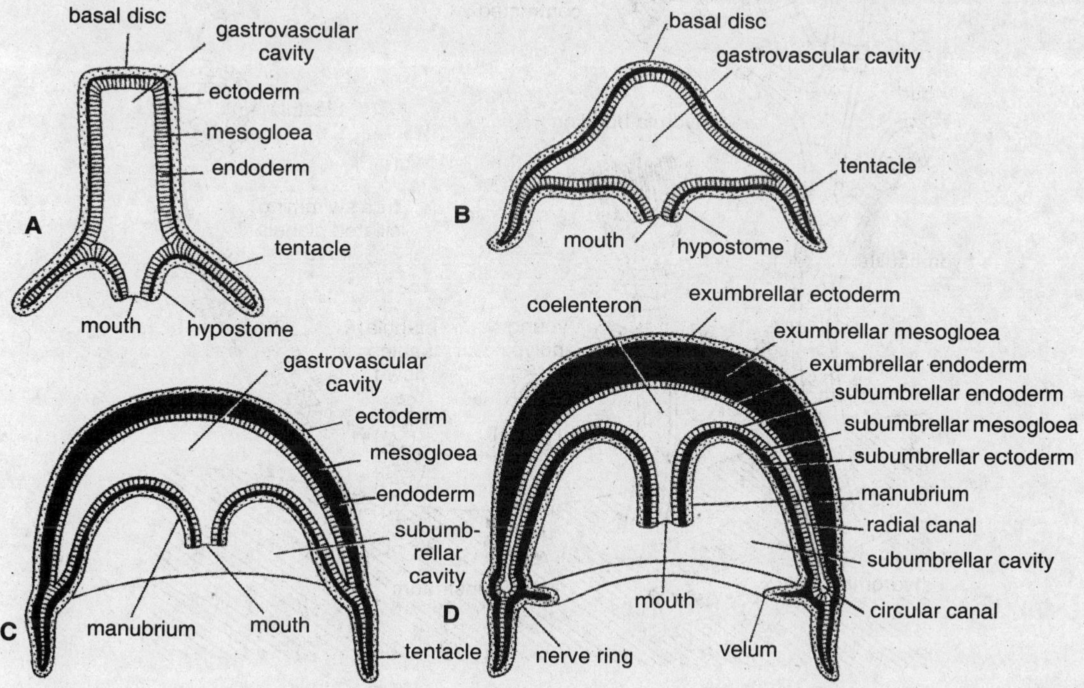

Fig. 17.11. *Obelia*. Diagram showing the derivation of the medusa from the polyp. A—Polyp in L.S., B—L.S. polyp with extended tentacular region; C—L.S. of bell; D—Vertical section of medusa.

17.4 REPRODUCTION AND LIFE CYCLE

Obelia colony propagates by the asexual method of budding producing hydranths (polyps) and blastostyles. These represent asexual generation. The blastostyles are specialized reproductive zooids forming **medusae** by budding. Medusae constitute the sexual or the reproductive individuals possessing four **gonads**. These gonads may be either male or female.

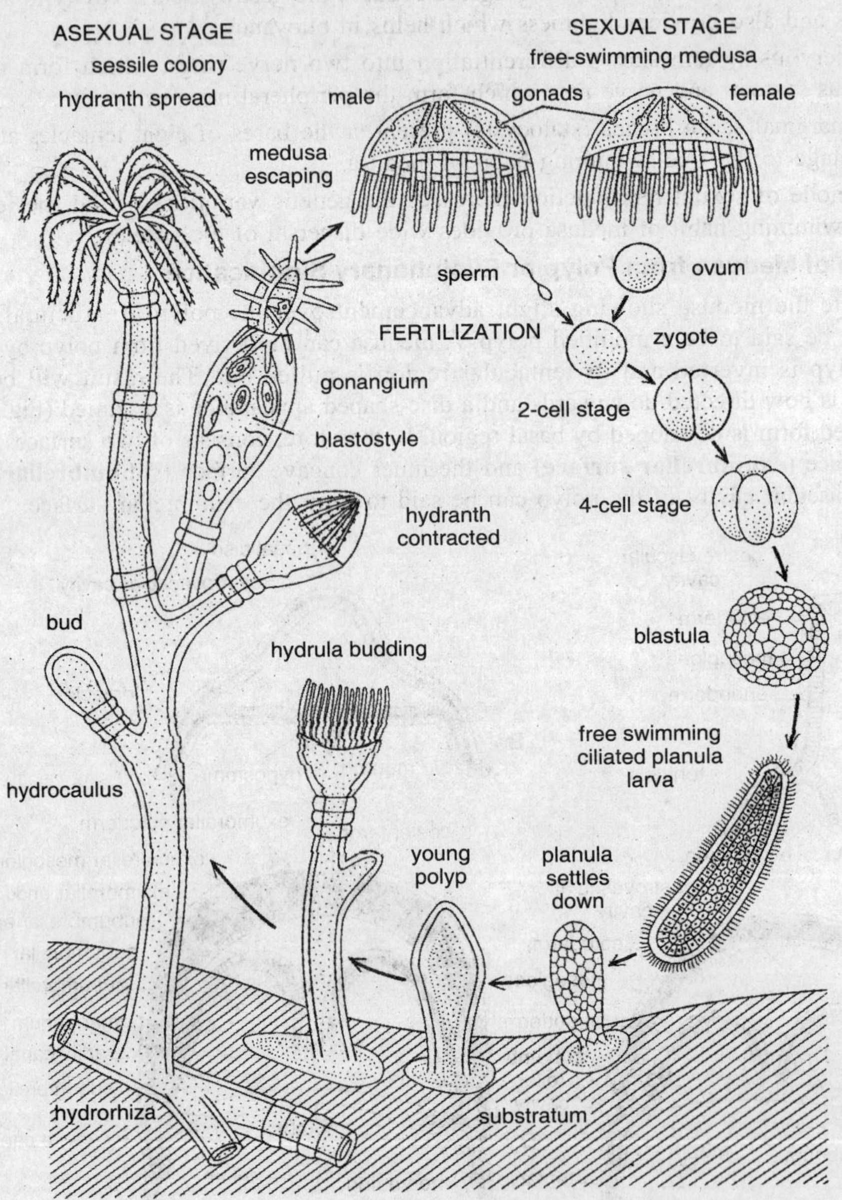

Fig. 17.12. *Obelia*. Stages in the development and life history.

Gonads

Gonads are situated on the sub-umbrellar surface of medusa, one in the middle of the course of each radial canals. They mature (*i.e.*, undergo meiosis and other steps of gametogenesis) after the medusae escape from the gonotheca. Each gonad is an oviod, knob-like body of undifferentiated interstitial cells between the epidermis (ectoderm) and gastrodermis (endoderm). Epidermis is continuous with that of the sub-umbrella and gastrodermis is continuous with that of radial canals (Fig. 17.9). The space between two layers is filled with a mass of **sex cells,** which become differentiated into haploid ova or sperms in female and male medusae respectively. On maturation of gametes, the endodermal/gastrodermal wall surrounding the gonad ruptures and the germ cells are set free in sea water (Fig. 17.12).

Fertilization. Fertilization is of cross-fertilization type. It takes place either in sea-water or the sperms may be carried by water currents to female medusae, where they fertilize the ova. The parent medusae die soon after liberating the gametes.

Development

The zygote or fertilized egg undergoes equal **holoblastic** division resulting in a solid ball of cells, the **morula**. Morula is transformed into **blastula** by the process of blastulation. The blastula is hollow, having a central cavity called **blastocoel**, which is surrounded by a single layer of cells, the blastomeres. During gastrulation, the blastocoel gets filled with the cells which are budded off from the wall of blastula. Thus, a solid gastrula (**stereogastrula**) is formed. Its outer most layer is known as **ectoderm** and inner mass of cells as **endoderm**. The gastrula elongates, the ectodermal cells develop cilia and endoderm develops **enteron** by a process of splitting or **delamination**. This ciliated gastrula represents the **planula larva**. It leads a free-swimming existence.

After leading an active free-swimming existence, the planula larva settles down, attaches itself by its anterior end to solid substratum and undergoes **metamorphosis**. Its anterior end forms **basal disc** for attachment, while the posterior or distal end forms **manubrium** with a **mouth** surrounded by a circlet of **tentacles**. The larva now closely resembles a polyp or hydra, hence called **hydrula**. The basal part of hydrula is elongated into hydrorhiza from which vertical and lateral branches are budded off and a complex branching *Obelia* colony is formed. This colony, later on, develops medusa buds on the blastostyles to start sexual generation.

The occurrence of free-swimming medusa and ciliated larva in a life-history of a fixed or sessile organism such as obelia, is of great significance; as it helps in the dispersal of organisms and prevents overcrowding of the species.

Alternation of Generation and Metagenesis

In the life-cycle of *Obelia*, there is a regular alternation of the colonial and sessile **hydroid phase** and the solitary and free-swimming **medusoid phase**. Hydroid phase represents the **asexual** phase and lacks gonads; it multiplies asexually. The medusoid phase represents **sexual** phase, the medusae contain gonads, but never produce medusae. They reproduce sexually by forming gametes (eggs or sperms) which after fertilization and embryogenesis produce new hydroid colonies, as can be shown by the following diagram:

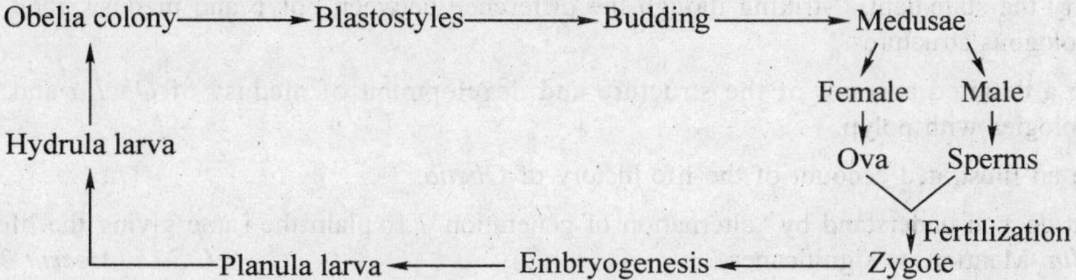

This phenomenon of alternation of sexual generation with asexual generation of life cycle of *Obelia* was termed as **alternation of generation**.

Hyman seems to be right in holding that the concept of alternation of generation is not valid in animals (*e.g.*, *Obelia*). It truely applies to plants, notably the bryophytes and pteridophytes, in which the haploid sexual generation called the **gametophyte** alternates with the diploid asexual generation, called the **sporophyte**. In contrast, in *Obelia* the apparent asexual and sexual generations, namely the polyps and medusae, respectively, are both **diploid**. In case of *Obelia*, only *the gametes are haploid* while the sexual gonad-bearing medusoid form and the hydroid form are both diploid. Further, the sexual generation of *Obelia* (*e.g.*, medusa) does not arise from a single-haploid cell. Thus, it becomes clear that there is no alternation of diploid asexual and haploid sexual generations in the life cycle of coelenterates. *Obelia*, instead shows alternation of fixed and free-swimming generations. This is advantageous for the *Obelia* as it prevents overcrowding and assures wide dispersal of the species.

It has been suggested that the hydroid colony represents **developmental** or **juvenile** stage as it does not bear gonads. Thus, the complete life cycle from egg to egg is one generation. The budding off of a medusae is merely a step in the attainment of sexual reproduction. Hence, instead of describing the phenomenon as alternation of generation, it is termed **metagenesis**, meaning postponement of the power of generation (*i.e.*, sexual reproduction).

QUESTIONS

Long Answer Questions

1. With the help of diagrams describe the structure of polyp and medusa of *Obelia*.
2. Give an account of various types of zooids found in the *Obelia* colony.
3. Why *Obelia* is called a trimorphic? Describe different types of zooids in *Obelia* colony.
4. What is polymorphism? Explain it with reference to the structure of *Obelia*.
5. What is polymorphism? Describe the various types of zooids found in the *Obelia* colony.
6. Describe the structure of various zooids found in the *Obelia*. List the functions of zooids of *Obelia* and *Physalia*. (*C.C.S.U. Meerut 2008*)
7. Describe the structure and function of medusa.
8. Give habit and habitats of *Obelia*. What do you understand by "division of labour"? Clarify it fully with reference to *Obelia*.
9. Compare the structure of polyp and medusa and add a note on their evolutionary significance.
10. Describe the form and structure of a hydranth in *Obelia*. Compare it with a medusa. How can the latter be derived from the former?
11. Enumerate the differences and similarities between polyp and medusa? Show how medusa can be derived from polyp? How does medusa show advance over polyp?
12. Give a comparative account of polyp and medusa.
13. Justify the statement, "striking though the difference between polyp and medusa, they are strictly homologous structures."
14. Give a detailed account of the structure and development of medusa of *Obelia* and discuss its homologies with polyp.
15. Give an illustrated account of the life history of *Obelia*.
16. What do you understand by "alternation of generation"? Explain the same giving the life history of *Obelia*. Mention its significance. (*C.C.S.U., Meerut 2004, 2005*)
17. What is alternation of generation in animals? Explain it with reference to the life history of *Obelia*.

18. Explain what is meant by metagenesis by referring to the life-history of *Obelia*?

19. Describe the structure and life cycle of *Obelia*.

Short Answer Questions

1. Describe the structure of medusa of *Obelia*. Compare it with hydranth. (*C.C.S.U., Meerut 2007*)

2. Describe the structure of a polyp in *Obelia*. Compare it with a polyp. (*C.C.S.U., Meerut 2006*)

3. Draw the well labelled diagrams of the following. No description is required.

(*i*) Structure of *Obelia*

(*ii*) V.S. hydranth of *Obelia*

(*iii*) Medusa of *Obelia*

(*iv*) Life cycle of *Obelia*

(*v*) Life history of a hydrozoan animal

Very Short Answer Questions

1. Name the cells that contain myonemes in *Obelia*.

2. What are the functions of polyp of *Obelia*.

3. Tentacles of *Obelia* have a solid core of gastrodermis. Is it true?

4. Which part of a nematocyst is discharged when cnidocil is stimulated?

5. What stimulates a nematoblast?

6. Which layer of cells is missing in the velum?

7. What term is used for the alternation of generation in *Obelia*.

8. Name the various kinds of zooids found in *Obelia* colony.

9. Medusa of *Obelia* has tetramerous radial symmetry. Is it so?

10. How many sense organs/statocysts occur in medusa of *Obelia*.

11. What is gonangium?

12. What are statocysts?

13. What is planula larva?

Fill in the Blanks

1. .... forms a trimorphic colony.

2. A zooid without mouth and tentacles is called .... in *Obelia* colony.

3. Polyp of *Obelia* is protected in ....

4. Perisarc forms .... around the blastostyles.

5. Majority of the cells in epidermis of *Obelia* are ....

6. The nematoblasts contain a poisonous fluid called ....

7. The .... are the unspecialised reserve cells of *Obelia*.

8. The muscle processes of the gastrodermal cells act as the .... muscles.

9. Mesogloea of *Obelia* is secreted by ....

10. *Obelia* has both .... and .... digestion.

Multiple Choice Questions

1. *Obelia* is a hydroid of
(*a*) fresh water (*b*) marine water
(*c*) brackish water (*d*) all the above

2. The stolon of *Obelia* colony is known as
(*a*) hydranth (*b*) hydrorhiza
(*c*) hyporhiza (*d*) root

3. Medusa stage of cnidarians is advanced to polyp because
(*a*) it has great power of contractility (*b*) it has marginal sense organs
(*c*) it is motile (*d*) all of these

4. Coenosarc is
(*a*) living (*b*) non-living
(*c*) inner tubular, living (*d*) non-cellular

5. Medusa of *Obelia* is
(*a*) herbivores (*b*) carnivores
(*c*) omnivores (*d*) insectivorous

6. The number of tentacles in young medusa of *Obelia* is
(*a*) 16 (*b*) 32
(*c*) 64 (*d*) 128

7. In *Obelia*, medusae arise from
(*a*) polyps (*b*) hydrorhiza
(*c*) hydrocaulus (*d*) blastostyles

8. The nematocysts are discharged on the stimulation of
(*a*) lasso (*b*) cnidocil
(*c*) nucleus (*d*) thread tube

9. In medusae of *Obelia*, statocysts occur at the bases of
(*a*) per-radial tentacles (*b*) inter-radial tentacles
(*c*) ring canal (*d*) all these canals

10. In medusae of *Obelia* gonads developed under the
(*a*) inter-radial canals (*b*) ad-radial canals
(*c*) per-radial canals (*d*) all these canals

11. Sexual zooid in the life history of *Obelia* is
(*a*) gonozoid (*b*) hydranth
(*c*) medusa (*d*) blastostyle

12. The free-swimming larva of *Obelia* is
(*a*) planula (*b*) amphiblastula
(*c*) parenchymula (*d*) scyphistoma

13. Gonotheca of *Obelia* is a protective covering of
(*a*) polyp (*b*) medusa
(*c*) blastostyle (*d*) hydrocaulus

14. Statocyst function as the organs for
(*a*) hearing (*b*) equilibrium
(*c*) light perception (*d*) smell

ANSWERS

Very Short Answer Questions

1. Epitheliomuscular cells.
2. Feeding and defence.
3. Yes.
4. Thread tube.
5. Mechanical and chemical stimulation.
6. Gastrodermis.
7. Metagenesis.
8. Polyps, blastostyles and medusae.
9. Yes.
10. Eight.
11. Gonangium is a gonozooid and non-feeding zooid of *Obelia* colony. It comprises a medusae bearing blastostyle and gonotheca.
12. Statocysts are minute fluid-filled ectodermal sacs. They contain a movable particle of $CaCO_3$ called **statolith**. Statocysts help medusae in maintaining equilibrium (balance) while swimming.
13. It is ciliated, free-swimming larva of *Obelia* and other cnidarians.

Fill in the Blanks

1. *Obelia*
2. Blastostyle
3. Hydrotheca
4. Gonotheca
5. Epitheliomuscular cells
6. Hypnotoxin
7. Interstitial cells
8. Circular
9. Epidermis and gastrodermis
10. Extracellular (or intercellular), intracellular.

Multiple Choice Questions

1. (*b*)
2. (*b*)
3. (*d*)
4. (*c*)
5. (*b*)
6. (*a*)
7. (*d*)
8. (*b*)
9. (*d*)
10. (*c*)
11. (*c*)
12. (*a*)
13. (*c*)
14. (*b*).

18

Polymorphism in Cnidaria

Polymorphism (Gr., *poly* = many + *morph* = form) is a phenomenon in which an organism has more than one type of individuals. It is more evident in class Hydrozoa and Anthozoa of Cnidaria/Coelenterata.

Coelenterates are basically colonial forms. Individuals of a colony of coelenterates are called **zooids**. These zooids are of two basic types: polyps and medusae.

1. Polyps. Polyps are cylindrical, sessile and fixed forms. These have mouth surrounded by tentacles and a wide gastrovascular cavity. These are the **nutritive zooids** of the colony, so, feed the colony. Polyps are also called **gastrozooids** or **trophozooids.**

2. Medusae. Medusa has an umbrella-like body with marginal tentacles and a centrally located mouth on a concave side. They bear gonads and help in sexual reproduction. Therefore, these are called **sexual zooids** or **gonozooids**. They are free swimming and derived from polyps.

Hydromedusae. The medusae of class Hydrozoa are called **hydromedusae**. They are of following types:

1. Trachymedusae. Bell margin of these medusae are not scalloped. These medusae contain long manubrium or pseudomanubrium. In them gonads are present on radial canals, *e.g.*, *Aglantha* (Fig. 18.1A).

2. Narcomedusae. Bell margin of these medusae are scalloped. These medusae contain reduced radial canals and lack manubrium. In them gonads occur on the floor of gastric cavity, *e.g.*, *Cunina* (Fig. 18.1B).

3. Anthomedusae. These medusae are taller, retain the tetraradiate form, have gonads on the manubrium and usually have ocelli, *e.g.*, *Sarsia* (Fig. 18.1C), *Pennaria* (Fig. 18.1D) and other gymnoblasts.

4. Leptomedusae. These medusae are flatter, may have lost their tetraradiate symmetry by branching of the radial canals, have gonads borne on these canals. They have statocysts on the bell margin, *e.g.*, *Obelia* (Fig. 18.1G), *Tima* (Fig. 18.1F), *Aequora* (Fig. 18.1E) and other leptomedusae.

18.1 TYPES OF POLYMORPHISM

According to the types of zooids, coelenterates may be of following types:

1. Dimorphic having two types of zooids.
2. Trimorphic having three types of zooids.
3. Polymorphic having more than three types of zooids.

1. Dimorphism

The coelenterate colonies bearing two types of individuals or zooids are called **dimorphic**. Such colonies and the phenomenon is termed **dimorphism**. For example, *Bougaivillea*, *Pennatula*, etc., consist of two types of zooids: 1. **Polyps**, the gastrozooids or trophozooids and 2. **Medusae**, the sexual zooids.

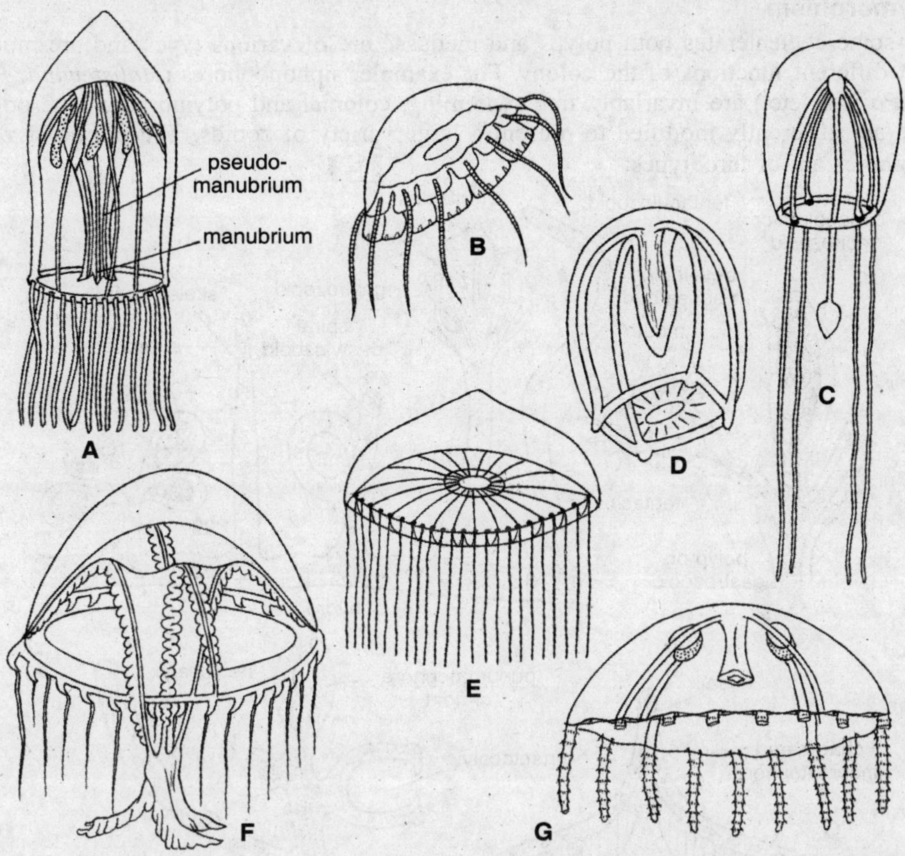

Fig. 18.1. Hydromedusae. A—Trachymedusa (*Aglantha*); B—Narcomedusa (*Cunina*); C—Anthomedusa (*Sarsia*); D—Anthomedusa (*Pennaria*); E—Leptomedusa (*Aequora*); F—Leptomedusa (*Tima*); G—Leptomedusa (*Obelia*).

Pteroids, *Corallium*, *Pennatula*, etc., of class Anthozoa show dimorphism and have two types of polyps: (1) **Autozooids** or **gastrozooids** feed the colony. (2) **Siphonozooids** circulate water current through the gastrovascular cavities of colony and bear gonads too. They lack a medusa stage in their life cycle.

2. Trimorphism

The coelenterate colonies bearing three types of zooids are called **trimorphic colonies** and the phenomenon is termed **trimorphism**. For example, *Obelia* colony has—

(*i*) Polyps or gastrozooids

(*ii*) Blastostyles or blastozooids or gonozooids which bear medusae

(*iii*) Medusae or sexual zooids.

In *Millipora*, *Plumularia*, etc., the zooids are of following three types:

(1) Gastrozooids. These are the polyps or nutritive zooids.

(2) Dactylozooids. These are the protective zooids with long and knobbed tentacles.

(3) Medusae. These are gonads-bearing sexual zooids.

3. Polymorphism

In some coelenterates both polyps and medusae are of various types and are modified to carry out different functions of the colony. For example, siphonophores (*Halistemma*, *Physalia*, *Velella*, *Porpita*, etc.) are invariably free swimming, colonial and polymorphic. The polyps and medusae are differently modified to present a large variety of zooids. The **polypoid zooids** of siphonophores are of three types:

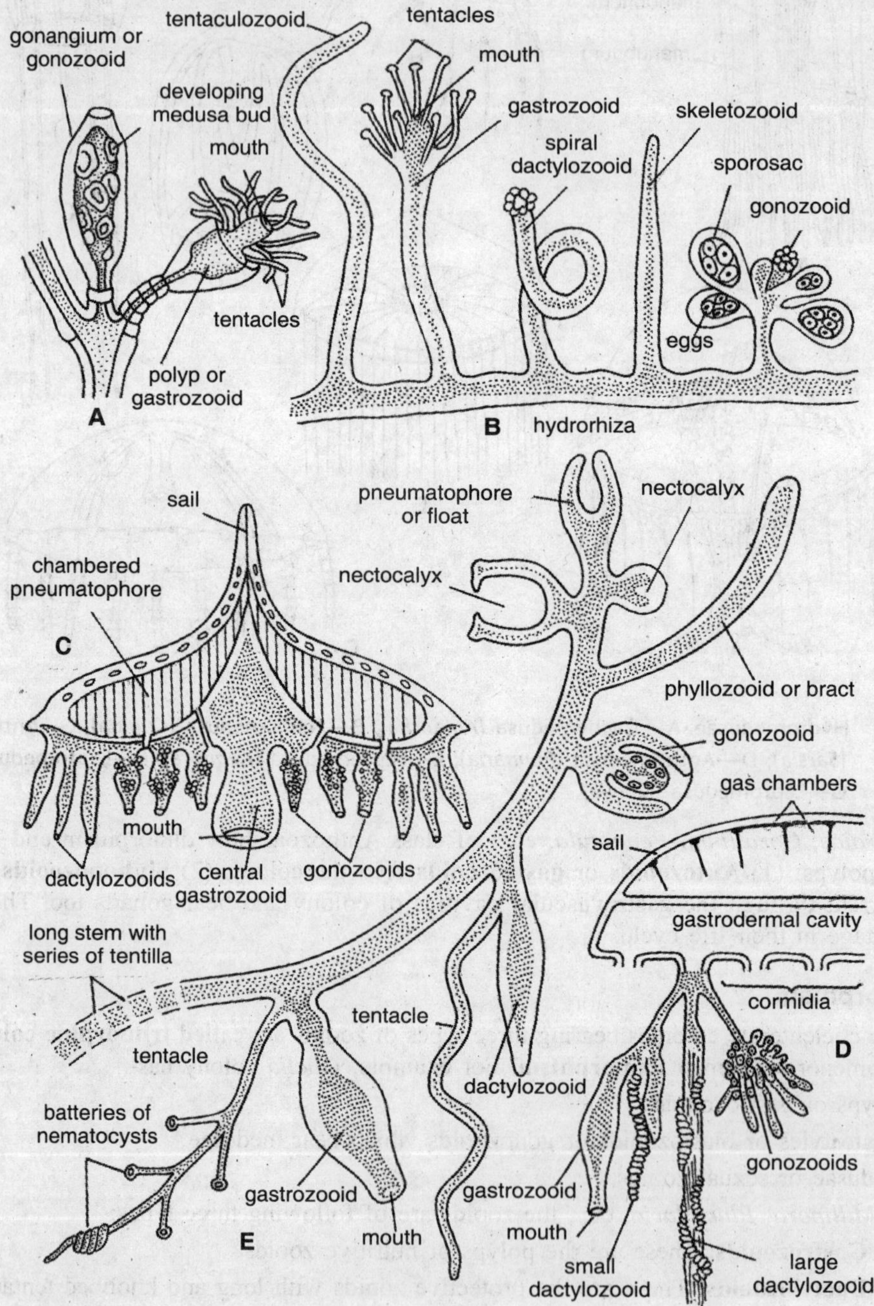

Fig. 18.2. Polymorphic colonies of cnidaria. A—*Obelia*; B—*Hydractina*; C—*Velella*; D—A single cormidium of *Physalia*; E—Generalized calycophoran Siphonophora showing different zooids.

(*i*) **Gastrozooids** or feeding polyps are each with a mouth and a long tentacle.

(*ii*) **Dactylozooids** or protective polyps are without mouth and usually with a long unbranched basal tentacles. They bear many nematocysts and are also called **palpons**, **tasters** or **feelers**.

(*iii*) **Gonozooids** or reproductive polyps lack mouth and tentacles. They reproduce **asexually** by budding and form medusae. In *Physalia*, gonozooids are branched stalks, called **gonodendra**, and bear grape-like clusters of gonophores or medusae (Fig. 18.2).

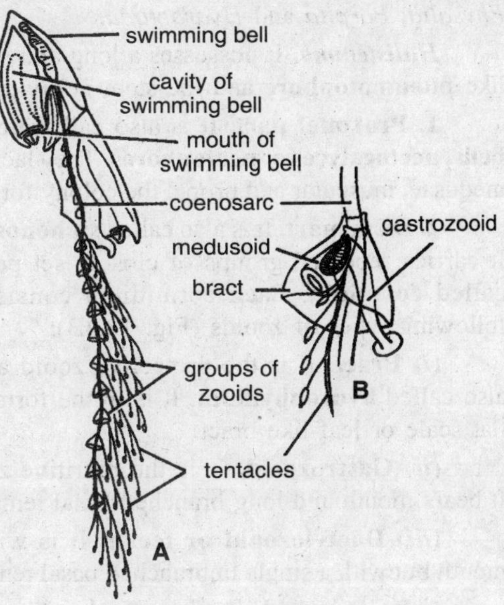

Fig. 18.3. *Diphyes*. A—Entire colony; B—Single group of zooids.

Similarly, the medusoid zooids are of following four types:

(1) Nectophores. They are also called **nectocalyces, nectozooids** or **swimming bells**. These are free-swimming zooids with swimming bells (Fig. 18.3), velum, four radial canals and a circular canal. Nectophores do not possess mouth, manubrium, tentacles and sense organs. They serve as the loco-motory organs of the colony.

(2) Pneumatophores or floats. These are bladder-like or vesicle-like structures filled with gas. Each pneumatophore represents an inverted medusa bell; it is devoid of mesogloea and consists of an external exumbrellar wall.

(3) Hydrophyllia. These are also known as **bracts** or **phyllozooids**. These are shield-like, leaf-like, helmet - shaped or prismatic covers. They are studded with nematocysts, serving for protection.

(4) Gonophores. They may occur singly on separate stalk or in clusters on polypoid gonozooids. They represent buds of immature or sessile medusae and are plate-like or bell-shaped. Gonophores are dioecious (Fig. 18.4) but the colonies are hermaphroditic bearing both types of gonophores in the same or separate clusters.

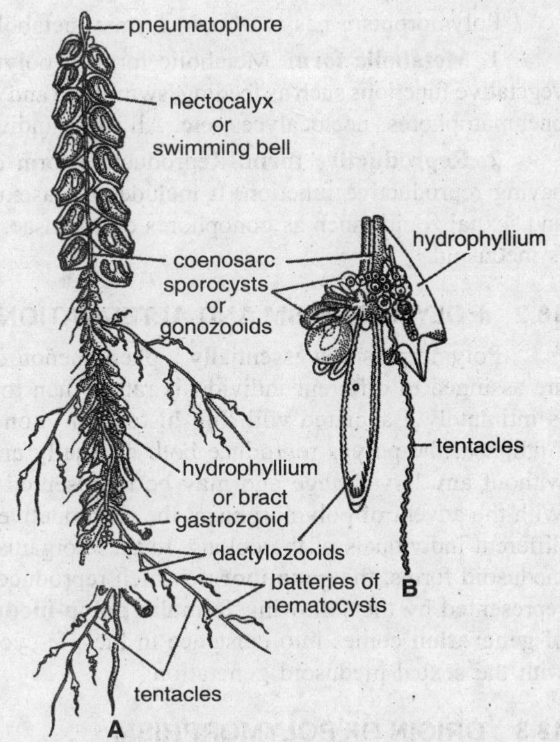

Fig. 18.4. *Halistemma*. A—Entire colony; B—One cormidium enlarged.

Polymorphism in *Halistemma*

Coelenterates of the order Siphonophora represent the most specialized hydrozoa attaining the highest degree of polymorphism. The composition and arrangement of various zooids is very different. The diversity is so great that each individual is to be studied separately. The best known polymorphic siphonophores are *Halistemma*, *Velella*, *Physalia*, *Porpita* and *Hydractinia*.

Halistemma. It possesses a long and slender **stem.** Its upper end is expanded into a bubble-like **pneumatophore** and the stem is divided into following two parts (Fig. 18.4):

1. Proximal part. It is also called **nectosome**. It bears closely set bell-shaped **swimming bells, nectocalyces** or **nectophores**. They lack mouth, tentacles and sense organs. They are modified medusae, muscular and propel the colony forward.

2. Distal part. It is also called **siphonosome.** It carries repeated groups of closely set polyps, called **cormidia**. Each cormidium consists of following types of zooids (Fig. 18.4B):

(*i*) Bract. It is the protective zooid and is also called **hydrophyllium**. It is in the form of a flat scale or leaf-like bract.

(*ii*) Gastrozooid. It is the nutritive zooid. It bears mouth and long branched basal tentacles.

(*iii*) Dactylozooid or feeler. It is without mouth but with a single unbranched basal tentacle.

(*iv*) Gonozooid. It is a reproductive zooid. It has either testes or ovaries.

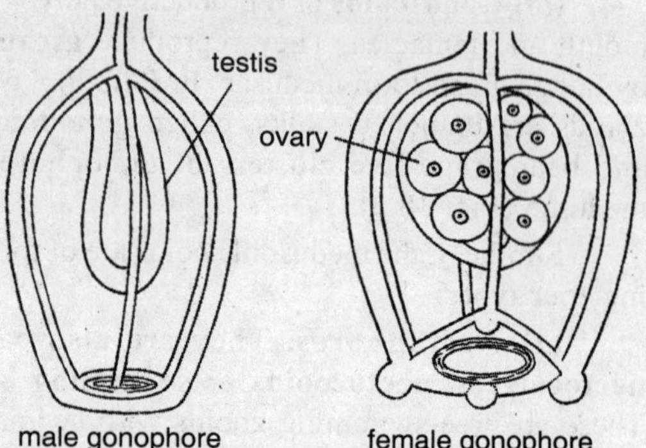

Fig. 18.5. Siphonophora. Male and female gonophores.

Forms of Polymorphism

Polymorphism has two main forms: metabolic and reproductive.

1. Metabolic form. Metabolic form of polymorphism is represented by the zooids having vegetative functions such as feeding, swimming and defence. It includes gastrozooids, dactylozooids, pneumatophores, nectocalyces, etc. All these individuals are polypoid.

2. Reproductive form. Reproductive form of polymorphism is represented by the zooids having reproductive function. It includes the asexual zooids, such as blastostyles or gonozooids and sexual zooids such as gonophores or medusae. The gonozooid is polypoid while the medusae is medusoid.

18.2 POLYMORPHISM AND ALTERNATION OF GENERATION

Polymorphism is essentially a phenomenon of "division of labour", *i.e.*, different functions are assinged to different individuals rather than to the organs of one individuals. Polymorphism is intimately associated with life history. In monomorphic forms such as *Hydra* and the class Anthozoa, the polyps reproduce both asexually and sexually, so that, life-cycle remains simple, without any larval stage and may be represented by the following, formula: **polyp-egg-polyp.** With the advent of polymorphism, the reproductive power of the organism are divided among the different individuals of the colony. In these organisms the polyps reproduce asexually to produce medusoid forms, the **gonophores**, which reproduce sexually to form polyp. The life-cycle then is represented by the following formula: **polyp-medusa-egg-planula-polyp**. Thus, the **alternation of generation** comes into existence in the life-cycle. The asexual polypoid generation alternates with the sexual medusoid generation.

18.3 ORIGIN OF POLYMORPHISM

Regarding the origin of polymorphism, there are many views. According to one view the original coelenterate was a polyp and through specialization the sexual function was relegated to secondarily developed medusoid form and this led to alternation of generation. According to another view, the ancestral coelenterate was a medusoid form, while the polypoid generation represent a persistent larval form, thus, leading to polymorphism.

Theories of Origin of Polymorphism

A number of theories have been put forward by various zoologists; some of them are as follows:

1. Polyorgan theory. This theory was put forward by **Huxley**, **Eschscholtz** and **Metschnikoff.** According to this theory the component zooids are organs of a single medusoid individual. Certain organs of these zooids such as manubrium, tentacles and umbrella have become multiplied independently and have assumed different forms to perform different functions.

2. Polyperson theory. This theory was put forward by **Leuckart, Vogt** and **Gegenbaur.** According to this theory a polymorphic individual is a colonial form of highly specialized polyps with the power to produce medusa. Their diversified organisms or zooids have grouped together to perform different functions. This theory maintains that the part of a polymorphic colony are either polyp or medusae but the primitive zooids of the colony are of polyp type.

QUESTIONS

Long Answer Questions

1. Write an essay on polymorphism in Coelenterata.
2. What is polymorphism? Describe it in the Coelenterata.
3. Select one example from Siphonophora and explain its polymorphic nature. Give a well labelled diagram of polymorphic coelenterate and comment upon its salient features.
4. Describe the polymorphism in *Velella* and *Physalia.*

Short Answer Questions

1. Describe different types of polymorphism found in Cnidaria.
2. Give an account of polymorphism in *Halistemma.*
3. Describe the polypoid zooids of siphonophores.

Very Short Answer Questions

1. Name four types of hydromedusae.
2. What is gonodendron?
3. Name the polymorphic colony with a rhomboidal float.
4. Give the function of nectocalcyes.
5. *Hydractinia* develops four types of zooids. What are their names?
6. Mention the theories advanced to explain polymorphism in cnidarians.
7. Which type of medusae are found in *Obelia*?

Fill in the Blanks

1. Three grades of polymorphism are dimorphic, ... and polymorphic.
2. Examples of dimorphic colonies are .... and *Pennatula.*
3. Two types of zooids in *Pennatula are* .... and ....
4. Common trimorphic colonies are *Obelia* and ....
5. Polymorphic colonies include .... , *Physalia* and ....
6. *Physalia* colony is kept afloat by .... that contains ....
7. A group of zooids in a polymorphic colony is called ....
8. Nectocalyces are modified ....

Multiple Choice Questions

1. Polyp helps in
   (*a*) reproduction (*b*) nutrition
   (*c*) excretion (*d*) respiration
2. Polymorphism is shown in
   (*a*) hydrozoa (*b*) anthozoa
   (*c*) scyphozoa (*d*) actinozoa
3. Division of labour between the several individuals of a colony is called
   (*a*) vital activity (*b*) labour division
   (*c*) polymorphism (*d*) poly-function
4. Gastrozooids, gonozooids and dactylozooids are the characteristic of .... colony
   (*a*) dimorphic (*b*) trimorphic
   (*c*) polymorphic (*d*) all the above
5. Modification of polyps are
   (*a*) gastrozooids (*b*) dactylozooids
   (*c*) gonozooids (*d*) all the above
6. Which zooid is generally motile?
   (*a*) polyp (*b*) medusa
   (*c*) both (*d*) none
7. Modification of medusa
   (*a*) phyllozooid (*b*) nectophore
   (*c*) pneumatophore (*d*) all the above
8. Cnidom is
   (*a*) battery of nematoblasts (*b*) group of nematoblasts
   (*c*) armed thread (*d*) unarmed thread
9. A cormidium of *Physalia* consists of
   (*a*) tentacles (*b*) zooids
   (*c*) both (*a*) and (*b*) (*d*) tentacles, zooids and pneumatophores
10. Pneumatophore lacks sail in
   (*a*) *Physalia* (*b*) *Velella*
   (*c*) *Porpita* (*d*) all the above

ANSWERS

Very Short Answer Questions

1. Trachymedusae, Narcomedusae, Anthomedusae and Leptomedusae.
2. A branched blastostyle (gonozooids).
3. *Velella*.
4. Locomotion of colony.
5. Gastrozooids, dactylozooids, gonozooids and sporosacs.
6. Polyorgan and polyperson.
7. Leptomedusae.

Fill in the Blanks

1. Trimorphic
2. *Bougainvillea*
3. Autozooids, siphonozooids
4. *Millepora*
5. *Halistemma*, *Velella*
6. Pneumatophore, gas
7. Cormidium
8. Medusae.

Multiple Choice Questions

1. (*b*) **2.** (*a*) **3.** (*c*) **4.** (*b*)
5. (*d*) **6.** (*b*) **7.** (*d*) **8.** (*b*)
9. (*c*) **10.** (*c*).

19 Corals and Coral Reefs

19.1 CORALS

Corals are deposits of lime or $CaCO_3$ formed by coelenterates. They are formed of exoskeleton secreted by the polyps of some colonial coelenterates. They may be solid and massive or branched like the trees. Most of them belong to the class Anthozoa comprising about 2500 living 5000 fossil species.

Types of Corals

1. Hydrozoan corals. They include *Millepora* (order Milleporina) and *Stylaster* (order Stylasterina). They are colonial and secrete massive calcareous exoskeleton. The skeleton is secreted by a modified epidermis called **calicoblastic layer**. Living within the skeleton are following two types of polyps: 1. **Large gastrozooid** or **nutritive zooids**, and 2. **Small dactylozooids** or **protective zooids**. These corals are found with other corals, which constitute coral reefs.

2. Anthozoan corals. They are of following types:

(*i*) Horny corals. These belong to subclass Octocorallia and order Gorgonacea. They are branched and tree-like. The skeleton is formed of horny proteinaceous material, **gorgonin,** along with calcareous spicules which surround the polyp, *e.g.*, *Gorgonia.*

(*ii*) Soft corals. These belong to subclass Octocorallia and order Alcyonacea. They are colonial polyps having endoskeleton of separate spicules, embedded in a massive mesogloea, *e.g.*, *Alcyonium* (Dead man's fingers, Fig. 19.1).

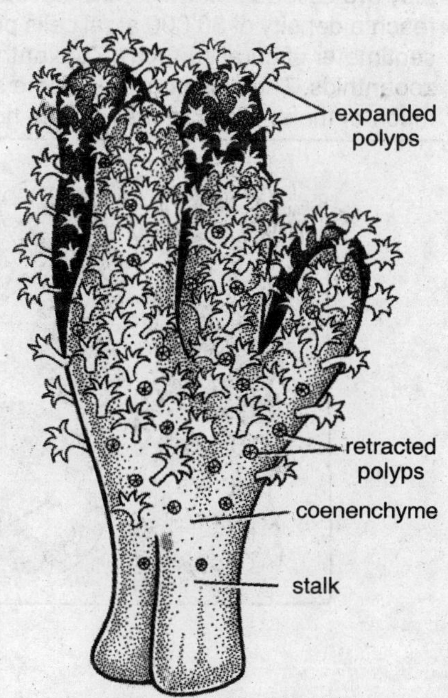

Fig. 19.1. *Alcyonium.*

(*iii*) Stony corals. These belong to subclass Octocorallia and order Stolonifera. The skeleton is made up of calcareous spicules and stained red with iron salts. The skeleton consists of vertical tubes connected by lateral platforms, *e.g.*, *Tubipora* (organ-pipe coral).

(*iv*) Blue coral. These belong to subclass Octocorallia and order Helioporacea. In its colony of polyps, the calcareous spicules form a massive skeleton. The larger cavities on the surface of skeleton contain polyps, *e.g.*, *Heliopora* (blue coral).

(*v*) True corals. These belong to subclass Hexacorallia and order Madreporaria. They form true corals. Majority of them are colonial. The colonies have different forms. These are the principal builders of the coral reef.

Organization of Corals

1. Coral skeleton. The exoskeleton of a coral is known as **corallite**. It is calcareous and secreted by the epidermis. In a colonial coral, the coralite of individual polyps fuse together to form a skeletal mass which is called the **corallium**.

Each corallite is cup-shaped depression with a **basal plate** or **theca**. The cavity of the cup contains a number of vertical radiating ridges or septa, called **sclerosepta**. The inner end of sclerosepta are fused to form an irregular skeletal mass or the **columella**.

2. Coral polyp. They are small anthozoan polyp measuring about 1 cm in length. These are cylindrical without pedal disc, and oral disc bears a cycle of six tentacles. The pharynx is without siphonoglyphs. The mesenteries are restricted to the upper part of the polyp. The muscles are poorly developed. At night and on cloudy days, the polyps expand and feed on plankton.

Box 19.1 Zooxanthellate Corals

One of the most noteworthy evolutionary achievements of cnidarians is their close relationship with unicellular photosynthetic partners. The relationship is widespread and occurs in many shallow-water cnidarians. The symbionts of freshwater hydrozoans (*e.g.*, *Chlorhydra*) are single-celled species of green algae (Chlorophyta) called "zoochlorelle". In marine cnidarians, the protists are unicellular cryptomonads and dinoflagellates called "zooxanthellae" (probably many genera including *Zooxanthella*; Fig. 19.2). These algae may be capable of living free from their hosts, but very little is known about their natural history (see **Brusca** and **Brusca**, 2003). The algae typically reside in the host's gastrodermis or epidermis, although some cnidarians harbour extra-cellular zooxanthellae in mesogloea. It is usually the algal symbionts that give cnidarians their green, blue-green, or brownish colour. Corals that are reef-builders (called **hermatypic corals**) typically harbour zooxanthellae (*i.e.*, they are called **zooxanthellate corals**). Resident populations of zooxanthellae in these corals may reach a density of 30,000 algal cells per cubic millimeter of host tissue (or from 1 to 2 × 10^6 cells per centimeter of coral surface). Zooxanthellae also occur in many tropical gorgonians, anemones, and zooanthids. They also occur in some scyphozoans, and in some cases they are thought to produce much of the energy required by the host (*e.g.*, *Cassiopea*, *Linche* and *Mastigias*).

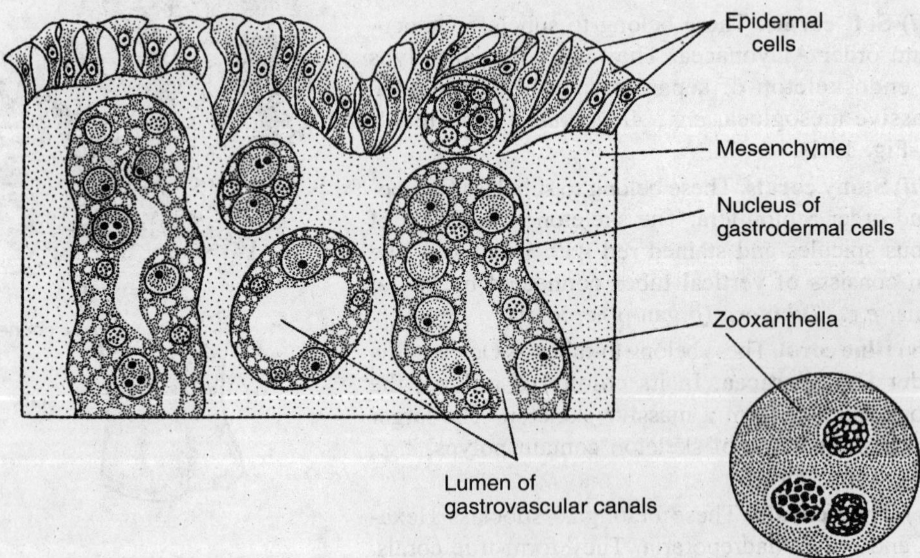

Fig. 19.2. An octocoral with zooxanthellae distributed throughout the gastrodermis (schematic section) (after Brusca and Brusca, 2003).

3. Formation of corals. The coral polyp develops from a planula which secretes a skeletal rudiment or **prototheca** with the help of ectodermal and basal plate. Radial folds of the larva secrete septa and a rim is built up around the polyp. Meanwhile, further skeletal material is added into the gaps between the septa. The septa alternate with the mesenteries of a living coelenterate.

Role of mutualistic zooxanthellae in coral formation. Corals and zooxanthellae (Fig. 19.2) exist in a **mutualistic** relationships: corals cannot live, or are handicapped without them (**Muscatine**, 1974). This restricts coral distribution to areas where zooxanthellae can thrive. Corals die after 18 days in total darkness, but they are completely carnivorous and expel zooxanthellae when starved. Zooxanthellae probably seek corals to satisfy their own nitrate and phosphorus requirements and in return help in the skeleton formation of coral. They help to dispose of H_2CO_3 released as by-product of the formation of $CaCO_3$ (**Rinkevich** and **Loya**, 1984).

19.2 CORAL REEFS

Coral reefs are limestone formations, formed of exoskeletons of stony corals (*Madreporaria*) and heavily charged with organic material, especially in upper strata where growing corals occur. Calcareous algae, foraminiferans, milleporines and other calcareous Alcyonaria contribute to reef.

Coral reefs are the ridges of calcium carbonate in the sea formed by the remains of skeleton and shells of animals.

Distribution

Coral reefs grow in warm, shallow waters and limited to continental and island shores in tropical and subtropical zones. They can not survive in water having temperature less than 22°C and have luxuriant growth in shallow water having a depth less than 11 metres. Coral reefs inhabit Maldives Island of Indian Ocean, Wake Island, Marshal Island, Fizi Island of Pacific Ocean, Whitesunday Island, Bahoem Island regions and the famous Great Barrier reefs of Australia.

Kinds of Coral Reefs

Coral reefs are considered to be of following three types (Fig. 19.3).

(*i*) Fringing reefs

(*ii*) Barrier reefs

(*iii*) Atolls

(1) Fringing reefs. The coral reefs are termed as **fringing reefs** which lie close to the shores of some volcanic island or part of some continent. A fringing reef extend from the shore a few metres to 1/4 mile out and consists of the **reef edges** or **front**, where the most active coral growth occurs and a slightly lower, more or less flat surface, the **reef flat**. The reef flat between the front and the shore is composed largely of coral sand and mud, dead corals and other debris usually giving a scarlet to brown green hue. The outer single beach of island may get merged into the seaward flat of the reef which already consists of **seaward slope** on which corals grow down to 20 fathoms or there may be a varying area of **boulder zone** in between. Sometime between island and boulder zone may be enclosed an **inner flat** which is hollowed out to give rise to shallow lake.

(2) Barrier reefs. Barrier reefs are same as the fringing reef but they are located some distance from the shore. They are separated from land by stretch of water or **lagoon** which may be 1/2 a miles to 10 mile or more and 60 to 1,000 feet deep.

The most notable example of barrier reef is **Great Barrier Reef** of Australia, extending over 1200 miles long and up to 90 miles from shore.

(3) Atoll or Atoll reef. They are also referred as **coral island** or **lagoon island**. They are irregular, more or less circular or horse-shoe shaped reef not enclosing any island but encircling lagoon.

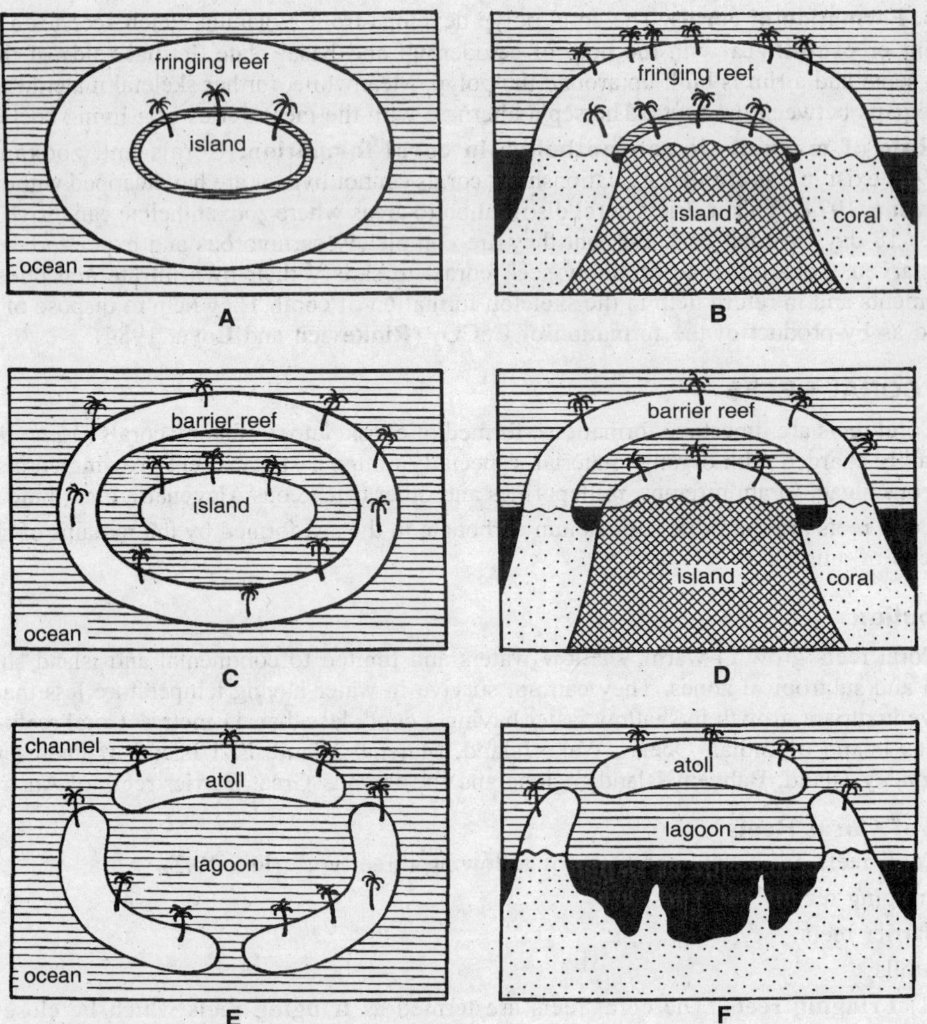

Fig. 19.3. Coral reefs. A—Fringing reef; B—Fringing reef in section; C—Barrier reef; D—Barrier reef in section; E—Atoll reef; F—Atoll reef in section.

The term atoll is derived from the Maldives language, where each government district is termed as "*atolu*" and the governor the "*atoluveri*". His province is a circular reef broken by many channels. The encircled lagoons varies from a few miles to 40 to 50 miles. The outer side of the reef slopes off rather steeply into the depth of the ocean.

Theories of Reef Formation

It is difficult to explain the great vertical thickness often attained by the coral reefs, since the reef building corals are littoral in habit and cannot grow below 150 feet. Many theories have been put forward to explain coral-reef formation. A few important ones are as follows:

1. Darwin-Dana theory of subsidence. In 1831, Darwin noticed that the place where coral reefs are formed now, a sinking of land had taken place in the past. Therefore, he assumed that the corals started their growth as fringing reefs around the sloping shores of an island in shallow tropical sea. Subsidence of the sea-floor then started in region of the reef followed by the upward and outward growth of the coral. The sinking island became smaller and smaller and finally disappeared and the barrier reef became an atoll.

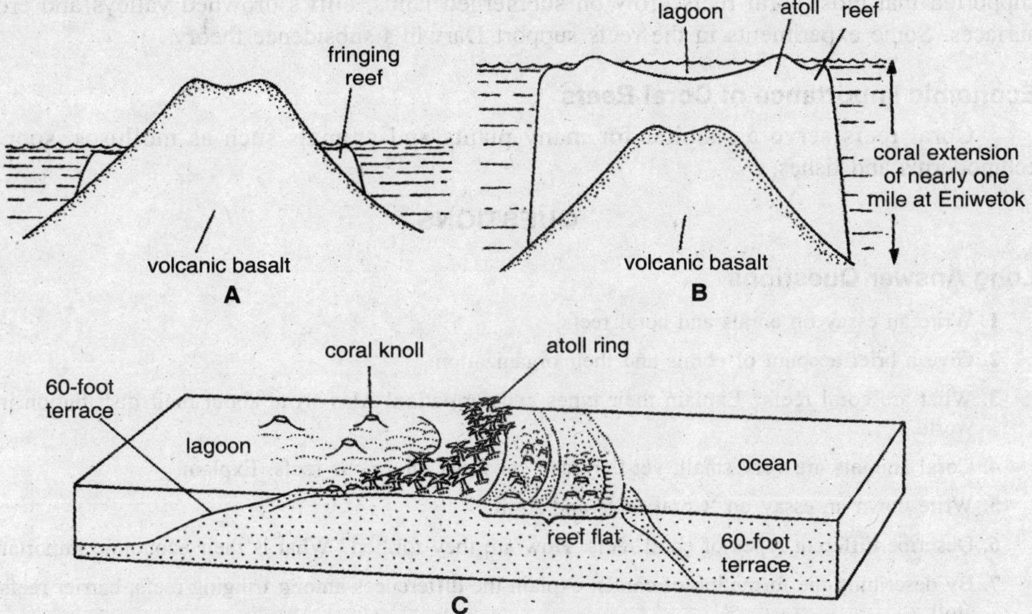

Fig. 19.4. Formation of an atoll. A—Fringing reef around an emergent volcano; B—Continuous deposition of coral as the volcanic cone subsides, leading to formation of a great coralline cap; the emergent part of the cap is the atoll; C—Section through part of the atoll.

2. Submerged bank theory. The recent scientists assumed that the coral grow to form reefs on flat pre-existing surface during or after their submergence. This is brought about by the erosion and denudation of an island both above and below water. An island is replaced completely by submerged plateau and builders will produce an atoll.

3. Daly glacial control theory. This theory is put forward by **Daly.** He assumed that during last glacial period, the widely extending polar ice caps, over a mile in thickness were formed. This lowered the sea level by at least 150 feet and exposed several flat platforms of sand and mud deposition cut out by the action of water currents during interglacial period. Due to the low temperature of island, the growth of corals are inhibited. When the ice melted, the sea level rose and more and more waters covered the platform. The sea temperature having risen to over 10°C became suitable for coral growth. This theory is based on accurate knowledge and close calculations for the very uniform depth of coral lagoon.

4. Super-Murray solution theory. According to **Sir John Murray** mounds are formed upon the sea bottom by deposition of limestone skeletons of many animals (such as shells of molluscs and starfish and deep sea corals) when they reach about 50 fathoms, the corals grow on them and reach the water surface. A barrier reef results at the outer edge and an atoll is produced by the dissolution of the inner coral rocks. The mound ultimately became flattened like plateau.

5. Struchbury's volcanic crater theory. According to **Struchbury**, the coral reefs are formed at the crater of volcano in the Pacific ocean. After formation of lagoon at the same region, the corals had continuous growth. Ultimately they came out from the water surface. But this theory is not entirely satisfactory.

The submerged bank theory and Daly's theory which supplement each other are at present most favoured although Darwin's idea still finds much support. Several lines of evidence have supported that most coral reefs grow on submerged lands, cliff's drowned valleys and eroded surfaces. Some experiments in the reefs support Darwin's subsidence theory.

Economic Importance of Coral Reefs

Coral reefs serve as habitat for many plants and animals such as molluscs, sponges, echinoderms and fishes.

QUESTIONS

Long Answer Questions

1. Write an essay on corals and coral reefs.
2. Give a brief account of corals and their organization.
3. What are coral reefs? Explain their types and formation. Also write about their distribution in the world.
4. Coral animals are very small, yet they can construct enormous reefs. Explain.
5. Write down an essay on 'Coral reefs and Atoll'.
6. Describe different types of coral reefs. How are they formed? What is their economic importance?
7. By describing the formation of corals, explain the differences among fringing reefs, barrier reefs and atoll.

Short Answer Questions

1. Write a short note on corals.
2. Coral reef.

Very Short Answer Questions

1. Name two hydrozoan corals.
2. Mention one soft coral and one horny coral.
3. Name the three types of coral reefs.
4. From which coral red 'munga' is obtained.
5. Mention an example of a solitary coral.
6. What are corals?

Fill in the Blanks

1. Two main types of coral are .... corals and anthozoan corals.
2. The major reef builders are .... corals.

Multiple Choice Questions

1. The exoskeleton of the colony is
   (*a*) corallum (*b*) columella
   (*c*) corallite (*d*) none of these
2. Corallite is
   (*a*) an animal (*b*) an individual
   (*c*) a cnidarian (*d*) skeleton of a solitary coral

3. Organ-pipe coral is
   (a) *Astrea* (b) *Tubipora*
   (c) *Fungia* (d) *Meandrina*
4. True corals are coelenterates belonging to
   (a) Hydroidea (b) Madreporaria
   (c) Pennatulacea (d) Coenothecalia
5. A coral island with a central shallow lake is known as
   (a) coral reef (b) lagoon
   (c) atoll (d) corallite
6. Coral islands are
   (a) Fiji (b) Maldives
   (c) Laccadives (d) all of these
7. Coral reefs are confined to the
   (a) tropical seas (b) temperate seas
   (c) polar seas (d) river deltas
8. A fringing reef is usually located
   (a) very near the shore (b) far away from the shore
   (c) in deep sea (d) in cold sea
9. Bramuda is a
   (a) fringing reef (b) coral island
   (c) atoll reef (d) barrier reef
10. Most notable example of barrier reef is
   (a) West Indies (b) Great Barrier reef
   (c) Florida keys (d) none of these

ANSWERS

Very Short Answer Questions

1. *Millepora* and *Stylaster*.
2. *Alcyonium* and *Gorgonia*.
3. Fringing reef, barrier reef and atoll.
4. *Corallium rubrum*.
5. *Fungia*.
6. Corals are colonial, polypoid marine cnidarians living in the skeleton secreted by themselves. This skeleton is calcareous or horny and is commonly known as coral. Some coral-building cnidarians are 1. *Millepora*; 2. *Tubipora*; 3. *Gorgonia*; 4. *Alcyonium*; 5. *Fungia*.

Fill in the Blanks

1. Hydrozoan 2. Madreporarian

Multiple Choice Questions

1. (a) 2. (d) 3. (b) 4. (b)
5. (c) 6. (d) 7. (a) 8. (a)
9. (b) 10. (b).

20

Phylum Ctenophora

20.1 GENERAL CHARACTERS

1. Ctenophores (Gr., *ktenes* = combs; *ophora* = carrying) are commonly called **sea walnuts, sea gooseberries** or **comb jellies.**
2. Marine, solitary, luminescent, pelagic and free-swimming animals.
3. There are only 50 species of Ctenophora. Some are abundant in coastal waters and many are oceanic.
4. Acoelomate and triploblastic animals having biradially symmetrical med-usa-like body.
5. Body delicate, transparent, gelatinous and unsegmented.
6. Body shape is usually pear-shaped, cylindrical, spherical flat or ribbon-like.
7. Body wall consists of three layers — outer **epidermis**, inner **gastrodermis** and middle thick **mesogloea** or **mesoderm** with scattered cells and muscle fibres.
8. Ctenophores perform locomotion by eight — vertical (meridional) rows of ciliary **comb plates**.
9. They are predatory (carnivorous) animals feeding on other planktonic forms (*e.g.*, copepods). Many species use in food-captur-ing their two long, branc-hed and retractile **tentacles** bearing special adhesive cells called **colloblasts** (Fig. 20.1).

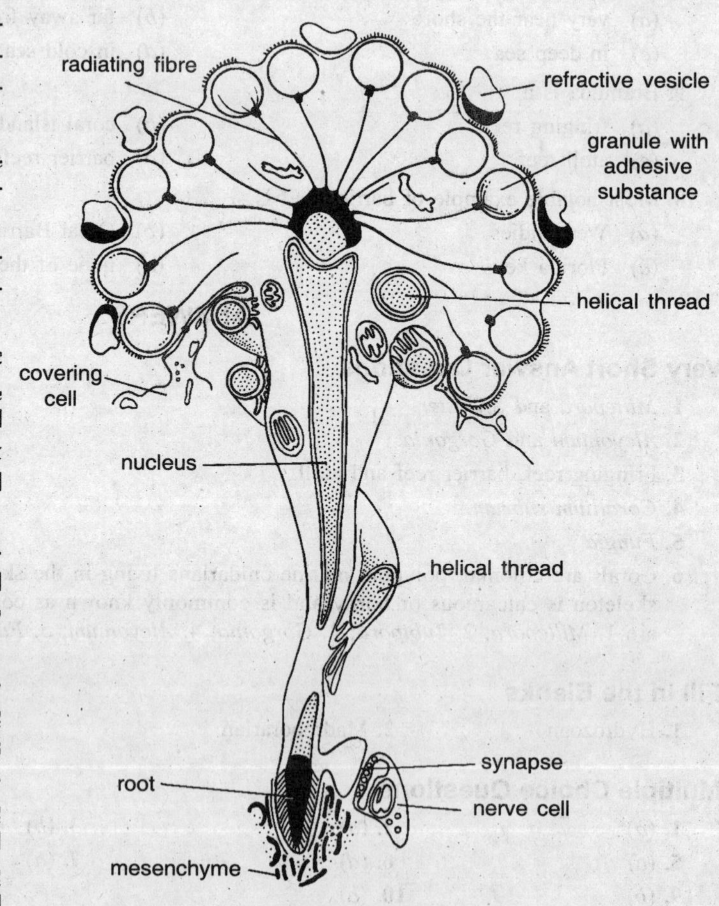

Fig. 20.1. Colloblast.

10. Nematocysts are entirely absent.
11. Digestive system consists of a **mouth**, a ciliated **pharynx, stomach** (infundibulum) and a system of gastrovascular canals.
12. Skeletal, excretory, circulatory and respiratory systems are absent.
13. Nervous system is diffuse having a subepidermal nerve network and aboral end bears a sense organ the **statocyst**.
14. They are **hermaphrodites**, *i.e.*, male and female gonads are found in the same organism.
15. Gonads are endodermal in origin.
16. Development is **indirect** type and highly **determinate**, producing a **cydippid larva**.
17. No alternation of generation occur.
18. **Regeneration** and **paedogenesis** are common.

Box 20.1

The tentacular epidermis of a ctenophore possess peculiar adhesive cells called **colloblasts**. A colloblast cell is somewhat pear-shaped, with the narrowed end anchored in the tentacular mesogloea and having a synaptic junction with a nerve cell. An intracellular helical thread winds around the long axis of the cell and at its distal end gives rise to a large number of small radiating fibres. Each radiating fibre terminates in a granule filled with an adhesive mucoid material, which is liberated on contact with prey.

20.2 CLASSIFICATION

The phylum ctenophora is divided in **two** classes:

Class 1. Tentaculata

Ctenophores of this class are with tentacles. Class Tetaculata is divided into following **four** orders:

Order 1. Cydippida

1. Body cylindrical or oval.
2. Two long tentacles are branched and retractile.
3. Anal pore is absent.

Examples. *Pleurobranchia, Hormiphora, Mertensia.*

Order 2. Lobata

1. Laterally compressed oval body.
2. Two large oral lobes present.
3. Two main tentacles reduced and without sheaths.

Examples. *Mnemiopsis, Bolinopsis.*

Order 3. Cestida

1. Ribbon-like body.
2. Two main tentacles are reduced and lie in a sheath.
3. Numerous lateral tentacles in a groove.
4. Comb plates in four rows but rudimentary.

Examples. *Cestum, Velamen.*

Order 4. Platyctenea

1. Body greatly flattened.
2. Two well developed tentacles with sheaths are present.
3. Comb plates are reduced or absent in adult.
4. Body is adapted for creeping.

Examples. *Ctenoplana*, *Coeloplana*.

Class 2. Nuda

Ctenophores of this class are without tentacles. It includes **single** order.

Order 1. Beroidea

1. Body conical or ovoid and laterally compressed.
2. Tentacles and oral lobes absent.
3. Mouth large and muscular and stomach (stomodeum) voluminous.
4. Meridional canals are highly anastomosing.
5. Voracious feeders feeding on other ctenophores.
6. Cydippid larva is lacking.

Examples. *Beroe*.

20.3 AFFINITIES OF CTENOPHORA

The evolutionary position of ctenophores is very obscure. They have long been considered related to cnidarians (coelenterates). However there are many fundamental similarities. The triploblastic origin of their tissues and determinate cleavage are similar to some bilateral phyla. It suggests that Ctenophora exhibit following affinities:

(*i*) Resemblance with Coelenterata

1. Radial symmetry.
2. Body has tissue grade organization. General lack of organ system.
3. Tentacles and mesogloea are present.
4. Coelom absent.
5. Presence of gastrovascular cavity.
6. Diffuse nervous system.
7. Respiratory system absent.
8. Nephridia absent.
9. Presence of statocyst.
10. Endodermal origin of gonads.

Due to above similarities the ctenophores are considered by many zoologists to be a class of Coelenterata.

(*ii*) Differences with Coelenterata (Cnidarians)

1. Absence of nematocysts.
2. Tentacles oppositely placed; it suggests a biradial symmetry.
3. Presence of stomodaeum.
4. Presence of eight locomotory ciliated comb plates.
5. Presence of colloblasts over tentacles.

6. Sense organ (statocyst) is aboral in position.
7. Presence of mesenchymal muscle cells.
8. Presence of a definite organization of digestive system with anal pores.
9. Development is of determinate type.

(*iii*) Advancement over Coelentrates (Cnidarians)

1. Mesenchymal muscle cells.
2. Presence of three germ layers.
3. Apical sense organs.
4. Complete digestive organs.
5. Gonoducts are present.
6. Cleavage/development of determinate type.

Since ctenophores exhibit many striking characteristics of their own so that it appears more justified to treat them as a separate phylum rather a class of phylum Coelenterata (Cnidaria).

(*iv*) Affinities with Platyhelminthes

The *Platyctenea* is considered to be a connecting link between Ctenophora and Bilateria. Ctenophora exhibit following similarities with turbellarians (Platyhelminthes).

1. Body contain cilia (*e.g.*, comb plates, ciliated pharynx).
2. Dorso-ventrally flattened body.
3. Crawling mode of life.
4. Origin of so-called mesoderm is more or less similar.
5. Ctenophora exhibits both radial as well as biradial symmetries.
6. Similar cleavage and gastrulation.
7. Gelatinous mesenchyme with muscle fibres and cells.

The view that the primitive Bilateria have evolved through *Platyctenea* has not been accepted.

Position of Ctenophora in Animal Kingdom

Ctenophores have certain characteristics in common with the Cnidarian (coelenterates). There is no evidence about its derivation from any existing group of Cnidaria. It is believed that Ctenophora is to be an offshoot from the ancestral medusoid coelentrata (*Hydractena*), which was probably more or less spherical animal with a concentration of cilia along their meridional rows, which later develop into comb plates. Such an offshoot gave rise to not a higher form (*i.e.*, ctenophores represent a blind evolutionary offshoot).

20.4 TYPE STUDIES

1. *Cestum*

Classification

Phylum	–	**Ctenophora**
Class	–	**Tentaculata**
Order	–	**Cestida**
Genus	–	***Cestum***

Habit and Habitat. *Cestum* (Fig. 20.2) is commonly called **venus's girdle**. It occurs in warmer seas specially the Mediterranean. It is a marine, solitary, free-swimming and bioluminescent ctenophore.

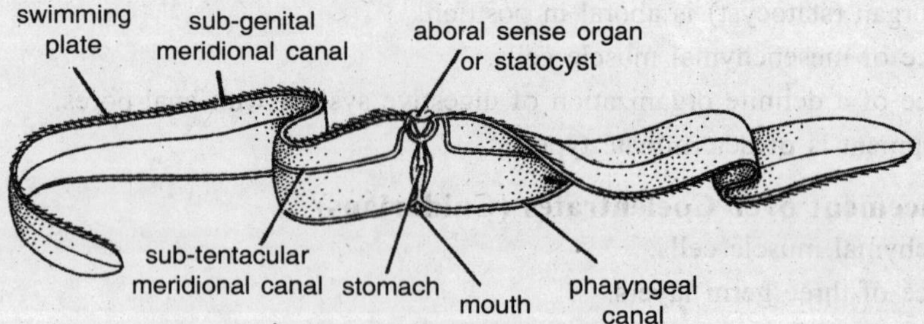

Fig. 20.2. *Cestum*.

Structure. The greatly elongated ribbon-like body is about 45 cm long and 3 cm wide and looks like a transparent celluloid belt. *Cestum* is green, blue or violet in colour. There are present eight aboral comb-plates, four of which are long and other four are rudimentary. At the middle of the oral surface of the body is the mouth. Two subtentacular canals run along the middle of each half of the body and two pharyngeal canals run along the oral surface. The base of two main tentacles are large and are enclosed in sheaths. Many small tentacles spring from grooves and extend in the whole length of oral edge. At the middle of aboral surface sensory statocyst is present. Locomotion takes place by means of long comb plates and muscular undulations.

2. *Ctenoplana*

Classification

Phylum	–	**Ctenophora**
Class	–	**Tentaculata**
Order	–	**Platyctenea**
Genus	–	***Ctenoplana***

Habit and Habitat. *Ctenoplana* (Fig. 20.3) is a small marine solitary and planktonic ctenophore. It is found in the coastal waters of Indo-China and Japan.

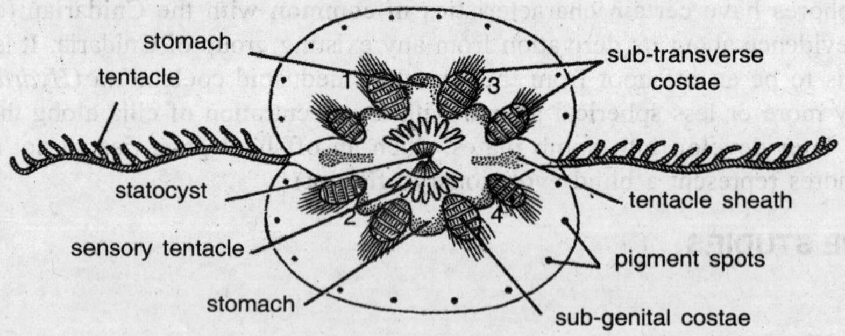

Fig. 20.3. *Ctenoplana*.

Structure. The body of *Ctenoplana* is dorso-ventrally flattened, disc-shaped and measuring about 6 mm in diameter. Colour of body on dorsal surface is mostly olive green, brown or reddish. The mouth lies in the centre of ventral or oral surface. It contains two pinnate and retractile tentacles having sheaths. In the centre of the body on the dorsal surface is a statocyst surrounded by a ring of small ciliated sensory tentacles. Eight rows of comb plates are connected to the statocyst by ciliated furrows. They help in creeping movement of *Ctenoplana*.

3. *Coeloplana*

Classification

Phylum	–	**Ctenophora**
Class	–	**Tentaculata**
Order	–	**Platyctenea**
Genus	–	***Coeloplana***

Habit and Habitat. *Coeloplana* is marine and solitary ctenophore. It is found in water of Red Sea and the coast of Japan.

Structure. The body of *Coeloplana* (Fig. 20.4) is oval and dorso-ventrally flattened. It is elongated in the tentacular plane and measures about 60 mm in length and is transparent or pink with various shades. The comb rows are absent, though the traces of ciliated furrows exist. Mouth is ventral in position. Irregularly branched alimentary canal is present. In *Coeloplana*, locomotion takes place by creeping. Statocyst is dorsal in position. There are 12 to 60 erectile papillae found and they are the diagnostic characters of the animal. There are eight meridional canals.

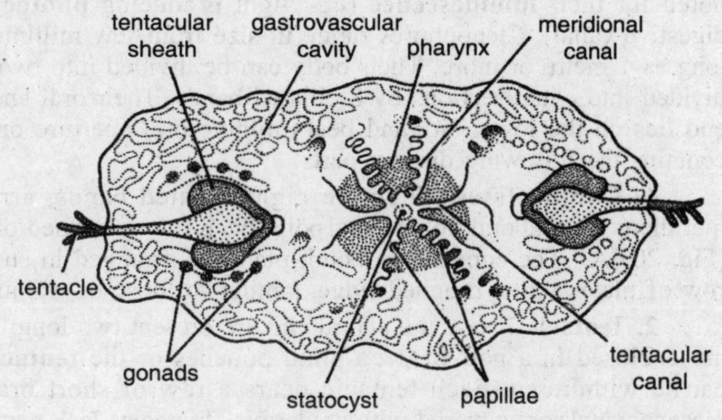

Fig. 20.4. *Coeloplana*.

4. *Pleurobrachia*

Classification

Phylum	–	**Ctenophora**
Class	–	**Tentaculata**
Order	–	**Cydippida**
Genus	–	***Pleurobrachia***

Habit and Habitat. *Pleurobrachia* is commonly called **comb jelly**. It is a marine and pelagic form. It is found all over the world.

Structure. The body of *Pleurobrachia* is pear-shaped and transparent (Fig. 20.5). It measures 2 cm in length and possesses two very long (15 cm) tentacles. Tentacles are solid, highly extensile and retractile into sheaths or lateral pits, called **tentacular pouches**. There also occur eight longitudinal rows of comb plates for the locomotion. A long tubular pharynx from oral pore or **mouth** opens into stomach. The stomach extends into a system of gastrovascular canals. The sense organ, called **statocyst**, occurs at the aboral side. *Pleurobrachia* uses its colloblast cells to capture the prey. It is hermaphrodite, *i.e.*, both male and female sex organs occur in the same organism.

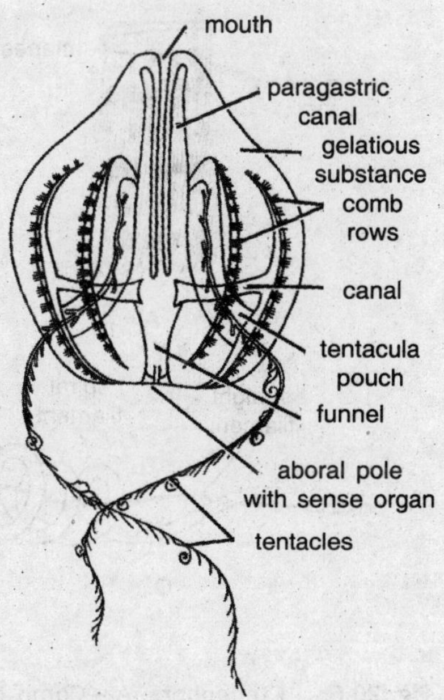

Fig. 20.5. *Pleurobrachia*.

III. General Morphology and Life History of Ctenophora

Ctenophores are commonly known as **comb jellies** or **sea walnuts**. The phylum takes its name from two Greek words: *ktenes* = combs and *ophora* = carrying.

Definition. Ctenophora is a small phylum of marine animals. These are biradially symmetrical radiates having transparent gelationous body without nematocysts but with special lasso cells (colloblasts) and with eight locomotory comb-like ciliary plates.

Structure. The generalized ctenophore is pear-shaped, roughly spherical in appearance (Fig. 20.5). Comb rows may be tinged with white orange or purple in colour. Ctenophores are noted for their **luminescence** (*i.e.*, light producing **photocytes** occur in walls of meridional digestive canal). Ctenophores range in size from few millimetres to 20 cm. Few species are as long as 1 metre or more. Their body can be divided into two hemispheres. Body can be further divided into equal sections by 8-ciliated bands. Their oral and aboral ends are distinct. The oral end lies on the lower side and bears minute oral aperture or **mouth**. The body of ctenophores contains the following distinct parts:

1. Comb plates. They are eight ciliated bands, arranged on the sides of body like meridians from aboral pole to oral pole, Each row is formed of cilia called **ctenes** or **comb plates** (Fig. 20.6A). The combs in a comb-plate are arranged in succession one behind the other. The row of plates forms roughed ridge, similar to those of walnuts and helps in locomotion.

2. Tentacles. Near the aboral end are present two long, solid and branched tentacles. These are enclosed in a pair of deep blind pouches or the **tentacular sheaths** into which tentacles can be withdrawn. Each tentacle bears a row of short branches or pinnae. It is formed of mesenchymal core covered with epidermis. Tentacles lack nematocysts but bear peculiar adhesive cells, called **colloblasts.**

3. Colloblast cell. It is modified epidermal cell (Fig. 20.6C) and contains the following parts: (*i*) **Head**. It is hemispherical and lies on the surface of epidermis. Its surface is covered with secretory granules that produce an adhesive secretion to be utilised for entangling and killing the prey. (*ii*) **Straight filament**. It connects the head to the mesenchyme core of tentacles. (*iii*) **Spiral filament**. It is spring-like and coiled around the straight filament. The colloblast lacks the mechanism of penetration.

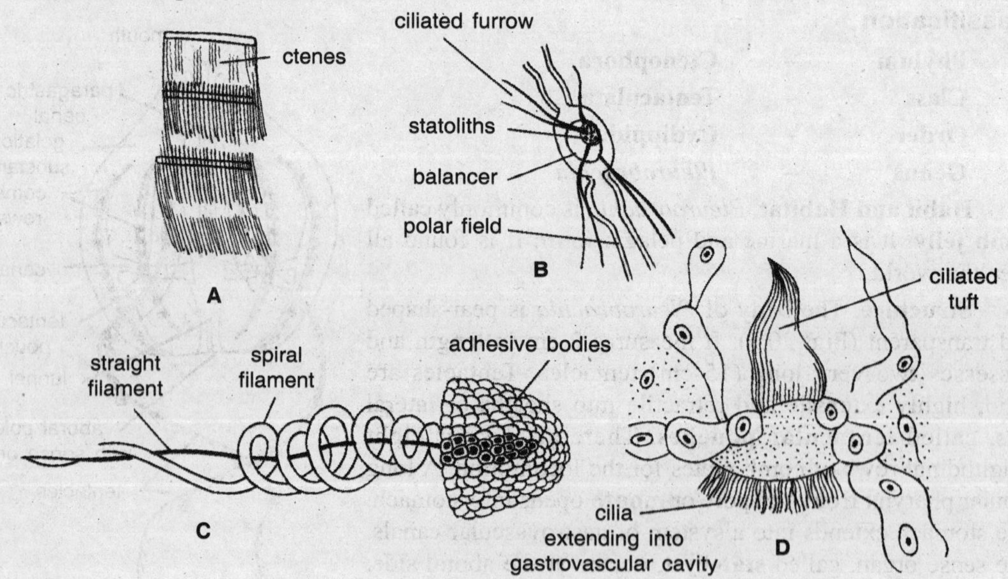

Fig. 20.6. Ctenophora. A—Comb plates or ctenes, highly magnified; B—Aboral apical sense organ; C—Colloblast; D—Rosette organ.

Body wall. The body wall is **triploblastic** and is formed of (*i*) **Epidermis**. It is outer layer, formed of syncytial cuboidal epithelium. Many types of specialized cells such as sensory, glandular, mucous and pigment cells are present in between these cells. (*ii*) **Mesogloea** or **mesoderm**. It is jelly-like layer found between epidermis and gastrodermis. It is very thick and contains amoebocytes and muscle fibres. Muscles are smooth, branched and anastomosing forming network. (*iii*) **Gastrodermis**. It is inner layer which lines the stomach and gastrovascular canals. It is formed of tall vacuolated and low ciliated cells.

Locomotion. The row of ciliated bands forms roughened ridges and helps in locomotion, *i.e.*, swimming or creeping.

Digestive system. Digestive system of ctenophores includes

(*i*) **Mouth.** It is a slit-like aperture, situated at the lower end of body.

(*ii*) **Pharynx or stomodaeum.** Mouth leads into pharynx which is a flattened chamber. Wall of pharynx is folded and is formed of long ciliated epidermal and glandular cells.

(*iii*) **Stomach.** Pharynx leads into a wide stomach or infundibulum which is connected with a system of five **gastrovascular canals**. These extend through the mesogloea in a definite arrangement up to the aboral pole.

Food, feeding and digestion. Food of ctenophores includes copepods and phytoplanktons. Food is usually caught by surface mucus and carried to mouth by cilia or entangled by colloblast on the tentacles. The tentacles are effective food traps and when they contract food is wiped off on the "lips" around mouth. In *Beroe*, which is cannibalistic, the mouth is apressed against a prey ctenophore. Then with a single gulp, the victim is swallowed.

Food is initially digested rapidly in pharynx and partly digested food enters the stomach. From there it is circulated through the gastrovascular canals where it is digested **intracellularly**.

Respiration and excretion. Ctenophores lack special respiratory organs; they exchange gases through their general body surface.

Their nitrogen is excreted in the form of **ammonia** which diffuses to the medium through body and gut surfaces. Most probably **cell rosettes** (Fig. 20.6D) are involved in excretion, osmoregulation and food distribution in mesogloea.

Nervous system. The nervous system is diffuse type. It is formed of a general sub-epithelial nerve plexus with multipolar ganglionic cells and neurites throughout the surface.

Reproductive system. Ctenophores are **hermaphrodites**. Gonads are situated in the wall of meridional canals in the form of bands. The gonads are derived from the endoderm. The mature gametes are discharged into the cavity of meridional canals. These are ejected out through mouth.

Development. In ctenophores fertilization is external. The zygote undergoes two meridional cleavages forming four blastomeres. It undergoes third vertical cleavage and results in the formation of a curved plate of eight cells which are arranged in two rows. These cells divide to produce eight **micromeres** and eight **macromeres**. Micromeres give rise to **ectoderm** and macromeres form **endoderm**. The mesogloea and muscles are derived from certain micromeres given off at the oral side of the embryo during gastrulation. Some classical authors have held that these are ectodermal cells, but many modern authors consider these cells to represent the **mesoderm**.

By rapid divisions, micromeres form a sheath of small cells over macromeres which also form a sheath of cells. As a result of invagination **gastrula** is formed. Four inter-radial bands of micromeres appear by the rapidly dividing cells that give rise to comb rows. Ectoderm at the oral end invaginates to form stomodaeum from which gastrovascular cavities arise by the active growth of endodermal cells. Two tentacles and their sheaths are formed by the ectoderm. At this stage, the embryo escapes as **cydippid larva** which is characteristic larval type for all ctenophores except the beroids.

QUESTIONS

Long Answer Questions

1. Give an account of the general characteristics of phylum Ctenophora and classify if giving important examples.
2. Describe the general characters and affinities of phylum Ctenophora.
3. Give an account of morphological organization and life history of Ctenophora.

Short Answer Questions

1. How do ctenophores differ from other coelenterates?
2. Write a short note on Ctenophora.
3. Classify the following animals given their distinguishing characters.
   (*i*) *Cestum* (*ii*) *Ctenoplana*
   (*iii*) *Coeloplana* (*iv*) *Pleurobrachia* (or *Hormiphore*)

Very Short Answer Questions

1. Name the larva of *Pleurobrachia*.
2. Combs of *Hormiphora* are composed of free cilia. Is it true?
3. The canals of the digestive tract are bilaterally arranged in *Pleurobrachia*. So you agree?
4. Where is the statocyst located in comb jellies?
5. What is the function of colloblasts?
6. Name the two classes of phylum *Ctenophora*.
7. Give the popular name of *Cestum*.
8. To which class does *Beroe* belong?
9. Give the classification of *Cestum*.

Fill in the Blanks

1. *Pleurobrachia* is popularly called ....
2. The larva of comb jelly is named ....
3. Statocyst lies at the .... pole in the sea gooseberry.
4. *Pleurobrachia* has .... meridional swimming plates.
5. Tentacles of comb jelly bear special cells, the ....
6. The ctenophores have .... symmetry.
7. Special adhesive cells, called .... , are the characteristics of ctenophores.
8. The ctenophores have .... level of organisation.
9. The ctenophores have .... egg, ... cleavage and .... gastrulation.
10. The larva of ctenophora is called .... larva.

Multiple Choice Questions

1. Venus's girdle is
   (*a*) *Ctenoplana* (*b*) *Cestum*
   (*c*) *Coeloplana* (*d*) *Beroe*
2. Mitre jelly-fish is
   (*a*) *Metridium* (*b*) *Beroe*
   (*c*) *Pleurobrachia* (*d*) *Ctenoplana*

3. One of the following Ctenophora is without tentacles
(*a*) *Coeloplana* (*b*) *Ctenoplana*
(*c*) *Beroe* (*d*) *Pleurobrachia*

4. Eight comb-like plates for locomotion are found in
(*a*) Arthropoda (*b*) Platyhelminthes
(*c*) Mollusca (*d*) Ctenophora

5. Development is indirect in Ctenophora and larva is know as
(*a*) planula (*b*) miracidium
(*c*) cydippid (*d*) ephyra

6. Select the creeping Ctenophora
(*a*) *Ctenoplana* (*b*) *Coeloplana*
(*c*) *Deiopea* (*d*) *Hormiphora*

7. Mnemiopsis belongs to
(*a*) Cydipidda (*b*) Platyctenea
(*c*) Lobata (*d*) Beroidea

8. Colloblasts are found in
(*a*) *Obelia* (*b*) *Aurelia*
(*c*) *Pleurobrachia* (*d*) *Metridium*

9. Digestive tract is complete in
(*a*) sea anemone (*b*) comb jelly
(*c*) *Obelia* (*d*) tapeworm

10. Symmetry in Ctenophora is
(*a*) spherical (*b*) radial
(*c*) biradial (*d*) bilateral

11. Ctenophores are
(*a*) monoecious (hermaphrodite) (*b*) dioecious
(*c*) viviparous (*d*) devoid of sex organs

ANSWERS

Very Short Answer Questions

1. Cydippid.
2. No.
3. Yes.
4. At aboral pole.
5. Capturing prey.
6. *Tentaculata and Nuda.*
7. Venus's girdle.
8. Nuda.
9. Order Cestida, Class Tentaculata, Phylum Ctenophora.

Fill in the Blanks

1. Sea gooseberry **2.** Cydippid **3.** Aboral **4.** Eight
5. Colloblasts **6.** Biradial **7.** Colloblasts **8.** Tissue
9. Lecithal, determinate, epibolic **10.** Cydippid.

Multiple Choice Questions

1. (*b*) **2.** (*b*) **3.** (*c*) **4.** (*d*)
5. (*c*) **6.** (*a*) **7.** (*c*) **8.** (*c*)
9. (*b*) **10.** (*c*) **11.** (*a*).

21

Phylum Platyhelminthes: Characters, Classification and Types

Platyhelminthes, (Gr., *platys* = flat; *helmins* = worms), *i.e.*, the flatworms are dorsoventrally flattened, triploblastic, bilaterally symmetrical acoelomates with blind sac body plan and organ grade of body organization. These primitive eumetazoans are without definite anus, circulatory, skeletal or respiratory systems but with protonephridial excretory system and mesenchyma (parenchyma) filling the space between various organs of body (see Box 21.1).

The term Platyhelminthes was first proposed by **Gaugenbaur** (1859). Flatworms help biologists to understand animal evolution.

Box 21.1 Primitive characters of flatworms

Phylum Platyhelminthes contains about 13,000 species and has following three primitive characters:

1. Absence of circulatory system.
2. The use of cilia rather than muscles for locomotion.
3. Presence of protenephridia.

These features are now believed to be the result of small body size.

21.1 GENERAL CHARACTERS

Phylum Platyhelminthes is characterized by the following characters:

1. Multicellular body shows **organ grade organization**.
2. Platyhelminthes are bilaterally symmetrical and dorso-ventrally flattened.
3. They are **triploblastic** worms, *i.e.*, their body is derived from three embryonic germ layers namely ectoderm, mesoderm and endoderm.
4. Body shape varies from moderately elongated flattened shape to long flat ribbons and leaf-like.
5. Most parasitic flatworms are white and colourless; some of them derive colour from the ingested food. The free-living forms are brown, grey, black or brilliantly coloured.
6. Animals have soft bodies and are unsegmented.
7. Anterior end of body is differentiated into the so called **head**.
8. Exoskeleton and endoskeleton are absent and hard parts of body such as spines, thorns, hooks, teeth, etc., are formed of scleroproteins.
9. Body is covered with syncytial or single layered, partly ciliated epidermis (*e.g.* Turbellaria). In parasitic forms (trematodes and cestodes) epidermis is absent and the body is covered with cuticle.

10. Great variety of adhesive secretions, organs of attachment and adhesion (*e.g.*, hooks and suckers) are present. Body surface of flatworms bears two kinds of glands: 1. **Cyanophilous glands** (*e.g.*, frontal glands) which secrete slime; 2. **Eosinophilous glands** (*e.g.*, glanduloepidermal adhesive organs) which secrete adhesive substances.
11. The **mouth** is typically the only opening to the digestive tract when the latter is present. The round or slit-like mouth appears at the site of blastopore (called **protostomia**). It may occur anywhere on the midventral surface.
12. A true body cavity or coelom is absent; so they are called **acoelomates**. Space between the organs is filled with a loose mesodermal tissue called **parenchyma** or **mesenchyma**.
13. Parenchyma helps in internal transport of materials from cell to cell distribution. Parenchymal tissue fluid is typically circulated as muscular movements squeeze and distort the parenchyma.
14. Digestive system is totally absent in Acoela and tapeworms but in other flat worms it consists of mouth, pharynx and blind and highly branched intestine. It is incomplete due to absence of **anus**.
15. No special organ for respiratory exchange or transport occurs. Respiration is **aerobic** in turbellarians but **anaerobic** in parasitic flatworms (see Box 21.2).

Box. 21.2

In turbellarians, the aerobic respiration releases energy mainly by a cyanide-sensitive system, *i.e.*, after dehydrogenation, hydrogen is transferred by a cytochrome-oxidase system to the final acceptor oxygen. A small cyanide-resistant fraction is probably maintained by iron-free oxidases.

Anaerobic respiration of parasitic flatworms involves glycogen degradation and formation of carbon dioxide and various organic acids. Different species release different combinations of acids, which indicate diversity among the array of respiratory enzymes.

16. Platyhelminthes are **ammonotelic**, releasing their nitrogenous waste in the form of ammonia.
17. Excretory system consists of single or paired **protonephridia** with **flame cells** or **solenocytes**. Protonephridia are absent in Acoela (see Box 21.3).

Box. 21.3

A **flame cell** is a hollow cell that has a tuft of vibratile cilia. The protonephridium has traditionally been thought of as a primitive precursor to more complex structures of excretion seen in higher invertebrates (**Wilson** and **Webster**, 1974). Recent work indicates, however, that rather than being primitive, protonephridia are a unique solution to the filtration of extracellular fluid in a body that has no coelom or blood-vascular system (**Ruppert** and **Smith**, 1985). In this case, the solenocytes must **actively transport** selective materials out of the body fluids into the protonephridial weir (=river dam).

18. Flatworms contain a primitive ladder-like nervous system. The main nervous system consists of a pair of **cerebral ganglia** or **brain** and one to three pairs of longitudinal **nerve cords** connected to each other by transverse commissures.
19. Turbellarians have a variety of sensory organs such as **photoreceptors** (for light reception), **chemoreceptors** (for food recognition), **tangoreceptors** (for touch reception), **rheoreceptors** (for orientation to water currents) and **statocysts** (for orientation to gravity). Sense organs are greatly reduced in parasitic forms.
20. Sexes are mostly united, *i.e.*, **hermaphrodite** or **monoecious**.
21. Asexual multiplication takes place by **fission** in many freshwater Turbellaria. Endogenous or exogenous budding is common in some tapeworms.
22. Flatworms have a remarkable power of **regeneration**.
23. Reproductive system is complex and highly evolved in most of the forms. Flatworms produce two types of eggs: 1. **Endolecithal** or **entolecithal ova** with yolk reserves stored in the egg cytoplasm; 2. **Ectolecithal ova** are free of yolk but encapsulated with special **yolk cells** containing the needed food reserve (Fig. 21.1).

24. Fertilization is **internal**. It is cross-fertilization in Turbellaria and Trematoda and self-fertilization in Cestoda.

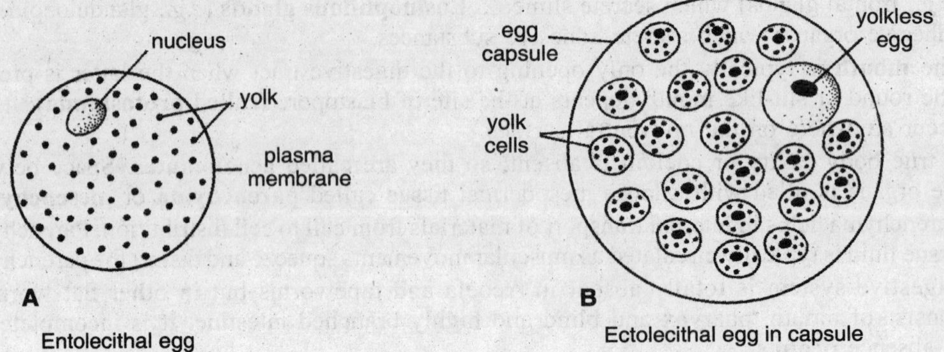

Fig. 21.1. A—Entolecithal (yolk-containing) egg; B—Ectolecithal egg.

25. Development may be direct (Turbellaria) or indirect. Life cycle is complicated and involves one or more hosts. **Digenetic life cycle** (*e.g.*, *Fasciola*) involves a **primary** or **definitive host** and an **intermediate host. Monogenetic life cycle** involves only one host.
26. Phenomena of **parthenogenesis** and **polyembryony** are common in trematodes and cestodes. Alternation of generation or **metagenesis** is also found in some cases (*e.g.*, *Fasciola*).
27. The flatworms are either **free-living** or **ecto-** or **endocommensals** or **endoparasites**.

21.2 OUTLINE CLASSIFICATION

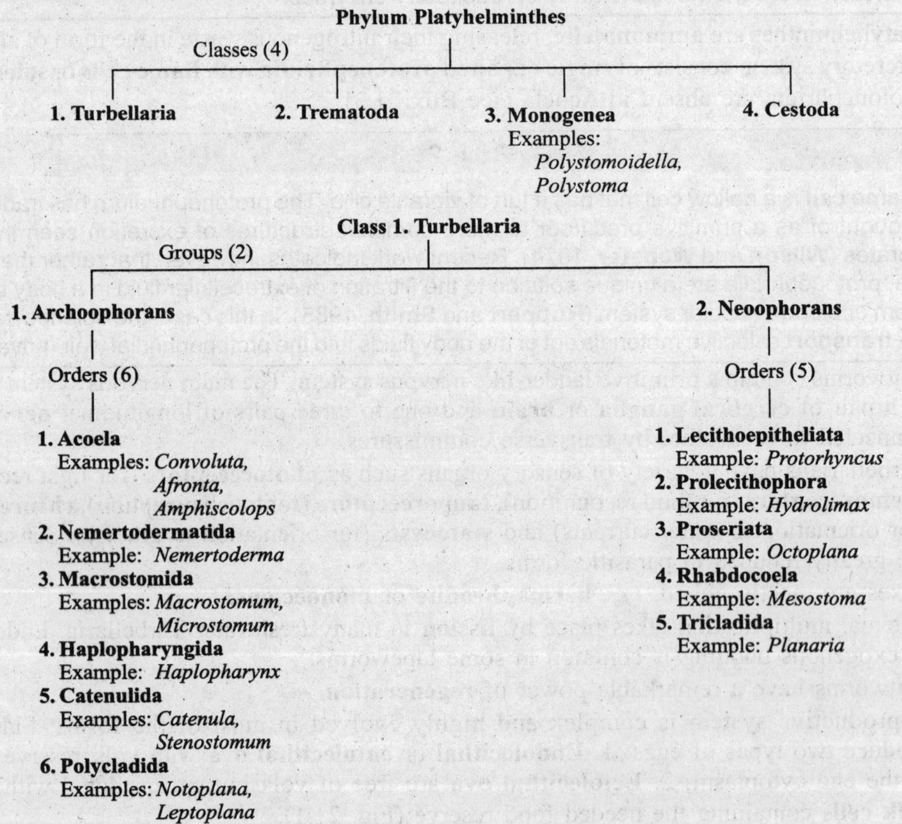

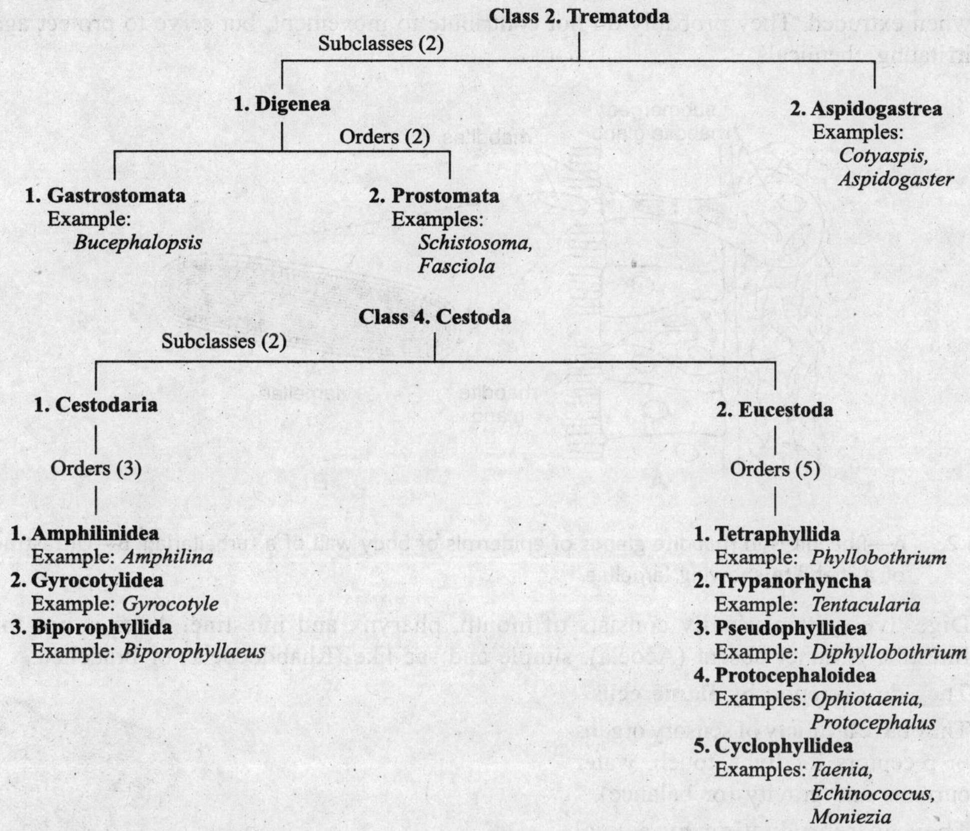

21.3 CLASSIFICATION

Phylum Platyhelminthes is divided into **four** classes of worms. Three classes are entirely parasitic and include the Trematoda, the Monogenea and the Eucestoda. The fourth class, the Turbellaria, is free-living and is certainly the group from which arose the ancestors of three parasitic classes.

Class 1. Turbellaria

(L., *turbella* = a little string)

1. Mostly free-living and carnivorous ciliated flatworms; a few are parasitic.
2. They are either aquatic, may be freshwater or marine, or terrestrial. Most aquatic forms are bottom dwellers or benthic, living in sand or mud.
3. They are mostly small but range in size from microscopic to more than 60 cm long (*e.g.*, the Lake Baikal triclad, *Rimacephalus arecepta*).
4. Body is unsegmented, dorsoventrally flattened and leaf-like, rounded or oval.
5. Body is covered with a cellular or syncytical **epidermis.**
6. Terrestrial species have strong epidermal ventral cilia and weak or no dorsal cilia. Epidermis of the ventral surface contains many unicellular **cyanophilous** or slime glands. Slime secretions tend to lubricate the surface; they increase water viscosity and provide resistance against which the cilia can thrust.
7. Suckers are absent, but adhesive organs are abundantly present.
8. Epidermal and subepidermal glands also form **rhabdites**. A rhabdite is a rod-shaped secretion, microscopically composed of successive lamellae (Fig. 21.2). Rhabdites dissolve rapidly

when extruded. They probably do not contribute to movement, but serve to protect against irritating chemicals.

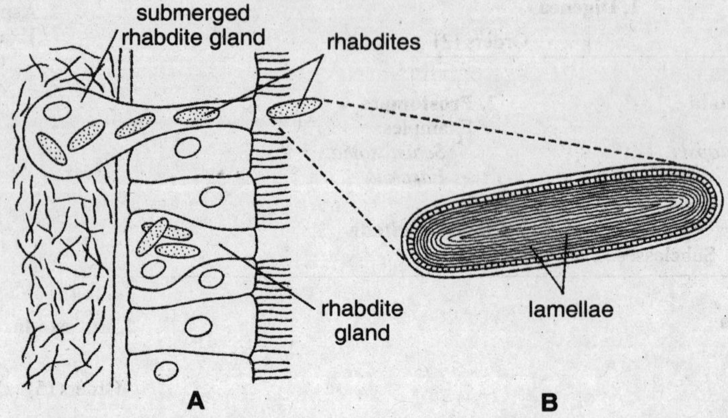

Fig. 21.2. A—Rhabdites in rhabdite glands of epidermis of body wall of a turbellarian; B—Ultrastructure of a rhabdite showing lamellae.

9. Digestive system usually consists of mouth, pharynx and intestine. Anus is not found. Intestine is either absent (Acoela), simple and sac-like (Rhabdocoela) or branched.
10. They do excretion by **flame cells**.
11. They have a variety of sensory organs or receptors for light, touch, water currents and gravity (or balance).
12. They do reproduction by sexual and asexual means. Asexual reproduction occurs by **paratomy** (or strobilation), **architomy** (= fission) and **parthenogenesis**. They have a remarkable power of regeneration.
13. Mostly hermaphrodite. Life cycle is simple. Development is **direct** in most turbellarians, but some polyclads produce free-swimming larvae that feed on plankton (**planktotrophic**). These larvae are of two types: **Goette's larvae** (having four arms or lobes) and **Mueller's larvae** (having eight arms) (Fig. 21.3). The larvae become metamorphosed into juvenile.

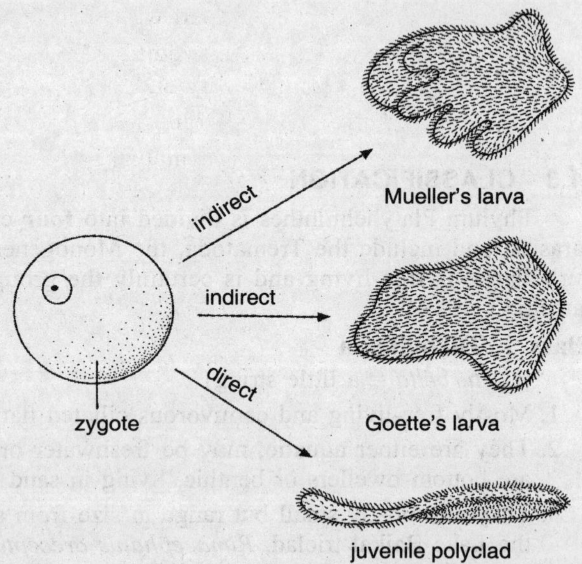

Fig. 21.3. Direct and indirect development of turbellarians.

14. About 3000 species of turbellarians have been described.

Class 2. Trematoda

(Gr., *trematodes* = having pores)

1. They are commonly called **flukes**.
2. They are **ectoparasites** or **endoparasites**.

3. Body unsegmented, dorsoventrally flattened and leaf-like.
4. Their body wall is without epidermis and cilia; instead it is covered with a thick cuticle.
5. They contain well developed suckers.
6. Their incomplete digestive system lacks anus but contains mouth, pharynx and branched intestine.
7. They are mostly **monoecious** or **hermaphrodite**.
8. They contain single ovary and two or many testes.
9. Their life history may be simple or complicated including many larval forms.

Class 3. Monogenea

1. This class includes flatworms which are mostly ecto- or endoparasites of aquatic vertebrates such as fishes, amphibians and reptiles (turtles).
2. Their body is dorsoventrally flattened and composed of a head, trunk and haptor.
3. Oral sucker is either absent or weak. Instead of it a pair of adhesive glands (head organs) are present.
4. Posterior end of body is provided with large posterior attachment organ called **opisthaptor** or **haptor**. Haptor bears **hooks** and **suckers**, allowing the parasite to cling tenaciously to the skin of fast-moving host.
5. Excretory pores are paired and situated anteriorly on the dorsolateral sides. In some species a urinary bladder is associated with each pore.
6. They have complex hermaphrodite reproductive system. Most contain a single circular or oval testis and a single ovary and extensive vitellaria.
7. Male and female gonopores are usually separate.
8. During copulation, there is a mutual exchange of sperm being passed into single or paired vagina. Males use a protrusible penis-like structure bearing hooks in copulation.
9. Their life cycle includes only **one host** (there is no intermediate host).
10. One egg (by way of a free-swimming ciliated larva called **oncomiracidium**) gives rise to only one adult worm. Hence the name **monogenea**, meaning "one generation."

    Examples. *Polystomoidella oblongum* (parasite in urinary bladder of turtles); *Benedenia melleni* (ectoparasite of fish); *Polystoma integerrinum* (endoparasite of urinary bladder of frogs and toads); *Dactylogyrus vastator* (ectoparasite on the gills of freshwater fishes).

Class 4. Cestoidea (Cestoda)

(Gr., *kestos* = girdle + *eidos* = form)

1. All are highly specialized endoparasites of vertebrates, commonly called **tapeworms**.
2. Body is flattened and ribbon-like.
3. Body is without epidermis and cilia but covered by a syncytial tegument (cuticle) having microvilli-like projections called **microtrichia**. Microtrichia plays a vital role in active transport of food molecules of various carbohydrates and amino acids.
4. Body of tapeworms is differentiated into **scolex, neck** and **body** or **strobila**.
5. Scolex is provided with adhesive structures (hooks, suckers) which are absent in subclass Cestodaria.
6. Neck is very short and narrow. It is **proliferative** giving rise to the body or strobila. Neck keeps on adding new proglottids to body by mitotic growth followed by transverse constriction.

7. Strobila consist of linearly arranged segment-like sections called **proglottids**. The youngest proglottids are at the anterior end and they increase in size and maturity as they are displaced toward the posterior end of strobila.
8. Mouth, digestive tract and sense organs are absent.
9. Excretory system consists of **protonephridia** with typical terminal **flame cells**.
10. Each mature proglottid contains one or two sets of complete hermaphroditic reproductive system.
11. Life cycle is complicated usually involving two or more hosts.
12. Embryos and larvae possess hooks.

CLASS 1. TURBELLARIA

Class Turbellaria is divided into following **two** major groups:

Group I. Archoophoran turbellarians

(*arch* = primitive + *oophoro* = ovary)

1. They have most primitive organization.
2. Vitellaria (yolk glands) are absent.
3. Eggs are entolecithal.
4. Cleavage is spiral.

Group II. Neoophoran turbellarians

(*ne* = new + *oophoro* = ovary)

1. They have an advanced level of organization.
2. Vitellaria (yolk glands) are present.
3. Eggs are ectolecithal.
4. Spiral cleavage is greatly modified.

Group I. Archoophoran turbellarians

This group includes the following **six** orders:

Order 1. Acoela

1. Minute marine turbellarians measuring less than 2 mm in length.
2. Mouth and sometimes a simple pharynx are present, but no intestine or body cavity.
3. Protonephridia are absent.
4. Gonads are often not bounded by a cellular wall.
5. Oviducts are absent.
6. Statocyst is with one statolith.
7. Usually free-living found under stones or on bottom of mud; some live in the intestine of various echinoderms.

Examples. *Convoluta, Afronta, Amphiscolops, Ectocotyla, Paratomella.*

Order 2. Nemertodermatida

1. Small marine flatworms similar to Acoela.
2. Have an epithelial digestive tract.
3. Statocyst with two statoliths.
4. They produce uniflagellate sperm.

Example. *Nemertoderma.*

Order 3. Macrostomida

1. Small (0.8 to 11 mm in length) marine and freshwater species.
2. Digestive system with a simple pharynx and a simple sac-like ciliated intestine.
3. Nervous system with one pair of ventrolateral nerve cords.
4. Testis is compact; penis armed with a stylet.
5. Yolk glands are absent.
6. Asexual reproduction with chains of zooids.

Examples. *Macrostomum, Microstomum.*

Order 4. Haplopharygida

1. Small marine species similar to macrostomids.
2. They have a proboscis.
3. A temporary anus is present.

Example. *Haplopharynx.*

Order 5. Catenulida

1. Mostly small freshwater species.
2. Digestive system contains a simple pharynx and ciliated sac-like intestine.
3. Excretory system consists of a single median protonephridia.
4. Statocyst sometimes present, usually with one statolith.
5. Gonads are unpaired. Single testis is in the form of compact mass. Penis unarmed. Male gonopore is dorsal above the pharynx.
6. Yolk glands absent. No female gonoduct.
7. Asexual fission occurs with the formation of chains of zooids.

Examples. *Catenula, Stenostomum.*

Order 6. Polycladida

1. Exclusively marine flatworms of moderate body size, averaging 3 to 20 mm in length.
2. Body usually greatly flattened and more or less oval in shape.
3. A pair of anterior marginal or dorsal tentacles may be present.
4. Many are brightly coloured.
5. Pharynx is tubular or plicate (= bell-shaped).
6. Intestine is elongated and centrally located, with many highly branched diverticula.
7. Nervous system consists of many radially arranged nerve cords.
8. Eyes are numerous.
9. Testes and ovaries are numerous and scattered.
10. They feed on coelenterates and store their undischarged nematocysts in their skin as means of self-defence.

Examples. *Notoplana, Leptoplana, Gnesioceros, Stylochus, Prostheceraeus, Pseudoceros.*

(**Note.** Order Polycladida is often divided into two suborders: 1. **Acotylea** (*e.g.*, *Notoplana*) and 2. **Cotylea** (*e.g.*, *Thysanozoon, Yungia*).

Group II. Neoophoran turbellarians

This group includes the following **five** orders:

Order 1. Lecithoepitheliata

1. Marine or freshwater species.
2. Mouth and pharynx are at the anterior end of body, intestine is simple.
3. Ovary produces eggs which are surrounded by follicle-like yolk cells.
4. Female ducts are simple or none.
5. Penis is with cuticular stylet.

Example. *Prorhynchus*.

Order 2. Prolecithophora

1. Usually small sized.
2. Marine and freshwater species.
3. They contain a plicate or bulbous pharynx and a simple intestine.
4. Ovary produces eggs and follicle-like yolk cells.

Examples. *Plagiostomum, Hydrolimax*.

Order 3. Proseriata

1. Small-sized marine turbellarians.
2. Statocyst with one statolith.
3. Pharynx is plicate and tubular but gut is not branched.
4. Female reproductive system consists of separate ovaries and yolk glands.

Examples. *Octoplana, Monocelis, Nemertoplana*.

Order 4. Rhabdocoela

1. It is a large group of small (less than 3 mm) marine, freshwater and terrestrial forms.
2. Digestive tract is complete having simple bulbous pharynx. Intestine unbranched sac-like without diverticula.
3. Excretory system with protonephridia.
4. One pair of nerve cords.
5. Eyes are usually present.
6. They contain one or two gonads, yolk glands present or absent.
7. Free-living, commensal or parasitic forms.

This order is divided into the following three suborders:

Suborder 1. Typhloplanoida, Example. *Mesostoma*.

Suborder 2. Dalyellioida, Examples. *Anoplodiera, Syndesmis, Kronborgia*.

Suborder 3. Kalyptorhynchia, Examples. *Gyratrix, Gnathorhynchus*.

Order 5. Tricladida

1. Relatively large (measuring from 2 to 60 cm in lenght) marine, freshwater and terrestrial turbellarians.
2. Pharynx is plicate, tubular and posteriorly directed.
3. Intestine (=gut) has three branches.
4. Eyes present in most species.
5. Protonephridia form paired lateral network.
6. Male reproductive organs consist of two to numerous testes; a penis papilla present.

7. Female reproductive organs consist of a pair of ovaries with yolk glands and a copulatory bursa.
8. Gonophore single.

This order is divided into following **three** suborders:

Suborder 1. Maricola. Examples. *Bdelloura* (It is commensal on the book gills of horseshoe crabs), *Ectoplana.*

Suborder 2. Paludicola. Examples. *Planaria, Dugesia, Dendrocoelum, Procerodes, Phagocata, Polycelis, Procotyla.*

Suborder 3. Terricola. Examples. *Bipalium, Geoplana, Orthodemus.*

CLASS 2. TREMATODA

The class Trematoda is divided into **two** subclasses: 1. Digenia and 2. Aspidogastrea.

Subclass I. Digenia

1. They include flukes which are endoparasites in vertebrates (such as fishes, amphibians, reptiles, birds and mammals) and invertebrates.
2. They contain two to four hosts in their life cycle. The **first intermediate host** is typically gastropod snail, **second intermediate host** is usually an arthropod and the **definitive host** is a vertebrate.
3. Adult digeneans range in size from approximately 0.2 mm to 6.0 cm in length.
4. They are typically dorsoventrally flattened, but some are thick and fleshy, whereas others are long and thread-like.
5. Most flukes contain two suckers which lack hooks. **Oral sucker** occurs around the mouth and aid in feeding. **Ventral sucker** or **acetabulum** is found on mid-ventral or posterior side of body and prevent dislodgement.
6. Their body is covered with a nonciliated cytoplasmic syncytium, the **tegument**. The tegument plays a vital role in the physiology of digeneans (Box 21.4).

Box 21.4

Tegument of Digenia provides protection, especially against host's enzymes in gut-inhabiting species. Nitrogenous wastes diffuse to the exterior through the tegument and it is the site of gas exchange. In endoparasites, the tegument as well as gut absorbs **glucose** and some **amino acids.** The **protein synthesis** involved in digenean egg production and in asexual reproduction in developmental stages (polyembryony) places especially heavy demands on the amino acid supply.

7. Excretory pore (nephridiopore) single at the posterior end of body. The protonephridia excrete water and waste metabolites such as unwanted iron from haemoglobin in blood-feeding flatworms.
8. Vagina is absent; uterus is usually long having many shelled eggs.
9. The eggs are **ectolecithal**, *i.e.*, the yolk is supplied by yolk cells, which also carry the material needed in egg capsule formation.
10. Life cycle of flukes is complicated involving many larval stages such as **miracidium, sporocyst, redia** and **cercaria**.
11. Life cycle includes one or more intermediate hosts.
12. Life cycle of flukes always includes at least two infective stages which account for the name Digenea, meaning "two generations". Two or more hosts are infected before the life cycle is completed.

Subclass 2. Aspidogastrea

1. They are endoparasites in the gut of fish and reptiles and in the pericardial and renal cavities of bivalve molluscs.
2. Oral sucker is absent.
3. Anterior end of body is without paired adhesive structures.
4. The entire ventral surface of body acts as an **adhesive organ** which is either a single septate sucker or a longitudinal row of suckers. This organ lacks hooks.
5. The digestive tract contains a single intestinal caecum.
6. They contain a single nephridiopore.
7. Male reproductive system has only one testis.
8. Their life cycle is simple involving no alternation of hosts.
   Examples. *Cotyaspis, Aspidogaster*.

Subclass 1. Digenia

The subclass Digenia is divided into following **two** orders:

Order 1. Gasterostomata

1. Adults are parasite of aquatic vertebrates, mainly fishes.
2. Mouth is ventral.

Examples. *Bucephalopsis, Bucephalus*.

Order 2. Prostomata

1. Adults are parasites on all vertebrates.
2. Mouth is subterminal or terminal in position.

Examples. *Schistosoma, Fasciola, Opisthorchis* (or *Clonorchis*) *sinensis* (Chinese liver fluke), *Echinostoma, Paragonimus* (lung fluke).

CLASS 3. CESTOIDEA (CESTODA)

Class Cestoda is divided into following **two** subclasses:

Subclass 1. Cestodria

1. Coelomic or intestinal parasites of vertebrates (mainly fish and rarely reptiles).
2. Body is undivided, *i.e.*, no strobila (**monozoic**, *i.e.*, body consists of a single differentiated unit).
3. Scolex is absent.
4. Alimentary canal is absent.
5. Only one set of monoecious reproductive organs is present.
6. Development is indirect including **lycophore larva** bearing ten hooks.

Subclass 2. Eucestoda

1. Endoparasitic forms in gut (intestine) of vertebrate hosts.
2. Body is usually very elongated ribbon-like.
3. Body is divided into scolex, neck and strobila (**polyzoic**). Strobila contains many proglottids.
4. Scolex with hooks and adhesive suckers.
5. Each proglottid contains one set of reproductive organs.
6. Embryo (larva) with six hooks.

Subclass 1. Cestodaria

This subclass is divided into following **three** orders:

Order 1. Amphilinidea

1. Coelomic endoparasites of fishes.
2. Body flattened, oval or elongated.
3. Scolex is absent.
4. Suckers are absent.
5. Anterior end of body bears protrusible proboscis and frontal glands.
6. Male and female (= vaginal) pores are located posteriorly.
7. Uterus is very much coiled opening near the anterior end.

Example. *Amphilina.*

Order 2. Gyrocotylidea

1. Intestinal endoparasites of fishes.
2. Body is elongated and flattened.
3. An adhesive sucker and a posterior rosette-shaped adhesive organ present.
4. Anterior end of body bears eversible proboscis.
5. Male, vaginal and uterine pores are together situated in the anterior half of the body.
6. Uterus is short and straight and runs directly to pore.

Example. *Gyrocotyle* (an intestinal parasite of *Chimaera* fish).

Order 3. Biporophyllida

1. Endoparasites in the intestine of sharks.
2. Anterior end of body contains an eversible proboscis.
3. Uterus is sac-like.
4. Male and female genital pores open in a common atrium.

Example. *Biporophyllaeus.*

Subclass 2. Eucestoda

This subclass is divided into more than 12 orders; of which main **five** orders are the following:

Order 1. Tetraphyllida

1. Intestinal endoparasites of elasmobranch fishes.
2. Four ear or leaf-like outgrowths or sessile suckers (called **bothridia**) are present on the scolex. They contain hooks.
3. Testes lie anterior to ovaries.
4. Vitelline glands are scattered in two lateral bands.
5. Cirrus (or penis) is armed with spines or hooks.
6. Common genital atrium is marginal.

Example. *Phyllobothrium.*

Order 2. Trypanorhyncha

1. Parasitic in the spiral valve of digestive tract of elasmobranch fishes.
2. Body is moderate-sized.
3. Scolex with four sucking grooves or bothria and four protrusible spiny proboscides.

4. Vitellaria (yolk glands) is present in a continuous layer in cortical parenchyma.
5. Testes extend behind the ovary posteriorly.
6. Gonophores lateral; uterus opens ventrally.

Example. *Tentacularia.*

Order 3. Pseudophyllidea

1. Intestinal parasites of teleost (or bony) fishes and terrestrial vertebrates.
2. Scolex with one terminal or two lateral groove-shaped bothria.
3. Most proglottids in a strobila are at the same stage of development.
4. Testes numerous, follicular and scattered in the mesenchyma of proglottids. Ovary bilobed.
5. Vitellaria (yolk glands) are follicular and numerous.
6. Genital aperture (gonopore) is midventral.

Example. *Dipyllobothrium.*

Order 4. Protocephaloidea

1. Intestinal parasite of freshwater fishes, amphibians and reptiles.
2. Scolex is mobile and contains four cup-shaped suckers.
3. Common genital atrium is marginal.
4. Ovary is bilobed and uterus is with many lateral diverticula.

Examples. *Ophiotaenia, Proteocephalus.*

Order 5. Cyclophyllidea

1. Parasites of intestine of reptiles, birds and mammals.
2. Large-sized tapeworms.
3. Scolex bears four large cup-shaped **suckers** often with an apical **rostellum** having two rows of hooks.
4. Protonephridial excretory system consists of four longitudinal vessels.
5. Ovary is two to many lobed; uterine opening absent.
6. Gonopores on one or both margins.
7. Vitellarium (yolk gland) is single and compact.

Examples. *Taenia, Echinococcus, Moniezia.*

Remaining orders of Subclass Eucestoda their examples are as follows:

Order 6. Diphyllida.

Example. *Echinobothrium.*

Order 7. Spathebothriidea.

Example. *Spathebothrium.*

Order 8. Caryophillidea.

Example. *Caryophyllaeus.*

Order 9. Aporidea.

Example. *Gastrotaenia.*

Order 10. Nippotaeniidea.

Example. *Nippotaenia.*

Order 11. Lecanicephalidea.
Example. *Lecanicephalum*.

Order 12. Disculiceptidea.
Example. *Disculiceps*.

Box 21.5 Recent Classification of Platyhelminthes (Brusca and Brusca, 2003)

I. **Class Turbellaria**
1. **Order Acoela**, *e.g.*, *Convoluta*.
2. **Order Catenulida**, *e.g.*, *Catenula*.
3. **Order Haplopharyngida**, *e.g.*, *Haplopharynx*.
4. **Order Lecithoepithelia**, *e.g.*, *Prorhynchus*.
5. **Order Macrostomida**, *e.g.*, *Macrostomum*, *Microstomum*.
6. **Order Nemertodermatida**, *e.g.*, *Nemertoderma*.
7. **Order Polycladida**, *e.g.*, *Notoplana, Planocera*.
8. **Order Prolecithophora**, *e.g.*, *Plagiostomum*.
9. **Order Proplicastomata**, *e.g.*, *Proplicastomata jenseni*.
10. **Order Proseriata**, *e.g.*, *Octoplana*.
11. **Order Rabdocoela**
   (*i*) **Suborder Dalyellioidea**, *e.g.*, *Callostoma*.
   (*ii*) **Suborder Typhloplanoida**, *e.g.*, *Typhlorhynchus*.
   (*iii*) **Suborder Kalyptorhynchia**, *e.g.*, *Cheliplana*.
   (*iv*) **Suborder Temnocephalida**, *e.g.*, *Temnocephala*.
12. **Order Tricladida**, *e.g.*, *Bipalium, Dugesia* (Formerly *Planaria*; see **Brusca** and **Brusca**, 2003), *Geoplana*.

II. **Class Monogenea**
1. **Subclass Monopistocotylea**, *e.g.*, *Polystoma, Silurodescoides vistulensis* (**Rastogi** *et al.*, 2008).
2. **Subclass Polyopistocotylea**, *e.g.*, *Diplozoon*.

III. **Class Trematoda**
1. **Subclass Digenea**, *e.g.*, *Fasciola*, *Schistosoma*, *Paragonimus westermani* (lung fluke), *Fasciolopsis buski* (intestinal fluke).
2. **Subclass Aspidogastrea**, *e.g.*, *Aspidogaster, Cotyaspis*.

IV. **Class Cestoda**
1. **Subclass Cestodaria**, *e.g.*, *Amphilina*, *Gyrocotyl*.
2. **Subclass Eucestoda**, *e.g.*, *Moniezia, Taenia, Echinococcus granulosus, Dipylidium caninum* (dog tapeworm), *Hymenolepis nana* (dwarf tapeworm of human beings).

21.4 TYPE STUDIES

1. *Echinococcus Granulosus*

Classification

Phylum	–	**Platyhelminthes**
Class	–	**Cestoda**
Subclass	–	**Eucestoda**
Order	–	**Cyclophyllidea**
Genus	–	***Echinococcus***
Species	–	***granulosus***

Habit and Habitat

Echinococcus granulosus is commonly called **dog tapeworm** or **hydatid worm**. It is pathogenic intestinal endoparasite of dogs, cats, foxes and wolves (**primary hosts**). It is common in Africa, Australia, New Zealand, Iceland, Holland and South America. Certain herbivores such as sheep, goats, cattle, camels, horses, kangaroos, deers, pigs and rabbits are **intermediate hosts** of *E. granulosus*. Humans are only **incidental intermediate hosts** of dog tapeworm and in them larva develops.

Size and Shape

The adult dog tapeworm measures only 3 to 6 mm having flat body.

Structure

The body of *E. granulosus* is differentiated into head or scolex, a narrow neck and three or four proglottids. **Scolex** is pear-shaped having four ovoidal **suckers** and a retractile **rostellum** which is armed with a double crown of 28 to 50 hooklets (Fig. 21.4). **Neck** is the narrow part of the scolex joining the body. **First proglottid** is small and immature containing rudiments of structures (sex organs). **Second proglottid** is mature and hermaphroditic containing a single set of male and female genital organs. Male reproductive system consists of spherical testes, vas deferens and cirrus. Female reproductive system comprises ovaries, oviduct, vitellaria, ootype, uterus and vagina. **Third** or **last segment** is gravid, elongated and biggest. It contains branched uterus filled with eggs under different stages of development (Fig. 21.4). Uterus bursts before or after the proglottid is separated from the worm.

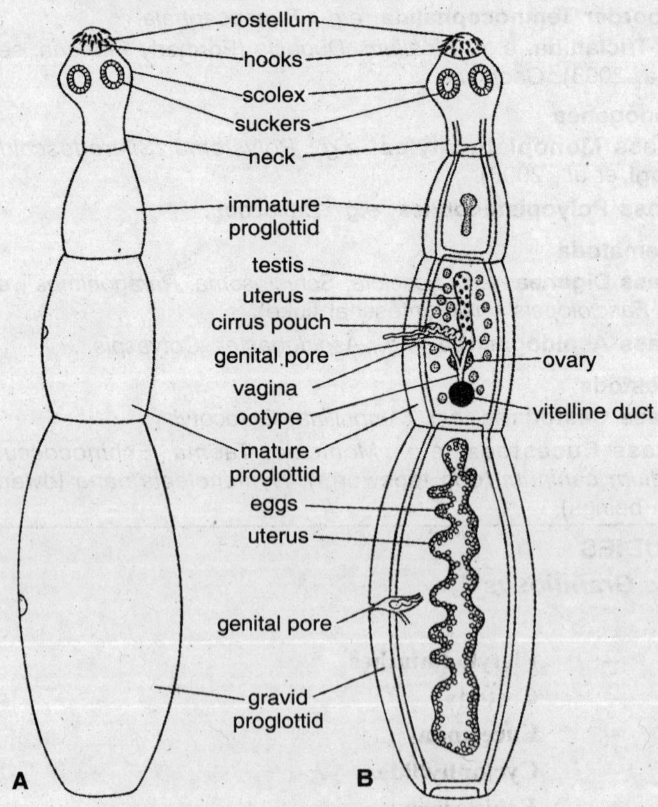

Fig. 21.4. *Echinococcus granulosus.*

Life Cycle

Eggs or gravid proglottids escape with the faeces of the primary host (dog). Inside uterus or on the ground eggs develop into **onchospheres** containing six-hooked embryos or **hexacanths**. They may be ingested by the intermediate hosts; in the intestine of which shell of onchospheres dissolves and releases hexacanths which enter mesenteric veins and make their way into various organs, specially lungs and liver. The young larva changes into a hollow **bladder** containing two layers—outer cuticular **ectocyst**, inner germinal layer or **endocyst**. The host forms an enveloping fibrous **cyst wall**, called **pericyst**. It is now called a **hydatid cyst** (Fig. 21.5).

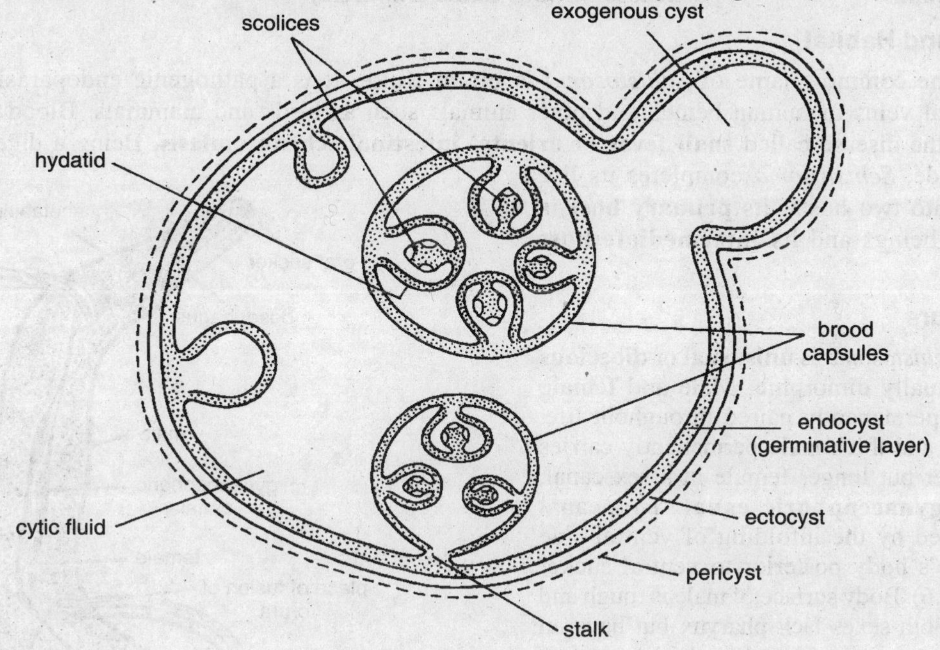

Fig. 21.5. *Echinococcus granulosus*. Hydatid cyst in cross section.

Development of the cyst is slow and after months or even years the bladder wall produces hollow **brood capsules** which remain attached by slender stalks. A cyst often throws out blind outgrowths forming **exogenous buds**. Budding is repeated within the brood capsules giving rise to the daughter buds or **hydatids** (**endogenous budding**). Invagination in hydatids gives rise to small scoleces with suckers and hooks. There may be as many as two million scoleces in a single large hydatid. Brood capsules under pressure burst to liberate tiny hydatids in the cystic fluid. Hydatid upon reaching the final host is capable of giving rise to an adult worm. This happens when dogs feed raw flesh of infected sheep and goats around mutton shops.

Human beings receive infection of dog tapeworm from the pet dogs through the tongue of dogs while they are licking the hands of children or the eggs are caught under the fingernails while caressing the pets. Infected meat of sheep and goat is also the source of human infection.

Hydatid cysts (bladder worms) have a colourless toxic fluid which may be up to many litres (one litre in a cyst of 15 cm diameter). Older cysts also have a granular deposit (called **hydatid sand**) consisting of brood capsules and free scolices. They become life-threatening when growing in such places as brain, liver or eye and release their toxic fluid into the host causing severe reaction (e.g., eosilophilia). Bladder worm cysts can be removed only by surgery.

2. *SCHISTOSOMA*

Classification

Phylum	–	**Platyhelminthes**
Class	–	**Trematoda**
Subclass	–	**Digenia**
Order	–	**Prosotoma**
Genus	–	***Schistosoma* (old name *Bilharzia*)**

Habit and Habitat

The common name of *Schistosoma* is **blood fluke**. It is a pathogenic endoparasite of intestinal veins of human beings and other animals such as birds and mammals. Blood fluke causes the disease called **snail fever** or **oriental intestinal schistosomiasis**. Being a digenetic trematode, *Schistosoma* completes its life cycle into two hosts: its **primary host** is human beings and its **intermediate host** is snail.

Structure

Schistosoma is **unisexual** or **dioecious** and sexually dimorphic. Male and female remain permanently paired throughout life. A short and thick male permanently carries a slender but longer female in a sex canal, called **gynaecophoric canal**. This canal is formed by the infolding of ventral side of male's body posterior to ventral sucker (Fig. 21.6). Body surface of male is rough and spiny. Both sexes lack pharynx but have an anterior and a ventral sucker; the acetabulum of male is better developed than female. **Male reproductive system** contains four to seven testes which open independently into vasa efferentia, all opening into a common vas deferens. The vas deferens opens to exterior by a gonopore located below the ventral sucker. **Female reproductive system** consists of an elongated ovary (or germarium), vitelline gland (vitellarium), ootype, Mehli's glands and short and coiled seminal receptacle. A tubular uterus arises from the ootype.

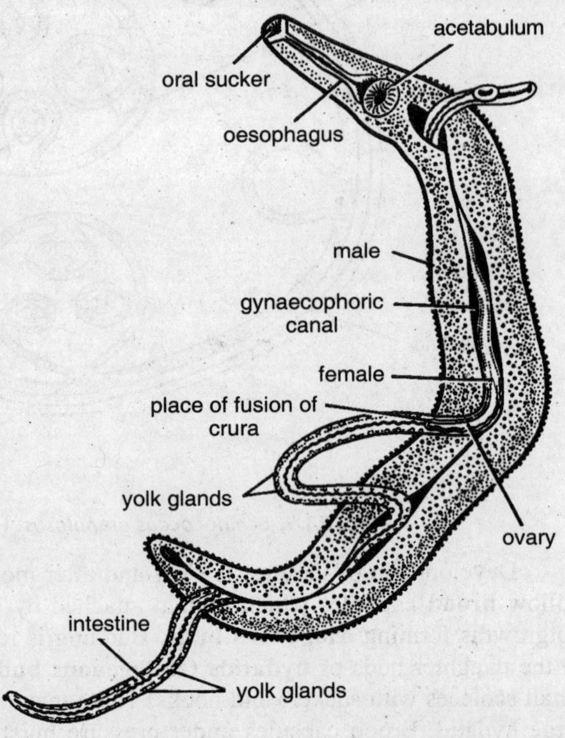

Fig. 21.6. *Schistosoma* (male and female).

Box 21.6

Following three species of *Schistosoma* are parasitic to human beings:

1. *S. haematobium* is found in Africa, Palestine, Iraq and Portugal in human pelvic veins; its egg capsule has a sharp terminal spine; its intermediate host is *Balinus* (snail).
2. *S. mansoni* is found in Africa and tropical America in the veins near the iliocaecal junction, its capsule has a lateral spine (Fig. 21.7); the intermediate host is *Planorbis*.
3. *S. japonicum* is found in Japan, China and Philippines in the hepatic portal and mesenteric veins; it is also parasitic in dogs, cats, cattle, horses and pigs; its egg capsule is small containing a rudimentary lateral spine; its intermediate host is *Oncomelania*.

Life Cycle

Life history of *Schistosoma* was worked out first by **Theodor Bilharz** in 1851 and later by **Wall** in 1941. When laying the eggs, the female protrudes from the groove of the male worm and deposits the eggs in small intestine venules of host. Eventually they break through by spines and enzymes into the lumen of the intestine and are released in the faeces of the host. If defecation occurs in the water, the thin walled egg capsules hatch, and the **miracidia** escape. The miracidia larva are ciliated and well supplied with sensory receptors; they seek out a particular species of freshwater snails (see Box 21.6). After penetration in the intermediate host, they transform into **sporocyst**, which eventually gives rise to a second generation of sporocysts. The sporocysts produce **cercariae** without an intermediate radia stage. Cercariae contain forked tails and are called **furocercariae** (Fig. 21.7). The cercariae leave the molluscan host, swim freely in water and on contact with human skin (when walking barefooted in water), penetrate, using enzymes of penetration glands and muscular boring movements. Cercariae develop into tailless young flukes, called **schistosomules** which are carried by the bloodstreams first to lungs, then to the liver, and finally to the veins of intestine or bladder. During this period the schistosomules gradually transform into adults. (**Note.** The eggs of *Schistosoma haematobium* occur in the vein of urinary bladder and they leave the primary host via the bladder and urine.)

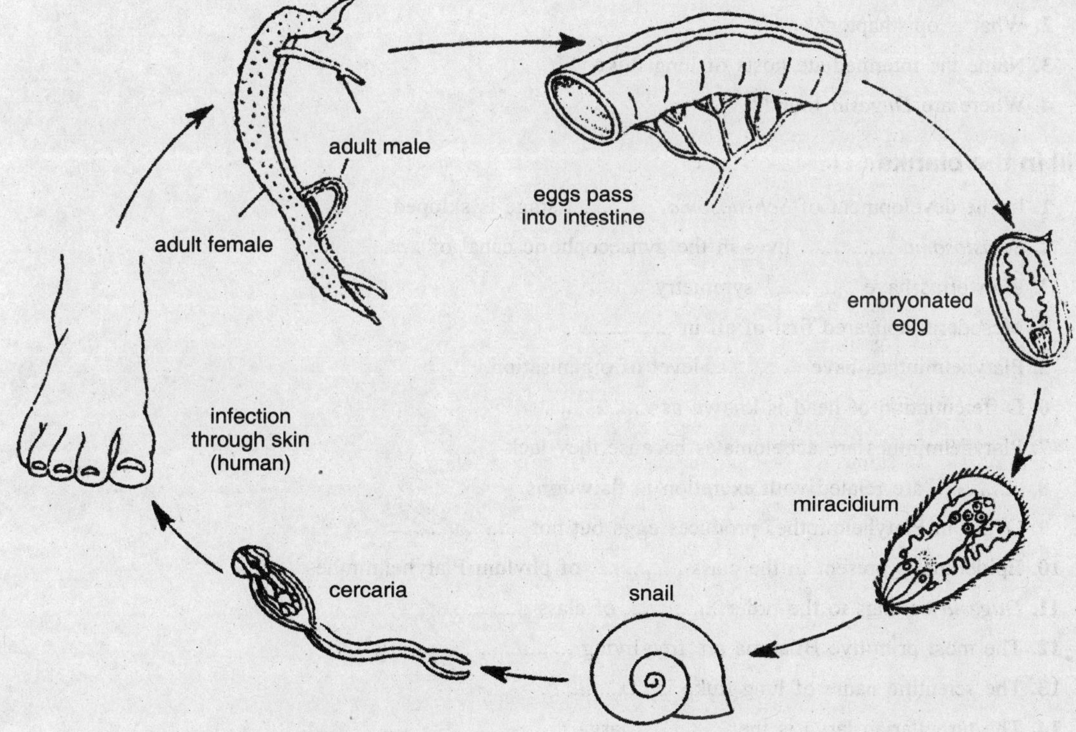

Fig. 21.7. Life cycle of *Schistosoma mansoni*.

Infection of *Schistosoma* causes headache, pain in back and legs, dysentery, daily fever, liver disorders, anaemia, eosinophilia, damage to urinogenital tract (**haematuria** which is a disorder of kidney or urinary track with discharge of blood). The disease can be prevented by sanitary control of water. It can be treated by tartar emetic or sodium antimony tartrate.

QUESTIONS

Long Answer Questions

1. Mention the distinguishing characters of phylum Platyhelminthes. Classify giving characters and examples of different classes.
2. Classify phylum Platyhelminthes giving diagnostic characters and examples of each class.

Short Answer Questions

1. Classify the following animals, giving their distinguishing characters:
   (*i*) *Planaria* (*Dugesia*)
   (*ii*) *Echinococcus granulosus*
   (*iii*) *Schistosoma* (Blood fluke)
   (*iv*) *Fasciola*
   (*v*) Tapeworm

Very Short Answer Questions

1. Name the primary host of blood fluke.
2. What is opisthaptor?
3. Name the intermediate hosts of lung fluke.
4. Where are *Dugesia* found?

Fill in the blanks

1. In the development of *Schistosoma* .... stage is skipped.
2. *Schistosoma* .... lives in the gynaecophoric canal of ....
3. Flatworms have .... symmetry.
4. Mesoderm appeared first of all in ....
5. Platyhelminthes have .... level of organisation.
6. Differentiation of head is known as ....
7. Platyhelminthes are acoelomates because they lack ....
8. .... are related with excretion in flatworms.
9. Ovary in Platyhelminthes produces eggs but not ....
10. Epidermis is present in the class .... of phylum Platyhelminthes.
11. *Dugesia* belongs to the order .... of class ....
12. The most primitive Bilateria are free-living ....
13. The scientific name of lung fluke is ....
14. The turbellarian larva is the .... larva.

Multiple Choice Questions

1. Flatworms are
   (*a*) diploblastic (*b*) triploblastic
   (*c*) monoblastic (*d*) none of these

2. The tissue in Platyhelminthes between viscera and body wall is called

(*a*) coelom (*b*) parenchyma

(c) mesoderm (*d*) choanoderm

3. Flame cells help in

(*a*) excretion (*b*) digestion

(*c*) respiration (*d*) reproduction

4. Coelom in flatworms is

(*a*) well developed (*b*) poorly developed

(*c*) schizocoel (*d*) absent

5. Sucker is absent in

(*a*) Turbellaria (*b*) Trematoda

(*c*) Cestoda (*d*) all of these

6. *Schistosoma* is a parasite found in

(*a*) lungs (*b*) intestine

(*c*) liver (*d*) blood

7. Alimentary canal is absent in the animals of class

(*a*) cestoda (*b*) apoda

(*c*) gastropoda (*d*) arachnida

8. *Echinococcus* is generally known as

(*a*) pig tapeworm (*b*) sheep tapeworm

(*c*) human tapeworm (*d*) dog tapeworm

9. Flatworms are characterised by presence of

(*a*) protonephridia (*b*) renette cells

(*c*) nephridia (*d*) all of these

10. *Polystomum* is endoparasite of

(*a*) frog (*b*) fish

(c) toad (*d*) all of these

11. Opisthaptor is

(*a*) anterior sucker (*b*) posterior sucker

(*c*) hook (*d*) spine

12. Rhabditis are present in the epidermal cells of

(*a*) *Dugesia* (*b*) *Taenia*

(*c*) *Fasciola* (*d*) *Schistosoma*

13. Parasitic animals have well developed

(*a*) respiratory system (*b*) reproductive system

(*c*) digestive system (*d*) nervous system

14. *Paragonimum* lives in

(*a*) intestine (*b*) blood

(*c*) lungs (*d*) brain

15. Cysticercus larva is formed in the development of

(*a*) *Dugesia* (*b*) *Fasciola*

(*c*) *Enterobius* (*d*) *Taenia*

ANSWERS

Very Short Answer Questions

1. Human being.
2. It is a posterior adhesive organ in monogenetic trematodes.
3. Freshwater snail and crab.
4. Freshwater.

Fill in the Blanks

1. Redia	**2.** Female, male
3. Bilateral	**4.** Platyhelminthes or flatworms
5. Organ-system	**6.** Cephalisation
7. Coelom	**8.** Flame cells
9. Yolk	**10.** *Turbellaria*
11. Tricladida, Turbellaria	**12.** Platyhelminthes or flatworms
13. *Paragonimus westermani*	**14.** Muller's

Multiple Choice Questions

1. (*b*)	**2.** (*b*)	**3.** (*a*)	**4.** (*d*)
5. (*a*)	**6.** (d)	**7.** (*a*)	**8.** (*d*)
9. (*a*)	**10.** (*d*)	**11.** (*b*)	**12.** (*a*)
13. (*b*)	**14.** (*c*)	**15.** (*d*).	

22

Fasciola Hepatica (Liver fluke) (Phylum Platyhelminthes)

22.1 CLASSIFICATION

Phylum	–	**Platyhelminthes**
Class	–	**Trematoda**
Order	–	**Digenea**
Genus	–	***Fasciola***
Species	–	***hepatica***

22.2 HABIT AND HABITAT

Fasciola hepatica lives as an endoparasite in the bile ducts of sheep. *Fasciola* is commonly called **liver fluke** on account of its leaf-like flattened shape and its habit to live in liver of the host. Besides sheep, it has been reported from the domestic animals such as buffaloes, goat, ass, horse, etc., and results in heavy mortality in them. *Fasciola* completes a part of its life cycle in snail. It has a worldwide distribution.

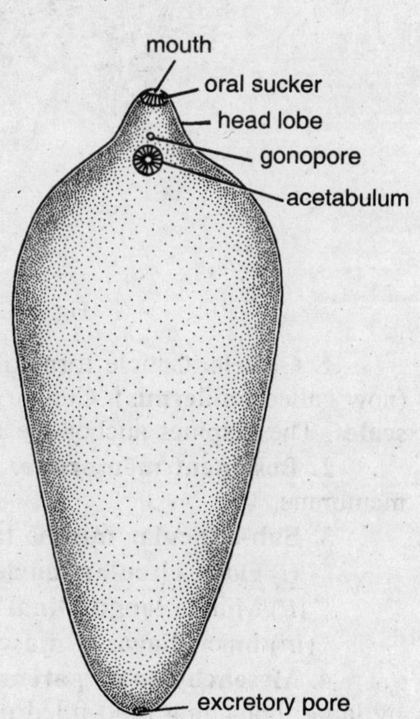

Fig. 22.1. *Fasciola hepatica*. Adult worm in ventral view.

22.3 MORPHOLOGY

Shape. Body of *Fasciola* is dorso-ventrally flattened, leaf-like and oval.

Size. The size of *Fasciola* varies from 15 to 30 mm in length and 10 to 15 mm in width. The broadest part of the body is about 1/3 distance from the anterior end.

Colouration. Generally the body of *Fasciola* is pinkish with brownish or blackish margins.

Structure

Anterior part of *Fasciola* is broad and rounded while posterior part is narrower and bluntly pointed or tapering. The anterior end of body is produced into a small conical projection, known as **oral cone, cephalic cone** or **head lobe.** The distal free end of the head lobe is perforated by a minute circular **mouth opening,** surrounded by a muscular cup-like depression, the **oral sucker** having radially arranged muscles. On the ventral surface, a little behind the head lobe, is found a large cup-shaped sucker, the **posterior sucker, ventral sucker** or **acetabulum.** It is a glandulo-muscular organ with which the animal is attached firmly to the walls of bile passages. A sucker works primarily on the pressure vacuum principle.

The entire body of *Fasciola* is covered with scales. There are following four pores on its external surface:

1. Mouth. It lies at the anterior end of body.

2. Excretory pore. It is situated at the posterior end of the body.

3. Gonopore. Closely in front of the acetabulum on the ventral surface there is a small opening, called **gonopore.** Through it, a curved muscular organ called **cirrus** or **penis** may be found protruding in live specimens.

4. Opening of Laurer's canal. During the breeding season a temporary opening is formed on the dorsal surface of body, it is known as opening of Laurer's canal.

22.4 BODY WALL

The body wall of *Fasciola hepatica* consists of following parts (Fig 22.2).

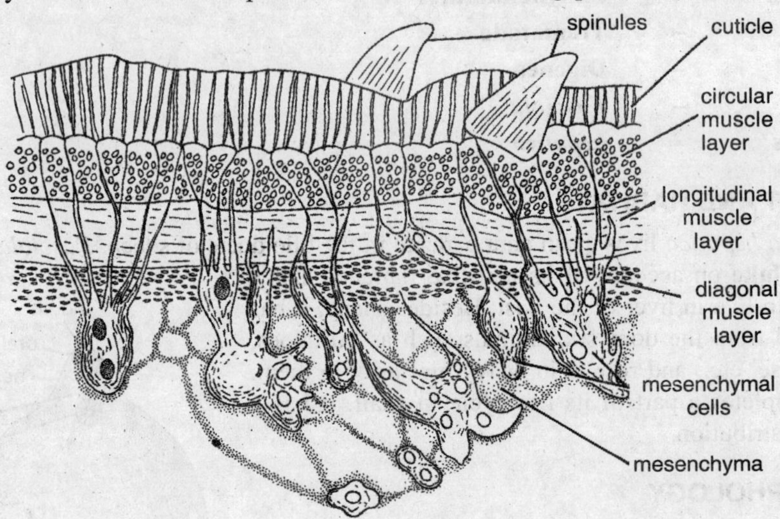

Fig. 22.2. *Fasciola hepatica*. T.S. through body wall.

1. Cuticle. Cuticle forms the outer most layer of body wall. It is of homogeneous nature (now called **epidermis**). Cuticle is rough and produced into minute spines or **spinules** or blunt **scales**. These spines anchor the fluke to the bile duct of the host.

2. Basement membrane. Just below the cuticle is present a thin and delicate basement membrane.

3. Sub-cuticular muscle layers. These layers are arranged in three groups:

(*i*) Outer **circular mucle** fibres.

(*ii*) Middle **longitudinal muscle** fibres.

(*iii*) Inner **diagonal muscle** fibres or **oblique** muscles. Muscles fibres are smooth type.

4. Mesenchyme or parenchyma. The interior of the body between the organs is occupied by large, branching fluid-filled parenchymatous cells.

5. Gland cells. Numerous unicellular gland cells are found embedded in the muscular layer. These are **ectodermal** cells which have sinked into parenchyma and are connected by protoplasmic projections.

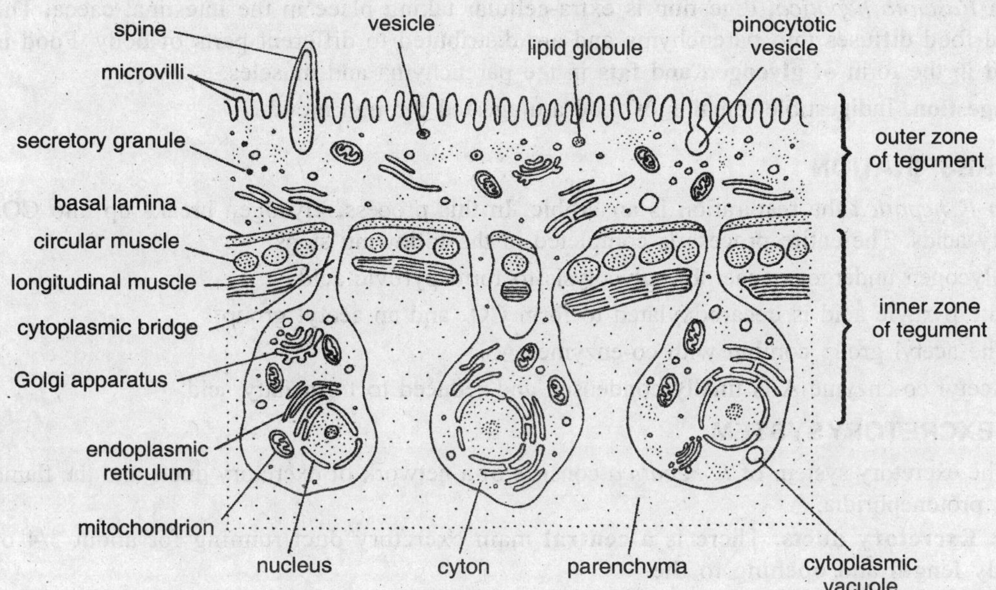

Fig. 22.3. Section through the tegument and body wall of a generalized digenean.

22.5 DIGESTIVE SYSTEM

The digestive system of *Fasciola* is complex, extensive and highly branched but without anus (*i.e.*, it is **incomplete**). It consists of following parts (Fig. 22.4):

1. Mouth. It is situated on the ventral side at the apex of head lobe and surrounded by oral sucker. Mouth leads into pharynx through the buccal cavity.

2. Oral chamber. Mouth opens into a short but funnel-shaped oral chamber.

3. Pharynx. Pharynx is short, rounded and highly muscular structure. Its walls are thick and provided with pharyngeal glands. The cavity of pharynx is small and narrow.

4. Oesophagus. The pharynx communicates with a short narrow oesophagus joining the intestine.

5. Intestine. Intestine divides into two branches or **intestinal caeca** or **crura** each running on one side to the posterior end and ending blindly. Intestinal caeca give out numerous branched **diverticula** in order to carry food to all parts of the body since there is no circulatory system. The median diverticula are short and lateral ones are long and branched.

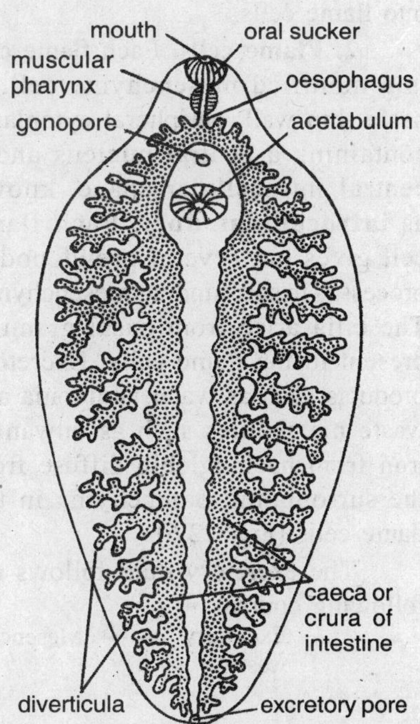

Fig. 22.4. *Fasciola hepatica.* Digestive system.

Digestion

Fasciola hepatica sucks bile and blood, lymph and cell debris from the wall of the host's bile duct by the oral sucker. The muscular pharynx acts as a suctorial apparatus. Amino acids, glucose and fructose directly diffuse into body of the fluke through general body surface from the surrounding fluid of the host.

In *Fasciola hepatica*, digestion is extra-cellular taking place in the intestinal caeca. The digested food diffuses into parenchyma and are distributed to different parts of body. Food is reserved in the form of **glycogen** and **fats** in the parenchyma and muscles.

Egestion. Indigestible remains of food are egested through mouth.

22.6 RESPIRATION

In *F. hepatica* the respiration is anaerobic. In this process, glycogen breaks up into CO_2 and fatty acids. The entire process is completed in the following steps:

1. Glycogen undergoes anaerobic glycolysis to form pyruvic acid.
2. The pyruvic acid is decarboxylated to form CO_2 and an acetyl group.
3. The acetyl group combin with co-enzymes A.
4. Acetyl co-enzyme A is finally condensed and reduced to form fatty acid.

22.7 EXCRETORY SYSTEM

The excretory system of *F. hepatica* consists of a network of excretory ducts and the flame cells or protonephridia.

1. Excretory ducts. There is a **central** main excretory duct running for about 3/4 of the body length and opening to the exterior through **excretory pore**, which is present at the posterior end of the body. The main canal through its course gives off side branches. All these branches divide and redivide into numerous minute capillaries which end into flame cells.

2. Flame cells. Each flame cell is a modified mesenchyme cell. It has a thin wall, peripheral cytoplasm containing a central nucleus and a central intracellular space, known as **intracellular duct**. Each flame cell gives out several pseudopodial processes in surrounding mesenchyme. The cilia arise from basal granules present in the flame cells. Excretory products such as water, ammonia and waste metabolites such as unwanted iron from haemoglobin diffuse from the surrounding parenchyma in the flame cells (Fig. 22.6).

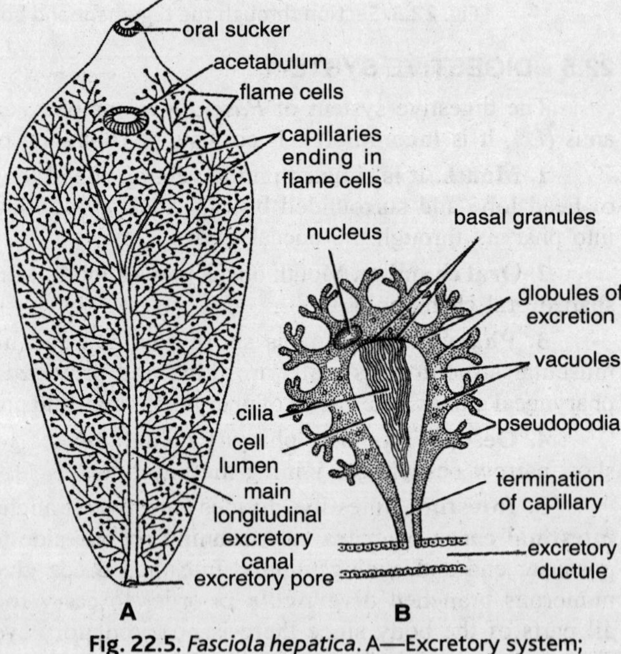

Fig. 22.5. *Fasciola hepatica*. A—Excretory system; B—A flame cell.

The excretory fluid follows the following course:

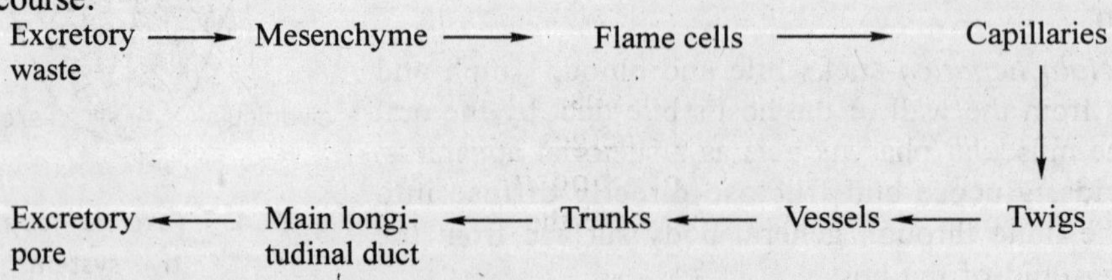

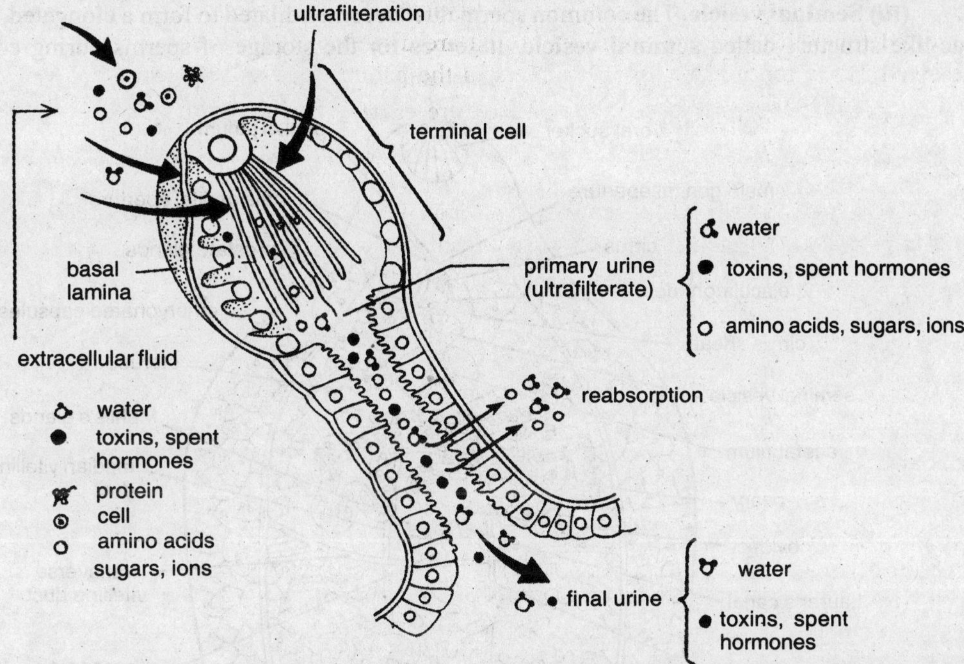

Fig. 22.6. Protonephridial excretion is a two-step process involving ultrafiltration and modification to form a final urine. Ultrafiltration occurs across the wall of a terminal cell, the flame cell. The force driving fluid across the terminal cell wall results from the action of flagella or cilia.

22.8 NERVOUS SYSTEM

The nervous system (Fig. 22.7) is well developed and consists of (*i*) A pair of cerebral ganglia. (*ii*) A pair of dorsal, a pair of ventral and a pair of lateral **nerve cords** are given off from the cerebral ganglia. (*iii*) A nerve ring exists around the oesophagus and connects the cerebral ganglia. (*iv*) Nerves are given out from the cerebral ganglia to the head lobe and to the hinder part of the body.

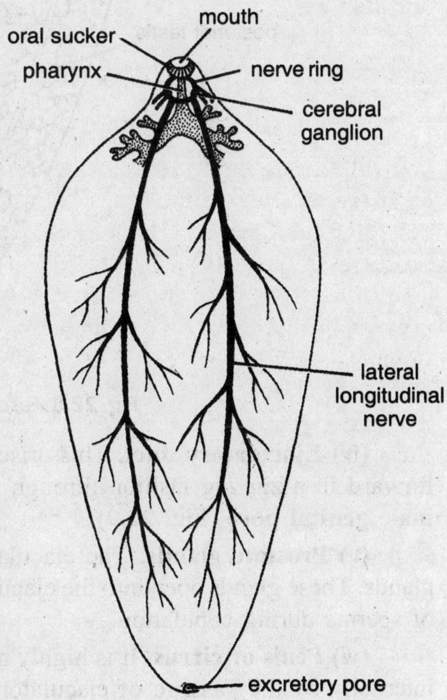

Fig. 22.7. *Fasciola hepatica*. Nervous system.

22.9 REPRODUCTIVE SYSTEM

Fasciola hepatica is **hermaphrodite** as both male and female sex organs occur in every individual. The system is studied under the following two headings:

1. Male Reproductive System

The male reproductive system consists of testes, vasa deferentia, seminal vesicle, ejaculatory duct, prostate glands, penis or cirrus and genital atrium (Fig. 22.8).

***(i)* Testes.** These are two in number, greatly ramified and tubular, situated in the middle part of body, one behind the other.

***(ii)* Vasa deferentia.** From each testis, a narrow delicate sperm duct or **vas deferens,** emerges out and runs forward to meet the fellow of opposite side forming the common sperm duct just below the ventral sucker.

(*iii*) Seminal vesicle. The common sperm duct becomes dilated to form a elongated, muscular, sac-like structure called **seminal vesicle**. It serves for the storage of sperms during copulation.

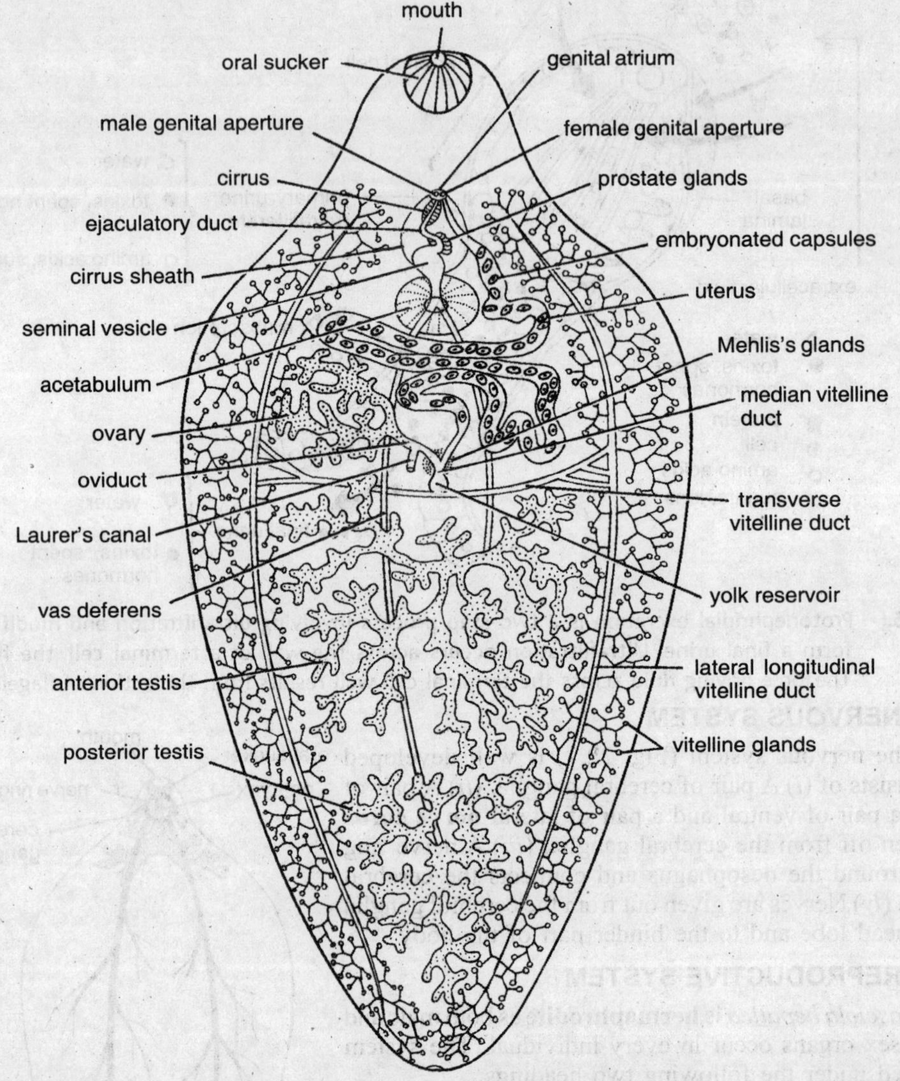

Fig. 22.8. *Fasciola hepatica*. Reproductive system.

(*iv*) Ejaculatory duct. This arises from seminal vesicle as a fine convoluted tube and runs forward in a zig-zag fashion through the cirrus to open into the **genital chamber** through the **male genital pore** (Fig. 22.9).

(*v*) Prostate glands. The ejaculatory duct is surrounded by numerous unicellular prostate glands. These glands open into the ejaculatory duct. Their alkaline secretion helps in free movement of sperms during copulation.

(*vi*) Penis or cirrus. It is highly muscular, elongated and cylindrical structure. It is traversed internally by the passage of ejaculatory duct. Cirrus opens into the **male genital aperture** and can be everted out and drawn in through the genital pore and thus helps in copulation.

The cirrus, seminal vesicle and prostate glands are enclosed in a bag-like cirrus-sheath.

2. Female Reproductive System

Female reproductive system consists of ovary, oviduct, uterus, vitelline glands, Mehlis's glands and Laurer's glands (Fig. 22.8).

***(i)* Ovary.** Ovary is single, tubular and highly branched, situated on the right side of body in front of testes. It occupies the anterior one-third of the body.

***(ii)* Oviduct.** All the branches of ovary open into short and narrow tube called **oviduct.** The oviduct travels down obliquely and opens into the median vitelline duct.

***(iii)* Uterus.** From the junction of oviduct and median vitelline duct arises a wide convoluted **uterus** having fertilized shelled **eggs** or **capsules**. Uterus runs forward to open into the genital atrium on the left side by **female genital pore**. The terminal part of uterus has muscular walls and is called **metraterm**; it helps in ejecting out the eggs. Metraterm also sometimes receives the cirrus during copulation.

***(iv)* Vitelline glands.** On each side of the body from one end to the other, there are the **yolk glands, vitelline glands** or **vitellaria** in the form of numerous minute vesicles or follicles. The yolk glands produce albuminous yolk and shell material for the eggs. A fine duct arises from each follicle or yolk gland and after uniting with similar ducts from the other follicles opens into the **lateral longitudinal vitelline ducts.** The two longitudinal ducts are connected together by a **transverse vitelline duct** placed above middle part of the body. The transverse vitelline duct is swollen in the centre to form the **yolk reservoir** or **vitelline reservoir.** From the yolk reservoir, a **median vitelline duct** starts and runs forward to join the oviduct (Fig. 22.9).

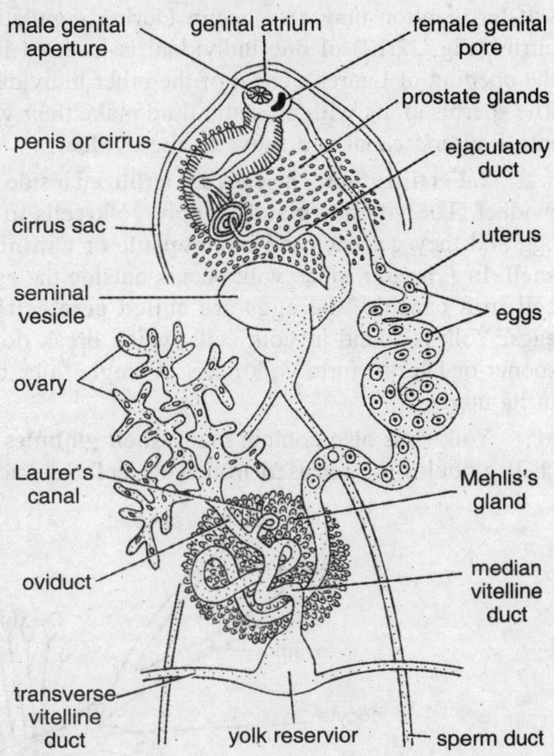

Fig. 22.9. *Fasciola hepatica*. Details of male and female reproductive organs in anterior region.

***(v)* Mehlis's glands or shell glands.** A mass of numerous unicellular shell or Mehlis's glands is found situated around the junction of median vitelline duct, oviduct and uterus (Fig. 22.10). The secretion of Mehlis's glands lubricates the passage of eggs in uterus; it probably also hardens the egg shells and activates spermatozoa (see Box 22.1).

***(vi)* Laurer's canal.** During the breeding season a temporary opening develops on the dorsal body surface. It arisers from the oviduct and acts as a vagina and is known as Laurer's canal.

Box 22.1

The junction of oviduct and median vitelline duct is swollen to form **ootype** in certain flukes such as *Fasciola indica*. In ootype, the parts of an egg are assembled and the eggs are shaped. An ootype is not present in *Fasciola hepatica*.

22.10 LIFE HISTORY

Life cycle of *Fasciola hepatica* is completed in two hosts, so it is called a **digenetic** trematode. The **primary host** of *F. hepatica* is sheep or cattle, while the **secondary host** is a snail of the genus *Limnaea*. The life cycle of sheep liver fluke is complicated by the introduction of a number of larval forms. This type of life-cycle is said to exhibit **polyembryony** and can be studied under the following headings:

1. Copulation. Copulation in *F. hepatica* occurs in the bile duct of sheep. Copulation is mutual and cross fertilization is the general rule. Sometimes self-fertilization may also occur. During copulation cirrus (Fig. 22.12) of one individual is inserted into the opening of Laurer's canal of the other individual. The sperms along with prostatic fluid make their way into Laurer's canal and move to the oviduct.

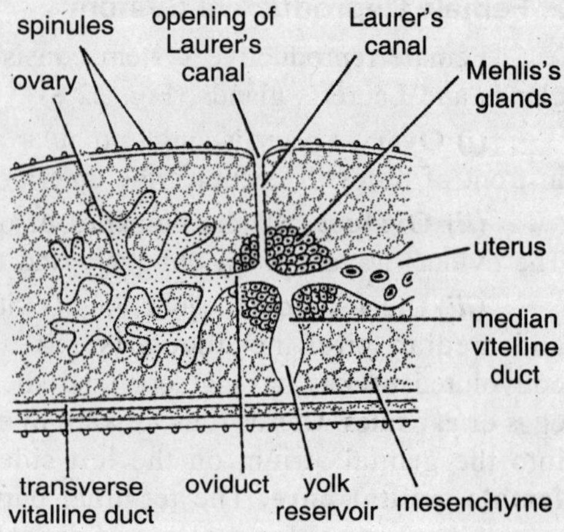

Fig. 22.10. *Fasciola hepatica*. T.S. body through Laurer's canal showing ovary and other female sex organs.

2. Fertilization. The eggs are fertilized inside the oviduct. The vitelline glands supply yolk cells to an egg and they get enclosed in a **capsule** or **chitinous shell.** In *Fasciola*, since yolk occurs outside the eggs cell in a capsule, the eggs are called **ectolecithal eggs.** Yolk is found in yolk cells which break down sooner or later to form yolk mass. A single fluke contains as many as 3,000—4,000 egg capsules in its uterus.

Yolk cells also contain many **shell globules** which are made of proteins and a phenol. The shell globules are released from yolk cells and form two-layered shell of egg capsule.

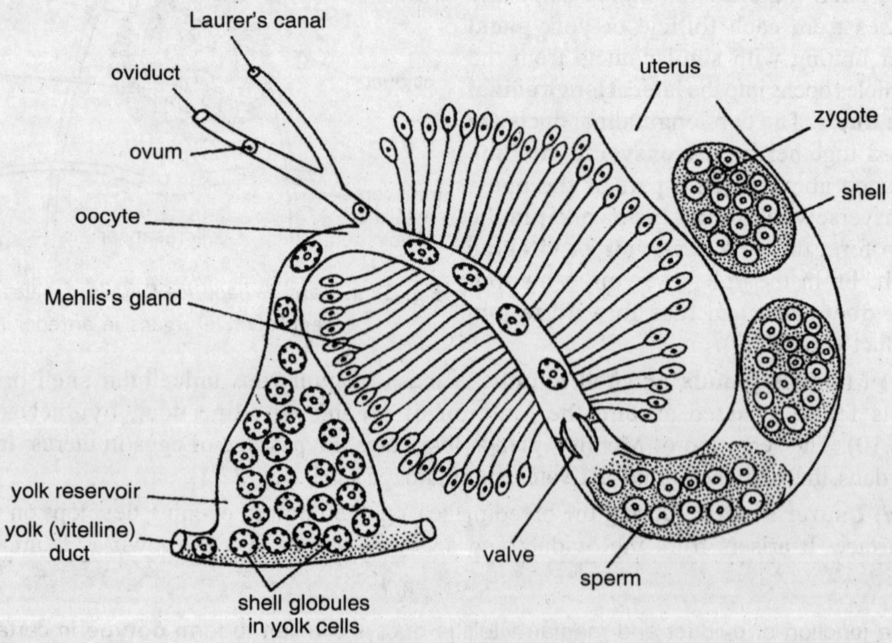

Fig. 22.11. *Fasciola hepatica*. The region of the ootype.

3. Eggs. The fertilized eggs are light brown in colour and oval in shape measuring about 130 to 150 μm in length and 63 to 90 μm in width. Each egg along with many yolk cells is surrounded by a shell or capsule which is made up of a double layer of lipoprotein and contains an **operculum** at one end. Such eggs are called **operculate.** Egg capsules come out of the gonopore into the bile duct of the sheep, they reach the intestine and are passed out with the faeces. The capsules which fall in water or damp places start development at about 75°F temperature.

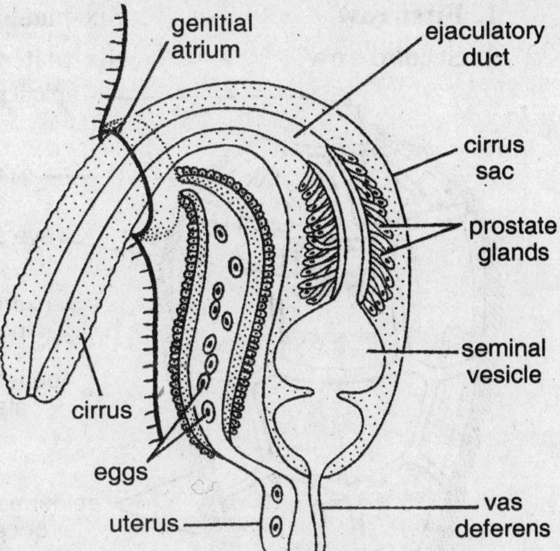

Fig. 22.12. *Fasciola hepatica*. Cirrus protrudes through gonopore.

4. Development. It includes following embryonic and larval stages (*i.e.*, it is of indirect type).

***(i)* Cleavage.** Cleavage or segmentation of the fertilized egg is started when the capsule is inside the uterus and continues on the damp ground. The fertilized egg is divided into a small **propagatory cell** and a larger **somatic cell** (Fig. 22.13). The somatic cell divides and forms the **ectoderm** of the larva. The propagatory cell divides and form two cells, one cell forms the **endoderm** and **mesoderm** of the larva. The other cell forms a mass of **germ cells** at the posterior end of the larva. This method of development is repeated during formation of different larval stages during the life history of fluke. Anyhow, within 9 to 19 days a fully formed ciliated larva, called **miracidium** is

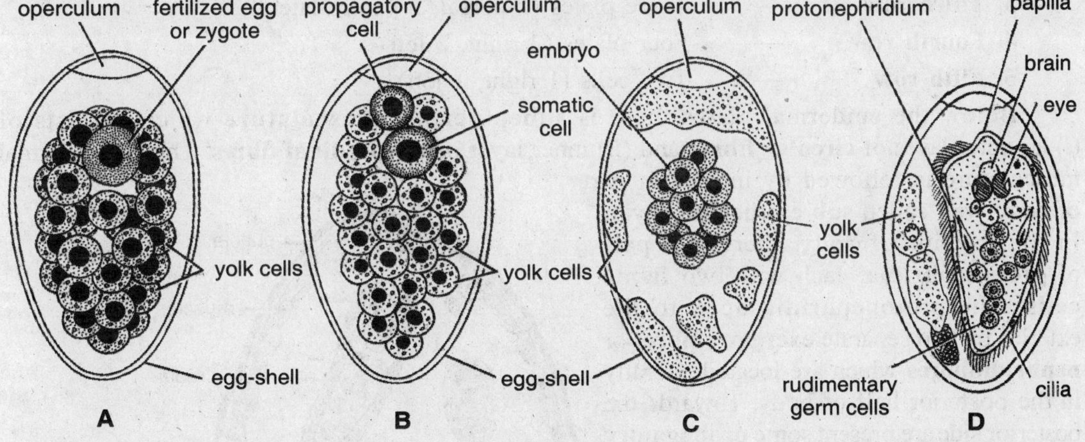

Fig. 22.13. *Fasciola hepatica*. Cleavage and larva formation. A—Fertilized egg; B—Two cell stage; C—Many cell stage (morula); D—Miracidium.

hatched out of the capsule by forcing off the operculum. The miracidium larva produces a proteolytic enzyme which erodes the lower surface of the operculum.

***(ii)* Miracidium.** Miracidium is a minute (0.13 mm long), oval, elongated and ciliated free-swimming larva. Its anterior end is broader than the posterior end and is produced into an **apical lobe** or **apical papilla.** It bears openings of (*i*) a pouch-like multinucleated **apical gland** and (*ii*) a pair of unicellular **penetration glands** or **cephalic glands** (Fig. 22.14).

Except the apical lobe, entire body of miracidium larva is **ciliated** and is covered with 2 closely fitted hexagonal **epidermal plates** which are arranged in five rows or tiers. Each tier has following arrangement of plates:

1. **First row** — Six plates (2 dorsal, 2 ventral and 2 lateral)
2. **Second row** — Six plates (3 dorsal, 3 ventral)

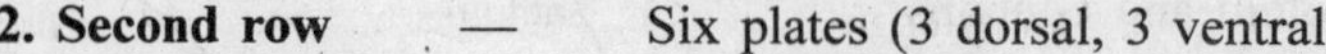

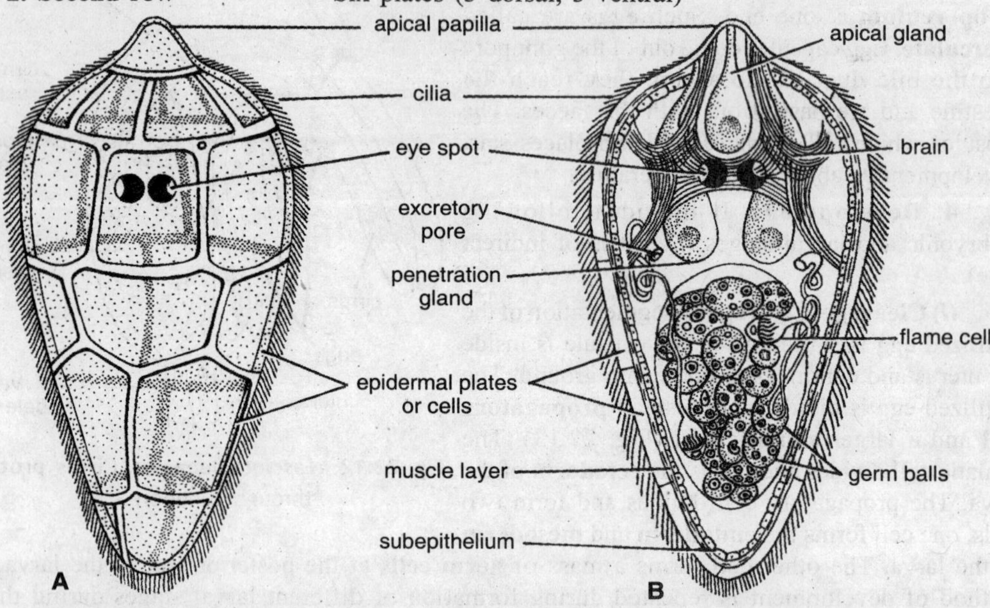

Fig. 22.14. *Fasciola hepatica*. Miracidium larva. A—External structure; B—Internal structure.

3. **Third row** — Three plates (1 dorsal, 2 ventro-lateral)
4. **Fourth row** — Four plates (2 right, 2 left)
5. **Fifth row** — Two cells (1 right, 1 left)

Below the epidermal plates there is **subepidermal musculature** which consists of (*i*) an outer layer of **circular fibres** and (*ii*) inner layer of **longitudinal fibres.** The subepidermal musculature is followed by innermost part of body wall called **sub-epithelium layer.** In the interior of body, there is a pair of **protonephridia,** each with two **flame cells.** The protonephridia open to the exterior by two separate **excretory pores** or **nephridiopores** which are located laterally in the posterior half of body. Towards the posterior side are present some **propagatory** or **germ cells,** some of which may have divided to form **germ balls** which are developing embryos.

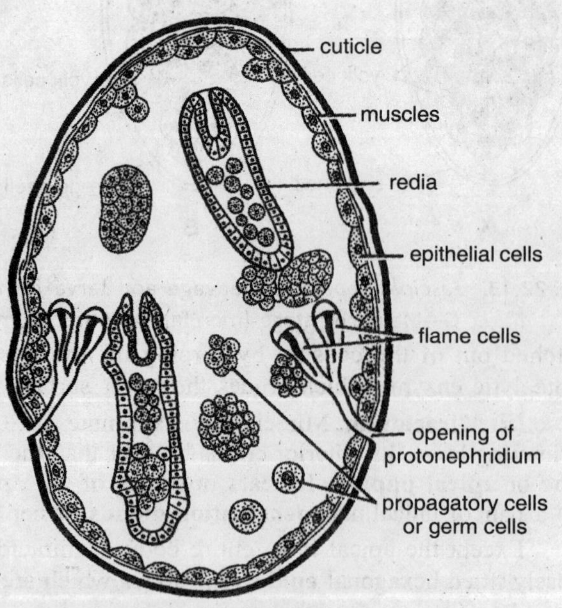

Fig. 22.15. *Fasciola hepatica*. Sporocyst.

A **brain** is present below the epidermal plates (or cells) of second row; it is associated with several nerve fibres. Above the brain is pair of X-shaped **eye spots.** The miracidium does not feed, it swims about in water. It dies in eight hours unless it can reach a suitable **intermediate host,** which is some species of amphibious snail of genus *Limnaea*. After getting a suitable host the miracidium adheres to it by its apical papilla and enters the pulmonary sac of the snail. From the pulmonary sac, the larva penetrates into the body tissues with the aid of penetration glands and finally reaches the snail's digestive gland. In intermediate host, miracidium casts off its ciliated epidermis, loses its sense organs and the larva swells up and changes its shape to become next larval stage called **sporocyst.**

***(iii)* Sporocyst (Fig. 22.15).** This is second larval stage of *Fasciola*. It is an elongated germinal sac about 0.7 mm long and covered with a thin cuticle. Below the cuticle, there are circular and longitudinal muscles and some mesenchyme cells. The hollow interior of sporocyst has a pair of protonephridia, each with two flame cells; it also has germ cells and germ balls. Sporocyst is called as a **living cyst.** The sporocyst moves about in the host tissues and its germ cells divide and passes through embryogenesis (such as morula, blastula, gastrula, etc.) to give rise to third larval stage, called **redia larva.**

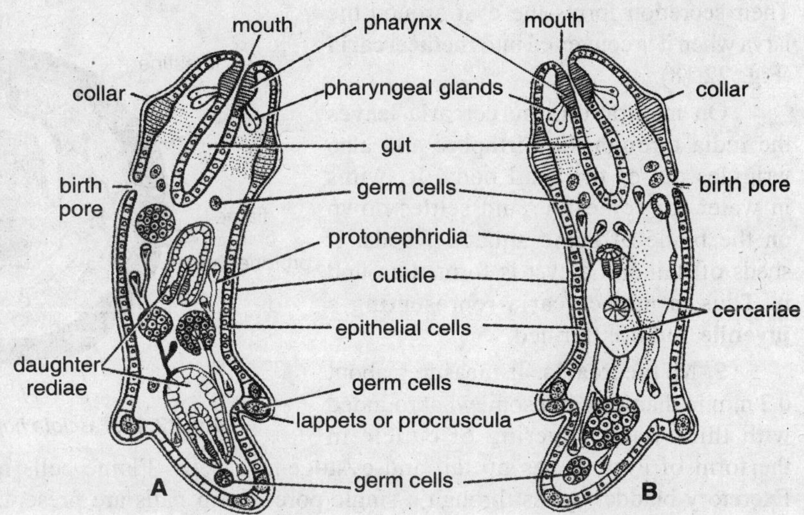

Fig. 22.16. *Fasciola hepalica*. A—Redia with daughter rediae, B—Redia with cercariae.

A sporocyst forms 5 to 8 radiae which are set free by rupture of the wall of sporocyst, feed on host's tissues and migrates to the liver of snail.

***(iv)* Redia larva.** It is an elongated (about 1.3 to 1.6 mm in length) cylindrical sac-like larva having two ventral processes, called **procruscula** or **lappets,** near the posterior end of body (Fig. 22.16). The body wall of redia larva contains cuticle, mesenchyme and muscles. Near the anterior end of body occurs a ring-like muscular swelling called **collar.** Collar is used in locomotion of larva. Behind the collar is present the **birth pore.** Redia has a small anterior **mouth,** a muscular **pharynx** with unicellular **pharyngeal glands** and a small sac-like **gut** or **intestine.** There is a pair of **protonephridia,** each with two pairs of **flame cells** (anterior and posterior) which open to exterior through a single **nephridiopore** on each side. The interior space of body is filled with loose parenchyma in which are found groups of **germ balls** and **germ cells.** The germ cells of redia give rise during summer months to a second generation of **daughter rediae,** but in winter they produce the fourth larval stage, the **cercaria larva.**

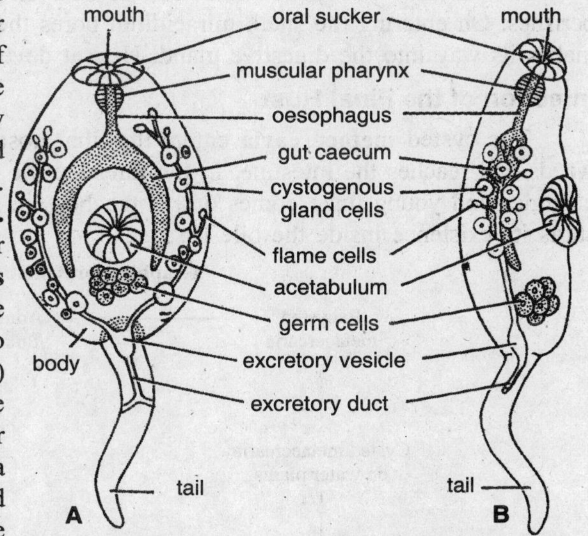

Fig. 22.17. *Fasciola hepatica*. A—Cercaria in ventral view; B—Cercaria in lateral view.

4. Cercaria larva (Fig. 22.17). Its body is small (0.5 mm long) rounded, oval or flattened with a long contractile tail. Its body surface is covered with backwardly directed **spines.** The anterior end of cercaria larva bears mouth, muscular pharynx, oesophagus and a bifid intestine. The mouth is surrounded by **oral sucker. A ventral sucker** or **acetabulum** is also present. There are numerous flame cells and the excretory ducts which unite to form the **bladder** or excretory pore situated at the base of the tail. Cercaria larva also contains germ cells and the rudiment of reproductive organs.

A number of unicellular **cystogenous glands** are situated below the larval body wall. Their secretion forms the **cyst** around the larva when it is converted into **metacercaria** (Fig. 22.18).

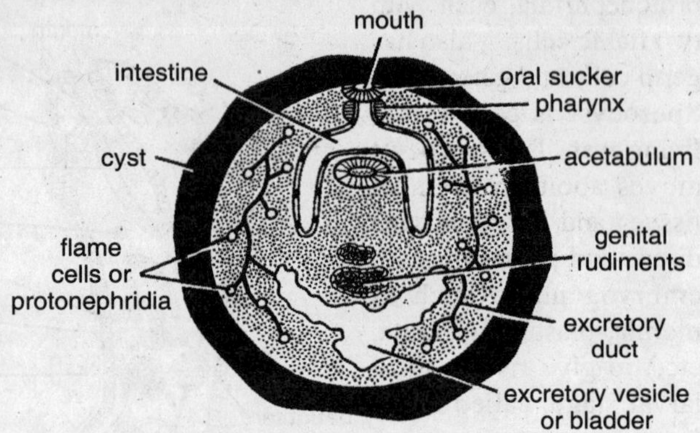

Fig. 22.18. *Fasciola hepatica*. Metacercaria.

On maturation, the cercaria leaves the redia through the birthpore and also wriggles out of the snail body. It swims in water for some time and settles down on the blade of some aquatic weeds. It sheds off tail and a cyst is formed around it. Thus, a metacercaria representing a **juvenile fluke** is formed.

5. Metacercaria. It measures about 0.2 mm in diameter and somewhat rounded with thick outer covering of cuticle in the form of **cyst.** It has no tail and cystogenous cells. Flame cells have increased in number. Excretory bladder opens through a single pore. Germ balls are present. Metacercaria can survive for weeks on grass and near water.

Infection of the Secondary Host

Miracidium swims freely in the water and searches for the intermediate host, the snail of genus *Limnaea* or *Planorbis*. Unable to find the host within a period of 24–30 miracidium perishes. On entering the snail, miracidium bores the pulmonary chambers, throws off cilia and makes its way into the digestive gland. Here it develops into **sporocyst larva.**

Infection of the Final Host

The cysted **metacercaria** enters the final host, the sheep, when it grazes on the aquatic weeds and reaches the intestine. In the intestine the cyst is dissolved by the action of digestive enzymes and young fluke comes out. It reaches the liver through the hepatic portal system and starts its existence inside the bile passage.

Graphic representation of the life cycle

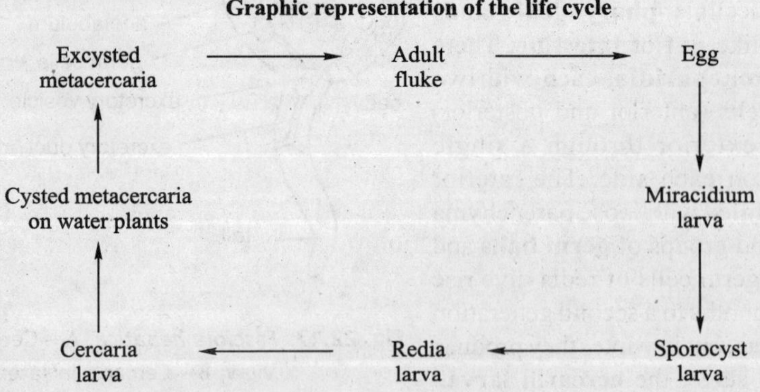

Significance of different larval stages in the life history of *Fasciola*

The different larval stages in the life cycle of *Fasciola* has following significance:

1. Alternation of generation or metagenesis. In the life cycle of *Fasciola*, a sexually reproducing adult phase alternates with asexually reproducing larval phase. Such alternation of two types of generations is called **metagenesis.**

2. Heterogamy. It was first thought (**Grobben** 1882) that the germ cells in different larva were eggs which developed into next stage by **parthenogenesis.** This type of sexual reproduction is called **heterogamy.** This hypothesis has now been totally rejected.

3. Polyembryony. According to this theory a single egg produces several embryos, itis called **polyembryony.** In the life cycle of *Fasciola* the division product of propagatory cells are called **germ cells** in the larval stage. The redia, daughter redia and cercaria are produced by **germ cells** by simple asexual multiplication. Thus, zygote gives rise to several larval stages, this shows polyembryony.

4. Extended metamorphosis. According to some, the asexual multiplication of larval forms is an extended metamorphosis due to parasitism.

5. Advanced larval stages. Body cavity, sense organs and locomotory organs are examples of advanced characters: They are absent in adult but present in larval stages. Thus, miracidium and cercaria are more advanced than the adult. The adult fluke has degenerated in many respect to suit its parasitic mode of life.

6. Complex life cycle. The presence of several larval stages has made the life-cycle of *Fasciola* quite complicated. Its life cycle is **digenetic,** *i.e.*, includes two hosts—**primary host** (sheep) and **secondary** or **intermediate host** (freshwater snail).

7. Over production. Rate of reproduction of an organism is directly proportional to the chances of death it faces at various phases of its life cycle. *Fasciola* being an endoparasite faces several threats to its survival and therefore its reproductive power is enormous.

A single fluke can produce more than 20,000 eggs in its life cycle. Each sporocyst produces 5–8 radia larvae. Each of these redia produces 8–12 daughter radia larvae and each of these radia produces 14 to 20 cercaria larva. Thus a single egg of *Fasciola* is capable of producing about 1,000 to 2,000 adults. But in nature it does not happen, only a few survive.

8. Chances of death (or mortality). The presence of a number of larval forms and a prolonged life-cycle, both increase the chances of destruction. These are:

(*i*) Dropping of eggs along with feces of sheep in dry places kills all eggs.

(*ii*) Failure of miracidium larvae to find a correct intermediate host within 24 hours kills the larvae.

(*iii*) Death of infected snail terminates the life cycle.

(*iv*) Drying of ponds kills the miracidium and cercaria.

(*v*) The sheep does not eat the plants containing metacercaria.

(*vi*) The pH and temperature of the pond water may be unsuitable for the life of larvae.

(*vii*) If the sheep swallows cercaria before encystment, the larvae are killed by the gastric juices.

To overcome the risk of disintegration of the race the parasite possesses great fertility and has evolved 'the asexual multiplication of larva'. Therefore, the extended and complicated life-history with many larval phases is an adaptation to parasitic mode of life.

22.11 ECONOMIC IMPORTANCE

1. Infection. The sheep, goat and cattle get infected with *Fasciola hepatica* by eating grass and other plants which contain **metacercaria.**

2. Pathogenesis. *Fasciola hepatica* causes serious damage to the bile duct and liver of its host. It causes the disease called **liver rot** or **fascioliasis.** The infected sheep develops the following symptoms: acute anaemia, appetite declines, loss of weight, enlargement of liver and fever. The sheep becomes weak and lazy, milk production is reduced, wool falls, eyes become yellowish and jaw muscles swell. Heavy infection of *Fasciola* causes death of the host within a few weeks.

3. Treatment. Anti-helminth drugs such as **emetin hydrochloride, phenothiazine, carbon tetrachloride, tetrachloroethane,** etc., are used to cure the sheep.

4. Preventive measures. Spread of infection of *Fasciola* can be prevented by

(*i*) Killing heavily infected sheep.

(*ii*) Introducing ducks in the ponds which eat up the intermediate host of *Fasciola*, the snail.

(*iii*) Drying the ponds.

(*iv*) By removing all vegetation around the ponds.

(*v*) By adding copper sulphate in the pond. It kills the snails.

22.12 PARASITIC ADAPTATIONS IN *FASCIOLA HEPATICA*

Adaptations can be defined as modification of an organism or its parts that make it more fit for existence under the conditions of its environment. **Parasitic adaptations** can be defined as structural, physiological and behavioural modifications developed by the organism in order to lead a parasitic mode of life.

I. Structural Adaptations

A. Morphological adaptations

1. Flat and thin body enables *Fasciola* to live in narrow spaces. It minimizes the danger of injury in narrow spaces.
2. Adhesive organs are of two types and of highly muscular nature. They occur in the form of **anterior sucker** and **ventral sucker.** By the help of these suckers, parasite attaches itself to the host's tissues.
3. In *Fasciola*, there is no distinction of body into head, trunk and tail. Body is also without any appendages, since these might prove harmful in endoparasitic mode of life.
4. Resistant tegument or the cuticle is an important parasitic adaptation. The parasite remains unaffected by alkaline pH of the bile.
5. The spines in the tegument protect the body.

B. Anatomical adaptations

1. There are no locomotory organs. As *Fasciola* is endoparasitic, it does not need to make excursion for food and safety.
2. Terminal mouth is surrounded by a muscular oral sucker which helps in feeding.
3. The pharynx is muscular and suctorial. It is adapted to suck blood, bile, etc.
4. The intestine of *Fasciola* is highly branched. Its branches supply the digested food to all parts of the body. Thus, intestine has taken over the work of circulatory system.
5. *Fasciola* feeds on already digested food so the anus is not required. Thus, alimentary canal is incomplete and intestine ends blindly.
6. Endoparasite respires **anaerobically**, as there is almost no oxygen in the bile passage of host.
7. Excretory system is well developed. It consists of extensively branched system of excretory ducts ending into flame cells. This facilitates the collection of excretory products from the interior of body.

8. Nervous system is poorly developed and sense organs are lacking, since the endoparasite does not react with the external environment, and there are least changes inside the body of host.
9. Male and female reproductive systems are well developed. The highly branched ovary and testes are capable of producing large number of gametes.
10. The eggs are covered with resistant capsule.
11. Bisexuality offers full chances of fertilization.

II. Physiological Adaptations

1. *Fasciola* is well adapted to the osmotic concentration of the bile in which it lives.
2. Respiration is anaerobic.
3. *Fasciola* can absorb certain amino acids through its body wall (tegument) as extra amino acids are needed for protein synthesis in the eggs and larval production.
4. Osmoregulation is not needed since the osmotic concentration of body fluids of *Fasciola* is almost equal to that of the host's bile.

III. Behavioural Adaptation

Fasciola hepatica is a hermaphrodite animal but in it cross-fertilization is the rule. Copulation is mutual and reciprocal. The cirrus of one partner is inserted into the vaginal canal of the other partner and sperms are ejected. After copulation both flukes lay fertilized eggs. In this way the rate of reproduction is doubled.

IV. Modifications in Life cycle

1. Very large number of fertilized eggs are produced.
2. The eggs are protected by the capsular covering.
3. The germ cells in a larva produce several new larvae asexually. This phenomenon is called **paedogenesis.**

QUESTIONS

Long Answer Questions

1. Give a detailed account of the structure of *Fasciola* and comment upon its parasitic adaptations.

2. Where is *Fasciola* found? Give an account of its structure.

3. Give the structure and life history of *Fasciola.*

4. Give an account of morphological features of *Fasciola* and point out how these features help them to lead a parasitic mode of life?

5. Give an account of the reproductive organs of *Fasciola hepatica.*

6. Describe the reproductive organs of *Fasciola.*

7. Describe the life-history of *Fasciola hepatica.*

8. Give a brief account of life cycle of *Fasciola* and explain the significance of different larval phases in its life history.

9. Give an account of the life history of *Fasciola hepatica.* Discuss its economic importance.

10. Name the various larval stages of liver fluke and discuss any two in detail.

11. What do you understand by alternation of generations or digenetic life cycle? Explain the same giving life history of *Fasciola hepatica.*

12. What do you understand by polyembryony? Explain this phenomenon with reference to the life-cycle of a digenetic parasite.

13. Name the primary and secondary hosts of *Fasciola hepatica* and describe various larval stages of this parasite.

14. Name one platyhelminth parasite of domestic cattle? State the mode of infection, pathogenic systems and control measures of parasite.
15. What is adaptation? Explain with reference to the structure and life-history of *Fasciola*.
16. Describe parasitic adaptations of *Fasciola hepatica*.

Short Answer Questions

1. Write about the parasitic adaptations of liver flukes.
2. Write short notes on the following:
   (*i*) Body wall of *Fasciola hepatica*
   (*ii*) Vitelline glands of liver flukes
   (*iii*) Miracidium larva.
3. Draw a well labelled diagram of the following:
   (*i*) Reproductive system of *Fasciola hepatica*
   (*ii*) Cercaria larva
   (*iii*) Metacercaria larva.

Very Short Answer Questions

1. Describe the flame cell in *Fasciola hepatica*. (*C.C.S.U. Meerut 2004*)
2. Name three sets of muscles found in the body wall of liver fluke?
3. What is the food of *Fasciola hepatica*?
4. How can you explain the lack of sense organs in liver fluke?
5. Name the connective tissue that fills spaces between the organs in liver fluke.
6. What is metraterm?
7. What is the name of excretory cells of liver fluke?
8. What is the function of flame cells?
9. Miracidium larva of *Fasciola* metamorphoses into the second larval stage. Name that larva.
10. Name the larval forms produced in a sporocyst of liver flukes.
11. Which larval forms are produced in a radia (*a*) in winter and (*b*) in summer?
12. Is miracidium more advanced than the adult liver fluke?
13. What is the encysted stage in the life history of *Fasciola* called?

Fill in the Blanks

1. *Fasciola hepatica* causes the .... disease in sheep and goats.
2. Liver fluke lives in the large .... of liver in sheep and goat.
3. The secondary host of liver fluke is ....
4. *Fasciola* has two suckers: oral and ....
5. Cuticle of *Fasciola hepatica* is living and called ....
6. Spaces between the viscera are filled up by .... in the liver fluke.
7. .... of liver fluke is incomplete.
8. Respiration of liver fluke is ....
9. The main function of flame cell is ....
10. In liver fluke, ammonia is mainly excreted through the ....
11. *Fasciola hepatica* has one ovary and .... testes.
12. One liver fluke may produce about .... capsule.

Multiple Choice Questions

1. Trematodes are usually described as

(*a*) blood flukes (*b*) flukes
(*c*) worms (*d*) flatworms

2. *Fasciola hepatica* is

(*a*) free living (*b*) ectoparasite
(c) endoparasite (*d*) ecto and endo-parasite

3. *Fasciola hepatica* lives in the

(*a*) intestine of sheep (*b*) liver of sheep
(*c*) spleen of sheep (*d*) pancreas of sheep

4. The causative agent of liver rot in sheep is

(*a*) *Dugesia* (*b*) *Fasciola*
(*c*) *Taenia* (*d*) *Schistosoma*

5. *Fasciola hepatica* occurs in

(*a*) bile duct of sheep (*b*) liver of sheep
(*c*) liver of human being (*d*) human bile duct

6. Liver fluke is not affected by enzymes of host because of

(*a*) cuticle (*b*) suckers
(*c*) epidermis (*d*) hooks

7. The secretion of vitelline glands

(*a*) facilitates the movement of sperms during copulation
(*b*) forms the shell around the egg
(*c*) is deposited around the fertilized egg
(*d*) none of above

8. Laurer's canal is present in

(*a*) *Fasciola* (*b*) *Taenia*
(*c*) *Ascaris* (*d*) Cockroach

9. Secretion of Mehlis's glands serve to

(*a*) lubricate the uterus (*b*) harden the capsule
(*c*) activate the sperm (*d*) does all these

10. Liver fluke has

(*a*) self fertilization (*b*) cross fertilization
(*c*) none of these (*d*) both of these

11. Primary host of *Fasciola* is

(*a*) human being (*b*) pig
(*c*) sheep (*d*) dog

12. The intermediate host in the case of liver fluke is

(*a*) snail (*b*) human being
(*c*) pig (*d*) fly

13. Larva of *Fasciola hepatica* that bores into snail is

(*a*) cercaria (*b*) hexacanth
(*c*) miracidium (*d*) redia

14. In *Fasciola,* miracidium develops into the next stage inside

(*a*) *Bulinus* (*b*) *Limnaea truncatula*
(*c*) *Pila globosa* (*d*) *Planorbis*

15. Which of the following larval stage of *Fasciola* does not produce several larvae?

(*a*) sporocyst (*b*) secondary redia
(*c*) primary redia (*d*) cercaria

16. In *Fasciola*, germ balls of redia give rise to
    (*a*) daughter sporocyst (*b*) daughter cercaria
    (*c*) daughter redia (*d*) metacercaria
17. The last infective larval form of *Fasciola hepatica* is
    (*a*) miracidium (*b*) sporocyst
    (*c*) metacercaria (*d*) redia
18. The first larval form in liver fluke is
    (*a*) redia (*b*) metacercaria
    (*c*) miracidium (*d*) cercaria

ANSWERS

Very Short Answer Questions

1. Flame bulbs or flame cells are protonephridia and form the osmoregulatory and excretory system of *Fasciola hepatica*. They are bulb-like bodies found distributed in the mesenchyma (= parenchyma) throughout the body of *Fasciola*. The flame cells are characteristic structures, each has a thin elastic wall with branched pseudopodia-like processes, a nucleus and an intracellular cavity having many long cilia arising from basal granules. In living condition, the cilia vibrate like a flickering flame, hence, the name flame cell. The undulating flagellar movement of cilia produces the pressure difference (force) needed to drive the **ultrafilteration** at the level of flame cell.
   The cavity of each flame cell continues into the capillary-like **collecting tubule.** Removal of excess water, ions, fatty acids are some important functions of the flame cells. According to **Brusca** and **Brusca** (2003) nitrogenous waste of the flukes is in the form of ammonia which is excreted, mainly across the tegument and protonephridia (*i.e.*, flame cells) are primarily osmoregulatory.
2. Circular, longitudinal and diagonal.
3. Bile, blood and cell fragments.
4. This is an adaptation for endoparasitic mode of life of liver fluke.
5. Parenchyma.
6. It is last muscular part of uterus of liver fluke.
7. Flame cells (*i.e.*, protonephridia).
8. Osmoregulation and excretion.
9. Sporocyst.
10. Rediae.
11. Secondary rediae, Cercaria.
12. Yes.
13. Metacercaria.

Fill in the Blanks

1. Liver-rot
2. Bile ducts
3. Freshwater snail, *Limnaea*
4. Ventral (acetabulum)
5. Tegument
6. Parenchyma
7. Alimentary canal
8. Anaerobic
9. Osmoregulation
10. Tegument
11. Two
12. 500,000

Multiple Choice Questions

1. (*b*)	**2.** (*c*)	**3.** (*b*)	**4.** (*b*)	**5.** (*a*)	**6.** (*a*)
7. (*c*)	**8.** (*a*)	**9.** (*d*)	**10.** (*b*)	**11.** (*c*)	**12.** (*a*)
13. (*c*)	**14.** (*b*)	**15.** (*d*)	**16.** (*c*)	**17.** (*c*)	**18.** (*c*).

23

Taenia Solium (Phylum Platyhelminthes)

23.1 CLASSIFICATION

Phylum	–	**Platyhelminthes**
Class	–	**Cestoda**
Order	–	**Cyclophyllidea (Taenioidea)**
Genus	–	***Taenia***
Species	–	***solium***

Taenia solium is commonly called as **pork tape worm.** It is an intestinal parasite (= endoparasite) of human beings. It has a cosmopolitan distribution.

23.2 HABIT AND HABITAT

Taenia solium is a **digenetic endoparasite,** *i.e.*, its life history is completed in two hosts, human beings is the **primary host** and pig is the **secondary host.** *Taenia* is found attached to intestinal mucosa of humans by its scolex, while the rest of body lies free. It is most common in the pork eating population of tropical and subtropical regions where pork is utilized as food without being properly cooked. It may live upto 25 years in the human host.

23.3 STRUCTURE

1. Shape. The body of tapeworm is long, narrow, dorso-ventrally flattened and ribbon-like (Fig. 23.1). It has two surfaces — a dorsal and a ventral.

2. Size. The size of adult tapeworm varies from 3–5 metres, but few are recorded to attain a length of about 8 metres.

3. Colouration. It is usually white in colour, but sometimes grey, yellow or cream coloured.

4. Segmentation. The body of tapeworm is differentiated into many segments, which are called **proglottids** and arranged in a chain. But this is not true segmentation, so is called **pseudosegmentation** (see Box 23.1).

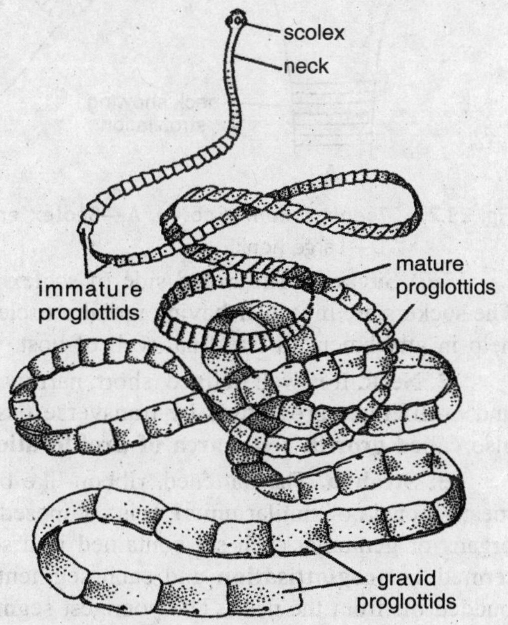

Fig. 23.1. *Taenia solium*. An adult tapeworm.

Box 23.1 Proglottids are not true segments.

The proglottids are not metameric segments like those of annelids, arthropods and other animals. In fact, these are independent, self-contained units, each with a complete set of reproductive organs both male and female and a part of excretory and nervous systems. They are formed by the process of strobilation at the zone of proliferation situated anteriorly behind the scolex. Each proglottid survives briefly after breaking away from the strobila.

5. Body divisions. The body of *Taenia* is well adapted for parasitic mode of life. It is divided into three parts: scolex, neck and body or strobila.

1. Scolex. The narrow anterior end of the body is knob-like and of the size of pin head. It is known as **scolex** (errorneously called **head**). It is a four-sided pear-shaped structure distinguished into following three parts (Fig. 23.2):

***(i)* Rostellum.** It is an anterior round prominence which arises from the middle of scolex. It is slightly retractile.

***(ii)* Hooks.** Hooks are present at the base of rostellum. They are curved, pointed chitinous structures. Hooks are 28 in number and are of two different size. They are arranged in two circles: larger hooks in inner circle and smaller hooks in outer circle. The smaller hooks alternate with the larger ones. Each hook is made of a **base,** a **handle** directed towards the apex and a conical **blade** directed outwardly. Hooks help in the attachment of scolex with the intestinal wall.

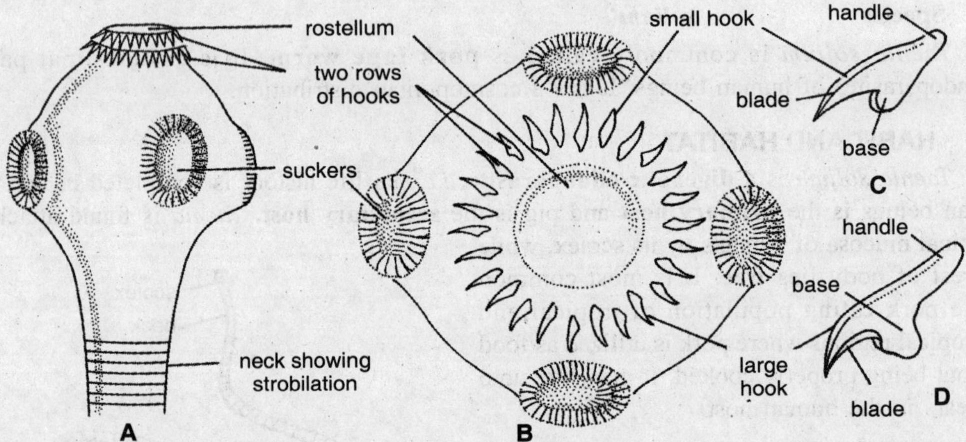

Fig. 23.2. *Taenia solium*. Scolex. A—Scolex enlarged; B—Frontal view of scolex; C—Small hook; D—Large hook.

***(iii)* Suckers.** The broad side of scolex contains four cup-like pockets known as **suckers.** The suckers are muscular having radial muscles. Each sucker measures 0.5 mm in diameter. They help in attachment to intestinal wall of host.

2. Neck. It is well defined, short, narrow region behind the scolex. Neck grows continuously and proliferates proglottids by transverse fission or asexual budding. Therefore, this region is also called **growth zone, area of proliferation** or **budding zone.**

3. Strobila. The flattened, ribbon-like body after the neck is known as **strobila** (strobilus means a chain of similar units). It is composed of linear series of 800 to 1000 sets of reproductive organs or genitalia, each set contained in a segment. This linear repetition of genital organs is termed as **proglottisation** and each segment is known as **proglottid,** since the segments are budded off from the neck. The youngest segments are towards the neck and the oldest segments are posterior most. The proglottids internally remain connected together by muscles, excretory vessels and nerve cords. The proglottids are of following three types (Fig. 23.1):

***(i)* Immature proglottids.** These are newly formed and youngest proglottids. Their width is more than the length. They lie next to neck region and include 200 anterior proglottids. In them the reproductive organs are either absent or in stage of development.

***(ii)* Mature or reproductive proglottids.** They are squarish 400 to 600 proglottids which are located after the immature proglottids. They are **hermaphrodite** containing fully developed male and female reproductive organs. About 100 to 150 anterior mature proglottids contain only the male reproductive system.

***(iii)* Gravid or ripe proglottids.** They are the oldest proglottids which are situated towards the posterior end of strobila. They include about 150 to 200 proglottids. They are longer than broad. Gravid proglottids contain a large, highly **branched uterus** which is packed with fertilized eggs and developing embryos. The rest of reproductive organs have already disappeared from them after performing their function.

The posterior most 5 to 6 proglottids are regularly cut off from the body and are expelled out of host body along with its feces. This process of separation of proglottids is known as **apolysis.**

Differences in a proglottid and a body segment

In a tapeworm (*Taenia solium*), each of the segment is virtually identical, differing mainly in size. Each segment, or **proglottid,** is derived from the **scolex** (head) of the tapeworm. The older proglottids are pushed back by new proglottids formed at the margin of the scolex and the most recent segment. This formation process in tapeworms constitutes a type of asexual reproduction. The proglottids are actually buds. When the gravid proglottids, swollen with self-fertilized eggs, arrive at the posterior end of the tapeworm, they break off, and surviving eggs eventually reach a new host.

In annelids, segmentation is part of a unitary body plan, and the segments are not by them-selves reproductively competent. Such structures like nephridia and setae do recur, but many other structures extend through the entire body of the worm and do not exist as separate structures in each segment. Some examples are the circulatory system and the reproductive system. Also, the digestive tract is not segmented but exists as a continuous one-way tube with varying specialization occurring along its length. Thus segmentation in Annelida is an advancement that facilitates development. It continues as a **morphogenetic** (showing a hereditary pattern of shape) feature in Arthropoda as well.

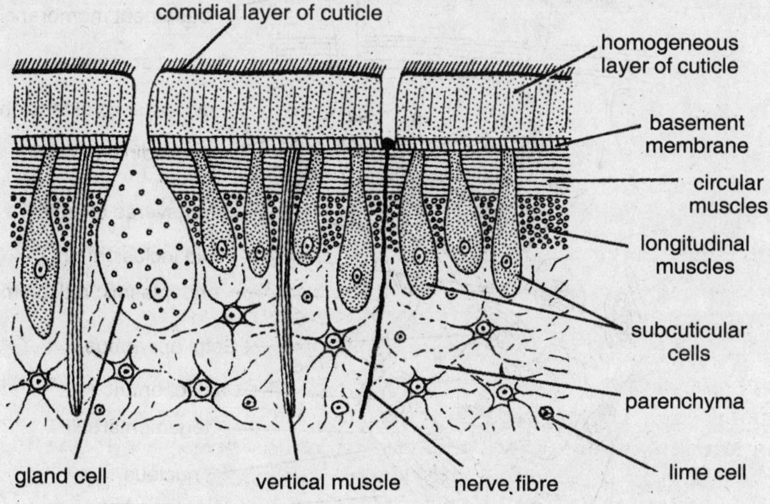

Fig. 23.3. *Taenia solium*. T.S. of body wall.

Body wall

According to classical light microscopic studies (Fig. 23.3) body wall of *Taenia* consists of following three structures: cuticle, muscular layer and parenchyma.

1. Cuticle. It forms the outermost, thick, waxy and enzyme resistant covering of body. It is composed of protein impregnated with calcium salts and is secreted by sub-cuticular cells. Cuticle is further divided into three layers: (*i*) **Comidial layer.** It is outermost layer and is produced into hair-like processes. (*ii*) **Homogeneous layer.** It is middle thick layer. (*iii*) **Basement membrane.** It is innermost thin layer.

Box 23.2 Basement membrane

Basement membrane or **basal lamina** is a dense membraneous layer of connective tissue that separates the layer of epithelial cells from underlying **lamina propria** (*i.e.*, highly vascular layer of connective tissue). It is secreted by epithelial cells and helps in communication between epithelium and connective tissue.

2. Subcuticular muscles. Beneath the cuticle is a layer of **circular muscles** followed by **longitudinal muscles.**

3. Parenchyma. It is fluid-filled spongy-network of mesenchyme cells. It forms a **packing** substance around the internal organs and subcuticular muscle layer. Except at the two sides in each proglottid, the parenchyma is divided into (*i*) outer **cortical region,** (*ii*) inner **medullary region.** Embedded in parenchyma are found few **gland cells, lime cells** and a large number of **cuticular cells.** Parenchyma also contains **parenchymal muscles** which help in bending movement of proglottids.

Ultrastructure of body wall. According to the electron microscopic studies of **Threadgold** (1984), there is no cuticle and no cellular epidermis in the body wall of *Taenia*. Instead of them it contains tegument and surface coat (Fig. 23.4).

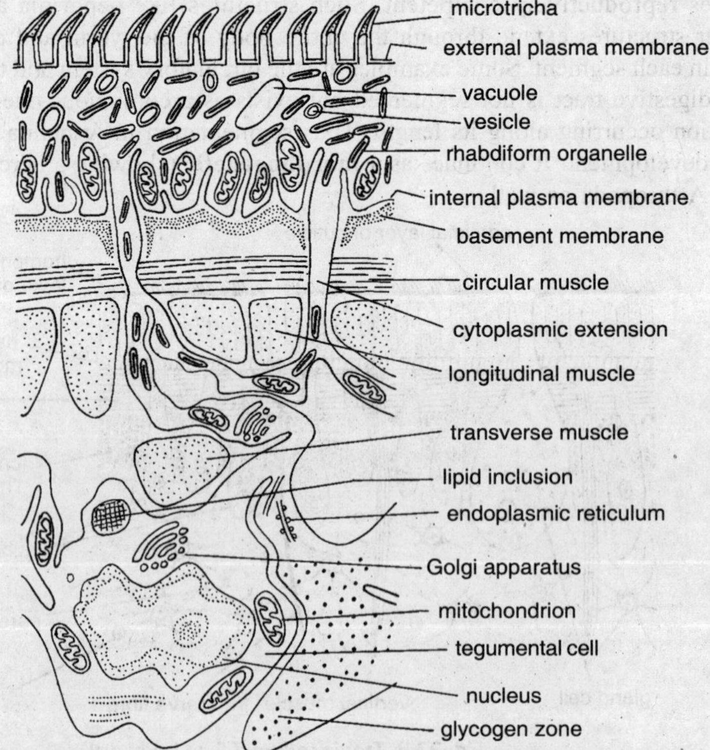

Fig. 23.4. Diagrammatic representation of electron microscopic structure of body wall of tapeworm.

Tegument and surface coat. Entire body of *Taenia* is covered with a cytoplasmic but non-cellular or **syncytium** layer. It is known as **tegument.** It varies in thickness from 1 to 30 µm. The tegument is bounded by **outer** and **inner plasma membranes.** The outer plasma membrane gives out microvilli-like folds, called **microtriches** (singular: **microthrix**). The long, smooth and flexible microtriches increase the surface area. Each of them contains an electron-dense apical tip. **Hooks** of rostellum are also formed by the tegument. The cytoplasm of tegument contains all typical cellular organelles such as mitochondria, Golgi apparatus, endoplasmic reticulum, ribosomes, glycogen granules, lipids and vacuoles. It also contains some specific structures such as **tegumental discs, dense bodies** and **multilaminar bodies.** These are required for the growth of tegument and for replacement of surface coat.

The tegument of adult tape worm does not contain **nucleus** but it is connected with nucleus-containing **subtegumental cells.** The **inner plasma membrane** of tegument is connected with the plasma membrane of finger-like extensions of parenchymal cells called **cyton** or **subtegumental cells.** Some of these cells contain stored glycogen.

The tegument is covered with a fuzzy **surface coat** which is a highly dynamic structure. It is composed of carbohydrates and mucopolysaccharides. It is renewed in every few hours. Different components of the surface coat are synthesized by the endoplasmic reticulum and Golgi apparatus of subtegumental cells and are transported by their ductules to the upper parts of the tegument.

Thus tegument is a metabolically active layer and serves the following functions:

1. Absorption of nutrients.
2. Osmoregulation and excretion.
3. Protection against the effects of digestive enzymes of host and host's immune system.

23.4 NUTRITION

Taenia solium lacks digestive organs. Digested food containing glucose, amino acids, glycerol, etc., from the host intestine is absorbed through the general body surface by diffusion. This is called **saprozoic nutrition.** Reserve food is stored as glycogen and lipids in the parenchyma.

23.5 RESPIRATION

Organs of respiration are also lacking in *T. solium*. The respiration is of **anaerobic** type. During respiration glycogen is broken into CO_2 and fatty acids/organic acids with the release of energy.

23.6 EXCRETORY SYSTEM

The excretory system is well developed and mostly helps in maintaining the **hydrostatic pressure** (water balance). Thus, it is involved in **osmoregulation.** Excretory system (Fig. 23.5) consists of following parts:

1. Longitudinal vessels. There are two **dorsal** and two **ventral vessels** extending into strobila and are linked in scolex by **nephridial plexus.** The four vessels run through proglottids (Fig. 23.5). The dorsal vessels are found in the anterior part of strobila while the ventral vessels extend its entire length. Both ventral vessels are connected except in last part of proglottid by **transverse excretory canal.** Last proglottid forms a pulsatile **caudal vesicle** opening out by an excretory pore (Fig. 23.6).

2. Secondary vessels and capillaries. Secondary vessels are formed by the union of several **capillaries.** These are ciliated while longitudinal vessels are lined by cuticle and are non-ciliated.

3. Flame cells. These are irregular bodies with granular nucleated cytoplasm and a central funnel-shaped space. Flame cells are found all over in the parenchyma. Many long flagella arise near the nucleus and flicker constantly like the flame inside the intracellular space. This space is continous with the space of excretory vessels through connecting vessels.

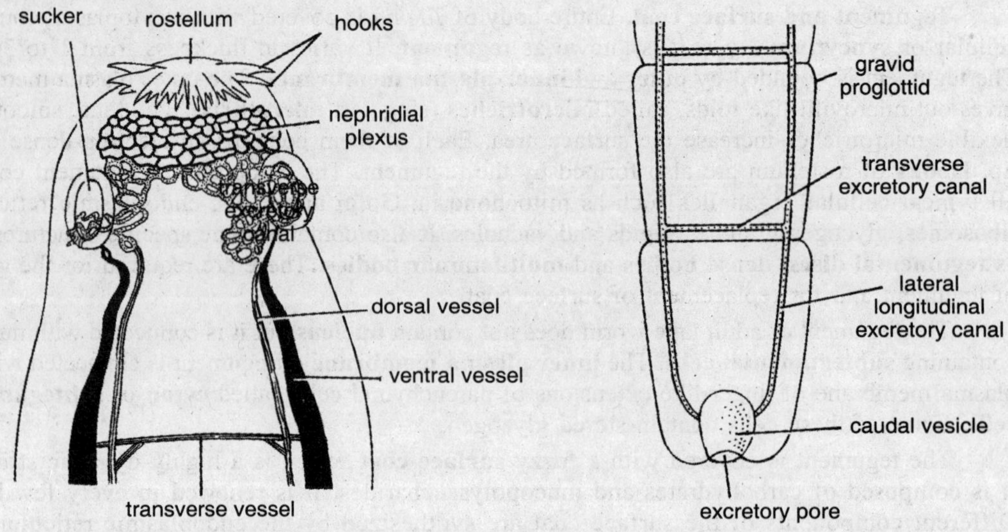

Fig. 23.5. *Taenia solium*. Excretory system of anterior part of body.

Fig. 23.6. *Taenia solium*. Excretory pore.

Function. Metabolic waste products such as ammonia, fatty acids, organic acids, etc., are removed by the excretory system of *T. solium*. Excretory system is also **osmoregulatory** in function.

23.7 NERVOUS SYSTEM

Nervous system of *Taenia* is not advanced. It consists of two **cerebral ganglia** in the scolex. These ganglia are connected by a thick transverse nerve band and by the **dorsal** and **ventral commissures** to form the **brain complex.** The brain complex is connected with another **rostellar nerve ring** having a pair of **rostellar ganglia,** in the rostellum. The suckers and rostellum are supplied by slender nerve from cerebral and rostellar ganglia. From the posterior side of brain complex, five pairs of **longitudinal nerve cords** arise and travel in the strobila. Out of these only two, the **lateral longitudinal nerves** are best developed and run along the entire length of strobila. They lie just outside the longitudinal excretory vessel and are connected in each proglottid by a **ring connective** situated below the transverse excretory canal.

23.8 REPRODUCTIVE SYSTEM

Taenia solium is **hermaphrodite.** Each mature proglottid contains a complete set of male and female reproductive organs (Fig. 23.7).

1. Male reproductive organs. The male reproductive organs are testes, vasa efferentia, vas deferens, cirrus or penis and genital atrium.

***(i)* Testes.** Testes are small, numerous, round (follicular) bodies scattered throughout in the dorsal part of the proglottid.

***(ii)* Vasa efferentia.** These are numerous minute ducts which arise from testes and collectively open into the vas deferens.

***(iii)* Vas deferens.** It is the common sperm duct. It is a long coiled tube and arises from the middle of the proglottid and transversely either to the left or right to open into the genital atrium.

***(iv)* Cirrus and cirrus sac.** The distal end of the vas deferens becomes muscular forming cirrus sac and modified into protrusible **cirrus** or **penis.** The cirrus is beset with spines, bristles or hooks and opens into the genital atrium through male genital pore.

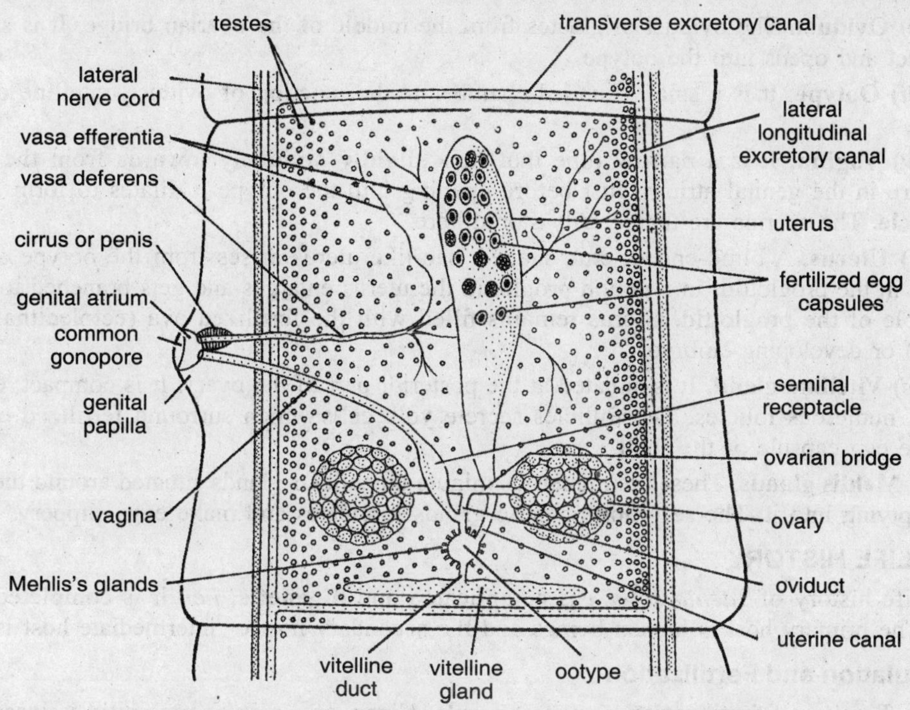

Fig. 23.7. *Taenia solium*. A mature proglottid showing male and female reproductive organs.

(*v*) Genital atrium. It is a cup-shaped structure, situated in the middle of proglottid at its lateral margin. Genital atrium is a common chamber and opens to the outside through a common **gonopore.** The gonpore is situated at the tip of a tiny protuberance, called the **genital papilla.** Genital papillae with gonopore occur alternately on the right and left margin of mature proglottids.

2. Female reproductive organs. The female reproductive system consists of ovary, oviduct, ootype, vagina, uterus, vitelline gland and Mehlis's glands (Fig. 23.7).

(*i*) Ovary (Germarium). It is a bilobed structure situated ventrally in the posterior part of the proglottid. Both the lobes of ovary are connected by a transverse ovarian bridge or **isthmus.** Each lobe of ovary is formed of radially arranged follicles. Each follicle produces yolkless ova.

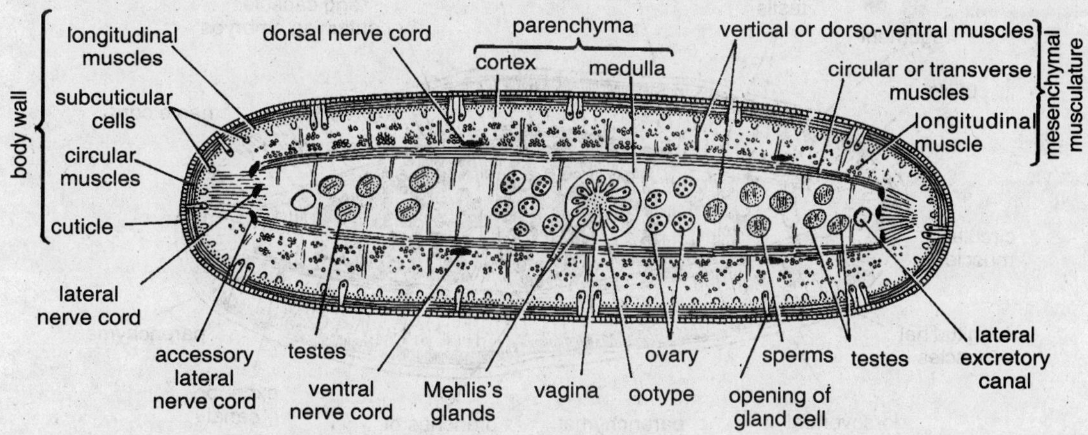

Fig. 23.8. *Taenia solium*. T.S. mature proglottid passing through ootype.

***(ii)* Oviduct.** The oviduct originates from the middle of the ovarian bridge. It is short but wide duct and opens into the ootype.

***(iii)* Ootype.** It is a small rounded chamber at the junction of oviduct, vitelline duct and uterus.

***(iv)* Vagina.** It is a narrow tube that runs slightly obliquely inwards from the **female gonopore** in the genital atrium. Just before opening into the ootype it dilates to form **seminal receptacle.** The sperms are temporarily stored here.

***(v)* Uterus.** A blind-ended, club-shaped, sac-like uterus arises from the ootype and runs forward in the proglottid. In a gravid proglottid the uterus enlarges and gets branched to occupy the whole of the proglottid. Uterus remains filled with the fertilized ova (ectolecithal ova in capsule) or developing embryos.

***(vi)* Vitelline gland.** It is situated at the posterior part of the ovary. It is compact, elliptical mass of numerous folicles. The follicles secrete yolk cells which surround fertilized eggs and form the egg capsule or the shell.

7. Mehlis glands. These are numerous minute unicellular glands situated around the ootype and emptying into it. The secretion of these glands lubricates and make eggs slippery.

23.9 LIFE HISTORY

Life history of *Taenia solium* is complicated and **digenetic,** *i.e.*, it is completed in two hosts. The primary host is human beings and the secondary host or intermediate host is pig.

1. Copulation and Fertilization

In *Taenia* **self-fertilization** occurs as a rule. Cirrus of a mature proglottid is inserted into the vagina of the same proglottid and sperms are transferred into the vagina. From the vagina the sperms come to lie in the seminal receptacle from where they fertilize the ova in oviduct.

Cross-fertilization due to **copulation** between different proglottids of the same tapeworm is also very common. *Taenia*, in fact, is **protandrous** and its testes mature prior to ovaries (in the anterior most mature segments). Thus, anterior mature proglottids can copulate with posterior most proglottids due to folding of the strobila.

2. Capsule Formation

In the ootype, fertilized ovum, egg or **zygote** becomes associated with a yolk-cell secreted by the vitelline gland. Materials of the yolk-cell forms a hard, resistant **shell** around the zygote and

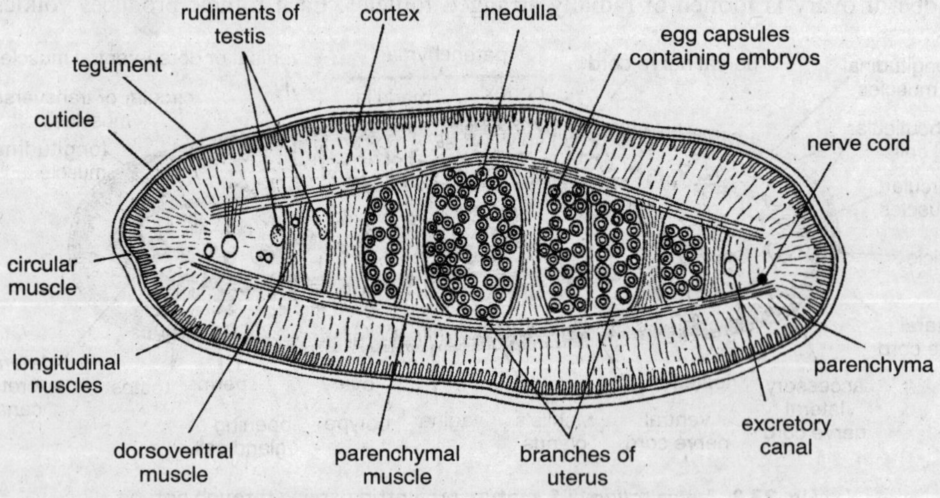

Fig. 23.9. *Taenia solium*. T.S. gravid proglottid.

yolk cell. The structure thus formed is called **capsule** which passes into the uterus. The secretion of Mehlis's gland facilitates the passage of capsule in the uterus. The uterus grows in size and becomes branched as it receives more and more (egg) capsules. Now the proglottid is known as **ripe** or **gravid** (Fig. 23.10). Soon the gravid proglottid breaks off (apolysis) and passes out of the host's intestine with his/her faeces during defaecation.

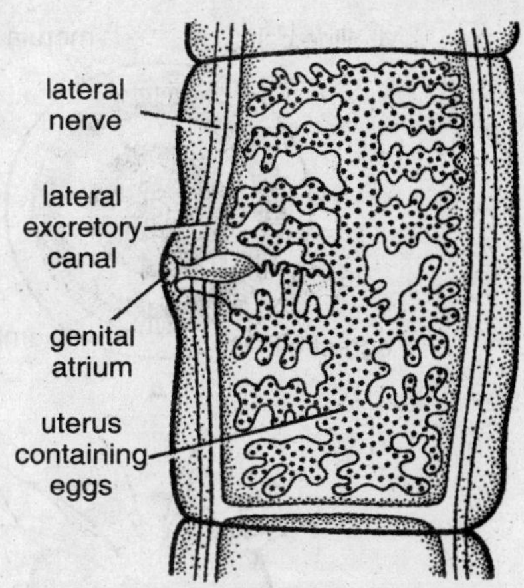

Fig. 23.10. *T. solium*. A gravid proglottid showing branched uterus.

3. Fertilized Eggs

The egg capsule is very small (about 40 microns in diameter). It contains a zygote and a large yolk cell and remains surrounded by a hard and resistant egg shell (Fig. 23.11B).

4. Development

In *T. solium* development is **indirect** and includes a single larval stage.

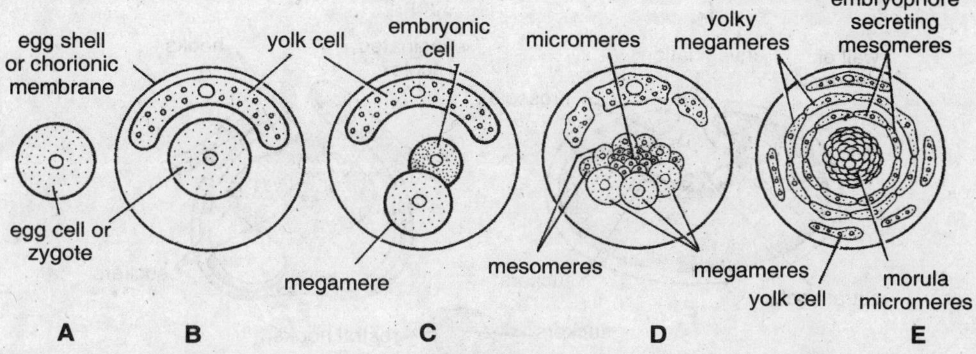

Fig. 23.11. T. solium. Early development. A—Zygote; B—Capsule; C—First cleavage; D—Embryo with micromeres, mesomeres and megameres; E—Morula.

***(i)* Cleavage and morula formation.** Cleavage starts when the egg capsule is in the uterus (inside the intestine of host). The first division is **unequal.** It produces a large **megamere** and a small **embryonic cell.** The megamere divides repeatedly to form many megameres. The embryonic cell divides repeatedly and produces two types of cells—large **mesomeres** and small **micromeres** (Fig. 23.11D). The micromeres form a ball of cells, called **morula,** in the centre. Mesomeres form an envelope around the morula. The megameres form an outer covering around the mesomere envelope. The megameres absorb yolk from the yolk cell and supply nourishment to the developing embryo. With time the yolk cell reduces in size, breaks up and gradually disappears.

***(ii)* Formation of hexacanth and onchosphere.** As development proceeds, the middle mesomere layer forms a thick, hard, cuticularized and striated **shell** or the **embryophore** around the morula. A basement membrane is also formed beneath the embryophore. The morula forms the **embryo** proper. Embryo develops six (= 3 pairs) chitinous **hooks.** These hooks are secreted by **onchoblast** cells of the embryo. This six-hooked embryo is called **hexacanth.** It also possesses a pair of **penetration glands** in between hooks. Hexacanth is surrounded by two **hexacanth membranes** (Fig. 23.12). The hexacanth embryo, hexacanth membranes, basement membrane, embryophore and shell together is known as **onchosphere** (Fig. 23.12B).

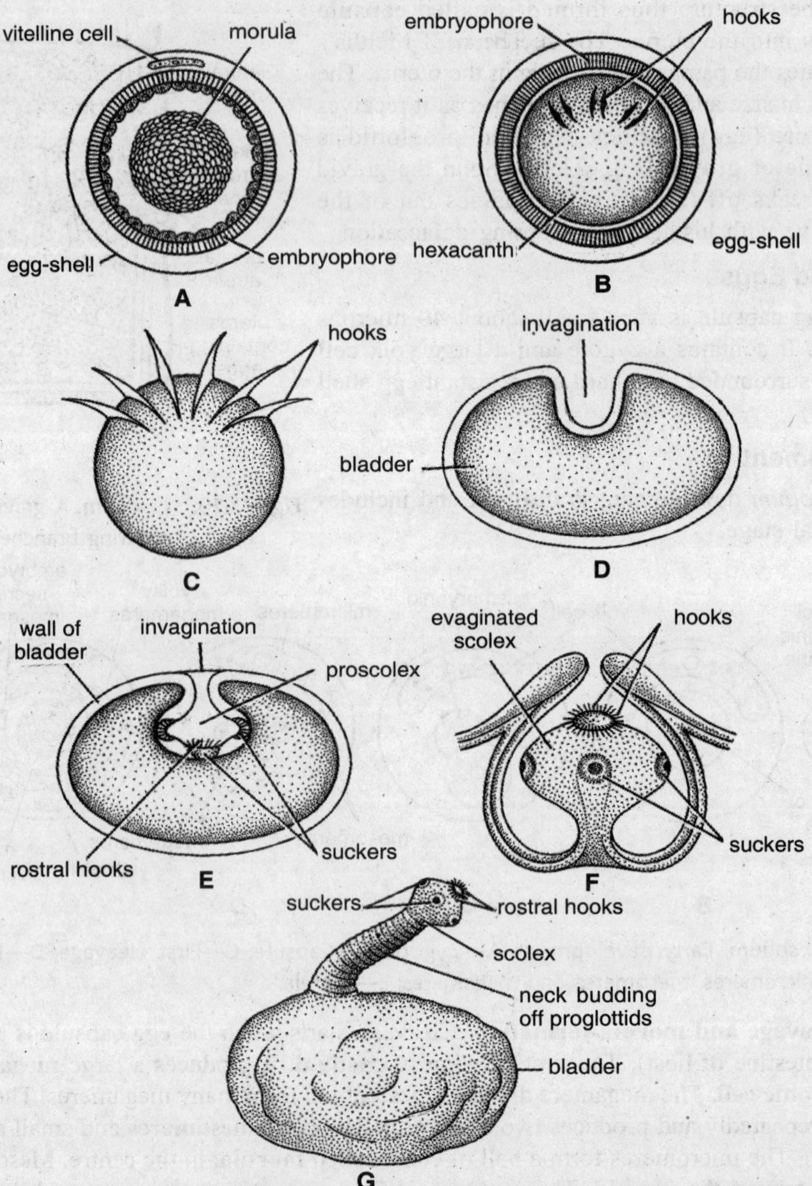

Fig. 23.12. *Taenia solium*. Stages in life cycle. A—Young onchosphere; B—Mature onchosphere; C—Free hexacanth; D—Bladderworm with invagination; E—Bladderworm with proscolex; F—Bladderworm with evaginated scolex; G—Cysticercus with neck budding off proglottids.

By the time the onchospheres are formed, the uterus in the gravid proglottid develops 7 to 13 branches on each side and fills the entire proglottid. The uterus of a gravid proglottid contains 30,000 to 40,000 onchospheres.

5. Infection of the Secondary Host (Pig)

The oncospheres which occur either in dislocated gravid proglottids or lie freely in the host's stool are eaten by the pigs (due to pig's **coprophagous** feeding habit). On reaching the stomach of the secondary host, the capsule (shell) and embryophore are digested by the gastric juices and the hexacanth is released. The hexacanth starts boring through the intestine wall with the help of its paired penetration glands. Hooks help in anchorage of hexacanth with intestinal wall. Hexacanth loses its onchospheric membranes and hooks in the lumen of intestine and enters the submucosal blood vessels. It is carried around by the blood in the host body. Eventually, these come out of blood capillaries to settle in the striated muscles of such organs as the limbs, neck, tongue, heart, etc. Once settled in muscles, each embryo absorbs a large amount of watery fluid (plasma) of host's blood and grows to a spherical, pea-sized sac-like cyst, called **bladderworm** or **cysticercus larva.** The cysticercus of *Taenia solium* is called **cysticercus cellulosae.** The flesh of pig or pork containing many cysticerci appears white spotted resembling something like that of measles, hence, it is characteristically called **measly pork.** Thus, the pig becomes infected with larvae of *Taenia solium*.

6. Development of Bladderworm or Cysticercus Larva

The hook-less hexacanth gradually becomes encysted by the connective tissue of the host (pig). All cells of cysticercus larva fall into two layers around the fluid-filled central cavity. Cells of the outer layer coalesce to form a syncytial **tegument**. The inner layer of cells is called **mesenchyme** or germinal layer. At the future anterior end (opposite to the side where hooks were present) the wall of cysticercus thickens and then invaginates into the cavity (Fig. 23.12D). Four suckers, hooks and rostellum develop upon the knob-like invagination which thus assumes the form of an inverted scolex, called **proscolex**. Formation of cysticercus or bladderworm with proscolex is completed in host muscles in about 4 months after ingestion of eggs or onchospheres.

The cysticerci do not develop further, their further development takes place only when they are eaten up by the primary host. Thus fully formed cysticerci or bladderworms are infective to human hosts. These usually survive (in dormant state) in the flesh for several years.

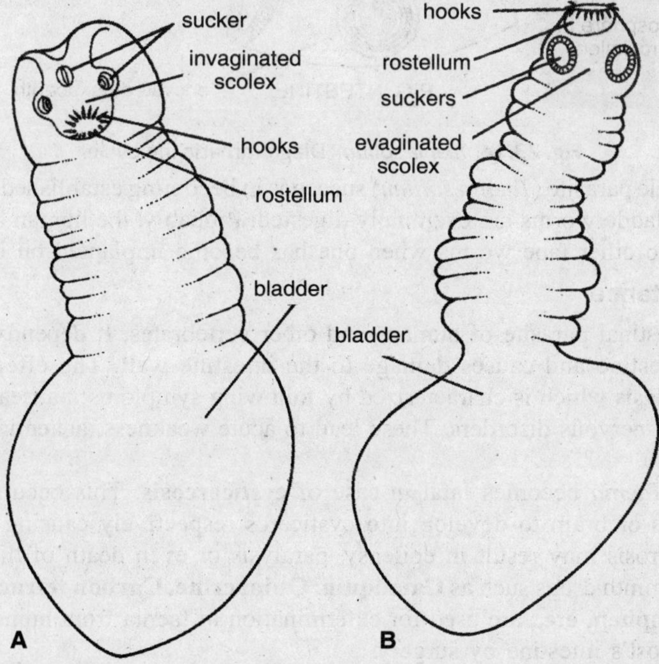

Fig. 23.13. *Taenia solium*. Cysticercus larva. A—Invaginated scolex; B—Evaginated scolex.

7. Infection of the Primary Host (Human)

Pork-eating people get the infection of *Taenia solium* by eating raw or imperfectly cooked measly pork. In their intestine, the outer cyst wall is digested off, the proscolex of a bladder worm everts out and anchors to the intestinal wall by the help of suckers and hooks. The bladder is progressively degenerated. The neck begins to produce proglottids by budding, In about 10 to 12 weeks, the parasite attains adult condition with gravid proglottids. Complete life cycle of *T. solium* has been depicted diagrammatically in Fig. 23.14.

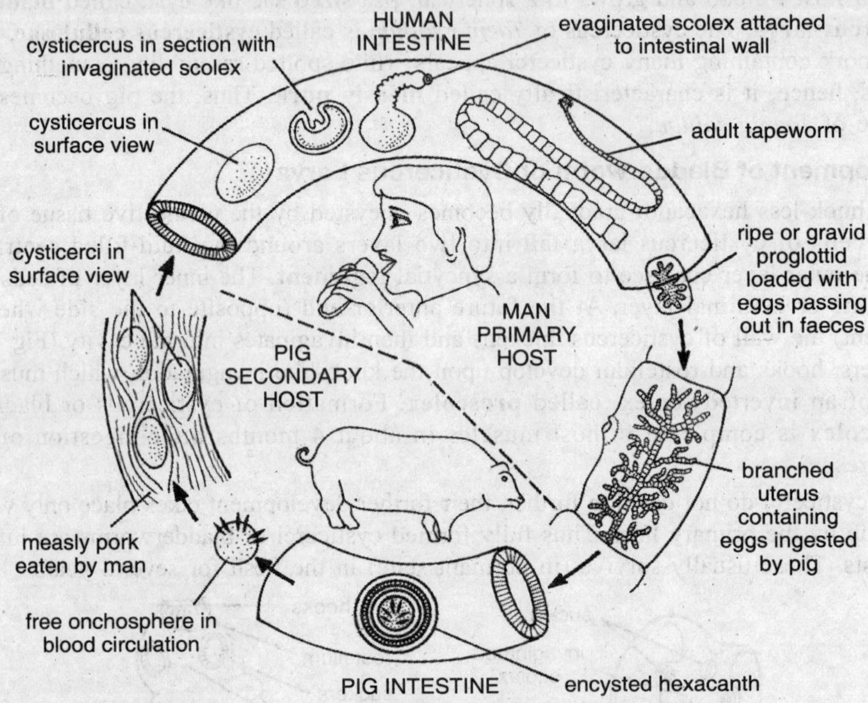

Fig. 23.14. *Taenia solium*. Diagrammatic life cycle.

Usually a single parasite (*Taenia solium*) succeeds in becoming established in human intestine; all other ingested bladderworms are eventually digested. Probably, the human host develops some kind of immunity to other tape worms when one has become implanted on its intestinal wall.

Economic Importance

Taenia is intestinal parasite of humans and other vertebrates. It depends upon the digested food from host intestine and causes damage to the intestine wall. The effect is occurrence of disease called **taeniasis** which is characterized by following symptoms: nausea (feeling to vomit), abdominal pain and nervous disorders. These lead to acute weakness, anaemia due to indigestion and eosinophilia.

Infection of *Taenia* becomes fatal in case of **cysticercosis**. This occurs when hexacanths invade eye muscles or brain to develop into cysticercs respectively causing blindness or brain necrosis. Brain necrosis may result in epilepsy, paralysis or even death of the host.

Some antihelminth drugs such as **Camoquin, Quinacrine, Carbon tetrachloride, Antiphen, Yomesun, Dichlorophen, etc.,** are used for extermination of taenia from human intestine. Scolex is detached from host's intestine by surgery.

Preventive measures against infection of *Taenia solium* include proper cooking, sterilization of nightsoil and personal hygiene.

23.10 PARASITIC ADAPTATIONS OF *TAENIA SOLIUM*

Parasitism is an association between two organisms (one is **host** on which other called **parasite** lives) for nourishment and shelter without any compensation for the host. Therefore, parasite tends to cause harm to host. Sometimes it proves so harmful that it ultimately results in the death of the host. The parasitic flatworms such as *Taenia solium* which is an intestinal endoparasite of humans have undergone profound adaptations to suit its parasitic mode of life. These adaptations are called **parasitic adaptations** and are discussed under the following two headings:

1. Morphological adaptations
2. Physiological adaptations

A. Morphological adaptations

(*i*) External form.

1. Body of *Taenia* is covered with thick tegument, provided with scales, afford suitable protection to the parasite.
2. Body is dorsoventrally flattened and ribbon-like or tape-like.
3. The anterior end of the body or the scolex is modified into a holdfast or region of attachment. It has developed suckers and hooks to ensure a firm hold of the host tissue.
4. Neck region possesses power of adding new segments to the body so that the animal grows throughout its life.
5. Body consists of large number of body segments (proglottids).

(*ii*) Internal organization.

1. Elaborate organs for nutrition are not needed because parasite can absorb the readily available digested and semi-digested food of the host.
2. In *Taenia*, respiratory organs are absent, since it has **anaerobic** type of respiration.
3. Locomotory organs and sense organs are absent. Need for quick "response to stimuli" is associated with free active life.
4. Each mature proglottid contains a complete set of reproductive organs. The posterior gravid segments contain only the fertilized eggs in the branched uterus and the developing embryo. These have power of detachment.
5. Hermaphroditism is of distinct advantage as

(*i*) It ensures copulation even when few individuals are present.

(*ii*) After copulation both the individuals lay eggs.

(*iii*) In the absence of a companion (*e.g.*, *Taenia solium*), the parasite can reproduce off-spring.

(*iv*) Germinarium (ovary) is highly branched, so that it can produce large number of eggs.

6. Eggs are covered with resistant capsule.
7. Uterus can store numerous fertilized eggs.

B. Physiological Adaptations

(*i*) Protective mechanisms.

1. Walls of gut of host secretes mucus, which form a protective clothing around the parasite.
2. Parasite secretes antienzymes to neutralize digestive enzymes of the host.

(*ii*) Anaerobic respiration. The worm respire anaerobically by breaking down glycogen.

***(iii)* Osmoregulation.** It maintains an equilibrium with the osmotic pressure of the host. The pressure inside the parasite and outside remains the same so that there is no difficulty in exchange of water.

***(iv)* High fertility.** In its life-time, *Taenia* produces eggs in millions.

Adaptations in the Life History

1. To cover the risk of finding individuals of different sexes in one host, the parasite becomes hermaphroditic and performs self-fertilization.
2. Each mature proglottid produces about 30 to 40 thousand eggs. Such an over-production of offsprings is necessary to maximize the chances of reaching host and ensure the perpetuation of race.
3. The complicated life-history is a remarkable feature in the parasite. Secondary or intermediate host the pig is introduced to ensure parasite's dispersal from one primary host to another.

Retrogressive and Progressive Parasitic Modifications

Parasitic adaptations of *Taenia* can also be classified into following two types:

1. Retrogressive modifications of *Taenia*. Besides the loss of surface epidermis and cilia as in flukes, the tapeworms exhibit advanced parasitic adaptation by a complete loss of digestive organs. Tapeworms obtain digested nutrients from the chyme (= partly digested food present in duodenum) of the host through body surface by absorption and pinocytosis and through surface pores. No locomotory organs and sense organs occur in adult tapeworms or their larvae, because no part of their life-cycle is passed outside host bodies.

2. Progressive modifications of *Taenia*. Besides having the protective tegument in place of surface epidermis, as in flukes, the tapeworms show advancement in parasitic adaptations by a better development of attachment organs. However, the most important advancement is shown by tapeworms over the flukes in multiplication of reproductive organ and great increase in their capacity to produce eggs. For example, an adult *Taenia* forms 7 to 8 proglottids each day throughout its life and each proglottid has a complete set of genitalia (*i.e.*, male and female reproductive systems), producing 30 to 40 thousand eggs.

23.11 COMPARISON OF IMMATURE, MATURE AND GRAVID PROGLOTTIDS

Table 23.1 Differences between immature, mature and gravid proglottid of *Taenia*.

	Immature proglottids		*Mature proglottids*		*Gravid proglottids*
1.	Present in the anterior part of the body next to neck region.	1.	Present in the middle part of the body.	1.	Present in the posterior part of the body.
2.	Their width is more than the length.	2.	Length and width of each proglottid is almost the same.	2.	These are long than broad.
3.	The reproductive organs are either absent or in stages of development.	3.	Reproductive organs are fully developed with mature male and female reproductive organs.	3.	After fertilization of the eggs the male and female reproductive organs are degenerated but uterus exists.
4.	No uterus.	4.	Uterus is small and unbranc-hed.	4.	Uterus becomes enlarged and highly branched and fills the entire space of the proglottid.
5.	—	5.	Uterus is without eggs.	5.	Uterus contains fertilized eggs and developing embryos.

23.12 COMPARISON OF *FASCIOLA* AND *TAENIA*

Table 23.2. Comparison of structure and life-cycle of *Fasciola hepatica* and *Taenia solium*.

	***Fasciola hepatica* (Sheep liver fluke)**		***Taenia solium* (Pork tapeworm)**
	I. Habit, Habitat and Distribution		
1.	Endoparasitic and digenetic.	1.	Endoparasitic and digenetic.
2.	Adults live in bile passages of sheep and goat **(primary host).**	2.	Adults live in the intestine of humans **(pri mary host).**
3.	Larval stages live in a freshwater snail, either *Limnaea* or *Planorbis* sp. **(secondary host).**	3.	Larval stages are found in pig (**intermediate** or **secondary host**).
	II. Structure		
4.	Body is bilaterally symmetrical, dorsoventrally flattened oval and **leaf-like.**	4.	Body is bilaterally symmetrical, dorsoventrally flattened, elongated, **tape** or **ribbon-like.**
5.	Its size varies from 1.2 to 5 cm in length and 1.2 cm in width.	5.	It measures 3 to 6 metres in length.
6.	Pinkish or dark brown in colour.	6.	Opaque white or cream (pale yellow) in colour.
7.	Anterior end of body contains a conical projection, called **head lobe.** Body is **unsegmented.**	7.	Body is segmented or divided into knob-like **scolex** at anterior end of body, **neck** and **strobila.** The strobila contains a chain of about 1000 proglottids (segments).
8.	Adhesive organs include oral or anterior sucker around the mouth and a ventral sucker. There are no hooks.	8.	Scolex acts as an adhesive organ; it bears four cup-shaped suckers (one dorsal, one ventral and two lateral) and two rows of hooks for this purpose.
9.	The neck is absent.	9.	Neck is present and possesses the power of proliferation. It adds new proglottids to strobila throughout life.
10.	Proglottids absent.	10.	Proglottids are of three types: immature, mature and gravid proglottids.
11.	Body aperture consists of a mouth, genital aperture, an excretory pore and a temporary opening for Laurer's canal.	11.	Body aperture consists of a common gonopore in each proglottid and one or two excretory pores.
	III. Body Wall		
12.	The body is covered with a cytoplasmic syncytium or the tegument. It contains backwardly directed spines or scales and numerous microvilli.	12.	Body is covered by tegument like the fluke. Tegument contains numerous long microvilli-like fold or microtriches. There are no spines or scales in the tegument. The hooks of scolex are formed by the tegument.
13.	Musculature consists of circular, longitudinal and diagonal muscles.	13.	Musculature consists of circular and longitudinal muscles.Diagonal muscles are absent.
14.	Parenchymal musculature is absent.	14.	Parenchymal musculature consists of longitudinal, circular and dorsoventral muscles.
	IV. Digestive System		
15.	Present but incomplete (anus is absent). It consists of mouth, pharynx, oesophagus and bifurcated intestine with diverticulae.	15.	Absent.

	Fasciola hepatica (Sheep liver fluke)		*Taenia solium* (Pork tapeworm)
16.	Mode of nutrition is holozoic; it sucks its food (bile, blood and cellular debris) by muscular and suctorial pharynx.	16.	Mode of nutrition is **saprozoic;** it absorbs its food through tegument.
	V. Respiratory System		
17.	Lacks definite organ for respiration. Respiration is anaerobic.	17.	Same as in *Fasciola*.
	VI. Nervous System		
18.	It contains a nerve ring bearing a pair of **cerebral ganglia** and a **ventral ganglion** surrounding the oesophagus. From the nerve ring **three** pairs of longitudinal nerve cords run posteriorly.	18.	It contains a pair of **cerebral ganglia** connected by a transverse nerve band and a ganglionated **ring commissure** in the scolex. Five pairs of longitudinal nerves which travel into the sotrbila present.
	VII. Degenerated Organs		
19.	Skeletal, locomotory, circulatory organs and organs of special senses are absent. However, eyes are present in miracidium larva.	19.	These organs are also absent in *Taenia*.
	VIII. Excretory System		
20.	It consists of numerous **flame cells** and a median **excretory duct** wich opens to outside by a single **excretory pore**.	20.	It consists of **flame cells** and four main **excretory ducts**, only two of these extend up to the last proglottid and open through a **common excretory pore**. When the last proglottide is lost by apolysis, the two exceretory ducts open separately.
	IX. Reproductive System		
21.	*Fasciola* is **monoecious** or **hermaphrodite.**	21.	*Taenia* is also **monoecious** or **hermaphrodite.**
22.	Only one set of male and female reproductive organs is found in the body of each organism.	22.	One organism has many sets of male and female reproductive organs, one set of them occurs in each mature proglottid.
	(A) Male Reproductive System		
23.	Male reproductive system consists of following organs: (*i*) Testes (*ii*) Vasa efferentia (*iii*) Seminal vesicle (*iv*) Cirrus and cirrus sac (*v*) Ejaculatory duct (*vi*) Prostate glands.	23.	Male reproductive system consists of following organs: (*i*) Testes (*ii*) Vasa efferentia (*iii*)Vas deferens (*iv*) Cirrus and cirrus sac.
24.	There is a pair of large, tubular and highly branched **testes** situated in the middle of the body one behind the other.	24.	There are numerous, small follicular (rounded) **testes** found scattered throughout the proglottid.
25.	There are two **vasa defferentia** arising one from each testis and running almost straight.	25.	**Vas deferens** is single and convoluted. It is formed by the union of fine ducts the **vasa efferentia,** arising from the testes.

	Fasciola hepatica (Sheep liver fluke)		*Taenia solium* (Pork tapeworm)
26.	Both vasa differentia run forwards and join to form a pear-shaped and muscular **seminal vesicle,** from which **ejaculatory duct** arises and enters into the muscular **penis** or **cirrus.**	26.	Single vas deferens runs transversely outwards and opens into the muscular penis; **seminal vesicle** and **ejaculatory duct** are absent.
27.	Ejaculatory duct is surrounded by many unicellular **prostate glands.**	27.	**Prostate glands** are absent.
28.	A **cirrus sheath** covers the cirrus, seminal vesicle and prostate glands.	28.	Cirrus sheath or sac surrounds the cirrus only.
	(B) Female Reproductive System		
29.	Female reproductive system consists of following organs: (*i*) Ovary (*ii*) Oviduct (*iii*) Uterus (*iv*) Laurer's canal (*v*) Vitelline glands and yolk reservoir (*vi*) Mehlis's glands (*vii*) Genital opening	29.	Female reproductive system consists of following organs: (*i*) Ovary (*ii*) Oviduct (*iii*) Ootype (*iv*) Uterus (*v*) Vagina (*vi*) Seminal receptacle (*vii*) Vitelline glands (*viii*) Mehlis's glands (*ix*) Genital opening
30.	**Ovary** is single, tubular and highly branched. It is situated on the right side infront of the testes.	30.	**Ovary** is single but **bilobed,** both lobes are connected by a transverse ovarian bridge or connected by a transverse ovarian bridge or **isthmus.** It is situated in the posterior part of proglottid.
31.	An **oviduct** extends from the ovary downwards to meet the median vitelline duct.	31.	An **oviduct** extends from the middle of **isthmus** and opens into ootype which receives vitelline duct.
32.	**Ootype** is absent.	32.	**Ootype** is present.
33.	**Vagina** and **seminal receptacle** are absent.	33.	A slender and tubular **vagina** extends from the genital atrium which expands near the oviduct to form seminal receptacle.
34.	A blind **Laurer's canal** extends from the oviduct to the dorsal side. It opens temporarily at the time of copulation and acts as vagina.	34.	Laurer's canal is absent.
35.	**Vitelline glands** are numerous, rounded vesicular gland scattered on the lateral sides of the body. **Yolk reservoir** is present.	35.	**Vitelline gland** is single, large, lobulated situated in posterior part of each proglottid below the ovary. Yolk reservoir is absent.
36.	**Uterus** is a long, convoluted and tubular structure. It opens into the genital atrium by female gonopore.	36.	**Uterus** is straight, blind and club-shaped sac.
37.	Uterus enlarges to store the capsule but does not branch.	37.	Uterus enlarges and develops lateral branches to store the capsules.

	Fasciola hepatica (Sheep liver fluke)		*Taenia solium* (Pork tapeworm)
38.	**Mehlis's glands** are numerous found around the junction of oviduct, uterus and median vitelline duct.	38.	Mehlis's glands are found around the ootype.
39.	Reproductive organs persist even after capsule formation.	39.	All reproductive organs except uterus degenerate after capsule formation; such proglottids are called gravid.
	(X) Life History		
40.	Usually two different flukes copulate so that **cross fertilization** occurs but sometimes self fertilization also occurs.	40.	Usually **self fertilization** occurs but interproglottid cross fertilization is also common.
41.	A fertilized egg along with yolk cell is packed into a **capsule.** Many capsules are stored in uterus and are laid in bile passage of the sheep. From there these capsules finally reach to the outside with host's faeces. Capsules are with operculum.	41.	Many capsules, each containing a fertilized egg and a yolk cell, are stored in the uterus. The uterus enlarges and becomes branched. Gravid proglottids packed with capsules are detached from the body of the tapeworm (apolysis) and reach to the exterior with the host's (humans) faeces. Outside the host's body, the proglottids break to release capsules. Capsules are without operculum.
42.	Development starts after capsules are laid and continue to develop if dropped at suitable places of temperature, moisture, etc.	42.	Development starts after capsule formation in the gravid proglottid.
43.	The life-history includes following five larval stages. (*i*) Miracidium (*ii*) Sporocyst (*iii*) Redia (*iv*) Cercaria (*v*) Metacercaria	43.	The life-history includes two larval stages: (*i*) Hexacanth stage (*ii*) Bladderworm stage
44.	Miracidia and cercaria are free swimming stages found in water.	44.	No free-living larval stage. The hexacanth larva gets freed in the alimentary canal of secondary host.
45.	The larva multiply asexually or parthenogenetically (polyembryony occur).	45.	No parthenogenesis and polyembryony occur.
46.	Sheep gets infection by ingesting metacercaria with vegetation.	46.	Human beings get infection by eating improperly cooked pork infected with cysticerci (*i.e.*, measly pork).
47.	Several flukes can live in a single host.	47.	Only one tapeworm can live in one host.
48.	Highly pathogenic, causing **facioliasis** or **liver rot.**	48.	It is also highly pathogenic causing **taeniasis** and **cysticercosis.**

In the light of above discussion it can be concluded that *Taenia solium* is better adapted and more specialized to the parasitic mode of life.

QUESTIONS

Long Answer Questions

1. Give an account of the structure of *Taenia solium*.
2. Describe the systematic position and life history of *Taenia solium*.
3. Give an account of the reproductive organs of *Taenia*.
4. Give an illustrated account of the reproductive organs of *Taenia solium*. What is meant by gravid proglottid?
5. Give an illustrated account of mature proglottid of *Taenia solium*.
6. Differentiate immature, mature and gravid proglottid.
7. Describe the life history of *Taenia*.
8. Describe life history and economic importance of *Taenia*.
9. What is parasitism? Explain how an animal adapts itself to the parasitic mode of life as evidenced by the structure and life history of *Taenia*.
10. Give an account of the parasitic adaptation of *Taenia solium*.
11. How does the segmentation of an annelid worm differ from the repeating units of the body of a tapeworm?
12. What is adaptation? Explain with reference to the structure and life-history of *Teania solium*.
13. Indicate the specialization, adaptations and degeneration of *Taenia* consequent to its parasitic mode of life.
14. Describe how the flatworms are adapted to their parasitic mode of life. Write a note on how *Taenia* infection can be prevented?
15. Compare general plan of life in *Fasciola* and *Taenia* and state which in your opinion is more suited to parasitic mode of life.
16. Describe the reproductive organs of *Taenia solium* and compare it with those of *Fasciola hepatica*.

Short Answer Questions

1. Write short notes on the following:
   (*i*) Scolex
   (*ii*) Strobila
   (*iii*) Gravid proglottid
   (*iv*) Onchosphere
   (*v*) Cysticercus
   (*vi*) Measly pork
   (*vii*) Taeniasis
   (*viii*) Economic importance of *Taenia*
2. Draw well labelled diagrams of the following:
   (*i*) *Taenia solium*
   (*ii*) Scolex
   (*iii*) A mature proglottid of *Taenia* or Reproductive organs of *Taenia*.

Very Short Answer Questions

1. What is the difference between hexacanth and onchosphere?
2. Describe structure and significance of cysticercus.
3. Where is Mehlis's gland found? What is its function?
4. Flesh is digested in the stomach of human beings but why *Taenia* is not digested?
5. What is scolex of *Taenia*?

6. What is flame cell?
7. Two species of *Taenia* commonly infect humans. Name these species.
8. Why *Taenia solium* is regarded as a digenetic parasite?
9. What is measly pork?
10. Where are the proglottids formed in a tapeworm?
11. What is the name of the organ of attachment in tapeworm?
12. How many suckers are there on the scolex of tapeworm?
13. Name the muscles present in the tapeworm.
14. Name the excretory organs of flatworms.
15. By which process is the excretory matter formed in *Taenia*?
16. Give two functions of tegument of a tapeworm.
17. What is an ootype?
18. How many vitelline glands occur in *Taenia*?
19. Name the larval stages found in *Taenia*.

Fill in the Blanks

1. Pork-eating humans are infected with a tapeworm named ....
2. *Taenia* .... infects the beef eating humans.
3. Tapeworm has a digenetic life cycle like ....
4. Nutrition in *Taenia* is ....
5. Symmetry of tapeworms scolex is ....
6. Tapeworm is attached to the intestinal wall with .... and suckers.
7. New proglottids are proliferated from the .... of tapeworm.
8. The outer layer in the tapeworm's body wall is called ....
9. *Taenia solium* has about .... proglotids of which about .... are gravid.
10. A hook of tapeworm consists of .... , guard and ....
11. The secondary host in the life history of *Taenia* is ....
12. The proglottid of *Taenia solium* which contains nothing but fertilized eggs, is called ....
13. In *Taenia solium*, the proximal portion of the oviduct that ends into the vagina is called ....
14. The first larval stage in the life history of a tapeworm is ....
15. The sac-like stage of the developing tapeworm is ....
16. The disease in human beings caused by infection of second stage larva of *Taenia solium* is known as ....

Multiple Choice Questions

1. *Taenia saginata* is commonly known as
   (*a*) pork tapeworm (*b*) beef tapeworm
   (*c*) human tapeworm (*d*) none of these
2. Sucker in tapeworm are meant for
   (*a*) attachment to intestinal wall (*b*) catching food
   (c) ingestion of food (*d*) all of these
3. In human beings, *Taenia solium* is found in
   (*a*) duodenum (*b*) large intestine
   (*c*) small intestine (*d*) rectum
4. Segments of tapeworm are called
   (*a*) scolex (*b*) cysticercus
   (*c*) proglottids (*d*) onchospheres

5. In *Taenia*, scolex bears in the middle a prominent
(*a*) cup shaped sucker (*b*) head
(*c*) rostellum (*d*) recurved hooks

6. The number of hooks in *Taenia solium*
(*a*) 32 (*b*) 28
(*c*) nil (*d*) unlimited

7. The outer layer in the body wall of tapeworm is
(*a*) noncellular cuticle (*b*) epithelial
(*c*) syncytial (*d*) none of above

8. *Taenia* lacks
(*a*) excretory system (*b*) digestive system
(*c*) reproductive system (*d*) all of these

9. The excretory product of *Taenia* is
(*a*) ammonia (*b*) fatty acid
(*c*) urea (*d*) ammonia and fatty acid

10. Tapeworm has
(*a*) one testis and many ovaries (*b*) many testes and two ovaries
(*c*) one testis and a two-lobed ovary (*d*) many testis and a two-lobed ovary

11. A proglottid is called **gravid proglottid** when it has
(*a*) both male and female reproductive units well developed
(*b*) only female units well developed
(*c*) only male reproductive units well developed
(*d*) branched uterus filled with fertilized eggs

12. Apolysis is a process in *Taenia* for
(*a*) regular detachment of gravid proglottids from main body
(*b*) embryonic development
(*c*) ingestion
(*d*) infection to human beings

13. Pig acquires infection of *Taenia solium* by ingestion of
(*a*) oncosphere (*b*) hexacanth
(*c*) cysticercus (*d*) none of these

14. In which form does *Taenia* pass through faecal matter in human beings?
(*a*) oncosphere (*b*) hexacanth
(*c*) cysticercus (*d*) young tapeworm

15. Cysticercus in the larva of
(*a*) *Scypha* (*b*) *Obelia*
(*c*) *Taenia* (*d*) *Fasciola*

16. The infective stage of *Taenia* in human being is
(*a*) hexacanth (*b*) onchosphere
(*c*) bladderworm (*d*) none of the above

17. The tapeworms are found in the
(*a*) voluntary muscles of human beings (*b*) voluntary muscles of pig
(*c*) liver of pig (*d*) body of snail

18. Cysticercosis is a disease caused by accidental infection of
(*a*) hexacanth of *Taenia*
(*b*) bladderworm of *Taenia*
(*c*) encysted cercaria of *Schistosoma*
(*d*) eggs of *Enterobius*

ANSWERS

Very Short Answer Questions

1. **Hexacanth** is a six-hook containing early embryonic stage of *Taenia solium*. **Oncosphere** is a tapeworm embryo that is the earliest differentiated stage of cyclophyllidean tapeworm. The oncosphere contains hexacanth embryo enclosed inside hexacanth membrane, basement membrane, embryophore and egg-shell (Fig. 23.12B).
2. Inside the secondary host of *Taenia solium*, the pig, the hexacanth grows in size and gradually becomes encysted by the connective tissue of the host. A central cavity appears in the hexacanth embryo due to breakdown of the cells. The cavity enlarges and becomes filled with a fluid. This fluid-filled cavity is called a **bladder** and the parasite is now called **bladder-worm** or **cysticercus** (Fig. 23.12E) Its wall is contractile. The wall thickens as a knob. The knob develops **rostellum,** four **suckers** and **hooks** and thus form the **scolex** or **proscolex**. The cysticercus measures 6 to 18 mm in diameter. The cysticerci do not develop further in the secondary host. These may remain in dormant condition for years. These develop further only when eaten by the primary host (the human beings).
3. Mehlis gland is one of the small unicellular glands around the ootype of tapeworm. Each gland empties into ootype and its secretion tends to lubricate and make eggs slippery.
4. Thick tegument of body wall of *Taenia* is enzyme resistant. Tegument also secretes antienzymes to nullify the effect of host's digestive enzymes.
5. Scolex is the head of *Taenia* having a muscular pad with hooks and two pairs of lateral suckers.
6. Flame cell is hollow cell that contains the terminal branches of excretory vessels in flatworms such as liver fluke and tapeworm. It is involved in osmoregulation and excretion.
7. *Taenia solium*, *T. saginata*.
8. When two hosts are involved in the life cycle of an endoparasite, it is called **digenetic.** *Taenia solium* is an endoparasite and digenetic as it completes its life cycle in two hosts, *i.e.*, human being is the primary host and pig is an intermediate or secondary host.
9. The pig flesh (or the pork) infected with cysticerci of *Taenia solium* is called measly pork due to its spotted appearance.
10. Neck.
11. Scolex.
12. Four
13. Longitudinal, transverse and dorsoventral.
14. Protonephridia (flame cells).
15. Ultrafilteration and reabsorption.
16. Intake of nutrients by active transport, evasion of host's immune response.
17. Sac at the junction of oviduct, uterus and vitelline duct.
18. One.
19. Hexacanth, Cysticercus.

Fill in the Blanks

1. *Taenia solium*
2. *saginata*
3. Liver fluke
4. Saprozoic
5. Radial
6. Hooks
7. Neck
8. Tegument
9. 850, 150 to 250
10. Blade, handle
11. Pig
12. Gravid proglottid
13. Uterus
14. Oncosphere
15. Cysticercus
16. Cysticercosis.

Multiple Choice Questions

1. (*b*)	2. (*a*)	3. (*c*)	4. (*c*)	5. (*c*)	6. (*b*)
7. (*c*)	8. (*b*)	9. (*d*)	10. (*d*)	11. (*d*)	12. (*a*)
13. (*a*)	14. (*a*)	15. (*c*)	16. (*c*)	17. (*b*)	18. (*b*).

24

Phylum Nematoda: Characters, Classification and Types

24.1 ASCHELMINTHES

Gegenbaur (1859) created a group **Nemathelminthes** to place some pseudocoelomate animals. However, **Grobben** (1910) introduced the term **Aschelminthes** in place of Nemathelminthes to avoid confusion. The Aschelminthes (Gr., *askes* = cavity; *helmins* = worm) are a heterogenous assemblage of marine and freshwater animals, such as gastrotrichs, rotifers and roundworms (nematodes). Although previously recognized as classes within the phylum Aschelminthes, each of eight former classes now designates a distinct phylum. The informal name aschelminthes, however, is still a convenient term for reference for entire assemblage.

Definition

Aschelminthes have in common a small body size, a pseudocoelom, bilateral symmetry, a cuticle (except rotifers and acanthocephalans), adhesive tubes; a lack of complete segmentation, a digestive tract with an anterior oral and a posterior anal opening, no circulatory or respiratory systems, usually with separate sexes, and with a protonephridial system of flame cells or flame bulbs (except nematodes and tardigrades). Nuclear numbers within species are constant.

General Characters

1. Most free-living aschelminthes are tiny worm-shaped animals, ranging from microscopic size to a centimeter in length, and their small body size is correlated with several familiar design features. For example, a well developed coelom and blood vascular system are absent.
2. Cilia are used for locomotion in several groups, but in those that have a thick cuticle, muscular form of locomotion predominates.
3. Protonephridia are the typical excretory organs.
4. Body is composed of relatively few cells — some aschelminths have as few as 1000 somatic cells. The rotifer *Epiphanes senta*, for example, has 958 nuclei (or cells) and the male of the roundworm *Caenorhabditis elegans* has 1031 cells.
5. Many species have an invariant and genetically fixed number of cells—a phenomenon called **eutely**. In eutelic animals, mitosis commonly ceases following embryonic development and growth continues only through increase in the size of cells.
6. Well formed **head** is absent, but **mouth** and sense organs are present.
7. An external cuticle is present in some phyla but absent in others.
8. In general, those that have a thin flexible cuticle or that lacks a cuticle altogether have antagonistic **circular** and **longitudinal muscles,** acting on a hydrostatic skeleton. Circular

musculature is absent in aschelminthes that have thick cuticle and the action of longitudinal musculature is antagonized by the elastic cuticle.

9. Many species have many **adhesive glands** which open to outside of the body via projecting cuticular tubes.
10. The digestive tube is often a complete tube bearing a mouth and anus. A specialized pharynx is always present.
11. Most aschelminthes are **pseudocoelomate animals,** meaning that they contain a fluid-filled cavity unlined by coelomic epithelium and which arose as a persistent embryonic blastocoel. This feature however has exceptions: (1) since many aschelminths have no body cavity at all, that is, they are **acoelomate animals,** *e.g.*, gastrotrichs, most nematodes and kinorhynchs, all are acoelomate protostomes; (2) a few aschelminths, such as rotifers and large roundworms, do indeed have an expanded, fluid-filled, connective-tissue cavity, the **pseudocoel;** (3) in acanthocephalans and tardigrades, in which the body cavity is voluminous; body cavity does not develop as a persistent blastocoel (a blastocoel is absent in the development of these taxa), but rather forms anew as an enlarged space in the connective tissue, exactly as haemocoel develops in arthropods and molluscs.
12. Pseudocoelomic fluid functions as hydrostatic skeleton and as a medium for internal transport, but a heart is absent and body movement circulates the fluid irregularly.
13. Nematodes, rotifers and tardigrades that inhabit the high stress environments of water film around soil particles and mosses are capable of remarkable degree of cryptobiosis (see Box 24.1).
14. Most aschelminths are dioecious, cleavage is strongly determinate and development may be direct or indirect.

Box 24.1 Cryptobiosis

Metazons that inhabit mosses, lichens, temporary ponds and damp soils periodically face the threat of dehydration, and other physical extremes. Often, such animals have evolved reproductive adaptations, such as resistant eggs, which allow their progeny to survive the extreme conditions. This same ability to resist and survive adverse conditions occurs in the adults of some remarkable species of roundworms, rotifers and tardigrades. As their surrounding environment worsen, the animals enter a death-like state of suspended animation, called cryptobiosis (meaning "hidden life") and are revived when conditions again become favourable. Cryptobiotic animals can remain dormant for a period of months to years–some for even as long as a century before reviving. While dormant they survive dessication and temperature extremes that seem antagonistic with life. There are records of cryptobiotic animals surviving temperatures ranging from 150°C (for a few minutes) to –200°C (for days). Clearly the ability to become cryptobiotic is advantageous to animals that live in briefly uninhabitable environments.

To successfully enter cryptobiosis, the animal must dry slowly, usually over several days, as would be typical in natural habitats. As the animals dry, they change their body shape to rninirnize surface area perhaps to retard the rate of desiccation. Many nematodes tightly coil their bodies, rotifers retract head and foot and adopt a spherical shape, and tardigrades contract into the shape of a barrel.

Cryptobiosis is truly a death-like state. Metabolism is almost immeasurably low, 0.01% of normal, or is entirely absent, and the water content of body decreases to less than 1%.

In the absence of metabolism and near total dehydration, cryptobiotic animals are essentially dead but revivable. If metabolic processes actually cease, then somehow the structural order essential to life must be preserved in a sort of stable crystallized condition until water returns and life resumes. Two molecules — **glycerol** and the diasaccharide **trehalose**, appear to be important in the formation and maintenance of the crystalline state. Both are synthesized as the animals slowly enter cryptobiosis. Glycerol is believed to protect the tissues against oxidation and to replace water bound to biological macromolecules. Trehalose replaces water in membranes. In substituting for water, the glycerol and trehalose act as spacers to preserve macromolecular organization during dehydration.

Classification

The Aschelminthes include the following **eight** phyla (**Ruppert** and **Barnes,** 1994):

Phylum 1. Gastrotricha. Freshwater and marine species; glide over bottom on ventral cilia; temporarily anchor by means of adhesive tubes that contain duo-gland organs; mostly feed on bacteria which they ingest by a sucking pharynx; hermaphrodite, internal fertilization, direct development; 40 species.

Examples. *Macrodasys*, *Chaetonotus*.

Phylum 2. Nematoda. Marine, freshwater and terrestrial; have long, tapered and cylindrical bodies; pharynx is muscular sucking type; teeth or stylets are present in plant-feeding nematodes, mostly are animal and plant parasites; 1200 species of roundworms.

Examples. *Ascaris*, *Toxocara*, *Entrobius*, *Trichinella*, *Wuchereria*, *Loa*, *Rhabditis*.

Phylum 3. Nematomorpha. Adults are free-living occur in freshwater or damp soil; larva are parasitic on arthropods; dioecious, internal fertilization; 320 species of horse-hair worms.

Examples. *Gordius*, *Nectonema*.

Phylum 4. Rotifera. Cuticle absent; ciliated corona present; pharynx modified into a special grinding structure, called **mastax;** mastax bears trophi; 1500 species.

Examples. *Philodina*, *Notommata*, *Conochilus*, *Collotheca*, *Brachionus*.

Phylum 5. Acanthocephala. Parasites of gut of aquatic and terrestrial vertebrates; retractile proboscis bearing curved spines; cuticle and digestive tract are absent; food is absorbed through general body surface; 1150 species of spiny- headed worms.

Examples. *Acanthocephalus*, *Polymorphus*.

Phylum 6. Kinorhyncha. Chitinous cuticle and body wall are segmented; body includes 13 segments or joints; **head, neck** and eleven segmented **abdomen;** mouth is at the end of a spiny, anterior, retractile head; dioecious; development direct; molting occurs with growth; 150 species.

Examples. *Echinoderes*, *Pycnophyes*.

Phylum 7. Loricifera. (It is most recently discovered phylum of animals; it was described in 1983 by Danish zoologist **R. M. Kristensen**). Minute marine species with cone-shaped anterior region (introvert) that can be retracted into lorica-encased abdomen; about 9 species.

Examples. *Nanaloricus*.

Phylum 8. Tardigrade. Tiny bear-like inhabitants of beaches, soils, mosses and lichens; body bears four pairs of telescoping stubby legs, each of which is tipped with claws or adhesive discs; a pair of pierching stylets is used in combination with a sucking pharynx for feeding; dioecious, direct development; molting occurs with growth; 600 species of water bears.

Examples. *Batillipes*, *Macrobiotus*.

Phylogeny

The aschelminths assemblage appears to represent multiple evolutionary lines. One line includes the gastrotrichs, nematodes and perhaps, the nematomorphs; another may include the rotifers and acanthocephalans. Loriciferans may be linked to the priapulids and kinorhynchs. Tardigrades appear to be arthropod relatives.

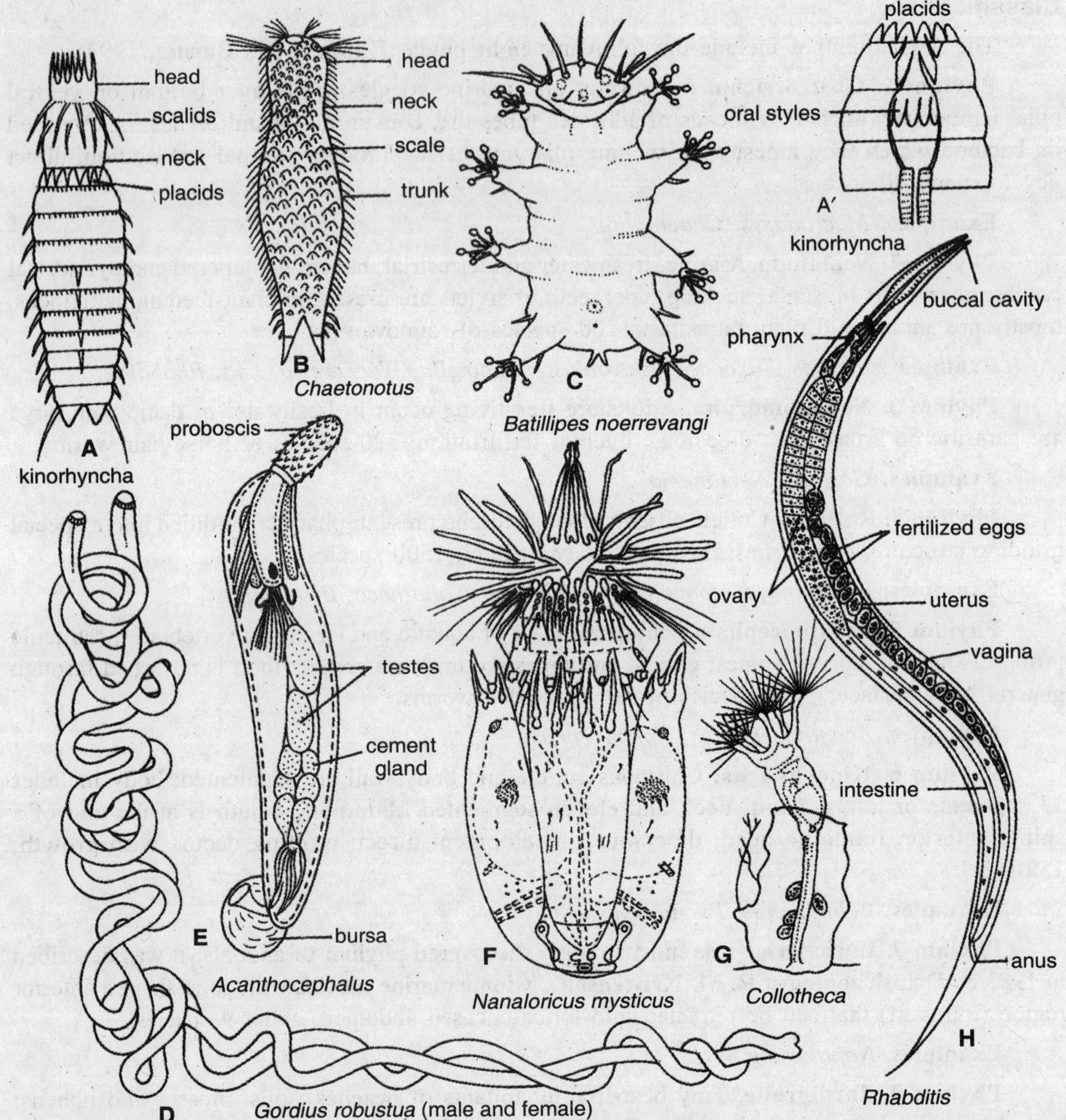

Fig. 24.1. Some Aschelminths.

24.2 PHYLUM NEMATODA

General Characters

1. Nematoda (Gr., *nema*=thread; *eidos*=form) include free-living (found in sea, freshwater and soil) or parasitic roundworms which are widely distributed.
2. The size and form of nematodes are adapted for living in interstitial spaces. Body of round worms are slender, elongated, cylindrical and tapering at both the ends. Most nematodes are microscopic in size.
3. Body is unsegmented, bilaterally symmetrical and not divisible into regions.

4. Mouth is located at somewhat rounded anterior end and is surrounded by lips and about 18 sensilla.
5. A **caudal gland** or **spinneret** is typical of many free-living nematodes.
6. Body wall consists of thick cuticle, epidermis (hypodermis) and longitudinal muscles. Circular muscles are completely absent.
7. Body is triploblastic. **Pseudocoel** is small or voluminous and filled with a fluid containing a variety of organic metabolites, including haemoglobin in some species but no circulating cells. A few **phagocytic cells** are permanently attached to the walls of the cavity and are important in internal defense.
8. Muscle fibres are arranged into four quadrants.
9. Coelom, circulatory system and respiratory system and cilia are absent.
10. Digestive system is well developed in nematodes. Nematodes are either carnivorous, phytophagous or microphagous (= feeding on diatoms, algae, fungi and bacteria).
11. Excretion of ammonium ions is performed by general body surface, excretory canal system or by excretory gland cells (called **renette**). Flame cells are absent.
12. Most nematodes move forward and rearward using sinuous eel-like undulations of the body. These undulations are in the dorsoventral plane and are produced by the alternate contraction of dorsal and ventral longitudinal muscles.
13. Sense organs are poorly developed, represented by small papillae and a pair of **amphids** on each side of head and a pair of **phasmids** at the posterior end.
14. Sexes are separate. Sexual dimorphism is well marked. In male, the tail is curved and tail bears a pair of pointed, curved, blade-like **copulatory spicules.** Sperms lack flagella and are amoeboid.
15. Fertilization is internal; parthenogenesis is common. Cleavage is strongly determinate. Development is direct.

Outline Classification

Phylum Nematoda

Classes (2)

1. Apasmida (Adenophorea)

Orders (6)

1. **Enoploidea**
   Examples. *Enoplus, Anticoma*
2. **Mermithoidea**
   Example. *Mermis, Allomermis*
3. **Dorylaimoidea**
   Examples. *Dorylaimus*
4. **Desmoscolecoidea**
   Example. *Desmoscolex, Tricoma*
5. **Monohysteroidea**
   Examples. *Monohystera, Siphonolaimus*
6. **Chromdoroidea**
   Examples. *Cromadora, Richtersia*

2. Phasmida (Secernentea)

Orders (10)

1. **Rhabditoidea**
   Examples. *Rhabditis, Heterodera*
2. **Rhabdiasoidea**
   Example. *Rabdias, Strogyloides*
3. **Oxyuroidea**
   Examples. *Oxyuris, Entrobus*
4. **Ascaroidea**
   Example. *Ascaris, Neoascaris*
5. **Strongyloidea**
   Examples. *Ancylostoma, Necator*
6. **Spiruroidea**
   Examples. *Rictularia, Thelazia*
7. **Dracunculoidea**
   Examples. *Dracunculus, Micropleura*
8. **Filarioidea**
   Examples. *Wuchereria, Loa*
9. **Tricuroidea**
   Examples. *Trichuris, Trichinella*
10. **Dioctophymoidea**
    Examples. *Dioctophyme, Hystrichis*

Detailed Classification

Phylum Nematoda is divided into **two** classes on the basis of presence or absence of phasmids.

Class 1. Aphasmida (Adenophorea)

1. Phasmids (caudal sensory organs) are absent.
2. Amphids (anterior sensory organs) are not pore like.
3. Cephalic organs, setae and hypodermal glands present.
4. Epidermal cells uninucleate. Cuticle four layered.
5. Caudal adhesive glands are present.
6. Usually two testes are present in males.
7. Most species are free-living; some are parasitic.

Class 2. Phasmida (Secernentea)

1. Phasmids are present.
2. Amphids open to outside by pores.
3. Cephalic sensory organs are present. Setae or papillae are not found in females.
4. Epidermal cells uni- or multinucleate; cuticle two or four layered.
5. Males contain single testis. Copulatory bursa is present in males.
6. Caudal adhesive glands are absent.
7. Most species are parasitic; free-living species mainly inhabit soil.

Class 1. Aphasmida

Class Aphasmida is divided into following **six** orders:

Order 1. Enoploidea

1. Mainly free-living and marine and of moderate size.
2. Anterior end with 6 labial papillae and with 10 to 12 sensory bristles.
3. Amphids are pocket-like, elongated or pore-like (called **cyanthiform**).
4. Cuticle is not ringed but with cuticular bristles.
5. Pharynx is a simple straight tube. Intestine ends in a blind pouch.

   Examples. *Enoplus*, *Anticoma*, *Synonchus*, *Triphla*, *Trilobus*.

Order 2. Mermithoidea

1. Long, slender and large sized juveniles are parasitic in invertebrates (mainly insects). Adults are free-living (in soil or water).
2. Anterior end with 6 labial papillae and no sensory bristles.
3. No buccal cavity is present, pharynx is long and intestine ends in a blind pouch. Adults do not feed, but live on food stores acquired during the parasitic existence.
4. Cuticle smooth and without bristles.
5. Amphids cyanthiform or reduced.

   Examples. *Mermis*, *Agameris*, *Paramermis*, *Allomermis*.

Order 3. Dorylaimoidea

1. Small sized, free-living nematodes found in soil and freshwater.
2. Anterior end with two circlets of labial papillae of 6 and 10 each.
3. Cuticle is smooth—without rings and bristles.

4. Buccal cavity with a hollow protrusible **spear** which is mounted on pharynx and used when sucking juices from plants and animal food organisms.
5. Rear part of pharynx is enlarged.
6. Amphids are sickle-shaped.

Examples. *Dorylamius*, *Tylencolaimus*.

Order 4. Desmoscalecoidea

1. Small sized marine nematodes.
2. Anterior end with four sensory bristles.
3. Head is armoured and distinct from body.
4. Amphids are crescent-shaped or hemispherical.
5. Cuticle is heavily ringed with prominent bristles throughout or in restricted areas.
6. Buccal cavity is reduced.

Examples. *Desmoscolex*, *Tricoma*.

Order 5. Monohysteroidea

1. Small sized, free-living, mostly marine and some are freshwater and some inhabit soil.
2. Anterior end with 4, 6, 8, or many sensory bristles.
3. Cuticle smooth or slightly ringed with scattered bristles.
4. Buccal capsule in the shape of a long tube.
5. Amphids are circular.

Examples. *Monohystera*, *Siphonolaimus*, *Cylindrolaimus*.

Order 6. Chromadoroidea

1. Small sized, free-living and mostly marine.
2. Anterior end with six lips and 10-papillae.
3. Amphids are spiral.
4. Ornamented cuticle is ringed and contains bristles.
5. Pharynx with a posterior bulb; usually with teeth in buccal cavity.

Examples. *Chromadora*, *Richtersia*, *Spilophorella*.

Class 2. Phasmida

Class Phasmida includes the following **ten** orders:

Order 1. Rhabditoidea

1. Medium sized worms. Free-living or parasitic in animals or plants.
2. Cuticle is smooth or ringed.
3. Pharynx with two bulbs.
4. Sensory bristles in the form of papillae arranged in two rings.
5. Amphids are reduced.
6. Phasmids are present in free-living forms.
7. Caudal glands are absent.
8. Males with copulatory spicules.

Examples. *Rhabditis*, *Heterodera* (lives in gall on the roots of plants), *Diplogaster*, *Tylenchus*, *Anguina*.

Order 2. Rhabdiasoidea

1. Medium sized worms. Parasite of vertebrates.
2. Cuticle is smooth.
3. Pharynx with a reduced bulb.
4. Life cycle includes both parasitic and free-living stages.
5. Parasitic stage is represented by either a protandrous hermaphrodite or a parthenogenetic female.

   Examples. *Rhabdias* (lung parasite of frog), *Strongyloides* (intestinal parasite of human beings).

Order 3. Oxyuroidea

1. Small sized, pin-shaped parasitic worms.
2. Mouth is surrounded by 3–6 simple lips.
3. Pharynx contains a posterior bulb, containing valves.
4. Female contain long, slender tail.
5. Males with one spicule or two equal spicules.

   Examples. *Entrobus* (Human pinworm), *Oxyuris*, *Heterakis*, *Histomonas*.

Order 4. Ascaroidea

1. Large sized, endoparasitic in digestive tract of vertebrates.
2. Mouth is surrounded by three prominent lips.
3. Pharynx is without bulb-like enlargement. Buccal capsule is absent.
4. Tail is ventrally coiled in males. Tail in female is blunt.
5. Males have two nearly equal sized copulatory spicules.

   Examples. *Ascaris*, *Neoascaris*, *Toxocara*.

Order 5. Strongloidea

1. Moderate sized parasitic worms (*e.g.*, intestine of human beings).
2. Mouth without lips.
3. Pharyngeal bulb absent. Buccal capsule is well developed and has teeth.
4. Males contain a expanded **copulatory bursa** which is supported by 13 muscular rays.
5. Females contain **ovijector**.

   Examples. *Necator* (commonly called **American Killer**), *Ancylostoma*, *Strongylus*.

Order 6. Spiruroidea

1. Moderate sized parasites.
2. Mouth with two lateral lips.
3. Pharynx has anterior muscular and posterior glandular portion; it lacks bulb.
4. The sensory papillae are present on the head. These are arranged in two circles of 6 and 4 or 8.
5. Males lack bursa but contain unequal and dissimilar copulatory spicules.
6. Males with spirally coiled tails.

   Examples. *Thelazia*, *Rictuillria*, *Oxyspirura*.

Order 7. Dracunculoidea

1. Large sized parasites of vertebrates.
2. Mouth without lips and surrounded by 4 or 8 papillae.

3. Pharyngeal bulb is absent. Pharynx is muscular anteriorly and glandular posteriorly.
4. Viviparous.
5. Filiform and equal sized copulatory spicules are present; bursa is absent.
6. Females are large in size. Males are quite small.

Examples. *Dracunculus* (Guinea worm), *Philometra*, *Micropleura*.

Order 8. Filarioidea

1. Filiform, slender, moderate to large-sized worms.
2. They are parasites of body fluids and tissues of vertebrates, require a blood sucking insect as intermediate host.
3. Six labial papillae are present.
4. Mouth is without lips.
5. Pharynx is divided into an anterior muscular and posterior glandular parts; bulb is absent.
6. Males are smaller than females. Males have coiled tails.
7. In males, the bursa is absent and copulatory spicules are dissimilar and unequal.
8. In females, vulva is present near the anterior end.

Examples. *Wuchereria* (filaria), *Loa* (eye worm).

Order 9. Trichuroidea

1. Moderate sized parasitic worms of vertebrates.
2. Anterior part of body is filiform.
3. Mouth is without lips.
4. Pharynx is simple and slender and without bulb.
5. Males are without copulatory apparatus or contain an eversible spicules or cirrus.

Examples. *Trichinella*, *Trichuris*, *Capillaria*.

Order 10. Dioctophymoidea

1. Mouth is without lips but is surrounded with 6, 12, or 18 papillae.
2. Pharynx is elongated and without bulb.
3. Female has single ovary.
4. Male contains only one or no copulatory spicule.

Examples. *Dictyophyme*, *Hystrichis*, *Eustronglides*.

24.3 TYPE STUDIES

1. *Rhabditis*

Classification

Phylum	–	**Nematoda**
Class	–	**Phasmida**
Order	–	**Rhabditoidea**
Genus	–	***Rhabdits***
Species	–	***elegans***

Habit and Habitat

The genus *Rhabditis* contains numerous species normally found in soil, organic matter or water and frequently in faeces of human beings and animals. *Rhabditis pellio* have, on few occasions, been found living in human vagina and larvae escaping in urine. *R. hominis* have been detected from stools of human beings and animals.

Structure

Adult male and female measures 2.0 and 1.7 mm in length respectively. Excretory system is H-shaped. Reproductive system is J-shaped. Male *Rhabditis* possesses single tubular testis and female has two ovaries. Female also contains oviducts, seminal receptacle and short uteri. Life cycle is simple. Adults breed as protandric hermaphrodites.

3. *Trichinelia spiralis*

Classification

Phylum	–	**Nematoda**
Class	–	**Phasmida**
Order	–	**Trichuroidea**
Genus	–	***Trichinella***
Species	–	***spiralis***

Habit and Habitat

Trichinella is commonly called **trichina worm**. It is an endoparasite of human beings and other vertebrates such as pigs and rats. The adult worm lives in the duodenal and jejunal mucosa of the intestine but embryos become encysted in striated muscles (*e.g.*, diaphragm, intercostals, etc.). *T. spiralis* is cosmopolitan in distribution but mostly in Europe and America.

Structure

They are slender worms. Their body is pointed anteriorly and fleshy posteriorly (Fig. 24.2). Females are 3 to 4 mm long and males are 1.5 mm long.

The long pharynx has an anterior muscular part and posterior glandular part. Pharynx contains large granular cells, called **stichosome.** Males are with a pair of conical appendages at the posterior end. They have no copulatory spicules.

Mode of Infection

The fertilized female bores through the intestine into spiral lymph spaces and gives birth to juveniles (about 10,000 at a time). They can develop further into the adult worm only if taken up by another rat or pig. Infection normally passes from rat to rat and pig to pig.

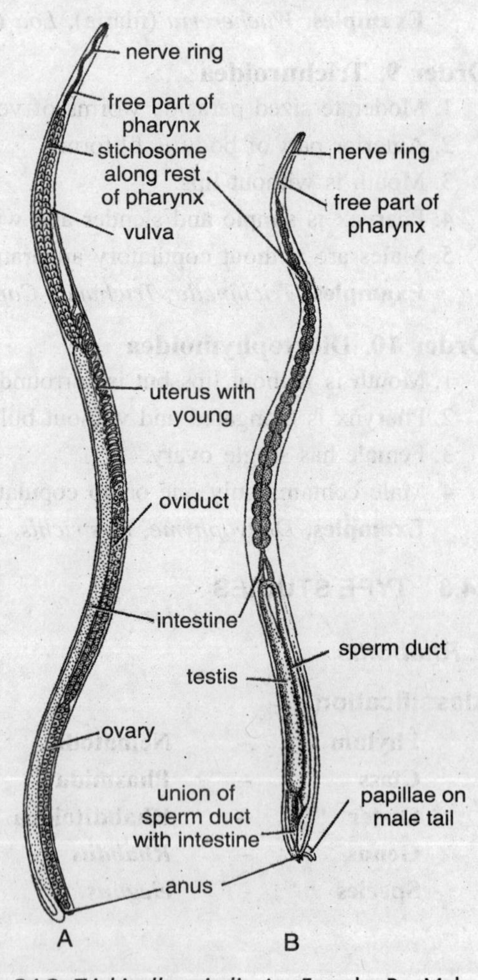

Fig. 24.2. *Trichinella spiralis*. A—Female; B—Male.

The pigs become infected through the ingestion of infected pork scraps (offal) from their swill. Inside the pig these undergo the same process of development and form lemon-shaped cysts (Fig. 24.3) in the muscles. The infected pig muscles containing larvae when eaten up by human beings cause infection.

Pathogenicity

Trichinella causes a very serious and dreaded disease known as **trichinosis** which is characterized by **myocarditis** or **myositis.** It may cause paralysis of the muscles. Only 5 per cent of heavy infection end fatally and the death is usually due to the paralysis of cardiac muscles.

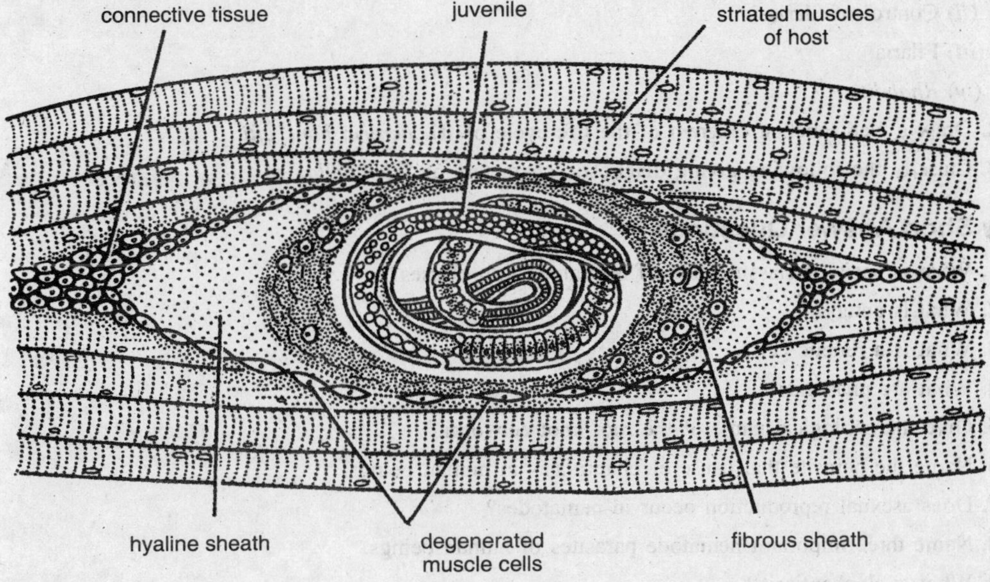

Fig. 24.3. *Trichinella spiralis.* Cyst in the striated muscle of pig.

Treatment

Daily doses of 2 to 3 gms of piperazine citrate are effective in removing the adults of *Trichinella* from human beings.

QUESTIONS

Long Answer Questions

1. Describe general characters of Aschelminthes and classify this group up to phyla/classes.

2. Give an outline classification of phylum Nematoda up to orders.

3. Give distinguishing features of phylum Nematoda. Classify up to order giving character and example of each.

4. Classify Nematoda or Nemathelminthes mentioning the characters of its orders. Cite an example of each order.

5. Give zoological name of Filarial worm. In India which animal is its vector, give its zoological name. What is meant by the periodicity of microfilaria? Describe pathogenicity, treatment and prevention of disease caused by worm.

6. Describe the morphology of the adult *Wuchereria bancrofti.* Mention the disease caused by it and the methods of its control.

7. Give an account of the structure, life history and mode of infection of *Wuchereria bancrofti*.
8. Give an account of the life-cycle of *Wuchereria bancrofti*, adding a note on day and night periodicity.
9. Describe sequential stages in life cycle of *Wuchereria bancrofti*.

Short Answer Questions

1. Write short notes on the following:
   (*i*) Elephantiasis
   (*ii*) Control of filariasis
   (*iii*) Filaria
   (*iv*) *Rhabditis*
2. Draw a well labelled diagram of the life cycle of *Wuchereria bancrofti*.
3. Classify the following giving diagnostic characters of *Trichinella*.

Very Short Answer Questions

1. What is basis of classification of Nematoda into classes?
2. What is eutely?
3. What is a renette cell?
4. What are phasmids?
5. Give the zoological name of the root knot eelworm.
6. Name the parasite that cause elephantitiasis in human beings.
7. Does asexual reproduction occur in nematodes?
8. Name three important nematode parasites of human beings.
9. What is elephantiasis?

Fill in the Blanks

1. Nematodes are the first to have a body cavity. It is called ....
2. In .... epidermis is syncytial.
3. Body wall musculature of roundworms consists of ....
4. A nematode muscle cell consists of .... part and cytoplasmic part.
5. Alimentary canal is .... in nematoda.
6. Sexes are.............. in nematodes.
7. Nerve cords in nematodes are located in ....
8. The excretory tubes of nematodes lie in the lateral .... chords.
9. Cloaca is found in .... roundworms.
10. The phylum Nematoda is divided into two classes: .... and Phasmida.

Multiple Choice Questions

1. The eyeworm of human beings is
   (*a*) *Wuchereria* (*b*) *Trichuria*
   (*c*) *Loa loa* (*d*) None

2. Nematode epidermis consists of
(*a*) cuticle (*b*) epidermis
(c) muscular layer (*d*) all of these

3. Number of cells in the various organs of the body is fixed in
(*a*) flatworms (*b*) segmented worms
(*c*) roundworms (*d*) walking worms

4. Gonads in male roundworms consists of
(*a*) one pair of testes (*b*) two pairs of testes
(*c*) three pairs of testes (*d*) unpaired testes

5. The nematode *Wuchereria bancrofti* belongs to order
(*a*) Ascaroidea (*b*) Strongyloidea
(*c*) Filarioidea (*d*) Oxyuroidea

6. Filarial worm is transmitted by
(*a*) mosquito (*b*) tse-tse fly
(*c*) sandfly (*d*) kissing bug

7. *Enterobius vermicularis* inhabits the human
(*a*) large intestine (*b*) caecum
(*c*) appendix (*d*) all of these

8. Trichinosis is caused by
(*a*) *Trichuris* (*b*) *Trichinella*
(*c*) *Ancylostoma* (*d*) None of these

9. Which of the following is a free living nematode?
(*a*) *Trichuris* (*b*) *Trichinella*
(*c*) *Rhabditis* (*d*) *Wuchereria*

10. Microfilaria is the larva of
(*a*) tape worm (*b*) pin worm
(*c*) round worm (*d*) filarial worm

11. The place where adult *Wuchereria bancrofti* lives is the
(*a*) human blood vessel (*b*) human lymph gland
(*c*) muscle of *Culex* (*d*) salivary gland of *Culex*

12. Elephantiasis spreads through the vector
(*a*) *Anopheles* (*b*) *Culex*
(*c*) *Musca domestica* (*d*) Flat worm

13. Organ system grade of body organisation is present in
(*a*) porifera (*b*) platyhelminthes
(*c*) aschelminthes (*d*) cnidaria

14. Intermediate host of *Wuchereria bancrofti* is
(*a*) male *Anopheles* (*b*) female *Anopheles*
(*c*) female *Culex* (*d*) male *Culex*

15. Which of the following is a digenetic nematode parasite?
(*a*) *Fasciola hepatica* (*b*) *Taenia solium*
(*c*) *Wuchereria bancrofti* (*d*) All of these

16. Transmission of *Wuchereria* to new host occurs in
(*a*) larval stage (*b*) encyst stage
(*c*) juvenile stage (*d*) none of the above

ANSWERS

Very Short Answer Questions

1. Phylum Nematoda is classified into two classes on the basis of presence or absence of phasmids, the sense organs. Two classes of Nematoda are: Aphasmida and Phasmida.
2. Constancy of the number of cells or nuclei in the various organs and also in body, *e.g.*, nematodes.
3. An excretory cell.
4. Phasmids are unicellular glandulo-sensory structures in the tail region of most nematodes.
5. *Meloidogynae marioni.*
6. *Wuchereria bancrofti.*
7. No.
8. *Wuchereria, Ancylostoma, Ascaris.*
9. **Elephantiasis** or **filariasis** is a disease of human beings caused by filaria worm *Wuchereria bancrofti.* This round worm inhabits the lymphatic vessels of human beings and infection is spread through *Culex* mosquito. This disease leads to the excessive swelling of legs, scrotum and breasts as a result of blockage of lymphatic circulation by the parasite. Most effective drug for elephantiasis is Hetrazan (diethylcarbmazine).

Fill in the Blanks

1. Pseudocoel	**2.** Nematoda	**3.** Longitudinal	**4.** Fibrillar
5. Complete	**6.** Separate	**7.** Epidermis	**8.** Epidermal
9. Male	**10.** Aphasmida.		

Multiple Choice Questions

1. (*c*)	**2.** (*d*)	**3.** (*c*)	**4.** (*d*)
5. (*c*)	**6.** (*a*)	**7.** (*d*)	**8.** (*b*)
9. (*c*)	**10.** (*d*)	**11.** (*b*)	**12.** (*b*)
13. (*c*)	**14.** (*c*)	**15.** (*c*)	**16.** (*c*).

25

Ascaris (Phylum Nematoda)

25.1 CLASSIFICATION

Phylum – **Nematoda**
Class – **Phasmida**
Order – **Ascaroidea**
Genus – ***Ascaris***
Species – ***lumbricoides***

25.2 HABIT AND HABITAT

Ascaris lumbricoides is the intestinal parasite (endoparasite) of human beings. It is also found in the intestine of apes, pigs, cattle, sheep, etc. It remains free in intestine and feeds on the partly digested food (chyme) of the host.

Though it is found in all parts of the world, its infection is much more common in India, China, Korea, Phillippines and Pacific islands.

25.3 STRUCTURE

Body of the *Ascaris* is large, cylindrical and tapers at both ends. It is pinkish white or yellowish white in colour. Sexes are separate; sexual dimorphism is well marked (Table 25.1). The female is larger measuring 20–40 cm in length and 4–6 mm in diameter. Male is smaller and measures 15–30 cm in length and 2–4 mm in diameter (Fig. 25.1). The tail of male is sharply curved and contains 50 pairs of preanal and 5 pairs of post-anal genital papillae. In female, the tail is straight.

Table 25.1. External features showing sexual dimorphism in male and female Ascaris.

	Male		*Female*
1.	Body is smaller, 15–30 cm long and 2–4 mm thick.	1.	Body is larger; 20–40 cm long and 4–6 mm thick.
2.	Tail end is curved ventrally.	2.	Tail end is straight.
3.	Anus and genital pore are common (cloaca).	3.	Anus and genital pore are separate (cloaca absent).
4.	A pair of needle-like and protrusible penial setae are present in cloaca.	4.	No penial setae are present.
5.	Preanal and postanal papillae are present upon ventral body surface in cloacal region to help in copulation.	5.	No such papillae are present.

At the anterior end three broad **lips** surround the **mouth** (Fig. 25.1C). Each lip contains fine **denticles** or **teeth** on its inner margin and **sensory papillae** on outer surface.

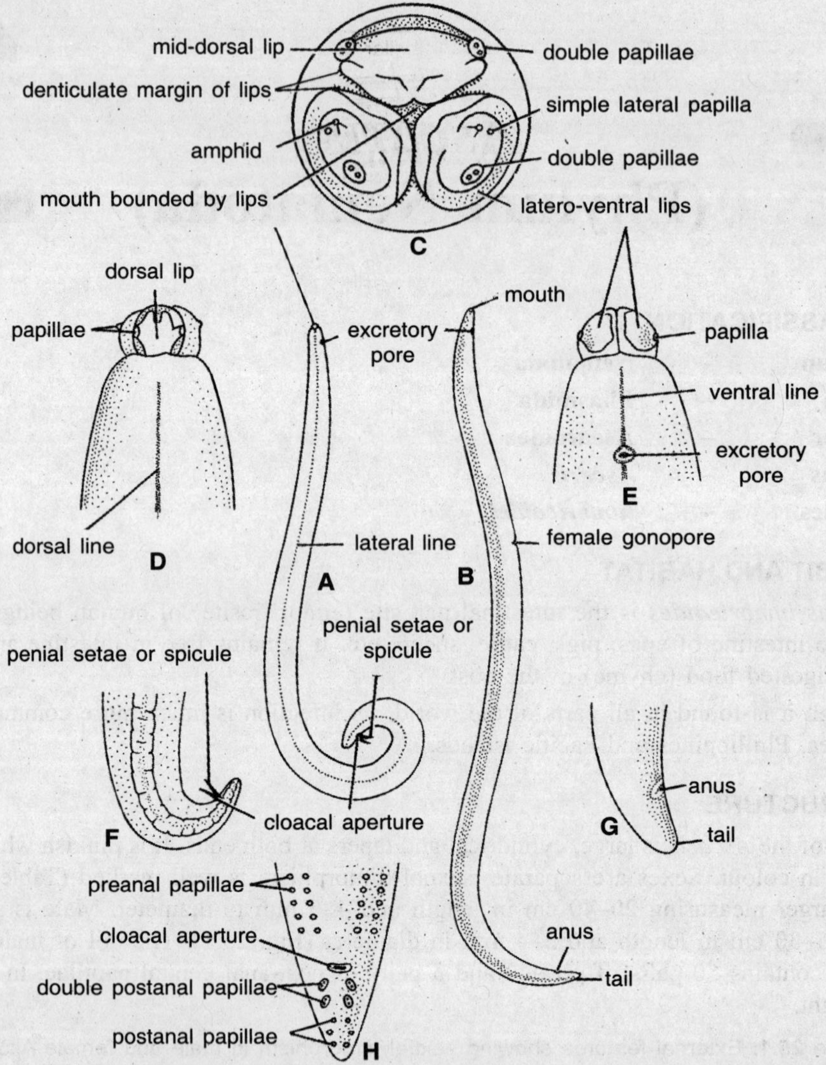

Fig. 25.1. *Ascaris lumbricoides*. A—Male; B—Female; C—Surface view of mouth and lips; D—Anterior end in dorsal view; E—Anterior end in ventral view; F—Posterior (tail) end of male, G—Tail end of female, H—Posterior end of male in ventral view showing papillae.

25.4 BODY WALL

The body wall of *Ascaris* consists of following three main layers (Fig. 25.2):

1. Cuticle;
2. Epidermis or hypodermis; and
3. Muscular layer.

1. Cuticle. The body of *Ascaris* is covered with a thick and tough non-cellular **cuticle** having transverse striations. Cuticle is a complex (Box 25.1) exoskeleton-like structure which provides protection to the body from mechanical injuries and also provides enough flexiblity and elasticity to body wall. Cuticle is followed by a thin-layer of **basement membrane.**

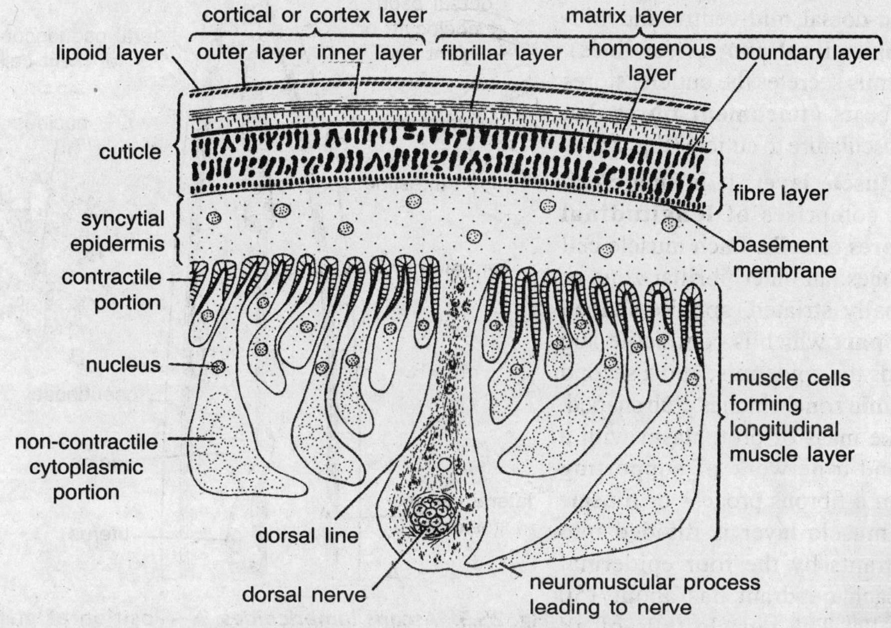

Fig. 25.2. *Ascaris lumbricoides*. T.S. body wall.

Box 25.1 Cuticle of Nematodes

The cuticle of body wall of nematodes is quite complex. It has **three** main layers, which contain **collagen** as well as other compounds and an **epicuticle** (Fig. 25.3). The thin epicuticle which may exhibit **quinone** (a golden yellow crystalline organic compound obtained from benzene and its homologs) **tanning,** cover the outer cuticular layer (cortex) externally. Quinone (and polyphenols) make epicuticle resistant to the digestive enzymes of the host.

The **outer layer** of cuticle is typically annulated (ringed) and sometimes divided into **outer cortex** and **inner cortex**.The **median layer** variedly has a uniform granular structure **os struts, skeletal rods, fibrils** or **canals,** depending on the species. The **basal layer** may be striated or laminated or may contain collagen fibres in a crossed helical arrangement.

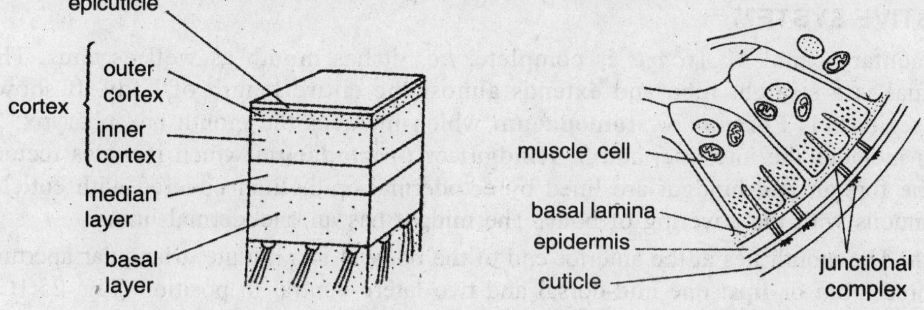

Fig. 25.3. Different parts of cuticle of nematodes.

Fig. 25.4. T.S. body wall of a nematode showing attachment of muscular layer with cuticle.

In nematodes, **growth** is accompanied by four moulting of the cuticle. Beginning at the anterior end, the old cuticle separates from the underlying epidermis and a new cuticle is secreted at least in part. The old cuticle is shed, sometimes in fragments. Moulting does not occur after the worm becomes adult, but the cuticle continues to expand as the animal grows.

2. Epidermis. The epidermis is syncytial and bulges into the pseudocoel to form mid-dorsal, mid-ventral and mid-lateral **longitudinal chords** (or lines). The epidermis secretes the cuticle, stores nutrients, bears **attachment fibres** that link the musculature to cuticle (Fig. 25.4).

3. Muscle layer. It is single-cell thick and comprises of **longitudinal muscle fibres** or **cells.** Each muscle cell has two zones: an outer fibrillar zone of longitudinally striated, spindle-shaped **muscular part** which is contractile and lies towards the epidermis; and a second **protoplasmic zone** which is club shaped, bladder-like mass of protoplasm with a nucleus and a network of supporting fibrils from a fibrous process or **muscle tail.** The muscle layer is divided into four quadrants by the four epidermal chords. Each quadrant has about 150 muscle cells. The muscle tails of all the muscle cells of two dorso-lateral groups join the dorsal nerves located in the mid-dorsal chords. Likewise, muscle tails of the two ventro-lateral muscle cell groups join the ventral-nerve cord located in the midventral longitudinal chord.

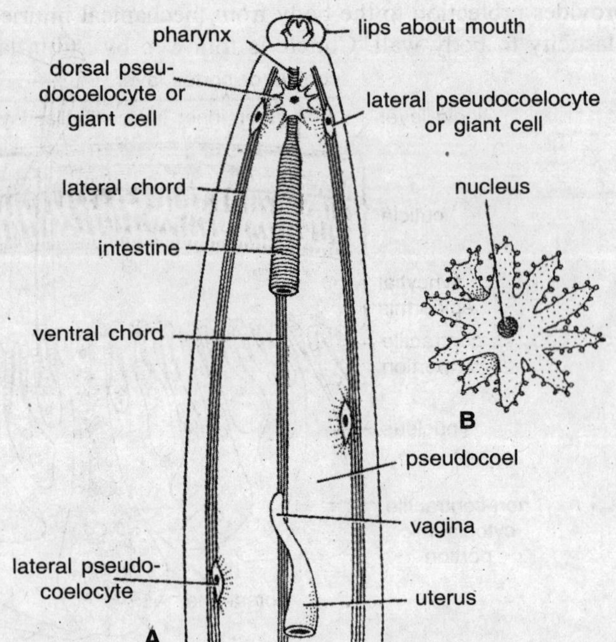

Fig. 25.5. *Ascaris lumbricoides*. A—Position of giant cells or pseudocoelomocytes (=pseudocoelocytes) in pseudocoel; B—single pseudocoelomocyte enlarged.

Pseudocoel

There is a space enclosed by the body wall, called **pseudocoel.** It is not lined by coelomic epithelium. The pseudocoel contains a clear fluid, fibrous tissue and four giant cells-called **pseudocoelomocytes** (Fig. 25.5) . Food, oxygen and waste products are transported by the pseudocoelomic fluid.

25.5 DIGESTIVE SYSTEM

The alimentary canal of *Ascaris* is complete, *i.e.,* it has mouth as well as anus. The alimentary canal is a straight tube and extends almost the entire length of body. It shows three distinct regions: 1. **Foregut** or **stomodaeum** which includes the mouth and pharynx; 2. **Midgut** which includes the intestine; and 3. **Hindgut** or **proctodaeum** which includes rectum (Fig. 25.6). The foregut and hindgut are lined by ectodermal epithelium covered with cuticle, which is continuous with the covering of body. The midgut has an endodermal lining.

1. Mouth. The mouth lies at the anterior end of the body. it is a minute, triangular aperture bounded by three **labia** or **lips**: one mid-dorsal and two latero-ventral in position (Fig. 25.1C). The lips have finely denticulate margin (*i.e.*, lips have fine teeth on their margins) and bear sensory **papillae**. The dorsal lip has two double papillae, while each latero-ventral lip has one double papilla, one simple papilla and a reduced sense organ called **amphid**. Amphid is a olfactory chemoreceptor.

2. Pharynx. The terminal mouth leads into cylindrical thick-walled **pharynx** or **oesophagus** which has a posterior swelling called end bulb which is provided with valves. The pharynx has muscular walls having radial muscle fibers which dilate its lumen (Fig. 25.6B). The pharynx has three large branched gland cells which open by cuticular ducts into the lumen; these are in fact, **pharyngeal** or **oesophageal glands**. The lumen or cavity of the pharynx has three deep longitudinal grooves lined by cuticle and in the transverse section the lumen of pharynx appears **triradiate**, connective tissue fibres arise from each three internal grooves and go to the cuticle covering the pharynx, they maintain the triradiate shape of the lumen (Fig. 25.7).

3. Intestine. The pharynx opens posteriorly into a thin-walled dorso-ventrally flattened **intestine**, which extends almost the entire length of the body. It is formed of a single layer of columnar epithelial cells lined externally by a thick layer of cuticle. The free inner margin of each cell is produced into several finger-like projections, the **microvilli**. They form a kind of tightly packed **brush border** (Fig. 25.6C, D). There is no muscle layer in the wall of intestine.

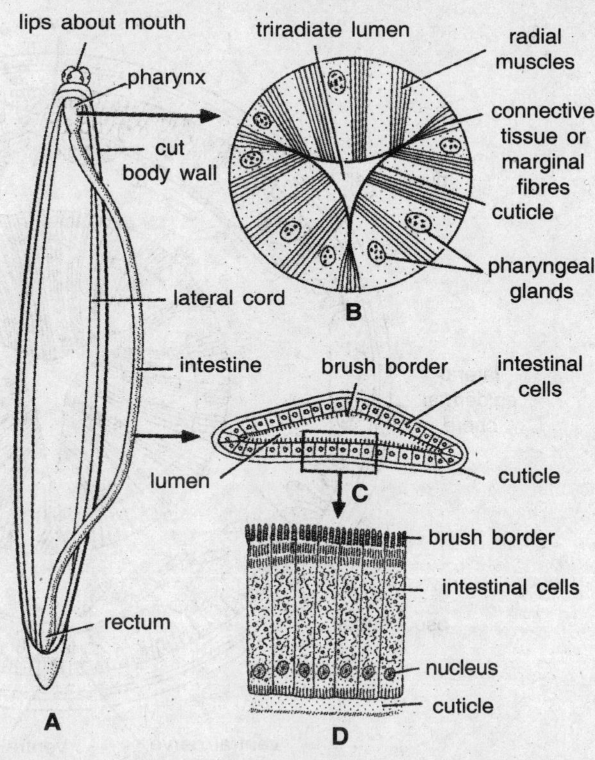

Fig. 25.6. *Ascaris.* A—Alimentary canal; B—Pharynx in T.S.; C—Intestine in T.S.; D—A portion of intestinal wall showing brush border.

4. Rectum. The intestine is followed by a short, relatively narrow chamber, called **rectum.** Its wall consists of tall columnar cells and lined internally by cuticle and externally by few muscle layers. In male, the rectum opens out by **cloaca** because it receives the ejaculatory duct but in female, the rectum opens out by a transverse aperture, the **anus** (Fig. 25.1G, H). The anal aperture is guarded by anterior and posterior lips and is provided with a few special dilator muscles running from the rectum to the body wall, called **depressor ani.** Their contraction from time to time causes the faecal matter to be discharged out. The rectum also has large unicellular **rectal glands**; three in the female and six in the male.

Food, Feeding and Digestion. Food of *Ascaris* comprises blood, tissue exudes and partly digested fluid food (chyme) from the host's intestine. Ingestion of food is accomplished by sucking action of pharynx. Sucking action of the pharynx is due to its periodic dilation by contraction of its radial muscles. Movement of the food down the pharynx and intestine is brought about partly by the pumping action of pharynx and partly by the contraction of the body wall. Absorption of food occurs in the intestine. Here the food diffuses through the intestine wall into the fluid filling the pseudocoelom. From the latter, the food is absorbed by the tissues and organs in which assimilation takes place. The unabsorbed food forms faeces in the rectum. The depressor ani muscles opens the anus (cloaca aperture in the male) periodically and the faeces is thrown out by the pressure developed in the intestine by pumping action of the pharynx.

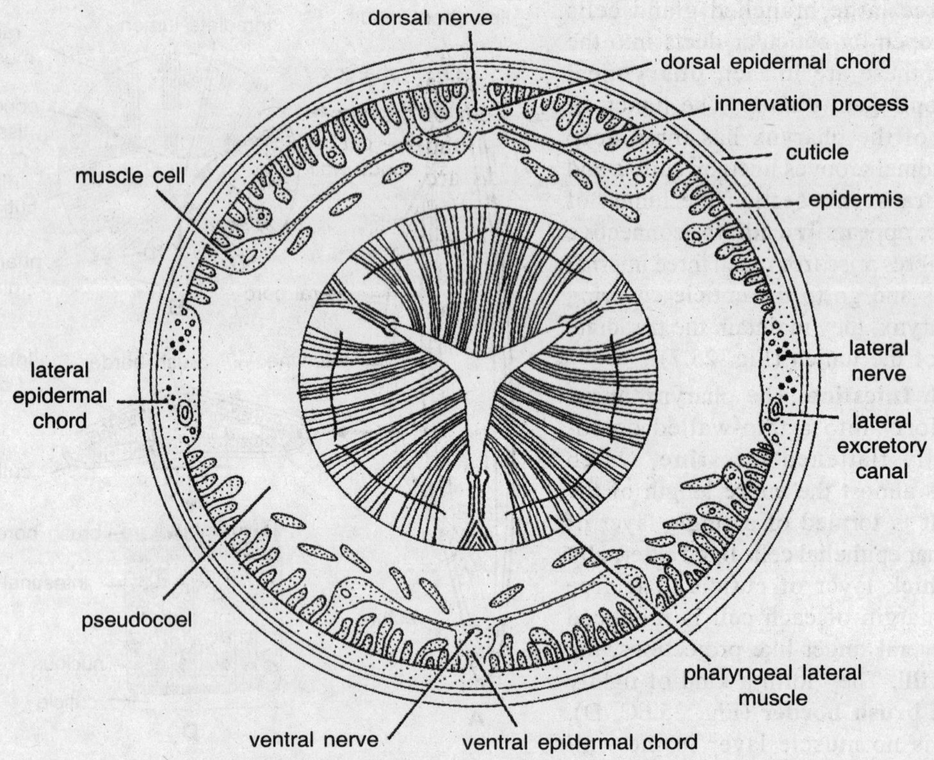

Fig. 25.7. *Ascaris lumbrircoides*. T.S. body through pharynx.

Some digestion in *Ascaris* is known to occur in the intestine. It starts in an extracellular manner in the intestine lumen and completed intracellularly. For the former, the pharyngeal gland cells and intestinal cells secrete enzymes, such as **proteases, amylase** and **lipase**. For intracellular digestion, the intestinal cells capture food particles into food vacuoles and are digested by lysosomal enzymes. Excess of food is stored mainly as glycogen and partly as lipids in the epidermis, intestinal epithelium and non-muscular parts of muscle cells. In *Ascaris*, intestine also synthesize **yolk** which is exported via pseudocoelomic fluid to the ovary for devloping oocytes.

25.6 RESPIRATION

In *Ascaris*, respiration is normally **anaerobic**, because the intestinal contents are deficient in free oxygen. In such respiration, glucose breaks down into carbon dioxide and a variety of fatty acids, mainly **valerianic acid**. It is these fatty acids, mixed perhaps with traces of other substances, which give a characteristic smell, rather like a tinned pineapple to the living *Ascaris* (see **Dhami** and **Dhami**, 2001). Energy released during respitation is utilised by the worm in its vital processes, such as locomotion, sucking of food, growth, reproduction, etc. Carbon dioxide leaves the body by diffusion through the skin.

25.7 EXCRETORY SYSTEM

Excretory system of *Ascaris* is H-shaped and consists of two slender **longitudinal excretory canals** (Fig. 25.8), one embedded in each lateral epidermal chord of body wall. A little behind the anterior end of body, both excretory canals are connected by a bridge-like network of several short and slender **transverse canals** located ventrally to oesophagus (pharynx). Both longitudinal canals extend still forwards for a short distance from the connecting bridge. Thus, a H-shaped excretory system is formed. A short **terminal duct** (called **common canal**) also extends forward from one side of transverse canal and opens out by a ventromedian **excretory pore.** The whole system of canals is formed of a single highly elongated and complicated **"giant cell"** in which the canals are formed as intracellular excavations. This cell is called **renette cell.** The nucleus of this cell is located upon the left longitudinal canal close to transverse canal.

Ascaris is both **ammonotelic** and **ureotelic,** *i.e.*, it excretes both ammonia and urea. Probably, the hydrostatic pressure of pseudocoelomic fluid causes ultrafiltration of excretory fluids from it into the excretory canals. In fact, the excretory system of *Ascaris* is osmoregulatory also.

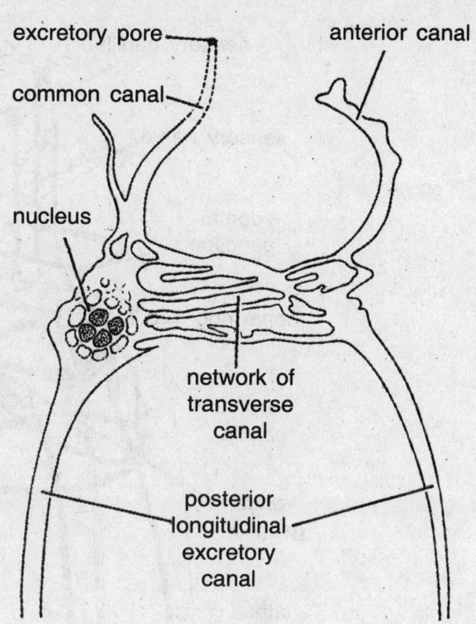

Fig. 25.8. *Ascaris lumbricoides*. Excretory system.

25.8 NERVOUS SYSTEM

The nervous system of *Ascaris* (Fig. 25.9 and 25.10) is well developed and complicated and is **intraepithelial** in position, being located in the epidermis, pharynx and rectum. Its nerve cells are constant in number, form, location and course of their fibers. They are 162.

Nervous system of *Ascaris* includes central nervous system and sympathetic nervous system.

1. Central nervous system. It includes a **circumpharyngeal nerve ring** round the pharynx. It is composed mainly of nerve fibers with very few nerve cells. From its anterior side, circum-pharyngeal nerve ring gives off six **papillary** or **sensory** nerves that proceed to the sensory papillae of the lips. Two of these nerves are dorso-lateral, two lateral and two ventro-lateral in position. Each papillary nerve bears a small **papillary ganglion** just beyond its origin from the nerve ring. Associated with the nerve ring are seven ganglia. These include a median **dorsal ganglion**, a pair of **subdorsal ganglia**, a pair of **lateral ganglia**, each subdivided into six smaller ganglia; and a pair of **large ventral ganglia.** From these ganglia of the nerve ring there proceed backward eight nerves: a **mid dorsal nerve** that passes through the dorsal epidermal chord, a **midventral nerve** that passes through the ventral epidermal chord, a pair of **dorsolateral nerves** that run just above the lateral chords; a pair of **lateral nerves** that extend through the lateral epidermal chords and a pair of **ventrolateral nerves** that run just below the lateral epidermal chords. Of these the ventral nerve is most prominent and is ganglionated throughout its length. Some authors have regarded it as **ventral nerve chord** of *Ascaris.* Posteriorly the nerve chord joins a **preanal ganglion** which sends nerves to the tail. The dorsal and ventral nerves are connected at intervals by asymmetrical **dorsoventral connectives** throughout their length. the ventral and lateral nerves are also connected at intervals by **ventrolateral connectives.**

One of the six ganglia formed by subdivision of each lateral ganglion is known as **amphidial ganglion**. It sends forward an **amphidial nerve** to the amphid of its side.

The genital papillae of the male worm are innervated from special **bursal nerves** formed one on either side near the lateral nerve.

2. Sympathetic nervous system. *Ascaris* also possesses two sympathetic system, namely, the **pharyngeal** and **rectal.**

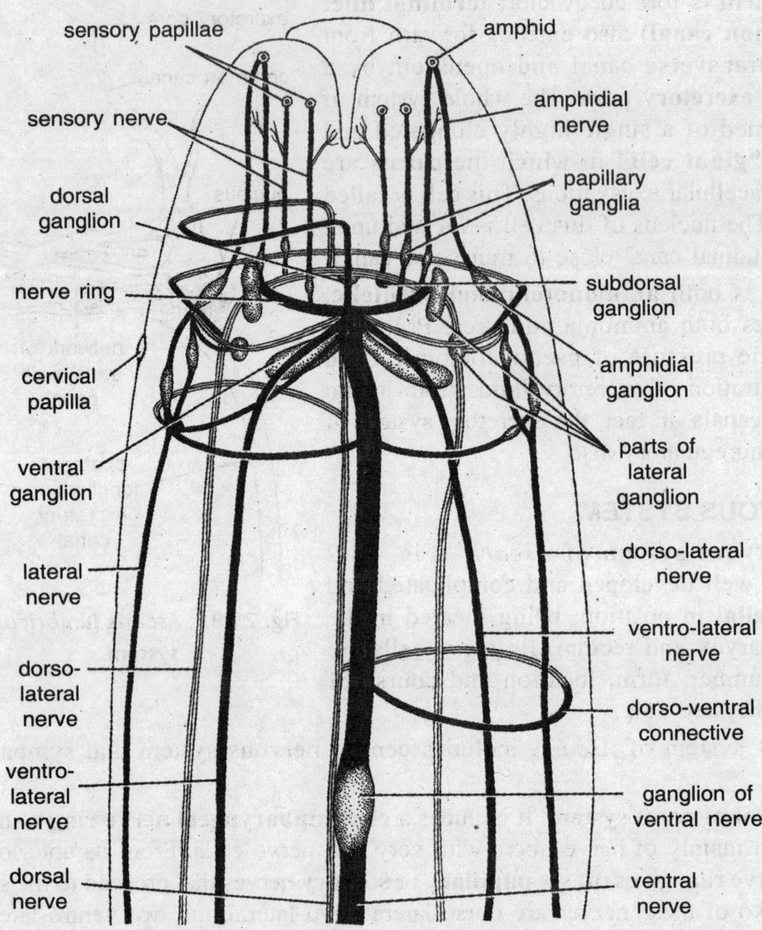

Fig. 25.9. *Ascaris lumbricoides*. Ventral view of anterior end showing central nervous system.

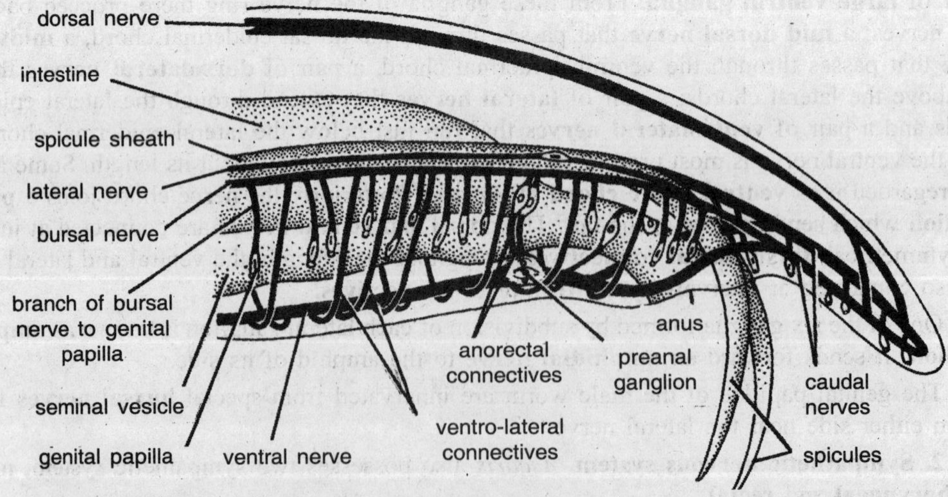

Fig. 25.10. *Ascaris lumbricoides*. Lateral view of posterior end showing nervous system.

25.9 SENSE ORGANS

The receptors or sense organs of *Ascaris* include papillae, amphids and phasmids. Receptors of the nematodes are of taxonomic importance.

I. Papillae. Papillae are primarily **tactile** sensory organs. They occur on the lips and on the sides of the anterior end in both the sexes and in front and behind the cloacal aperture in the male. The dorsal lip has two double papillae, each consisting of a **labial papilla** on the median side and a **cephalic papilla** on the outer side. Each ventrolateral lip has a labial papilla near its upper margin and a double papilla near its lower margin. The double papilla, like the one of the dorsal lip, comprises a labial papilla toward the medial side and a cephalic papilla toward the outer side. Thus, the lips bear six labial and four cephalic papillae. The papillae present on the sides of the anterior end are known as the **cervical papillae**, and those in front and behind the cloacal aperture in the male are termed as the **preanal** and **postanal genital papillae**. Structure of four main types of papillae is as follows:

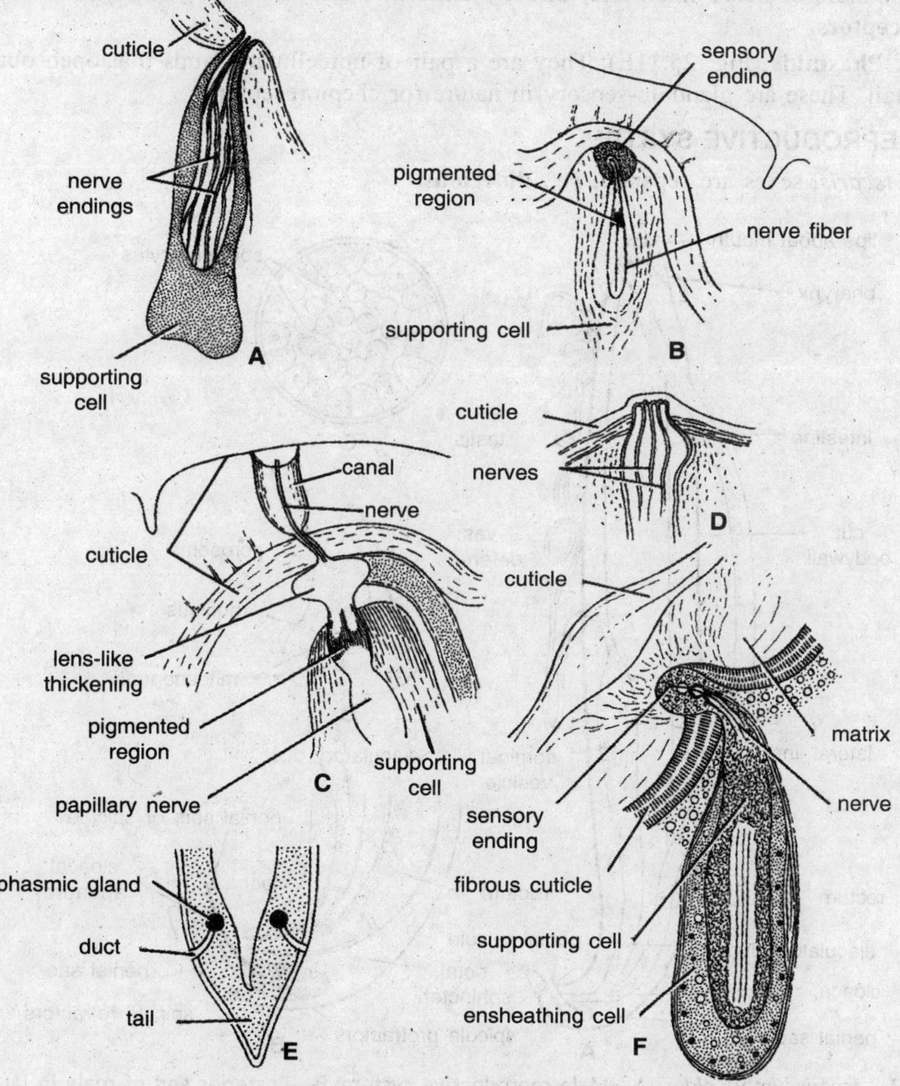

Fig. 25.11. *Ascaris lumbricoides*. Sense organs or receptors. A—Amphid; B—Labial papilla; C—Cephalic papilla; D—Genital papilla; E—Phasmid; F—Cervical papilla.

1. Labial papilla (Fig. 25.11B). It consists of a stout nerve fibre with a bulb-like ending under the cuticle and surrounded by a supporting cell. It also acts as a gustatory or taste organ.

2. Cephalic papilla (Fig. 25.11C). It also consists of a stout nerve fiber surrounded by a supporting cell, but the nerve fiber has just beneath the cuticle a lens-shaped expansion from which a canal extends to the surface.

3. Cervical papilla (Fig. 25.11F). It also consists of a nerve fiber having a bulb-like ending under the cuticle and surrounded by a supporting cell. The supporting cell is surrounded by a ensheathing cell.

4. Genital papilla (Fig. 25.11D). It consists of three nerve fibers embedded in a supporting cell and ending distally in a narrow canal that opens out by a pore.

II. Amphids (Fig. 25.11A). There are two amphids, one on each ventrolateral lip little above the labial papilla. An amphid consists of a supporting cell opening out by a pore and having a bundle of nerve fibers that directly unite into one fibre. The amphids are **olfactory chemoreceptors.**

III. Phasmids (Fig. 25.11E). They are a pair of unicellular glands that open out on the sides of tail. These are glandulo-sensory in nature (or chemoreceptors).

25.10 REPRODUCTIVE SYSTEM

In *Ascaris*, sexes are separate, *i.e.*, **dioecious.**

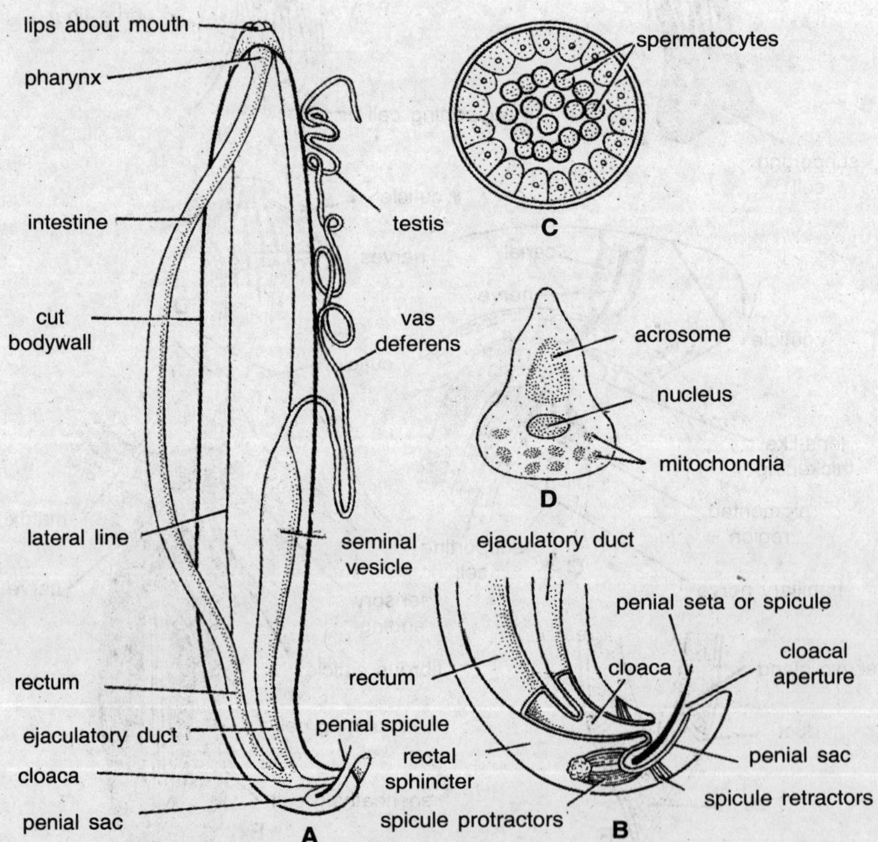

Fig. 25.12. *Ascaris lumbricoides.* A—Male reproductive system; B—Posterior end of male in lateral view showing cloaca and spicule; C—T.S. of vas deferens; D—Amoeboid sperm.

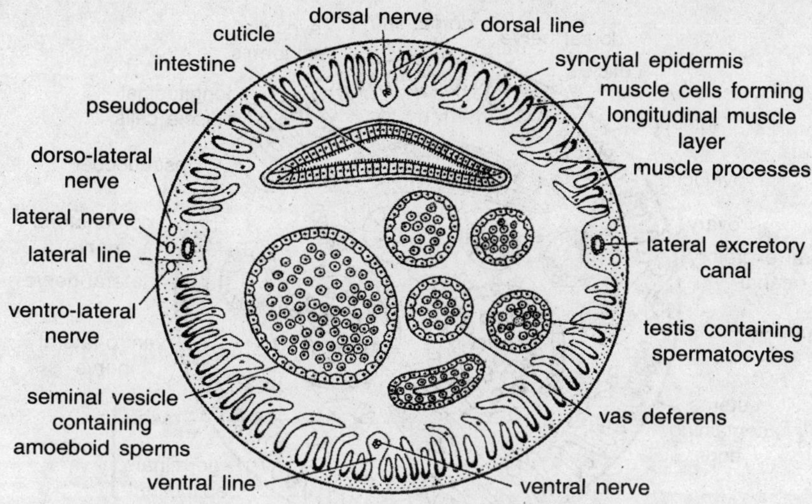

Fig. 25.13. *Ascaris lumbricoides*. T.S. of a mature male.

1. Male Reproductive System

The male reproductive system contains a single, slender, coiled, tubular **testis, vas deferens, seminal vesicle, ejaculatory duct** and a sac containing two penial spicules, which are inserted into the vulva of the female during copulation. Ejaculatory duct opens into the cloaca.

2. Female Reproductive System

The female reproductive system consists of a pair of long, slender, thread-like **ovaries, oviducts** and **uteri.** The two uteri join to form a short single **vagina** which opens to outside through **vulva** (Fig. 25.14). In some nematodes the end part of the uterus or the end part of the vagina, when present, forms muscular ovejectors which by peristaltic movement forces the eggs one by one through the gonopore or vulva.

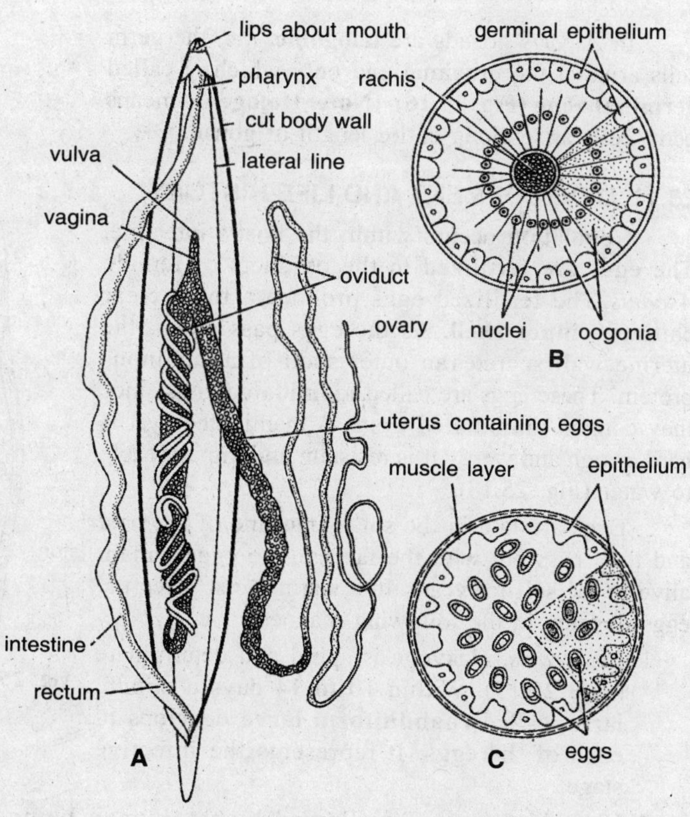

Fig. 25.14. *Ascaris lumbricoides*. A—Female reproductive system; B—T.S. of ovary; C—T.S. of uterus.

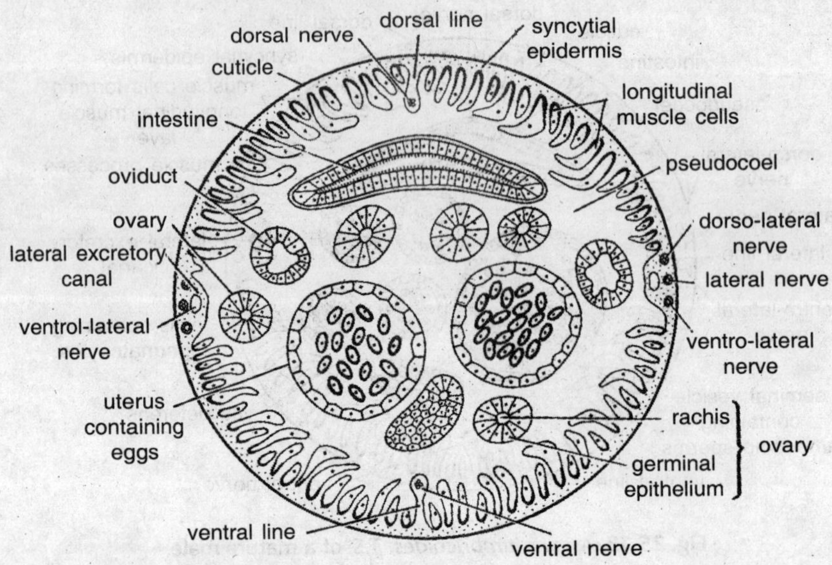

Fig. 25.15. *Ascaris lumbricoides*. T.S. of a mature female.

In *Ascaris* gonads are **telogonic,** *i.e.*, the germ cells arise at the proximal end only which is called **germinal zone** (Fig. 25.16). (**Note. Hologenic** means germ cells arise along entire length of gonads).

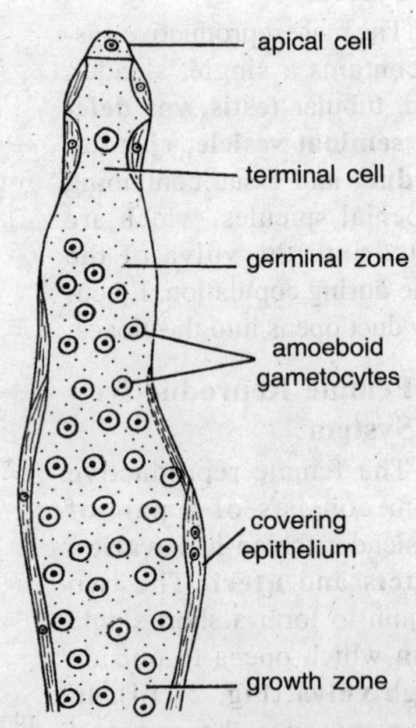

Fig. 25.16. *Ascaris lumbricoides*. Telogonic origin of sex cells in the proximal end of gonad.

25.11 DEVELOPMENT AND LIFE HISTORY

Copulation occurs within the host's intestine. The eggs are fertilized in the oviducts of female *Ascaris*. The fertilized eggs produce a thick clear chitinous **inner shell.** As the eggs pass down, the uterine wall secretes an outer shell of albuminous protein. These eggs are called **mamillated eggs** since they contain **mamillated cysts.** A mamillated cyst is hard, rough and warty. It is resistant and impermeable to water (Fig. 25.18).

Eggs are laid in the small intestine of the host and they pass out with the faeces. The eggs remain alive in the soil for years. In a warm moist place the eggs develop in the following manner:

1. In *Ascaris*, **cleavage** is spiral and determinate (Fig. 25.17). Within 10 to 14 days a minute larva, called **rhabditiform larva** develops in each of the eggs. It represents the infective stage.
2. If such eggs are swallowed by the human beings along with contaminated food or water, the larvae hatch and burrow into the veins or lymph vessels of the intestinal wall.

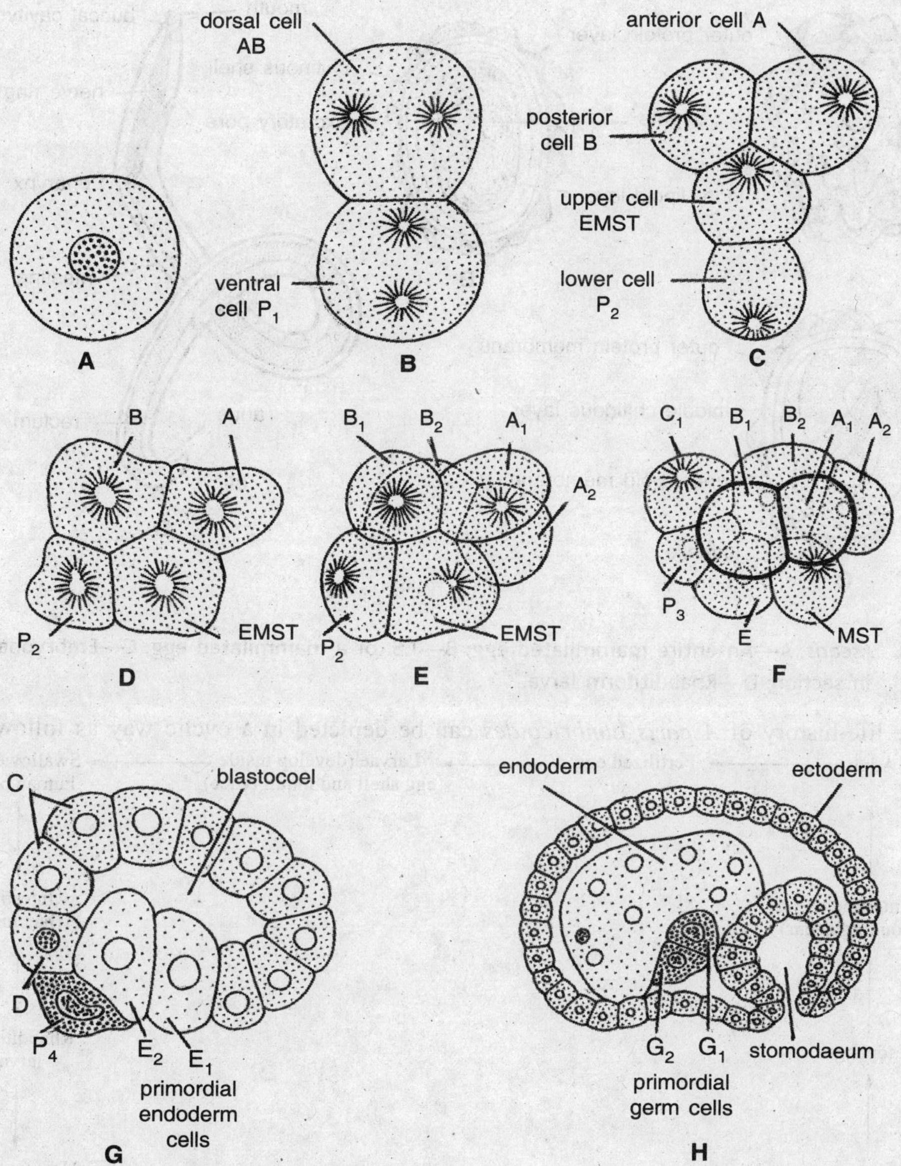

Fig. 25.17. *Ascaris.* A—Various stages of spiral and determinate cleavage. A—Fertilized egg; B—Two-cell stage; C—Four-cell stage (T-shaped); D—Four-cell stage (rhomboidal); E—Six-cell stage; F—Eight-cell stage; G—Median sagittal section through blastula; H—M.S. section of gastrula.

3. These larvae reach the capillaries of the lungs through the heart. In a few days they break into the air passages and move through the trachea, oesophagus and stomach to the intestine again, where they grow to adult stage.
4. In the 60 to 75 days they grow into adult males and females. The length of life span of the parasite in the host average only 9 months to a year.
5. During these migrations the larva moults for four times and grows.

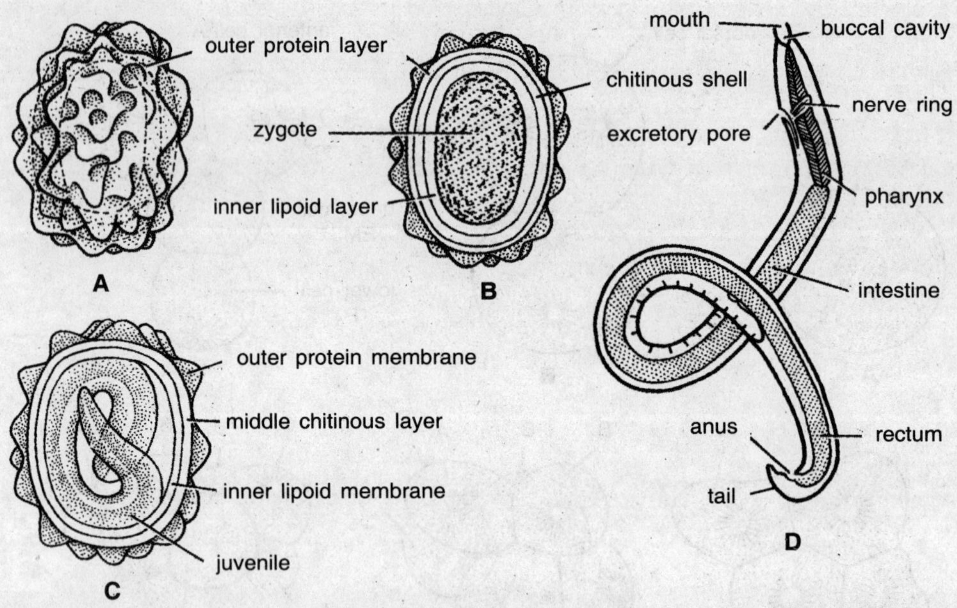

Fig. 25.18. *Ascaris*. A—An entire mammilated egg; B—T.S. of a mammilated egg; C—Embryonated egg in section; D—Rhabditiform larva.

The life-history of *Ascaris lumbricoides* can be depicted in a cyclic way as follows:

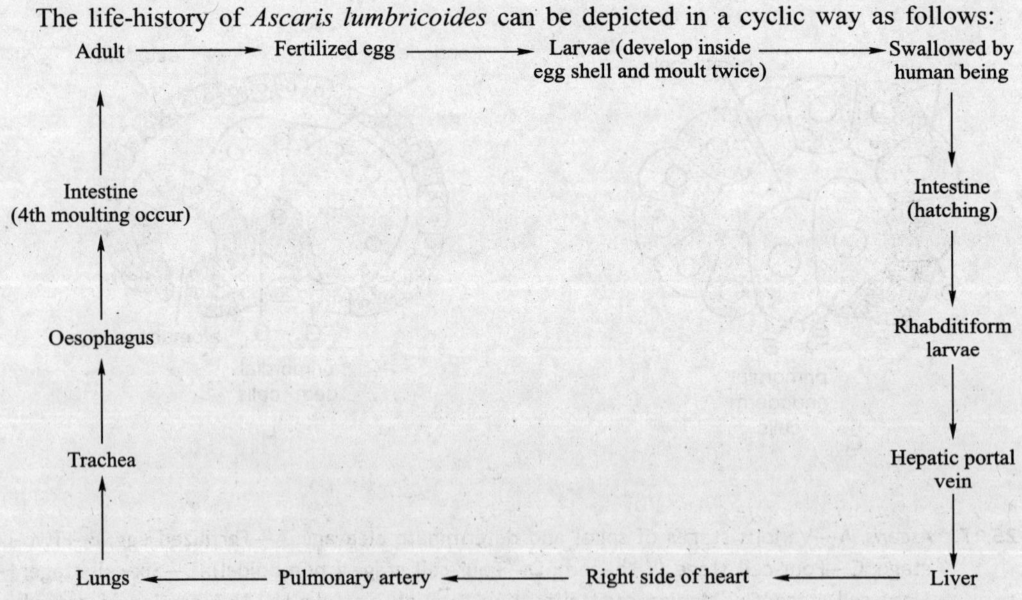

25.12 ECONOMIC IMPORTANCE

As many as 500 to 5000 *Ascaris* may be present in a single host. Both the larval stages and adults cause a variety of disorders in human beings. The larval stages are more harmful.

1. Harmful effects of larval stages. In the intestine of the host, larvae cause internal bleeding, muscular spasm, fever and anaemia. Passage of larvae through the lungs causes local inflammation and pneumonia like symptoms.

2. Adult stage. Adult worm may cause fever, colic pain, indigestion, loss of appetite, insomnia, tumours, intestinal ulcers, diarrhoea, vomiting, convulsions and eosinophilia.

Thus, *Ascaris* is **pathogenic** and causes the disease called **ascariasis.** Sometimes the worms migrate upto mouth or nose causing serious illness or death.

25.13 TREATMENT AND PREVENTION

Ascaris can be killed by taking following drugs: oil of chenopodium with tetrachloroethylene, piperazine, helmacid, tetramezole, dithiazanine, hetrazin, etc. By adopting strict sanitary and hygienic measures alone can effectively check the incidence of infection of *Ascaris.*

25.14 PARASITIC ADAPTATIONS OF *ASCARIS*

1. Its body is long and narrow for easy accomodation in the intestine of the host (human beings).
2. Since *Ascaris* feeds upon the chyme of the host, it possesses gustatory receptors concentrated upon the lips around the mouth.
3. Its cuticle is structurally and chemically well-adapted for protection against injurious effects of chyme of the host. For example, *Ascaris* produces enzyme inhibitors or **anti-enzymes** to protect itself from host's digestive enzymes.
4. The muscle cells of body wall of *Ascaris* contract against the pressure of pseudocoelomic fluid to bring about undulations of body that counteract the force of peristalsis of host intestine.
5. Digestive system of *Ascaris* is simple, because almost predigested nutrients are available to the worms in the chyme of the host.
6. The pharynx is adapted as a sucking and pumping organ to suck in the chyme and pump it into the intestine.
7. Special organs for locomotion and respiration are lacking. *Ascaris* respires anaerobically.
8. Nervous system and sensory organs are relatively simple.

QUESTIONS

Long Answer Questions

1. Give the structure and life-history of Ascaris.
2. Describe the structure and function of body wall of *Ascaris.*
3. Give an account of nervous system of *Ascaris.*
4. Describe sexual dimorphism. Describe the life cycle of *Ascaris* in detail.
5. Give an account of the life cycle of *Ascaris*. How will you control Ascariasis?
6. What are the symptoms of Ascariasis? Describe the body layers and reproductive system of *Ascaris.*
7. Describe male and female reproductive organs of *Ascaris.*

Short Answer Questions

1. Write short notes on the following:

   (*i*) Excretion and osmoregulation in *Ascaris.*

   (*ii*) Ascariasis.

2. Draw a labelled diagram showing
   (*i*) T.S. of female *Ascaris;*
   (*ii*) T.S. of male *Ascaris.*

Very Short Answer Questions

1. How many epidermal chords occur in *Ascaris*?
2. Name the animals having intracellular excretory system.
3. Name the parts of the gut lined with cuticle.
4. How many longitudinal nerves *Ascaris* has? Where do they lie?
5. Name the three layers enclosing an egg of *Ascaris*.
6. Pseudocoel is the persistent blastocoel. Is it so?
7. What is the function of pharynx in *Ascaris.*
8. Reproductive organs of *Ascaris* are didelphic. What does it mean?
9. Why is the young of *Ascaris* called juvenile and not larva.
10. What are the accessory structures that are present in penial sacs of *Ascaris*?

Fill in the Blanks

1. Name of human ascarid is .... *lumbricoides* variety ....
2. .... shows distinct sexual dimorphism.
3. Hind end of .... *Ascaris* is hooked/curved.
4. Mouth of *Ascaris* is surrounded by .... lips.
5. .... of *Ascaris* is syncytial.
6. Body cavity of *Ascaris* is called ....
7. Phenomenon of having fixed number of cells in the body is termed ....
8. Alimentary canal is .... in roundworms.
9. Amphids of *Ascaris* are ....
10. The sperms of *Ascaris* are ....

Multiple Choice Questions

1. *Ascaris* inhabits the host's
   (*a*) blood (*b*) lungs
   (*c*) intestine (*d*) liver
2. Male *Ascaris* has
   (*a*) curved/hooked hind end (*b*) anal papillae
   (c) monorchic gonad (*d*) all of these
3. Body wall of *Ascaris* lacks
   (*a*) muscles (*b*) longitudinal muscles
   (c) circular muscles (*d*) both (*b*) and (*c*)
4. Digestion in *Ascaris* is
   (*a*) intercellular (*b*) intracellular
   (*c*) both (*a*) and (*b*) (*d*) neither (a) nor (b)

5. Juvenile of *Ascaris* undergoes .... molts.

(*a*) two (*b*) three

(*c*) one (*d*) four

6. In *Ascaris*, the reserve food is stored as glycogen and fats in

(*a*) intestine (*b*) muscles

(*c*) hypodermis or epidermis (*d*) all of these

7. *Ascaris* is

(*a*) host (*b*) bisexual

(*c*) dioecious (*d*) digenetic

8. In *Ascaris*, eggs are fertilized in

(*a*) oviduct (*b*) uterus

(*c*) vagina (*d*) outside human body

9. The lips of *Ascaris* are

(*a*) denticulate (*b*) smooth

(*c*) cartilaginous (*d*) bony

10. Amphids upon the lips of *Ascaris* are

(*a*) organs of smell and chemoreceptors (*b*) secrete saliva

(*c*) sense organs (*d*) adhesive organs

11. Which type of excretory canal system is present in *Ascaris*?

(*a*) C-shaped (*b*) J-shaped

(*c*) H-shaped (*d*) W-shaped

12. The tail end of male *Ascaris* is

(*a*) straight (*b*) thickened

(*c*) coiled (*d*) curved

13. In *Ascaris* penial setae arise from

(*a*) cloaca of male (*b*) anus of female

(*c*) vulva of female (*d*) none of these

14. Infective stage of *Ascaris* for new host is

(*a*) I stage juvenile (*b*) embryonated egg

(*c*) III stage juvenile (*d*) larva after II moult

15. The stage hatched from the ingested egg of *Ascaris* is called

(*a*) bladder worm (*b*) hexacanth

(*c*) maggot (*d*) rhabditiform larva

16. Embryonated egg is

(*a*) shelled egg (*b*) shelled zygote

(*c*) shelled II stage juvenile (*d*) II stage juvenile

ANSWERS

Very Short Answer Questions

1. Four (*i.e.*, one dorsal, one ventral and two lateral).

2. Roundworms (nematodes).

3. Pharynx (stomodaeum) and rectum (proctodaeum).

4. Eight, epidermis.
5. Proteinous (albuminous), chitinous membrane and lipoid membrane; such eggs are called **mammiliated eggs**.
6. Yes.
7. Ingestion, movement of food, digestion.
8. *Ascaris* has a double set of genital organs.
9. As it resembles adult in all respects, but sexually immature.
10. Penial setae.

Fill in the Blanks

1. *Ascaris, humanis*
2. *Ascaris*
3. Male
4. Three
5. Epidermis.
6. Pseudocoel
7. Eutely
8. Complete
9. Chemoreceptors
10. Amoeboid.

Multiple Choice Questions

1. (*c*)
2. (*d*)
3. (*c*)
4. (*c*)
5. (*d*)
6. (*c*)
7. (*c*)
8. (*a*)
9. (*a*)
10. (*a*)
11. (*c*)
12. (*d*)
13. (*a*)
14. (*b*)
15. (*d*)
16. (*c*).

26

Ancylostoma (Phylum Nematoda)

26.1 CLASSIFICATION

Phylum	–	**Nematoda**
Class	–	**Phasmida**
Order	–	**Strongyloidea**
Genus	–	***Ancylostoma***
Species	–	***duodenale***

Ancylostoma is commonly called **hookworm** of human beings. There are two main species of hookworms — (*i*) *Ancylostoma duodenale* is the "**old world hookworm**"; (*ii*) *Necator americanus* is the "**new world hookworm**" or **American hookworm**.

Box 26.1 Homologous and Heterologous Hosts

Though primarily a human parasite, *Ancylostoma duodenale* can be induced experimentally to parasitize monkeys, cats and dogs. An animal which normally or naturally harbours a parasite is called a **homologous host,** and an animal that is artificially or experimentally infected with a parasite is termed a **heterologous host.**

26.2 HABIT AND HABITAT

A. duodenale is a dangerous nematode parasite of human beings which occur in Asia (including India), Africa, Europe, South America, etc. It lies attached to the wall of intestine (*i.e.*, jejunum of human beings and feeds on blood, lymph and mucous membrane. The infective juveniles (= larvae) find their way in the human percutaneously from the soil contaminated with faeces in which they live.

26.3 STRUCTURE

Adult *Ancylostoma duodenale* are small and cylindrical in shape. Sexes are separate; the male is about 8 mm in length and 0.4 mm in diameter, while female is generally longer about 12.5 mm in length and 0.6 mm in diameter.

The anterior end of worms of both sexes is slightly bent dorsally (hence, the name hookworm) and has a large **buccal capsule** having six **cutting plates** or **teeth** (4 hook-like on the ventral surface and 2 knob-like triangular plates called **sharp lancets**, on the dorsal surface). The posterior end of female worm tapers bluntly in a short post-anal tail, while that of male is expanded and umbrella-like (Fig. 26.1). This expanded structure is called **copulatory bursa** and surrounds the cloaca. Bursa has two large lateral lobes with six **muscular rays** in each and small median lobe. Each ray is given a name as dorsal ray, externo-dorsal ray, postero-lateral, medio-lateral, externo-lateral, latero-ventral, ventro-ventral, etc. A pair of **cephalic gland** (lie on both sides of pharynx) open in the buccal cavity. Its secretion contains enzymes for digesting the intestinal mucosa of the host and other substances (such as **acetylcholine esterase**) which have a role in attachment of the worm to the host's intestinal wall.

The body of *Ancylostoma* is covered externally by **cuticle**, it is followed internally by the **epidermis** and **musculature** which is directed longitudinally. Its body cavity is **pseudocoel**.

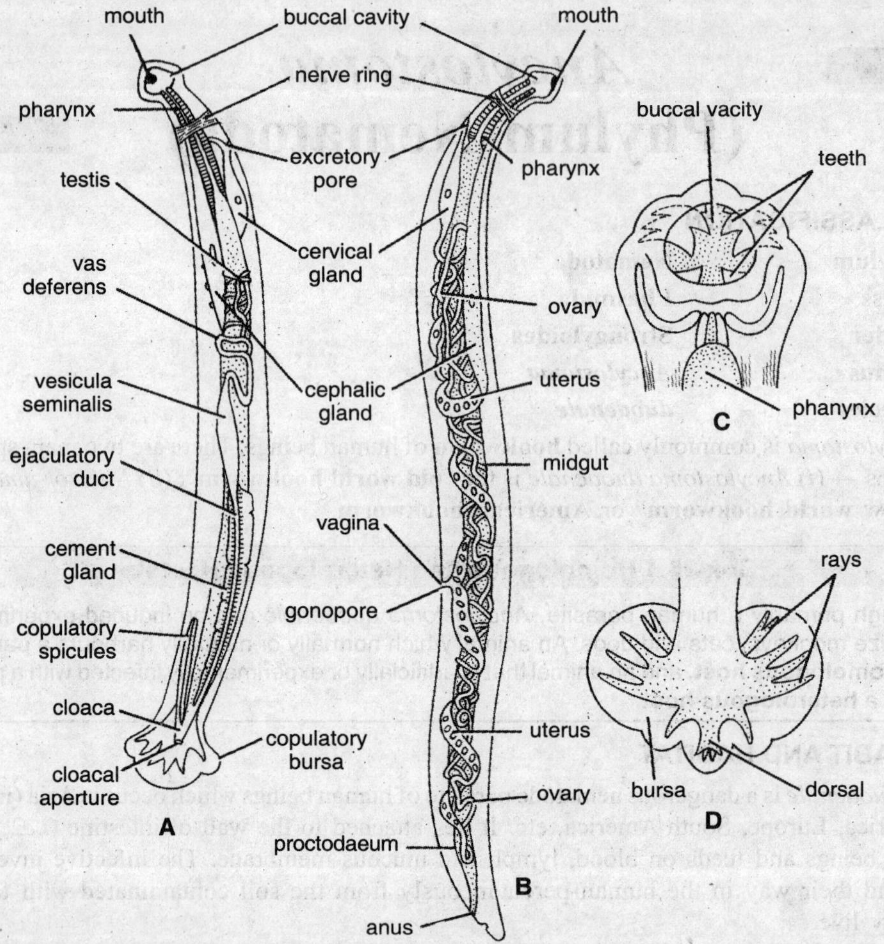

Fig. 26.1. *Ancylostoma duodenale*. A—Adult male; B—Adult female; C—Anterior end; D—Posterior end.

26.4 DIGESTIVE SYSTEM

Alimentary canal of *Ancylostoma* is tubular and very simple. It consists of the mouth, buccal capsule, muscular pharynx having a triradiate lumen lined by cuticle, oesophageal bulb, intestine, rectum and cloaca in male but anus in female. There are five glands connected with the digestive system, one of the is called **oesophageal gland** and secretes an anticoagulant secretion which prevents the clotting of blood so that the worms can suck blood from the host.

The food of this holozoic and blood-sucking nematode includes blood, lymph, tissue exudates and pieces of intestinal mucous membrane. *Ancylostoma* tends to live firmly anchored by drawing a bit of the human intestinal lining into its mouth. In the process of feeding, the tiny teeth and cutting plates of the buccal capsule make small wounds in the intestinal mucosa through which the food, blood, lymph is sucked by the pumping action of pharynx. After feeding it leaves a bleeding wound and moves to another location. An adult worm is said to suck nearly 0.8 ml of blood in a day from the host causing severe **anaemia**. Digestion of food is however, completed in the intestine.

It is held that the red blood corpuscles ingested by the *Ancylostoma* worm provide oxygen needed in the worm's metabolism, as the small intestine of the host has very little oxygen. Other systems of *Ancylostoma* such as **respiratory system, excretory system, nervous system** and **receptors** are like those of *Ascaris lumbricoides*.

26.5 REPRODUCTIVE SYSTEM

As already has been described, in *Ancylostoma* sexes are **separate** and **sexual dimorphism** is well distinct.

Male reproductive system comprises a single, tubular, thread-like **testis** twisted around the intestine in the middle of the body. Testis continues posteriorly in **vas deferens** which finally opens into an elongated, swollen, sac-like **seminal vesicle.** The seminal vesicle soon tapers to forms a narrow passage called the **ejaculatory duct** which opens into the cloaca.

Female reproductive system comprises of two much highly twisted tubules, the **ovaries**. One ovarian tubule is placed anteriorly and other posteriorly from a little behind the middle of body. Both the ovarian tubules continue into the **oviducts** and, thus, the two oviducts open into elongated and dilated **seminal receptacles**, each continues into muscular **uterus**. Thus, the two uteri (one from anterior and other from posterior) meet little behind the middle of the body to form a short tubular **vagina** which opens out by **vulva** or the **gonopore** situated at junction of the posterior and middle third of the body.

26.6 LIFE CYCLE

Ancylostoma duodenale is **monogenetic**. There is no intermediate host and human being is its main host.

1. Copulation. Copulation occurs in the intestine of the host. During copulation, male and female assume a Y-shaped figure and the copulatory bursa of male is applied on the vulva of the female and sperms are transferred. Sperms reach the seminal receptacle of female, where fertilization of eggs occur. Fertilized eggs are then pushed into uteri. About 9000 fertilized and uncleaved eggs are laid by female per day in the intestine of host and are passed out with faeces. Cleavage of eggs is started in the soil.

2. Development in soil. Under favourable environmental conditions such as moisture, oxygen and temperature (20 to 30°C), the fertilized egg divides, develops and hatch within 48 hours into the **first stage juvenile** or **rhabditiform larva**. This larva is 250 μm long and possesses the mouth, buccal capsule, elongated pharynx, bulb-like oesophagus (resembling with the nematode *Rhabditis*) and intestine. The larva feeds on bacteria and other debris in faeces or moist soil and grows rapidly. At the end of second day, it undergoes first moult (*i.e.*, shed its cuticle and grows) and changes into second stage larva, called the **strongyliform larva**. At about 5th day, it undergoes second moult to become third stage larva called **filariform larva** or **juvenile.** This is the infective, nonfeeding stage. The skin of second stage larva is retained as a protective sheath. The larva lives in the 1 cm of the top soil. It remains alive for several weeks under favourable conditions.

3. Infection of a new host. Infection of human host occurs when the filariform larvae or juveniles come in contact with human's skin. They cast off their cuticular sheath and penetrate the skin of human host. The anterior end of the larva is provided with oral spears by which it penetrates the soft skin of feet and hands, generally through the hair follicles.

4. Larval migration inside host. On reaching the subcutaneous tissues the larvae enter the lymphatics and small venules. Ultimately they reach the venous circulation and are carried through the right heart into the pulmonary capillaries, where they break through the capillary walls to enter into the alveolar spaces. They then migrate in the bronchi → trachea → larynx and crawl over the epiglottis to the back of pharynx and are finally swallowed. During its migration, when larva reaches oesophagus, its **third moulting** occurs and a terminal buccal capsule is formed. The time taken in this migration is about 10 days.

Finally the growing larvae (3 to 5 mm long) settle down in the small intestine and undergo **fourth** and **final** moult to become adult. In about 3 to 4 weeks time they become sexually mature to repeat the life history again. The average **life span** of adult worm in the human intestine is about five years.

26.7 PATHOGENICITY (ECONOMIC IMPORTANCE)

The hookworms are the most dangerous parasitic nematodes because they hold on to the intestinal villi and suck blood and body fluids of the host by their muscular pharynx. They also cut holes in the intestinal mucosa and leave bleeding gaps. It results in severe **anaemia.** In children, where incidence of infection is very great, they retard the physical and mental growth. Some toxin secreted by the glands of head region of worms cause stomachache; food fermentation, diarrhoea, constipation, dyspnea (= laboured respiration), palpitation of heart, eosinophilia, ill health and the patient may finally collapse.

During penetration of larvae in the skin local irritation is caused resulting in inflammation

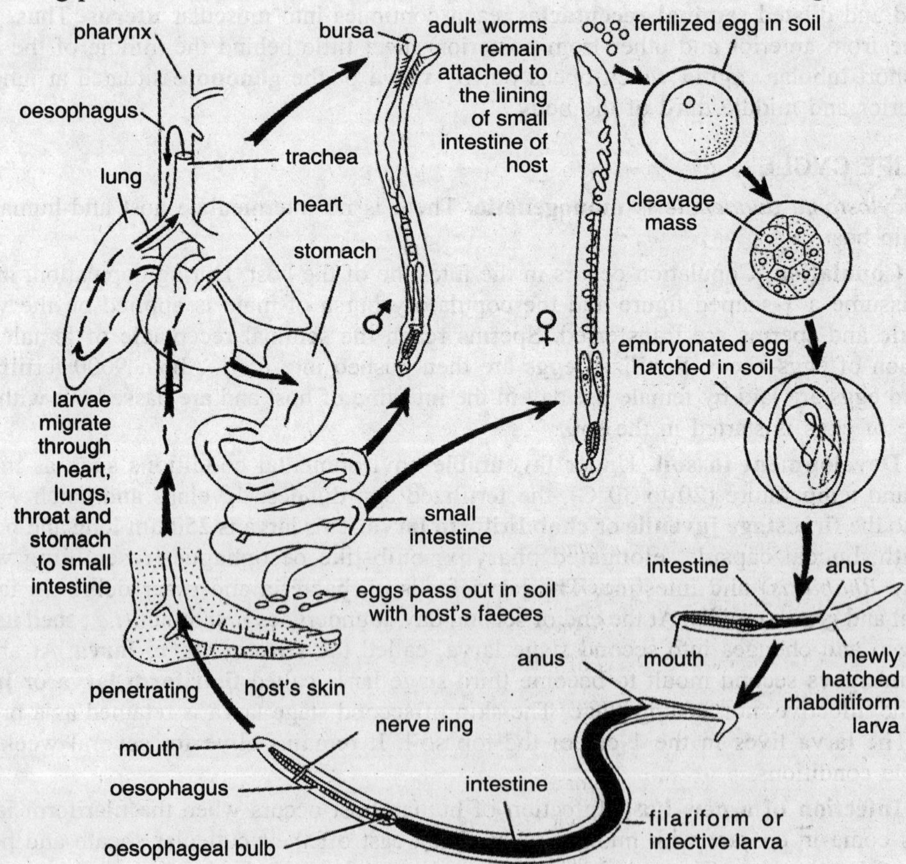

Fig. 26.2. *Ancylostoma duodenale*. Life cycle.

of the surrounding tissues; these may result into tiny sores. This condition is commonly called **miner's itch, foot itch, toe itch, dew itch** or **ground itch**. The migratory larvae in lungs may cause haemorrhage and bronchial pneumonitis (*i.e.*, coughing, sore throat and blood sputum).

26.8 TREATMENT

Drugs like **carbon tetrachloride, thymol, oil of *Chenopodium*, hexylresorcinol,** etc., are generally used effectively to control the infection of *Ancylostoma*. **Tetrachloroethylene** is an effective drug but it is dangerous to give it to alcoholic patients. **Thiabendozole, albendozole** and **pyrantel** are more effective and less toxic drugs.

Prevention and Control

Safe disposal of all human excreta is the most important prevention and control measure. Wearing of hand gloves and shoes by people working in fields of hookworm infested areas helps in avoiding catching of hookworm infection. Children should be directed to keep their hands and nails clean.

QUESTIONS

Long Answer Questions

1. Describe the structure and life history of *Ancylostoma*.
2. Give the habit, habitat, classification, characters and economic importance of hook worm.
3. Describe the life cycle or *Ancylostoma*. (*CCSU, Meerut 2004, 2005*)

Short Answer Questions

1. What is the economic importance of hook worms? How do they infect human beings?
2. Write a note on copulatory bursa and penial spicules of *Ancylostoma*.
3. Describe the development of larva of *Ancylostoma* in soil.
4. Write short note on rhabditiform larva.
5. Draw labelled diagrams of male and female *Ancylostoma*. (*Garhwal 2001*)

Very Short Answer Questions

1. Give the zoological name of the old world hookworm.
2. What is the food of hookworm?
3. What is extracorporeal digestion?
4. Give the popular name of *Necator americanus*.
5. What is the name of a chitinous thickening on the dorsal wall of the cloaca in the male hookworm?
6. Name the main excretory matter of *Ancylostoma*.
7. Are the female reproductive organs of hookworm monadelphic or didelphic?
8. Name the larvae forms of *Ancylostoma*.
9. Are the hookworm larvae free-living or parasitic?
10. How do hookworm larvae infect human beings?
11. What is the name of second stage larva of *Ancylostoma*?
12. Name new world worm that causes necatorasis in human beings.

Fill in the Blanks

1. The popular name of *Ancylostoma duodenale* is ....

2. *Ancylostoma* belong to the order ....

3. Copulatory bursa is present in .... hookworm.

4. .... *Ancylostoma* is longer than ....

5. *Ancylostoma duodenale* or old word hookworm feeds on .... , ... , and pieces of intestinal mucosa of the host.

6. Excretory matter of the hookworm is mainly ....

7. .... helps the male hookworm in holding the female in copulation.

8. The eggs of *Ancylostoma* have a .... layered protective coat.

9. First stage larva of hookworm is called .... larva.

10. Rhabditiform larva of hookworm feeds on ....

Multiple Choice Questions

1. Strongyliform larva occurs in the life history of
(*a*) *Ascaris* (*b*) *Fasciola*
(*c*) *Taenia* (*d*) *Ancylostoma*

2. *Ancylostoma* is a
(*a*) flat worm (*b*) round worm
(c) segmented worm (*d*) none of these

3. *Ancylostoma* belongs to the order
(*a*) Filarioidea (*b*) Strongyloidea
(*c*) Oxyuroidea (*d*) Ascaroidea

4. Excretory organs of the hookworm are
(*a*) flame cells (*b*) green glands
(*c*) renette glands (*d*) nephridia

5. No secondary host is found in the life history of
(*a*) *Fasciola* (*b*) *Taenia*
(*c*) *Ancylostoma* (*d*) *Plasmodium*

6. Characteristic symptoms of ancylostoiasis is
(*a*) gastro-intestinal disturbance (*b*) anaemia
(*c*) nervous disorder (*d*) all the above

7. Which of the following penetrates through skin and enters the intestine?
(*a*) *Ascaris* (*b*) tape worm
(*c*) hook worm (*d*) *Amoeba*

8. Diagnosis of *Ancyclostoma* infection in human beings is made by examination of
(*a*) stool (*b*) urine
(*c*) blood (*d*) saliva

9. The host of *Ancyclostoma duodenale* is
(*a*) human being (*b*) mosquito
(*c*) snail (*d*) pig

10. *Ancyclostoma duodenale* is found in
    (*a*) intestine of human beings (*b*) stomach of human beings
    (*c*) intestine of pig (*d*) liver of human beings
11. Maximum people are infected by
    (*a*) *Ascaris* (*b*) *Taenia*
    (*c*) *Trypanosoma* (*d*) *Ancylostoma*

ANSWERS

Very Short Answer Questions

1. *Ancylostoma duodenale.*
2. Blood, lymph and pieces of intestinal mucous membrane.
3. Digestion outside the body.
4. New world hookworm.
5. Gubernaculum.
6. *Ammonia.*
7. Didelphic.
8. Rhabditiform, strongyliform and filariform.
9. Free-living.
10. By boring through the skin.
11. Strongyliform.
12. *Necator americana.*

Fill in the Blanks

1. Old world hookworm 2. Strongyloidea 3. Male
4. Female, male 5. Blood, lymph 6. Ammonia
7. Copulatory bursa 8. Three 9. Rhabditiform
10. Bacteria and organic debris.

Multiple Choice Questions

1. (*d*) 2. (*b*) 3. (*b*) 4. (*c*)
5. (*c*) 6. (*d*) 7. (*c*) 8. (*a*)
9. (*a*) 10. (*a*) 11. (*d*).

27

Wuchereria Bancrofti *(Filaria)* (Phylum Nematoda)

27.1 CLASSIFICATION

Phylum	–	**Nematoda**
Class	–	**Phasmida**
Order	–	**Filarioidea**
Genus	–	***Wuchereria***
Species	–	***bancrofti***

It is commonly known as *Filaria bancrofti*.

27.2 HABIT AND HABITAT

Wuchereria is a digenetic nematode which lives in the lymphatic vessels and lymph nodes of human. Male and female worms copulate in the lymph glands.

27.3 STRUCTURE

The body of filaria is thread-like tapering, transparent and often creamy-white in colour (Fig. 27.1). The female is usually larger than the male, measuring 8 to 13 cm in length and 0.2 to 0.3 mm in thickness. The male is only 2.5 to 4 cm long and 1 mm broad. The tail end of the male is curved and carries two spicules of unequal length. In female, tail is abruptly pointed. The male and female remain coiled together and are separated with difficulty.

The female is ovoviviparous and lays eggs which contain well developed larvae. The hatched larvae are called **microfilariae**.

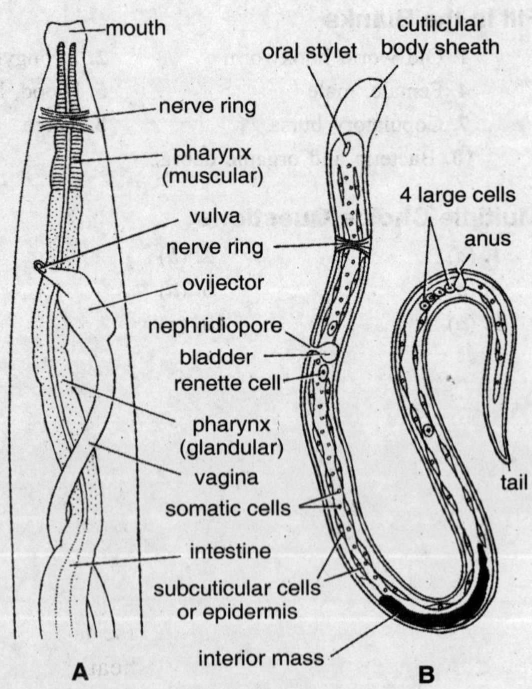

Fig. 27.1. *Wuchereria bancrofti.* A—Anterior part of female; B—Microfilaria.

27.4 LIFE CYCLE

The life cycle of *Wuchereria bancrofti* is completed in two hosts (Fig. 27.2). The primary host is human being and secondary host is mosquito—*Culex pipiens* in India.

Life cycle in human beings. A microfilaria has a colourless and transparent body with blunt anterior end and rather pointed tail end. Its body is covered in a hyaline sheath followed by **cuticle** which is lined by flattened subcuticular cells or **epidermis** and an inner column of cytoplasm containing nuclei. It contains future mouth or oral stylet, nerve ring band, nephridiopore, renette cells, a dark coloured inner mass and future anus (Fig. 27.1B). Microfilaria are active forms and can move both with and against the blood streams. These larvae are unable to develop further in the human body, unless they are taken up by the intermediate host. If these microfilariare are not sucked up by the mosquito they die within 70 days.

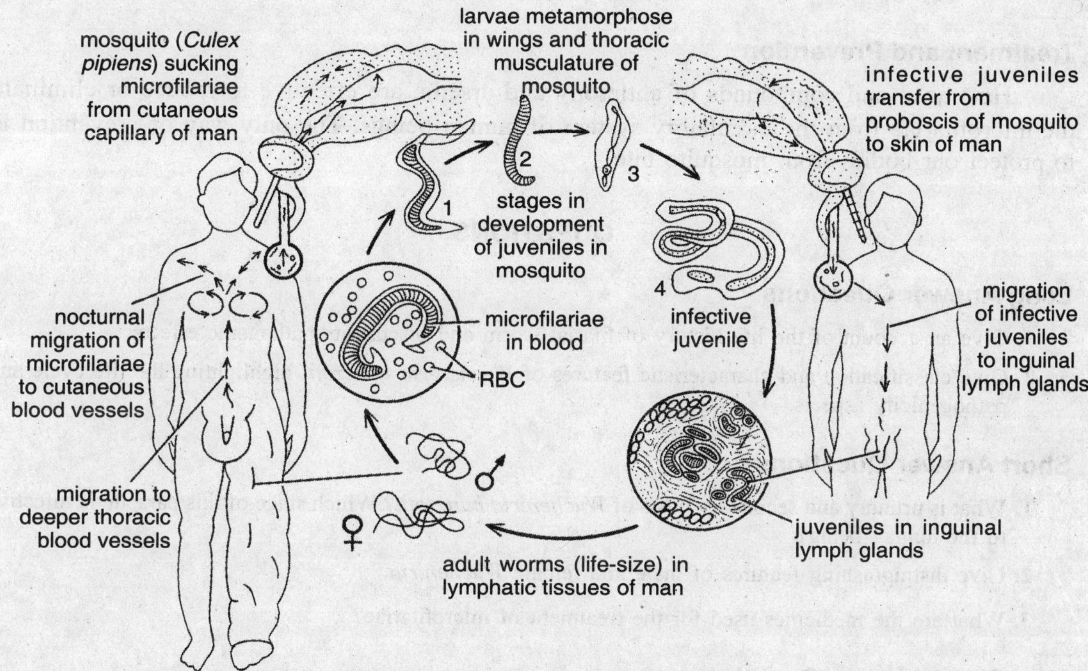

Fig. 27.2. *Wuchereria bancrofti*. Life cycle.

Periodicity of microfilariae. The microfilariae of oriental countries such as India and China exhibit **nocturnal periodicity.** They appear in peripheral circulation periodically at night, only generally between 10 pm and 4 am, but disappear during rest of day. It is believed that nocturnal periodicity of microfilariae is related with the nocturnal feeding habit of their secondary host, *Culex pipiens*.

Life cycle in mosquito. The microfilariae are sucked up by the mosquito (*Culex pipiens*) along with blood meal. These larvae are collected in the anterior part of the stomach where these cast off the hyaline sheath, penetrate the gut wall and within an hour or two migrate to the thoracic muscles and change into **first stage larvae.** They undergo moulting once or twice and form **second stage larvae.** On tenth or eleventh day the metamorphosis is completed. The tail, the digestive system, body cavity and genital organs are developed.

This is the **third stage larvae** measuring 1500 to 2000 μm in length and 18 to 20 μm in diameter. These larvae are infective to humans. They become inactive and come to lie in the labium of the mosquito. When the mosquito bites the warm and moist skin of human beings, the larvae creep out of the labium to the human skin. They penetrate human skin and finally come to settle down in the lymphatics.

27.5 PATHOGENICITY

The disease caused by the infection of *W. bancrofti* is, in general, referred to as **wucheriasis** or **filariasis.** Accumulation of microfilaria worms blocks the human lymphatic system which results in an inflammation of lymph glands. In some cases, the obstruction of lymph vessels may result in **elephantiasis,** resulting in enormous swelling and excessive growth of the affected part such as arm, leg, scrotum or mammae.

Treatment and Prevention

Heterazan and compounds of antimony and arsenic are effective to reduce or eliminate the microfilariae from the circulatory system of human beings. The only way of prevention is to protect our bodies from mosquito bite.

QUESTIONS

Long Answer Questions

1. Give an account of the life history of filarial worm and discuss its pathogenic effects.
2. Give classification and characteristic features of *Wuchereria bancrofti* highlighting the life cycle and pathogenicity aspects.

Short Answer Questions

1. What is primary and secondary hosts of *Wuchereria bancrofti*? Which stage of this parasite is infective to the human being?
2. Give distinguishing features of male and female *Wuchereria.*
3. What are the medicines used for the treatment of microfilariae?

Very Short Answer Questions

1. Give the name of insect vector that is involved in the transmission of elephantiasis.
2. What is elephantiasis?

Fill in the Blanks

1. Human disease called elephantiasis is caused by ....
2. In India and China, the secondary host of *Wuchereria bancrofti* is ....
3. The primary host of nematode *Wuchereria bancrofti* is the ....

Multiple Choice Questions

1. The place where adult *Wuchereria bancrofti* dwells

   (*a*) human blood vessels (*b*) human lymph glands

   (*c*) muscles of *Culex* (*d*) salivary gland of *Culex*

2. The vector of elephantiasis in India is

(*a*) *Culex* mosquito
(*b*) female mosquito
(*c*) female *Anopheles* mosquito
(*d*) infected female *Culex pipiens*

3. Microfilaria is found in the infection of

(*a*) *Fasciola*
(*b*) *Taenia*
(*c*) *Wuchereria*
(*d*) *Schistosoma*

4. Viviparity generally occurs in

(*a*) *Wuchereria*
(*b*) *Enterobius*
(*c*) *Ancylostoma*
(*d*) *Trichuris*

ANSWERS

Very Short Answer Questions

1. Infected female *Culex pipens* mosquito.

2. Elephantiasis is a disease in which there is a massive swelling and enlargement of the skin and subcutaneous tissue of limbs, scrotum or mammary gland due to obstructed flow of lymph. It occurs due to infection of nematode worm of genus *Wuchereria bancrofti.*

Fill in the Blanks

1. *Wuchereria bancrofti* **2.** *Culex pipiens* **3.** Human beings.

Multiple Choice Questions

1. (*b*) **2.** (*d*) **3.** (*c*) **4.** (*a*).

28

Parasitic Adaptations in Platyhelminthes (Flatworms) and Nematodes

Hickman *et al.*, (1996) have defined *parasitism* as the condition of an organism living in or on another organism (the host) at whose expense the parasite is maintained. It is a sort of destructive symbiosis. Also, they have defined *adaptation* as an anatomical structure, physiological process, or behavioural trait that evolved by natural selection and improves an organism's ability to survive and leave descendants.

28.1 DEFINITION

Parasitism is an association of two organisms of different sizes and species, in which smaller organism is benefited and the larger organism is harmed. The organism which is benefited is known as **parasite**, and that which suffers, is referred to as **host**. The host can live without the parasite, but the parasite cannot survive without the host.

28.2 PERKS OF PARASITES

Parasite gets nourishment from the host. It feeds on tissues or body fluids of the host or utilizes the food ingested by the host. Internal parasites, the endoparasites, may have additional advantages, such as protection, lack of competition and ensured survival.

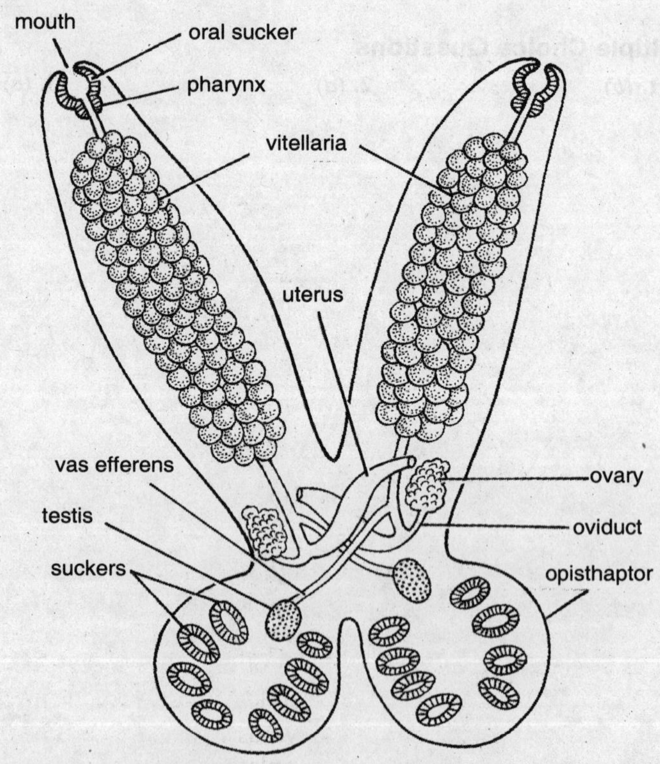

Fig. 28.1. *Diplozoon*. Two individuals are permanently united in the form of a cross.

28.3 TYPES OF PLATYHELMINTH PARASITES

The helminth parasites are parasites for the whole of their life, so are known as the **permanent parasites.** Some live on the surface of body of host and are called **external parasites** or **ectoparasites.** *Diplozoon indicum* is such an ectoparasite (Fig. 28.1). It inhibits the gills of freshwater fishes such as *Barbus* and feed on the blood of the host.

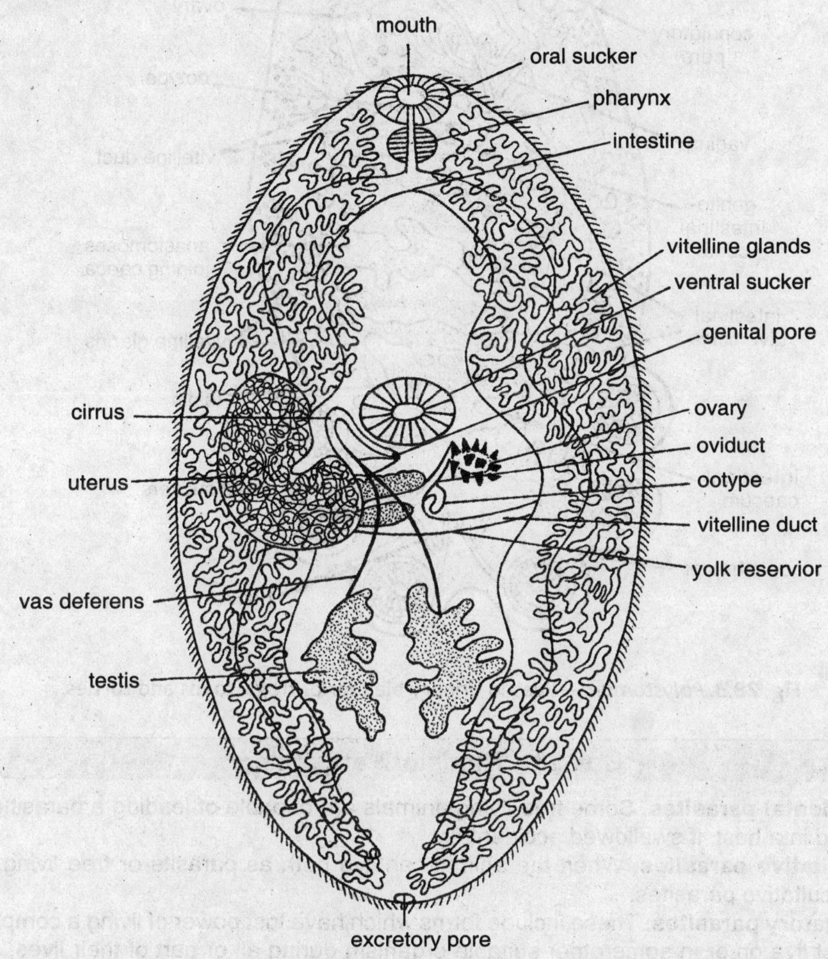

Fig. 28.2. *Paragonimus westermani*—the lung fluke.

Many flatworms live inside the body of the host and are called **internal parasites** or **endoparasites.** Such endoparasites may occur in the intestine as *Taenia solium* (Fig. 23.1), lungs as *Paragonimus westermani* (Fig. 28.2), urinary bladder as *Polystomum integerrinum* (Fig. 28.3), liver as *Fasciola hepatica* (Fig. 22.1) or blood as *Schistosoma* (Fig. 23.3).

Some parasitic platyhelminthes have a simple life history with a single larval form and no intermediate host. They are known as **monogenetic parasites**, *e.g.*, *Polystomum*, *Diplozoon.* Many helminth parasites have a complex life history with several larval forms that inhabit one or more intermediate hosts, *e.g.*, *Paragonimus, Taenia, Fasciola*, etc. The host for the adult stage of the parasite is called **definitive host** or **primary host** and that for the larval forms as the **intermediate host.**

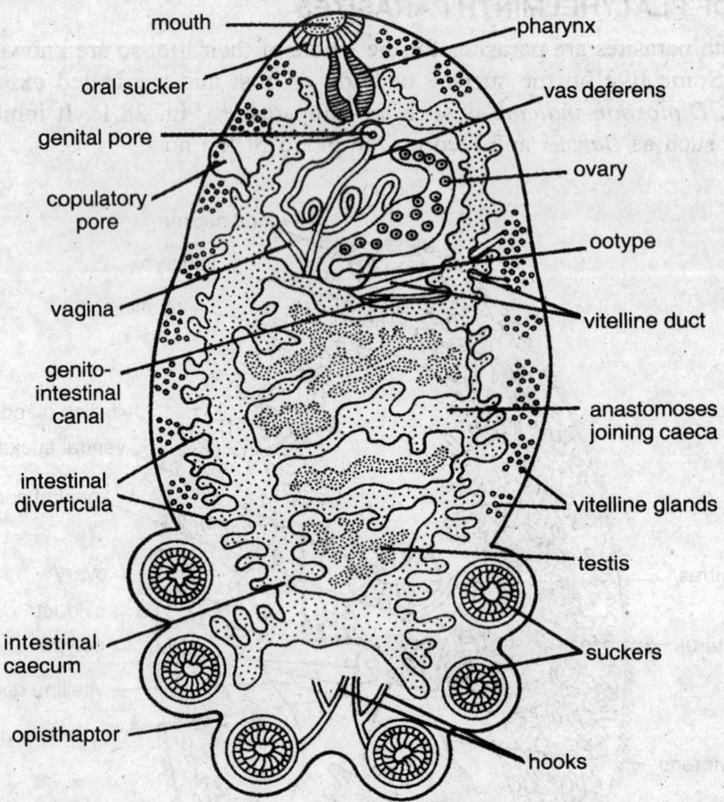

Fig. 28.3. *Polystomum*—lives in urinary bladder of frogs, toads and turtles.

Box 28.1 Kinds of Parasitism.

1. Accidental parasites. Some free-living animals are capable of leading a parasitic life for a certain period in a host, if swallowed accidentally.

2. Facultative parasites. When the animal can live both as parasite or free living, they are known as facultative parasites.

3. Obligatory parasites. These include forms which have lost power of living a completely free life, and must live on or in some other suitable organism during all or part of their lives, otherwise they die.

4. Ectoparasitism. In ectoparasitism, the animals live on the external surface of the body of host and totally depend for food on the host.

5. Endoparasitism. In endoparasitism, the animals live inside the body of host and depends entirely for the food on the host.

28.4 DIFFICULTIES OF PARASITES

Parasites face certain difficulties in the establishment and maintenance of a parasitic relationship with the host. A primary problem for the parasites is that of locating and infecting new hosts. Hosts often have defenses and respond to the invading helminth parasites as to the proto-zoan, bacterial and viral parasites. Any tissue damage provokes inflammation and phagocytic reactions of leucocytes (white blood cells). The host's immune system is also stimulated by the protein antigens on the parasites' surface. The parasites which overcome these defenses of the host may be walled off in the host's body. Hydrochloric acid and proteases (protein digesting enzymes) present in the vertebrate host's stomach are effective barriers which few sturdy parasites can cross. Low level of oxygen in the gut contents and high body temperatures of birds and mammals are also barriers for the parasites. Parasites have adaptations to overcome all the barriers which are posed by their hosts.

28.5 MODES OF INFECTION IN PLATYHELMINTH PARASITES

Parasitic flatworms infect new host in one of the following ways:

1. By swallowing. Most of the parasitic flatworms enter the host through the mouth as encysted larvae. Access to the mouth is effected with leaves, vegetables and fruits, or with improperly cooked meat (beef, pork, fish) of the intermediate hosts.

2. Boring through skin. The cercariae of blood flukes, such as *Schistosoma*, penetrate the human body through the skin from water of rivers, canals, tanks, etc.

3. Direct contact. Larvae of the gill-inhabiting ectoparasite *Diplozoon*, fix on the gills on hatching from the capsules, that are also attached to the gills by coiled filaments. Larva of *Polystomus* clings to the internal gills of a tadpole and metamorphoses into a young fluke that migrates to the urinary bladder during the metamorphic changes of the tadpole into a frog or toad.

28.6 ADAPTATIONS OF PARASITIC PLTYHELMINTHES

The parasitic flatworms have both morphological and physiological adaptations for their parasitic mode of life.

I. Morphological Adaptations of Parasitic Flatworms

The morphological parasitic adaptations of the flatworms are of following three types:

1. Loss of those organs which are useful only for free existence; such organs remain unnecessary for the parasitic life.
2. Continuation of pre-existing features found beneficial in parasitic life also.
3. Development of new structures in response to new needs of parasitic life.

1. Loss of organs. The organs lost by parasitic flatworms include locomotory, digestive and sense organs.

***(i)* Loss of locomotory organs.** The parasites have a abundant supply of readymade food and they have not to face the competition with its fellows. As a consequence, they do not need locomotory organs which have been lost. Tapeworms lack locomotory organs (cilia) throughout their life. Flukes, however, retain cilia in the free-living stage (miracidium) to move about in search of a snail.

***(ii)* Loss of digestive organs.** Endoparasitic flatworms of human intestine (tapeworm) are surrounded by digested food ready for absorption and assimilation. Therefore, they have lost the alimentary canal (the digestive organs) and have taken to saprozoic nutrition. Long and flat body of a tapeworm provides a large surface to facilitate this mode of nutrition.

***(iii)* Loss of sense organs.** The parasitic flatworms, particularly endoparasites, live in a uniform environment with practically no changes to be perceived. Their sense organs have therefore disappeared.

2. Continuation of preexisting features. The preexisting features continued in the endoparasitic flatworms include flat body and shell around the eggs.

***(i)* Thin, flat body.** The thin and flat body of the endoparasitic flatworms is beneficial in following three ways: 1. It enables the parasites to fit in narrow spaces in the host; 2. It quickens the elimination of carbon dioxide by diffusion; 3. It facilitates distribution of food digested in gut (*Fasciola*) or absorbed through the skin (*Taenia*).

***(ii)* Shell around eggs.** A fertilized egg, alongwith one or more yolk cells, is enveloped by a resistant shell. The structure thus formed is called the **capsule**. The capsules leave the host and lie in the open, where the shell protects the egg or the developing embryo from injury, desiccation, high and low temperature, etc. This adaptation ensures survival of the parasite outside the host.

3. Development of new structures. The new organs developed by the parasitic flatworms include adhesive organs, tegument, sucking pharynx and cyst wall around the infective larvae.

***(i)* Adhesive organs.** Adhesive organs have been developed in the endoparasitic flatworms to check dislodging of them from the host. Thus, once established within or on a host, it is advantageous for a fluke to stay more or less in one place. In that regard, nearly all of them are equipped with external organs for temporary or permanent attachment (Fig. 28.4, Fig. 28.5).

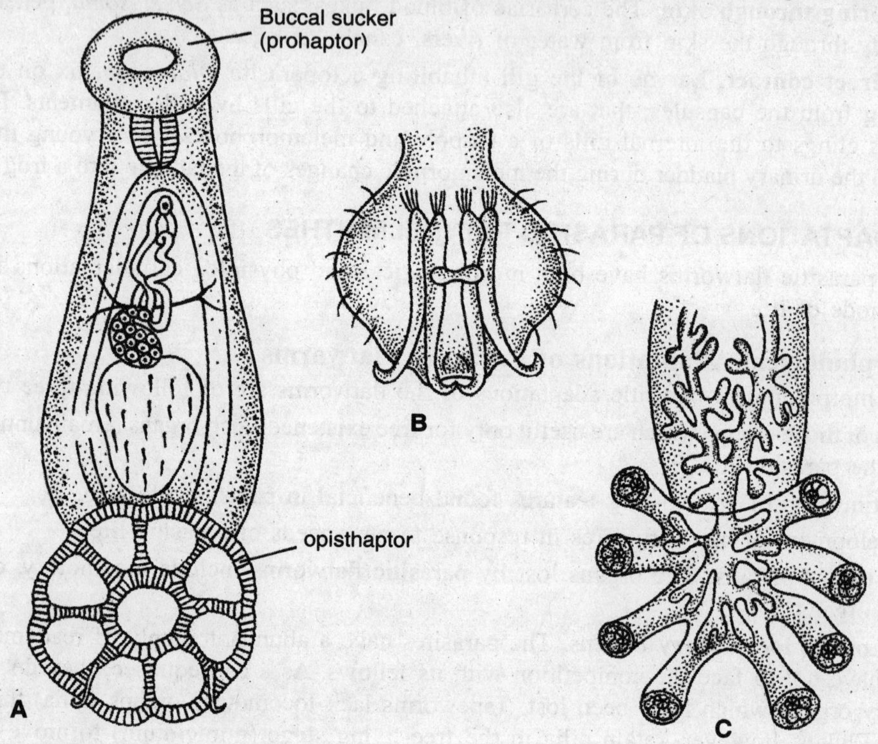

Fig. 28.4. Some attachment organs of monogenetic flukes. A—*Anoplocotyloides papillate*; B, C—Opisthaptor from monogenetic flukes.

Monogenetic flukes typically have an anterior and a posterior adhesive organs called the **prohaptor** and the **opisthaptor**, respectively. The prohaptor consists of a pair of adhesive structures, one on each side of the mouth, bearing suckers or simple adhesive pads. The opisthaptor is usually the major organ of attachment, and includes one or more well developed suckers with hooks and claws.

The digenetic fluke possess two hookless suckers. One, the **oral sucker**, surrounds the mouth, and the other, the **acetabulum** (*e.g.*, *Fasciola hepatica*). These suckers are usually supplied with adhesive gland cells, although the well developed ones operate mainly on suction produced by muscle action. The aspidogastrean flukes lack an oral sucker but have a large subdivided ventral sucker (Fig. 28.5).

Adult tapeworms do not move around much, but they are capable of muscular undulations of the body. They remain fixed to the host's intestinal wall by the scolex (in eucestoda) or adhesive organs (in cestodaria). The tip of the scolex in many cestodes (*e.g.*, *Taenia*) is equipped with a movable hook-bearing **rostellum**. In others (*e.g.*, *Cephalobothrium*) the anterior end bears a protrusible sucker, or adhesive pad, called a **myzorhynchus**. The rest of the scolex bears various suckers and sometimes hooks or spines. There are three categories of adhesive suckers upon which classification of cestodes is partially based.

***(a)* Bothria.** They are elongate, longitudinal grooves on the scolex. They possess weak muscles but are capable of some sucking action. Bothria occur as a single pair in order Pseudophyllides (*e.g.*, *Diphyllo-bothrium*).

***(b)* Bothridia.** Members of order Tetraphyllidea (*e.g.*, *Acanthobothrium*) bear four symmetrically placed **bothridia** around the scolex. These foliose structures are often equipped with suckers on their anterior ends.

***(c)* Acetabula.** The third and most familiar type of attachment structures on the scolex are true suckers or **acetabula.** They are characteristic of order Cyclo-phyllidea (*e.g.*, *Taenia*). Acetabula are identical in structure and are probably homologous to the acetabula of digenetic trematodes. There are usually four aceta-bula, placed symmetrically around the circumference of the scolex.

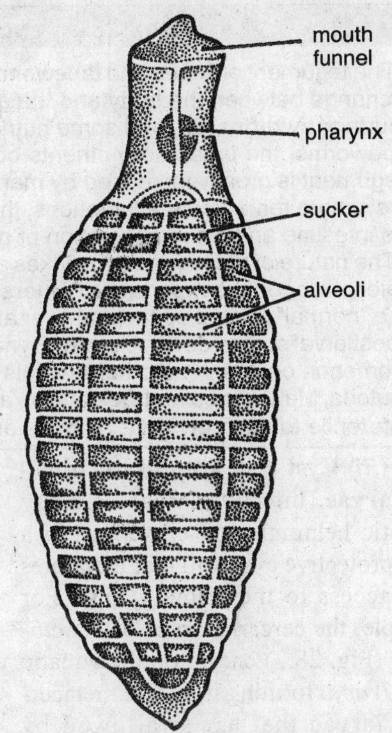

Fig. 28.5. *Aspidogaster* (a trematode). It is a endoparasite of freshwater mussel and lives in pericardial and renal cavities of the host.

***(ii)* Tegument.** The parasitic flatworms have lost epidermis and developed a thick tegument which is resistant to the host's digestive enzymes and antitoxins. This ensures survival of the parasites in their environment (Box 28.2).

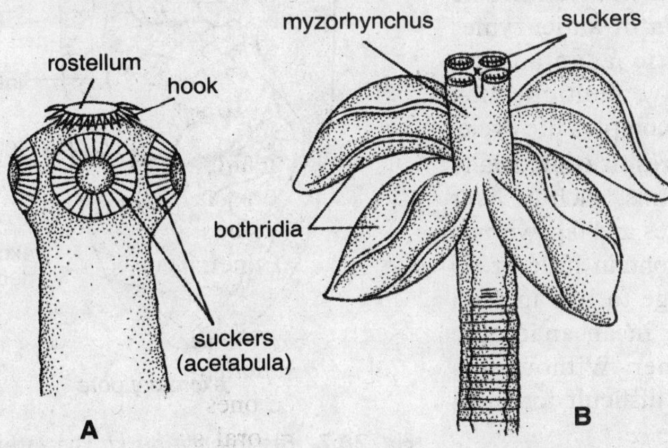

Fig. 28.6. Scoleces of two eucestodes. A—"Typical" scolex with rostellum, hooks and suckers (*Taenia solium*); B—Complex scolex with suckered myzorhynchus and leaflike bothridia (*Myzophyllobothrium*).

Box 28.2 Tegument: A key synapomorphy

The tegument of flukes and tapeworms not only provides some protection but is an important site of exchange between the body and the environment. Gases and nitrogenous wastes move across this surface by **diffusion**, and some nutrients, especially amino acids, are taken in by **pinocytosis**. In tapeworms, the uptake of nutrients occur solely across the body wall, and the surface area of the tegument is greatly increased by many tiny folds called **microtriches** (Fig. 23.4.). As one of the nature's more remarkable adaptations, these folds may interdigitate with the internal microvilli of the host's intestine and aid in absorption of nutrients.

The nature of the tegument in flukes and tapeworms is viewed by some zoologists as unique and of major phylogenetic importance (**Ehlers** 1984, 1986, 1995). The larvae of these parasitic flatworms have a "normal" ciliated epithelium over at least part of their bodies. However, this epidermis is shed, and postlarval stages develop a new, syncytial body covering, called **neodermis**, or **tegument**. This phenomenon occurs in no other animals and should be viewed as a **key synomorphy** uniting the Trematoda, Monogenea and Cestoda as a monophyletic taxon that **Ehlers** (1985) called **Neodermata** (in reference to the "new skin" of these animals) (**Brusca** and **Brusca**, 2003).

(*iii*) Cyst wall around the infective larvae. Infective larvae of many parasitic helminths get enveloped in a thick protective cyst wall that facilitates their access to the primary host. For example, the cercaria larvae of *Fasciolopsis* (Fig. 28.7) encyst on the aquatic plant *Trapa*, forming infective metacercaria larvae that are swallowed by human beings with *Trapa*. Hexacanths of *Taenia solium* encyst in muscles of pig, forming cysticerca. These encysted larvae are taken by human being with undercooked or raw pork.

II. Physiological Adaptations of Parasitic Flatworms

The physiological adaptations of parasitic flatworms include anaerobic respiration, secretion of antienzymes and mucus, osmotic adaptability, chemotaxis, excessive multiplication (high fecundity) and complex life cycle.

1. Anaerobic respiration. Many endoparasitic flatworms, such as liver flukes, intestinal flukes and tapeworms, live in media that contain little or no oxygen. They manage to live in such media by respiring in an anaerobic (anoxybiotic) manner. Without this ability, it would be difficult for them to survive.

2. Secretion of antienzymes and mucus. The intestinal flatworms secrete antienzymes which neutralize the

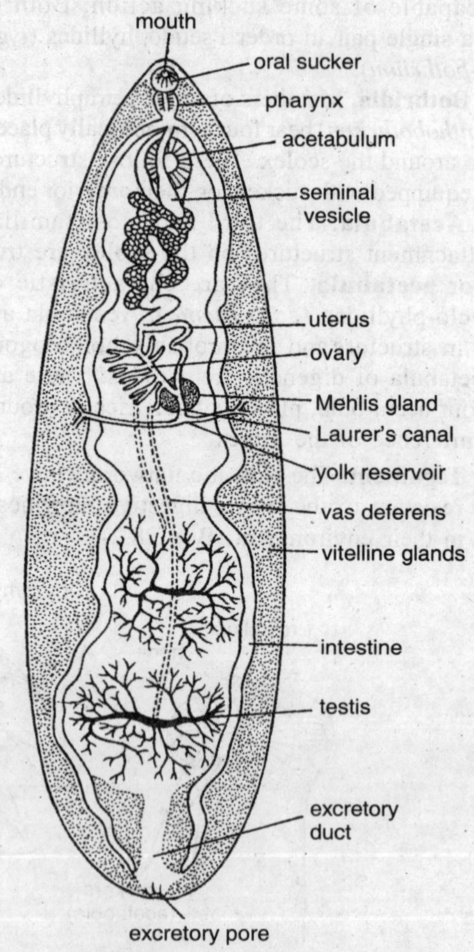

Fig. 28.7. *Fasciolopsis buski*. The intestinal fluke of human beings. Its intermediate host is snail *Planorbis* and *Segmentina*.

digestive enzymes of the host. Tapeworms stimulate host's intestine to produce a large quantity of mucus, that surrounds the parasite to further reduce the action of host's enzymes. Lime cells in the body wall of the tapeworms, neutralize the acids formed in the host's gut (Box 28.3).

Box 28.3

One study showed that *Hymenolepis diminuta* (a common tapeworm of rats and mice) releases proteins (*i.e.*, antitoxins or enzyme inhibitors) that appear to inhibit trypsin activity. The tapeworm can also regulate the pH of its immediate environment to about 5.0 by excreting **organic acids**; this acidic output also may inhibit the activity of trypsin (**Uglem** and **Just**, 1983).

3. Evasion of host's immune system. Some endoparasitic flatworms simply avoid host's immune system by dwelling at the sites, such as the gut, that are largely out of the reach of the immune system of the host. Blood flukes conceal themselves immunologically by absorbing the host's antigens on to their surface so that the host's immune system does not recognize them as "foreign" or "nonself".

4. Osmotic adaptability. Endoparasitic flatworms seem to possess a considerable osmotic adaptability, as they can successfully live in different media. Liver fluke of sheep, for example, passes through the stomach, body cavity and liver before reaching the bile duct.

5. Chemotaxis. Parasitic flatworms must be having marked chemotaxis that enables them to find their way to their final habitat in the host's body.

6. Excessive multiplication (Fecundity). The endoparasitic flatworms face a problem of migration from one host to another. Many are likely to perish in this adventure. Numerous capsules are, therefore, produced by the endoparasite, so that enough survive to continue the race. For example, one liver fluke (*Fasciola hepitica*) may lay about 50,000 capsules. A single tapeworm (*Taenia solium*) can live for about 30 years, and every year, it sheds about 2500 gravid proglottids, each containing 30,000 to 40,000 oncospheres. Formation of several larvae from one zygote in flukes and budding in the hydated cysts in *Echinococcus*, further add to multiplication. Bisexuality also increases the reproductive rate, as all individuals lay eggs (Box 28.4).

Box 28.4

High fecundity is a general rule among parasites and the flukes and tapeworms are no exceptions. The dangers of complex life cycles and host location result in extremely high mortality rates that must be offset by increased zygote production or asexual processes. Flukes may produce as many as 100, 000 times as many eggs in free-living turbellarians (see **Brusca** and **Brusca**, 2003).

Fecundity means productiveness of an individual. It is the innate potential reproductive capacity of the individual organism, as denoted by its ability to form and separate from the body the mature germ cells, *i.e.*, capable of producing many offspring.

7. Complex life cycle. The parasitic flatworms generally have complex life cycle due to the formation of several larval forms that inhabit one or more intermediate hosts. This adaptation brings about dispersal and reduces the period of exposure to the hazards of external environment.

28.7 HOST-PARASITE RELATIONSHIP

A successful parasite usually does not kill its host because the death of the host means the end of parasite also. The parasite or its progeny (eggs or larvae) must escape before the host dies. In many cases, a host can tolerate a small population of parasites without any serious consequences. If the effect of a parasitic infection is manifested in the host, the condition is called **disease.** The parasitic disease may damage host cells and tissues, and may block gut, blood vessels or bile ducts. The parasites may produce wastes or substances which have a toxic or allergic effect on the host.

The disease-producing parasites are called **pathogenic parasites** or **pathogens**. Here we can consider two serious human diseases caused by *Taenia solium*: taeniasis and cysticercosis.

1. Taeniasis. Normally only one *Taenia solium* (tapeworm) is found in the intestine of human being. It is because the presence of a tapeworm provides immunity (called **premunition**) to the host against fresh infection.

Adult tapeworm may cause mechanical injury by blocking the intestinal canal and by injuring the mucous membrane of the intestine. Symptoms of **taeniasis** include abdominal pain, increased appetite, weakness, loss of weight, nausea, difficult breathing, digestive disturbances, dizziness, insomenia, restlessness, false sensation and occasional convulsions and epileptic fits.

2. Cysticercosis. This disease is caused by cysticercus larvae. It is much more dangerous than taeniasis. It is caused by eating food contaminated with the oncospheres of tapeworm. It is also caused when the gravid proglottid is destroyed in the intestine of human beings and the released oncospheres and eggs are carried into stomach by reverse peristalsis. It is called **self-infection**. The hexacanth may reach to any part of the body including eyes, heart, spinal cord, brain and bones and form cysticercus cysts. Brain infection leads to epileptic fits, convulsions, violent headaches, giddiness, local paralysis, vomiting and brain tumours. Cysts in the bones cause spontaneous fractures. Symptoms result both from toxins produced by the parasite and mechanical pressure on adjacent tissues caused by the cysts.

(***Note:*** For parasitic adaptations of Nematodes refer Chapter 25).

QUESTIONS

Long Answer Questions

1. What are structural and functional adaptations of platyhelminthes to parasitic mode of life. *(Awadh 2001)*

2. What do you understand by parasitism and parasitic adaptations? Discuss parasitic adaptations in platyhelminthes.

Short Answer Questions

1. Write short notes on:

(*i*) Tegument

(*ii*) Adhesive organs of platyhelminthes.

Very Short Answer Questions

1. Where does *Schistosoma* live?

2. Name the intestinal fluke of human beings.

3. Name a tapeworm having two intermediate hosts?

4. Give the habitat of *Paragonimus*.

5. What is a capsule?

6. Write the name of the disease caused by *Schistosoma mansoni.*

7. Which organ is affected by infection of *Diphyllobothrium*.

Fill in the Blanks

1. Absorptive surface area of tapeworm is increased by the development of ....

2. Parasite gets food and .... from the host.

3. The host containing the adult stage of the parasite is called .... or .... host.

4. An intermediate host harbours the .... stages of the parasite.

5. *Diplozoon* is a .... parasite.

6. .... is a digenetic parasite.

7. *Enterobius vermicularis,* the common parasitic platyhelminth of human beings is found in ....

8. There is only single host in the life history of ....

9. Tapeworm body cannot be dissolved in the human digestive system because it has ....

10. When an intermediate host is not found, eggs and larvae of tapeworm ....

Multiple Choice Questions

1. Among the following, an ectoparasite is

(*a*) *Fasciola* (*b*) *Paragonimus*

(*c*) *Diplozoon* (*d*) *Taenia*

2. An unisexual fluke is

(*a*) *Schistosoma* (*b*) *Opisthorchis*

(c) *Fasciola* (*d*) *Paragonimus*

3. Locomotory and receptor organs are totally missing in

(*a*) liver fluke (*b*) lung fluke

(*c*) intestinal fluke (*d*) tapeworm

4. Flatworms have

(*a*) ectolecithal egg (*b*) endolecithal egg

(*c*) both a and b (*d*) neither a nor b

5. Cysticercus is the larva of

(*a*) sponges (*b*) cnidarians

(*c*) flukes (*d*) tapeworms

6. Tapeworm is

(*a*) anoxybiotic (*b*) oxybiotic

(*c*) aerobic (*d*) none of these

7. Four suckers present in *Taenia solium* is about

(*a*) sucking food (*b*) attachment

(*c*) crushing food (*d*) all of these

8. The parasite showing alternation of generations and alternation of hosts in its life cycle is

(*a*) *Fasciola* (*b*) *Ascaris*

(*c*) *Wuchereria* (*d*) *Taenia*

ANSWERS

Very Short Answer Questions

1. Blood capillaries of bladder and intestine.

2. *Fasciolopsis buski.*

3. *Diphyllobothrium latus* (*Note.* Primary host = human beings; Secondary hosts = a crustacean the cyclop and a fresh water fish).

4. Lungs.

5. Egg plus one or more yolk cells enclosed in a shell.

6. *Schistosoma mansoni* is a blood fluke of humans which cause severe dysentery and anaemia.

7. Human intestine.

Fill in the Blanks

1. Microtriches	**2.** Shelter (protection)	**3.** Definitive, primary	**4.** Young (larvae)
5. Monogenetic	**6.** *Fasciola*	**7.** Caecum and large intestine	**8.** *Polystomum*
9. Cuticle	**10.** Are destroyed.		

Multiple Choice Questions

1. (*c*)	**2.** (*a*)	**3.** (*d*)	**4.** (*a*)
5. (*d*)	**6.** (*a*)	**7.** (*b*)	**8.** (*a*).

29

Parasitic Nematodes of Humans and Plants

29.1 NEMATODES AND HUMAN DISEASES

The human diseases caused due to infestation with the nematode worms constitute **nematodiasis.** Some of the common type of nematode related diseases of human beings are the following:

1. Ascariasis

Ascariasis is a very common disease of human beings caused by the large-sized nematode (roundworm) *Ascaris lumbricoides*. This roundworm is found in the intestine especially of children. In human beings the disease follows the ingestion of embryonated eggs with contaminated food (raw vegetables, fruits) and drinking water. Among children playing in the contaminated soil, there is also hand to mouth transfer of eggs by dirty fingers. Ascariasis is characterized by indigestion, protuberance of abdomen, loss of appetite and intestinal obstruction, insomnia, appendicitis, gastric ulcers, diarrhoea, liver disorders, etc. The larvae while pursuing a migratory course enter the lungs and their heavy infection affects the lungs (causing bronchitis). Ascariasis can be successfully treated by oral administration of piperazine citrate syrup, alkendazole, mebendazole, santonin and oil of chenopodium.

2. Ancylostomiasis

Ancylostomiasis is caused by two hookworms—*Ancylostoma duodenale* and *Necator americanus* (American hookworm). Both the hookworms are parasites within human intestine. They are most frequent in rural areas. Infection occurs, when human beings walk bare-foot on the faecally contaminated soil. Their filariform larvae penetrate directly through host's skin when they come in contact with it. These hookworms cause itching in skin, haemorrhage and bronchial pneumonitis in the lungs in the larval condition and the adults produce weightloss, constipation, diarrhoea, dyspnea (labored breathing), anaemia, nervous disorders, etc. Ancylostomiasis can be treated by following drugs — tetrachloroethane and carbon tetrachloride.

3. Enterobiasis

Enterobiasis (or pinworm disease) is caused by *Enterobius vermicularis* commonly called **pin-worm** or **seat-worm.** Adult pinworms live in caecum, colon and vermiform appendix of human beings (especially children). The females migrate out through colon and rectum and lay large number of eggs in the skin folds near the anus, causing intense itching. These eggs may infect the entire family through clothes, furniture, air and dust in the room. The symptoms of enterobiasis include severe itching around anus, loss of appetite, sleeplessness, bed-wetting, grinding of teeth, nausea and vomiting. Piperazine and enteric-coated capsules of gentian violet are most effective drugs in the treatment of pinworms.

4. Trichuriasis

Trichuriasis (or whipworm disease) is caused by *Trichuris trichura* and *T. ovis* (Fig. 29.1), commonly called **whip-worm.** The adult worms live in caecum and vermiform appendix of human beings. Host is infected when the embryonated eggs are swallowed with food or water. It causes abdomi-nal pain, mucous diarrhoea, often with blood, streaked stool and loss of weight. It may also show symptoms of acute appendicitis. Trichuriasis can be treated by drugs such as stibazium idodide, deftarsone, thiabendazole and mebendazole.

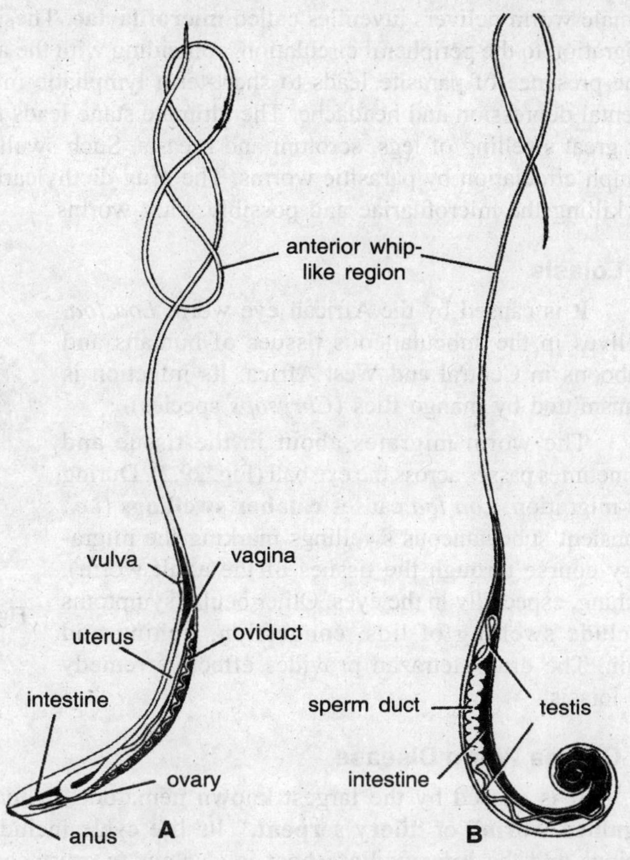

Fig. 29.1. *Trichuris ovis*. A—Female; B—Male.

5. Trichinosis

Trichinosis is caused by *Trichinella spiralis*. It is one of the smallest nematodes infecting human beings. Trichinosis is common and widespread in Europe and America. Humans become infected by *Trichinella* by eating raw or under-cooked meat (pork) containing its encysted larvae. Trichinosis is characterized by eosinophilia, hae-morrhage, muscular pain, thrombosis and restlessness. The drug thiabendazole is found to be effective in the treatment of trichinosis.

6. Strongyloidiasis

This disease is caused by the nematode *Strongyloides stercoralis*, commonly called **thread-worm**. These worms invade the lining (mucous epithelium) of intestine of human beings. The female lays large number of eggs that pass out in faeces and develop into ***rhabditiform larvae***. These then transform into infective ***filariform larvae*** that penetrate the skin of the barefooter, enter blood vessels and pass to the lungs, where they invade alveoli. They ascend trachea, descend the oesophagus and mature to become pathenogenic females in the small intestine. Invading larvae cause dermatitis. Larvae migrating through lungs tend to produce cough, haemoptysis (= expectoration of blood in sputum) and dysponea. Severe infection of the instestine causes vomiting, diarrhoea and constipation. Its hyperinfection results in ulceration, oedoma, congestion, fibrosis and severe inflammation of the intestine. This disease can be treated by the drug thiabendazole.

7. Filariasis or Elephantiasis

Filariasis is caused by *Wuchereria bancrofti,* commonly called the filaria worm. The adult worm inhabits lymphatic vessels of human beings, most frequently those in the lymph nodes, testes and epididymis. Infection is spread through *Culex* mosquito. Following copulation the viviparous female worm delivers juveniles called **microfilariae**. These exhibit a distinct periodicity of larval migration to the peripheral circulation, coinciding with the activity of mosquito (intermediate host). The presence of parasite leads to short-term lymphatic inflammation marked by pain and fever, mental depression and headache. The ultimate stage leads to **elephantiasis** which is characterised by great swelling of legs, scrotum and breasts. Such swellings take place due to blockage of the lymph circulation by parasitic worms. The drug diethylcarbamazine (Hetrazan) is found effective in killing the microfilariae and possibly adult worms.

8. Loiasis

It is caused by the African eye worm *Loa loa*. It lives in the subcutaneous tissues of humans and baboons in Central and West Africa. Its infection is transmitted by mango flies (*Chrysops* species).

The worm migrates about in the tissue and sometimes passes across the eye ball (Fig. 29.2). During its migration, *Loa loa* causes **calabar swellings** (*i.e.*, transient subcutaneous swellings marking the migratory course through the tissues of the adult worm), itching, especially in the eyes. Other ocular symptoms include swelling of lids, congestion, itching and pain. The drug Hetrazan provides effective remedy of loiasis.

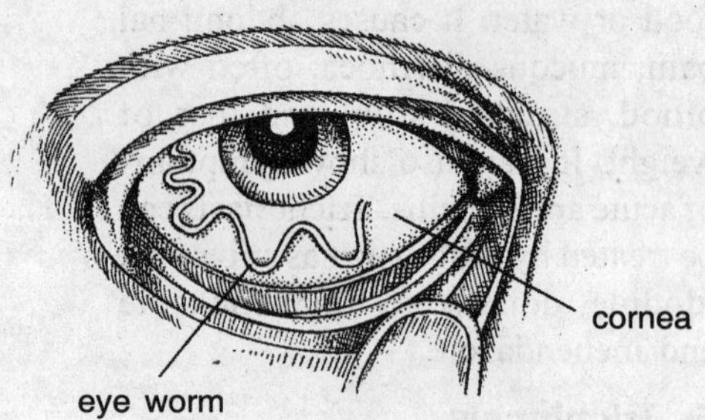

Fig. 29.2. Eye worm *Loa loa* in the cornea of human eye.

9. Guinea Worm Disease

It is caused by the largest known nematode *Dracunculus medinensis*, commonly called **"guinea worm"** or **"fiery serpent."** Its life cycle includes two hosts: **primary host** is human beings and the **intermediate host** is *Cyclops* (a crustacean). Guinea worm lives in the general viscera (coelom) and subcutaneous tissues of arms, legs and shoulders of human beings. It also infests dogs, foxes, racoons and minks. It is commonly found in dry regions of India, Arabia, Tropical Africa, South America, Persia, Turkey, Guyana and China. In India, it is found in Andhra Pradesh, Tamil Nadu, Rajasthan and foot hills of the Himalayas.

The female guinea worm is about 1 mm in diameter and up to 120 cm in length. After a period of development in the body cavity and connective tissue of the host, the gravid female migrates to the subcutaneous tissue and produces an ulcerated opening to the exterior. If the ulcerated area of the host comes in contact with water, larvae are released. After a short free living stage, the larvae are ingested by species of *Cyclops* and continue their development in the host's haemocoel. When definitive host (human being) swallows *Cyclops* in drinking water, the nematode larvae are released and penetrate the intestinal wall to reach the coelom or subcutaneous tissue.

Dracunculus secretes toxic substances that cause asthma, eosinophilia, itching, vomiting; diarrhoea, and giddiness. The worm can be surgically removed but the ancient method, still practised, is to slowly wind them out on a small stick (Fig. 29.3). A safe extraction takes 10 to 11 days. Intramuscular injection of phenothiazine emulsified in olive oil into a number of places close to worm kills the worm. Infection can be prevented by drinking filtered water.

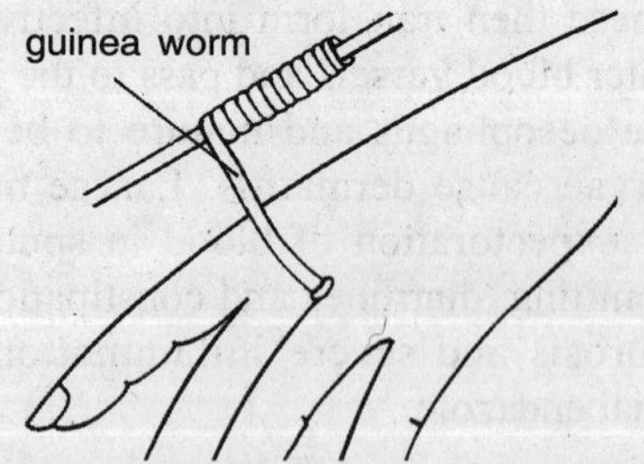

Fig. 29.3. A guinea worm being removed from the ulcerated opening of arm by slow winding on a matchstick.

Box 29.1

The dracunculid and its removal from the human is the basis of the **caduceus** (Fig. 29.4) or symbol of the medical profession. The symbol, however, has been misinterpreted as a staff (wand) with snakes instead of a nematode wrapped around a stick.

Fig. 29.4. Caduceus.

29.2 PLANT NEMATODES

Nematodes parasitic in plants attack all types of commercial crops such as grasses, flowers, vegetables, fruits and trees (see Table 29.1). Because of the damage they can cause to crop plants, many of these plant nematodes are of great economic importance.

Table 29.1 Indian examples of plant nematodes and their host plants.

	Phytoparasitic nematodes	*Host plants*
1.	*Heterodera avenae*	Wheat, barley, oat, maize, sorghum
2.	*Heterodera cajani*	Pigeon - pea
3.	*Heterodera rostochiensis*	Potato
4.	*Meloidogyne incognita*	Tomato, egg-plant
5.	*Meloidogyne javanica*	Okra, tomato
6.	*Meloidogyne hapla*	Red clover, potato
7.	*Rotylenchulus reniformis*	Castor, coffee, papaya, tomato
8.	*Tylenchulus semipenetrans*	Citrus
9.	*Rotylenchus robustus*	Carrot
10.	*Ditylenchus dipsaci*	Onion
11.	*Trichotylenchus astriatus*	Grapewine
12.	*Trophurus similis*	Coconut palm
13.	*Criconema punici*	Pomegranate
14.	*Tylenchorhynchus phaseoli*	Leguminous crop
15.	*Tylenchorhynchus vulgaris*	Maize
16.	*Hirschmanniella shamimi*	Paddy
17.	*Hirschmanniella indica*	Paddy
18.	*Hirschmanniella oryzae*	Rice

Types of Phytoparasitic Nematodes

1. Ectoparasites of plants. Worms feed on the external cells of plants by puncturing the cell wall with oral stylets and sucking out the cell contents. *Pratylenchus*, for example, is an ectoparasite of plant roots.

2. Endoparasites of plants. Worms in juvenile stage enter the plant body and feed on the living cells; producing gall-like structures or causing tissue death. Reproduction takes place within the host, and the new generation of juvenile migrates to other plants. Cyst-forming nematode, *Heterodera,* and root-knot nematode, *Meloidogyne*, have many such species.

3. Phytoparasitic juveniles and zooparasitic female adults. The female worm produces juveniles within a plant-feeding insect host. When the insect punctures the plant tissue, the juvenile worms enter the plant and remain as endoparasites. When mature and after copulation, the female enters the larva of the insect host, which lives on the same plant. The larval host metamorphoses into an adult and deposits a new generation of juvenile worms. *Heterotylenchus aberrans*, a parasite of onion flies, illustrate this type of life cycle. Likewise, adult females of *Tylenchinema oscinellae* live in the haemocoel of fruit flies. Their larvae enter the fly. Larval flies and larval nematodes develop independently on the oat plant and as the nematode mature, the females enter the fly larvae after mating, while the male dies.

General Characters of Plant Nematodes

1. All plant nematodes belong to orders **Tylenchida** and **Aphelenchida**.
2. Phytophagous/phytoparasitic nematodes are small and the plants on which they feed are relatively huge.
3. In plant parasitic nematodes, the buccal cavity carries a long cuticular **oral stylet** which can be protruded from the mouth and used in puncturing plant cells and then the contents of plant cells are sucked out. Oral stylet is formed either by the modified lining of the buccal cavity (**stomatostyle**) or sometimes from a highly modified tooth (**odontostyle**).
4. Plant nematodes generally penetrate inside the host plants through their roots.
5. Feeding of phytoparasitic endoparasites (*e.g.*, *Nacobbus*, *Meloidogyne*) cause hyperplasia and hypoplasia resulting in a **gall**, inducing **giant nutritive cells** that allow the adult worms to be sedentary. **Hypoplasia** means the conditions of arrested development (*i.e.*, poor growth and production). **Hyperplasia** means excessive production of cells in a part.
6. Some plant nematodes show high degree of host-specialization (*e.g.*, *Heterodera*).
7. The exudations (*e.g.*, root leachates) from host plants are attractive to phytoparasitic nematodes; juveniles can find their way toward host plants by this means (*e.g.*, *Heterodera*).
8. Plant parasitic nematodes also exhibit a wide variety of **life history** patterns. In some cases, the males and females live freely in the soil and the juveniles move into and out of root on which they feed and are not restricted to one feeding site. But there are others which penetrate deeply into the plant, become sedentary and undergo sexual reproduction, the resultant juveniles move off into the soil and find new plants.

Examples of Some Plant Nematodes

1. *Meloidogyne marioni* (Root-knot Eelworm)

Root-knot is one of the oldest known diseases of plants. One of earliest references to the root galling attributable to these nematodes is that of **Berkely** (1855) who described and illustrated galls found on the roots of cucumber in English green house (Fig. 29.5). **Chitwood** (1949) reconfirmed the generic name *Meloidogyne*.

The second-stage juveniles of *Meloidogyne* enter the host plant and pierce and suck out the contents of cells. Their feeding causes hyperplasia and hypoplasia resulting in a **gall**, inducing **giant nutritive cells** that allow the adult worms to become **sedentary**. Juveniles mature in the gall. Male nematodes swell as do females, but in the fourth stage moulting they metamorphose into a vermi- form stage. They embark from the home gall into soil to seek other infected plants, where they fertilize the female and die. The females become more and more plump, losing all resemblance to the cylindrical larvae and several hundred ova form within them (Fig. 31.6B). Like *Heterodera*, females of *Meloidogyne* become semiendoparasitic, so that only the head and anterior end of their body remain embedded in host tissues and eggs are deposited outside the host.

In addition to cause direct damage to host plants, *Meloidogyne* is involved in spreading the infection of certain fungal diseases such as *Fusarium* wilt of cotton.

2. *Heterodera* (Cyst-forming Nematode)

This genus was created by **Thome** (1949). It includes all the true **cyst-forming nematodes**, *i.e.*, those nematodes in which some or all of the eggs are retained within the swollen body of the female; her body wall after her death becomes oxidized and tanned to form a tough, resistant or protective container or lemon- or pear-shaped **cyst** for the developing eggs. *Heterodera* also exhibits following two peculiar features which are not known in plant parasitic nematodes: 1. It has a high degree of host-specialisation; 2. Its hatching process depends on stimulatory substances (root leachates) emanating from the host roots. Thus, members of *Heterodera* are highly adapted to a plant parasitic existence. Species of Heterodera are best suited to relatively cool climates.

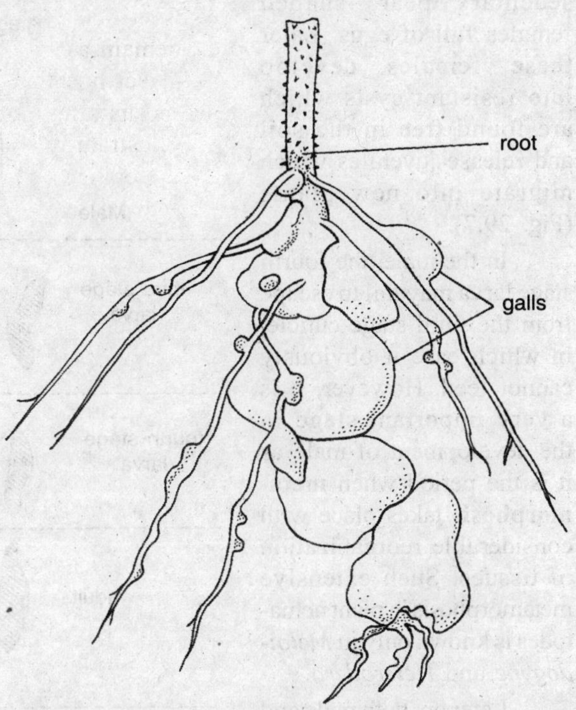

Fig. 29.5. Galls of root-knot nematode ('vibrio') on cucumber root.

Both male and female juveniles of *Heterodera* penetrate the host plant (via roots). In the female, all larval stages are separated by distinct complete moulting and the third and fourth stage larvae have well developed **oral stylets.** They continue to grow and enlarge to become

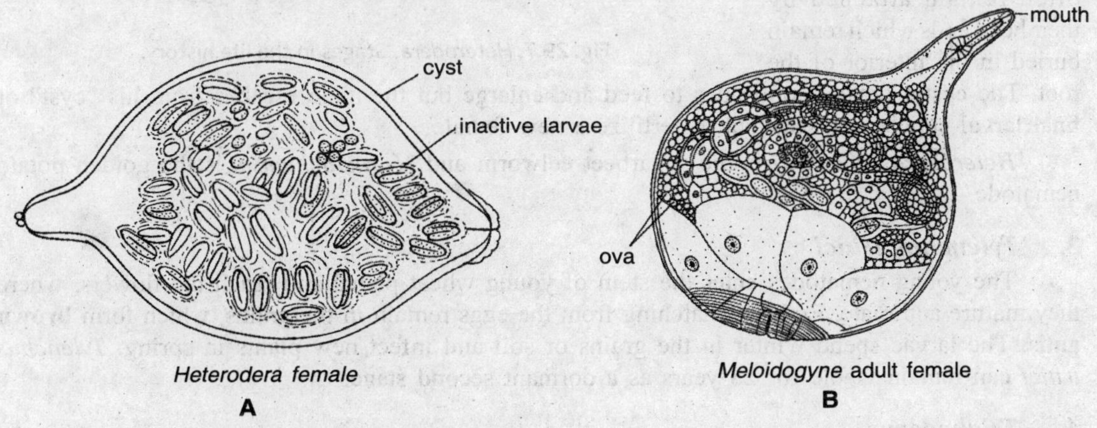

Fig. 29.6. A—*Heterodera* female after conversion into cyst filled with inactive larvae. B—*Meloidogyne*, adult female with maturing ovaries.

sedentary pear -shaped females full of eggs. Later these females develop into resistant cysts which are found free in the soil and release juveniles which migrate into new plants (Fig. 29.7).

In the male, the fourth stage larva may fail to escape from the third stage cuticle, in which case it obviously cannot feed. However, it is a very important stage in the development of male as it is the period when **metamorphosis** takes place with considerable reorganization of tissues. Such extensive metamorphosis in plant nematodes is known only in *Meloidogyne* and *Heterodera*.

Later on, both male and female of *Heterodera* become secondarily **semi-endoparasitic**: the outer root tissues rupture under the pressure of the enlarging parasites within, so that these burst through to the surface. However, they often remain attached by their head ends which remain buried in the interior of the root. The cystic female conti-nues to feed and enlarge but the male escapes from his "cyst" or final larval cuticle and seek out to fertilize a new female.

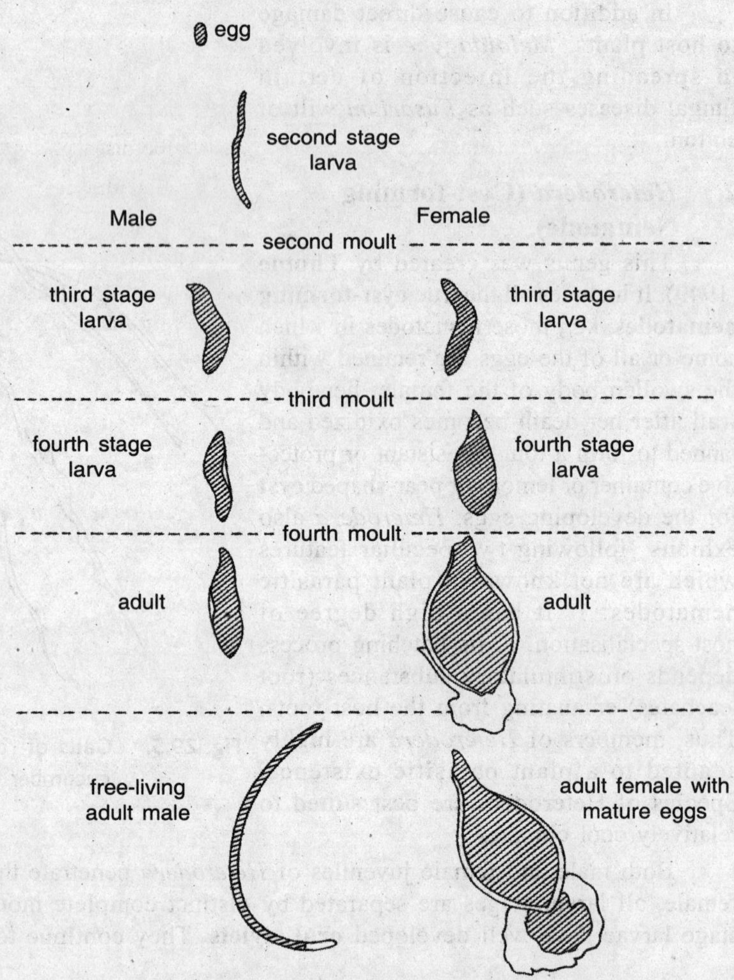

Fig. 29.7. *Heterodera*. Stages in the life history.

Heterodera schachtili is the sugarbeet eelworm and *H. rostochiensis* is the golden potato nematode.

3. *Tylenchus tritici*

The young nematodes enter the stem of young wheat plants and reach the flowers, where they mature and mate. Embryos hatching from the eggs remain in the grains, which form **brown galls**. The larvae spend winter in the grains or soil and infect new plants in spring. *Tylenchus tritici* can remain viable for 28 years as a dormant second stage.

4. *Trichodorus*

It is called the **stubby root nematode**. It is ectoparasite on vegetables, fodder, grain and fruit crops. Its infection checks the growth of root tip, producing stubby roots.

QUESTIONS

Long Answer Questions

1. Write an essay on nematodes and disease.
2. Write names of human nematode parasites. Describe diseases caused by any three of them.
3. Name any three nematodes which are harmful to man and describe the life-history of one nematode studied by you.
4. Discuss the economic importance of Nemathelminthes.
5. Write early history of plant nematodes. Describe four phytophagus nematodes giving their common names and plants they cause damage. Add a note on their control.
6. Name the common plant nematodes. Give a brief account of any one of them.

Short Answer Questions

1. Suggest some measure to control phytoparasitic nematodes.
2. Write short notes on the following:
   (*i*) Plant parasitic nematodes
   (*ii*) Root gall
   (*iii*) Root-knot nematode

Very Short Answer Questions

1. How human beings get infection of *Ascaris*?
2. What protects the gut nematodes from digestive enzymes?
3. Name a nematode that is parasitic in the adult stage and free-living in the young stage.
4. Which parasitic nematode is introduced into the human body by a mosquito?
5. Name any nematode parasite that occurs in the human muscles.
6. Cite one example of the order Oxyuroidea.

Fill in the Blanks

1. The vector of elephantiasis is ....
2. *Dracunculus medinensis* (guinea worm of humans) is a parasitic worm which invades ....
3. Nematode parasites that attack plants are called ....
4. .... are parasites which infect animal.
5. The nematode parasite that lives in the lymph is ....
6. The nematode .... attacks wheat plants.
7. The nematode parasite of horse is ....

Multiple Choice Questions

1. The intermediate host of guinea worm is
   (*a*) house fly (*b*) *Cyclops*
   (*c*) mosquito (*d*) none of the above
2. *Dracunculus medisensis* is a parasite of
   (*a*) body tissue (*b*) body fluids
   (c) gut (*d*) lungs

3. The common parasite of human large intestine is

(*a*) *Ascaris* (*b*) *Trichinella*

(c) *Enterobius* (*Oxyuris*) (*d*) *Ancylostoma*

4. *Enterobius vermicularis* inhabits the human

(*a*) large intestine (*b*) caecum

(*c*) appendix (*d*) all these

5. Trichinosis is caused by

(*a*) *Trichuris* (*b*) *Trichinella*

(*c*) *Ancylostoma* (*d*) none of these

6. *Trichuris trichura* is commonly known as

(*a*) *Ascaris* (*b*) pinworm

(*c*) whipworm (*d*) roundworm

7. The scientific name of pinworm or seatworm is

(*a*) *Trichinella* (*b*) *Enterobius*

(*c*) *Ancylostoma* (*d*) *Wuchereria*

8. The transmission of infection caused by *Loa loa* is through

(*a*) mango flies (*b*) *Anopheles*

(*c*) *Culex* (*d*) housefly

ANSWERS

Very Short Answer Questions

1. By swallowing eggs of *Ascaris* in water or food.

2. Thick cuticle of body wall.

3. *Ancylostoma duodenale.*

4. *Wuchereria bancrofti.*

5. *Trichinella spiralis.*

6. *Enterobius* (= *Oxyuris*) *vermicularis.*

Fill in the Blanks

1. *Culex* **2.** Subcutaneous tissue

3. Phytophagus **4.** Zoophagus

5. *Wuchereria bancrofti* **6.** *Tylenchus tritici*

7. *Parascaris equonium* (= *Ascaris megalocephala*).

Multiple Choice Questions

1. (*b*) **2.** (*a*) **3.** (*c*) **4.** (*d*)

5. (*b*) **6.** (*c*) **7.** (*b*) **8.** (*a*).

30

Minor Phyla

30.1 PHYLUM ROTIFERA

Rotifera (L., *rota*=wheel; *ferre*=to bear) is also called **Rotatoria**. It represents a **pseudocoelomate** minor phylum. Rotifera includes about 1800 species of minute (or microscopic), bilaterally symmetrical, unsegmented and aquatic animals. Rotifers are commonly called "**wheel animalcules**" because their anterior end bears a **ciliated crown** or **corona** which on beating gives the appearance of a rotating wheel.

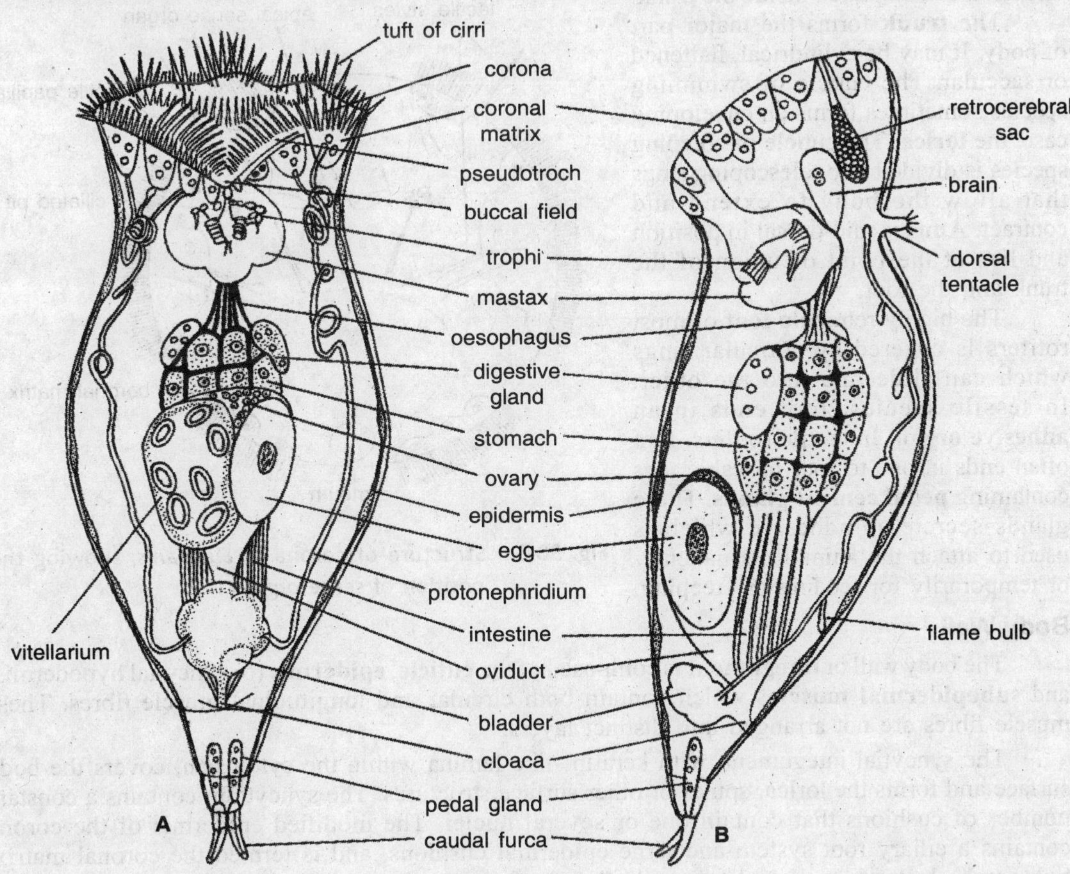

Fig. 30.1. Rotifer body plan (*Epiphanes senta*). A—Ventral view; B—Lateral view.

Habit and Habitat

Rotifers are cosmopolitan in distribution. They are found in fresh and marine water as well as in moist terrestrial habitats among mosses and lichens. Some float passively, others creep or swim actively, some are sessile and live in cuticular tubes and a few form colonies. A few rotifers are parasitic.

Structure

1. Size and shape. Rotifers are the **smallest metazoans**, approximately of the size of ciliate protozoans. Most species measure 0.1 to 1 mm in length. Their body is composed of about 1000 cells.

Shape of rotifers is highly variable. Bottom dwellers are worm-like; epizoic (attached) species are **flower-like** and floating species are usually **globular**.

2. Body plan. The body is divided into **head**, **trunk** and **tail** or **foot**. A **neck** may separate the head and trunk.

The head bears a retractile crown or disc, called **corona** or **trochal disc** that surrounds mouth. The corona of primitive, creeping species is composed of an evenly ciliated **buccal field** arranged around the mouth. Swimming rotifers usually have a reduced buccal field. In some rotifers, the corona bears two rings (bands or circlets) of cilia. The outer or posterior band is called **cingulum** and inner or anterior **trochus**. The cilia beats in a circular manner (metachronally) to give an impression of a revolving wheel. Corona is used in locomotion, feeding and respiration.

The head bears a central **mouth**, a pair of **eyes** and other sensory projections (Fig. 30.2). The head can be pulled inside the trunk.

The **trunk** forms the major part of body. It may be cylindrical, flattened or saccular. The cuticle of swimming species sometimes forms an enveloping case, the **lorica**. The cuticle of creeping species is divided into telescoping rings that allow the body to extend and contract. **Anus** is mid-dorsal in position and lies at the point of union of the trunk and the foot.

The highly retractile **foot** of most rotifers is covered by cuticular rings which can slide one into the other. In sessile species, foot ends in an **adhesive organ**. In motile rotifers, foot often ends in one to four movable toes containing **pedal cement glands**. These glands secrete an adhesive which is used to attach the animal permanently or temporarily for feeding or creeping.

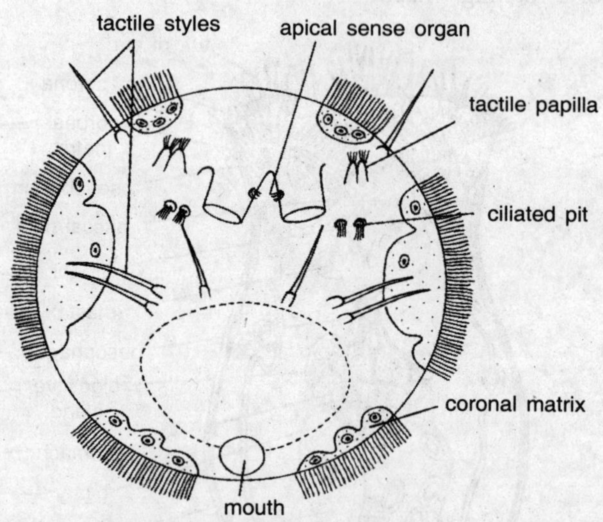

Fig. 30.2. Structure of corona of *Euchlanis*, showing the position of sense organs.

Body Wall

The body wall or **integument** is composed of the **cuticle**, **epidermis** (or syncytial hypodermis) and **subepidermal muscles** which contain both circular and longitudinal muscle fibres. These muscle fibres are not arranged into distinct layers.

The syncytial integument, with keratin -like lamina within the cytoplasm, covers the body surface and forms the lorica, spines or other surface structures. The syncytium contains a constant number of cushions that contain one or several nuclei. The modified epidermis of the corona contains a ciliary root system and large epidermal cushions, and is termed the coronal matrix. Sub cerebral glands, pedal glands and other surface glands are epidermal in origin but extend below the epidermis.

Body Cavity

It is a spacious **pseudocoel**. It is filled with perivisceral fluid containing loose syncytial branching network of amoeboid cells. It helps in internal transport. Circulatory system is absent. Respiration is mainly aerobic, sometimes anaerobic.

Locomotion

Rotifers can move about either by swimming or crawling. The beating of the coronal cilia moves the animal about in swimming. Pelagic forms maintain their buoyancy with fat globules. Leech-like crawling movement of rotifers is performed by the help of foot and its associated glands.

Digestive System

Rotifers are either **suspension feeders** or **raptorial** (=living on prey). Gut is typically developed in most rotifers. **Mouth** is ventral and is surrounded by corona. Sphincter and dilator muscles may open and close the mouth. **Pharynx** is called **mastax**. It is a muscular chamber containing a set of jaws, formed of cuticularized trophi. Jaw structure is complex (Fig. 30.3). A median **fulcrum** and pairs of **rami**, **unci** and **manubria** make up the major parts. Pharyngeal jaws of rotifers are most important distinctive structures; they are used in the identification of a species. The mastax helps in food capture and its maceration.

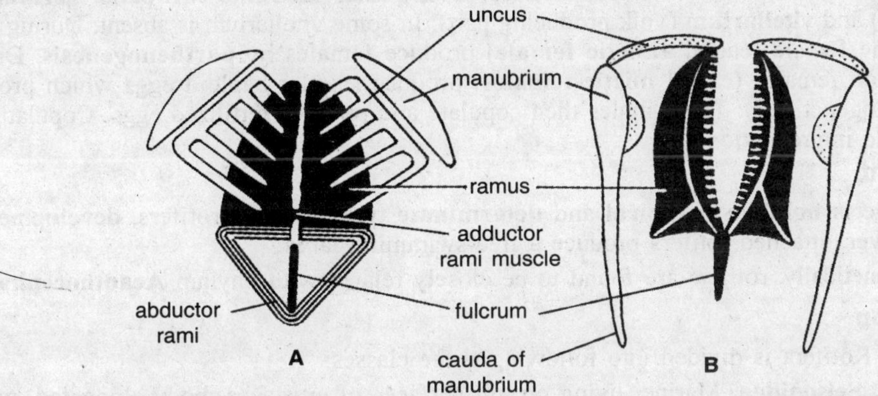

Fig. 30.3. Two types of trophi in a rotifer. A—Malleate type; B—Forcipulate type.

Pharynx opens into **oesophagus** which leads into **stomach**. Stomach is thick-walled, ciliated sac composed of constant number of cells. A delicate mesh of circular and longitudinal muscles covers its surface and a pair of syncytial **gastric glands** open into stomach. Gastric glands secrete enzymes for **extracellular digestion**. A thin-walled, syncytial and straight **intestine** is sometimes ciliated. A posterior **cloaca** receives the gut, protonephridia and oviducts.

While **feeding**, a suspension feeder remains attached to a piece of debris, etc. Rapid beating of the cilia draws a current of water towards the mouth. The water carries protozoans and algae into the mouth. The predatory rotifers shoot out their armed mastax to catch the prey. The mastax acts as pincers or forceps. **Digestion** is both extracellular and intracellular. Intracellular digestion occurs in syncytium of stomach. Carbohydrate digestion is limited. Digestion of proteins and lipids is very rapid.

Excretory System

Rotifers have two syncytial **protonephridial** tubules which bear terminal **flame bulbs** (=cells). Posteriorly the protonephridia drain into cloaca by way of common duct or a **urinary bladder**. Protonephridial system is a water-regulating device, also maintaining salt concentration of pseudocoel. Rotifers easily tolerate dehydration and freezing.

Nervous System

It is bilaterally symmetrical. A bilobed **brain** gives rise to a pair of longitudinal ventrolateral nerve trunks. Several ganglia are connected to the nerve trunks. Peculiar features of nervous system of rotifers include: (1) Clear separation of sensory and motor elements; (2) Exhibition of the phenomenon of constancy of cells (called **eutely**), *i.e.*, brain cells and nerve cells are constant in number and position.

Rotifers contain a variety of sense organs. Examples of sensory organs include photoreceptive **eyes** (*e.g.*, cerebral eyes, coronal eyes or lateral eyes), chemoreceptive **ciliated pits**, tactile **antennae** and stiff bristles (styles).

Cyclomorphosis

Rotifer phenotypes can be environmentally controlled. The form of amictic females gradually changes during the year in response to environmental conditions, a process called **cyclomorphosis**.

Reproductive System

Rotifers reproduce sexually. Sexes are separate. Males are generally small and degenerate. They lack the digestive and excretory systems. They live only for few days. Rotifers produce two types of sperms: **ciliated sperms** and **rod-shaped sperms**. The rod-shaped sperms probably help to pierce the cuticle of female integument. Some rotifers have cirrus or penis.

Females contain single or paired ovaries. Ovary is divided into two parts: **germarium** (ovary proper) and **vitellarium** (yolk producing part). In some vitellarium is absent. During most of the year the female (called **amictic female**) produce females by **parthenogenesis**. During breeding season females (called **mictic females**) produce smaller haploid eggs which produce males parthenogenetically. These males then copulate and produce fertilized eggs. Copulation is by hypodermic impregnation.

Development

Cleavage is **holoblastic, spiral** and **determinate** type. In most rotifers, development is **direct**. However, attached rotifers produce a free-swimming larva.

Phylogenetically, rotifers are found to be closely related with phylum **Acanthocephala**.

Classification

Phylum Rotifera is divided into following **three** classes:

Class 1. Seisonidea. Marine, living on gill surfaces of crustacea, body elongated, with a long neck, jaws (trophi) as forceps; small corona, paired gonads, without vitellaria, males present.

Example. *Seison.*

Class 2. Monogononta. Swimming or sessile, males with testis, females with one ovary (or germovitellarium).

Examples. *Conochilus*, *Hexarthra*, *Collotheca*, *Notommata*, *Brachionus*, *Keratella*, *Asplanchna*, *Gastropus*.

Class 3. Bdelloidea (Digononta). Swimming or creeping forms, with retractable anterior end, mastax adapted for grinding and contains one pair of teeth, males absent, females with two ovaries or germovitellaria).

Examples. *Philodina roseola*, *Rotaria*, *Embata*.

30.2 PHYLUM SIPUNCULA

Phylum **Sipuncula** or **Sipunculoidea** (L., *sipunculus*=a little siphon) is a **coelomate minor phylum** and includes 320 species. Sipunculans are commonly called **peanut worms** because they resemble a peanut when fully contracted (they do so when disturbed).

Habit and Habitat

Sipunculans are widespread bottom-dwelling marine animals that are mainly found in shallow water (littoral zone) of warmer seas, but also occur in polar waters and abyssal depths. They are sedentary creatures that burrow in the sand or mud (*e.g.*, *Sipunculus*) or occupy any protected den of appropriate size, whether a rocky crevice, coral reef or an annelid tube or a molluscan shell (*e.g.*, *Phascolion*) or even foraminiferan shell. Sipunculans are mostly detritus or deposit feeders.

Structure

1. Shape and size. Sipunculans are light brown coloured worms that measure 2 mm to more than 72 cm in length. Most are less than 10 cm.

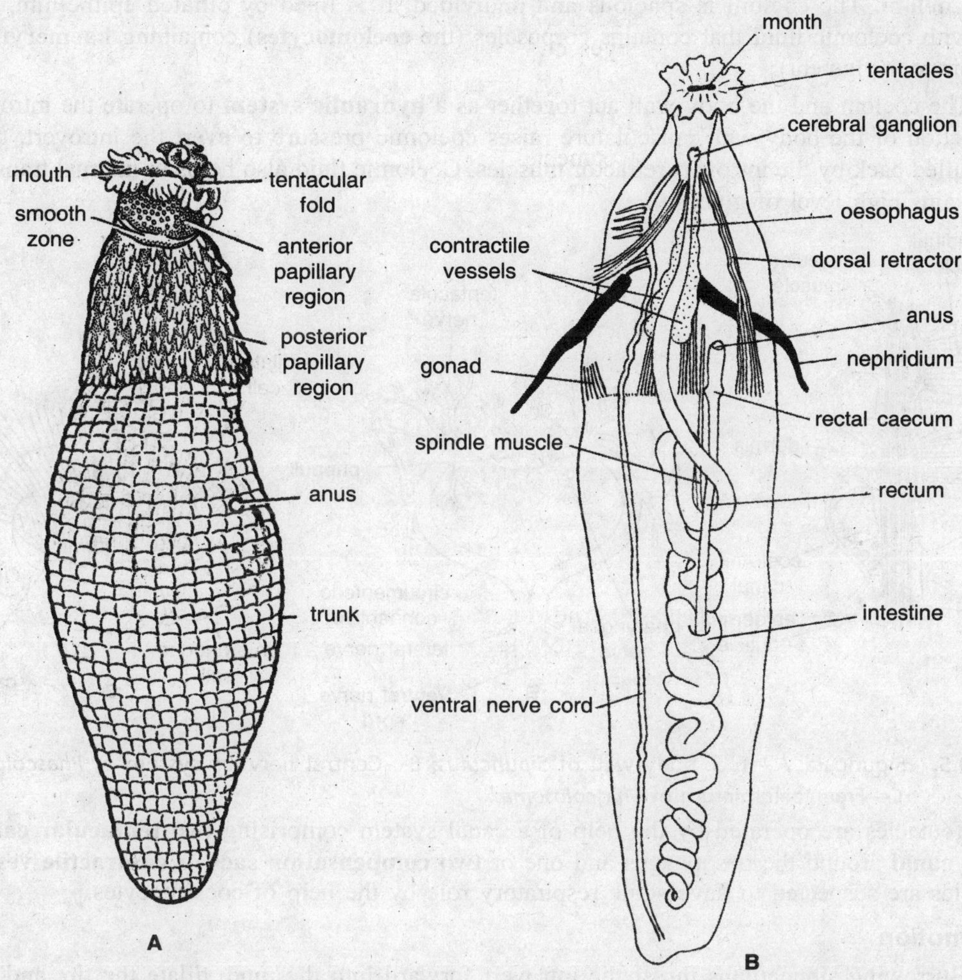

Fig. 30.4. *Sipunculus*. A—External structure; B—Internal structure.

2. Body plan. Sipunculans have a very simple external form. The body is cylindrical and vermiform without segmentation, appendages, spines or parapodia. It is divided into anterior long or short iridiscent **introvert** and posterior **trunk**.

The introvert represents head. It is narrow, funnel-shaped anterior part of body that can be involuted inside the trunk. **Mouth** lies at the tip of introvert and is surrounded by a pleated **tentacular fold** which looks like a circlet of tentacles. Tentacles are hollow, ciliated, conical, filiform or digitiform processes that bear deep **ciliated groove** on their surface. A smooth zone, called **collar**, may be found immediately behind the tentacles. Minute hooks or spines and sensory or glandular papillae often cover the surface of the introvert. The **trunk** is cylindrical bears only sensory and glandular papillae. **Anus** is mid-dorsal, at the anterior end near the base of introvert.

Body wall. Body wall of sipuncula is of annelid type in histology. It is composed of following layers: (1) **cuticle;** (2) **epidermis;** (3) **dermis** which is formed of thin connective tissue; (4) **circular muscles;** (5) **longitudinal muscles.** The circular and longitudinal muscles of *Sipunculus* can be seen through the body wall giving it a graded appearance.

Coelom. The coelom is spacious and undivided. It is lined by ciliated epithelium. It is filled with coelomic fluid that contains corpuscles (the **coelomocytes**) containing **haemerythrin** (a respiratory pigment).

The coelom and the body wall act together as a **hydraulic system** to operate the introvert. Contraction of the body wall musculature raises coelomic pressure to evert the introvert, to be later pulled back by the introvert retractor muscles. Coelomic fluid also helps in internal transport (it contains high level of glucose).

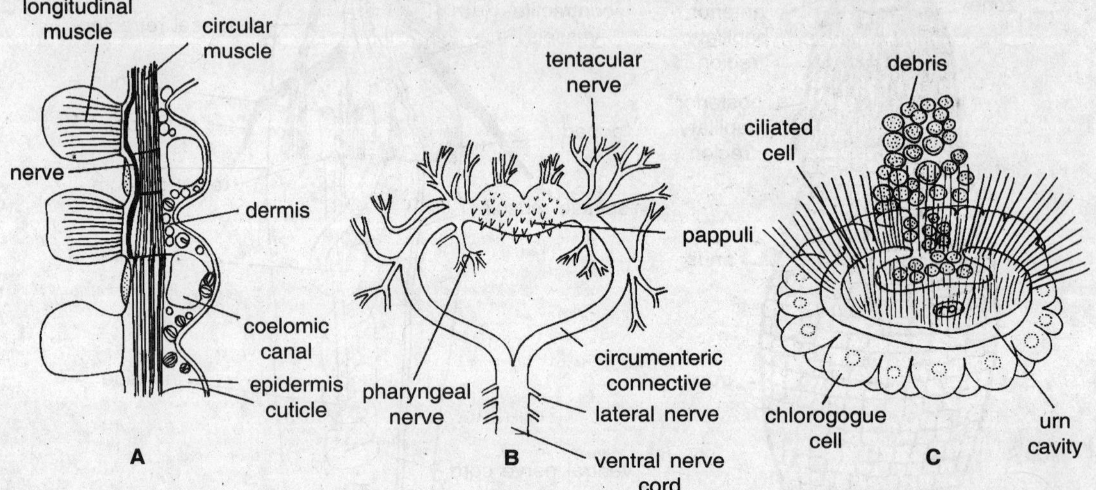

Fig. 30.5. Sipuncula. A—T. S. body wall of *Sipunculus*; B—Central nervous system of *Phascolopsis*; C—Free coelomic urn of *Phascolosoma*.

Tentacles are operated by the help of a canal system comprising two **tentacular canals**, a **ring canal** around the oesophagus and one or two **compensation sacs** or **contractile vessels**. Tentacles are suspected to have some respiratory role by the help of coelomocytes.

Locomotion

Burrowing sipunculans thrust the introvert forward into the mud, dilate the tip, and then contract the retractor muscles. This pulls the body forward.

Digestive System

Alimentary canal includes a large **mouth**, a short muscular **pharynx**, slender **oesophagus**, recurved and coiled **intestine**, **rectum** and **anus**. Most sipunculans being deposit feeders eat sand and mud and subsist on whatever digestible organic material that may yield. Diatoms, protozoa, larvae and small particles of organic detritus provide most of their nourishment. Digestion and absorption occur in the descending arm of intestine and ascending arm is involved in faeces formation.

Excretory System

The excretory organs are pair of large, saccate, brownish **metanephridia** with open nephrostomes and nephridiopores. The brownish glandular walls of the nephridia excrete nitrogenous waste (*e.g.*, ammonia) into the tubule.

Some sipunculans have curious peritoneal structures known as **coelomic urns** (Fig. 30.5C). Each urn is like a vase capped with a ciliated cell. Urns may be found fixed to the peritoneum or they may get detached and swim in the coelomic fluid. The free urns collect excretory material and form either brown bodies or trailing aggregates. The latter are discharged through the nephridia. Annelid-like chlorogogue tissue remains attached with urns.

Nervous System

It is annelidan type with the brain (**cerebral ganglion**) lying above the oesophagus, a pair of **circumoesophageal connectives** and a single **ventral nerve cord**. Ventral nerve cord lacks segmentation. Sensory organs include a pair of pigmented spots or ocelli embedded in the brain and sensory cells in **nuchal organs** or **ciliated pits** at the end of introvert.

Reproductive System

Sexes are separate (dioecious) but no sexual dimorphism. Ovaries and testes are simple masses of cells shed into the coelom. Ripe ova and sperms leave the body through nephridia. Fertilization is external.

Development

Cleavage is spiral. Development is **indirect** producing a typical **trochophore larva**. Phylogenetically sipunculans are probably related to the annelids.

Classification

Phylum Sipuncula is divided into following two classes:

Class 1. Sipunculidea. Tentacles encircling mouth; introvert hooks are simple.

Example. *Sipunculus.*

Class 2. Phascolosomatidae. Tentacles are confined to arc surrounding dorsal nuchal organ; introvert hooks recurved.

Examples. *Aspidosiphon*, *Phascolion.*

30.3 PHYLUM ECHIURA

Phylum Echiuroidea or **Echiura** (Gr., *echis* = an adder; *aura* = tail) is a small coelomate minor phylum including about 140 species. Echiurans are commonly called **spoon worms.**

Habit and Habitat

Echiurans are marine, bilaterally symmetrical, unsegmented, cylindrical and worm-like animals. They lack setae and parapodia. They are widely distributed in the shallow coastal waters of warm

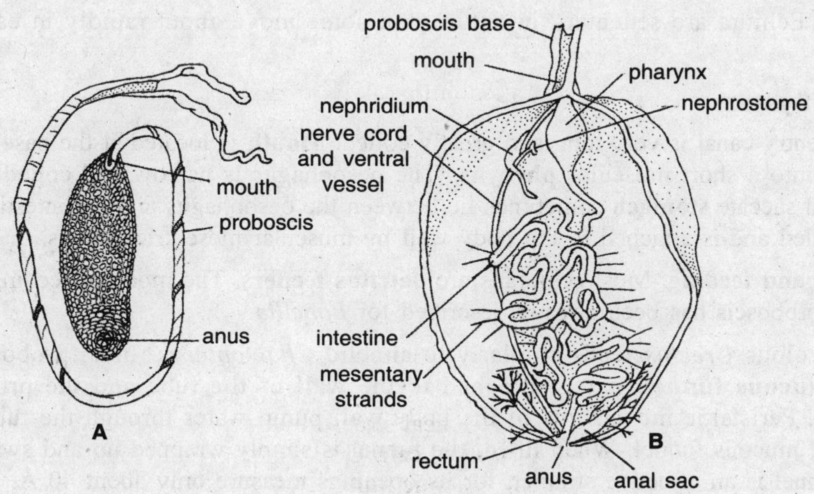

Fig. 30.6. *Bonellia viridis.* A—External structure; B—Internal structure.

and temperate seas. Echiurans are mostly bottom dwellers. Most of them dig and live in U-shaped mucus-lined tubes or burrow in sand or mud. Some live under stones, rocks and in coral crevices. Most echiurans have cryptic habits, dwelling in burrows and crevices, only thrusting out the remarkable proboscis as a food finder.

Structure

Shape, size and colour. Bilaterally symmetrical body of echiurans is cylindrical, ovoid or T-shaped. They are generally grey, red, yellow or green in colour. They are about 25 mm in length. But *Ikeda,* found in Japan, contains a trunk of 40 cm and a 1.5-m-long proboscis.

Body plan. Body of echiurans is differentiated into a cylindrical **trunk** and a prominent anterior **proboscis**. The proboscis is the centre for chemoreception and for food procurement. It is a large, flattened, spoon-like extension of the head. It contains brain, so is considered homologous to the prostomium of the annelids. Proboscis is very extensible. For example, in *Bonellia viridis* the proboscis is about 7 cm long when contracted; it can be extended fully to as much as 1.5 m almost transparent thread. However, like the Sipuncula, proboscis cannot be retracted into the trunk. Proboscis secretes a mucus cover and contains a **ciliated groove** on its ventral surface. Minute food organisms and small particles (detritus) stick to mucus and are conveyed to the mouth by the groove and muscular movements of the proboscis itself. The **mouth** lies at the base of the proboscis.

Trunk is a simple bag of guts and gonads. Its surface is either smooth or covered with **papillae** which are usually arranged in rings or are irregularly distributed. A pair of annelid-like hooked **setae** are present on the ventral surface near the anterior end of trunk. Some echiurans such as *Echiurus* possess one or two circles of **anal setae** around the posterior end of the trunk.

Body Wall and Coelom

Body wall of echiurans is highly muscular and similar to that of annelids and sipunculids. It consists of **cuticle**, **epidermis**, **dermis**, layers of **longitudinal** and **circular muscles. Coelom** is spacious and lined by ciliated coelomic epithelium. It contains a coelomic fluid which has amoebocytes and haemocytes containing haemoglobin. **Coelomic urns** are absent.

Locomotion

Most Echiura are sedentary in nature, but some move about rapidly in captivity (*i.e.*, aquarium).

Digestive System

Alimentry canal is very long and greatly coiled. **Mouth** is located at the base of proboscis and opens into a short muscular **pharynx**. The oesophagus is narrow and coiled. A muscular **gizzard** and saccate **stomach** sometimes lie between the oesophagus and the intestine. **Intestine** is long, coiled and is attached to the body wall by muscular mesentric strands.

Food and feeding. Most echiurans are **detritus feeders**. The mode of feeding via ciliated groove of proboscis has been already described for *Bonellia.*

Tubicolous *Urechis* feeds similarly to annelid *Chaetopterus*. In it, proboscis secretes a slimy, **mucous funnel** that is attached to the wall of the tube and the proboscis base (Fig. 30.7). Peristaltic movements of the body wall pump water through the tube, pulling it through the mucous funnel. When filled, the funnel is simply wrapped up and swallowed, The mucous funnel is an effective strainer, for its openings measure only about 40 Å.

Excretory System

Echiurans do excretion by sipunculid-like **metanephridia** which vary widely in number. For example *Bonellia* has one, whereas *Ikeda* has hundreds. Most species, however, have from one to three pairs of metanephridia. A pair of modified nephridia or **anal vesicles** open into rectum. These are contractile, simple or branched. They contain many **ciliated funnels** which open into coelom, Anal vesicles remove waste from the coelomic fluid and discharge them into the rectum.

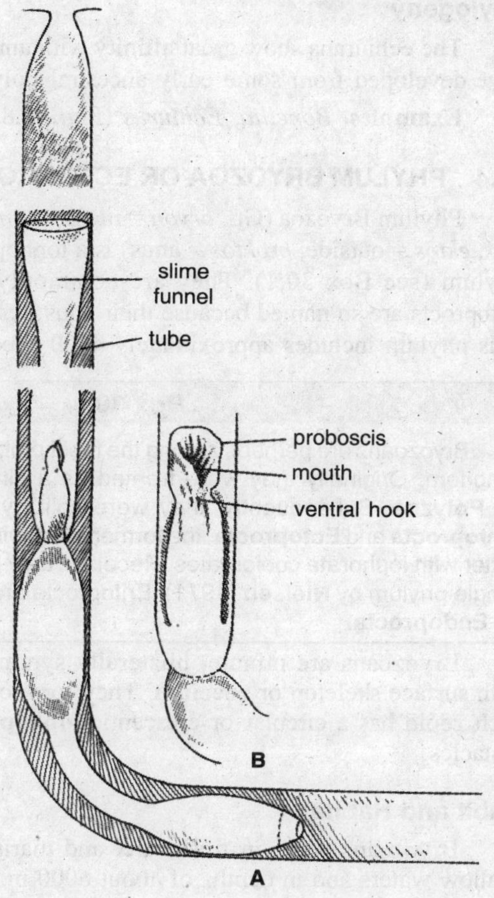

Fig. 30.7. *Urechis caupo*, in its tube with mucus funnel.

Respiration

Haemoglobin in the coelomocytes appears to serve primarily for oxygen storage during period of low oxygen availability, such as when the tide is out.

Circulatory System

It is simple and contains **dorsal vessel** and **ventral vessel** both connected by contractile **circum-intestinal** and **circum-oesophageal vessels.**

Nervous System

It is simple and built on the annelid pattern, but it contains no ganglia. The circumenteric ring is drawn out into a long loop that follows the proboscis margin and connects with ventral nerve cord. Echiurans lack special sense organs.

Reproductive System and Development

Sexes are separate and sexual dimorphism is well marked in some echiurans, *e.g., Bonellia.* The trunk of *Bonellia viridis* measures 8 cm while the male is only 1 to 3 mm in length. Male *Bonellia* is ciliated and has no proboscis. Its alimentary canal is reduced and without mouth and anus.

Gonads are specialized regions of ventral mesentery. Gametes are shed into the coelom before they reach maturity. After they have ripened, they escape through the nephridia. **Fertilization** occurs externally in ocean except in *Bonellia* and *Hamingia*. In these genera, the ova are fertilized in the nephridia and remain there during development. **Cleavage** is spiral type and development is **indirect type** producing free-swimming **trochophore** larva like the annelids (this larva shows metameric segmentation).

In case of *Bonellia*, any larva that settles by itself develops into female; it becomes sexually mature in one year. A larva that settles near to or in contact with female becomes a male (**Jaccarini** *et al.*, 1983). Under the influence of hormonal secretion, the development of the male is retarded and he remains a small, ciliated organism with many juvenile characteristics. He may cling to proboscis of the female for a time, but eventually enters the female mouth and takes up residence in a fold of the nephridium. Male becomes sexually mature in one or two weeks and lives the rest of his life as an **endocommensal**; a female may contain up to 20 males.

Phylogeny

The echiurans show great affinity with annelids. It seems that they have developed from some early ancestral polychaete.

Examples. *Bonellia, Echiurus* (Fig. 30.8), *Urechis, Ikeda.*

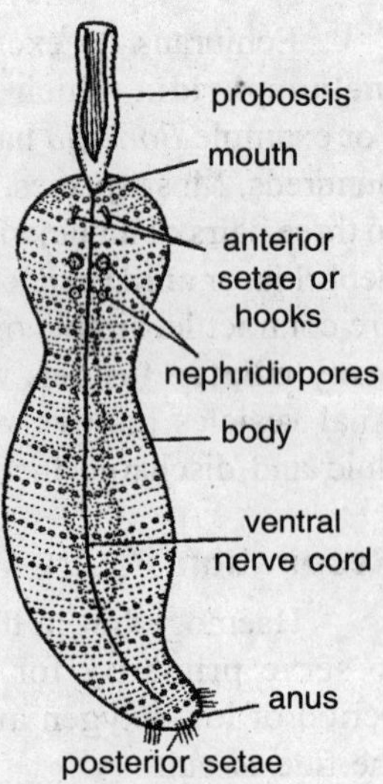

Fig. 30.8. *Echiurus.*

30.4 PHYLUM BRYOZOA OR ECTOPROCTA

Phylum Bryozoa (Gr., *bryon* = moss; *zoon* = animal) or Ectoprocta (Gr., *ektos* = outside, *proktos* = anus) is a lophophorate coelomate minor phylum (see Box 30.1). They are commonly called **moss animals.** Ectoprocts are so named because their anus lies outside the lophophore. This phylum includes approximately 4000 species.

Box 30.1

Bryozoans are perhaps among the most problematic of all in the animal kingdom. Originally they were treated as a single group, the **Bryozoa** or **Polyzoa**. Subsequently, they were split by **Hyman** into two phyla, **Entoprocta** and **Ectoprocta**, the former allied with pseudocoelomates, the latter with lophorate coelomates. Recently, they have been rejoined into a single phylum by **Nielsen** (1971). Entoprocta are also called **Kamptozoa** or **Endoprocta**.

Bryozoans are minute, bilaterally symmetrical sessile, colonial unsegmented coelomates with surface skeleton or **zoecium**. They form colonies of **zooids** by budding. On its anterior end, each zooid has a circular or crescentic **lophophore** bearing a series of slender ciliated post-oral tentacles.

Habit and Habitat

Bryozoans occur in freshwater and marine habitats everywhere in the world. They live in shallow waters and in depths of about 6000 m. They are also abundant in both polar and tropical seas. Their colonies resemble the colonies of coelenterates and look like sea-weeds. Bryozoan colonies are found attached to rocks, shells, sea-weeds, pilings and to other animals.

Structure

1. Size, shape and colour. An individual (zooid) measures about 0.5 mm in length. A large colony may contain as many as 2,000,000 zooids. A colony or **zoarium** of ectoproct may measure one to 50 centimetres or more in diameter. A zooid is box-like, oval or cylindrical. Mostly their colonies form a thin and flat sheet, called **sea mats**. The colonies occur in many shapes, such as encrusting, foliose, branching and globular. Colour of most bryozoan colonies is white or pale, some are red, brown, orange, blue or violet.

2. Body plan. Zooid is box-like, vase-like or tubular and is enclosed in a nonliving cuticular covering, called **zoecium**. The opening of zoecium to exterior is known as **orifice** and is covered with a lid, called **operculum**. Lophophore of zooid protrudes out of the orifice.

Each zooid contains two types of tissues: (1) **Cystid** which includes body wall tissues; (2) **Polypide** which includes business part of animal such as lophophore and the internal organs. Externally, the body of zooid is differentiated into **introvert** and **trunk**. Introvert is the anterior part of the zooid. Its free distal end bears **mouth** which is surrounded by a circlet of ciliated **tentacles**. The disc bearing mouth and tentacles is called **lophophore**. It may be circular or horse-shoe shaped. The latter bears more tentacles than the former type of lophophore. Tentacles drive the food towards mouth and also serve as tactile and respiratory organ. In some forms (*e.g.*, *Lophopus*) a ciliated lobe, called **epistome**, overhangs mouth on the anal side.

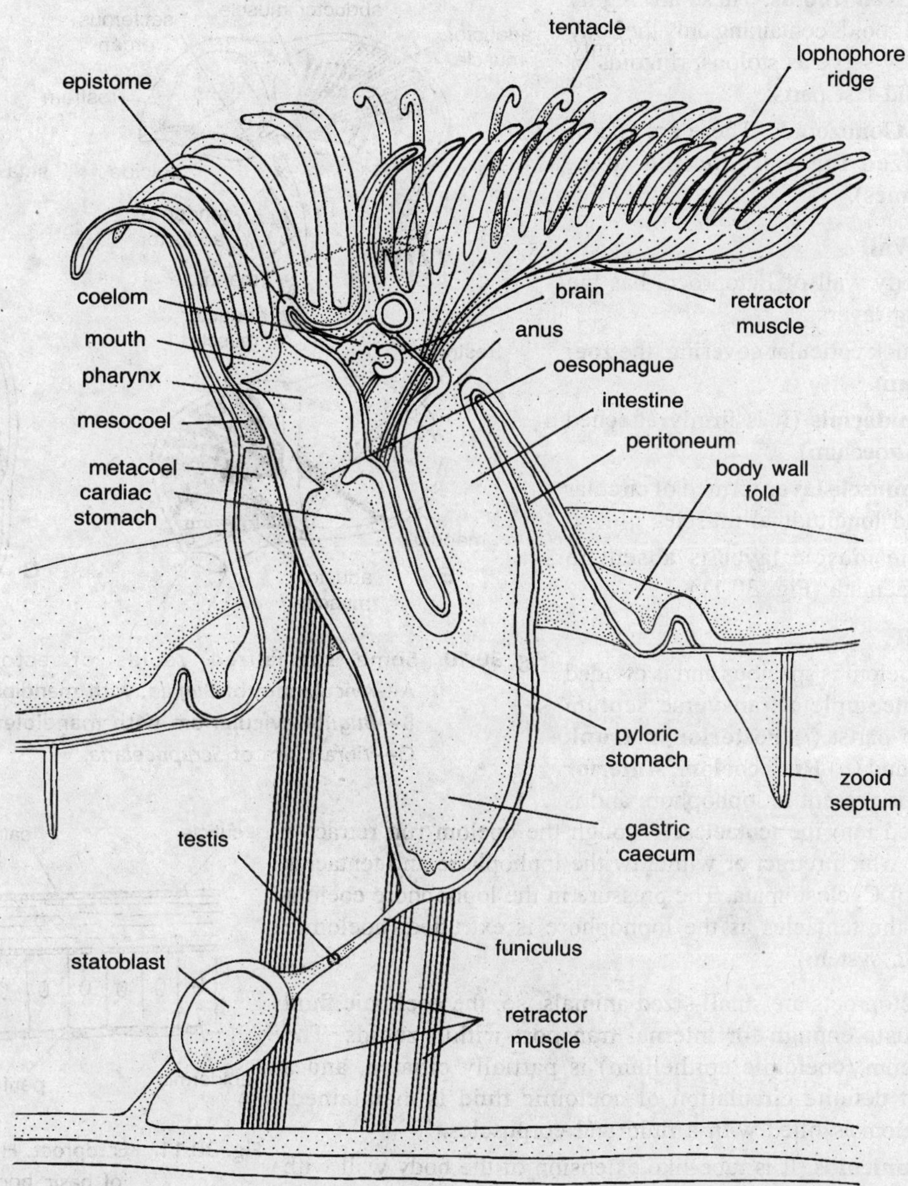

Fig. 30.9. *Cristatella*. Anatomy.

Polymorphism. Bryozoan colonies are polymorphic. They contain following types of zooids:

1. Autozooids. These are most numerous complete feeding zooids containing lophophores.

2. Vibracula. They are a type of defensive zooids in which the operculum becomes modified into a long movable bristle. These sweep away debris and settled larvae.

3. Avicularia ('little bird'). They resemble the head of a little bird. They contain flexible neck and a pair of strong beak-like pincers (Fig. 30.10). They are used to repel or crush intruders (settling larvae).

4. Kenozooids. These are highly modified zooids containing only the body wall. They serve as stolons, rhizoids or other hold-fast parts.

5. Gonozooids. These are zooids specialized for reproduction (*e.g.*, cyclostomes).

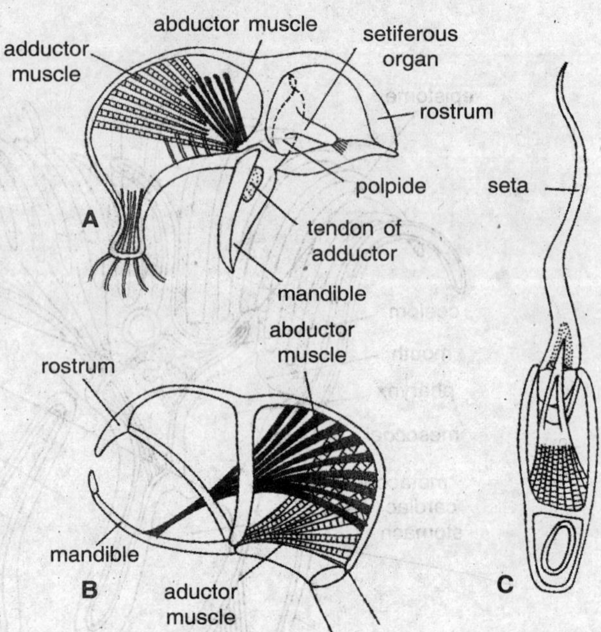

Fig. 30.10. Some specialized zooids of ectoprocta. A—Avicularium of *Bugula*, with mandible open; B—*Bugula* avicularium with mandible closed. C—Vibraculum of *Scrupocelaria*.

Body Wall

Body wall of Ectoprocta has the following layers:

(*i*) Thick cuticular covering, the **zoecium**.

(*ii*) **Epidermis** (It is firmly attached to zoecium).

(*iii*) A **muscle layer** formed of circular and longitudinal muscles.

The muscle layer is absent in Gymnolaemata (Fig. 30.11).

Coelom

Coelom is spacious and is divided by an incomplete transverse septum into two parts: (*i*) **Posterior** or **trunk coelom** and (*ii*) **Ring coelom. Anterior coelom** is present in lophophore and is prolonged into the tentacles. Through the coelom run retractor muscles which retract or withdraw the lophophore and tentacles in case of Cyclostomata. The pressure in the lophophore coelom inflates the tentacles as the lophophore is extruded (coelomic hydraulic system).

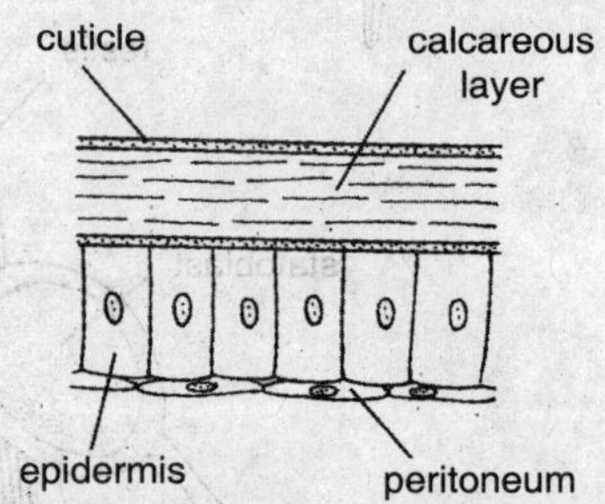

Fig. 30.11. Ectoproct. Histology of basic body wall.

Ectoprocts are small-sized animals, so, the coelomic fluid is adequate enough for internal transport within zooids. The peritoneum (coelomic epithelium) is partially ciliated, and a constant definite circulation of coelomic fluid is maintained. The coelom is filled with a fluid and corpuscles.

Funiculus. It is tube-like extension of the body wall with muscle fibres which are attached to the stomach. In phylactolaemates and cyclostomates, funiculus is found to contain fluid-filled cylinders with wall of peritoneal cells and also longitudinal muscles. Funiculus is a homologue of the circulatory system of brachiopods and phoronids.

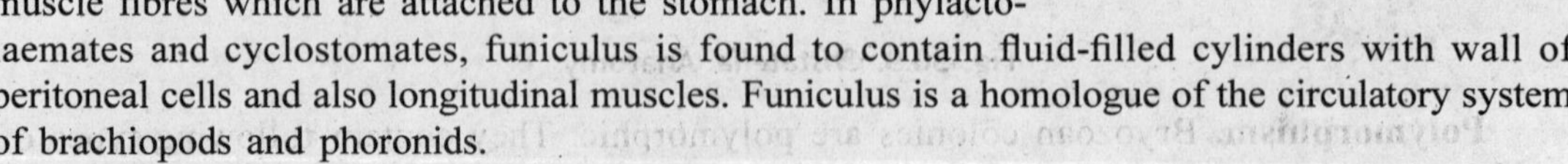

Special **respiratory** and **excretory organs** are absent.

Digestive System

Alimentary canal of Ectoprocta is V-shaped. **Mouth** is surrounded by lophophore. **Pharynx** is short, broad, ciliated and muscular. **Oesophagus** is long, narrow and delicately muscled. **Stomach** is differentiated into three parts: **cardia, caecum** and **phylorus**. The caecum forms the bottom of U and the **rectum** is sharply set off from the pylorus. A thick-walled **gizzard** containing grinding denticles occurs at the base of oesophagus in case of some Ctenostomata.

Bryozoans filter their planktonic and detritus food particles from the surrounding water by altering the direction of beat of the impinging cilia. Ciliated currents are caused by the lophophore. The cilia on the lateral surfaces of the tentacles pull water down and out between the tentacles, directing detritus particles toward the mouth. Large particles are rejected by retraction or muscular movements of the tentacles and small particles are delivered to the mouth by ciliary currents at the lophophore base.

Digestion is very rapid and wholly extracellular. Starch is digested slowly. Proteins are hydrolyzed in stomach and absorbed there. Fat globules are phagocytized by stomach epithelial cells.

Respiration

The exchange of respiratory gases occurs at the body surface, no doubt largely through the lophophore.

Excretory System

Coelomic leucocytes collect the wastes, which are ejected out either through the lophophore or along with **brown bodies** through the alimentary canal (called **storage excretion**). Sometimes waste materials become incorporated into a body that eventually be shed or be retained in an inert condition.

Nervous System

There are no special sense organs but sensory cells are provided for reception of **tactile, chemical** and **water current stimuli.**

The nervous system is simple. In gymnolaemates, there occurs a **nerve ring** around the pharynx. A small cephalic ganglion, with peripheral neurons and central neuropile, is attached to the ring and lies within the lophophoral coelom. Two motor and two sensory fibres pass to each tentacle from the ring.

Reproductive System

Freshwater ectoprocts reproduce asexually and sexually. The **asexual reproduction** takes place by four methods: (*i*) budding; (*ii*) hyperhacula, (*iii*) brown bodies and (*iv*) by statoblast formation. Budding is also used in the growth of the colony.

Statoblasts (Fig. 30.12). Buds of statoblasts appear on the funiculus. They grow and differentiate into a dark, resistant covering and a mass of yolky, undifferentiated germinal cells. Statoblasts are of three types: (1) **Floatoblasts** which contain air-filled cells in the capsule wall; (2) **Sessoblasts** which are attached to the zoecium wall by cement; (3) **Siphonoblasts** with spiny projections. Statoblasts are liberated when polypide degenerate. They are resistant and protect the species during the winter. In spring, they germinate to form a new zooid when the environmental conditions become favourable. Statoblasts are used in species identification.

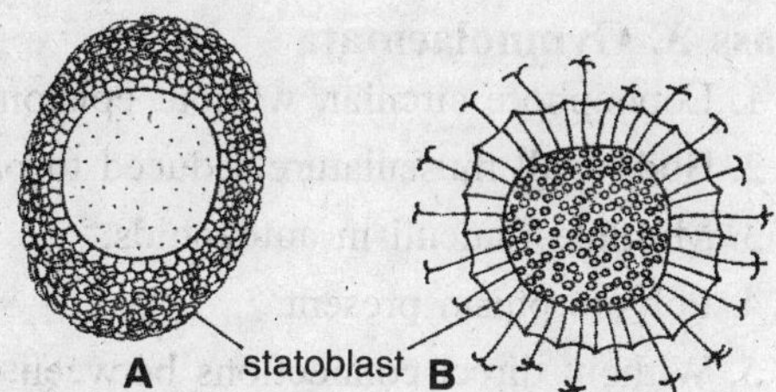

Fig. 30.12. Statoblast. A—of *Plumatella*; B—of *Cristatella*.

Sexual reproduction. Most marine ectoprocts are protandric hermaphrodites. The **gonads** develop in the peritoneum of the coelom, on the body wall, gut or funiculus. All phylactolaemates are hermaphroditic with **ovaries** on the body wall and **testes** on the funiculus. Eggs and sperms are shed into the coelom where fertilization takes place. The fertilized eggs pass out of the coelom through a special pore. However, in many bryozoans the eggs are brooded, either in the coelom, or in a modified chamber (the **ovicell** in *Bugula*) of the zooid. Development is **indirect**. They produce a modified larva, either plankton feeding **cyphonautes larva** (*e.g.*, *Frustrella*) or **coronate larva** (*e.g.*, *Bugula*) that subsists on yolk and develops in brood chamber. The larva gets attached with substratum and metamorphose to produce first zooid of new colony. The zooid undergoes budding and forms the colony.

Phylogeny

Ectoprocta show affinities with three invertebrate phyla namely—Phoronida, Brachiopoda and Endoprocta.

Classification

Phylum Ectoprocta is divided into following **three** classes:

Class 1. Phylactolaemata

1. Zooids with horse-shoe shaped lophophore.
2. Epistome present.
3. Body wall with musculature.
4. Zoecium incomplete and not calcified.
5. Coelom of adjacent zooids is continuous.
6. Polymorphism of zooids is absent.
7. Exclusively freshwater species.

Examples. *Lophopus, Cristatella, Plumatella, Pectinatella, Stolella, Fredricella (Fig. 30.13 B), Plumatella.*

Class 2. Stenolaemata

Class Stenolaemata includes one living order.

Order Cyclostomata

1. Tubular, fully calcified zoecium having terminal, typically circular orifice with operculum but the orifice closed by perforated membrane.
2. Coelom divided by membranous sac into inner and outer regions, used in lophophore extrusion.
3. Polymorphism common, with formation of gonozooids, kenozoids and nannozooids.
4. Many young formed by polyembryony from single embryo in brood.
5. Exclusively marine species.

Example. *Crisia* (Fig. 30.13 C).

Class 3. Gymnolaemata

1. Lophophore circular, without epistome.
2. Body wall musculature reduced to parietal muscle.
3. Multiple funiculi in autozooids.
4. Polymorphism present.
5. Without direct connections between coeloms of adjacent zooids.
6. Cyphonautes larva.

Class Gymnolaemata is divided into following **two** orders:

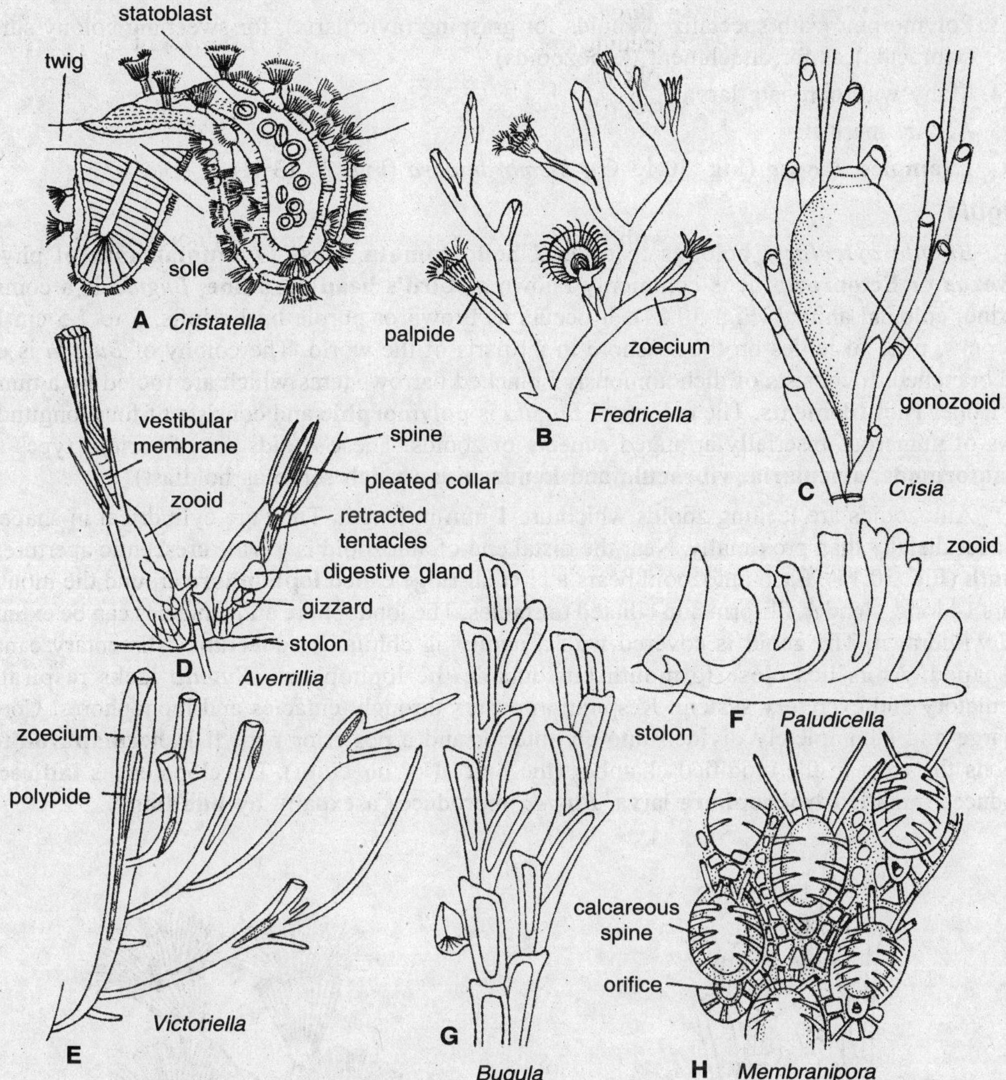

Fig. 30.13. Ectoprocta. A—*Cristatella*; B—*Fredricella*; C—*Crisia*; D—*Averrillia*; E—*Victoriella*; F—*Paludicella*; G—*Bugula*; H—*Membranipora*.

Order 1. Ctenostomata

1. With simple, flexible zoecium of chitinous material having terminal or subterminal orifice, without closing plate.
2. Only two types of zooids: feeding zooids and stolonic zooids.
3. Most are marine, some freshwater.

Examples. *Averrillia*, *Victoriella*, *Paludicella* (Fig. 30.13F).

Order 2. Cheilostomata

1. With membranous or calcareous, box-like zoecium.
2. Zoecium with subterminal opening and typically, with hinged lid, operculum.

3. Polymorphic with specialized zoöids for grasping (avicularia), for sweeping colony surface (vibracula) or for attachment (kenozooids).
4. Many with coronate larva.
5. All are marine.

Examples. *Bugula* (Fig. 30.13 G), *Membranipora* (Fig. 30.13 H).

Bugula

Bugula avicularia belongs to order **Cheilostomata**, class **Gymnolaemata** of phylum **Bryozoa** or **Ectoprocta**. It is commonly known as **bird's head coralline**. *Bugula* is a common marine, colonial animal (Fig. 30.14). It occurs as brown or purple bushy tufts, 5 to 7.5 cm long on rocks, piles of jetties on the seashore in all parts of the world. The colony of *Bugula* is erect and branched. It consists of dichotomously branched narrow stems which are rooted by a number of slender **root filaments**. The colony of *Bugula* is **polymorphic** and consists of four longitudinal rows of numerous biserially arranged **zooecia** of zooids. These zooids are of various types such as **autozooids, avicularia, vibracula** and **kenozooids** (which serve as holdfast).

Autozooids are feeding zooids which are 1 mm in length. They are cylindrical in shape but broader distally than proximally. Near the distal end of autozooid is a wide crescentic aperture, the **mouth** (Fig. 30.15). Each autozooid bears a circular ridge called **lophophore** around the mouth; it bears 14 long, slender, filiform and ciliated **tentacles**. The lophophore and tentacles can be extended and withdrawn. The zooid is covered with a nonliving chitinous **zooecium**. Alimentary canal is U-shaped. **Anus** lies close to mouth but outside the lophophore. *Bugula* lacks respiratory, circulatory and excretory system. Respiration occurs through tentacles and lophophores. Coelom is large and incompletely divided into an anterior and a posterior part. It is **hermaphrodite**. It broods the eggs in the modified chamber (the **ovicell** or **ooecium**). Development is **indirect**. It produces modified **trochophore** larva. *Bugula* reproduces asexually by **budding**.

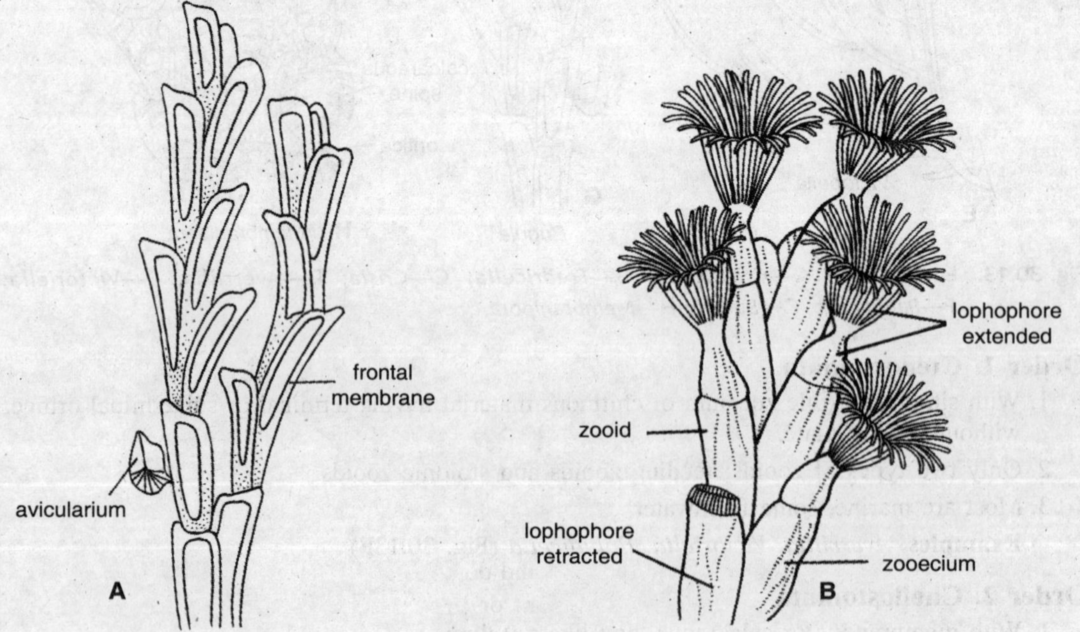

Fig. 30.14. *Bugula* colony A—A part of colony; B—Colony with extended lophophore.

Avicularia consists of a **peduncle** which is attached to the zooids of colony, a **head** which is, in fact, the modified zooecium (zoecium) and a **mandible** that is modified operculum (Fig. 30.15). The reduced parts of the polypide and cystid can be seen in the interior. The most conspicuous internal elements are the opening (**abductor**) and closing (**adduc-tor**) muscles. Avicularia snap viciously when disturbed; they grasp and hold small organisms that visit the colony. Thus, they prevent fouling of the colony by other sessile organisms.

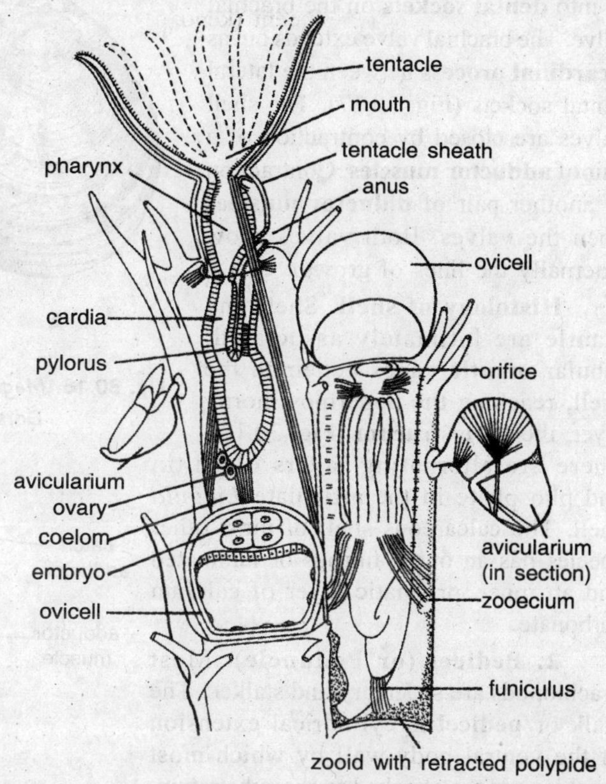

Fig. 30.15. *Bugula*. Structure of two zooids. Right zooid with retracted polypide; left zooid with extended polypide.

30.5 PHYLUM BRACHIOPODA

Phylum Brachiopoda (Gr., *brachion* = arm; *podos* = foot) is a group of lophophore-bearing coelomate animals which resem-ble bivalve molluscs in possessing a mantle and a calcareous bival-ved shell. However their two shell valves are **dorsal** and **ventral;** dorsal valve being smaller than the ventral one. Brachiopods are commonly called **lamp shells**. The phylum Brachiopoda includes about 45,000 extinct species and 350 living species.

Phylum Brachiopoda made its appearance in Cambrian period (about 590–505 million years ago) and reached its peak of evolutionary development during Ordovician period.

Habit and Habitat

Brachiopods are entirely marine and found from the intertidal zone to the deep sea. Most species live attached to the rocks but some brachiopods such as *Lingula* live in burrows in sand or mud. They ate ciliary feeders.

Structure

Brachiopods range from 5 mm to 8 cm in length. Their shape may be oval, spherical, flattened or elongated. Some species contain transparent shells. Their body colour may be yellowish grey, orange, pink or red. The body includes the following parts:

1. Shell. The body is bilaterally symmetrical and dorsoventrally flattened. It is enclosed in a **shell** which is formed of two **valves**. One is **dorsal** or **brachial valve** and other is **ventral** or **pedicle valve** (Fig. 30.16). Dorsal valve is smaller and fits over the larger ventral valve. The apex of the ventral valve forms a beak which bears a hole. It looks like a Roman lamp. In inarticulate brachiopods, the two valves are held together by muscles only. In articulate brachiopods, interlocking processes form a **hinge**. Such a hinge occurs at the posterior margin of the valves and consists of hinge teeth on the pedicle valve which fit into **dental sockets** on the brachial valve. The brachial valve extends out as a **cardinal process** between the lateral dental sockets (Fig. 30.17). The shell valves are closed by contraction of a pair of **adductor muscles**. Contraction of another pair of **diductor muscles** open the valves. Both valves show externally the lines of growth.

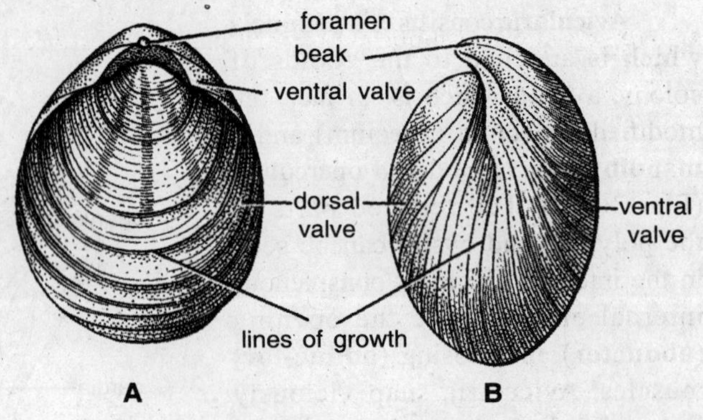

Fig. 30.16. *Magellania* (an articulate brachiopod) shell. A—Dorsal; B—Lateral.

Histology of shell. Shell and mantle are intimately associated; tubular **mantle caeca** penetrate the shell, reaching the uppermost horny layer, the **periostracum** (Fig. 30.18). There are alternating layers of chitin and phosphate in the articulate *Lingula* shell. The calcareous shell of most other species has an outer, fibrous or laminated and an inner, prismatic layer of calcium carbonate.

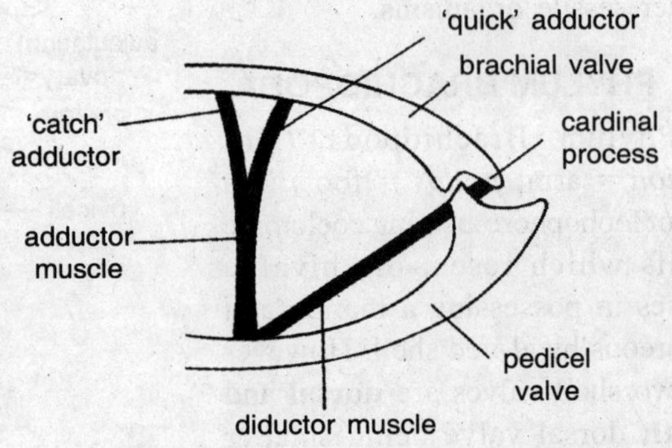

Fig. 30.17. Articulate brachiopod shell. Hinge arrangement.

2. Pedicel (or Peduncle). Most brachiopods are sedentary and stalked. The stalk or **pedicel** is cylindrical extension of the ventral body wall by which most brachiopods are attached to the substratum. Pedicel may be long, fluid-filled and muscular (in Inarticulata) or short and without muscles (in Articulata). The pedicel has been permanently lost in some brachiopods such as *Crania* (Inarticulata) and *Lacazella* (Articulata).

Inarticulates (*Lingula*) can contract the pedicel and achieve some evasive movements by moving deeper into sediments (Fig. 30.21).

3. Mantle. The two mantle lobes or folds surround the body, line the valves of the shell and enclose a large **mantle cavity**. Mantle folds are extensions of body wall. The body proper of brachiopods occupies only the posterior part of the mantle cavity. Anterior part of mantle cavity is occupied by the lophophore. The edge of the mantle bears prominent, stiff **setae**, formed by single chaetoblasts. Setae are protective and sensory in function.

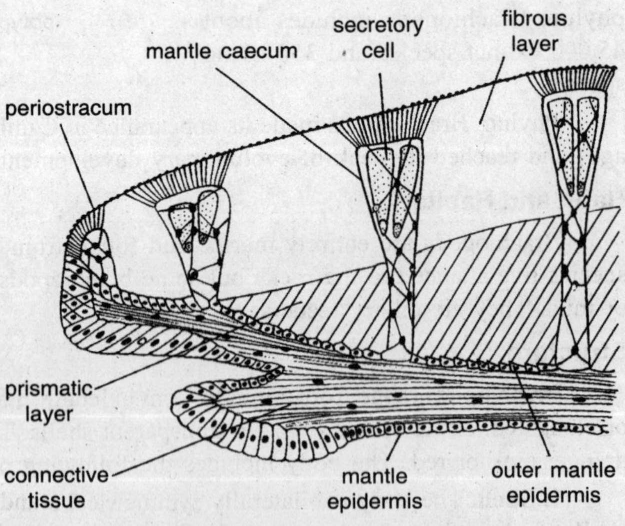

Fig. 30.18. L.S. through mantle and shell valve of an articulate brachiopod. Note caeca that extend into calcareous and fibrous parts of shell, conferring characteristic porous texture.

4. Lophophore. The body wall forms a discoid or horse-shoe shaped fold, the **lophophore**, around the mouth. Each arm of lophophore is coiled in complicated way. Each arm bears a row of numerous ciliated **tentacles**. Coiling of lophophore greatly increases the food collecting area.

The inner surface of brachial or dorsal valve is often sculptured to accommodate the lophophore, and articulates bear the **brachidium** or **lophophore** skeleton (Fig. 30.19).

Body Wall and Coelom

The body wall consists of an outer layer of epithelium, a layer of connective tissue and a muscle layer.

Coelom is large and divided incompletely into two lateral halves by a longitudinal mesentery. Each half is again divided into a **protocoel**, a **mesocoel** and a **metacoel** by two incomplete septa. Branches of mesocoel extend into tentacles, and branches of the metacoel extend into mantle.

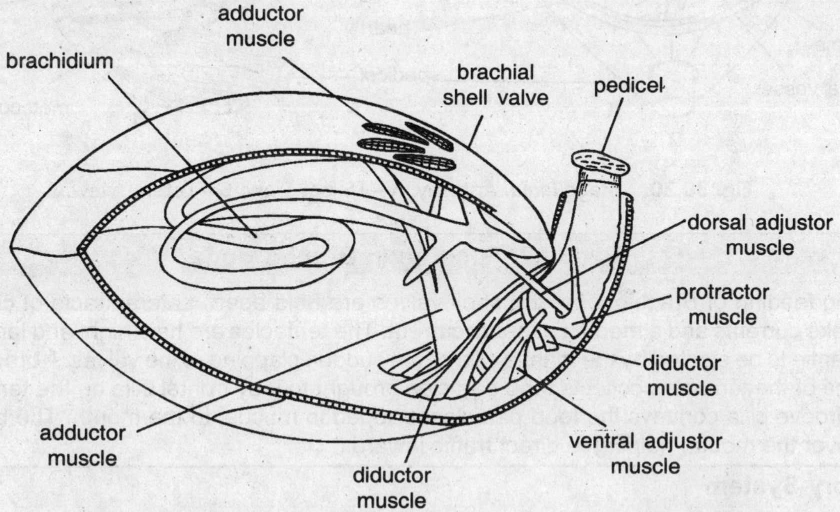

Fig. 30.19. *Magellania*. Muscles, muscle scars, brachidium and pedicel.

Coelom is filled with the coelomic fluid which contains several types of corpuscles, including haemerythrin containing corpuscles. Peritoneal cilia or flagella circulate the coelomic fluid. The fluid moves rapidly in the narrow passageways of the mantle (*i.e.*, tubular coelomic channels called **mantle canals**) and tentacles and was once thought to be blood.

Digestive System

Alimentary canal is V-shaped in Articulata (*Magellania*) having mouth, oesophagus, stomach with paired gastric glands or liver and a narrow and blind intestine (Fig. 30.20). In articulata, the intestine is looped or coiled having **anus** which opens anteriorly in the mantle cavity (metacoel). Gastric glands are probably sites of intracellular digestion and absorption (*e.g.*, *Lingula*).

The inarticulates subsist on particulate matter and are especially adept at sorting edible and inedible particles as they enter the tentacle field (see Box 30.2). They require an anus to void wastes. Articulates derive most of their nutrition from the absorption of dissolved and colloidal organic nutrients in sea water (*e.g.*, polar and subpolar areas). The gut, since is dealing with a low-waste food source, lacks an anus.

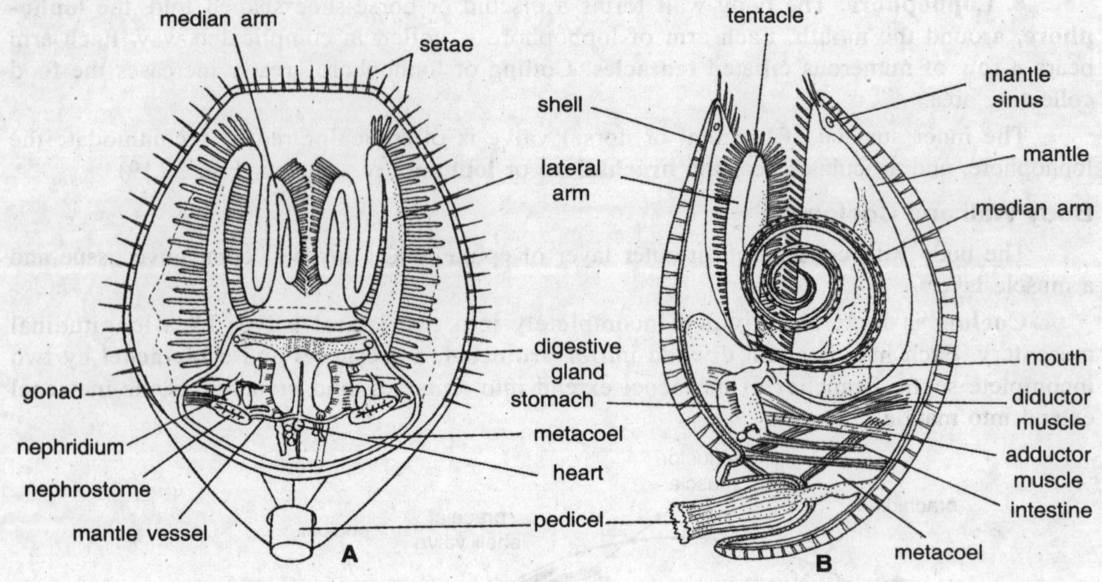

Fig. 30.20. *Magellania*. Antomy. A—Dorsal view; B—Lateral view.

Box 30.2 Feeding in Brachiopoda

During feeding of Brachiopoda, the shell valves are held apart. Lateral tracts of cilia produce lateral intake currents and a median exhaust current. The tentacles are held high, and large particles drop to mantle to be ejected by the mantle cilia or the sudden clapping of the valves. A brachial groove at the base of the tentacles collects food particles brought to it by frontal cilia on the tentacles. The brachial groove cilia conveys the food particles, trapped in mucus, to the mouth. The brachial fold projects over the mouth, helping to direct traffic inward.

Circulatory System

Colourless blood containing respiratory pigment haemerythrin in cells, circulates in an **open type** circulatory system. The contractile vesicle or **heart** lies dorsally to the stomach. It is divided into two **anterior branches**, each communicating with an extensive sinus around the gut and continuing up to the lophophore. Heart bifurcates posteriorly to form **dorsal** and **ventral mantle vessels**. Ventral mantle vessel also communicates with sinuses in the nephridia and gonads. The primary function of circulatory system of Brachiopods is circulation of nutrients.

The blood and coelomic channels communicate extensively through tissue spaces.

Respiratory System

Brachiopods respire by the help of tentacles, mantle lobes and mantle caeca of shell. They have low rates of oxygen consumption. A respiratory pigment, haemerythrin, occurs in blood and coelomic fluid cells. It does not take up oxygen as readily as haemoglobin, but gives it up very readily.

Excretory System

Excretory system includes one or two pairs of **metanephridia**. Their ruffled and flagellated **nephrostomes** open into the metacoel and the **nephridiopores** into the mantle cavity.

Lingula is mainly **ammonotelic**, although enzymes to produce urea are present (**Hammen,** 1977). Particulate nitrogenous material of coelomic fluid is taken up by **coelomic phagocytes** and by **peritoneal cells** that detach and disintegrate. Streams of particles move to nephridia through nephrostomes and are excreted out through nephridiopores. The excreted particles remain suspended in mucus strands.

Nervous System

It consists of a **nerve ring** around the oesophagus and several nerves. The nerve ring bears a **small ganglion** on its dorsal side and a **larger ganglion** on the ventral side.

Sensory Organs

Mantle margin is the most important site of sensory reception. Mantle setae transmit tactile stimuli to these sensory receptors.

Reproductive System

Sexes are separate. Four pairs of gonads are present. Sex cells or gametes are released into coelom; from there they depart through the nephridia. Fertilization is external.

Development

Cleavage is **holoblastic** and usually of **radial** type. Plankton-feeding larvae occur in Inarticulata; they develop from less yolky eggs. Yolk-subsisting larvae occur in Articulata; they develop from very yolky eggs.

Classification

Following **two** classes of brachiopods may be recognized according to whether or not the valves of the shell are joined together by a hinge.

Class 1. Inarticulata (or Ecardines)

1. Shell is without hinge held together by muscles only.
2. Lophophore is without skeletal support.
3. Digestive tract is complete with anus.

This class includes following **two** orders:

Order 1. Lingulida

1. With pedicel valve ventral but both shell valves are modified to permit pedicel passage.
2. Shell mineral is calcium phosphate.

Examples. *Lingula, Glottida.*

Order 2. Acrotretida

1. Without pedicel, or with pedicel emerging from notch or foramen in pedicel valve.
2. Shell minerals are calcium carbonate and/or calcium phosphate.

Examples. *Crania, Discinisca.*

Class 2. Articulata (or Testicardines)

1. Shell valves are hinged by interlocking processes.
2. Pedicel is emerging through pedicel valve.
3. Shell mineral is calcium carbonate.
4. Lophophore is generally with skeletal support.

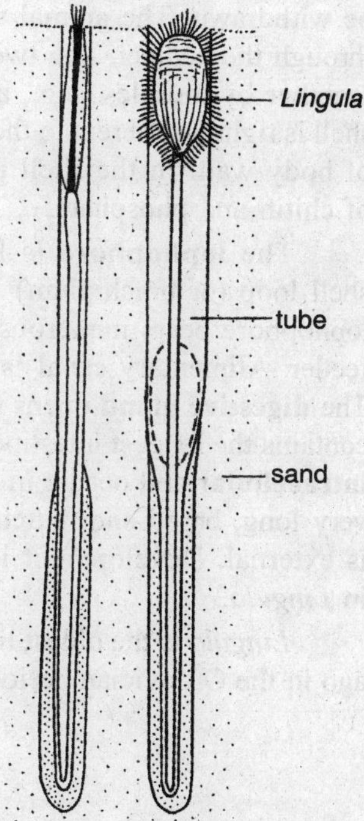

Fig. 30.21. *Lingula anatifera* in their tubes in the sand; the dotted line in the right tube indicates position of the body when retracted.

5. Digestive tract is incomplete (no anus is present).

Class Articulata includes following **two** orders:

Order 1. Rhynchonellida

Example. *Rhynchonella*

Order 2. Terebratulida

Examples. *Magellania, Terebratulina.*

Lingula

Lingula contains about 12 species and belongs to order **Lingulida**, class **Inarticulata** and phylum **Brachiopoda**. *Lingula* is marine and lives in vertical burrows in the bottom of sand (Fig. 30.21). It is found in the tropical and subtropical waters of the Indo-West Pacific area. Japanese eat *Lingula.*

The body of *Lingula* is covered by a **shell** (Fig. 30.22). Shell consists of two almost equal and weakly convex **valves**. Shell valves are not united by a hinge. A long **pedicel** passes through the proximal end of two valves. The pedicel is extensible and can be withdrawn. The animal remains attached to the bottom of tube through the pedicel. The two shell valves of *Lingula* are connected together by muscles (*e.g.*, adductor, protractor, diductor, etc). The shell is tightly adherent to the two **mantle folds** which are extensions of body wall. In the shell of *Lingula*, there are alternating layers of chitin and phosphate.

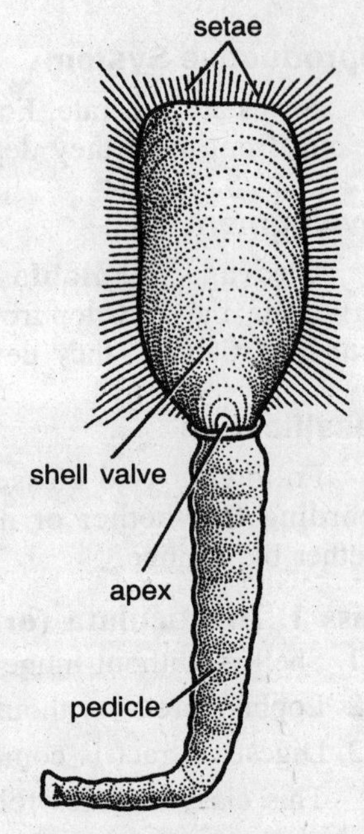

Fig. 30.22. *Lingula*. External structure.

The **lophophore** is long and spirally coiled. There is no shell loop (or brachidium) for the support of the lophophore. The lophophore bears numerous ciliated **tentacles**. Lingula is a ciliary feeder. Alimentary canal is U-shaped and complete having **anus**. The **digestive gland** opens into the stomach by four ducts. *Lingula* contains the longest intestine of all brachiopods. Digestion is chiefly **intracellular** and occurs in the digestive gland. **Metanephridia** are very long, broad and flattened. Sexes are separate and fertilization is external. Development is indirect. There is no metamorphosis in *Lingula.*

Lingula is the oldest living multicellular animal on earth. It appeared about 400 million years ago in the Ordovician period. Since then it is surviving unchanged and is called a **"living fossil".**

QUESTIONS

Long Answer Questions

1. Describe general characters of phylum Rotifera.

2. Describe in brief the general characters of phylum Sipunculida.

3. Describe general characters of phylum Echiura or Echiuroidea.

4. Describe general characters of phylum Bryozoa.

5. Describe general characters and classification of phylum Brachiopoda.

(*C.C.S.U. Meerut 1999, 2001*)

Short Answer Questions

1. Write a short note on *Sipunculus*.
2. Write a note on Bryozoa.
3. Write a short note on *Bugula*.
4. Write a short note on *Lingula*.
5. Draw a well labelled diagram of *Bonellia*.
6. Write a note on Brachiopoda.

Very Short Answer Questions

1. Give the popular name of the rotifera.
2. Cite one example of phylum Rotifera.
3. What is corona of rotifers?
4. Explain cyclomorphosis in context with rotifers.
5. How does introvert operate in *Sipunculus*?
6. What is common name of brachiopods?
7. What is the common name of animals of Phylum Ectoprocta?
8. What is the popular name of sipunculans?
9. Give the common name of Echiuroidean animals?

Fill in the Blanks

1. Rotifers are commonly called ....
2. Rotifers are either suspension feeders or ....
3. Phylum Rotifera includes three classes: Seisonidea, ... and ....
4. Peanuts worms are the ....
5. Echiura are commonly called ....
6. Avicularia are the zooids of ....
7. Larva of *Bugula* is called ....
8. Brachiopoda are commonly called ....
9. *Lingula* is called .... fossil.

Multiple Choice Questions

1. *Sipunculus* is commonly known as
   (*a*) nutworm (*b*) peanut worm
   (*c*) peacock worm (*d*) clam worm
2. Colonies of zooids of Ectoprocta are called
   (*a*) lamp shells (*b*) sea mats
   (*c*) slipper animalcules (*d*) none of these
3. *Magellania* belongs to phylum
   (*a*) Rotifera (*b*) Ectoprocta
   (*c*) Brachiopoda (*d*) Acantocephala
4. Brachiopods are
   (*a*) acoelomate (*b*) coelomate
   (*c*) schizocoelomate (*d*) pseudocoelomate

5. Lophophore acts as a
   (*a*) food catching apparatus (*b*) locomotory organ
   (*c*) digestive gland (*d*) none of these
6. Brachiopods are
   (*a*) filter feeders (*b*) ciliary feeders
   (*c*) mucous feeders (*d*) none of these
7. Ectoprocta (Bryozoa) form colonies known as
   (*a*) sea-mats (*b*) sea-colonies
   (c) sea-huts (*d*) none of these
8. Complete ectoproct colony is called
   (*a*) zooid (*b*) zoarium
   (c) corona (*d*) none of these
9. The zoological name of Bird's Head Coralline is
   (*a*) *Bugula avicularia* (*b*) *Brachionus rubens*
   (*c*) *Echinorhyncus gigas* (*d*) *Oligacanthorhynchus*
10. In Ectoprocts asexual reproduction takes place by
    (*a*) fragmentation (*b*) binary fission
    (*c*) budding (*d*) none of these
11. The larva of *Bugula* is
    (*a*) cyphonant (*b*) pilidium
    (*c*) nauplius (*d*) zoa
12. Rotifers are also known as
    (*a*) slipper animalcule (*b*) wheel animalcule
    (*c*) ladder animalcule (*d*) none of these
13. In *Brachionus rubens*, pedal glands, secrete
    (*a*) adhesive material (*b*) saliva
    (*c*) both (a) and (b) (*d*) none of these
14. *Brachionus rubens* belong to phylum
    (*a*) Acanthocephala (*b*) Rotifera
    (*c*) Entoprocta (*d*) Ectoprocta
15. *Pedicellina* belongs to
    (*a*) Entoprocta (*b*) Nemertinea
    (*c*) Acanthocephala (*d*) Ectoprocta
16. Larvae form of phylum Nemertinea is
    (*a*) pilidium (*b*) nauplius
    (*c*) miracidium (*d*) zoea

ANSWERS

Very Short Answer Questions

1. Wheel bearers (wheel animalcules).
2. *Philodina roseola*.
3. The head of rotifers or wheel animalcules consists of a characteristic retractile ciliary crown of disc called **corona** or **trochal disc**. It has two rings of cilia which beat in a circular manner and appear as rotating wheel.

4. In the phenomenon of cyclomorphosis, the phenotypic form of amictic female (*i.e.*, female producing the eggs parthenogenetically in case of rotifers) changes gradually in response to environmental conditions during the year.
5. The body of *Sipunculus* consists of **introvert** and **trunk**. The introvert may be withdrawn into the trunk. In this process the coelom and body wall together work as a **hydraulic system** to operate the introvert. The introvert is everted by the contraction of body wall musculature which raises the coelomic pressure. It is retracted by the retractor muscle of body wall.
6. Lamp shells.
7. Moss animals.
8. Peanut worms.
9. Spoon worms.

Fill in the Blanks

1. Wheel animalcule **2.** Raptorial **3.** Monogononta, Bdelloidea
4. Sipunculoidea **5.** Spoon worm **6.** Bryozoans
7. Coronata **8.** Lamp shells **9.** Living.

Multiple Choice Questions

1. (*b*) **2.** (*b*) **3.** (*c*) **4.** (*b*)
5. (*a*) **6.** (*b*) **7.** (*a*) **8** (*b*)
9. (*a*) **10.** (*c*) **11.** (*a*) **12.** (*b*)
13. (*a*) **14.** (*a*) **15.** (*a*) **16.** (*a*).

PRINCIPLES OF ECOLOGY
(CC–II)

1

Definition and History of Ecology

1.1 DEFINITION OF ECOLOGY

Ecology, is the study of the interrelations of organisms with their total environment, physical and biological (Box 1.1). The term **ecology** (*oekologie*) is derived from two Greek words–*oikos* = means 'house' or 'place to live' and–*logos* means 'a discussion or study'. Literally, ecology is the study of organisms 'at home', in their native environment. There is some controversy about the author of who coined the term ecology and first used it in the literature. The term was first of all introduced by **Reiter** in 1868, but because the German biologist **Ernst Haeckel** (1869) first fully defined this term and made an extensive use of this term in his writings, he is therefore, usually credited for the coinage of the term ecology (see **C.B. Knight**, 1965). However, **Kormondy** (1969) tried to give credit for the first use of the term ecology to **H.D. Thoreau** in 1858.

Box 1.1 Some Other Definitions of Ecology

Ecologists continue to work toward a deeper definition of ecology as follows :

1. The science of community (**Clements** 1916).
2. The scientific natural history (**Charles Elton** 1927).
3. The scientific study of the distribution and abundance of organisms (**Andrewartha** 1961).
4. Interaction of form, functions and factors (**R. Misra** 1967).
5. Study of ecosystems (**Margalef** 1968).
6. Study of animals and plants in relation to their habit and habitat (**Colinvaux** 1970).
7. The study of structure and function of nature (**Odum** 1971).
8. The study of the adaptation of organisms to their environment (**Emlen** 1973).
9. The scientific study of the relationships between organisms and their environments (**McNaughton and Wolfe** 1979).
10. The scientific study of the interactions that determine the distribution and abundance of organism (**Krebs** 1985).
11. The study of the patterns of nature and how those patterns come to be and how they change in space and time (**Kingsland** 1985).
12. The study of the principles which govern temporal and spatial patterns for assemblages of organisms (**Fenchel** 1987).
13. The study of the relationships between organisms and the totality of the physical and biological factors affecting them or influenced by them (**Pianka** 1988).
14. The study of organisms in relation to the surroundings in which they live (**Chapman** and **Reiss** 1995).

The term *ecology* is derived from the same root word as *economics* which means management of the household. Ecology then could be considered as the economics of nature. **Wells, Huxley** and **Wells** (1939) commented that "Ecology is really an extension of economics of the whole world of life". In fact, some economic concepts, such as resource allocation, cost-benefit ratios and optimization theory, have found place in ecology.

Ecologists are persons who study ecology. They are always aiming to understand how an organism fits into its environment. The environment is of supreme importance to an organism and its ability to exist in the environment where it lives will determine its success or failure as an individual.

Environmental Biology

The subject of ecology is gaining more popularity with its name as "**environmental biology**". However, the environmental biology is not a new discipline but simply an extension of ecological approach which *stresses the study of the environment in its totality with special emphasis on the welfare of man and his comfort as the man has been considered the most evolved and an intelligent component of the ecosystem* (**K.C. Agarwal,** 1999).

1.2 HISTORICAL BACKGROUND OF ECOLOGY

The roots of ecology lie in natural history, which is as old as man himself. The vague beginning of ecology may be traced back to prehistoric man, who utilized environmental information to hunt food; fishing, trap animals, find edible vegetation, and locate shelter to survive the hardships imposed by nature. An increased knowledge of the importance of environmental conditions led quite naturally to religious rituals, worship of weather gods, and rain dances, all of which are quite prevalent in different social groups and tribes of Indian civilization. The establishment of agriculture increased the need to learn about the practical ecology of plants and domestic animals.

Further, the science of ecology has had a gradual but irregular development during the recorded history of human civilization. Consequently, the historical background of ecology can be traced as follows:

(a) Ecology in Indian Classics

Early Indian philosophers and thinkers had some knowledge of ecology as has been revealed by the Indian classical writings, in the Vedic, Upanishadic, Pauranic, and Epic literature. **Charaka** considered the important factors of *Vayu* (air and gases), *Jala* (water), *Desha* (topography) and *time* in regulating the life of plants (see **R. Misra**, 1962). The great classical Indian Poet **Kalidasa** has displayed his ecological faculty in his classics–*'Ritu Sanghar'*, *'Meghdoot'* and *'Abhigyan Shakuntalam.'*

(b) Early Greek's Ideas of Ecology

The early Greek philosophers too were well aware of the importance of environmental studies. One of the papers entitled *Airs, waters and places* of **Hippocrates** stressed on the need of ecological background for medical students. According to him–"whoever wished to investigate medicine properly, should proceed thus: in the first place to consider the seasons of the year, and what affects each of them produces........" In the fourth century B.C. **Aristotle** made references about the habits of animals and environmental conditions prevailing in certain areas, in his writings on natural history. One of the Aristotle's friend and associate **Theophrastus** (370–250 B.C.) who is regarded as the first true ecologist, wrote about the association of plant communities and the relation of plants to each other and the non-living environment.

(c) Ecology in 18th Century

After a considerable long gap of many centuries, the first naturalist who made a serious attempt to systematise knowledge concerning relations of animals to environment was **Buffon**

(1707–1788). Though such works as *Historia animalium* by **Konard Gesner** (1519–1565); 14 volumes of collected works of natural history by **Ulisse Aldrovandi** (1522–1605); the natural histories of fishes and birds by **Pierre Belon** (1517–1564); *The plants* by **Andrea Caesalpino** (1519–1603), the botanical work of **Gaspard Bauhin** (1550–1624); demographical studies about human ecology by **Graunt** (1664); studies of the sexuality of plants by **Rudolph Camerarius** (1665–1721); the works of English naturalist **John Ray** (1627–1705); studies on natural history of insects by French naturalist **Rene Reaumur** (1683–1757) and the works of many other naturalists doubtlessly laid the foundation upon which **Buffon** built his remarkable work. **Buffon** considered that animals and plants developed adaptations which enabled them to favourably respond to their environment and termed this **environmental induction**. He touched on many of our modern ecological problems and recognised that population of man, other animals and plants are subjected to the same processes. **Buffon** also explained how the great fertility of every species was counter-balanced by innumerable agents of destruction. He, thus, dealt with problems of population regulation.

Anton van Leeuwenhoek, best known as a pioneer microscopist of the early 1700s, also pioneered the study of "food chains" and "population regulation", which are two important areas of modern ecology.

Malthus published one of the earliest controversial book on demography. In his *Essay on Population* (1798) he calculated that although the number of organisms can increase geometrically (1,2,4,8,16,3(b)......); their food supply may never increase faster than arithmetically (1, 2, 3, 4, 5....). Such a great disproportion between these two powers of increase led **Malthus** to infer that reproduction must eventually be checked by food production. Many workers questioned the ideas of **Malthus**. For example, in 1841 **Doubleday** brought out his true law of population. To contradict the **Malthus** idea, he believed that whenever a species was threatened, nature made a corresponding effort to preserve it by increasing the fertility of its members. Human populations that were undernourished had the highest fertility; those that were well fed had the lowest fertility. **Doubleday** explained these effects by the oversupply of mineral nutrients in well-fed populations.

Further, many of the early developments in ecology came from the applied fields of agriculture, fisheries, and medicine. For example, work on the insect pests of crops has been one of the important sources of ideas. The regulation of population size of obnoxious insects is a basic problem that has been under study. In 1762 the mynah bird was introduced from India to the island of Mauritius to control the red locust. Due to such **biological control**, by 1770 the locust threat was a negligible problem. **Forska** wrote in 1775 about the introduction of predatory ants from nearby mountains into date palm orchards to control other species of ants feeding on the palms in South Western Arabia. In subsequent years an increasing knowledge of insect parasitism and predation led to many such introductions all over the world in the hope of controlling introduced and native agricultural pests.

Similarly, medical work on infectious diseases such as malaria around the 1890's gave rise to the study of epidemiology and interest in the spread of disease through a population. Before malaria could be controlled adequately it was necessary to know in detail the ecology of mosquitoes. Likewise, the production ecology had its beginnings in agriculture, and **Egerton** (1969) has traced this back to the eighteenth-century botanist **Richard Bradley**. Bradley recognized the fundamental similarities of animal husbandary and production, and he proposed methods of maximizing agricultural yields (and hence profit) for vineyards, trees, poultry, rabbits and fish.

(d) Ecology in 19th Century

In 1807, **Humboldt** after extensive travelling to tropical and temperate South America, discussed the geographical distribution of plants and animals in relation to climate. **Edward**

Forbes in 1844 described the distribution of animals in British coastal waters and part of the Mediterranean sea, and he wrote of zones of different depths which were distinguished by the associations of species they contained. He noted that some species are found only in one zone and that other species have a maximum of development in one zone but occur sparsely in other adjacent zones. Mingled in are stragglers that do not, fit in zonation pattern. **Forbes** recognized the dynamic aspect of the interrelations between these organisms and their environment. As the environment changes, one species might die out, another might increase its abundance. English naturalist, **St. George Jackson Mivart** coined the term **hexicology** around 1859 and in 1894 he defined it as "devoted to the study of the relations which exist between the organisms and their environment as regards to the nature of the locality they frequent, the temperatures and the amounts of light which suit them, and their relations to other organisms as enemies, rivals or accidental and involuntary benefactors." **Saint-Hilaire** (1859) used the term **"ethology"** for the study of relations of organisms to environment and outlined a volume, which he never wrote, on instincts, habits, and other ecological matters. The term ethology, though was not accepted by early ecologists, but this term is now used as an important ecological discipline dealing with animal behaviour. In 1869, **Haeckel** proposed **"oecologie"** for the relations of organisms to organic and inorganic environments. The modern term *ecology* is derived from the word *oecologie*. The observations of **Priestley** and **Scheele** that plants produce oxygen led to an understanding of the inter-relationships of plants and animals. The discovery that green plants and animals use carbon dioxide and water to form organic matter for use by animals and result in the release of oxygen and water as wastes, provided new understanding of ecological food chains. **Justus von Liebig** (1803–1873) initiated the idea of carbon and nitrogen cycles in nature and his work (1840) led to **ecoclimatology** and **physiological ecology**. Further, out of such studies in aquatic environment developed the field of **ecological energetics**. **Louis Agassiz** (1807–1873) initiated the research in marine ecology by establishing many laboratories for marine ecological research. **Karl Mobius** (1877) introduced the word **"biocoenosis"** to designate a group of organisms as an ecological unit. **Charles Darwin** (1809–1882) and **Alfred R. Wallace** (1823–1915) have increased our knowledge of island or insular life with their characteristically different types of environmental conditions. **Hensen** (1887) in a classical paper on "*The Lake as a Microcosm*" suggested that the species assemblage in a lake was an organic complex and that by affecting one species we exerted some influence on the whole assemblage. Thus each species maintains a 'community of interest' with the other species, and we cannot limit our studies to a single species. **Forbes** believed that there was a steady balance of nature, which held each species within limits year after year, even though each species was always trying to increase its number. The Russian pioneer ecologist **V.V. Dokuchaev** (1846–1903) and his chief disciple **G.F. Morozov** (who specialized in forest ecology) placed great emphasis on the concept of the **"biocoenosis"**, a term later expanded by Russian ecologist to **"geobiocoenosis"**, a term equivalent to ecosystem.

Schroter (1896) introduced the terms **"autecology"** and **"synecology"**. **Herdman** (1896) indicated the economic possibilities of scientific agriculture. Danish botanist **Warming** (1895, 1909) raised questions about the structure of plant communities and the associations of species in these communities. In 1899, **H.C. Cowles** described *plant succession* on the sand dunes at the Southern end of Lake Michigan.

Thus, as a recognized distinct field of biology, the science of ecology dates from 19th century, and only in the mid 20th century has the word become the part of the general vocabulary.

(e) Ecology in the 20th Century

At the beginning of the twentieth century ecology was a young, but an established science and such eminent ecologists as **Wasmann** (1901), **Dahl** (1901) and **Wheeler** (1902) were discussing whether **Saint Hilaire's** ethology or **Haeckel's** ecology should be used to designate the science of relations of organisms to environments. The latter term has gradually come into general usage. **Case** (1905) discussed environmental conditions of past geological periods and has continued to make contributions in that field. Two botanists, **Clements**, in his "*Research Methods in Ecology*" (1905) and subsequent writings, and **Warming** (1909, 1925) in his "*Oecology of Plants*" have done much to clarify ideas and contribute terms. **Adams** (1913) gave an excellent review of significant ecological literature, urged the usefulness of ecological surveys, and discussed methods to be used in making them. **Shelford** (1912) demonstrated excellent examples of ecological succession and has made notable contributions concerning animal communities (1913) and environmental factors (1929). In 1920 the German limnologist **Thienemann** introduced the concept of trophic levels in terms of producers and consumers. The theoretical approaches in population dynamics of **Lotka** (1925) and **Volterra** (1926) stimulated the experimental approaches by biologists. **Gleason** (1926) studied the development and dynamics of plant communities. **Shelford** (1913, 1937), **Adams** (1909) and **Dice** (1943) in America and **Elton** (1927) in England investigated the interrelations of plants and animals. In 1935 **A.G. Tansley** first proposed the term **ecosystem.**

Gause (1935) investigated the interactions of predators and prey and the competitive relationships between species. At the same time **Nicholson** studied interspecific competition. In 1954 **Andrewartha** and **Birch** and **Lack** provided a broader foundation for the study of regulation of populations. The discovery of the role of territory in bird life by **H.E. Howard** in 1920 led to further studies by **Nice** in the 1930s and 1940s. Out of such studies came the field of **behavioural ecology**. In the 1940s and 1950s **Lorenz** and **Tinbergen** developed concepts of instinctive and aggressive behaviour. In 1962, **Wynne-Edwards** of England explored the role of social behaviour in the regulation of populations.

In the field of ecological energetics, in 1920 the German limnologist **Thienemann** introduced the concept of trophic levels in terms of producers and consumers. Two American limnologists, **Birge** and **Juday** in the 1940s, through their measurements of the energy budgets of lakes, developed the idea of primary production. Out of their studies came the trophic-dynamic concept of ecology. Introduced by **Lindemann** in 1942, this concept marked the beginning of modern ecology. Out of **Lindemann's** study came further pioneering work on energy flow and energy budgets by **Hutchinson** and **H.T.** and **E.P. Odum** in the 1950s in America. Early work on the cycling of nutrients was done by **Ovington** (1957) in England and Australia and by **Rodin** and **Bazilevig** (1967) in Russia.

One of the interesting aspects of development of ecology during the first third of 20th century remains that a tendency towards increased specialization in ecology has occurred as is evidenced by the establishment of such disciplines as **palaeoecology**, a study of environmental conditions and life as it existed in past ages. Pollen analysis, radioactive dating, and paleontology have aided the palaeoecologists. Other specialized fields of ecology which took their birth during this time, are **zoogeography**, the scientific study of the geographic distribution of animals; **oceanography**, the study of the biotic and physical conditions existing on oceans, bays, and estuaries, and **limnology**, the study of the living and non-living components of inland waters. With the establishment of specialized disciplines of ecology, the value of application of ecology to such fields as agriculture, forestry, wildlife conservation, fisheries management, and pest control became apparent, and, thus, a new discipline called **applied ecology** came into its existence.

Further, in the second decade of 20th century great emphasis was laid on statistical studies on populations, sampling techniques, and community studies. During this period several important texts published–**Lotka's** *Elements of Physical Biology;* **Elton's** *Animal Ecology;* **Volterra's** *Animal Ecology;* **Gause's** *The Struggle for Existence,* and **Bodenheimer's** *Problems of Animal Ecology.* Certain distinguished ecologists of 20th century were **Clements**, **Cowles**, **Shelford**,

Pearse, Oosting, A.M. Woodbury, Hutchinson, Deevey, Lindeman, F.E. Smith, Slobodkin, Hairston, T. Park, O.Park, Emerson, MacArthur, G.L. Clark, C.B. Knight, H.T. and **E.P. Odum, R. L. Smith, F.J. Vernberg; C.J. Krebs, R. Misra, M.S. Mani**, etc.

Development of ecology during 20th century have been greatly stimulated by organization of ecologists. The British Ecological Society was founded in 1913 and this society sponsored two journals namely *Journal of Ecology* (started in 1913) and *Journal of Animal Ecology* (started in 1932). Following the British, the Ecological Society of America was established in 1915 which also sponsored two publications, *Ecology* in 1920 and *Ecological Monographs* in 1931.

Growth of Ecology in India

In India, ecological studies were started in 1940's. But a strong foundation was laid during 1960's when some of the today's well known ecologists such as **J.S. Singh, S.S. Raman, R.S Ambasht, P.S. Ramakrishnan, P.D. Sharma** and others were trained at Banaras Hindu University under able guidence of **Prof. R. Mishra** (Box 1.2). In the field of animal ecology important contributions have been made by **S.L. Hora, M.S. Mani, S.M. Das, S.Z. Qasim, T.N. Ananthkrishnan, Madhav Gadgil, Ishwar Prakash, M.C. Dash,** etc.

Box 1.2

Two pioneer centres of ecological research were established in 1930's in India, first by **R. Misra** at the Department of Botany, Banaras Hindu University, and second by **F.R. Bharucha** at the Institute of Science, Bombay. From 1937 onwards, R. Misra did extensive work at Banaras (Varanasi) on the ecology of grasslands, wastelands, playgrounds, lakes and ponds, low-lying lands, riparian eroded slopes and ecotypic differentiation. In 1950, Prof. Misra established another Ecological Research Institute at Sagar University (M.P.). In 1956, **Misra** moved to Varanasi and developed an internationally recognized centre of Ecological Investigation. At this centre, in between 1955-1962, investigations were carried on in the diverse areas such as ecophysiology (**V.Kaul**), root-soil relationships of grasslands (**S.S. Raman**) and riparian lands (**R.S. Ambasht**), ecotypic differentiation (P.S. Ramakrishnan) and reproductive capacity of herbage species (**H.R. Sant** and **V.G. Nelvigi**). During 1963-1971, Banaras centre of ecology carried mainly autecological work on the medicinal plants and the weeds (**R.S. Tripathi**), grassland productivity (J.S. Singh) and forest litter decomposition and productivity (**K.P. Singh**).

1.3 BRANCHES OF ECOLOGY

The science of ecology is often divided into autecology and synecology. **Autecology** deals with the study of the individual organism or an individual species. In this, life histories and behaviour as a means of adaptation to the environment are usually emphasized. **Synecology** deals with the study of groups of organisms which are associated together as a unit (*i.e.,* community). Thus, if a study is made of the relation of a white oak tree (or of white oak trees in general) to the environment, the work would be autecological in nature. If the study concerned the forest in which the white oak lives, the approach would be synecological (**Odum, 1971**). While autecology is experimental and inductive, the synecology is philosophical and deductive (**Smith, 1974**).

Synecology is often further subdivided into **aquatic** and **terrestrial ecology**. The aquatic ecology includes **freshwater ecology, eustuarine ecology** and **marine ecology**. Terrestrial ecology, subdivided further into such areas as **forest ecology, grassland ecology, cropland ecology** and **desert ecology**, is concerned with terrestrial ecosystems—their microclimate, soil chemistry, nutrient and hydrological cycle and productivity.

Further, **demecology** is that branch of ecology which deals with the ecology of populations. Early ecologists have recognized two major subdivisions of ecology in relation to plants and animals—**plant ecology** and **animal ecology**. But when it was found that in the ecosystem plants and animals are very closely associated and interrelated, then, both of these major subdivisions of ecology into plant ecology and animal ecology became vague. Besides these major subdivisions the ecology has been classified in the following branches according to the level of organization, kind of environments or habitats and taxonomic position:

1. Habitat ecology. It deals with the study of different habitats of the biosphere. According to the kind of habitat, ecology is subdivided into **marine ecology**, **freshwater ecology**, and **terrestrial ecology**. The terrestrial ecology too is further subdivided into **forest ecology, cropland ecology, grassland ecology**, etc., according to the kind of study of its different biomes.

2. Ecosystem ecology. It deals with the analysis of ecosystem from structural and functional point of view including the interrelationship of physical (abiotic) and biological (biotic) components of environment.

3. Conservation ecology. It deals with methods of proper management of natural resources such as land, water, forests, sea, mines, etc., for the benefit of human beings.

4. Production ecology. It is the modern subdivision of ecology which deals with the gross and net production of different ecosystems such as freshwater, sea water, cropfields, orchards, etc., and tries to do proper management of these ecosystems so that maximum yield can be obtained from them.

5. Radiation ecology. It deals with the study of gross effects of radiations and radioactive substances over the environment and living organisms.

6. Taxonomic ecology. It is concerned with the ecology of different taxonomic groups and eventually includes following subdivisions of ecology–**plant ecology, insect ecology, invertebrate ecology, vertebrate ecology, microbial ecology** and so on.

7. Human ecology. It deals with the study of relationship of man with his environment.

8. Space ecology. It is a modern subdivision of ecology which remains concerned with the development of partially or completely regenerating ecosystems for supporting life of humans during long space flights or during extended exploration of extra-terrestrial environments.

9. Systems ecology. The systems ecology is the most modern branch of ecology which is concerned with the analysis and understanding of the function and structure of ecosystem by the use of applied mathematics such as advanced statistical techniques, mathematical models and computer science.

10. Resource ecology. It emphasizes upon the judicious management of all resources including the plants, animals, water, soil and the minerals.

1.4 RELATIONSHIP OF ECOLOGY WITH OTHER DISCIPLINES

Modern ecology is a multidisciplinary science which depends on a variety of disciplines such as **physics, chemistry, mathematics, statistics, meteorology, climatology, geology, geography, economics, sociology, agriculture science, forestry, horticulture, genetics, physiology**, etc. All these disciplines have helped in the better understanding of many ecological principles.

For instance, meteorological and climatological data for certain geographic localities allow for a more implied interpretation of results. A basic knowledge of forestry can be valuable for a forest ecologist to understand forest type distribution, floristic composition and prevalent environmental factors. Likewise, statistical data helps in interpreting the reasons for activity, population increase, migrations, probability of ecological events occurring in a particular area, sampling techniques and reliability of results. Paleontology (geology) provides information about the ancestral organisms and environmental situations prevalent in past. Evolution and genetics are utilized to interpret the reason for organic changes when linked with environmental conditions, establishment of new populations and species, environmental effects on genetic populations and species and, so on. Such interdisciplinary approaches to ecology, consequently have given rise to following subdivisions of ecology:

(a) Ecological genetics. An ecologist recognised a kind of genetic plasticity in the case of every organism. In any environment only those organisms that are favoured by the environment survive. The branch of ecology dealing with genetics in relation to ecology is called **ecological genetics** or **genecology.**

(b) Palaeoecology. It deals with the movements of biotic elements based on palaentological evidence, which provides information about ancestral organisms and environmental conditions existing in the past.

(c) Ecophysiology. The factors of environment have a direct bearing on the functional aspects of organisms. The ecophysiology deals with the survival of populations as a result of functional adjustments of organisms with different ecological conditions of the ecosystem.

(d) Chemical ecology. It deals with the adaptations of animals or preferences of particular organisms such as insects to particular chemical substances.

5. Pedology. It is a branch of terrestrial ecology and it deals with the study of soils, in particular their acidity, alkalinity, humus content, mineral contents, soil type, etc., and their influence on the organisms.

6. Ecogeography. It deals with the study of the role of the environment in animal distribution. It is related with **biogeography** which is concerned with the structural and functional relations of living organisms in space, which form the immediate environment of the individuals as well as populations. **Ecofloras** and **ecofaunas** are the lowest units of which a biogeographic flora or fauna is made up of.

7. Ecological energetics. It deals with energy conservation and its flow in the organisms within the ecosystem. In it thermodynamics has its significant contribution.

Because of its far-flung involvements with so many fields, ecology is often regarded as a generality rather than a speciality. Visualizing this fact, an ecologist, **A. Macfadyen** (1957) wrote in his book *Animal Ecology: Aims and Methods*—*"The ecologist is something of a chartered libertine. He roams at will over the legitimate preserves of the plant and animal biologist, the taxonomist, the physiologist, the behaviourist, the meteorologist, the geologist, the physicist, the chemist, and even the sociologist; he poaches from all these and from other established and respected disciplines. It is indeed a major problem for the ecologist in his own interest, to set bounds to his divagations"*.

Ecological Interfacing

Because ecology is a broad, multilevel discipline, it interfaces well with traditional disciplines that tend to have more narrow focus. During the past decade there has been a rapid rise of interface fields of study accompanied by new societies, journals, symposium, volumes, books and new careers. A new interface discipline **ecological economics** is beginning to bridge the gap between ecology and economics. Others that are receiving a great deal of attention, especially in resource management are **agroecology, biodiversity, conservation ecology, ecological engineering, ecosystem health, ecotoxicology, environmental ethics** and **restoration ecology** (**Odum** and **Barrett** 2005).

1.5 SIGNIFICANCE OF ECOLOGY FOR HUMAN BEINGS

Man himself is an organism within an environment. Like other animals man is influenced by the physical features of his environment. He is absolutely dependent upon other species for his food, clothing, medicine, and other similar aspects and he has to adjust to other individuals of his own species. Therefore, the basic laws of ecology apply well to him and its fundamental knowledge is must for man for his own existence on this planet (Earth).

Man almost always has a modifying influence, and without proper regulation he often has a destructive effect. For instance, by applying certain ecological principles to such fields as agriculture, biological surveys, game management, pest control, forestry, horticulture, and fishery biology, he has received tremendous economic gains. Ecological knowledge is found critically important for intelligent conservation whether in relation to soil, forest, wildlife, water supply or fishery resources. Further, though man has been known to control his environment successfully to meet his needs, indiscriminate control of different pathogens or pests such as bacteria, fungi, weeds, insects, rodents, etc., through use of different chemical-poisons or pesticides (bacteriocides, fungicides, insecticides, etc.), the release of massive quantities of radioactive debris, **xenobiotics** (*i.e.*, chemical compounds synthesized by humans which are not naturally found in living organisms and cannot normally be metabolised (broken down) by them), discarded chemicals and industrial wastes into rivers result in atmospheric and aquatic pollution, which may have short and long term ecological effects. Further, rapid growth of urbanization and fast rate of multiplication of human population have resulted in fatal threat of scarcity of wildlife, food, open space, and of survival.

There are certain other ecological problems. Agriculture and now forestry are concentrating upon monoculture–**single species ecosystems**–in spite of the difficulties and dangers associated with unnaturally simplified ecosystems that lack a diversity of species. Over much of the world, especially in the grasslands, we continue to disrupt the energy balance through overgrazing and end up with eroded mountain sides, silt-clogged streams and lakes, and a scarcity of water. Therefore, future of human life on earth demands more knowledge about the ecosystems and other ecological problems.

Outcomes of Environmental Education

W.P. Cunningham and **B.W. Saigo** (1995) have enlisted the following four advantages of **environmental literacy** and **environmental education.**

(a) The natural context. An environmentally educated person understands the scientific concepts and facts that underlie environmental issues and the interrelationships that shape nature.

(b) The social context. An environmentally educated person understands how human society is influencing the environment, as well as the economic, legal and political mechanisms that provide avenues for addressing issues and situations.

(c) The valuing context. An environmentally educated person explores his or her values in relation to environmental issues, from an understanding of the natural and social contexts, the person decides whether to keep or change those values.

(d) The action context. An environmentally educated person becomes involved in activities to improve, maintain or restore natural resources and environmental quality for all.

Recently, ecologists called for broad environmental literacy in which every citizen is fluent in the principles of ecology and has working knowledge of the basic grammar and underlying syntax of environmental wisdom.

1.6 ECOLOGY AND ECONOMICS

Ecology or study of the environmental house includes all organisms in it and all the functional processes that make the house habitable. Thus, literaly, ecology is the study of " life at home" with emphasis on "the totality or pattern of relations between organisms and the environment" (see **Odum** and **Barrett** 2005).

Like the ecology, the word **economics** is also derived from the Greek root *Oikos*. As *nomics* means "management", economics translates as **"the management of the household"** and accordingly, ecology and economics should be companion disciplines (Box 1.3). Unfortunately,

Box 1.3

Before the term ecology came into use perhaps term *"bionomics"* was used for all or part of that we now study in ecology. In the words of **Wells**, **Huxley** and **Wells** (1939) *"ecology is really an extension of economics to the whole world of life"*. **Clarke** (1954) stated that economics and sociology might be thought of as the *"ecology of man"*, in a broad sense. According to the famous chipko movement leader **Sunderlal Bahuguna**, who was awarded the Swedish Right Livelihood Award for 1987, *ecology is permanent economy.*

many people view ecologists and economists as opponents with antithetical (*i.e,* opposed to each other) visions. Table 1.1 attempts to illustrate perceived differences between economics and ecology.

Table 1.1 A summary of perceived differences between economics and ecology (**Odum** and **Barrett** 2005)

	Attribute	Economics	Ecology
1.	School of thought	Cornucopia*	Neo-Malthusian
2.	Currency	Money	Energy
3.	Growth form	J-shaped	S-shaped
4.	Selection pressure	r-selected	*K*-Selected
5.	Technological approach	High technology	Appropriate technology
6.	System services	Services provided by economic capital	Services provided by natural capital
7.	Resource use	Linear (disposal)	Circular (recycling)
8.	System regulation	Exponential expansion	Carrying capacity
9.	Futuristic goal	Exploration and expansion	Sustainability and stability

***Cornucopia**. The horn of plenty, represented as over flowing with fruits and symbolize peace and prosperity. In Greek mythology, the goat on whose milk Zeus was bred, one of whose horns, *the horn of Amalthea*, is broken off by the god, became the cornucopia or horn of plenty.

Zeus. In Greek mythology, the supreme deity, ruler of the celestial realm, son of Kronos and Rhea and husband of Hera, he is identified with the Roman Jupiter.

Ecological Economics

Ecology was of practical interest early in human history. In primitive society, all individuals needed to know their environment—that is, to understand the forces of nature, in fact, coincided with the use of fire and other tools to modify the environment. Because of technological achievements, humans seem to depend less on the natural environment for their daily need, many of us forget our continuing dependence on nature for air, water and indirectly, food, not to mention waste assimilation, recreation and many other services supplied by nature. Also economic systems, of whatever political ideology, value things made by human beings that primarily benefit the individual, but they place little monetary value on the goods and services of nature that benefit us as society. Until there is a crisis, humans tend to take natural goods and services for granted, we assume they are unlimited or somehow replaceable by technological innovations, even though, we know that life necessities such as oxygen and water may be recycled but not replaceable.

As long as the life-support services are considered free, they have no value in current market systems (**Odum** and **Odum** 2000).

Natural Capital and Economic Capital

Capital means wealth owned by a person or organisation or invested, lent, or borrowed. **Natural capital** is defined as the benefits and services supplied to human societies by natural ecosystems, or provided "free of cost" by unmanaged natural systems. These benefits and services include purification of water and air by natural processes, decomposition of wastes, maintenance of biodeversity, control of insect pests, pollination of crops, mitigation (*i.e*, lessening) of floods, and provision of natural beauty and recreation, among others (**Daily** 1997).

Economic capital is defined as the goods and services provided by humankind, or the human workforce, typically expressed as the gross national product (GNP). **Gross national product** is the total monetary value of all goods and services provided in a country during one year. Natural Capital is typically quantified and expressed in units of energy, whereas economic capital is expressed is monetary units (Table 1.1). Only in recent years has there been an attempt to value the worlds ecosystem services and natural capital in monetary terms. **Costanza d'Arge,** *et al.* (1997) estimated this value to be in the range of 16 to 54 trillions U. S. dollars per year for the entire biosphere, with an average of 33 trillions dollars per year. Thus, it is wise to protect natural ecosystems, both ecologically and economically, because of the benefits and services they provide to human societies.

1.7 ECOLOGY, ENVIRONMENTAL BIOLOGY AND ENVIRONMENTAL SCIENCE

Currently ecologists differentiate the following three terms: ecology, environmental biology and environmental science (see **Singh** 2005; **Cain** *et al*, 2011).

1. Ecology

An organism has its identity as an individual and also as a component of a population, community, ecosystem and biosphere. But at all levels of organization, (see Chapter 6), there is sufficient interaction between the organism and its environment. The study of this interaction (or relationship) between organism and its environment (both abiotic and biotic) is called **ecology**.

Ecology is both a pure and applied science and ecologists work on a variety of spatial and time scales. The **fundamental ecology** deals with four higher levels of organization–individual organisms, populations (individuals of same species), communities (an assemblage of different populations) and ecosystems (the community together with its physical and biotic environment). The **applied ecology** attempts to make practical applications of ecological understanding to predict, manage and control. In ecology, the unit of study is a natural ecosystem.

Thus, ecological studies focus their attention on how individual organisms interact with their environments, or how ecological populations or communities interact with their environment. Study of how organisms are adapted to respond to temperature or other factors of their physical environment is an example of such a focus. This approach is called **ecophysiology** (Fig. 1.1)

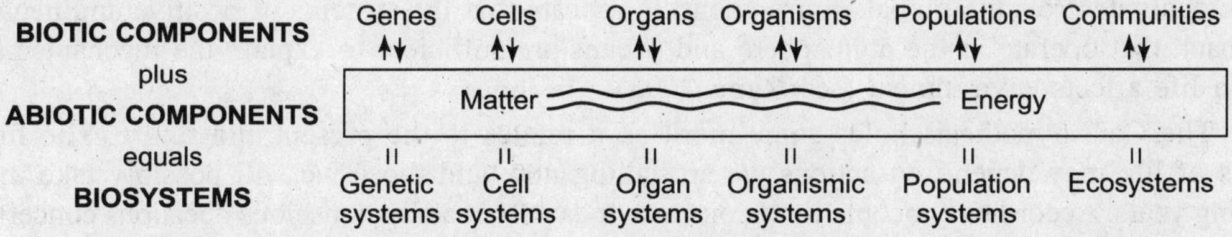

Fig. 1.1 Ecological levels-of-organization spectrum emphasizing the interaction of living (biotic) and nonliving (abiotic) components.

2. Environmental Biology.

The term "environmental biology" is often used for "ecology". These two words depict the same things which may appear like the two sides of the same coin. While ecology is deeply rooted in biology and in the classical ecology often the focus is mainly on the "ecosystem studies" (energy flow and productivity), but in environmental biology the focus is on the "biological organisms of environment" and their ecophysiological adaptations and interactions. For example, if we use the phrase "environmental biology of fish", it will mean how the fish species of a particular environment (pond, lake, river, estuary, ocean) are interacting with the environmental factors, both biotic and abiotic; which of these factors govern their distribution and abundance, and how these factors have influenced their adaptations and life history strategies. We also know that it is not only environment and its physical and biotic factors which affect the organism but the organism itself is also able to modify the environment to a certain extent (called **Gaia Hypothesis**). This hypothesis states that Earth's atmosphere would not support life without regulation by the totality of life in the biosphere (**Lovelock** and **Margulis** 1974; **Lovelock** 1988; **Mann** 1991). According to these environmental biologists, we should return to the ancient Greek idea of 'Gaia' or idea of 'mother earth' as the sustainer.

3. Environmental Science

Ricklefs and **Miller** (2000) consider "environmental science' yet another subfield of ecology that focuses on the specific effects of humans on the environment. According to them, the study of environmental science integrates the principles of Ecology, Chemistry, Physics and other sciences with Economics, Politics and Ethics. Thus, environmental science is an amalgamation of all the disciplines of science and social sciences. Environmental science attempts to solve the major environmental problems with the help of interdisciplinary and transdisciplinary approaches (**Jantsch** 1970).

Environmental scientists often study the action and fate of human-made substances, such as detergents and chemical pesticides, in the biosphere. This study is called **ecotoxicology**. **Ecotoxicologists** often use fish and other animals as models to study the action of toxic compounds (**Misra** 1982; **Mukherjee** and **Pankajakash** 1995; **Feisthaur** *et al*., 2000).

Since all the three terms, i.e, ecology, environmental biology and environmental science cannot be defined without reference to environment, they show too much overlapping which is inevitable and unavoidable (see **H.R. Singh** 2005).

1.8 GAIA HYPOTHESIS

Gaia or **gaea** is the term used for Greek goddess which is mother of Earth. In Greek mythology gaia is the mother and wife of Uranus and mother of Titans (see Webster Comprehensive Dictionary, International Edition 1996). Gaia hypothesis suggests that life manipulates the environment for the betterment of life. The planet Earth works as a large organism capable of self–maintenance. In fact, gaia hypothesis is a series of several hypotheses. The first of them suggests that life greatly affects planetary environment. The second idea asserts that life affects the environment for the good of life. A third hypothesis of gaia is that life deliberately and consciously controls the global environment. It appears that the systems of positive and negative feedback that operate in the atmosphere and oceans are sufficient to explain the mechanisms by which life affects environment (see **Rana** 2003).

The Gaia hypothesis holds some merit as it relates to the present and future. The future status of life may depend on actions we are taking now and those we will possibly take in the coming years. Accordingly people living on earth today are making conscious decisions concerning the future of planet.

1.9 SOME ECOLOGICAL MAXIMS

Connections in nature form the basis for the first of eight **ecological maxims** (A maxim is a short statement expressing a general truth or rule of behaviour), namely, "you can never do just one thing" (Table 1.2; **Cain** *et al.*, 2011). This maxim is meant to suggest that all actions have multiple effects because events in nature are interconnected. Overall, although the natural world may not be as predictable or as tightly woven as early ecologists may have thought, species are

Table 1.2. Eight ecological maxims (Source. Cain, Bowman and Hacker, 2011)

1. You can never do just one thing.

Organisms interact with one another and with their physical environment. As a result, events in nature are connected, and what affects one organism or place can affect others as well.

2. Everything goes somewhere.

There is no "away" into which waste materials disappear.

3. No population can increase in size forever.

There are limits to the growth and resource use of every population, including our own.

4. There is no free lunch.

An Organism's energy and resources are finite and increasing inputs into one function (such as reproduction) results in a trade-off in which there is loss for other functions (such as growth).

5. Evolution matters.

Organisms evolve or change over time. It is a mistake to view them as static. Evolution is an ongoing process because organisms continually face new challenges from change in both the living and non-living components of their environment.

6. Time matters.

Ecosystems change over time. When we look at the world as we know it, it is easy to forget how past events may have affected our present, and how our present action may affect the future.

7. Space matters.

Abiotic and biotic environmental conditions can change dramatically from one place to another, sometimes across very short distances. This variation matters because organisms are simultaneously influenced by processes acting at multiple spatial scales, from local to regional to global.

8. Life would be impossible without species interactions.

Species depend on one another for energy, nutrients and habitat.

connected to one another. For some people, the fact that events in nature are interconnected provides an ethical imperative (urgent thing) to protect natural systems. A person who feels an ethical obligation to protect human life, for example, may also feel an ethical obligation to protect the natural systems on which human life depends.

QUESTIONS

Long Answer Questions

1. Define the term ecology, autecology and synecology. *(Rohilkhand 2001)*
2. What do you mean by ecology? Describe its subdivision and relationships with other branches of sciences?
3. Define the term ecology, autecology and synecology.
4. Differentiate between ecology and environmental biology.
5. Describe historical development of ecology. Add a note on growth of ecology in India.
6. Describe relations of ecology with other disciplines and its scope.
7. In recent years scientists are giving much emphasis on the study of ecology. Why?

Short Answer Questions

1. Write short note on Palaeoecology. *(Avadh 1999)*
2. Differentiate between the following:
   (i) Ecology and Environmental Biology
   Hits. refer 1.1
   (ii) Autecology and synecology *(Avadh 1999)*

Very Short Answer Questions

1. Give a modern definition of ecology.
2. What is Gaia hypothesis?
3. Define economics.
4. What is economic capital?
5. What is ecosystem?

Fill in the Blanks

1. Schrotev (1896) introduced the term .... and ....
2. .... investigated the interactions of predators and prey and the competitive relationship between species.
3. The term ecology was coined by ....
4. J.V. Liebig initiated the idea of .... and .... cycle in nature.
5. The term ecosystem was proposed by .... in 1935.

Multiple Choice Questions

1. Ecology deals with the study of
   (*a*) microorganisms (*b*) economic differences of countries
   (*c*) insects (*d*) relationship between living and nonliving
2. Ecology is the study of relationship of
   (*a*) organism and environment (b) man and environment
   (c) soil and water (*d*) husband and wife
3. The study of interaction between living organisms and their environment is called
   (*a*) phytogeography (*b*) ecology
   (*c*) ecosystem (*d*) phytosociology

4. The term "Oekologie"(ecology) was coined by

(*a*) Linnaeus (*b*) Harvey
(*c*) Haeckel (*d*) Odum

5. The term ecology was proposed by

(*a*) Reiter (*b*) Aristotle
(*c*) Brown (*d*) Khorana

6. Ecology takes into account

(*a*) environmental factors (*b*) plant adaptations only
(*c*) effect of plants on environments (*d*) all the above

7. Study of interrelationship between a species and its environment is called

(*a*) forest ecology (*b*) autecology
(*c*) synecology (*d*) resource ecology

8. Autecology refers to the study of the ecology of

(*a*) community (*b*) a species
(*c*) animal population (*d*) an individual

9. The study of an entire community in relation to its environment is called

(*a*) autecology (*b*) synecology
(*c*) resource ecology (*d*) space ecology

10. Synecology is the study of

(*a*) community in relation to environment (*b*) water
(*c*) individuals (*d*) environment

11. Who gave the terms "Autecology" and "Synecology"?

(*a*) Schroter (*b*) A.G. Tansley
(*c*) Karl Mobius (*d*) Alexander von Himboldt

12. 'Biocenosis' is the study of

(*a*) living (*b*) environment
(*c*) interaction between living and nonliving (*d*) none of these

13. The term 'Biocenosis' was proposed by

(*a*) Tansley (*b*) Warming
(*c*) Karl Mobius (*d*) None of the above

14. Science which deals with the management of plants, animals, soil, water and minerals is

(*a*) autecology (*b*) synecology
(*c*) resource ecology (*d*) phytosociology

15. The study of plant community structure is called

(*a*) ecology (*b*) phytosociology
(*c*) autecology (*d*) ecosystem

16. Cytological study of species growing in different environments is termed

(*a*) gene ecology (*b*) gynaecology
(*c*) ecotypes (*d*) cytoecology

17. Pedology is the study of

(*a*) soil (*b*) locomotion of animals
(*c*) rocks (*d*) crop disease

18. Predator-prey relation was given by
(*a*) Gause G.F.
(*b*) Lotka and Voltera
(*c*) Holling C.S.
(*d*) Kitching J.A.

19. Name of the famous plant ecologist
(*a*) M.S. Swaminathan
(*b*) Birbal Sahni
(*c*) Ramdeva Mishra
(*d*) Jagdish Chandra Bose

20. The word ecology was given by (*Rohilkhand 2001*)
(*a*) E. Haeckel
(*b*) H. Reiter
(*c*) C. Elton
(*d*) Taylor

21. Fundamental ecology deals with
(*a*) organism and population
(*b*) communities
(*c*) ecosystems
(*d*) all of them

ANSWERS

Very Short Answer Questions

1. Ecology is a branch of science dealing with interactions and relationships between organisms and the environment; the study of goods and services provided by natural ecosystems, including the integration of these non-market services with the economic market (Odum and Barrett 2005).
2. Gaia hypothesis was formulated by James Lovelock in 1968 and it holds that organisms, especially microorganisms, have evolved with the physical environment to provide control (self-regulation) and to maintain conditions favorable to life on Earth.
3. Economics is the branch of science dealing with the goods and services provided by humankind, encompassing the integration of market services with non-market services provided by natural ecosystems.
4. Economic capital is goods and services provided by humankind or the human workforce, typically expressed as gross national product (GNP).
5. Ecosystem means all the organisms in a given area as well as the physical environment in which they live.

Fill in the Blanks

1. Autecology, synecology **2.** Gause **3.** E. Haeckel **4.** Carbon, Nitrogen
5. A.G. Tensley.

Multiple Choice Questions

1. (*d*) **2.** (*a*) **3.** (*b*) **4.** (*c*)
5. (*a*) **6.** (*d*) **7.** (*b*) **8.** (*d*)
9. (*b*) **10.** (*a*) **11.** (*a*) **12.** (*c*)
13. (*b*) **14.** (*c*) **15.** (*c*) **16.** (*d*)
17. (*a*) **18.** (*b*) **19.** (*c*) **20.** (*a*)
21. (*d*).

2

Autecology and Synecology

2.1 COMPARISON OF AUTECOLOGY AND SYNECOLOGY

As discussed earlier in the introductory chapter, both plant and animal ecology are artificially divided into two parts, synecology and autecology. A brief comparison of these two ecological subdivisions will explain the basic significance of each of them. **Autecology** is concerned with the study of the interrelations of individual organisms with the environment. The ecological study of a number of species (population community), ecosystem is called **synecology** (Fig. 2.1).

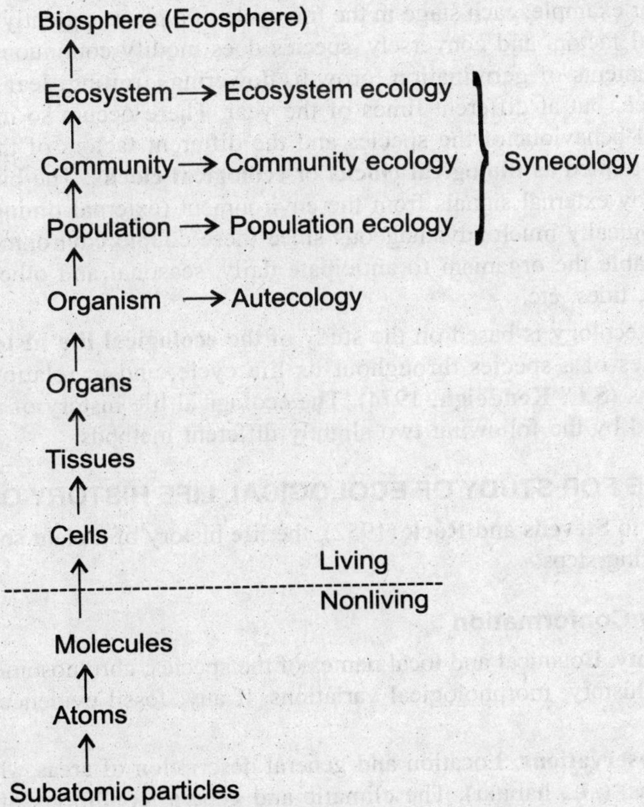

Fig. 2.1 Levels of organization (a classical view) and two main divisions is of ecology.

Both have developed independently, although the knowledge of both of them is necessary for the understanding of the individual, population or ecosystem. Autecology is experimental and inductive. Synecology is philosophical and deductive. Autecology, because it is concerned with the relationship of an organism to one or more variables such as light, temperature, moisture, or salinity, is easily quantified and subject to experimental design both in the laboratory and in the field. Synecology is largely descriptive and not easily subject to experimental design. Autecology has borrowed techniques from physics and chemistry. Synecology, on the other hand, has entered a strong experimental phase in recent years, with the development of such tools as computers and radioactive tracers (**R.L. Smith**, 1977).

Autecology has great significance for the following economic biological disciplines: economic botany which includes agriculture, horticulture, silviculture and forestry; and economic zoology that includes pisciculture, prawn fishery, pearl culture, animal husbandary, apiculture, sericulture, etc. It is also important for soil conservation and wild life **conservation**. Before one proceeds to cultivate any organism of economic or any other significance, he must be well aware to all the requirements of that species for its best possible growth in the environment. Further, autecology gives an idea of distribution, adaptation, speciation, etc., of a particular species. For instance, one finds in nature so many species living in various habitats. Some species are confined to limited areas due to limited characteristics of adaptation, whereas others may grow in different habitat couditions due to having a wide range of ecological adaptations. To find out the answers of some such questions, one has to study autecology of the species.

There exists an intimate interrelationship between the individual organism and its surrounding environment. For example, each stage in the life cycle of a plant is greatly influenced by a number of environmental factors and conversely, species does modify continuously the environment. In fact, the requirements of germination, growth, flowering, fruiting, leaf fall, etc., are met with at the same place, but at different times of the year. There occurs so much synchronisation in the phenological behaviour of the species and the different factors of the environment that the plants are often termed as **biological clocks** or **ecological clocks**. The biological clocks are said to be regulated by external signals from the environment (external timing hypothesis), and such clocks are ecologically much advantageous since these couple environmental and physiological rhythms and enable the organism to anticipate daily, seasonal and other periodicities in light, temperature and, tides, etc.

In fact, autecology is based on the study of the **ecological life history** which is concerned with the activities of a species throughout its life cycle, and in relation to its adjustments to natural conditions (**S.C. Kendeigh**, 1974). The ecological life history of a given plant or animal species is studied by the following two slightly different methods:

2.2 OUTLINE FOR STUDY OF ECOLOGICAL LIFE HISTORY OF THE PLANT

According to **Stevens** and **Rock** (1952), the life history of a plant species should be studied under the following steps:

1. Introductory Conformation

1. Taxonomy. Botanical and local names of the species; chromosome number; geographical distribution and history; morphological variations, if any; fossil evidences, centre of origin and migration route.

2. Field observations. Location and general description of areas where plams grow under natural conditions, (*i.e.*, habitat). The climatic and general conditions of the habitats in which plants grow.

2. Ecological Relations

1. Natural distribution. General distribution, altitude limits, effect of slope, lakes, low lying areas, etc.

2. Soil relations. Type of soil, humus coments, water holding capacity, wilting coefficient, pH range and other edaphic factors.

3. Climatic relations. Light (intensity, duration and quality) and temperature, wind and soil water, etc., which influence the vegetative growth of plant.

4. Plant associatiom. Inter-as well as intraspecific competitions at different growth stages.

5. Modification of the species. Correlation between plant variation and changing environmental conditions, development of ecotype, biotypes, etc.

6. Phenology. Time of seedling appearance, time and rate of vegetative growth, time of flowering, fruiting, seed maturity, and fruit dispersal, etc.

3. Regeneration or Developmental History

This depends mainly upon the average seed output, viability of seeds, seed dormancy, reproductive capacity, seed dispersal, seedling growth, vegetative propagation, vegetative growth and reproductive growth.

(i) Seed output. Collection of seeds, date, habit and weather of seed collection, weight of seed and average production condition of seeds; percentage of seed production and seed germination. Average seed output of a species is calculated as follows:

$$\text{Average seed output} = \frac{\text{Total number of seeds}}{\text{Number of plants from which seeds are collected}}$$

(ii) Seed dispersal. Fruits, bulbs, bulbils, spores, shoot apices, and seeds are generally carried away from parent plants by such natural agencies as animals, wind and water. Thus availability of these dispersal agents at a proper period of life cycle is a very significant factor for successful dispersal of seeds.

(iii) Viability of seeds. The seeds usually have a longer period in their life before they lose their capacity to germinate. This period is called the viability period. Seeds are stored for long time in soil, water or mud to withstand adverse environmental conditions. The viability of seeds lying in the soil is generally influenced by depth, water content, temperature and microbial population of soil.

(iv) Seed dormancy. Methods of break the seed dormancy.

(v) Seeds germination and reprodutive capacity. Normally all the seeds produced by a plant do not germinate due to various reasons. Reproductive capacity of any species indicates its pressure upon the environment. Species with high reproductive capacity is presumed to have better chances of its survival and dispersal. Reproductive capacity of a species is calculated as follows:

$$\text{Reproductive capacity} = \frac{\text{Average seed output} \times \text{percentage germination}}{100}$$

Light, temperature, water, and levels of oxygen and carbon dioxide concentrations are the chief environmental factors affecting the seed germination. From his extensive survey of the reproductive capacity of flowering plants, **Salisbury** (1946) concluded that seed size is determined by the length of time for which the seeding needs to be supported by nutrient reserves in the seed before it becomes photosynthetically self-supporting. **Garrett** (1973) extended Salisbury's conclusions to fungi with reference to spores of some fungi causing leaf spots, macroconidia and chlamydospores of root-infecting fungi (*Fusarium* sp.), mycelial strands and rhizomorphs of fungi infecting tree roots and sclerotia of pathogenic root-infecting fungi.

(vi) Seedling growth. Seedling represents the juvenile stage of plants. Seedlings of trees in forests, annuals, shrubs, climbers, etc., differ in their requirements of seedling establishment, especially in light conditions, water relations, soil characteristics and other environmental parameters.

Extremes of environmental factors such as light, temperature, humidity, pathogens and birds and grazing animals adversely affect the establishment of seedlings.

(vii) Vegetative growth. The vegetative growth is influenced by various environmental, mostly edaphic and aerial factors, such as the intensity, duration and quality of temperature, light, water, pH, etc. In grasses and some weeds, the vegetative growth, as the length of shoot, root depth, number of nodes, internode length, number and size of leaves, stomatal frequency, thickness of cuticle on leaf, etc., are affected by the environmental conditions. In other plants, vegetative growth include study of root system, root-shoot ratio, at different growth stages and arrangement, type, shape, variation, surface leaf area, chlorophyll, etc., in relation to environment in various growth stages.

(viii) Reproductive growth. It includes the flowering, pollination and fruiting of a species. Most of the terrestrial plants, for their successful growth, reproduce sexually, *i.e.*, flower and fruit. Various environmental factors influence the flowering, pollination and fruiting of a plant species. Different species differ in their time of flowering and their light and temperature requirements for flowering. Various characteristics of flowers affect pollination and agencies involved in the process. Plant species also differ in structure, and number of fruits, time of their formation and agents which damage their fruits. However, aquatic plants generally reproduce by vegetative means.

4. Growth and Dry Matter Accumulation

Measurement of Net Assimilation Rate (NAR), Relative Growth Rate (RGR), Leaf Area Index (LAI), Net primary production, biomass, energy accumulation pattern, phytochemical composition and accumulation pattern with reference to nitrogen, phosphorus, and other nutrients.

5. Economic Importance of the Plant Species

(For further details of autecology see **R.F. Daubenmire's** *Plants and Environment*: *A Text Book of Plant Autecology* (1959); and **R. Misra's** *Ecology workbook* (1968).

2.3 OUTLINE FOR STUDY OF ECOLOGICAL LIFE HISTORY OF THE ANIMAL

In case of animals, the ecological life histories usually proceed with, first, analysis of the behaviour adjustments needed for the survival of the mature animal; then of its reproductive behaviour and lastly, of the development of behaviour and physiological adjustments of the young animal. A complete ecological life history of an animal that is needed for the autecological study includes the following items:

1. Phylogenetic and geological history.
2. Geographic and habitat distribution with an analysis of adjustments to the physical environment and of biotic interrelations within the community.
3. Variation in population, through time and in space.
4. Changes in seasonal activities and physiological states: breeding, migration, hibernation.
5. Food and enemies.
6. Parasites and diseases.
7. Reproductive potential, mortality, rate of population turn-over.
8. Requirements for reproduction: home range, territory, nestsite, nesting materials, etc.
9. Breeding behaviour: mating, nesting, etc.
10. Development of offspring: rate, stages, generations per year, etc.

Thus, to study the ecological life history of a **plant** or **animal** species each stage of the life cycle is studied in detail both in field and cultures grown under controlled or known environmental conditions. Then, observations and data are put in proper sequence and the whole picture of autecology is drawn.

QUESTIONS

Long Answer Questions

1. Define autecology and show, why it is necessary to study?
2. Compare autecology and synecology.
3. How will you study the autecology of a plant species?
4. Write in brief outline about the method of study of the life history of an animal species.

Short Answer Questions

1. Write short notes on the following: Autecology; Synecology; Life history and Economic significance of autecology.

Very Short Answer Questions

1. Define biosphere.
2. What are three subdivisions of the biosphere?
3. What is the biological clock?

Fill in the Blanks

1. The ecological study of one species is called ....
2. The ecological study of a number of species (population, community, ecosystem) is called ....
3. The term biosphere was first used by the Austrian geologist .... in 1857.
4. .... (1970) has defined biosphere as that part of the Earth in which life exists.
5. The three subdivisions of the biosphere are: .... , ... and ....

Multiple Choice Questions

1. Ecological study of the interrelations of individual organism with the environment is called
   (*a*) synecology (*b*) autecology
   (*c*) ecosystem ecology (*d*) none of these

ANSWERS

Very Short Answer Questions

1. Biosphere is that part of the Earth in which life exists (**Hutchinson** 1970).
2. Atmosphere, Lithosphere, Hydrosphere.
3. Biological clock is an endogenous, physiological mechanism that keeps time independent of external events, enabling organisms to respond to daily, lunar, seasonal, or other periodicities.

Fill in the Blanks

1. Autecology 2. Synecology 3. Eduard Suess 4. Hutchinson
5. Atmosphere, Lithosphere, Hydrosphere.

Multiple Choice Questions

1. (*b*).

3

Levels of Organisation, Hierarchy and Biosphere

3.1 INTRODUCTION

An evident unit of life in nature is the individual organism. Linguistically as well as biologically, the root of the word '*organism*' is organisation: a characteristic **structural order**. Each kind of living organism possesses a certain degree of organisation which enables them to perform different life activities. In fact, close examination of any aspect of the living world reveals that everything has a number of interrelated component parts. For example, a human being possesses in his body various organ systems such as digestive system, circulatory system, respiratory system, etc., for sustaining the life of an individual. Each organ system, in its turn, contains numerous different organs. For example, the digestive system is built up of such organs as mouth, buccal cavity, oesophagus, stomach, intestine, rectum, anus, liver, pancreas, etc. Each organ too is made up of different kind of tissues. Thus, stomach possesses epithelial tissues, muscles, glandular tissues and so on. Even each kind of tissue contains numerous cells, as muscles of stomach possess muscle cells.

Further, each cell contains different cellular organelles such as mitochondria, endoplasmic reticulum, Golgi bodies, lysosomes, nucleus and so on. Each organelle is built up of different kinds of macromolecules such as polypeptides, polysaccharides, lipids and nucleic acids (DNA and RNA). Each kind of organic macromolecule is built up by several simple molecules (monomers), such as polypeptides contain many molecules of amino acids. Each molecule contains atoms of different elements, such as a molecule of amino acid consists of atoms of many elements like carbon, nitrogen, hydrogen. Lastly, each atom has its own structural components in the form of subatomic particles such as electrons, protons and neutrons. Thus, in the living world, there exist different separate levels of organisation. In non-living world too we may find different levels of structures of matter.

Perhaps the best way to delimit modern ecology is to consider the concept of **levels of organization**, visualized as an **ecological spectrum** (Fig. 3.1) and as an extended **ecological hierarchy** (Fig. 3.2). **Hierarchy** means an arrangement into a graded series. Interactions with the

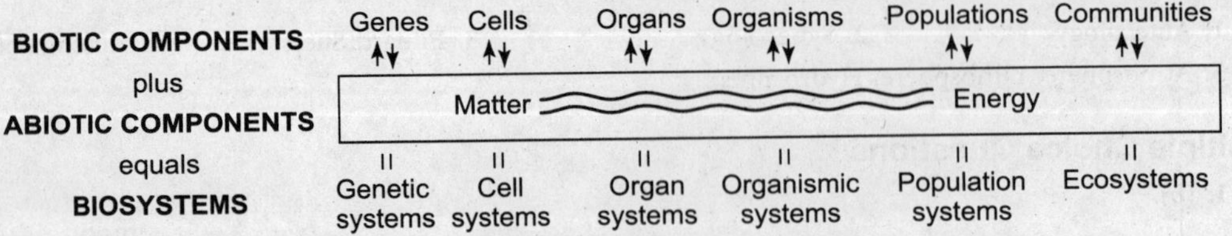

Fig. 3.1 Ecological levels of organization spectrum emphasizing the interaction of living (biotic) and nonliving (abiotic) components (after odum and Barrett 2005).

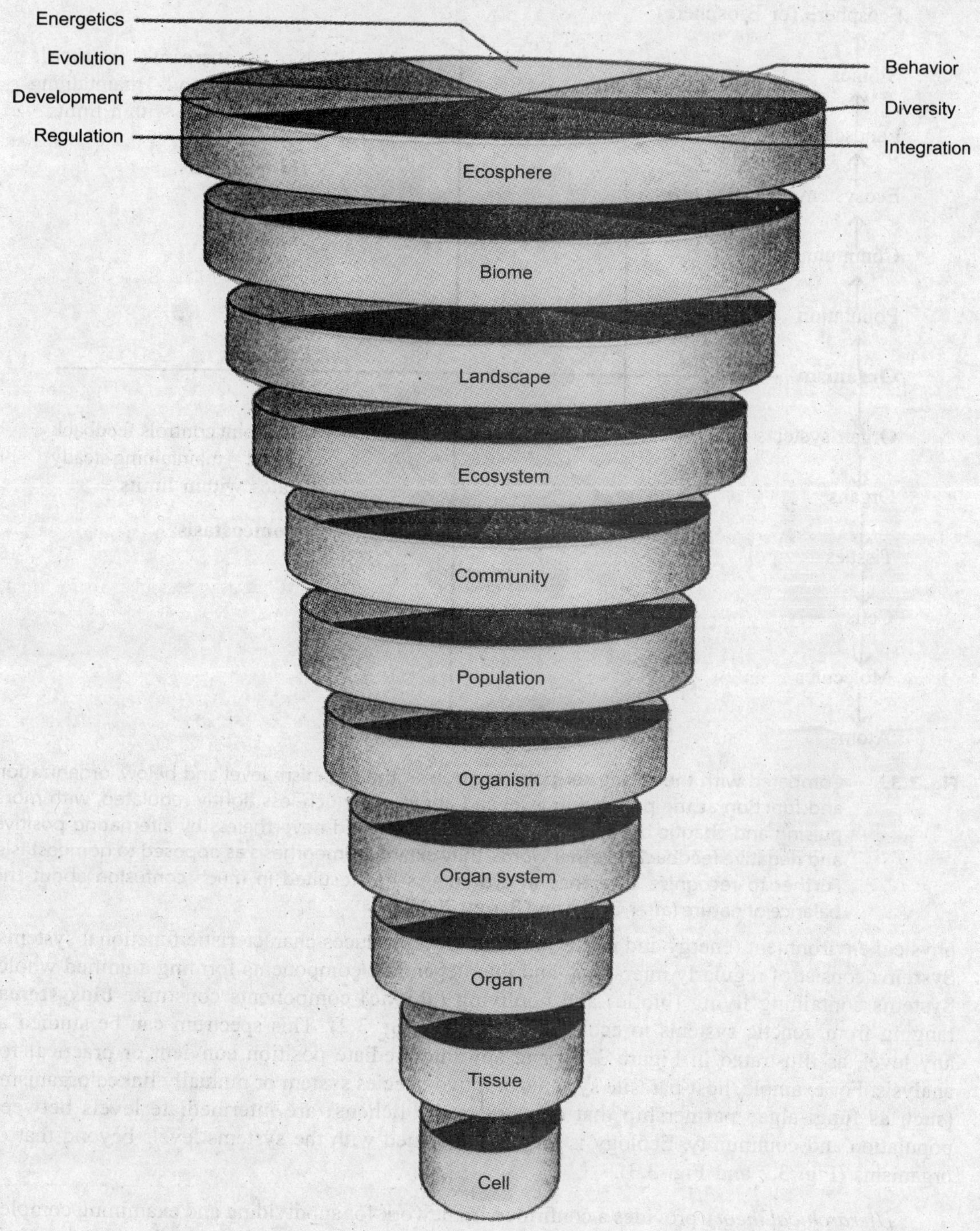

Fig. 3.2 Ecological levels of organization hierarchy. Model suggested by Barrett *et al.* (1997) has seven processes or functions which are depicted as vertical components of eleven integrative levels of organization (afte Odum and Barrett 2005).

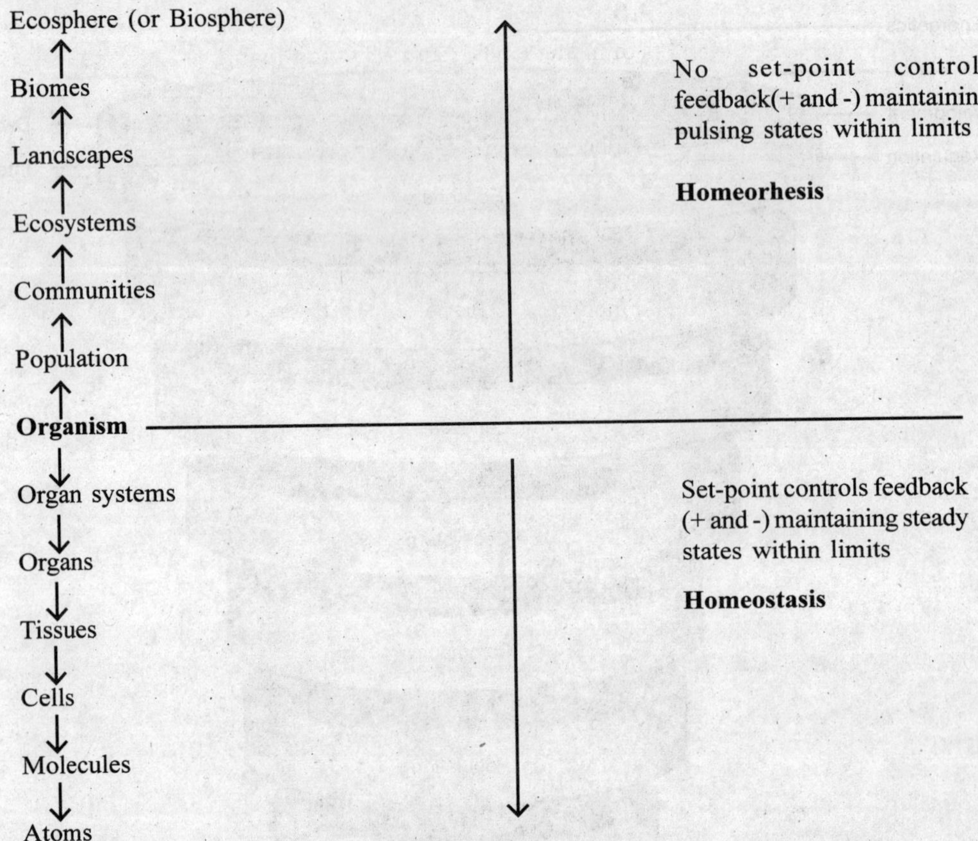

Fig. 3.3. Compared with the strong set-point controls at the organism level and below, organization and function at the population level and above are much-less tightly regulated, with more pulsing and chaotic behaviour, but they are controlled nevertheless by alternating positive and negative feedback–in other words, they exhibit homeorhesis as opposed to homeostasis. Further to recognise difference in cybernetics has resulted in much confusion about the balance of nature (after Odum and Barrett 2005).

physical environment (energy and matter) at each level produces characteristic functional systems. **Systems** consist of regularly interacting and interdependent components forming a unified whole. Systems containing living (biotic) and nonliving (abiotic) components constitute **biosystems**; ranging from genetic systems to ecological systems (Fig. 3.2). This spectrum can be studied at any level, as illustrated in Figure 3.1, or at any intermediate position convient or practical for analysis. For example, host-parasite systems or a two-species system or mutually linked organisms (such as fungi-algae partnership that constitutes the lichens) are intermediate levels between population and community. Ecology is largely concerned with the systems levels beyond that of organisms (Fig. 3.2 and Fig. 3.3).

Hierarchical theory provides a confirmed framework for subdividing and examining complex situations or extensive gradients, but it is more than just a useful rank-order classification. It is **holistic approach** to understand and deal with complex situations and is an alternative to the **reductionist approach** of seeking answers by reducing problems to lower-level analysis (**Ahl** and **Allen** 1996).

3.2 HIERARCHY OF LEVELS OF ORGANISATION

The smallest structural units of matter (living as well as non-living) are **subatomic particles**. The next larger units are **atoms**, each of which consists of subatomic particles. Atoms in turn form still more complex combinations called **chemical compounds** (molecules). Chemical compounds are variously joined together as even more elaborate units or **complexes of compounds** (macromolecules). These units are representing successively **higher levels of organisation of matter**. They form a pyramid or hierarchy, in which any given level contains all lower levels as components and is itself a component of all higher levels. For example, atoms contain subatomic particles as components and atoms are themselves components of molecules.

All structural levels upto and including those of macromolecules are encountered both in the non-living and the living world. For example, water and table salt are two chemical compounds which are found in non-living and living matter. Examples of complexes of compounds in the non-living world are rocks which are composed of several types of compounds. In living world, complexes of compounds often occur as microscopic and submicroscopic bodies called **organelles**. But even in their most elaborated and complicated forms, complexes of compounds (macromolecules like proteins, lipids, carbohydrates, nucleic acids, etc.) or organelles (*i.e.*, mitochondria, endoplasmic reticulum, chloroplasts, etc.) cannot qualify as living units. To reach the level of life, we must go to the next higher structural level, that of cells.

A cell is a specific combination of organelles. It is usually a microscopic bit of matter organised just complexly enough to contain all the necessary apparatus for the performance of metabolism and self-perpetuation. A cell in effect represents the least-elaborate known structure that can be fully alive. It follows that a living organism must consists of at least one cell. In fact, **unicellular** (single celled) organisms like Monera (Bacteria and blue-green algae) and Protista (unicellular eukaryotic organisms like Protozoa and algae) constitute the majority of living creatures on earth. All other organisms are **multicellular**, each composed of up to hundred of trillions of joined cells.

Further, within a multicellular organism, several distinct levels of organisations can be distinguished. The simplest multicellular types contain comparatively small number of cells. If all such cells are more or less alike, the organism is often referred to as a **cellular colony**. If two or more different groups of cells are present each such group usually forms a **tissue**. Thus, there occur cellular colony in bacteria, two-tissue colony in slime-molds and three-tissue colony in *Volvox*. In structurally more complex organisms, there are not only several tissues, but some of the tissues also joined-further as one or more units called **organs**. The most complex organisms contains not only many tissues and organs, but also groups of organs united as one or more **organ systems**. Thus, living organisms exhibit at least five levels of structural complexity: the single-celled form, the colonial form, the organisms with tissues, the type with organs and the type with organ systems.

Thus, whether of one cell or of trillions cells, the individual is the relevant unit with regard to physiological organisation, total metabolism, responsiveness, reproduction, growth and development, and certain other essential phenomena of living things. The individual is the separate and objectively concrete biological unit. However, several still higher forms of life can be distinguished beyond the level of the individual organism.

An individual cannot live in isolation in nature. For its survival and for perpetuation of its race, an individual often associates with other individual of its own kind. All the similar individuals of a given locality which potentially form a single sexually interbreeding group, form a next higher level of organisation called **population**. Since Mendelian laws of heredity apply to the transmission of genes among the individuals of a population, so, it is also termed as **Mendelian population** by the geneticists. The size of a population may vary, but it is usually considered to be a local group which is called **local population** or **dame**. Each individual member of a dame has an equal chance of mating with any other member of the opposite sex. For example, all individuals of *Paramecium* in a fresh-water pond form a dame or local population or population.

A group of similar dames or of actually interbreeding natural populations is called **species**. Thus, a species is a group of like individuals that live together and interbreed to produce fertile offspring. Thus all corn plants on earth, all bull frogs, all the human beings, each represent a species.

Many groups of populations belonging to different species form the next higher level of organisation, the **biotic community**. For example, different species of micro-organisms, plants and animals inhabiting a fresh water pond along the population of *Paramecium* form a biotic community. A biotic community together with the physical space in which it exists represents an ecological unit, the **ecosystem**. For example, a fresh water pond with its biotic community represents an ecosystem.

Referring again to Figure 3.2, the next level in the ecological hierarchy is the **landscape**, a term originally referring to a painting and defined as "an expanse of scenery seen by the eye as one view" In ecology, landscape is defined as *a heterogeneous area composed of a cluster of interacting ecosystems that are repeated in a similar manner throughout* (**Forman and Gordon** 1986). A **watershed** is a confirmed landscape-level unit for large-scale study and management because it usually has identifiable natural boundaries. Watershed is the catchment or drainage basin of a river; it includes total area above a given point on a stream or river that constitutes water to flow at that point. **Biome** is a large regional or subcontinental system characterized by a major vegetation type or other identifying landscape aspect, as, for example, the Temperate Deciduous Forest biome or the Continental Shelf Ocean Biome. A **region** is a large geological or political areas that may contain more than one biome-for example, the regions of Midwest, the Appalachians Mountains, or the Pacific Coast. The largest and most nearly self-sufficient biological system is the **ecosphere** or **biosphere**, which includes all the living organisms of Earth interacting with the physical environment as a whole to maintain a self-adjusting, loosely controlled pulsing state (homeorhesis). Biosphere is a huge ecosystem which contains living components in the form of biological world (living matter) and non-living or abiotic components in the form of earth or soil (**lithosphere**), water (**hydrosphere**) and air (**atmosphere**).

3.3 DEDUCTION OF CERTAIN BASIC PRINCIPLES FROM THE LEVELS OF ORGANISATION

From the hierarchy of levels in the organisation of non-living and living matter, following generalizations can be made:

1. Each level of organisation has fewer units than the preceding lower level. Thus, there are fewer communities than species and fewer cells than organelles.

2. Each level of organisation is structurally more complex than lower ones. It includes the complexity of all the lower levels but has some additional special features of its own. This fact supports the **principle of emergent properties** by showing the emergence of new properties at each level of organisation when parts are organised to form a living system.

3. In any hierarchy of levels, a jump from one level to the next can often be achieved only at the expense of energy. It takes energy to build atoms into chemical compounds, and it takes energy to create cells out of chemical complexes. Indeed once a higher living level has been created, energy expenditures must usually continue thereafter to maintain that level. For example, if the energy supply to the cell, the organ, or the organism is stopped, death and decomposition soon follow and reversion to lower levels occur.

The Emergent Property Principle

An important outcome of hierarchical organization is that as component or subsets, are combined to produce larger functional wholes, new properties emerge that were not present at the level below. Accordingly, an **emergent property** of an ecological level or unit cannot be predicted from the study of the components of that level or unit. Another way to express the same concept is **nonreducible property**- that is, a property of the whole not reducible to the sum of the properties of the parts. Though findings at one level help in the study of the next level, they never completely explain the phenomena occurring at the next level, which must be itself studied to complete the picture. Following two example may explain the concept of emergent properties:

1. When hydrogen and oxygen are combined in a certain molecular configuration, water is formed–a liquid with properties utterly different from those of its gaseous components.

2. When certain algae and coelenterate animals evolve together to produce a coral, an efficient nutrient cycling mechanism is created that enables the combined system to maintain a high rate of productivity in waters with very low nutrient content. Thus, the fabulous productivity and diversity of coral reefs are emergent properties only at the level of the reef community.

Salt (1979) suggested that a distinction be made between emergent properties, as defined previously, and **collective properties**, which are summations of the behaviour of components. Both are properties of the whole, but the collective properties do not involve new or unique characteristics resulting from the functioning of the whole unit. *Birth rate* is an example of population level collective properties, as it merely a sum of the individual births in a designated time period, expressed as a fraction or percent of the total number of individuals in the population. New properties emerge because the components interact, not because the basic nature of the components is changed. Parts are not "melted down", as it were, but integrated to produce unique new properties. It can be demonstrated mathematically that integratives hierarchies evolve more rapidly from their constituents than non-hierarchical systems with the same number of elements; they are also more resilient in response to disturbance. Theoretically, when hierarchies are decomposed to their various levels of subsystems, the latter can still interact and reorganize to achieve a higher level of complexity.

Transcending Functions and Control Processes

Whereas each level in the ecological hierarchy can be expected to have unique emergent and collective properties, there are basic functions that operates at all levels. Examples of such **transcending functions** are *behaviour, development, diversity, energetics, evolution, integration* and *regulation* (Fig. 3.2). Some of these (energetics, for example) operate the same throughout the hierarchy, but others differ in *modus operandi* at different levels. Nature selection evolution, for example, involves *mutations* and other direct genetic interactions at the organism level but indirect coevolutionary and group selection processes at higher levels.

It is especially important to emphasize that although **positive** and **negative feedback controls** are universal, from the organism level down, control is **set point**, in that it involves very exacting *genetic, hormonal and neural* controls on growth and development, leading to what is often called **homeostasis** (Box 3.1). As noted on the right-hand side of Figure 3.3, there are no set-point controls above the organism level (no chemostats or thermostats in nature). Accordingly, feedback control is much looser, resulting in pulsing control than steady states. The term **homeorhesis** has been suggested for this pulsing control. In other words, there are no equilibriums at the ecosystem and ecosphere levels, but there are **pulsing balances**, such as between production and respiration or between oxygen and carbon dioxide in the atmosphere. Failure to recognize this difference in **cybernetics** (the science dealing with mechanisms of control or regulation) has resulted in much confusion about the realities of the so-called "balance of nature" (**odum** and **Barrett**, 2005).

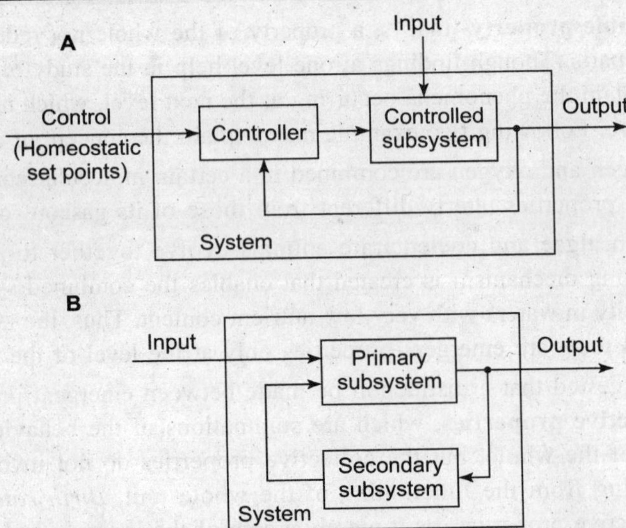

Fig. 3.4. Feedback control systems. A—Model appropriate for human-made autonomic control systems and homeostatic, goal-seeking organismic systems; B—Model appropriate for non–teleological systems, including ecosystems, where control mechanisms are internal and diffuse, involving interactions between primary and secondary subsystems.

Box 3.1 Homeostasis and Homeorhesis

The science of cybernetics, as founded by **Norbert Wiener** (1948), embraces both inanimate and animate controls. Mechanical feedback mechanisms are often called **servomechanisms** by engineers, whereas biologists use the phrase **homeostatic mechanisms** to refer to organismic systems. **Homeostasis** (from *homeo*=same, and stasis= standing) at the organism level is a well-known concept in physiology, as outlined in **Walter B.Cannon's** book, *The Wisdom of the Body* (1932). In servomechanisms and organisms, a distinct mechanical or anatomical "controller" has a specified "set point"(Fig. 3.4 A). **Homeostasis** is the tendency of a system to resist change and maintain itself in a state of stable equilibrium. For example, in the familiar household heating system, the thermostat controls the furnace, in a warm blooded animal, a specific brain center controls body temperature, and genes tightly control the growth and development of cells, organs and organisms. There are no thermostats or chemostats in nature, rather, the interplay of material cycles and energy flows along with subsystem feedbacks in large ecosystems, generates a self-correcting **homeorhesis** (Fig. 3.4 B). **Homeorhesis** (Greek rhesis = flow or pulsing; maintaining the flow) is the tendency of a system to maintain itself in a pulsing state of equilibrium. **Waddington** (1975) coined the term *homeorhesis* to denote the evolutionary stability or preservation of a system's flow or pulsing process as a pathway of change through time. The goal of homeorhesis is to keep systems altering in the same manner as they have altered in the past (**Naveh** and **Lieberman** 1984). Control mechanisms operating at the ecosystem level include microbial subsystems that regulate the storage and release of nutrients, behavioural mechanisms and predator-prey subsystems that control population density, to mention just a few examples.

3.4 BIOLOGICAL IMPORTANCE OF INCREASING COMPLEXITY AND ORGANISATION

The trend of increasing complexity and organisation from lower level toward higher one produces following new properties: **cooperation, division of labour** and **specialization.**

One property which arises at each new higher level of organisation is united, integrated function. Disunited structure means independent function and by extension, **competition**; while united structure means joint function and by extension, **cooperation**. For example, the cells can remain structurally independent and can compete for space and raw materials. But if they form a multicellular unit, they surrender their independence and become a cooperative, integrated system.

The fundamental advantage of cooperation is **operational efficiency**: the cooperating whole is more efficient in performing the functions of life than its lower-level components separately and competitively. For example, separate cells must expend more energy and materials for survival than if that same number of cells were integrated as a tissue.

One underlying reason for this difference is that, in the integrated unit, duplication of effort can be avoided. In a set of separate cells, for example, every cell is exposed to the environment on all sides and must, therefore, expend energy and materials on all sides to cope with the impact of the environment. However, if the same cells are grouped together as a compact tissue only the outermost cells are in direct contact with the environment, and inner cells then need not to channel their resources into protective activities.

Besides avoiding duplication of efforts, cooperative groupings make possible continuity of efforts. For example, an unicellular form must necessarily carry out all survival functions in its one cell. Sometimes, the performance of even one of these functions requires most or all of the capacities of all the cell. Thus, the entire cell surface must often serve as gateway for entering nutrients and departing wastes. And all parts of the cell may have to participate directly in locomotion or in feeding, for example.

By contrast, in multicellular organisms continuity of a given effort becomes possible through **division of labour**. The total task of survival can be divided up into several subtasks, and each of these can become the continuous responsibility of particular cells only. Some cells might function in feeding, continuously so, and others in locomotion, again continuously. The division of labour becomes so pronounced that many or most cells are permanently limited in functional capacity: they can perform only certain jobs and no others. For example, mature nerve cells can conduct nerve impulses only and are quite unable to reproduce or move. Thus, different cells of multicellular organisms exhibit **specialization** for particular set of functions.

3.5 THE BIOSPHERE

The entire world basically can be broadly classified into **abiotic world** and **biotic world**. The abiotic world can be further divided into three packages: the **atmosphere** (air), the **lithosphere** (earth) and the **hydrosphere** (water). To these three ecological components we can add the biological world or biotic world, the **biosphere**. All four subdivisions thus represent four major global components of the world ecosystem. These four spheres continuously exchange matter with one another, but obviously there are differences in the relative and total amounts of the natural elements in each organism take in inorganic nutrients from the soil, water and air; they eliminate their wastes, and their bodies are rendered into inorganic molecules once again. Water and minerals shuttle among the air, land and water with an occasional pause in an organism.

Thus, the biosphere is that part of the earth in which life exists (**Hutchinson**, 1970). More specifically, the sum of those portions of the hydrosphere, lithosphere and atmosphere into which life penetrates is the biosphere. However, together with the geological, chemical features of the totality of our habitats, these (air, water, earth, and organisms) are sometimes grouped under the term **ecosphere** (**Cole**, 1958). Biosphere or ecosphere may be thought of as a biochemical system capable of capturing, converting, storing, and utilizing the energy of the sun. Approximately three hundred thousand species of green plants and micro-organisms are recognised as **primary producers** which utilize inorganic elements and compounds to synthesize the organic materials of life. Their productivity is consumed by more than a million other species of organisms which convert this organic storehouse into animal form, adding to the beauty, and value of the biosphere as well as its complexity. Still other species, primarily bacteria and fungi, accomplish the recycling process by returning plant and animal wastes and residues to inorganic form so the process may be renewed. In this cyclic process of life, many elements are shared from a common global pool and are converted from inorganic to organic form and back again. Examples of such elements which commonly shuttle among the air, water, earth and organisms are carbon, oxygen, hydrogen, nitrogen, phosphorus, sulphur, sodium, potassium, calcium, magnesium, iron, manganese, cobalt, copper, and zinc.

The oldest known photosynthetic organisms are at least two billion years old, and throughout this vast period of time, the entire biosphere has matured and evolved as a self-regulatory system with internal checks and balances. It has developed a remarkable degree of homeostasis and ecological control within its various components. In forthcoming chapters; we will consider in more detail the process of photosynthesis or primary production, the cyclic passage of elements, the flow of energy through living organisms, the growth and development of populations, and the diversity and maturation of communities and ecosystems.

In physical terms, the biosphere is a relatively thin and incomplete envelope covering most of the world. It represents a mosaic of different biotic communities from simple to complex, aquatic to terrestrial, and tropical to polar. It does not exist in the extremities of the polar regions, the highest mountains, the deepest ocean troughs, the most extreme deserts, or the most highly polluted areas of land and water. Its total thickness, including all portions of the earth where living organisms can exist, is less than 26 kilometres. Its zone of active biological production, in terms of photosynthesis, is much narrower, and varies from a few centimetres to over 100 metres. This zone would, for instance, be only a few centimetres in muddy or turbid water, whereas in very clear ocean water, it could be more than 100 metres in thickness. On land, the zone of biological production might be only a few millimetres in a desert or rock environment, whereas it might again be more than 100 metres in a sequoia or tropical rain forest. Living organisms can exist, of course, beyond the range of active biological production; some insects or birds may be airborne to altitudes above 20,000 feet, and viable spores, seeds of plants, and micro-organisms may be found in the atmosphere and mountain tops above 25,000 feet. In the ocean depths, many animals can exist well below a thousand feet—one hydra-like animal has been photographed at a depth of 15,900 feet in the south Atlantic. However, in both extreme altitude and extreme depth, the organisms depend upon the much thinner zone of active biological production, that portion of the system which converts the energy of sunlight into the chemical and physical energy of living organisms.

3.6 FROM BIOSPHERE TO ECOLOGICAL NICHE

The biosphere can be divided into a series of **biogeographical regions**, each inhabited by distinctive species of animals and plants. These organisms are able to move freely from place to place within each region, but not from one region to another, this being prevented by various natural barriers like river, lake, sea, mountain, desert, etc. Such biogeographical regions include Eurasia, South America, Africa and Australia.

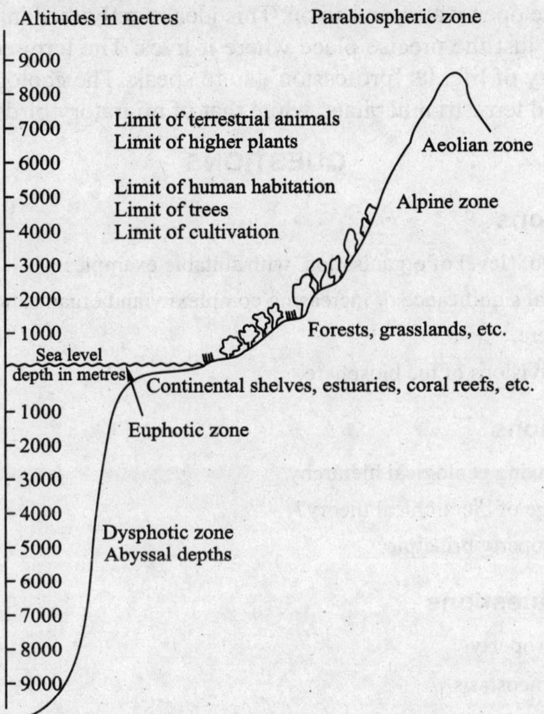

Fig. 3.5. Vertical extent of biosphere. The parabiospheric zone represents altitudes where only dormant forms of life, such as bacterial and fungal spores, exist. The euphotic zone in aquatic environments is the zone of active photosynthesis (after Southwick, 1976).

Although they may possess a different fauna and flora, two such geographical regions may be virtually identical in their climatic conditions, and this is reflected in similar adaptations shown by their respective inhabitants. These equivalent areas, cutting right across different continents, are lumped together as **biomes**. Each biome has its unique set of conditions, and each supports a particular type of flora and fauna. Such areas include the tundra of the Arctic and Antarctic regions; grasslands, deserts, and the different kinds of forests (coniferous, temperate and tropical). The oceans can be regarded as a single biome (the marine biome), though coral reefs are generally regarded as a separate biome. Within each biome only those organisms with the necessary adaptations for surviving the physical conditions are found. Thus coniferous forest and tundra biomes only have organisms capable of withstanding long periods of extreme cold. Desert organisms, on the other hand, must be able to cope with intense heat and drought. Marine organisms must be able to thrive in salt water, and so on.

Biomes are subdivided into small units called **zones**, each with its own particular set of physical conditions. For example, a forest biome can be divided into **ground zone** and **canopy zone**; the desert into **surface** and **subterranean zones** and the marine biome into **surface, abyssal** and **intertidal (littoral) zones**. Individual organisms may be specially adapted to live in one of these different zones, in some cases providing them with a means of avoiding the extremes of climatic conditions met with in the biome as a whole. For example, many desert animals avoid the heat of the mid-day sun by burrowing into the subterranean zone.

Within each biome are numerous **habitats**, specific localities each with a particular set of conditions and an appropriately adapted community of organisms. Typical habitats include freshwater ponds, slow-flowing streams, rock pools, and hedgerows.

Within a particular habitat an organism is generally confined to a restricted situation. For example, in a slow-flowing stream, crayfish live in burrows in the river-bank, flatworms and shrimps under stones, fishes in the open water, and so on. This idea is embodied in the term **ecological niche** which means more than just the precise place where it lives. The term ecological niche embraces the organism's entire way of life, its 'profession', so to speak. The ecological niche of amphibians includes both aquatic and terrestrial habitats, while that of migratory birds span several biomes.

QUESTIONS

Long Answer Questions

1. Explain the concept of 'level of organisation' with suitable examples.
2. What is the biological significance of increasing complexity and organisation in living beings?
3. Describe the biosphere.
4. Describe different divisions of the biosphere.

Short Answer Questions

1. Draw a diagram showing ecological hierarchy.
2. What is the advantage of hierarchical theory?
3. Explain emergent property principle.

Very Short Answer Questions

1. What is emergent property?
2. Define the term homeostasis.
3. What is homeorhesis?

Fill in the Blanks

1. The part of the environment of Earth in which living organisms are found is called ....
2. Hierarchy means an arrangement into a .... series.
3. Term homeorhesis was coined by .... in 1975.
4. Term Homeostasis was coined by .... in 1932.

Multiple Choice Questions

1. All the elements of life-support system are
   (*a*) interconnected (*b*) interrelated
   (*c*) interdependent (*d*) all the above

ANSWERS

Very Short Answer Questions

1. Emergent property refers to properties at various levels of organization that cannot be derived from lower-level systems studied in isolation.
2. Homeostasis is the tendency of a system to resist change and maintains itself in a state of stable equilibrium.
3. Homeorhesis is the tendency of a system to maintain itself in a pulsing state of equilibrium.

Fill in the Blanks

1. Biosphere 2. Graded 3. Waddington 4. W.B. Cannon.

Multiple Choice Questions

1. (*d*).

4

Environment: Atmosphere, Hydrosphers and Lithosphere

Each and every living organism has its specific surrounding, medium or environment to which it continuously interacts and remains fully adapted. The **environment** is the sum total of physical and biotic conditions influencing the responses of the organisms. The life supporting environment of planet earth—the **biosphere** is composed of following three chief media— **air**, **water** and **soil**, which are the components of three major sub-divisions of the biosphere— **atmosphere**, **hydrosphere** and **lithosphere**, respectively (Box 4.1).

These media are not completely isolated from each other, however, some of the atmospheric gases are dissolved in all natural waters, and some moisture is present almost everywhere in the atmosphere. Each of these media can be discussed separately in the following manner:

Box 4.1

The solid (lithosphere) and liquid (hydrosphere) portions of earth form the **geosphere**. Above the geosphere lies the atmosphere, at the interface between these two regions is found almost all of the biosphere or zone of life (**Joseph A. Angelo, Jr., 1999**).

4.1 ATMOSPHERE (AIR)

The multilayered gaseous envelope surrounding the planet earth is called **atmosphere**. The atmosphere remains in contact with all the major types of environment of earth, interacting with them and greatly affecting their ability to support life. It filters sunlight reaching the earth, affect climate, and is a reservoir of several elements essential for life.

Various Zones of Atmosphere

The atmosphere is divided into five distinct layers or zones: troposphere, stratosphere, mesosphere, ionosphere and exosphere (Fig. 4.1). The **tropopause** separates the stratosphere from the troposphere and **stratopause** separates mesosphere from stratosphere.

1. Troposphere. It is the lowest region of atmosphere which subjects to differential heating, temperature inversions and convection currents and which extends from the surface of the earth up to a height of 8 to 10 km at polar latitudes (poles), 10 to 12 km at moderate latitudes and 16 to 20 km at the equator. For the organisms, troposphere forms a most important zone of atmosphere. Many important climatic events such as cloud formation, lightening, thundering, thunder storm formation, etc., all take place in troposphere. In this zone the percentage concentration of different gases in air does not vary with an increase in height. But the water vapour content in air depends upon the weather (*e.g.,* part of the troposphere over an ocean carries more moisture than that over a land surface) and it decreases sharply with an increase in height as does the air temperature. Air temperature in this zone gradually decreases with height at the rate of about 6.5^{0}C per km (more specifically 5^{0}C per km in the lower troposphere and 7^{0}C per km in the upper troposphere). In fact, towards the upper layers of troposphere, the temperature may decrease up to – 60^{0}C. Upper region of the troposphere has a narrow boundary called the **tropopause** which has a constant temperature.

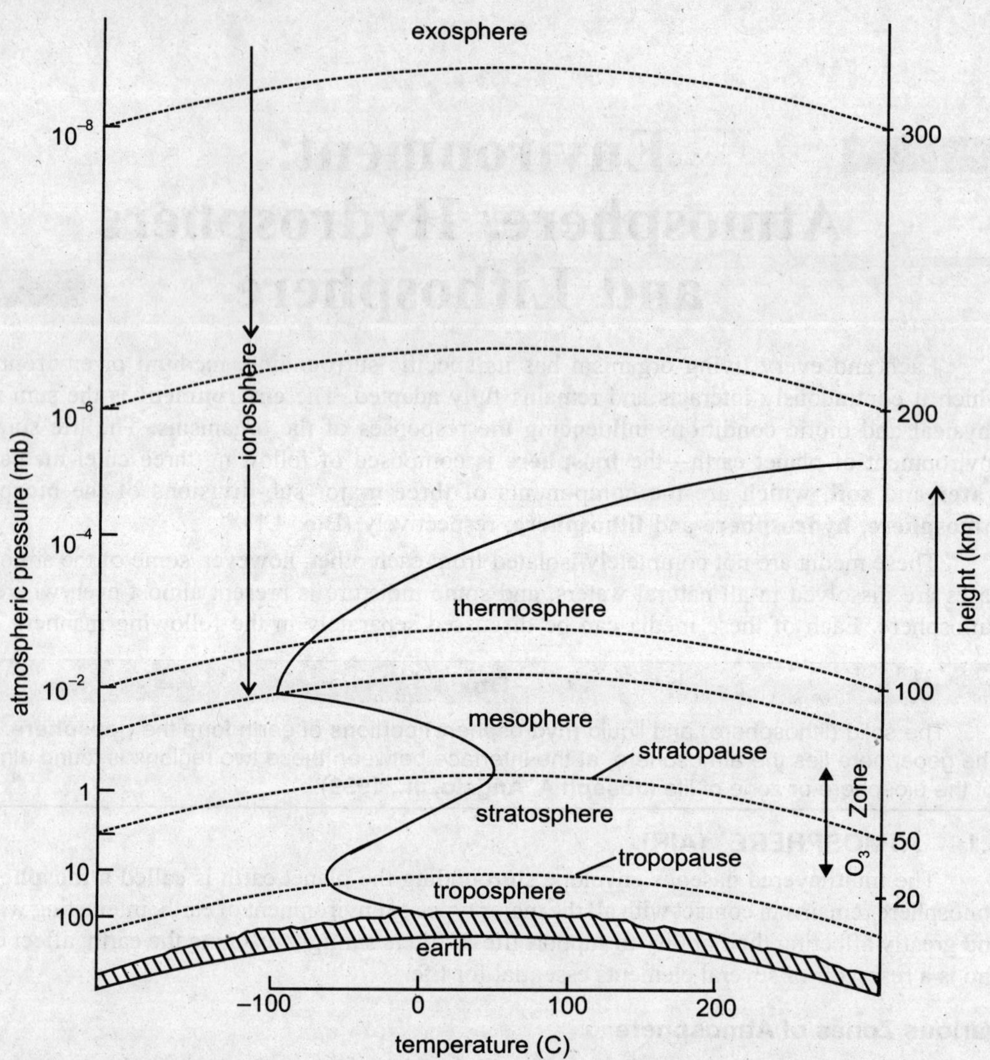

Fig. 4.1. Diagrammatic sketch showing the principal zones of atmosphere and variations in temperature and pressure in them.

Further, the non-uniform heating of the ground surface by sun's radiations produces ascending and descending air currents, which cause turbulence and mixing of air masses vertically. Moreover, the average air pressure at the earth's surface is 1,014 millibars (*viz.,* 1 millibar= 1/1000 bar; 1 bar = 1.019 kg per cm^2 close to 1 atmosphere, 1.332 millibars=1mm of mercury (Hg). At an altitude of 5 km, the air pressure is half that at the surface; at 11 km it is 225 millibars and at 17 km it is only 90 millibars.

2. Stratosphere. Next to troposphere is the second zone, called **stratosphere**, which is about 30 km in height. This zone is free from clouds and aeroplanes usually fly in its lower zone. The temperature of stratosphere increases up to 90^{0}C and such an increase in temperature is due to ozone formation under the influence of ultraviolet component of sunlight. Such a layer of ozone is called **ozonosphere**. In ozonosphere, the sunlight ionize oxygen to ozone by photochemical dissociation. The ozonosphere completely absorbs solar radiation, ultraviolet radiation from the sun and also a lot of the solar infra-red, thus, becoming warmer than adjacent layers above or below. There is a serious threat to this ozone layer now due to the harmful effects of gaseous pollutants. A big hole (*i.e.,* thinning of the ozone layer) has occurred in it, above the antarctic region. Upper layers of stratosphere form **stratopause.**

3. Mesosphere. Stratosphere is followed by next zone called **mesosphere** which is 40 km in height. In mesosphere, temperature shows again a decrease up to –80°C. Upper layers of this zone form the **mesopause.**

4. Ionosphere. The remaining part of atmosphere above the mesosphere, up to the height of about 300 km above earth's surface, is called **ionosphere**. Ionosphere contains several layers of ionized air. Thus, most of the gaseous components which become ionized under the influence of radiant energy, remain as ions. Ionosphere reflects short radio waves, making telecommunication possible over long distances.

5. Exosphere. The exosphere is the outer fringe of earth's atmosphere and outer space begins after it. The air density is very low in this zone; hydrogen being dominant element of it.

Air

The gaseous mixture of troposphere, is utilized by most organisms in respiration to liberate energy from food during oxidation and is called air. In atmosphere, about 95 per cent of the total air is present up to the height of about 20 km above earth's surface and the remaining 5 percent in the rest, of about 280 km height. In the gaseous mantle, there is found a mixture of gases in different proportions (Table 4.1). Of these various gases nitrogen and oxygen are the major components of air.

Table 4.1. Composition of atmospheric air.

Component	% by volume	% by weight
Nitrogen (N_2)	78.09	75.54
Oxygen (O_2)	20.93	23.14
Argon (A)	0.93	1.27
Carbon dioxide (CO_2)	0.032	0.46
Miscellaneous	0.02	0.02

Miscellaneous component of air includes traces of hydrogen, ozone, radon, helium, neon, krypton, xenon, sulphur dioxide, hydrogen sulphide, ammonia, methane, etc.. Besides these gases, air may carry suspensions of liquids such as water in clouds and solids such as dust from the ground (soil), smoke from fires or salt from ocean spray. Air also contains microorganisms (viruses, bacteria, etc.), pollen grains and fungal spores, all forming biological constituents of the atmosphere.

In general, the composition of these gases in air and in other media on earth, such as water or soil, is in equilibrium with the atmosphere. In special ecological habitats such as the anaerobic regions of deep lakes or sand-mud flats, however, the composition of gases of air is altered dramatically. Slight differences occur in the atmosphere at different latitudes and, at places where gases are entering or leaving the atmosphere, such as volcanoes, fires, smelters, cities, metropolitan areas, and vegetation.

Physiologic-ecologic inter-relationships of gases and animals. The atmospheric composition is not always the most important factor to organism; rather, it is the partial pressure of a gas, especially oxygen and CO_2, which influences the existence of life, as illustrated by altitudinal studies on organisms. The **partial pressure** of a gas is the product of the total barometric pressure times the concentration of gas in dry air.

The solubility characteristics of oxygen and carbon dioxide in water are different, for carbon dioxide is about 200 times more soluble than oxygen. Both temperature and salinity have evident effects on the solubility of these two gases; with either increased temperature or increased salinity, there is a decrease in their solubility in water as has been shown in Table 4.2.

Table 4.2. Coefficients of saturation of CO_2 in water at different temperatures and salinity (ml/litre in equilibrium with 760 mm Hg).

Temperature, °C	Salinity, %		
	0	28.91	36.11
0	1715	1489	1438
12	1118	980	947
24	782	695	677

In the process of photosynthesis, green plants use CO_2 and release O_2 during the day time; oxygen is used in respiration by all the organisms all the time. Photosynthesis, thus, regulates the oxygen and carbon dioxide balance in nature. Nitrogen which is abundantly present in the atmosphere is not used directly by plants or animals except by some bacteria and blue green algae. If gases are to be functionally important to an organism, there must be a mechanism whereby gases can enter and leave the body. Apparently the passage of oxygen across a membrane is accomplished by **diffusion**. The rate of diffusion can be calculated by **Fick's Law**, which is based on concentration coefficient, amount of surface thickness of the membrane, and time, and is expressed as the diffusion coefficient. The diffusion coefficient value is not the same for different animal tissues. For example, some representative values of diffusion coefficient of O_2 are : muscle, 0.000014; connective tissue, 0.000011; water, 0.000034; air, 11.0; chitin, 0.000013. Not only is the diffusion rate different in various tissues, but the behaviour of each gas also differs. Carbon dioxide, for example, diffuses through water and animal tissues 25 times faster than oxygen.

Air as a medium for living organism. Air is not an easy and suitable medium to support life (biota) and actually no organism ever originated in air, though, certain aquatic and terrestrial organisms have become secondarily adapted for aerial existence. Air has so much less **buoyancy** than water that organisms emerging from water or land are immediately subjected to a stronger pull of **gravity** which holds them to the earth. They are also exposed to **evaporation** of water from their bodies, which threatens loss of water from the protoplasm and death by desiccation. Changes in **temperature** are much more drastic in air and there is great danger of chilling or overheating. Further, periods of light exposure are much longer and much more intense in air than in water or upland. The problem of **mineral supply** becomes acute in the air.

Biota of the air. Only a few microorganisms, plants and animals have invaded the air. The most common examples of aerially adapted animals are insects, flying fish (*Exocoetus volitans*), flying frog, flying lizard (*Draco volans*), flying phalangers (*Petaurus*), bats (*Eptesicus*), and birds.

4.2 HYDROSPHERE (WATER)

The oceans, rivers, streams, lakes, ponds, pools, polar ice caps, water vapour, etc., form the **hydrosphere**. About three-fourth of the earth's surface (75 pre cent) is covered with hydrosphere, the main component of which is **water**. Water is one of the most unusual natural compounds found on earth, and it is also one of the most important. The water remains in solid (snow), liquid (water) and gaseous (water vapour) forms. Life on earth began in the seas, and water in some form or the other is absolutely essential for the maintenance of all life. Water is one of the main agent in pedogenesis (soil formation) and is the medium for several different ecosystems. It permeates the atmosphere and the outer layers, of lithosphere and has uneven distribution on earth, so that, some of the great ocean depths are approximately six or seven miles (9750 meters). Further water in its two forms, salt water and freshwater, forms two chief aquatic environments—namely **marine environment** and **fresh water environment** of earth. The oceans holding marine environment are two and one half times more extensive than land and provide over 300 times the living space, since they are habitable throughout their entire depth by certain groups of organisms. Water is obviously heavier than air which imparts a greater buoyancy to the aquatic medium enabling organisms to float at variable levels. The most unique features of water concern its physical properties.

Physical Properties of Water

Henderson (1913, 1924) listed in his stimulating book, *The Fitness of the Environment,* the following characteristics of water which are favourable to biological systems:

1. A tremendous quantity of water exists on earth in three forms: gaseous, liquid, and solid. Not only is 75 per cent of the earth's surface covered by water but the atmosphere also contains an abundance of aqueous vapour, and the polar region is ice-covered. Moreover, at least 60 per cent of active protoplasm is water.

2. Water is an extremely inert substance in relation to most other chemical substances.

3. It has unique thermal properties such as heat capacity, latent heat and higher freezing point.

(a) Heat capacity. Water has high heat capacity and it can withhold large amounts of heat. Because of the high heat capacity of water, oceans and lakes tend to maintain a relatively constant temperature, and therefore, the temperature of the biosphere is relatively stable. This property of water is functionally important to animal life. For example, a 165 lb (about 75 kg) man at rest produces sufficient heat to raise his body temperature more than 32°C, but if the heat capacity of his body were to correspond to that of many other substances, this amount of heat production would raise his temperature 100–150°C.

(b) Latent heat of melting and evaporation. The latent heat of melting is the number of calories required to convert 1 gram of solid at the freezing point into 1 gram of liquid at the same temperature. This value is about 90 for water, and means that the amount of heat necessary to melt ice is the same as that required to raise the temperature of the resulting ice water to 80°C. The latent heat of evaporation is defined as the number of calories required to change 1 gram of liquid into vapour. For water this value is 536; therefore, as much heat is required to boil away 1 gram of water as to raise the temperature of 536 grams through 1°C. These properties of water are important not only because they moderate the temperature of the biosphere, but also because they play a basic role in the evaporation of water and its precipitation as rain and as dew in the **hydrological (water) cycle**.

(c) Thermal conductivity. Although water is a poor thermal conductor compared to metals, among the common liquids it is excellent : for example, the conductivity value for silver is 1.10, for water, 0.0125, for alcohol, 0.00048, and for benzene, 0.00033.

(d) Expansion before freezing. The relationship between temperature and density, or mass per unit volume, is very unusual. When water is cooled from room temperature, it contracts, becoming denser until it reaches a maximum density at 3.94°C. If cooled further, it begins to expand again. At the freezing point (0°C) it expands markedly, unlike almost all other substances. Thus, ice will always float on the top of a lake or stream, and it is very unusual for an aquatic ecosystem ever to freeze solid, unless it is very small.

4. No other compound compares to water as a **solvent**. So many different substances can be dissolved in it that it is known as the **universal solvent**. More things, in fact, can be dissolved in water than in any other liquid. This is especially true for inorganic chemicals which split, or dissociate to form electrically charged entities termed **ions**. Ionization influences most electrical phenomena and many chemical phenomena of solutions. It is probable that all natural elements are soluble in water, at least in trace amounts, and that they are all found in natural water at some place or other on the earth's surface. In addition, many organic chemicals are water-soluble. Thus, water is the main medium by which chemical constituents are transported from one part of an ecosystem to the other. It is the only medium by which these constituents can pass from the abiotic portion of the ecosystem into the living (biotic) portion. Even in the driest of terrestrial environments, nutrient materials pass into the roots of plants in aqueous solution; when air is breathed by animals, oxygen is dissolved in water at the surface of the lung before it can cross the mucous membrane and be absorbed by the blood.

5. Water has the greatest **surface tension** of all common liquids, except mercury. The role of surface tension is most obvious in the way it allows certain things, such as pollen, dust, and water striders, to remain at the surface of a water body even though they are denser than the water. More important, however, the high surface tension of water allows soils to contain a significant amount of water through capillary attraction and to make it available to terrestrial plants.

Visualizing these unique physical properties of water, **Henderson** (1913) concluded that no other known substance could be substituted for water as a basic abiotic environmental factor. Now, ecologists have known certain other important physical properties of water which affect life in someway. Some of the additional important physico-chemical properties of water are the following:

(i) Viscosity. Water is a fairly viscous liquid. Animals that live and move in water need to be much more streamlined than those that move through air, because the resistance to motion in a viscous medium is high. But at the same time, the viscosity of water allows organisms to swim using relatively simple movements. Further, high viscosity of water protects the aquatic animals and plants from the mechanical disturbances.

(ii) Buoyancy. Water is buoyant medium. Organisms can exist in it without specialized supportive structures such as those that are needed by organisms that inhabit terrestrial environments.

(iii) Transparency. Water is a transparent medium. Its transparency enables the penetration of light to the depths where it is ultimately absorbed. Different wavelengths are absorbed at different depths. The long heat waves are stopped near the surface. Shorter waves with more energy penetrate successively farther. The ultraviolet rays penetrate beyond 100 meters. The zone up to which light rays penetrate is called as **photic zone** and below this zone there is complete darkness and organisms that require light cannot live. The transparency of water is greatly affected by the presence of suspended particles, phytoplankton, etc., which absorb light and so penetration of light in turbid water is less.

(iv) Pressure. Organisms living at sea level experience a pressure of about 15 psi, which is defined as 1 atm (760 mm of Hg). Pressure increases with increased depth of water at the rate of one atmosphere (1 atm) for every 10 meters of descent. Organisms inhabiting the floor of deep-sea areas at depth of 10,500 meters are exposed to hydrostatic pressure of about one ton per square centimeter. Pressure influences solubility, ionic dissociation, and surface tension, and water is slightly compressible with increased pressure.

(v) Salinity. Salinity has been defined as "the total amount of solid material in grams contained in one kilogram of the water, when all the carbonate has been converted into oxide, bromine and iodine replaced by chlorine and all organic matter completely oxidized." All types of natural waters contain various amounts of different salts (ions) such as Na, K, Mg, Cl, SO_4, PO_4, CO_3, HCO_3, NO_3, etc., and all these salts are responsible for the **saltiness**, **salinity** or **salt content** of water (Table 4.3). The salinity of marine water is rather constant being about 3.5%. The salinity of fresh water varies greatly. Some salt lakes may have a salinity of 25% to 30% which greatly restricts life in them.

Table 4.3. Comparison of some of the principal ions found in different kinds of water.

Water	Na	K	Ca	Mg	Cl	SO_4	CO_3	Total per litre
1. Soft water	0.016	–	0.01	0.0005	0.019	0.007	0.012	0.065
2. Hard water	0.021	0.016	0.62	0.014	0.041	0.025	0.119	0.301
3. Sea	10.56	0.30	0.40	1.27	18.98	2.65	0.71	34.85
4. Great salt lake	65.54	3.76	0.065	4.47	110.08	13.04	–	197.51

Salinity of water acts as an important limiting factor for the distribution of a number of species of plants and animals. Certain animals such as, spider crab, *Maia*, etc., can tolerate only narrow fluctuations in salinity of water and are known as **stenohaline** animals. While some animals such as *Mytilus, Aplysia*, etc., can withstand wider ranges of salinity and are called

euryhaline animals. However, there are certain animals, such as, *Anguilla, Salmon,* etc., which are both stenohaline and euryhaline.

Chemical Properties of Water

Water consists primarily of a single compound, H_2O. It is a universal solvent and most chemical compound ionize are readily in water and provide many radicals and considerable versatility in the rearrangement of chemical substances. It has following chemical properties :

(a) Solubility of gases in water. Most gases dissolve readily in water, most notably those that are essential for life. The concentration of any gas in water generally varies between zero and a theoretical maximum or **saturation** (Table 4.4). The latter is the amount of gas that can be dissolved in water when the atmosphere and the water are in equilibrium with one another. Except for water falls and very turbulent streams, the water in natural ecosystem is seldom in equilibrium with the atmosphere. A gas may show a **deficit**, if it is being utilized in the ecosystem faster than it is going into solution across the air water interface, or it may be **supersaturated**, if it is being produced in the ecosystem faster than it is being released from solution across the interface. The concentration of important gases may vary widely in any ecosystem, both horizontally and vertically.

Table 4.4. Comparison between equilibrium concentrations of important gases in the atmosphere and in water.

Gas	Atmospheric concentration	Saturation in water
Oxygen	210 cc./1. (21%)	7 cc./1. (32.9%)
Nitrogen	780 cc./1. (78%)	14 cc./1. (65.7%)
Carbon dioxide	0.3 cc./1. (0.03%)	0.3 cc./1. (1.4%)

The saturation level of any gas in water depends on several variables, most notably temperature, salinity (see Table 4.2), the concentration of the gas in the atmosphere, and its relative solubility in water. The greater the concentration of a gas in the atmosphere, the greater its concentration in the water will tend to be, depending on its relative solubility in water.

1. Oxygen. One of the most critical factor in an aquatic environment is the amount of oxygen in the water because most living organisms (excepting anaerobic forms) require this gas for respiration. In contrast to atmosphere, the oxygen becomes limiting factor for aquatic animals as the saturation concentration of oxygen in water is governed by temperature and salinity. As is evident in Table 4.5, the lower the temperature, the greater the oxygen retaining capacity of water, whether it is freshwater or sea water.

Table 4.5. Comparison of the saturation concentration of oxygen in freshwater and salt water environment with varying temperature.

Water	Temperature, °C	Saturation point in millilitre per litre
Freshwater	0	10.27
Saltwater	0	8.08
Freshwater	30	5.57
Saltwater	30	4.52

2. Nitrogen. Nitrogen is significantly less soluble in water than oxygen. But because it constitutes 78 per cent of the atmosphere, it still accounts for about 65 per cent of the dissolved gases at equilibrium. It is fairly inert chemically and does not react with water, although some bacteria, fungi, blue-green algae, and so on, can use it to satisfy their nitrogen requirements

(Nitrogen fixation); and other bacteria can produce it through reduction of nitrate under conditions of very low oxygen concentration.

3. Carbon dioxide. The decomposition of organic matter and the respiratory activity of aquatic plants and animals produce carbon dioxide. This gas is one of the essential raw materials necessary for photosynthetic activity by green plants. Carbon dioxide combines chemically with water to produce carbonic acid (H_2CO_3), which influences the hydrogen ion concentration (pH) of water. Carbonic acid dissociates to produce hydrogen (H^+) and bicarbonate (HCO_3^-) ions. The bicarbonate radical may undergo further dissociation forming more hydrogen (H^+) and carbonate ($CO_3^=$), as represented in the following reactions:

$$\underset{\text{atmospheric}}{CO_2 + H_2O} \underset{\text{air-water interface}}{\rightleftarrows} \underset{\text{dissolved}}{CO_2 + H_2O} \rightleftarrows \underset{\text{carbonic acid}}{H_2CO_3} \rightleftarrows \underset{\text{bicarbonate}}{H^+ + HCO_3^-} \rightleftarrows \underset{\text{carbonate}}{2H^+ + CO_3^=}$$

The amount of free or uncombined carbon dioxide in water is of ecological importance : it governs the precipitation of calcium in the form of calcium carbonate ($CaCO_3$). Calcium precipitates when temperature and salinity are high and the amount of uncombined carbon dioxide is low. This means more carbonate ($CO_3^=$) is present to combine with the calcium cation (Ca^{++}). These conditions exist in shallow tropical waters, where evaporation is high. This raises the salinity and photosynthetic activity of plants and reduces the quantity of free carbon dioxide in water. The precipitation of calcium carbonate in tropical areas as the Bahamas explains the preponderance of thick calcareous shells of shallow water tropical molluscs, plankton and algae. In deep oceanic water, temperature is low and there are no photosynthetic plants, consequently, the carbon dioxide content of the water is high. Deep water fauna (molluscs, crustaceans) possess very fragile skeleton because the precipitation of calcium carbonate is minimum.

4. Hydrogen sulphide. The deeper layers of many bodies of water, including ponds, lakes, and some estuaries, may contain significant amounts of the toxic gas, hydrogen sulphide, which is released by decaying organic matter. If concentration of the gas builds up, all life but anaerobic bacteria is excluded from the area (*e.g.,* deeper strata of Black sea).

Effect of Factor of Aquatic Environment on Aquatic Organisms

The aquatic environment is subject to water movements ranging from small vertical circulations to strong currents. The streams have a unidirectional movement and in seas the movement is reversible. Many aquatic animals have accordingly taken to sedentary or sessile lives depending on water movements. Radial symmetry is a characteristic of such animals. Transformation from a sessile to a locomotive existence favours a bilateral symmetry. The currents of water often abrade (= rub off) the inhabiting flora and fauna and varied modifications are encountered to withstand this abrasive action. Thick scales, strong shells and many attachment devices such as the holdfasts and suckers all are the results of this environmental stress. The ability to breathe air dissolved in water, at times even resorting to anaerobic existence, the modification of various senses to respond to stimuli characteristic of aquatic environments, the phenomenon of osmoregulation, and above all the phenomenon of external fertilization are other remarkable physiological adaptations to live in an exclusively aquatic medium.

Water and Ecological Adaptations

Water makes up a large proportion of the bodies of plants and animals, whether they live on land or in water. Active cytoplasm holds about 70–90 per cent of water. It has several important physiological properties. There exists a strong relationship between the water status of soil, plant and atmosphere. The rooting zone of the soil (zone of soil in which the water absorbing organs, roots, root hairs are present), the plant body and the lower layer of atmosphere behave as a continuum, called **SPAC** or **soil plant atmosphere continuum** in relation to water transfer (**Phillip**, 1966). Solar radiation is the primary energy source for the water transport process in the SPAC. On the other hand, animals obtain water (i) by drinking (ii) by absorbing it through their skin from contact with some damp ground, (iii) directly from their food or (iv) from water produced by metabolism. The method of obtaining water and the relation to the supply of liquid water as well as resistance to the drying effects of the surrounding atmosphere are important in determining the distribution of animals.

The scarcity or abundance of water brings about adaptations in living organisms. Plants which grow in areas where water is available in plenty, are classified as **mesophytes** and terrestrial animals living under such conditions are called **mesocoles**. Plants growing in water are called **hydrophytes**, while animals that live in the aquatic environment are called **aquatic animals** or **hydrocoles**. Some plants can grow in ecosystems where water is scarce and where the day temperature is very high. These plants are called **xerophytes** and the animals living in such xeric conditions are called **desert animals** or **xerocoles**. Xerophytes living in physiologically dry soils, *i.e.,* saline soils with high concentrations of salts such as NaCl, Mg Cl_2 and Mg SO_4, are called **halophytes**. Based on their specific habitat, halophytes can be further classified into **lithophilous**, **psammophilous**, **pelophilous** and **helophilous** plants growing on rock and stones, sand, mud and swamp, respectively. Helophilous helophytes include mangroves of sea shores of Bombay such as *Rhizophora mucronata* and *Sonneratia*.

1. Hydrophytes and hydrocoles and their adaptations. Hydrophytes include : (a) **free-floating hydrophytes** (*e.g., Wolffia, Lemna, Spirodella, Azolla, Eichhornia crassipes, Salvinia* and *Pistia*), (b) **rooted hydrophytes with floating leaves** (*e.g., Trapa, Nelumbo, Nymphaea, Marsilea,* etc.), (c) **Submerged floating hydrophytes** (*e.g., Ceratophyllm, Utricularia, Najas,* etc.), (d) **rooted submerged hydrophytes** (*e.g., Hydrilla, Chara, Vallisneria,* etc.), (e) **rooted emergent hydrophytes** or amphibious plants (*e.g., Sagittaria, Ranunculus*, etc.). **Tenagophytes** are amphibious plants—they grow in water bodies as well as in water logged soil.

The hydrophytes grow in hydric conditions and show the following general adaptive features: They possess poor mechanical, absorbing, conductive and protective tissues. They also contain an extensive development of air spaces (aerenchyma) in the tissues. Roots are either absent (*e.g., Wolffia*) or poorly developed (*e.g., Hydrilla*). Roots may not have root hairs, root cap (instead of root cap *Eichhornia* has root pockets) and vascular tissue. Roots of hydrophytes are generally fibrous and adventitious, when present. The stem of hydrophytes is weak, slender and spongy. In some it is like a horizontal rhizome covered with mucilage, while it may be hard, as in *Nelumbo*. The aerial leaves may be broad but the submerged leaves are thin, long or ribbon-shaped. Stomata are completely absent in submerged leaves (*e.g., Anacharis*), but in floating forms, stomata are confined only to the upper surfaces of leaves as in *Nymphaea.*

Aquatic animals or hydrocoles in general exhibit an elongated stream-lined body having a compressed head, body and tail. Hydrocoles include fishes, sea turtles, mammals such as whales, and many others. There are also amphibious forms such as frog, toad, crocodile, etc., and many birds which visit water bodies either for reproduction or for collection of food.

2. Xerophytes and xerocoles and their adaptations. Xerophytes grow in conditions of water scarcity, high temperature, strong winds, high transpiration rate and evaporation higher than precipitation. The soil is very dry and porous. The essential adaptations of xerophytes involve increased water absorption by roots, storing of water and retardation of transpiration. Thus, in search of water, roots of xerophytic trees may go very deep in the soil and have extensive root hairs to absorb it. The roots of plants such as *Calotropis procera, Ficus,* and *Acacia nilotica* may go as deep as 10 to 16 metres and may reach the water table. As a consequence, these plants survive in deserts or arid conditions even if their rate of transpiration is higher. The storage of water is facilitated either by modifications of leaves, as in *Mesembryanthemum* and in the malacophyllous xerophytes (in which leaves contain turgescent parenchymatous cells) as *Aloe, Begonia, Bryophyllum, Agave, Yucca,* etc., or modification of stems, as in cacti such as *Opuntia* (phylloclade) and *Euphorbia.* In some xerophytes the water is stored in their roots as in *Asparagus* and *Ceiba parvifolia.* All these xerophytes are called **succulents** because they possess thick, fleshy, water storage organs such as stems, leaves and roots. Non-succulent xerophytes such as *Calotropis*, *Prosopis, Acacia, Zizyphus, Casuarina, Nerium, Saccharum* and *Pinus* possess other sort of xerophytic adaptations, *viz.,* extensive root system, high osmotic pressure and other modifications in the leaves. Reduced transpiration is achieved by decreasing the leaf surface, as in *Casurina, Acacia* and *Asparagus* or by modifying the leaves into spines and barbed bristles, as in cacti, or by having thick, leathery, thick cuticle or wax-coat bearing leaves with well-developed hypoderma and sunken stomata to reduce transpiration, as in *Calotropis* and *Nerium.* Halophytes (*e.g.,* mangroves) resemble xerophytes and have high osmotic pressure; succulent organs; thin, evergreen, small leathery leaves with water storing tissues and thick cuticles and special air-breathing roots called **pneumatophores.**

Different animals have evolved the following adaptive features to live in arid environment :

1. Nocturnal life style. Most desert animals are nocturnal and seek shade or burrow deep in the soil in the day time to avoid excessive heat and dryness. Some xerocole rodents passively lose heat through conduction by pressing their bodies against the burrow walls.

2. Deceptive colouration. Desert animals are usually grey, brown or red matching with the colour of the sand or rock.

3. Suspended animation. Certain animals, usually with simpler organization, such as rotifers, nematodes, tardigrades, desert snails, etc., retain their vitality in long dry environment. Other forms (frogs, toads, etc.) **aestivate** during droughts and are active during moist season of the year.

4. Fast movement. Desert animals move much faster than other land animals, since they have to travel long distances in search of food and water.

5. Migration. Many birds and mammals of arid zones migrate when water becomes scarce or as a result of drought or for other reasons, the food supply is less.

6. Heat loss by radiation. Animals such as jack-rabbits (*Lepus*) and fox (*Vulpes velox*) have large ears that reduce the need of water evaporation to regulate the body temperature. Their ears function as efficient radiators to the cooler desert sky, which on clear days may have a radiation temperature 25°C below than that of the animal body. By seeking shade and sitting in depressions, *Lepus* could radiate 5 kcal/day through its two large ears (400 cm^2).

7. Impervious skin. The drier habitats (deserts, etc., are invaded by only those animals which contain a thick impermeable body covering. Such integuments occur in many insects, birds and mammals. Some mammals such as human beings, apes and horses lose much water (and salt) through sweat glands in heat regulation. Most rodents and some ruminants such as antelope nearly or completely lack sweat glands. Moist skinned forms (most amphibians and earthworms), certain mites and soft-bodied insects are restricted to swamps, stream margins, moist soils and other similarly damp places.

8. Upturned nostrils. Desert animals have nostrils directed upwards; this may provide a protection from clogging by wind blown sand.

9. Water from food and from metabolism. Most herbivores and carnivores (frog, toad, lizards, snakes) live on the moisture obtained with food. Many insects utilize the high water contents of plants to meet their water requirements. In fact, most animals make use of water released during metabolism when fats and carbohydrates are broken.

Kangaroo rat, which seals its burrow by day to keep its chamber moist, can live throughout year without drinking water. It obtains its water from its own metabolic processes and from hygroscopic water in its food.

10. Internal lungs or tracheal system. Mode of respiration has some correlation with water. Crustaceans, with their gills covered by a water-retaining carapace, carry with them a liquid environment for their gills. The scaly body covering of a fish may be practically impermeable to water and exchange of gases may be limited to gills and gut. Internal lungs, whether in pulmonate snails, land isopods, spiders or higher vertebrates (amphibians, reptiles, birds and mammals) together with the internal tracheal system of insects are water saving. Much water is lost in breathing even in animals having internal lungs.

11. Dry excretion. A further water-saving device is the excretion of concentrated, relatively dry nitrogenous and faecal waste materials (*e.g.,* cow, donkey). Water-saving insects, reptiles and birds excrete nitrogenous waste as solid uric acid.

Camel and African antelope, *Oryx* provide good examples of xerocoles or drought-resistant animals. Camel can go without food and water as long as 10 days at a stretch. For water conservation, camel has following adaptations: (1) When camel drinks water, it can take in up to 50 litres in one gulp and this water is evenly distributed all over its body in tissues and not in specific pockets or organs (*e.g.,* in stomach) as the common misconception holds. (2) Camel excretes highly concentrated urine. Its dung also contains very little water, compared to the dung of the donkey or the cow (both of which are xerocoles). (3) It perspires very little and its breathing rhythm is very slow. (4) Camel can withstand dehydration up to 25 per cent of body weight and it loses water from body tissues rather than from blood. (5) Body temperature of camel is labile, dropping to 33.8^{0}C overnight and raising to 40.6^{0}C by day, at which point it begins to sweat. Due to such thermoregulation, the amount of water loss by perspiration and in other ways is greatly reduced. (6) Camel accumulates its fat in the hump rather than all over body. This speeds heat flow away from the body and its thick coat prevents the flow of heat inward towards the body.

Snow as Habitat

Snow which is the solid state of liquid water, is an integral factor in the physical environment. It is important particularly in the higher latitudes and on the higher parts of mountains where snow cover may last throughout the year. Its significance is increased by the large amount of land surface in northern Eurasia and North America. Snow covers much of the available food for many animals and is a factor causing migration, both of birds and mammals, towards areas that are somewhat free from its direct influence. Certain mammals (rodents, etc.) tunnel through the snow near the underlying mosses and get access of food and often protection from winter cold. In the Tundra, the non-hibernating lemmings (*Dicrostonyx*) breed during winter beneath the snow. Snow prevents the locomotion of mammals and of running birds, and animals with long feet only are well adapted for moving on snow cover.

4.3 LITHOSPHERE (SOIL) (EDAPHIC FACTORS)

The earth is a cooled, spherical, solid planet of solar system, which rotates on its axis and revolves around the sun at a certain constant distance. The solid component of earth is called **lithosphere**. The lithosphere is multilayered including following main layers— crust, mantle, and **outer** and **inner core**. The core is the central fluid or vapourised sphere having diameter of about 2500 kms from the centre and is possibly composed of Nickel-Iron. The **mantle** extends about 2900 kms above the core. This is in molten state. The outer most solid zone of the earth is called **crust** which is about 8 to 40 kms above the mantle. The crust is very complex and its surface is covered with **soil** supporting rich and varied biological communities, for living organisms find in the soil an environment providing food, shelter, and concealment from predators.

Soil

The word soil is derived from a Latin word *'solum'* meaning earthy material in which plants grow. The science which deals with the study of soil is called **Soil Science, Pedology** (*pedos = earth*) or **Edaphology** (*edaphos = soil*). The study of soil is significant for us due to various reasons. Soil is a natural habitat for microorganisms, plants and animals. Its knowledge is helpful in practices of agriculture, horticulture, forestry, etc., such as cultivation, irrigation, artificial drainage, and use of fertilizers. Pedology is important also for geology, petrology, mineralogy, palaeobotany and palaeozoology.

The soil can be defined as the weathered surface of the earth's crust which is mixed with organic material and in which microorganisms live and plants grow. Soil consists of the inorganic materials (the mineral matter) derived from parent rocks; the organic materials (the humus) derived from dead organisms; the air and water occupying the pores between the soil particles which are loosely-packed (the soil water and the soil atmosphere); small organisms (the biological systems) such as bacteria, fungi, algae, protozoans, rotifers, nematodes, oligochaetes, molluscs and arthropods, and higher plants which live in it. Soil is the ultimate source of all food production since plants form the base of all ecological pyramids and plants grow in soil and derive nutrition from them. It provides mechanical anchorage to plants, besides serving as a reservoir of food materials and water. It is the site where nutrient elements are brought into biological circulation by mineral weathering. The soil harbours the bacteria which incorporate atmospheric nitrogen into the soil. Plant's roots occupy a considerable portion of the soil; tie the vegetation to the soil; and pump water and its dissolved minerals to other parts of the plant for photosynthesis and other biochemical processes.

Further, a **mature soil** is that state of soil that has assumed the profile features (*i.e.,* succession of natural layers) characteristic of predominant soils on the smooth uplands within the general climatic and botanic regions in which it is found.

Soil formation or Pedogenesis

Soil is a stratified mixture of inorganic and organic materials, both of which are decomposition products. The mineral constituents of soil are derived from some parent material, the **soil forming rocks** by fragmentation or weathering, while, organic components of soil are formed either by decomposition (or transformation) of dead remains of plants or animals or through metabolic activities of living organisms present in the soil. Before discussing the soil-forming processes, let us describe in brief the nature of soil forming rocks as follows:

Soil forming rocks. Basically, there are following three kinds of soil forming rocks:

1. Igneous rocks which are formed due to cooling of molten megma or lava, *e.g.,* granite, diorite and basalt.

2. Sedimentary rocks which are formed by deposition of weathered minerals which are derived from igneous rocks, *e.g.,* shales, sandstone and limestone.

3. Metamorphic rocks which are formed by change of pre-existing rocks (*e.g.,* igneous or sedimentary rocks) through heat and pressure, *e.g.,* gneiss, schist, slate, quartzite and marble.

Chemistry of minerals of soil-forming rocks. Rocks are the chemical mixture of numerous kinds of minerals. The chemical nature of certain most common and abundant minerals of soil-forming rocks has been listed in Table 4.6.

Table 4.6. Chemical composition of some common soil minerals.

Minerals	Chemical constituents
A . Sand and silt minerals	
1. Quarts or silica	SiO_2
2. Feldspars	
a. Orthoclase	$K_2Al_2Si_6O_{16}$
b. Plagioclase	$NaAlSi_3O_8$
c. Calcium feldspar	$CaAl_2Si_2O_8$
3. Micas	
a. Muscovite	$K(OH)_2Al_2(AlSi_3)O_{10}$
b. Biotite	K, Mg, Fe, Al silicate
4. Pyroxene	(Mg, Fe) SiO_3
5. Amphibole	$(Mg, Fe)_7 (Si_4O_{11})_2 (OH)_2$
6. Olivine and serpentine	$(Mg, Fe)_2 SiO_4$
7. Calcite; magnesite; and dolomite	$CaCO_3$, $MgCO_3$; and ($CaCO_3$, $MgCO_3$)
8. Iron oxides	
a. Haematite	Fe_2O_3
b. Magnetite	Fe_3O_4
c. Limonite	FeO (OH), xH_2O
B. Clay minerals	
1. Kaolin	Al_2O_3, $2SiO_2$, $2H_2O$
2. Montmorillonite	$(Ca,MgO)Al_2O_3,5SiO_3,5H_2O$

Process of Soil Formation

The processes which are involved in the formation of mature soil can be studied under the following heads:

A. Weathering of Soil Forming Rocks

Soil formation is started by **disintegration** or **weathering** of parent rocks by some physical, chemical and/or biological agents, because of which the soil-forming rocks are broken down in small particles called **regoliths**. Regoliths are the basic materials which under the influence of various other pedogenic processes finally develop into mature soil.

(a) Physical weathering. The physical weathering agents are primarily climatic in character, exerting a mechanical effect on the substratum with the result that fragments are pulverised into progressively decreasing particle sizes (*i.e.,* regoliths). Such climatic weathering of rocks does not cause any chemical transformation of rock-minerals and commonly occurs in deserts, in high altitudes, in high latitudes, and in localities with marked topographic relief and sparse vegetation cover. The agents which are involved in climatic weathering of rocks, are **temperature, water, ice, gravity** and **wind**.

The **temperature** causes break down of those rocks which have heterogeneous structure, due to the fact of differential expansion and contraction coefficient of materials composing the rocks. Minerals composing the rocks have got different degrees of expansion. These minerals expand in the high temperature of day and contract when the temperature falls. The differential expansion and contraction of different minerals of rocks set up internal tensions and produce cracks in the rocks and consequently, the rocks weather into finer particles.

In its liquid state **water** causes mechanical weathering of rocks by following methods: (i) **Rain water**. Natural water falling either in the form of rain drops or hail storm on the surface of rock with beating effect bring about abrasion of massive rocks into smaller particles. (ii) **Torrent water**. Rapidly flowing water rolls the heavy rock masses such as rock boulders along the bottom of stream and grinds them into finer particles. (iii) **Wave action**. The wave actions are most effective in sea shores. The rapidly striking water waves dislocate solid particles of varying diameters from sea shore rocks and the debris is then settled at the sea bottom to form **marine soil**. Water also acts as a mechanical carrier. Nearly 5 billion tons of mineral matter is annually carried away in solution from land to sea.

In its freezing and ice-melting states water causes rock-weathering by **frost action** and **glacier formation**. Water in the form of frost or ice, is an extremely effective physical weathering agent of rocks. It seeps into rock crevices, freezes due to sudden fall of temperature of rocks, expands about nine per cent of its original volume, exerts a pressure (expansion force) of approximately 150 ton/ft^2 and eventually cracks the rock into smaller pieces. Likewise, in summer when ice at mountain tops starts melting and glaciers (huge sliding masses of ice) move downwardly on the slope, then during the glacier movement, the rock over which they move is gradually worn down to produce fine particles which are deposited as **drift** or **till**, when glacier finally retreats.

Gravitation weathering action is most effectively demonstrated by land slides and rock slippages caused by earth quakes and faulting during which the rock is fragmented by abrasion and the forces of impact. Lastly, the rapid stormy **wind** carrying suspended sand particles causes the abrasion of exposed rock. It acts like a mechanical carrier in moving the particles over the surface of earth as **dunes** or **drifts** and in transporting large quantities of fine suspended particles long distances.

(b) Chemical weathering. The physical weathering produces a greater surface area of rock exposed to the **chemical weathering**, which occurs simultaneously with physical weathering and continues much beyond that. During chemical weathering, a chemical transformation or decomposition of parental mineral materials into new mineral complexes or secondary products occurs. For example, primary minerals that contain aluminium and silicon, such as feldspar, are converted to secondary minerals such as clay. Because chemical weathering requires the presence of moisture and air as essential factors, therefore, chemical weathering is not effective in deserts. It includes reactions such as solution, hydrolysis, oxidation, reduction, carbonation and hydration.

(c) Biological weathering. Though the surface of bare rock is unsuitable for many forms of life, even then a number of microorganisms (bacteria, protozoans, fungi, nematodes, etc.), lichens and mosses can gain a foothold. These early colonizers transform the rock into a dynamic system, storing energy and synthesizing organic material. Their activities alter the mineral composition as well as the physical structure of the rock. For example, lichens are present in the initial stages of biological succession and their growth may cause cracking or flaking, exposing greater area of rock to further weathering. The lichens and mosses extract mineral nutrients such as P, S, Ca, Mg, K, Na, Fe, Si, and Al, from the rock. These elements are combined with organic complexes, and eventually return to the developing soil when the vegetation decomposes.

Products of Weathering and Soil types

The soil which is formed by weathering of soil-forming rocks is called **embryonic** or **primary soil**. The **secondary** or **transported** soils are those which are carried to other places by carriers such as gravity, water, glacier and wind. Soil material transported from one area to another is known as **loess** or **eolian**; that transported by water as **alluvial, lacustrine** (lake soil) and **marine deposits**; that transported by gravity as **colluvial soil** and that transported by sand storms as **sand dunes**. In a few places soil material comes from accumulated organic matter as **peat**. The soil which develops *in situ* above parent bedrock is called **residual** or **sedentary soil**.

B. Mineralization and Humification

The weathering of inorganic soil matter is followed by decomposition of organic matter, which starts at an early stage of pedogenesis, and continues up to much later stages. During early stages of pedogenesis, the organic contents of the embryonic soil are not very high, as the vegetation and its associated fauna are not richly developed. Also, the products of organic composition frequently occur in combination with those of inorganic decomposition as complexes, not easily separable into component parts in these initial stages. However, as both the size of soil particles and the spaces between them become smaller with the result that the water holding capacity increases. This together with the increasing amount of plant nutrients and organic material, allows the soil to support higher plant life, such as grasses; the protective covering effect and binding action of root systems so provided, promotes a greater stability of soil structure. Thus, soil development begins under some influence of plants. They root, draw nutrients from mineral matter, reproduce and die. Their roots penetrate and further breakdown the regolith. The plants pump up nutrients from its depth and add them to the surface, and in doing so recapture minerals carried deep into the material by weathering processes. By photosynthesis plants capture the sun's energy and add a portion of it as organic carbon to the soil each year. This energy source, the plant debris, enables bacteria, fungi (such as *Rhizoctonia solani, Armiliaria mellea,* etc.), earthworms and other soil organisms to colonize the area.

The breakdown of organic debris into humus is accomplished by decomposition and finally mineralization. Higher organisms in the soil—millipedes, centipedes, earthworms, mites, springtails and others— consume fresh material and leave partially decomposed products in their excreta. This is further decomposed by microorganisms, the bacteria and fungi, into various compounds of carbohydrates, proteins, lignins, fats, waxes, resins and ash. These compounds are broken down into simpler products such as carbon dioxide, water, minerals and salts. This latter process is called **mineralization**. The residual amorphous, incompletely decomposed black coloured organic matter which undergoes mineralization is called **humus**. The process of humus formation is called **humification**. **Muller** (1879, 1884) has recognized two kinds of humus–mor and mull. **Mor humus** is acidic and supports an abundant fungal growth and low number of soil bacteria. Fungal mycelia may help to bind together particles of humus and decomposition litter into matted layers and in a well developed mor, three such layers can be distinguished, namely a surface **litter** or **L layer**, comprising undecomposed leaves and twigs, below which is the **fermentation** or **F layer**, in which decomposition has proceeded toward the development of humus, and beneath this, the **humus** or **H layer**, in which degraded humus fractions accumulate. Mor has low calcium content and developed on sandy soils under conifers. **Mull humus** is neutral or slightly alkaline and contains rich microflora of bacteria. It lacks distinct layering of the mor, largely because the presence of calcium compound favours the development of a rich earthworm fauna which promotes a greater mixing of organic and mineral materials. The mull humus develops in brown forest soils under tree species having a relatively high calcium content, such as alder, elm, bass wood, aspen and few conifers (red cedar). Intermediate between the two types of humus is the '**moder**' which has richer and varied fauna.

In order to emphasize the part played by living plants and animals in soil formation, **Taylor** (1930) proposed the following formula:

$$S = M\,(C+V+VA+A)\;t+D$$

where, A signifies animals ; C = climate; D = erosion or deposition ; M = parent material, S = soil; t = time and V = vegetation.

Role of earthworms in soil formation and soil fertility. Earthworms play a significant role in soil formation. They may burrow two metres into the soil, make numerous transverse furrows and produce **worm casts** by eating the soil and soil organic matter (litter). These casts are very rich in nitrogen, water soluble aggregates and mineral substances and increase soil fertility. The burrows facilitate aeration and increase water holding capacity of the soil. **Dash** and **Patra** (1979) have calculated 77 tonnes of dry weight of earthworm cast production per hectare per year in an Indian grassland site. It is estimated that soil turn over due to earthworm action provides a stone-free layer about 15 cm deep on the surface.

C. Formation of Organo-Mineral Complexes

During final stage of pedogenesis, colloidal particles which are formed due to weathering, humification and mineralization accumulate and may aggregate into **crumbs** or **concretions**. Some colloidal humus particles may become associated with mineral particles to form organo-mineral complexes. According to **Wallwork** (1970) the crumbs or organo-mineral complexes are formed by following two mechanisms— **electro-chemical bonding** and **cementing**. In electrochemical bonding method of crumb formation, aggregation of negatively charged colloidal clay and/or humus particles of water molecules and metallic ions, particularly calcium, takes place. The cementing mechanism of crumb formation involves the action of substances absorbed on the surface of soil particles which effectively glues them together.

The crumbs increase the total pore space in the soil, allowing good aeration and drainage. Eventually, a characteristic profile develops under the influence of climate, vegetation, parent material and the activities of the soil communities. The mature soil, thus, becomes a complex system of living and non-living materials not inert but active.

Soil Profile

Soil profile is the term used for the vertical section of earth crust generally up to the depth of 1.83 meter or up to the parent material to show different layers or horizons of soil for the study of soil in its undisturbed state. It is made up of a succession of horizontal layers or **horizons**, each of which varies in thickness, colour, texture, structure, consistency, porosity, acidity and composition. A **pedon** is the smallest three-dimensional volume of soil needed to give full representation of horizontal variability of soil.

In general, soils have following four horizons— an **organic** or **O-horizon** and three **mineral (A,B,C) horizons**. Some workers recognized a **D-horizon**, in which rocks are in active weathering state, in between C- and R-horizons. **R-horizon** is the consolidated bed rock on which a soil profile rests. A- and B-horizons form the **true soil** or **solum**. Each horizon of soil profile is further subdivided. Horizon subdivisions are indicated by arabic numbers, *e.g.*, O_1, O_2, A_1, A_2, etc. Different horizons of soil-profile have following characteristics (Fig. 4.2):

1. O-horizon. The O-horizon, once designated as **L**, **F**, **H**, or Ao and Aoo, is the surface layer forming above the mineral layers and composed of fresh or partially decomposed organic material, as found in temperate forest soils. It is usually absent in cultivated soils and grasslands. O-horizon contains both kinds of humus (*e.g.*, mull and mor) and is subjected to the greatest changes in soil temperatures and moisture conditions and contains most organic carbon. O-horizon and upper part of A horizon is the region where life is most abundant. The O horizon is divided into following two sub-layers:

(i) O_1 (Aoo) region. It is the uppermost layer which consists of freshly-fallen dead leaves, branches, flowers and fruits, dead remains of animals, etc. All these do not show evident breakdown.

(ii) O_2 (Ao) region. Below the O_1 region is the O_2 layer of partly decomposed organic matter. The process of decomposition of the litter is started in O_2 region. Thus, organic matter is found under different stages of decomposition and microorganisms such as bacteria, fungi, actinomycetes are frequently found in it. Upper layers contain detritus in initial stage of decomposition, in which material can be faintly recognized, whereas the lower layers contain fairly decomposed matter, the **duff**.

2. A-horizon. It is characterized by major organic matter accumulation, by the loss of clay, iron and aluminium and by the development of organo-mineral complexes, granular crumbs or platy structures. The A-horizon is divided into following two sub-layers:

(i) A_1 region. This region is dark and rich in organic matter and is called **humic** or **melanized region**. The amorphous, finely-divided organic matter here becomes mixed with the mineral matter, which is now known as **humus** and is dark brown or black-coloured.

(ii) A_2 region. It is light-coloured region where the mineral particles of large size as sand (silica) are more with little amount of organic matter. In areas of heavy rainfall, the mineral elements and organic chemicals are rapidly lost downwards in this region, making it light-coloured. A_2 region is, thus, also called **podsolic** or **eluvial zone** or zone of leaching.

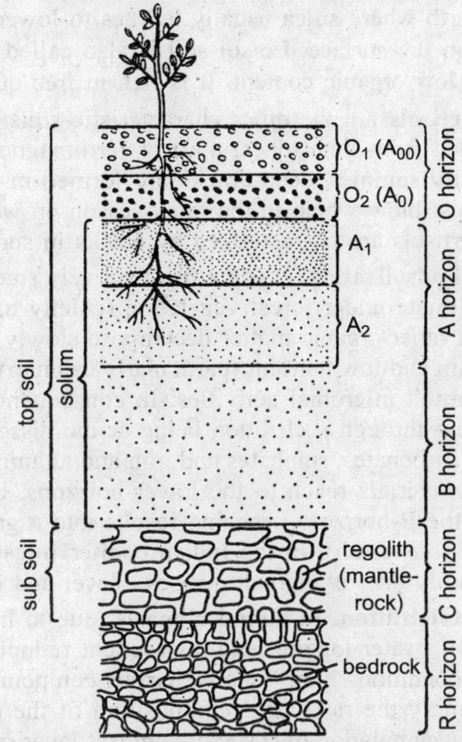

Fig. 4.2. A generalized (hypothetical) diagram of the soil profile to show principal horizons.

3. B-horizon. It lies below A-horizon and also called **sub-soil** or **illuviation** or **illuvial zone**, since, the nutrients received from A-horizon due to leaching are accumulated in this region. B-horizon is dark-coloured and coarse textured due to the presence of silica rich clay, organic compounds, hydrated oxides of aluminium, iron, etc. It is poorly developed in dry areas. B-horizon can be divided into B_1 (A_3), B_2 and B_3 regions, depending upon the stages of soil development in the area.

4. C-horizon. Below B-horizon and above the surface of weathered parent rock, is the zone of regolith or C-horizon. It is a light coloured horizon containing weathered parent material.

5. R-horizon. Below all these horizons may lie the **R-horizon**, which is the parent, unweathered bedrock. The percolated soil water tends to collect at the surface of the bedrock.

Climate and Soil Types

Variation in climatic factors and rock properties, therefore, lead to different types of soil development in different climatic conditions, as follows :

1. Laterization. Laterite (L. *latus*=brick) soils are reddish brown in colour and are found in the warm and humid (=heavy rainfall) tropics. In the process of laterization, iron and aluminium oxides which are resistant to decomposition, do not leach down but remain in the surface soil. Silicic acid (*i.e.,* colloidal form of silica) on the other hand leaches to lower horizons. Laterites are comparatively less fertile soils and are found in Australia, India, some warm parts of Europe and North America.

Lotsol is a wider term than laterite and currently used to refer to soils of warm humid belt of the earth where silica usually leaches to lower horizons while oxides of iron and aluminium remain on the surface. Lotsol soil is also called **oxide** and **ferrasol** and is poor in bases and contains low organic content. It is rich in free quartz grains and kaolinite clays.

Vertisols are the other characteristic soils where there is a long dry season. These soils are formed from limestones, marl or ferromagnesium rocks and rich in montmorillonite clays. During dry summer, deep cracks are formed in vertisols and these cracks get filled with clay due to dust blow which in the rainy season on wetting swells so much as to form small bumps. Thus, vertisols are characterized by cracks in summer and bumps in rains.

2. Podsolization. It is the most actively studied soil forming process. It takes place in cold humid climate under forest vegetation (chiefly of genera of Ericaceae and conifers) where leaf litter and other organic matter decompose slowly, chiefly through fungal activity. Litter contains high lignin and low nutrient (particularly calcium) contents; it is also rich in phenolic compounds which inhibit microbial activities. In consequence, such a litter is acidic in nature. The water percolating through such litter, being acidic dissolves out with its minerals and humus contents (such as carbonates, sulphates and iron and aluminium compounds) from A-horizon of soil. These leached materials reach to the lower horizons, being collected in the form of a hard, distinct layer in the B-horizon. Leaching results into a grey ashlike surface called **podsol** (Russian *pod* = under, *zole* = ash). In the Indian conifer forests of the Himalayas true podosols are not met with, mainly due to alternate phases of wet and dry seasons.

3. Gleization. In humid climates, due to high ground level or retention of surface water in the soil, water-logging and consequent reducing conditions are fairly common. Under such reduced conditions firstly due to ferrous compounds, the soil colour becomes blue-grey or grey, and secondly the rates of decomposition of the organic matter are slow. These together result into the accumulation of a sticky compact layer of blue-grey colour at the bottom of B- horizon. This process in called **gleization** and the soil as **gleys**. Gleization also takes place in ice covered regions of arctics (also tundra) where soil is not saline.

4. Melanization. This process is very common in regions of low humidity where humus is formed from the organic matter. As a result of melanization, the black humus along with water is mixed in the A-horizon of the soil which turn dark-coloured. Such a blackish brown top soil is called **chernozem** and it has a minimum leaching, so it is rich in organic detritus.

5. Calcification. It is a common process found in North India, where calcium in the form of calcium bicarbonate dissolves in water and leaches down escaping adsorption around dry particles. The leached $CaCO_3$ gets precipitated at the depths of about half to three metres, depending on the quantity of rainfall and depth of infiltration. The precipitated calcium remains in the form of nodules or *kankar* and form a hard pan. This gradually reduces soil fertility. In the Northern India the kankar nodules are unearthed and are used in the building of roads and in the preparation of lime. Calcification is typical of grassland and a layer of calcium carbonate generally occurs at the base of B horizon. Sometimes, a layer of more soluble calcium sulphate (gypsum) is also present.

There are some other types of soils such as a hydromorphic, halomorphic and azonal. **Hydromorphic soils** are formed on poorly drained water filled regions with poor aeration, *i.e.,* they are highly reducing soils in which ferric ions are reduced to ferrous. But in dry seasons the soil becomes aerated and ferrous is again converted into ferric. The hydromorphic soils of tropical paddy fields are called **grey soil**. **Halomorphic soils** are characterized by the presence of high salt contents on the surface. Salinity and alkalinity are commonly found in big patches in the North Indian states resulting into the formation of *usar* or *reh* soils. In fact, such soils result due to upward capillary movement of dissolved salts from the rising water table in too frequently irrigated and fertilized regions, or from nutrient rich water received through stream flow from the surroundings during the periods of high rainfall. On evaporation, salts in different forms are left on the surface. Grazing animals often lick such salt deposits to fulfil their metabolic needs. **Azonal soils** lack horizon B and the horizon A is also thin. Among azonal soils, the common are **lithosols** found on mountain tops where parent rock is quite hard and resistant to weathering. **Regosols** are formed from volcanic deposits.

Morphology of Soil

In the field, differences among soils and among horizons within a soil are primarily reflected by variations in texture, arrangement, structure and colour.

1. Texture of soil. The texture of a soil is determined by the production of different sized soil particles. Mineral fractions or particles of soil are called "**soil fractions**" or "**soil separates**". The soil particles have been classified into **gravel**, **sand**, **silt** and **clay** on the basis of their size differences. Gravel consists of coarse particles larger than 2.0 mm. Sand ranges from 0.02 to 2.0 mm in diameter and is easily seen and feels gritty. Silt consists of particles from 0.002 to 0.02 mm in diameter, which scarcely can be seen by the naked eye, feels and looks like flour. Clay particles range below 0.002 mm in diameter, are too fine to be seen even under the ordinary microscope, and are colloidal in nature. Clay controls the most important properties of soils, including plasticity and exchange of ions between soil particles and soil solution.

Most soils are mixture of these various particles. Based on the proportions of the various particles contained in them, soils can be grouped into 5 textural groups: **(1) Coarse textured soils** are loose, consist mainly of sand and gravel. They retain very little moisture and supply some plant nutrients. **(2) Moderately coarse soils** include sandy loam to very fine sandy loam. **(3) Medium textured soils** are mixture of sand, silt and clay, high enough to hold water and plant nutrients. **(4) Moderately fine textured soils** are high in clay. They are moderately sticky and plastic when wet, they may form a crust on the surface if organic matter is low. They have a high moisture-holding capacity. **(5) Fine textured soils** contain more than 40 per cent clay, may be sticky and plastic when wet and hold considerable water and plant nutrients, but may have restricted internal drainage. All these 5 textural soil-groups include 12 textural classes (see Table 4.7) which have been recognized on the basis of relative percentage of soil particles (*i.e.,* sand, silt and clay).

Table 4.7. Textural classes of soils.

Soil classes or Textural names	Range in relative percentage of soil separates		
	Sand	Silt	Clay
1. Sandy soil	85-100	0-15	0-10
2. Loamy sand	70-90	0-30	0-15
3. Sandy loam	43-80	0-50	0-20
4. Loam	23-52	28-50	7-27
5. Silt loam	0-50	50-88	0-27
6. Silt	0-20	8-10	0-12
7. Sandy clay loam	45-80	0-28	20-35
8. Clay loam	20-45	15-53	27-40
9. Silty clay loam	0-20	40-73	27-40
10. Sandy clay	45-65	0-20	35-45
11. Silt clay	0-20	40-60	40-60
12. Clay	0-45	0-40	40-100

2. Structure of soil. Soil separates (sand, silt and clay) are held together in clusters or shapes of various sizes, called **aggregates** or **peds**. The arrangement of these peds in earth crust is called **soil structure**. Like texture, there are many types of soil structure. Soil peds may be classified as

granular, **crumb-like**, **blocky plate-like** or **platy**, and **prismatic** or **columnar**. The properties of different soil peds have been tabulated in Table 4.8.

Table 4.8. Characteristics of peds.

	Types of ped	Properties
1.	Granular	Small, spherical and non-porous.
2.	Crumb-like	Small, porous and spheroidal.
3.	Agranular blocky	Block-like with sharp ends, one end may be pointed.
4.	Sub-granular blocky	Block-like but bounded by other aggregates.
5.	Platy	Plate-like, sometimes plates are overlapped.
6.	Prismatic or columnar	Prism-like but without rounded surface.

Structureless soil can be either single-grained or massive soil aggregates (peds) tend to become larger with increasing depth. Structure is influenced by texture, air, moisture, organic nutrients, micro-organisms, root-growth and soil chemical status.

3. Soil colour. Soils exhibit a variety of colour. Soil colour may be inherited from the parental material (*i.e.,* **lithochromic**) or sometimes it may be due to soil forming processes (**acquired** or **genetic colour**). Though colour has little direct influence on the functions of a soil, but is important in the identification of soil type. In temperate regions dark-coloured soils generally are higher in organic matter than light coloured ones. Well-drained soil may range anywhere from very pale brown to dark brown and black, depending upon the organic matter-content. Red and yellow soils are the result of iron oxides, the bright colours indicating good drainage and good aeration. Other red soils obtain their colour from parent material and not from soil-forming processes. Well-drained yellowish sands are white sands containing a small amount of organic matter and such colouring matter as iron oxide. Red and yellow colours increase from the cool regions to the equator. Quartz, kaolin, carbonates of lime and magnesium, gypsum and various compounds of ferrous iron give whitish and grayish colours to the soil. The grayest are permanently saturated soils in which the iron is in the ferrous form.

Soil colour and temperature. Soil colour influences the soil temperature. The dark coloured soils absorb heat more readily than light coloured soils. **Ramdas** *et al.,* (1936) showed that black cotton soils of Poona absorbed 86 per cent of the total solar radiations falling on the soil surface as against 40 per cent by the grey alluvial soil.

Physical Properties of Soil

Soil possesses many characteristic physical properties such as density, porosity, permeability, temperature, water and atmosphere, each of which can be studied under following separate headings:

1. Soil density. Average density of soil is 2.65 gms. per ml. Density of soil varies greatly depending upon the degree of weathering.

2. Porosity. The spaces present between soil particles in a given volume of soil are called **pore spaces**. The percentage of soil volume occupied by pore space or by the interestitial space is called **porosity** of the soil. Porosity of soil depends upon the texture and structure compactness and organic contents of soil. Porosity of the soil increases with the increase in the percentage of organic matter in the soil.

The pore spaces are of two types– **(1) Micro-pore spaces** (capillary pore spaces) and **(2) Macro-pore spaces** (non-capillary pore spaces). Capillary pore spaces can hold more water and restrict the free movement of water and air to a considerable extent, whereas macro-pore spaces have little water holding capacity and allow free movement of moisture and air in the soil under normal conditions.

3. Permeability of soil. The characteristic of soil that determines the movement of water through pore spaces is known as **soil permeability**. Soil permeability is directly dependent on the pore size, therefore, it is higher for the loose soil with large number of macro-pore spaces than it is for compact soil with numerous micro-pore spaces.

4. Soil temperature. Soil gets heat energy from different sources such as solar radiation, decomposing organic matter, and heat formed in the interior of earth. The temperature of soil is affected by its colour, texture, water content, slope, altitude of the land and also by climate and vegetational cover of the soil. Evaporation of water from soil makes it cooler. Black soils absorb more heat than white soils. Sandy soils absorb more heat and radiate it out quickly at night than clay or loam soils.

The soil temperature greatly affects the physico-chemical and biological processes in the soil. For example, the germination of seeds, normal growth of roots and biological activity of soil-inhabiting micro-and macro-organisms which require proper and specific temperature.

5. Soil water. In soil, water is not only important as a solvent and transporting agent but in various ways it maintains soil texture, arrangement and compactness of soil particles and makes soil habitat livable for plants and animals. It comes in soil mainly through infiltration of precipitated water (dew, rain, sleet, snow, and hail) and irrigation. Soil water is classified into following types— gravitational and ground water, capillary water, hygroscopic water, combined water and water vapour.

(i) Gravitational water. In a well-saturated soil, the accessory (extra) amount of water displaces air from the pore spaces between soil particles and percolates downwardly under gravitational influence and finally it is accumulated in the pore spaces. This accumulated excess water of large soil spaces is called **gravitational water**. When this gravitational water further percolates down and reaches to the level of parent rock, it is called **ground water**. Both kinds of these soil water are ecologically important in the leaching of nutrients.

(ii) Capillary water. The water which is held by capillary forces (*i.e.,* surface tension and attraction forces of water molecules) in smaller soil channels, after the gravitational water and ground water have been drained, is called **capillary water**. Capillary water occurs as a thin-film around soil particles in the capillary spaces and represents the normal available water to the plants. It remains in soil for long periods and carries with it nutrients in solution. Humus has more capillary water than soil minerals.

(iii) Hygroscopic water. Soil particles retain some water so firmly that the plants cannot absorb this. Such soil water is called hygroscopic water.

(iv) Combined water. Combined water is the water of chemical compounds held by chemical forces of molecules such as hydroxides of silicon, iron and aluminium. It is of no ecological significance.

(v) Water vapour. Some soil water occurs as moisture or water vapours in the soil atmosphere.

Further, the total amount of water present in the soil is called **holard**. The quantity of water that plant-roots can absorb out of holard is called **chresard** and that amount of soil water which cannot be absorbed by plant-roots is called **echard**. Moreover, there are some terms that reflect the water status of soil and are generally used for comparative studies of different soils.

(a) Soil water potential. This is an expression of the total reduction of water potential in the soil, due to mineral matrix, solubility, external pressure and gravitation effects.

(b) Field capacity. When a soil holds all the water it can, but no gravitational water, it is said to be at its field capacity. Field capacity is generally defined as the water content of an undisturbed soil (per cent oven-dry weight) after it is saturated by rainfall and drainage of gravitational water has completely stopped. Field capacity of soil may be taken as the total amount of capillary, hygroscopic and combined water plus water vapour.

(c) Moisture equivalent. It is defined as the water content (per cent oven-dry weight) retained by the undisturbed soil.

(d) Water holding capacity (or storage capacity). This is the extent to which a soil can hold capillary water against gravity. It is equal to field capacity less hygroscopic water.

(e) Hygroscopic-coefficient. It is defined as the water vapour (%) absorbed by unit weight of dry soil, when placed in an atmosphere completely saturated with water vapours.

(f) Permanent wilting percentage (or wilting coefficient). It is the amount of water (per cent oven-dry weight) that remains in a soil when permanent wilting is present in the plants growing in soil.

The status of soil water is also represented in terms of **capillary potential** (= metric potential). The relationship that exists between soil water content and water potential is usually determined by a pressure membrane apparatus, called **tensiometer**.

6. Soil atmosphere. Gases found in pore spaces of soil profiles form the **soil atmosphere.** The soil atmosphere contains three main gases namely O_2, CO_2, and N_2. Soil air differs from atmospheric air in having more of moisture and CO_2 and less of O_2. The soil atmosphere is affected by temperature, atmospheric pressure, wind, rainfall, etc. Loam soils with humus contain a normal proportion of air and water (about 34 per cent air and 66 per cent water) and therefore are good for majority of crops.

Coarse-textured or well-structured soils contain higher gaseous diffusive transfer rates than fine-textured or poorly-structured soils under **wet conditions**. This is because they contain many large pores which remain gas-filled. The water-filled pores of the fine-textured soils form potent barriers of gaseous diffusion; O_2 diffuses ten thousand times more slowly in water than in gas. In **dry soil**, the situation is reversed, as the fine textured soils have a greater total pore space and provide a larger gas-filled cross sectional area for diffusion. Soil aeration is very important in growth of roots, seed germination and microbial activity. Poor soil aeration suppresses root hair development, and may reduce rate of absorption of water and nutrients.

7. Soil solution. There exists a weak solution of various salts, alongwith other liquids and gases in the soil mass. This soil solution contains almost all the essential minerals. Complex mixture of minerals such as carbonates, sulphates, nitrates, chlorides, and organic salts of Ca, Mg, Na, K, etc., are found dissolved in water. The chemical nature of soil solution depends on the nature of parent matter, chemical nature of organic matter and climatic factors and other factors involved in pedogenesis. The soil solution is the primary source of inorganic nutrients for plant roots. Soils with optimal concentration of various nutrient solutes are called **eutrophic**, whereas those with suboptimal concentration of these nutrient salts are called **oligotrophic**.

Chemical Properties of Soil

Soil is a mixture of various inorganic and organic chemical compounds and exhibits certain significant chemical properties, all of which can be discussed as follows :

(a) Inorganic elements and compounds of soil. The chief inorganic constituents of soils are the compounds of following elements—A1, Si, Ca, Mg, Fe, K and Na. Soil also contains smaller amounts of compounds of following inorganic elements—B, Mn, Cu, Zn, Mo, Co, I, F, etc.

(b) Organic matter of soil. The chief organic component of soil is **humus** which chemically contains amino acids, proteins, purines, pyrimidines, aromatic compounds, hexose sugars, sugar alcohols, methyl sugars, fats, oils, waxes, resins, tannins, lignin and some pigments. Further humus is black coloured, odourless, homogeneous complex substance.

(c) Colloidal properties. As soil is composed of crystalloids and colloids, therefore, it exhibits all the physico-chemical properties which are related with these two soil particles. Colloids for example exhibit absorption, electrical properties, coagulation, Tyndal phenomenon, Brownian movement, dialysis, etc.

(d) Soil pH. Many chemical properties of soils centre around soil reaction. As regards their nature, some soils are **neutral**, some are **acidic** and some **basic**. The acidity, alkalinity and neutrality of soils are described in terms of **hydrogen-ion concentrations** or **pH** values. A pH value of 7.0 indicates neutrality, a value above this figure (7.1–14.00) indicates alkaline condition and a value below it (0–6.9) indicates acid conditions. Normally, the pH value of soils lies between 2.2 and 9.6. In India, acidic soils (pH below 5.5 to 5.6) occur in the high rainfall areas of Western ghats, Kerala, Eastern Orissa, West Bengal, Tripura, Manipur and Assam. The saline, alkaline or basic soils (called *'Usar'*, contain pH upto 8.5) of India, occur in U.P., West Bengal, Punjab, Bihar, Orissa, Maharashtra, Madras, M.P., A.P., Gujarat, Delhi and Rajasthan.

Some pH values are ecologically significant for the plants; plants regarded as **calcicoles** usually occur in soil with pH 6.5, whereas **calcifuges** occur in soil with pH value below 3.8–4.0. Soils above pH 6.5 are generally cation-saturated (those containing free $CaCO_3$ called **calcareous soils**) while soil below pH 3.8-4.0 contain a considerable content of exchangeable hydrogen. These limits of pH value are also reflected in the nature of the soil's organic matter: raw humus or mor which is associated with soils below of pH 3.8, while mull is characteristic of the more cation saturated soils of pH 4.8–5.0 and above.

Highly acidic and highly saline or alkaline soils often remain injurious for plant growth, micro-organisms, etc. Soil pH strongly affects the microbial activities, as at below pH 5.0 bacterial as well as fungal activities are reduced. Neutral or slightly acidic soil, however, remains best for the growth of majority of plants.

Soil as Habitats for Animals

Soil is used by various animals as a hideout from enemies, to escape desiccation and extremes of temperature; as material for abodes, and as a highway. Clay is cold, dense, poorly aerated and though it holds much water, does not give up water readily to plants and animals. Sand is dry, loose (cave in readily) and has a variable temperature. Rock heats and cools quickly, has little available water and practically no air, and is usually too resistant to allow penetration by burrowing animals. Alkaline soils are "**physiologically dry**" because the salts present hold the water so that it is not readily available to organisms. Humus has a great water capacity and gives up water readily, is usually well-aerated and furnishes nitrogen as well as mineral salts and water. It remains rich in biota. Peaty soils are raw humus and are uninhabitable for most animals because they are poorly aerated and have an acid reaction. In general, the character of the vegetation cover serves as a good index of the habitability of soils for animals; conditions that make plants flourish are usually favourable for animals.

Soil Fauna and Soil Flora

Soil supports a wide array of organisms of different body-sizes and taxonomic groups. Generally, soil organisms are classified into three major groups namely **microfauna** and **microflora**, **mesofauna**, and **macrofauna**. Mesoflora and macroflora because occur above the surface of soil (land-surface), therefore, are excluded from this discussion :

1. Microfauna. It includes animals with body size within the range of 20μm to 200μm. It includes all Protozoa and small-sized mites, nematodes, rotifers, tardigrades and copepode Crustacea. Soil inhabiting **protozoans** such as amoeba, ciliates, zoomastigine flagellates occur near the surface soils, while the testate forms such as *Thecamoeba, Euglypha* and *Difflugia,* have a wider vertical distribution. The common terrestrial **polyclad** is *Bipalium*. The **nematodes** such as *Rhabditis, Diplogaster, Tylenchus, Heterodera, Aphelenchoides, Mononchus, Pratylenchus, Xiphinema* and *Criconemoides,* abound by as much as 1-3 million in raw humus soils to 20 million/m^2 in grassland soils.

2. Microflora. The microflora of soil includes bacteria, soil fungi, soil actinomycetes, blue green algae and algae. In soil, microflora bacteria form about 90 per cent of the total population. Fungi and algae together represent one per cent and actinomycetes cover only 9 per cent.

Soil bacteria grow fairly well in the neutral soils richly supplied with organic nutrients. Soil inhabitant bacteria fall into categories namely—**autotrophic bacteria** and **heterotrophic bacteria**. The autotrophic bacteria derive their energy from the oxidation of simple carbon compounds or from inorganic substances and their carbon from the atmospheric CO_2. The common autotrophic bacteria of soil are nitrifying bacteria, hydrogen bacteria, sulphur bacteria, iron bacteria, manganese bacteria, carbon monoxide bacteria and methane bacteria. Most of soil bacteria are heterotrophic bacteria depending upon the organic matter of soil for their energy source and are primarily concerned with the decomposition of cellulose, and other carbohydrates, proteins, fats and waxes. They bring about mineralization of organic matter of soil and release considerable amount of nitrogen, phosphorus and other nutrients for plants. The common nitrogen-fixing bacteria of soil are *Rhizobium* (occurs in root nodules of leguminous plants); *Azotobacter* and *Clostridium pasteurianum* (the latter two are free occurring in soil).

Majority of **soil fungi** are found in acidic soils. They may be parasitic, saprophytic and symbiotic. **Parasitic fungi** of soil infect roots of plants and cause plant diseases such as cotton root rot and many kinds of wilts, rusts, blights and smuts. Certain wilt-forming fungi produce toxins which are harmful, for example, *Fusarium lini,* which causes wilt of flax (*Alsi*) and secretes HCN and *Fusarium udum*, a fungus causing wilt of pigeon pea (*Arhar*) secretes fusaric acid in the roots of host plants. However, certain parasitic fungi produce growth stimulating substances for host plant. *Fusarium* sp., for example, have been found to secrete gibberellin and gibberellic acid ($C_{19}H_{22}O_6$). **Symbiotic fungi** of soil live on the roots of certain plants and both fungus and plants are benefited. **Saprophytic fungi** depend on dead organic matter of soil and derive the energy from decomposition of the latter. They break down cellulose, lignin, and gum, sugars, starch, protein, etc., into simple substances to be utilized by higher plants as nutrients.

Actinomycete fungi prefer saline soils and bring about the decomposition of organic matter such as cellulose. They produce a variety of antibiotics of great economic significance for human beings. The most important blue green algae of soil are those which fix nitrogen in soil. *Anabaena, Nostoc, Microcystis, etc.,* are important nitrogen fixing blue green algae of soil. They also make soil aggregates because of having mucilage.

3. Mesofauna. Mesofauna includes animals with body size within the range 200μm to 1 cm. The micro-arthropods Acari (mites) and Collembola (spring tails) are important members of this group which also include the larger nematodes, rotifers, and tardigrades, together with most of the isopods, Arachnida (spiders), Chelognathi (pseudoscorpions), Opiliones (harvestmen), Enchytraeidae (pot-worms), insect larvae and small millipedes (Diplopoda), isopods and molluscs.

Among **annelids** the microscopic enchytraeids are represented by *Enchytraeus fridericia* and *Achaeta lumbricellus*, which are more abundant in organic soils and forests than in grasslands. *Oniscus, Porcellio,* and *Armadillidum* are the most common isopods (**crustacean**) of the tropics in the humid zone. Among the soil **arachnids**, mites are the most predominant. Mites flourish in moist organic soils and certain mites such as *Galumna, Cepheus, Hemorobates* occur in lichens and mosses. Certain mites, such as *Schelorbates* and *Brachychthornus* live in humus. The mites are saprophagous, predatory and phytophagous and are involved in the process of organic decomposition and its resultant processes. Certain arachnids such as scorpions, *Thelyphonus,*

Galeodes; and some spiders are crepuscular, hiding under rocks or in crevices in soil and in loose litter, and has no ecological significance in decomposition like other arachnids. Many **opiliones** or harvestmen occur in forest litter, frequently preying upon soil organisms.

Besides mites, only the pseudoscorpions or **chelognathi** occur in surface soils and most decaying vegetations. Of the xerophil litter inhabitants are *Stenatemnus indicus, Dhanus indicus, Fealla indica* and the hygrophil inhabitants living in the litter and under stones are–*Comsaditha indica, Tyrannochthonius madrasensis, Tyrannochithonius chelatus* and *Hygrochelifer indicus* (**Murthy**, 1964), feeding on Collembola, enchytraeids, etc.

The common millipedes or diplopodes of forest soil which are chief decomposers of soils are *Spirostreptus, Thyropygus, Glomeris, Arthrosphaera, Polydesmus, Julus,* etc. **Tardigrades** or bear animalcules occur in surface layer of moist soils in grassland being represented by the *Macrobiotus* and *Hypsibius.*

Among **insects**, apterygote Collembola form numerically the most important groups of soil insects. Other insects such as Dermaptera, Psocoptera, Dictyoptera, Isoptera, Coleoptera, a few Hymenoptera and some Diptera also occur in soil, sometimes as juveniles. The termites such as *Reticulotermes* and *Odontotermes* are important soil-dwellers of tropics and play an important role in the break up of organic materials and their mixing up with mineral soils. Among the Hymenoptera, ants are the most important soil dwelling forms. Among Collembola, Onchuridae, Isotomidae, Poduridae and Entomobrydae are richly represented in the soil both in number and species composition. Large-sized Collumbola such as *Tomocerus, Entomobrya* and *Orchesella* occur in surface layer, while the smaller *Onychurus, Tullbergia,* etc., occur in deeper layers of soils.

Among Diplura, *Anajapyx, Japyx* and *Campodea* are often found in small numbers in moist soils under stones and in humus. Proturans are more abundant than Diplura and very common in moist forest and grassland soils which abound in species of *Eosentomon, Acerentomon* and *Acerentulus.*

4. Macrofauna of soil. Macrofauna of soil includes those animals whose body size is greater than 1 cm. Here belong the majority of Lumbricidae, the Mollusca, the largest insects and arachnids and the soil-dwelling vertebrates.

Earthworms usually occur in abundance in alkaline and moist soils and sparse in acid soils. They have been acknowledged for their influence on the process of decomposition of organic materials, breaking up litter fragments and mixing them thoroughly with mineral soils resulting in the formation of organic soils. Some of the common Indian annelidan species of soil are *Megascolex, Pheretima, Octochaetus, Drawida* and *Moniligaster*. Among chilopods the carnivorous *Scolopendra* and *Lithobius* are common in moist soils feeding on leaf litter inhabitants.

Among soil vertebrates, following animals are well adapted for burrowing life in soils–*Ichthyophis, Cacopus systema, Breviceps* (Amphibia), *Sphenodon, Uromastix,* limbless lizards and snakes (Reptilia), *Talpa, Dasyurus, Notoryctes* and various insectivores and rodents (Mammalia).

Adaptations of soil animals. Animals which are adapted for digging the burrows and for subterranean mode of life are called **fossorial animals**. These animals may dig either for their food or simply for retreat. Zoologically they are primitive, defenseless and unambitious animals. They have following adaptations: 1. The body contour is either cylindrical (*e.g., Ichthyophis,* limbless lizards, snakes, earthworms, *Scolopendra,* etc.), or spindle-shaped or fusiform (e.g., *Talpa, Echidna,* etc.), so as to offer least resistance to subterranean passage. 2. The head tapers anteriorly to form a sort of snout for burrowing. 3. The tail is short or vestigial. 4. The eyes tend to become vestigial as they are of no use in dark habitat. 5. The external ears also tend to disappear since they would be obstructing in burrowing. 6. For digging, many structures may be found in different fossorial animals, *e.g.*, hands are well adapted for digging. In the insect *Gryllotalpa*, the fore-legs are modified for digging purpose.

The ground-dwelling animals which may be **cursorial** (running), such as ostrich, rhea, ungulates, wolves, cats, bears, hyaenas, etc., **saltatory** (jumping) such as rodents, rabbits, wallabies, kangaroos, etc., or **graviportal** (heavy) such as turtles, armadillos, elephants, hippopotami, etc., exhibit different kinds of adaptations for different kinds of soils. For example, if the soil is firm and hard, the large animals inhabiting the ecosystem tend to have small hooves or paws; if the soil is wet and spongy, they tend to have broad hooves or paws.

QUESTIONS

Long Answer Questions

1. What are the physico-chemical properties of air and describe biota of atmosphere.
2. What are the physico-chemical properties of water and discuss the effect of water on the life of organism.
3. What are physio-chemical properties of soil. Describe biota of soil. Write a short note on the edaphic factor.

Short Answer Questions

1. Differentiate between Hydrosphere and Lithosphere.
2. Write a short note on soil profile.
3. Describe physico-chemical properties of water.
4. Describe adaptive features of animals of arid environment.

Very Short Answer Questions

1. What is chemical composition of Humus?
2. Give an example of physiologically dry soil.
3. Give an example of graviportal ground-dwelling animal.
4. What is lithosphere?
5. What are the physical weathering agents of soil formation?

Fill in the Blanks

1. The life supporting environment of planet Earth is called ....
2. Ozonosphere exist in the .... of the atmosphere.
3. .... animals can tolerate only narrow fluctuations in the salinity.
4. .... are amphibious plants.
5. *Aloe* is a .... xerophytes.
6. Edaphology is science which deals with ....
7. The total amount of water present in the soil is called ....

Multiple Choice Questions

1. Ozonosphere is a part of
   (*a*) stratosphere (*b*) mesosphere
   (c) thermosphere (*d*) ionosphere
2. Which one of the following layers of the atmosphere has maximum density?
   (*a*) stratosphere (*b*) troposphere
   (*c*) mesosphere (*d*) thermosphere
3. The largest amount of freshwater on our planet is in
   (*a*) polar ice caps and glaciers (*b*) underground
   (*c*) rivers (*d*) lakes and streams

4. Root cap is absent in
(*a*) lithophytes
(*b*) hydrophytes
(*c*) xerophytes
(*d*) mesophytes

5. In submerged hydrophytes functional stomata are found
(*a*) on upper surface of leaf
(*b*) on lower surface of leaf
(*c*) on both surfaces of leaf
(*d*) no where on the plant

6. Gaseous exchange in submerged hydrophytes takes place through
(*a*) lenticels
(*b*) stomata
(*c*) hydathodes
(*d*) general surface of plants by diffusion

7. Diatoms are commonly
(*a*) benthos
(*b*) in the upper surface of water in lakes and oceans
(*c*) saprophytic
(*d*) parasitic

8. In an aqueous environment microscopic animals and plants are collectively known as
(*a*) commensals
(*b*) herbivores
(*c*) fauna and flora
(*d*) planktons

9. The abyssal zone in an ocean is characterized by
(*a*) presence of sunlight and producers
(*b*) absence of sunlight and living organisms
(*c*) absence of sunlight
(*d*) absence of sunlight and presence of consumers and decomposers

10. Among the plants listed below point out a plant that does not fit into ecological group represented by other plants
(*a*) *Euphorbia*
(*b*) *Vallisneria*
(*c*) *Acacia*
(*d*) *Aloe*

11. Sunken stomata are present in
(*a*) xerophytes
(*b*) hydrophytes
(*c*) mesophytes
(*d*) epiphytes

12. Extreme xerophytic conditions are found in
(*a*) *Brassica*
(*b*) argemone
(*c*) cactus
(*d*) *Capparis*

13. Which group of plants has to face physiological dryness
(*a*) halophytes
(*b*) lithophytes
(*c*) hydrophytes
(*d*) glaciation

14. Plants growing in xerophytic conditions have
(*a*) extensively developed shoot system
(*b*) extensively developed root system, small and thick leaves
(*c*) large flowers
(*d*) extensive air spaces

15. Which one of the following is the most characteristic feature of a xeric environment
(*a*) extremes of temperatures
(*b*) high rate of evaporation
(*c*) low atmospheric humidity
(*d*) low precipitation

16. Mangrove vegetation is found in
(*a*) Sunderbans
(*b*) Kullu valley
(*c*) Western ghats
(*d*) Dehradun valley

17. *Acacia arabica* is a
(*a*) halophyte
(*b*) mesophyte
(*c*) xerophyte
(*d*) hydrophyte

18. *Eichhornia crassipes* is a
(*a*) desert plant (*b*) parasite
(*c*) water plant (*d*) terrestrial plant

19. Measurement of the rate of O_2 consumption in unit volume of water over a period of time is done to find out
(*a*) biogas generation (*b*) biological oxygen demand
(*c*) biosynthetic pathway (*d*) fermentation

20. Actively moving or swimming animals in aquatic systems are
(*a*) benthos (*b*) phytoplankton
(*c*) nektons (*d*) zooplanktons

21. Carbon dioxide contents in atmospheric air is about
(*a*) 0.034% (*b*) 0.34%
(*c*) 3.34% (*d*) 6.5%

22. The solid components of earth comprise
(*a*) hydrosphere (*b*) lithosphere
(*c*) atmosphere (*d*) ionosphere

23. Plants which are attached to the stones and rocks are
(*a*) halophytes (*b*) aerophytes
(*c*) lithophytes (*d*) psammophytes

24. Edaphic factors are related to
(*a*) human (*b*) animal
(*c*) temperature (*d*) soil

25. Pedology is the study of
(*a*) rocks (*b*) soils
(*c*) crop disease (*d*) locomotion of animals

26. Development of soil from parent rock is termed as
(*a*) edaphic factor (*b*) pedogenesis
(*c*) edaphotropism (*d*) edaphic climax

27. The most common source of water for plants from soil is
(*a*) gravitation water (*b*) hygroscopic
(*c*) rain water (*d*) capillary water

28. Property of soil based on the size of its particles is termed
(*a*) colour (*b*) texture
(*c*) field capacity (*d*) water holding capacity

29. Soil particles arranged in order of increasing size are
(*a*) clay, sand and silt (*b*) silt, sand and clay
(*c*) sand, silt and clay (*d*) clay, silt and sand

30. The pH of fertile soil is usually around of
(*a*) 2–3 (*b*) 6–7
(*c*) 8–10 (*d*) 11–12

31. The soil A_O layer consists of
(*a*) undisintegrated material (*b*) litter
(*c*) mineral substances (*d*) none of the above

32. Water logging occurs in

(*a*) clay soil (*b*) loamy soil
(*c*) gravel soil (*d*) sandy soil

33. Plants die from prolonged water logging because

(*a*) soil nutrients become very dilute
(*b*) root respiration stops
(*c*) cell saps in the plants become too dilute
(*d*) nutrients leak down due to excess water

34. Nitrogen is a critical element in ecosystem because

(*a*) it is abundant is atmosphere (*b*) nitrogen fixation comes by microbes
(*c*) it is an essential plant element (*d*) it is very labile element

35. Humus is

(*a*) totally decomposed organic matter in the soil
(*b*) partially decomposed organic matter in the soil
(*c*) totally decomposed organic matter in a pond
(*d*) partially decomposed organic matter in a pond

36. Humus is an example of

(*a*) crystalloid (*b*) soil structure
(*c*) organic colloids (*d*) none of them

37. Insectivorous plants grow in soil which is deficient in

(*a*) N_2 (*b*) Ca
(*c*) Mg (*d*) Water

38. Plants which live on saline soil are known as

(*a*) lithophytes (*b*) halophytes
(*c*) mesophytes (*d*) hydrophytes

39. Free living aerobic nitrogen fixing bacterium of soil is

(*a*) *Azotobacter* (*b*) *Rhizobium*
(*c*) *Clostridium* (*d*) All of these

40. Which of the following soils is suited for fossorial animals such as earthworms?

(*a*) loam and humus (*b*) clay
(*c*) silt (*d*) sand

41. The least porous soil among the following is

(*a*) sandy soil (*b*) loam soil
(*c*) clay soil (*d*) fine gravel

42. Biosphere is made up of

(*a*) organism
(*b*) organism + lithosphere
(*c*) organism + lithosphere + atmosphere
(*d*) organism + lithosphere + atmosphere + hydrosphere

43. The factors of soils are studied under the heads: (*Rohilkhand 2001*)

(*a*) climate (*b*) biotic
(*c*) topographic (*d*) edaphic

44. Animals which live at the bottom of sea are called (*CCSU Meerut 2014*)

(*a*) diatoms (*b*) benthos
(*c*) neuton (*d*) plankton

ANSWERS

Very Short Answer Questions

1. Humus contains the following chemicals: amino acids, proteins, purines, pyrimidines, aromatic compounds, hexose sugar, sugar alcohols, methyl sugars, fats, oils, waxes, resins, tannins, lignin and some pigments.
2. Alkaline soils.
3. Elephants.
4. The solid component of earth is called lithosphere.
5. Temperature, water, ice, gravity and wind.

Fill in the Blanks

1. Biosphere	**2.** Stratosphere	**3.** Stenohaline	**4.** Tenagophytes
5. Succulent	**6.** Soil	**7.** Holard.	

Multiple Choice Questions

1. (*a*)	**2.** (*b*)	**3.** (*a*)	**4.** (*b*)
5. (*d*)	**6.** (*d*)	**7.** (*b*)	**8.** (*d*)
9. (*d*)	**10.** (*b*)	**11.** (*a*)	**12.** (*d*)
13. (*a*)	**14.** (*b*)	**15.** (*d*)	**16.** (*a*)
17. (*c*)	**18.** (*c*)	**19.** (*b*)	**20.** (*c*)
21. (*a*)	**22.** (*b*)	**23.** (*c*)	**24.** (*d*)
25. (*b*)	**26.** (*b*)	**27.** (*d*)	**28.** (*b*)
29. (*d*)	**30.** (*b*)	**31.** (b)	**32.** (*a*)
33. (*b*)	**34.** (*b*)	**35.** (*b*)	**36.** (*c*)
37. (*a*)	**38.** (*b*)	**39.** (*a*)	**40.** (*a*)
41. (*c*)	**42.** (d)	**43.** (*d*)	**44.** (*b*).

5

Abiotic Environmental Factors

5.1 LAW OF MINIMUM AND LAW OF TOLERANCE

An organism is seldom, if ever, exposed solely to the effect of a single factor in its environment. On the contrary, an organism is subjected to the simultaneous action of all factors in its immediate surroundings. However, some factors exert more influence than do others, and the attempt to evaluate their relative roles had led to the development of the law of the minimum by **Justus von Liebig** in 1840.

1. Law of Minimum

The adaptiveness of an organism to its environment is exhibited by its ability to function between upper and lower limits within a range of environmental conditions. The role of environmental limitations in the response of organisms to their environment was recognized in 1840 by the German organic chemist **Justus von Liebig.** In his book "*Organic chemistry and its application to Agriculture and Physiology*", Liebig described plant chemistry and growth and noted. "*The crops of a field diminish or increase in exact proportion to the diminution or increase of the mineral substances conveyed to it in manure*".

What **Liebig** was saying is that each plant requires certain kinds and quantities of nutrients. If one of these mineral substances is absent, the plant fails to grow or dies. *If the substance is present in minimal quantities only, the growth of the plant will be minimal.* This statement became known as the **law of minimum.**

Leibig had determined that *phosphorus* was not only important to agriculture, but that if it was present in small amounts the plant growth was poor and that in its absence there was no plant growth at all. Thus growth was limited by the amount of phosphorus present.

As originally visualized, Liebig law of minimum applied to inorganic nutrients with the passage of time, the concept expanded to include a wide spectrum of physical factors as well, temperature and rainfall (*e.g.,* a minimum of either or both) being among the most apparent.

The law of minimum, strictly applied, carries with it some restrictions. It applies only under steady-state conditions with all other resources (= factors) being in excess of needs. If the quantity of the limiting resource increase, so that the rate of growth of the plant increases, then that substance is no longer limiting, but another might become so, as the increased growth places more demand on other resources. Further, the concentration or availability of another nutrient may modify or influence the availability or use of the minimum one. For example, if availability of *phosphorus* is low, *molybdenum,* another essential plant nutrient needed in very small amounts, may be sufficient for the plant's needs. If the availability of phosphorus increases to a point that it stimulates additional plant growth, molybdenum will become the limiting element. This concept applies not only to nutrients but also to other factors of the environment.

2. Law of Limiting Factors

Too little of a nutrient or other resource, such as light, may be harmful; but two much of a good thing can be just as bad, and thus limiting. **F. F. Blackman** (1905) advanced this concept. He pointed out that *a maximum quantity of a resource tolerated by an organism would limit response as well.* Blackman's observation became known as the **law of limting factors** (Box 5.1).

Box 5.1

Blackman noted that "in treating physiological phenomena, assimilation, respiration, growth, and the like, which have a varying magnitude under varying external conditions of temperature, light, supply of materials, etc., it is customary to speak of three cardinal points, the **minimal** condition below which the phenomenon ceases altogether, the **optimal** condition at which it is exhibited to its highest observed degree and the **maximal** condition above which it ceases again".

In this same paper, **Blackman** considered the case of a given chloroplast engaged in photosynthesis noting that there are *five* controlling factors: the amounts of CO_2, H_2O and chlorophyll present; intensity of available radiant energy; and temperature in the chloroplast. This led him to conclude that "*when a process is conditioned as to its rapidity by a number of separate factors, the rate of the process is limited by the pace of the 'slowest' factor*".

3. Law of Tolerance

Organisms may be limited in their growth and their occurrence not only by too **little** of an element or too **low** an intensity of a factor but also by too **much** of the element or too **high** intensity of the factor. For example, carbon dioxide is necessary for the growth of all green plants, small increase in concentration of carbon dioxide in the atmosphere will, under certain circumstances, increase the rate of plant growth, but very considerable increases become toxic. Likewise, small additions of arsenic to the human diet actually have a tonic effect, further increase in the dosage, however, soon proves fatal.

The idea that factors could be limiting at their maximum as well as minimum quantities was incorporated in **law of tolerance** formulated by **V.E. Shelford** in 1913. This law postulates that each ecological factor to which an organism responds has maximum and minimum limiting effect between which lies a range or gradient that is now known as the **limits of tolerance.**

Between the **lower** and **upper** limits of tolerance lies a broad middle sector of a gradient which is called the **zone of compatibility**, the **zone of tolerance,** the **biokinetic zone** or the **zone of capacity adaptation**. The region at either end of the zone of compatibility is called the **lethal zone** or the **zone of resistance** or **zone of intolerance.** The zone of compatibility too includes a broad **range of optimum** and narrow **zones of physiological stresses** in between the range of optimum and lethal zones.

Upper and **lower limits of tolerance** are intensity levels of a factor at which only half of the organisms can survive. These limits are sometimes difficult to determine, as for example with low temperature, organisms may pass into an inactive, dormant, or hibernating state from which they may again become functional when the temperature rises above a threshold. At high temperatures, there may be similar inactivation or aestivation before the lethal level is attained. Even without dormancy occurring, there are normally **zones of physiological stresses** before the limits of tolerance are reached.

The law of tolerance can be illustrated as a bell-shaped tolerance curve (Fig. 5.1). The ordinate or *x* axis represents the gradient or range of a particular environmental factor. The

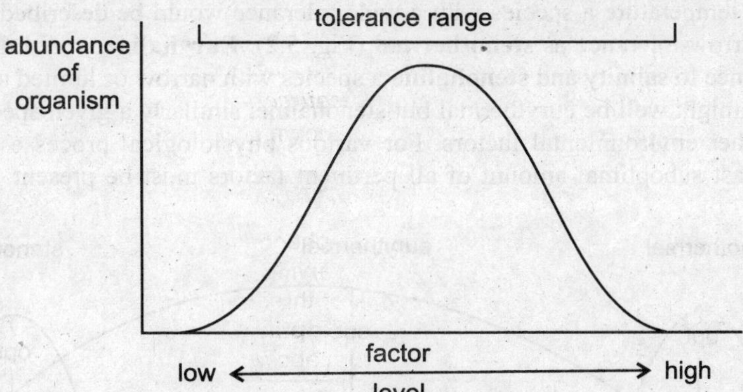

Fig. 5.1. The law of tolerance.

abscissa or *y* axis represents the response of the species or individuals of a species on that gradient. The upper or middle part of the curve embraces the optimal state for reproduction or fitness. Descending parts of the graph represent conditions under which individuals grow but posses lower fitness, survive but not reproduce and fail to survive.

Some subsidiary principles to the law of tolerance may be stated as follows:

1. Organisms may have a wide range of tolerance for one factor and narrow range for another.

2. Organisms with wide ranges of tolerance for all the factors are likely to be most widely distributed.

3. However, tolerance limits and range of optima may be shifted, displaced or expanded. When conditions are not optimum for a species with respect to a particular environmental factor, the limit of tolerance may be reduced with respect to other factors. For example, according to **Penman** (1950), when soil nitrogen is limiting, the resistance of grass to drought is reduced.

4. The limits of tolerance for reproductive individuals, seeds, eggs, embryos and larvae are usually narrower than for nonreproductive adult plants or animals. For example, adult blue crabs and many other marine animals can tolerate brackish water or fresh water with a high chloride content, but their larvae cannot live in such waters and hence they cannot reproduce in rivers and are not established there permanently.

4. Factor Interaction

The following terms (Table 5.1) are used to indicate the relative extent to which organisms can tolerate variation in environmental factors. The prefix *steno*—means that the species, population, or individual has a narrow range of tolerance and the prefix *eury*—indicates that it has a wide range.

Table 5.1. Terminology for ranges of certain abiotic environmental factors.

Stenothermal–Eurythermal	Pertaining to temperature
Stenohaline–Euryhaline	Pertaining to salinity
Stenohydric–Euryhydric	Pertaining to water
Stenophagic–Euryphagic	Pertaining to food
Stenobathic–Eurybathic	Pertaining to depth
Stenoecious–Euryecious	Pertaining to niche or habitat selection

In case of temperature a species with a wide tolerance would be described as **eurythermal** and one with narrow tolerance as **stenothermal** (Fig. 5.2). **Euryhaline** would describe a species with wide tolerance to salinity and **stenohaline** a species with narrow or limited salinity tolerance. A given species might well be eurythermal but stenohaline; similarly a given species has different tolerances to other environmental factors. For various physiological processes to be optimally functional, at least suboptimal amount of all pertinent factors must be present.

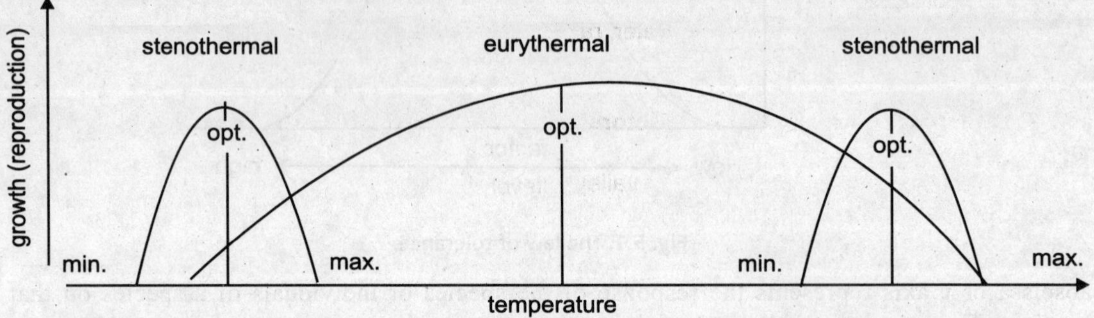

Fig. 5.2. The ranges of temperature tolerance in stenothermal and eurythermal animals.

The conclusion that organisms with a wide range of tolerance to many factors are likely to be very widely distributed should not be surprising. What may not be immediately self-evident is that when one factor becomes less than optimum the limit of tolerance for other factors may become limiting. This is called **combined concept of limiting factors (Odum).**

Age, season, and physiological or behavioural state affect tolerance. The stage of life that is usually least tolerant to all factors is the egg or very young. For example, salmon eggs hatch and the young begin development in freshwater; they can be described as being *stenohaline.* Adult salmon by contrast, are *euryhaline,* spending the majority of the adult period in salt water. Also in 83.5 per cent of 119 studies, increased temperature during rearing of a wide variety of animals, plants, protists and a bacterium led to reduced size at a given stage of developments (**Atkinson** 1994).

5.2 TYPES OF ABIOTIC ENVIRONMENTAL FACTORS

The distribution, abundance, growth and reproduction of the organisms comprising the individual members of populations are controlled by certain **environmental** or **ecological factors**. An environmental factor is any external force, substance or condition which surrounds and affects the life of an organism in any way. The abiotic components are characterized by physical and chemical factors such as light, temperature, rainfall, pressure, pH, the content of oxygen and other gases, and so on. These factors exhibit diurnal, nocturnal, seasonal, and annual changes. Abiotic environmental factors are customarily classified as follows:

Physical Factors of Environment

1. Temperature
2. Light
3. Water
4. Humidity (Moisture)
5. Current and pressure
6. General weather conditions
7. Soil (edaphic factor)
8. Microclimate
9. Fire

Box 5.2

According to an alternative classification, abiotic environmental factors have been classified as follows:

1. **Climatic factors**
   (i) Light;
   (ii) Temperature;
   (iii) Water (including atmospheric water, rainfall or precipitation, soil moisture, etc.);
   (iv) Atmosphere (gases and wind);
   (v) Fire.
2. **Topographic** or **physiographic factors**
   (i) Altitude;
   (ii) Direction of mountain chains and valleys;
   (iii) Steepness and exposure of slopes.
3. **Edaphic factors** (soil formation, physical and chemical properties of soil, nutrients).

5.3 TEMPERATURE

Temperature is one of the essential and changeable environmental factor. It penetrates into every region of the biosphere and greatly influences all forms of life by exerting its action through increasing or decreasing some of the vital activities of organism, such as behaviour, metabolism, reproduction, ontogenetic development (*viz.,* embryogenesis and blastogenesis) and death. Temperature is a universal influence and is frequently a limiting factor for the growth or distribution of animals and plants. Normal life activities go on smoothly at a specific temperature or at a specific range of temperatures. This is called the **optimum temperature** or the optimum range of temperature. Organisms react to any rise or fall of the optimum temperature range and biotic communities more often encounter alterations only due to extremes of temperature (*viz.,* **minimum** or **maximum** temperatures). The interaction of temperature with certain other abiotic environmental factors such as humidity, etc., cause into many other climatic changes which influence the living organisms in one way or another.

Nature of Temperature

Temperature is a measure of the intensity of **heat** in terms of a standardized unit, and is commonly expressed as degrees on either the **Fahrenheit** or **Celsius scale**. Heat is a form of energy, and as such is necessary for the very existence of life. Heat may be received or lost by animals as molecular vibrations transmitted from one part of an environment to another by radiation, convection, or conduction. It may be received as radiant energy, of course largely from the sun. Heat may be produced by mechanical work as in Hawk moth, the heat is generated by vibrating the wings before take-off; or by chemical reactions.

Heat Budget

The temperature at the earth's surface is governed by the brightness of the sun; the constancy of brightness of sun has remained virtually unchanged for about 3 billion years. In fact, average terrestrial temperatures now probably do not differ radically from those at the earth's beginning (**Schwarzschild**, 1967). The total amount of heat entering the biosphere from the sun must balance the amount lost per unit time if temperatures are to remain unchanged, since the flow of geothermal heat from the interior of the earth is small by comparison and probably has been negligible for at least the past 500 million years. The estimate of this energy flow is referred to as the **heat budget** (**Vernberg** and **Vernberg,** 1970). Heat budget has been projected for the total surface of the earth as well as for special environments.

Temperature Fluctuations in Different Environments

Environmental temperatures fluctuate both daily and seasonally. Different environments such as freshwater, marine and terrestrial environments, are subjected to varied responses to fluctuating temperature. Temperature fluctuations are comparatively less in the aquatic environment than in the terrestrial environment. The increase in depth of aquatic medium often increases the temperature fluctuations. There exists a distinct difference in the response of living organisms of the freshwater and sea to temperature fluctuations because of the presence of dissolved salts in sea water. The minimum temperature in the sea is –3^{0}C, while in freshwater pond it never goes below 0^{0}C. The maximum temperature of ocean generally goes up to 36^{0}C, but in shallow pools of freshwater and tide of pools littoral zone, temperature may go higher. In deeper bodies of water, heating and cooling are restricted to the surface strata. But the deeper layers also get a lot of heat as a result of what is usually termed **vertical circulation**, wherein, due to circulatory movement of water, surface waters are brought to the deeper regions, and vice versa. Studies on the vertical changes of temperature have led to the hypothetical classification of the freshwater media into three strata. The superficial layer of freshwater is constantly stirred by wind and is called **epilimnion**. It is the layer of warmer water and its temperature may rise up to 22^{0}C during summers. The stagnant water of the bottom constitutes the **hypolimnion**. The hypolimnion has temperature of 5^{0}C to 9^{0}C. In between epilimnion and hypolimnion occurs an intermediate zone called **thermocline** or **metalimnion**, which has rapid vertical temperature changes (Fig. 5.3 B). During winter, the temperature of epilimnion of a freshwater lake becomes 0^{0}C and the lake becomes ice covered. The process of differentiation of freshwater habitat into these three strata is called **thermal stratification**.

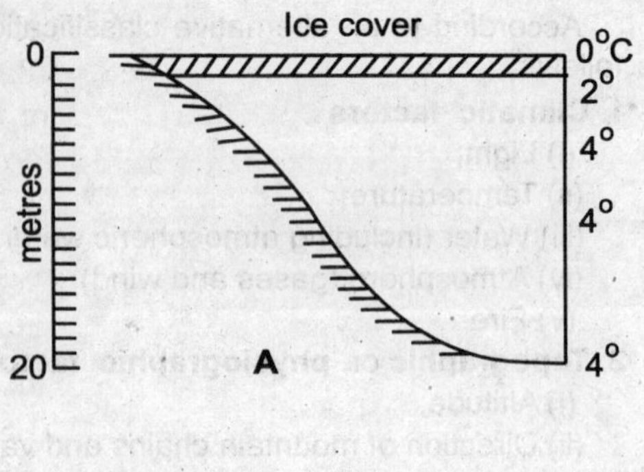

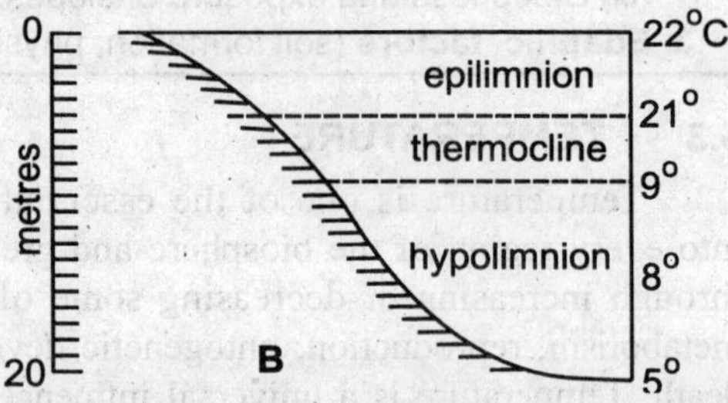

Fig. 5.3. Thermal stratification of a deep lake of the temperate region, A–during winter, B–during summer.

In the terrestrial environment the seasonal and daily fluctuations in temperature are varied and marked. The lowest temperature recorded for any land mass is –70^{0}C (Siberia in 1947). Higher temperatures may likewise go often 85^{0}C as in certain deserts at noon. However, the water in hot springs and geysers may approach 100^{0}C, and even higher temperatures occur occasionally in the very special situation presented by volcanic areas. In Rajasthan the highest temperature exceeds 50^{0}C. On land, diurnal variations of temperature are quite staggering. The fluctuation between day and night temperatures may be 17^{0}C as in ordinary land masses or 40^{0}C as in deserts. The Thar Desert of South Rajasthan (India) shows a diurnal range of 20–30^{0}C for all seasons. Difference in latitudes often causes variation in the annual temperature cycles with visible effects on the organisms. Further, with the increased altitude for every 150 meters, the decrease in temperature of 1^{0}C takes place.

Range of Temperature Tolerance

Life in this universe exists within a range of—200 to 100^{0}C. Though normal life persists within narrow temperature limits of about –10 to + 50^{0}C. Individual species survive in a smaller range and are active within even narrower limits. Many nematodes, rotifers and tardigrades have shown to withstand cooling to –272^{0}C without ill effects. Larvae of chironomids and certain other

Diptera have been found to thrive at temperatures near 55^{0}C. Praying mantis is reported to live in bare grounds at a temperature of 62^{0}C in deserts. Some algae and bacteria are reported to live in hot springs at 88^{0}C. Non-photosynthetic bacteria inhabiting hot springs can actively grow at temperatures greater than 90^{0}C (**Bott** and **Brock**, 1969). On the other hand, some arctic algae can complete their life cycles in places where the temperatures barely rise above 0^{0}C. Further, the eggs of the acanthocephalan *Macracanthorhynchus hirudinaceus* have been known to withstand temperatures from –10 to –45^{0}C for about 140 days, and desiccation at temperatures up to 39^{0}C for a period of 265 days.

The organisms (microbes, plants and animals) which can tolerate very large fluctuations in temperature are called **eurythermal organisms**. For example, cyclops, toad, wall lizard, grass snake, humans, etc., are the eurythermal animals. The organisms which can tolerate only a small variation in temperature are termed **stenothermal organisms**. The common stenothermal animals are fishes, snails, coral reefs, etc. In organisms all metabolic processes necessary for life start at a certain **minimum temperature** and increase with rise in temperature until they reach the maximum level at a temperature called **optimum temperature**. Further rise in temperature beyond optimum brings about decrease in metabolic rate, until it ceases at a temperature called **maximum temperature**.

Thus, the favourable temperature range for any particular species is determined by the prevailing temperature at which normal physiological activities of the animals take place, as in the distribution of the rotifer *Keratella procura* (**Nayar**, 1970). This rotifer species is known to appear in the ponds of Pilani, Rajashthan (India), when the temperature is below 24^{0}C and to disappear when it rises above 24^{0}C, the frequency of distribution reaching a peak during months of October to March, with the fall in temperature.

Poikilotherms and Homeotherms

An animal's response to wide ranges of temperature is influenced by its physiology. All invertebrates, lower chordates (hemichordates, urochordates and cephalochordates), cyclostomes, fishes, amphibians and reptiles have no internal mechanism for temperature regulation, and their body temperatures vary with the surrounding environmental temperatures. Such animals are called **cold blooded, ectothermic** or **poikilothermic animals**. The temperature range that a poikilothermic animal tolerates can be correlated closely with the environmental temperatures normally encountered in nature: tropical animals cannot withstand low temperatures; polar animals cannot withstand high temperature; temperate-zone animals survive a wide temperature range. Many poikilothermic animals show a rather precise discrimination of temperature. If the temperature decreases or increases appreciably, these animals may tolerate and adjust to change, may avoid the change by seeking a less exposed environment or may become inactive. Some animals respond by becoming dormant during periods of extreme temperatures. Such dormancy is called a **estivation** at high temperatures and **hibernation** at low temperatures. For example, most poikilotherms become inactive when the temperature of their surroundings goes below 8^{0}C or rises to 42^{0}C.

A few exceptional poikilotherms, especially insects, certain amphibians and reptiles, exercise a degree of thermoregulation by either physiological or behavioural mechanisms. For example, Hawk-moths can rise the temperature of their flight muscle to 32^0–36^{0}C by vibrating the wings before take-off and gregarious butterfly larvae may raise their temperature 1½–2^{0}C when clustered together. Locust and grasshoppers may increase their temperature 10^{0}C by basking sideways in the sun (Box 5.3). Ants move their larvae to warm or cool places within the nest and bees maintain temperatures within their hives between 13^0 and 25^{0}C by fanning with their wings to evaporate water droplets when it is too hot, or releasing body heat through increased metabolic activity, when too cold.

Box 5.3 Heliothermism

Semiterrestrial frogs such as bull frogs and green frogs, are able to exert great control over their body temperature. By basking in the sun (called **heliothermism**), frogs can raise their body temperature as much as 10°C above ambient temperature. Because of associated evaporative water losses, such amphibians must either be near water or partially submerged. Forms that live near water also do evaporative cooling through the skin to reduce body heat loads.

When temperature drops, lizards bask in the sun to achieve the desired body temperature; once this thermal level is attained, they will divide their time between sun and shade to maintain it. Poikilotherms such as desert dwellers that live in environments where the temperature is apt to be very high are often nocturnal and, thus, avoid the highest temperatures of the day. Some poikilotherms, both vertebrates and invertebrates, lower their body temperatures slightly by evaporative cooling. In frogs and reptiles, evaporative cooling can occur through the skin or via the respiratory tract by **panting**. Among the invertebrates, evaporative cooling has been reported in tropical intertidal-zone animals.

In contrast to poikilotherms, birds and mammals can, within limits, maintain constant body temperatures, regardless of temperature variations of air and water. Such animals are termed **warm blooded**, **homeothermic** or **endothermic animals**. The life processes are adjusted to function at the animals's normal temperature, averging a little less than 38^{0}C in mammals and 3 to 4^0 higher in birds. If its temperature control fails, the animal dies. The homeothermic animals are able to maintain the constancy of body temperature by a combination of several factors: (1) a thermoregulating centre in the brain (hypothalamus); (2) insulation; (3) a peripheral vascular response to ambient temperature; and (4) metabolic compensation.

Among these factors, **insulation** is highly important particularly in enabling large arctic animals to withstand very low temperature. If the temperature is lowered, the oxygen consumption rate of tropical mammals increases, whereas that of arctic mammals remains basal. This response pattern is due to differences in insulation; some arctic animals are fur-insulated, whereas others are insulated with a layer of blubber. Many small animals, however, are not well insulated and must seek heat retention by other means, often by huddling together.

Homeothermic animals living in very hot environments cannot tolerate greatly elevated body temperatures and they utilize methods that facilitate heat transfer to the environment, including an increased peripheral blood flow and surface cooling by sweating or panting. Birds may accomplish this by rearranging their plumage so that more skin is exposed. There is however, no reduction in metabolic rate. Animals subjected to high temperature may exhibit diurnal patterns of behaviour—that is, they may reduce locomotor activity during the heat of the day or may move into the shade to avoid direct sunlight. Some animals such as monotremes and marsupials have a limited power of temperature regulation, they are called **heterothermic animals**; *e.g.*, the pigmy mouse and the little pocket mouse, which respond to temperature extremes by estivating or hibernating, and others such as the humming bird, experience a nocturnal drop in temperature.

Plants too can be divided into the following three categories on the basis of their heat tolerating capacity: **megatherms**, **microtherms** and **mesotherms**. Megatherm plants occur in warm habitat (*e.g.* desert vegetation). Microtherm plants occur in cold habitat, (*e.g.*, plants of high altitudes). Mesotherm plants are the plants of the habitat which is neither very hot nor very cold (*e.g.*, aquatic plants). **Hekistotherms** include alpine vegetation which tolerate very low temperatures.

Effect of Temperature on Plants and Animals

Temperature has been found to affect the living organisms in various ways. Some of well-studied effects of temperature on living organisms are the following:

1. Temperature and cell. The minimum and maximum temperatures have lethal effects on the cells and their components. If too cold, cell proteins may be destroyed as ice forms, or as water is lost and electrolytes become concentrated in the cells; heat coagulates proteins (**Lewis** and **Taylor**, 1967).

2. Temperature and metabolism. Most of metabolic activities of microbes, plants and animals are regulated by varied kinds of enzymes and enzymes in turn are influenced by temperature, consequently increase in temperature, up to a certain limit, brings about increased enzymatic activity, resulting in an increased rate of metabolism. For instance, the activity of liver arginase enzyme upon arginine amino acid, is found to increase gradually, with the simultaneous increase in the temperature from 17⁰C to 48⁰C. But an increase in temperature beyond 48⁰C is found to have an adverse effect on the metabolic rate of this enzymatic activity which retards rapidly.

In plants, the absorption rate is retarded at low temperature. Photosynthesis operates over a wide range of temperature. Most algae require lower temperature range for photosynthesis than the higher plants. The rate of respiration in plants, however, increases with the rise of temperature, but beyond the optimum limit high temperature decreases the respiration rate. The rate of respiration is doubled (like in animals) at the increase of 10⁰C above the optimum temperature, provided other factors are favourable (**Vant Hoff's law**). However, optimum temperature for photosynthesis is lower than that for respiration.

3. Temperature and reproduction. The maturation of gonads, gametogenesis and liberation of gametes takes place at a specific temperature which varies from species to species. For example, some species breed uniformly throughout the year, some only in summer or in winter, while some species have two breeding periods, one in spring and other in fall. Thus, temperature determines the breeding seasons of most organisms.

Temperature also affects fecundity of animals. **Fecundity** of an animal is defined as its reproductive capacity, *i.e.,* the total number of young ones given birth during the life time of the animal. For example, females of the insect, acridid *Chrotogonus trachypterus* became sexually mature at 30⁰C and 35⁰C than at 25⁰C, and the highest number of eggs per female was laid at temperature of 30⁰C. The number of eggs decreased from 243 to 190 when the temperature was raised to 30–35⁰C (**Grewal** and **Atwal**, 1968). Likewise, in grasshopper species—*Melanoplus sanguinipes* and *Camnula pellucida* when reared at 32⁰C produce 20-30 times as many eggs than those reared at 22⁰C. On the other hand, the fecundity of certain insects such as cotton stem weevil (*Pempherulus affinis*) was found to decline with an increase in temperature beyond 32.8⁰C (**Ayyar** and **Margabandhu**, 1941).

4. Temperature and sex ratio. In certain animals the environmental temperature determines the sex ratio of the species. For example, the sex ratio of the copepod *Macrocyclops albidu* is found to be temperature dependent. As the temperature rises there is a significant increase in number of males. Similarly in plague flea, *Xenopsylla cheopis*, males outnumber females on rats, on days when the mean temperature remains in between 21–25⁰C. But the position becomes reverse on more cooler days.

5. Temperature and ontogenetic development. Temperature influences the speed and success of development of poikilothermic animals. In general, complete development of eggs and larvae is more rapid in warm temperatures. Trout eggs, for example, develop four times faster at 15⁰C than at 5⁰C. The insect, chironomid fly *Metriocnemus hirticollis*, requires 26 days at 20⁰C for the development of a full generation, 94 days at 10⁰C, 153 days at 6.5⁰C, and 243 days at 2⁰C (**Andrewartha** and **Birch**, 1954). However, the seeds of many plants will not germinate and the eggs and pupae of some insects will not hatch or develop normally until chilled. Brook trout grows best at 13°C to 16°C, but the eggs develop best at 8°C.

6. Temperature and growth. The growth rate of different animals and plants is also influenced by temperature. For example, the adult trouts do not feed much and do not grow until the water is warmer than 10°C. Likewise, in the oyster *Ostraea virginica*, the length of the body increases from 1.4 mm to 10.3 mm when temperature is increased from 10°C to 20°C. Sea urchin *Echinus esculentus* shows maximum size in warmer waters. Corals flourish well in those waters which contain water below 21°C.

7. Temperature and colouration. The size and colouration of animals are subject to influence by temperature. In warm humid climates many animals such as insects, birds and mammals bear darker pigmentation than the races of some species found in cool and dry climates. This phenomenon is known as **Gloger rule**. In the frog *Hyla* and the horned toad *Phrynosoma*, low temperatures have been known to induce darkening. Some prawn (crustacean invertebrates) turn light coloured with increasing temperature. The walking stick *Carausius* has been known to became black at 15°C and brown at 25°C. Another good example is provided by Australian grasshopper which is dark in colour at low temperature and absorbs solar radiation, thus, heating up rapidly (called **basking heliotherms**). As the temperature rises above a set point, further absorption is reduced by the cuticle lightening in colour. This colour change is believed to be a direct response **shown in pigment cells** to body temperature.

8. Temperature and morphology. Temperature also affects the absolute size of an animal and the relative proportion of various body parts (**Bergman's rule**). The two parameters, *i.e.,* surface area (for heat loss) and volume (of metabolically active tissues), are inversely proportional to each other. For this reason animals living in cold regions tend to be large, for example, polar bears and whales, whilst animals living in hot climates are generally smaller, for instance insectivorous mammals. Tiger too decreases in its body size with distance from the poles.

There are certain exceptions to Bergman's rule, but the organisms concerned have adaptations favouring survival in these regions. For example, small mammals in temperate and arctic regions have a large appetite enabling them to maintain a high metabolic rate. They have small extremities to reduce heat loss and are forced to hibernate in winter. Large mammals living in hot regions, such as elephant and hippopotamus, have the opposite problems. The elephant has extremely large ears which are well supplied with blood and flapping of these ears encourage heat loss by radiation and convection. The hippopotamus lacks sweat glands and adopts a similar behavioural response to temperature as crocodile in that it moves between land and water in an attempt to minimise the effects on its body of changes in the temperature.

Further, the extremities of organism such as tail, ears and legs of mammals often appear to be shorter in colder climate (**Allen's rule**). Mice reared at 31° to 33.5°C have longer tails than those of the same strain reared at 15.5° to 20°C. Another example of Allen's rule can be observed in closely related species of fox (Fig. 5.4). Likewise the races of birds with relatively narrow and more acuminate (*i.e.,* tapered to a slender point) wings tend to occur in colder regions, while those in warmer climates tend to be broader (**Rensch's rule**). Temperature also influences the morphology of certain fishes and is found to have some relation with the number of vertebrae (**Jordon's rule**). Cod which hatches off New Foundland at a temperature between 4° and 8°C has 58 vertebrae, while that hatches East of Nantucket at a temperature between 10° and 11°C has 54 vertebrae.

9. Temperature and cyclomorphosis. The relation between seasonal changes of temperature and body form is manifested in a remarkable phenomenon termed **cyclomorphosis** exhibited by certain cladocerans such as *Daphnia* during the warm months of summer. These crustaceans show a striking variation in the size of their helmet or head projection between the winter and summer month. The helmet develops on the *Daphnia* head in spring, it attains its maximum size in summer and disappears altogether in winter to provide usual round shape to the head.

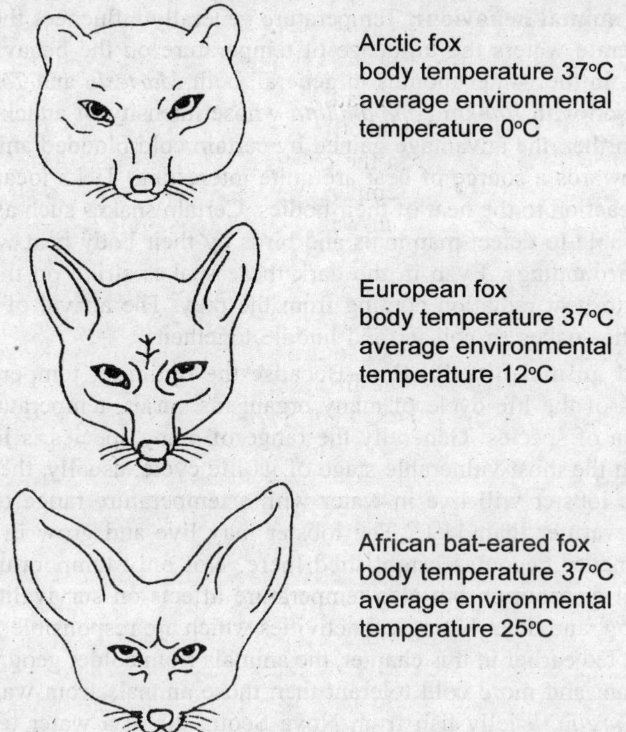

Fig. 5.4. Variation in ear length shown by three species of fox which each occupies a different geographical region. This is an example of Allen's rule.

Such a kind of cyclomorphosis in the terms of size of the helmet is clearly showing a correlation to the degree of warmth of different seasons. These prolongations of the helmet have been interpreted as an adaptation aiding floatation since the buoyancy of water becomes reduced as the temperature increases (the **buoyancy hypothesis**). According to another interpretation (*viz.*, **stability hypothesis**), the helmet acts like the rudder and gives greater stability to the animal. Besides temperature such structural polymorphism can be caused by other environmental factors including the food.

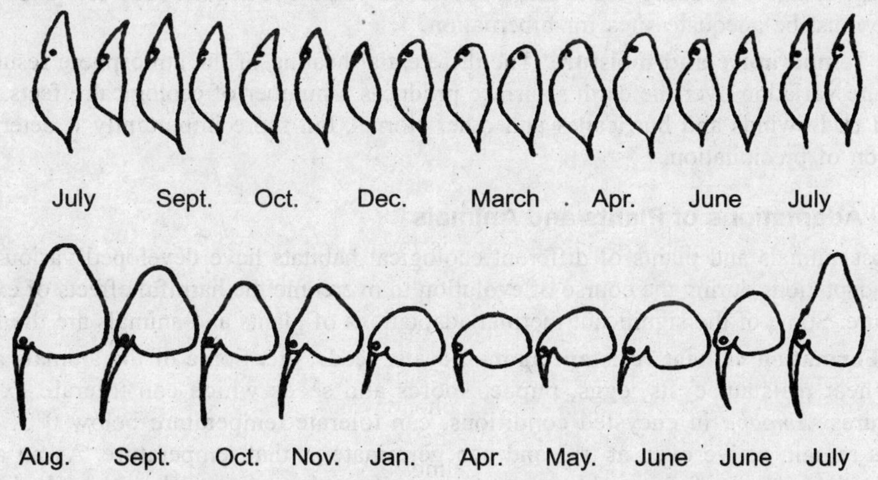

Fig. 5.5. Cyclomorphosis in *Daphnia cucullata* due to seasonal change in temperature.

10. Temperature and animal behaviour. Temperature generally influences the behavioural pattern of animals. In temperate waters the influence of temperature on the behaviour of wood borers is great. For example, in the winter months in general, both *Martesia* and *Teredo* occur in smaller numbers in comparison with *Bankia campanulata* whose intensity of attack is maximum during the winter months. Further, the advantage gained by certain cold blooded animals through thermotaxis or orientation towards a source of heat are quite interesting. Ticks locate their warm blooded hosts by a turning reaction to the heat of their bodies. Certain snakes such as rattle snake, copper heads, pit vipers are able to detect mammals and birds by their body heat which remains slightly warmer than the surroundings. Even in the dark these snakes strike on their prey with a remarkable accuracy, due to heat radiation coming from the prey. The arrival of cold weather in temperate zones causes the snakes to coil up and huddle together.

11. Temperature and animal distribution. Because the optimum temperature for the completion of several stages of the life cycle of many organisms varies, temperature imposes a restriction on the distribution of species. Generally the range of many species is limited by the lowest critical temperature in the most vulnerable stage of its life cycle, usually, the reproductive stage. Although the Atlantic lobster will live in water with a temperature range of 0^0 to 17^0C, it will breed only in water warmer than 11^0C. The lobster may live and grow in colder water but a breeding population never becomes established there. Not only temperature affects on breeding in the geographical distribution but also temperature affects on survivality (*i.e.,* lethal effect of temperature), feeding, and other biological activities which are responsible in geographic distribution of animals. As noted earlier in this chapter, the animals from colder geographic regions are generally less heat tolerant and more cold tolerant than those animals from warmer regions; for example, members of *Aurelia,* a jelly fish from Nova Scotia die at a water temperature of 29^0–30^0C, while *Aurelia* from Florida can tolerate temperatures up to 38.5^0C. Thus, lethal limit of temperature may regulate the range of distribution of *Aurelia.*

Terrestrial invertebrates, particularly arthropods generally are distributed in all thermal environments where life is found. Many arthropods that have invaded the colder areas have one stage in their life cycle which is very resistant to cold, enabling them to over-winter until warmer weather returns. Birds and mammals are also adapted to live in nearly all thermal environments. The distribution of amphibians and reptiles, however, is limited to the relatively warmer thermal climates. **Hock** (1964) has listed three factors that limit the invasion of reptiles into cold environments: the daily ambient (= atmospheric) temperature must be high enough and long enough to allow breeding and to allow adults and youngs to acquire food for "overwintering" and there must be adequate sites for hibernation.

12. Temperature and moisture. The differential heating of the atmosphere resulting from temperature variation over the earth's surface produces a number of ecological effects, including local and trade winds and hurricanes and other storms, but more importantly it determines the distribution of precipitation.

Thermal Adaptations of Plants and Animals

Most animals and plants of different ecological habitats have developed various kinds of thermal adaptations during the course of evolution to overcome the harmful effects of extremes of temperature. Some of the significant thermal adaptations of plants and animals are the following:

1. Formation of heat resistant spores, cysts, seeds, etc. Some of the animals and plants produce heat resistant cysts, eggs, pupae, spores and seeds which can tolerate extremes of temperatures. *Amoeba* in encysted conditions, can tolerate temperature below 0^0C. Similarly, *rye* seeds remain active even at 0^0C and can germinate at that temperature. As an adaptation against frost the starch of plants changes to fats or oils in the autumn. The fatty oils diminish the freezing points and, thus, increase the power of resistance in plants against frost. Many leaves, that grow in the coldest lands, store fats. Pentosans mucilage and pectic substances which have high moisture retaining power are abundant in many plants. They decrease the danger of plants from desiccation during extremes of heat and save them from death.

2. Removal of water from tissue. Dried seeds, spores and cysts avoid freezing because there remains no liquid in them that can freeze. Due to removal of water from seeds, the cold resistance of seeds of certain plants increases up to the extent that their exposure for 3 weeks to 190^{0}C, does not diminish their germinating capacity.

3. Dormancy. Dormancy includes two already discussed phenomena namely **hibernation** and **aestivation**. During both kinds of dormancies metabolic rate becomes reduced, body temperature becomes low and heart beat rate is also reduced.

4. Thermal migration. Thermal migrations occur only in animals. The journeys taken by animals that enable them to escape from extremely hot or cold situations are referred to as **thermal migrations**. For example, desert animals move to shaded places to avoid burning heat of noon and some animals such as desert reptiles and snakes become nocturnal to avoid heat of the day. The frogs, toads, other amphibians, turtles, etc., make short trips into or out of water (or moist places) and this provides desired cooling and warming to the animal.

5.4 LIGHT AND RADIATIONS

The radiant energy from the sun is the basic requirement for the existence of life on the earth. This source of energy is of fundamental importance to the photosynthetic production of food by plants and as mentioned previously, the heat budget of the world is dependent on solar radiation. Although we generally, think only in terms of visible light, the sun emits other radiations of different wavelengths — cosmic rays, gamma rays, X-rays, ultraviolet rays, infra-red rays, heat waves, spark discharges, radar waves, radio waves, slow electro-magnetic waves. Biologists have been primarily interested in only three regions near the centre of the electromagnetic spectrum— (1) the infra-red, (2) the visible light and (3) ultraviolet regions.

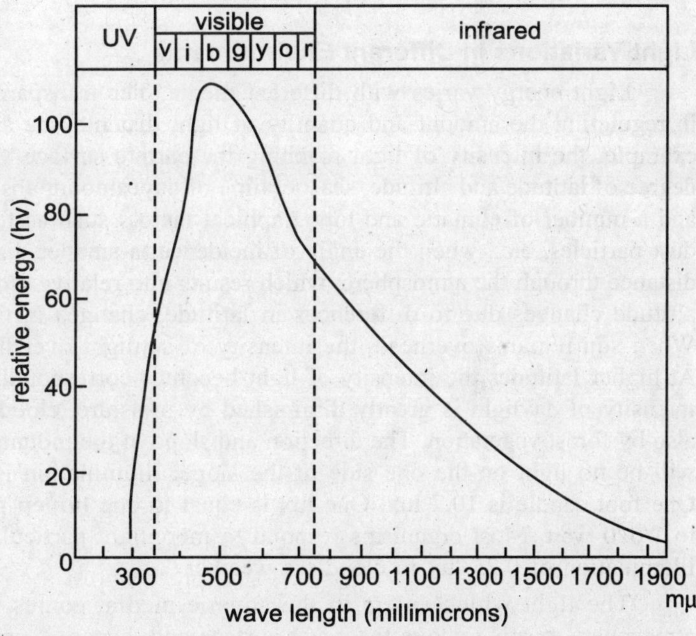

Fig. 5.6. Spectral distribution of ultraviolet light, visible light and infra-red radiation at the earths surface.

The **infra-red wavelengths**, which are the longest of these three, are not visible to the human eye; they contribute to the warmth of the earth at the high altitudes in the terrestrial atmosphere. **Visible light**, is only a small fraction of the radiation spectrum and contains the frequency of wavelengths ranges from 390 to 700 millimicrons (mμ). It is made up of a series of colours ranging from violet through indigo, blue, green, yellow, orange and red, all constituting the **visible spectrum**. Light energy, thus, reaches the earth as electro-magnetic waves of solar radiation with tremendous velocity and supplies most of the warmth the earth receives from the sun and also supplies the main source of energy which is utilized in photosynthesis of plants, oriented and rhythmic behaviour of animals, bioluminescence, periodicities of occurrence and periods of inactivity. Unlike temperature, light is a non-lethal ecological factor and it has a specific direction in its flow. The wavelength of **ultraviolet light** from the sun is shorter than that of visible light, and it produces the upper levels of the earth known as the ionosphere.

Light Receptors of Animals

Since animals depend on light for orientation, diurnal migrations and synchrony of rhythmic activities, light reception probably is the most important sensory modality in the exploration of the environment. It is not surprising, therefore, that receptors for light are common to almost all animals: in those with better developed photoreceptors, light greatly influences behaviour. Light receptors may be well-defined organs such as the vertebrate eyes, the compound eyes of Crustacea and insects, the simple eyes or ocelli of other arthropods and invertebrates, and the dermal light receptors.

Many homeotherm animals with normal vision do not commonly respond to ultraviolet or infra-red rays, since their retinas are not stimulated. They also do not properly distinguish between the different colours of the visible spectrum. Dogs, cats, hamsters, and oppossums are colour blind, while horned cattle cannot identify red light. Horses, deer, sheeps, pigs and squirrels cannot distinguish close to red and green. Primates can distinguish colours. Some insects such as honey bee can see ultraviolet radiation.

Light Variations in Different Environments

Light energy varies with different media. The transparency of air and water is important in regulating the amount and quantity of light that may be available in particular habitats. For example, the intensity of light reaching the earth's surface varies with the angle of incidence, degree of latitude and altitude, season, time of day, amount absorbed and dispersed by atmosphere and a number of climatic and topographical factors such as fog, clouds, suspended water drops, dust particles, etc., when the angle of incidence is smaller, light rays have to travel by a longer distance through the atmosphere, which results into relative reduction in intensity. Likewise, sun's altitude changes due to differences in latitude, changes in the season and in the time of day. When sun remains overhead, the intensity of sunlight over the earth's surface will be greatest. At higher latitudes the intensity of light becomes correspondingly reduced. The **illumination** or intensity of daylight is greatly diminished by moisture, clouds, and dust in the atmosphere and also by forest vegetation. The direction and slope of the mountain also affect light intensity. There will be no light on the one side of the slope. Illumination is measured in **lux** or **foot candle**. One foot candle is 10.7 lux. One lux is equal to one **lumen** per square meter. A lumen is equal to 1/620 watt. Most organisms respond to moonlight particularly on a full moon, which has an illumination of 0.25 lux in cloudless weather.

The light which enters in the aquatic media, comes from sun by passing through the atmosphere existing above the water surface and, hence, is subjected to all kinds of atmospheric factors like that of terrestrial media. About 10 per cent of the sunlight which falls over the water surface, is reflected back and rest 90 per cent of that pass downward in the water and is modified in respect to intensity, spectral composition, angular distribution (refraction) and time distribution. The phytoplankton, zooplankton, suspended organic and inorganic particles either reflect or absorb the light rays. Further, in water there is a selective absorption of light at various depths. The longer light rays are absorbed near the surface and in general the shortest light rays penetrate deepest. Thus, long infra-red rays are absorbed in the upper layers of water (about 4 metres); red and orange rays are completely absorbed up to the depth of 20 metres; yellow rays penetrate up to 50 metres and green and blue rays penetrate up to 80 to 100 metres deep. Violet and ultraviolet rays penetrate beyond 100 metres and no light ray penetrates beyond 200 metres depth. Depending upon the penetration of light, oceans are divided into **euphotic zone** (up to 50 metres depth), **disphotic zone** (up to 80 to 200 meters depth) and **aphotic zone** (below 200 meters of depth). In the ocean, algae are distributed according to length of light rays that their colours are best suited to absorb and to utilize. Photosynthesis in deeper waters occurs with blue and green rays, which are absorbed by the brown and red pigments of red algae (*Phyllophora*). The red and blue-green algae use pigments of photocyanin and photoerythrin for photosynthesis. Each plant performs optimum photosynthesis in complementary colours, *e.g.,* green pigment for red rays, brown pigment for green rays, and red pigment for blue rays. Further, since the photosynthesis is a function of the illumination, a variation in the amplitude of which depends on the earth's rotation, altitude, season, presence or absence of clouds and dust and so on.

Effect of Light on the Plants

Light energy influences almost all the aspects of plant life directly or indirectly. Thus, it controls plant's structure, form, shape, physiology, growth, reproduction, development, local distribution, etc. On the basis of light factor certain ecologists have classified plants into **sciophytes, shade loving**, or **photophobic plants** which have best growth under lower intensities of light, and **heliophytes** or **photophilous plants** which have best growth in full sunlight. The light factor affects following aspects of plant life :

A. Direct effects of light on plants. Light affects directly the following physiological processes of the plants— 1. It is an essential factor in the formation of **chlorophyll** pigment in chlorophyllous plants. 2. It has a very strong influence on the number and position of **chloroplasts**. The upper part of the leaf which receives full sunshine has larger number of chloroplasts which are arranged in line with the direction of light. In leaves of plants which grow under shade, chloroplasts are very few in number and are arranged at right angle to the light rays, thus, increasing the surface of light absorption. 3. Light has its most significant role in **photosynthesis**. During photosynthesis, the green plants which are the "primary producers" of an ecosystem, synthesize their carbohydrate food from water and CO_2 in the presence of sunlight. Thus, during photosynthesis, the solar radiant energy is transformed into the chemical or molecular energy which remains stored in chemical bonds of carbohydrates and this chemical energy is utilized by other chlorophyllous and non-chlorophyllous parts of plants, all animals, fungi, bacteria and viruses in their different life activities. The rate of photosynthesis is greater in intermittent light than in the continuous light. At high intensity of light a photo-oxidation of chlorophylls and other enzymes takes place, which consequently reduces the rate of synthesis of carbohydrates and proteins. However, high intensity of light results in the formation of **anthocyanin** pigments. It is for this reason alpine plants have beautifully coloured flowers. 4. Light inhibits the production of **auxins** or **growth hormones** as a result of which it influences the shape and sizes of plants. Plants grown in insufficient light or in the total darkness, produce maximum amount of growth hormones, as a result of which they are elongated with weak pale yellow stems with very few branches. 5. Light influences certain chemical compounds of plants which affect the differentiation of specialized tissues and organs. 6. Leaf structure is influenced by the intensity of light. 7. The development of flowers, fruits and seeds is greatly affected by light intensity. Diffused light or reduced light promotes the development of vegetative structures and causes delicacy. For example, vegetative crops such as turnips, carrots, potato and beets give highest yield in regions with high percentage of cloudy days. Intense light favours the development of flowers, fruits and seeds. 8. Duration of light is also very important. Actual duration or length of the day (**photoperiod**) is a significant factor in the growth and flowering of a wide variety of plants. The controlling effect of **photoperiod** is called **photoperiodicity**. According to the response to length of photoperiods, the plants have been classified into following three groups:

(i) Short day plants. Which bloom when the light duration is less than 12 hours per day *e.g., Nicotiana tabacum* (tobacco), *Dahlia variabilis, Chrysanthemum indicum, Cosmos bipinnatus, Cannabis sativa* (hemp), etc.

(ii) Long day plants. Which bloom when the light duration is more than 12 hours per day, *e.g., Allium cepa* (onion), *Beta vulgaris* (beet root), *Daucus carota* (carrot), *Papaver somniferum* (opium poppy), *Vicia faba* (broad bean), *Brassica rapa* (turnip), *Avena sativa* (oat), *Secale cereale* (rye), *Sorghum vulgare* (sorghum), etc.

(iii) Day neutral plants. Which show little response to length of day light, *e.g., Cucumis sativus* (cucumber), *Gossypium hirsutum* (cotton), *Solanum tuberosum* (potato), etc.

9. Light also affects the movement in some plants. The effect of sunlight on the plant movement is called **heliotropism** or **phototropism**. The stems elongate towards light (positive phototropism) and the roots are negatively phototropic. The leaves grow transversely to the path of light. 10. The seeds when moist are very sensitive to light. In some cases the **germination of seeds** is delayed in light. 11. Light is an important factor in the distribution of plants. Some plants grow in full sunlight, while others prefer to grow in the shades.

B. Indirect effects of light on plants. Light affects opening and closing of stomata, influences the permeability of plasma membrane and has heating effect. All these in turn affect **transpiration** which in turn affects **absorption** of water. Light affects **respiration** of plants indirectly, as in the presence of light the respiratory substrates are synthesized. In many plants the respiratory rate increases with the increase in the light intensity (*e.g., Canna, Nerium, Bougainvillea*). However, in certain plants respiration rate is decreased slightly in intense light.

Effects of Light on Animals

Light affects divergent aspects of animal's life. It influences cellular metabolism, growth, pigmentation, locomotion, reproduction, ontogenetic development, and also controls the periodicity and biological clocks of animals. Some of its significant effects can be discussed as follows:

1. Effect of light on protoplasm. Though the bodies of most animals remain protected by some sort of body covering which save animal tissues from the lethal effects of solar radiations. But, sometimes sun rays penetrate such covers and cause excitation, activation, ionization and heating of protoplasm of different body cells. Ultraviolet rays are known to cause mutational changes in the DNA of various organisms.

2. Effect of light on metabolism. The metabolic rate of different animals is greatly influenced by light. The increased intensity of light results in an increase in enzyme activity, general metabolic rate and solubility of salts and minerals in the protoplasm. Solubility of gases, however, decreases at high light intensity. Cave dwelling animals are found to be sluggish in their habits and to contain slow rate of metabolism.

3. Effect of light on pigmentation. Light influences pigmentation in animals. Cave animals lack skin pigments. If they are kept out of darkness for a long time, they regain skin pigmentation. The darkly pigmented skin of human inhabitants of the tropics also indicate the effect of sunlight on skin pigmentation. The skin pigment's synthesis is dependent on the sunlight. Light also determines the characteristic patterns of pigments of different animals which serve the animals in sexual dimorphism and protective colouration. Animals that dwell in the depths of the ocean where the environment is monotonous, though are pigmented, but do not show patterns in their colouration.

4. Effect of light on animal movements. The influence of light on the movement of animals is evident in lower animals. Oriented locomotory movements towards and away from a source of light is called **phototaxis**. Positively phototactic animals such as *Euglena, Ranatra*, etc., move towards the source of light, while, negatively phototactic animals such as planarians, earthworms, slugs, copepodes, siphonophores, etc., move away from the source of light.

The light directed growth mechanisms are called **phototropisms** which occur in sessile animals. Phototropisms also include responsive movement of some body part of some active animals to the light stimulus, such as the movement of flagellum of *Euglena* towards light and movements of polyps of many coelenterates.

The velocity or speed of the movement of certain animals is also regulated by light. It has been observed that animals when responding to light reduce their velocity of movement and these movements which are non-directional are called **photokinesis**. Photokinesis may be a change in linear velocity (**rheokinesis**) or in the direction of turning (**klinokinesis**). During photokinesis when only a part of the body of an animal deviates away from the source of light, the reaction is termed **photoklinokinesis**. Larvae of *Musca domestica* show such movements. When animals are confronted with two lights of equal brightness they move towards or away to a position that is distanced between the two lights. This is termed **phototropotaxis**. Attraction of males towards the flash of the female is called **telotaxis**. Movement of animals at a constant angle towards the source of light is called **light compass reaction** or **celestial orientation**.

Celestial orientation. Some organisms, particularly arthropods, birds and fish, utilize their time sense as an aid to find their way from one area to another. To orient themselves, the animals use the sun, moon, or stars as a compass. To do this, they utilize both their biological clock and observations on the azimuthal position of the sun in relation to an established direction. The **azimuth** is the angle between a fixed line on the earth's surface and a projection of the sun's direction on the surface. Using the sun as a reference point involves some problems for animals because the sun moves. The target angle changes throughout the day. But animals which use the sun as a reference, correct their orientation somehow. Such celestial orientation has been observed in fishes, turtles, lizards, most birds, and such invertebrates as ants, bees, wolf spiders and sand hoppers.

5. Photoperiodism and biological clocks (Biorhythms). During evolution, organisms have acquired a variety of endogenous rhythms, their periods are matched with the rhythmic events in the environment. A **rhythm** is a recurring process which is wave like in character, because maximum and minimum states appear at identical intervals of time. The time taken between two maxima (peaks) or two minima (troughs) is called a **period** or **cycle** and consists of two phases, a rise and fall in the biological process. The **amplitude** is the range of fluctuations from an average value. The response of different organisms to environmental rhythms of light and darkness is termed **photoperiodism**. Each daily cycle inclusive of a period of illumination followed by a period of darkness is called the **photoperiod**. The term **photophase** and **scatophase** are sometimes used to denote the period of light and the period of darkness respectively.

Some rhythms of organisms are matched to **24 hours** cycle of light and dark (**circadian**), other to the **12.4** or **24.8 hours** tidal cycle (**circatidal**), the **29 days** lunar cycle (**circalunar** or **circasynodic**), the **yearly seasons** (**circanular**) or the time between successive spring low waters (**14.7 days**; **semilunar** or **circasyzygic**). Prefixing word *circa* (which means about) with most types of biological rhythms is very necessary because all these internal clocks are only approximately matched, *e.g.,* 24 hours cycle or rhythm does not mean 24 hours by wrist watch, it means about 24 hour or circadian (L. *circa* = about + *di*(*em*) = day). This term was coined by an American ethologist **Franz Halberg** in 1959.

Now it is believed that all animals (and also plants) possess an internal (endogenous) and automatic **clock** that controls the rhythm of behaviour and keeps it going with the help of internal stimuli. This clock is entrained (*i.e.,* set and reset) by external environmental stimuli, called **zeitgebers** or **entrainers** such as day length, lunar phases, tides, temperature, humidity, etc. The biological clocks show following general characteristics: 1. They keep almost the same timings at high and low temperatures, *i.e.,* they are unaffected by metabolic inhibitors which are known to block biochemical reactions. 3. They are controlled by the physiological processes of body. 4. They keep normal cycle even in the absence of environmental cues and are self sustained in nature. 5. They function through nerves (brain) and hormones.

6. Effect of light on reproduction. In many animals (*e.g.,* birds) light is necessary for the activation of gonads and in initiating annual breeding activities. The gonads of birds are found to become active with increased illumination during summer and to regress during shorter periods of illumination in winter.

7. Effect of light on development. Light in some cases (*e.g., Salmon* larvae) accelerates development, whereas, in other (*e.g, Mytilus* larvae) it retards it.

Further, occasionally the output of sunlight is increased by the development of **sunspots**. As a result of this excess energy is radiated to space and this naturally increases the output of solar energy near the earth. A direct consequence of this is the greater evaporation of water which results in cloud formation to prevent more exposure to sunshine and thus to equalize temperature and modify climate.

5.5 HUMIDITY OF AIR

Atmospheric moisture in the form of invisible vapour is known as **humidity**. The humidity of air is expressed in terms of values of **relative humidity** which is the amount of moisture in air as percentage of the amount which the air can hold at saturation at the existing temperature. Relative humidity is measured by the instrument called **psychrometer** or by paper strip **hygrometer** or a **thermo-hydrograph**. Humidity is greatly affected by intensity of solar radiation, temperature, altitude, wind exposure, cover and water status of soil. High temperature increases the capacity of the air to retain moisture and results in lower relative humidity. Low temperatures result in higher relative humidity by decreasing the capacity of air for moisture and reaches 100 per cent or saturation point. Further cooling results in condensation of vapour into water and this temperature/moisture point is called the **dew point**. Daily variations in relative humidity values depend upon the type of habitat conditions. In plains and deserts, it may show variations during day, whereas in oceanic islands there is little variation, being same throughout the year.

Humidity plays an important role in the life of plants and animals. It affects the life processes such as transpiration, absorption of water, etc. The rate of evapo-transpiration is modified by the factors as saturation deficit (*i.e.,* degree of wetness or dryness), temperature and wind velocity. Parasitic fungi becomes abundant in moist weather and reduces primary production. Some plants such as orchids, lichens, mosses, etc., make direct use of atmospheric moisture. In fungi, and other microbes, humidity plays an important role in germination of spores and subsequent stages in life cycle.

Precipitation (Rainfall)

The moisture falling on an area in liquid, vapours or frozen form is termed as **precipitation**. Thus, precipitation includes all moisture that comes to earth in the form of rain, snow, hail and dew. Precipitation is the chief source of soil water. The water available to plants and animals from soil comes as a result of rainfall. Due to **water cycle** or **hydrological cycle**, there occurs an interchange of water between the earth's surface and the atmosphere. In this cycle following, two important events are involved: **precipitation** and **evapo-transpiration**. The ecofactor of precipitation depends upon season, wind, air pressure and temperature.

Precipitation occurs as a result of the cooling and condensation of water vapour at high altitudes. The low temperature at high altitudes cools the air, which gets saturated and loses its water-holding capacity. As the temperature starts falling, the water vapour condenses and falls as rain due to gravity. Depending on the environmental conditions, precipitation falls as hail, snow or rain. In winter the ground temperature falls and as a result, atmospheric vapour gets condensed as **dew** or **frost** on the surfaces of objects, plants, animals, soils, etc. Dew becomes an important source of moisture to plants in the winter season. **Drizzle** involves minute drops appearing as to float in air. **Rain** is the drops of liquid water, which are larger than drizzle and also heavier. **Snow** is the moisture as solid state. **Sleet** is the form of small grains or pellets of ice, whereas **hail** consists of balls or lumps of ice.

Light drizzle is of little importance as very little moisture penetrates the soil because much of it evaporates rapidly. Snow is injurious to plants, breaking tender branches, flowers and fruits. Hail and sleet also cause similar damage. Some sedges grow in snow patches. Of all the above forms of precipitation, the rain is most important. It is the source of soil water and also affects humidity of atmosphere. The total precipitation of the world is about 4.46×10^{20} g per year and of this amount 0.99×10^{20} g fell on land and the rest on oceans (**Hutchinson**, 1957). In India, rains are caused by monsoons; **monsoon** is a special pattern of moist air movement prevalent in India. During monsoon, the air masses moving from the Arabian sea to the West coasts of India and from the Bay of Bengal to the eastern part of country, become extremely moist. About 45 per cent of the water available during annual precipitation flows into river, 20 per cent percolates in the ground and the remaining 35 per cent is lost by evaporation. Gentle steady rains are most effective because much of it penetrates the soil. Torrential rains on the land are most disastrous because they lead to flooding, soil erosion, destruction of vegetation and of animals.

The quantity, duration and intensity of rainfall profoundly regulate the vegetation of any place. For example, in tropical areas with heavy rainfall throughout the year, vegetation mainly include **evergreen forests**. In countries, with heavy rainfall during winter and low during summer, there are present **sclerophyllous forests**. The plants are shrubs, stunted in height, with leathery, thick, evergreen leaves. The areas with heavy rainfall during summer and low during winter are characterized by the presence of **grasslands**. The regions where rainfall is scanty, are seen with deserts and xerophytic vegetation.

The average annual rainfall of India is about 117 cm, the highest in the world. However, there exists great variations between the different regions in India with regard to rainfall. Thus, the average annual rainfall in Assam and the north-east is about 250 cm. In Cherrapunji in Meghalaya, the annual average of rainfall is about 1,100 cm, whereas in Jaisalmer in Rajasthan it is only 20 cm. The distribution of rainfall over India depends largely on the position of hills and mountains and the forest cover. Thus, monsoon winds strike the Western Ghats from a southwest direction and shed most of their water on the windward side. Bombay being on the windward side, receives 200 cm of rainfall annually during the monsoon, while, Pune being on the leeward side (*i.e.,* side of anything away from wind) receives only about 75 cm. If the rainfall to evaporation ratio is zero or less, deserts develop. Grasslands develop when this ratio is more than 0.2 and less than 1 and forests develop when the ratio is more than 1 (around 1.6 to 2). Often in forests, the floor receives rain from trees in the form of drips from tree leaves due to accumulation of condensed water on winter mornings.

5.6 CURRENTS AND PRESSURE

Sea water is never static. Waves, currents and tides are the regular features of sea water. Most world-wide water currents occur at the surface and at great depths of sea water. These currents (both horizontal and vertical) determine the interchange between the surficial and deep water masses as well as horizontal movements.

Currents in water not only influence the concentration of dissolved gases and nutrients, but act directly as limiting factors. Thus, in freshwater ecosystem, the differences between a lotic and a lentic community are mainly due to great difference in the current factor. Currents have direct influence on the distribution of many species by transporting them from place to place, especially the small holoplanktonic forms. Currents probably also provide a means of navigation.

The strong moving current of air is called **wind**. It is an important ecological factor of the atmosphere affecting variously the plant life on flat plains, along sea coasts and at high altitudes in mountains. Wind is directly involved in transpiration, in causing several types of mechanical damage and in dissemination of pollen, seeds and fruits. Wind also modifies the water relations and light conditions of a particular area. The movement or velocity of wind is affected by such factors such as temperature, atmospheric pressure, geographical features (including topography), vegetation masses and position with respect to sea shores. Air moves from a region of high pressure to low pressure. The pressure differences are mainly due to differential heating of atmosphere. The equatorial regions receive more heat than north or south regions, thus, low pressure occurs at lower latitude. The air generally moves from poles towards equator. Winds result in various physical, anatomical and physiological effects on plants: 1. **Physical effects** such as breakage and uprooting; deformation; lodging or flattening of herbaceous plants such as wheat, maize, sugarcane, etc.; abrasion of buds of plants (by wind-carried soil particles), erosion and deposition of soil around plant roots; plant injury due to salt spray along sea coasts. 2. **Anatomical** and **physiological effects** such as formation of dense, reddish xylem called **compression wood** on the compressed side of wind deformation, desiccation due to increased rates of evaporation and transpiration which are caused by strong winds and **dwarfing** of trees on sea coasts, arctic or alpine timberline due to prolonged dehydration and consequent loss of turgidity under the influence of drying winds.

In the ocean, **hydrostatic pressure** is of importance because of tremendous gradient from surface to the bottom. In the deepest part of the sea the pressure reaches 1000 atmospheres. Many animals can tolerate a wide range of pressure (**eurybathic**) when their bodies do not contain free air or gas. If these are present, **gas embolism** (embolism means stopping up of a vein or artery, as by a blood clot or gas depressing bubble) occurs. Generally great pressure of ocean exerts a depressing effect, so that the pace of life is slower.

5.7 GENERAL WEATHER CONDITIONS

Animals in temperate climate deal with fluctuating weather conditions that are not met with in tropical countries. Seasonal changes influence habits and distribution of animals. Many of the physical factors such as temperature and light are correlated with seasonal changes in weather. Some animals in the temperate regions can tolerate a wide range of temperature fluctuations. Many, however, are killed at extremely low temperatures. But their races are preserved by their spores, eggs and larval forms, which can withstand such extreme conditions. The hazards of winter are faced by animals in various ways such as migration, hibernation and change of food habits. Due to this reason food chains in winter differ greatly from those in summer.

5.8 MICROCLIMATE

Until now, we have discussed the influence of different ecological media and of different abiotic environmental factors on the living organisms. All these factors form the macroclimate which is measured by the meterologists. But the climate in which animals actually live is very much different from the gross environmental conditions. This immediate environment of animal is called **effective climate** or **microclimate**. Microclimate (ecoclimate) can be defined as the effective climate upon which life of an organism depends. According to Geiger (1950), the characteristics of microclimate may or may not be similar to the regional or gross climate. Microclimate relates to such of the habitats of animals as those living in the soil within crevices of barks of trees, inside sheaths of grasses, within burrows, in ant's nests, beneath logs of wood or rock, within caves, on vegetation, in sand dunes, within plant galls and a variety of habitats which are occupied by endoparasites. Woodlice, centipedes and millipedes, for example, remain restricted to dark microhabitats during the daytime to prevent rapid water loss. When temperature falls and relative humidity increases at night, these animals emerge from the daytime microhabitats.

Concept of microclimates is particularly important in relation to moisture because relative humidity varies greatly within short distances in irregular habitat. For example, in swampy woodland the rate of evaporation near the damp soil and tops of trees varies markedly. This type of vertical change in moisture and also vertical gradients in light and other factors contributes toward the establishment of a characteristic vertical layering in vegetation and also in arboreal life of the forests. Among animals, the difference in evaporation rate with distance above the ground is of particular importance to soft bodied insects. They move higher in the vegetation at night but during the day they are forced by excess evaporation to move to lower levels.

Burrows of desert animals provides a good example of microclimate. Their physical conditions are very much different from those of the region as a whole. When *Dipodomys* (kangaroo rat) retreats within its burrow, it breaths air two or five times more humid than that outside. If this animal remains out of its burrow continuously, the rate of evaporation from its lungs exceeds the rate at which it obtains water, and the resulting water loss eventually causes death. The temperature in the desert burrow has been found to be as low as 28°C when its value at the same time for the surface of the soil was 77°C or more.

5.9 FIRE

Fire is an interesting ecofactor. Fire is of a common occurrence in natural vegetation all over the world; it is more common in drier habitats than the wet. Lightening is the commonest natural cause of fire initiation. Our earth's surface is hit by lightening every second in one or another part of the globe and many of these are of great magnitude. Other causes of fire are abrassive effects of falling rocks or of dried plant material such as bamboos, or spontaneous combination of very dry and hot material or by volcanic activities. Most forest fires are now man-caused, *i.e.,* by incendiarists (such as poachers), debris burners, smokers, campers, short-circuiting of high-tension electric lines, and nearby railway lines. In the Meghalaya region, tribals and even modern farmers practice the *Jhum* cultivation by periodically slashing the forest, burning them and raising crops on the ash enriched soil for a few years and moving to another segment of forest as the earlier one becomes eroded and nutrient deficient.

Types of Fire

Fires are generally classified as (i) **Ground fires** which develop in such conditions where organic matter (litter) accumulates richly as heaps and they catch fire which generally smoulder for longer periods. Thus, in dry litter, fire is rapid and extinguishes quickly, while in somewhat moist litter, the fire is slow and with its heat the inner parts of litter-heap also get dried and fire continues for a longer period. Ground fires are flameless and subterranean and kill almost all herbaceous plants rooted in the burning material except some woody species (shrubs and trees). (ii) **Surface fires** which sweep over the ground surface rapidly and their flames consume the litter, living herbaceous vegetation and shrubs and also scorching the tree bases if comes in contact. (iii) **Crown fires** which are most destructive, burning the forest canopy and are common in dense woody vegetations. They spread in the top layers from the canopy of one tree to the other and so on. Canopy fires produce a temperature up to 704.5^{0}C, killing the trees, shrubs and herbs. However, in moist soil surface, the underground plant parts and buried seeds escape death.

Effects of Fire

Fire has direct (*e.g.,* lethal) as well as indirect effects on plants and wild-life. Some well confirmed indirect effects of fire on plants are as follows:

1. Fire causes injury to some plants, resulting large scars on their stems. Such scars may serve as suitable avenues of entry of parasitic fungi and insects.

2. Fire arrests the course of succession and modify the edaphic environment very much.

3. Fire brings about distinct changes of such ecofactors as light, rainfall, nutrient cycles, fertility of soil, litter and humus contents of soil, pH, water holding capacity and soil fauna (earthworms, nematodes, arthropods, etc.) Soil fungi are reduced while bacteria increase due to post-fire changes in the soil.

The microclimate too is greatly changed due to addition of ash, loss of shade, loss of raindrop interception, accelerated erosion, etc.

4. Fire plays an important role in the removal of competition for surviving species. **Fire tolerant plant** species generally increase in abundance at the expense of those killed by fire (**fire-sensitive plants**) due to considerable reduction in competition and possibly due to alteration in other conditions.

5. Some plants such as *Populus tremuloides* and *Epilobium angustifolium* are stimulated to growth by fires. A number of such grasses as *Aristida stricta, Cynodon dactylon, Paspalum notatum*, are stimulated by fire to produce large quantities of seeds. In some grasses and legumes, the seeds would germinate only after these get fire treatment (*e.g.*, seeds of *Themeda, Heteropogon, Andropogon, Rheus, Tephrosia* and *Stipa*).

6. Some fungi, mainly some ascomycetes, grow in soils of burnt areas. Such fungi are known as **pyrophilous** (*e.g., Pyrophilous confluens*).

Adaptations to Fire

In frequent fire prone areas certain plants develop the following adaptations:

1. Certain trees, particularly conifers such as *Pinus* and *Larix* and dicots such as *Quercus* develop fire resistant bark with insulating effect against heat. These trees also have tall trunk with the crown restricted to upper zone only. This helps in escaping the destructions against surface and ground fires.

2. In some plants leaves are fire-resistant due to poor contents of such compounds such as resin or oil, and may check surface fires.

3. Some plants as *Pinus rigida* and species of *Eucalyptus* have adventitious or latent axillary buds which may develop into new branches. Similarly *Betula papyrifera* and *Vaccinium* sp., may develop new shoots after fire kills the older ones.

4. *Epilobium anguistifolium* acts as a fire indicator species. It grows in patches and in dormant condition. In case of fire, these plants rapidly grow while other plants die due to fire.

5.10 CHEMICAL FACTORS OF ENVIRONMENT

Chemical factors of environment can be enumerated as follows:

1. Atmospheric gases. As described in chapter 2, air consists mainly of *oxygen* (about 20 per cent), *carbon dioxide* (about 0.03 per cent), *nitrogen* (about 70 per cent), *water vapour* (in varying amounts, depending on climatic conditions) and *inert gases* (minute traces). Except the inert gases, all these gases of air serve as metabolites and each of them circulates in a global cycle (*i.e.,* biochemical cycles).

Oxygen of air forms most important factor for living organisms. All terrestrial animals use the oxygen of the atmosphere. **Carbon dioxide** of the air is used by green plants in photosynthesis. In fact, atmospheric carbon dioxide is the only source of carbon which is the basic constituent of all organic compounds. The carbon dioxide content of the atmosphere is replenished not only through respiration but also through other sources such as burning of fuels, forest fires, decay, volcanic eruption, and underground biota. **Nitrogen** is required by the organisms for the synthesis of proteins, nucleic acids and other nitrogenous organic compounds. As the aerial nitrogen is inert chemically, it cannot be used in gaseous form in majority of organisms, so it has to be fixed. The nitrogen cycle depends upon at least four different kinds of bacteria called **nitrogen fixers, decay causers, nitrifiers** and **denitrifiers.** Blue green algae (cyanobacteria) also play a crucial role in nitrogen fixation in aquatic environments. Thus, there is a regular circulation of nitrogen through the air, soil, plants, animals and microorganisms.

2. Dissolved gases. Dissolved oxygen content of the water is a significant factor and is often affected by various other factors such as temperature, carbon dioxide and presence or absence of aquatic vegetation and decaying organic matter. The oxygen supply of water comes mainly by diffusion from the air and from photosynthesis by aquatic plants (hydrophytes). In aquatic environment the dissolved oxygen often becomes a limiting factor and its concentration at different depths has a great bearing on the presence and abundance of different organisms. Requirement of oxygen of aquatic organisms varies to great extent. Most of the fish are **oxykinetic,** *i.e.,* they are made restless by lack of oxygen and stimulated to move away to better aerated water. Bloodworms, however, can live at oxygen concentration less than 2mg/1.

Carbon dioxide is readily soluble in water. Unlike oxygen, carbon dioxide enters into chemical combination with water to form carbonic acid (H_2CO_3) which in turn reacts with available lime stones to produce carbonates and bicarbonates. These compounds serve as a reservoir pool of carbon dioxide in the ocean. They are not only a source of nutrients but also act as buffers in keeping the hydrogen ion concentration of aquatic environments near the neutral point. These oceanic reservoirs also tend to regulate the amount of carbon dioxide in the atmosphere. Moderate increases in carbon dioxide in water seem to speed up photosynthesis and the development process of many organisms. High carbon dioxide concentration becomes a limiting factor to animals because such high CO_2 concentrations are usually associated with low oxygen concentrations. Fishes respond vigorously to high CO_2 concentration and may be killed when water contains too much of CO_2.

3. Hydrogen ion concentration or pH. In aquatic environments the pH of the water is a function of the dissolved CO_2 content, which in turn, is decreased by photosynthesis and increased by respiration. Formerly pH has been regarded to be a significant limiting factor, but now it is considered to have a limited importance in the distribution of animals. However, some animals prefer alkalinity and others acidity, but many forms can endure a wide range of pH. For example, tapeworms can live in pH 4 to pH 11, or very acidic to very narrow alkaline medium, whereas others, such as certain species of *Euglena,* live and flourish in water that has a pH range of 2 to 8. Some mosquito larvae thrive well in water with a pH of less than 5 and will not live in an alkaline medium.

Fish seem to adjust to wide range of pH. Molluscs may be more sensitive to acidic pH, because their shells of calcium carbonate are corroded by acids. This is reason why water with a pH of 6 or less contains a few molluscs. Soils and waters of low pH (acidic) are quite frequently deficient in nutrients and low in productivity. Thus among other things the pH of soils determines its fauna and productivity.

4. Nutrients. Life is dependent not only on energy but also on the availability of some 20 chemical elements required in life processes of most organisms and an additional 10 or so required by different species, usually in trace amounts. Although carbohydrates can be photosynthesized from water and carbon dioxide, the more complex organic substances require additional components either in ample amounts, as in case of nitrogen and phosphorus, or in trace amounts, as in case of zinc and molybdenum. Further, photosynthesis and other metabolic reactions in both plants and animals occur in the presence of enzymes that themselves contain a variety of elements in trace amounts.

All of the thirty-two elements that have been identified as nutrients are among the lightest elements. Twenty-eight of the first thirty-five elements in the periodic table are essential for plant and/or animal life; the remaining four are still regarded as light elements, with atomic number of 38 (strontium), 42 (molybdenum), 50 (tin) and 53 (iodine).

Those elements which are required in relatively large amounts are generally referred to as **macronutrients,** each consituting 0·2 per cent or more of dry organic weight. **Micronutrients** are those elements needed in trace amounts, each constituting less than 0·2 per cent of dry organic weight.

1. Macronutrients. Macronutrients can be considered into two groups: 1. those that constitute more than 1 per cent each of dry organic weight—**carbon, oxygen, hydrogen, nitrogen** and **phosphorus**; and 2. those that constitute 0·2 to 1 per cent of dry organic weight — **sulphur, chlorine, potassium, sodium, calcium, magnesium, iron** and **copper.** Of the total number of atoms in the human body, hydrogen constitutes 63 per cent, oxygen 25·5 per cent, carbon 9·5 per cent and nitrogen 1·4 per cent (**Frieden** 1972).

Function of macronutrients. The role of the *first group of macronutrients* is well known: *oxygen* (*O*) as building block of protoplasm and the primary agent in biological oxidation; *carbon* (*C*) as the building block of all organic compounds; *hydrogen* (*H*) as an acceptor of oxygen; *nitrogen* (*N*) as the building block of protein and nucleoprotein, as well as a component of enzymes and *phosphorus* (*P*) as a constituent in photosynthesis, proteins and nucleoproteins.

The role of *second group of macronutrients* is also quite distinct: *sulphur* (*S*) as a basic constituent of protein; *chlorine* (*Cl*) is required in minute amounts by plants, but in greater quantities in animals as a key factor in maintaining the acid-base balance; *calcium* (*Ca*), a major component of cell walls, needed for normal mitotic divisions and root-tip development and for a variety of functions in animals including the maintenance of acid-base balance, contraction of muscles, and in vertebrates, arthropods and molluscs, the key element in providing structural support (bones, exoskeletons and shells); *magnesium (Mg)* critical to the structure of chlorophyll and energy transfer from ATP to ADP; *iron (Fe)* required in electron transfer in photosynthesis and the key element in haemoglobin; *copper (Cu)*, present in chloroplasts, influencing the rate of photosynthesis.

2. Micronutrients. The known micromutrients include **aluminum (Al), arsenic (As), boron (B), bromine (Br), chromium (Cr), cobalt (Co), fluorine (F), gallium (Ga), iodine (I), manganese (Mn), molybdenum (Mo), nickel (Ni), selenium (Se), silicon (Si), strontium (Sr), tin (Sn), titanium (Ti), vanadium (V)** and **zinc (Zn).** In fact, some of these micronutrients may be macronutrients for some species, whereas some of the second group of macronutrient, for example, **sodium** and **chlorine** in plants may be micronutrients.

The list of micronutrients has recently been expanded with the inclusion of four new essential trace elements (*e.g.,* arsenic, nickel, silicon and vanadium) identified in animals in the 1970s (**Mertz 1981**). At least one, nickel, was the first essential trace element identified in plants nearly 30 years ago, when its presence was determined to be necessary for development and functioning of root nodules in soyabeans (**Eskew** *et al.,* 1983).

Function of micronutrients. Generally many micronutrients play key roles in enzymatic activity. Some, however, are involved in a wide variety of roles: *boron* is involved in cell division, translocation of water in plants and carbohydrate metabolism, among a number of other functions. *Aluminum* is necessary for normal development of some ferns, *silicon* is the structural element of diatom tests and *cobalt* is required in root nodules for proper functioning of nitrogen-fixing bacteria. *Selenium* deficiency, primarily affecting juveniles such as black-tailed deer, results in increased mortality during the neonatal and preweaning period (**Flueck,** 1994).

Nutrient interaction. The presence or absence of an essential nutrient may adversely hit the availability or activity of another nutrient. **Howarth** and **Cole** (1985) showed that *sulphate* inhibit the assimilation of *molybdenum* by photoplankton, thus, making it less available in sea water than it is in freshwater. Molybdenum is required in the conversion of elemental nitrogen to nitrate (nitrogen fixation), the form in which it can be used by plants, as well as in nitrate assimilation. Thus a low levels of molybdenum requires a greater expenditure of energy for these nitrogen processes in sea water than in freshwater and may explain the relatively low levels of nitrogen in coastal marine ecosystems.

5. Food. The availability and the kind of **organic food** also plays an important role in the distribution of free living organisms. Many animals have a varied diet and use the food that they get (**omnivores**). Other animals are plant feeders (**herbivores**) or flesh feeders (**carnivores**). Aquatic animals feed on particular phytoplankton or zooplankton, or on small or large animals of their own or of their kinds. The food related inter-relation between animals, plants and microorganisms form food chains in all habitats.

In fact, the presence of an animal in a particular medium is determined to a great extent by its organic food. Animals restricted to seasonal foods are sometimes forced either to starve or to migrate to more favourable places.

5.11 HOMEOSTASIS

Daily and seasonally, most organisms are confirmed with a changing external environment. In spite of it, they have to maintain a fairly constant internal environment. They need some means of regulating internal conditions such as body temperature, water balance, pH and amount of salts in fluids and tissues, relative to external ones. They take in substances from the environment and use them in cellular chemical reactions; but they also have to release in the environment both excessive intake of substances and metabolic waste products to maintain a fairly constant state (Box 5.4).

Box 5.4

In the living world constancy is not a static affair but a **dynamic phenomenon**. Forces tend to strengthen a particular trait or raise the concentration of some material are balanced against forces tending to weaken the trait or reduce the concentration. A dynamic "equilibrium" maintains a balance within all organisms. This balance, steady-state or homeostasis is the singlemost important feature unifying all aspects of physiology (function) at the level of cell, organism and even population.

Thus, the maintenance of a relatively constant internal environment in a varying external environment is called **homeostasis.** The term *homeostasis* was first used by **Walter B. Cannon** (1871-1945), an American physiologist who was particularly interested in the movements of the digestive tract and their control (Box 5.5).

Box 5.5

Historically, **Hippocrates** hinted to the belief that disease, a disturbance in health, is righted (cured) by natural forces. This notion of Greek physician, who first formulated the physician's oath that bears his name, recognizes a natural tendency to oppose abnormality, a resistance to pathological change. However, it was **Claude Bernard** (1813-1878) who is recognized as the father of the concept of homeostasis. This French physician developed physiology as an experimental science in a series of elegant laboratory investigations during the mid-nineteenth century. His writings at that time stressed the importance of constancy in the internal environment. He was aware of the fact that every cell in the body is bathed in a fluid that is both nourishing and sustaining. The totality of these tissue fluids, which are intimately associated with the blood and lymph, makes up the **internal environment.** Bernard stated that the *constancy of the internal environment is necessary for free life.* The term homeostasis was first used by **Cannon** in 1920.

Homeostasis has been studied most intensively in multicellular animals, particularly in vertebrates. However, it is operative at all levels of life. Homeostasis is accomplished by means of morphological, physiological and behavioural adaptations. It is most highly developed in birds and mammals. The main regulators of homeostasis in mammals are the liver and kidney.

Mode of operation of homeostasis. Feedback systems within an animal bring about physiological and behavioural adjustments that minimize deviations in internal conditions. Each system has three basic components: **receptors**, which detect changes in conditions; **integrators**, which evaluate these changes and **effectors,** which work to bring the body back to its original state. The feedback is usually **negative**, *i.e., a change in one direction produces a response in the opposite direction.* Thus, an increase in body temperature causes effector activity that decreases body temperature.

Receptors are sensory cells that respond to changes in such internal features as body temperature, carbon dioxide concentration in the blood, blood pressure and muscle stretch. Receptors convert information about these changes into electro-chemical impulses that travel to brain.

Integrators are special areas of the brain that receive information from receptors, evaluate it, and then send commands to the effectors capable of correcting the deviation from normal. Two major homeostatic centres of the vertebrate brain are the **hypothalamus** and the **medulla oblongata.**

Effectors are muscles and glands. They act in ways that return the body to its original, more suitable state after some deviation from normal. Muscles contract and relax to produce movements of skeleton, circulatory system and internal organs. Glands synthesize and release special materials, of which some act directly to promote homeostasis and others act indirectly to alter the activities of certain target cells.

Negative feedback operates, for example, to maintain body temperature. Receptors in skin detect changes in skin temperature and neurons in the hypothalamus detect changes in blood temperature. This information is sent to special areas within the hypothalamus, where it is interpreted and evaluated. The hypothalamus then sends commands to appropriate effectors. *When the temperature is above normal,* arteries in the skin dilate, which brings warm blood to the surface, and **sweat glands** release sweat, which evaporates and removes heat from skin. *When the temperature is below normal,* arteries in the skin constrict, preventing warm blood from reaching the surface, and skeletal muscles **shiver,** forming heat from excess cellular respiration, while the adjustments are taking place, the hypothalamus continues to receive information about body temperature and to adjust its commands in response to the information it receives (Box 5.6).

Box 5.6 Physiology of Homeostasis

In humans, normal body temperature or set point is 37°C (98.6°F). When the environmental temperature drops, sensory mechanisms in the skin detect it and send a message to the *hypothalamus* in the brain. The hypothalamus (which behaves much like a **thermostat**) sends messages to higher brain centres that result in **voluntary acts,** such as putting on coat or moving closer to heat *i.e.,* behavioural response (Fig. 5.7). It also sends a message to the **autonomic nervous system** that triggers **shivering**, an involuntary muscular exercise producing more heat. The hypothalamus stimulates the **adrenal gland** to release the hormone **epinephrine** into the blood stream. This hormone speeds up the conversion of liver *glycogen* to *glucose*. At the same time the hypothalamus stimulates the **pituitary gland** to release a thyroid stimulating hormone, causing the **thyroid** to increase the release of its **thyroxine** hormone. Both hormones stimulate body cells to increase their respiratory activity (at the level of mitochondria), producing body heat. As the body temperature increases, the hypothalamus detects the change and slows down heat production. Functioning by negative feedback, the hypothalamus maintain an optimal body temperature.

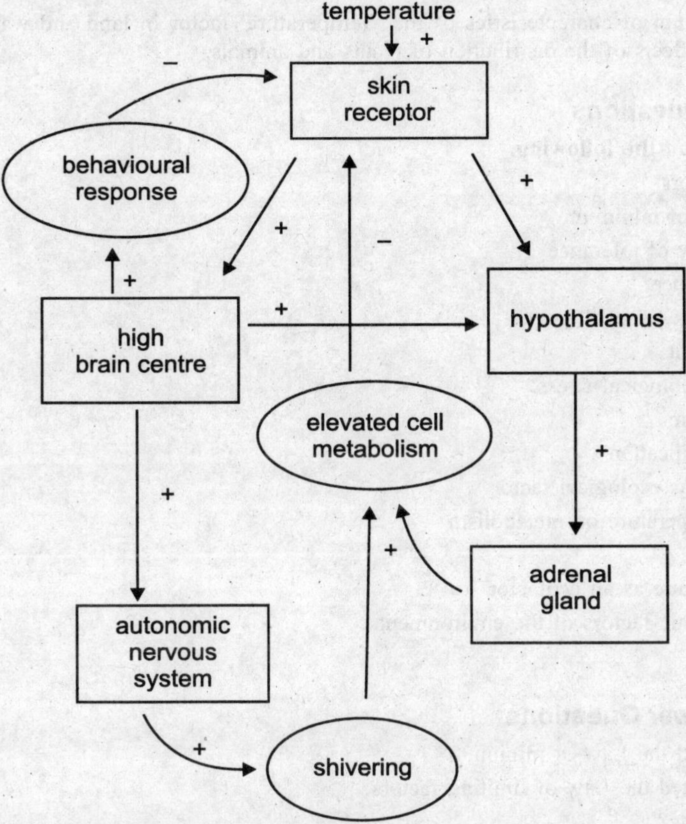

Fig. 5.7. Homeostatic control of human body temperature.

QUESTIONS

Long Answer Questions

1. Give an account of Leibig's law of minimum and Shelford's law of tolerance and the combined concept with examples.
2. Describe the role of limiting factors in ecology.
3. What are limiting factors? Describe the limiting factors of a terrestrial ecosystem. (*Kerala 2001*)
4. Define Shelford's law of tolerance.
5. What are the various abiotic factors of environment? Describe the role of "temperature" as an important ecological factors.
6. Describe the different abiotic factors governing any ecosystem.
7. Describe the role of climatic factors of the environment.
8. Describe the importance of light, temperature and water as ecological factors.
9. Discuss in detail the significance of light and temperature as abiotic factors.
10. What are abiotic factors? Discuss the role of light as ecological factors.
11. Discuss in detail the role of light and temperature on metabolism of aquatic animals.
12. What are environmental factors? Describe the effect of temperature on animals.
13. What are environmental factors? Describe the effects of light on animals.
14. Give an account of the abiotic factors affecting the animal life of the environment.

15. Discuss soil as physical environmental factor.
16. Give an account of characteristics of the 'Temperature' factor in land and water ecosystems and describe its effects of the distribution of plants and animals. (*Rohilkhand 2001*)

Short Answer Questions

Write short notes on the following.

1. Limiting factors
2. Liebig's law of minimum
3. Shelford's law of tolerance
4. Law of tolerance
5. Abiotic factors (of ecology)
6. Threshold limit
7. Zone of physiological stress.
8. Photoperiodism
9. Thermal stratification
10. Temperature as ecological factor
11. Effect of temperature on metabolism
12. Humidity
13. Role of moisture as an ecofactor
14. Role of climatic factors of the environment.
15. Homeostasis

Very Short Answer Questions

1. Who proposed the Law of minimum?
2. Who formulated the Law of limiting factors?
3. What is abiotic?
4. What is Liebig law of minimum?
5. What is limiting factor?
6. Define limits of tolerance.
7. Define Shelford law of tolerance.
8. What is tolerance?
9. What is torpor?
10. Define poikilotherm.

Fill in the Blanks

1. The law of tolerance was formulated by .... in 1913.
2. .... is dormancy in animals during periods of drought.
3. Organism that uses metabolic energy to maintain a relatively constant body temperature is called ....
4. .... include alpine vegetation which tolerate very low temperature.
5. .... are animals which do not have an internal mechanism for temperature regulation.
6. .... is the zone of rapid temperature change in a lake beneath the epilimnion and above the hypolimnion.

Multiple Choice Questions

1. Nonliving components of the environment including soil, water, air, light, nutrients, etc., are called
   (*a*) biotic factors (*b*) abiotic factors
   (*c*) limiting factors (*d*) ecological factors

2. Light loving plants are known as
(*a*) heliophytes (*b*) xerophytes
(*c*) lithophytes (*d*) none of the above

3. Which of the following habitats would be ideally suited to sciophytes?
(*a*) ponds and lakes (*b*) dry and sunny place
(*c*) moist and shady place (*d*) desert

4. Light has significant effect on
(*a*) vision (*b*) metabolism
(*c*) locomotion (*d*) reproduction

5. Temperature changes in the environment affect most of the animals which are
(*a*) poikilothermal (*b*) homoisothermal
(*c*) aquatic (*d*) desert living

6. Plants growing in extreme cold soil are
(*a*) oxylophytes (*b*) halophytes
(*c*) psychrophyte (*d*) psammophytes

7. Which of the following pairs belong to the category of cold blooded animals
(*a*) frogs and snakes (*b*) lizards and rats
(*c*) snakes and birds (*d*) birds and monkeys

8. Dormancy in animals through a drought or dry season is called
(*a*) hibernation (*b*) aestivation
(*c*) diapause (*d*) none of these

9. Warm, oxygen-rich upper layer of water in a lake or other body of water is called
(*a*) thermocline (*b*) epilimnion
(*c*) hypolimnion (*d*) all of these

10. Which law of evolution says that warm-blooded animals of hot and humid areas have melanin pigment in abundance?
(*a*) Dollo's law (*b*) Cope's law
(*c*) Gloger's law (*d*) Gause's law

11. The animals of cold countries have relatively shorter and not fully developed ears, eyes, hairs and other phenotypic characters. This is known as
(*a*) Allen's law (*b*) Bergamann's law
(*c*) Cope's law (*d*) Dollo's law

12. The direct and dominating factor of plant vegetation is
(*a*) light (*b*) water
(*c*) temperature (*d*) carbon dioxide

13. Which one of the following is a direct ecological factor
(*a*) soil structure (*b*) wind
(*c*) altitude (*d*) temperature

14. Cold-blooded animals are those
(*a*) which are cold to touch
(*b*) which have cold blood
(*c*) which live in cold environment
(*d*) which cannot regulate their body temperature

15. The factor most important for living organisms is
(*a*) water (*b*) temperature
(*c*) humidity (*d*) light

16. Aquatic animals which can tolerate only a narrow range of salinity are said to be
(*a*) euryhaline (*b*) stenohaline
(*c*) poikilothermy (*d*) catadromus

17. Euryhaline species denotes an animal
(*a*) without osmoregulation
(*b*) which lives in saline water
(*c*) which has a wide range of saline tolerance
(*d*) which has a wide range of temperature tolerance

18. The factor governing the structure of earth's surface is
(*a*) topographic (*b*) edaphic
(*c*) biotic (*d*) temperature

19. Which of the following is not a climatic factor?
(*a*) humidity (*b*) rainfall
(*c*) snow fall (*d*) substratum

20. Circadian rhythm is the metabolic or behavioural rhythm of
(*a*) 24 minute cycle (*b*) 24 hour cycle
(*c*) 24 day cycle (*d*) 24 month cycle

21. Law of limiting factors in photosynthesis was given by
(*a*) Calvin (*b*) R. Hip
(*c*) Blackman (*d*) Arnon

22. The idea of homeostasis was first presented in 1859 by
(*a*) Claude Bernard (*b*) Julian Huxley
(*c*) Charles Darwin (*d*) Louis Pasteur

23. Homeostasis occurs in
(*a*) all organisms to some extent (*b*) multicellular organisms only
(*c*) animals only (*d*) birds and mammals

24. UV rays are (*CCSU Meerut 2014*)
(*a*) electromagnetic (*b*) non-ionizing radiations
(*c*) physical mutagens (*d*) both (*b*) and (*c*)

ANSWERS

Very Short Answer Questions

1. Justus von Liebig in 1840.

2. F.F. Blackman in 1905.

3. Nonliving components of an ecosystem, such as water, air, light, and nutrition.

4. Concept, first stated by Baron J. von Liebig in 1810, that the essential materials or resource most closely approaching the minimum need tend to be the limiting one.

5. Resource that limits the abundance, growth and distribution of an organism or species, is called a limiting factor.

6. Upper and lower limits to the range of particular environment factors (such as light or temperature) with in which an organism or species can survive.

7. Law, proposed by V.E. Shelford in 1911, stating that the presence and success of an organism or species depends on both the maximum and minimum resource or set of conditions.

8. Tolerance is the ability to survive stressful environmental conditions.

9. Torpor is the state of dormancy in which endotherms drop their lower critical temperature or function at the expense of another.

10. Poikilotherm is an animal (organism) whose body temperature varier (alters) directly with the temperature of its environment.

Fill in the Blanks

1. V.E. Shelford	**2.** Aestivation	**3.** Homeotherm or endotherm
4. Hekistotherm	**5.** Ectotherms	**6.** Thermocline.

Multiple Choice Questions

1. (*b*)	**2.** (*a*)	**3.** (*c*)	**4.** (*d*)
5. (*a*)	**6.** (*c*)	**7.** (*a*)	**8.** (*b*)
9. (*b*)	**10.** (*c*)	**11.** (*a*)	**12.** (*c*)
13. (*d*)	**14.** (*d*)	**15.** (*a*)	**16.** (*b*)
17. (*c*)	**18.** (*a*)	**19.** (*d*)	**20.** (*b*)
21. (*c*)	**22.** (*a*)	**23.** (*a*)	**24.** (*d*).

6

Biotic Environmental Factors (Ecological Relationships)

Organisms do not exist alone in nature but in a matrix of other organisms of many species. Many species in an area will be unaffected by the presence or absence of one another (This is often termed **neutralism**), but in some cases two or more species will interact. The evidence for such interaction is quite direct : populations of one species are different in the absence and in the presence of a second species. Under biotic or living environment are studied interactions of organisms with the members of their own species and with those of other species. Relationships between the organisms of the same species or between conspecific animals is known as **intraspecific relationships** and that between two different species is known as the **interspecific relationships.**

6.1 INTRASPECIFIC RELATIONSHIPS/INTERACTIONS

Individuals of a community are related to one another through their cooperative or disoperative coactions. Their relationship may be beneficial or harmful, affecting a single animal, or a group of animals, or even the whole community. The intraspecific relationships are influenced by following factors:

1. Reproduction. All organisms possess an inherent property to reproduce and to exist by tiding over the resistance of the environment. The rate of reproduction and the ability to survive depends on the **biotic potential** of a species. If the biotic potential is high, it will be able to reproduce and to survive in order to continue its race. Reproduction is a means to increase the number of individuals and to produce individuals that may have new combinations of characteristics (phenotype) that differ from those of their parents and that may enable them to survive in a particular environment. In sexual reproduction of unisexual organisms, two or more individuals are always involved. Reproductive activities are usually cyclic or seasonal in nature and are often correlated with specific environmental changes.

In different organisms, rate of reproduction also varies greatly. In some animals it is enormous. In all cases, however, *over production* leads to **competition** for food, space to live and also for mate.

2. Competition for food. Over production of animals leads to competition for food among them. As the food of a particular species is usually common, this competition is natural to occur. In the absence of food in its environment, an animal may eat the members of one's own species (**Cannibalism**; Box 6.1). This phenomenon is quite common among animals, both small and large. It has been observed among ants, termites, beetles, scorpions, snakes etc. In facing the competition for food mighty individuals get the benefit while weaker forms have to suffer from the lack of food. In the latter case, they may perish or sometimes leave their environment for another place. Thus, migration may occur due to lack of food at a particular place.

Box 6.1

Cannibalism or intraspecific predation is found in a wide variety of animals and functions as a means of interference competition, limiting population size before the resource itself becomes limited (**Fox** 1975).

Migration involving the seasonal movement to and from given areas, has a very significant regulatory role, both in decimating effects of the migration movement itself and in avoiding an otherwise limiting environment during particular seasons.

3. Competition for space (shelter). Over production of animals also leads to competition for space among them. However, in many cases large number of animals of the same species can live side by side in small space without showing any sign of interaction. For example, this happen in many species of leaf-eating insects. But in higher animals competition for space is more fierce and results in hostility. The competition for space has given rise to the phenomenon of **territorial behaviour.**

Territorial behaviour. Territoriality is a space oriented behaviour and is evident in many vertebrates, with its greatest expression in birds and mammals. An individual or a group, stakes out a geographic area, which it then defends in some manner. The extent of the territory is itself determined by a number of factors *e.g.,* vigour of the holder, amount of competition by would -be intruders and so on. If the size of the territory is consistent with a sufficient food supply, the holder of the territory has greater chance of survival and reproduction than do those excluded from the territory. Holding of territory involves intense agonistic behaviour (*e.g.,* dogs). Sometimes it is achieved by a conspicuous visual or acoustical display (*e.g.,* frogs).

4. Competition for mate. Among higher animals such as birds or mammals, competition for mates is also common and results in such interactions as fight or even death of an individual. In red deer, for example, males fight for harems of females. In certain birds, for example, peacocks, males (which are polygynous) compete without fighting and females choose the one which they see as most attractive. During the courtship, the peacock takes numerous forms such as fluffing out of ornamental plumage, fanning and erecting the tail and dancing in front of the females (*i.e.,* 4 to 5 hens forming a drove or harem). In small water birds, such as bronze-winged jacanas (*Metopidius indicus*) competition occurs between females for males.

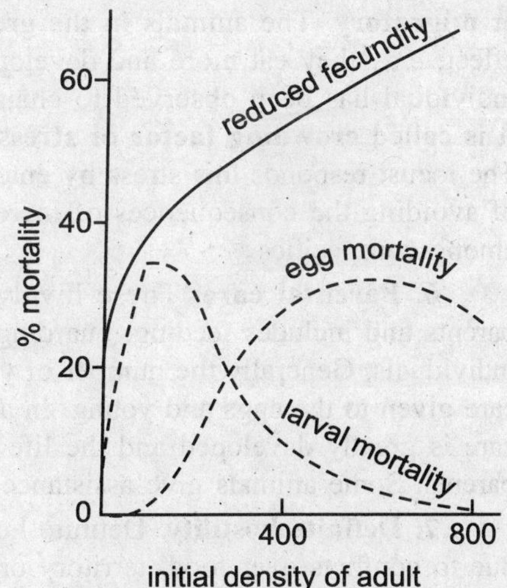

Fig. 6.1. Intraspecific competition between azuki bean weevils change as population density is increased in laboratory cultures. Fecundity is reduced and mortality among younger shows more complex and contrasting patterns.

Intraspecific competition is responsible for levelling off or fluctuations around a certain density of a population. Exact mechanism do vary in different species. Intraspecific competition is worked out in the beetle, the azuki bean weevil (*Callosobruchus chinensis*) which feed on stored legume seeds. Fig. 6.1 shows results from a series of experiments begun with different numbers of adult weevils. Thus, at high density, female fecundity is reduced because overcrowding leads to fewer successful matings due to mechanical interference between individuals. These conflicts are the main factors which result in the decrease in population density. Minor effects include increased egg mortality and increased level competition for food at high density. The egg mortality is due to adult knocking eggs from the sites at which they were laid.

Dash and **Hota** (1980), **Misra** and **Dash** (1984) and **Mohapatro** and **Dash** (1984) studied the density effect on the growth and metamorphosis of three species of tropical anuran tadpoles *Rana tigrina* (= *tigerina*), *Polypedates* (= *Rhacophorus*) *maculatus* and *Bufo stomaticus.* For each of these tadpoles, they found the competition for food due to crowding which affects growth rate and the threshold size for metamorphosis. In very high density populations (with severe competition) some larvae were not able to attain the threshold size and could not metamorphose.

5. Social life and communal relationship. In many animals such as social insects, on the basis of morphological differentiation and division of labour "castes" are formed, *e.g.*, reproductive individuals (quens and drones), workers and soldiers. The highly developed societies of ants, bees, wasps and termites are maintained for years or even decades. Sexual relationships and care of the young are the simplest and most common forms of social behaviour within a species.

Other types of communal relationships which may be temporary or permanent and serving a variety of functions may also develop. Their familiar examples include **school** of fish, **flocks** of birds and **herds** of ungulates. In herds, for example, the individual enjoys (1) greater protection from predators by joint defensive action, (2) the benefits of communal food-finding, (3) other benefits. Some groups contain one of these factors in particular and animals may join together particularly for *migrating, hunting, sleeping, hibernating, building* or *breeding.*

Group effect and phase phenomenon. In insects when some individuals are separated from their society, they live for shorter time than their conspecifics left in the group. A similar effect has been observed in many fishes, birds, rodents and even apes. This phenomenon has been described as **group effect. Uvarow** (1921) had used the term **phase phenomenon** for group effect like situation in insects. He discovered that migratory locusts of a species (*Schistocerca gregaria*) can form two different types of individuals: **solitary** or **non-migratory** and **gregarious** or **migratory**. The animals in the gregarious phase show certain characteristics of the group effect, *e.g.,* they eat more and develop more rapidly. In some animals such as field mouse, the individual has been observed to change its behaviour when the population density increases. It is called **crowding factor** or **stress syndrome** and also occur in some birds and mammals. The locust responds this stress by entering the migratory phase. Mass migrations offer a chance of avoiding the consequences of increased density competition and other negative interactions among conspecifics.

6. Parental care. These involve protection given to the helpless young ones by their parents and includes feeding, guarding, nursing, keeping them warm and transporting immature individuals. Generally the number of youngs produced is inversely proportional to the amount of care given to the eggs and young. In advanced animals such as birds and mammals, the parental care is greatly developed and the life of young is very much protected and influenced by their parents. Some animals give assistance to young ones of another species, but this is quite rare.

7. Definite hostility. Definite hostility occurs among the members of a particular species due to conflicts over food, territory or due to conflicts over mates. If a worker bee of one hive tries to enter another hive by mistake, it is immediately attacked and killed. Similar examples can be observed among other social insects. Definite hostility over territories or in selection of mates is usually seen among some reptiles, and in many birds and mammals.

8. Other factors. Besides the above factors, there is a compensation among factors and as a result subpopulations may develop. As they differ slightly in habits or needs, the competitions can be reduced within a species by the development of subpopulations. This sort of adaptation plays a central role in evolution.

6.2 INTERSPECIFIC/RELATIONSHIPS INTERACTIONS

The interactions between species may have positive or negative results. Following six general types of interactions have been described:

A. Positive interactions (Symbiosis)
1. Mutualism;
2. Commensalism;
3. Protocooperation.
B. Negative interactions (Antagonism)
4. Exploitation;
(i) Social parasitism;
(ii) Parasitism;
(iii) Predation.
5. Amensalism and antibiosis;
6. Competition.

Box 6.2 Consortism

When a plant or animal comes into close relationship with another organism during its existence, it exhibits **consortism** (**Woodbury**, 1953). Consortism is classified into four types according to the benefit derived by the **consortes:** 1. If it is **voluntary** with mutual benefit but not asymmetrical or damaging to either consorte, it is **mutualism.** 2. **Commensalism** is a voluntary asymmetrical consortism in which one organism is benefited with no appreciable benefit or damage to the partner. 3. **Parasitism** is an involuntary consortism with progressive asymmetrical damage to one for the benefit of the other. 4. **Predation** is a nonsymbiotic consortism with damage to one for the benefit of the other. Thus, consortism includes both harmful and beneficial coactions and may occur between two animals, two plants or plant and animal.

A. Positive Interactions

In case of positive interactions, populations help one another and either one or both the species are benefited. This benefit may be regarding the food, shelter, substratum or transport. Further, such an association may be continuous or transitory, obligate or facultative and the two partners may be in close contact (*i.e.*, their tissues remain intermixed with each other) or one of them may live within a specific area of the other or attached to its surface. Moreover, for the positive interactions, how does one determine that what is the benefit to the species ? Individuals that participate in the interaction should have a higher relative fitness than those that do not participate. At the population level, the growth rate of a population containing mutalistic individuals should be higher than that of a population lacking such individuals. Positive interactions may be of the following types:

I. Mutualism. Mutualism is an obligatory positive interspecific interaction that is strongly beneficial to both species. Such mutually beneficial interactions are more common in the tropics than elsewhere. In this case, both of the species derive benefit and there exists a close and often permanent and obligatory contact which is more or less essential for survival of each. In mutualism, two populations enter into some sort of physiological exchange and resulted in coevolution of both species. Some common examples of mutualistic association are the following:

1. Pollination by animals. Certain insects such as bees, moths, butterflies, etc., and birds derive food from the nectar, pollen or other plant products, and in return bring about cross pollination (Fig. 6.2A). To ensure the success of this function, various structural adaptations have occurred in both plants and animals, leading to coevolution.

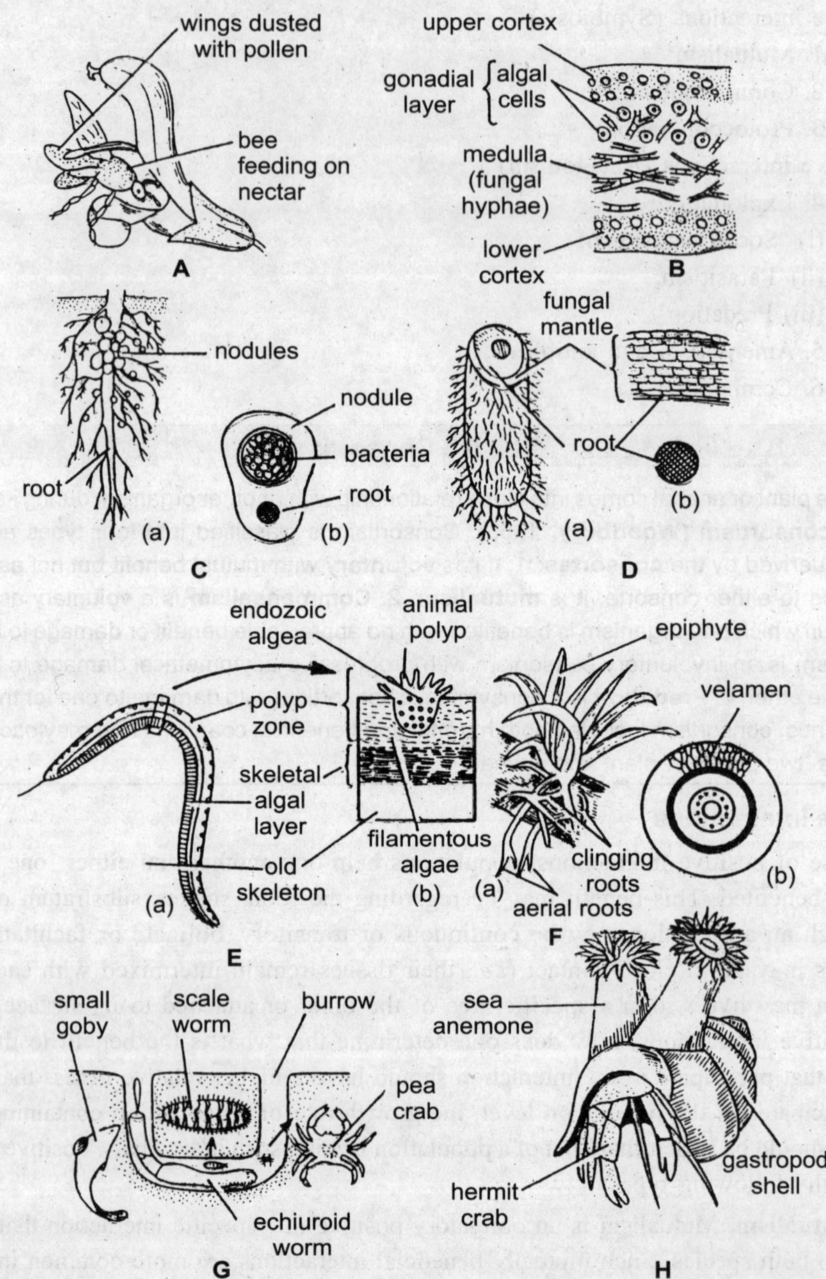

Fig. 6.2. Positive interspecific interactions. A—Mutalism, a bee feeding on nectar and also is dusted with pollen; B—Mutalism, a part of T.S. of lichen thallus showing algal and fungal partners; C—Mutalism, (a) root system of a legume showing nodule, (b) T.S. through the root and nodule; D—Mutalism, mycorrhizae, *i.e.*, association between root of higher plant and fungal hyphae, (a) A part of mycorrhizal root coverd by fungal hyphae, (b) A part of mycorrhizal root showing fungal mantle, E—Mutualism, a relationship between the polyp (animal) and several algae, (a) Cross section of "head" of coral (polyp) showing algal layer, (b) A part of C.S. of coral head magnified; F—Commensalism, epiphyte orchid growing perched on the branch of another plant, (b) T.S. of aerial root of orchid showing velamen; G—Commensalism, a diagrammatic sketch showing three commensals, the small goby, scale worm and pea crab, living in the burrow of the echiuriod worm; H—Protocooperation, association between sea anemone and hermit crab.

2. Dispersal of fruits and seeds. Birds and mammals are of great importance as agents of plant distribution. Seeds, fruits, even entire plants become attached to feathers or fur or ingested seeds are eaten and eliminated unharmed with the faeces.

3. Mutual defence in ants and acacias. A mutualistic system of defence has been achieved by the swollen-thorn acacias and their ant inhabitants in the New World Tropics. The ants (*Pseudomyrmex* sp.) depend on acacia (*Acacia* sp.) tree for food and a place to live and the acacia depends on the ants for protection from herbivores and neighbouring plants. Swollen-thorn acacias (*e.g., Acacia cornigera*) have large, hollow thorns in which ants live (Fig. 6.3). The ants feed on modified leaflet tips, called **Beltian bodies**, which are the primary source of protein and oil for the ants, and also on enlarged nectaries, which supply sugars. Swollen-thorn acacias maintain year-around leaf production, even in the dry season, to provide food for the ants. The acacia ants continually patrol the leaves and branches of the tree and immediately attack any herbivore that attempts to eat acacia leaves or bark. The ants also bite and sting any foreign vegetation that touches an acacia, and they clear all the vegetation from the ground beneath the acacia tree. Some of the species of ants that inhibit acacia thorns are obligate acacia ants and live nowhere else.

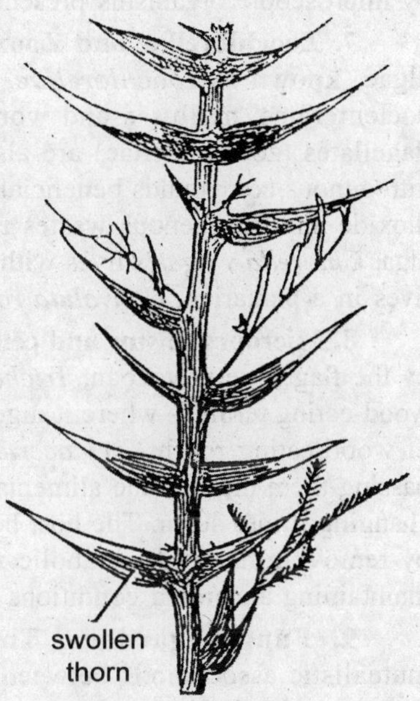

Fig. 6.3. Swollen thorns of *Acacia cornigera* on a lateral branch. Each thorn is occupied by 20-40 immature ants and 10 to 15 workers ants. All the thorns on the tree are occupied by one ant colony.

4. Lichens. They form the examples of mutualism where contact is close and permanent as well as obligatory. The body of lichens is made up of a matrix formed by a fungus, within the cells of which an alga (*e.g., Graphis, Parmelia* and *Cladonia*) is embedded (Fig. 6.2B). The fungus makes available the moisture and minerals to the algae, which prepare food by photosynthesis. In nature, neither of the two can grow alone independently. Lichens tend to grow abundantly on bare rock surfaces.

5. Symbiotic nitrogen fixation. Here a bacterium *Rhizobium* forms nodules in the roots of leguminous plants and lives symbiotically with the host (Fig. 6.2 C). Bacteria get a protective space to live in and derive prepared food from the roots of higher plants and in return fix gaseous nitrogen, making it available to the plants. The leguminous plants use this nitrogen in the protein synthesis and so legumes are rich in proteins. Nitrogen fixation-like association also occurs in root nodules of *Alnus, Alopecurus, Casuarina, Cycadacease, Myrica, Podocarpus, etc.*, and leaves of about 400 species of non-leguminous plants.

6. Mycorrhizae. In mycorrhizal associations, tree roots become infested with fungal hyphae (Fig. 6.2D). The fungi derive their food from the tree roots and in return their hyphae supply water and minerals that they absorb from the soil much like the root hairs of trees. It is believed that the fungus also regulates the pH and sugar level for a good growth of roots in acidic soils (*e.g.,* conifers). Mycorrhizae may be on the surface of roots (**ectotrophic**) or inside between the cells of the roots (**endotrophic**). Ectotrophic mycorrhizae are common in nature on pines, oaks, hickories and beech and endotrophic ones occur in red maple and are common in roots and other tissues of many orchids and members of Ericaceae.

The other similar root associations are met with actinomycetes (actinorhizal) in form of nodules and with some blue-green algae (cyanobacteria) such as *Anabaena* and *Nostoc* forming coralloid roots of *Cycas.* Both of these kinds of associations are connected with nitrogen fixation by microscopic organisms present in the roots.

7. Zoochlorellae and Zooxanthellae. Some unicellular photosynthetic plants, especially algae, known as *zoochlorellae*, live symbiotically in the outer tissues of certain sponges, coelenterates, molluscs and worms (Fig. 6.2E). Some brown or yellow cells, and probably flagellates (zooxanthellae) are also present. Algae are photosynthetic and produce oxygen and nitrogenous compounds beneficial to hosts and in exchange, they obtain materials such as carbon dioxide and nitrogenous wastes released by metabolism of host animals. The unicellular green alga, *Chlorella vulgaris* lives within the gastrodermal cells of *Hydra.* Likewise, alga *Zoochlorella* lives in a planarian, *Convoluta roscoffensis.*

8. Microorganisms and cellulose digestion. Interspecific mutualism is nicely demonstrated by the flagellate protozoan, *Trichonympha,* an obligate anaerobe in the gut of several species of wood-eating termites where it digests cellulose. *Trichonympha* also occurs in the alimentary canal of wood-eating roach *Cryptocerus.* The termite and roach reduce the wood to small fragments, passing them through the alimentary canal to hind gut, where the protozoans digest the cellulose, changing it into sugar. The host benefits the protozoa by removing harmful metabolic waste products and maintaining anaerobic conditions in the intestine.

9. Fungus gardens. There exist certain mutualistic associations between insects and fungi. For example, the insect-fungus-galls of gall insects (*e.g.,* midge, *Lasioptera*) are lined by a fungus which is parasitic on the host plant. It is believed that the female midge deposits spores of fungus at the time she lays her eggs in the plant or when insect sucks plant sap. Gall fungus seems to assist the insect indirectly by partly breaking down the tissues so that the insect can digest it (Fig. 6.4). Besides these gall insects, there are various cases of insect gardeners such as ambrosia beetles, tropical termites and American ants which do gardening of fungi in their nests to get a regular supply of their food directly or indirectly.

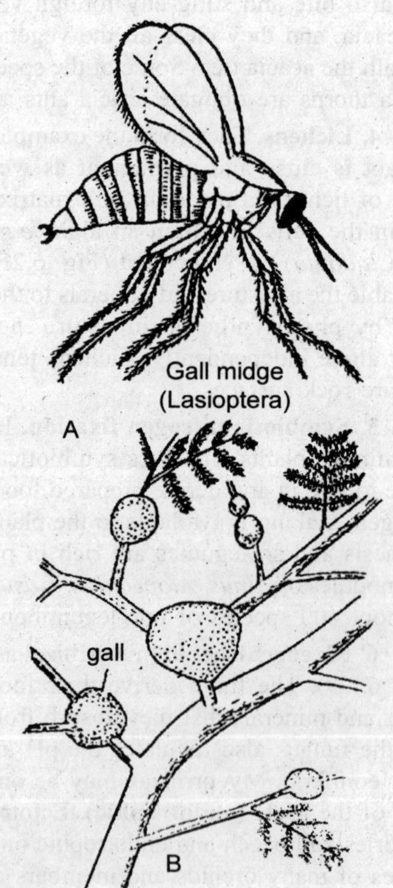

Fig. 6.4. A—Gall midge; B—Plant galls.

II. Commensalism. Commensalism defines the coaction in which two or more species are associated and one species at least, derives benefit from the association, while the other associates are neither benefited nor harmed. The concept of commensalism has been broadened in recent years, to apply to coactions other than those centering on food; thus, in it cover, support, protection, and locomotion are now frequently included. Some common examples of commensalism are the following:

1. Lianas. These plants are common in tropical rain forests where light at ground level is scarce because of the dense and multistoreyed growth of vegetation. Lianas are vascular (woody) plants rooted in the ground but which climb up with the support of other trees and reach almost to the top of the forest canopy. The woody stem of lianas is closely attached to the supporting tree but it is not involved with it in any nutritional relationship. Thus, with much economy of mechanical tissues, lianas are able to get better light. On the basis of the type of device used for climbing their supports, lianas are classified as **leaners, thorn lianas, twiners** or **tendril lianas.** In North Indian tropical deciduous forests *Bauhinia vahili* is a common liana. Other common lianas are species of *Ficus* and *Tinospora.*

2. Epiphytes. Epiphytes are the plants growing on other plants (Fig. 6.2F). They use other plants only as support and not for water or food supply. Epiphytes differ from the lianas in that they are not rooted into the soil. They may grow on trees, shrubs or larger submerged plants. They grow either on the trunks or leaves. Certain common examples of epiphytes are orchids, bromeliads, hanging "mosses", *Usnea* and *Alectoria*. Epiphytes are most common in tropical rain forests. They derive their supply of moisture and nutrient from the frequent rains and debris accumulated in bark crevices. Dust is also a source of the nutrients. Roots of epiphytes often store water in a special tissue, called **velamen**. For meeting water requirements, in the epiphyte *Dischidia*, the leaves fold to form a jug-like structure having a narrow mouth where water accumulates; roots of the epiphyte enter and grow in these specialized cavities of leaf. Many epiphytes develop a thick network of roots upon which wind blown dust accumulates and provides the necessary edaphic environment.

3. Epizoans. Some plants grow on the surfaces of animals. For example, some green algae grow on the long, grooved hair of sloth. *Basicladia* (Cladophoraceae) grows on the backs of fresh water turtle.

4. Epizoite. Likewise, ectocommensals or **epizoite** animals are associated with another animal for the purpose of anchorage and protection. Many small animals, for example, become attached to the outside of larger ones, such as the protozoans *Trichodina* and *Kerona* on *Hydra,* vorticellids (*Vorticella, Epistylis*) or various other aquatic organisms, branchiobdellid annelids on crayfish, the barnacles, which attach themselves to the backs of whales and shells of horse-shore crabs and so on. The remora fishes attach themselves to the bellies of sharks, swordfish, tunny, barracuda, or sea turtle by means of a dorsal fin highly modified into a suction disk on the top of the head. They are of small size and are not burdensome to the host. The host benefits the remora, for when the host feeds, the scraps of food floating back are swept up by the remora.

5. Commensals may also be internal. For example, many harmless protozoans occur in the intestinal tract of mammals, including human beings. Some microorganisms such as bacterium *Escherchia coli* is found in human colon. The pea crab *Pinnotheres ostreum*, lives as a commensal in the mantle cavity of certain sea mussels; the crab steals food collected by the host mollusc but does little if any other known injury. The hermit crab normally lives inside gastropod shell. The pitcher of the pitcher plant found in bogs furnishes a breeding site or home for certain species of midge flies, mosquitoes, and tree toads. Many kinds of microorganisms, plants and animals (such as small crabs, shrimps, polychaets, ophiuroids, ascidians, etc.) live in the canal system of sponges.

The nest of one species often furnishes shelter and protection for other species as well. Ant nests may contain guest species of various other insects; some birds place their nests close to wasps, bees or ants for the protection offered by these insects. Woodchuck burrows are used also by rabbits, skunks and racoons, especially in winter.

6. Commensalism-like associations also exist between a variety of microorganisms and higher plants. For example, the zone of soil around the roots of higher plants characterized by intense microbial activity, called the **rhizosphere**; the surface proper of the roots growing in soil—the **rhizoplane**; the boundary layer of air over the green leaves with active microbes—the

phyllosphere; and the leaf surface—**phylloplane**, constitute important ecological niches, where occur rich populations of microorganisms including fungi, actinomycetes and bacteria. The living roots and leaves of higher plants are known to continuously diffuse various metabolic products, mainly sugars and amino acids. These are sources of nutrition for microorganisms present therein.

III. Protocooperation. Protocooperation is less extreme sort of interaction than mutualism in which the interaction is clearly beneficial to both species, allowing the equilibrium population levels of both to be higher than they otherwise would be. However, it is not obligatory for either species. Examples of protocooperation are many, including the relationship between water moccasins and large birds such as herons and ibises on several islands off the west coast of Florida. The birds nest in the lower branches of relatively unprotected trees, while the snakes congregate around the bases. This protects the birds from tree climbing predators such as racoons. In turn, the snake feed, in part, on fish dropped by the birds and the occasional baby birds that fall out of the nest.

Another good example of protocooperation is of coelenterate, sea anemone (*Adamsia palliata*) attached to the empty molluscan shell harbouring hermit crab (*Eupagurus prideauxi*; Fig. 6.2H). The sea anemone is carried by the crab to fresh feeding grounds and crab in turn is said to be protected from its enemies by sea anemone. Some ecologists prefer to include this association under commensalism.

B. Negative Interactions

In case of negative interactions, one or both species are harmed in any way during their life period. Some authors such as **Clarke**, 1954, prefer to call these types of interactions as **antagonism**. The negative interactions include the following three broad categories :

1. Exploitation. In exploitation, one species harm the other by making its direct or indirect use for support, shelter or food. It is of following types :

(i) Social parasitism. Social parasitism describes the exploitation of one species by another, for various advantages. It is a kind of parasitism in which the parasite foists the rearing of its young into the host. Social parasitism in various stages of development is found among some higher vertebrates and insects. For example, there occurs an **egg parasitism** in two species of birds— old world cuckoos and the brown headed cowbirds of North America, both of which do not build nests of their own, rather they deposit their eggs in nests of other species, abandoning eggs and young to the care of foster parents. Similarly, there occurs a **brood parasitism** in between two Indian birds—Indian koel and crow. The Indian koel (*Eudynamys scolopacea*), like many of its cuckoo cousins (Family Cuculidae), neither incubates its eggs nor feed its young. But these functions are secretly forced upon the house crow (*Corvus splendens*) and jungle crow (*Corvus macrorhynchus*). One species of ant attacks foraging workers of another species and snatches away the food they are transporting or the robber species may deliberately rob another nest for food. Some species of ants make slaves of the workers of other species. Such kind of dependency of one species on another also occurs in other social insects such as termites, wasps and bees. Social insects are apparently the only animals other than humans to have succeeded in domesticating other species, and of cultivating plants particularly fungi for food.

Kleptoparasitism. It is the piracy of food by one individual from another. It is common in hawks and gulls, skuas and wading birds. Kleptoparasitism is most common where potential hosts are aggregated in breeding colonies and feeding groups.

(ii) Parasitism. Parasitism is a kind of harmful coaction (disoperation) between two species. It is the relation between two individuals wherein one individual called **parasite** receives benefit at the expense of other individual called **host**. Parasitism is mainly a food coaction, but the parasite derives shelter and protection from the host, as well. A parasite usually parasitizes a host which is larger in body size than it. Further, a parasite does not ordinarily kill its host, atleast not until the parasite has completed its reproductive cycle. However, a host may die due to some secondary infection or suffer from stunted growth, emaciation, or sterility. The balance between parasite and host is upset if the host produces antibodies or other substances which hamper normal development of the parasite. In general the parasite derives benefit from the relation while the host suffers harm. But this does not mean that all parasites are harmful. The former category constitutes the **pathogenic parasites.**

Classification of parasites. Parasites exhibit a vast diversity in ways and adaptations to exploit their hosts (Box 6.3). The parasites may be **viral parasites** (*e.g.,* bacterial, plant and animal viruses), **microbial parasites** (*e.g.,* bacteria, protozoa, fungi, etc.), **phytoparasites** (*e.g.,* plant parasites) and **zooparasites** (*e.g.,* animal parasites such as platyhelminthes, nematodes, arthropodes, etc.). They may parasitize microorganisms, plants and animals. They may occur on the outside of the host (**ectoparasites**) or live within the body of the host (**endoparasites**). The endoparasites usually live in the alimentary tract, body cavities, various organs or blood or other tissues of host. Ectoparasites may be parasitic only in the immature stages—the hairworm larvae, parasitic in aquatic insects; only the adult parasitic— fleas, on birds and mammals; or both larvae and adults may be parasitic—the blood sucking lice and flies, biting lice, mites, and ticks that occur on birds, mammals, and sometimes reptiles, and the monogenetic trematodes on fish. Similar relation occurs in endoparasites but in them usually all the stages of life are parasitic—entozoic amoebae, trichomonad flagellates, opalinid ciliates, sporozoans, pentastomids, nematodes, digenetic trematodes, acanthocephalans, cestodes, and some copepodes.

Box 6.3 Microparasites and Macroparasites

Parasites may be divided into **microparasites** and **macroparasites.**

1. Microparasites include the viruses, bacteria and protozoa. They are small in size, have a short generation time, multiply rapidly in the host, tend to produce immunity and spread by direct transmission. They are usually associated with dense population of the host.

2. Macroparasites. They are larger in size and include parasitic worms, lice, ticks, fleas, rusts, smuts, fungi and other forms. Macroparasites have a comparatively long generation time, rarely multiply directly in the host, are persistent with continual reinfection and spread by **direct** and **indirect transmission.**

Further, endoparasites rely on various means of transport from one host to another so that their survival, range of distribution, and life cycle are unaffected. For the endoparasite, the interior of its host is its **microhabitat** and **microenvironment**, while the outside environment is the **macroenvironment**. The host acts as a buffer between the parasite and the outside environment. Changes in the microenvironment directly influence the parasite while alterations in the macroenvironment influence it indirectly. Thus, mutual adaptations and tolerance between the host and parasite are important factors facilitating successful transmission of endoparasites, and climatic fluctuations tend to alter the physiology of the parasite, host or vector. Mutual adaptations are the parasite's abilities to establish and maintain itself in a favourable location within the host and exist in a proper form for sufficient length of time enabling subsequent transport to another host and continuation of species.

Animals may also be parasitic to plants. Nematodes infest the roots of plants. Wasps or gnats form galls on plants such as oaks, hickories, willows, roses, goldenrods, and asters. Mites stimulate formation of witches brooms in hackberry. A variety of insects the larvae of which are leaf miners, wood borers, cambium feeders, and fruit eaters should be included here.

Lastly, parasites may be **full-time** (or **permanent**) **parasites** or **part-time** (or **temporary**) **parasites**. Mosquitoes and bugs that suck the blood of their hosts are temporary parasites. Some temporary parasites spend only a part of their life cycle as parasites. For example, glochidium larva of *Anodonta* (freshwater mussel) attaches itself to the body of the fish by means of its hooks and penetrates inside the fish integument to remain buried there for several weeks and finally emerges out as the young mussel to lead an independent existence. Permanent parasites, however, spend their life completely on other organisms. The common examples of permanent parasites are—*Plasmodium, Entamoeba histolytica,* and other protozoan pathogens, different platyhelminthes, nematodes, arthropods, etc.

Parasitic adaptations. Hosts are the habitats of parasites. The problem faced by parasites is to gain entrance into and escape from the host. The life cycles of parasites revolve about these two problems. While studying parasitism as a biotic factor influencing the life and activities of parasites and hosts we often encounter manifold adaptations, both offensive and defensive, being developed by the host, as well as by the parasite. An endoparasite entering a host often meets with the antibodies or phagocytic cells produced by the host. The ectoparasites, and endoparasites have following parasitic adaptations :

1. In parasitic animals, a reduction of organs of special sense of nervous system, and of locomotory organs occur.

2. Most ectoparasites develop some clinging organs such as hooks, suckers, etc., to get attached with the body of their hosts. They also develop special piercing and sucking organs to suck the blood of animals or sap of plants. Certain blood-sucking ectoparasites such as leeches, blood-sucking mosquitoes, etc., contain certain anticoagulant enzymes in their salivary secretions.

3. Most endoparasites exhibit anaerobic respiration, high rate of reproduction, parthenogenesis, hermaphroditism, polyembryony, intermediate host and a complicated life-cycle. Some endoparasites, such as tapeworm, have become so adapted to the host that they no longer require a digestive system. They simply absorb their food directly through their body wall. Further, parasites that live within the bodies of plants and animals possess cuticles or develop cysts resistant to the digestive enzymatic action of the host.

4. Many parasites pass their entire existence in a single host; others require one, two, even three intermediate hosts. It is of ecological significance that both primary and intermediate hosts of a parasite occur in the same habitat or community. Even then the hazards to successful passage from one host to another are so great and mortality so high that enormous quantities of offspring are produced to ensure that at least a few individuals will complete the cycle.

5. Parasites are transferred from one host to another by active locomotion of the parasite itself; by ingestion, as one animal sucks the blood of or eats another; by ingestion as an animal takes in eggs, spores, or encysted stages of the parasite along with its food or drinking water; as a result of bodily contact between hosts; or by transportation from host by way of **vectors**. For example, the bacteria that cause tularemia in humans are carried from rabbit to rabbit by ticks. Human beings contact the disease when they handles infected rabbits but the incidence of infection is greatly reduced in the autumn when cold weather forces the ticks (*i.e.,* the vectors of bacteria) to leave the rabbits and go into hibernation.

Host specificity of parasites. Certain parasites, such as copepodes, infect a wide range of hosts and they are commonly present in various invertebrates and fishes. Most parasitic genera, however, are adapted to hosts of one phylum only. For example, each order of birds possesses its own particular species of tapeworms. This is true even when several order of birds live in the same habitats as do, for instance, grebes, loons, herons, ducks, waders, flamingoes and cormorants. Species of flagellate protozoans that occur in termite's alimentary canals are largely host-specific. Some species of gall wasps attack only one species of oak. Where a single species parasitizes two or more host species, the shape and structure of the gall formed around the egg and larva on both hosts is essentially similar. When several insects are found on the same oak, each kind of parasite produces its own characteristic gall form. The parasite specific gall formation is found to be related with parasite-specific enzyme.

Effects of parasite on the host : disease. Parasites may not cause immediate mortality but they cause damage to body structures, should it become excessive, may cause death. Because of these parasitically caused anatomical damages, the finely adjusted balance of different vital processes of host's body become disturbed and host is said to be diseased. There are a number of causes of diseases. Parasites are one; physiological stress, nutritional deficiency and poisoning are others. Some of the common parasitic agents of disease and consequent mortality of animals are following:

1. Viruses are the potent agents of several disastrous diseases of plants and animals including humans. For instance, some viruses are the agents of hoof and mouth disease in deer, spotted fever in rabbit, encephalitis and distemper in foxes.

2. Bacteria may produce localized inflammatory changes in tissue, enter the blood stream, or produce powerful poisons known as **toxins**. They cause a variety of diseases, notably tularemia, paratyphoid, and tuberculosis among birds and mammals, as well as other diseases in lower types of organisms.

3. Fungus spores of *Aspergillus* may be drawn into the lungs of ground-feeding birds, where they germinate and grow, causing **aspergillosis disease**. Fungus may also develop on the external surface of animals (*e.g.,* ringworm causing fungus).

4. Protozoan parasites are especially important in the alimentary tract and in the blood. A sporozoan species of *Eimeria* damages the walls of the intestine in upland game birds and causes **coccidiosis disease**; *Toxoplasma* becomes encysted in the brain of rodents; *Leucocytozoon* is a common blood parasite of waterfowl and game birds.

5. Worm parasites, such as tapeworms, nematodes, and acanthocephalans may wander through the host's body doing mechanical injury as well as destroying and consuming tissues. The host may respond by forming a fibrous capsule or cyst around an embedded parasite.

6. External parasites such as ticks, fleas, lice, mites, and flies do not commonly produce serious mortality by themselves, but they are often vectors, transmitting protozoa, bacteria and viruses from one animal to another. Heavy infestations of external parasites however lower the vitality of an animal and cause diseases of fur or feathers.

7. Nutritional deficiency in vitamins or minerals, or improper balance among carbohydrates, proteins, and fats may produce malformations, lack of vigour, or even death.

8. Food poisoning, **botulism**, occurs when certain food become contaminated with the toxins released by the bacterium *Clostridium botulinum.*

9. **Physiological stress** is a term that has come to be applied to changes produced in the body non-specifically by many different agencies which may accompany any disease. Effects of stress include loss of appetite and vigour, ache and pains, and loss of weight. Internally, physiological stress causes following abnormalities, such as, acute degeneration of the lymphatic organs, decrease of the blood eosinophils, enlargement and increased secretory activity of the adrenal cortex, and a variety of changes in the chemical constitution of the blood and tissues.

Parasitism in the plants. There are some parasitic vascular plants. Species of *Cuscuta* (doddar) which are **total stem parasites**, grow on other plants on which they depend for nourishment. Young stem twines around the host stem (Fig. 6.5) and it develops adventitious roots which finally penetrate the stem of host, establishing relationship with its conducting elements. The specialized roots are called **haustoria**. Other examples of such association are **total root parasites** as *Orabanche, Conopholis, Epifagus, Balanophora* and *Striga* which are found on roots of higher plants. *Orabanche* affects the yield of mustard and brinjal and is parasitic to the roots of Solanaceae, Cruciferae and Gramineae. *Rafflesia* is root parasite of *Vitis*. Partial root parasites belong to family Loranthaceae (*Viscum album, Loranthus* sp.) They grow rooted in branches of host tree. Examples of partial root parasites include *Santalum album* and *Thesium*. Fungi, bacteria, mycoplasmas, rickettsias and viruses are other common examples of plant parasites. Thus, fungi such as rusts and smuts are parasitic to many crop plants and reduce their production.

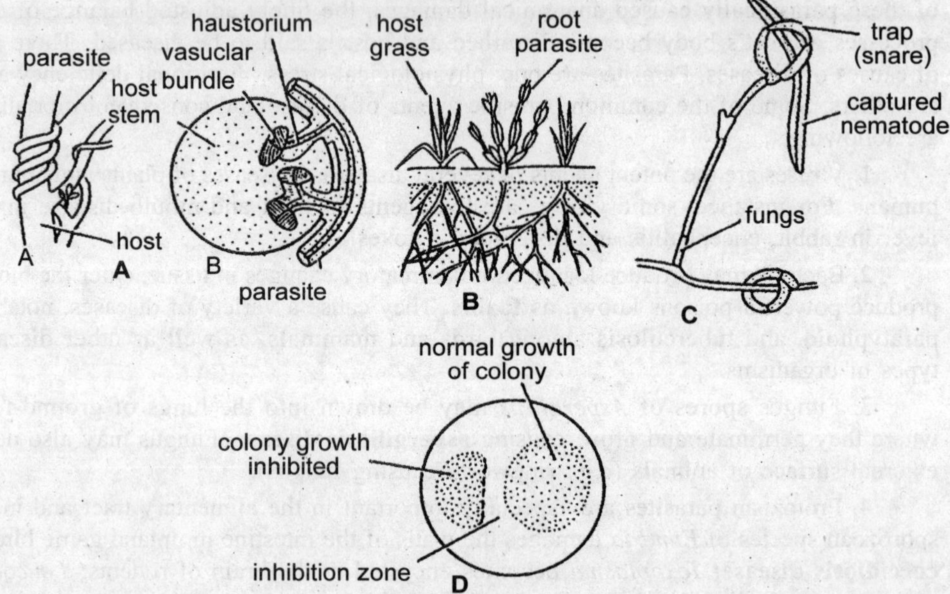

Fig. 6.5. Negative interactions. A—Parasitism, total stem parasites : a–*Cuscuta* twinning the host stem, b–T.S. of stems of host and parasite to show haustoria. B—A partial root parasite with its haustoria attached to the roots of host. C—Predation, a part of the mycelium of a nematophagous fungus (*Arthrobotrys dactyloides*) capturing the nematode as its prey. D—Antibiosis, inhibition of growth of one microbe colony by other colony.

Parasitism may occur within the species. For example, in higher plants the growing pollen tube may be considered as a male plant parasitic on the tissues of the stigma and style of the flower (female plant).

Parasitodism. Some Diptera and Hymenoptera deposit their eggs in the immature stages of other insects; the larvae on hatching feed on the host until they are fully grown. In such type of parasitism in which host generally dies of the larval destruction before the larva emerges and the parasitoid larva generally lives in spite of the host's death, is called **parasitodism**. It stands in between parasitism and predation. The parasitoid may in turn be infested with **hyperparasitoids**. In Chicago, *Samia cecropia*, a saturniid moth, suffers the destruction of nearly 23 per cent of its cocoon by an ichneumonid parasitoid, *Spilocryptus extrematis*, which deposits an average of 33 eggs inside of each cocoon or on the surface of larva. The host larva dies in a few hours after the parasitoid hatches and the ichneumonid larva moves about freely, feeding on the cuticle and burrowing into the tissues to drink the body fluids. *S. extrematis* is infested by another hyperparasitoid, *Aenoplex smithii* which causes about 13 per cent destruction of *S.extrematis* cocoons. A chalcidid, *Dibrachys boucheanus*, feed both upon *S. extermatis* and as a tertiary parasitoid, upon *A. smithii.* A pycnidial fungus, *Cicinnobolus cesatii* is found as hyperparasite on a number of powdery mildew fungi.

(iii) Predation. Predation occurs when members of one species eat those of another species. Often, but not always, this involves the killing of the prey. Most of the predatory organisms are animals, but there are some plants (carnivores) also. Four types of predation have been recognized:

Herbivores are animals that prey on green plants or their seeds or fruits; often the plants eaten are not killed but may be damaged. Typical predation occur when **carnivores** prey on herbivores or on other carnivores. **Insect parasitism** is another form of predation in which the insect parasite lays eggs on or near the host insect, which is subsequently killed and eaten. Finally, **cannibalism** is a special form of predation in which the predator and prey are from the same species.

Predation is an important ecological process from three points of view. One effect of predation on a population may be to restrict distribution or reduce abundance. If the affected animal is a pest, predation appears useful. If it is a valuable resource such as caribou, the predation seems undesirable.. Second, along with competition, predation is the second major process which affects community structure. Third, predation is a major selective force and many adaptations we see in organisms, such as warning colouration, have their explanation in predator-prey coevolution (Evolutionary changes in two or more interacting species are called **coevolution**).

(A) Predation in animals. A predator animal tends to be larger than its prey. The main components of predation are predator and prey. A successful **predator** has following characteristics:

1. The hunting ability of a predator remains well developed.

2. By their hunting activities predators can be regarded as specialized or generalized. **Specialized predators** are those adapted to hunt only a few species. They are forced to move where the vulnerability of a staple prey item drops to a point where the predator population cannot support itself. For example, Peale's falcon shows a marked preference for ducks and pheasants. Deer exhibits a pronounced preference for certain species of browse. **Generalized predators**, not so restricted in diet, adjust to other food sources. The horned owl and buted hawk have a large range of collective prey available. Foxes can shift to vegetable and carrion diet, should conditions require it.

3. Hunting ability and success of predator involve the development of a searching image on the part of the predator. Once it has secured a palatable item of prey, the predator finds it progressively easier to find others of the same kind.

4. Though a predator may have strong preference for a particular prey, it can turn in the time of relative scarcity to an alternate, more abundant species that provides more profitable hunting. For example, if rodents are more abundant than rabbits and quail, foxes and hawks will concentrate on them instead of game animals.

5. Habitat preferences or overlapping territories can bring predator and prey into close contact, increasing prey risks. For example, predatory rainbow trout in Paul Lake, British Columbia, moves into the shoals when their prey, the red side shiner, are most heavily concentrated there.

6. Age, size, and strength of prey influence the direction that predation takes. Predators select food on the basis of size. Mountain lions, for example, avoid attacking large healthy elk, which they cannot successfully handle, and concentrate instead on deer and young or feeble elk (**Hornocker**, 1970).

7. Predators hunt only when it is necessary for them to procure food. The searching rate of predator is influenced by the speed of the predator relative to the speed and escape reactions of the prey, to the distance at which predators first notice and attack the prey, and to the proportion of attacks that result in successful capture.

Like the predators, the prey has its certain defensive specializations. The prey risk is determined by density of prey population, availability of food and protective cover (concealment place), movement, activity, habits, size, age, strength and escape reactions of prey. Besides these striking morphological features, prey species often develop the following two antipredator defence strategies: aposematic colouration and group living.

(i) Aposematic colouration. Many animals have evolved distastefulness as a means of predator defence. Toxins may be obtained from food or synthesized by the prey species. Many of these toxic prey species feature bright colours, termed **aposematic colouration** and these colours are usually regarded as a signal to predators. For example, Monarch butterflies advertise their noxiousness to blue jays by their large size and distinct colouration.

(ii) Group living. Group living is normally not advantageous for the individuals. Then why do birds form flocks and ungulates go about in herds ? One possible reason for living in groups is to reduce predation losses. There are three main advantages a prey organism can obtain by living in a group. First, early detection of predators may be facilitated if many eyes are looking. Prey in groups may, thus, be able to spend more time feeding and less time looking for potential predators. Second, if the prey are not too much smaller than the predator, several of them acting together may be able to deter the predator from attacking. One possible example of this is mobbing in birds (*e.g.*, crow). Primate troops will also mob predators in some circumstances and musk-oxen threatened by wolves gather into a circular defence formation. Third, if the predator is still able to attack a group, it must select one individual in the group to capture. By fleeing in confusion, the prey may confuse the predator, who may not be able to concentrate on any one individual. Alternatively, by being in the centre of a group, an individual may reduce its chances of being eaten.

Regulating effect of predation. It is commonly concluded that if the predator–prey interaction causes a reduction in the prey population, it is deterimental in same fashion to the prey. This belief has led to extensive "predator control" efforts conducted in the name of wildlife conservation. But the coevolution of species within natural ecosystems has in fact led to a dynamic balance between the populations in a community, so that the population sizes of predator and prey species are inter-regulated by feed back mechanisms that effectively control the population of both.

(B) Predation in plants. Plants provide the following examples of their predation:

1. Carnivorous fungi. A number of fungi such as species of *Dactylella, Dactylaria, Arthrobotrys, Zoophagus,* etc., capture insects, nematodes and other worm-like animals. Such fungi use specialized structures such as **traps** or **snares** (Fig. 6.5C) which are formed on their mycelia to capture the nematodes.

2. Carnivorous plants. A number of highly specialized plants such as *Nepenthes, Utricularia, Drosera, Dionaea, Darlingtonia, Sarracenia,* etc., are dependent for part of their nitrogen requirements upon small animals such as insects. These **insectivorous plants** have some specialized structures and mechanisms to attract, trap and ingest insects and then secrete some proteolytic enzymes to "digest" the protein contents of the insect body.

(C) Plants as prey (food) of animals: Browsing, grazing and scraping. Herbivores such as cattle and wild herbivores kill the plants and use unharvested herbs, shrubs or even trees as their food and sometimes pose great problem of management of natural and artificial vegetations (forests and grasslands). Different plants receive varying degree of damage as a result of browsing and grazing. Many insects and ruminants browse lightly over the vegetation. cattle, camels, goats, etc., frequently browse the tender shoots of shrubs and trees and sheep graze the grasses. Generally, annuals suffer more due to grazing than the perennials, shrubs are damaged less than herbs.

Grazing has a profound effect on the composition, structure and physiognomy of vegetation. Heavy grazing reduces the photosynthetic parts so much that many plants succumb or their population decreases. This is true for palatable species. The unpalatable species avoided by grazing animals multiply and increase in number in absence of competition due to decrease in the population of palatable species. In a pasture with *Cynodon dactylon* and *Eragrostis amabilis* as dominant grass species, **Senapati** and **Dash** (1981) reported that a grazing pressure of 1.3 livestock/acre caused the accumulation of higher proportion of plant biomass (68 per cent) underground compared to

46 per cent in an ungrazed (*i.e.,* protected) pasture. Plant diversity was more in the grazed plot and about 18 per cent of NP (= nitrogen and phosphorus) was removed by grazing. Grazing also stimulated higher secondary production by soil invertebrate animals, particularly oligochaetes, but it also increased mortality due to hoof action and trampling.

In *Dichanthium annulatum*, there is a considerable loss of seeds and herbage due to grazing. The reduced reproductive capacity due to loss of seeds is compensated by a remarkable adaptation of increased reproductive buds on rhizomes (**Ambasht** and **Maurya**, 1970). The indirect effect of grazing is on the rate of soil erosion. This is certainly the greatest menace of grazing aspect of biotic factor. The decreased shoot cover ultimately affects root growth and exposes the soil surface to erosion due to rain and run off water and also to wind erosion.

Ambasht (1974) has indicated that grazing involves the large scale consumption of seeds, due to their high nutritive value, which results in poor successive crops, unless the plants adapt to vegetative propagation. Rats selectively remove seeds from grasses and store them in their burrow in large quantities. Fruit-eating birds and animals have some adavantageous effects also. Some seeds fail to germinate unless they have passed through the gut of birds or other animals. The seed coat in these seeds are hard and impermeable to water or oxygen. The seed coat is softened and often digested by the animals and the rest of the seeds come out and germinate quickly.

Further, in the outskirts of Indian forests, villagers tend to scrape grasses to feed their cattle in summer. **Scraping** is usually done right at the ground level. Completely exposed and loosened top soil is blown by strong summer wind. This is followed by rainy season, during which initial heavy showers result in great soil erosion.

2. Amensalism. Amensalism is a comprehensive term in biotic interactions to denote the adverse or depressing effect of one of the competing population or species while the other, remains stable. Thus, amensalism is a situation in which one population definitely inhibits the other while remaining unaffected itself. By so modifying the environment, the organism improves its own chance of survival. Ammensalism which commonly involves some type of chemical interaction, by which the organisms of one species affects the well-being and growth of individuals or the population biology of other, is called **allelochemic**. The inhibitor substance may be inorganic chemicals such as acids or bases which are produced by pioneering organisms (inhibitor species) and which reduce the competition for nutrients, light, and space, between the amensal species and inhibitor species. The production of relatively simple organic toxins is another source of chemical inhibitors. These toxins inhibit seedling growth in the vicinity. This may affect succession of plant species, especially important in the early stages (**Muller**, 1966). A third type of inhibitor chemical is the **antibiotic**–a potent antimicrobial agent. It is a substance produced by an organism, which, in low concentrations, can inhibit or kill the growth of another organism. For example an antibiotic called **penicillin** is secreted by *Penicillium notatum* and *P.chryogenum* and **griseofulvin** (an antibiotic for ringworm-fungs/pathogen by (*P.griseofulvum*).

The majority of inhibiting chemicals are produced as secondary substances by plants and released into the soil through the roots or leaf wash. The suppression of growth through the release of chemicals by a higher plant is known as **allelopathy**. Thus, allelopathy (*allelon* = each other + *pathy* = suffering) means chemical control of distribution among plants. For example, walnut tree produces a non-toxic substance, juglone which is found in its leaves, fruits and other tissues. When the leaves or fruits fall on the ground, **juglone** is released to soil, where it is oxidized to a substance that inhibits the growth of certain under story species and garden plants such as heaths and broad-leaf herbs, and favours others such as bluegrass and blackberries. In Southern California, *Salvia leucophylla* leaves emit some volatile oils which reach the soil surface and inhibit the germination of seeds of other species and the inhibitory effects persist for several months. Similar allelopathic effects are also reported to be exerted by *Adenostoma fasciculatum* (**Christensen** and **Muller**, 1975). **Datta** and **Chakrabarti** (1982) prepared weed plant extracts such as of *Croton bonplandianum* and *Clerodendrum viscosum* in water and tested their effects under different dilutions on seed germination and growth of crops as pea, mustard, rice and lettuce. *C. bonplandianum* inhibits germination of seeds of above crops except of pea at high concentration. They found that the active ingredient in *C.bonplandianum* leaves showing allelopathic potential are **abscisic** and **phaseic acids**, while in *C.viscosum* inhibiting mustard seed germination may be a **terpene**. *Clerodendrum* is also reported to exert allelopathic potential on several weeds such as *Abutilon indicum, Amaranthus spinosus, Cassia tora,* etc.

Agriculturists have recognized the action of **smother crops** as weed suppressors. The smother crops include barley, rye, sorghum, millet, sweet clover, alfalfa, soyabean and sunflower. Their inhibition of weed growth was assumed to be due to competition for water, light or nutrients. However, **Overland** showed that barley (*Hordeum vulgare*) inhibited the germination and growth of several weeds, even in the absence of competition for nutrients or water. He found that the adverse effect of barley is partly due to the secretion by its roots of chemicals (alkaloid) that reduce growth and germination of nearby weeds.

The term **antibiosis** generally refers to the complete or partial inhibition or death of one organism by another through the production of some substance's or environmental conditions as a result of metabolic pathway. These substances and/or conditions are harmful (antagonistic) to other organisms. Antibiosis is more commonly referred to secretions by microorganisms that check the growth of others. Production of chemicals that are antagonistic to microbes—the **antibiotics** is well known. Antibiotics produced by bacteria, fungi, actinomycetes, and lichens are widespread in nature and may be one of the reasons why bacteria pathogenic to humans cannot multiply well in soils. *Penicillium*, a fungus found in soil produces antibiotic substances that check the growth of a large variety of bacteria. A number of antibiotics such as penicillin, have been used extensively in human medicine.

Antagonistic substances are also reported in some algae, as for example in culture of *Chlorella vulgaris*. Some substance accumulates which inhibits the growth of the diatom, *Nitzschia frustrulum*. Substances produced by senescent cultures of *Chlorella* and of the diatoms, *Navicula* and *Scendesmus* inhibit the filter-feeding of *Daphnia* in laboratory. Pond "blooms" of blue-green algae especially *Microcystis* are known to produce toxins such as hydroxylamine which causes death of fish and cattle. In marine waters, populations of some microbes, popularly known as **red tide**, cause catastrophic destruction of fish and other aquatic animals.

Allelochemic effects are of great variety in both plants and animals: repellants, escape substances, suppressants, venoms, inductants, counteractants, attractants, signals, stimulants, autotoxins, autoinhibitors, and so on. **Pheromones** (ectohormones) are chemical messages between members of a species and are especially important in reproductive behaviour, social regulation and recognition, alarm and defense, territory or trail marking, food location, and so on. Many of these effects are beneficial to the individual, while other serve competitive purposes.

3. Competition. The presence of other organisms may limit the distribution of some species through **competition**. Such competition can occur between any two species that use the same sorts of places. Note that two species do not need to be closely related to be involved in competition. For example, birds, rodents and ants may compete for seeds in desert environments. Competition among animals is often over food, water and mates. Animals may also do competition for space, *i.e.,* nesting sites, wintering sites, safer sites from predators. Plants can compete for light, water, nutrients, or even for pollinators and/or attachment sites. Competition is an important process affecting the distribution and abundance of plants and animals.

There are two different types of competition, defined as follows:

(a) Resource competition (also called **scramble competition**) occurs when a number of organisms (of the same or of different species) utilise common resources that are in short supply.

(b) Interference competition (also called **contest competition**) occurs when the organisms seeking a resource harm one another in the process, even if the resource is not in short supply.

Gravity of interspecific competition depends on the extent of similarity or overlap in the requirements of different individuals and the shortage of supply in the habitat. Such sort of competition may have the following effects on the population:

(i) Regulation of population size. The role of competition in regulating size of the population remains directly effective by causing mortalities through *fighting, nest destruction* and *loss of food supplies.* It also results from increased predation and dispersal into other regions.

(ii) Segregation of species into different niches. Due to interspecific competition a species is confined to its particular niche. In the absence of competition, there occurs an expansion of the species beyond the usual limits of its niche. Thus, interspecific competition causes the segregation of species into different niches.

(iii) Speciation. Interspecific competition also acts as the process of evolutionary differentiation between population, in which one older species may split into two or more new species (*i.e.,* speciation).

(iv) Effect on biological variation. In nature the selective effect of competition affects the extent of variation that is maintained in species population. Such an effect will be related to taxonomic nature of the competing populations. If the habitat is occupied by a species population composed of interbreeding but geographically or ecologically distinct segments, such as *subspecies* or *ecotypes,* competition between them will tend to extend variation within the individual populations until it approximates to the variation within the habitat. Thus, overall tolerance range of the species will widen or competition will tend to widen the biological variation already present in the segments.

On the other side, if the habitat is occupied by competing independent species populations, gene flow will accordingly be very much restricted. In this case, the pressure of interspecific competition will select from each species population those genomes that impart a special adaptation to a particular and limited range of the environment. Here the competition will tend to narrow the biological variation present in each of the competing species.

(v) Exploitation. Competitive interaction may cause exploitation. Laboratory studies have provided various examples of such exploitation. When two flour beetles, *Tribolium confusum* and *Tribolium castaneum* are kept together, the former drives out the latter in competition. Thus, **exploitation** develops when two species are competing for the same resource. However, when their cultures were infested with *Adelina*, a sporozoan parasite that may kill the beetles, it was found that *Adelina* have significant effect on *T. castaneum* but not on *T. confusum.*

These experiments clearly demonstrated that the out come of competition is not constant but rather is affected both by *extrinsic factors* such as weather and parasites and by *intrinsic factors* such as genetic composition (genotype) of the competing populations.

(vi) Interference. It occurs when two species are not competing in the exploitation sense, but when their coexistence in the same habitat modifies some behavioural pattern of one or the other. For example, when the smell or mere sight of a male mouse of a different species may cause recently fertilized females of some species to abort. Similarly, when brook trout and rainbow trout occupy the same stream, they treat each other as if they were the same species. The largest fish occupy the deepest and darkest parts of the stream and the smaller ones are forced to live in less favourable parts.

Gause Principal

G. F. Gause (1934) has proposed that *no two forms can share exactly the same ecological niche for an idefinite period of time, eventually one will replace the other.* Thus, each mode of life is filled by just one species; niches may overlap, but a single niche is inhabited by only one form.

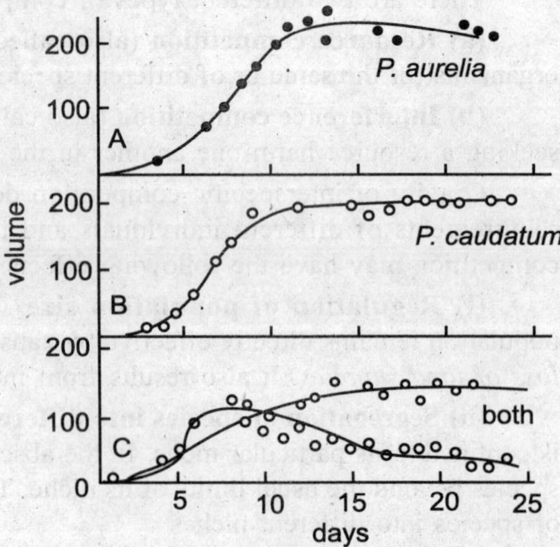

Fig. 6.6. Population growth and interspecific competition between two species of *Paramecium* in laboratory cultures. When cultured separately (A) and (B), the two species approximate to logistic growth. However, in mixed cultures (C) *P. aurelia* survives and *P.caudatum* dies out.

This principle (of Gause) supports directly the conception of competition and so is known as the **competitive exclusion principle (Hardin** 1960). The following experimental data have supported the Gause's principle:

1. Gause himself performed a classic experiment to demonstrate his principle by using two species of the same genus of a ciliated protozoan, *Paramecium aurelia* and *Paramecium caudatum.* By growing two organisms in similar vessels, containing the same fluid culture medium and food (*i.e.,* a kind of bacterium) and under similar conditions of temperature and light, Gause made sure that the *niches occupied by the two populations were similar.* Fig. 6.6A and B show the growth curves and population densities when the two protozoans are grown in separate culture and Fig. 6.6 C, when both protozoans grown in the same culture. In the latter case, *P. caudatum* was eliminated. In isolation, *P. aurelia* grew more rapidly than *P. caudatum* (Fig. 6.6 A) indicating that *P. aurelia* uses the available food more rapidly and efficiently than does *P. caudatum* (Fig. 6.6 B). Therefore, when both occupy the same niche, one would expect *P. aurelia* to be more successful competitor. The results are consistent with the competitive exclusion principle predicted by mathematical theory (Box 6.4).

Box 6.4

A.J. Lotka and **V.Volterra** in 1920 developed theoretical or mathematical models to explain prey-predator interactions which produce oscillatory patterns. They also discussed the problem of competition. If two different species populations require a common resource for living and if this resource is potentially or actually limiting, then the two species are said to be in competition for it. These models predict the following three types of outcomes:

1. One species only survives, it being the one with the greatest negative effect on its competitor. Growth of the surviving population to its carrying capacity is slower than if the second population had been absent.

2. Both species coexist indefinitely. This occurs when interspecific competition is less intense than intraspecific one in both species. Neither population reaches the carrying capacity it would have in the absence of other species.

3. The species beginning at higher density persists, and the other is eliminated. This is a special case when the populations have equally negative effects on the growth of each other, but interspecific competition is stronger than intraspecific one.

2. An experimental study of interspecific competition was carried out by **Thomas Park** (1962) with flour beetles in a very simple homogeneous habitat, a jar of flour or wheat bran. He found that when two different species of the same genus (*i.e., Tribolium castaneum* and *Tribolium confusum*) were placed in a jar of flour, invariably one species was eliminated sooner or later while the other continued to live.

3. Experiments of **Frank** (1957) with *Daphnia* have also given comparable results with algae as food in the culture, *Daphnia pulicaria* eliminates the closely related *Daphnia magna.*

Lotka–Volterra Predator-Prey Model

In the 1920s, **Alfred Lotka** and **Vito Volterra** independently represented the dynamics of predator-prey interactions with what is now called the **Lotka–Volterra Predator-Prey Model:**

$$\frac{dN}{dt} = rN - aNP$$

$$\frac{dP}{dt} = baNP - mP \qquad \text{(Equation 6.1)}$$

In these equation, N represents the number of prey individuals and P represents the number of predator individuals. The equation for change in the prey population over time (dn/dt) assumes

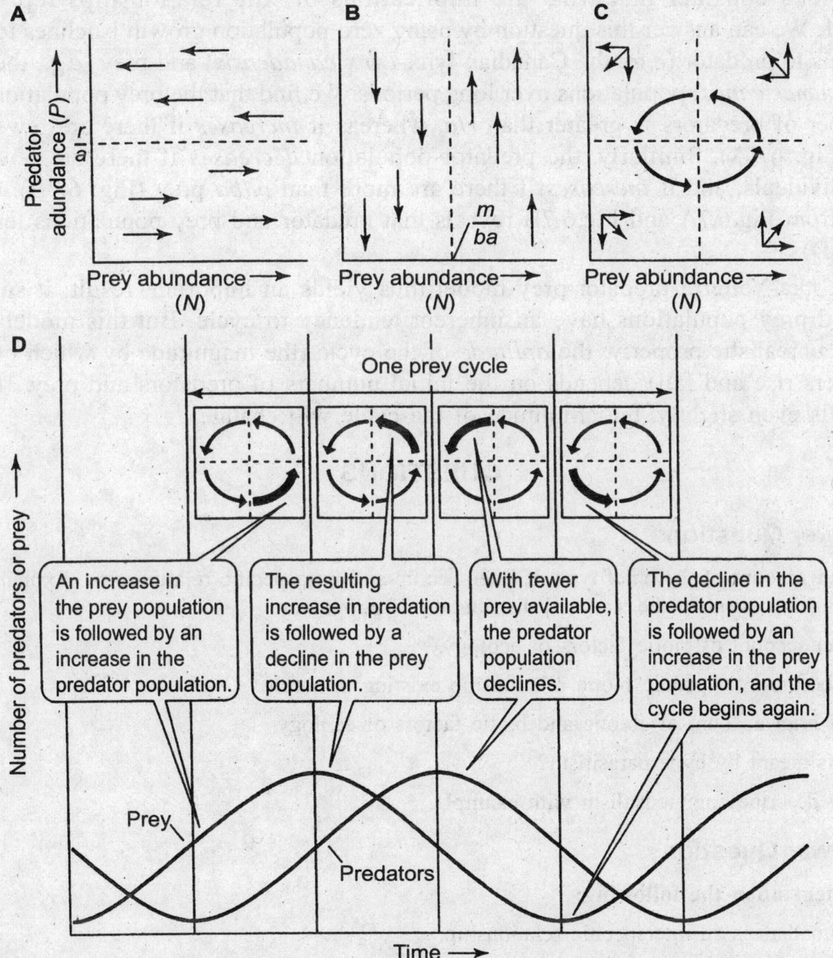

Fig. 6.7. The Lotka- Volterra predator-prey model produces population cycles. A- Considering the prey population first, the abundance of prey does not change when $dN/dt = 0$, which occurs when $P = r/a$ (see equation 6.1). B- Similarly, considering the predator population, the abundance of predators does not change when $dP/dt.=0$, which occurs when $N = m/ba$. Combining the results in parts (A) and (B) suggests that predator and prey populations have an inharent tendency to cycle. These cycles are shown here in two ways: C- by plotting the abundance of prey, and D- by plotting the abundance of both predators and prey versus time, the four inset diagrams in D plot the abundance of predators versus the abundance of prey. In D, note the predator abundance curve is shifted ¼ of a cycle behind the prey abundance curve (after Cain *et al.*, 2015).

that when predators are absent (*P*=O), the rate at which they kill prey depends in part on how frequently predators and prey encounter one another. This frequency is expected to increase with the number of prey (*N*) and with the number of predators (*P*), so a multiplicative term (*NP*) is used in the equation for *dn*/*dt*. The rate at which predators kill prey also depends on the efficiency with which predators capture prey; this capture efficiency is represented by the constant *a*, so that the overall rate at which predators remove individuals from the prey population is *aNP*.

Predators starve when there are no prey. Thus, the equation for change in the predator population over time (*dP*/ *dt*) assumes that in the absence of prey (*N*=O), the number of predator decreases exponentially with a mortality rate of *m* (*i.e.*, *dp*=*dt*=-*mP*).When prey are present ($N \neq 0$), individuals are added to predator population according to the number of prey that are killed (*aNP*) and the efficiency with which prey are converted into predator offspring (represented by the consant *b*). Thus, the rate at which individuals are added to the predator population is *baNP*.

Now let's consider that what are implications of the relationships represented by Equation 6.1. We can answer this question by using zero population growth isoclines to determine what happens to predator (*e.g*., the Canadian lynx-*Lynx canadensis*) and prey (*e.g*., the snowshoe hare (*Lepus americanus*) populations over long periods. We find that the prey population *decreases* if the number of predators is greater than *r*/*a*, whereas it *increases* if there are fewer than *r*/*a*, predators (Fig. 6.7A). Similarly, the predator population *decreases* if there are fewer than *m*/*ba* prey individuals, and it *increases* if there are more than *m*/*ba* prey (Fig. 6.7B). Combining the results from Fig.6.7A and Fig.6.7B reveals that predator and prey populations tend to cycle (Fig. 6.7C, D).

The Lotka-Volterra predator-prey model thus yields an important result: it suggests that predator and prey populations have an inherent tendency to cycle. But this model also has a curious and unrealistic property: the *aplitude* of the cycle (the magnitude by which predator and prey members rise and fall) depends on the initial numbers of predators and prey. If the initial number shifts even slightly, the amplitude of the cycle will change.

QUESTIONS

Long Answer Questions

1. Give an account of the chief type of interspecific and intraspecific relationships. Explain How these relationships maintain the size and distribution of population.
2. Give an account of biotic factors of ecology.
3. Describe the interspecific biotic relationship existing in nature.
4. Give a brief account of abiotic and biotic factors of ecology.
5. What is meant by hyperparasitism?
6. Briefly describe commensalism with example.

Short Answer Questions

1. Comment upon the following:

(*i*) Mutualism is an interspecific relationship.

(*ii*) Competition is a positive interaction between the two organisms.

(*iii*) Coexistence means living together of different members of the same species in the same environment.

2. Write short notes on the following:

(*i*) Parasitism (*ii*) Symbiosis
(*iii*) Mutulasim (*iv*) Predation
(*v*) Antibiosis (*vi*) Antagonism

(*vii*) Interspecific relationship (*viii*) Botic factors
(*ix*) Gause's principal

3. Differentiate between the following pairs.
(*i*) Commensalism and mutualism (*ii*) Predator and parasite
(*iii*) Parasitism and symbiosis (*iv*) Commensalism and amensalism
(*v*) Interspecific and intraspecific relationship (*vi*) Symbiosis and commensalism
(*vii*) Commensalism and parasitism *(Rohilkhand 2001)*

4. Write short notes on any two biotic factors of ecology.

Very Short Answer Questions

1. What are Lotka-Volterra equations?
2. Define biotic.
3. Define competition.
4. What is competitive exclusion principle?
5. What is mutualism?
6. Give examples of mutualism.
7. What is cannibalism?
8. Define commensalism.
9. What is protocooperation.
10. Enlist types of negative interactions of populations.
11. What is amensalism?

Fill in the Blanks

1. Lichen is the example of ....
2. Symbiotic nitrogen fixation takes place by .... of root nodules of .... plants.
3. In .... associations, tree roots become infested with fungal hyphae.
4. Kleptoparasitism occurs in .... and gulls.
5. Griseoflavin is an .... for .... ; it is produced by .... *griseofulvum.*
6. Allelopathy means .... of distribution among plants.
7. Gause principle states that no two species with same ecological requirements can coexist. It is also called .... principle.

Multiple Choice Questions

1. Biotic factors are
(*a*) chemical factors of soil which affect life
(*b*) physical factors of soil which affect life
(*c*) all living organisms which influence other organisms
(*d*) factors of atmosphere which affect life

2. Occurrence of zoochlorellae in the body wall of *Hydra* is an example of
(*a*) predation (*b*) a food chain involving parasite
(*c*) commensalism (*d*) mutualism or symbiosis

3. Mycorrhiza is the phenomenon of
(*a*) parasitism (*b*) symbiosis
(*c*) antagonism (*d*) endoemism

4. The relationship between water fern *Azolla* and cyanobacterium *Anabaena* is
(*a*) symbiosis (*b*) competition
(*c*) scavenging (*d*) commensalism

5. In *a* population where competition between individuals is severe then the distribution is likely to be
   (*a*) random (*b*) uniform
   (*c*) irregular (*d*) non-random

6. Competition for food, light and space is most severe in
   (*a*) closely related species growing in the same area or in the same niche
   (*b*) closely related species growing in different habitat
   (*c*) distally related species growing in the same habitat
   (*d*) distally related species growing in different habitat

7. Mutualism is an
   (*a*) interaction favourable to both but obligatory
   (*b*) interaction favourable to both but not obligatory
   (*c*) interaction favourable to one and harmful to other
   (*d*) interaction favourable to one and neutral to other

8. A relation which is harmful to both the population
   (*a*) neutralism (*b*) competition
   (*c*) parasitism (*d*) antibiosis

9. Which one of the following interspecific interactions are antagonistic relationships
   (*a*) parasitism and neutralism (*b*) competition and antibiosis
   (*c*) competition and predation (*d*) parasitism and predation

10. Which of the following statement is incorrect
   (*a*) sea-anemone on hermit crab is an example of protocooperation
   (*b*) mutualism, protocooperation, commensalism cannot be included under symbosis
   (*c*) lichens, an association of fungs and algae is an example of mutualism
   (*d*) those epiphytes which use other plants for support only and not for water or food supply are examples of commensalism

11. The tick bird on a rhinoceros would exemplify:
   (*a*) commensalism (*b*) antibiosis
   (*c*) protocooperation (*d*) neutralism

12. Which of the following is considered as negative interaction between two populations
   (*a*) predation (*b*) mutualism
   (*c*) commensalism (*d*) protocooperation

13. In commensalism
   (*a*) one population is benefitted and other is not affected
   (*b*) one population is benefitted and other is inhibited
   (*c*) one population is benefitted and other is affected adversely
   (*d*) both the populations are benefitted

14. The animal which preys the other living species is known as
   (*a*) cannibalistic (*b*) predator
   (*c*) parasite (*d*) primary consumer

15. The association between *Sacculina* and crab is referred to as
   (*a*) symbiotic association (*b*) mutualistic association
   (*c*) commensalic association (*d*) parasitic association

16. An association where one is benefitted and the other (host) is neither benefitted nor adversely affected is called as
   (*a*) mutualism (*b*) protocooperation
   (*c*) ammensalism (*d*) commensalism

17. The type of interaction between two species where protozoan (*Trichonympha*) lives in the gut of termite for cellulose digestion, is referred to as
(*a*) commensalism (*b*) protocooperation
(*c*) mutualism (*d*) neutralism

18. The relation between algae and fungi in a lichen is best described as
(*a*) symbiosis (*b*) parasitism
(*c*) convenient (*d*) commensalism

19. Which of the following indicates exploitation of one population by the other
(*a*) parasitism (*b*) competition
(*c*) antagonism (*d*) antibiosis

20. One of the following relation which is useful to one population and harmful to other population
(*a*) competition (*b*) predation
(*c*) commensalism (*d*) mutualism

21. One of the following relation is useful to one population and harmful to other population
(*a*) commensalism (*b*) amensalism
(*c*) mutualism (*d*) parasitism

22. Which one of the following is an example of mutualism between plant and animal
(*a*) lichen made up of alga and fungus
(*b*) *Hydra* having Zoochlorella in gastrodermal cells
(*c*) bacterium *Rhizobium* and root nodules of leguminous plant
(*d*) none of these

23. Intraspecific competition is competition among
(*a*) species (*b*) individuals of a population
(*c*) individuals of a community (*d*) different populations

24. Which of the following is a negative interaction between two populations
(*a*) symbiosis (*b*) mutualism
(*c*) commensalism (*d*) antibiosis

25. One of the following is a positive interaction between two populations
(*a*) antagonism (*b*) competition
(*c*) predation (*d*) symbiosis

26. According to the Competitive Exclusion Principle developed by the Russian biologist **G. F. Gause** two species cannot remain in the same community if they have the same
(*a*) taxonomy (*b*) niche
(*c*) habitat (*d*) food requirements

27. The effect of inter-specific competition on niches is to make them
(*a*) smaller (*b*) larger
(*c*) change location (*d*) more triangular

28. *Cuscuta* is an example of
(*a*) complete stem parasite (*b*) complete root parasite
(*c*) partial root parasite (*d*) partial stem parasite

29. Parasitic castration occurs in
(*a*) *Fasciola* (*b*) *Nereis*
(*c*) *Sacculina* (*d*) *Palaemon*

30. A sucker fish (*Remora*) on a shark would exemplify
(*a*) parasitism (*b*) commensalism
(*c*) mutualism (*d*) neutralism

31. Symbiosis is expressed by
(*a*) soil (*b*) fungi
(*c*) lichens (*d*) virus

32. The term antibiosis means (*Rohilkhand 2001*)
(*a*) inhibition/death of one organism by another through the production of some substances in environmental conditions
(*b*) production of antibiosis (*c*) production of antigen
(*d*) both (*a*) and (*b*)

ANSWERS

Very Short Answer Questions

1. Lotka-Volterra equations belong to a model developed in the 1920s by Alfred Lotka, an American mathematician and Vito Volterra, an Italian mathematician, based on the logistic equation, expressing interspecific competition (such as predator-prey relationships).
2. Biotic refers to the living components of an ecosystem.
3. Competition is relationship between two species that is mutually deterimental to both populations.
4. It is the principle stating that no two species can permanently occupy the same ecological niche.
5. Mutualism is a positive interaction. It is a relationship between two species in which the growth and survival of both populations benefit.
6. Pollination by animals, lichens, symbiotic nitrogen fixation, mycorrhizae, zoochlorellae, cellulose digestion in termites.
7. Intraaspecific predation.
8. Commensalism is positive interspecific interaction. It is a relationship between two species in which one population is benefited but the other is not affected, *e.g.*, lianas, epiphytes, epizoans, epizoites, etc.
9. Protocooperation is relationship between two species, in which both population benefit from the relationship, but the association is not obligatory, frequently termed facultative cooperation, *e.g.*, sea anemone attached to the empty molluscan shell harbouring hermit crab.
10. Exploitation such as parasitism, predation, amensalism and competition.
11. Amensalism is a relationship between two species in which one population is inhibited and the other not affected, *e.g.*, antibiotics.

Fill in the Blanks

1. Mutualism **2.** *Rhizobium*, leguminous **3.** Mycorrhizae **4.** Hawks
5. Antibiotic, ringworm, *Penicillium*. **6.** Chemical control
7. Competitive exclusion principle.

Multiple Choice Questions

1. (*c*)	**2.** (*d*)	**3.** (*b*)	**4.** (*a*)
5. (*d*)	**6.** (*a*)	**7.** (*a*)	**8.** (*b*)
9. (*b*)	**10.** (*a*)	**11.** (*c*)	**12.** (*a*)
13. (*a*)	**14.** (*b*)	**15.** (*d*)	**16.** (*d*)
17. (*c*)	**18.** (*a*)	**19.** (*a*)	**20.** (*b*)
21. (*d*)	**22.** (*b*)	**23.** (*b*)	**24.** (*d*)
25. (*d*)	**26.** (*b*)	**27.** (*a*)	**28.** (*a*)
29. (*c*)	**30.** (*b*)	**31.** (*c*)	**32.** (*a*).

7

Population (Population Ecology)

The term population has its origin in the Latin word *populus*, meaning people. In ecology, a **population** may be defined as a group of organisms of the same species occupying a particular space. Thus, we may speak of the deer population of Corbett National Park, the deer population of Kaziranga Wild Life Sanctuary, the human population of Delhi, or the human population of India. The ultimate constituents of the population are **individual organisms** that can potentially interbreed. The populations may be subdivided into **demes** or local populations, which are groups of interbreeding organisms, the smallest collective unit of a plant or animal population. Individuals in demes, thus, share a common gene pool. The boundaries of a population both in space and in time are vague and in practice are usually fixed arbitrarily by the investigator.

Box 7.1

The population has various **group characteristics**, which are statistical measures that cannot be applied to individuals. These group characteristics are of three general types. The basic characteristic of a population is its **size** or **density** which is affected by four primary population parameters such as **natality** (births), **mortality** (deaths), **immigration** and **emigration** (Fig. 7.1). In addition to these attributes, one can derive secondary characteristics of a population, such as its **age distribution**, **genetic composition**, and **pattern of distribution** (distribution of individuals in space). These population parameters result from a summation of individual characteristics. Thus, group attributes of population are different from those of an individual, for instance an individual cannot have a density, it cannot have a rate of birth or death, although it occupies a place, is born and dies. The group attributes of a population with respect to rates of birth, death, reproduction, etc., form the science of **demography**. However, when they are studied from a ecological point of view, this forms a **population ecology** or **democology**. In other words, population ecology is the study of individuals of the same species where the processes such as aggregation, interdependence between individuals, etc., and the various factors governing such processes are emphasized.

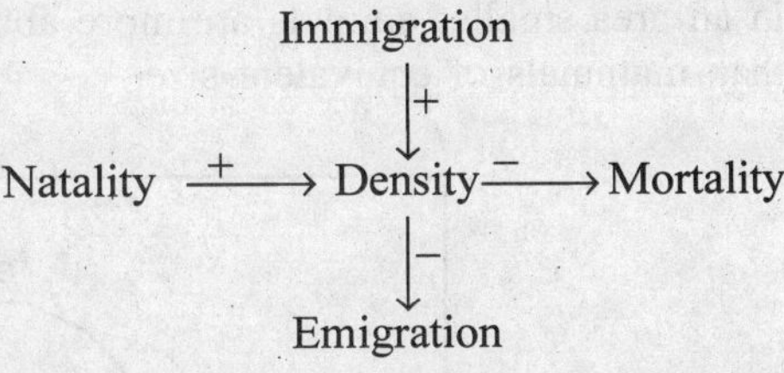

Fig. 7.1. The primary population parameters.

The ecological study of populations include the following three main aspects:

1. Population characteristics;
2. Population dynamics;
3. Regulation of population.

7.1 POPULATION CHARACTERISTICS

The population has the following characteristics :

A. Population Size and Density

Total **size** of population is generally expressed as number of individuals in a population. The population size (N) at any given place is determined by the processes of **birth** (B), **death** (D), new arrivals from outside or **immigration** (I) and going out or **emigration** (E). Therefore, change in population size between an interval of time Nt + 1 is Nt (initial stage) + B – D + I – E.

Population density is defined as numbers of individuals per unit area or per unit volume of environment. Larger organisms as trees may be expressed as 600 trees per hectare, whereas smaller ones such as phytoplanktons (as algae) as 2 million cells per cubic metre of water. In terms of weight it may be 100 kilograms of fish per hectare of water surface. Density may be **numerical density** (number of individuals per unit area or volume) or **biomass density** (biomass per unit area or volume). When the size of individuals in the population is relatively uniform, as mammals, birds or insects, then density is expressed in terms of number of individuals (numerical density). But, when the size of individuals is variable, which is true for fishes, trees or mixed populations (biomass density). To measure density of a population in terms of measurement of biomass which represents wet weight, dry weight, volume and carbon and nitrogen weight. Population density is also measured as **abundance** or absolute number of population. Generally, in an area smaller animals are more abundant than larger animals. But birds are less abundant than mammals of equivalent size.

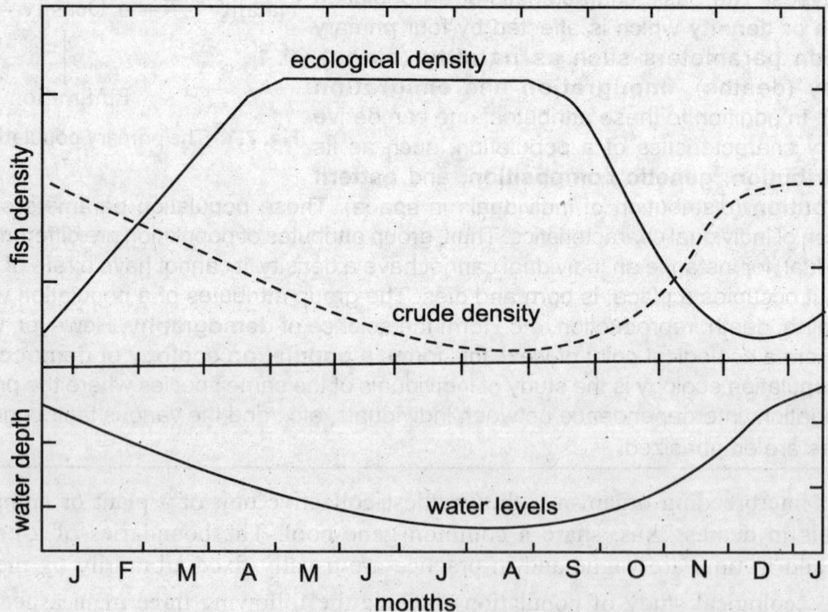

Fig. 7.2. Crude density and ecological density of fish prey in relation to the breeding of stork predator.

Further, since the pattern of distribution of organisms in nature is different it becomes important to distinguish between crude density and ecological density. **Crude density** is the density (number or biomass) per unit total space. **Ecological density** (also called **specific** or

economic density) is the density (number or biomass) per unit of habital space, *i.e.,* available area or volume that can actually be colonised by the population. In nature, organisms grow generally clumped into groups and rarely are uniformly distributed. For example, in plant species such as *Cassia tora* and *Oplismenus burmanni* individuals are found more crowded in shady patches, and few in other parts of same area. Thus, density calculated into total area (shady as well as exposed) would be crude density, whereas the density value for only shady area (where the plants actually grow) would be ecological density.

Fine distinction between the crude density and ecological density becomes more clear by the Fig. 7.2, which is based upon **Kahl's** (1964) study on fish density in a water body. As shown in the figure the crude density of small fish in the area as a whole goes down as the water level drops during the winter dry season, but the ecological density in the contracting pool of water increases as fish are crowded into smaller and smaller water area. Thus, with changing time, water level and total number of fish, the two densities become different.

B. Patterns of Population Dispersion (or Spatial Distribution)

Individuals of a population arrange themselves in a manner that is specific for each population and these arrangements in space appear to be of great significance in the studies of the dynamics of the ecosystems. In nature, due to various biotic interactions and influence of abiotic factors, the following three basic population distributions can be observed: random distribution, uniform or regular dispersion, and clumped dispersion (Fig. 7.3).

C. Age Structure

In most types of populations, individuals are of different age. The proportion of individuals in each age group is called **age structure** or **age distribution** of the population. Age distribution of the population influences natality (birth rate) and mortality (death rate). Reproductive ability is determined by age of the females and, hence, natality denotes the numerical representation of young ones produced in unit time by the same age group. Similarly chances of death are more towards the earlier and later periods of life span and so mortality is under the control of age. Therefore, the ratio of the various age groups in a population determines the current reproductive status of the population and helps in anticipating its future. From an ecological point of view there are three major **functional** or **ecological ages** (age groups) in any population. These are (i) **pre-reproductive** (or juvenile or dependent phase), (ii) **reproductive** (or adult phase) and (iii) **post-reproductive** (or old age). These categories were proposed by **Bodenheimer** (1958) and help in visualizing the status of a population (Fig 7.4).

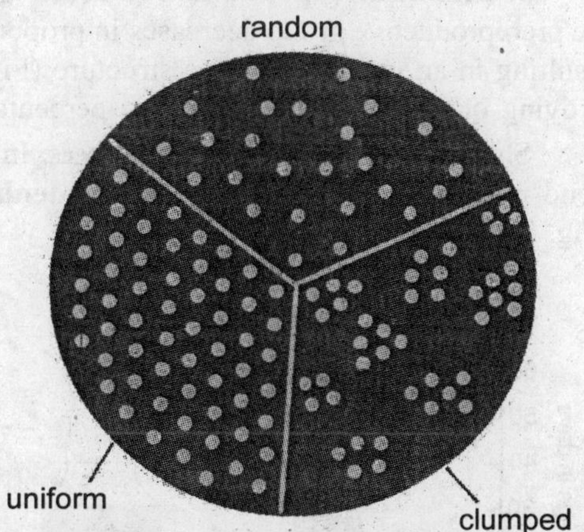

Fig. 7.3. Three basic patterns of population dispersion.

1. Expanding population. In a rapidly growing population (Fig. 7.4 A) birth rate is high and population growth is exponential, as in case of housefly, *Paramecium,* yeast and algae.

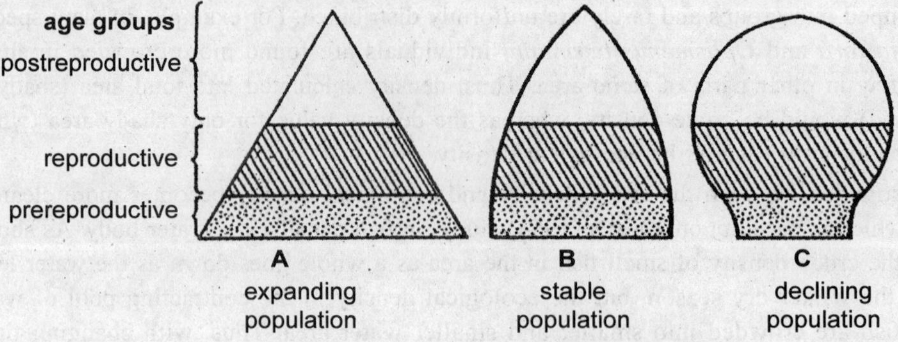

Fig. 7.4. Age structure in different types of population.

Under these conditions each successive generation will be more numerous than the preceding one; as a result a pyramid-shaped (triangular) age structure results (such graphic forms are called **age pyramids**; Box 7.2).

Box 7.2

An age pyramid is a vertical bar graph in which the number or proportion of individuals in various age ranges at any given time is shown from youngest at the bottom of the graph to oldest at the top. It is very useful in monitoring the commercial exploitation of food species by humans among other things.

2. Stable population (stationary population). As the rate of growth slows and stabilizes, the prereproductive and reproductive age groups become more or less equal in size; the post-reproductive group remain as the smallest. The graphic representation of this stabilized population is that of a **bell shape** (Fig. 7.4 B).

3. Diminishing population (Declining population). If the birth rate is drastically reduced, the prereproductive group decreases in proportion to the reproductive and postreproductive groups, resulting in an **urn-shaped** age structure (Fig. 7.4 C). This is representation of a population that is dying off. It also indicates a low percentage of young individuals.

Shift in age structure in honey bees. In a study of a hive of honey bees, where a seasonal trend in the population takes place, **Bodenheimer** observed all the three shifts in age structure (Fig. 7.5).

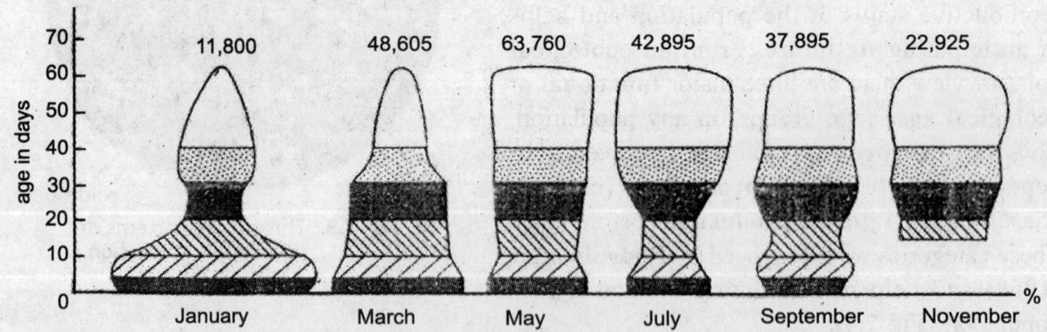

Fig. 7.5. Changes in age structure in a hive of honey bees over one season.

There is a change from rapidly growing population, with the triangular shape in January, to a stable population, represented by a bell shape in March, and a progressive dying off represented by a series of urn shapes from July through November.

Species differences in age structure. The relative proportion of the life cycle occupied by each of the "ecological ages" differ from species to species. In humans, for example, the *prereproductive period* (birth to about age 15) occupies 21 per cent of the mean life expectancy, the *reproductive period* (ages 15 to 45) occupies about 42 per cent and the *postreproductive period* occupies about 37 per cent. In rat, these proportions are about 25, 20 and 55 per cent, respectively.

Examples of age structure in animals

1. In contrast to relatively short prereproductive stages of humans and rats, many insects spend well more than half their life cycle in that stage. Some dragon flies spend 2 years in the egg and larval stages and survive hardly 4 weeks as an adult. Thus, they are being capable of reproduction for as little as only 1 or 2 days of that brief life span.

2. Many birds and mammals have a very short postreproductive period. The chaparrel black-tailed deer, for example, is capable of reproduction until death at an age of 10 years. Because the 1 year olds are not reproductive, the age structure in that population is a diminishing one with a peculiar pyramidal urn shape: about 42 per cent population is in the prereproductive category (age 0-1); 58 per cent is in reproductive segment, equally divided into the age group 1-3 and age group 3-10.

3. In their study of human louse, **Evans** and **Smith** (1952) determined that the stable age distribution consisted of 5.69 per cent adults, 26.43 per cent larval and 67.88 per cent eggs. Similar age distribution exists in the grain-and rice-weevil.

Examples of age structure in plants

1. Among plants, a number of **annual plants** start flowering and set their first seeds when very young. This situation is particularly true in extremely demanding and unpredictable environments, such as deserts.

2. Other annual plants developing in stable, predictable environments have relatively long prereproductive (vegetative) stage whose change into reproductive phase is often timed by **photoperiod**, that is the relative proportion of daylight and darkness.

3. In **biennial plants** at least 1 year of prereproductive growth occurs before seed is produced and under stressful conditions flowering may be delayed for several years.

4. Among herbaceous **perennial plants,** the prereproductive period appears to be quite variable: from 2 years in carpenter weeds (*Prunella vulgaris*) to 8 years in water avens (*Geum rivale*). Among bulbous species, the prereproductive period is from 1 (dahlias and gladiolus) to 4 to 7 years (daffodils, tulips).

5. In woody perennial plants, conifers have earlier reproductive ages for a given length of life than angiosperms. Several conifers that survive 200 years reproduce before age of 10, whereas most angiosperms with life span of over 200 years spend at least 20 years in the juvenile period. In general, the ratio of life span of prereproductive stage among angiosperms is 10:1. *Plants with short prereproductive periods have short longevity and those with long prereproductive periods have long longevity and a long reproductive period.*

D. Natality

Population increases because of natality. **Natality** is a broader term covering the production of new individuals by birth, hatching, germination, or fission. The **natality rate** may be expressed as the number of organisms born per female per unit time. In human population, the natality rate is equivalent to the 'birth-rate'.

At this stage, one should be clear regarding two essential aspects of the reproduction which forms main component of natality: 1. **Fertility** is the physiological notion which indicates that *an organism is capable of breeding.* 2. **Fecundity** is an ecological concept that is based *on the numbers of offspring produced during a period of time.* Fecundity is of two types—**potential fecundity** and **realized fecundity**. For example, the realized fecundity rate for an actual human population may be only one birth per 8 years per female in the childbearing ages. Whereas the potential fecundity rate for human is one birth per 9 to 11 months per female in the childbearing ages.

Natality is of following two types:

1. Maximum natality. (Also called **absolute**, **potential** or **physiological natality**). It is theoretical maximum production of new individuals under ideal conditions which simply means that there are no ecological limiting factors and that reproduction is limited only by physiological factors. Hence, absolute natality is constant for a species population. This is also called **fecundity rate.**

2. Ecological natality. (Also called **realized natality** or simply **natality**). It refers to population increase under an actual, existing specific condition. Thus, ecological natality takes into account all possible existing environmental conditions. This is also called **fertility rate.** Ecological natality varies with the size and composition of the population, and with environmental conditions. It is expressed as—

Δ Nn = production of new individuals

Δ Nn/ Dt = the absolute natality rate (B)

Δ Nn/NDt = the specific natality rate (b) (*i.e.,* natality rate per unit of population)

where, N = the reproductive part (mature female) of the population or initial number of organisms or total population

n = new individuals in the population

t = time

Δ = delta : a change in value.

Further, the rate at which females produce offsprings is determined by the following three population characteristics:

1. **Clutch size** or the number of young produced on each occasion;
2. the time between one reproductive event and the next, and
3. the age of first reproduction.

Thus, natality usually increases with the period of maturity and then falls again as the organism gets older (Fig. 7.8). But there are some trees which continue to increase fruit production as they get older. Natality patterns differ in tropical and temperate populations.

E. Mortality

Mortality means the rate of death of individuals in the population. It is a negative factor for population growth. Like natality, mortality may be of following types :

1. Minimum mortality. It is also called **specific** or **potential** mortality. Minimum mortality represents the theoretical minimum loss under ideal or non-limiting conditions. It may be constant for a population.

2. Ecological mortality. It is also called **realised mortality**. It is the actual loss of individuals under a given environmental condition. Ecological mortality is not constant for a population and varies with population and environmental conditions, such as predation, disease and other ecological hazards.

All organisms have a **physiological** or **potential longevity** representing the age (called **old age**) up to which the organisms can live under ideal conditions without stresses or diseases and death occurs purely due to aging or **senescence**.

Like natality, mortality may be expressed as the number of individuals dying in a given period (death per time), or as specific rate in terms of units of the total population or any part there of. The death rates vary among species and are correlated with rates of reproduction (natality). The death rate of a species is influenced by a number of factors such as destruction of nests, eggs or young by storms, wind, floods, predators, accidents, and desertation of parents, but of fundamental importance is the number of young that are born in relation to the carrying capacity of the habitat. When more youngs are born than the habitat can support, the surplus must either die or leave the area. Mortality rate is usually expressed as a percentage of individuals dying within a given time. The mortality rate in many species varies from one age level to another; thus, a mean death rate has only general significance. In birds, however, the death rate is nearly constant once they become adult, and it is then apparent that it varies inversely with adult longevity. In adult penguins, pelicans, shore birds, gulls, and swifts, the annual mortality rate is commonly between 12 and 30 per cent; in herons, hawks, and owls it is about 30 per cent; in ducks, doves, and song birds it is between 40 and 68 per cent, while in gallinaceous birds it is the highest, 60 to 80 per cent. As mortality varies positively with age in the majority of organisms, the specific mortalities at particular ages can be illustrated in the form of life table.

Vital index and survivorship curves. A birth-rate ratio (100 × births/death) is called **vital index**. For a population the surviving individuals are more significant for a population than the dead ones, the death rate can be represented by **survivorship curves** or **life tables**, both of which provide an estimate of organisms surviving at various ages.

(i) Survivorship curves. The pattern of mortality with age is best illustrated by survivorship curves which plot the numbers surviving to a particular age. There are following three types of survivorship curves (Fig. 7.6) which represent the different nature of survivors in different types of populations:

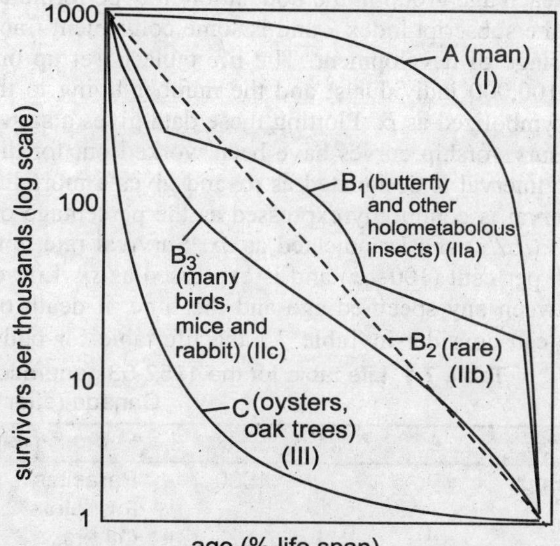

Fig. 7.6. Different types of survivorship curves plotted on the basis of survivors and age. Examples of each has been given within the brackets.

(a) Highly convex curve. This type of curve (pattern I) in the figure is the characteristic of the species in which the mortality rate of the population is low until near the end of the life span. Thus, such species tend to live throughout their life span, with low mortality rate in young and adult. Many species of large animals as deer, mountain sheep, man and small rotifers show such curves. Among various large-sized perennial plants, generally the plant die after reaching reproductive phase within a period of old age.

(b) Highly concave curve. This type of curve (pattern III) is the characteristic of such species where mortality rate is high during the young stages. Some birds, blacktail deer, oysters, shell fish, oak trees, short-lived weedy annuals, etc., exhibit pattern III type curve.

(c) Diagonal straight-line curve. This type of curve (pattern IIb) indicates an age-specific constant survivorship, *i.e.,* a constant rate of mortality occurs at every age. Some animals such as hydra, gull, American robin, etc., exhibit this type of curve.

In fact, no population in the real world has a constant age-specific survival rate throughout the whole life span. Thus, a slightly concave or sigmoid curve (pattern IIc) is characteristic of many birds, mice and rabbits. In them, the mortality rate is high in the young but lower and almost constant in the adult (1 year or older). In some holometabolous insects (*i.e.,* insects with complete metamorphosis), such as butterflies, the survival rate differs in successive life-history stages and the curve becomes the **stair-step** type survivorship curve (curve IIa), the initial, middle and final steep segments represent the egg population and short lived adult stages and the two middle flatter segments represent the larval and pupal stages which exhibit less mortality. Some earthworms also exhibit stair-step type survivorship curve. Lastly, crowding (high density) in certain populations such as black tail deer population may influence the shape of survivorship curve (**Taber** and **Dasmann**, 1957).

The survivorship curve of human population is highly convex. This has become possible because of increased medical care, better hygiene, improved nutrition, and so on. The average life span of an Indian male has gone up from 30 years (before 1947) to 58 years now. But maximum longevity has not gone up very much in any society.

(ii) Life tables. Species differ widely in the number of young produced each year, in the average age to which they live and in their average rate of mortality. When sufficient facts about a species are known, a **life table** that tabulates the vital statistics of mortality and life expectancy for each age group in the population may be formulated. In such tables, age is usually represented by the subscript index x and is some convenient fraction of a species mean life span, such as year or stage of development. The life table is set up on the basis of an initial cohort of 100, 1000, or 100,000 individuals; and the number living to the beginning of each successive age interval is symbolized as lx. Plotting these data gives a **survivorship curve** for the species and a number of survivorship curves have been worked out for different species. The number dying with each age interval is designated as dx and gives a mortality rate. The rate of mortality during each age interval is commonly expressed as the percentage of the number at the beginning of the interval 100 (dx/lx) and is indicated as qx. Survival rate is the difference between the mortality rate and 100 per cent (100–qx) and is expressed as sx. Life expectancy (ex) is the mean time that elapses between any specified age and the time of death of all animals in the age group. For example we can consider in Table 7.1 the life table for budworms of Canada.

Table 7.1. Life table for the 1952-53 generations of the spruce budworm in New Brunswick, Canada (after **Morris** and **Miller**, 1954).

x	lx	Factors responsible for dx	dx	qx
Eggs	1000	Parasites	17	2(—)
		Predators	86	9(—)
		Others	6	1(—)
			Total = 109	12
Instar I	891	Dispersal, etc.	428	48
Hibernacula	463	Winter	79	17
Instar II	384	Dispersal	242	63
Instars III-VI	142	Parasites	51	36
		Disease	3	2
		Birds	20	14
		Others	61	43
			Total = 135	95
Pupae	7	Parasites	0.6	8
		Predators	0.7	10
		Others	1.3	18
			Total = 2.6	36
Moths	4.3	Total for generation	= 995.7	99.5

Uses of life tables. Life tables are useful for computing the average longevity of a population, for showing the age composition of a population, for indicating critical stages in the life cycle at which mortality is high, for showing differences between species, for showing the success of the same species in different biotopes, for furnishing information of value in game and fish exploitation (yield) and in control of pests.

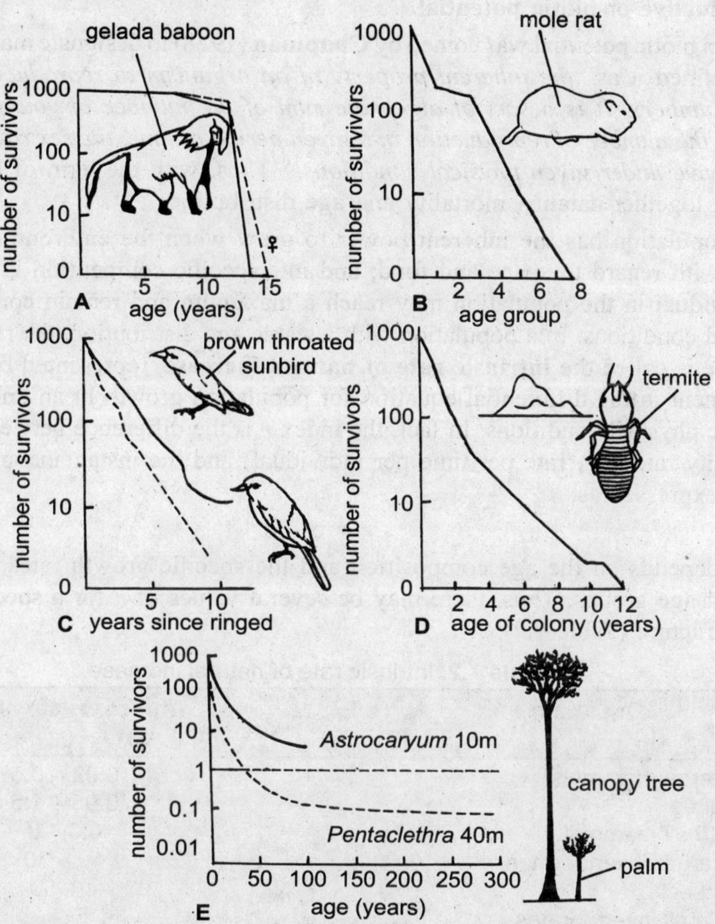

Fig. 7.7. Some examples of survivorship curves of real population. A—Convex curve in gelada baboons, (montane Ethiopia); B—Diagonal curve of mole rat (Nairobi, Kenya); C—Diagonal curve of brown-throated sunbird (forests of south-east Asia); D—Concave curve of whole termite colonies (Nigerian savanna); E— Concave curves of two rainforest trees (solid line is for understory palm in Mexico; broken line for a canopy tree in Costa Rica).

F. Biotic Potential

Charles Darwin (1859) had shown in his natural selection theory that all organisms reproduce many more offspring than survive. Thus, he estimated that a single pair of elephants would have over 19 million offspring alive after 750 years. **Kormondy** (1978) mentioned that a single pair of English sparrows after being introduced into USA could give rise to 275, 716, 983, 698 descendants in ten years. **Howard** estimated that the female *Musca domestica* on the average produces 120 eggs at a time and there are seven generations per year. Assuming that all survive one year and all the females (50 : 50 sex ratio) produce each generation, then at the end of the seventh generation, there would be more than six trillion flies. **Dash** and **Kar** (1990) found that *Lepidochelys olivacea* (the olive ridley sea turtle) female lays, on the average 110 eggs in the Orissa Coast. Assuming a 50 : 50 sex ratio, a female after one egg-laying session is replaced by 55 newborn females. A female lays eggs many times in her life time. If all the offspring survive, one female can be replaced by a few hundred females. All these examples indicate the potential of organisms to reproduce. This reproductive capacity of the organism is called **reproductive** or **biotic potential**.

The term biotic potential was coined by **Chapman** (1928) to designate maximum reproductive power. He defined it as "*the inherent property of an organism to reproduce to survive, i.e., to increase in numbers. It is a sort of algebraic sum of the number of young produced at each reproduction, the number of reproduction in a given period of time, the sex ratio and their general ability to survive under given physical conditions.*" Thus, with the term of biotic potential, one is able to put together natality, mortality and age distribution.

Each population has the inherent power to grow when the environmental conditions are not limiting with regard to space and food; and interspecific competition is absent. The growth rate per individual in the population may reach a maximum and remain constant under specific environmental conditions. In a population with a stable age distribution, the reproductive capacity or growth rate is called the **intrinsic rate of natural increase**, represented by the symbol r. The r is the exponent in the differential equation for population growth in an unlimited environment under specific physical conditions. In fact, the index r is the difference between the instantaneous specific natality rate (*i.e.*, rate per time per individual) and the instantaneous specific death rate and may be expressed as:

$$r = \text{b} - \text{d}$$

The r depends on the age composition and the specific growth rates due to reproduction of component age groups. Thus, there may be several values of r for a species depending upon population structure (Table 7.2):

Table 7.2. Intrinsic rate of natural increase.

	Organism	Approximate biotic potential (r) per year
1.	Large mammals	0.02 – 0.5
2.	Birds	0.05 – 1.5
3.	Small mammals	0.3 – 8
4.	Larger invertebrates	10 – 30
5.	Insects	4 – 50
6.	Small invertebrates (including large protozoans)	30 – 800
7.	Protozoa and unicellular algae	600 – 2,000
8.	Bacteria	3,000 – 20,000

Under natural conditions biotic potential is a rare phenomenon, since environmental conditions do not permit unlimited growth of any population. The difference between the maximum r (biotic potential) and the rate of increase which occurs in an actual laboratory or field conditions is often taken as a measure of the **environmental resistance**, which is the sum total of environmental limiting factors which prevent the biotic potential from being realized. Environmental resistance is often low when a species is first introduced into a new territory so that the species increases in number at a fantastic rate as when the rabbit was introduced into Australia and English sparrow and Japanese beetle were brought into the United States. But as a species increases in number the environmental resistance to it also increases, in the form of organisms that prey upon it or parasitize it, and the competition between the members of the species for food and living space, and consequently, population decreases. An equilibrium will be reached either by decreasing the birth rate (natality) or by increasing the mortality rate.

7.2 POPULATION DYNAMICS

Populations have characteristic pattern of increase which are called **population growth forms**. Such growth forms represent the interaction of biotic potential and environmental resistance. The study of population dynamics is done by three approaches (1) mathematical models, (2) laboratory studies and (3) field studies.

The growth is the most fundamental dynamic feature that a species population displays. Populations characteristically increase in size in a sigmoid, S-shaped or logistic fashion. When a few organisms are introduced into an unoccupied area, the growth of the population is at first slow (**positive acceleration phase**), then becomes very rapid (**logarithmic phase**) and finally slows down as the environmental resistance increases (**the negative acceleration phase**) until an equilibrium level is reached around which the population size fluctuates more or less irregularly according to the constancy or variability of given environment. The level beyond which no major increase can occur represents the saturation level or **carrying capacity**. The carrying capacity or equilibrium density is represented by the letter ***K***. It is often useful to define the maximum rate of growth of the population. This parameter, generally termed the **intrinsic rate of natural increase**, is symbolized r_0 and represents the growth rate of a population that is infinitely small. Accordingly such type of population growth can be described by following **logistic equation**:

$$dN/dt = r_0 N (K - N) /K$$

where r_0 = innate capacity of population to increase (birth rate without resource limitation), N = population size and K = highest population density that can be maintained in real environment, *i.e.*, at carrying capacity.

There are two main types of population growth forms. (1) J-shaped and (2) S-shaped or sigmoid forms (Fig. 7.8). The growth forms are due to the nature of species and prevailing environmental conditions. In **J-shaped curve** there is a rapid increase in density with the passage of time (called **exponential growth**). The density values when plotted against time give a J-shaped growth curve and at the peak the population growth ceases abruptly due to environmental resistance. For example, the population growth curve in human populations and growth of yeast, *Drosophila* and rabbit under laboratory conditions show an initial slow rate and then it accelerates and finally slows giving the growth curve which is **sigmoid** or **S-shaped**. The peak constant level represented by K or upper level (called **asymptote**) of the sigmoid curve is called the **maximum carrying capacity**. It marks the limit to which the environment can support the population.

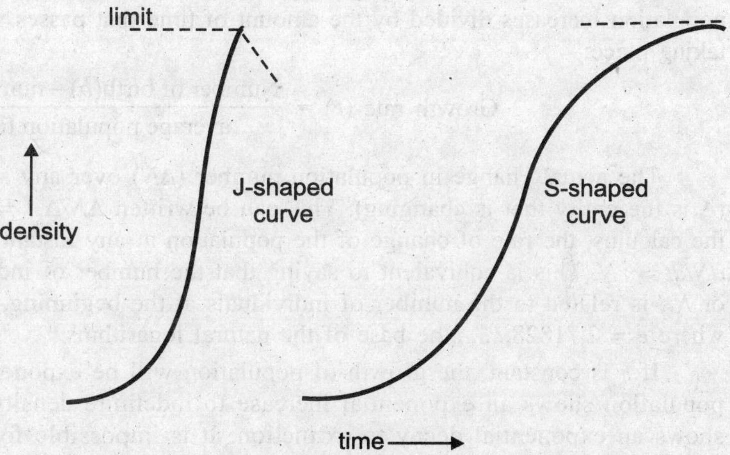

Fig. 7.8. J-shaped and S-shaped population growth curves.

Plant Population Dynamics

In many respects, plant populations behave like the animal population, but, they have some unique features such as follows : Most higher plants are **modular** organisms, developing from a single zygote but producing an indeterminate number of repetitive structures, called **modules** vegetatively. A clump of herbs, grasses or trees may be product of one zygote. Plants cannot move to mate or disperse. Thus, they have evolved means as gravity, wind, water flow or animals for dispersal of pollen, seed, vegetative parts, etc. The **seed population** present in the soil for different species are referred to as **seed bank** or **seed pool** (**Silvertown**, 1987). All these seeds do not germinate or all the seedlings do not establish. Some die due to environmental stresses and this is called **environmental sieve** which allows only the stronger individuals to survive. In most cases, the seeds germinate in batches and seedlings of one lot is known as **cohort**. In this way, from a huge seed bank through ecological selection, cohorts are formed and these in turn result into adult population. This process is referred to as **recruitment**. Further, a plant may originate from a vegetative part, called **ramete** (or **tiller**) or from seed called **genet**. Thus, ramete and genet form two levels of population structure. The term **clone** is normally used to designate the population derived from ramete of the same parent plant.

The 3/2 thinning law. Most aspects of growth of population are density related. One important generalisation applied is **3/2 thinning law**. If we plot the relationship between the dry weight and density of shoots (known number of individuals of) in plant population, the line relating weight of each individual to density has a slope of –1.5 (or – 3/2). The slope would be –1, if increasing density has been exactly compensated by reduction in weight of individuals. Thinning is normally inversely density dependent, but does not always occur if the growth of the plants is extremely plastic. The 3/2 is universal and applied well in a wide variety of plants from mosses to trees. Exact reason of its occurrence is still not clear.

Growth Rate of Population

The rate of growth of a population is expressed as the number of individuals by which the population increases divided by the amount of time that passes while this population increase is taking place:

$$\text{Growth rate } (r) = \frac{\text{Number of birth}(b) - \text{number of deaths } (d)}{\text{average population time internval}}$$

The actual change in population number (ΔN) over any span of time (Δt) is equal to rN (Δ is the entity that is changing). This can be written $\Delta N/\Delta\, t = rN$ or, using the symbology of the calculus, the rate of change of the population at any instant time (dN/dt) can be expressed $dN/dt = rN$. This is equivalent to saying that the number of individuals at any arbitrary time t, or Nt, is related to the number of individuals at the beginning, N_0, by the equation $Nt = N_0e^{rt}$, where $e = 2.71828$........,the base of the natural logarithms.

If r is constant, the growth of population will be exponential. If r is positive ($b>d$), the population shows an exponential increase to indefinite density and if r is negative ($b<d$), it shows an exponential decay to extinction. It is impossible for a population to change at an exponential rate indefinitely. However, there are many cases in which conditions are such that b is substantially larger than d for a period of time, following which conditions change so that d becomes much larger than b. The responses of populations to variations of this sort is an exponential "**population explosion**" during favourable conditions, followed by a "**crash**" when conditions change. Diatom populations in Lake Michigan, USA, for example, undergo such exponential increases at different times of years, triggered by variations in abiotic factors within the lake, followed by equally rapid declines.

7.3 POPULATION DISPERSION

Populations have a tendency to disperse, or spread out in all directions, until some barrier is reached. Accordingly, population dispersion is the movement of individuals into or out of the population area. It takes three forms : **emigration**—one-way outward movement; **immigration**—one-way inward movement; and **migration**—periodic departure and return. Dispersion supplements natality and mortality in shaping population, growth form and density, and also it plays a significant role in the distribution of plants and animals even to the areas previously unoccupied by the members of the population. Most types of population dispersion occur due to a number of reasons such as for obtaining food, avoiding predators, preventing overcrowding, result of action of wind and water, environmental factors as light and temperature, breeding behaviour, physiological reasons as secretion of some hormone or for interchange of genetic material between populations.

Emigration

Emigration under natural conditions occurs when there is overcrowding in the migratory locust, lemming, grouse, snowy owl, snowshoe rabbit, Arctic fox, gray squirrel and occasionally in other species. This is generally regarded as an adaptive behaviour that regulates the population on a particular site and prevents over-exploitation of the habitat. Further, it leads to occupation of new areas elsewhere. By dispersing into new localities, there is opportunity gained for interbreeding with other populations leading to more genetic heterozygosity and adaptability.

It is, of course, population pressure that is responsible in large part for the dispersal of the young and extension of ranges into new areas. Under normal conditions adult animals, especially among the higher vertebrates, are well established on their territories and the youngs are forced to seek homes elsewhere. Among insects, there is a relation between emigration and inherited behaviour tendencies. Individual tent caterpillars, both larvae and adults, differ innately in the extent to which they show activity even within same colony. In the development of populations of excessive size, spread of infestations of the insect into new regions is largely by the more active individuals. The outbreak finally terminates when the proportion of sluggish individuals comes to predominate in the population (**Wellington**, 1966). Continuous emigrations are rare and when they occur, result in **depopulation**. Equilibrium of populations is maintained in such circumstances by enhancing the reproductive ability as well as by decreased mortality among the populations.

Immigration

Immigration leads to a rise in population level, causing an **overpopulation** which may lead to an increase beyond the carrying capacity. These immigrations result in increased mortality among the immigrants or decreased reproductive capacity of the individuals. Both emigration and immigration are initiated by weather and other abiotic and biotic environmental factors.

Migration

Migration is a peculiar kind of population dispersion which involves the mass movement of entire population. This can occur only in mobile organisms and best developed in insects such as desert locust, *Schistocerca gregaria,* migratory locust, *Locusta migratoria*, butterfly, *Danaus plexippus* and in the migratory dragonflies, *Libellula quadrimaculata* and *Pantala flavescens*; in fishes as salmons and eels, in birds and in certain mammals. Most two-way migratory movements are rhythmic processes of population and regular periodicity is a common feature. Very often, environmental periodicities control these migratory movements, as for example day and night rhythms, lunar periods, tides and changing seasons. The monarch butterflies, *Danaus plexippus*, travel very long distances and their migrations are found to be pathed every year through their same conventional routes, and their migratory movements are initiated by the oncoming winter and the return trip being influenced by spring.

In most cases, migration of population may occur for food, shelter, or reproduction. Better utilization of uninhabited or hitherto untouched habitats and their resources are the greatest benefits derived from the migratory movements. However, during migration of population, mortality of numerous individuals may occur due to different ecological hazards such as temperature fluctuations, scarcity of food, predation, etc. Anyhow, migration has certain benefits for populations—as it enables wider dispersion of populations; it avoids intraspecific competition for food, shelter or any other means.

7.4 REGULATION OF POPULATION SIZE

The inherent tendency of all animal populations is to increase in number. But this increase in number is, however, not infinite since the carrying capacity of the environment always imposes a restriction upon it. Thus, after reaching the carrying capacity level, the population density tends to fluctuate above and below this level and such fluctuation in population between upper and lower limits tend to give some stability to the population. The nature of the processes that regulate the numbers of plants and animals is a major problem of population ecology.

Many groups of animals and plants are provided with unique and intrinsic self-regulatory mechanism such as failure of reproduction and self-inflicted mortality for controlling the size of the population. Populations grow when natality exceeds mortality, and they decline when mortality exceeds natality. However, limitation of animal number in a population is brought about by the action and interaction of two basic regulatory processes namely **density independent** and **density dependent factors**. Density independent factors are the extrinsic factors which tend to regulate the density of a population under different conditions, appearing to act on the population and inflict loss of individuals irrespective of the population density. Variations in space or cover, favourable weather and food occur independently of population densities and may cause drastic changes in the abundance of animals. Such ecological or environmental factors influence negatively or positively all the individuals of a population irrespective of density.

The density dependent factors are intrinsic or biotic factors and they depend on coaction between individuals within the same population or between populations of different species. Density-dependent factors may stabilize populations at an asymptote, the level of which is determined by the carrying capacity of the environment. Some of the important density dependent factors are competition, reproductivity, predation, emigration and disease. The combination of factors or any specific factor involved in the density dependent action may vary from species to species. Further these species have been known to display the property of intercompensation. According to intercompensation, if there is a change in the environment to relieve the population from the pressure of an existing effect, then the population increases till it reaches a level when the second effect takes over. Thus, if the predators that normally keep herbivorous animals down are removed, the population of herbivores may increase to become overcrowded and result is starvation. A supply of abundant food at this junction would make the individuals susceptible to diseases due to the intensity of crowding (**Wilson** and **Bossert**, 1971). Generally, the population fluctuations controlled by extrinsic factors tend to be irregular and correlated with the variation in one or more major physical limiting factors such as temperature, food, water, etc., while fluctuations of populations controlled by intrinsic factors exhibit regularity and population cycles.

Population Cycles

Populations are said to be cyclic when they alternatively erupt and subside in a more or less uniform manner between high and low levels of density. Different animals exhibit population cycles different times. The best established cycles of population density of fluctuations are those of periodicities of 3-4 years and 9-10 years. The **3-4 year cycles** are most commonly observed in many birds such as snowy owl, willow ptarmigan, capercaillie, blackgame, hazel grouse etc.; mammals such as lemmings, voles, arctic foxes, etc., and fishes. The **9-10 years cycles** of population density fluctuations have been observed in birds such as ruffed-grouse, sharp-tailed grouse, etc., and mammals such as snowshoe rabbit, muskrats, Canada lynx, etc. Among invertebrates, insect pests of coniferous forests in Germany fluctuate in periods variously from 6 through 18 years. *Asterias forkesi* is found to contain periodicity of 14 years.

Attempts to explain these vast fluctuations in numbers on the basis of climatic changes have been unsuccessful. At one time it was believed that these were caused by sunspots, and the sunspots and lynx cycles do appear to correspond during the early part of the nineteenth century. However, the cycles are of slightly different lengths and by 1920 were completely out of phase, sunspot maxima corresponding to lynx minima. Attempts to correlate these cycles with other periodic weather changes and with cycles of disease organisms have been unsuccessful.

The snowshoe hares die off cyclically even in the absence of predators and in the absence of known disease organisms. The animals apparently die of "shock" characterized by low blood sugar (hypoglycemia), exhaustion, convulsions and death symptoms which resemble the "alarm responses" induced in laboratory animals subjected to physiological stress. Visualizing this similarity, **J.J. Christian** (1950) proposed that their death, like the alarm response is the result of some upset in the adrenal-pituitary system. As *population density increases*, there is increasing physiological stress on individual hares owing to crowding and competition for food. Some individuals are forced into poorer habitats where food is less abundant and predators more abundant. The physiological stresses stimulate the adrenal medulla to secrete **epinephrine** or **adrenaline** which stimulates pituitary via the hypothalamus to secrete more **ACTH** (adrenocorticotropic hormone). This in turn, stimulates the adrenal cortex to produce corticosteroids, an excess or imbalance of which produces the alarm response or **physiological shock.**

In the latter part of the winter of a year of peak abundance, with the stress of cold weather, lack of food and the onset of the new reproductive season putting additional demands on the pituitary to secrete gonadotropins, the adrenal-pituitary system fails, becomes unable to maintain its normal control on carbohydrate metabolism, and low blood sugar (hypoglycemia), convulsion and death follows.

According to **Kendeigh** (1974) several intrinsic factors such as disease, predation, food factor and natural selection affect animal cycles in addition to several extrinsic factors such as weather, solar radiation, and so on.

7.5 FECUNDITY

Species vary greatly in the characteristic number of generations, broods or litters produced per year, and in the size of them. Protozoans often divide so rapidly that they produce a new generation every few hours. Plankton organisms being less fecund, may produce a new generation every few days. Many vertebrates breed but once a year, some large animals only once every two or three years. Several species of small birds and mammals have two or more broods per year. Rodents may continue to breed throughout the winter under favourable environmental conditions, so that their reproductive potential is enormous.

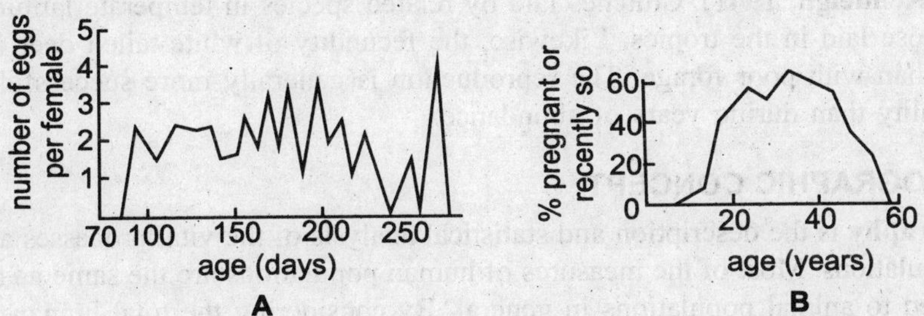

Fig. 7.9. Normally fecundity of female animals is low at the onset of the sexual maturity, then reaches a lengthy plateau, followed by a decline in old age. A—Milkweed bug (*Oncopeltus unifasciatellus*) cultured in laboratory; B—African elephants (*Loxodonta africans*) in a National Park, Uganda.

The maximum size of a litter is determined by the physiological and morphological characteristics of the species. In mammals, which produce viviparous young, the size of uterus and body cavity as well as the number of mammary glands for suckling the young after birth acts as limiting factors for the size of the litter. In birds there is a limit on the number of eggs that one individual can cover and successfully incubate. However, in species that do not take care of their eggs after laying, the number produced may be limited only by the energy resources of the parent.

The number of eggs or young produced per litter is correlated inversely with the amount of attention (*viz.,* parental care) that they require. When parental care is altogether lacking, invertebrates may lay 1,000 to 500,000,000 eggs at one maturation; where there is some protection afforded by brood pouches, 100 to 1,000 eggs may be laid; with a high degree of brood protection, 1 to 10 or more eggs may be laid. Mammals seldom have more than a dozen young in a single litter and in larger species usually only one. Characteristic clutch size among birds varies from 1 to 15; rarely, 20.

Further, there is a limit on the size of the brood or litter that adult warm blooded animals can successfully feed and raise to maturity. There is no advantage, for example, for starlings to have broods larger than five. In larger broods, each individual receives less food, and, hence, has less vigour and weight on leaving the nest. In such weak individuals, mortality increases either before fledging or in subsequent months (**Perrins**, 1965). In those species that feed their young in the nest, the clutch size has evolved through natural selection to the greatest number that can be hatched and raised successfully through efforts of the adults (**Lack**, 1967). In those species whose young leave the nest and feed themselves at hatching, the clutch size depends in large part on the capability of the female to mobilize energy in her body to produce eggs of a particular size (**Ryder**, 1970). The variability in clutch and litter size for most species allows them to take advantage of temporarily improved conditions.

Box 7.3 Semelparous and Iteroparous Organisms

Clutch size is also influenced by the number of times the parents can reproduce. In organisms (*e.g.,* humans) that have a good chance of surviving to produce several clutches, selection favours a smaller clutch size so that the parent is not exhausted, and its life threatened, by the one reproductive event (called **interoparous organisms**). In contrast, organisms such as Pacific salmon, that typically have only one clutch in lifetime, produce as many offspring at one time as possible, often using up their own body tissues in the efforts; selection does not favour conserving resources to ensure the survival of these parents (called **semeleparous organisms**).

The size of brood, litter or clutch is also regulated by weather. For example, the clutches laid by birds during periods of hot weather are usually smaller than those laid when temperature is moderate (**Kendeigh**, 1941). Clutches laid by related species in temperate latitudes tend to be larger than those laid in the tropics. Likewise, the fecundity of white-tailed deer is higher with good forage than with poor forage. The reproduction is generally more successful after periods of high mortality than during years of abundance.

7.6 DEMOGRAPHIC CONCEPT

Demography is the description and statistical analysis of the vital processes and movement of human populations. Most of the measures of human populations are the same as those we have already applied to animal populations in general. By considering the total human population of world one finds that the generalised population formula is applicable; population equals biotic potential minus environmental resistance. According to usage by the demographers biotic potentials can be expressed in terms of **fecundity,** which is defined as the theoretical or maximum capacity of the individual or population to reproduce. **Fertility** is the actual reproductive performance of a given population. Fecundity in human population has been stable for a long time, but fertility, being dependent upon ecological and social factors, has varried greatly.

Human **birth rate** (natality) is the fertility expressed as the average number of live births per thousand population per year. Since not all age groups in a population have equal fertility, demographers often use more specialised measures of birth rates that relate to age groups.

Human **mortality rate** is the average number of individuals who die per thousand population per year. Again, demographers often use more sensitive measures, for different mortality rates. Both the very young and the very old segments of the population have higher mortality rates than are found in other age groups.

Human population shows uneven or clumping pattern of distribution on earth. The **density** of human population in a village, district, city, province, country or any area can be obtained by dividing the total number of persons living in the given region by the total land area of that region. The average number of people per square unit of land area tells us how dense or sparse is the population in a given area. The average population density of the world is calculated about 27 persons per square kilometre.

The information gained from the study of human populations is known as **vital statistics.** These have many applications in our modern society. For example, life insurance is based upon the computation of **actuarial tables** which express the incidence of death by sex, age, and cause. Some general findings resulting from an analysis of vital statistics are as follows:

1. Birth rates are higher in rural areas and among farming people than in cities. Furthermore, the larger the city, the lower the birth rate.

2. Within a given region, such as city or a farming area, the lower economic classes have the higher birth rates.

3. Birth rate decrease during economic depressions and increase during times of prosperity.

4. More males than female are born.

5. Social and economic factors which produce low birth rates also produce low mortality rates. Mortality rate are lower in times of depression, in higher economic groups and in city dwellers.

6. Women have lower mortality rates than do man.

7. Both marriage and divorce rates increase during time of prosperity.

8. Both married men and married women live longer than single persons.

7.7 IMPACT OF HUMAN POPULATION INCREASE

Consequences of over-population can be quite devastating. To satisfy the needs of an increased population, a country has to increase its agricultural and industrial production and rate of urbanization is also have to increase. India being an agricultural country, two-third of its population (72.05 per cent) is engaged in farming. Since the cultivator requires land, he is forced to encroach upon forests. Simultaneously, the population of cities and town is swelled due to immigration from the villages. Consequently, to accommodate more and more people, the rate of urbanization is increased resulting in the reduction of farm land around the cities for the construction of houses, colonies, schools, hospitals, markets, roads and industries. Because of reduced availability of agricultural land, production of various crops has suffered and the amount of food proves insufficient to feed the increasing population. Besides the space and the food, fuel is also becoming short in supply. To meet the ever-increasing needs of energy of huge population, new dams are constructed for the power generation and for that forests are being destroyed and we are left with scant land under forests. Deforestation has resulted in soil erosion, desertification, recurring flood, droughts, and ever-shrinking rivers due to reduction in the size of catchment-basin. Thus, power, transport, health care, education and many other human needs have failed to keep pace with the population explosion.

QUESTIONS

Long Answer Questions

1. Write an essay on population ecology.

2. Distinguish between natality and mortality of population.

3. What is population equilibrium? How is it maintained by a biotic community.

4. Explain the meaning of term "population". What is population density? What are various methods for its determination?

5. What are the structural characteristics of a population with reference to dispersion and growth patterns? What are the important physical and biological limiting factors of population growth.

6. How does population affect the environment?

7. Describe important aspects of population dynamics.

8. Write an essay on "population fluctuations".

9. What is meant by population dispersion?

10. Explain briefly the factors controlling the population density.

11. Give an account of the attributes of population.

12. Give an account of the chief types of interspecific and intraspecific relationships. Explain how these relationships maintain the size and distribution of populations.

13. What are factors that govern population?

14. State any two factors involved in "biological" equilibrium.

Short Answer Questions

1. Write short notes on the following:
    1. Fecundity
    2. r-selection and K-selection
    3. Birth rate of population
    4. Age distribution
    5. Demographic concept
    6. Impact of human population increase
    7. Catastrophic control of population
2. Differentiate between the following:
    (*i*) Natality and mortality rate
    (*ii*) Population and community

Very Short Answer Questions

1. What is population?

2. What is population density?

3. Define mortality?

4. What is natality?

5. What is migration?

6. Define fecundity.

7. What is population dynamics?

8. What is population structure?

9. What is demography?

10. What do you mean by census?

11. Define age distribution.

12. What is survivorship curve?

13. What is life table?

14. Define biotic potential.

15. Define density dependence.

16. What is density independence.

Fill in the Blanks

1. The .... is a group of organisms of the same species occupying a particular place.
2. The population size at any given place is determined by the process of .... , ... , ... and emigration.
3. Age distribution of population is also refer to as ....
4. Bell shaped age pyramid occurs when population is ....
5. Natality depends on two aspects of reproduction: .... and ....
6. .... means number of young produced on each occassion.

Multiple Choice Questions

1. A population is a group of

(*a*) species in a community
(*b*) individuals in a species
(*c*) communities in an ecosystem
(*d*) individuals in a family

2. Population density is the

(*a*) number of species per community
(*b*) number of communities per ecosystem
(*c*) number of individuals per species
(*d*) number of individuals per species per unit area or volume

3. Population dispersion is the

(*a*) mixing of two populations
(*b*) movement away from a natal site
(*c*) spatial distribution of individuals
(*d*) movement from one fixed point to another and back again

4. The geographic limits within which a population exists is its

(*a*) niche
(*b*) habitat
(*c*) range
(*d*) territory

5. Exponential growth occurs when there is

(*a*) a sexual reproduction only
(*b*) sexual reproduction only
(*c*) a fixed carrying capacity
(*d*) no inhibition from crowding

6. Logistic growth occurs when there is

(*a*) a fixed carrying capacity
(*b*) a sexual reproduction only
(*c*) sexual reproduction only
(*d*) no inhibition from crowding

7. A bacterium can divide every 20 minutes. Beginning with a single individual, how many bacteria will there be in the population if there is exponential growth for 3 hours

(*a*) 18
(*b*) 440
(*c*) 512
(*d*) 1024

8. Beginning with a single bacterium, how large will the population be if there is logistic growth, with a carrying capacity of 18, for a period of 3 hours?
(a) 18
(b) 440
(c) 512
(d) 1024

9. The carrying capacity of a population is determined by its
(a) population growth rate
(b) birth rate
(c) death rate
(d) limited resource

10. Organisms with very high intrinsic growth rates have
(a) long generation times
(b) short generation times
(c) no courtship behaviours
(d) no carrying capacities

11. Intraspecific competition is competition among
(a) species
(b) individuals of a population
(c) individuals of a community
(d) populations and their regulatory factors

12. Intraspecific competition is strongest when the
(a) species overlap in their distributions
(b) populations overlap in their ranges
(c) population is at its carrying capacity
(d) reproductive rate is at its maximum

13. Density-dependent factors utilize a population near its
(a) niche
(b) carrying capacity
(c) ancestral range
(d) ancestral habitat

14. Demography is the study of statistics that affect
(a) geography
(b) family pedigrees
(c) population growth
(d) local concentrations

15. The age structure of a population is its
(a) relative number of individuals at each age
(b) number of newborns each year
(c) number of young reaching puberty each year
(d) relative number of deaths at each age

16. A life table is composed of
(a) mortality records based on age
(b) records of reproductive events in a life time
(c) lists of current ages of individuals in a population
(d) a collection of pedigrees from the society

17. Life tables of humans are constructed most often by
(a) medical doctors
(b) life-insurance companies
(c) geographers
(d) psychologists

18. A semelparous organism has
(a) several clutches per lifetime
(b) one clutch per lifetime
(c) several mates per lifetime
(d) one mate per lifetime

19. An iteroparous organism has
(a) several clutches per lifetime
(b) one clutch per lifetime
(c) several mates per lifetime
(d) one mate per lifetime

20. Which of the following organism is iteroparous?
(a) pacific salmon
(b) annual plant
(c) bacterium
(d) human

21. Which of the following organism is semelparous?

(*a*) pacific salmon (*b*) annual plant

(*c*) bacterium (*d*) human

22. The population growth rate *r* is inversely related to

(*a*) clutch size

(*b*) number of clutches per lifetime

(*c*) generation time

(*d*) age structure

23. A population in which each female produces *a* single large clutch very early in life is

(*a*) *k*-selected

(*b*) *r*-selected

(*c*) iteroparous

(*d*) a relict of earlier conditions on earth

24. Population density is affected by

(*a*) natality (*b*) mortality

(*c*) immigration and emigration (*d*) all of the above

ANSWERS

Very Short Answer Questions

1. Population is a group of individuals of the same species living in a given area or habitat at a given time.

2. The number of individuals of a species per unit area (or volume) is its **population density**.

3. Mortality is death of individuals in a population.

4. Natality is the ability of a population to increase by reproduction; OR production of new individuals in a population.

5. Migration is periodic departure and return of the same individuals in a population.

6. Fecundity is number of disseminules (egg or seeds) produced by an organism.

7. Population dynamics refers to the quantitative variations of growth, reproductive rate, mortality rate, fluctuations in number, age distribution, etc.

8. The distribution of the ages of individuals in a population and the way those individuals are distributed over the environment describe population structure.

9. The study of birth, death and movement rates that give rise to population dynamics known as demography.

10. The complete count of individuals in an area is called a census.

11. Ratio of each age group (prereproductive, reproductive and postreproductive) in a population is called **age distribution.**

12. Survivorship curve is a graphic description of the pattern of survival of individuals in a population from birth to the maximum age attained by each individual.

13. Life table is the tabulation of mortality and survivorship schedules of a population based on an initital cohort (*i.e.*, group of individuals of the same age class).

14. Maximum reproductive potential of an organism is called **biotic potential.**

15. Density dependence is 1. regulation of population size or growth by mechanisms whose effectiveness increases as population size increases; or 2. effect on pupulation size as a function of density.

16. Density independence is 1. regulation of population growth not tied or related to population density; or 2. change in population numbers is independent of population size.

Fill in the Blanks

1. Population **2.** Birth, death, immigration **3.** Age structure

4. Stable **5.** Fertitity, fecundity **6.** Clutch size.

Multiple Choice Questions

1. (*b*)	**2.** (*d*)	**3.** (*c*)	**4.** (*c*)
5. (*d*)	**6.** (*a*)	**7.** (*c*)	**8.** (*a*)
9. (*d*)	**10.** (*b*)	**11.** (*b*)	**12.** (*c*)
13. (*b*)	**14.** (*c*)	**15.** (*a*)	**16.** (*a*)
17. (*b*)	**18.** (*b*)	**19.** (*a*)	**20.** (*d*)
21. (*a*)	**22.** (*c*)	**23.** (*b*)	**24.** (*d*).

8 Community, Niche, Ecotone, Edge Effect and Bioindicators

8.1 BIOTIC COMMUNITY

A **biotic community, ecological community** or **community** is a naturally occurring assemblage of plants and animals living in the same environment and interacting in many ways, including mutualism, predation and competition (**R. L. Smith 1996**). Thus, community is the assemblage of species populations that has the potential of interaction. More precisely, a community may be defined as an interactive assemblage of species occurring together within a particular geographical area, a set of species whose ecological function and dynamics are in some way interdependent (**Putman** 1994). These interactions include critically determining competitive interactions and feeding relationships as well as delicate display such as dependence of plants on animals for pollination and seed dispersal or of animals on plants for meeting habitat requirements. The range and nature of these interactions are indeed myriad.

The approach of botanists and zoologists to community studies is quite different. While the zoologist is mainly concerned with the functional relationships within a community, involving both the plants and animals, the botanists are generally concerned with community structure and the changes that undergo in time and space.

Characteristics of a Community

Like a population, a community has a series of characteristics such as follows:

1. Species diversity. Various species of plants and animals live in a community and exhibit species richness or species diversity. The study of species diversity is an essential component of community study. For animal communities, a study of age structure and growth pattern is important, while for plant communities, floristics, study of taxonomy, life forms such as herbs, shrubs, climbers, trees are important. Since seasonal changes occur in the appearance of plant structure and growth, periodicities and phenology (= seasonal succession in natural communities) are significant parameters.

2. Growth form and structure. The type of the community is described by major categories of growth forms (*e.g.,* trees, shrubs, herbs, mosses, etc.) These different growth forms determine the stratification, or vertical layering of the community.

3. Dominance. Among several species present in a community, a few exert a major controlling influence by virtue of their size, numbers, or activities. These are called **ecological dominants** or **dominant species**.

4. Relative abundance. Different populations in a community exist in relative proportions and this idea is called **relative abundance**.

5. Trophic structure. Who eats whom? The feeding relations of the species in the community will determine the flow of energy and materials from plants to herbivorous animals to carnivorous animals.

6. Emergent and collective properties. Just as populations have attributes beyond those of the individuals that comprise them, so a community is more than the sum of its parts. The community has attributes called **emergent properties** which is a term coined by **Begon** *et. al.,* in 1986. He explained these with analogy of a cake: a cake has emergent properties of texture and flavor not apparent by considerations of its ingredients. The nature of the community is influenced by the interactions of species and their adaptations to the physical environment. These adaptations and interactions are reflected in such attributes of community as structure, dominance and niches.

Salt (1979) considered emergent properties of community to be merely **collective properties.** An example of collective properties is the age structure of a population, which is the result of the combined ages of all the individuals making up a population. Likewise, community respiration can be regarded as a collective community characteristic since it is an aggregation of the totality of respiration carried on by the individuals in the community.

7. Individualistic and organismic communities. Historically, plant ecologists debated whether plant communities are tightly integrated units (**super organisms**) or chance assemblage of species with similar physiological tolerance. The first view was called **organismic community** and proposed by **F. Clements** (1916) and supported by **S. A. Forbes, C. Allee** and **E. Odum.** The second view was called **individualistic community** and was proposed by **Henry Gleason** (1926).

Types of Communities

A community may be **autotrophic** in the sense it includes photosynthetic plants and gains its energy from the sun. Other communities, such as those in springs and caves are **heterotrophic.** They depend upon the input of fixed energy, such as organic material, from the outside. Autotrophic communities usually contain a number of heterotrophic micro-communities.

Communities have been classified by different ecologists from different view points. In terms of the general growth, composition, shape, etc., of vegetation, and organisms associated with them, communities may be classified as **forests**, **deserts**, **grasslands**, **tundra** and so on. Likewise, according to the amount of water in the habitat, communities may be divided as **hydrophytic** in predominantly aquatic habitats, **mesophytic** in moderately moist soils and **xerophytic** in arid or dry conditions. Communities growing in condition of abundant light are called **heliophytic** and those growing in shade are called **sciophytic**. **Clements** (1916) recognized the fact that plant communities are not always the same at any place and he classified the communities on two parallel lines: One in the process of change which are called **seral communities** and the others are called **stable** or **climax communities.**

Further, the **global community** is an enormous mass of life, comprising all the plants and animals in the world. The global community is further divided into: **continental communities** and **oceanic communities**. Since due to great variability in climatic factors, an exhaustive study in such vast areas is practically impossible, therefore, communities are often studied as **biotic province**. A biotic province can be defined as a considerable geographic area, over which the climate is relatively uniform though often modified by physiographic features (**Dice**, 1952). Since biotic province is an abstract community, so based on ecological criteria, the associations are studied as **concrete communities**. A concrete community can be defined as a specific area which can be observed directly and which is an assemblage of plants and animals that actually exists and from which some ecological data can be collected.

Lastly, a **stand** is the largest concrete community, for example, a particular forest, river, swamp, meadow or lake that can be seen, observed, measured and worked over by the ecologist.

A **microstand** is a small localized area within stand. Thus, each individual plant or animal with its associated parasites, epiphytes and commensals are good examples of the microstands.

Composition of community

Composition of community comprises the following parameters:

1. Size. Communities may be large or small. Larger one extends over areas of several thousands of square kilometers, as forests, others such as deserts, etc., are comparatively smaller with dimensions in hundreds of kilometers, and still others such as meadows, rivers, ponds, rocky plateaus, etc., covering a more restricted area. Very small-sized communities are the groups of microorganisms in such microhabitats as leaf surface, fallen log, litter, soil, etc.

2. Number of species. The number of species and population abundance in communities vary greatly. **Charles Elton** (1927) calculated that the communities of certain British rivers included 131 species of invertebrates, in addition to fish, amphibians, populations of algae, protozoa, bacteria, and rooted aquatic plants. In contrast to this, he found a community on a sandy beach consisting of only five species of invertebrates. A meadow (=a rich pasture ground especially beside a stream) on clay near Oxford, England included 93 species of invertebrates in the soil and on above ground vegetation. In India too, various ecologists recently have estimated the number of species in the composition of various Indian communities. **M.C.Dash** and his coworkers (1980, 1981 and 1984) studied the composition of a biotic community in some tropical pastures of Sambalpur, India and found that it consisted of 22 grass and other herbaceous species (the primary producers), 5 species of earthworms, 25 of test-bearing protozoa, 15 of nematodes and 7 of microarthropods. The team of **Dash** (1980, 1984) reported 9 species of grasses and other herbaceous species, 1 of earthworms, 11 of testate protozoa and 16 of nematodes in a tropical hill ecosystem of Sambalpur. At one site of river Ib, Orissa, the community consisted of 1 species of rooted plant, 32 of phytoplanktons, 20 of zooplanktons and 7 of fish (**Kar** *et al.,* 1987; **Dash** *et. al.,* 1988).

3. Dominants. In each community there occurs diverse species. All these species are not equally important but there are only a few over topping species which by their bulk and growth modify the habitat and control the growth of other species of the community, thus, forming a kind of characteristic nucleus in the community. These species are called the **dominants**. Generally in most of the communities, only a single species is particularly conspicuous and dominant and in such case the community is called by the name of dominant species, *e.g.,* spruce forest community. In other cases, there may be more than one dominants in a community, *e.g.,* oak-hickory forest community.

According to **Clements** and **Shelford** (1939), dominance is most commonly expressed in the reactions of an organism on its habitat. Community dominants sustain the full impact of the climate or the environment but modify this effect for other organisms within the community by tempering light, moisture, space, and other conditions. Only those other organisms that find these modified physical conditions tolerable can exist within the community. Furthermore, dominants are ordinarily the most prominent species in the community, make up its greatest mass of living material (**biomass**), and serve as the major source of food, substrate, and shelter for the animals that are present. In a forest community, trees are dominant. They decrease light intensity, increase the relative humidity, intercept precipitation, monopolise most of the moisture and nutrients in the soil, decrease wind velocity and furnish shelter and food for animals. Grasses play a similar, though less conspicuous, role in prairie communities; sedges, rushes and cattails in marsh communities: sagebrushes in the arid habitat of the Great Basin; mussels and barnacles on a rocky seashore; and so forth.

4. Ecological amplitude. The range of environmental conditions which a taxon can tolerate is called **ecological amplitude**. The composition of a biotic community in any habitat is dependent upon the frequency of environmental conditions in that habitat and the ecological amplitude of species populations. Thus, the climate and other abiotic and biotic conditions of a habitat determine the type of community which survives and develops.

Characters Used in Community Structure

Structure of plant communities can be determined by quantitative and qualitative features as follows:

A. Quantitative Structure of Plant Communities

Quantitative characters of the community include such characters as frequency, density, abundance, cover and basal area, etc.

1. Frequency. Various species of the community are recorded by different phytosociological methods, by taking any sampling unit such as **transect**, **bisect**, **quadrat**, **point centre** or **point quarters**, etc. Frequency is the number of sampling units (as %) in which a particular species occurs. Thus, the frequency of each species is calculated in the following way:

$$\text{Frequency} = \frac{\text{Number of sampling units in which the species occurred}}{\text{Total number of sampling units studies}} \times 100$$

After determining the percentage frequency of each species, various species are distributed among **Raunkaier's** (1934) five frequency classes depending upon their frequency values (Table 8.1).

Table 8.1. Raunkaier's frequency classification of plant communities.

Frequency class	Species with frequency range (%)
A	0–20
B	21–40
C	41–60
D	61–80
E	81–100

The **relative frequency** (**RFR**) is determined by the use of the following formula (data collected using the quadrant method).

$$\text{Relative frequency (RFR) of the species} = \frac{\text{Number of occurrence of a species}}{\text{Number of occurrence of all species}} \times 100$$

Since environmental conditions vary, the frequency values differ in different communities. **Raunkaier** observed the following relationships between different frequency classes: A = 53, B = 14, C = 9, D = 8 and E = 16. He generalized his findings in the **law of frequency**, which showed that

$$A > B > C \gtreqless D < E$$

Such a relationship occurs only in undisturbed communities or communities where disturbance is minimum.

2. Density. Density represents the numerical strength of a species in the community. The number of individuals of the species in any unit area is its density. It gives an idea of degree of competition. Samples are taken by the help of different types of samples and sampling techniques (**Southwood**, 1966) and estimation are made of the numerical strength of a species population and of the total community. The data can be multiplied with an appropriate factor to express the total number of organisms per unit of area. The density of a species is calculated as follows:

$$\text{Density} = \frac{\text{Total number of individuals of the species in all the sampling units}}{\text{Total number of sampling units studied}}$$

The **relative density (RDE)** of a species is calculated in the following way:

$$\text{RDE} = \frac{\text{Total number of individuals of the species in all quadrates}}{\text{Number of individuals of all species in all quadrates}} \times 100$$

3. Abundance. This is the number of individuals of any species per sampling unit of occurrence. Abundance is calculated as follows:

$$\text{Abundance} = \frac{\text{Total number of individuals of the species in all sampling units}}{\text{Number of sampling units in which the species occurred}}$$

4. Cover and basal area. The above ground parts of plants (such as leaves, stems and inflorescence) cover a certain area— if this area is demarcated by vertical projections, the area of the ground covered by the plant canopy is called **canopy cover**, **foliage cover** or **herbage cover**. Canopy cover is a good measure of the herbage availability and is estimated by chart quadrat, line intercept or point frame.

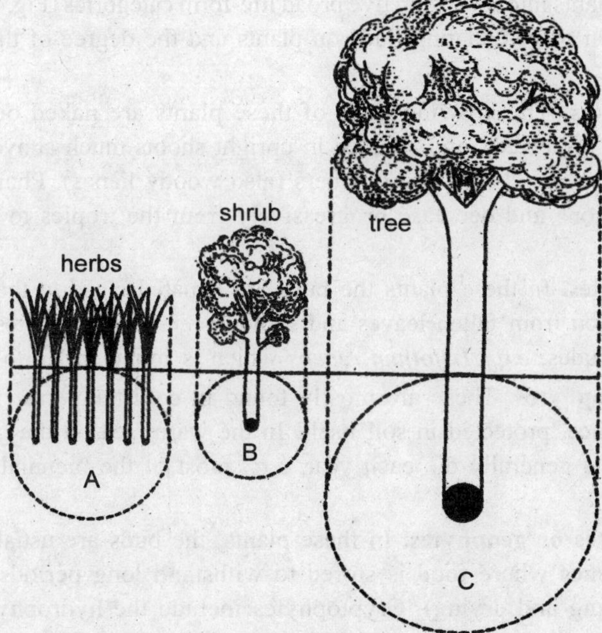

Fig. 8.1. Diagram showing relationship between herbage cover and basal area in different types of growth forms (herbs, shrubs and trees) of plants.

Basal area refers to the ground actually penetrated by all stems and is readily seen when the leaves and stems are clipped at the ground surface. It is either measured 2.5 cm above ground or actually on the ground level. Basal area of a plant is measured by callipers, line-interception or point-centered quadrat method. It is one of the chief characteristics to determine dominance.

Since plants differ in their growth form, so the relationship between herbage cover and basal area differs in different types of plants as shown in Figure 8.1.

For the measurement of crown cover, the diameter is taken at ground level, across the canopy perimeter passing by the main stem. The cover is expressed as a percentage. Based on their cover values, species are grouped into the following six groups : Group 1 (less than 5%), Group 2 (5–25%), Group 3 (25–50%), Group 4 (50–75%), Group 5 (75–95%) and Group 6 (95–100%).

B. Qualitative Characteristics of Plant Communities

These involve parameters such as physiognomy, phenology, stratification, abundance, sociability or gregariousness, vitality and vigour, life form (growth form), etc. All these parameters can be described and may be grouped in point scales.

1. Physiognomy. Physiognomy is the general appearance of vegetation as determined by the growth form of dominant species. Such a characteristic appearance is often expressed by single term. For example, observation of a plant community having trees and some shrubs as the dominants, clearly forced us to conclude that it is a forest. Similarly on the basis of appearance of a community, it may be a grassland, desert, etc.

2. Life-form (Growth form). Ecologists generally use **Christen Raunkiaer's** classification (1934) of plant life forms. A **life-form** is the sum of the adaptation of the plant to the community (it is determined by the stature of the plant species, their spread and characters of life-form). **Raunkiaer** classified plants into following five broad life-form categories (Fig. 8.2); his classification is based on the position of perennating buds on plants and the degree of their protection during adverse conditions.

(i) Phanerophytes. The growing buds of these plants are naked or covered with scale (*i.e.,* they are not well protected), and situated in upright shoots much above the ground surface. Phanerophytes include trees, shrubs and climbers (also woody lianas). Phanerophytes are found mostly in tropical regions and decrease progressively from the tropics to the temperate to the polar regions.

(ii) Chamaephytes. In these plants the buds are situated close to the ground surface and these buds get protection from fallen leaves and snow cover. Chamaephytes commonly occur in high altitudes and latitudes, *e.g., Trifolium repens* which is found in North America.

(iii) Hemicryptophytes. These are mostly found in cold temperate zone. Their buds are hidden under soil surface, protected in soil itself. In the warm season the growth of aerial parts is marked. Their shoots generally die each year, *e.g.,* most of the biennial and perennial herbs including grasses.

(iv) Cryptophytes or geophytes. In these plants, the buds are usually buried in the soil or in bulbs and rhizomes where food is stored to withstand long periods of adverse climatic conditions (*i.e.,* freezing and drying). Cryptophytes include the hydrophytes (buds remaining under water), helophytes (marsh plants with rhizomes under the soil) and geophytes (terrestrial plants with underground rhizomes or tubers).

(v) Therophytes. These are seasonal plants, completing their life cycle in a single favourable season, and remain dormant throughout the rest unfavourable period of year in the form of seeds. Therophytes are commonly found in dry, hot or cold environments (*i.e.,* deserts).

(vi) Epiphytes. Plants growing on other plants, their roots remain up in the air.

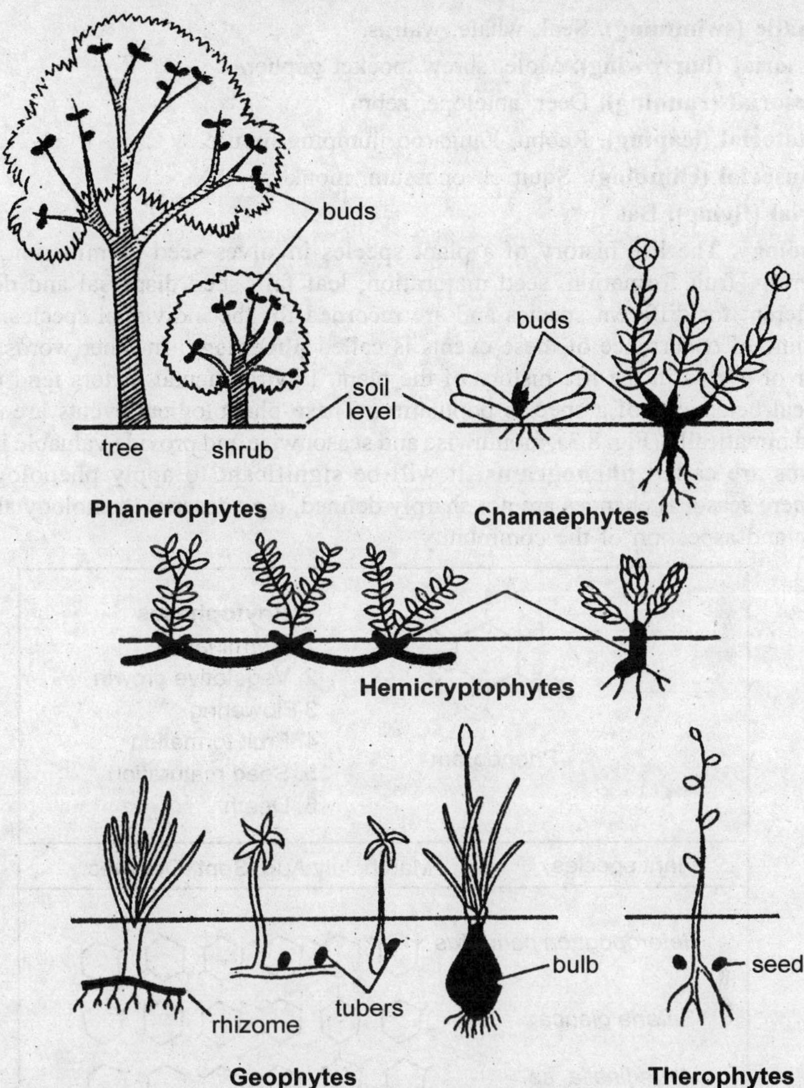

Fig. 8.2. Diagrammatic representation of Raunkiaer's life-forms. The solid black regions survive during adverse periods of the growing season. Note the successive increasing degree of protection of renewal or perennial buds, organs or seeds from phanerophytes to therophytes.

Life-forms of animals. There have been several attempts to classify the life-forms of animals, but no definite system has resulted (**Remane**, 1952; **Krivolutskii**, 1972). The major life-forms more often agree with their taxonomy than do plants, but some life-forms include representatives from several different taxonomic groups. These can be recognized **encrusting forms**, such as the freshwater bryozoan *Plumatella* and some sponges; **coral forms**; **radiate forms**, such as coelenterates and echinoderms; **bivalve forms**; **snail forms**; **slug forms**; **worm forms**; **crustaecean forms**; **insect forms**; **fish, snake, bird** and **four-footed forms**. Each of these major types may be divided into narrower structural or behavioural types, for example, **Osburn** *et al.,* (1903) subdivided four-footed mammalian forms into following types:

(i) **Aquatic (swimming).** Seal, whale, walrus.

(ii) **Fossorial (burrowing).** Mole, shrew, pocket gopher.

(iii) **Cursorial (running).** Deer, antelope, zebra.

(iv) **Saltatorial (leaping).** Rabbit, kangaroo, jumping mouse.

(v) **Scansorial (climbing).** Squirrel, opossum, monkey.

(vi) **Aerial (flying).** Bat.

3. Phenology. The life history of a plant species involves seed germination, vegetative growth, flowering, fruit formation, seed maturation, leaf fall, seed dispersal and death. These events are different for different species and are recorded for the individual species. A study of the date and time of occurrence of these events is called **phenology**. In other words, phenology is the calender of events in the life history of the plant. Environmental factors tend to influence the phenological behaviour of a species population. These phenological events are customarily recorded diagrammatically (Fig. 8.3), monthwise and seasonwise and provide valuable information. These diagrams are called **phenograms**. It will be significant to apply phenology to those ecosystems where seasonal changes are not sharply defined, *e.g.*, deserts. Phenology also includes the periodicity and aspection of the community.

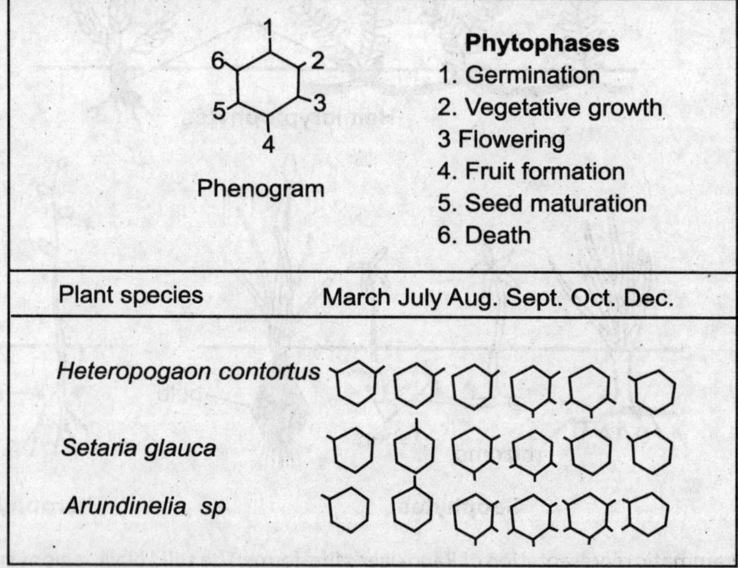

Fig. 8.3. A phenogram showing phenological events of three Indian herbaceous species.

4. Stratification. Stratification of communities is the way in which plants of different species are arranged in different vertical layers in order to make full use of the available physical and physiological requirements.

The stratification in a community may be resulted by the following factors: 1 Specific tolerations and adaptations. 2. Chemical reactions taking place between the by-products and physical stratification. 3. Other organisms take up temporary or permanent residence as a direct response to the presence of initial residents. The stratification in community may be horizontal or vertical and is different in aquatic and terrestrial habitats.

1. Horizontal Stratification

A community may be divided horizontally into **subcommunities** which are units of homogeneous life forms and ecological relations. This horizontal division constitutes the **zonation** in the community. Figure 8.4 shows the zonation of different distinct vegetational types on a mountain. Latitudinal as well as altitudinal zonations of vegetation have also been shown in relation with climatic factors.

In shallow ponds, the zonation is very little. However, in deep ponds and lakes, there may be recognized three zones, *viz.,* **littoral zone, limnetic zone** and **profundal zone** (Fig. 8.5). In each zone, organisms differ from each other. Further, in case of animal communities of rocky sea shore, there are following four distinct zones: 1. **Balanus zone**. This zone includes the animals such as *Balanus* and *Patella* which live on exposed rocks. 2. **Littoral zone** or **intertidal zone**.(*i.e.,* zone between high and low tides). This habitat develops in the sheltered niche beneath loose stones and contains a variety of crustaceans and collembola (springtail). Lower on the shore, where the tides always flood are living beneath the stone some annelids and nemertine worms (*e.g., Lineus*). Adjacent to low water mark are certain species of *Neries,* shore crabs, etc. 3. Holes and cracks in the rocks are occupied by various annelid worms, crustaceans, small sea-cucumbers and rock-boring bivalves. 4. There are rock-pools that enclose those animals which escape direct current of water.

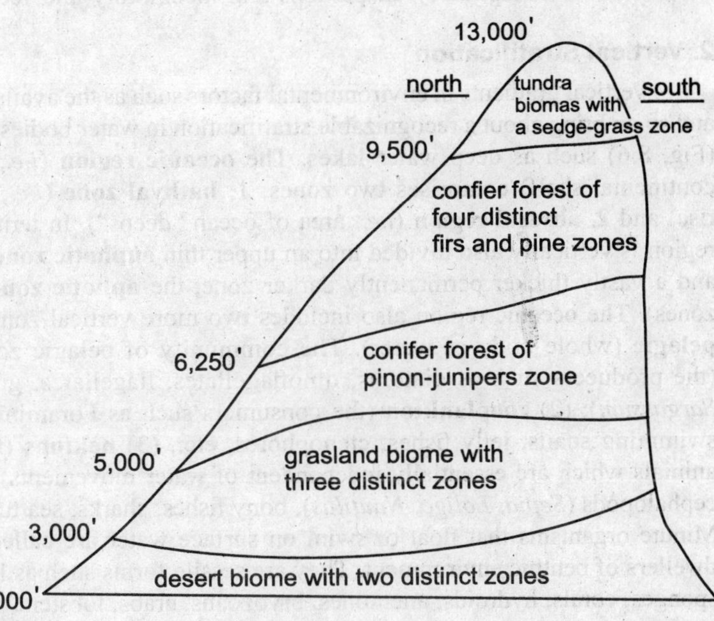

Fig. 8.4. Horizontal vegetational zones (*i.e.,* zonation) on one of the mountains of Western North America.

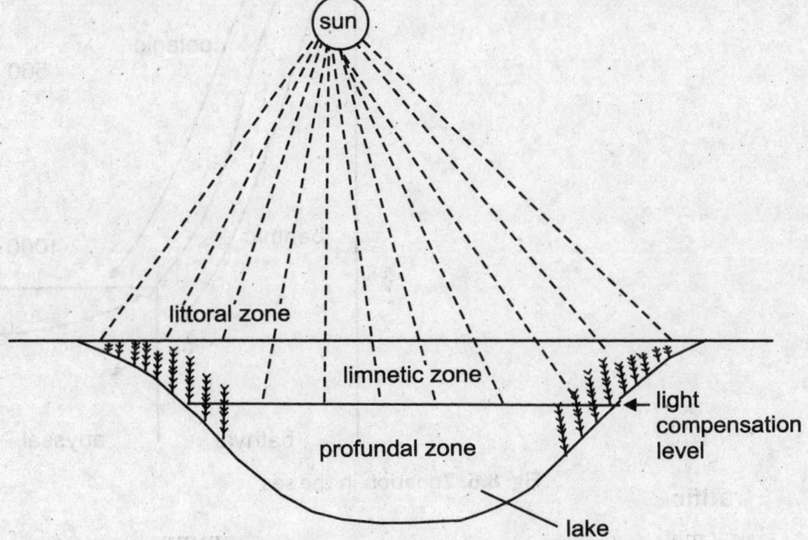

Fig. 8.5. Three major zones of a freshwater lake.

Sandy-shore dwelling marine animals are in great part burrower such as spinoid (*Prionospio, Aricia*), annelids (*Glycera, Nerina*), archiannelids (*Protodrilus*) and crustacean (*Emerita*, *Albunea*). Populations which inhabit this substrata adjust to both sand and mud particles. This becomes evident in their respiratory adaptations and locomotory and feeding behaviour.

2. Vertical Stratification

Vertical gradients in environmental factors such as the availability of sunlight, the temperature and so on bring about a recognizable stratification in water bodies, particularly in marine ecosystem (Fig. 8.6) such as deep water lakes. The **oceanic region** (*i.e.,* the region of open sea beyond continental shelf) comprises two zones: 1. **bathyal zone** (*i.e.*, region of continental slope and rise) and 2. **abyssal region** (*i.e.,* area of ocean "deeps"). In terms of light penetration, the ocean region is vertically also divided into an upper thin **euphotic zone** (= the light compensation zone) and a vastly thicker permanently darker zone, the **aphotic zone** (including bathyal and abyssal zones). The oceanic region also includes two more vertical zones: the **benthic** (bottom) and the **pelagic** (whole body of water). The community of pelagic zone includes (1) **phytoplankton** (the producers such as diatoms, dinoflagellates, flagellates, green algae, brown algae such as *Sargassum*); (2) **zooplankton** (the consumers such as Foraminifera, Radiolaria, annelid worms, swimming snails, jelly fishes, ctenophores, etc., (3) **nektons** (these are free swimming marine animals which are essentially independent of water movements). Nektons are consumers such as cephalopods (*Sepia, Loligo, Nautilus*), bony fishes, sharks, sea turtles, whales, seals and sea birds. Minute organisms that float or swim on surface water are called **neustons**. **Benthos** are bottom dwellers of benthic environment. They are sessile forms such as limpets, chitons, mussels, oysters, sponges, corals, hydroids, anemones, bryozoans, crabs, lobsters, some echinoderms and flat fishes.

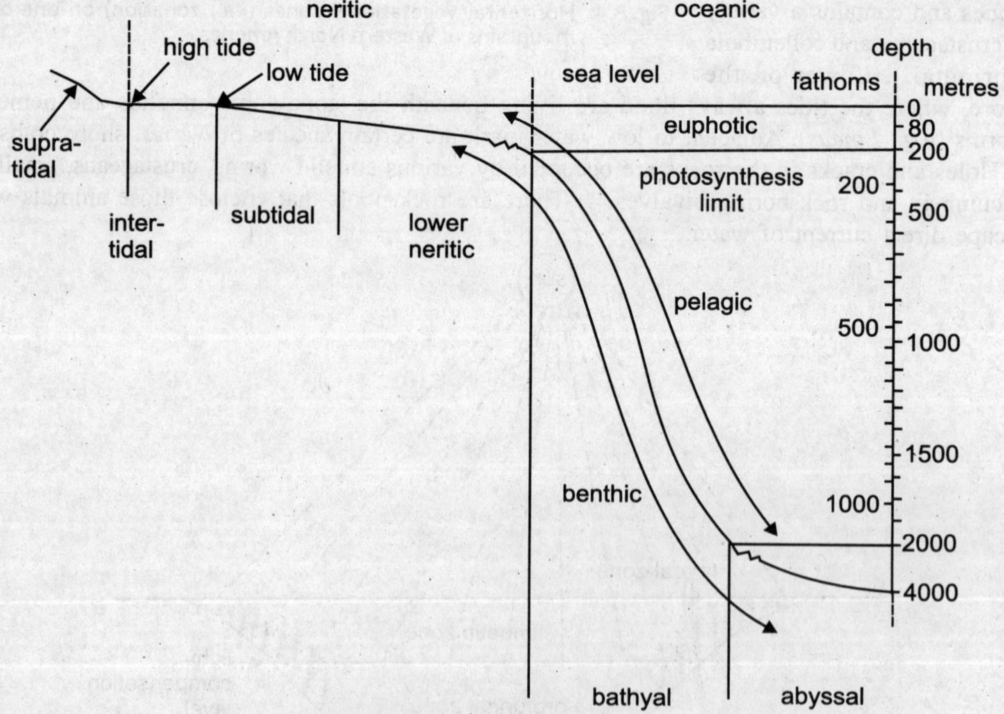

Fig. 8.6. Zonation in the sea.

Vertical stratification in communities is also observed in terrestrial ecosystems. For example, in a **grassland community**, the euphotic zone consists of the **herbaceous stratum** consisting of grasses and other herbs as producers and rodents, cattle and insects as consumers. The **subterranean stratum** consists of different layers of soil containing litter, humus, mineral fraction of soil, plant roots, rhizomes and organisms such as bacteria, fungi, protozoa, microarthropods, insect larvae and, pupae, oligochaetes and molluscs. In **forest community**, however, the vertical stratification is most complicated and includes following five subdivisions: 1. **subterranean subdivision**; 2. **forest floor**; 3. **herbaceous vegetation**, 4. **shrubs** and 5. **trees** (Fig. 8.7). In some **tropical rain-forests** three more subdivisions exist in the forest floor such as leaf litter layer (**L layer**), partially decomposed leaf litter layer (**F layer**) and a humus layer. Thus, based upon the light and relative humidity requirements, we find stratification in above ground parts.

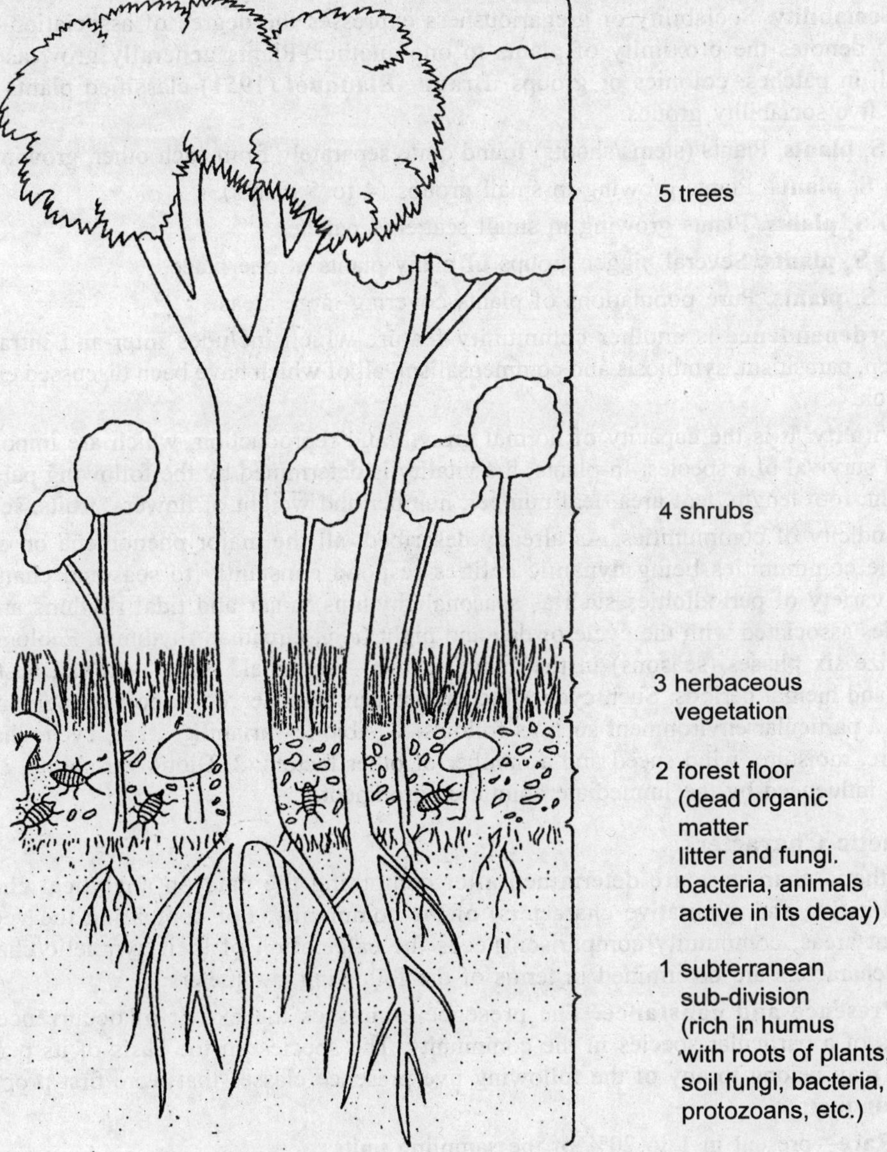

Fig. 8.7. Vertical stratification in a complex forest community.

Ecological potential of a zone or strata determines the distribution of community. **Fenton** (1947) has classified these communities as (1) **Micro-biota** which include bacteria, algae, fungi, protozoans, rotifers, nematodes and mites. (2) **Meso-biota** which include many nematodes, centipedes, millipedes, spiders, insects, etc. (3) **Macro-biota** which include plant roots, earthworms, toads, lizards, snakes, rodents and large mammals including humans. He also classified these communities according to their feeding habits such as **chemophagous, ectophagous, entophagous, predators** and **shelterers.**

5. Abundance. Plants are not found uniformly distributed in an area. They are found in smaller patches or groups, differing in number at each place. Abundance is divided into following five arbitrary groups, depending upon the number of plants: 1. **very rare**, 2. **rare**, 3. **common**, 4. **frequent**, and 5. **very much frequent**.

6. Sociability. Sociability or gregariousness expresses the degree of association between species. It denotes the proximity of plants to one another. Plants generally grow as isolated individual, in patches, colonies or groups. **Braun -Blanquet** (1951) classified plants into the following five sociability groups:

(i) S_1 plants. Plants (stems/shoots) found quite separately from each other, growing singly.

(ii) S_2 plants. Plants growing in small groups (4 to 5 plants).

(iii) S_3 plants. Plants growing in small scattered patches.

(iv) S_4 plants. Several bigger groups of many plants at one place.

(v) S_5 plants. Pure populations of plants covering large areas.

Interdependence is another community feature which includes inter-and intraspecific competition, parasitism, symbiosis and commensalism, all of which have been discussed elsewhere in this book.

7. Vitality. It is the capacity of normal growth and reproduction, which are important for successful survival of a species. In plants, the vitality is determined by the following parameters: stem height, root length, leaf area, leaf number, number and weight of flowers, fruits, seeds, etc.

Periodicity of communities. As already described, all the major phenomena on earth are cyclic. The communities being dynamic entities respond constantly to seasonal changes and exhibit a variety of periodicities such as seasonal rhythms, lunar and tidal rhythms and many periodicities associated with the cycle of day and night (*e.g.*, circadian rhythms). Ecologists tend to recognize six phases (seasons) in the annual cycle—prevernal, vernal, extival, aestivation, autumnal and hiemal periods. Such cyclic fluctuations can be observed in both abiotic and biotic factors of a particular environment such as follows: 1. Abiotic variability , *i.e.*, cycle changes in temperature, moisture, wind speed and a number of other factors. 2. Biotic variability, *i.e.*, flora and fauna influenced by the immediate abiotic environment.

C. Synthetic Characters

Synthetic characters are determined after computing the data on analytical characters (*i.e.,* quantitative and qualitative characters) of the community. For comparing the vegetation of different areas, community comparison needs the calculation of their synthetic characters. Synthetic characters are determined in terms of the following parameters:

1. Presence and constance. The presence expresses the extent of occurrence of the individuals of a particular species in the community. The species on the basis of its percentage frequency may belong to any of the following five presence classes that were first proposed by **Braun Blanquet**:

I. Rare—present in 1 to 20% of the sampling units.

II. Seldom present—present in 21–40% of the sampling units.

III. Often present—present in 41–60% of the sampling units.

IV. Mostly present—present in 61–80% of the sampling units.

V. Constantly present—present in 81–100% of the sampling units.

A species with high frequency is considered constant. The term constancy is used when equal sample areas are taken in each stand, and presence is used when the area of the sample varies from stand to stand.

2. Fidelity. Fidelity or "faithfulness" is the degree with which a species is restricted in distribution to one kind of community. Such species are sometimes known as **indicators**. The species have been grouped into the following fidelity classes which were first formulated by **Braun Blanquet**:

(i) Fidelity 1 or Strangers. Rarely occurring species which either have arrived from other communities or are relicts from earlier stages of succession.

(ii) Fidelity 2 or Indifferents (or **ubiquitous).** Species which occur in any community without exhibiting any preference to any particular kind of community.

(iii) Fidelity 3 or Preferentials. Species which occur in several kinds of communities but are predominant in one.

(iv) Fidelity 4 or Selectives. Species occurring frequently in one kind of community and sometimes also in others. Sometimes preferential and selective species are considered together as **characteristic**.

(v) Fidelity 5 or Exclusive or True. Species which are completely restricted to one kind of community.

3. Dominance. It is used as a synthetic as well as analytical character (**Daubenmire**, 1959). The number of organisms sometimes may not give correct idea of the species. If one bases his conclusion on number, a single or few trees in a grassland, or few grasses in a forest should be of little value. But if he considers the species on the basis of area occupied or weight (biomass), the situation may be different. Thus, in dominance, cover is included as an important character. The single tree in grassland may occupy fairly a large area and may have much mass. Relative dominance (cover; RDO) is calculated in the following way:

$$\text{Relative dominance (cover, RDO)} = \frac{\text{Dominance (cover) of the species}}{\text{Total dominance (cover) of all the species}} \times 100$$

4. Importance value index (IVI) and polygraph construction. This index is used to determine the overall importance of each species in the community structure. In calculating this index, the percentage values of the relative frequency, relative density and relative dominance are summed up together and this value is designated as IVI or importance value index of the species. The IVI provides an idea of the sociological structure of a species in its totality in the community but does not indicate its position separately with regard to other aspects, such as frequency, density, and so on. In consequence, plant ecologists often use **polygraphs** to show the individual and combined aspects of the position of each species in the structure of the community.

The method of polygraph preparation involves drawing a circle and dividing it into four equal segments, by drawing two lines at right angles to each other passing through its centre. Three of the radii are divided into 100 equal parts from the centre to the circumference and the fourth is divided into 300 equal parts to mark the IVI value. On the 0–100 scale in the three radii, the values of relative frequency, relative density and relative dominance are marked. Fig. 8.8 shows some of the polygraphs (or phytographs) which are generally used by plant ecologists (*e.g.,* **Dani** and **Behera**, 1990).

Animal ecologists use an **importance value** (IV) as follows (**Sajise** *et al.,* 1976):

$$IV\ (\%) = \frac{Wa}{Wt} \times 100$$

where Wa is the average dry weight (dry weight in grams per square meter per month) of the species and Wt is the total dry weight of all species of the community (of the area).

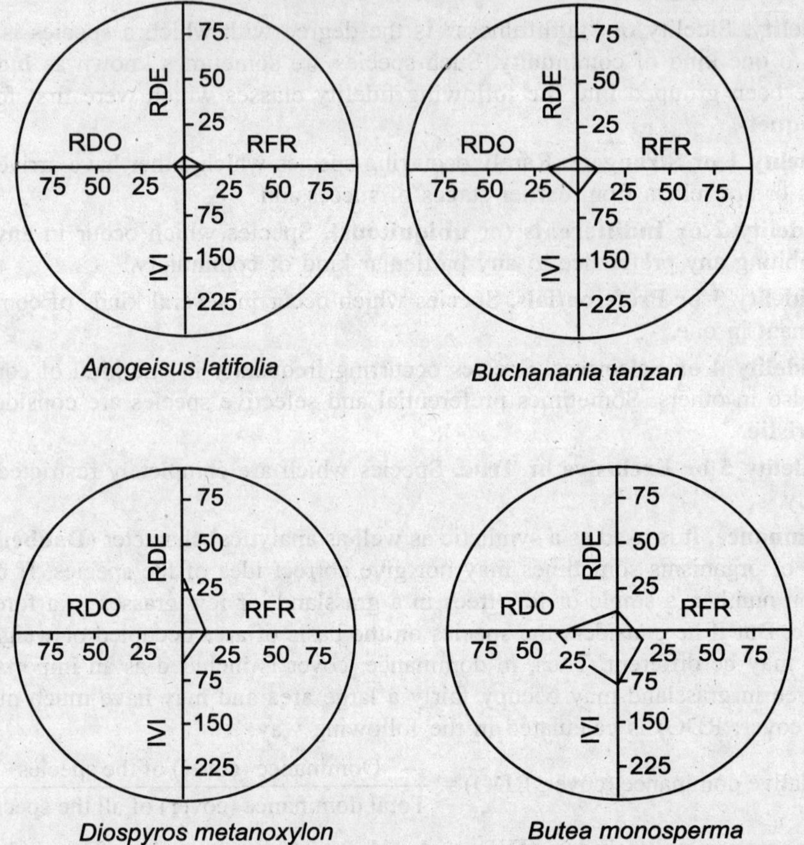

Fig. 8.8. Polygraphs prepared by **Dani** and **Behera** (1990) to show the sociological characters of individual species in a montane deciduous forest ecosystem (foot hill) of Sambalpur, India. IVI = importance value index, RDO = relative dominance, RDE = relative density, RFR = relative frequency.

Community Stability

One of the most important considerations about communities is their stability, or lack of variation in time. There are two discrete senses in which this term is used. The first, **community stability**, refers to the degree of fluctuation in population size of the populations comprising the community. The second, **environmental stability**, refer to the fluctuations in the abiotic factors of the ecosystem. The two types of stability are closely related, and stable communities are generally to be found in stable environments, while unstable environments are more often characterized by unstable communities.

The global range of variation in community stability is extraordinarily broad. Some communities such as the tropical rainforest or the coral reef, appear exceedingly stable, and the densities of their populations are constant over long periods of time. Others, such as the tundra, are extremely variable, and they are made up of populations that fluctuate widely in abundance.

Much of this fluctuation is due to variations in environmental factors which the community cannot avoid. However, genetically determined adaptations that have arisen through natural selection also play a part in determining the responses of a population to environmental fluctuations.

Further, the structure of community in which a population is found, has a major role in determining the degree of environmental variability to which the species is subjected. There are at least two mechanisms by which community structure affects the stability or variability of a community. The first mechanism involves the relationship between community structure and microenvironment. Many organisms are protected from the full impact of environmental fluctuations because of the presence of other members of the community. For example, the floor of a conifer forest is protected from winds, evaporation, and temperature shifts, making it a much less harsh and variable environment for organisms than the forest canopy or an open field adjacent to the forest. The second mechanism by which community structure affects community stability is community diversity. But about the correlation of community stability and community diversity, many ecologists have expressed their doubts (**Hairston** *et al.,* 1968).

8.2 HABITAT AND NICHE

Any position specially adapted to its occupants is called **niche.** The term *niche* was for the first time used by **Joseph Grinnel** (1971) to explain microhabitats of California thrashers. According to him, "*niche is the ultimate distributional unit, within which each species is held by its structural and instinctive limitations......no two species in the same general territory can occupy for long identically the same ecological niche.*" The **area** of a species or any other taxon refers to the total extent of its geographical range of dispersal; this can be plotted on a map. The **habitat** of a species described in a single word or phrase is the totality of abiotic factors to which the species is exposed in this area. Thus, one can talk of marine habitats, coastal habitats, marsh habitats, forest habitats, disturbed habitats or even dry habitats.

In many cases, however, the habitat of the species is highly specialized. Certain species of leaf miners, for example, live only in the upper photosynthetic layer of the leaves of certain species of plants, while other species live in the lower cell layer. These patterns of location are so consistent that it is clear that the habitats of the species are different. Subdivisions of the environment on this scale are commonly called **microhabitats**. Thus, the leaf constitutes a microhabitat for leaf miners within the total forest, and the different cell layers of leaf constitute different microhabitats within the leaf for different species of leaf miner. Conditions within the leaves are quite different from the general conditions in the forest. The specific environmental variables in the microhabitat of a population is called **microenvironment** or the **microclimate** .

Though habitats and microhabitats indicate the places where organisms of a community live, but, if one wishes to understand how a species fit into the fabric of the ecosystem, it is not enough to merely describe the creature's habitat or microhabitat, he must also understand its functional role within the community—"address" of the organism but its "profession" as well. This includes what it eats, what it does, where it lives—everything about it that influences the community. Thus, the ecological niche is the property of the community and it represents the place of the species in the formal community structure.

In fact, the ecological niche is an inclusive term that involves not only the physical space occupied by an organism, but also its functional role in the community (*i.e.,* trophic position occupied) and its position in environmental gradients including other conditions of existence. These three aspects of ecological niche are generally designated as (i) the **spatial** or **habitat niche** (the physical space occupied); (ii) the **trophic niche** (functional role, *i.e.,* trophic position), and (iii) the multidimensional or **hypervolume niche** (position in the environmental gradients).

1. Spatial or Habitat Niche

This type of niche pertains to the physical space occupied by an organism. **O′ Neill** (1967) has cited a good example of spatial niche in the distribution of seven species of millipedes in forest floor of a maple-oak forest. All the seven species occur in the same general habitat, *i.e.,* forest floor and all belong to the same basic trophic level (*i.e.,* all are decomposers in their role as detritivores). Each species of millipedes predominates in its own specific different microhabitat. In fact, there exist several gradients in decomposition stage from the centre of log to the position underneath the leaf litter, and these gradients constitute distinct microhabitats in the same general habitat, the forest floor. Thus, *Euryurus erythropygus* predominates in the heartwood at centre of logs, *Pseudopolydesmus serratus* in superficial wood of logs, *Narceus americanus* in the outer surface of the logs beneath the bark, *Scytonotus granulatus* under log but on the log surface, *Fontaria virginiensis* under the log but on ground surface, *Cleidogonia caesioannularis* within leaves of litter, and *Abacion lacterium* beneath litter on ground surface.

2. Trophic Niche

In this case two species live in the same habitat but they occupy different trophic niches, because of differences in food habits. For example, two aquatic bugs, *Notonecta* and *Corixa* live in the same pond but occupy different trophic niches. The former is an active predator that swims about grasping and eating other animals, but the latter feeds mainly on decaying vegetation.

Finches of Galapagos Islands in America provide another good example of trophic niche. These birds belong to three genera, *viz., Geospiza* (ground finche), *Camarhynchus* (tree finche) and *Certhidia* (warbler finch). These all birds live in the same general habitat, but differ in terms of their trophic position. For example, in tree finches, one of the species, *Camarhynchus crassirostris* has a parrot-like beak and is basically vegetarian, feeding on buds and fruits. Rest of the tree finches are insect eaters, three of which feed on insects of different sizes. Another species, *C. heliobates* feeds on insects in mangrove swamps. The most interesting is wood-pecker finch, *C. pallidus* which climbs tree in search of insects in the cracks of bark. Although, it has wood-pecking beak, it lacks long extensive tongue of woodpecker.

3. Multifactor or Hypervolume Niche

This concept, where niche is considered as an abstract n-dimensional inhabitated hypervolume, was developed by **G.E.Hutchinson** (1965) of Yale University. Suppose that we measure the range of some environmental factor (*e.g.,* temperature) over which a particular species can live and reproduce (in effect, its range of tolerance) and we put this on a graph. Suppose that we then do the same for another environmental factor (*e.g.,* humidity) and put this on the second axis of the graph. The space that is enclosed will represent the niche of the species. If the effects of two variables (or factors) (*i.e.,* temperature and humidity) are independent, the space would be two-dimensional box (Fig. 8.9 A). However, in fact, temperature and relative humidity are not independent in their biological effects. Thus, tolerance to higher temperature may be related with increase in relative humidity. Under such condition of interacting variables or factors the niche instead of being the box-like would be an oval in shape (Fig. 8.9 B). Now suppose that the tolerance to levels of a third factor (*i.e.,* available phosphorus) was affected by interactions with both temperature and relative humidity. We would now have a niche representing three variables or three-dimensional, volumetric figure (Fig. 8.9 C). If there is present a fourth variable, the space enclosed would be a hyper-volume with four dimensions. Since there are a large number (n) of other environmental factors, both abiotic and biotic, that affect the population, the niche is n-dimensional hypervolume, an abstraction since one is able to draw only with respect to three dimensions. This is the **fundamental niche** of the species.

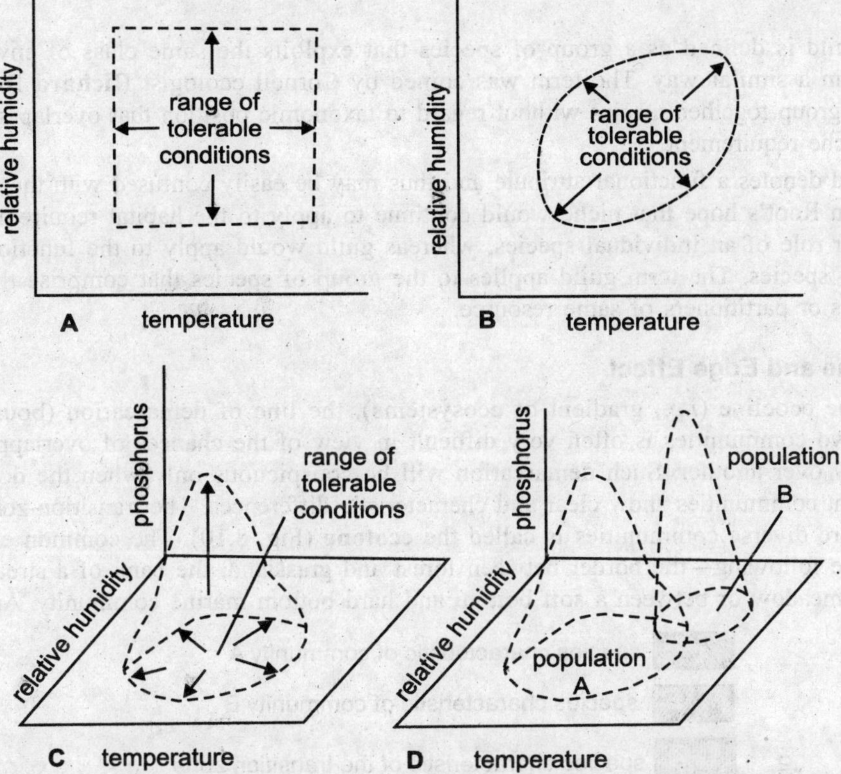

Fig. 8.9. Multifactor or hypervolume niche. A—A two-dimensional niche with independent variables or factors; B—A two-dimensional niche with interdependent factors; C— A three-dimensional niche; D—Two populations with overlapping niches.

When the fundamental niches of two species overlap them, the two species are competing. Competitive exclusion would be called into play only in the complete overlapping of n-dimensional niches (Fig. 8.9 D). Thus, this concept refers to the totality of abiotic and biotic factors to which a given species is uniquely adapted.

Advantages of Niche Segregation

The major advantage which organisms gain by occupying different niches is escape from continuous intense competition. In fact, the niche occupied is favourable to the species physically in furnishing suitable substratum and microclimate, although many species have the ability to live elsewhere where competition is not involved. Automatic segregation of a species into its niche through inherited behaviour patterns or imprinting avoids the great expenditure of energy and loss of time that would be required if this segregation had to be worked out anew each year or each generation. Segregation into niches also avoids confusion of activities between organisms in the community and permits a more orderly and efficient life cycle on the part of each species. Furthermore, the segregation of each species into different niches permits the occupancy of the area by a large number of species, since they will better divide the available resources between them. Similarly, the more distinct the niche of a species is, the more it can avoid conflict with its neighbours and lead a life that is orderly, productive and quite efficient (**Kendeigh**, 1974).

1. Guilds

A guild is defined as a group of species that exploits the same class of environmental resources in a similar way. The term was coined by Cornell ecologist **Richard Root** (1967). This term group together species without regard to taxonomic position that overlap significantly in their niche requirement.

Guild denotes a functional attribute and thus may be easily confused with the term niche. It had been Root's hope that niche would continue to apply to the habitat requirement, due to function or role of an individual species, whereas guild would apply to the functional role of a group of species. The term guild applies to the group of species that comprise the potential competitors or partitioners of same resource.

2. Ecotone and Edge Effect

In the ecocline (*i.e.*, gradient of ecosystems), the line of demarcation (boundary-line) between two communities is often very difficult in view of the chances of overlapping of one community over another. Such demarcation will be conspicuous only when the dominants of the adjacent communities show clear and characteristic differences. The transition zone between two or more diverse communities is called the **ecotone** (Fig. 8.10). The common examples of ecotone are following—the border between forest and grassland, the bank of a stream running through a meadow or between a soft bottom and hard bottom marine community. An ecotone,

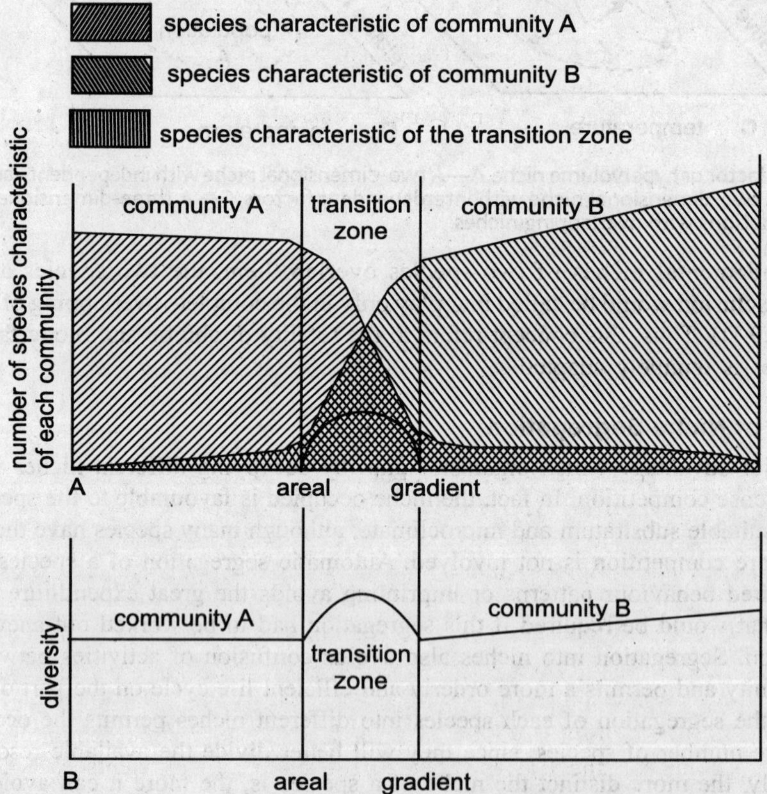

Fig. 8.10 Diagram of the species distribution across a transition zone or ecotone between two communities, labelled A and B.

thus, presents conditions that are intermediate to the communities which are on either side of it. The ecotones may be narrow or very wide, extending to large areas. The community on either side of the ecotone may have a typical structure, but the ecotone is strikingly different. The ecotone has a higher diversity than either of the main communities, a diversity that is not directly controlled by the climate or further fundamental environmental factors, but because of the migrations of individuals of different species-populations from both communities. Further, a number of special populations can become adapted specifically to the ecotone, even when both of the major communities are too simple or are otherwise unsuited for successful colonization by the species. This potential for the ecotone to act as habitat for species found in neither major community is termed as the **edge effect**. A common example of the edge effect in action can be seen in those species of owl that live in or near ecotones between forests and grassland. They depend on forest trees for nesting, and they do their hunting in the grassland, where they depend on field rodents for food. In man-made communities such as agricultural fields, the transition zone (ecotone) between the field and the forest may act as a refuge for animal species formerly found in the plowed area, as well as for other plants such as weeds. Ecotones of this type are also the prime habitat of many species of insect, game bird, and mammal.

3. Ecotypes

Organisms are not just 'slaves' to the physical environments, they adapt themselves and modify the physical environment so as to reduce the limiting effects of temperature, light, water, and other physical conditions of existence. Such **factor compensation** is particularly effective at the community level of organization, but also occurs within the species. Species with wide geographical ranges almost always develop locally adapted populations called **ecotypes** that have different genetically based tolerance ranges and optima to correspond to the environments within which they normally live (**Odum**, 1971; **Clapham Jr**., 1973). The term ecotype was coined by a Swedish botanist, **Gote Turesson** in 1912 and in 1922 he defined it as the *product arising as a result of genotypical response of an ecospecies to a particular habitat*. Ecotypes are especially common in plants, but are also shown by animals and microorganisms.

The ecotypes of the plant species though remain genetically distinct, but are interfertile and are put into one taxonomic species. The different ecotypes of a particular species may differ in their edaphic, biotic, or microclimatic requirements. In India **Misra** and **Rao** (1948) first time reported ecotypic differentiation in *Lindenbergia polyantha.* Ecotypes have also been reported in following Indian plant species as *Euphorbia hirta, Euphorbia thymifolia, Cassia tora, Xanthium strumarium,* etc.

Among animals, ecotypes have been reported in jelly fish, *Aurelia aurita* (**Bullock**, 1955), European oyster, *Ostrea edulis* (**Korringa**, 1957), and frog. Some frogs, for example, undergo rapid development at lower temperatures at one extreme of their range of distribution, while others show a more retarded development rate at higher temperatures at another extreme of their range.

4. Community metabolism

Once a basic idea of community structure is gained it becomes necessary to ask the question how does a community function? The study of functional organisation or community metabolism is fundamentally a study of food relationship within the community. Such studies involve investigations on the flow of energy and cycling of nutrients within a community. Through the **food web** one can make the graphic depiction of the actual flow of food materials from population to population through the community.

8.3 BIOLOGICAL INDICATORS

Indication means pointing out. **Biological indication** or **bioindication** can be defined as the use of living matter as qualitative measures of the environmental status. Biological indication makes use of relatively easy, available and observable reactions of living matter as **indicators** (**Kannan** 1997). The living things that are used in the process of indication of environmental quality are called **bioindicators** or **biological indicator.** Bearers of life such as macromolecules, organelles, cells, tissues, organs, organisms, population, community, ecosystem and biomes can all be used for bioindication.

Advantages of Biological Indicators

1. When there is a change in the normal pattern of a specific biological process or response, bioindicators could be the first to indicate that the health of the enviroment is suffering from some sort of stress.

2. Whereas chemical and physical methods of analysis demand more time, money and space, bioindicators could prove useful for extrapolation of observed results, with limited number of chemical and physical measurement.

3. Observable differences in a specific biological process or response among various geographical regions, are likely to bring into distinct focus the differences in pollution stress. When bioindicators point out a gradient in pollution stress, it could well be verified by suitable chemical methods.

4. Lastly, when there is multitude of factors influencing the quality of environment, bioindicators effectively exhibit their combined effects all at the same time.

Disadvantages of Bioindicators

Inherent uncertainties associated with method of bioindication can make them unsuitable if they are to be used as the sole process to evaluate pollution stress or environmental change.

Classification of Biological Indicators

Biological indicators can be classified into two main categories:

1. **Biological indicators** (also called **ecological indicators, ecosensors** or **ecological indicator species**).

2. **Biological accumulators** (also called **chemical monitoring species** or **collector organism**).

In this chapter, only biological indicators will be dealt with.

Example of Bioindications and Biological Indicators

Organisms, chiefly plant, species, communities or even systems serve as a measure or index (indicator) of the environment. If plants serve as indicators, they are called **plant indicators.** Each response of plant is the effect of some factor or factor complex (interacting factors) acting as a cause and is, therefore, the indication of this factor. It is thus evident that *every plant is a product of the conditions under which it grows and is, therefore, a measurement of environment.* Dominant species in an area are most important indicators, as they receive the full impact of the habitat for over longer periods. Consequently, plant communities are more reliable indicators than individual plants. Plants are indicators of conditions, processes and uses of enviornment.

Further, large plant species serve as better indicators than small species. Usually "*steno*" (*steno* = narrow) species serve much better indicators than "*eury*" (*eury* = broad) species. Some of the evident cases where plants, animals and microbes serve as biological indicators of some characteristic types of environmental conditons are as follows:

1. Indicators of potential productivity of land. Forests serve as good indicators of productivity of land. For example, vegetative growth of tress such as species of *Quercus.* (*Q. marilandica* and *Q. stellata*) is comparatively poor on lowland or sterile sandy soil than the normal soil in which they grow under natural conditions. Plants as *Chrozophora rottleri, Heliotropium supinum* and *Polygonus plebejum* grow better in low-lying lands.

2. Indicators of agriculture. Native vegetation of a particular region is the safe criterion of agricultural possibilities. Thus, plants growing under natural conditions provide more correct information on capabilities of land for the crop growth than those obtained through meteorological data of soil.

3. Indicators of climate. Plant communities characteristic of a particular region provide information on the climate of that area. For instance, **evergreen forests** indicate high rainfall in winter as well as in summer; **sclerphyllous** vegetation indicates heavy rainfall in winter and low during summer; **grasslands** indicate heavy rains during summer and low during winter; **xerophytic vegetation** indicates a very low or no rainfall in the year.

4. Indicators of soil type and other soil characteristics. Many plant species are helpful in indicating the soil type. Thus, luxuriant growth of some taller and deeply rooted grasses such as *Psoralea* indicate a sandy loam type of soil, whereas presence of grasses such as *Andropogon scoparium* indicates the iron-rich soil in the area. Certain plants as *Shorea robusta, Cassia obtusifolia, Geranium* sp., and *Impatiens* sp., indicate proper aeration of soil. Grasses as *Saccharum tridentata, Kochia vesrita, Salicomia uthahensis* and *Salicomia rubra* indicate saline soils. *Woodfordio foribunda* and *Chloris virgata* indicate high lime content of the soil.

Certain plants as *Capparis spinosa* and *Carissa spinaum* indicate intense soil erosion. *Zizyphus rotundifolia* is an indicator of soil formation. *Enicostema liltorale* indicate well drained soil. *Saccharum spontaneum* indicate poorly drained soil. In Africa, *Accacia glandulefera* is used as indicator of ground water. *Argemone mexicana* indicates recently disturbed soil or flooded soil.

5. Indicators of minerals. Some plants grow in metalliferous soils. For example, pines and junipers grow in uranium rich soil; *Viola calamine* grows only in soils rich in zinc in Belgium and England; *Stellaria setacea* grows in mercury rich soils in Spain and *Equisetum plebejum* grows in gold rich soil (in India). *Silene cobalicola* indicates cobalt rich soil and *Waltheria indica* indicate presence of copper, lead and zinc in the soil (in India).

6. Indicators of petroleum deposits. Some animals as the protozoan *Fusilinds* indicate petroleum deposits in the area.

7. Indicators of fire. Some plant such as *Agrostis hiemalis, Epilobium spicatum, Pinum contorta, Populus termuloides, Pteris aquilina* and *Pyronema confluens* (fungus) dominate in areas destructed by fires. *Pteridum* sp., in particular indicates burnt and highly disturbed coniferous forests.

8. Indicators of grazing. Four different categories of animals are indicators of grazing. Horses and cattle prefer shrubs; goats are grazers of non-grassy herbs and sheep are grazers of the pasture land.

9. Indicators of over grazing. Annual weeds and short lived perennials as *Amaranthus, Chenopodium* and *Polygonum, etc.,* grow better in overgrazed areas. Frequent visits of the areas by animals as cattle, horses, sheep, goats, etc., also indicate that the area is under intense grazing.

10. Indicators of water conditions. Burrowing may fly (*Hexangenia* sp.) indicates proper oxygen regimes in the water. *Hydrilla verticillata* and *Ceratophyllum demersum* indicate hard water. *Atriplen* sp., *Salsola* sp. and *Saueda* sp., indicate saline water.

11. Indicators of pollution. Plants as *Utricularia, Chara* and *Wolffia* prefer to grow in polluted waters. Bacterium *Escherichia coli* is an index organism of water pollution. Presence of diatoms in water indicate pollution by sewage. Maggots and chironomid larvae also increase in number in polluted water.

Emigration and disappearance of certain fish as *Catla catla, Labeo gonius, Labeo bata, Labeo rohita, Notopterus,* etc., from the fresh water body indicate industrial pollution in water.

QUESTIONS

Long Answer Questions

1. Define biotic community and describe in detail its different main characters.

2. What do you understand by community? Discuss its structure, stratification and periodicity.

3. Write an essay on animal communities or community structure.

4. Discuss in detail the community and its organization.

5. What is stratification? Explain the phenomenon in a grassland, a pond or a lake.

6. Comment upon the following:

(i) An ecological niche of any given habitat is usually occupied by more than one species.

(ii) Habitat and niche means posession of habitat by an organism.

7. What is meany multidimensional niche?

8. Write an account of habitat and ecological niche.

9. What do you known about the biological indicators of aquatic pollution.

Short Answer Questions

1. Write short notes on the following:

(i) Ecological niche

(ii) Niche

(iii) Niche and habitat

2. Write short notes on the following:

(i) Guilds

(ii) Ecotone and Edge effect or Ecotone

3. Write short notes on the following:

(i) Biological indicators

(ii) Ecological indicators

(iii) Indicator species

(iv) Ecosensors

4. Answer the following question must not exceed 20 words.

(i) Differentiate between Ecotone and Edge effect.

(ii) Define edge effect.

Very Short Answer Questions

1. Define ecotypes.

2. Who coined the term ecotype?

3. Name the zone of ocean characterized by absence of sunlight and living organisms.

(*C.C.S.U. Meerut 2003*)

4. Name the water zone below 200 meters. (*C.C.S.U. Meerut 2003*)

5. Name the deepest part of lake where no light reaches. (*C.C.S.U. Meerut 2004*)

6. What is alpine tundra?

7. What is edge effect?

8. Define edge species.

9. Define stratification.

Fill in the Blanks

1. .... includes all the populations inhabiting a specific area at the same time.

2. Zone of transition from one type of community or ecosystem to another is called ....

3. Functional role of a species in a biotic community or ecosystem is called ....

4. .... have buds close to ground surface and are hidden under soil surface.

5. Zebra is a .... animal form.

6. Benthos are mostly .... animal forms.

7. Forest communities have .... stratification.

Multiple Choice Questions

1. The niche of a population is the

(*a*) set of conditions and resources it uses

(*b*) place where it lives

(*c*) geographical area it covers

(*d*) set of interactions it has with other populations

2. The habitat of a population is the

(*a*) set of conditions and resources it uses

(*b*) place where it lives

(*c*) geographical area it covers

(*d*) set of interactions it has with other populations

3. According to the Gause's competitive exclusion principle, two species cannot remain in the same community if they have the same

(*a*) taxonomy (*b*) niche

(*c*) habitat (*d*) food requirements

4. All the populations in a given area are defined as

(*a*) population (*b*) ecosystem

(*c*) biotic community (*d*) neither of these

5. Microclimate is

(*a*) specific climatic condition to which an organism is exposed in its habitat

(*b*) climate during a particular season

(*c*) climate during a given short period

(*d*) neither of three

6. The transition zone between two ecosystems is termed (*CCSU Meerut 2014*)

(*a*) ecad (*b*) ecotone

(*c*) ecoline (*d*) none of them

ANSWERS

Very Short Answer Questions

1. Ecotypes are subspecies or local populations adapted to a particular set of environmental conditions.

2. Gote Turesson (1912), a swedish botanist coined the term ecotype.

3. Hadal zone (below the depth of 5000 meters).
4. Bethyal or aphodal zone.
5. Aphodal zone.
6. Tundra-like conditions found above tree lines of high mountains.
7. Edge effect is a response of plants and animals to the site where two or more communities or ecosystems meet (typically creating an increase in biotic diversity along the edge site).
8. Species that inhibit edge or boundary habitats; species that uses edges for production are survivorship purpose.
9. Stratification is the layering of water in oceans and lakes due to differences in water density and temperature with depth.

Fill in the Blanks

1. Community 2. Ecotone 3. Niche 4. Chamaephytes
5. Cursorial 6. Sessile 7. Vertical.

Multiple Choice Questions

1. (*a*) 2. (*b*) 3. (*b*) 4. (*c*)
5. (*a*) 6. (*b*).

9

Ecological Succession

When stripped of its original vegetation by fire, flood, or glaciation, an area of bare ground does not remain devoid of plants and animals. Beginning with plants, area is rapidly colonized by a variety of both plant and animal species that subsequently modify one or more environmental factors. This modification of the environment may in turn allow additional species to become established. The transitional series of communities which develop in a given area are called **sere** or **seral stages**, while the final stable and mature community is called the **climax**. The development of the community by the action of vegetation on the environment leading to the establishment of new species is termed **succession**. Succession is the universal process of directional change in vegetation during ecological time. It can be recognized by the progressive change in the species composition of the community. Retrogression in community development does not occur unless succession is disturbed or halted by fire, grazing, scraping or erosion.

H. C. Cowles (1899) in his study of vegetation development on sand dunes of Lake Michigan first of all explicitely described the process of plant succession. However, prior to him two other plant ecologists attempted to study succession. These were **Henry Thoreau** (1860) who described plant succession following logging in New Zealand and **Anton Kerner** (1863), who wrote about plant succession in the Danube Basin in central Europe. Later **F. E. Clements** (1916) developed a descriptive theory of succession and advanced it as a general ecological concept. Clement's theory of succession dominated ecology for many years, almost becoming a dogma, because it provided an orderly, logical explanation for the development of plant communities.

9.1 CAUSES OF SUCCESSION

Since succession involves a series of complex processes, so there exist many causes of its occurrence. Ecologists have recognized the following three primary causes of succession:

1. Initial or Initiating causes. These are climatic as well as biotic in nature. The climatic causes include factors such as erosion and deposits, wind, fire, etc., which are caused by lightening or volcanic activity. The biotic causes include various activities of organisms. All these causes produce the bare areas or destroy the existing populations in an area.

2. Ecesis or Continuing causes. These are processes as migration, ecesis, aggregation, competition, reaction, etc., which cause successive waves of populations as a result of changes, chiefly in the edaphic (soil) features of the area.

3. Stabilising causes. These include factors such as climate of the area which result in the stabilisation of the community.

Trends of Succession

Trends of changes during ecosystem development from a young to mature stages include the following features:

1. A continuous change occurs in the kinds of plants and animals.

2. An increase in the diversity of species takes place. The general appearance of the community or the physiognomy keeps on becoming more and more complex as the succession proceeds.

3. There is a progressive increase in the amount of living biomass and dead organic matter. Such an increase occurs in gross as well as net primary production in the initial and seral stages. Thus, there is more biomass accumulation, gradually reaching a huge biomass structure in the climax.

4. Green pigment (Chlorophyll) go on increasing during the early phase of primary succession. The ratio of yellow/green pigments remains around 2 in the early stages and increases to 3 to 5 in the climax stage. Pigment diversity also increases.

5. The community respiration increases but the P/R (*i.e.,* Production/Respiration) ratio remains more than 1 in the seral stages. The huge living biomass respires a lot in the climax stage and the P/R ratio equals 1 (*i.e.,* P/R = 1). Thus, in the early stages P>R and in the climax stage P = R.

6. The food chain relationships become more complex as succession proceeds.

7. Nutrients in the young stage are allocated mostly in the soil, but as the seral stages advance, nutrients get allocated more in the vegetation and less in soil. Further the nutrient cycling becomes more closed or intrabiotic with an efficient cycling mechanism whereas in the young stage the nutrients easily leak out from the system, *i.e.,* the cycling is more of an open type.

8. The role of detritus becomes progressively more and more important.

9. The quality of the habitat gets progressively modified to a more mesic condition from either too dry or too wet condition, in the early seral stage.

10. The niche specialization increases, *i.e.,* different functions are more effectively performed by specialist species in mature seral stage, whereas in early stage many functions are performed but less efficiently by a few species.

11. The life cycle of mature community species are longer and more complex.

12. The importance of macroenvironment becomes less in later stages.

13. Relationship among component species becomes more and more mutualistic or helpful even though there exist enough competition and allelochemic activities to prevent invasion of outside elements.

14. Dispersal of seeds and propagules is by wind in young stage, while by animals in mature stage.

9.2 TYPES OF SUCCESSION

Based on different criteria, there are following kinds of succession :

1. Primary succession. If an area in any of the basic environments (such as terrestrial, freshwater or marine) is colonized by organisms for the first time, the succession is called **primary succession**. Thus, primary succession begins on a sterile area (an area not occupied previously by a community), such as newly exposed rock or sand dune where the conditions of existence may not be favorable initially.

2. Secondary succession. If the area under colonization has been cleared by whatsoever agency (such as burning, grazing, clearing, felling of trees, sudden change in climatic factors, etc.) of the previous plants, it is called **secondary succession**. Usually the rate of secondary succession is faster than that of primary succession because of better nutrient and other conditions in area previously under plant cover.

3. Autogenic succession. After the succession has begun, in most of the cases, it is the community itself which, as a result of its reactions with the environment, modifies its own environment and, thus, causing its own replacement by new communities. This course of succession is known as **autogenic succession**.

4. Allogenic succession. In some cases replacement of one community by another is largely due to forces other than the effects of communities on the environment. This is called **allogenic succession** and it may occur in a highly disturbed or eroded area or in ponds where nutrients and pollutants enter from outside and modify the environment and in turn the communities.

5. Autotrophic succession. It is characterized by early and continued dominance of autotrophic organisms such as green plants. It begins in a predominantly inorganic environments and the energy flow is maintained indefinitely. There is gradual increase in the organic matter content supported by energy flow.

6. Heterotrophic succession. It is characterized by early dominance of heterotrophic organisms such as bacteria, actinomycetes, fungi and animals. It begins in a medium which is rich in organic matter such as small areas of rivers, streams; these are polluted heavily with sewage or in small pools receiving leaf litter in large quantities.

7. Induced succession. Activities such as overgrazing, frequent scraping, shifting cultivation or industrial pollution may cause deterioration of an ecosystem. Agricultural practices are retrogression of a stable state to a young state by man's deliberate action.

8. Retrogressive succession. It means a return to simpler and less dense or even denuded form of community from an advanced or climax community. In most cases, the causes are allogenic, *i.e.,* forces from outside the ecosystem become severe and demanding. For example, most of our natural forest stands are degrading into shrubs, savanna or impoverished desert-like stands by the severity of grazing animals brought from surrounding villages. Excessive removal of wood, leaf and twig litter also leads to retrogressive succession.

9. Cyclic succession. It is of local occurrence within a large community. Here *cyclic* refers to repeated occurrence of certain stages of succession whenever there is an open condition created within a large community.

Further, in ecological literature, there are mentioned still so many other kinds of succession, depending mainly upon the nature of the environment where the process of succession has begun. Thus, it may be a **hydrosere** or **hydrarch** when succession occurs in regions where water is in plenty, as ponds, lakes, streams, swamps, bogs, etc., a **mesarch** when there are present adequate moisture conditions; and a **xerosere** or **xerarch** when there is present minimum amount of moisture, *e.g.,* deserts, rocks, etc. Sometimes, few more categories are recognized such as **lithosere** – succession initiating on rocks, **psammosere** –succession initiating on sand and **halosere** – succession occurring in saline water or saline soil.

9.3 GENERAL PROCESS OF SUCCESSION

The entire process of primary autotrophic succession is completed through the following sequential steps:

1. Nudation. This is the development of a bare area without any form of life. Exposure of new surface may occur due to several causes such as landslide, erosion, deposition, or other catastrophic agency. These causes of nudation are of three main types:

(i) Topographic. Due to soil erosion by gravity, water or wind, the existing community may disappear. Other topographic causes include deposition of sand, landslide, volcanic activity and other factors.

(ii) Climatic. Glaciers, dry period, hails and storm, frost, fire, etc., may also destroy the community.

(iii) Biotic. Humans form a most important biotic factor; they are responsible for destruction of forests, grasslands for industry, agriculture, housing, etc. Other factors are disease epidemics due to fungi, viruses, etc., which destroy the whole population.

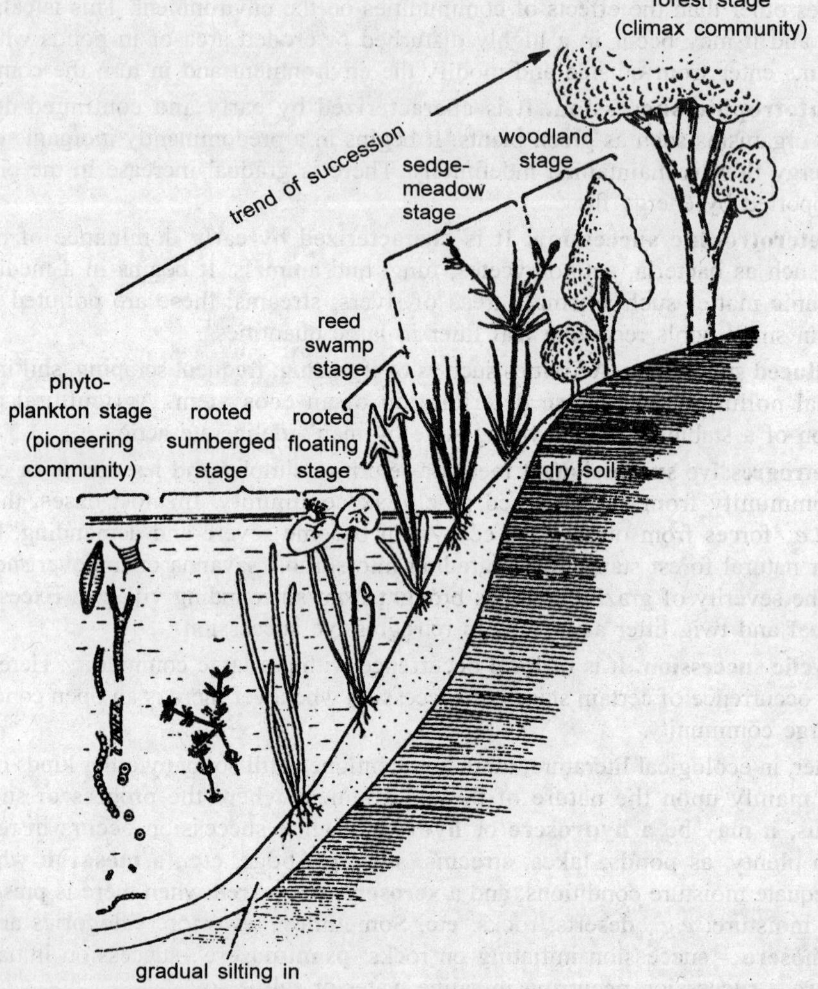

Fig. 9.1. Diagrammatic representation of succession through autogenic processes. Seral communities modify the environmental complex which in turn changes the vegetation and the process continues until the dynamic equilibrium is reached. Note the pioneer, seral and climax communities.

2. Invasion. Invasion is the successful establishment of a species in a bare area. The species actually reaches this new site from any other area. Invasion includes the following three steps:

(i) Migration (dispersal). The seeds, spores, or other propagules of the species reach the bare area. This process is called **migration** and is generally brought about by air, water, etc.,

(ii) Ecesis (establishment). After reaching to new area, the process of successful establishment of the species, as a result of adjustment with the conditions prevailing there, is known as **ecesis**. Ecesis is followed by full-scale **colonization**. In plants, after migration, seeds or propagules germinate, seedlings grow, and adults start to reproduce.

(iii) Aggregation. Colonization by successive offspring and new migrants help increase the population, a process called **aggregation**. Thus, after ecesis, as a result of reproduction, the individuals of the species increase in number and they come close to each other. Plants or autotrophic organisms which are the first to colonize and aggregate are called **pioneers**. The pioneer communities are likely to be more dynamic and have low-nutrient requirements and to take minerals in comparatively more complex forms. They are small-sized and make less demand from environment.

3. Competition and coaction. Due to aggregation of a large number of individuals of the species at the limited place, there develops **competition** (*i.e.,* interspecific and intraspecific competition) for space and nutrition. Individuals of a species affect each other's life in various ways and this is called **coaction**. The species which fail to compete with other species are ultimately discarded. The reproductive capacity, wide ecological amplitude, etc., help the species withstand the competition.

4. Reaction. Reaction includes mechanism of the modification of the environment through the influence of living organisms on it. Due to this very significant stage, changes take place in soil, water, light conditions, temperature, etc., of the environment. As a result of reaction, the environment is modified and becomes unsuitable for the existing community which sooner or later is replaced by another community (seral community). The whole sequence of communities that replaces one another in the given area is called a **sere**, and different communities constituting the sere are called **seral communities**, **seral stages** or **developmental stages**.

5. Stabilization (Climax). Finally, there occurs a stage in the process, when the final terminal community becomes more or less established for a longer period of time and it can maintain itself in equilibrium with the climate of the area. The final community is not replaced and is known as **climax community** and the stage as **climax stage**.

9.4 SOME EXAMPLES OF SUCCESSION

1. Hydrosere. A good example of succession is the **hydrarch succession** or **hydrosere** (Fig. 9.2), in which a pond and its community are converted into a land community. In the initial stage, phytoplankton [*e.g.,* some blue green algae (cyanobacteria), green algae (*e.g., Spirogyra, Oedogonium*), diatoms and bacteria] are the pioneer colonizers. They are consumed by zooplankton (*e.g.,* protozoans such as *Amoeba*, *Paramecium*, *Euglena*, etc.), fish such as blue gill fish, sun fish, large mouth, etc. Gradually these organisms die and increase the content of dead organic matter in the pond. This is utilized by bacteria and fungi, and minerals are released after the decomposition. The nutrient-rich mud then supports the growth of rooted hydrophytes such as *Hydrilla, Elodea, Vallisneria, Ceratophyllum,* etc., in the shallow water zone. This submerged stage is also inhabited by the animals such as dragon flies, may flies and crustaceans such as *Asellas, Gammarus, Daphnia, Cypris, Cyclops,* etc. The hydrophytes die and are decomposed by mircroorganisms, thus, releasing nutrients. In addition to this, due to silting, the water depth of the pond is reduced and at the margin of the pond grow rooted floating vegetation (*i.e.,* the plant species whose leaves reach the water surface and roots remain in the mud). Plants such as *Nelumbo nucifera, Trapa, Monochoria,* etc., grow in these conditions. In floating stage faunal living space is increased and diversified. Hydras, frogs, salamanders, gill-breathing snails, diving beetles (*Dysticus*) and host of new insects capable of utilizing the under surfaces of floating leaves appear. Some turtles and snakes also invade the pond.

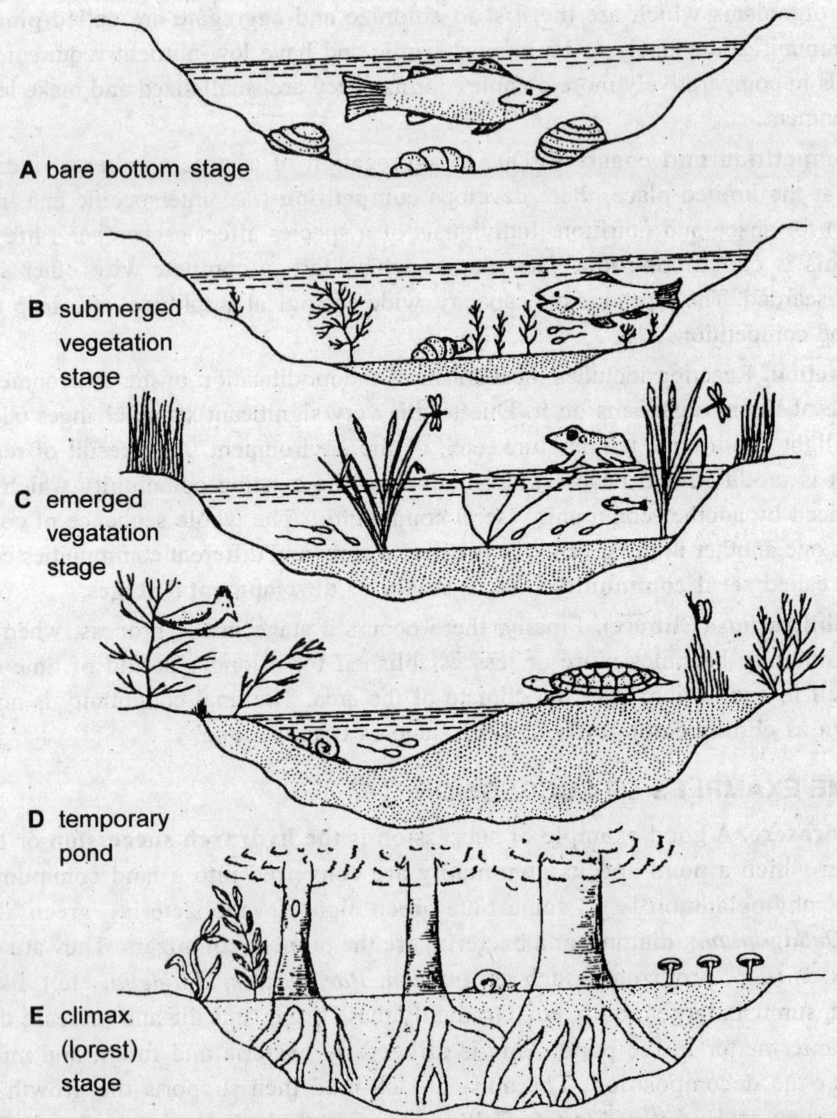

Fig. 9.2. Diagram showing ecological succession from a pond to a climax forest.

Gradually the water depth in the pond decreases due to evaporation and the deposition of organic matter and the concentration of the nutrients increases. Free-floating plants such as *Azolla, Lemna, Pistia, Wolffia, Spirodella,* etc., increase in number because of the high nutrient availability. Gradually their dead parts fill up the pond ecosystem, resulting in the further build up of the substratum.

At this stage, the pond becomes a swampy ecosystem. The reed swamp species (such as *Scirpus* or bulrushes, *Typha* or cattail, *Phragmites* or reed grass, *Rumex,* etc.) and sedges (*e.g., Carex, Juncus, Cyperus,* etc.) invade the pond and the latter are gradually replaced by mesic communities as the water depth is reduced greatly. Gradually land plants, such as, shrubs (*Salix, Cornus*) and trees (*Populus*, *Almus*) invade ending in the climax community such as deciduous forest (Fig. 9.2). In association with the changes in water depth and vegetation, the aquatic fauna also change and ultimately gets replaced by land animals.

Thus, possible trend of succession in the aquatic environment is as follows:

Climax community
↑
Open scrub land = Deciduous forest
↑
Terrestrial communities
↑
Mesic communities
↑
Reeds and sedges
↑
Free floating and rooted plants
↑
Rooted and aquatic plants
↑
Phytoplankton
↑
Pond ecosystem

Climax communities vary from place to place. For example, in low lying lands in some parts of Kashmir, the climax community has trees such as *Salix* (**Ambasht,** 1988). In Indian upland plateaus, the climax woody species consist of *Diospyros, Butea* and *Zizyphus* and ground vegetation of *Eragrostis, Sporobolus, Bothriochloa,* etc. Climax vegetation of lowlands and valleys which provide a mesic environment includes *Terminalia, Ficus, Sterculia, Salix,* etc.

2. Xerosere. Xerosere or xerarch succession begins on exposed parent rocks (**lithosere**) or dry sand (**psammosere**). A lithosere (Fig. 9.3) involves the following stages: crustose lichens stage (pioneers) → foliose lichens stage → moss stage → herbs stage → shrub stage → forest stage (climax stage). Thus, pioneer plants are lichens, mosses and *Selaginella*, which help in soil formation by accelerating erosion. In course of time grasses, annuals and herbaceous vegetation grow on the soils deposited on rocks.

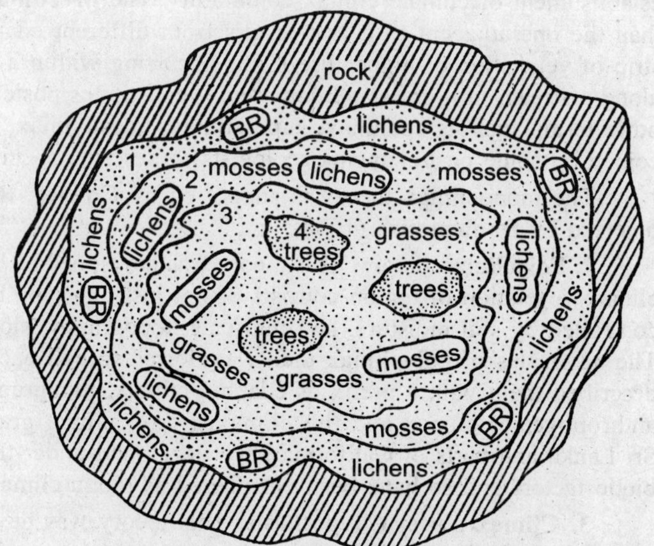

Fig. 9.3. Diagrammatic representation of different plant communities of a lithosere appearing on a rock. The various zones from outside towards the centre of rock are (1) lichens (pioneers), (2) ring of mosses, (3) grasses (broad zone) and (4) trees (seedlings). Note the pioneer community of lichens around the outer edge and more advanced stage of trees located in the centre. BR=bare rock.

Like the hydrosere, the lithosere involves successive changes in animal life. The pioneer animals of the lichen stages are few species of mites, ants and spiders. These animals are exposed to harsh environment such as extreme fluctuations in temperature. During the moss stage, many new species of mites, small spiders, tardigrades and springtails invade the community. The herb stage is characterized by nematodes, mites, collembola, ants and various insect larvae. During the shrub and forest stages great qualitative and quantitative modifications occur in the fauna. Thus, there occur numerous kinds of animals such as slugs, snails, wire worms, millipedes, centipedes, mites, ants, sow bugs, springtails, amphibians such as salamanders, frogs, etc., reptiles such as turtles, skinks and other lizards, snakes, birds such as flycatcher and grouse, and mammals such as mole, mouse, shrews, squirrels, chipmunk and fox.

9.5 CONCEPT OF CLIMAX

The concept of climax has since long been a subject of much controversy and discussion. There are following three theoretical approaches to the climax:

1. Monoclimax theory. This theory is developed largely by **Frederick Clements**. This theory recognizes only one climax, determined solely by climate, no matter how great the variety of environmental conditions is at the start. All seral communities in a given region, if allowed sufficient time, would ultimately converge to a single climax. The whole landscape would be clothed with a uniform plant and animal community. All other communities than the climax are related to the climax by successional development and are recognized as subclimax, disclimax, preclimax, postclimax and so on. A **subclimax** is a stage in succession of forests just preceding the climate climax community. The **disclimax** is the particular type of vegetation maintained in an area as a result of recurrent disturbance, chiefly the biotic, thus preventing a successful establishment of climate climax community. The **preclimax** is a vegetation of lower-life forms than the one adjacent to it and results from different edaphic conditions. The **postclimax** is a strip of vegetation of higher life-forms occurring within a climate climax, for example, a forest along a stream in a grassland community constitutes postclimax community. Later on numerous other terms such as coclimax, superclimax, quasiclimax, anticlimax, pseudoclimax, etc., were coined by some post-clements ecologists to describe specific situations.

The monoclimax theory is supported by **Cowles**, **Ranganathan**, and **Puri**, but strongly objected by **Daubenmire** (1968).

2. Polyclimax theory. This theory was developed by **Tansley**. This theory considers that the climax vegetation of region consists of not just one type but a mosaic of vegetational climaxes controlled by soil moisture, soil nutrients, topography, slope exposure, fire and animal activity. The advocates of polyclimax theory preferred to call each stable community as a climax and described them with a prefix as edaphic climax, topographic climax, biotic (zootic, grazing or anthropogenic) climax and fire climax. For example, grassland communities of central India, Sri Lanka and parts of California have developed under the influence of fire, grazing and other biotic factors and so have been considered as biotic climaxes by **Misra**, **Pandeya** and **Holmes**.

3. Climax pattern hypothesis. This theory was developed by **Whittaker**, **Mclntosh** and **Sellack**. According to this theory, the composition, species structure and balance of a climax community are determined by the total environment of the ecosystem and not by one aspect, such as climate alone. Involved are the characteristics of each species population, their biotic interrelationships, availability of flora and fauna to colonize the area, the chance dispersal of seeds and animals, and the soils and climate. The pattern of climax vegetation will change as the environment changes. Thus, the climax community presents a pattern of populations that corresponds with and changes with the pattern of environmental gradients, intergrading to form ecoclines. The central and most widespread community in the pattern is the prevailing or climatic climax. It is the community that most clearly expresses the climate of the area.

9.6 MODELS OF SUCCESSION

The processes and mechanisms involved in succession have been the subject of considerable study, controversy and review. Following models have been proposed to interpret the successional process:

1. Pioneering concepts. F. E. Clements (1916) viewed succession as process involving several phases, to each of which he gave his own terminology. Succession began with the development of a bare site, called **nudation** (disturbance). Nudation was followed by **migration,** the arrival of propagules. Migration was followed by the establishment and initial growth of vegetation. **Clements** called this phase **ecesis** (growth). As vegetation became well established, grew and spread, various species began to compete for space, light and nutrients. This phase Clements called **competition.** It was followed by **reaction,** the self-induced (autogenic) effects of plants on the habitat. The outcome of this reaction was the replacement of the plant community by another, resulting in the persistence of one species complex or community, or **stabilization.** Clement's theory of succession dominated ecology for years.

Ecologists remain critical of Clement's climax concept. To Clements, the climax was an assemblage of vegetation that belonged to the highest type of vegetational community possible under the prevailing climate. The climax was able to reproduce itself. During succession, each seral stage so modifies the environment that plants of that stage eventually can no longer exist there. Instead they prepare the site for the replacement of plants of next stage. The process continues until the vegetation arrives at the self-reproducing climax. This marks the end of succession.

Henry Gleason (1917) challenged Clements view of succession. He regarded a plant association or climax not as a superorganism, but as community consisting of individual species that respond independently to environmental conditions. Succession resulted from these individual responses of different species. Plants involved in succession were those that arrived first on the site and were able to establish themselves under prevailing environmental conditions. Competitive and other interaction among species, the plant modification of the microenvironment, and the ability of the plants to exploit nutrients, moisture and other environmental inputs on the site determine the final outcome.

2. Holistic concept. This approach was proposed by **E. P. Odum** (1969, 1983). It views succession as an orderly transition from one seral stage to another driven by self-induced modification of the environment. Thus, holistic concept views succession as driven by changes in attributes between youthful and mature systems. Succession involves changes in the emergent properties (such as nutrient flow, biomass accumulation and species diversity) that arise out of organization of the ecosystem.

Success begins with the development stages of short-lived, shade-intolerant plant species and terminates with a mature stage dominated by long-lived, shade tolerant species. Young stages so modify the environment that the existing community is replaced by a more mature one, better able to exploit the changed environment.

3. Reductionist concept. Reductionists reject the holistic concept. They argue that successional changes can be explained by **population dynamics** (especially **competition, regeneration, mortality**), **physiology** and **life history strategies** (**Bazzaz** 1979; **Peet** and **Christensen** 1980; **Noble** and **Slatyer** 1980).

Reductionist approach to succession emphasizes the role of population dynamics and life history patterns. Succession in its earliest stages begins with an open site that is eventually colonised by early successional species, variously called **pioneer, opportunistic,** or **fugitive** (*r* species). On primary sites, propagules of colonizing species arrives by means of wind, water or animals. On secondary sites (*i.e.,* in case of secondary plant succession), residual propagules, *i.e.,* seeds lying dormant in the soil, roots and rhizomes, as well as dispersed seeds are important in colonization. The colonizers establish themselves and grow.

Early colonists soon lose their temporary dominance in part because they have to renew their photosynthetic biomass each year to the taller, more vigorous growth of later stage plants, which carry over biomass from year to year. These plants gradually assume dominance and place a much greater demand on resources such as nutrients, light and moisture. As availability of resources decreases and demand increases, competition within and among populations of plants becomes more intense. Mortality, reflected in self-thinning, increases. As the population declines, resources are released for use by plants of still later stages (**Peet** 1981). Species characteristic of later stages of succession grow more slowly and are longer period of time because much of their production goes into storage and maintenance. These species are mostly specialists, adapted to a narrow range of environmental conditions in which the plants either hoard resources or use them more effectively.

4. Connel-Slatyer models. Connell and **Slatyer** (1977) proposed following three models of succession. These tend to emphasise some points of Clemen's original theory.

(i) Facilitation model. This model is autogenic. Changes are brought about from within by the organisms themselves. Early stage species modify the environment and prepare the site for later stage species, facilitating their success.

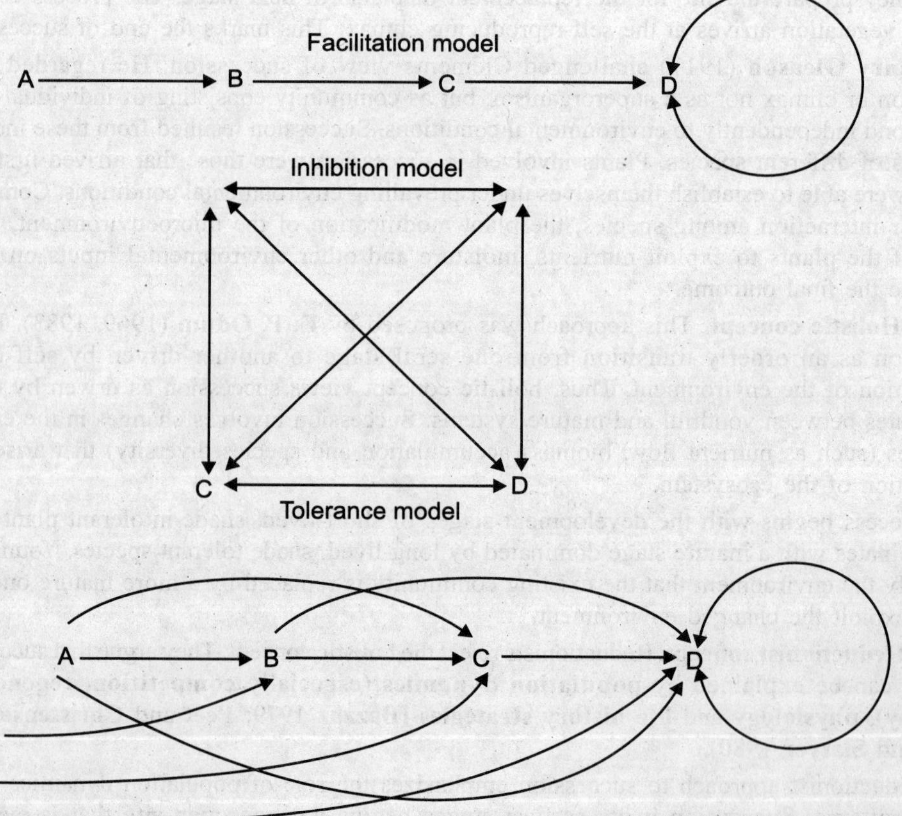

Fig, 9.4. Three models of succession proposed by Connell and Slayter. Four species are represented by A, B, C, and D. An arrow indicates "is replaced by".

(ii) Inhibition model. This model is purely competitive. No species is competitively superior to another. This site belongs to those species that become established first and are able to hold their position against all invaders. These plants make the site less suitable for both early and late successional species. As long as they live, they maintain their position. The ultimate winners, however, are the long-lived plants. Such succession is not orderly and is less predictable.

(iii) Tolerance model. It is intermediate between two models and involves the interaction of life history traits, especially competition. It suggests that later successional species are neither inhibited nor aided by species of earlier stages. Later stage species can invade a site, become established and grow to maturity in the presence of those preceding them. They can do so because these later species have a greater *tolerance* for a lower level of resources than the earlier ones. Such interactions lead to communities composed of those species which are most efficient in exploiting resources either by interference competition or by using sources unavailable to other species.

5. Mechanistic models. A mechanistic approach involves experimental and computer simulation studies of succession. It considers both direct process by which an interaction, such as competition, occurs and how the physiology, morphology and life history traits of individual species influence these processes.

The **resource-ratio model** (**Tilman** 1987, 1988) and the **individual-based plant model** (**Huston** and **Smith** 1987) are more mechanistic. According to these models, the pattern of species replacement involves competition and species-specific life-history traits and physiological properties of plants interacting under changing environmental conditions, particularly light, nutrients and moisture. Such environmental changes are brought about in part by individual plants.

QUESTIONS

Long Answer Questions

1. What is ecological succession? Explain the phenomenon in a pond.
2. What do you understand by ecological succession. Describe various patterns of succession.
3. Write an essay on "Biotic succession" in nature.
4. Discuss the importance of biotic diversity in ecological succession.
5. What is succession? Describe the basic types of succession and add a note on climax community.
6. Write a short note on 'Ecological Succession'.
7. What do you understand by climax? Describe monoclimax and polyclimax theory. Write briefly on climax community.
8. Describe various models to explain the process of ecological succession.
9. Define population. How does it differ from a community? Explain the reason why in an ecological succession one type of community is replaced by another type of community and also give main features of climax community.

Short Answer Questions

1. Write short notes on the following:

   (*i*) Xerosere

   (*ii*) Hydrosere

   (*iii*) Secondary Succession

Very Short Answer Questions

1. Define succession.
2. What are pioneers?

3. What is pioneer stage?

4. Define seral stage.

5. What is sere?

6. What do you understand by climax stage?

7. Define the polyclimax concept.

8. What is hydrarch?

9. What is hydrosere?

10. What is halosere?

11. What is xerach and xerosere?

Fill in the Blanks

1. .... (1885) was first to use the term succession for the orderly changes in communities.

2. Father of successional theory is .... (1916).

3. Succession initiating on sand is called ....

4. .... is development of the bare area without any form of life.

5. Invasion step of succession includes ...., ... and aggregation.

6. In succession, the final community which is not replaced is called ....

7. In hudrosere, the pioneer colonizers include .... , ... , diatoms and ....

Multiple Choice Questions

1. Succession involves a turnover of species during
(*a*) the season of the year (*b*) ecological time
(*c*) microevolutionary time (*d*) macroevolutionary time

2. Which of the following organism will be the first to colonize a bare rock?
(*a*) annual plant (*b*) biennial plant
(*c*) perennial plant (*d*) lichen

3. The kind of climax community in an area depends mostly on the areas
(*a*) pool of available colonists (*b*) soil organism
(*c*) climate (*d*) bedrock

4. Primary succession takes much longer than secondary succession, because it involves
(*a*) development of the soil
(*b*) development of a seed bank
(*c*) colonization by organisms that are farther away
(*d*) none of the above

5. Early colonists in a primary succession are usually lichens, legumes and plants with mycorrhizae, which suggests that colonization is easier when an organism has a
(*a*) competitive edge (*b*) parasitic association
(*c*) commensal association (*d*) mutualistic association

6. The stability established in succession (*CCSU Meerut 2014*)
(*a*) in initiation stage (*b*) in migration stage
(*c*) in seral stage (*d*) in climax stage

7. Xerosere initiates from (*CCSU Meerut 2014*)
(*a*) rocks (*b*) water
(*c*) both the above (*d*) none of the above

8. The correct sequence of general process of succession is (*CCSU Meerut 2014*)

(*a*) nudation → invasion → competition → reaction → stabilisation

(*b*) nudation → competition → invasion → reaction → stabilisation

(*c*) invasion → nudation → competition → reaction → stabilisation

(*d*) nudation → competition → invasion → reaction → stabilisation

ANSWERS

Very Short Answer Questions

1. Succession is replacement of one community or seral stage by another.

2. The first organisms that invade a bare area are called pioneers.

3. Pioneer stage is an initial seral stage of a sere characterizad by early successional plant species (typically annuals).

4. Seral stage is one of the successional stages in a sere.

5. Sere is series of successional stages at a particular site leading to a mature, climax community.

6. Climax is the final mature, stable, self-maintaining and self-sustaining stage of vegetational development in a climatic unit.

7. Polyclimax concept is the theory that the final seral stage of succession is controlled by one of several local environemntal forces or conditions such as soil, fire, or climate.

8. Succession beginning in aquatic habitat (such as pond, lake or marsh) is called hydrarch.

9. The entire sequence of communities that appear during ecological succession in aquatic habitat is called hydrosere.

10. The succession beginning in saline water is called halosere.

11. The ecological succession beginning in dry situations are called **xerarch** and the different stages of the sere constitute xerosere.

Fill in the Blanks

1. Hult	**2.** Clements	**3.** Psammosere	**4.** Nudation
5. Migration, ecesis	**6.** Climax stage	**7.** Cyanobacteria, green algae, bacteria.	

Multiple Choice Questions

1. (*b*)	**2.** (*d*)	**3.** (*c*)	**4.** (*a*)
5. (*d*)	**6.** (*d*)	**7.** (*a*)	**8.** (*a*).

10

Ecosystem: Structure and Function

10.1 CONCEPT OF ECOSYSTEM

Ecosystem is composed of a community and the physical environment it occupies. The central theme of ecosystem concept is that living organisms of a community not only interact among themselves but also have functional relationship with their nonliving environment (Box 10.1). Ecosystem consists of three basic components — **producers, consumers** (including decomposers) and **inorganic** and **organic substances** – which exchange **energy** and **nutrients.**

Box 10.1

A system consists of two or more components that interact and it is surrounded by an environment with which it may or may not interact. One characteristics of any system is organization, that is, a unified group of components forming a systemized whole. **O'Neill** et al. (1986) have recognized the following properties in an ecological system or ecosystem.

1. *Ecosystems exist independently of specific components, e.g.,* an individual tree may die, but the forest's organization persists.
2. *The components of an ecosystem are interdependent, e.g.,* when removed from its colony, a social insect does not often survive.
3. *An ecosystem has a function, e.g.,* the component parts each have functions that together produce a function of the whole.
4. *Ecosystem is active and something dynamic, past or present, is implied, e.g.,* change occurs or has occurred.
5. *A sliding scale of organization exists, e.g.,* two population may independently coexist in an area or they may be intertwined in a complex relationship.

The word *ecosystem* is relatively recent term. A British ecologist **Sir Arthur Tansley** (1935) invented the word to apply to a whole community of organisms and its environment as one unit. For many years, ecologists had realised that a habitat and the community it contains, are a single working system. They applied many terms to this concept, but apart from the term ecosystem, only one earlier word/term **biocoenose** (**Mobius,** 1877), is frequently used, especially in Europe and Russia.

Kinds of Ecosystem

An ecosystem can be natural or artificial, temporary or permanent and large or tiny. Thus, various constituent ecosystems of the biosphere fall into the following categories:

1. Natural ecosystems. These types of ecosystems operate by themselves without any major interference by man. Based upon the particular kind of habitat, these are further classified as:

(i) Terrestrial ecosystems such as forests, grasslands, deserts, a single log, etc.

(ii) Aquatic ecosystems which may be further distinguished as follows:

(a) Fresh water ecosystems. These may be **lotic** (running water as spring, brook, stream or river) or **lentic** (standing water as lake, pond, pool, puddle, ditch, swamp, etc.) **ecosystems.**

(b) Marine ecosystems. These include salt water bodies which may be deep bodies as an ocean or shallow ones as a sea or estuary.

2. Artificial ecosystems. These are also called **human-made** or **human-engineered ecosystems.** These are maintained artificially by human beings where, by addition of energy and planned manipulations, natural balance is disturbed regularly, *e.g.,* croplands such as sugarcane, maize, wheat, rice-fields; orchards, gardens, villages, cities, dams, aquarium and manned spaceship (Box 10.2).

Box 10.2 Complete and Incomplete Ecosystems

A portion of land, a lake or a river can be considered as a **complete ecosystem** in which green plants produce organic matter which is consumed by herbivore. Herbivores are in turn eaten by carnivores while decomposers mineralize the exudates and remains of the living organisms.

Abysmal depth of oceans, where no light penetrates represent **incomplete ecosystem**. There are no plants and hence there is no primary production. The organic matter or energy to support the living organisms in the system is obtained by dead remains, exudates and debris which sink down from above. Likewise our cities/metropolitans are man-made but incomplete ecosystem, in which all life activities depends on the food or energy imported from the countryside.

Structure of Ecosystem

The structure of an ecosystem is basically a description of the species of organisms that are present, including information on their life histories, populations and distribution in space. It is a guide to who's who in the ecosystem. It also includes descriptive information on the non-living (physical) features of environment, including the amount and distribution of nutrients. An ecosystem typically has two major components:

A. Abiotic or Non-living Components

Abiotic components of the ecosystem comprise three sort of components: (1) **Climatic condition** and **physical factors** of the given region such as air, water, soil, temperature, light (*i.e.,* its duration and intensity), moisture (relative humidity), pH, etc. (2). **Inorganic substances** such as water, carbon (C), nitrogen (N), sulphur (S), phosphorus (P) and so on, all of which are involved in cycling of materials in the ecosystem (*i.e.,* biogeochemical cycles). The amount of these inorganic substances, present at any given time in an ecosystem, is designated as the **standing state** or **standing quality**. (3). **Organic substances** such as proteins, carbohydrates, lipids, humic substances, etc., present either in the biomass or in the environment, *i.e.,* **biochemical structure** that link the biotic and abiotic components of the ecosystem.

B. Biotic or Living Components

In the trophic structure of any ecosystem, living organisms are distinguished on the basis of their nutritional relationships, which are discussed as follows:

1. Autotrophic component. Autotrophic (*auto* = self; *trophic* = nourishing) component of ecosystem includes the **producers** or **energy transducers** which convert solar energy into chemical energy (that becomes locked in complex organic substances such as carbohydrate, lipid, protein, etc.) with the help of simple inorganic substances such as water and carbon dioxide and organic substances such as enzymes. Autotrophs fall into following two groups: (i) **photoautotrophs** which contain green photosynthetic pigment **chlorophyll** to transduct the solar or light energy of sun, *e.g.,* trees, grasses, algae, other tiny phytoplanktons and photosynthetic bacteria and cyanobacteria (= blue green algae). (ii) **Chemoautotrophs** which use energy generated in oxidation - reduction process, but their significance in the ecosystem as producers is minimal, *e.g.,* microorganisms such as *Beggiatoa,* sulphur bacteria, etc.

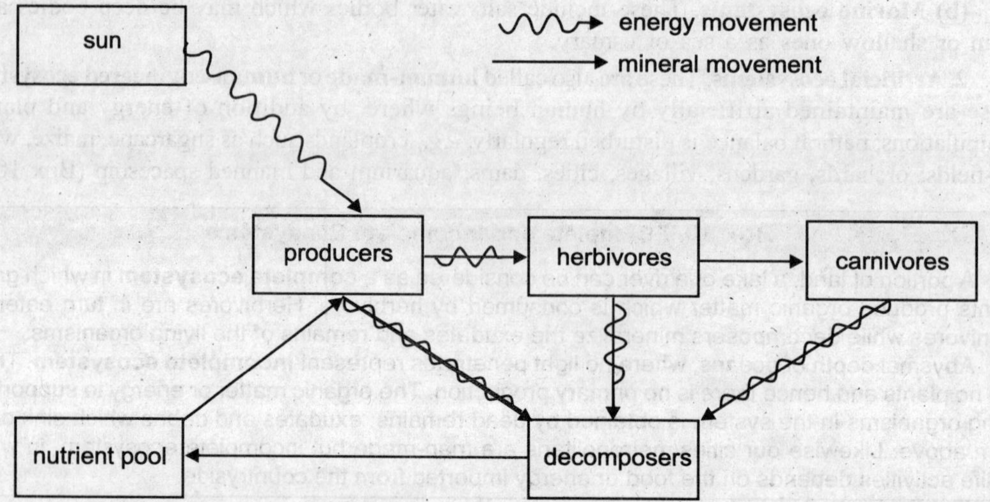

Fig. 10.1. A simple model of energy and mineral movement in ecosystems. Note that energy flow is unidirectional and noncyclic, whereas nutrient movement is cyclic.

2. Heterotrophic component. In the heterotrophic (*hetero* = other; *trophic* = nourishing) organisms predominate the activities of utilization, rearrangement and decomposition of complex organic materials. Heterotrophic organisms are also called **consumers**, as they consume the matter built up by the producers (autotrophs). The consumers are of following two main types:

(a) Macroconsumers. These are also called **phagotrophs** (phago = to eat) and include mainly animals which ingest other organisms or chunks of organic matter. Depending on their food habits, consumers may either be **herbivores** (plant eaters) or **carnivores** (flesh eaters). Herbivores live on living plants and are also known as **primary consumers**, *e.g.,* insects, zooplanktons and animals such as deer, cattle, elephant, etc. **Secondary** and **tertiary consumers**, if present in the food chain of the ecosystem, are carnivores or omnivores, *e.g.,* insects such as preying mantis, dragon flies; spiders and large animals such as tiger, lion, leopard, wolf, etc. Secondary consumers are the carnivores which feed on primary consumers or herbivores. Carnivores are, often, recognized as carnivore order - 1 (C_1), carnivore order - 2 (C_2) and so on, depending on their food habits.

Ticks and mites, leeches and blood-sucking insects (mosquito, bed-bug) are dependent on herbivores, carnivores and omnivores.

(b) Microconsumers. These are also called **decomposers, reducers, saprotrophs** (*sapro* = decompose), **osmotrophs** (osmo = to pass through a membrane) and **scavengers. Wiegert** and **Owen** (1971) have coined the term, **biophages** for heterotrophic decomposers which feed on the dead organic matter. Microconsumers include microorganisms such as bacteria, actinomycetes and fungi. Microconsumers breakdown complex organic compounds of dead or living protoplasm, absorb some of the decomposition or breakdown products and release inorganic nutrients in the environment, making them available again to autotrophs or producers. Some invertebrate animals such as protozoa, oligochaeta such as earthworms, etc., use the dead organic matter for their food, as they have the essential enzymes and, hence, can be classified as decomposer organisms. Some ecologists believe that microorganisms are **primary decomposers**, while invertebrates are **secondary decomposers**.

The disintegrating dead organic matter is also known as **organic detritus** (Latin word *deterere* means to wear away). By the action of **detritivores** (=decomposers), the disintegrating detritus result into particulate organic matter (POM) and dissolved organic matter (DOM) which play important role in the maintenance of the edaphic environment.

Functions of Ecosystem

When we consider the function of an ecosystem, we must describe the flow of energy and the cycling of nutrients. That is, we are interested in things like how much sunlight is trapped by plants in a year, how much plant material is eaten by herbivores, and how many herbivores are eaten by carnivores. Thus, the producers, the green plants, fix radiant energy and with the help of minerals (such as C, H, O, N, P, Ca, Mg, Zn, Fe, etc.) taken from their edaphic (soil) or aerial environment (the nutrient pool) they build up complex organic matter (carbohydrates, fats, proteins, nucleic acids, etc.). Some ecologists prefer to refer to the green plants as **converters** or **transducers**, since in their view, the most popular and prevalent term 'producer' from energy view point is somewhat misleading. Their view point is that green plants produce carbohydrates and not energy and since they convert or transduce radiant energy into chemical form, they must be better called **converters** or **transducers**. The two ecological processes of energy flow and mineral cycling involving interaction between the physico-chemical environment and the biotic communities, may be considered the 'heart' of ecosystem dynamics. In an ecosystem, energy flows in non-cyclic manner (**unidirectional**) from sun to the decomposers via producers and macroconsumers (herbivores and carnivores), whereas the minerals keep on moving in a cyclic manner (Fig. 10.1).

10.2 POND ECOSYSTEM

The freshwater pond as a whole represents a complete self-maintaining and self-regulating ecosystem. The **pond** can be defined as a body of shallow standing water (*i.e.*, lentic aquatic system) characterized by relatively quite waters having numerous plants, animals and microorganisms. In the pond ecosystem all the four basic components (*i.e.*, abiotic factors, producers, consumers and decomposers) of an ecosystem are well represented (Fig. 10.2).

1. Abiotic components. The abiotic components of pond ecosystem consist of two things: the **materials** (matter) and **energy.** The materials are inorganic and organic substances. **Inorganic substances** of pond ecosystem include water, minerals (calcium, sulphur, etc.) atmospheric gases (carbon dioxide, oxygen, nitrogen), salts (*e.g.*, phosphates, nitrates and chlorides of sodium, potassium and calcium). **Organic substances** of pond ecosystem include amino acids, humic acid and other products of decay of living things. The quantity of abiotic materials such as minerals present at any given time in an ecosystem is termed as the **standing state** (similar to the **standing crop** which refers to the quantity of biotic components). A small portion of these nutrients circulates through their death and decay returning to the soil and the environment. The process is called **mineral circulation** or the **biogeochemical cycle.** A great portion of materials is held reserve in the environment as well as the organisms themselves.

The **energy** is in the form of light, heat and stored energy in chemical bonds of organic material. While the materials circulate, the solar energy is continuously trapped by the green plants during the process of photosynthesis on one hand and lost in space through respiration or loss of heat by all types of organisms, on the other. The energy flows through the pond ecosystem on an one-way traffic. For the proper maintenance of the ecosystem, the energy must be regularly added and transferred from organism to organism (*i.e.*, food chain).

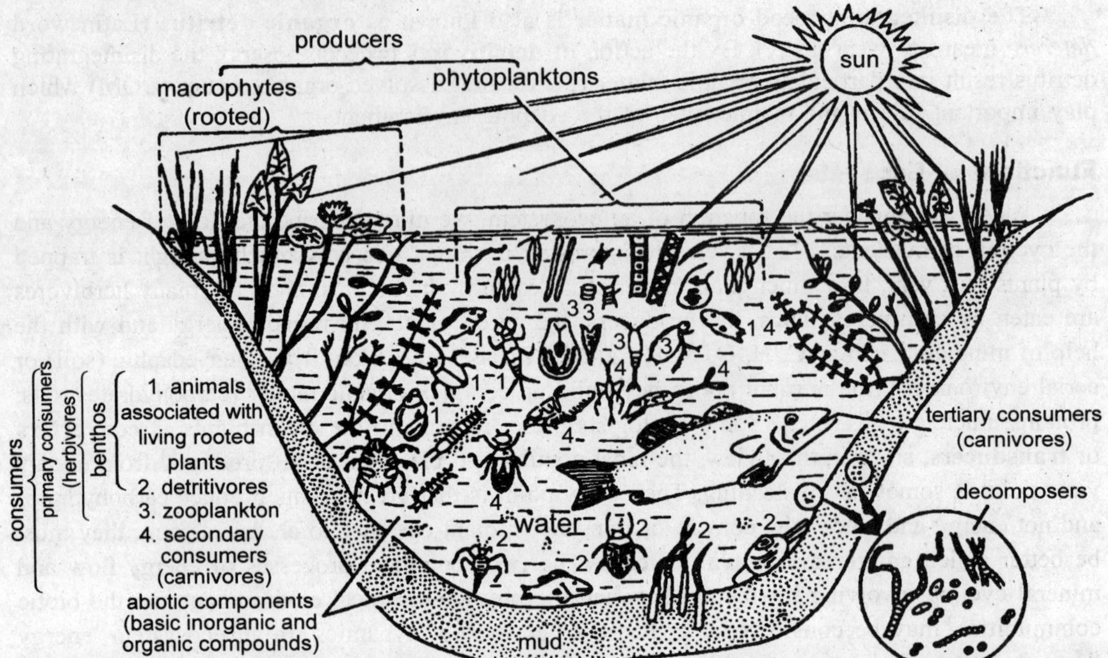

Fig. 10.2. A pond ecosystem showing its basic structural units – the abiotic (inorganic and organic compounds) and biotic (producers, and consumers – herbivores, carnivores and decomposers) components.

2. Biotic components. They include various organisms which are classified into the following types:

(a) Producers. These are photoautotrophic green plants and photosynthetic bacteria. The producers fix radiant energy of sun and with the help of minerals derived from water and mud, they manufacture complex organic substances as carbohydrates, proteins and lipids. Producers of pond are of following types:

(i) Macrophytes. These include mainly the rooted large-sized plants which comprise three types of hydrophytes: partly or completely submerged, floating and emergent aquatic plants. The common plants are species of *Trapa, Typha, Eleocharis, Sagittaria, Nymphaea, Potamogeton, Chara, Hydrilla, Vallisneria, Utricularia, Marsilea, Nelumbo,* etc. Besides these plants, some free floating forms also occur in the pond ecosystem, *e.g., Azolla, Salvinia, Wolffia, Eichhornia, Spirodella, Lemna,* etc.

(ii) Phytoplankton. These are microscopic (minute), floating or suspended lower plants (algae) that are distributed throughout the water, but mainly in the photic zone. Most of them are filamentous algae such as *Spirogyra, Ulothrix, Zygnema, Cladophora* and *Oedogonium.* There also occur some chlorococcales *e.g., Chlorella, Closterium, Cosmarium, Eudorina, Pandorina, Pediastrum, Scendesmus, Volvox,* Diatoms, *Anabaena, Gloeotrichia, Microcystis, Oscillatoria, Chlamydomonas, Spriulina,* etc., and some flagellates.

(b) Macroconsumers. They are phagotrophic heterotrophs which depend for their nutrition on the organic food manufactured by producers, the green plants. Macroconsumers are of following three types:

(i) Herbivores (Primary consumers). These animals feed directly on living plants (producers) or plant remains. They may be large or minute in size and are of following two types: 1. **Benthos** which are the bottom dwelling forms such as fish, insect larvae, beetles, mites, molluscs (*e.g., Pila, Planorbis, Unio, Lamellidens,* etc.), crustaceans, etc. 2. **Zooplankton** which feed chiefly on phytoplankton and are chiefly the rotifers as *Brachionus, Asplanchna, Lecane,* etc., although some protozoans as *Euglena, Coleps, Dileptus,* etc., and crustaceans such as *Cyclops, Stenocypris,* etc., are also present in the pond.

Besides these small-sized herbivores, some mammals such as cow, buffaloes, etc., also visit the pond casually and feed on marginal rooted macrophytes. Some birds also regularly visit the pond to feed on some hydrophytes.

(ii) Carnivore order-1 (Secondary consumers). These carnivores feed on the herbivores and include chiefly insects, fish and amphibians (frog). Most insects are water beetles which feed on zooplanktons; some insects are the nymphs of dragonflies which feed upon aquatic insects.

(iii) Carnivore order-2 (Tertiary consumers). These are some large fish as game fish that feed on the smaller fish and, thus, become the tertiary (top) consumers.

(c) Decomposers. They are also called microconsumers, since they absorb only a fraction of the decomposed organic matter. They bring about the decomposition of dead organic matter of both producers (plants) as well as macroconsumers (animals) to simple forms. Decomposers help in returning of mineral elements again to the medium of the pond and in running biogeochemical cycles. Decomposers of pond ecosystem include chiefly bacteria, actinomycetes and fungi. Among fungi, species of *Aspergillus, Cephalosporium, Cladosporium, Pythium, Rhizopus, Penicillium, Thielavia, Alternaria, Trichoderms, Circinella, Fusarium, Curvularis, Paecilomyces, Saprolegnia,* etc., are most common decomposers in water and mud of the pond.

10.3 PRODUCTIVITY OF ECOSYSTEM

The productivity of an ecosystem refers to the rate of production, *i.e.,* the amount of organic matter accumulated in any unit time. It is of following types:

1. Primary productivity. It is defined as the rate at which radiant energy is stored by photosynthetic and chemosynthetic activity of producers. Primary productivity is of following types:

(i) Gross primary productivity. It refers to the total rate of photosynthesis including the organic matter used up in respiration during the measurement period. GPP depends on the chlorophyll content. The rate of primary productivity are estimated in terms of either chlorophyll content as chl/g dry weight/unit area or photosynthetic number, *i.e.,* amount of CO_2 fixed/g chl/hour.

(ii) Net primary productivity. It is the rate of storage of organic matter in plant tissues in excess of the respiratory utilization by plants during the measurement period.

Primary production is measured by following methods — harvest method, oxygen measurement method (or light or dark method), oxygen diurnal curve method, carbon dioxide measurement method (enclosure method), the aerodynamic method, the pH method, radioisotope method, chlorophyll estimation method.

2. Secondary productivity. It is the rate of energy storage at consumer's levels—herbivores, carnivores and decomposers. Consumers tend to utilise already produced food materials in their respiration and also convert the food matter to different tissues by an overall process. So, secondary productivity is not divided into 'gross' and 'net' amounts. Due to this fact some ecologists such as **Odum** (1971), prefer to use the term **assimilation** rather than production at this level – the consumers level. Secondary productivity, in fact, remains mobile (*i.e.,* keeps on moving from one organism to another) and does not live *in situ* like the primary productivity.

3. Net productivity. It is the rate of storage of organic matter not used by the heterotrophs or consumers; *i.e.,* equivalent to net primary production minus consumption by the heterotrophs during the unit period as a season or year, etc.

Autochthonous and Allochthonous Production

The overall energy budget of the ecosystem reflects the balance between income and expenditure, just as it does in a bank account. The ecosystem gains energy through photosynthetic assimilation of light by green plants and the transport of organic matter into the system from outside. Sources of latter are referred to as **allochthonous inputs** (Gr., *Chthonous* = of the earth; *allos* = other); local production is called **autochthonous.**

G.W. Minshall (1978) surveyed the relative amounts of autochthonous and allochthonous inputs in aquatic systems: large rivers, lakes and most marine ecosystems are dominated by autochthonous production; allochthonous imports were most important in small streams and springs under the closed canopies of forests; life in caves and abyssal depths of the oceans to which no light penetrates, subsists, entirely on energy transported in from outside.

Examples of Productivity of Ecosystems

The productivity of ecosystems varies widely over the globe. It is strongly influenced by temperature and rainfall. This fact is exemplified by net productivity (*i.e.,* accumulation of biomass) of following ecosystems:

1. The most productive terrestrial ecosystems are tropical forests with high rainfall and warm temperatures; their net productivity ranges between 1000 and 3500 g/m^2/yr.

2. The net productivity of temperate forests, where rainfall and temperature are lower, range between 600 and 2500 g/m^2/yr.

3. Shrublands have net production of 700 to 1500/g/m^2/yr.

4. Desert grasslands produce about 200 to 300 g/m^2/yr.

5. Net production of deserts and tundra range between 100 to 250 g/m^2/yr.

6. Net production of open sea is generally quite low. Thus, the productivity of the North Sea is about 170 g/m^2/yr and of the Sargasso sea is 180 g/m^2/yr.

7. Swamps and marshes, ecosystems at the interface of land and water, have net productivity ranging from 900 to 3300 g/m^2/yr.

8. Estuaries, because of inputs of nutrients from rivers and tides, and coral reefs, because of inputs from changing tides, have net productivity between 1000 and 2500 g/m^2/yr.

9. Among agricultural ecosystems sugarcane has a net productivity of 1700 to 1800 g/m^2/yr, hybrid corn 1000 g/m^2/yr and some tropical crops 3000 g/m^2/yr.

10.4 FOOD CHAINS

In an ecosystem one can observe the transfer or flow of energy from one trophic level to other in succession. A **trophic level** can be defined as the number of links by which it is separated from the producer, or as the *n*th position of the organism in the food chain. Thus, trophic level represents nutrient status or functional level of an organism in any food chain. A **food chain** is an energy sequence of links or trophic levels that starts with a species that eats no other species and ends with a species that is eaten by no other species (**Kormondy,** 1996). Thus, the patterns of eating and being eaten forms a linear chain called **food chain** which can always be traced back to the producers. Thus, primary producers trap radiant energy of sun and transfer that to chemical or potential energy of organic compounds such as carbohydrates, proteins and fats. When a herbivore animal eats a plant (or when bacteria decompose it) and these organic compounds are oxidized, the energy liberated is just equal to the amount of energy used in synthesizing the substances **(first law of thermodynamics)**, but some of the energy is heat and not useful energy **(second law of thermodynamics)**. If this animal, in turn, is eaten by another one, along with transfer of energy from a herbivore to carnivore a further decrease in useful energy occurs as the second animal (carnivore) oxidizes the organic substances of the first (herbivore or omnivore) to liberate energy to synthesize its own cellular constituents. Such transfer of energy from organism to organism sustains the ecosystem and when energy is transferred from individual to individual in a particular community, as in a pond or a lake or a river, we come across the food chains. The number of steps in a food chain are always restricted to four or five, since the energy available decreases with each step. For example, in a typical food chain of an Indian river, a diatom may be eaten by a copepod which is eaten by a small fish, which forms the food source of large fish and so on.

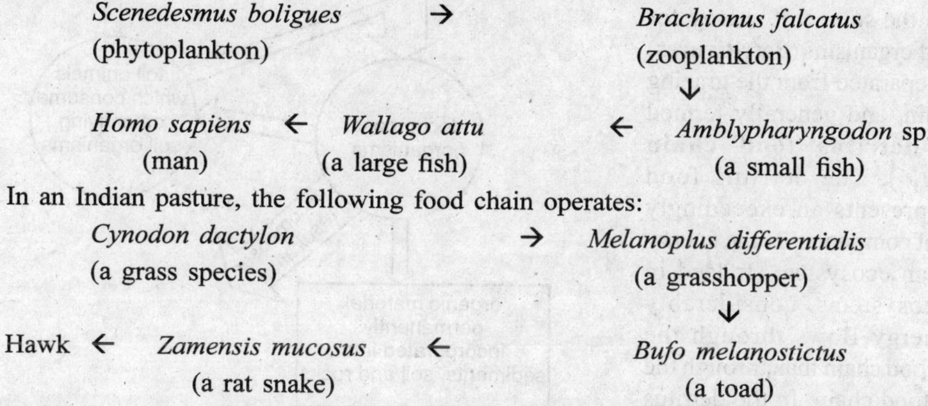

Many direct or indirect methods are employed to study food chain relationships in nature. They include gut content analysis, use of radioactive isotopes, precipitin test, etc.

In nature, basically two types of food chains are recognized—grazing food chain and detritus food chain.

1. Grazing food chain. This type of food chain (Fig. 10.3) starts from the living green plants, goes to grazing herbivores and on to the carnivores. Ecosystems with such type of food chain are directly dependent on an influx of solar radiation. Thus, this type of food chain depends on autotrophic energy capture and the movement of this energy to herbivores. Most of the ecosystems in nature follow this type of food chain. These chains are very significant from energy standpoint. The phytoplanktons → zooplanktons → fish sequence or the grasses → rabbit → fox sequence are the examples of grazing food chain. Further the producer → herbivore → carnivore chain is a **predator chain**. **Parasitic chains** also exist wherein smaller organisms consume larger ones without outright killing as the case of the predators.

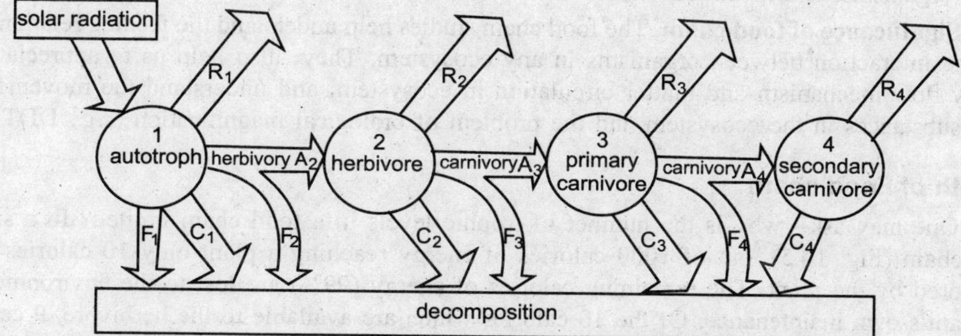

Fig. 10.3. Diagrammatic representation of a grazing food chain. Showing inputs and losses of energy at each trophic level. Trophic levels are numbered and used as subscripts to letters indicating energy transfer. A–assimilation of food by the organisms at the trophic level; F–energy lost in the form of faeces and other excretory products; C–energy lost through decay and R–energy lost to respiration.

2. Detritus food chain. The organic wastes, exudates and dead matter derived from the grazing food chain are generally termed **detritus**. The energy contained in this detritus is not lost to the ecosystem as a whole; rather it serves as the source of energy for a group of organisms (**detritivores**) that are separated from the grazing food chain, and generally termed as the **detritus food chain** (Fig. 10.4). The detritus food chain represents an exceedingly important component in the energy flow of an ecosystem. Indeed in some ecosystems, considerably more energy flows through the detritus food chain than through the grazing food chain. In the detritus food chain the energy flow remains as a continuous passage rather than as a stepwise flow between discrete entities. The organisms of the detritus food chain are many and include algae, bacteria, slime molds, actinomycetes, fungi, Protozoa, insects, mites, Crustacea, centipedes, molluscs, rotifers, nematodes, annelid worms, and some vertebrates. Some species are highly specific in their food requirements and some can eat almost anything. Many Protozoa, for instance, need certain specific organic acids, vitamins, and other nutrients before they can thrive; on the other hand, the guts of small Collembola (a group of tiny soil insects) have been reported to contain decaying plant material, fungal fragments, spores, fly pupae, other Collembola, parts of decaying earthworms, and cuticle from their own faecal casting. In contrast to the grazing food chain, in which energy storage is entirely within the tissues of living organisms, energy storage for the detritus food chain may be largely external to the organisms, and in the detritus itself.

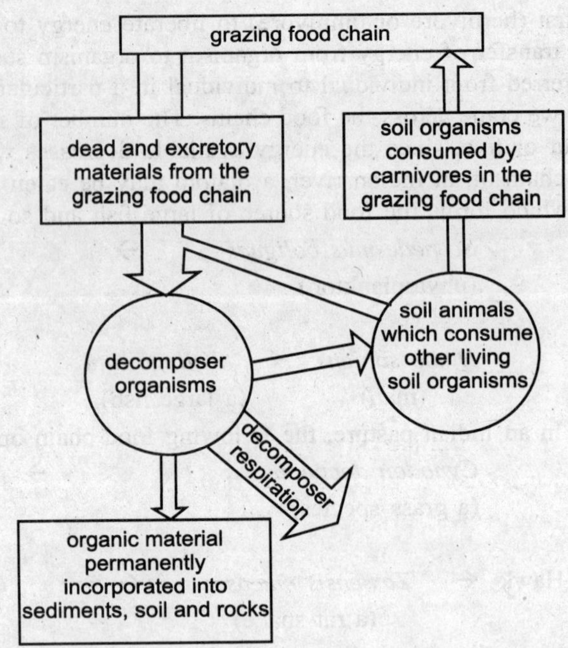

Fig. 10.4. Diagrammatic representation of the detritus food chain, showing energy transfers between it and the grazing food chain, as well as energy losses to the detritus food chain.

Significance of food chain. The food chain studies help understand the feeding relationships and the interaction between organisms in any ecosystem. They also help us to appreciate the energy flow mechanism and matter circulation in ecosystem, and understand the movement of toxic substances in the ecosystem and the problem of biological magnification (*e.g.,* DDT).

Length of Food chain

One may ask—why is the number of trophic levels in a food chain limited? In a simple food chain (Fig. 10.5), out of 1000 calories of energy reaching a plant only 10 calories (1%) are stored by the plant. The remaining calories of energy (99%) are lost to the environment or for plant's own maintenance. Of the 10 calories which are available to the herbivore, 9 calories (99%) are lost at its level and only 1 calorie is passed down to the carnivore. Thus, at each trophic level in a food chain, a large portion of energy is used for its own maintenance and ultimately lost as heat. Consequently, organisms in each trophic level pass on less and less energy than they receive. This tends to limit the number of steps or trophic levels to four or five. The longer the food chain, the less is the energy available to the final member.

Following two hypotheses have been forwarded regarding the length of food chains:

1. Energetic hypothesis. This hypothesis was advanced by **Hutchinson** (1959). It states that food chains should be longer in ecosystems with higher primary productivity.

2. Dynamical stability hypothesis. This hypothesis was advanced from theoretical or mathematical models by **Pimn** and **Lawton** (1978). The gist of this hypothesis is that food chains should be longer in ecosystems that are not subjected to distrubance by catastrophic events such as fire, flooding, widespread disease, etc.

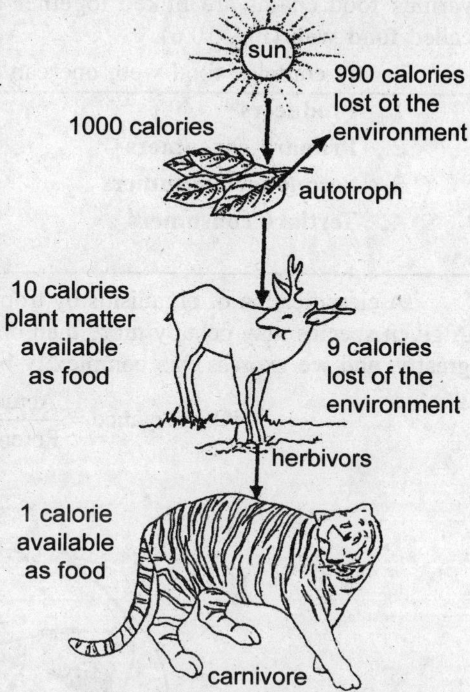

Fig. 10.5. Mode of energy flow through a simple food chain.

In their review of 113 community food webs from natural communities, **Briand** and **Cohen** (1987) found no evidence to support either the energetic or the dynamic stability hypothesis. They did conclude, however, that the spatial dimensions of an ecosystem are related to food chain length, with three dimensional environments having longer food chains than two dimensional ones. **Two-dimensional environments** are regarded as essentially flat such as a grassland, the tundra, a sea or lake bottom, a stream bed or the rocky intertidal zone; **three-dimensional environments** were regarded as solid, such as a forest canopy or the water column of the open ocean. In their survey, though the smallest food chain length was 2 species and the maximum 10, the more general trend is 3 to 5.

Recently the length of food chain is found to be limited by ecological efficiency. Thus, one can estimate the average length of food chains in a community from net primary production, average ecological efficiency and average energy flux (flow) of a top predator population. The energy available ($E(n)$) to a predator at a given trophic level [n (plants being level 1)] is equal to the product of the net primary production (NNP) and the intervening ecological efficiencies (Eff.). Thus,

$$E\,(n) = NNP\ Eff^{n-1}$$

where Eff is the geometric mean of the efficiencies of transfer between each level. This expression can be rearranged to give

$$n = 1 + \frac{\log\,(E(n)) - \log\,(NPP)}{\log\,(Eff)}$$

using this equation and rough estimates for the values on its right-hand side, one can calculate the average *number of trophic levels to be about 7 for marine plankton based system, 5 for inshore aquatic communities, 4 for grassland, and 3 for wet tropical forests* (**Rickleff** 1990).

10.5 FOOD WEB

In nature simple food chains occur rarely. The same organism may operate in the ecosystem at more than one trophic level, *i.e.,* it may derive its food from more than one source. Even the same organism may be eaten by several organisms of a higher trophic level or an organism may feed upon several different organisms of a lower trophic level. Usually the kind of food changes with the age of the organism and the food availability. Thus, in a given ecosystem various food chains are linked together and intersect each other to form a complex network called **food web** (Fig. 10.6).

In any complex food web, one can recognize several different trophic levels:

1.	**Producers**	Green plants	First trophic level
2.	**Primary consumers**	Herbivores	Second trophic level
3.	**Secondary consumers**	Carnivores, insectivores	Third trophic level
4.	**Tertiary consumers**	Higher carnivores, insect hyperparasites	Fourth trophic level

A classification of organisms by trophic levels is one of function and not of species as such. A given species may occupy more than one trophic level. The complexity of food web can vary greatly, and we express this complexity by a measure called the **connectance** of the food web:

$$\text{Connectance} = \frac{\text{Actual number of a interspecific interactions}}{\text{Potential number of interspecific interactions}}$$

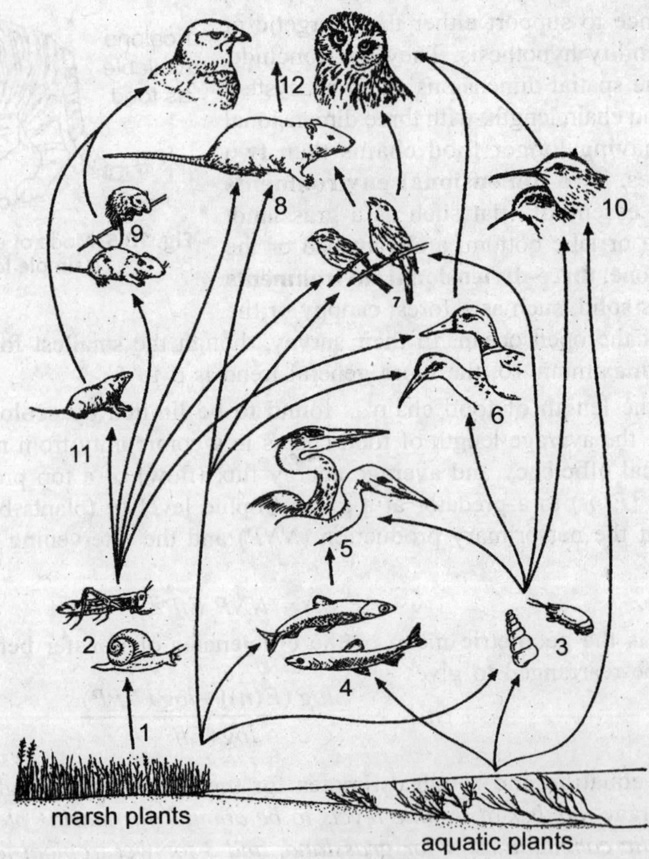

Fig. 10.6. A midwinter food web in a salt marsh in San Francisco Bay area. Producer organisms are marsh and aquatic plants (the basal trophic species). Marsh plants (1) are eaten by terrestrial herbivorous invertebrates (the intermediate species) represented by grasshopper and snail (2) and by herbivorous marine and intertidal invertebrates (3) Fish, represented by smelt and anchovy (4) feed on vegetable matter from both ecosystems. The fish in turn are eaten by first-level carnivores, represented by the great blue heron and common egret (5) Continuing through the web are the following omnivores (*i.e.*, intermediate trophic species): clapper rail and mallard duck (6) Savanna and song sparrows (7) Norway rat (8) California vole and salt marsh harvest mouse (9) least and western sandpipers (10) The shrew (11). is a first-level carnivore. Top carnivores (*i.e.*, top trophic species) are marsh hawk and short-eared owl (12).

Trophospecies. Since several species may occur in each of the trophic levels of food web, the collection of organisms that feed on a common set of organisms (prey) and are fed on by another common set of organisms (predators) are referred to as a **trophic species** (**Briand** and **Cohen** 1984) or **trophospecies.** These species are classified as basal, intermediate and top species. The **basal species,** which include both primary producers and detritus, are at the bottom of food web and feed on no other species. The **top species**, occupying the apex of the food web, are ones on which no other species feed. The **intermediate species** may feed on more than one trophic level (omnivores). Predators are species that feed on other species in the food web, and prey are species that are fed upon by some other species.

Generalization about food web. Food chains, in general, tend to be short, rarely exceeding four links (**Pimm** 1982). No matter how productive an ecosystem, the number of links remains essentially the same. Increased productivity of an ecosystem does not support longer food chains, as often assumed. *It may support more species and thus more complexity,* but as the number of species in the web increases, the connectance decreases.

Further, in any food web there are four possible connectances (their proportion in each following categories remains almost same in every habitat):

1. Top and intermediate (35 per cent);
2. Different intermediates (30 per cent);
3. Intermediate and basal (27 per cent); and
4. Top and basal (8 per cent).

Thus, rarely does the top species link with basal. *What happens if a species is removed from or reduced in a food web?* According to recent experimental results, the structure of a food web often is *affected adversely* when a new or exotic species such as a carnivore or grazing herbivore, is introduced into an ecosystem, or when a key species is removed from the food web. Thus, if the species removed is a top predator, then the effect cascades down through the intermediate to the basal species. Some of the best examples come from experiments in fisheries and lake productivity (**Carpenter** *et. al.,* 1985). We can consider the case of bass which is a predatory fish and a top piscivore of food web. The bass reduces the biomass of vertebrate planktivores, allowing invertebrate planktivore biomass and large zooplankton herbivores to increase, and causing small herbivore zooplankton and primary production to decrease. Conversely, if the top predator (*i.e.,* bass) is reduced, vertebrate planktivores increase, invertebrate planktovire and large herbivorus zooplankton decreases and small herbivore zooplankton and primary production increase.

10.6 ECOLOGICAL PYRAMIDS

In the successive steps of grazing food chain–photosynthetic autotroph, herbivorous heterotroph, carnivorous heterotroph, decay bacteria–the number and mass of the organisms in each step is limited by the amount of energy available. Since some energy is lost as heat, in each transformation the steps become progressively smaller near the top. This relationship is sometimes called **"ecological pyramid"** and was first devised by **Charles Elton** (1927). The ecological pyramids represent the trophic structure and also trophic function of the ecosystem. In many ecological pyramids, the producer form the base and the successive trophic levels make up the apex.

Thus, communities of terrestrial ecosystems and shallow water ecosystems contain gradually sloping ecological pyramids because these producers remain large and characterized by an accumulation of organic matter. This trend, however, does not hold for all ecosystems. In such aquatic ecosystems as lakes and open sea, primary production is concentrated in the microscopic algae. These algae have a short-cycle, multiply rapidly, accumulate little organic matter and are heavily exploited by herbivorous zooplankton. At any one point in time the standing crop is low.

As a result, the pyramid of biomass for these aquatic ecosystems is inverted: the base is much smaller than the structure it supports.

Types of Ecological Pyramids

The ecological pyramids may be of following three kinds:

1. Pyramid of number. It depicts the number of individual organisms at different trophic levels of food chain. This pyramid was advanced by **Charles Elton** (1927), who pointed out the great difference in the number of the organisms involved in each step of the food chain. The animals at the lower end (base of pyramid) of the chain are the most abundant. Successive links of carnivores decrease rapidly in number until there are very few carnivores at the top. The pyramid of number ignores the biomass of organisms and it also does not indicate the energy transferred or the use of energy by the groups involved. The lake ecosystem provides a typical example for pyramid of number.

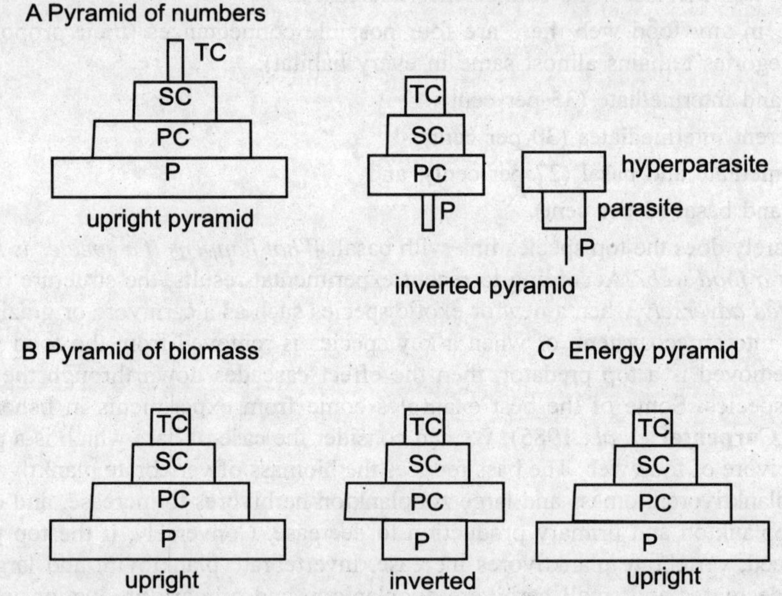

Fig. 10.7. Ecological or Eltonian pyramids: pyramid of numbers can be both upright and inverted, pyramid of biomass can also be both upright and inverted but the pyramid of energy is always upright (P=producers; PC=primary consumers (herbivores); SC=secondary consumers (carnivores); TC=tertiary consumers (carnivores).

2. Pyramid of biomass. The biomass of the members of the food chain present at any one time forms the pyramid of the biomass. Pyramid of biomass indicates decrease of biomass in each trophical level from base to apex. For example, the total biomass of the producers ingested by herbivores is more than the total biomass of the herbivores in an ecosystem. Likewise, the total biomass of the primary carnivores (or secondary consumer) will be less than the herbivores and so on.

3. Pyramid of energy. When production is considered in terms of energy, the pyramid indicates not only the amount of energy flow at each level, but more important, the actual role the various organisms play in the transfer of energy. The base upon which the pyramid of energy is constructed is the quantity of organisms produced per unit time, or in other words, the rate at which food material passes through the food chain. Some organisms may have a small biomass, but the total energy they assimilate and pass on, may be considerably greater than that of organisms with a much larger biomass. Energy pyramids are always slopping because less energy is transferred from each level than was paid into it. In cases such as in open water communities the producers have less bulk than consumers but the energy they store and pass on must be greater than that of the next level. Otherwise the biomass that producers support could not be greater than that of the producers themselves. This high energy flow is maintained by a rapid turnover of individual plankton, rather than an increase of total mass.

10.7 ENERGY FLOW IN ECOSYSTEMS

The behaviour of energy in ecosystem (*i.e.,* its entrance, relation within ecosystem and its dissipation into space) is termed as **energy flow** because of its unidirectional energy transformations **(K. C. Agarwal 1998)**.

1. Concept of energy. Energy is the capacity to do work. Biological activities require consumption of energy which ultimately comes from the sun. Radiant energy of sun (or **solar energy**) is transformed into **chemical energy** by the process of photosynthesis—this is stored in plant tissues and then transformed into **mechanical** and **heat** form of energy during metabolic activities. In the biological world, the energy flows from the sun to plants and then to all heterotrophic organisms, such as microorganisms, animals and human beings in the following manner:

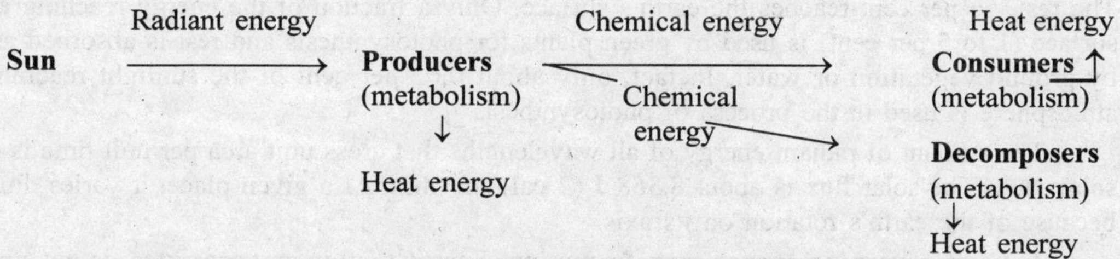

Mechanical energy has two forms, namely kinetic or free energy and potential energy. The energy a body possesses by virtue of its motion is called **kinetic energy**, and is measured by the amount of work done in bringing the body at rest. **Potential energy** is stored energy (the energy at rest) and becomes useful after conversion into kinetic energy. All organisms require a source of potential energy, which is found in the chemical energy of food. The oxidation of food releases energy which is used to do work. Thus, chemical energy is converted into mechanical energy. Food means material containing energy that organism can use. *Food is the means to transfer of both matter and energy in the living world.* Plants synthesize food with the help of solar energy and inorganic substances such as nutrients, CO_2 and H_2O in a biochemical process called **photosynthesis:**

$$6CO_2 + 6H_2O \xrightarrow[\text{Enzymes, etc.}]{\text{Solar energy}} \rightarrow \underset{\text{Glucose; food}}{C_6H_{12}O_6 + 6O_2}$$

$$\text{(Chlorophyll)}$$

2. Unit of energy. The unit of measurement of energy is **erg**; the work done in lifting 1 gram of weight to a height of 1 cm against the force of gravity is equal to 981 ergs. One crore ergs (10^7 ergs) is equal to one **Joule**. All forms of energy can be completely converted into heat energy. For a better and uniform expression in ecology, therefore, energy is measured not in terms of ergs but joules or units of heat measurement. Heat is measured in calories. One calorie is equal to the heat energy required to raise the temperature of 1 gram of water from 14.5^{0}C to 15.5^{0}C, and one calorie is equal to 4.2 joules or 4.2×10^7 ergs. One thousand calories (10^3) makes one kilo calories or a kilogram calories (Kcal or Cal). Now, there is a trend of expressing energy in ecological literature in terms of kilojoules.

3. Ecological energetics. Ecological energetics includes energy transformation which occur within ecosystems. In ecological energetics, we consider (i) quantity of energy reaching an ecosystem per unit of area (say a square metre) per unit of time (say one hour, day or year); (ii) quantity of energy trapped by green plants and converted to a chemical form (photosynthesis) and (iii) the quantity and path of energy flow from green plants to organisms of different trophic levels over a period of time in a known area (*i.e.,* energy flow from producers to consumers).

The energy used for all plant life processes is derived from solar radiations. A fraction, *i.e.,* about 1/50 millionth of the total solar radiation reaches the earth's atmosphere. Solar radiation travels through the space in the form of waves, their wavelength ranging from 0.03 A^0 to several kilometres. While most radiations are lost in space, those ranging from 300 mμ to 10 mμ and above 1 cm (radiowaves) enter the earth's outer atmosphere (which is about 28 km altitude). The energy reaching the earth's surface consists mainly of visible light (390–760 mμ) and infra-red component. On a clear day radiant energy reaching the earth's surface is about 10 per cent UV, 45 per cent visible and 45 per cent infra-red. Green plants absorb strongly the blue and red light (400 to 500 mμ and 600 to 700 mμ respectively).

About 34 per cent of the sunlight reaching the earth's atmosphere is reflected back (by clouds and dust), 10 per cent is held by ozone layer, water vapour and other atmospheric gases. The rest, 56 per cent reaches the earth's surface. Only a fraction of the energy reaching earth's surface (1 to 5 per cent) is used by green plants for photosynthesis and rest is absorbed as heat by ground vegetation or water. In fact, only about 0.02 per cent of the sunlight reaching the atmosphere is used in the process of photosynthesis.

The amount of radiant energy of all wavelengths that cross unit area per unit time is called **solar flux.** The solar flux is about 8.368 J (2 cal)/cm^2 min. At a given place, it varies diurnally because of the earth's rotation on its axis.

4. Laws governing energy transformation. Energy flow in any ecosystem is governed by two laws of thermodynamics. The **first law of thermodynamics** (called **Law of conservation of energy**) states that energy is neither created nor destroyed. It may change forms, pass from one place to another, or act upon matter, transforming it to energy in various ways. However, regardless of what transfers and transformations take place, no gain or loss in total energy of a system occurs. All of it can be accounted for. This law applies to respiration/burning **(exothermic reactions)** and photosynthesis **(endothermic reactions)**.

The **second law of thermodynamics** (called **Law of entropy**) makes an important generalization about energy transfer. It states that when energy is transferred or transformed, part of the energy is lost as waste; it assumes a form that cannot be passed on any further. For example, in an ecosystem as energy is transferred from one organism to another in form of food, a large part of energy is degraded as heat through metabolic activity, with a net increase in entropy. The remainder is stored as living tissue.

5. Concept of free energy, enthalpy and entropy. Free energy is that component of the total energy of a system which can do work under isothermal conditions. All physical and chemical processes proceed with a decline in free energy until they reach an equilibrium where the free energy of the system is at a minimum.

$\Delta G = \Delta H - T \Delta S$ (The sign Δ refers to a change in quantity)

Where ΔG = Change in the free energy of the system

ΔH = Change in **enthalpy** (*i.e.,* a change in the amount of energy in the form of heat liberated or absorbed by the system during physical or chemical changes)

ΔS = Entropy change of the system (**Entropy** is the name of a quantity in thermodynamics representing the degree of disorder in a physical system or the extent to which the energy in a system is available for doing work)

T = Absolute temperature

Thus, a decline in G is accompanied by an increase in $T \Delta S$. These are equal if there is no heat transfer between the system and the surrounding. If a reaction proceeds with a decline in free energy, we call it spontaneous.

6. Lindeman's trophic-dynamic concept. Lindeman (1942) made first attempt to show energy flow diagram showing energy flow through different trophic levels in Cedar Bog Lake of Minnesota (USA) and developed the concept of **energy** dynamics, called **trophic dynamics**. He pointed out, the amount of energy reaching each trophic level is determined by the net primary production (NPP) and the efficiency with which food energy is converted to biomass energy within each trophic step. Of the light energy assimilated by plants, 15 to 70 per cent is used for maintenance and, therefore, is unavailable to consumers. Most herbivores and carnivores are more active than plants and spend correspondingly more of their assimilated energy on maintenance. As a result productivity of each trophic level is usually no more than 5 to 20 per cent that of the level below it. The percentage transfer of energy from one trophic level to the next is called both the ecological efficiency and food chain efficiency.

7. Maintenance cost of secondary producers. In general, 55 to 75 per cent of the assimilated energy is spent on maintenance of secondary producers. Temperature, moisture conditions of the habitat and the type of species determine the maintenance cost. The drier and hotter the habitat, the higher is the maintenance cost, irrespective of species. For example, the average maintenance cost of some Indian earthworm species was found to be 6.48, 9.96 and 20.54 kJ/g dry tissue/month in the winter, rainy and summer seasons respectively in tropical pastures (**Dash**, 1987). The maintenance cost varies seasonally and is three times more in summer than in winter.

8. Assimilated energy and respiration energy. Once food is eaten, its energy follow a variety of paths through the organisms (Fig. 10.8). Regardless of an organism's source of food, what it digests and absorbs is referred to as **assimilated energy**, which supports maintenance,

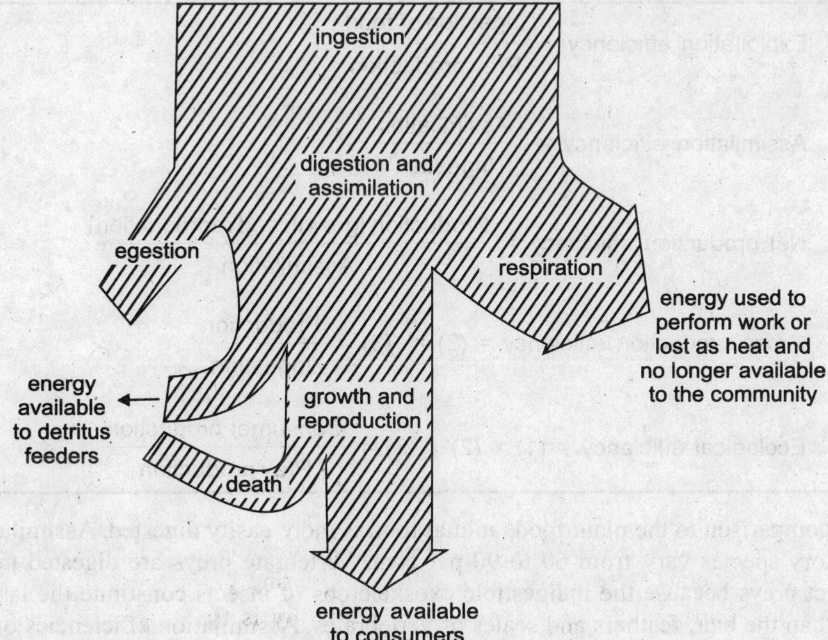

Fig. 10.8. Partitioning of energy within a link of the food chain.

builds tissues, or is excreted in unusable metabolic byproducts. The energy used to fulfil metabolic needs, most of which is lost as heat, is known as **respired energy**. A smaller fraction of assimilated energy is excreted in the form of organic, nitrogen –containing wastes (primarily ammonia, urea, or uric acid) produced when the diet contains an excess of nitrogen. Assimilated energy retained by the individual organism is available for the synthesis of new biomass (**production**) through growth and reproduction, which may then be consumed by herbivores, carnivores, and detritivores.

Further, not all food can be fully assimilated; hair, feathers, insect exoskeletons, cartilage and bone in animal foods, and cellulose and lignin in plant foods resist digestion by most animals. These materials are egested either by defecation or by regurgitation of pellets of undigested remains. Some egested wastes are substances that have been relatively unaltered chemically during their passage through an organism, but nearly all have been mechanically broken up into fragments by chewing and by contractions of the stomach and intestines, which makes them more readily usable by detritus feeders (**Edwards** and **Heath**, 1963).

9. Ecological efficiency. Ecological efficiency is the product of efficiencies with which organisms exploit their food resources and transform them into biomass which becomes available to the next higher trophic level. Because most biological production is consumed, **exploitation efficiency** is 100 per cent overall, and ecological efficiency depends on two factors: the proportion of assimilated energy incorporated in growth, storage and reproduction. The first proportion is called the **assimilation efficiency** and the second, the **net production efficiency** (Table 10.1). The product of the assimilation and net production efficiencies is the **gross production efficiency:** the proportion of food energy that is transformed into consumer biomass energy.

Net production efficiency for the plants is defined as the ratio of net to gross production. This index has been found to vary between 30 and 85 per cent, depending on habitat and growth form. Rapidly growing plants in temperate zones, whether trees, old-field herbs, crop species, or aquatic plants, have constantly high net production efficiencies (75 to 85 per cent). Similar vegetation types in the tropics exhibit lower net production efficiencies, perhaps 40 to 60 per cent respiration increases relative to photosynthesis at low latitudes.

Table 10.1. Definition of several energetic efficiencies.

1. Exploitation efficiency = $\dfrac{\text{Ingestion of food}}{\text{Prey production}}$

2. Assimilation efficiency = $\dfrac{\text{Assimilation}}{\text{Ingestion}}$

3. Net production efficiency = $\dfrac{\text{Production (growth and reproduction)}}{\text{Assimilation}}$

4. Gross production efficiency = (2) × (3) = $\dfrac{\text{Production}}{\text{Ingestion}}$

5. Ecological efficiency = (1) × (2) × (3) = $\dfrac{\text{Consumer production}}{\text{Prey production}}$

In comparison to the plant food, animal food is more easily digested. Assimilation efficiency of predatory species vary from 60 to 90 per cent. Vertebrate preys are digested more efficiently than insect preys because the indigestible exoskeletons of insects constitute the larger proportion of body than the hair, feathers and scales of vertebrates. Assimilation efficiencies of insectivorous vary between 70 and 80 per cent, whereas those of most carnivores are about 90 per cent.

The nutritional value of plant foods depends upon the amount of cellulose, lignin, and other indigestible materials present. Herbivores assimilate as much as 80 per cent of the energy in seeds, and 60 to 70 per cent of that in young vegetation (**Chew** and **Chew**, 1970). Most grazers and browsers (*e.g.,* cattle, elephants and grasshoppers) assimilate 30 to 40 per cent of the energy in their food. Millipedes, which eat decaying wood composed mostly of cellulose and lignin (and the microorganisms that occur in decaying wood), assimilate only 15 per cent (**O'Neil**, 1968).

Maintenance, movement and in warm-blooded animals, heat production require energy that otherwise could be utilized for growth and reproduction. Active warm-blooded animals (homeotherms) exhibit low net production efficiency—birds less than 1 per cent, small mammals with high reproductive rates up to 6 per cent. More sedentary, cold-blooded animals (poikilotherms), particularly aquatic species, channel as much as 75 per cent of their assimilated energy into growth and reproduction (**Welch**, 1968). The extreme high value approaches the biochemical efficiency of egg production and tissue growth , between 70 to 80 per cent in domesticated animals (**Ricklefs**, 1974).

The **gross production efficiency** (*i.e.,* biomass production efficiency within a trophic level) is the product of assimilation efficiency and net production efficiency. Gross production efficiencies of warm-blooded terrestrial animals rarely exceed 5 per cent, and those of some birds and large mammals fall below 1 per cent (**Turner**, 1970). Gross production efficiencies of insects lie within the range of 5 to 15 per cent, and those of some aquatic animals exceed 30 per cent (Fig. 10.9).

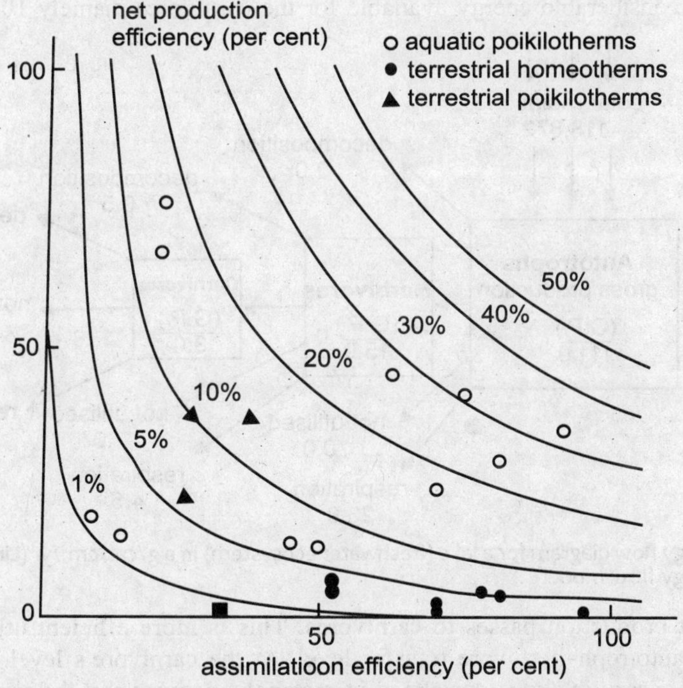

Fig. 10.9. Relationship between assimilation efficiency and net production efficiency for a variety of animals. Gross production efficiencies are indicated by the curved lines on the graph.

Gross ecological efficiency. From studies based on laboratory populations of algae, microcrustacea and hydra, **Lawrence Slobodkin** (1960) suggested that in the transfer of energy from one trophic level to the next the **gross ecological efficiency** is on the order **of 10 per cent.** Thus, if there are 100 calories of net plant production, only about 10 calories net production would be expected at the herbivore level and only 1 calorie at the carnivore level. Various studies to date, however, show a range in gross ecological efficiency from 5 to about 30 per cent, averaging about 10 per cent from producer to herbivore, and 15 per cent from herbivore to carnivore.

Models of Energy Flow

For explaining mode of energy flow in various types of ecosystems, ecologists have suggested the following models:

1. Single-channel energy flow models.

(a) Lindeman's model. This model was first suggested by **Lindeman** (1942) who died at the age of 26 years. As shown in Fig. 6.10, out of total incoming solar radiation (118,872 g cal/cm^2/yr), 118,761 g cal/cm^2/yr remain unutilised. Thus gross production (net production plus respiration) by autrophs is 111 g cal/cm^2/yr with an efficiency of energy capture of 0.10 per cent. It may also be noted that 21 per cent of this energy, or 23 g cal/cm^2/yr is consumed in metabolic reactions of autotrophs for their growth, development, maintenance and reproduction. It can be observed further that 15 g cal/cm^2/yr are consumed by herbivores that graze or feed on autotrophs — this amounts to 17 per cent of net autotroph production. Decomposition (3 g cal/cm^2/yr) accounts for about 3.4 per cent of net production. The remainder of plant material, 70 g cal/cm^2/yr or 79.5 per cent of net production is not utilised at all but becomes part of accumulating sediments of the total energy incorporated at the herbivore's level (*i.e.,* 15 g cal/cm^2/yr) 30 per cent or 4·5 g cal/cm^2/yr is used in metabolic reactions. Thus, there is considerably more energy lost via respiration by herbivores (30 per cent) than by autotrophs (21 per cent). Again there is considerable energy available for the carnivores, namely 10.5 g cal/cm^2/yr or 70

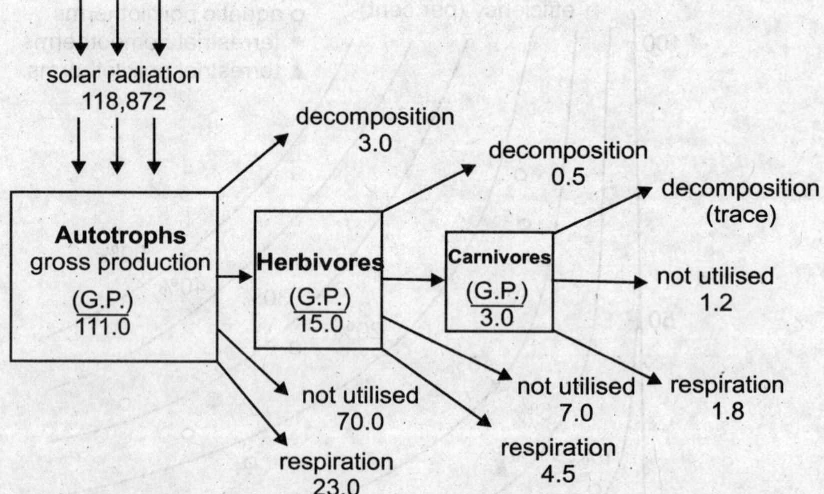

Fig. 10.10. Energy flow diagram for a lake (freshwater ecosystem) in a g/cal/cm^2/yr (Lindeman's single-channel energy flow model).

per cent of net production passes to carnivores. This is more efficient utilization of resources that occurs at autotroph–herbivore transfer level. At the carnivore's level about 60 per cent of the carnivore's energy intake is consumed in metabolic activity and the remainder becomes part of the not utilised sediments; only an insignificant amount is subject to decomposition yearly.

From this energy models two facts become clear:

1. There is one-way street along which energy moves (*i.e.,* unidirectional flow of energy). Thus, energy that is captured by the autotrophs does not revert back to solar input; that which passes to the herbivores does not pass back to the autotrophs. As it moves progressively through the various trophic levels it is no longer available to previous level. Thus due to one-way flow of energy, the system would collapse if the primary source, the sun, were cut of.

2. There occurs a progressive decrease in energy level at each trophic level. This is accounted largely by the energy dissipated as heat in metabolic activities and measured here as respiration coupled with unutilised energy.

(b) Box and pipe energy flow model. Fig. 10.11 presents a very simplified energy flow model of three trophic levels; it was suggested by **E. P. Odum** in 1953. In such single-channel energy flow model, "**boxes**" represent the trophic levels (*i.e.,* population mass or biomass) and the "**pipes**" depict the energy flow in and out of each level. Energy inflows balance outflows as required by the first law of thermodynamics, and energy transfer is accompanied by dispersion of energy into unavailable heat (*i.e.,* respiration) as required by the second law. From this model, it also becomes evident that the energy flow is greatly reduced at each successive trophic level from producers to herbivores and then to carnivores. At each transfer of energy from one level to another, major part of energy is lost as heat or other form. A closer examination of the diagram reveals that all the energy available to a trophic level is not consumed but a good portion of it is left unutilized. The herbivores do not feed upon roots, wood and barks of a plant and carnivores leave unused the bones, skin, hairs, etc.

Thus, of the 3000 Kcal of total light falling upon the green plants, approximately 50 per cent (1500 Kcal) is absorbed, of which only 1 per cent (15 Kcal) is converted at first trophic

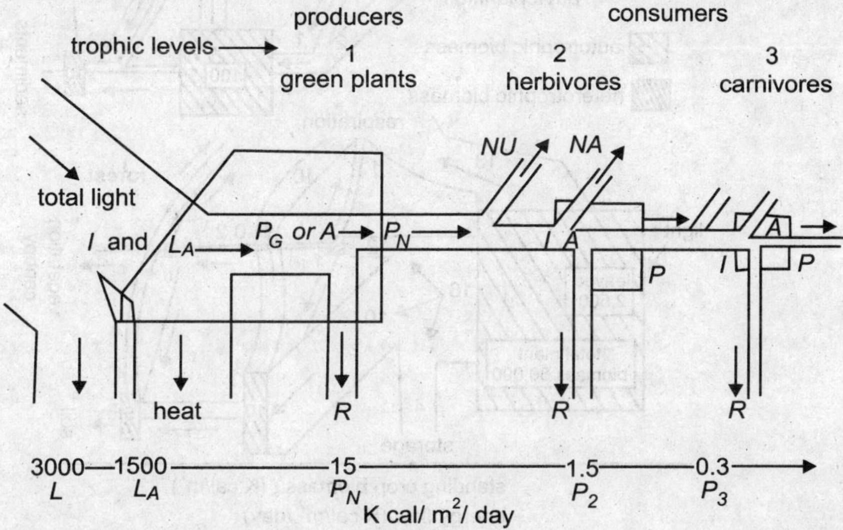

Fig. 10.11. A simplified energy flow diagram suggested by **E.P. Odum** in 1963 depicting three trophic levels (boxes numbered 1, 2, 3) in a linear food chain. I = total energy output, L_A = light absorbed by plant cover, P_G = gross primary production, A = total assimilation, P_N = net primary production, P = secondary production (consumer), N = energy not used (stored or exported), NA = energy not assimilated by consumers (egested), R = respiration.

level. Hence net primary production is merely 15 Kcal. Secondary productivity (P_2 and P_3 in Fig. 10.11) tends to be about 10 per cent at successive consumer trophic levels, (*i.e.,* herbivores and carnovores) although efficiency may be sometimes higher, as 20 per cent, at the carnivore level as shown (or P_3 = 0.3 Kcal) in the diagram.

2. Y-shaped or two channel energy flow models

These models were depicted by **E.P. Odum** in 1962. He pointed out that in nature two basic food chains are present in any system: 1. The **grazing food chain** beginning with green plant base going to herbivores and then to carnivores and 2. The **detritus food chain** beginning with dead organic matter acted by microbes, then passing to detritivores and their consumers (predators).

Fig. 10.12 shows two Y-shaped energy flow models. In each Y shaped model one arm represents the herbivore food chain and the other, the decomposer (detritus) food chain. The two arms differ fundamentally in the way in which they can influence primary producers in each model the grazing and detritus food chains are sharply separated. This figure contrasts the biomass energy flow relationships in the sea and forest. In the marine bay the energy flow via the grazing food chain is shown to be larger than via the detritus pathways, whereas reverse is shown for the forest in which 90 per cent or more of the net primary production is normally utilised in detritus food chain. Thus in marine ecosystem the grazing food chain is the major pathway of energy flow whereas in the forest ecosystem the detritus food chain is more important.

In grazing food chain herbivores feed on living plants and therefore, directly affect the plant populations. What they do not eat is available, after death, to the decomposers. As a result, decomposers are not able to directly influence the rate of supply of their food.

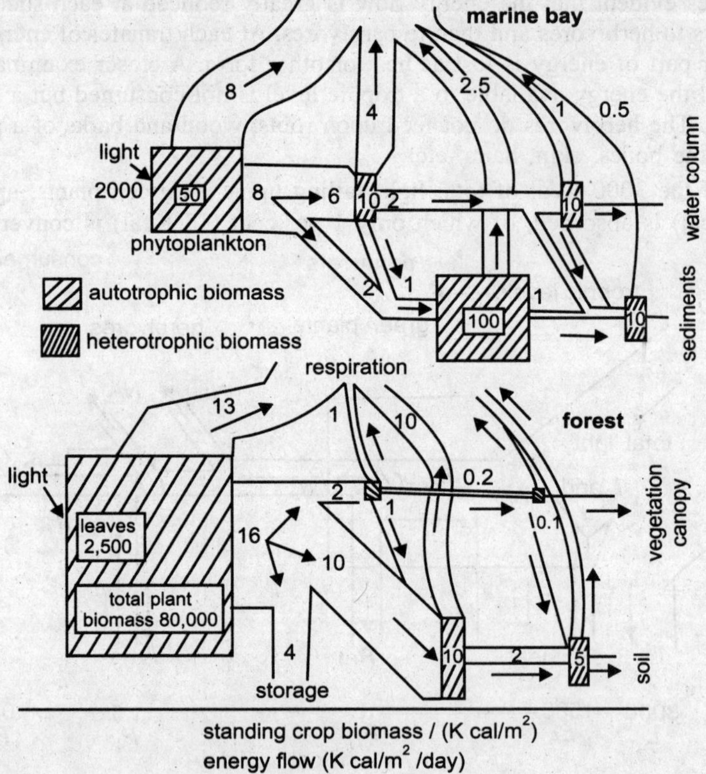

Fig. 10.12. A Y-shaped or two-channel energy flow diagram that separates a grazing food chain (water column or vegetation canopy) from a detritus food chain (sediments and in soil). Estimates for standing crops (shaded boxes) and energy flows compare a hypothetical coastal marine ecosystem (upper diagram) with a hypothetical forest (lower diagram).

E. P. Odum (1983) gave a generalized model of Y-shaped or two-channel energy flow which is applicable to both terrestrial and aquatic ecosystems (Fig. 10.13).

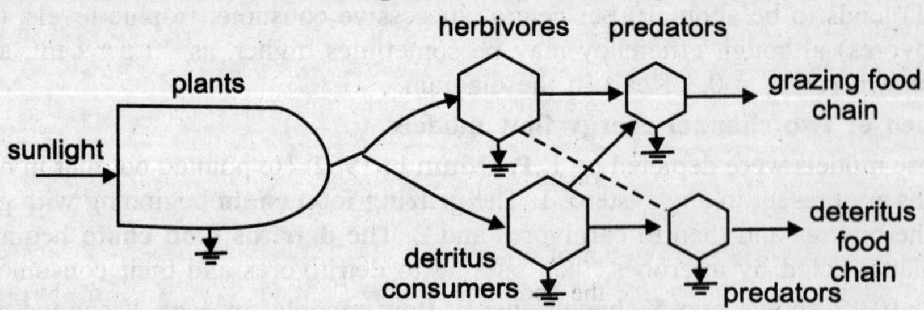

Fig. 10.13. Y-shaped energy flow model of Odum (1983) showing linkage between the grazing and detritus food chains.

The Y-shaped model is more realistic and practical working model than the single-channel model because (*i*) is confirms to the basic stratified structure of ecosystem; (*ii*) it separates the grazing and detritus food chains (*i.e.,* direct consumption of living plants and utilization of dead organic matter respectively) in both time and space, and (*iii*) that the microconsumers (absorptive bacteria, fungi) and macroconsumers (phagotrophic animals) differ greatly in size-metabolism relations. The important point in Y-shaped model is that the two food chains are not isolated from each other.

Odum's Universal Model

In 1968, **E. P. Odum** suggested a Y-shaped energy flow model (Fig. 10.14) which is applicable to any living components, whether a plant, animal, microorganism or individual, population or

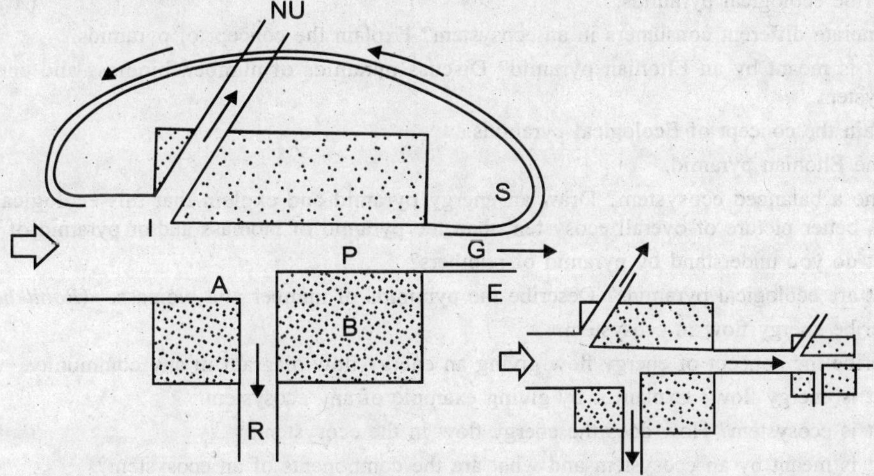

Fig. 10.14. Components of a "universal" model of energy flow. I = input or ingested energy; NU = not used; A = assimilated energy; P = production; R = respiration; B = standing crop biomass; G = growth and reproduction; S = stored energy; E = excreted energy.

a trophic group. It is called **universal model of energy flow.** In the figure, the shaded box labelled "B" represents the living structure or biomass of the component. The total energy input is indicated by "I", which is light for autotrophs and organic food for heterotrophs. Such a model may depict food chain as already shown in other Y-shaped energy flow systems, or the bioenergetics of an entire ecosystem. This model can be used in two ways: (*i*) It can represent a species population in which case the appropriate energy inputs and links with other species would be shown as a conventional species oriented food-web diagram, or (*ii*) the model can represent a discrete energy level in which case the biomass and energy channels represent all or parts of many populations supported by same energy source.

QUESTIONS

Long Answer Questions

1. What do you understand by ecosystem? Describe the various components of an ecosystem giving suitable examples.
2. Give an account of ecosystem structure.
3. Write an essay on ecosystem. (*Avadh 1999*)
4. Write all you know about concept of ecosystem. (*Kurukshetra 1999*)
5. Define the ecosystem. Describe the components of the ecosystem of small freshwater pond.
6. Describe pond ecosystem. (*Allahabad 1999*)

7. Discuss pond as an example of ecosystem.
8. Describe the structure of a freshwater ecosystem.
9. Discuss productivity of an ecosystem.
10. Differentiate between gross primary production and net primary production.
11. Describe food chain and various trophic levels.
12. What do you understand by food chain in an ecosystem. Give suitable examples.
13. What is food chain? *(Rohilkhand 2001)*
14. What are food chains? Describe various kinds of food chains involving fish.
15. Why the number of trophic levels of food chain is limited.
16. Describe food web. *(Avadh 1999)*
17. Describe ecological pyramids. *(Avadh 1999)*
18. Enumerate different consumers in an ecosystem? Explain the concept of pyramids.
19. What is meant by an Eltonian pyramid? Discuss pyramids of number, biomass and energy in an ecosystem.
20. Explain the concept of Ecological pyramids.
    Define Eltonian pyramid.
21. Define a balanced ecosystem? Draw an energy pyramid and explain that this ecological pyramid gives better picture of overall ecosystem than the pyramid of biomass and/or pyramid of number.
    What do you understand by pyramid of numbers?
22. What are ecological pyramids? Describe the pyramids of number and biomass. *(Rohilkhand 1999)*
23. Describe energy flow in ecosystem.
24. Describe the concept of energy flow giving an energy flow diagram in any community.
25. What is energy flow? Explain it by giving example of any ecosystem.
26. What is ecosystem? How does the energy flow in the ecosystem.
27. What is meant by an ecosystem and what are the components of an ecosystem?
    Discuss the process of energy flow in an ecosystem.
28. Give an account of flow of energy through different trophic levels.
29. Describe the transformation of energy in an ecosystem and discuss its significance in the substance of ecosystem.
30. Explain with the help of ecosystem pyramids " The trophic organization" of ecosystem. Also explain how energy flows through these systems.
31. What is energy flow? Explain it with reference to an aquatic system.
32. Describe different types of food chains and ecological pyramids.
33. What is food chain? Discuss different steps in typical food chain? Give an account of pyramids of number and biomass.
34. What is food chain? Describe the flow of energy in a freshwater lake.

Short Answer Questions

1. Write short note on Ecological pyramids.
2. Write a brief note on "pyramid of biomass for fresh water fish".
3. What is meant by ecological energetics?
4. What is transfer of food energy?
5. Write short note on flow of energy through an ecosystem. *(Avadh 1999)*
6. Write short note on the following:
   (i) Food chain *(Allahabad 1999)*
   (ii) Trophic levels
   (iii) Trophic levels in fish

7. Comment upon the following:
   (i) An ecosystem comprises of abiotic components only.
   (ii) An ecosystem comprises of biotic components only.
   (iii) A food chain consists of producer only.
   (iv) A food link consists of several interlocked food chains.
   (v) A food web consists of one food chain.
   (vi) A parasitic food chain exhibits an upright ecological pyramid.
8. Answer this question in only 20 words:
   The pyramid of biomass is upright in a terrestrial ecosystem.
9. Write short notes on the following:
   (i) Producer
   (ii) Detritivores
   (iii) Biomass
   (iv) Ecosystem
   (v) Food webs
10. Differentiate between the following:
   (i) Food Chain and food web
   (ii) Pyramid of numbers and pyramid of biomass
   (iii) Autotrophic and heterotrophic

Very Short Answer Questions

1. Who proposed the term ecosystem?
2. What is detritus?
3. Define the ecosystem.
4. What is plankton?
5. What is predator?
6. Define producer.
7. Define consumer.
8. What is the trophic pyramid?
9. Define the trophic level.
10. What is pyramid of biomass?
11. Define pyramid of energy?
12. What is pyramid of number?
13. Dercribe progressive efficiency.
14. Which food chain show inverted pyramid of number?
15. Which ecosystem possesses inverted pyramid of biomass?
16. Which ecological pyramid is never inverted.

Fill in the Blanks

1. The successive levels of .... constituting the link of the food chain are called ....
2. Ecosystem consists of three basic components: .... , ... and abiotic factors.
3. Macroconsumers of ecosystem are also called ....
4. Osmotrophs are also called .... and include .... , actinomycetes and ....

5. *Euglena* and *Cyclops* form the examples of .... of the pond ecosystem.
6. .... help in returning of mineral elements to the pond ecosystem.
7. The concept of pyramid of number was advanced by ....
8. Trophic-dynamic concept of the ecosystem was developed by .... in 1942.

Multiple Choice Questions

1. The ecological pyramid was first described by
   (*a*) Tansley (*b*) Smith
   (*c*) Charles Elton (*d*) Webster
2. At each trophic level of a food chain (pyramid) the energy not used or passed along is given off as a
   (a) matter (*b*) free energy
   (*c*) heat (*d*) water
3. What is the source of energy that flows through the living world?
   (*a*) photosynthesis (*b*) chemical bond
   (*c*) green plant (*d*) the sun
4. Energy and nutrients enter a community by way of the
   (*a*) producer (*b*) consumer
   (*c*) scavenger (*d*) detritivores
5. A sequence of species through which the organic molecules in a community pass is called
   (*a*) pyramid of energy (*b*) food chain
   (*c*) food web (*d*) nutrient cycle
6. A group of interconnected food chains is called a
   (*a*) pyramid of energy (*b*) complex food chain
   (*c*) food web (*d*) food cycle
7. Most food chains are composed of
   (*a*) 1 or 2 species (*b*) 3 or 4 species
   (*c*) 9 or 10 species (*d*) More than 16 species
8. About how much of the solar energy that falls on the leaves of a plant is converted to chemical energy
   (*a*) 1 per cent (*b*) 10 per cent
   (*c*) 30 per cent (*d*) 50 per cent
9. About how much of the chemical energy within producer tissues become chemical energy within herbivore tissue
   (*a*) 1 per cent (*b*) 10 per cent
   (*c*) 30 per cent (*d*) 50 per cent
10. An ecological pyramid of energy flow is often an inverted pyramid in which of the following system?
    (*a*) desert (*b*) ocean
    (*c*) tundra (*d*) rain forest
11. Herbivores are also known as (*CCSU Meerut 2014*)
    (*a*) autotrophs (*b*) primary consumers
    (*c*) secondary consumers (*d*) tertiary consumers
12. Which of the following is a biotic component of the ecosystem (*CCSU Meerut 2014*)
    (*a*) autotrophs (*b*) decomposers
    (*c*) saprophytes (*d*) all of the above

13. Insects feed on grass and frogs feed on insects, hence frog is a (*CCSU Meerut 2014*)
(*a*) decomposers (*b*) Secondary consumer
(*c*) primar consumer (*d*) none of the above

14. Inverted pyramid is found in (*CCSU Meerut 2014*)
(*a*) biomass pyramid of aquatic system (*b*) energy pyramid of grass land
(*c*) biomass pyramid of grass land (*d*) number pyramid of aquatic system

15. Which of the following is not an ecological pyramid? (*CCSU Meerut 2014*)
(*a*) pyramid of biomass (*b*) pyramid of food
(*c*) pyramid of number (*d*) pyramid of energy

16. Ratios (expressed as percentages) between energy flow at different points along food chain are (*CCSU Meerut 2014*)
(*a*) ecological niche (*b*) ecological magnification
(*c*) ecological efficiencies (*d*) ecological barriers

17. Which of the food chain directly depends on solar radiations? (*CCSU Meerut 2014*)
(*a*) grazing (*b*) detritus
(*c*) predator (*d*) parasitic

18. In the following which pyramid represents the number of the organisms in the following chain:
Grass ⟶ Insects ⟶ Bird (*CCSE Meerut 2014*)

(P) (Q)

(R) (S)

(*a*) S (*b*) R
(*c*) Q (*d*) P

19. Which of the following is not used for construction of ecological pyramids (*CCSU Meerut 2014*)
(*a*) rate of energy flow (*b*) weight of organisms
(*c*) structure of organisms (*d*) number of organisms

20. Pyramid of energy is inverted in which ecosystem (*CCSU Meerut 2014*)
(*a*) grass land (*b*) forest
(*c*) both of these (*c*) none of these

21. The term ecosystem was proposed by (*CCSU Meerut 2014*)
(*a*) S. A. Forbes (*b*) H.T. Odum
(*c*) A.G. Tansley (*d*) Sumner

22. Bacteria and fungi in a forest ecosystem are (*CCSU Meerut 2014*)
(*a*) Producers (*b*) decomposers
(*c*) primary consumers (*d*) secondary consumers

23. Who discovered the ecological pyramid? (*CCSU Meerut 2014*)
(*a*) Odum (*b*) Tansley
(*c*) Mobius (*d*) Charls Elton

24. Which one of the following is a tertiary consumer in an aquatic ecosystem? (*CCSU Meerut 2014*)
(*a*) *Cyclops* (*b*) *Utricularia*
(*c*) *Zygnema* (*d*) cod fish

ANSWERS

Very Short Answer Questions

1. A.G. Tansley (1935).

2. Detritus is freshly dead or partially decomposed remains of organism.

3. 1. All the organisms in a given area as well as the physical environment in which they live, form the ecosystem. 2. A discrete unit that consists of living and nonliving parts interacting to form an ecological system.

4. Small, often microscopic organisms that live suspended in water; although many plankton are mobile, none can swim strongly enough to overcome water currents.

5. An organism that kills and eats other organisms, referred to as predator.

6. An organism that can produce its own food by photosynthesis or chemosynthesis; also called primary producer or autotroph.

7. Consumer is an organism that obtains its energy by eating other organisms or their remains.

8. A common approach to conceptualizing trophic relationship in an ecosystem in which a stack of rectangles is constructed, each of which represents the amount of energy or biomass within one trophic level.

9. 1. Trophics level is the position in a food chain as determined by the number of energy transfer steps to that level (primary producer to secondary consumer, for example); or (2) Functional classification of organisms in an ecosystem according to feeding relationships.

10. Pyramid of biomass is a model or diagram that depicts the amounts of standing crop biomass at different trophic levels of an ecosystem.

11. The pyramid of energy is a model or diagram that depicts the rates of energy flow through the different trophic levels of an ecosystem.

12. Pyramid of numbers is a model or diagram that depicts the number of organisms present at each trophic level of an ecosystem; frequently termed as the Eltonian pyramid, attributed to Charles Elton, a British ecologist.

13. The relationship between the energy consumed at two successive trophic levels is called **progressive efficiency**. **Lindeman** described it for the first time.

14. Parasitic food chain, *e.g.*, tree ecosystem.

15. Pond or lake ecosystem.

16. Pyramid of energy.

Fill in the Blanks

1. Energy flow, trophic levels **2.** Producers, Consumers **3.** Phagotrophs
4. Microconsumers, bacteria, fungi **5.** Zooplankton
6. Decomposers **7.** Charles Elton **8.** Lindeman.

Multiple Choice Questions

1. (*c*)	**2.** (*c*)	**3.** (*d*)	**4.** (*a*)
5. (*b*)	**6.** (*c*)	**7.** (*b*)	**8.** (*a*)
9. (*b*)	**10.** (*b*)	**11.** (*b*)	**12.** (*d*)
13. (*b*)	**14.** (*a*)	**15.** (*b*)	**16.** (*c*)
17. (*a*)	**18.** (*d*)	**19.** (*c*)	**20.** (*d*)
21. (*c*)	**22.** (*b*)	**23.** (*d*)	**24.** (*d*).

11

Biogeochemical Cycles

The total mass of all the organisms that have lived on the earth in the past 1.5 billion years is much greater than the mass of carbon and nitrogen atoms present. According to the **law of conservation of matter**, matter is neither created nor destroyed; obviously the carbon and nitrogen must have been used over and over again in the course of time. The earth neither receives any great amount of matter from other parts of the universe nor does it lose significant amount of matter to outer space. The atoms of each element as carbon, hydrogen, oxygen, nitrogen, phosphorus, calcium, and the rest are taken from the environment, made a part of some cellular component of an organism and finally, perhaps by a quite circuitous route involving several other organisms, are returned to the environment to be used over again. The cyclic movements of chemical elements of the biosphere between the organism and the environment are referred to as **biogeochemical cycles**, (**Vernadsky** 1934). "*Bio*" refers to living organisms and "*geo*" to the rocks, soil, air and water of the earth.

Organic and Abiotic Phases of Geochemical Cycles

Since an element is necessary for the maintenance of life, its movement through biotic communities (organisms) can be viewed in the terms of food chain. The flow of a chemical element through the food chain can be viewed as the **organic phase** of the biogeochemical cycle. Further, the biogeochemical cycles also include **abiotic phases** which are the functions of the chemistry of the elements in question. These abiotic phases are of critical importance to the ecosystem, as the major reservoirs for all nutrient elements are external to the food chains, and flow in the abiotic phases tends to be much slower than in the organic phase. The rapidity and direction of nutrient cycling through the abiotic phases determine not only the distribution of the element in the total environment, but also its availability of living systems. There are two classes of abiotic phases in biogeochemical cycles, a **sedimentary phase**, which is part of all cycles, and an **atmospheric phase**, which is possessed by some. In some cycles, such as nitrogen, the atmospheric phase is more important than the sedimentary. In others, such as phosphorus, the atmospheric phase is essentially non-existent. In still others, such as sulphur, both phases are present and their relative importance depends on other environmental factors. Biogeochemical cycles that have dominant atmospheric phases are often called **atmosphere-reservoir cycles**; those whose sedimentary phase is dominant are termed **sediment-reservoir cycles**.

Types of Biogeochemical Cycles

There are two types of biogeochemical cycles, the **gaseous** and the **sedimentary**. In gaseous cycles the main reservoir of nutrients is the atmosphere and the ocean. In sedimentary cycles the main reservoir is the soil and the sedimentary and other rocks of the earth's crust. Both involve biotic and abiotic agents, both are driven by the flow of energy and both are tied to the **water cycle**.

Types of Biogeochemical Cycles

A. Water or hydrological cycle

B. Gaseous cycles:

1. Oxygen cycle;
2. Carbon cycle;
3. Nitrogen cycle.

C. Sedimentary cycles:

1. Sulphur cycle;
2. Phosphorus cycle.

D. Nutrient cycle.

11.1 WATER CYCLE

Living organisms, atmosphere and earth maintain between them a circulation of water and moisture, which is referred to as **water cycle** or **hydrological cycle**. As we have already discussed in Chapter 2, water forms a very significant factor of environment and without the cycling of water, biogeochemical cycles could not exist, ecosystems could not function, and life could not be maintained. Water is important for an ecosystem for several reasons— it is the medium by which nutrients are introduced into autotrophic plants; it is an important part of living tissue, either as liquid water or as part of essential organic molecules; it serves as a means of thermal regulation for both plants and animals; it is the medium by which sediments — a prime source of mineral nutrients— are removed from or added to local ecosystem; it covers the great majority of the earth's surface, and is the dominant feature of all aquatic ecosystems.

The hydrological cycle is driven by solar energy and gravity. More than 80 per cent of the total insolation that is not lost immediately as electromagnetic radiation goes to **evaporate** water. The atmospheric water vapour produced by this means can then condense around particles of dust in the atmosphere, often called **nucleation particles**. The atmosphere possesses a limited capacity for holding water vapour, thus, the droplets formed by this means are heavy enough to fall as precipitation, under the influence of gravity. Eventually, the hydrologic cycle can be defined as an alternation of evaporation and precipitation, with the energy used to evaporate the water being dissipated as heat in the atmosphere as the water condenses.

Distribution of water in earth's surface. Water is not evenly distributed throughout the earth. Almost 95 per cent of the total water on earth is chemically bound into rocks and does not cycle. Of the remainder, about 97.3 percent is in the ocean, about 2.1 per cent exists as ice in the polar caps and permanent glaciers, and the rest is fresh water, present in the form of atmospheric water vapour, ground water, soil water, or inland surface water (Table 11.1).

The rate of cycling of water. The rate of cycling of water between surface and atmosphere is very rapid. The amount of water vapour in the atmosphere is sufficient, on the average, to cover the entire earth to a depth of 2.25 cm. But the average annual rainfall for the earth is about 81.1 cm (**Furon**, 1967) and in some places it ranges up to 1200 cm. This means that the average turnover time for atmospheric water is about 11.4 days, or that the equivalent of all the water vapour in the entire atmosphere falls as precipitation and is re-evaporated more than 32 times per year.

Table 11.1. Distribution of water in the earth's crust and surface.

A. Chemically bound water of rocks: Does not cycle		
1.	Crystalline rocks	250,000 × 10^{17} Kg.
2.	Sedimentary rocks	21,00 × 10^{17} Kg.
B. Free water : Moves via hydrologic cycle		
1.	Oceans	13,200 × 10^{17} Kg.
2.	Ice caps and glaciers	292 × 10^{17} Kg.
3.	Ground water to a depth of 4000 m	83.5 × 10^{17} Kg.
4.	Freshwater lakes	1.25 × 10^{17} Kg.
5.	Saline lakes and inland seas	1.04 × 10^{17} Kg.
6.	Soil moisture	0.67 × 10^{17} Kg.
7.	Atmospheric water vapour	0.13 × 10^{17} Kg.
8.	Rivers	0.013 × 10^{17} Kg.

Further, the distribution of evaporation and rainfall is quite uneven. If we compare the annual evaporation and precipitation over land and sea, we find that relatively more water precipitates on land than evaporates from land. This is fortunate from the point of view of terrestrial organisms. Even so, the amount of rainfall on the open ocean is proportionately greater than that on land, taking into account the relative percentages of the earth's surface covered by land and sea.

Nature of hydrological cycles. The hydrological cycle over the oceans is extremely simple—the water is evaporated from the surface of the ocean and water vapours form the clouds which when cool down precipitate the water as rain fall. But several routes are open to precipitation that falls on land—direct evaporation, transpiration, entry of water into ground water system and runoff. Consequently, the routes of hydrological cycles on land can be divided into following three main categories–the rapidly cycling portion, or **evapotranspiration**, which includes the evaporation and transpiration, the less rapidly cycling water, or surface runoff, and very slowly cycling ground water that seeps into the soil can end up in any one of these three categories.

1. Evapotranspiration. Evapotranspiration includes evaporation and transpiration. **Evaporation** refers to water that is evaporated directly from any surface other than a plant, such as a lake, soil surface, or animal skin. In most cases, the main effects of direct evaporation are to moderate the temperature of local area and to allow the hydrological cycle to continue. In some ecosystems, evaporation also leads to a concentration of salts in the water of soil which may be a critical environmental factor. **Transpiration** is water that evaporates from the surface of leaves of plants. Transpiration acts to move the biogeochemical cycles for all mineral nutrients that enter the food chain via the roots of plants.

2. Surface runoff. If transpiration is related to the mechanism of nutrient uptake, the gross movement of soluble and solid particles in the ecosystem is accomplished largely by runoff. Nutrients that have accumulated in sediments or soils can be eroded by streams and removed altogether from a local ecosystem, or soluble nutrients may be carried by soil seepage into surface waters, where they are removed from the area. Streams may carry sediment particles which can be chemically altered through additional weathering so that the nutrient elements they contain may be utilized by organisms. Finally moving water acts as an agent of **erosion** which removes soil and allows weathering of the underlying rock to make their nutrients available to plants.

3. Ground water. Ground water is water that saturates either sediment or rock below the water table. In general, i is not trapped by plants for transpiration and it is too deep to be

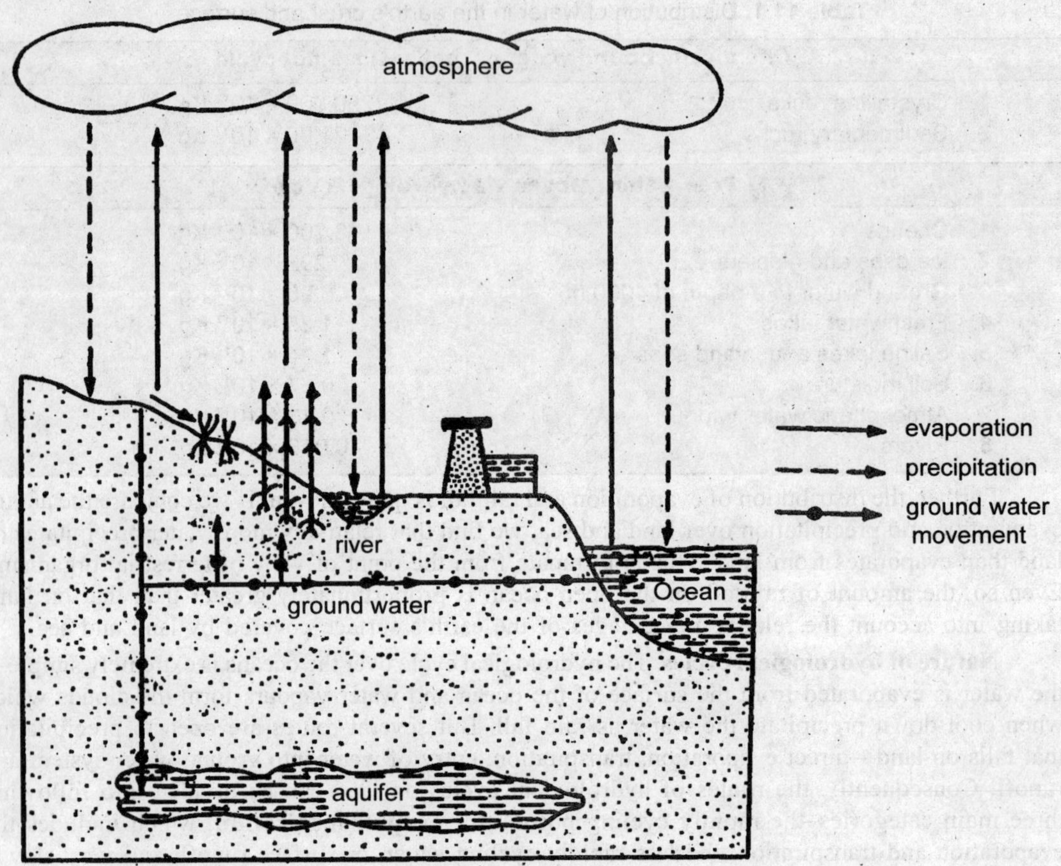

Fig. 11.1. The hydrological cycle.

directly evaporated from the soil surface. It is an exceedingly important reservoir for water which moves from one place to another under the influence of gravity. The area where the net water movement is from the surface into the ground water systems is termed as **catchment area**; areas where ground water reaches the surface and runs off are termed **springs**. A rock body through which ground water flows is called an **aquifer**. A well drilled into an aquifer that has sufficient hydrostatic pressure to force water up into it is called an **artesian well**.

The hydrological cycle on land, thus, includes evapotranspiration of water from earth's surface and leaf surface → formation of clouds → precipitation → surface runoff + accumulation of water as ground water → return of water to sea via streams or direct evaporation and cloud formation, and so on.

B. Gaseous Cycles

The gaseous geochemical cycles are of following types:

11.2 OXYGEN CYCLE

Oxygen (O_2), the byproduct of photosynthesis, is involved in the oxidation of carbohydrates with release of energy, carbon dioxide and water. Its primary role in biological oxidation is that of a hydrogen acceptor. The break-down and decomposition of organic molecules proceeds primarily by dehydrogenation. Hydrogen is removed by enzymatic activity from organic molecules in a series of reactions and is finally accepted by the oxygen, forming water. Though oxygen is necessary for life, but being very active chemically, molecular O_2 may be toxic to living body cells. Therefore, for the protection from toxic effects of molecular O_2, cells possess the cellular organelles called **peroxisomes** which mediate oxidative reactions resulting in the production of hydrogen peroxide which in turn is used through the mediation of other enzymes as an acceptor in oxidizing other compounds.

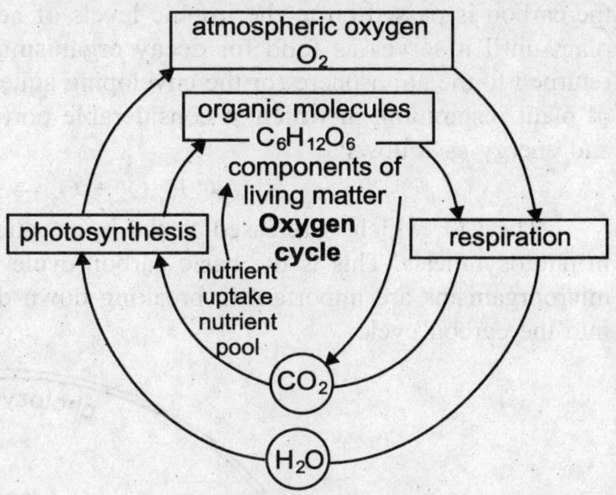

Fig. 11.2. The oxygen cycle.

The major supply of free oxygen which supports life occurs in the atmosphere (Fig. 11.2). There are two significant sources of atmospheric oxygen. One is the **photodissociation** of water vapour in which most of the hydrogen released escapes into outer space. The other source is **photosynthesis**, active only since life began on earth. Because photosynthesis and respiration are cyclic, involving both the release and utilization of oxygen, one would seem to balance the other, and no significant quantity of oxygen would accumulate in the atmosphere. However, at some time in the earth's history the amount of oxygen introduced into the atmosphere had to exceed the amount used in the decay of organic matter and that tied up in the oxidation of sedimentary rocks. Part of the atmospheric oxygen represents that portion remaining from the unoxidized reserves of photosynthesis–coal, oil, gas, and organic carbon in sedimentary rocks. The amount of stored carbon in the earth suggests that 150×10^{20}g of oxygen has been available to the atmosphere, over 10 times as much as now present, 10×10^{20}g (**Johnson**, 1970). The main non-living (abiotic) oxygen pool consists of molecular oxygen, water, and carbon dioxide, all intimately linked to each other in photosynthesis and other oxidation-reduction reactions, and all exchangeable in such compounds as nitrates and sulphates utilized by organisms that reduce them to ammonia and hydrogen sulphide.

The cycling of oxygen is very complex. As a constituent of CO_2, it circulates freely throughout the biosphere. Some carbon dioxide combines with calcium to form carbonates. Oxygen combines with nitrogen compounds to form nitrates, with iron to ferric oxides, and with many other minerals to form various other oxides. In these states oxygen is temporarily withdrawn from circulation. In photosynthesis the oxygen freed is split from the water molecule. This oxygen is then reconstituted into water during plant and animal respiration. Some part of the atmospheric oxygen that reaches the higher levels of the troposphere is reduced to ozone (O_3) by high energy ultraviolet radiation.

11.3 CARBON CYCLE

The carbon being a basic constituent of all organic compounds and a major element involved in the fixation of energy by photosynthesis, is so closely tied to energy flow that the two are inseparable (Fig. 11.3). The source of all the fixed carbon both in living organisms and fossil deposits is carbon dioxide CO_2, found in the atmosphere and dissolved in the waters of the earth. During photosynthesis, carbon from atmospheric CO_2 is incorporated into the production of the carbohydrate, glucose, $C_6H_{12}O_6$, that subsequently may be converted to other organic compounds such as polysaccharides (sucrose, starch, cellulose, etc.), proteins and lipids. All the polymeric organic compounds containing carbon are stored in different plant-tissues as food and from them the carbon is passed on to the trophic levels of herbivores or phytoparasites, or retained by the plant until it serves as food for decay organisms (*viz.,* decomposers). Some of the carbon is returned to the atmosphere (or the enveloping aqueous medium) in the form of CO_2 a by-product of plant respiration, in which, a considerable portion of glucose is oxidized to yield CO_2, H_2O and energy as follows:

$$1/6C_6H_{12}O_6 + O_2 \rightarrow CO_2 + H_2O + \text{Energy}$$

The CO_2 which is released as the by-product of plant respiration is again used by plants in photosynthesis. This is the basic carbon cycle which is simple and complete. Decomposing microorganisms are important in breaking down dead material with the release of carbon back into the carbon cycle.

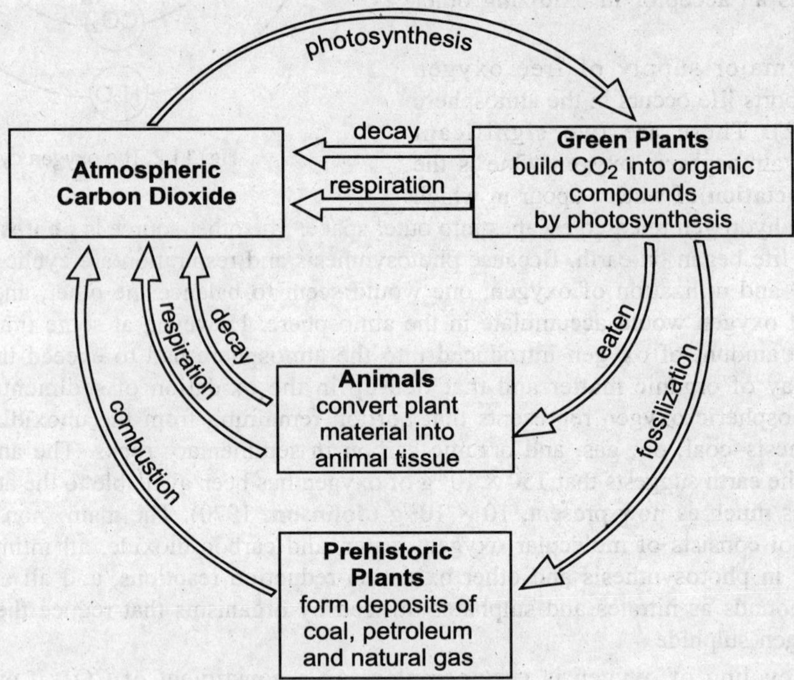

Fig. 11.3. The carbon cycle.

Similarly, carbon taken up by herbivores or phytoparasites may travel a number of routes. It may be incorporated into protoplasm (assimilation) and stored until the organism dies, where upon it is utilized by decomposers; it may be released through animal respiration; it may serve as live food for other organisms; or finally it may be stored in the environment as CO_2. Similar fates await carbon at the carnivore trophic levels. In fact, all the carbon of plants, herbivores, carnivores and decomposers is not respired but some is fermented and some is stored. The carbon compounds that are lost to the food chain after fermentation, such as methane, are readily oxidized to carbon dioxide by inorganic reactions in the atmosphere. As for the storage of carbon in sediments, just as deposition works to store materials, erosion may uncover them, and inorganic chemical weathering of rock can oxidize the carbon contained there. Some carbon is permanently stored in sediments and not uncovered by weathering; it may be replaced by carbon dioxide released from volcanoes and other similar examples of intense geological activity. In modern age, humans have greatly increased the rate at which carbon is passing from sedimentary form to carbon dioxide. The combustion of fossil fuels is a significant means of recycling sedimentary carbon much faster than natural weathering.

Small portion of carbon, especially in the sea, is found not as organically fixed carbon, but as carbonate ($CO_3^=$), especially calcium carbonate ($CaCO_3$). $CaCO_3$ is very commonly used for shell construction by such animals as clams, oysters, some protozoa, and some algae. Carbon dioxide reacts with water to form carbonate in the following three step reaction :

$$CO_2 + H_2O \rightleftarrows \underset{\text{Carbonic acid}}{H_2CO_3} \leftrightarrows H^+ + \underset{\text{Bicarbonate}}{HCO_3^-} \leftrightarrows 2H^+ + \underset{\text{Carbonate}}{CO_3^=}$$

The exact amount of each of these constituents in the water depends on the pH of the water. Organisms such as clams can combine bicarbonate or carbonate with calcium dissolved in the water to produce calcium carbonate. After the death of the animal, this calcium carbonate may either dissolve or remain in sedimentary form.

Certain control mechanisms are inherent in the carbon cycle. The rate of carbon utilization is depends on its availability. If excessive amounts of carbon are taken up in any one phase of the cycle, other phases of activity may be inhibited or slowed down. For example, if the pH of water is alkaline, more carbon is tied up in a carbonate and less is in solution. This removal of carbon in solution would upset the equilibrium established between the atmospheric and the dissolved CO_2 and the net effect would be a movement of CO_2 into solution until equilibrium is reached.

Peculiarities of carbon cycle. Though carbon-cycle exhibits basic similarity with other biogeochemical cycles, yet it is unusual in that the organic phase is not essentially a complete cycle within itself. The organic (biotic) and atmospheric (abiotic) phases, however, are so closely intertwined that the rapid cycling typical of the organic phase is present. The multiplicity of paths along which carbon can flow is typical of biogeochemical cycles in general, and provides a well-buffered system with adequate feedback mechanisms to insure an adequate supply of the carbon. It is significant that all phases of the cycle yield carbon dioxide at some time, and carbon dioxide is the raw material for them. Thus, despite its relative low concentration in the atmosphere (0.03 per cent), carbon in a form in which it can be used by living organisms is virtually always present.

11.4 NITROGEN CYCLE

Nitrogen is an essential constituent of different biologically significant organic molecules such as amino acids and proteins, pigments, nucleic acids and vitamins. It is also the major constituent of the atmosphere, comprising about 79 per cent of it. The paradox is that in its gaseous state, N_2 is abundant but is unavailable to most life. Before it can be utilized it must be converted to some chemically usable form.

To be used biologically, the free molecular nitrogen has to be fixed and fixation requires an input of energy. In the first step molecular nitrogen, N_2 has to be split into two atoms: $N_2 \rightarrow 2N$. The free nitrogen atoms then must be combined with hydrogen to form ammonia, with the release of some energy:

$$2N + 3H_2 \rightarrow 2NH_3$$

This fixation comes about in two ways. One is by high-energy fixation such as cosmic radiation, meteorite trails, and lightning that provide the high energy needed to combine nitrogen with oxygen and hydrogen of water. The resulting ammonia and nitrates are carried to the earth in rain water. The second method of nitrogen-fixation which contributes about 90 per cent of fixed nitrogen of earth, is biological. Some bacteria, fungi, and blue-green algae can extract molecular nitrogen from the atmosphere and combine it with hydrogen to form ammonia. Some of this ammonia is excreted by the nitrogen-fixing organism, and, thus, becomes directly available to other autotrophs. Some of these nitrogen-fixing organisms may be free-living, either in the soil (*e.g.*, bacteria — *Azotobacter* and *Clostridium*) or in water (*e.g.*, blue-green algae—*Nostoc, Calothrix* and *Anabaena*) and produce vast quantities of fixed nitrogen. In other cases, certain symbiotic bacteria of genus *Rhizobium*, although unable to fix atmospheric nitrogen themselves, can do this when in combination with cells either from the roots of legumes (*e.g.*, peas, beans, clover and alfalfa) and of other angiosperms such as *Alnus, Ceanothus, Shepherdia, Elaeagnus* and *Myrica*, or from the leaves of African genera of *Rubiaceae* and *Pavetta.* The bacteria invade the roots or leaves and stimulate the formation of root-nodules or leaf-nodules, a sort of harmless tumors. The combination of symbiotic bacteria and host cells remains able to fix atmospheric

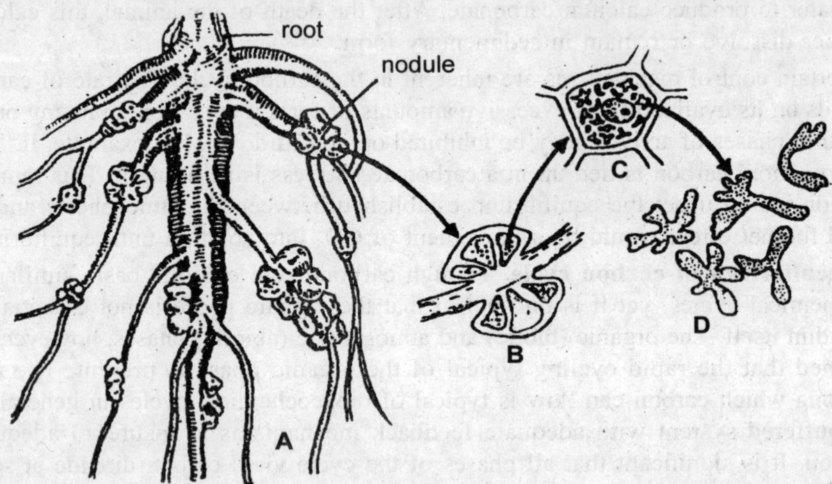

Fig. 11.4. Root-nodules of a legume plant. A—Legume root with root nodules. B—T.S. of root nodule. C—Single cell of nodule. D— Nitrogen-fixing bacterium, *Rhizobium*.

nitrogen and for this reason legumes are often planted to restore soil fertility by increasing the content of fixed nitrogen. Nodule bacteria may fix as much as 50 to 100 kilograms of nitrogen per acre per year, and free soil bacteria as much as 12 kilograms per acre per year. Further both free soil bacteria (*Azotobacter* and *Clostridium*) produce ammonia as the first stable product and like the symbiotic bacteria, they require molybdenum as an activator and are inhibited by an accumulation of nitrates and ammonia in soil.

Certain lichens (*Collema tunaeforme* and *Peltigera rufescens*) were also implicated in nitrogen fixation (**Henriksson**, 1971). Lichens with nitrogen-fixing ability possess nitrogen-fixing blue green species as their algal component.

Nitrogen fixed by symbiotic and non-symbiotic microorganisms in soil and water is one source of nitrogen. Another source is organic matter. The nitrogenous wastes and carrion of animals are degraded by the detritus organisms, nitrogen is converted to the amino form (*e.g.*, L-Alanine). The amino group ($-NH_2$) is liberated from organic molecules to form ammonia; this process is called **deamination**. Certain specific bacteria, most notably of the genus *Nitrosomonas*, can oxidize ammonia to nitrite (NO_2) by the reaction.

$$2NH_3 + 3O_2 \rightarrow 2NO_2^- + 2H_2O + 2H^+$$

This reaction takes place in the soil, in lake or sea water or sediments, and whenever ammonia is being released and oxygen is present. As fast as nitrite is produced, other bacteria, such as *Nitrobacter*, can combine nitrite with oxygen to form nitrate (NO_3) by the reaction:

$$2NO_2^- + O_2 \rightarrow 2NO_3^-$$

Both of these reactions which are performed by two nitrifying bacteria — *Nitrosomonas* and *Nitrobacter* are the parts of a single biological process called **nitrification**. In nitrification process, thus, ammonia is oxidized to nitrate and nitrite yielding energy. This energy is used by the bacteria to make their organic materials directly from carbon dioxide and water. Nitrate can be taken up by autotrophs at the beginning of food chain.

Under certain circumstances, nitrate is either not produced in the nitrogen cycle or it is degraded before it can be utilized by autotrophs. Degradation of nitrate is called **denitrification**, and may be important when oxygen concentration is low. Denitrifying bacteria such as *Pseudomonas* can use the energy of the nitrate ion to drive their metabolism, and in so doing, they break the nitrate down to nitrite, ammonia, or molecular nitrogen:

$$C_6H_{12}O_6 + 12NO_3^- \rightarrow 12NO_2 + 6CO_2 + 6H_2O$$
$$C_6H_{12}O_6 + 8NO_2 \rightarrow 4N_2 + 2CO_2 + 4CO_3^= + 6H_2O$$
$$C_6H_{12}O_6 + 3NO_3^- \rightarrow 3NH_3 + 6CO_2 + 3OH^-$$

If denitrification is significant in an ecosystem, nitrite is transitory and is also degraded into either ammonia or molecular nitrogen.

Cycling of nitrogen in the ecosystem. The sources of inputs of nitrogen under natural conditions are the bacterial fixation of atmospheric nitrogen, addition of inorganic nitrogen in rain from such sources as lightning fixation and fixed "juvenile" nitrogen from volcanic activities, ammonia absorption from the atmosphere by plants and soil, and nitrogen accretion from windblown aerosols, which contain both organic and inorganic forms of nitrogen (Fig. 11.5). In terrestrial ecosystems, nitrogen, largely in the form of ammonia and nitrates is taken up by plants, which convert it into amino acids and proteins. Animals (primary macro-consumers) may eat the plants and utilize the amino acids from the plant proteins in the synthesis of their own proteins and other cellular constituents. When animals and plants die, the decay bacteria convert the nitrogen of their proteins and other compounds into ammonia. Animals excrete several kinds of nitrogen-containing wastes—urea, uric acid, creatinine, and ammonia and the decay bacteria converts these wastes to ammonia. Ammonia may be lost as gas to the atmosphere, may be acted upon by nitrifying bacteria, or may be taken up directly by plants. The nitrates may be utilized by plants, immobilized by microbes, stored in decomposing humus, or leached away. This material is carried to streams, lakes, and eventually the sea,

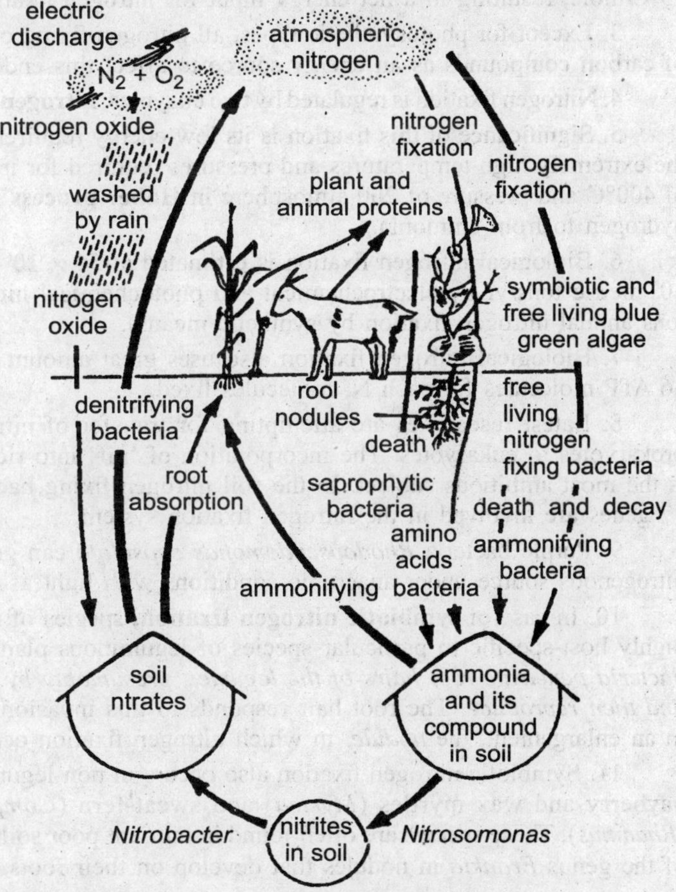

Fig. 11.5. The nitrogen cycle.

where it is available for use in aquatic ecosystems. There nitrogen is cycled in a similar manner, except that the large reserves contained in the soil humus are largely lacking. Life in the water contributes organic matter and dead organisms that undergo decomposition and subsequent release of ammonia and ultimately nitrates. In aquatic ecosystem atmospheric nitrogen is fixed by numerous blue-green algae.

Under natural conditions nitrogen lost from ecosystems by denitrification, volatilization, leaching, erosion, wind blown aerosols, and transportation out of the system is balanced by biological fixation and other sources. Both chemically and biologically, terrestrial and aquatic ecosystems constitute a dynamic equilibrium system in which a change in one phase affects the other.

Box 11.1 Nitrogen Fixation

1. Fixation of nitrogen first requires the activation of molecular nitrogen by spliting it in two atoms of free nitrogen ($N_2 \rightarrow 2N$). It is an energy-requiring step in biological nitrogen fixation that consumes 160 C/mole.

2. The combination of free nitrogen with hydrogen to form ammonia ($N \rightarrow NH_3$) releases 13 C/mole, resulting in a net energy input for nitrogen fixation of 147 C/mole.

3. Except for photosynthetic types, all nitrogen-fixing organisms require an external supply of carbon compounds as an energy source to affect this endothermic reaction.

4. Nitrogen fixation is regulated by two enzymes, **nitrogenase** (**Moffat** 1990) and **hydrogenase.**

5. Significance of this fixation is its low-energy requirement. This situation is in contrast to the extremely high temperatures and pressures required for industrial fixation (*i.e.,* a temperature of 400°C and pressure of 200 atmosphere in **Haber process** in which nitrogen is combined with hydrogen to from ammonia.

6. Biological nitrogen fixation is estimated at 54×10^6 metric tons/yr (In contrast to 7.6×10^6 metric tons/yr by electrochemical and photochemical nitrogen fixation and 30×10^6 metric tons annual nitrogen fixation by synthetic means).

7. Biological nitrogen fixation also uses great amount of energy, namely between 12 and 36 ATP molecules for each N_2 molecules fixed.

8. Latest researches are attempting for transfer of nitrogen-fixing genes (the 'nif') from prokaryotes to eukaryotes. The incorporation of 'nif' into rice by means of genetic engineering is the most ambitious subject. In the soil nitrogen fixing bacteria *Klebsiella pneumoniae,* about 17 genes are involved in the nitrogen fixation system.

9. Purple bacteria *Rhodopsendomonas capsulata* can grow with dinitrogen gas as the sole nitrogenous source under anaerobic conditions with light as energy source.

10. In case of **symbiotic nitrogen fixation,** species of root-nodule bacteria, *Rhizobium* are highly host-specific to particular species of leguminous plants (peas, clover, beans, etc.). *These bacteria penetrate root hairs of the legumes, presumably by lysing the cell wall and membrane and then reproduce.* The root hair responds to this invasion by a differential growth, resulting in an enlargement, the *nodule,* in which nitrogen fixation occurs.

11. Symbiotic nitrogen fixation also occurs in non-leguminous plants such as alder (*Alnus*), bayberry and wax myrtles (*Myrica*) and sweat-fern (*Comptonia peregrina*) and buckethron (*Rhamnus*). These species are often found in nutrient poor soils that harbour actinomycete bacteria of the genus *Frankia* in nodules that develop on their roots.

12. Economically significant symbiotic nitrogen fixation also occurs in aquatic ecosystems. Thus, the water ferm *Azolla* along with its blue-green algae symbiont *Anabaena* is grown extensively in subtropical or tropical rice fields in China as a fertilizer crop. Symbiosis of diatom (*Rhizosolenia catracanei* and *R. imbricata*) and bacteria (*Richelia intracellularis*) in the pelagic

zone in the eastern Pacific ocean form another example of symbiotic nitrogen fixation in aquatic ecosystem.

13. Among the nonsymbiotic and free-living nitrogen fixers are bacteria of both aerobic and anaerobic types and blue green algae. **Aerobic** nitrogen-fixing bacteria, such as *Azobacter*, are widely distributed in soils as well as fresh and marine waters; the same is true for anaerobic forms, such as *Clostridium.*

14. Four strains of bacteria from fresh water and mud are the first **anaerobic** and cellulolytic bacteria known to use cellulose as an energy source for nitrogen fixation (**Leschine** *et al.,* 1988).

15. Cyanobacteria of genus *Anabaena, Gleotrichia* and *Oscillatoria* (= *Trichodesmium*) account for high rate of nitrogen fixation in aquatic ecosystems.

C. Sedimentary Cycles

Mineral elements required by living organisms are obtained initially from inorganic sources. Available forms occur as salts dissolved in soil water or lakes, streams, and seas. The **mineral cycle** varies from one element to another, but essentially it consists of two phase; the **salt-solution phase** and the **rock phase**. Mineral salts come directly from the earth's crust by weathering. The soluble salts then enter the water-cycle. With water they move through the soil to streams and lakes and eventually reach the sea, where they remain indefinitely. Other salts are returned to the earth's crust through sedimentation. They become incorporated into salt beds, silts and limestones; after weathering they again enter the cycle.

Plants and animals, the living components of the ecosystems, fulfil their mineral requirements from mineral solutions in their environments. Other animals acquire the bulk of their minerals from plants and animals they consume. After the death of living organisms the minerals are returned to the soil and water through the action of the organisms and process of decay.

There are different kinds of sedimentary or mineral cycles, depending on the kinds of elements, but following two cycles are very significant for a ecosystem:

11.5 SULPHUR CYCLE

Sulphur, like nitrogen, is an essential part of protein and amino acids and is characteristic of organic compounds. It exists in a number of states–elemental sulphur, (S), sulphides, sulphur monoxide, sulphite and sulphates. Of these three are important in nature : elemental sulphur, sulphides and sulphates.

The sulphur cycle (Fig. 11.6) is both sedimentary and gaseous (*i.e.,* it includes gaseous phase and sedimentary phase). The sedimentary phase of sulphur cycle is long-termed and in it sulphur is tied up in organic and inorganic deposits. From these deposits, it is released by weathering and decomposition, and is carried to terrestrial and aquatic ecosystems in a salt solution. Atmospheric (gaseous) phase of sulphur-cycle is less pronounced and it permits circulation on a global scale.

Sulphur enters the atmosphere from several sources—the combustion of fossil fuels, volcanic eruption, the surface of the oceans and gases released by decomposition. Initially sulphur enters the atmosphere as hydrogen sulphide, H_2S, which quickly oxidizes into another volatile form, sulphur dioxide, SO_2. Atmospheric sulphur dioxide, soluble in water, is carried back to earth in rainwater as weak sulphuric acid, H_2SO_4. Whatever its source, sulphur in a soluble form, mostly as inorganic sulphate ($SO_4^{=}$) is absorbed through plant roots, where it is incorporated into certain organic molecules, such as some amino acids (*e.g.,* cystine) and proteins. From the producers the sulphur in amino acids is transferred to the consumer animals, with excess being excreted in the faeces.

Excretion and death carry sulphur in living material back to the soil and to the bottoms of ponds, lakes, and seas where the organic material is acted upon by bacteria of detritus food chain.

Within the detritus food chain, the sulphydryl group (–SH) of amino acids (*e.g.*, L-cysteine) is separated from the rest of the molecule as hydrogen sulphide (H_2S) by most decomposing fungi (*e.g.*, *Aspergillus, Neurospora*) and bacteria as a normal part of the degradation of proteins. In an aerobic environment, the hydrogen sulphide is oxidized to sulphate by bacteria specially adapted to perform this conversion:

$$H_2S + 2O_2 \rightarrow SO_4^{=} + 2H^+$$

The sulphate produced then can be reused by the autotrophs. In anaerobic environments, such as bottom of certain lakes, it is impossible to oxidize sulphide by this means, because the process of oxidation requires oxygen. But if infrared radiation is present in these environments, there are photosynthetic bacteria that can use it to manufacture carbohydrates and oxidize sulphide either to elemental sulphur or to sulphate:

$6CO_2 + 12H_2S + hv \rightarrow C_6H_{12}O_6 + 6H_2O + 12S$ (Green-sulphur bacteria, *Chlorobium*)

$6CO_2 + 12H_2O + 3H_2S + hv \rightarrow C_6H_{12}O_6 + 6H_2O + 3SO_4^{=} + 6H^+$ (Purple-sulphur bacteria, *Chromatimum*)

Elemental sulphur can also be utilized by other bacteria to form sulphate. If oxygen is present, the reaction is quite rapid.

$$2S + 3O_2 + 2H_2O \rightarrow 2SO_4^{=} + 4H^+$$

Under anaerobic conditions, elemental sulphur can still be oxidized to sulphate by certain bacteria if nitrate is present:

$$6NO_3^- + 5S + 2CaCO_3 \rightarrow 3SO_4^{=} + 2\ CaSO_4 + 2CO_2 + 3N_2$$

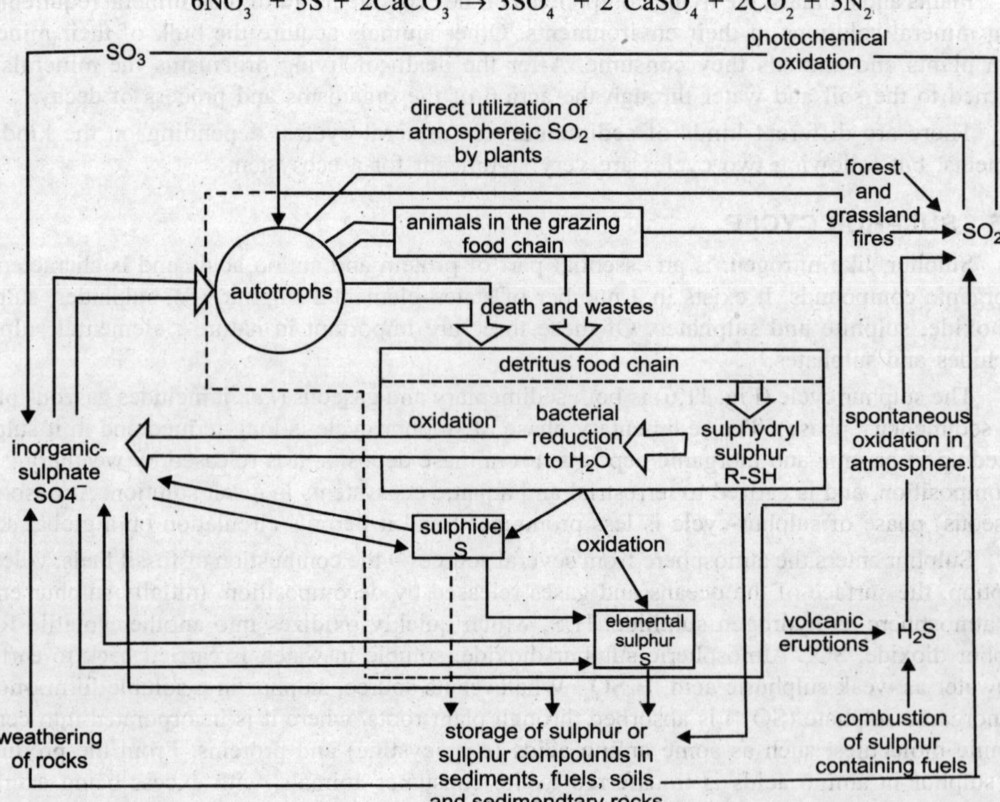

Fig. 11.6. Sulphur cycle.

None of these bacterial reactions is unidirectional; under certain conditions, sulphate can also be reduced either to sulphide or to elemental sulphur by bacteria. This series of reactions operating within the organic phase of the sulphur cycle provides a rather finely tuned mechanism for regulating the availability of sulphur to autotrophs.

The sulphur is removed from the organic phase in the form of elemental sulphur which is insoluble and accumulates in sediments. If iron is present in the sediment, it can combine with sulphide to form iron sulphides, all of which are highly insoluble:

$$Fe^{++} + S^{=} \rightarrow FeS$$

Fe(ionic) + 2S(ionic) → FeS_2 (ferrous sulphide or pyrite)

FeS_2 is highly insoluble under neutral and alkaline conditions and is firmly held in mud and wet soil. Some ferrous sulphide is contained in sedimentary rocks overlying coal deposits. Exposed to the air in deep and surface mining, the ferrous sulphide oxidizes and in the presence of water produces ferrous sulphate and sulphuric acid:

$$2FeS_2 + 7O_2 + 2H_2O \rightarrow 2FeSO_4 + 2H_2SO_4$$

$$12FeSO_4 + 3O_2 + 6H_2O \rightarrow 4Fe_2(SO_4)_3 + 4Fe(OH)_3$$

In this manner sulphur in pyrite rocks, suddenly exposed to weathering by man, discharges heavy slugs of sulphur, sulphuric acid, ferric sulphate and ferrous hydroxide into aquatic ecosystems. These compounds destroy aquatic life and cause acidic water.

11.6 PHOSPHORUS CYCLE

Phosphorus cycle has no atmospheric phase (Fig. 11.7). It occurs naturally in environment as phosphate (PO_4^-, or one of its analogues, HPO_4^- or $H_2PO_4^-$), either as soluble inorganic

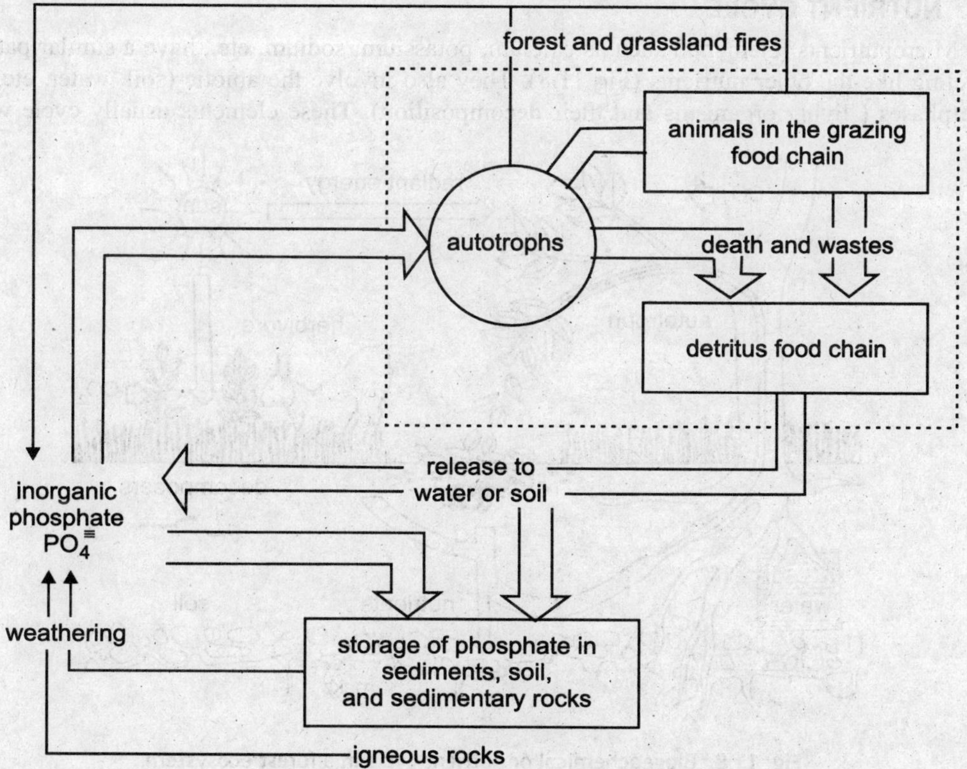

Fig.11.7. Phosphorus cycle.

phosphate ions, as soluble organic phosphate (*i.e.,* as a part of a soluble organic molecule), as particulate phosphate (*i.e.,* as part of an insoluble organic or inorganic molecules) or as mineral phosphate (*i.e.,* as part of a mineral grain as found in a rock or sediment). The ultimate source of phosphate in the ecosystem is crystalline rocks. As these are eroded and weathered, phosphate is made available to living organisms, generally as ionic phosphate. This is introduced into autotrophic plants through their roots, where it is incorporated into living tissues. From autotrophs, it is passed along the grazing food chain in the same fashion as nitrogen and sulphur, with excess phosphate being excreted in the faeces. An extreme example of faecal phosphate is the tremendous guano deposits built up by birds on the desert west coast of South America. Phosphates can also be released as particulate matter from forest and grassland fires.

In the detritus food chain, as large organic molecules containing phosphate are degraded, the phosphate is liberated as inorganic ionic phosphate. In this form it can be immediately taken up by autotrophs, or it can be incorporated into a sediment particle, either in the soil of a terrestrial ecosystem or in a sediment of an aquatic ecosystem. The sedimentary phase of phosphorous cycle remains comparatively slow than the organic phase.

Thus, the geochemical cycles of different chemical substances are closed : the atoms are used over and over again. To keep the cycles going does not require new matter but it does require **energy**, for the energy cycle is not a closed one. Further, the patterns of flow, both of energy and of chemical substances, are of great significance. The simpler patterns involve energy, as the sources of energy are external to the ecosystem, and flow is unidirectional through it. Chemical substances, on the other hand, are finite and have their origin inside the ecosystem, thus, they must continuously cycle within the system.

11.7 NUTRIENT CYCLE

Micronutrients (elements) such as calcium, potassium, sodium, etc., have a similar pathway of cycling like the other nutrients (Fig 11.8). They also involve the abiotic (soil, water, etc.) and biotic phases (living organisms and their decomposition). These elements usually cycle within

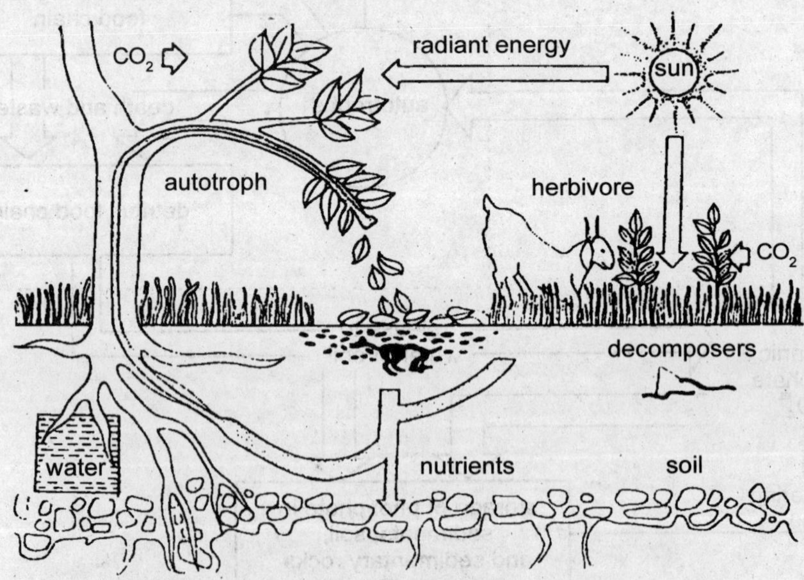

Fig. 11.8. Biogeochemical or nutrient cycle in a forest ecosystem.

the same ecosystem and in small quantities globally. Usually, the rocks are their reserve pools and the weathering process enriches the soil and water with these elements. Though, the dust of these nutrients may be blown away by the wind and carried across ecosystems and some of their amount is lost in the runoff to lakes and oceans and then to deep sediments. For example, in a tropical rain forest on average, more than 80% of total store of these micronutrients is found in the soil, with progressively less in vegetation, litter and the fauna. Though, the concentration of trace elements tends to increase along the food chains in the sequence such as soil – plants – herbivores – carnivores and litter – detritivores. Thus, concentration of iron in a rain forest were 6 ppm in soil, 218 ppm in vegetation, 3054 ppm in herbivores and 3387 ppm in the carnivores. Many of these micronutrients become toxic to plants in high concentration.

The nutrient concentration of the leaf litter of some plants have been tabulated in Table 11.2.

Table 11.2. Concentration of nutrients (in %) in the leaf litter of some plant species (Source : **Vyas** and **Galley**, 1975 ; **Rajvanshi** and **Gupta**, 1985).

	Species	Calcium (Ca)	Magnesium (Mg)	Potassium (K)	Sodium (Na)	Nitrogen (N)	Phosphorus (P)
1.	*Terminalia arjuna*	4.42	0.37	0.54	0.06	0.85	0.24
2.	*Terminalia tomentosa*	3.73	0.54	0.40	0.06	0.93	0.52
3.	*Tectona grandis*	2.54	0.25	0.24	0.20	0.78	0.18
4.	*Diospyros melanoxylon*	1.90–2.10	0.51–0.69	0.57–0.90	0.05–0.10	0.56–0.61	0.008–0.02
5.	*Butea monosperma*	1.85	0.56	0.75	0.06	2.26	0.25
6.	*Shorea robusta*	1.03–1.58	0.36–0.48	0.26–0.59	0.05–0.08	0.71–0.81	0.41–0.50
7.	*Dalbergia sissoo*	1.02	0.96	0.25	0.07	2.23	0.025

QUESTIONS

Long Answer Questions

1. Give an account of the biogeochemical cycles and their importance.

2. Describe different types of biogeochemical cycles. Discuss their significance. (*Purvanchal 1999*)

3. What are biogeochemical cycles? Describe the following cycles in nature.

(*i*) Oxygen cycle; (*ii*) Carbon cycle; (*iii*) Nitrogen cycles; (*iv*) Phosphorus cycle.

4. Describe with the help of examples, the different types of biogeochemical cycles.

5. What do you understand by biogeochemical cycles? Discuss carbon cycle. (*CCSU Meerat 2005*)

6. Write an essay on biogeochemical cycles.

7. Give an account of hydrological cycles.

8. Explain water cycle in the ecosystem. (*Kerala 2001*)

9. Describe carbon cycle in nature. (*Kanpur 1998; Kurukshetra 1999*)

10. What do you mean by biogeochemical cycles? Give an account of nitrogen cycle in nature with the help of a diagram.

11. What is the role of nitrogen fixing bacteria in nitrogen cells.

12. Describe phosphorus cycle in environment.

13. What is meant by a sedimentary biogeochemical cycles. Describe the phosphorus cycle in nature.

14. Describe nitrogen and phosphorus biogeochemical cycles with suitable examples.

15. Give an account of carbon cycle in nature. What is the relationship between oxygen, water and carbon cycle in nature.

Short Answer Questions

1. Write a short note on carbon cycle. (*Avadh 1999*)

2. Write short notes on the following:

(*i*) Biogeochemical cycle. (*Allahabad 1999*)

(*ii*) Water cycle

(*iii*) Biological water cycle (*Himachal 1999*)

(*iv*) Nutrient cycle (*Himachal 1999*)

(*v*) Nitrogen cycle

3. Differentiate between the following:

1. Gaseous and sedimentary types biogeochemical cycles. (*Kurukshetra 1999*)

2. Nitrogen cycle and phosphorus cycle.

Very Short Answer Questions

1. What is biogeochemistry?

2. Define biogeochemical cycle.

3. How many types of biogeochemical cycles exist in nature?

4. Define the carbon cycle?

5. What is sedimentary cycle?

6. Define the nitrogen cycle.

7. Give an example of sedimentary cycle.

8. Sulphur is found in nature in gaseous as well as in sedimentary form. Yet its cycle is classified under sedimentary cycle. Why?

9. Which biogeochemical cycle has no gaseous phase?

10. What is nitrification?

11. Why fertilizers contain nitrogen?

12. When does global concentration of carbon dioxide remain lowest?

13. When is the concentration of carbon dioxide remain highest?

14. What is the reason of fluctuation in atmospheric CO_2 with change in season of Northern Hemisphere?

15. Nitrogen is almost 80 percent in atmosphere. Yet it often acts as a limiting factor. Explain.

Fill in the Blanks

1. The cyclic movements of chemical elements of the biosphere between the organism and the environment are refered to as ....

2. *Azotobacter* and *Clostridium* are free-living .... of the soil.

3. .... can oxidize ammonia in to nitrite (NO_2).

4. *Pseudomonas* is an example of .... bacteria.

5. Symbiotic .... occurs in non-leguminous plants such as *Alnus*.

6. Sulphus cycle is both .... and ....

7. .... cycle has no atmospheric phase.

Multiple Choice Questions

1. The main nitrogen reservoir in the biosphere is the
   (*a*) atmosphere (*b*) ocean
   (*c*) organism (*d*) rocks
2. Which of the following kingdoms contain species that can fix nitrogen?
   (*a*) monera (*b*) protista
   (*c*) fungi (*d*) plantae
3. Which of the following categories of organisms can convert ammonium or nitrate into amino acids?
   (*a*) producers (*b*) primary consumers
   (*c*) primary detritivores (*d*) scavengers
4. The most common mutalistic association between a nitrogen-fixing organism and a plant is between bacteria of genus *Rhizobium* and plants in the
   (*a*) composite family (*b*) rose family
   (*c*) legume family (*d*) grass family
5. The largest reservoir of sulphur in the biosphere is the
   (*a*) atmosphere (*b*) ocean
   (*c*) organism (*d*) rocks
6. The phosphorus cycle is unusual in that it is entirely
   (*a*) within aquatic systems (*b*) within terrestrial ecosystems
   (*c*) sedimentary (*d*) gaseous

ANSWERS

Very Short Answer Questions

1. (1) Biogeochemistry is the branch of science that focuses on movement of elements or natrients through organisms and their environment; (2) The study of natural cycles of elements and their movement through biological and geological compartments.
2. The cyclic pathway through which chemical elements travel from the environment into the organisms and from organisms into the environemnt regularly and continuously are called biochemical cycle.
3. Two types: (a) Gaseous biogeochemical cycles and (b) Sedimentary biogeochemical cycles.
4. Carbon cycle involves movement of carbon, C, between the atmosphere, hydrosphere, and biosphere and transformations (such as photosynthesis and respiration) between its different chemical forms.
5. Sedimentary cycle is any global cycle where geological processes, such as the weathering of existing rock, erosion, and sedimentation, dominate or originate the cycle; calcium and potassium exemplify this type of cycle.
6. (1) Nitrogen cycle is description or model depicting the movement of nitrogen-containing compounds as they cycle among the atmosphere, soil, and living matter or (2) Movement of nitrogen, N, among the almosphere, biosphere, and hydrosphere, including transformations between different chemical forms.
7. Phosphorus cycle, sulphus cycle, calcium cycle or potassium cycle.
8. The residence time of sulphur in the atmosphere is very small and its main reserve pool is found in the soil. Soil is considered under sedimentary cycle.
9. Phosphorus cycle.
10. The process of conversion of ammonia into nitrates is called nitrification.
11. There is intense competition between plants for available nitrogen. The fertilizers containing nitrogen supply nitrogen to the plants to fulfil their requirements.

12. During northern hemisphere's summers.

13. During northerh hemishers's winters.

14. There is more land in the Northern Hemispere than in Southern Hemisphere, therefore, more vegetation. The vegetation has a maximum photosynthetic activity during the summer, reducing the global amount of CO_2. During the winter vegetation respire more CO_2 than it uses for photosynthesis, causing a global increase in the CO_2 gas.

15. Plants cannot fix almospheric nitrogen themselves. All nitrogen available to plants comes from nitrogen fixing bacteria and in aquatic environments throgh cyanobacteria. There is acute competition between plants for this available nitrogen. Thus, nitrogen acts as a limiting factor.

Fill in the Blanks

1. Biogeochemical cycles **2.** Nitrogen-fixing **3.** *Nitrosomonas* **4.** Denitrifying
5. Nitrogen fixation **6.** Sedimentary, gaseous **7.** Phosphorus.

Multiple Choice Questions

1. (*a*) **2.** (*a*) **3.** (*a*) **4.** (*c*)
5. (*d*) **6.** (*c*).

12

Aquatic Ecosystems: Freshwater Communities

The **habitat approach**, describes the distinctive features of the major habitats and their subdivisions, how they are organized, the organisms present in each and the ecological role of these organisms in that region (*i.e.,* the identity of the major producers, consumers and decomposers). According to **Elton** (1949) habitat is definable as an area possessing uniformity of physiography, vegetation, climate or any other quality the investigator assumes important. Four major habitats can be distinguished: **marine**, **estuarine**, **freshwater** and **terrestrial**. No plant or animal is found in all four major habitats and, indeed, no animal or plant is found everywhere within any one of these. According to recent ecological trend only two major habitats or more specifically ecosystems namely **aquatic ecosystems** and **terrestrial ecosystems**, are recognized.

12.1 AQUATIC ECOSYSTEMS

Liquid water covers about three quarters of the earth's surface either as oceans or as freshwater. Virtually all these waters contain life in one form or other; hence, aquatic ecosystems would be important for their sheer volume, if for nothing else. But aquatic ecosystems are historically the source of life on earth. Even now, tiny single-celled marine plants are the main source of the earth's oxygen supply, because most photosynthetic oxygen is derived from these plants. In addition, aquatic ecosystems are simpler in many ways than terrestrial ecosystems. The reason for this is that the omnipresent environmental factor which sets the tone for all aquatic systems, regardless of their biotic complexity, is water. It is the medium within which all aspects of the ecosystem coexist, both living and non-living; it is the source of all nutrients for aquatic life, including the gaseous nutrients such as oxygen and carbon dioxide; it is the medium by which organic and inorganic wastes and sediments are distributed throughout the ecosystem. The amount of light energy reaching the community is determined by the way in which light is absorbed by water; the heat properties of water determine much of the circulation patterns within the ecosystem and have a major control over the structure of the aquatic community that exists in any one place.

Subdivisions of Aquatic Ecosystems

Global aquatic system fall into two broad classes definable by salinity, or amount of material dissolved in water—the **freshwater ecosystems** and **salt-water ecosystems**. The latter may include inland brackish water, as well as marine and estuarine habitats. In fact, estuary represents a transitional zone between a river and the sea and it contains dissolved solid content intermediate between those of fresh and marine waters. Freshwater ecosystems are conveniently divided into two groups—**lentic**, **standing** or **still water habitats** and **lotic** or **running water habitats**. Both can be considered on an environmental gradient. The lotic habitat follows a gradient from springs to mountain brooks to streams to rivers. The lentic habitat involves a gradient from lakes to ponds to bogs, swamps and marshes. The study of the physical, chemical and biological properties and feature of freshwaters is referred to as **limnology,** litreally *the study of marshes.*

12.2 PHYSICO-CHEMICAL NATURE OF FRESHWATER

Freshwater ecosystems cover about 2 per cent of the Earth's surface, an area of about 2·5 million km^2. Although referred to as *freshwater,* a number of lakes contain a variety of chemicals referred to collectively as **salts.** The volume of inland saline waters (containing more than 3 grams of salts per litre) is estimated to be 104×10^3 km^3 in contrast to a total volume of 125×10^3 km^3 for nonsaline waters (**Burgis** and **Morris** 1987).

Box 12.1

With a volume of 23×10^3 km^3, a maximum depth of 1620 m (deepest lake in world) and mean depth of 740 m, **Lake Baikal** in Siberia, Russia, contains about one-fifth of the earth's freshwater. The surface area of **Caspian Sea**, an inland salt lake between Europe and Asia, of 371×10^3 km^2 makes it *largest single freshwater lake,* though it does contain saline water. At 83×10^3 km^2 **Lake Superior** of Canada, is the *second largest lake on earth*, and the largest in North America.

Although the **Nile River** at 6,693 km is *longest river* on earth; the **Amazon**, the next longest at 6299 km, carries the *largest volume of water,* delivering an average of 643×10^6 m^3 of water every hour to the Atlantic ocean, 60 times the flow of the Nile. Others of the longest rivers in world, in descending order, are the **Chang** in China (5525 km), **Ob-Irtysh** in the Russia (5569 km; 3680 km long Ob + 5569 km long Irtysh), **Huang** (or **Yellow**) in China (4671 km) and the **Congo** in Africa (4664 km).

The freshwater of both kinds—lentic and lotic, has low percentage of dissolved salts and is subjected to the influence of a wide array of physical and chemical factors. The rise and fall of these factors very frequently affect the fauna, altering their number and diversity. Some of the important factors of freshwater environment are following:

1. Pressure, Density and Buoyancy

The pressure imposed on a lake-dwelling organism is the weight of the column of water above it plus the weight of the atmosphere. In all freshwater environments maximum pressure is much less than in the ocean, and organisms appear to adjust to them readily. The absence of animal life from deep water is ordinarily a consequence of low oxygen supply, or low temperature, rather than pressure.

The density of water varies inversely with temperature and directly with the concentration of dissolved substances. Water is most dense at approximately 4^{0}C and becomes progressively less dense as it cooled below +4^{0}C. Ice also expands markedly the colder it gets. It is because the coldest water is at the surface in winter that ice forms there, rather than at the bottom of a lake. In summer, the coldest waters of deep lakes are at the bottom. Dissolved salts increase the density of water; the density of most inland water-bodies is much less than that of the ocean. However, when great evaporation occurs in a lake having no outlet, the lake may come to contain a higher percentage of salts (*i.e.,* hypersalinity) than the ocean. The few species capable of living in these very salty lakes (*e.g.,* Great Basin of USA) include some algae and Protozoa, the brine shrimp *Artemia gracilis* and the immature stages of two brine flies, *Ephydra gracilis* and *E. hians.*

According to the **law of Archimedes** the buoyancy of an object is equal to the weight of the water it displaces. Buoyancy varies with the density of water, and is influenced by the factors that affect density. Viscosity, the measure of the internal friction of water, varies inversely with temperature and also influences buoyancy. Most aquatic organisms keep stations by swimming movements or have special adaptations to decrease the specific gravity of the body and take advantage of any turbulence in the water. For this purpose freshwater aquatic organisms have some swimming adaptations, clinging organs (in case of animals of lotic habitat) or following adaptations—absorption of large amount of water to form jelly-like tissues; storage of gas or air bubbles within the body; formation of light-weight fat deposits within the body or oil droplets within the cell; increase of surface area in proportion to body mass, which increase frictional resistance. When an organism so equipped dies, the special mechanisms quickly cease to function, and it sinks to the bottom.

2. Temperature

The unique thermal properties of water are best demonstrated by freshwater environment. Diurnal and seasonal variations of temperatures are very much evident in these environments than in marine environments. A diurnal variation range of 4.8–5.0°C has been recorded by **Sreenivasan** (1964) in a tropical pond, with an average depth of 3.0 metres. In shallow water habitats, difference between day and night temperatures remain more conspicuous. For example, in a polluted moat with an average depth of 1.5 metres, the lowest night time temperature was 26.6°C, the highest day time temperature was 32.°C with a variation of 5.4°C (**Sreenivasan**, 1964). However, the Kodaikanal lake in South India showed a diurnal variation of only 2.8°C (**Sreenivasan**, 1964). Flowing lotic waters of streams and rivers lack such wide fluctuations in temperature.

Thermal stratification. The lentic water of lakes and ponds undergo thermal stratification phenomenon according to the seasons. Thermal stratification has been reported most frequently in the lakes of tropical countries such as Java, Sumatra, etc. In fact, according to their temperature relations, lakes have been classified into three types : (1) **Tropical lakes** in which surface temperatures are always maintained above 4°C, (2) **Temperate lakes** in which surface temperature vary above and below 4°C, and (3) **Polar lakes** in which surface temperatures never go above 4°C. The seasonally regulated thermal stratification of lentic habitats has a significant influence on their inhabitant biotic communities. Decreasing temperatures often cause a fall in metabolism, resulting in a lower rate of food consumption. The extremes of lower and higher temperature have lethal effects on the aquatic organisms. So fluctuations in temperature of aquatic media regulate the breeding periods, initiate hibernation, gonadial activitation and a number of other biological phenomena such as thermally oriented migration, etc., of freshwater biota. On the basis of their ability to tolerate thermal variations, most freshwater organisms are **stenothermic** with a narrow range temperature tolerance, but some are **euthermic** with a wide range of temperature tolerance. For example, the stenothermic oligochaets includes **steno-minimothermal forms** (narrow range of temperature, *e.g., Aeolosoma, Megascolex mauritii*), **steno-maximothermal** (*e.g., Dero limosa*) or **stene-optimothermal forms** (*e.g., Branchiodrilus semperi and B. menoni*) (**Sitaramiah**, 1966).

3. Light

Light influences freshwater ecosystems greatly. The freshwaters often have a lot of suspended material. While affording protection to the light sensitive species, these substances more often obstruct the light that normally reaches the water. The degree of such obstruction of light influence the productivity of the freshwater ecosystems. A shallow lake receives light to its very bottom resulting in an abundant growth of vegetation both phytoplankton and rooted vascular plants. These plants in living or dead states form nice food for consumers of grazing food chain or organisms of detritus food chain, respectively. The running water contains little plant or animal plankton not due to the lack of sunlight but because of the action of the currents in washing it away. Further, light controls the orientation and changes in position of attached species and their nature of growth and it also causes the diurnal migration of planktonic species of freshwater.

4. Oxygen

Chemically pure water is biologically uninhabitable and all freshwaters contain an array of chemical substances. The oxygen, which is a most essential chemical component of life processes, remains dissolved in freshwaters. The aquatic environments which remain in close proximity with atmosphere contain an abundance of oxygen, that reaches the water either by direct diffusion or by movements of water such as wave action or water circulation. Lotic (moving) water of streams and rivers often have a high percentage of oxygen. Aquatic plants supply water with oxygen that is formed as a product of photosynthesis. Rooted vegetation of shallow water zones and floating phytoplankton of open waters also produce oxygen. The amount of photosynthetically produced oxygen remain high at warm temperatures and at greater light intensities.

The oxygen level in a tropical pond exhibits diurnal variation—it remains at peak between 14.00 and 17.00 hours of day (**Sreenivasan**, 1964). Oxygen contents of a freshwater body are depleted in numerous ways. Primarily oxygen is utilised in the respiration of organisms and decomposition of dead organisms in the aquatic environment. While photosynthesis remains restricted to the surface layer of water containing phytoplankton and exposed regions of rooted vascular plants, respiration and decomposition occur at all levels. In stagnant pools with a lot of decaying vegetation oxygen content often reaches a stage of complete depletion. The reduction in dissolved oxygen is magnified by the release of many gases as end products of decomposition or by the mixing up of waters of low oxygen content reaching the habitat as an inflow. Aquatic animals with very few exception, *i.e.,* those that breath air, utilize the oxygen dissolved in water. Certain freshwater inhabitants such as many anaerobic bacteria and insect larvae of chironomids perform anaerobiosis and requires no oxygen.

5. Carbon Dioxide

Aquatic vegetation and phytoplankton require carbon dioxide for photosynthetic activity. The carbon dioxide of freshwater environments is produced as the end product of respiration and of decomposition. Carbon dioxide also diffuses directly from the atmosphere and is readily dissolved in water to result in carbonic acid (H_2CO_3) which affects the pH of water. It is also present in the freshwater as carbonates and bicarbonates of calcium, magnesium and other minerals. The growing plants and lime deposition bacteria and other animals may cause a depletion in carbon dioxide resources. Photosynthesis is the major cause for its drain. The high saturation levels of O_2 and CO_2 have been found to have toxic effects on aquatic biota.

6. Other Gases

Streams and lakes contaminated by sewage and stagnant pools with decaying vegetation show an abundance of the gas, hydrogen sulphide which is a decomposition product. This gas is highly toxic to living organisms and results into complete denudation of bottom fauna. Methane and carbon monoxide are other toxic gases which are the products of decomposition. Nitrogen, hydrogen, sulphur dioxide and ammonia are some of the other gases which are found dissolved in freshwaters.

7. Dissolved Salts and Salinity

Freshwater being efficient solvent contains many solutes in solution, but even then its salt contents remain under 1/5 per cent than marine water which contains about thirty-five parts per thousand (‰) dissolved salts. Different dissolved salts reach the water by erosion, inflow and decay of aquatic forms. Dissolved substances have peculiar significance for floating aquatic vegetation and phytoplankton, since these organisms do not depend on the substratum for the supply of nutrients. Compounds of nitrogen, phosphorus and silicon are most important substances found dissolved in freshwater. Nitrates, nitrites and ammonium salts are essential for the food of aquatic vegetation such as algae and water weeds. Nitrate always remains available due to nitrogen cycle occurring between nitrogen fixing bacteria and nitrogen consuming plants. Ammonium salts in excess have a lethal effect on the fauna. Dissolved silicates of freshwater are readily utilized by diatoms and sponges in constructing their body structures such as shell in case of diatoms and spicules in case of sponges.

All freshwater environments also contain small amounts of phosphorus which more often acts as a limiting factor. Utilization of phosphorus by plankton during the periods of abundance may result in a total elimination of other plants that require the element (phosphorus). Many other elements such as calcium, magnesium, manganese, iron, sodium, potassium, sulphur, and zinc are found dissolved in water and influence the fauna variously. Iron being a growth promoting element for plants exists as the compound of oxygen (ferrous oxide) or sulphur (ferrous sulphide) in different freshwater bodies. Its influence is often modified by the pH of water. Calcium is an essential element for plants. The abundance and scarcity of carbonate of calcium determine the faunal composition. Deposition of calcium carbonate in water called **marl** is produced by the activity of plants. External coverings of arthropods and the shell of molluscs and tubes of some worms need calcium carbonate. Snails are found to develop a heavy shell if the waters in which they lived contained excess of calcium. Bryozoans, sponges, and cladocerans prefer an increased calcium content.

Due to low salinity of freshwater, animals face the problem of **osmoregulation**. Because the salt concentration of body fluids of animals remains higher than the freshwater, hence, the water continuously tend to enter the body which should be readily removed. Most aquatic animals (*e.g.,* Protozoa and fishes) have the means to excrete extra amount of water of body by osmoregulation. For this purpose Protozoa employ contractile vacuoles and other multicellular invertebrates and chordates use excretory organs, such as nephridia, kidney, etc.

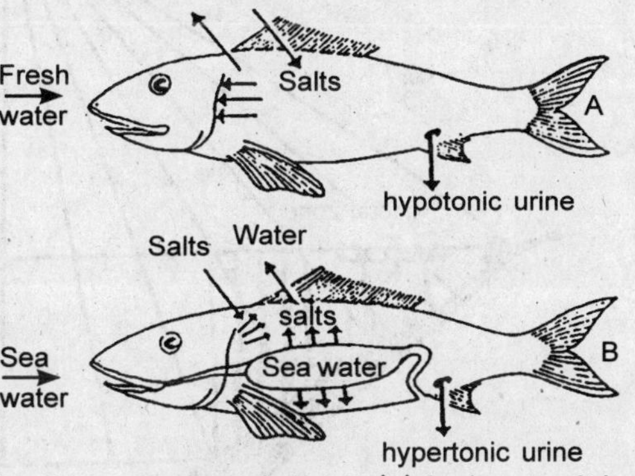

Fig. 12.1. Osmoregulation in (A) fresh water fish and (B) marine fish.

8. Hydrogen Ion Concentration or pH

In freshwater environments pH is a determining factor for the biota by becoming a limiting factor. The pH value of different freshwater bodies may fluctuate seasonally and annually. The pH of surface waters and deeper waters exhibit marked differences. **Sreenivasan** (1968) has reported a marked pH variation of 2.2 units between surface water and deeper water in the Sandynulla reservoir in the Nilgiris (India). Though pH range is species specific, yet lower aquatic forms in general showed little reaction to alterations in pH, while higher aquatic organisms (*e.g.,* fishes) responded quickly to little pH variations.

12.3 LENTIC ECOSYSTEMS

Lentic (*lenis,* calm; standing water) ecosystems include all standing water (freshwater) habitats such as lakes, ponds, marshes, swamps, bogs, meadows, moats, etc.

1. Lakes and Ponds

Lakes are inland depressions containing standing water (lentic habitat). They may vary in size from small ponds of less than a hectare to large seas covering thousands of square kilometres. They may range in depth from a few centimetres to over 1666 metres. **Ponds**, however, are considered as small bodies of standing water so shallow that rooted plants can grow over most of the bottom. Most ponds and lakes have outlet streams and both are more or less temporary features on the landscape because their filling is inevitable.

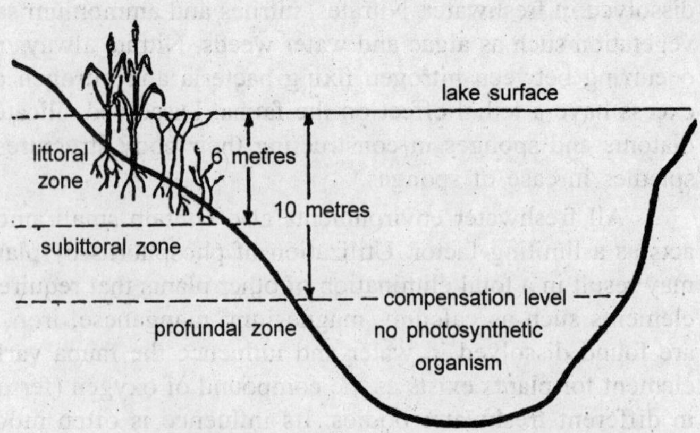

Fig. 12.2. Different zones of a deep freshwater lake.

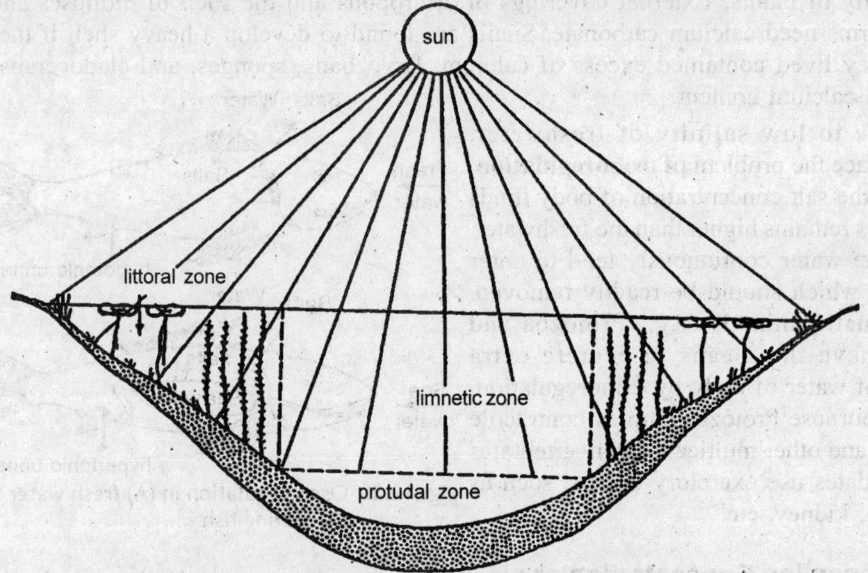

Fig. 12.3. Different zones of a freshwater pond.

The aquatic habitats of lake and pond remain vertically stratified in relation to light intensity, wave length absorption, hydrostatic pressure, temperature, etc. In a lake, for example there are three well recognized horizontal strata namely ; **(i)** Shallow water near the shore forms the **littoral zone**. It contains upper warm and oxygen rich circulating water layer which is called **epilimnion**. The littoral zone includes rooted vegetation. **(ii) Sublittoral zone** extends from rooted vegetation to the non-circulating cold water with poor oxygen zone, *i.e.,* **hypolimnion**. **(iii) Limnetic zone** is the open water zone away from the shore. It is the zone upto the depth of effective light penetration where rate of photosynthesis is equal to the rate of respiration. **(iv) Profundal zone** is the deep-water area beneath limnetic zone and beyond the depth of effective light penetration. **(v) Abyssal zone** is found only in deep lakes, since it begins at about 2,000 metres from the surface. Ponds have little vertical stratification. In them littoral zone is larger than the limnetic zone and profundal zone. In a small pond the limnetic and profundal zone are not found. Lakes, thus, differ from ponds in having relatively larger limnetic zone and profundal zone than littoral zone.

Further, lentic water of a lake or pond is also classified on the basis of the depth of light penetration enabling photosynthesis into **trophogenic zone** (includes littoral plus sublittoral zones), and a **tropholytic zone** (upper part of profundal zone). The former is often distinguished by abundant plant growth and dependent fauna, while the latter denotes a general absence of vegetation and harbours mostly saprobes. In between the two zones is the **compensation level** which forms a boundary between two zones. It exhibits perfect equilibrium between respiration and photosynthesis.

Thermal Stratification of a Lake

In temperate latitudes, the lake ecosystem shows marked seasonal temperate changes which may be described as follows:

1. Winter. During winter the coldest water form ice at 0°C (32°F) and floats at the lake's surface. The water at increasing depth below the ice is progressively warmer and more dense. The heaviest water, at the bottom of lake, has a temperature of 4°C and throughout winter the water remains relatively stable.

2. Spring. Following the melting of ice, the surface water slowly warms to 4°C. At this point the water column is nearly **isothermal**, *i.e.,* all the water is of uniform temperature and density. Hence, the strong spring winds cause marked stirring, which results in a complete mixing of water, dissolved oxygen and nutrients from the lake surface to lake bottom, a phenomenon known as the **spring overturn** or **spring turnover**. As the spring progresses, however, the surface water naturally become warmer and lighter than the water at lower levels. As a result, the lake becomes **thermally stratified** into following three zones (Fig 12.4).

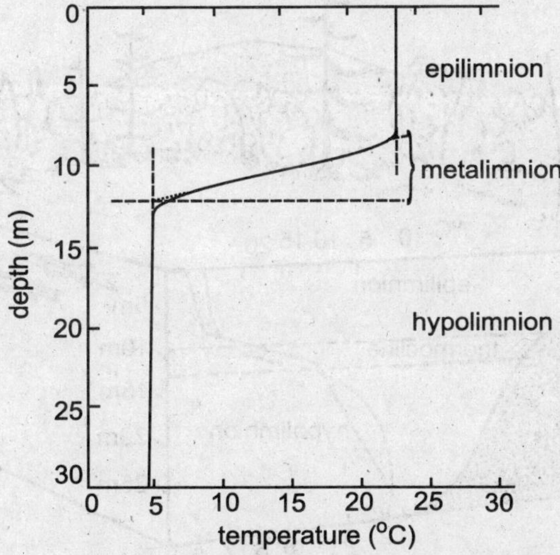

Fig. 12.4. Typical thermal stratification of a lake into three strata.

(i) Epilimnion. The upper stratum, which usually has the highest dissolved oxygen concentration and is characterized by a temperature gradient of less than 1°C per metre of depth, is the **epilimnion** (which litreally means the "upon lake" or the "upper lake"). Epilimnion contains more or less uniformly warm, circulating and fully turbulent water.

(ii) Hypolimnion. The lower stratum of water characterized by a temperature gradient of less than 1°C per metre of depth is the **hypolimnion** (which litreally means "lake below" or "bottom lake"). Hypolimnion contains more dense, cooler and relatively quite water.

(iii) Metalimnion. It is the transitional stratum of marked thermal change between the epilimnion and hypolimnion. The middle layer of the lake, characterized by a temperature gradient of more than 1°C per metre of depth is called the **thermocline**. The term 'thermocline' refers to the plane or surface of maximum rate of decrease of temperature in the metalimnion. The depth at which the thermocline forms is not fixed, being determined by *the degree of solar heating, the transparency of the water,* and the *morphological features of the lake.* However, wherever the thermocline forms, it effectively divides the lake into two distinct layers, the upper epilimnion and the lower hypolimnion. These two divisions of lake have great effects on the ecology of lake.

Thus, epilimnion is well lit and oxygenated and with sufficient high temperature to promote algal productivity and hence to support zooplankton and fish. When nutrients are in ample supply, algae growth is accelerated and bloom may occur. By contrast, hypolimnion is cold, dark and becomes progressively deoxygenated as the decaying remains of organisms (*i.e.,* detritus) sinks down from the epilimnion. Conditions in the hypolimnion may become so extreme that anoxia follows after which the biological productivity becomes minimum. Under less extreme conditions the epilimentic material provides an energy source for benthic invertebrates. The sinking of dead algae and zooplankton from the epilimnion not only contributes to the potential deoxygenation of the hypolimnion but also prevents immediate recycling of nutrients. Nutrient depletion may become so high that algal growth is limited.

3. Summer. Except that the lake is very clear and allows photosynthesis, the hypolimnion often becomes depleted of oxygen in summer. This happens because of the **biological oxygen demand (BOD)** of bacterial decomposers, the reduced photosynthetic activity and the minimal mixing with upper waters as a result of density differences.

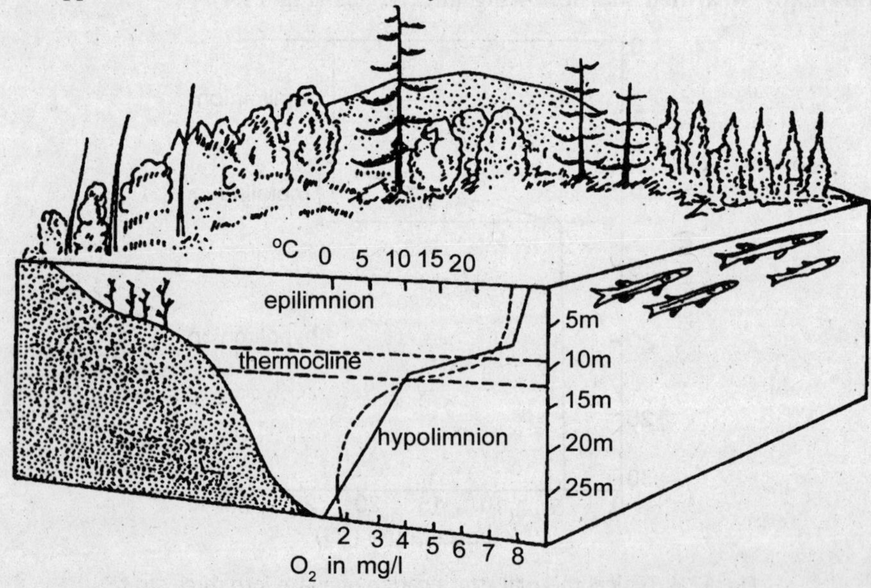

Fig. 12.5. Distribution of oxygen and temperature in mid-summer in a lake ecosystem. Depletion of oxygen in hyplimnion is depicted in the figure.

4. Autumn. At the end of thermal stratification the surface water of epilimnion slowly cools as a result of *conduction, evaporation* and *convection.* Now the lake attains temperature uniformity from top to bottom. The lake's water at this stage, since, contains uniform density, it becomes well mixed by wind and wave action, a phenomenon known as the **fall overturn** or **fall turnover.** As a result, the nutrients, dissolved oxygen and plankton becomes uniformly distributed.

5. Winter. As the winter approaches, the lakes get colder until the water attains a uniform temperature of 40°C at which it has maximal density. As the surface water cools below this temperature it becomes lighter. Ultimately surface water may freeze at 0°C. During the winter season, the ice cover forms on the surface and in such ice bound lakes there exists an **inverse stratification** of water temperature, with the coldest water (ice) at the surface and the warmest water (4°C) on the bottom.

Lakes undergoing complete circulation in spring and autumn separated by thermal summer stratification and winter inverse stratification are called **dimictic lakes.**

Kinds of lakes. Different classifications of lakes on the basis of physical factors, productivity, etc., exist. Based on temperature **Hutchinson** (1957) classified lakes into **dimictic**, **monomictic** and **polymictic** lakes. The dimictic lakes exhibit two overturns every year, while monomictic lakes may be **cold monomictic** and **warm monomictic**, the former being characterized by a circulation only during summer, while the latter has a complete circulation in winter as well. Polymictic lakes present circulations throughout the year. Based on the humic acid contents the lakes of world have been classified into **clear water lakes** and **brown water lakes**, brown water contains high humus content. Clear water lakes may be divided into two types—the **oligotrophic type**, whose water is poor in nutritive plant material and shows nearly equal distribution of oxygen during summer and winter months and whose mud bottom contains little organic material; and the **eutrophic type**, which is rich in nutrients. At greater depth below the thermocline in summer, eutrophic lakes show a considerable reduction in oxygen content and their mud bottom is composed of typical muck. Eutrophic lakes are polluted lakes and support the growth of the microorganisms. With regards to above features, an intermediate type of lake is called **mesotrophic lake.**

Special types of lakes. There also occur the following five types of lakes:

1. Dystrophic lakes. Lakes that receive large amounts of organic matter from surrounding watersheds, particularly in the form of humic materials that stain water brown, are called **dystrophic lakes.** Such lakes have low planktonic production but have highly productive littoral zones, particularly those that develop bog flora. Littoral vegetation adds both dissolved and particulate organic matter to the dystrophic lakes.

2. Deep ancient lakes. Lake Baikal in Russia is the most famous of ancient lakes. It is the deepest lake in the world. This lake is often nicknamed as **"Australia of Freshwater"**, since it contains various unique animals. It contains about 384 species of arthropods, 98 per cent of which are **endemic** (*i.e.,* found no where else). Likewise, 81 per cent of its 36 species of fish are also endemic.

3. Desert salt lakes. These lakes exist in sedimentary drainages in arid climates where evaporation exceeds precipitation, *e.g.,* Great Salt Lake of Utah, USA. Biotic community of these lakes comprises only those species which can tolerate high salinity, *e.g., Artemia* (Brine shrimp).

4. Desert alkali lakes. These lakes occur in igneous drainages in arid climates, *e.g.,* Pyramid lake, Navada, USA. High pH and concentration of carbonates are the characteristic features of these lakes.

5. Volcanic lakes. Acid or alkaline lakes associated with volcanic regions are called **volcanic lakes**, *e.g.,* Japanese and Philippine lakes. In these lakes exist the extreme chemical conditions.

Types of ponds. Ponds too are of the following four types:

(i) Flood plain ponds. These are ponds which are found in stream flood plains. These are quite productive habitats due to accumulation of organic matter in them.

(ii) Temporary ponds. These ponds remain dry for most part of the year but support a unique community. Organisms in such ponds must be able to survive in a dormant stage during dry periods or be able to move in and out of ponds.

(iii) Artificial ponds. These ponds are the result of damming of a stream or basin by humans. They are artificially managed habitats.

(iv) Beaver ponds. Ponds constructed by beavers (=amphibious rodents) are called beaver ponds. The beaver often does not construct pond but live in holes in the bank of streams and, thus, becomes essentially the stream animal.

Origin of lakes and ponds. Lakes and ponds arise in many ways. For example, North American lakes were formed by glacial erosions and deposition and a combination of the two. Glacial abrasions of slopes in high mountain valleys engraved basins, which filled with water from rain and melting snow to produce **tarns,** *i.e.,* small mountain lakes. Retreating valley glaciers left behind crescent-shaped ridges of rock debris, which dammed up water behind them. Lakes are also formed by the deposition of silt, driftwood, and other debris in the beds of slow-flowing streams. Craters of extinct volcanoes may fill with water and landslides can block off streams and valleys to form new lakes and ponds. Further humans intentionally create artificial lakes by damming rivers and streams for power, irrigation, and water storage, or by constructing small ponds and marshes for water, fishing and wild life. Human-made lakes are often called **impoundments**. They may be a 'close' type or 'open' type depending on the discharge of water into it. Sandynulla is a closed impoundment, while Pykara is an open impoundment (**Sreenivasan,** 1957). Impoundments differ from natural lakes in thermal and oxygen variations and are often characterized by a low percentage of bottom dwellers or benthos.

Ponds occur in most regions of adequate rainfall. Temporary ponds, which are often shallow and quick drying contain a characteristic lentic environment. Permanent ponds contain water throughout the year, while flood-plain ponds are formed when a stream shifts its position, leaving the formed bed isolated as a body of standing water.

Biotic Communities of Lakes and Ponds

Different organisms of the lentic environment can be ecologically classified based on whether they are dependent on the substratum or free from it. Organisms depending on the substratum are called **pedonic forms** and those that are free from it are the **limnetic forms**. Further, the aquatic organisms may also be classified into following groups depending upon their sizes and habits:

1. Neuston. These are unattached organisms that live at the air-water interface. They may include floating plants such as duckweed, as well as many types of animals. Animals that spend their lives on top of the air-water interface, such as water striders, are termed **epineuston (supraneuston)**, while others, including insects such as diving beetles and back swimmers, which spend most of their time on the underside of the air-water interface and obtain much of their food from within water, are termed **hyponeuston (infraneuston)**.

2. Plankton. These are forms which are found in all aquatic ecosystems except for fast-moving rivers. They are small plants and animals whose powers of self-locomotion are so limited that they cannot overcome currents. Thus, their distribution is controlled largely by the currents in their ecosystems. Most plankton (phytoplankton and zooplankton) can move a bit, however, either, to control their vertical distribution or to seize prey. Certain animal plankton or zooplankton are exceedingly active and move relatively great distances considering their small size, but they are so small that their range is still controlled largely by currents and such planktons are also known as **nektoplankton** (Table 12.1).

Table 12.1. Classification of plankton based on various criteria.

Criteria	Terminology	Characteristics
1. Quality	1. Phytoplankton (i) Phytoplankton proper (ii) Saproplankton (iii) Zooplankton	Plant plankton Photosynthetic plankton Non-photosynthetic plankton (Bacteria and fungi) Animal plankton

Criteria	Terminology	Characteristics
2. Size	1. Macroplankton (mesoplankton) 2. Net plankton 3. Nanoplankton (microplankton)	Larger units of plankton, visible to the nacked eye Plankton secured by the plankton net Very minute plankton, not secured by plankton net'
3. Environmental	1. Limnoplankton distribution 2. Rheoplankton (potamoplankton) 3. Heleoplankton 4. Haliplankton 5. Hy palmy roplankton	Lake plankton Running water plankton Pond plankton Salt-water plankton Brackish-water plankton
4. Life history	1. Euplankton (holoplankton) 2. Meroplankton	Permanent plankton Temporary planktonic phase
5. Origin	1. Autogenic plankton 2. Allogenic plankton	Plankton produced locally Plankton introduced from other localities
6. Content	1. Euplankton 2. Pseudoplankton	True plankton Debris mingled in plankton

3. **Nekton.** Nektonic animals are swimmers and are found in all aquatic systems except for fastest-moving rivers. In order to overcome currents, these animals are relatively large and powerful; they range in size from the swimming insects of quiet water, which may be only about 2 mm long, to the largest animal that has ever lived on earth, the blue whale.

4. **Benthos.** The benthos include the organisms living at the bottom of the water mass. They occur virtually in all aquatic ecosystems. The bentho organisms living above the sediment-water interface are termed **benthic epifauna** and those living in the sediment itself are termed **infauna.**

Box 12.2

Hutchinson (1967) has divided the benthic community based on the niche they occupy into following groups:

1. **Rhizobenthos** are rooted in the substratum but well extended into the aquatic phase.
2. **Haptobenthos or periphyton** are attached to a immersed solid surface.
3. **Herpobenthos** are growing or moving through mud.
4. **Psammon** are growing or moving through sand.
5. **Endobenthos** are boring into solid substrate.

Neustons of lentic aquatic environment. The water surface of lake or pond contains certain free-floating hydrophytes such as *Wolffia, Lemna, Spirodella, Azolla, Salvinia, Pistia* and *Eichhornia*. These plants remain in contact with water and air, but not soil.

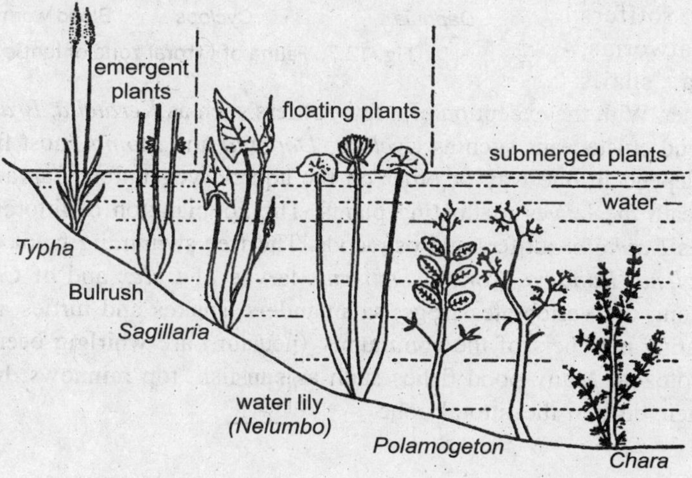

Fig. 12.6. Flora of the lentic habitat.

Biota of littoral zone. Lentic aquatic life is most prolific in the littoral zone. The littoral zone of a lake remains rich in pedonic flora especially up to the depth to which effective light penetration is possible facilitating the growth of rooted vegetation. At the shore proper are the **emergent vegetation** which remains firmly rooted in the shore substratum but their tops with thin chlorophyll bearing regions are exposed. Certain rooted emergent plant species of littoral zone are *Ranunculus, Monochoria, Cyperus* and *Rumex.* Interspersed with these plants are the cattails (*Typha*), bulrushes (*Scirpus*), arrowheads (*Sagittaria*) and pickerelweeds. Slightly deeper are the rooted plants with floating leaves such as the water lilies–*Nymphaea, Nelumbo, Aponogeton, Trapa* and *Marsilea.* Still deeper in the Jittoral zone of lake are the agile thin-stemrned water weeds, rooted but totally submerged. Such a submerged vegetation includes plants such as *Elodea, Vallisneria, Myriophyllum, Isoetes, Hydrilla, Chara, Potamogeton, etc.*

The phytoplankton of littoral zone of lake is chiefly composed of diatoms (*e.g., Navicula, Cyclotella*), blue green algae (*e.g., Microcystis, Oscillatoria*), green algae (*e.g., Cosmarium, Staurastrum*) and holophytic flagellates. **Sreenivasan** (1970) has observed the abundance of marine dinophycean flagellate *Ceratium hirudinella* in some impoundments of India.

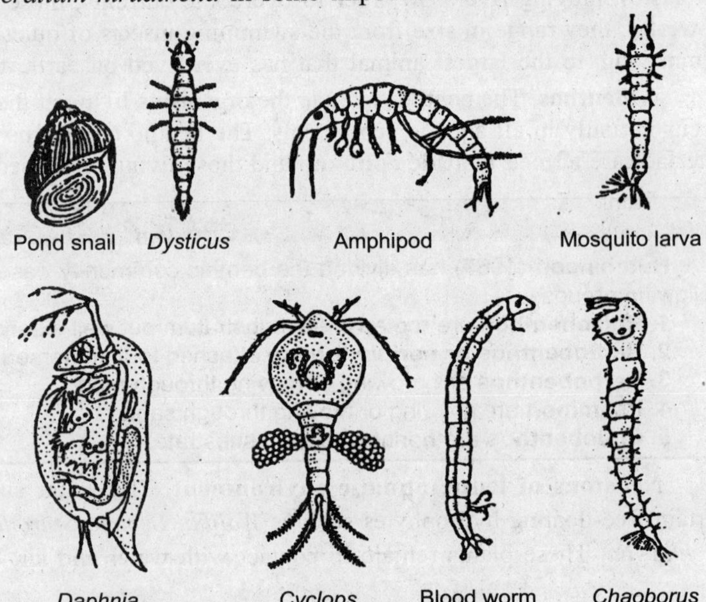

Fig. 12.7. Fauna of littoral zone of lentic habitat.

The littoral zone also contains great concentration of animals, which remain distributed in recognizable communities. In or on the bottom are various dragonfly nymphs, crayfish, isopods, worms, snails and clams. Other animals live in or on plants and other objects projecting up from the bottom. These include protozoans such as *Vorticella, Stentor*; larvae of *Dytiscus, Laccotrophes*, leach-like *Glossophonia*, climbing dragonfly, damsel fly nymphs, rotifers (*Brachionus*), flatworms, Bryozoa, Hydra, snails (*Lymnaea*) and other. With the exception of a few rotifers such as *Keratella, Brachionus, Diurella,* and *Trichocerca* and crustaceans such as *Cyclops, Daphnia* and *Moina,* most freshwater animals of lentic habitat depend in one way or other on the aquatic vegetation. Larvae of *Chironomous* are found underneath the leaves of floating plants. The zooplankton of littoral zone consists of water fleas such as *Daphnia,* rotifers and ostracods. The free swimming fauna (Nekton) includes *Paramecium, Euglena, Ranatra, Corixa, Dytiscus,* larvae of *Culex* and of *Chaoborus, Gerris, Gyrinus,* etc. Among the vertebrates frogs, salamanders, snakes and turtles are the nektons of littoral zone. Floating members of the community (neuston) are whirlgig beetles, water spiders and numerous Protozoa. Many pond fishes such as sun-fish, top minnows, bass, pike and gar spend much of their time in the littoral zone.

Biota of limnetic zone. Limnetic zone is the region of rapid variation, with the water level, temperature and oxygen content varying from time to time. In certain Himalayan lakes which have a glacial origin, change in water level is a conspicuous phenomenon, the onset of summer bringing forth a rapid and sudden rise in water level. Various protozoans which are capable of encystment during adverse ecological conditions, tardigrades such as *Macrobiotus,* rotifers such as *Rotaria, Philodina,* copepods, snails and frogs occupy the limnetic zone. Sedentary and slow moving forms are excluded from this zone because of predators and lack of permanent substratum for attachment. Many microscopic plants (*Volvox, Euglena*) and fishes also occupy this zone. The limnetic zone has autotrophs in abundance.

Biota of profundal zone. The deep (profundal) life consists of bacteria, fungi, clams, blood worms (larvae of midges), annelids and other small animals capable of surviving in a region of little light and low oxygen. In the profundal zone, autotrophs cannot produce food, and the main source of energy is detritus that rains out of the limnetic zone. All the organisms of this zone are heterotrophs, either as detritus feeders or as carnivores. Most of our largest lake fishes inhabit the dark waters of hypolimnion for most of their lives. In addition to detritus from the limnetic zone, food washed into lakes by rivers can settle out and serve as food for profoundal animals.

Benthos of lake bottom. The lake bottom in young lakes may be of the original rock; in older lakes, it will have been covered with sediment to form a uniform substrate of mud or sand. The benthos community includes several species of insect larvae, including those of small mosquito such as midges, burrowing mayflies, clams, snails, and tubeworms.

12.4 LOTIC ECOSYSTEMS

Running-or moving-water or lotic ecosystems include rivers, streams, and related environments. They are remarkably variable, ranging in size from Ganga, Yamuna, Hindon, Kali Nadi, Sutlej, Gomti, Brahmaputra, Narmada, Mahanadi, Kaveri, Krishna, Godavari, etc., to the trickle of a small spring. They vary from raging torrents and waterfalls to rivers whose flow is so smooth as to be almost unnoticeable. A given river varies considerably over its length, as it changes from a mountain brook to a large river. This is most noticeable in the abiotic factors of the environment, but all features of the ecosystem vary in response. The most obvious factors which distinguish a stream or river from a lake are the **unidirectional** and **relatively rapid flow** of the water (Box 12.3).

Box 12.3

Freshwater streams such as **springs, rivulets, creeks, brooks**, etc., and rivers change over their course being narrow, shallow and relatively rapid to become increasingly broad, deep and slow moving. The degree of each feature depends on the streambed fall—that is, the vertical drop over its length. **Waterfalls** are not common features of lotic ecosystems.

Characteristics of Lotic Ecosystems

Moving waters differ in the three major aspects from lakes and ponds : current is a controlling and limiting factor; land-water interchange is great because of the small size and depth of moving water systems as compared with lakes and oxygen is almost always in abundant supply except when there is pollution. Temperature extremes tend to be greater than in standing water. Besides these, the most distinctive features of moving water ecosystems are those related to their motion, *i.e.,* the **rate of flow** and the **stream velocity**. The rate of flow or discharge refers to the volume water passing a given observation point during a specific unit of time; it is measured in units such as m^3/sec., ft^3/sec., or acre-feet/sec., one acre-foot is equivalent to a volume of water 1 acre in area by 1 foot deep. It tends to increase steadily growing down stream, as tributaries join with the main river. The **velocity of flow** is the speed at which the water moves, and is measured in m./sec., ft./sec., or mi./hr. Velocity is variable, but it also tends to increase downstream with increasing discharge. A factor that may be even more important to the biotic community is the **turbulence**, or the irregularity of the motion of the water particles. Perfectly even flow, in which water particles move parallel to one another, is called **laminar flow**. It is contrasted with **turbulent flow**, in which the movement of water particles is highly irregular. In highly turbulent water the erosive power is great, the sheer forces at the water-sediment interface at the bottom of the river are powerful, and the amount of oxygen incorporated into the water is very high. In streams with laminar flow the erosive power is lower, the sheer forces at the bottom of the stream are lower, and relatively less oxygen is incorporated into the water than turbulent waters.

Zonation and Communities of Stream

Streams and rivers have a **longitudinal zonation** or **profile** (Fig. 12.8) in contrast to horizontal in lakes and ponds. **Illies** and **Botosaneanu** (1963) have recognized two main subdivisions of a river course:

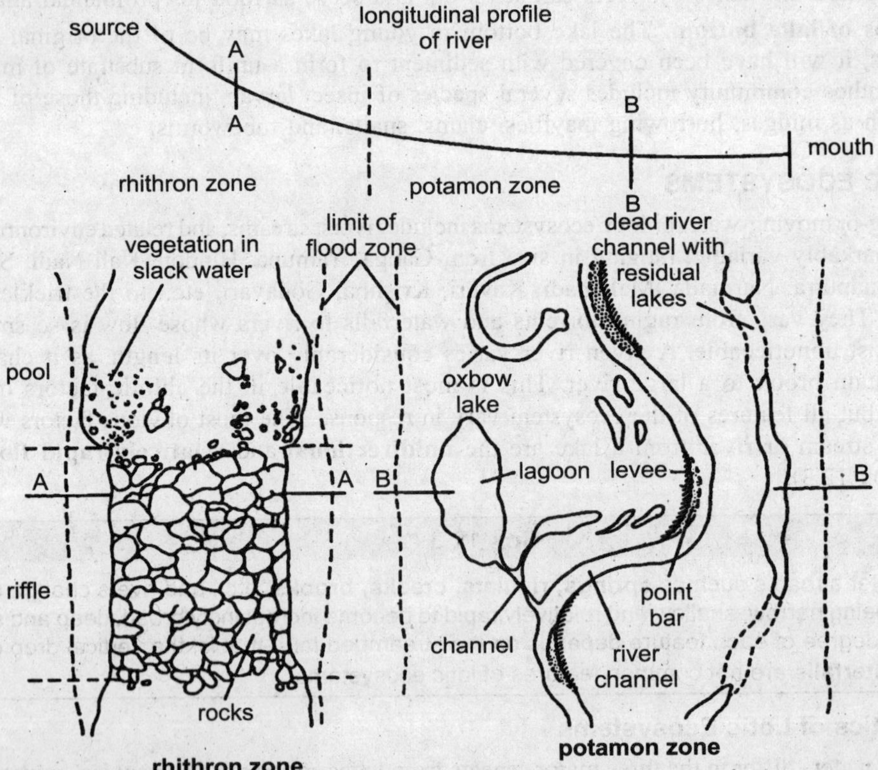

Fig. 12.8. Diagrammatic representation of a river course showing rhithron (A–A) and potamon (B–B) zone.

A. Rhithron, the steep and torrential upper course;

B. Potamon, the flat, slow-following lower course.

A. Rhithron zone. It is characterized by the presence of (1) steep, narrow, shallow and turbulent **riffles** or **rapids** and (2) flatter, wider, deeper and quite reaches, termed pools (Fig. 12.8). Intermediate areas of moderate current often found in larger streams and rivers are termed **runs.** Riffles have high, turbulent flow, coarse bottom of boulders, rocks or pebbles and limited attached vegetation.

Box 12.4

Rhithron is characterized by fast or swiftly flowing streams, having velocity of flow 50 cm/sec or higher. At this velocity the current will remove all particles less than 5 mm in diametre and will leave behind a stony bottom.

A riffle is a stretch of shallow, choppy water caused by shoal, rocky obstruction lying beneath the surface of a river. Pools have slow flow, bottom of somewhat finer material and some rooted vegetation. Low temperatures and high value of dissolved oxygen are significant characteristics of the rhithron zone. The environment of pools differ in chemistry, intensity of current and depth. Pools are sites of decomposition. They are the catch basins of organic materials, for here the velocity of the current is reduced enough to allow a part of the load to settle out. Pools are sites for free carbon dioxide production during summer and fall.

Communities of rhithron. Riffles are the sites of primary production in the stream. Here aufwuchs or periphyton assume dominance. The **aufwuchs** (German word proposed by **Ruttner,** 1953) consists chiefly of diatoms (called **microphyta**), blue-green algae, green algae (*e.g., Cladophora;* they are called **mesophyta**) and water moss. Extensive strands of algae grow over rocks and rubble on the streambed and form a slippery covering. Many small algae species are **epiphytes** growing on top of or in among other algae.

The outstanding feature of much of this algae growth is its **ephemeral nature**. Scouring action of water and debris it carries tears away larger growth, epiphytes and all, and sends the algae downstream. As a result, there is a constant contribution from upstream to the downstream sequence.

Rhithron also contains zooplankton, nektons and benthos. The **zooplankton** of rhithron zone mainly consists of ciliate protozoans (*e.g., Vorticella*), rotifers (*e.g., Asplancha, Brachionus, Keratella, Philodina*) and crustaceans (*e.g., Daphnia, Bosmina, Cyclops, Diaptomus, Gammarus pulex*).

There are also many insect species adapted to life on the stream bottom (*i.e,* **benthos**) where they shelter among or cling to the rubbles or bedrocks. The following insects are commonly encountered in the rhithron:

1. Nymphs of mayflies (*e.g., Baetis rhodani, Rhithrogena, Ecdyonurus, Ephemerella, Caenis, Ameletus, Epeorus;* order **Ephermeroptera**).

2. Larvae of caddisflies (*e.g., Hydropsyche, Rhyacophila, Philopotamus, Drusus, Limnephilus, Agraylea;* order **Trichoptera**).

3. Larvae of stoneflies (*e.g., Nemura, Protonemura, Perlodes, Diura, Arcynopteryx, Perla, Isoperla, Leuctra, Brachyptera;* order **Plecoptera**).

4. Larvae of two-winged flies (*e.g., Simulium* (black fly), *Chironomus riparius* (bloodworm), *Antocha* and *Liponeura*; order **Diptera**).

5. Larvae of dragon flies (*e.g., Libellula*; order **Odonata**).

Some worms, *e.g.,* flatworms (*Dugesia*); roundworms (*Dolichodorus*); annelid worms (*Tubifex,* leeches as *Haemopis, Glossiphonia,* and *Erpoddela*); molluscs (*e.g., Bithynia, Dreissena*) and crustacea (*e.g., Gammarus*) may also occur among the benthic forms.

The nektons of the rhithron zone include water bugs (order **Hemiptera** *e.g., Notonecta, Micronecta, Gerris, Corixa*); beetles (order **Coleoptera**, *e.g., Dytiscus*) and fish. The resident fish species in rhithron zones are entirely **rheophilic** and belongs to two main groups: Firstly, there are those species which live among rocks of the bottom and are found only in riffles. These fish are of small size and are adapted to grip or cling to the substrate (*e.g., Chiloglanis, Astroblephus, Glyptothorax, Pseudecheneis*). Other fish species such as *Mastacembelus* have pointed snout and eel like shape that enable them to get into the holes in the rocky bottom. The second group of fish are those species which are adapted to swim fast as to resist the current and some of them even move against it (*e.g., Barbus* (Punti), *Salmo, Barilius, Schizothorax,* etc.). Because of the severity of habitat, the diversity of resident species tend to be low.

Adaptations of rhithron community. Animals of rhithron zone of stream exhibit following adaptations for living in swift current.

1. Streamlined bodies. Almost all stream animals possess more or less egg-shaped body which offers minimum resistance to water flowing over it. Good examples of hillstream fish with streamlined bodies are trout (*Salmo*), *Barilius* and *Schizothorax* species.

2. Flattened bodies. Many stream animals have flattened bodies which enable them to find shelter under stones, *e.g.,* insects such as *Epeorus, Rhithrogena* (mayflies, order Ephemeroptera) and fishes such as *Noemacheilus* (Cobitidae) and *Pseudecheneis* (Sisoridae).

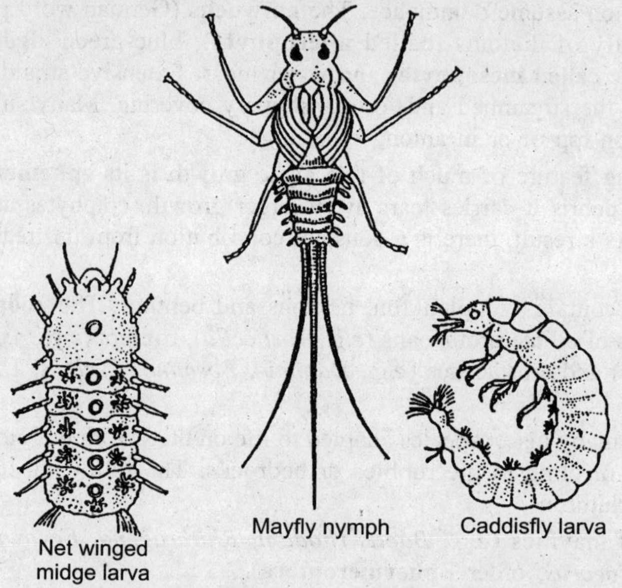

Fig. 12.9. Some inhabitants of lotic environment.

3. Hooks and suckers. Hooks and suckers enable riffle animals to grip smooth surfaces. For example, *Bibicocephala* and *Liponeura* have a row of ventral suckers and *Baletora* (Indian loach which has one large ventral sucker; **Hora,** 1947).

4. Adhesive apparatus. Some hillstream fishes, *e.g., Glyptothorax, Garra* and *Pseudecheneis* possess a well developed adhesive apparatus on the ventral surface.

5. Sticky undersurfaces. Some snails and flatworms are able to adhere to the substrate by their sticky undersurfaces.

6. Positive rheotaxis. Stream animals orient themselves upstream and if capable of swimming movement move against the current (*rheo* current; *taxis,* arrangement). Hellgramites (an insect larva) are large and covered with spines. Their size makes them somewhat harder for current to sweep them away, while the spines help in holding the larvae in places between stones.

7. Positive thigmotaxis. Many stream animals have an inherent behaviour pattern to cling close to a surface or keep the body in close contact with the substratum (*thigmo,* touch or contact). Almost all the larvae of mayflies and stoneflies and many fishes exhibit this type of behaviour.

8. Permanent attachment. Some organisms of the lotic environment remain permanently attached to a firm substrate such as stone moss, log moss and leaf moss. Green algae *Cladophora,* encrusting diatoms such as *Nitzchia, Gomphonema, Melosira* and some aquatic mosses (*Fontinalis*) are good examples.

B. Potamon zone. The potamon zone of a river is the flat, slow-flowing lower course. It has both longitudinal and lateral zonation (Fig 11.8). Longitudinally, there is a repetition of differing habitats associated with the **meanders** (= tortuous or winding courses) and the **channel.** Laterally, there is the distinction between the **main channel** and its **floodplain.** The floodplain is normally an area of relatively flat land flanking the main channel. The channel at each band is deeper by the outer, concave bank where the current is fastest, whereas the inner convex bank consists of a sandy or muddy point bar. The plain itself contains many types of water body, some of which retain water throughout the inter-flood period. Because of deposition of silt, such features show a succession from open **lagoon,** through vegetation-lined **pools** and heavily vegetated **swamps** to dry land.

Thus, potamon zone is environmentally more complex and differ from the rhithron zone in the following features:

1. Geomorphology of the channel;
2. Physico-chemical qualities of water;
3. Nature of substratum;
4. Biota, including floating, emergent and submerged vegetation;
5. High bacterial density.

Communities of potamon. With increasing temperatures, decreasing current, depleting oxygen and accumulating bottom silt of potamon the resident communities include populations of the following groups:

1. Plankton. The occurrence of plankton is closely related to the flow conditions. During the floods planktonic organisms are rare whereas during the dry seasons algae blooms may develop within the lentic waters of the plain and also in the main channel. In short rivers plankton are generally confined to backwaters. In longer rivers the time taken for individual masses of water to travel downstream is sufficient to allow the development of plankton.

(a) Algae and diatoms. On the basis of their sensitivity to the pollutants, algae species of potamon may be categorised into following three groups (**Bilgrami** and **Datta-Munshi** 1985):

(i) Maximum pollution tolerant. This group includes following algae species: *Oscillatoria limosa, Microcystis aeruginosa, Chlorella vulgaris, Stigeoclonium tenue, Ankistrodesmus falcatus, Scenedesmus quadricauda, Synedra ulna, Navicula viridula, Euglena viridis, Phacus caudatus* and *P. viridis.* They are usually confined to the sites of urban and industrial discharge and can be used as **pollution indicators.**

(ii) Moderate pollution tolerant. This group includes following algae species: *Oscillatoria princeps, Phormidium uncinatum, Anabaena circularis, Closterium acerosum, Pediastrum duplex, Scenedesmus dimorphus, Gomphonema parvulum, Cymbella turgida, Navicula cuspidata* and *Synedra acus.*

(b) Zooplankton. Zooplankton of the potamon zone consists of members of protozoa, rotifera, cladocera and copepoda. Forms such as *Filinia longiseta, Brachionus angularis, B. calyciflorus, B. forficula, B. quadridentata, Keratella tropica* and *K. cochlearis* may be present in clean as well as moderately polluted waters. However, forms as *Brachinous rubens, Platiyas polyacanthus* and *Rotatoria rotatoria* are characteristics of polluted part of the potamon zone.

2. Nekton. Some insects (including beetles) and fish form the dominant nekton of potamon. Some animals which are not true nekton such as *Kachuga kachuga* (sail terrapin), *Kachuga tectum* (roofed-terrapin), *Trionyx gangeticus* (tortoise or soft-river terrapin), *Lissemys punctata* (Indian flashelled turtle), *Crocodilus porosus* (Muggar or Marsh crocodile), *Gavialis gangeticus* (Gharial), *Platanista gangetica* (gangetic dolphin) have also been reported from the Ganga and other rivers.

Fish communities of the potamon zone are represented by two main groups: (i) There are fish species adapted to resisting low dissolved oxygen concentration. Their adaptation may be in the form of **morphological** as accessory respiratory organs for using atmospheric oxygen as in *Channa* (*Ophiocephalus* or snakehead), *Clarias* (Mangri), *Erythrinus* and *Notopterus* (Chital); **physiological** as in *Carassius;* or **behavioural** as in cyprinodonts. The same species often have a capacity to tolerate high temperatures. (ii) Other group of fish includes the species which usually live in the main channel and are fast swimmers such as *Mystus seenghala, Labeo rohita* (Rohu, carp), *Catla catla* (Katla, carp), *Cirrhinus mrigala* (Mrigal, carp), and *Wallago attu.*

3. Benthos. The benthos (or bottom-dwelling forms) of the potamon is relatively poor as unstable mud bottoms, heavy situation and seasonal dessication do not favour well settled bottom-living organisms. The benthic forms may include some molluscs (*e.g., Limnaea, Bithynia, Corbiula*), Chironomids, Ephemeroptera, Plecoptera and worms and leeches.

4. Macrophytes. The presence of macrophytes is a characteristic feature of the potamon zone. Most aquatic macrophytes are angiosperms (flowering plants) but aquatic ferns, mosses, liverworts and even the large algae of the Charophyceae may be present.

The free floating macrophytes such as *Salvinia* (water fern), *Pistia* (water lettuce), *Eichhornia crassipes* (water cabbage) may be quite common. In temperate and subtropical parts much smaller free-floating plants such as the duckweed (*Lemna*) may be present in backwaters. Rooted macrophytes may have all or part of their vegetative and sexually reproductive parts above the water or may be completely submerged. Some examples of rooted macrophytes of streams are *Nymphaea* (water lily), *Potamogeton* (pond weed), *Ceratophyllum* (coontail), *Myriophyllum* (milfoil) and *Ranunculus* (water crowfoot). The eel grass (*Zostera*) and widgeon grass (*Ruppia*) are the common rooted macrophytes in estuaries.

12.5 NUTRIENT STATUS OF LAKE

A close relationship exists between land and water ecosystems. Primarily through the hydrological cycle one feeds upon the other. The water that falls on land runs from the surface or moves through the soil to enter streams, springs and eventually lakes. The water carries with it **silt, clay, organic matter** and **nutrients in solution**, all of which enrich aquatic ecosystems. Human activities including road building, logging, mining, construction and agriculture, add an additional heavy load of silt, organic matter and nutrients, especially **nitrogen** and **phosphorus.** Outcomes of all these factors result in *nutrient enrichment* of aquatic systems, called **eutrophication.** The term **eutrophy** (from the Greek "well nourished") mean a condition of being nutrient rich. The opposite of eutrophy is **oligotrophy,** the condition of being nutrient-poor. Both terms were introduced by the German limnologist **C. A. Weber** in 1907, for the development of peat bogs. **E. Naumann** later associated the terms with phytoplankton in lakes: **Eutrophic lakes** found in fertile lowland regions hold high population of phytoplankton; **Oligotrophic lakes,** typical of rock upland regions, contain little plankton.

1. Oligotrophic lakes. These lakes are characterized by a low surface-to-volume ratio; water that is clear and appears blue or blue-green in the sunlight; bottom segments are largely *inorganic* and a *high oxygen concentration through the hypolimnion* (Fig. 12.10). The nutrient content of the water is low; although *nitrogen* may be abundant, *phosphorus* is highly limiting. Low nutrient availability results from a low input of nutrients from external sources. This factor in turn causes a low production or organic matter, particularly phytoplankton. Low organic production results in low rate of decomposition and high oxygen concentration in the hypolimnion. These oxidizing conditions are responsible for low nutrient release from the sediments. The lack of decomposable organic substances result in low bacterial populations and slow rates of microbial metabolism.

Table 12.2. Difference between oligotrophic and eutrophic lakes.

Oligotrophic lake	Eutrophic lake
1. These lakes have changed little since the time of their formation, *i.e.*, geologically they are young.	1. They are geologically old as they have changed a lot since their formation.
2. Poor in nutrients	2. Rich in nutrients, especially of phosphate and nitrate ions.
3. Low primary productivity.	3. High primary productivity.
4. Deep basin.	4. Shallow basin.
5. Clear water.	5. Turbid water with unpleasant taste and odour.
6. Plankton population density is low although number of species may be large (characterized by species of such algae as Chlorophyta, Chyrophyta and Bacillariophyta).	6. Plankton population is denser but with few species (characteristic blue green algae are *Oscillatoria rubescens, Aphanizomenon flosaquae* and *Microcystis aeruginosa*).
7. Algal blooms are formed very rarely.	7. Rich in algal blooms.
8. No severe oxygen depletion; high oxygen demanding species of fish such as trout are present.	8. Dissolved oxygen is low, high oxygen-demanding species are replaced by carp and other organisms.

Although the number of organism in oligotrophic lake may be low but the diversity of species is often high. Fish life is dominated by members of salmon family.

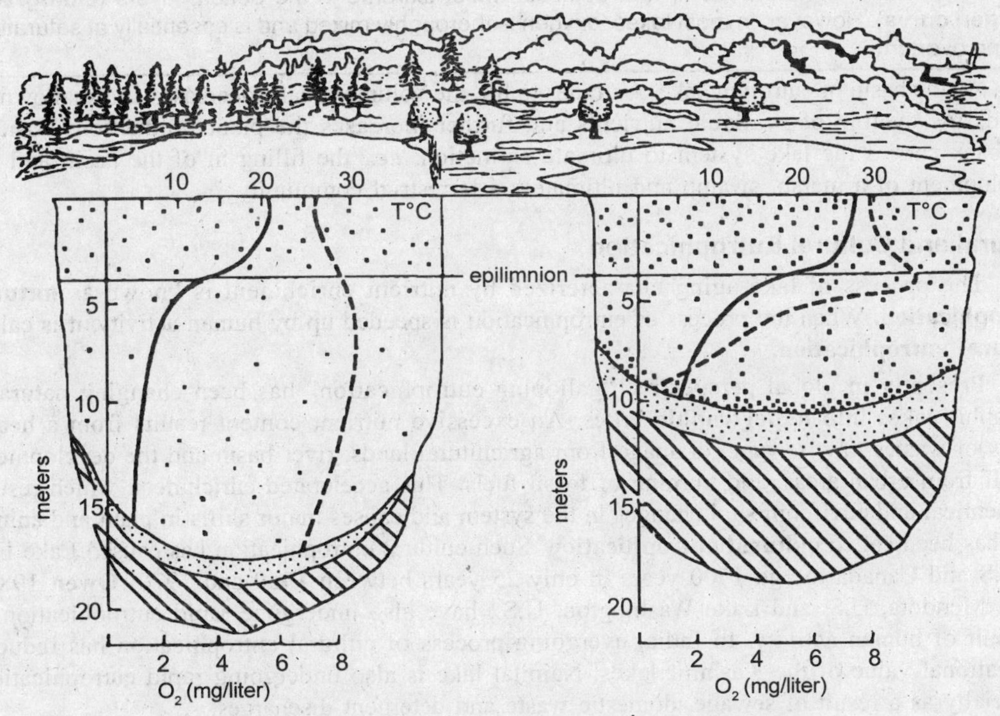

Fig. 12.10. Comparison of oligotrophic and eutrophic lakes.

2. Mesotrophic lake. When moderate amounts of nutrients are added to oligotrophic lakes, they are rapidly taken up and circulated. If nitrogen and carbon are in excess and phosphorus is limiting, the addition of phosphorus would stimulate growth. If nitrogen is limiting, the addition of that element would do the same. Increasing quantities of nutrients are added to the lake, it begins to change from oligotrophic to **mesotrophic** to eutrophic. Mesotrophic lake contains a moderate amount of nutrients. This change has been happening at an increasing rate to clear oligotrophic lakes around the world.

3. Eutrophic lake. A typical eutrophic lake (Fig. 12.10) has a high surface-to-volume ratio, that is, the surface area is large relative to depth. It has an abundance of nutrients, especially *nitrogen* and *phosphorus* that stimulate a heavy growth of algae and other aquatic plants. Increase photosynthetic production leads to increase regeneration of nutrients and organic compounds, stimulating further growth. Phytoplankton becomes concentrated in the upper layer of the water, giving it a murky (or darkened) green appearance. The **turbidity** reduces light penetration and restricts biological productivity to a narrow zone of surface water. Algae, inflowing organic debris and sediment, and the remains of rooted plant accumulate to the bottom, adding to the highly organic sediments. On the bottom of lake bacteria partially convert dead matter into inorganic substances. The activities of these decomposers deplete oxygen supply of the bottom sediments and deep water to a point where deeper parts of the lake are unable to support aerobic forms of life. Thus, aerobic decomposition ceases. The amount of oxygen needed for oxidative decomposition is greater than the amount of oxygen available. The needed amount is called **biochemical oxygen demand (BOD).**

Box 12.5

It is paradoxical that the greater the productivity of a lake the lower the oxygen concentration in its hypolimnion. For example, oligotrophic lake which is virtually sterile biologically has low production but its water remain well saturated with oxygen from top to bottom (**orthograde oxygen curve**). Likewise, a eutrophic lake has high productivity, but there is substantial oxygen depletion from top of the hypolimnion to the bottom, with depletion most extreme in the bottom layers (**clinograde oxygen curve**). However, in most lakes, oxygen is thoroughly mixed and is essentially at saturation during overturn.

As the basin of eutrophic lake continues to fill, the volume of lake decreases and shallowness speeds the cycling of available nutrients and further increases the plant production. Positive feedback carries the lake system to ultimate extinction, *i.e.,* the filling in of the basin and the development of a marsh, swamp and ultimately a terrestrial community.

Natural and Cultural Eutrophication

The process of lake-aging characterized by nutrient enrichment is known as **natural eutrophication.** When the process of eutrophication is speeded up by human activity, it is called **cultural eutrophication.**

Presently in global perspective "galloping eutrophication" has been changing naturally eutrophic lakes into **hypertrophic lakes.** An excessive nutrient content results from a heavy influx of wastes, raw sewage, drainage from agricultural lands, river basin and the development, runoff from urban areas, and burning of fossil fuels. This accelerated enrichment, which results in chemical and environmental changes in the system and causes major shifts in plant and animal life, has been called **cultural eutrophication**. Such cultural eutrophication has caused Lake Erie in U.S and Canada to age 1500 years in only 25 years between 1950 and 1970 (**Owen** 1985). Lake Mendota, U.S. and Lake Washington, U.S., have also undergone rapid eutrophication as a result of human activity. In India, evergoing process of cultural eutrophication has reduced recreational value of the Kashmir lakes. Nainital lake is also undergoing rapid eutrophication, especially as a result of sewage, domestic waste and detergent discharges.

Effects of Eutrophication

Various adverse effects of nutrient pollution (eutrophication) on a lake are the following:

1. Eutrophication destroys the aesthetics of the lake, rendering it repulsive to swimmers and sport enthusiasts. Algal blooms mar the beauty of lake.

2. The algal blooms impair water quality by giving it a bad taste and odour. If the lake is a source of drinking water, considerable expense may be involved in improving its quality.

3. Toxic gases (*e.g.,* hydrogen sulphide) that emanates from rotting algae has foul smell. Some of the blue-green algae (Cyanobacteria) release poisonous chemical which affects fish, birds, domestic animals and humans.

4. Dense algal blooms at the lake surface reduce penetration of sunlight to the lake bottom. As a result, the deeper waters contain less amount of dissolved oxygen (DO), which is further reduced by decomposing algae and organic matter accumulated at the lake bottom. The oxygen depletion often causes **winter kill** of fish of northern lakes.

5. The algal bloom and rooted weeds gradually reduce the lake size and also interfere with fishing, bathing, irrigation and the production of hydroelectric power.

6. As the clean water becomes turbid and dirty, the game fish such as the Trout and the Mahseer are replaced with trash fish.

7. The lake basin is gradually filled in by sediments (**sediment pollution**) making the lake shallow and in due course of time the lake dies or becomes "extinct" as a result of eutrophication.

Control of Eutrophication

The various measures suggested to stop eutrophication are as follows:

1. Limitation of nutrient input by treatment of the waste water before it is being discharged into a lake or river.

2. Reduction in the amount of nutrient solubulized in water through stimulation of bacterial multiplication which will lead to disruption of algal food-web.

3. Harvesting and removal of algal blooms to check recycling of nutrients into the water upon their death and decomposition.

4. Removal of dissolved nutrients from water physically or chemically: phosphorus can be removed by various method of precipitation (**Agarwal** 1999). Nitrogen can be removed by (i) biological nitrification and denitrification; (ii) air stripping of ammonia from an alkanized waste water; (iii) ion exchange; (iv) electrolysis and (v) reverse osmosis.

12.6 PRODUCTION IN LOTIC ECOSYSTEM

The lotic or flowing water system is open and largely **heterotrophic,** especially in headwater streams. A main source of energy and nutrients is **detrital material** from the outside. Much of this organic matter comes in the form of leaves and woody debris dropped from streamside vegetation, collectively called **coarse particulate organic matter** or **CPOM** (particles larger than 1mm). Another type or organic input is **fine particulate organic matter** of **FPOM** (material less than 1 mm), including leaf fragments, invertebrate faeces, and precipitated dissolved organic matter. A third input is **dissolved organic matter** or **DOM** (material less than 0.5 micron in solution). One source of DOM is rainwater dropping through overhanging leaves, dissolving the nutrient-rich exudates on them. Another is subsurface seepage, which brings nutrients leached from adjoining forest, agricultural and residential lands. Many streams receive inputs from dumping of industrial and residential effluents. Supplementing this detritus input is autotrophic production in streams by diatomaceous algae growing on rocks and by rooted aquatics such as water moss (*Fontinalis*).

The processing of this organic matter (*i.e.,* detritus) involves both physical and biological mechanisms. In fall, leaves drift down from overhanging trees, settle on the water, float downstream and deposit against banks, debris and stones. Soaked with water, the leaves sink to bottom, where they quickly lose 5 to 50 per cent of their dry matter as water leaches soluble organic matter from their tissues. Much of this DOM is either combined with detrital particles or precipitated to become part of FPOM. Another part is mixed with *microbial biomass.* Once softened, leaves and other debris are processed by a number of species of invertebrates, which are classified into following **functional groups:**

1. Microbes. Within one or two weeks, depending upon the temperature and the plant species involved, the surface of the leaves is colonized by **bacteria** and **fungi,** largely aquatic hyphomycetes. Fungi are more important on CPOM because large organic particles offer more surface for mycelial growth (**Cummins** and **Klug** 1970). Bacteria are associated more with FPOM. Microorganisms degrade cellulose and metabolize lignin. Their population form a layer on the surface of leaves and detrital particles that is much richer nutritionally than the detrital particles themselves (**Anderson** and **Cummins** 1979).

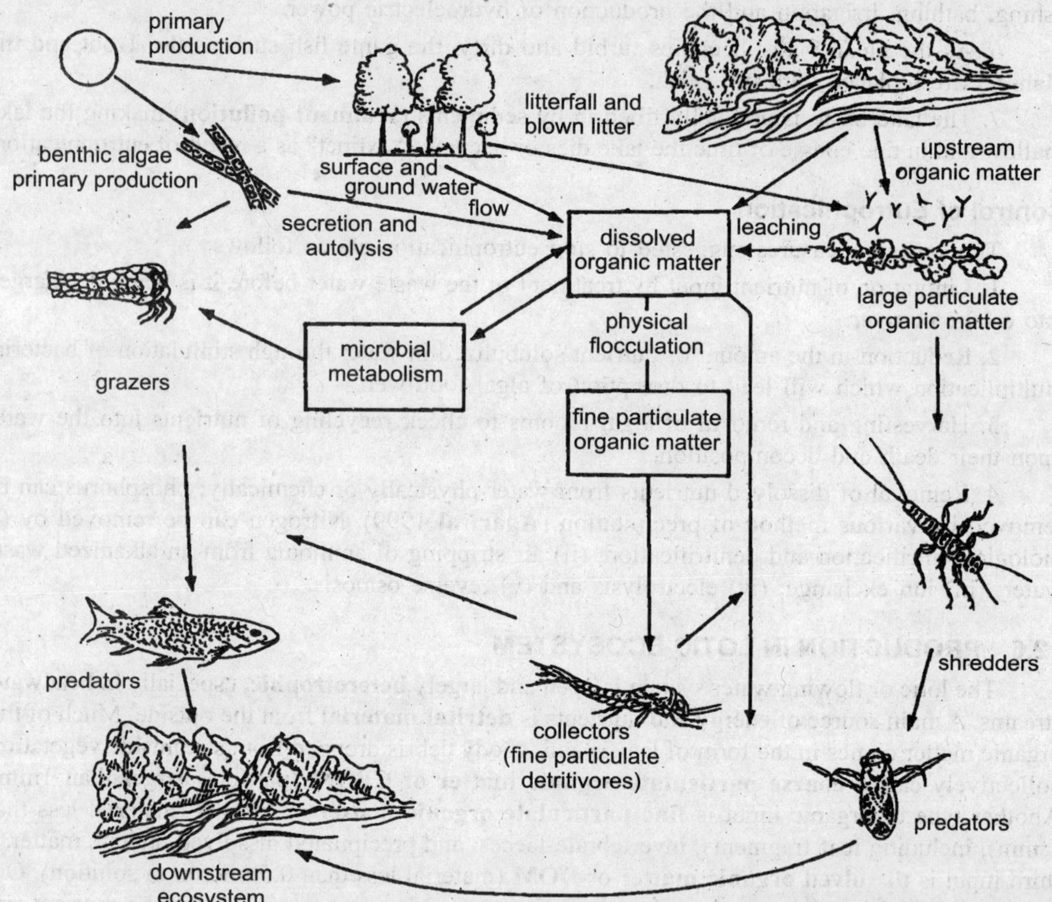

Fig. 12.11. The processing of leaves and other particulate matter and dissolved organic matter in lotic ecosystem by various functional groups such as grazers, shredders, collectors and predators.

2. Shredders. Leaves and other detrital particles are attacked by a major feeding group, called **shredders,** *i.e.,* invertebrates that feed on leaves and other large organic particles. Examples of shredders include the larvae of craneflies (Tipulidae) and caddisflies (Trichoptera), the nymphs of stoneflies (Plecoptera) and crayfish. Shredders breakdown the CPOM, feeding on the material not so much for the energy it contains but for the bacteria and fungi growing on it (**Cummins** 1974; **Cummins** and **Klug** 1979). In doing so they skeletonize the leaf by feeding on the softer portions. Shredders assimilate about 40 per cent of the material they ingest and pass off 60 per cent as faeces.

3. Collectors. Broken up by shredders and partially decomposed by microbes, the leaf material along with faecal material becomes part of the FPOM, which also includes some precipitated DOM. Drifting downstream and settling on the stream bottom, FPOM is picked up by another feeding group of stream invertebrates, called filtering collectors and gathering collectors. The **filtering collectors** include the larvae of blackflies (Simuliidae) with filtering fans and net-spinning caddisflies, including *Hydropsyche.* The **gathering collectors** include the larvae of midges which pick up particles from stream bottom sediments. Collectors obtain much of their nutrition from bacteria associated with fine detrital particles.

4. Scrapers. Whereas shredders and collectors feed on detrital material, another group feeds on algae coating of stones and rubble. They are the **scrapers,** which include the beetle larvae, popularly known as the **water penny** (*Psephenus* sp.) and mobile caddisfly larvae. Much of the material they scrape loose enters the drift as FPOM. Scrapers are adapted to maintain their position in the current either by flattening to avoid main force of flow or by weighting down with heavy mineral cases.

5. Piercers. They feed on mosses and filamentous algae and include mainly microcaddisflies.

6. Gougers. They are associated with woody debris and include invertebrates that burrow into waterlogged limbs and trunks of fallen trees.

7. Predators. Feeding on the detrital feeders and scrapers are predaceous insect larvae such as large stoneflies, the powerful dobsonfly larvae (*Corydalus cornutus*), stream salamander and fish such as sculpin (*Cottus* sp.) and trout.

Because of current, quantities of CPOM and FPOM and invertebrates tend to move downstream to form a traveling benthos called **drift.** Drift is so characteristic of streams that a mean rate of drift can serve as an **index of the production rate** of the stream. Drift is essential for the production process of downstream systems.

Total Dissolved Solids (T.D.S.)

Solids occur in water either in solution or in suspension. These two types of solids are distinguished by passing the water sample through a glass fibre filter. By definition, the **suspended solids** are retained on the top of filter and **dissolved solids** pass through the filter with the water. If the filtered portion of the water samples were placed in a small dish and then evaporated, the solids in the water would remain as a residue in the evaporating dish. This material is usually called TDS. The concentration of TDS is expressed in terms of mg/l. It can be calculated as follows:

$$\text{TDS} = \frac{(\text{A} - \text{B}) \times 1000}{\text{C}}$$

where A = weight of dish plus residue, mg

B = weight of empty dish, mg

C = volume of sample filtered

In drinking water, dissolved solids may cause taste problems. Hardness, corrosion, aesthetic problems may also accompany excessive TDS concentrations. In waste water analysis and pollution control, the suspended solids retained on the filter are of primary importance and are referred to as **TSS** (*i.e.,* total suspended solids). The TSS concentration can be computed using equation

$$TSS = \frac{(A-B)\times 1000}{C}$$

where A = weight of the filter plus retained solids.

B = weight of clear filter

C = volume of sample filtered

12.7 THE RIVER CONTINUUM CONCEPT

The composition of biological communities in streams and rivers changes with stream order (Fig. 12.12) and channel size. The **river continuum concept** was developed to describe these

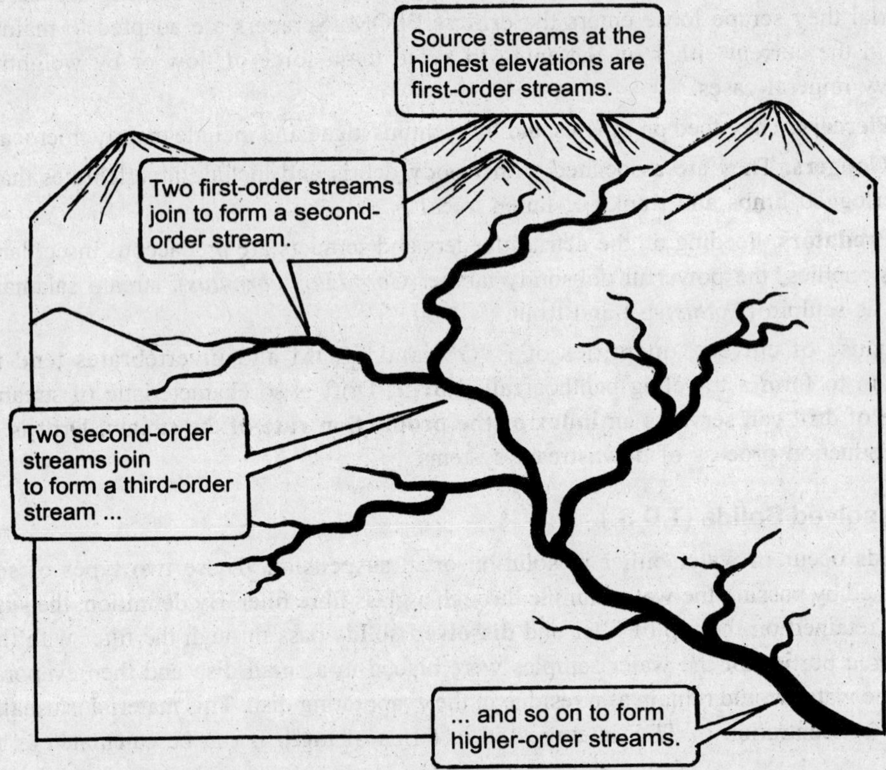

Fig. 12.12. Stream orders. Stream order affects environmental conditions, community composition and the energy and nutrient relationships of communities within stream (after **Cain** *et al*, 2011).

changes in both physical and biological characteristics of a stream (**Vannote** *et al.*, 1980). This conceptual model (Fig. 12.13 and Fig. 12.14) holds that as a stream flows downslope and increases in size, the input of detritus from the vegetation adjacent to stream (known as **riparian vegetation**) decreases relative to the volume of water, and the particle size in the stream bed decreases from boulders and coarse rock to fine sand. As a result, the importance of terrestrial vegetation as a food source for stream organisms decreases in downstream direction. Coarse terrestrial detritus is most important near the stream source, while the importance of fine organic matter, algae, and rooted floating aquatic vascular plants (known as **macrophytes**) increases downstream.

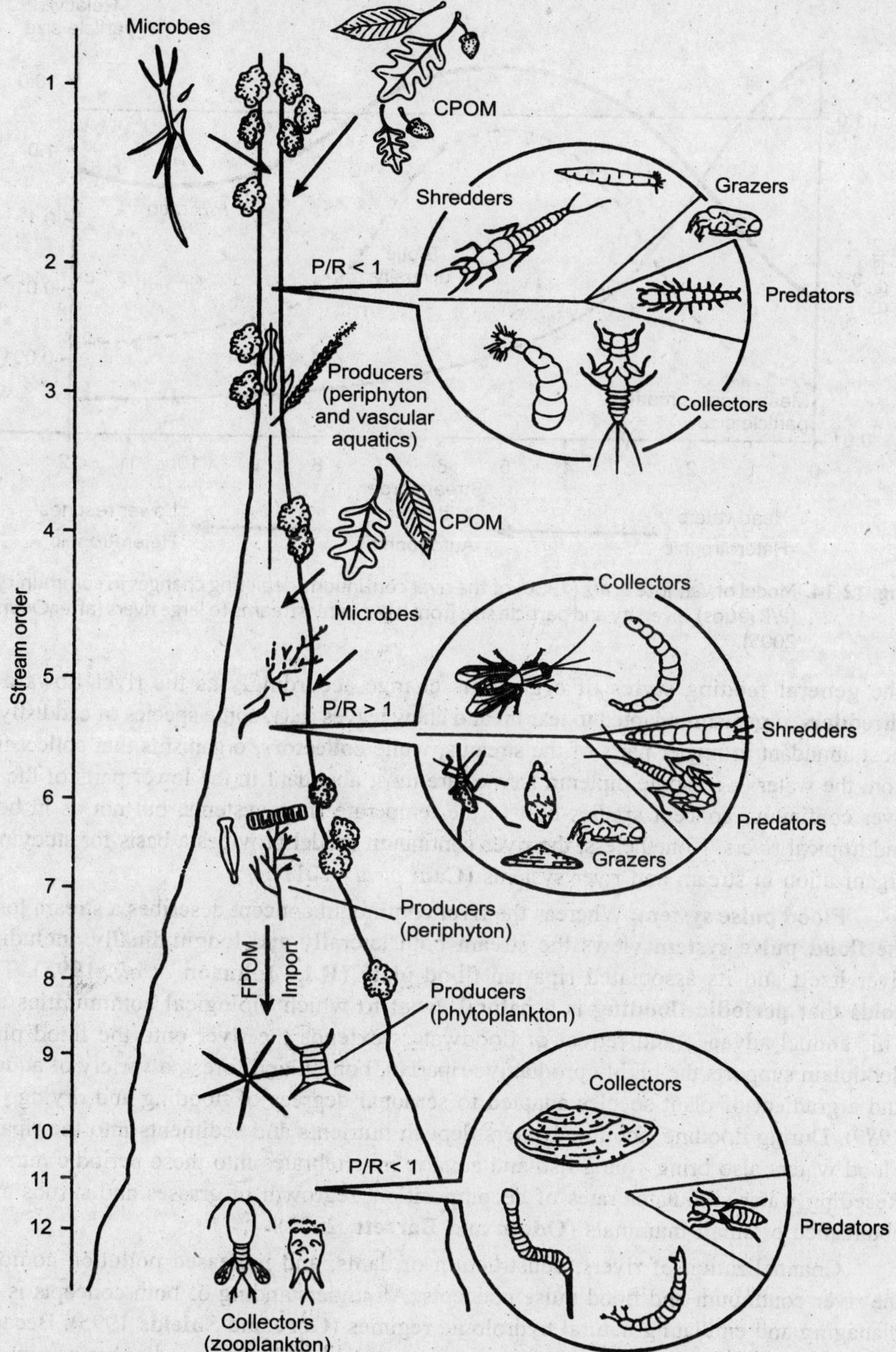

Fig. 12.13. **Cummins** (1977) overview of the river continuum, depicting stream order, organisms by feeding type, and changes in particulate matter. CPOM = coarse particulate organic matter; FPOM = fine particulate organic matter (after **Odum** and **Barrett**, 2005).

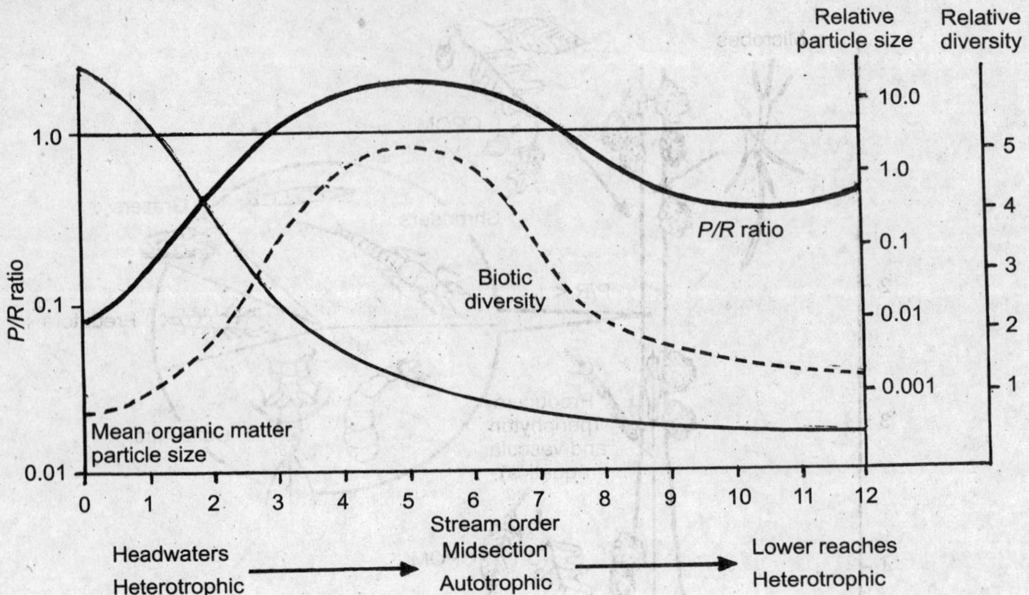

Fig. 12.14. Model of Vannote *et al.*, (1980) of the river continuum, depicting changes in community metabolism (P/R ratios), diversity and particle size from headwater streams to large rivers (after Odum and Barrett, 2005).

The general feeding styles of organisms change accordingly as the river flows downstream. **Shredders**, organisms adapted to tear up and chew leaves (*e.g.*, some species of caddisfly larvae), are most abundant in higher parts of the streams, while **collectors**, organisms that collect fine particles from the water (*e.g.*, some dipteran larvae) are most abundant in the lower parts of the stream. The river continuum concept applies best to the temperate river systems, but not as in boreal/ Arctic and tropical rivers. Nonetheless, the river continuum model provides a basis for studying biological organization in stream and river systems (**Cain** *et al.*, 2011).

Flood pulse system. Whereas the river continuum concept describes a stream longitudinally, the **flood pulse system** views the stream both laterally and longitudinally, including both the river itself and its associated riparian flood plain (**B.L. Johnson** *et al.*, 1995). This concept holds that **periodic flooding** is a natural event to which biological communities are adapted. This annual advance and retreat of floodwaters extends the river onto the flood plain. Thus, a floodplain supports the highly productive riparian, bottomland forest, a variety of aquatic habitats, and a gradient of plant species adapted to seasonal degrees of flooding and drying (**Junk** *et al., 1989*). During flooding, the floodwaters deposit nutrients and sediments into the riparian system. Flood waters also bring young fish and aquatic invertebrates into these periodic nursery grounds. Receding water stimulates rates of decomposition, regrowth of grasses and shrubs and pulses in abundance of small mammals (**Odum** and **Barrett**, 2005).

Channelization of rivers, construction of dams, and increased pollution compromise both the river continuum and flood pulse concepts. An understanding of both concepts is essential for managing and emulating natural hydrologic regimes (**Gore** and **Shields** 1995). Because so many rivers have one or more dams on them a new concept called **river discontinuum concept**, has emerged (see **Odum** and **Barrett**, 2005). In many places, there are too many dams, so dams on small rivers are being taking down.

Stream spiraling. In the biosphere, organisms are not just passive adapters to a gradient of changes in the physical environment. The concerted action of stream animals, for example, works to recycle and reduce the downstream loss of nutrients to the ocean. Aquatic insects, fish and other organisms collect particulate and dissolved organic material that are held and cycled through the food chain. **Stream spiraling** is the cycling of essential element (such as nitrogen, carbon and phosphorous) between organisms and available pools as they move downstream. In other words, stream spiraling is the process in which elements alternate between organic and inorganic forms as they move downstream.

12.8 HUMAN EFFECTS ON LOTIC ECOSYSTEM

Most fourth-and higher order rivers have been altered by the results of human activities, including **pollution**, increases in **inputs of sediments**, and introduction of non-native species. Streams and rivers have been used as conduits for the disposal of sewage and industrial wastes in most parts of the world inhabited by humans. These pollutants often reach levels that are toxic to many aquatic organisms. Excessive application of fertilizers to croplands results in runoff and leaching of nutrients into groundwater, which eventually reaches rivers. Inputs of nitrogen and phosphorous from fertilizers alters the composition of aquatic communities. **Deforestation** increases sediment inputs into nearby streams, which can reduce water clarity, alters benthic habitat and inhibit gill function in many aquatic organisms. Introduction of non-native species such as sport fishes (*e.g.,* bass and trout), have lowered the diversity of native species in both stream and lake ecosystems. Lastly the *construction of dams* on streams and rivers tremendously alters their physical and biological properties, converting them into still waters (*i.e.,* lentic ecosystems) (**Cain** *et al.,* 2011).

QUESTIONS

Long Answer Questions

1. Describe the ecological features and adaptations to freshwater habitat. Add a note on freshwater fauna.
2. Describe ecology of a lentic ecosystem.
3. Mention chief characteristics of fauna in a lotic habitat.
4. Define nutrient status of a lake. Describe oligotrophic, mesotrophic and eutrophic lakes.
5. What is trophic relationship? Write the trophic relationship in a freshwater community.
6. Describe production in a lotic ecosystem.
7. List the chemical abiotic factors of freshwater ecosystem. Describe the effects of total dissolved solids and total suspended solids on the productivity of freshwater ecology.
8. What do you mean by TDS and TSS? How do these factors affect freshwater ecosystem?
9. Define adaptation? Give an account of the various adaptations of aquatic animals.
   Describe adaptation of hill stream fishes.

Short Answer Questions

1. Write a short note on freshwater adaptations.
2. Write a short note on eutrophication.
3. Describe thermal stratification in lakes.
   Write short notes on the following:
   1. Lentic waters.

4. Differentiate between the following:
   (i) Spring overturn and fall overturn
   (ii) Limnetic zone and profound zone.
   (iii) Metalimnion and epilimnion
   (iv) Littoral and limnetic zone
   (v) Eutrophic and oligotrophic lakes

Very Short Answer Questions

1. What is lotic ecosystem?
2. Define lentic ecosystem?
3. What is limnology?
4. Define limnetic zone.
5. What is littoral zone?
6. Define the profundal zone.
7. What is river continuum concept?
8. Define riparian.
9. What is thermocline?
10. What is eutrophication?
11. What is epilimnion?
12. Define the flood pulse concept.
13. What is stream spiraling?

Fill in the Blanks

1. Freshwater ecosystems are divided into. .... and. .... ecosystems.
2. Lakes are inland depressions containing ....
3. Floating plants such as duckweed are the ....
4. All lakes form in land basin and so eventually fills with .... and become ....
5. Freshwater systems are characterized according to the speed of water flow. These are .... ecosystems of nonflowing water and .... ecosystem of flowing water.
6. A lake has three ecological zones; the .... zone of shallow water, where light penetrates to the bottom substrate; the .... zone of deep water, which lacks sufficient light for photosynthesis; and the .... zone, above the deep-water zone, which receives sufficient light for photosynthesis.
7. .... is the process in which elements alternate between organic and inorganic forms as they move downstream.
8. Flood pulse concept deals with changes in river, both .... and ...., during heavy rain and ....
9. Continuum of changes in .... structure, dominant .... and ecosystem processes along the length of a river system is the .... concept.

Multiple Choice Questions

1. Aquatic ecosystems cover approximately how much of the earth's surface?
(*a*) 20% (*b*) 40%
(*c*) 75% (*d*) 90%
2. The largest amount of freshwater on our planet is in
(*a*) polar ice caps and glaciers (*b*) underground
(*c*) rivers (*d*) lakes and streams
3. Which of the following aquatic ecosystems has very little primary productivity?
(*a*) river (*b*) estuary
(*c*) lake (*d*) marsh
4. Small, floating or weakly swimming plants and animals in freshwater and marine ecosystem are
(*a*) nekton (*b*) benthos
(*c*) plankton (*d*) all of these
5. What is plankton?
(*a*) sea scum
(*b*) floating vegetation
(*c*) litter of aquatics
(*d*) floating microscopic plants and animals
6. Most primary productivity in the surface waters of lake is by
(*a*) floating plants (*b*) red and brown algae
(*c*) zooplankton (*d*) phytoplankton
7. Organisms capable of swimming and resting on the surface of water are termed
(*a*) neuston (*b*) nekton
(*c*) plankton (*d*) benthos
8. Littoral form of animals are found
(*a*) in freshwater (*b*) in seawater near the shore
(*c*) in deep sea (*d*) at the bottom of the sea
9. Benthos are
(*a*) periphyton
(*b*) zooplankton
(*c*) epineuston
(*d*) bottom dwelling primary consumers
10. The swimming fish constitute
(*a*) neustons (*b*) nektons
(*c*) planktons (*d*) benthos
11. The dominant second trophic level in a lake ecosystem is
(*a*) zooplanktons (*b*) phytoplanktons
(*c*) benthos (*d*) nektons
12. Thermocline is
(*a*) movement of water due to temperature
(*b*) movement of water due to wind and temperature
(*c*) winter stratification
(*d*) summer stratification

13. Oligotrophic lakes
(*a*) are poor in nutrients
(*b*) are excessively mineralized
(*c*) are excessively loaded with organics
(*d*) have great biological diversity

14. Phosphate pollution is caused by
(*a*) phosphate rocks only
(*b*) agriculture fertilizers only
(*c*) sewage and phosphate rocks
(*d*) sewage and agriculture fertilizers

15. Sudden mass death of fishes from oxygen depletion is more likely in case of
(*a*) eutrophic lakes
(*b*) mesotrophic lakes
(*c*) oligotrophic lakes
(*d*) oxalotrophic lakes

16. When huge amounts of sewage is dumped into a river, the BOD (Biological Oxygen Demand) will
(*a*) remain unchanged (*b*) slightly decrease
(*c*) decrease (*d*) increase

ANSWERS

Very Short Answer Questions

1. The term lotic (from Latin *lotus* = washed) running water ecosystems such as streams and rivers.

2. The term lentic (from Latin *lenis* = calm) stands for standing water ecosystem such as lakes and ponds.

3. Limnology is the study of freshwater ecosysytem such as lakes.

4. The limnetic zone is the open water of a lake beyond the littoral zone.

5. The littoral zone contains rooted floating and emergent vegetation along the shore of a lake or pond.

6. The profound zone is the deep – water area of a lake that lies beyond the depth of effective light penetration.

7. River continuum concept is the model depicting a continuum of changes in physical structure, dominant organisms and ecosystem proceses along the length of a river system.

8. Riparian area refers to area along banks of rivers and streams.

9. Thermocline is 1. layer of water in thermally stratified lake where the temperature profile changes rapidly relative to the body of water as a whole or 2. zone of water in a thermally stratified lake between the epilimnion and hypolimnion.

10. Eutrophication is the processes of nutrient enrichment (typically of phosphates and nitrates) in aquatic ecosystems, resulting in increased primary productivity.

11. Epilimnion is warmer, oxygen-rich upper part of a lake when thermally stratified during summer.

12. Flood pulse concept is the description of changes in a river, both laterally and longitudinally (both of the river and associated riparian flood plain) especially during heavy rain or flooding (pulse) conditions.

13. Stream spiraling is the movement and cycling of essential elements (such as carbon, nitrogen and phosphorous) between organisms and available pools as they move downstream.

Fill in the Blanks

1. Lentic, lotic **2.** Standing water **3.** Neuston
4. Sediments, terrestrial; **5.** Lentic, lotic; **6.** Littoral, profundal, limnetic
7. Stream spiraling **8.** Laterally, longitudinally, flooding **9.** Physical, organisms, river continuum.

Multiple Choice Questions

1. (*c*) **2.** (*a*) **3.** (*a*) **4.** (*c*)
5. (*d*) **6.** (*d*) **7.** (*a*) **8.** (*a*)
9. (*d*) **10.** (*b*) **11.** (*a*) **12.** (*d*)
13. (*a*) **14.** (*d*) **15.** (*a*) **16.** (*a*).

13

Aquatic Ecosystems: Marine Communities

13.1 OCEANS AND SEAS

The great stretches of salt water are termed as the **oceans,** while the smaller areas are referred to as **seas** (Box 12-1). Marine ecosystem includes the salt water mass of seas and oceans that covers 70.8% of the earth's surface and contain about 97% of the planet's water; a volume of 1370 × 10^6 km^3. The average depth of oceans is 3730 m. Oceans strongly influence climate by acting as a reservoir of solar heat, thus, improving temperature extremes, and as the source of water which falls as precipitation (*i.e.,* hydrologic cycle). The **Pacific ocean** is the largest (179.7 × 10^6 km^2), coldest (average 3.36°C), least saline (34.6 parts per thousand) and deepest (average depth 4028m). The **Atlantic ocean** is warmer (average 3.73°C), more saline (average 34.9 parts per thousand), and relatively shallow (average depth 3310 m); it also has the longest coastline providing many natural harbours. The smallest and shallowest ocean is the **Arctic.** The **Antarctic (Southern) ocean** shows wide variations in temperature (–1.8—10°C). The **Indian ocean** covers 77 × 10^6 km^2 to an average depth of 3872m.

Box 13.1

There are five great oceans. These are **Pacific, Atlantic, Indian, Arctic and Antarctic oceans**. In spite of their vast size, all the oceans are connected with each other. In addition to the oceans, there are seas. Some of these, such as **Arabian seas** and the **Sargasso sea**, are parts of oceans. Other seas are surrounded by continents, such as the **Mediterranean sea**, **Red sea** and the **Black sea**, are the flooded edges of continents. The largest of world's sea is the **South China sea**.

The volume of surface area of marine environment lighted by the sun is small in comparison to the total volume of water involved. This and the dilute solutions of nutrients limits production. All the seas are interconnected by currents, dominated by waves, influenced by tides and characterized by saline waters. Not only the seashore and banks which are the homes of many organisms but the open ocean, many hundreds of miles away from land, supports plant and animal communities of great diversity and complexity.

13.2 PHYSICO-CHEMICAL NATURE OF MARINE ENVIRONMENT

In the marine environment, the most important physical factors which influence marine life are light, temperature, pressure, salinity, tides and currents.

1. Light. Light is a very significant factor in regulating the pattern of distribution of marine animals and it contributes significantly to organic production. The autotrophic primary producers exploit the light energy in the photosynthetic production of food for primary macro- consumers of marine ecosystem. The amount of light exploited in photosynthesis depends on intensity of light and turbidity of water. For example, a ten meters deep turbid coastal zone receives an equal amount of light energy as a hundred meters deep clear oceanic zone, but the maximum intensity at a ten meters coastal water is greater. Light determines diurnal migrations of marine organisms, and it also regulates colour pattern of marine animals. The deep sea fauna which lives in total darkness exhibit either colourlessness or uniform colouration. It is also somehow related with the development of visual sense organs as shown by absence of functional eyes in deep-sea animals.

2. Temperature. The range of temperature in sea is far less than that on land. Arctic waters at 27°F (–2.8°C) are much colder than tropical waters at 81°F (27.2°C), and currents are warmer or colder than the waters through which they flow. Seasonal and daily temperature changes are larger in coastal waters than on the open sea. The surface of coastal waters is the coolest at dawn and the warmest at dusk. In general, sea water is never more than 2^0 to 3^0 below the freezing point of freshwater or higher than 81°F (27.2°C). At any given place the temperature of deep water is almost constantly cold, below the freezing point of freshwater. Unlike freshwater, sea water does not have a density maximum at 4°C; rather it becomes continuously denser as it gets colder.

3. Pressure. Pressure in the ocean varies from 1 atm at the surface to 1,000 atm at the greatest depth. Pressure changes are many times greater in the sea than in terrestrial environments and have a pronounced effect on the distribution of life. Certain organisms are restricted to surface waters when the pressure is not so great, whereas other organisms are adapted to life at great depths. Some marine organisms, such as the sperm-whale and certain seals, can dive to great depths and return to the surface without difficulty.

4. Ocean floor. It can be studied under following two headings:

(i) Beach. Where the land meets the ocean or seas is called **seashore** or **beach** (Fig. 13.1). A beach extends from the farthest point where sand has been carried by wave action to the depth beyond which wave action does not have sufficient force to move the sand particles.

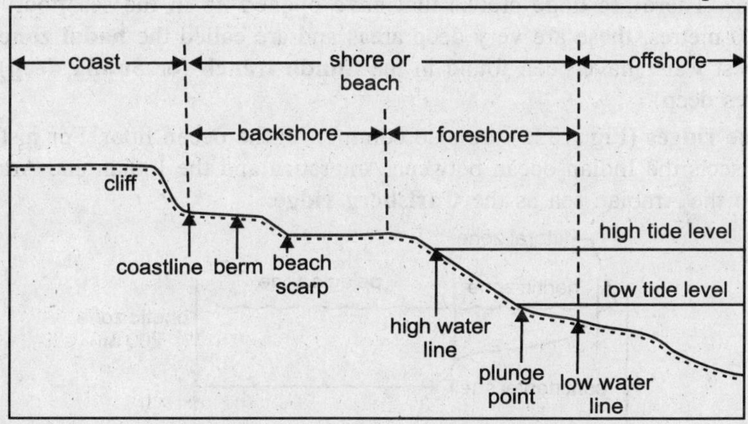

Fig. 13.1. The structure of a sea beach.

From the point where sand is stabilized to a certain extent, the **backshore** region starts and extends down to a sudden change in slope, known as the **beach scarp.** The backshore generally contains one or more horizontal sloping portions, called **berms.** From the scarp, the beach slopes seawards and is known as **foreshore.** The foreshore extends from the high water line to the low water line and may have a continuous slope or a **low-tide terrace.** Beyond the low-tide level is the **offshore,** which is always submerged.

(ii) Sea floor. Close to the beach the sea is mostly shallow, the bottom shelving from the shore to a depth of about 200 metres. This coastal projection of shallow sea-bottom is the

continental shelf (Fig. 13.2) and its seaward margin is called **continental edge.** Beyond the continental edge, the floor descends downward steeply and is called **continental slope.** The

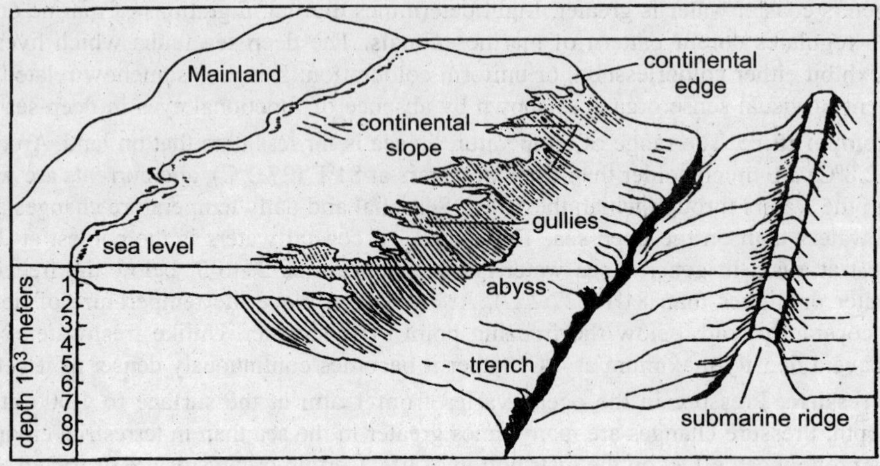

Fig. 13.2. Structure of sea floor.

slope plunges into the floor of ocean basin, often reaching a depth of 3000 to 6000 metres or even deeper in some places. The slope is often fissured by irregular **gullies** and steep-sided submarine **canyons.**

The continental slope levels out into an extremely deep underwater plain, called the **abyss.** The abyssal region lies between 2000 to 5000 metres of depth. There are also deep **trenches** under the oceans. There are huge cracks that have opened up in the sea floor. Trenches may drop below 6000 metres, these are very deep areas and are called the **hadal zone.** In the Indian ocean, the deepest water have been found in the **Sunda trench** (or **Sunda deep**) south of Java (*i.e.,* 7453 metres deep).

Submarine ridges (Fig. 13.2.) are also common in the ocean floor. For instance, the **mid-Indian ridge** bisects the Indian ocean between Antarctica and the Indian and Arabian peninsula and extends into the Arabian sea as the **Carlsberg ridge.**

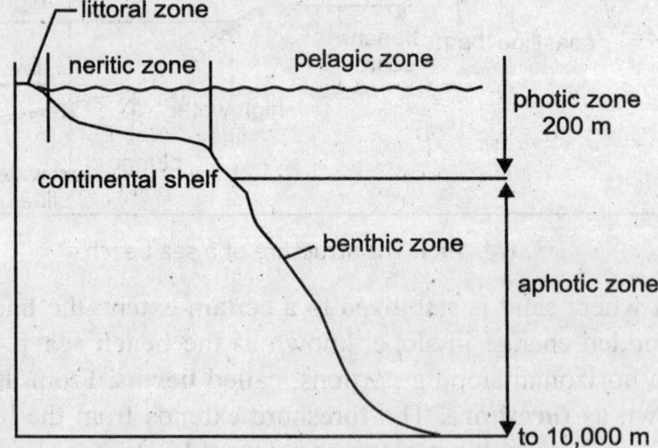

Fig. 13.3. Zonation of the ocean.

5. Zonation of marine environment. Just as lakes exhibit stratification and zonation, so do the seas. The major zones in marine ecosystem are the littoral, neritic, pelagic and benthic (Fig. 13.3).

The ocean itself is divided into two main divisions, the **pelagic**, or whole body of water, and the **benthic**, or bottom region (Fig. 13.3). The pelagic region is further divided into two provinces: the **neritic province** (near shore), water that overlies the continental shelf, and the **oceanic province**. In respect to illumination, the oceanic zone vertically contains two distinct zones: 1. **Photic zone** which extends from surface upto the depth of 200 metres and **aphotic zone** which extends from 200 metres to 6000 metres depth.

A. Pelagic division. On the basis of other ecological parameters, the oceanic province (or pelagic zone) includes following three vertical zones (Fig. 13.4).

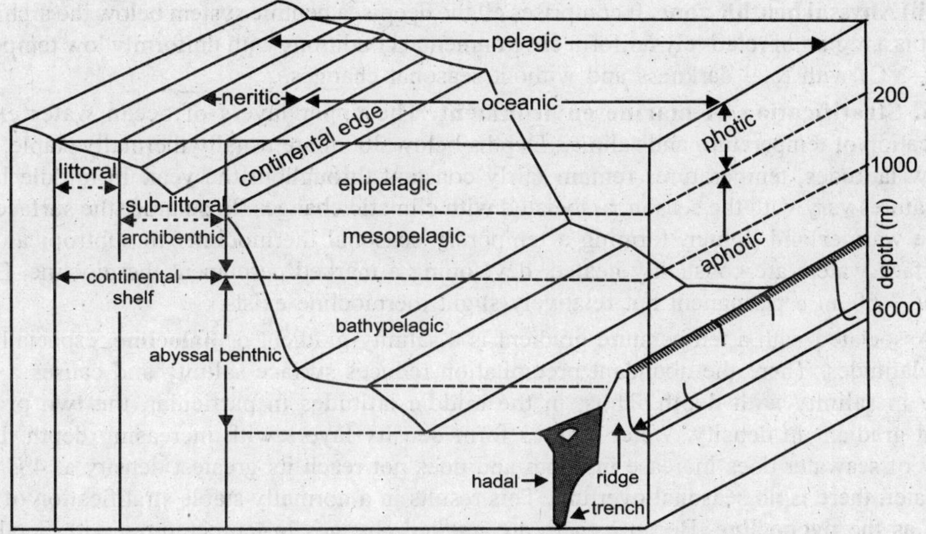

Fig. 13.4. Horizontal and vertical zonation of ocean.

1. Epipelagic zone. It extends vertically downward from the surface to a depth of 200 metres. In this zone, sharp gradients of illumination and temperature occur between the surface and deeper layers. In this zone, temperature gradients and thermocline is common. Marked diurnal and seasonal changes in light intensity and temperature occur in the epipelagic zone. The effect of water movement (in the form of waves) is prominent especially in the upper layers and determines the conditions in the layers beneath.

2. Mesopelagic zone. It extends from 200 metres to about 1000 metres. Wave action does not reach here and very little light is available. The temperature gradient is more even and gradual and there is very little seasonal variation. Mesopelagic zone also contains an oxygen-minimum layer along with maximum concentration of nitrates and phosphates.

3. Bathypelagic zone. It extends from about 1000 metres to the sea floor which is typically 3000 to 4000 metres in depth. Here darkness is virtually complete, except for bioluminescence. Temperatures are low but constant. Water pressure is, however, enormous.

B. Benthic division. This includes all the bottom terrain from wave-washed shoreline at flood tide to the ocean abyss. Benthic division includes the following two zones:

1. Littoral zone (or **Intertidal zone**). It extends from the high-tide level to a depth of 200 metres, that is, it includes the benthic region upto the continental edge. Littoral region is subjected to the physical violence of waves and tides and to fluctuations of temperature, moisture and light intensity. It has two subdivisions:

(i) Eulittoral zone. It extends from the wave-splashed high tide level to a depth of about 40 to 60 metres. Its lower border indicates lowest limit at which the more abundant attached plants can grow. Its upper limit indicates tidal or intertidal zone that is bounded by the high and low tide levels.

(ii) Sublittoral zone. It extends from the lower part of eulittoral zone down to a depth of 200 metres, or the edge of continental shelf.

2. Deep sea zone. It includes the benthic region from the edge of continental shelf to the ocean abyss which is usually at great depths. Deep sea zone has following two zones:

(i) Archibenthic zone. It extends from the sublittoral to a depth between 800 metres to 1100 metres.

(ii) Abyssal benthic zone. It comprises all the deep-sea benthic system below the archibenthic zone. It is a region of relatively uniform environmental conditions with uniformly low temperatures (5° to –1°C) with total darkness and without seasonal changes.

6. Stratification of marine environment. The upper layers of ocean water exhibit a stratification of temperature and salinity. Depths below 300m are usually thermally stable. In high and low latitudes, temperatures remain fairly constant throughout the year. In middle latitudes temperatures vary with the season, associated with climatic changes. In summer the surface waters become warmer and lighter, forming a temporary seasonal thermocline. In subtropical regions the surface waters are constantly heated, developing a marked permanent thermocline. Between 500 and 1,500m a permanent but relatively slight thermocline exists.

Associated with a temperature gradient is a salinity gradient or **halocline**, especially at the higher latitudes. There the abundant precipitation reduces surface salinity and causes a marked change in salinity with depth. Thus, in the middle latitudes in particular, the two produce a marked gradient in density. Water masses form density layers with increasing depth. Because density of seawater does increase in depth and does not reach its greatest density at 4°C as with freshwater, there is no seasonal overturn. This results in a normally stable stratification of density known as the **pycnocline**. Because there are marked changes in temperatures with depth in the open ocean and, thus, with density, the pycnocline is the deep zone which is oxygen rich, cold and comprises 80 per cent of the ocean.

7. Salinity. The sea is salty, with an average salinity of 35 parts of salt (weight basis) per 1000 parts of water or 3.5 percent, that is usually written as 35‰, *i.e,* parts per 1000 (in contrast the salinity of freshwater is 0.5%). The main salts are chloride, sulphates, bicarbonates, carbonates and bromides of sodium, magnesium, calcium and potassium. Of these salts, sodium chloride is present in maximum amount. The salinity of marine water fluctuates from place to place. The marine animal life has specific osmoregulatory adaptation for high saline sea water. The absence of many animal species in marine environment has been related to their inability to tolerate the high salt contents of sea water. Except few insects such as *Halobates* and shore collembolans such as *Isotoma* and *Sminthurus,* most insects do not occur in marine ecosystems.

8. Currents and tides. Sea water is never static and waves, currents, and tides are the regular features of sea water. All these phenomena are controlled by winds, cosmic forces and varying water densities. Most world-wide water currents occur at the surface and at great depths of sea water. These currents determine the interchange between the surficial and deep water masses, as well as horizontal movements. Both the horizontal and vertical movements of ocean water are significant, but for different reason. Within a given current, such as warm Gulf stream, for instance, the water mass retains its identity for great distances, and the warm-sea community ranges far north of what have been anticipated as its northernmost limits. In addition, the climate of terrestrial ecosystems is strongly affected by the nature of adjacent water masses. The prevailing winds blow across the ocean and are heated or cooled by the waters they traverse. In marine ecosystems, however, the oceanic current system controls the distribution of productivity of marine communities through vertical mixing of water masses. Because most of the ocean is permanently stratified, any essential nutrient that settles out of the upper layers, either as an inorganic mineral or as organic detritus, is not returned to the upper layers at the same location. This had led to such an improvisation of the surficial waters of the oceans that productivity in most parts of the open oceans is about the same as that of desert, because of the very low concentrations of nutrients, especially phosphorus. At the same time, the lowest waters of the ocean are nutrient-rich. There are certain regions in the world, notably at the west coasts of continents and in subpolar latitudes in both hemispheres, where deep currents rise to the surface. These are called **zones of upwelling**, and even though their waters are very cold, they are sites of intense production because of their relatively high concentration of nutrients. Nutrients are then cycled back to the rest of the ocean via surface currents.

Tides consist of a rise and fall of water level caused by the gravitational effects of the Moon and Sun. When, Earth, Moon and Sun are alinged (as at the time of a new and a full moon); the gravitational pull is the strongest, resulting in the highest, or **spring tides.** When the Moon and Sun are at right angles to each other, as at the time of the first and last quarters, their gravitational fields counteract, resulting in the smallest, or **neap tides.** Each of these tides occurs monthly. Tidal range varies from less than one-third of a metre to 15 metres in some enclosed bays.

9. Nutrients. The concentration of nutrients in the ocean is low. **Nitrates, phosphates** and other nutrients are measured in parts per billion in contrast to salts such as **sodium chloride,** which are measured in parts per thousand. After **hydrogen** and **oxygen,** which constitute about 96 per cent of the oceans elements, the *major elements* (those constituting 0·3 to 1·9 per cent) in descending order are **chlorine, sodium, magnesium, sulphur, calcium** and **potasium** (**Klemm** *et. al.*, 1990). The *minor elements* (·00013 to ·0067 per cent), also in descending order, are **bromine, carbon, nitrogen, strontium, boron, silicon** and **fluorine.** There is wide spectrum of *trace elements* (less than ·000063 per cent) from **argon** to **zirconium,** some 45 in all.

In the Indian ocean the amount of **carbonates** and **bicarbonates** vary from 30-50 mg/l and 160-185 mg/l respectively (Table 13.1).

Table 13.1. Chemical characteristics of the Laćcadives and Andaman seas (**Sengupta and Naqvi 1984; Mukherjee 1992**).

Parameters	Andaman sea	Laccadive sea
1. Salinity (‰)	35·8 – 36·5	35·0
2. Chlorinity (‰)	19·88 – 20·22	19·37
3. Oxygen (mg/l)	3·8 – 6·5	1–5
4. Free CO_2 (mg/l)	55	–
5. Carbonates (mg/l)	30	–
6. Bicarbonates (mg/l)	185	–
7. NO_3–N (µg–at/l)	0	0–1·0
8. PO_4–P (µg–at/l)	0·17–0·54	0·2–1·0
9. pH	7·5–8·5	8·2
10. Fe (µg/l)	–	25·3
11. Mn (µg/l)	–	12·0
12. SiO_2 (µg/at/l)	0–25	0–20
13. Ca (mg/kg)	412–434	425
14. Mg (mg/kg)	1268	1339
15. Oxygen minimum (ml/l)	< 0·5	< 0·5

10. Dissolved organic matter. Varying quantities of organic matter are present in solution in sea water. There is on average about 15 kg of organic matter beneath each square metre of ocean surface. The sources are diverse and include materials originating from the breakdown of tissues, excretory materials and organic secretions (exocrines). The sea may be considered to contain a mosaic of following secretions: **pacifenol, aplysin, sarcophene, gorgosterol, fucoxanthin, saxitonin.** Some of these secretions help in recognition and location of similar species, others are antipredatory keeping off potential predators, while still others are antifouling; restricting the growth of other species. Such secretions have a major impact in niche segregation and territoriality.

13.3 MARINE COMMUNITIES

The sea biota is not abundant but contains well marked diversity. Every major group of algae and almost every major group of animals can be found some where in the oceans. The only striking omissions are the vascular plants and insects, which are abundant in estuaries but which have few or no marine representatives. Communities of marine environment includes : (1) **producers** such as phytoplanktonic diatoms and dinoflagellates, green algae, brown algae, red algae and kelps; (2) **consumers** such as zooplanktons, benthos, nekton and neuston and bacteria. The marine biotic communities can be studied separately for different life zones of the ocean as follows:

A. Biotic Communities of Oceanic Region

The oceanic region or pelagic zone is less rich in species and numbers than the coastal areas, but it has its characteristic species. Many of these are transparent or bluish and since the sediment-free water of the open sea is marvelously transparent, these animals are nearly invisible. Animals that are too thick to be transparent frequently have smooth shiny and silvery bodies which make them invisible by mirroring the water in which they swim. Further, the animals of the pelagic zone of ocean encounter stable and uniform environmental conditions due to the continuity of the waters of the sea. This continuity of sea waters causes the individuals of the open ocean to stay away temporarily or permanently from any solid objects, and they are provided with different locomotor organs. However, the planktonic forms lack a locomotor organs. However, the planktonic forms lack a locomotor organ and float freely in the vast expanse of the sea.

(i) Biotic communities of pelagic zone. The pelagic plankton also called **epiplankton** are exceedingly diverse. **Pytoplankton** includes diatoms and dinoflagellates, which together produce most of the organic carbon in the sea (and most of the oxygen in the atmosphere), as well as other forms of golden-brown algae and flagellated green algae. Some seaweeds, such as the large brown algae *Sargassum*, may have a floating stage. The **zooplankton** of pelagic zone includes representatives of every major phylum and most minor phyla, either as permanent members of the plankton community (**holoplanktononic forms**) or as transients during their larval stages (**meroplanktonic forms**). Common among permanent planktonic forms are the Foraminifera and Radiolaria, arrow worm (*Sagitta*), certain annelid worms, swimming snails, jelly fishes and most abundant of all are the crustaceans such as shrimps, copepods and cladocerans. Among the temporary zooplanktons are the larvae of large animals from all marine environments and even some freshwater environments. For example, in the zooplankton of inshore water of Mandapam (South India) **Prasad** (1956) reported following animal species—flagellate protozoan *Noctiluca miliaris*; larval forms of coelenterates such as *Planula, Ephyra.*

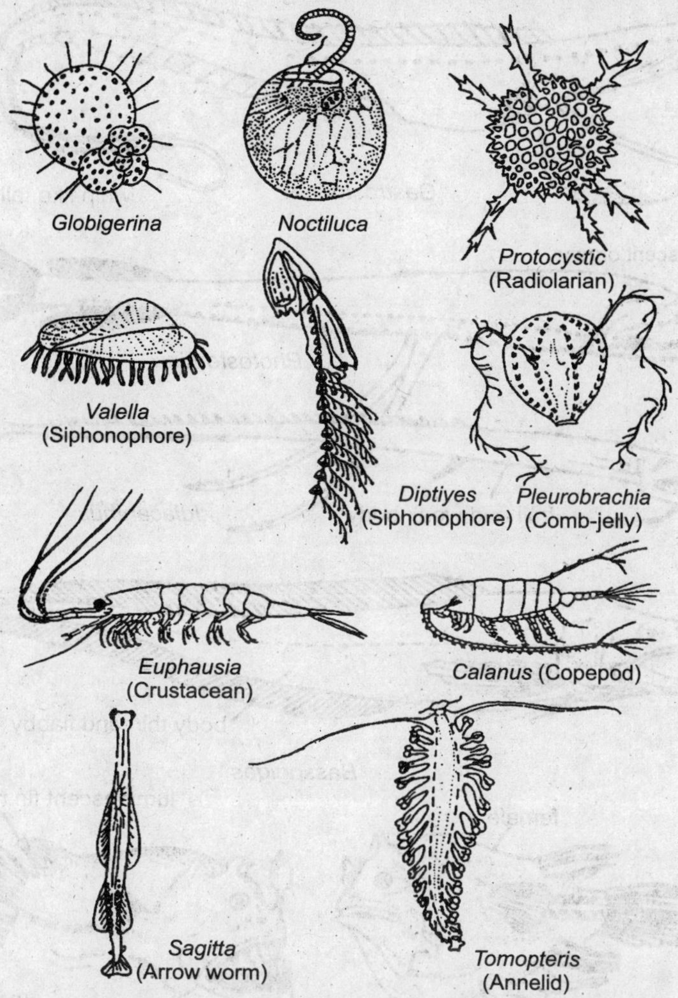

Fig. 13.5. Certain holoplanktons of pelagic zooplankton.

Semper's larva and *Cerianthus* larva; medusae of certain coelenterates such as *Bougainvillia, Obelia, Charybdea, Rhizostoma* and *Rhopalina;* ctenophores such as *Pleurobrachia* and *Beroe;* larvae of many polychaet annelids and annelids such as *Tomopteris, Alciopa* and *Autolytus,* several species of chaetognaths such as *Sagitta* and *Spadella*; heteropod and pteropod molluscs such as *Carinaria, Pterotrachea, Cliona, Creseis;* larvae of echinoderms such as bipinnaria, ophiopluteus, echinopluteus, auricularia; tornaria of *Balanoglossus*; species of copepods such as *Acartia, Calanopia, Labidocera, Centropages, Paracalanus, Canthocalanus, Acrocalanus, Eucalanus* and *Microsetella;* larvae of decapod *Lucifer*; protochordate species of Larvacea, Thaliaceae and Ascidiaceae such as *Thalia democratica, Jasis zonaria, Salpa cylindrica;* appendicularian—*Oilopleura diocia, O. parava* and *O.cephalocera* and also cephalochordate *Amphioxus.*

The largest animals in the pelagic region are the nekton. These include cephalopods such as squid and nautili among the invertebrates, as well as many marine vertebrates, such as bony fishes, sharks, sea turtles and whales. In addition, sea birds also feed on many of the same food types as these large carnivores. Air-breathing nekton, such as the turtles, and whales, are found

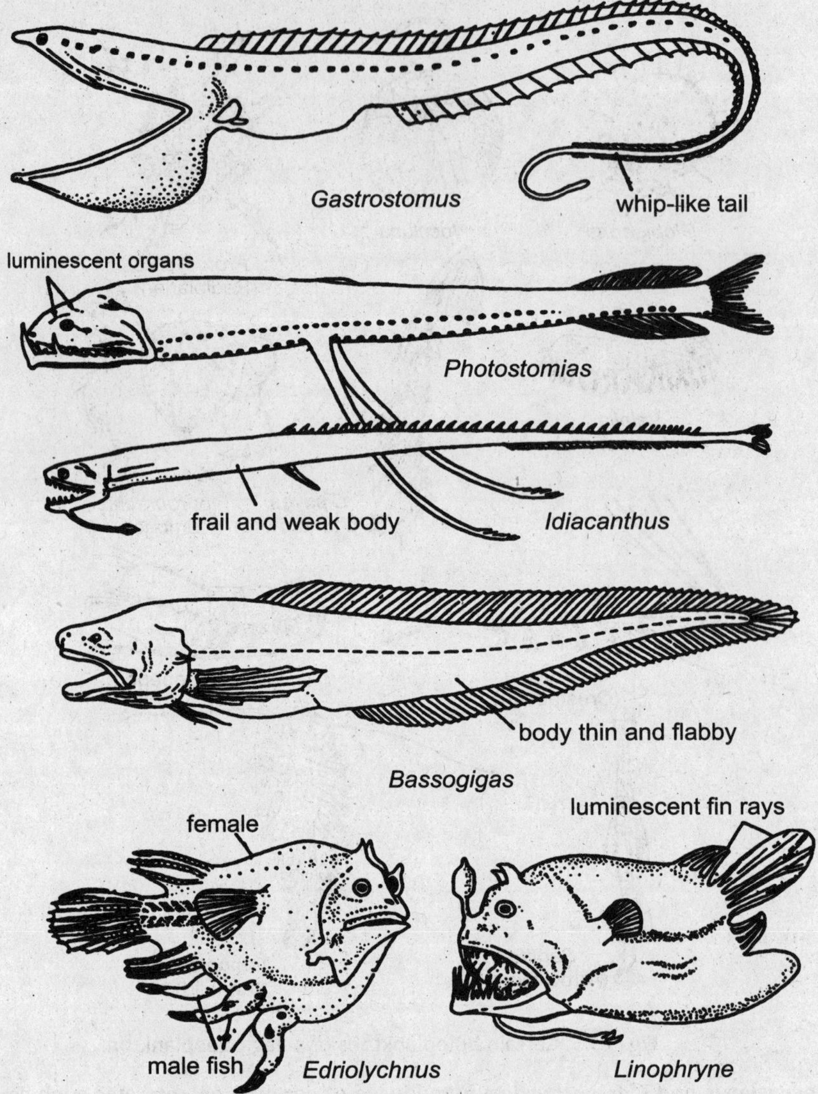

Fig. 13.6. Fishes of the deep water of the ocean.

mainly in the photic zone, but fishes extend from the sea surface to the bottom. Certain fishes such as tuna, shark, sardine, mackerel, herring, bonito, and anchovy live near the surface. Those fishes which occur at greater depths are often bizarre and unlike any fish found at the surface. They tend to be small— 15 to 20 m is large for the ocean deeps —and they are exceedingly dispersed. Many have luminescent appendages with which to lure prey, or mouth that looks too large for the rest of the body. Food is not plentiful in the deep waters of the sea, so these fishes must go for long periods of time without food, and then consume as much as they can when they have a chance.

(ii) Biotic communities of abyssal benthic zone. The abyssal benthic zone or deep sea of oceanic region is pitch dark and universal absence of light in this environment excludes the possibility of any growth of vegetation or other photosynthetic organisms. Consequently, most of the deep sea dwellers depend upon the detritus (dead body and excreta of surface forms) which sink to the sea bottom while a large number of species are active predators (carnivores).

However the deep ocean benthos are surprisingly diverse. Food is a little more plentiful on the bottom than it is in the deep waters of the ocean, because the bottom sediments are the ultimate resting place for all detritus raining out of the upper layers (oceanic surface and pelagic zones). The bottom of the sea is a soft ooze, made of the organic remains and shells of Foraminifera, Radiolaria, and other animals and plants. Sea cucumbers, brittle stars, crinoids (sea lilies), sea urchins, certain benthic fishes, and several types of crustaceans, as well as sea anemones, clams, and similar animals are all found at the bottoms of the deepest ocean trenches. The biomass of carnivorous brittle stars is often higher than that of the detritus feeders that serve as their food source; however, their metabolic rate is much lower. The great diversity of sea bottom fauna has been related with great constancy in the physical environment of the deep sea.

B. Biotic Communities of Continental Shelf

The communities of continental shelf are both richer and more diverse than those of open ocean. Diatoms and dinoflagellates are still the most productive phytoplanktons, but in the shallower regions green, brown, and red algae anchored to the bottom may be of great importance. Several kinds of these seaweeds are harvested from rocky shores as human food or for some commercial purpose in eastern countries, USA and India. The zooplankton of continental shelf are generally the same as in the pelagic region, but some purely pelagic species have been replaced by neritic species, and the overall diversity is somewhat higher. The temporary zooplanktons are much more abundant over the shelf region than in the open ocean.

The necktons of the **neritic sea** are both diverse and well known. The significant neckton species of the region are large squid, whales, seals, sea otters, and sea snakes. The most numerous necktonic forms, of course, are the fishes, which include many species of shark as well as herring-like species such as menhaden, herring, sardine and anchovy; cods and their relatives, such as haddock and pollack; salmon and sea trout; flounder and other flatfish such as sole, plaice and halibut; and mackerels, including tuna and bonito.

Communities of sea shore or intertidal zone. The littoral, intertidal or eulittoral zone, is the region of sea shore which exists between the high and low tide lines. The region of the high tide mark is called **supratidal** or **supralittoral zone**. The region of low tide is called **subtidal zone. Davenport** (1903) has divided sea shore or intertidal zone into **submerged zone** which is the portion of the shore below the low tide exposed by very low tides, a **lower beach** occupying the area between the mean low tide and mean high tide and an **upper beach** which is reached by the highest of high tides. The intertidal zone is the most variable zone in the entire sea. It is completely covered at high tide, and is completely uncovered at low tide except for tide pools. Any organism that is to survive in the intertidal must be either resistant to periodic desiccation or able to burrow to water level. For example intertidal polychaets escape desiccation by resorting to a tubicolous mode of life whereby they can resist desiccation by retaining sufficient moisture till the occurrence of the next wetting. Barnacles can survive desiccation by remaining in a state of suspended activity during dry periods. Many plants of sea shore contain jellylike substances such as agar, which absorb large quantities of water and retain it while the tide is out. Further, the majority of shore animals utilize oxygen dissolved in water and when the low tide exposes them during the periods of drought, they utilize the oxygen stored in the blood plasma, *e.g.,* nereid *Arenicola.* Barnacles achieve this by enclosing an air bubble between the shells.

One of the notable characteristics of this region is the ever present action of the waves, therefore, the organisms living on a sandy or rocky beach have had to evolve ways of resisting wave action. Many seaweeds have tough pliable bodies, able to blend with the waves without breaking, while the animals either are encased in hard calcareous shells, such as those of mollusks, Bryozoa, starfish, barnacles and crabs, or are covered by a strong leathery skin that can bend without breaking, like that of the sea anemone and octopus. Many sedentary shore animals (*e.g.,* sponges, tunicates, etc.) have special modes of adhesion which keep them firmly attached to the substratum. However, animals of sandy beaches escape wave action by taking burrowing mode of life, because, these habitats lack a hard substratum for the attachment of the animals.

The biota of intertidal zone also have to overcome the wide fluctuations in salinity and temperature. Thus, like the estuary, this variable environment is a zone in which an organism must be adapted to a broad range of environmental conditions. Like the estuary also, it is an area of very high productivity with a simple community, many of whose members may be exceedingly abundant. The diversity of biota of intertidal zone is determined by nature of substratum whether loose sand or mud or rocky coast. Ecologists have recognized three major types of intertidal zones or seashores, namely, **rocky**, **sandy** and **muddy**. **Eltringham** (1971) has included a fourth type of sea shore, the **pabble shore** is this classification. Each type of sea shore possesses specific biota each of which remains specifically adapted for its peculiar habitats. Here we will discuss only rocky shore and sandy beach habitats.

1. Biotic communities of rocky shore. The rocky shore presents solid substratum for the attachment of many sessile animals which often remain abundant here. The animals which are sedentary and inhabit rocky shore are limpets such as *Patella, Haliotis, Fissurella,* and oysters, barnacles, tunicates and bryozoans. Sessile organisms such as the sponges, the colonial hydrozoans and anthozoans such as *Gammaria* and *Zoanthus* attached to rocks. Certain animals such as some sponges, annelids and molluscs either bore the soft rocks like limestone or seek protection in the crevices of hard rocks. All these sedentary sessile animals are adopted for filter feeding.

Further, all these animals of rocky shore occur in successive zones. At the uppermost end is **a zone of bare rock** marking the transition between land and sea. Next is a **spray zone** with dark patches of algae on which the periwinkles (*Littoria*) graze. Below this is the zone called **barnacle zone** regularly covered by the high tide; rocks in this zone are encrusted with acorn barnacles, limpets and mussels (*e.g., Mytilus*). This zone is a vulnerable zone being exposed to the stress of wave action and active predation by starfishes and gastropods. The mussels act like hard substratum for tunicates, sponges and small mussles. The next one is the **zone of oysters** which is less vulnerable and includes green algae (*Enteromorpha* and *Ulva*) and barnacles. The next zone includes mussels, *i.e.,* boring bivalves, which remain within crevices with exposed siphons only to be protected from waves and predation. The chief species of borer bivalves are *Pholas, Martesia, Hiatella, Tridacna,* etc. Besides bivalves a variety of filter feeder bryozoans, and brachiopods, and tubicolous polychaets such as *serpulids, terebellids, sabellids* and *cirratulids* which live in tubes are the chief inhabitants of this zone. The subtidal zone includes red algae, brown algae (Laminarians and fucoids), sea anemone, sea urchin, corals, etc.

For example, the Okha coast of Gujarat in India is a typical rocky sea coast which has well defined fauna and flora (**Gopalakrishnan**, 1970). Dominant weeds of subtidal zone of Okha coast are *Sargassum, Dictyopteris, Graciliaria, Padina, Ulva, Polysiphonia,*etc. This subtidal zone also contains sea urchins, serpulid tubes of species *Pomatoleius crosslandi,* and mollusc species such as *Trochus radiatus, Cypraea arabica, Thais alveolata.* The upper limit of the dense weed growth of subtidal zone marks the lower end of eulittoral or intertidal zone, where occurs a dense cluster of *Crassostrea cucullata* and *Littorina* sp. The intertidal zone includes three subzones, *i.e.,* 1. the subzone of patchy weed growth; 2. the subzone of limpets such as *Sellana radiata* and barnacles and 3. the subzones of oysters and barnacles. The supratidal zone is inhabited by green algae and molluscs. Chlorophyceae such as *Enteromorpha, Ulva, Bryopsis, Codium, Halimeda, Caulerpa, Dictyosphaeria* and *Cladophora;* Phaeophyceae such as *Padina, Sargassum, Colpomenia, Iyengaria, Spathoglossum, Dictyopteris* and *Turbinaria*; Rhodophyceae such as *Gracilaria, Hypnea, Acanthophora, Polysiphonia, Cryptonemia* are the chief algae of the area. The common animals of this area are coelenterate species *Stoichactis, Ixalactis,* and *Gyrostama;* Nereid (*e.g., Eunice, Tubifex*); molluscs such as *Chiton, Nerita rumphii, Cypraea* *integerrimus, Pilumnus vespertilio,* seaweed grazer (nudibranch) molluscs, *Aplysia benedicti, Onchidium, Elysia grandifolia, Eolis, Doris* and floating mollusc *Janthina*, etc. The midlittoral reef community of Okha coast consists of fishes such as *Epinephelus, Petrachus*; the crab *Pilumnus vespertilio*, the annelid *Eurythoe complanata* and molluscs such as *Murex, Bursa*, etc. Rock crevices of intertidal zone harbour within them crabs such as *Atergatis, Grapsus, Pilumnus, Matuta, Charybdea, Gelasimus,* squillids (*e.g., Sesarma* and *Gonodactylus*); coelenterates (*e.g., Stoichactis, Ixalactis, Zoanthus, Lobophytum*; the hydroid species such as *Lytocarpus, Sertularia* and *Plumularia*); and annelids such as *Sepula, Polynoe, Eurythoe, Sabella,* etc.

(ii) Biotic communities of sandy shore. The sandy shore may be even more harsh than the rocky shore. It is subjected to all the extremes of the factors (temperature, salinity, turbidity, wave action, etc., plus inconvenience of a constantly shifting substratum. The last factor makes life on the surface almost impossible; life has retreated below the surface. Generally sandy beaches are characterized by gentler wave action. Because of the prolonged time taken for drying up, these beaches are suitable for animal life. Decaying seaweeds and dead remains of animals result in the addition of organic matter to the sand. The coastal subsoil water is the environment of a special interstitial brackish water fauna. Most of the interstitial organisms are small, elongate, vermiform with transparent bodies. They have no eye but possess well developed adhesive organs and sense organs. They are negatively phototactic and gregarious in habit. They may be herbivore, carnivore, omnivore or detritus feeder. Large animal forms which are few in number prefer coarse sand grains and smaller one which are very abundant, prefer finer sand. The principal animal groups of interstitial water are the Nematoda, Turbellaria, Annelida, Gastrotricha and Acraina. Certain animal groups such as Protozoa, occur occasionally. The top 20 cm of sand between low and midtide levels of the beach is occupied by the inhabitants such as Hydrozoa, Turbellaria, Nemertina, Rotifera, Archiannelida, Polychaeta, Ostracoda, Halacaridae and Nudibranchiata. Species of Gastrotricha, Kinorhyncha, and Isopoda occur at mid-tide levels preferably in deep layers of sand. Oligochaets are sporadic and are found at different levels. Harpactcoid copepods remain restricted to mid-tide level. Tardigrades occupy the region between mid-tide and high-tide. Thus, sandy beaches also possess characteristic successive zones or zonations of animals.

For example, according to studies of **Ganapathy** and **Rao** (1962) sandy beach fauna of Visakhapatnam coast though lacks or contains small number of tubicolous polychaets and Crustacea due to very unstable substratum, but include following animal species— spinoid *Prionospio krusadiensis, Aricia* and *Lumbriconeries*, all of which possess the capacity to secrete mucus and formation of cover of sand grain over the body for protection, *Glycera lancadivae, G. alba, Nerina bonnieri, Pisionidens indica, Pisione complexa* and archiannelids *Saccocirrus minor* and *Protodrilus* also occur. The crustacean fauna of this zone includes *Emerita asiatica, Albunea symnista* (burrowing), amphipod species *Harpinia* sp., isopod *Sphaeroma walkeri,* mysid shrimp *Gastrosaccus spinifer,* hermit crabs *Clibanarius aretheustus* living within the shells of *Thais* and *Cerithium*, the eight-oared swimming crab *Matuta victor*, the spider crab, *Philyra scabruescula, Ocypoda platytarsis* and *O.macrocera.* Lamellibranchs such as *Donax cuneatus* and *Pamphia textile,* burrowing snails (*e.g., Sinum neritoideum*), common gastropods (*e.g., Oliva gibbosa* and *Terebra* sp.), *Bullia vittata* are also found along the coast of Visakhapatnam. This beach also contains eel *Ophichtys orientalis* and siphonophore coelenterates such as *Physalia, Velella* and *Porpita*; and medusae such as *Aequoria* and *Chiropsalmus.*

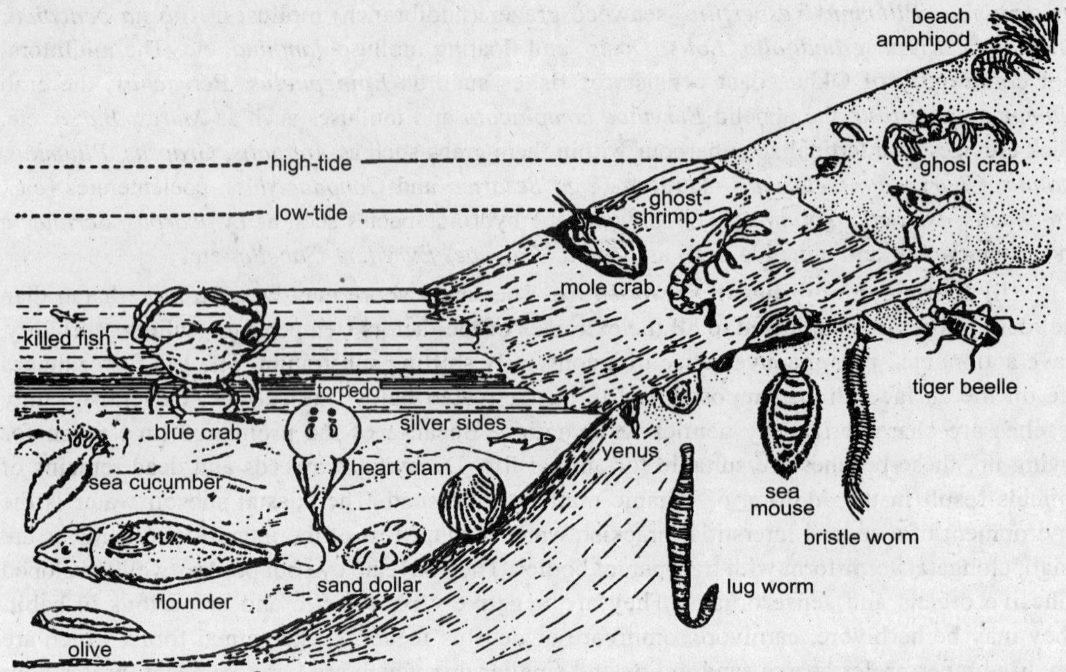

Fig. 13.7. Life on a sandy beach along the Atlantic coast.

13.4 ESTUARIES

Waters of all streams and rivers ultimately drain into the sea ; the place where this freshwater joins the salt water is called an **estuary**. An estuary (L. *aestus*—tide) is the wide lower tidal part of a river which is strongly affected by tidal action. Estuaries are the transitional zones (ecotones) between the sea and rivers and are the sites of unique ecological properties possessing a characteristic biological make up. They are semienclosed coastal bodies of water that have a free connection with the open sea and within which sea water is measurably diluted with freshwater from rivers. Examples of estuaries are river mouths, coastal bays, tidal marshes and water bodies behind barrier beaches. Not all rivers open into estuaries: some simply discharge their runoff into the ocean. Estuaries differ in size, shape and volume of water flow, all influenced by the geology of the region in which they occur. As the river reaches the encroaching sea, the stream carried sediments are dropped in the quite water. These accumulate to form **deltas** in the upper reaches of the mouth and shorten the estuary (Box 13.2). When silt and mud accumulations become high enough to be exposed at low tide, **tidal flats** develop, which divide and braid the original channel of the estuary. At the same time, ocean currents and tides erode the coast line and deposit material on the seaward side of the estuary, also shortening the mouth. If more material is deposited than is carried away, then **barrier beaches**, **island** and **brackish lagoons** appear.

Estuaries include **drowned river beds, fjords** (*i.e.,* U-shaped basin gouged by glaciers), **geological faults, river delta** and **shallow basins** enclosed by a chain of offshore bars or barrier islands.

Box 13.2 Delta

A delta is a concentration of sediment at a river mouth with splitting channels on the surface. The name comes from the ancient Greek capital letter Δ (delta) which roughly resembles with outline of Nile delta. Deltas form where a river entering an ocean, depositing more sediments than waves of current can remove. The lengthening of a channel across the sediments means that cross slopes are reduced. During times of flood, when the channel banks are over topped, some of the water is apt to take a short cut. Repeated channel splitting produces the typical morphological system of distributaries. The deltas are process response systems, the process being sedimentation.

For example, Ganges conveys annually to the Bay of Bengal more than 356,000,000 tons of sand and clay—an average of over 900,000 tons of silt per day. There are some rivers whose waters are more silt laden than those of the Ganges. The solid matter suspended in the Indus waters and discharged below Hyderabad, in Sindh is estimated to be around 1,000,000 tons per day. The Brahmputra carries down more silt than the Indus or Ganges.

Types of Estuaries

Estuaries are classified in various ways depending on diverse criteria such as geomorphology, water circulation and stratification and systems energetics. Thus, on the basis of geomorphology, estuaries include (1) drowned river valleys, (2) ford-type estuaries, (3) bar-built estuaries, (4) estuaries produced by tectonic processes, (5) river delta estuaries, etc. Based on water circulation and stratification, estuaries are classified as (i) highly stratified or salt-wedge estuaries, (ii) partially mixed stratified estuaries, (iii) completely mixed or vertically homogeneous estuaries and (iv) hypersaline estuaries. On the basis of the ecosystem energetics, the estuaries have been classified by **Odum** *et al.,* (1969) as (1) physically stressed systems of wide latitudinal range, (2) natural arctic ecosystems with ice stress, (3) natural temperate ecosystem with seasonal programming, (4) natural tropical coastal ecosystems of high diversity, and (5) emerging new systems associated with humans.

Physico-Chemical Aspects of Estuaries

Current and **salinity**, both very complex and variable, shape life in the estuary, where the environment is neither freshwater nor salt. Estuarine currents result from the interaction of a one-direction stream flow, which varies with the season and rain fall, with oscillating ocean tides, and the wind. The salinity of estuaries varies vertically and horizontally and fluctuates amazingly between 0.5–0.35 per cent. Due to low salinity freshwater has a lower density than sea water (± 1.00, as opposed to 1.03). Were there no tides in an estuary to mix fresh and salt-water the lighter freshwater would simply flow over the heavier sea water and dissipate in the ocean (Fig. 13.8). However, the tidal action acts as a plunger to thoroughly flush the estuary and mix the fresh and salt water. This has several results. The water level in the estuary fluctuates regularly, unlike that of river. For this reason the habitat that is covered at high tide and uncovered at low tide is a prominent one in the estuary, and has no analogue in any purely freshwater ecosystem. Further, the salinity is exceedingly variable and may change by a factor of ten over the course of a day at any location. At low tide, most of water passing through the estuary is fresh-river water, and the salinity is correspondingly low. At high tide, most of the

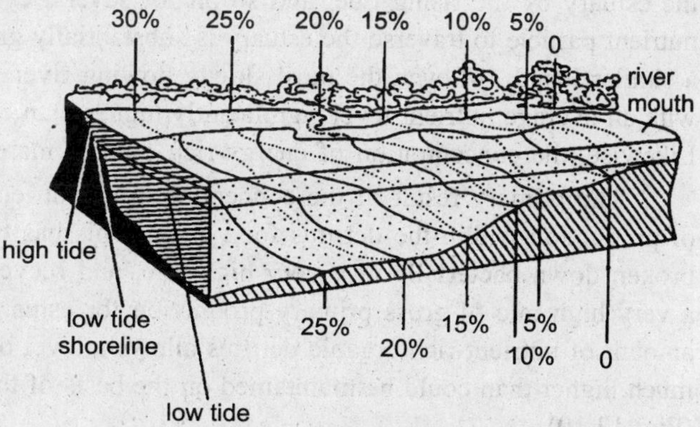

Fig. 13.8. Vertical and horizontal stratification of salinity from the river mouth to the estuary at both low and high tide. At high tide the incoming seawater increases the salinity toward the river mouth; at low tide salinity is reduced. Note also how salinity increases with depth, because lighter freshwater flows over denser salt water.

water may be of marine origin, and the salinity correspondingly high. Flow may be somewhat stratified, or mixing may be incomplete, and the salinity at any time may vary considerably from one location to another. The intensity of current, and, hence, of the degree of mixing, is a function of the intensity of the tides and of the rivers flow rate. In other words, an estuary is an exceedingly variable environment. Further, salinity of estuaries remains highest during the summer and during periods of drought, when less freshwater flows into the estuary. It is lowest during the winter and spring, when rivers and streams are discharging their peak loads. No stenohaline organism (such as Echinodermata, Cephalopoda, and other molluscs) could hope to survive in the estuaries. Likewise, no organism that could not tolerate strong currents and the turbid water that results from strong currents could live there.

The **temperatures** in estuaries fluctuate considerably diurnally and seasonally. Waters are heated by solar radiation and inflowing and tidal currents. High tide on the mud flats may heat or cool the water, depending on the season. The upper layer of estuarine water may be cooler in winter and warmer in summer than the bottom, a condition that, as in a lake, will result in a spring and autumn overturn.

Anyhow, all estuaries have high productivity. Estuary's high productivity is exceeded among aquatic communities only by coral reefs. The reason is that although an estuary may be a harsh environment in some ways it is not in others. Because the typical estuary is shallow and turbulent, the amount of dissolved oxygen tends to be fairly high although it may be low in bottom layers of those estuaries where water does not mix from top to bottom. More important, the tidal action acts to accumulate (concentrate) the nutrient and energy materials that wash in from upstream or, in some cases, that enter from the nutrient rich bottom waters of the sea (Box 13.3). The mechanism of nutrient concentration is very simple—particulate nutrient material enters the estuary at its upper end is carried seaward by the falling tide and is brought back through the estuary by the rising tide, and so on for several cycles. The length of time it takes from a nutrient particle to traverse the estuary is substantially greater than it would take for it to traverse a similar length of even the most slowly flowing river. Thus, the estuary acts as a nutrient trap, with an average nutrient level significantly higher than either the river or the sea that it connects. Likewise, the concentration of energy-rich organic materials remain high in estuaries.

The primary result of the concentration of nutrients and fixed carbon is a very high level of production within the detritus food chain. This has two aspects. First the nutrient material is broken down bacterially at a very high rate, and recycled back into soluble form. This allows a very high rate of gross primary production for estuarine plants. More important, perhaps, the amount of nutrient-rich organic detritus allows a level of productivity for detritus-eating animals much higher than could be maintained on the basis of the primary production by estuarine plants (Fig. 13.10).

Biotic Communities of Estuaries

The estuary is characterized by *low salinity, high turbidity* and shallow water. Organisms inhabiting the estuary are faced with two problems: 1. Maintenance of position and adjustment to changing salinity. The bulk of estuarine organisms are benthic and are securely attached to bottom, buried in the mud or lodged in crevices and crannies about sessile organisms. Motile inhabitants are mostly crustaceans and fish, largely young of species that spawn offshore in high salinity water. Phytoplankton are wholly at the mercy of the currents. Changing salinity dictates the distribution of life in estuary (Box 13.3).

Box 13.3

Mixing waters of different salinities and temperatures acts as a **nutrient trap.** Inflowing river waters more often fertilize the estuary with nitrogen and phosphorus. Nutrients and oxygen are carried into the estuary by tides. If vertical mixing takes place, these nutrients are not swept back out to sea, but circulate up and down among organisms, water and bottom sediments (Fig. 13.9).

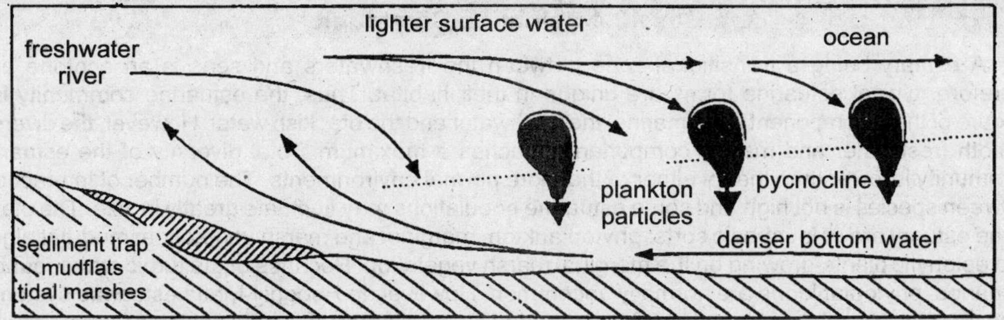

Fig. 13.9. Circulation of fresh and salt water in estuary creates a nutrient trap. The trap develops because of a salt wedge of intruding seawater on the bottom, a surface flow of lighter fresh water and a counterflow of heavily brackish water. This countercurrent serves to trap nutrients, recirculating them toward the tidal mass. The same countercurrent also sends phytoplankton up the estuary, repopulating the water.

Carikker (1967) has classified the regions of estuaries into the **head**, where freshwater enters the estuary, **upper**, **middle** and **lower reaches** with increasing range of salinities and the

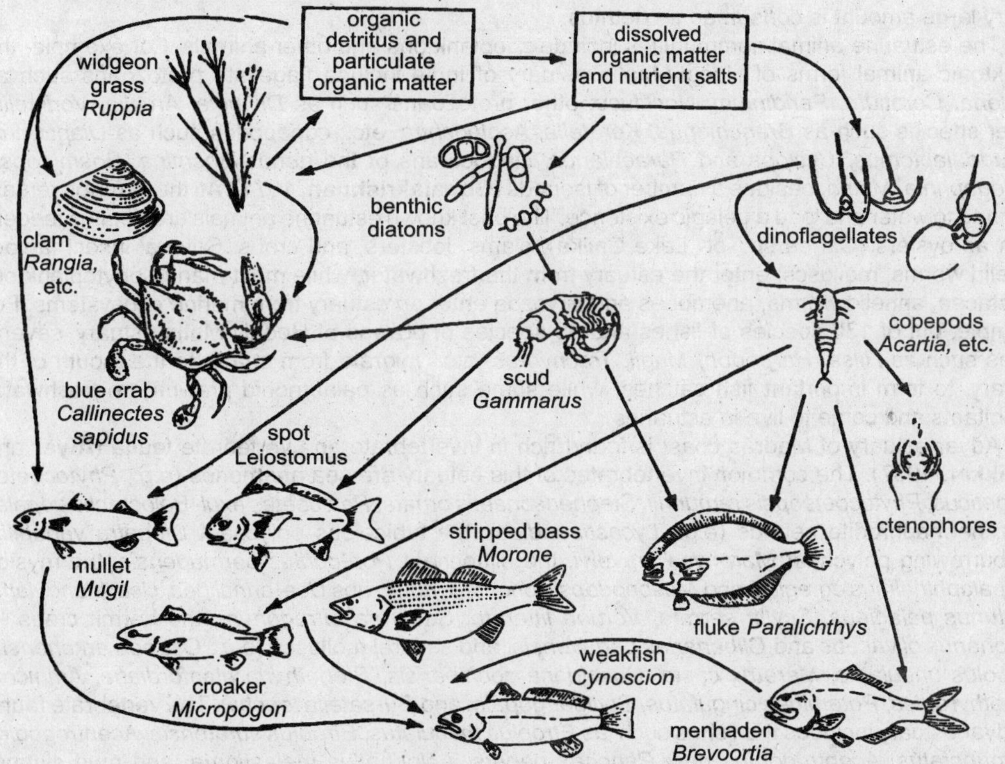

Fig. 13.10. Simplified estuarine food web based on an estuary (Pamlico River estuary in North Carolina, USA). The energy base is largely heterotrophic, supported by particulate and dissolved organic matter.

mouth with salinity nearly equal to the sea. He has also classified the animals inhabiting the estuarine region into **oligohaline** (0.5 to 5 per cent), **mesohaline** (5 to 18 per cent) in the upper reaches, **polyhaline** (18 to 25 per cent) in the lower reaches, 25 to 30 per cent in the middle and **euhaline** (30 to 50 per cent) in the mouth of the estuary. Communities of estuaries are a mixture of endemic species and those which come in from sea (Box 13.4).

Box 13.4 Estuarine Community

A estuary being a transitional zone between the freshwaters and seas is an ecotone and, therefore, typical estuarine forms are unique in their habitat. Thus, the estuarine community is a mixture of three components: the marine, the freshwater and the brackish water. However, the diversity of both freshwater and marine components reaches a maximum. Total diversity of the estuarine community is lower than that of either of the more normal environments. The number of interactions between species is not high and some estuarine populations may fluctuate greatly in size. The plants of the estuary are of four basic sorts: phytoplankton, marginal and marsh vegetation, mud-flat algae, and epiphytic plants-growing on the marginal marsh vegetation. Because of the turbid water found in estuaries, phytoplanktons are normally uncommon. However, in Hooghly-Matla estuarine system in West Bengal, **Gopalakrishnan** (1971) has reported an abundance of phytoplanktonic forms—several species of diatoms, *Synedra, Navicula, Rhizostoma, Fragilaria, Asteriobella, Biddulphia, Planktoniella, Hemidiscus, Chaetoceros, Cyclotella, Stephanodiscus, Triceratium*; several species of green algae such as *Pediastrum, Spirogyra, Eudornia, Tribonema, Closterium, Zygonema, Pandorina, Volvox, Chlorella;* and blue-green algae such as *Microcystis, Oscillatoria, Anabaena,* and *Trichodesmus*. Most estuarine algae are of marine origin. Further, the most significant estuarine plants are marsh grasses such as *Spartina, Salicornia,* and *Scirpus*, as well as some submerged filamentous algae such as *Cladophora, Chara* and *Enteromorpha*. Very few animals feed on these plants directly, but a very large amount is consumed as detritus.

The estuarine animal communities include zooplanktons and other animals. For example, the planktonic animal forms of Hoogly-Matla estuary of India include flagellate protozoans such as *Euglena, Ceratium, Peridinium, Noctiluca*; other protozoans such as *Difflugia, Arcella, Vorticella;* rotifer species such as *Branchionus, Keratella, Asplanchna*, etc.; copepodes such as *Diaptomus, Pseudodiaptomus, Cyclops* and *Paracalanus;* cladocerans of the genus *Bosmina, Bosminopsis Ceriodaphnia, Moina*, besides a number of isopods (**Gopalakrishnan**, 1971). All these forms remain confined to water and lead a pelagic existence. The best known estuarine animals are detritus feeders such as oysters (*Ostrea* sp., of Lake Chilka), clams, lobsters, and crabs. Several insect larvae, annelid worms, molluscs, enter the estuary from the freshwater, while most marine phytoplankton, Crustacea, annelid worms, anemones and Bryozoa enter an estuary from marine ecosystems. For example, out of 130 species of fishes and 30 species of prawns of Hoogly-Matla estuary, several fishes such as *Hilsa, Harpodon, Mugil, Trachyurus*, etc., migrate from sea in to the mouth of the estuary, to form important fish catches, while some such as palaemonid prawns are freshwater inhabitants and come to live in estuaries.

Adyar estuary of Madras coast is found rich in invertebrate and vertebrate fauna (**Aiyar** and **Panikkar**, 1937). The common invertebrates of this estuary are sea anemones (*e.g., Phytocoetes gangeticus, Phytocoeteopsis ramunnii, Stephensonaetis ornata, Pelocoetes exul, Bolocractis gopalai*, etc.); hermaphrodite nereids (*e.g., Lycastis indica*); the tubicolous polychaet *Diopatra variabilis*; the burrowing polychaet *Marphysa gravelyi*, the oligochaet *Pontodrilus bermudensis*; the mysids *Rhopalophthalmus egregius* and *Mesopodopsis orientalis*; the crabs *Uca annulipes, Uca triangularis, Neptunus pelagieus, Scylla serrata, Veruna litterata, Sesarma tetragonum*; the hermit crabs — *Clibonarius olivaceus* and *Clibanarius padavensis*, and several molluscs (*e.g., Ostraea arkanensis, Modiolus undulatus, Meretrix casta, Cuspidaria cochinensis, Stenothyria blanfordiana, Amnicola stenothyroides, Potamides cingulatus, Styliger gopalai* and *Bursatella leachii*). The vertebrate fauna of Adyar estuary includes the fishes such as *Etroplus maculatus, Etroplus suratensis, Acentrogobius virdipunctatus, Acentrogobius neilli, Panchax parvus, Aplocheilus melastigma,* and mud skipper *Periophthalmus koelreuteri* and snakes such as *Natrix piscator* and *Cerberus rhyncops.*

13.5 CORAL REEF

Coral reefs are the specialized ecosystems of ocean which are among the most productive of all ecosystems anywhere, with a diversity equalled only by tropical rainforests. The reef environment is very rugged one because it is constantly subjected to the pounding of the waves, and parts of the reef are often above water at low tide. But reef is also a very rich environment for those species that can withstand the physical pounding of water. The amount of oxygen is very high. During the day, it may reach 250 per cent of saturation because of the production of oxygen by algae in the reef structure. Likewise, the tremendous amount of CO_2 produced by the metabolism can be carried off before it becomes too high.

Reefs can be found wherever a suitable substrate exists in the lighted waters of the tropics, away from areas of continental sediments which would tend to bury the reef in mud or cold upwellings which would tend to lower the temperature too much (Box 13.5). They may form **barrier reefs** (*e.g.,* barrier reef in Australia which contains canal or lagoon between them and coast) adjacent to continents; a **tolls** or rings shaped or horse-shoe-shaped islands on the top of submarine heights, or **fringing reefs**, as in Mandapam, Appa Islands and Hare Islands.

Box 13.5

In the Indian ocean, corals are known to occur in the Gulf of Kutch, Ratnagiri, Malvan and Redi on west coast between Bombay and Goa; Angria bank, off Ratnagiri, Gaveshani bank, off Malpe, off Marmagoa harbour; off Visakhapatnam, Rameshwaram, Kanyakumari, Laccadives and Andman and Nicobar island.

In all modern reefs, the main reef structure is formed by madreporarian corals of following families—Poritidae (*Porites minicoiensis, Porites mannarensis*), Acroporitidae (*Acropora intermedia*), Astraeidae and Milliporidae. Secretions of calcareous algae go into the building up of reefs. Equally important are calcareous Bryozoa, lamellibranchs and gastropod molluscs and Foraminifera, which become significant elements in the total reef structure. Although the name 'coral reef' implies that corals should be the most abundant organisms, algae of many sorts comprise over three times the biomass of the corals. In addition, the calcareous algae are not only producers; their skeletons act to cement the entire reef to make it more resistant to the pounding surf.

Coral reefs normally exist in water with mean temperature 20.5^{0}C, but they can withstand very high temperatures. The growth of coral reef is influenced by salinity and reduced tides. Moderate water movement around the coral reefs help in the circulation of food.

Biotic Communities of Coral Reef

Coral reef ecosystems harbour an abundant and well diverse biota. The well known Krusadi Island is the coral island of gulf of Manaar. It's dominant corals are *Porites, Montipora, Poecilopora, Acropora, Madrepora, Favia, Symphylia, Psammocora, Platygyra, Merulina* and *Leptastrea.* In the Krusadi area there is a luxuriant growth of sea weeds such as *Sargassum, Caulerpa, Turbinaria, Spatoglossum, Gracilaria, Ulva, Enteromorpha, Chaetomorpha* and *Polysiphonia.* The abundant sponges of this reef are *Spirastrella incorstans* and *Callyspongia fibrosa.* The other inhabitants of this reef are following—hydroid coelenterates such as *Syncoryne, Sertularia, Eudendrium, Pennaria, Tubularia, Thyroscypus, Plumularia;* alcyonarians such as *Clavularia, Alcyonium, Lobophytum, Scelerophytum, Telesto,* zooantharians such as *Zoanthus, Germmaria;* polychaet annelids such as *Nereis, Polynoe, Eunice, Lepidonotus, Iphion, Eurythoe, Hesione*; sipunculid species such as *Dendrostomum, Aspidosiphon* and *Phascolosoma;* molluscs such as *Doris, Dendradoris, Acanthochiton, Ischnochiton, Haliotis, Emarginula, Trochus, Turbo, Nerita, Natica, Cypraea, Thais*; and echinoderms such as *Holothuria atra, H. scabra, H. edulis* and starfish *Proteaster lincki.*

Beautifully pigmented fishes such as *Holocentrus, Lutianus* sp., *Pomacanthoides, Chaetodontopsis, Chaetodon, Linophora, Thallasoma* and *Zanclus* and the eels *Gymnothorax undulatus* and *G. punctatus* are commonly frequent in the coral reefs. These inhabitants of coral reef community, when exposed to low tides are subjected to predation by large number of birds such as the crested tern *Sterna bergii*, golden plover *Pluvialis dominica fulva*, etc.

According to **Clapham, Jr.,** (1973) the great productivity of reefs is obviously caused by something more than the abundance of light and oxygen. He considers nutrient cycles which runs very quickly and efficiently between the components of reef ecosystem to be main cause of great productivity of reef ecosystems. In fact, no organic matter is incorporated permanently into the sedimentary system portion of reef. The conditions of rapid nutrient cycling allow a mechanism by which the natural paucity of nutrients in the ocean can be effectively circumvented. It may also allow a gradual increase in nutrient availability by incorporating the nutrients of the non-reef organisms that wash into the reef, while still continuing to prevent the reef's own nutrients from being lost to the deeper seas.

QUESTIONS

Long Answer Questions

1. Explain the physical and biological characters of various divisions of a sea.
2. Comment on the composition of fauna of oceanic ecosystem.
3. Describe in detail the adaptations of pelagic organisms in the marine environment.
4. What are characteristic physical factors which affect life in deep sea? Describe the 'deep sea fauna' and their adaptations? How far is it correct to call 'deep sea' a desert.
5. Write an essay on the deep sea fauna with special reference to their adaptation.
6. What is estuary? Describe physico-chemical properties of an estuary. Add a note regarding the estuarine production and food chains.
7. Describe coral reef as habitat of oceanic biota.

Short Answer Questions

1. Give a short account of zonation of marine environment.
2. Describe in brief biotic communities of the pelagic zone of ocean.
3. Write short notes on the following:
   (a) Estuaries;
   (b) Coral seef ecosystem.

Very Short Answer Questions

1. What are nekton?
2. What is neretic?
3. Define pelagic.
4. What is oceanography?
5. Define abyssal
6. What is estuary?
7. What are benthos?
8. What are coral reefs?
9. Eenlist three types of coral reefs.
10. What are mid-oceanic ridges?

Fill in the Blanks

1. Marine ecosystem covers .... of the earth's surface.
2. Five great oceans of the world are .... , Atlantic, ... , Arctic and Antarctic oceans.
3. Pelagic zone of the ocean is subdivided into three zones: .... , mesopelagic and ....
4. Littoral zone of ocean is also called ....
5. *Noctiluca miliaris* is .... of inshore water of Mandapam.
6. The largest animals in pelagic region of ocean are ....

Multiple Choice Questions

1. Animals living in the bottom of the aquatic systems is called
   (*a*) pelagic animals (*b*) benthos
   (*c*) bathyal (*d*) abyssal
2. The most stable natural ecosystem is
   (*a*) forest (*b*) desert
   (*c*) ocean (*d*) pond
3. The largest ecosystems in world are
   (*a*) grasslands (*b*) forests
   (*c*) great lakes (*d*) oceans
4. Littoral form of animals are found
   (*a*) in freshwater (*b*) in sea water near the surface
   (*c*) in deep sea (*d*) at the bottom of the sea
5. The most diverse communities are in the
   (*a*) abyssal zone (*b*) littoral zone
   (*c*) neritic zone (*d*) oceanic zone
6. Upper part of sea water mainly contains
   (*a*) nekton (*b*) neuston
   (*c*) plankton (*d*) nekton and neuston
7. Pelagic forms of animals are found
   (*a*) on the surface of the sea floating along the waves
   (*b*) in deep sea
   (*c*) in the soil
   (*d*) as parasites of mammals
8. Benthos are
   (*a*) periphyton
   (*b*) zooplankton
   (*c*) epineuston
   (*d*) bottom dwelling primary consumers
9. Most animals that live in deep oceanic waters are
   (*a*) producers (*b*) primary consumers
   (*c*) secondary consumers (*d*) detritivores
10. Benthoic animals are
    (*a*) deep dwellers in sea (*b*) floating organisms
    (*c*) submerged in sea (*d*) float on the sea surface

11. The abyssal zone of oceans is characterized by
(*a*) presence of sunlight and producers
(*b*) absence of sunlight and all living organisms
(*c*) absence of sunlight and presence of consumers and decomposers
(*d*) absence of sunlight but presence of producers

12. Plants of salty seashore wetlands are called
(*a*) heliophytes (*b*) halophytes
(*c*) saprophytes (*d*) hydrophytes

ANSWERS

Very Short Answer Questions

1. Nekton are the free-swimming organisms.

2. Neretic refers to regions of marine environments where land masses extend outward as a continental shelf.

3. Pelagic refers to the open sea.

4. Oceanography is the branch of ecology dealing with the biology, chemistry, geology, and physics of the ocean.

5. Abyssal refers to bottom waters of oceans.

6. Estuary (from Latin *aestus* "tide") is a partially enclosed embayment, such as river mouth or coastal bay, where freshwater and salt (sea) water meet and where tidal action is an important physical regulator and energy subsidy.

7. Benthos are bottom dwelling organisms that inhabit the bottom of rivers, lakes and the sea.

8. Coral reefs are colonial groups of cnidarians that secrete an external skeleton of calcium carbonate, usually in mutualistic relationship with algae.

9. Atolls, barrier reefs and fringing reefs.

10. Mid-oceanic ridges are undersea in which spreading tectonic plates create vents, hot sulphurous springs and seeps.

Fill in the Blanks

1. 70.8% **2.** Pacific, Indian **3.** Epipelagic, bathypelagic
4. Intertidal zone **5.** Zooplankton **6.** Nekton.

Multiple Choice Questions

1. (*b*) **2.** (*c*) **3.** (*d*) **4.** (*b*)
5. (*b*) **6.** (*c*). **7.** (*a*) **8.** (*d*)
9. (*d*) **10.** (*a*) **11.** (*c*) **12.** (*b*).

14

Terrestrial Ecosystems: Realms and Biomes (Forests)

Only about a quarter of the earth's surface is dry land, yet the complexity and variegation of terrestrial ecosystems are much greater than those of aquatic ecosystems. There is no unifying theme in terrestrial ecosystems analogous to the physical properties of water. Rather, the variety of climates, the diversity of all lithosphere, and heterogeneity of terrestrial biotic communities all conspire to give rise to such a variety of themes that they are woven together in nature to give an enormous diversity of biotic communities and terrestrial ecosystems.

14.1 PHYSICO-CHEMICAL NATURE OF TERRESTRIAL ECOSYSTEMS AND THEIR COMPARISON WITH AQUATIC ECOSYSTEMS

An aquatic system is essentially a **single-phase system**, where water sets the tone for entire habitat. A terrestrial ecosystem, on the contrary, is a **three-phase system**, where the characteristics of the habitat are a function of the atmosphere and climate, the soil, and the biotic community itself. The atmosphere is the source of oxygen for animals and carbon dioxide for plants. Both are sufficiently common and well-mixed in the atmosphere that they are essentially never limiting. Air as a medium of support is much less buoyant than water. Thus, an organism need relatively little support to survive in water, but it needs a fairly rigid skeleton to live and move on land. Climate in a terrestrial ecosystem is much more variable than it is in water. The most important aspects of climate are the temperature and water relations of ecosystems. These relations include mean annual temperature, amount of temperature fluctuation, annual rainfall, degree and time of fluctuation in rainfall and potential evapotranspiration. Rainfall tends to be somewhat irregular even in the most predictable terrestrial ecosystems, and in arid regions or during periods of drought, lack of water may be a source of extreme stress. The temperature is much more variable on land than in water. This is caused by the fact that aquatic ecosystems are heated by absorption of radiation by water, and water has a high specific heat; terrestrial ecosystems, on the other hand, are heated by the absorption of heat by soil, rock, and vegetation, whose specific heat are much lower (Table 14.1). A given quantity of heat energy absorbed by solid material can change its temperature two to five times as much as the same energy absorbed by water. Thus, heat is gained and lost rapidly by terrestrial ecosystems, with a resulting wide fluctuations of temperature, both diurnally and seasonally.

Table 14.1. Specific heat of several common substances.

1.	Freshwater	1.00
2.	Sea water	0.93
3.	Wet mud	0.60
4.	Moist sandy clay	0.33
5.	Solid rock	0.20
6.	Copper	0.09

In terrestrial ecosystems, the soil serves two major functions—it provides support for living organisms, and it is the source for all essential nutrients except for carbon, oxygen and hydrogen (for detail see Chapter 2). Even its supportive role is different from the way in which a lake sediment supports rooted vegetation. Because the entire weight of all living organisms is borne by the soil, there is no partition of support between the soil and the air. The role of soil as a source of nutrients is unique to terrestrial environments. It is the site of the entire detritus food chain and, thus, is central to the biogeochemical cycling of nutrient materials. Different types of soils have different properties that affect the availability of nutrients to plants, so the productivity of the terrestrial ecosystems is very closely tied to the chemistry of the soil. Further, in terrestrial biotic communities interactions between species are real but they are simple, involving relationships such as predation, mutualism, parasitism and the like. In contrast to aquatic ecosystems, living organisms of terrestrial ecosystems leave a permanent mark on the system. For example, the role of plants in breaking down and weathering of rock and building soil is much more evident than it is in almost all aquatic ecosystems. In addition, the succession that takes place in a terrestrial community is almost entirely a function of the organisms within the ecosystem, unlike aquatic successions such as lake eutrophication, which are dependent in large part on materials being washed in from outside.

Lastly, in each of these two ecosystems, namely aquatic and terrestrial ecosystems, several types of major subdivisions can be recognized : thus, one can distinguish freshwater, estuarine and marine aquatic ecosystems, and several major types of terrestrial ecosystems such as prairies, forests, tundra, etc. The former are distinguished on the basis of a major chemical difference (*i.e.,* salt content), the latter generally on the basis of the predominant type of vegetation (grass, tree, etc.).

14.2 CLASSIFICATION OF TERRESTRIAL ECOSYSTEMS

The earth surface— the continental land masses have been classified by biogeographers and ecologists into following regions:

(i) Biogeographic realms; (ii) Biomes.

(I) Biogeographic Realms or Regions

Careful studies of the distribution of plants and animals over the earth have revealed the existence of six major biogeographic realms, namely **Palaearctic realm, Nearctic realm, Neotropical realm, Ethiopian realm, Oriental realm** and **Australian realm,** each characterized by the presence of certain unique organisms. Each of these realms embraces a major continental land mass and each remains separated by oceans, mountain ranges, or desert.

(II) Biomes

Within these biogeographic realms and established by a complex interaction of climate, other physical factors and biotic factors, are large community units called **biomes**. A biome is a large community unit characterized by the kinds of plants and animals present. In each biome the **kind** of climax vegetation is uniform—grasses, conifers, deciduous trees—but the particular **species** of plant may vary in different parts of the biome. The kind of climax vegetation depends upon the physical environment and the two together determine the kind of animals present. The definition of biome includes not only the actual climax community of a region, but also the several intermediate communities that precede the climax community. Further, there is usually no sharp line of demarcation between adjacent biomes; instead each blends with the next through a fairly broad transition region termed as **ecotone**. There is, for example, an extensive region in Northern Canada where the tundra and coniferous forests blend in the tundra-coniferous ecotone. Ecologists have identified the following biomes:

1. Tundra Biome

Tundra presents the most common example of "fragile ecosystem". Tundra which means "marshy plain", lies largely north of latitude 60°N (*i.e.,* between the Arctic ocean and polar ice-caps and the forests to the south) and is characterized by the absence of trees, the presence of dwarfed plants, and an upper ground surface that is wet, spongy and uneven, or hummocky, as a result of freezing and thawing of this poorly drained land. Some five million kilometers areas of Tundra stretch across Northern America, Northern Europe and Siberia. Although there is variation from place to place within the biome, temperature, precipitation, and evaporation are characteristically low, the warmest months averaging below 10°C and the wettest with about 25 mm of precipitation. Despite the small amount of precipitation, water is usually not a limiting factor because the rate of evaporation is also very low.

The ground usually remains frozen except for the uppermost 10 to 20 cm, which thaw during the brief summer seasons. The permanently frozen deeper soil layer is called **permafrost**. The permafrost line which may exist at a depth of a few centimeters to several meters, is the ultimate limit of plant root growth, but the immediate control is the depth to which soil is thawed in summer. The rather thin carpet of vegetation of tundra biome includes few species : grasses and sedges are characteristic of numerous marshes and poorly drained areas, but large areas consist of grasses, sedges, mosses and lichens, with occasional occurrence of dwarf birches (*Betula*) and willows (*Salix*). Perhaps the most characteristic arctic tundra plant is the lichen known as " reindeer moss" (*Cladonia*). Tundra is also characterized by heath-plants such as *Erica* and other members of Ericaceae family. The net primary productivity is extremely low.

The animals that have adapted to survive in the tundra are caribou or reindeer, musk ox, the arctic hare, arctic fox, polar bear, wolves, lemmings, snowshoe hare, snowy owls, ptarmigans and during the summer, swarms of flies (Dipteran black flies), mosquitoes and a host of migratory birds (*e.g.,* water fowl).

Tundra biomes cover large areas of the arctic zone. There are two tundra biomes one in the palaearctic, and another in the nearctic region. Tundra-like areas, called the **alpine tundra** are also quite similar to some of Arctic tundra, but in differs the absence of permafrost and in the presence of better drainage and generally a longer growing season. Mosses and lichens are less prominent, flowering plants are also scanty.

2. High Altitude or the Alpine Biome

The region of mountain above the timber line contains a distinct flora and fauna and is referred to as the **alpine zone**. Alpine zone remains conspicuous in those mountains whose peaks reach up to the niveal or snow zone, as in the Himalayas. The alpine zone (zone which lies between timber line and snow zone) includes in the descending order, a **sub-snow zone** immediately below the snow zone, a **meadow zone** in the centre and a **shrub zone** which gradually merges into the timber zone. According to **Mani** (1957) snow zone of Himalayas lies over 5100 m above mean sea level and alpine zone exists at a height of 3600 m. From an ecological view point, the zone above the limits of tree growth (timber line) exhibits extreme environmental conditions which greatly influence the biota of this region.

The characteristic features of high altitude environment are following—a low air density, low oxygen and carbon dioxide contents and water vapour, high ozone content, high atmospheric transparency affording greater penetration of light, cold, snow cover, increased rate of desiccation, high wind velocity, insolation of high intensity during the day and rapid radiation during the night, the high glare from the sky and snow, high intensity of ionizing radiation and the exclusion of trees.

In the western Himalayas of Jammu and Kashmir, Himachal Pradesh and Uttar Pradesh and the eastern Himalayas of Assam and Arunachal Pradesh, **subalpine forests** dominated by trees of such species as *Abies, Pinus*, and so on, occur at altitudes of 2900 to 3500 meters, while **alpine scrub** forests are found at altitudes of 36,000 meters and more. These scrub forests contain the broad-leaved *Betula utilis*. Alpine zone of Himalayas is also characterized by a sparseness of animal groups, the scarcity being relatively important as far as tropical Indian, south Chinese, Indo-Chinese and Malayan derivatives are concerned. Important constituents of the fauna of this zone are cold-adapted palaearctic forms.

Many invertebrates of alpine zone are predatory and occur in lakes, streams and ponds. Among vertebrates fishes and amphibians are totally lacking and reptilian fauna is greatly reduced. However, pelobatid frog, *Aleurophryne mammata* is found to occur at 4500 meters of height in Himalayas and viviparous *Mabuya varia* has been reported to live at 4000 meters of height in Mount Kilimanjaro. Some of the representative vertebrates of Himalayan alpine zone are—crow, *Corvus corax*; snow partridge, *Lerwa nivicola*; snow leopard, *Felis unica* and *Felis lynx*; European leech marten, *Mustela foina*; the vole, *Lagomys*; Tibetan yak, *Bos grunnius*; Tibetan sheep, *Ovis hodgsoni*; pamir sheep, *Ovis poli*; ibex, *Capra sibirica*; markhor, *Capra falconeri*; and Persian wild goat, *Capra aegagrus.*

A large number of insects and arachnids remain best adapted to Himalayan alpine zone and at 6900 meters height. The torrential streams which are the products of melting snow contain many mayflies, stoneflies and caddisflies. Some of common insects of this zone are—stonefly *Rhabdiopteryx lunata*; grasshoppers such as *Bryodema* and *Gomphomastax*; apterous species *Conophyma*; *Dicranophyma*; *Spingonotus*; tettigonid *Hypsinomus fasciata*; earwigs such as *Anechura*; bugs such as *Dolmacoris, Tibetocoris,* and *Nysius ericae;* aquatic beetle *Amphizoa;* Carabids such as *Bembidion, Carabus, Calosoma, Harpalus, Trechus, Calathus, Bradytus, Broscus,* and *Dyschirius;* staphylinids such as *Atheta, Aleochara, Geodromicus, Oxypoda, Philonthus* and *Tachinus;* tenebrionids such as *Bioramix* and *Chianalus;* ants such as *Formica picea*; bumble bees such as *Lapidariobombus separandus;* butterflies such as *Calias, Argynnis* and *Parnassius*; the fly *Deuterophlebia*; collembolans such as *Entomobrya, Isotoma, Proistoma, Hypogastrura, Isotomurus, Tomocerus* and *Onychiurus* ; and numerous simulids blepharocerids, syrphids, anthomyids tachinids and sarcophagids.

3. Forest Biomes

The word **forest** is derived from Latin '*Foris*' meaning outside, the reference being to village boundary fence, and must have included all uncultivated and uninhabitated land. Today a forest is any land managed for the diverse purposes of forestry whether covered with trees, shrubs, climbers, etc., or not. The forest biomes include a complex assemblage of different kinds of biotic communities. Optimum conditions of temperature and ground moisture responsible for the growth of trees contribute greatly to the establishment of forest communities. In addition, 50 mm rainfall is a prerequisite for the trees. The nature of soil, wind and air currents determines the distribution (abundance or sparseness) of forest vegetation. Normally ecologists recognise among forest communities such features as their evergreen nature, whether deciduous or indeciduous, as well as their shape, whether broad-leaved as in temperate forests or more needle-like as in the conifers. On the basis of these features the forest biomes of the world have been classified into following biomes—**coniferous forest, tropical forest**, and **temperate forest**. All these forest biomes are generally arranged on a gradient from north to south or from high altitude to lower altitude.

(i) Boreal coniferous forest. Cold regions with high rainfall and strongly seasonal climates with long winters and fairly short summers are characterized by boreal coniferous forest which is transcontinental. For example, adjacent to the tundra region either at Latitude or high altitude is the **northern coniferous forest**, which stretches across both North America and Eurasia just south of the tundra (*i.e.,* Canada, Sweden, Finland, Siberia and Mussoorie). The term **taiga** is applied to the northern range of coniferous forests. This is characterized by evergreen plant species such as white spruce (*Picea glauca*), balsam fir (*Abies balsamea*) and pine trees (*Pinus resinosa, Pinus strobus*) and by animals such as snow shoe hare, the lynx, the wolf, bears, red fox, porcupines, squirrels, bats, amphibians such as *Hyla* and *Rana,* etc. Large size is common, both in the trees, which range upto some 40 m in height, and in vertebrates, which include the giants of several groups of animals, such as moose, caribou (wild reinder), elk, grizzly bear, wolverine, beaver, and several large species of birds. A wide variety of insects are also found. Species diversity is low, and pure stands of trees and shrubs are common.Understory trees are uncommon as a result of the continual low light penetration. Among common understory associates are orchids and ericaceous shrubs such as the blue berry. The thalloid mosses and lichens being very rich understory vegetation.

Boreal forest soils are thin podozols, and are rather poor both because the weathering of rock proceeds slowly in the cold environments, and because the litter derived from conifer needle is broken down very slowly and is not particularly rich in nutrients. These soils are acidic and mineral deficient as the result of the movement of a large amount of water through the soil. In the absence of a significant counter upward movement of evaporation, soluble essential nutrients such as calcium, nitrogen, and potassium are sometimes leached beyond the roots thereby leaving no alkaline-oriented cations to encounter the organic acids of the accumulating litter. The productivity and community stability of a boreal coniferous forest are lower than those of any other biome.

(ii) Temperate deciduous forest. The temperate forest biomes are characterized by a moderate climate and broad-leaved deciduous trees, which shed their leaves in fall, are bare over winter and grow new foliage in the spring. These forests are the characteristics of North America, Europe, Eastern Asia, Chile, part of Australia and Japan, with a cold winter and an annual rainfall of 75—150 cm and a temperature of 10—20°C. In these biomes the precipitation may be fairly uniform throughout the year. In India, at elevations of 9000' (=2743.2m)—12,000' (=3655.6m) in Himalayas occur temperate vegetation including pines, fir, yew and juniper trees with an undergrowth of scrubby rhododendrons.

Soils of temperate forests are podozolic and fairly deep. Trees are quite tall—about 40—50m in height and their leaves are thin and broad. The predominant genera of this biome are maple (*Acer*), beech (*Fagus*), oak (*Quercus*), hickory (*Carya*), basswood (*Tilia*), chestnut (*Castnea*), cottonwood (*Populas*), sycamore (*Platanus*), elm (*Ulmus*) and willow (*Salix*). In some locations, coniferous vegetation may be quite predominant and that incudes white pine (*Pinus strubos*), hemlock (*Tsuga canadensis*) and red cedar (*Juniperus virginianus*). There are not many epiphytes or lianas save for some species of mosses, algae and lichens growing on tree trunks, and a few vines, notably *Vitis,* the grape. The understory of shrubs and herbs in the deciduous forest is typically well-developed and richly diversified, with a considerable portion of the photosynthesis and flowering attuned to the short day of the spring season, prior to the leafing out of and consequent shading by the tree canopy. Accordingly, there are often two separate herb assemblages. One consists of spring flowers, which bloom before the trees have expanded their leaves and are gone by summer, and the other is adapted to the low light levels of the forest floor and lasts into the fall.

The animals originally present in temperate forests are deer, bears, squirrels, gray foxes, bobcats, wild turkey and wood packers. Other common animals of this region are invertebrates such as earthworms, snails, millipedes, Coleoptera and Orthoptera and vertebrates like amphibians such as newts, salamanders, toads, and cricket frogs; reptiles such as turtles, lizards and snakes; mammals such as racoon, opossum, pigs, mountain lion, etc., and birds such as horned owl, hawks, etc. All these animals and plants show a profound seasonality; some may even hibernate throughout the winter. The range of animal size and adaptations is wide; the largest animals include such forms as the deer and black bear. The dominant carnivores are large, including the wolf and mountain lion, although smaller carnivores such as fox and skunk are also common. Diversity of fauna is lower than in any of the rain forests and a few species seems clearly to be dominant.

(iii) Temperate evergreen woodland (Chaparral). Many parts of the world have a Mediterranean-type climate with warm, dry summers and cool, moist winters. These are commonly inhabited by low evergreen trees with small hard needles or slightly broader leaves. The most important area of tropical evergreen woodland in North America is the **'chaparral'** of the Pacific coast, the Mediterranean **'maquis'**, Spanish **'encinar'** and **'melle scrab'** on Australia's South coast are the same type of community. In such a woodland, trees are essentially lacking, although shrubs may range upto 3-4 m in height. Species diversity is roughly intermediate between that of a temperate deciduous forest and a drier grassland. Fire is an important factor in this ecosystem, and the adaptations of the plants enable them to regenerate quickly after being burned. The characteristic animals of temperate evergreen woodland or chaparral are mule deer, brush rabbits, wood rats, chipmunks, lizards, wemtits and brown towhees. Small-hooved cursorial ungulates are the dominant herbivores. Saltatorial (jumping) animals and many fast moving ungulates are also common in this fauna.

(iv) Temperate rainforests. The temperate rainforest is a colder ecosystem than any other rainforest. Such a forest has a definite seasonality, with both temperatures and rainfall varying throughout the year. Rainfall is high, but fog may be very heavy and actually more important as a source of water than rainfall. The diversity is much lower, both in plants and animals, in comparison to warmer rainforests, yet it remains still higher than other temperate forest types. The diversity is much lower, both in plants and animals, in comparison to warmer rainforests, yet it remains still higher than other temperate forest type. The dominant trees (canopies) are coast redwood (*Sequoia sempervirens*) of the Pacific coast of North America and the alpine ash (*Eucalyptus regnans*) of Australia and Tasmania, both of which reach more than 100m in height. Epiphytes and lianas are common but are not abundant like those of other rainforest. The animals of temperate rainforests are similar to those of deciduous forests, but show a somewhat higher diversity.

(v) Tropical rainforests. Tropical rainforests occur near the equator in Central and South America, Central and Western Africa (Congo, Zambesi river), Southeast Asia (parts of India and Malaysia), Malaya, Borneo, New Guinea and Northwest Australia. Tropical rainforests are among the most diverse communities on earth. Both temperature and humidity are high and constant. The annual rainfall which exceeds 200 to 225 cm is generally evenly distributed throughout the year.

The flora is highly diversified: a square mile may contain 300 different species of trees, a diversity unparalleled in any other biome. The extremely dense vegetation of the tropical rainforests remains vertically stratified with tall trees often covered with vines, creepers, lianas, epiphytic orchids and bromeliads. Under the tall trees is a continuous evergreen carpet, the canopy layer, some 25 to 35 meters tall. The lowest layer is an understory of trees, shrubs, herbs as ferns and palms, all of which become dense where there is a break in the canopy. Nearly all plants are evergreen, and those that do lose their leaves entirely do so at irregular intervals with no apparent regard to the climatic regime. The leaves of most plant are of moderate size, leathery and dark green in colour. Their roots are often shallow and have swollen bases or flying buttresses.

Soils of tropical rainforests are red latosols, and they may be exceedingly thick. The high rate of leaching makes these soils virtually useless for agriculture purposes, but if they are left undisturbed the extremely rapid cycling of nutrients within the litter layer (due to decomposition) can compensate for the natural poverty of the soil. It is the nature of the soil, both its potential for high leaching as well as its chemical composition that promotes a rock-like quality when exposed to air, that largely has prevented western style agriculture from being applied to the tropical forests.

Invertebrate density and abundance are very high in tropical rainforests, but while vertebrates are diverse, they are not as abundant as in many other communities. The common invertebrates of these forests are worms, snails, millipedes, centipedes, scorpions, isopods, spiders, insects, planarians and leeches. Among insects, heteropterans, orthopterans, blattids, mantids, phasmids, bees, termites and ants are most common. The common vertebrates of tropical rainforests are the arboreal amphibian *Rhacophorus malabaricus*; aquatic reptiles, chameleons, agamids, geckoes, and many species of snakes; many species of birds, social birds being predominant; and a variety of mammals. Nocturnal and arboreal habits are most common in many mammals such as insectivores, leopards, jungle cats, ant-eaters, giant flying squirrels, monkeys and sloths. But in New Guinea and Northern Queensland, where monkeys are absent, there are arboreal kangaroos (*Dendrolagus* sp.), despite the fact that the basic anatomy of the kangaroo is not particularly well-suited to arboreal life. Further, in the foot hills of the forest zone of peninsular India covered with dense tropical vegetation, we have the tiger (*Panthera tigris*), the elephant (*Elephas maximus*), samber deer (*Rusa unicolor*), muntjac (*Muntiacus muntjak*), the gaur (*Bibos gaurus*), the chital or spotted deer (*Axis axis*), and the swamp deer (*Rucervus duraucelli*) as the major ground dwelling mammals.

(vi) Tropical seasonal forests. Tropical seasonal forests occur in those regions whose total annual rainfall is very high, but segregated into distinct wet and dry periods. Tropical seasonal forests are found in Southeast Asia, Central and South America, Northern Australia, Western Africa and the tropical islands of the Pecific as well as India and Southeast Asia. In exceedingly wet tropical seasonal forests, commonly known as **monsoon forests**, the annual precipitation may be several times higher than that of the tropical rainforests. Trees may reach heights over 40 m, but are more commonly 20-30 m high. Stratification is of a relatively simple type with a single understory tree layer, canopy is deciduous and understory is evergreen. Teak is often a major large tree in the best-known tropical seasonal forests, those of India (central India) and southeast Asia. Bamboo is also an important climax shrub in these areas although in other areas it is important only in earlier stages of the succession.

(vii) Subtropical rainforests. In regions of fairly high rainfall but where temperature differences between winter and summer are less marked, as in Florida (USA), the broad-leaved evergreen subtropical biome is found. The vegetation includes mahogany, gumbo limbo, bays, palms, oaks, magnolias, tamarinds, all laden with epiphytes (of pineapple and orchid families), ferns, vines and strangler fig (*Ficus aureus*). This stratification is simpler, with only one understory tree horizon. All these plants tend to be evergreen, but may lose their leaves during the dry season. Animal life of subtropical forest is very similar to that of tropical rainforests.

Rainforests of India. In India, patches of rainforests are found in Kerala, Assam and the Gandhamardan hills of Orissa. They are of following three types:

(a) Moist tropical forests. They include the southern tropical wet evergreen forests found in Assam and West Bengal, the northern semi-evergreen forests of Assam and Orissa, and the southern tropical semi-evergreen forests of Andamans.

(b) Montanic subtropical forests. They include the northern subtropical broad-leaved wet hill forests of Assam and West Bengal, the southern subtropical broad-leaved hill forests of Orissa and Kerala and the subtropical pine forests of Uttar Pradesh, Himachal Pradesh, Assam, Manipur, etc. The *Chir* forests of UP and HP are also example of this type of forests.

(c) Montanic wet temperate forests. They include the forests of Kodaikanal and Udagamandalam in Tamil Nadu and Kerala, the northern wet temperate forest of the north-eastern region and West Bengal and the Himalayan moist temperate forests.

4. Tropical Savanna Biomes

Savannas are tropical grassland with scattered, drought resistant trees which generally do not exceed above ten meters in height and do not form a canopy. Thus, a savanna is an intermediate between a forest and a grassland. Savannas constitute extensive areas in Eastern Africa, which support the richest diversity of grazing mammals in the world, and also occur in Australia, South America and Asia. The climate is generally characterized by a rainy (May through October) and dry (November through April) season; in the llanos of Venezuela, for example, nearly 90 per cent of the annual rainfall of 130 cm falls during the rainy season. The latosol soils of savannas are nutrient poor owing a heavy leaching. Quite widely distributed in savannas are soils called laterite which when dried harden to rock-like consistency thereby precenting their use in western style agriculture. Although climate and soil are significant regulating factors in this biome, the controlling factor appears to be fire, which gives grass and certain species of trees a powerful advantage over other tree species. As a result of this latter factor, species diversity is quite low in comparison to adjacent tropical forests; in some situations a single species of both grass and tree may be dominant over large areas.

The trees of savannas are resistant to desiccation, and may be either deciduous or evergreen. Their leaves are often hard and drought-resistant. Grasses are the most conspicuous plants, and may reach heights of 1½–2m. Gigantism of certain animal groups in these tropical savannas is as clear as it is in the boreal forest including such giants as many antelopes, giraffes, elephants, buffalo and lions. In addition, a rich insectan fauna especially those with strong mandibles capable of mastication are conspicuous in this biome. Grasshoppers and termites are encountered in large numbers.

Savannas of India. The savannas of India are dominated by grasses and sal trees and the consumers are cattle, rodents, insects, jackals, hyaenas, etc. They are classified into the following types:

(1) High savanna. It occurs in the Brahmaputra valley and constitutes open stands of low branching trees which are usually 2 to 3 metres tall, such as *Syzigium cerasoideum* and *Emblica officinalis.* Its common grasses are *Imperata, Saccharum spontaneum, Ophiurus* and *Vetiveria.*

(2) Moist sal savanna. It occurs in the Gangetic plains and consists of open forests of sal (*Shorea robusta*) having tall grasses. Its common grasses are *Imperata cylindrica, Themeda arundinacea, Cymbopogon nardus, Erianthus* and *Apluda.*

(3) Low alluvial savanna woodlands. It occurs in the Gangetic plains and on riverine flats. The soil of this savanna is sandy and alluvial and contains patches of clay in depressions. The common trees of this region are *Dalbergia sissoo, Butea monosperma, Albizia, Adina cordifolia* and *Zizyphus procerum, Arundinella, Themeda gigantea* and *Erianthus* sp.

(4) Dry savanna. This region is found in Punjab, Haryana, Bihar, Orissa and eastern Tamil Nadu. It is characterized by trees which stand far apart, singly or in small groups such as *Acacia lenticularis, Emblica officinalis, Gardenia turgida, Crotalaria hirta* and *Pterocarpus marsupium.* The common and abundant grasses of this region are *Themeda triandra, T.quadrivalves, Apluda mutica* and *Arundinella setosa.*

(5) Saline alkalina scrub savanna. It occurs throughout the Indo-Gangetic Plain. The common tree of this region are the *Phoenix sylvestris, Acacia* sp., *Tamarix* sp., *Calotropis procera, C. giganta* and *Kochia indica.*

5. Grassland Biomes

The grassland biome is found where rainfall is about 25 to 75 cm per year, not enough to support a forest, yet more than of a true desert. The seasonality of grasslands is clear, both with respect to rainfall, which is concentrated in the summer and to temperature. Grasslands typically occur in the interiors of continents and include the **tall grass prairie**, **short grass prairie** or great plains and **arid grassland** of North America as well as the **steppes** of Eurasia (southern Russia, Siberia and Asia), **pusztas** in Hungary, the **veldt** of South Africa, **llano** in Venezuella, and the **pampas** of South America (Argentina). The grassland communities are open land communities with limited moisture conditions, irregular rainfall, sharp seasonal and diurnal variations and very high radiation. Vegetation of grassland is dominated by grasses, legumes and composites. Since tall trees or other thick vegetations are excluded from these communities there is a free movement of air. These winds carry the particles of sand or dust. These open habitats provide natural pasture for grazing animals (herbivore) which are excluded from predation by predators which hide in thick vegetation and prey upon them.

Grasslands are dominated by grasses, members of Family Poaccae (Gramineaa). Grasses are distinctive because their stems (called **culms**) produce narrow leaves that grow from their bases. This growth form allows grasses to bę grazed or moved with minimum mortality. Grasses are **sod-formers** or **bunch grasses** (Fig. 14.1). As the names imply, sod-formers develop a solid mat of grass over the ground and bunch grasses grow in bunches.

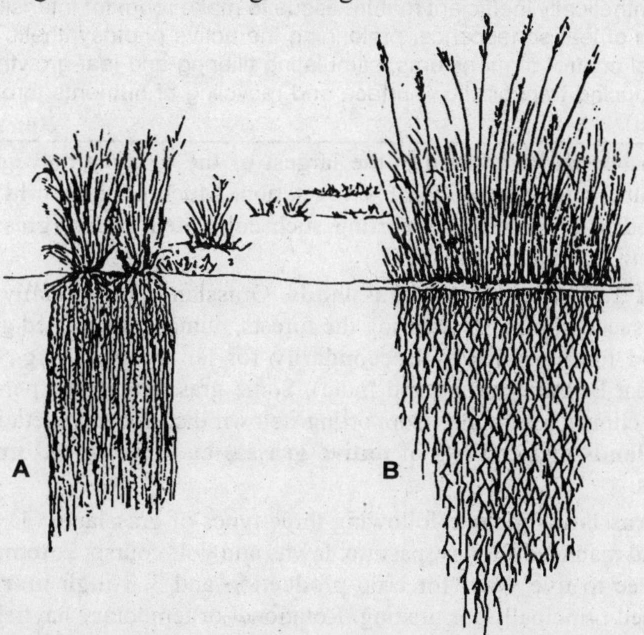

Fig. 14.1. Growth forms and root penetration (maximum depth of about 2.5 metres) of a sod grass (B) and a bunch grass (A).

Examples of typical bunch grasses include orchard grass (*Dactylis glomerata*), broom-sedge (*Andropogon virginicus*), crested wheat grass (*Andropyron desertorum*) and little bluestem (*Anderopogon scorpius*). These bunch grasses form clumps by the erect growth of all the culms and spread at the base by shoots that develop from underground stems (called **tillering**). The space between bunch grasses is often occupied by other plants, usually herbs.

Sod grasses include species of kentucky blue grass (*Poa pratensis*) and western wheat grass (*Agropyron smithii*), spread by underground stems.

Some grasses may be either sod or bunch depending upon the local environment. For example, bluestem (*Andropogon gerardii*) will develop a sod on a rich, moist soil and form bunches on a dry soil. Further, **cool season grasses**, such as blue grass, put on growth in the spring and becomes senescent in summer. **Warm season grasses** such as switch grass (*Panicum virgatum*) put on most growth and flower in summer.

Factors controlling the development of grasslands

There are various factors controlling the development of grassland such as follows:

1. **Climatic.** Particularly the length of the dry season.
2. **Edaphic.** Soil condition.
3. **Biotic.** Grazing by wild herbivores and domestic herds (Box 14.1)
4. **Pyrogenic.** Fire caused by natural events or by humans.

Box 14.1

Grasslands evolved under grazing pressure of ungulates since the Cenozoic. Grasses have adapted to grazing by compensatory growth for tissue damaged or removed by grazing. Such compensation involves 1. increased photosynthetic rates in residual tissues; 2. real location of nutrients and photosynthates from one part of the plant to another, especially from roots to stems; 3. removal of photosynthetically inefficient mature tissue to make room for intensities to stimulate new growth; 4. reduction of leaf senescence, prolonging the active photosynthetic period of remaining tissues; 5. hormonal control of meristems, stimulating tillering and leaf-growth; 6. conservation of soil moisture by reducing transpiration surface and recycling of nutrients through dung and urine (**McNaughton** 1979).

Types of grasslands. Grassland is the largest of the four major vegetational formation, accounting for about 24 per cent of Earth's vegetation. Much of the world's natural grasslands have been converted to croplands, supporting such cultivated grains (grasses) as corn, wheat, oats and barley, or to cultivated grasslands.

1. Cultivated and successional grasslands. Grasslands in normally forested region are either cultivated or successional. By clearing the forests, humans developed grasslands principally as a source of food for livestock and secondarily for landscaping (*e.g.,* eastern and central North America, Great Britain, Europe and India). Some grasslands have persisted for centuaries, becoming a sort of climax community supporting its own distinctive vegetation. These are called **successional grasslands** and consists of **native grasses** and **introduced grasses** along with an assortment of herbs.

Cultivated grasslands include following three types of grasslands: 1. **permanent**, in grass over seven years and managed for hay, pasture, lawns and golf course; 2. **temporary** or **rotational** ploughed every three to five years for crop production and 3. **rough marginal, unimproved, semi-wild lands** used principally for grazing. Rotational or temporary hayfields are dominated by two or three cultivated species usually two grasses and a forb as legume (*e.g.,* alfalfa *Medicago sativa;* red clover or *Trifolium pratense*). Such hayfields support denser and heavier growth than permanent successional grasslands.

2. Prairies. These are grasslands of central part of North America and includes tall grass prairie, mixed grass prairie, short grass prairie or plain, desert grasslands and palouse prairie.

3. Indian grasslands. These include following three types:

(i) Xerophilous grasslands. They are found in the dry regions of north-west India under semi-desert conditions and generally consists of the *Eragrostris Aristida–Andropogon–Cenchrus* association.

(ii) Mesophilous grasslands. These are termed as savannahs and include the extensive grassy banks, typical of moist deciduous forests of Uttar Pradesh. It has a *Saccharum–Vetiveria* association. The alpine meadow of the Himalayas and the grassy down of the Nilgiris represent temperate variations of this type.

(iii) Hygrophilous grasslands. These are termed as wet savannahs generally have an *Erianthus–Phragmites* association and are found in marshy areas.

Box 14.2 Distribution of Grasslands in India

1. Dichanthium–Cenchurus–Elionurus type grasslands. They occur in the plain of Punjab, Western Uttar Pradesh and Northern Rajasthan, in dry and somewhat arid conditions with an annual rainfall of about 750 mm. These grasslands have alluvial sandy to sandy loam soils. The perennial grasses in this community are *Dichanthium annulatum, Cenchrus ciliaris, Elionurus hirstus, Cynodon dactylon* (Bermuda grass, doob), *Sporobolus pallidus, Panicum turgidum* and *Heteropogon contortus.*

2. Sehima–Dichanthium type grasslands. They occur in most parts of tropical India, in the *Deccan* plateau extending to South Bihar and Eastern part of West Bengal. This type of grassland is common in the upper Damodar valley near Hazaribagh. The rainfall in this area is between 500 to 1375 mm. Summer temperatures are high, though winters may be milder. This association occurs on a number of soil types such as red and yellow, red gravelly and mixtures of various types of red and black. Common species of this grassland include *Sehima nervosum, Dichanthium annulatum, Chrysopogon montanous, Bothriochloa pertusa, Eremophogon foveolatus, Heteropogon contortus, Cymbopogon* spp., *Cynodon dactylon, Aristida setacea* and *Eragrostris coarctata.*

3. Phragmites–Saccharum type grasslands. These grasslands occur in the plains of Punjab, western Uttar Pradesh, Assam and Manipur states, in low lying poorly drained areas. They are found on sandy soils with alkaline to strongly acidic reactions. The characteristic species of these grasslands are *Phragmites kakra, Saccharum spontaneum* (thatch grass, kans), *Saccharum arundinaceum, Saccharum munja, Chrysopogon hexandra, Sporobolus indicus, Vetiveria zizanioides* (vetiver; khas) and *Imperta cylindrica.*

Structure. Grasslands possess three strata–roots, ground layer and herbaceous layer. The **root layer** is more distinct in grasslands than in any other major ecosystem. Half or more of the plants biomass is hidden in the soil. Most of the root biomass uniformly occupies the upper 16 cm of the soil profile and decreases with depth. The depth to which roots of grasses extend is great. For example, root of little bluestem reaches 1.3 to 1.7 metres and forms a dense mat to 0.8 metres. Roots of blue grama and buffalo grass penetrate vertically to one metre. In addition many grasses possess underground stems, or rhizomes, that serve both to propagate the plants and to store food. Forbs such as golden-rods and asters possess large woody rhizomes and fibrous roots that add to the root mat in the soil. The plant roots absorb nutrients from different depths in the soil, depending upon moisture.

The ground layer is dominated by **mosses** and plant with a **rosette-type growth form** such as dandelions. It is characterized by low light intensity during the growing season, temperatures cooler than ambient on the ground, and reduced wind flow. Even though the grass tops may move like waves of water, air close to ground is calm. However, conditions on grazed lands are different. Because the grass cover is closely cropped, the ground layer is subject to higher temperatures and greater wind velocity. The seasonally variable herbaceous layer consists of three or more strata.

Mulch. Grasslands accumulate a layer of mulch (or detritus) on the ground. The oldest layer consists or decayed and fragmented remains or fresh mulch. Leafy fresh mulch consists of dead, often standing, residual herbage (**thatch** of lawn owners). Three or four years must pass before natural grassland herbage will decompose completely. **Grazing** reduces mulch, as do fire and mowing. An accumulation of mulch has both positive and negative effects. It conserve soil moisture by reducing evaporation, but it also retards infiltration of precipitation. Mulch alters the microclimate and physiology of emerging shoots, so CO_2 uptake is reduced. By insulating the soil surface from solar radiation, mulch reduces root productivity and inhibits the activity of soil microbes and invertebrates. It provides nesting cover for grassland birds and offers cover for small mammals.

Since the precipitation—evaporation ratio is below one in the grassland, leaching is considerably less. The soils of the grasslands are rich, fertile **mollisoils,** *i.e.,* prairie soils (black-earths) and chernozems. Organic matter (humus) accumulates in the upper portion of the soil, rendering it dark; this upper portion remains neutral to slightly alkaline because of the continued replenishment of cations such as calcium and potassium through the upward movement associated with evaporation.

Typical animals of grasslands tend to be quite small, with the exception of a few very large cursorial herbivore mammals such as the bison and pronghorn in North America, the wild horse, ass, and saiga antelope of Eurasia and zebra and antelops of Southern Africa. The large herbivores are no where near as diverse as they are in savanna areas. Likewise, carnivores are relatively small such as are coyotes, weasels, badgers, foxes, ferrets, owls and rattle snakes. Rodents such as prairie dogs, rabbits, and ground squirrels are common. Most herbivores characteristically aggregate into herds or colonies; this aggregation provides some protection against predators. The characteristic birds of grasslands are prairie chickens, meadow larks, and rodent hawks. Further, saltatorial motion is widespread, not only in mammals such as rabbits and kangaroo rats, but also in insects, such as grasshoppers and crickets. This probably has to do with increasing rate at which animal can move through or over the tall grasses, as well as making it harder for a predator to catch the animal. Visibility in the grasslands in exceedingly good and the premium placed on efficiency including predators is very high.

The common insects of steppes are termites, locusts (*Locusta migratoria, Schistocerca gregaria* and *Melanoplus* sp.), bees, burrowing wasps, mutilid wasps, bumble bees and blister bees. The steppes also harbour an undisturbed reptilian fauna. Lizards and snakes are met with in large numbers and possess, remarkable diversity. Most of them are fossorial, insectivorous and carnivorous.

Box 14.3 Production in Grasslands

Grassland types, species composition, **productivity** and other attributes are directly related to rainfall and temperature. Primary production is lowest where precipitation is lowest and temperatures are high. It is highest where mean annual precipitation is greater than 80 cm and mean annual temperature is above 15°C. Production, however, is most directly related to precipitation. The greater the precipitation, the higher is production. This relationship reflects in part the reduction of water stress and the enhancement of nutrient uptake and in part differences in water use efficiency.

Grasslands have the highest index of productivity per unit of standing crop (20 to 55 per cent) among all terrestrial ecosystems. For example, the tundra during its short growing season has an index of 10 to 20 per cent and for the temperate deciduous forest, 8 to 10 per cent. It fact, a great amount of energy in woody ecosystems goes into the maintenance of high standing crops, which grasslands do not support. Efficiency of energy capture by grasslands is 0.75 per cent for net primary production. Light to moderate grazing increases that efficiency to 0.87 per cent.

In grasslands, although large herbivores are the most conspicuous **consumers**, invertebrates are the most important grazers. The overground biomass of invertebrates—including **plant consumers, saprovores** and **predators** —ranges from 1 to 50 g/m^2, while grazing mammals amount to about 2 to 5 g/m^2. Belowground invertebrates exceed 135 g/m^2. The major aboveground consumers are

grasshoppers and major belowground consumers are **nematodes**. Nematodes account for 90, 95 and 93 per cent of all belowground herbivory, carnivory and saprophagous activity respectively. They account for 46 to 67 per cent of root and crown consumption, 23 to 85 per cent of fungal consumption, and 43 to 88 per cent of belowground predation. Aboveground herbivores consume 2 to 7 per cent of primary production, while belowground herbivores consume 13 to 46 per cent.

Grassland Fauna

A grassland is a region where grasses are dominant but associated with relatively abundant trees in the wetter areas and with thorn bush scrub near the drier margins. India does not have a true grassland biome type. The grasslands present in the country owe their origin to the destruction of natural forests vegetation by humans as well as due to various other biotic pressures such as overgrazing. Thus grasslands are not climax formations but have developed secondarily.

Animal Life of Grasslands

All types of grassland support similar forms of animal life. Much of it exists within the strata of vegetation: roots, ground layer and herb cover. Invertebrates, particularly insects, occupy all strata at some time during the year. In winter insect life is confined largely to the soil, litter and grass crowns as pupae and eggs. In spring, soil occupants are mostly earthworms and ants. Ground and litter layers harbour scavenger carabid beetles and predaceous spiders. Since this layer provides only limited supports for webs, there are more hunters than web builders among spiders.

Life in the herbaceous layer varies as the strata become more pronounced from summer to fall. In this layer, invertebrate life is more diverse and abundant. Homoptera, Coleoptera,

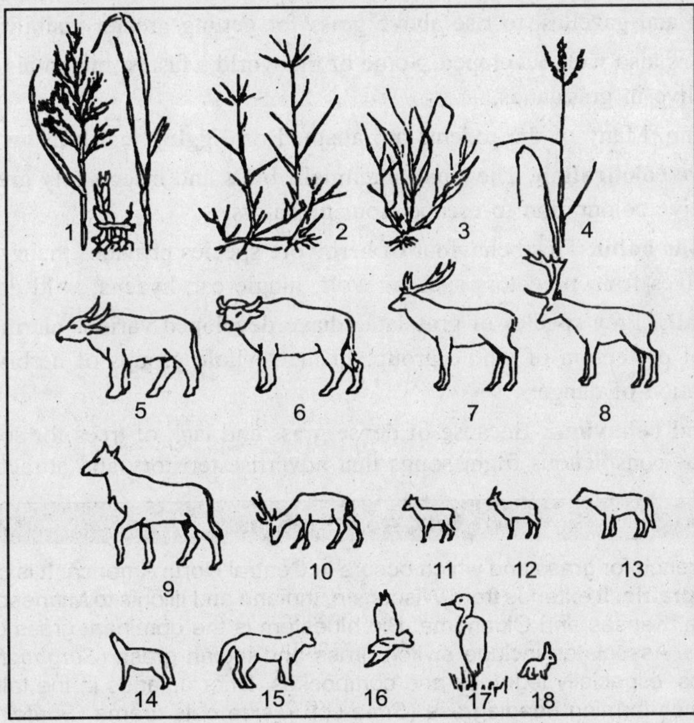

Fig. 14.2. Some plants and animals of Indian grasslands. 1. *Phragmites*, 2. *Cynodon*, 3. *Cenhrus*, 4. *Andropogon*, 5. Gaur, 6. Buffalo, 7. Sambar, 8. Barasingha, 9. Nilgai, 10. Cheetal, 11. Chausingha, 12. Chinkara, 13. Wild dog, 14. Jackal, 15. Jungle cat, 16. Hawk, 17. Egret, 18. Hare.

Orthoptera, Diptera, Hymenopetera and Hemiptera are all represented. Insects life reaches two highs during the year, a major peak in summer and a less well-defined one in fall.

Mammals are the most evident vertebrates of the grasslands, and the majority of them are herbivores. A large and rich ungulate fauna evolved on the grasslands (Fig. 14.2). The bison (*Bison bison*) and pronghorn antelope (*Antilocarpa americana*) of North America were surpassed only by the richer and more diverse ungulate fauna of the East African plains. In Australia, marsupial kangaroos are the grazing equivalent of placental ungulates. South America lacks an ungulate fauna. Their closest equivalent are the large grazing rodents that live in herds, the capybara (*Hydrochaeris hydrochaeris*). Today herds of cattle and flocks of sheep have replaced most of the native grazing hebivores.

Grassland birds are more limited in kind but include meadowlark, prairie chicken, long spur, horned lark, dickcissel, rodent-eating hawks and grasshopper sparrow (named for its song, not for feeding on grasshoppers, the most abundant arthropod of grassland).

Adaptations of grassland animals. Animals living in grasslands show the following adaptations:

1. Cursorial locomotion. The grassland ungulates are cursorial (*i.e.,* adapted for running or walking) and have developed long-range vision which is necessary to perceive the approach of danger from a distance.

2. Hopping or leaping (**Saltatorial motion**). It is a common method of locomotion of animals of grassland. Strong hind legs enable a variety of animals, such as grasshoppers, jumping mice, jack rabbits and gazelles, to rise above grass for getting greater visibility.

3. Speed. It is also well developed. Some of the world's fastest mammals such as antelopes and the cheetah, live in grasslands.

4. Burrowing. Many of the rodents are adapted to **digging** and **burrowing.**

5. Protective colouration. The small mammals, birds and insects rely greatly on their well developed protective colouration to escape from predators.

6. Gregarious habit. This behaviour of herbivore species provides them with a mechanism to protect themselves from predators such as wolf, jungle cat, hyaena, wild dog, etc.

7. Alarm calls. Prey species of grasslands have developed various alarm calls and signals for the purpose of protection of whole group. In fact, whole groups of herbivore start moving at the slightest notion of danger.

8. Territorial behaviour. Because of dense grass and lack of trees for song perches, some grassland birds has conspicuous flight songs that advertise territory and attract mates.

Box. 14.4 Prairies

Prairies is French for grassland which occurs in Central North America. It is of following types:

1. Tallgrass prairie. It extends from Wisconsin, Indiana and Illionis to Minnesota, eastern south Dekota, Nebraska, Kansas and Oklahoma. Big bluestem is the dominant grass of moist lowlands in tallgrass prairie. Associates include switch grass and Indian grass (*Sorghastrum nutans*) and a diversity of forbs, especially legumes and composites. Drier uplands in the tallgrass prairie are dominated by bunch-forming needlegrass (*Stipa* spp.), side oats grama (*Bouteloua curtipendula*) and the prairie dropseed (*Sporobolus heterolepis*). Much of this grassland is interspersed with trees and shrubs, especially along streams and lower slopes of rolling hills.

2. Mixed grass prairie. It extends from southern part of central Canada to Texas. It is an **ecotone** between the tallgrass prairie and the shortgrass prairie or plains. Mid-height grasses occupy lowlands and shortgrass the higher elevations. The grasses are largely cool season bunch species.

3. Shortgrass prairie or plain. It extends from Nebraska panhandle, Wyoming, easter Colorado to New Mexico and Texas. The shortgrass plain or **steppe** reflects a climate where the rainfall is slight and infrequent (25 to 43 cm in the west, 51 cm in the east), the humidity low, the winds high and the evaporation rapid. Being shallow-rooted, the short grasses use moisture in the upper soil layers, beneath which is a permanent dry zone into which the roots do not penetrate. Sod-forming blue grama (*Bouteloua gracilis*) and buffalo grass (*Buchloe dactyloides*) dominate the shortgrass plain. On wet bottomlands, switch grass, Canada wild rye (*Elymus candensis*) and western wheat grass (*Agropyron smithii*) replace blue grama and buffalo grass. Because of the dense sod, fewer forbs grow on plains.

4. Desert grassland. It extends from southeastern Texas to southern Arizona into Mexico. Desert grassland is similar in many respects to the shortgrass plain except that triple awn grass replaces buffalo grass. Composed largely of bunch grasses, the desert grasslands are widely interspersed with other vegetation types, such as mesquite (*Prosopis*) woodland. Here climate is hot and dry. Rain falls only during summer and winter in amounts that vary from 30 cm to 51 cm. Evaporation is rapid, upto 203 cm in a year.

5. Palouse prairie. This is inter-mountain region extending from southwest Canada to Oregon. Because this region is dominated by cool-season short bunch grasses and sage brush (*Artemisia* spp.), it is often called **shrub steppe.** Originally the dominant grasses of this prairie were blue bunch wheatgrass and blue bunch fescue. Being highly sensitive to grazing, these grasses were mainly replaced by cheatgrass, Kentucky blue grass and such forbs as sunflower and prickly lettuce.

Because annual precipitation in these central grasslands increases from west to east, the natural vegetation was of *shortgrasses,* such as blue grama and buffalo grass, in the western portions; *medium sized grasses,* such as needlegrass and slender wheatgrass, in the middle portions; and *tallgrassses*, mainly big blue stem in the eastern portions.

6. Desert Biomes

Deserts are the biomes formed in the driest of environments. Temperature may range from very hot as in **hot deserts** to very cold as in **cold deserts**. Major hot deserts of the world are situated near the tropics of cancer and capricon, with a rainfall of less than 10 mm. The most important hot deserts of world is the Sahara-Arabia-Gobi desert complex extending from Africa to Central Asia and contains highly irregular and very insignificant rainfall, and low humidity due to excessive evaporation. Fairly extensive hot deserts also occur in India (Sind-Rajasthan deserts; Box 14.5), South America (Chile), North America and Australia. The cold deserts occur at high elevations where the temperatures are low and rainfall scanty as the air losses all its moisture content as it ascends higher and higher. Cold deserts occur in Ladakh regions of Himalayas, Tibet, and Bolivia Arctic. The hot and cold deserts may also be distinguished by differences in plant population which are mostly succulent type (*e.g.,* cactus, palo verde trees, creosote bush, etc.) in hot deserts. Most cold deserts have sage brush.

Box 14.5 Sind-Rajasthan Desert

In India desert biome is found in the **Thar** deserts of Rajasthan including Rajputana, Ajmer, Jaipur, Jhelum and Sind. The Thar desert comprises about 2,30,000 hectares and consists of wind–blown sands, fixed and shifting sand dunes, interstand dunal area and rocky outcrop.

(i) **Climate.** In west of Aravali range climate is extreme, hot and dry (annual rainfall less than 300 mm) in summer (temperature reaches upto 48·9°C in Ganganagar) and cold and dry in winter (a temperature of — 3·7°C, Churu). The south eastern parts receive a fair amount of rainfall (380-600 mm) and therefore have a milder climate.

January temperature of southern Rajasthan is about 15-20°C, while that of northern Rajasthan is in the range of 10 to 15°C. Summer temperatures of the desert region may rise to about 49°C. Cold winds increase the chill factor at night while hot winds during the day increase the dessicating effects. Rainfall is highly irregular and long drought periods are typical. The ground surface is generally baked hard and most of the rains run off. Here evaporation is in excess of precipitation.

Soil. The desert soils of Rajasthan are typically sandy, gravelly and with low organic matter, low nitrogen and other plant nutrients. Soil pH varies from 8.0 to 9.0.

(ii) Vegetation. The vegetation in the desert is sparse and includes desert scrub. Desert vegetations have been divided into following three associations:

(1) Sand community (or ***Calligonum – Aerva – Panicum* association**). This type of association is found in the Barmer region and the habitat includes **sand dunes**, sandy plains and sand hills. The dunes are formed by the piling up of sand due to wind action. The leeward side being steep, vegetation occurs on the windward side. On well developed old sand dunes the vegetation shows an distinct *elevation zonation.* Beginning with the foot of the dune, the *Aerva* association is dominant, consisting of *Aerva tomentosa, Aerva pseudotomentosa, Aerva persica, Calatropis procera* (swallow wart, Ak), *Leptodenia pyrotechnia* and *Crotolaria burhia.* Near the crest of sand dune occur species as *Panicum targidum* and *Indigofera argentea,* while at the top of dune, pure associations of *Calligonum polygonoides* occur. Intermixed in the association are other plant species such as *Acacia jacquemontic, Capparis decidua* and other. *Cyperus arenarius* binds sand of dunes and arrests their spreading.

(2) Gravel community (or ***Tribulus – Heliotropium – Zizyphus* association**). This habitat is formed by the sorting action of the wind which carries away the sand and leaves the larger gravel in place. It has a more firm surface layer than the sand dunes, but has poor water-retention capacity. Typical gravel vegetation consists of plants with stiff woody branches lying prostrate on the ground. Typical plants of this community are *Tribulus terrestris* (land caltrops, Gokhru), *Heliotropium rariflorum* and *Zizyphus rotundifolia.* Associated species include *Leptadenia, Blepharis* and *Sericostoma.* Grasses such as *Eleusine compressa* and *Aristida lystricula* are also commonly found here.

3. Rock community (***Euphorbia–Capparis–Ziziphus* association**). Western side of Aravali has rocks as habitat. *Euphorbia caducifolia* is the most conspicuous and prominent plant in such habitats along with *Capparis decidua* and *Zizyphus jujuba* (Jujube, Ber). Other plants such as *Salvadora oleoides, Acacia senegal* and *Anogeissus pendula* are also commonly found in such habitats. The community supports a large number of climbers including *Sarcostemma, Convolvulus, Asparagus* (Satavar) and others.

Low abnormal precipitation coupled with soil and air temperatures that are extremely high by day and drop abruptly by night, low humidity, and high insolation are the major desiccating environmental factors to which desert vegetation and animals have adapted. Desert plants which tend to be shrubs are adapted to drought conditions through reduced leaf size and the dropping of leaves in dry conditions, both reducing water loss via evapotranspiration. The roots of most desert plants remain well developed and occur in the top meter of the soil to take maximum possible advantage of any rainfall. Further, the root hairs on many desert plants are ephemeral, drying back under drought conditions and thereby reducing potential water loss by osmosis. Yet other species are short-lived annuals that complete their life cycles during the short-moist period. In most hot deserts, there occur plants such as cacti, water-storing succulents such as acacias, euphorbias, cacti, prickly pears, etc., which are adapted by their protoplasmic colloids, which enable the accumulation of substantial water reserves, as well as by a reduced leaf surface, which prevents water loss via evapotranspiration.

The animals present in the desert are reptiles, insects and burrowing rodents. All these animals possess special morphological, physiological and ethological adaptations for deserts. In general, large animals are very uncommon except mule deer and some species of gazelle and all animals have cursorial, fossorial and/or saltatorial adaptations. Some desert animals are nicely adapted for high extremes of temperature. For example, the lethal temperature of different species of insects found to be following—for the canal spider *Galeodes granti* it is 50°C, for the *Gryllus domesticus* it is 40°C and for the forficulid *Labidura riparia* it is 38°C (Cloudsely-Thompson, 1962). Diurnal rhythms are perhaps the best method of avoiding the heat. While some desert plants close their petals at night, many blossom only at night. Some insects such as tenebrionid bettle *Akis spinosa* remain active during the day, and the centipede *Scolopendra clavipes* restricts its activities to night time.

Further, certain reptiles and certain insects are well adapted for survival in deserts because of their thick, impervious integuments and the fact they excrete dry waste matter. A few species of mammals have become secondarily adapted to the desert by excreting very concentrated urine. They avoid the sun by remaining in their burrows during the day. Kangaroo rat and pocket mouse, both are able to live without drinking water by extracting the moisture from the seeds and succulent cactus they eat. The camel and the desert birds (ostrich, etc.) must have an occasional drink of water but can go for long periods of time using the water stored in the body. Most insects of deserts are herbivores and as a correlation the number of small insectivorous lizards found in the desert is usually high.

Fauna of Indian deserts. Prakash (1974) have recognized four habitat types in the Indian desert based on land forms namely **aquatic**, **sandy**, **rocky** and, **ruderal habitats**. The aquatic habitats of Indian deserts which remain confined to the perennial lakes, are inhabited by various species of fishes such as *Labeo nigripinnis, Oxygaster clupeoides, Tor khurdee, Puntius amphibia, Noemacheilus denisonii.* Among amphibians of the Indian desert are a species of toad (*Bufo andersonii*) and five species of frogs (Indian bull frog *Rana tigrina,* cricket frog *Rana limnocharis* and burrowing frog *Rana breviceps* being dominant species). Among reptiles, there occur two species of testudines (Loricata), 18 species of lizards and 18 species of snakes. Of the lizards, some species such as *Acanthodactylus cantoris, Calotes versicolor, Uromastix hardwicki, Ophiomorus tridactylus* are predatory on the desert locust inhabiting localized areas in the Thar desert. The avian fauna of the Indian desert includes species such as painted partridge *Francolinus pictus pallidus,* grey partidge *Francolinus pondicerianus,* common quail *Coturnix coturnix,* black breasted quail *Coturnix coromandelicus,* rock bush quail, the little bustard quail *Lurnix sylvatica,* button quail *Turnix tanki,* great Indian bustard *Choriotes nigriceps,* florican *Sypheotides indica,* sandgrouses such as *Petrocles exustus erlangeri, Petrocles alchata, P. senegallus,* and *P. indicus indicus.* During winter numerous aquatic birds become abundant in Indian deserts. Such birds are white fronted goose *Anser albifrons,* wigeon *Anas penelope*, garganey *anas querguedrila,*red crested pochard *Netta fufina* and tufted duck *Aythya futigola.* Among the predominant predatory birds are two species of the vultures namely *Gyps bengalensis* and the white vulture *Neophron percnopterus.*

The mammalian fauna of Indian deserts includes many species, some of which have been tabulated in Table 14.2.

Table 14.2. Certain common mammals of Indian deserts.

Order	Species	Popular Name
1. **Chiroptera**	*Rasetius arabicus*	Fruit bat
	Pteropus giganteus giganteus	Fruit bat
	Rhinopoma kinneari	Rat-tailed bat
	Rhinopoma hardwicki	The lesser rat-tailed bat
	Taphozour perforatus	Tomb bat
	Megaderma lyra	Indian false vampire
	Rhinolophus lepidus lepidus	Horse-shoe-bat
2. **Insectivora**	*Hemiechinus auritus lepidus*	Long-ear hedgehog
	Paraechinus micropus	Long-ear hedgehog
	Suncus muinus sindensis	House Shrew
3. **Pholidota**	*Manus crassicaudata*	Scaly ant eater

Order	Species	Popular Name
4. Lagomorpha	*Lepus nigricollis caudatus*	Hare
	Lepus nigricollis dayanus	Hare
5. Rodentia	*Funambulus pennanti*	Northern palm squirrel
	Hystrix indica indica	Indian crested porcupine
	Rattus rattus	House rat
	Mus musculus rufescens	House mouse
	M. musculus bactrianis	House mouse
	Gerbillus gleadovi	Indian hairy footed gerbil
	Tatera indica	Gerbil
	Merriones hurrianae	Gerbil
6. Artiodactyla	*Sus scrofa cristata*	Wild boar
	Boselaphus tragocamelus	Blue bull
	Antilope cervicapra rajputani	Black buck
	Gazella gazella bennetti	Gazelle
7. Carnivora	*Canis lupus*	Wolf
	Canis aureus	Asiatic jackal
	Vulpes vulpes	Red fox
	Vulpes bengalensis	Bengal fox
	Lutra perspicillata	Indian otter
	Viverricula indica	Small Indian civet
	Herpestes aeropunctatus	Small Indian mongoose
	Felis chaus prateri	Jungle cat
	Felis libyes ornata	Jungle cat
	Panthera pardus sindica	Panther

All these mammals have well specialized adaptations for survival in thermal extremes and low humidity. Many of them possess diurnal rhythmic activities.

7. Wetland Biomes

Wetlands may be defined as submerged or water saturated lands (natural and artificial, permanent or temporary) with water that is static or flowing and fresh, brackish or saltish. They are characteristic transitional conditions between terrestrial land and deep water bodies and are represented by **swamps**, **marshes** (or palustrine lands), **fens** (*i.e.,* tract of marshy land), **peatlands**, **lagoons** (*i.e.,* a shallow lake, especially one near and communicating with the sea or river), **lakes**, etc. Wetlands are also represented by paddy fields, fishery ponds, *Trapa* (Singhara) cultivation ponds, riverine flood plains, lacustrine (L.*lacus*= a lake) marshy lands, marine backwaters and coastal wetlands. Wetlands are often seasonal, *i.e.,* with free overlying water in the rainy season, marshy and often almost dry in summer. Wetlands support specialized vegetation and fauna; they are suitable niche for fish and other aquatic fauna, as breeding and nursery ground for water fowls and as filters for sediments and pollutants. Flood plains and marshes are sometimes not easily accessible and, hence, serve as a secure place for many kinds of wild life. Sunderbans is one such habitat in the coastal belt of the Bay of Bengal. Since wetlands contain abundant water, they are quite productive. In tropics (*e.g.,* India), because of associated bright sunshine and warm condition their primary productivity becomes very high. Rice which is the major crop of India, is best grown in wetlands. Biomass of wetlands is well distributed between the above ground foliage and underground rhizomatous and cormous parts. *Typha* and *Phragmites* have very high production efficiency. *Euryale ferox*, a prized dry fruit crop is most extensively grown in wetlands of north Bihar. Wetlands are fast disappearing in many parts of world because of dumping of wastes and reclamation for agriculture, housing, forestation, etc. The flowing or lotic wetlands are most easily polluted by city sewage, industrial effluents and runoff pesticides and other agrochemicals applied in the drained upland crop fields.

8. Mangroves in India

Halophytes growing in the muddy swamps of the estuaries of tropical and subtropical regions and sea coasts flooded by rivulets and tides form the **mangroves**. Mangroves are, thus, the salt-tolerant forest ecosystems found mainly in tropical and subtropical inter-tidal regions. They comprise the swamps, forest land within and its water spread areas. In India, the total area of mangroves is about 6740 sq. km., which is nearly 7% of the world's mangroves. The mangrove areas of Sunderbans (W. Bengal) and Andman and Nicobar Islands constitute about 80 per cent of the Indian mangroves, the rest are scattered in the states of Andhra Pradesh, Tamil Nadu, Orissa, Maharashtra, Gujarat, Goa and Karnataka.

Mangroves are dense forests of low trees, especially adapted to (a) fixation, (b) respiration, (c) vivipary, (d) migration, and (e) xeric characters. In mangroves, the plant and animal diversity is very rich. Typical plant associations of Indian mangroves include *Avicennia officinalis, Avicennia alba, Aegiceras majus, Aegialitis rotundifolia, Acanthus ilicifolius, Kandelia rheedii, Bruguiera caryophyloides, Sonneratia griffithi* and *Carapa obovata.*

Mangroves are highly productive ecosystems. The net primary production is usually high. Secondary productivity and decomposition activity of the mangroves are also very high. The terrestrial consumers are snakes, lizards, turtles, jackals, wild boars, hyaena, tigers and so on. In the Bhittar Kanika sea coast of Orissa adjacent to the mangroves exists the largest sea turtle (*Lepidochelys olivacea*) rookery. Every year 300,000 to 500,000 female turtles come to the sea beach (arribada) in late December-January and again in March-April to lay about 50 million eggs (**Dash** and **Kar**, 1990). Birds are also the most colourful and conspicuous inhabitats of the mangroves. Mangroves of the Florida key support a fauna of mostly insects and spiders, along with scorpions, isopods and other arthropods (**Krebs**, 1985). In recent years, the Indian mangroves have been subjected to reckless exploitation and other biotic interference.

QUESTIONS

Long Answer Questions

1. Describe characteristics of terrestrial habitat and differences from the aquatic habitat.
2. What are biomes? Describe different biomes.
3. Discuss the characteristics of Forest ecosystems. *(Kerala 2001)*
4. What are physico-chemical conditions which the animals and plants in the desert biomes face? Discuss the adaptations undergone by desert animals. Which parts of the globe represent desert biomes?
5. Write an essay on the 'Grassland Fauna' with the special reference to their adaptations to the habitat.

Short Answer Questions

Write short notes on the following:

1. Pasture land
2. Prairies
3. Terrestrial ecosystem

Very Short Answer Questions

1. What is biome?
2. What is desert?
3. What are the mangroves?
4. Define the savanna.

5. What are wetlands?
6. Define the forest.
7. What is grassland?
8. Give an example of desert biome of India.

Fill in the Blanks

1. The most important factor controlling the type of biome found on land is the ________.
2. Up to 300 species of trees can be found within a single hectare of a ________ biome.
3. The trees of tropical deciduous forest lose their leaves during the ________ season. The trees of temperate deciduous forests lose their leaves during the ________ season.
4. Between latitudes 50° and 70°, whether a forest is coniferous or deciduous broad-leaved trees depends on the ________ of the ________.
5. The arctic summer is less than 6 weeks long, yet the productivity of arctic tundra plant is greater than that of alpine tundra plants, because the arctic ________ are ________ than in the alpine tundra.
6. Deserts form wherever ________ and ________ exceed precipitation, which typically occurs at ________ latitude.
7. Most desert animals avoid water loss by living in ________ and by being ________ during the summer.
8. The two North American biomes that support the largest herbivores are the ________ and the ________.

Multiple Choice Questions

1. Dry climate typically occur at
   (*a*) 0° latitude (*b*) 30° latitude
   (*c*) 45° latitude (*d*) 60° latitude
2. Wet climate typically occur at
   (*a*) 0° latitude (*b*) 30° latitude
   (*c*) 45° latitude (*d*) 90° latitude
3. Most animals are tree dwellers in a
   (a) tropical rainforest (*b*) temperate deciduous forest
   (*c*) thorn woodland (*d*) coniferous forest
4. Most of the small mammals and non-migratory birds are seed-eaters in a
   (*a*) tropical rainforest (*b*) temperate deciduous forest
   (*c*) thorn woodland (*d*) coniferous forest
5. Epiphytes are most abundant in a
   (*a*) tropical rainforest (*b*) temperate deciduous forest
   (*c*) thorn woodland (*d*) shrubland
6. Which of the following biomes has the deepest, richest soil?
   (*a*) tropical rainforest (*b*) temperate deciduous forest
   (*c*) tallgrass prairie (*d*) tundra
7. The arctic tundra is a mosaic of plant-covered land, bogs and shallow lakes, because
   (*a*) there is so much water from melting glaciers
   (*b*) water cannot drain downward
   (*c*) weathering of rocks occurs unevenly
   (*d*) there is so much snow fall in the winter

8. Nights are cool in desert, because
(*a*) the dry air does not radiate much heat back to the earths surface
(*b*) the elevation is always high
(*c*) so much heat is absorbed by the earth's surface during the day
(*d*) the air pressure is low

9. The tropical savanna differs from a temperate grassland in that it has
(*a*) more perennial plants (*b*) deeper soil
(*c*) scattered clumps of trees (*d*) a higher water table

10. Which of the following is the largest terrestrial community?
(*a*) biosphere (*b*) biome
(*c*) biomass (*d*) ecosystem

11. What ecological factors are most strong determinants of various biomes?
(*a*) light and wind (*b*) soil and wind
(*c*) temperature and precipitation (*d*) pH

12. Largest terrestrial community of an entire geographical belt over which plant and animal species may be found is called
(*a*) biome (*b*) habitat
(*c*) range (*d*) niche

13. What determines limits of a biome?
(*a*) type of soil and presence of barrier (*b*) altitude and latitude
(*c*) temperature and rainfall (*d*) all the above

14. Which of the following biomes is very delicate and recovers from any disturbance only very slowly?
(*a*) tundra (*b*) savanna
(*c*) mangrove (*d*) tropical forest

15. Deciduous forests have
(*a*) variety of leaves (*b*) broad-leaved trees
(*c*) narrow-leaved trees (*d*) variety of crocodile

16. In India coniferous forests are found in
(*a*) madhya pradesh (*b*) satpura hills
(*c*) himalyan region (*d*) rajasthan

17. Alpine forest occurs at an altitude of
(*a*) 3900-6000 m (*b*) 1900-3000 m
(*c*) 1000-1500 m (*d*) 500-1000 m

18. Temperate evergreen forests are found in
(*a*) himalayan ranges (*b*) western ghats
(*c*) aravali ranges (*d*) assam

19. Forest floors in general support a wide variety of arthropods. The predominant forms closely associated with forest litter are the
(*a*) insects (*b*) ants
(*c*) mites (*d*) spiders

20. Prairies and steppes and parts of the natural region are called
(*a*) tropical grasslands (*b*) temperate grasslands
(*c*) prairie grasslands (*d*) steppes grasslands

21. During extreme aridity, desert rat
(*a*) stores water (*b*) saves water
(*c*) uses metabolic water (*d*) does not use water

22. A natural region characterized by hot summer, warm winter and treeless vegetation is
(*a*) tropical desert (*b*) steppes grassland
(*c*) savanna grassland (*d*) temperate desert

23. Desert can be converted into grassland by
(*a*) oxylophytes (*b*) psammophytes
(*c*) halophytes (*d*) tropical trees

24. Mangrove vegetation is found in
(*a*) Kullu valley (*b*) Western Ghats
(*c*) Sunderbans (*d*) Dehradun valley

25. A region of severe winter and mild summer is known as
(*a*) noosphere (*b*) cultigen
(*c*) holard (*d*) taiga

26. Biomes describe (*Rohilkhand 2001*)
(*a*) desert vegetation
(*b*) several communities in an area represented by plants and animals
(*c*) communities represented by animals only
(*d*) communities represented by plants only

ANSWERS

Very Short Answer Questions

1. (1) Biome is large regional or subcontinental system characterized by a particular major vegetation type (such as a temperate deciduous forest); or (2) Biomes are distinguished by the predominant plants associated with a particular climate (especially temperature and precipitation).
2. Desert is a biome with less than 10 inches (25 centimeters) of rainfall per year, dominated by stem succulents, such as cacti, and desert shrubs that are frequently regularly spaced in their distribution.
3. (1) Mangroves are emergent woody plants that tolerate the salinity of the open sea; or (2) trees that dominate tropical intertidal forest.
4. Savanna is tropical grassland on which trees and woody shrubs are widely spaced.
5. Wetlands are the habitats that are perpetually or periodically flooded.
6. Forest is a biome in which dominant plants are the trees.
7. Grassland is a biome which has herbaceous vegetation dominated by grasses.
8. In India desert biome is found in Thar deserts of Rajasthan.

Fill in the Blanks

1. Climate; **2.** Tropical rainforest; **3.** Dry, cold; **4.** Length, winter;
5. Days, longer; **6.** Evaporation, transpiration, 30°;
7. Burrows, nocturnal; **8.** Tundra, grassland

Multiple Choice Questions

1. (*b*) **2.** (*a*) **3.** (*a*) **4.** (*d*)
5. (*a*) **6.** (*c*) **7.** (*b*) **8.** (*a*)
9. (*c*) **10.** (*b*) **11.** (*c*) **12.** (*a*)
13. (*b*) **14.** (*a*) **15.** (*b*) **16.** (*c*)
17. (*a*) **18.** (*b*) **19.** (*b*) **20.** (*b*)
21. (*c*) **22.** (*a*) **23.** (*b*) **24.** (*c*)
25. (*d*) **26.** (*b*).

15

Landscape Ecology

15.1 DEFINITION

Landscape ecology is an integrative field of study that weds ecological theory with practical application; addresses the exchange of biotic and abiotic materials among ecosystems; and investigates human actions as responses to and reciprocal influences on ecological processes (**Odum** and **Barrett** 2005).

Landscape ecology is a subdiscipline of ecology that emphasizes the causes and consequences of spatial variation across a range of scales. As such, **landscape ecologists** document observed spatial patterns – including those that occur across broad geographic regions- and study how those patterns affect and are affected by ecological processes (**Cain** *et al*, 2011). They are interested in the spatial arrangement of different landscape elements across Earth's surface. Examples of **landscape elements** include **patches** of forest surrounded by pastures or lakes scattered across a large region of northeast forest (USA). At smaller spatial scales, individual cerosote bushes in a desert or areas of a certain soil type could be considered landscape elements. But whatever these elements can influence what species live in an area, as well as the dynamics of ecological processes such as disturbance and dispersal.

Landscape ecology had its beginning in North America during the 1980s, when **Gary W. Barrett**, the Ecology Program Director at the National Science Foundation, recommended funding for a workshop held at Allerton Park, Piatt County, Illionois in April, 1983. This meeting served as a catalyst for annual meetings of the United States International Association for Landscape Ecology (IALE); the first meeting of IALE was held at the University of Georgia in January 1986. This meeting immediately followed by the first volume of the journal *Landscape Ecology*, published in 1987 with **Frank B. Golley** serving as editor-in-chief. The now classic book *Landscape Ecology* by **R.T.T Forman** and **M.Godron**, was published in 1986. Thus, the 1980s proved the decade when landscape ecology took root in North America.

The study of the causes and consequences of spatial patterns in the landscape is the cornerstone of the emerging science of landscape ecology. This emerging biological science has also resulted in new emerging properties and landscape scale understanding, such as the role of **gamma diversity** (number of species or other taxa occurring on a regional basis); quantifying the spread of **disturbance**; the importance of **source-sink dynamics**; species-specific responses of individuals to landscape element such as **corridors**; and rates of **biotic exchange** among ecosystem types.

To understand **landscape patterns** (such as heterogeneity in an agricultural landscape mosaic) and **processes** (such as eutrophication of a watershed), theory and application must be integrated into a holistic research and management approach. Integrated concepts include **hierarchy theory, sustainability, net energy, patch connectivity** and **cybernetic** regulatory mechanisms.

Landscape ecology has now gained general acceptance as a branch of modern ecology dealing with the interrelationship between humans and both natural landscapes and human-built tecno-landscapes. Landscape ecology provides a scientific basis such fields as **design planning, management, protection, conservation** and **restoration** and provides foundation for natural and human- dominated land management at the regional scale (**Hersperger** 1994). Landscape change during the course of history, not only because of ongoing natural processes (because of processes such as ecosystem development) but also because of *social, political* and *economic* processes occurring within these systems. Landscape ecology emphasizes these changing relationships and emphasizes landscape as a system and level of organization. The better one understand landscape-level patterns and processes, the better one will understand processes and phenomena occurring at the autecological level (organism) and synecological levels (population, community and ecosystem) as well.

15.2 TOOLS AND TECHNIQUES OF LANDSCAPE ECOLOGY

The advent of landscape ecological angle has been followed by, or perhaps driven by, a powerful assemblage of **tools** that permits us to monitor the environment in multiple dimensions and at many scales. For example, the emergence of **aerial photography** gave ecologists a ready means of looking at "the big picture". More recently, our approach to space has vastly expanded our ability to acquire images of Earth through **remote sensing** and has permitted the interpretation of many large-scale ecological patterns, such as global patterns of net primary production. The use of **geographic information system (GIS)** has become standard for landscape planning efforts, whether for urban development or for conservation.(Box 15.1) In the field handheld global positioning system (GPS) have permitted ecologists to document precise locations and integrate them with other landscape variables through GIS. **Radiotelemetry** has greatly enhanced our ability to follow animal movements and migration patterns, again with the help of GIS. Our ability to analyse all this information is constantly growing, thanks to better **computers** and new statistical methods of spatial analysis.

In the chapter of biodiversity (Chapter 17) it has been discussed that **habitat loss, fragmentation** and **degradation** are primary causes of the current declines in biodiversity. In this chapter, we will see how the tools and methods of landscape ecology are used to address this problem at the landscape and ecosystem scale.

Box 15.1 Geographic Information Systems (GIS)

Geographic Information Systems (GIS) are computer-based systems that allow the storage, analysis, and display of data pertaining to specific geographic areas. The data used in GIS are derived from multiple sources, including *aerial photographs, satellite imagery*, and *ground-based field* studies (Fig.15.1). Examples of such data include rainfall, elevation, and vegetation cover at specific locations. Each of these and many other variables could be used in a particular application of GIS-but whatever variables are used, the data are keyed to or referenced by spatial or geographic coordinates, so that they can be assembled into a multi-layered map.

Layers of mapped data can be put together in ways that help to address particular questions. We will illustrate this process with an approach often used in conservation biology, called gap analysis. The acronym GAP refers to the Gap Analysis program, a U.S. Geological Survey program whose mission is to help prevent biodiversity decline by identifying species and communities that are not adequately represented on existing conservation lands.

The lark bunting (Calamospiza melanocorys) is one such species. It depends on prairie habitat for its breeding grounds, but much of this habitat has been destroyed by humans. As a result, populations of lark bunting have been declining by an average of 1.6% per year over the past 40 years, making it a species of conservation concern (U.S. Fish and Wildlife Service 2008).

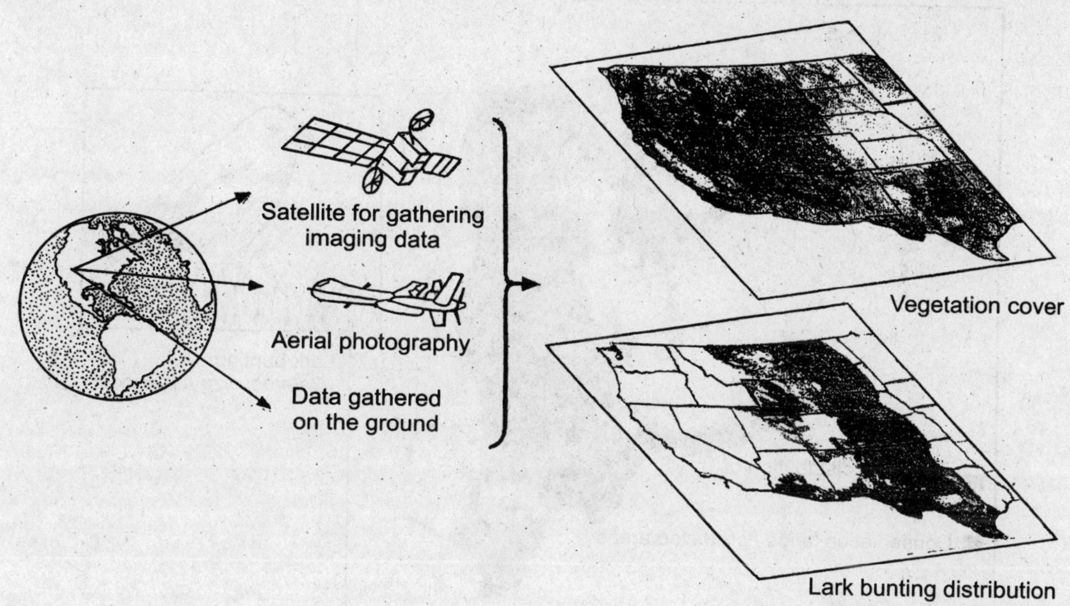

Fig.15.1. GIS integrates spatial data from multiple sources (Cain *et al*, 2011).

For the bird lark bunting and any other species, gap analysis is a two-step process. First, input data on vegetation cover (see the top GIS layer in Fig.15.1) and other environmental conditions required or preferred by the lark bunting is used to predict its geographic distribution (the second GIS layer in Fig. 15.1). Next the lark bunting's distribution is compared with a third GIS layer showing the locations of conservation lands. By combining these two layers, we can calculate that only a small percentage of the bird's distribution is protected (Fig. 15.2). Such information is critical to decisions about what lands should be protected to prevent future losses of biodiversity.

15.3 LANDSCAPE ELEMENTS

Landscape means the landforms of a region in the aggregate. A landscape is an area in which at least one element is spatially **heterogeneous** (varies from one place to another). Landscapes often include multiple ecosystems. They can be heterogeneous either in (1) what they are composed of –is the landscape composed of twelve different vegetative cover types or only three? – or (2) in the way their elements are arranged – are there many small patches arranged regularly cover the landscape, or are there a few large patches? Ecologists often refer to this composite (or pattern) of heterogeneous elements that make up a landscape as a **mosaic** (**Cain** *et al*. 2011).

The different ecosystems that make up a landscape are dynamic and continually interacting with one another. These interactions may occur through the flow of water, energy, nutrients, or pollutants between ecosystems, as from a forest ecosystem into a adjacent lake. There is also **biotic flow** between habitat patches in the mosaic as individuals or their gametes (*e.g.*, pollen) move between them (**Forman** 1995). For such movements to occur, patches of the same habitat type must be connected to one another, or the surrounding habitat (the **matrix**) must be of a

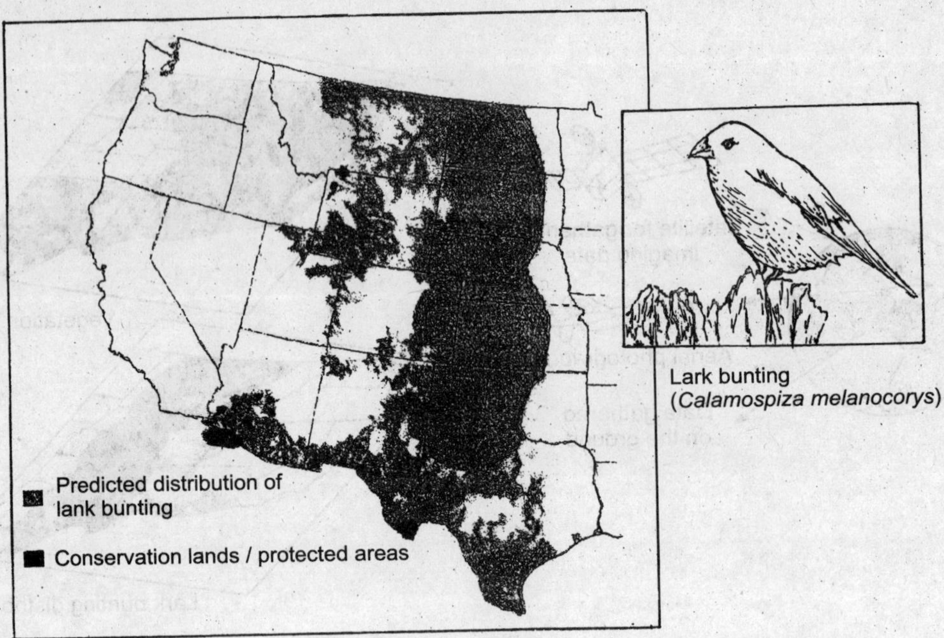

Fig.15.2. A conservation gap. Less than 3% of the lark buntings distribution is in protected areas (Cain *et al.* 2011)

type through which dispersal is possible (Fig.15.3). In Australia, for example, rats regularly leave patches of forest habitat to forage in adjacent macadamia (the edible nut of an Australian tree) plantation (a part of the surrounding matrix). As a result, nut losses along the plantation edges adjacent to forests are greater than along edges adjacent to grasslands or agricultural fields (**White** *et al.*,1997).

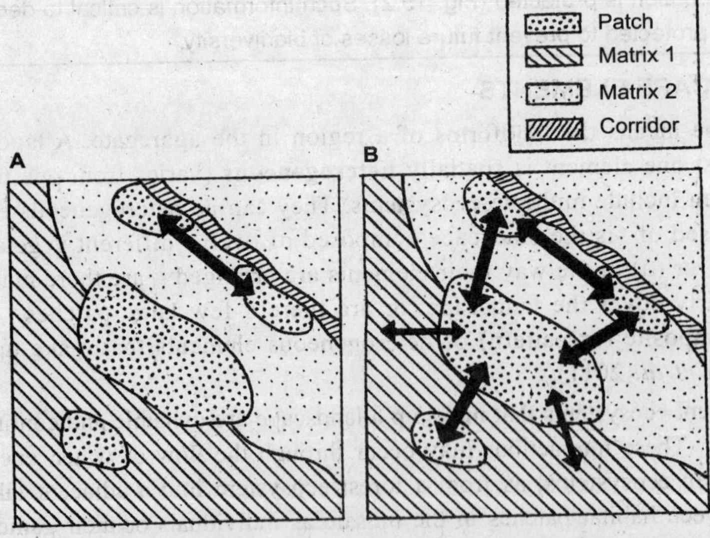

Fig.15.3. Movement across the landscape. Movements between adjacent landscape elements may occur frequently (thicker arrows) or rarely (thinner arrows). A-Exchange between patches of same type occurs frequently if a corridor that allows movement connects the patches. B-Exchange between patches of the same type occurs frequently, but exchange with the matrix occurs only rarely (after Cain *et. al.*, 2011).

The **landscape mosaic** is composed of three major elements: 1. Landscape matrices; 2. Landscape patches and 3. Landscape corridors:

1. Landscape matrix. It is a large area of similar ecosystem or vegetation types (such as agricultural, prairie, old field, or forest) in which landscape patches and corridors are embedded(Fig.15.3).

2. Landscape patch. It is relatively homogeneous area that differs from the surrounding matrix [such as forest patch or woodlot embedded in an agricultural matrix, or a meadow (an area of grassland) embedded in a subalpine forest]. A landscape patch differs from its surrounding matrix and may be referred to as a **low-quality patch** or a **high-quality patch** depending on its vegetative cover, plant quality (e.g., protein content) and species composition.

3. Landscape corridor. It is a strip of environment that differs from the matrix on either side and frequently connects-either naturally or by design—two or more landscape patches of similar habitat. A stream with its riparian vegetation is an example of a natural landscape corridor. The vegetation of corridors is often similar to the patches they connect but different from the surrounding landscape matrix in which they are embedded.

Types of corridors. Based on their origin the corridors are of following five types: (i) remnant corridors; (ii) disturbance corridors; (iii) planted corridors; (iv) resource (natural) corridors; and (v) regenerated corridors. The function of a corridor depends, among other factors, on its structure (both natural and human built), size, shape, type and geographic relation to the surroundings.

(i) Remnant corridors. They occur when most of the original vegetation is removed from an area, but a strip of native vegetation is left uncut. Remnant corridors include uncut vegetation along streams, steep land, railroad tracks, or property borders. Remnant corridors increase species diversity in the region, improve nutrient cycling, protect natural capital and provide habitat for *k*-selected edge species, which include many game species.

(ii) Disturbance corridor. It is produced by a linear disturbance through the landscape matrix. Disturbance corridors disrupt more homogeneous landscape but provide important habitat for native plants and animal "opportunistic" species adapted to disturbance or for species commonly found during the early stages of secondary succession. Power – line corridors that cut through the forest landscape provide an example of a disturbance corridor. Forest – interior species seldom use such corridors to nest or reproduce; however, forest-edge wildlife species may flourish in such corridors. Disturbance corridors may act as **barriers** to the movement of some species, but they provide **routes of dispersal** for other species, such as white-footed mouse (*Peromyscus leucopus*) and the eastern chipmunk (*Tamias striatus*); see **Handerson** *et al.*,1985).

Filter effect. Disturbance corridors can also act filters for some species but not for others. This filter effect can be minimized by providing gaps or nodes of matrix vegetation in the corridor, allowing certain species to cross while restricting others.

(iii) Planted corridors. These are strips of vegetation planted by humans for a variety of economic and ecological reasons. For example, thousands of miles (kilometers) of planted tree corridors were established in the treeless Great Plains of USA as part of the Shelter-belt Project during the 1930s to reduce wind erosion, to reduce snow drift and to provide wood and wildlife habitat. Planted corridors also provide excellent habitat for insectivorous birds and predaceous insects and they function as dispersal routes for small mammal species.

Planted corridors have also been used in agricultural landscape for a variety of ecological purposes. For example, **Kemp** and **Barrett** (1989) established grassy corridors within soybean (*Glycine max*) agroecosystems to retard the movement of adult potato leafhopper (*Empoasca fabae*). Moreover, *Nomuraea rileyi*, a fungal pathogen, infected significantly higher proportion of green cloverworms (*Plathypena scarba*) in plots divided by grassy corridors. This fungus is the primary *natural biotic control* agent of this major lepidopteran pest in the Midwest of the United States.

In North America planted hedgerows of Osage orange trees (*Maclura pomifera*) and multiflora rose (*Rosa multiflora*) have been planted by farmers and game managers for decades to provide firewood, fenceposts, and wildlife habitat, and to serve as barriers to livestock movement.

(iv) Resource corridors. These are narrow strips of natural vegetation that extend for long distances across the landscape (such as gallery forest along a stream). **Karr** and **Schloser** (1978) and **Lowrance** *et al.* (1984) described how vegetated stream corridors benefit the agricultural landscape by intercepting nutrient and sediment (soil) runoff from the croplands that otherwise end up in the streams contributing to *cultural eutrophication* problems (see Chapter 12 Freshwater Ecology). These strips not only improve water quality, but also reduce fluctuations in stream levels and help to conserve the natural biotic diversity within the agricultural landscape mosaic.

(v) Regenerated corridor. It results from the regrowth of a strip of vegetation in a landscape matrix. Hedgerows that develop along fences due to the natural processes of secondary succession provide an excellent example of a regenerated corridor. Birds are common inhabitants of the regenerated corridors; avian species also aid development and plant species composition of these corridors by providing a mechanism of seed dispersal. Although some animals, especially insect pests, do economic damage by feeding on crops adjacent to corridors, **Price** (1976) and **Forman** and **Baudry** (1984) found that regenerated corridors were often the source of natural enemies that colonize adjacent crops and aid biotic pest control. Forest-edge bird species and birds that forage in croplands, often nest in wooded regenerated corridors. These bird species also help to regulate insect species in agricultural crops. Red fox (*Vulpes vulpes*), white deer (*Odocoileus virginianus*) and wood–chucks (*Marmota monax*) frequently use regenerated corridors.

Importance of corridors. Landscape corridors form the important landscape elements that provide a means of enabling animal dispersal, reducing soil and wind erosion, allowing transfer of genetic information between patches, aiding in integrated pest management, and providing habitat for nongame species. Corridors however, may have negative as well as positive effects (for example, transmission of contagious diseases, spread of disturbances such as fire, and increased exposure of animals to predation, **Simberloff** and **Cox** 1987).

Grain. A component of spatial distribution related to dispersion is the concept of **grain** (**Pielou** 1974). Grain is the size of the smallest homogeneous unit of study (such as pixel in a digital image) and determines the resolution at which we view the landscape (**Cain** *et al.*, 2011).Regarding habitat or landscape composition, grain relates to the relationship of landscape patch size to an animal's mobility. A habitat or landscape patch is said to be **coarse grained** for a given animal species if its **vagility** (ability to move freely about) is low relative to the size of habitat patches. A habitat is considered **fine grained** for a species if that organism has high vagility relative size of habitat patches. For example, a wide ranging predator, such as red-tailed hawk (*Buteo jamaicensis*), will experience a mosaic of agroecosystem patches (woodlots) as a fine – grained habitat, whereas another species, such as white-footed mouse (*Peromyscus leucopus*), will likely experience a mosaic of agroecosystem by spending most of its time in a single patch (woodlot).

Extent refers to the area or time period encompassed by a study. Consider how differently we might describe the composition of a landscape depending on how we define its spatial extent (**Cain** *et al.*, 2011).

15.4 BIODIVERSITY AT THE COMMUNITY AND LANDSCAPE LEVELS

Descriptions of how populations and communities are arranged within a given geographical region at the landscape level have featured two contrasting approaches:

1. The **zonal approach**, in which discrete communities are recognized, classified and listed in a sort of checklist of community types.

2. **The gradient analysis approach,** which involves the arrangement of populations of community based on frequency distributions, similarity coefficients, or other statistical comparisons. The term **ordination** is frequently used to designate the ordering of species populations or communities. In general, the steeper the environmental gradient, the more probable is the physical environment, but also because boundaries are sharpened by competition and coevolutionary processes between interacting and interdependent species.

Deciding where to draw boundaries is not too much of a problem in ecosystem analysis as long as input and output environment are considered part of the system, no matter how the system is delimited, whether by natural features or by arbitrary lines drawn for convenience. **Strayer** *et. al*,. (2003) classified boundaries into four main classes of boundary traits: (i) **origin** and **maintenance**; (ii) **spatial structure** (such as geometric shape); (iii) **function** (such as transmissive or permeable); and (iv) **temporal dynamics** (such as age and history of the boundary). However, when one delineates biotic communities by their component species population, there is a problem (see Bioscience August 2003).

15.5 ISLAND BIOGEOGRAPHY

MacArthur and **Wilson** (1963,1967) first published the theory of island biogeography. Simply stated, the **island biogeography theory** holds that the number of species on an island is determined by the equilibrium between the immigration of new species and the extinction of those species already present.

Islands have fascinated biologists, geographers and ecologists since **Charles Darwin** visited the Galapagos Islands. It has also become apparent that landscape patches on the mainland likely function as island within the landscape mosaic. For example, the peaks of mountains (Andes Mountains of Ecuador), especially at approximately the same elevations, function as **terrestrial islands** regarding plant and animal community types. These patches, which vary in size –large and small –and distance -near and far-fit the theory of island biogeography. For example, a patch of forest may be located in a "sea" of agricultural cropland. The effect of patch size and isolation appears to have a pronounced influence on the nature and diversity of species within these landscape patches.

15.6 NEUTRAL THEORY

The **neutral theory** of biodiversity and biogeography (**Hubbell** 2001) is a generalization and extension of the theory of island biogeography. It is termed *neutral theory* because all species are treated as having identical vital rates on a per capita basis (*i.e.*, the same birth and death rates, the same rate of dispersal, and the same rate of speciation). Neutral theory applies to communities of organisms that are on the same trophic level and that compete for the same or similar limiting resources-that is, no species can increase in abundance or biomass without a matching decrease in the collective abundance or biomass of all other competing species. Neutral theory assumes that species are largely substitutable in their use of limiting resources, so that if one species happens to be absent from a community, other species will fully use the resource freed up by its absence. The neutral theory of biodiversity and biogeography extends the MacArthur Wilson theory by also predicting steady-state patterns for species' commonness and rarity (relative species abundances on islands or in local communities) but by incorporating speciation, a process not included in the original theory of island biogeography (see **Odum** and **Barrett** 2005).

Demographic stochasticity. Neutral theory also assumes that large landscape patterns are the result of random speciation, dispersal and random drift in population size of individual species, known as **demographic stochasticity**. **Sugihara** *et al*., (2003) and **Chase** and **Leibold** (2003) argued in support of **niche assembly theory** – the hypothesis that ecological communities are equilibrium assemblages of niche-differentiated, competing species, coexisting because each species is the most effective competitor in its own niche. Contrary to **MacArthur** and **Wilson** (1967) and **Hubbell** (2001), niche assembly theory asserts that dispersal is not very important in determining which species are present or absent from a particular ecological community.

15.7 TEMPORAL AND SPATIAL SCALE

Ecological processes vary in their effects or importance at different spatial and temporal scales. For example, biogeochemical processes may be relatively unimportant in determining local patterns but may have major effects on regional or landscape patterns. At the population and community levels, processes leading to population decline or reduced biodiversity may produce **extinction** at the local scale, but at the landscape level the same processes may appear only as spatial redistributions or alternations. The **concept of scale** encourages analyses at different levels of organization (different levels in the hierarchical system). For example, a landscape might appear to be heterogenous at one scale but quite homogeneous at another scale. Thus, when ecologists select a scale for investigation, they must understand how a change in temporal or spatial scale can affect patterns, processes and emergent properties across scales.

K.E.F.Watt (1973) in his textbook *Principles of Environmental Science* listed five categories of fundamental ecological variables that need to be understood in order to understand large temporal-spatial processes, including the interactions among these five categories of world resources. These resources (or variables) are **1. energy, 2. matter, 3. diversity, 4. Time** and **5. space.** Since 1973, numerous books and publications have focussed on the importance of time and space as fundamental world resources-indeed, temporal-spatial relationships support the emerging disciplines of **conservation biology**, **ecosystem health**, **landscape ecology** and **restoration ecology**, among others.

Figure 15.4 outlines how processes and patterns change at different temporal and spatial scales (**Urban** *et al*., 1987). Figure 15.4A shows short-term disturbances that involve increasing spatial but not temporal scales, ranging from tree falls to large-scale fires and floods. Figure 15.4B depicts forest processes that involve both increasing temporal and spatial scales, ranging from the role of local seed banks to speciation and extinction. Figure 15.4C exhibits environmental constraints, scaled from microhabitat conditions to glacial cycles; and Figure 15.4D illustrates changes in vegetation patterns, scaled from ground cover to major biomes. Each of these space-time relationships can best be understood when viewed as changes in temporal and spatial scales.

Both temporal and spatial scales are involved in the ecological/economic hierarchy of farming. The success of the crop depends not only on field conditions (small temporal-spatial scale), but also the sustainability of the whole farm (ability to profit), on water from the watershed, and on a market for a crop (large temporal-spatial scale). Problems such as pest control, stream eutrophication and economic constraints change with scale, ranging from small- scale (and frequently short term) problems at the field or single-crop level to large-scale (and frequently long-term) challenges at the landscape or national level. For example, a specific pesticide applied at the

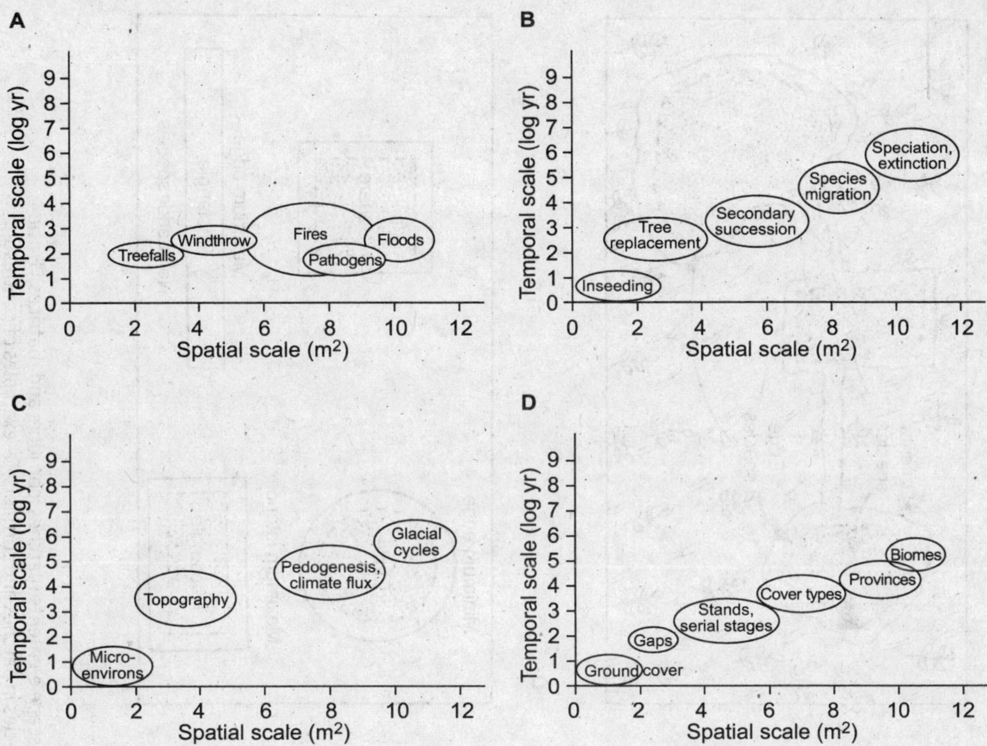

Fig.15.4 Landscape ecological examples of changes at different temporal spatial scales in A-disturbance regimes; B-forest processes, C-environmental constraints; and D-vegetation patterns (after **Odum** and **Barrett**, 2005).

field or crop level may control a particular insect species at a critical time during the growing season, whereas only the implementation of large- scale, ecologically based pest management programs at the landscape level improve the problem on a long-term basis (**NAS** 2000; **Odum** and **Barrett** 2000).

The impact at natural disturbances or human-originated anxieties on population dynamics and community structures is also a matter of scale, involving an assessment of size of the area affected and the duration and intensity of the anxiety (**Connell** 1979).

15.8 LANDSCAPE GEOMETRY

Just as the size and quality of landscape patches and corridors (amount of vegetation cover and food quality, for example) affect ecological processes and plant and animal abundances, the **geometry and configuration of landscape elements** affect ecological processes at the population and community levels. It is now widely recognized that the size and shape of landscape patches influence biotic diversity, home range, size and shape, animal dispersal behaviour and species abundance. Recent investigations have addressed the role that landscape geometry plays in plant and animal survivorship source-sink dynamics. Future investigations need to focus on the role of **landscape geometry** (the study of the shapes, patterns and configurations of landscape elements) and **landscape architecture** (patch stratification, "soft versus hard" edge, and three dimensional use of habitat space) if ecologists are to more fully understand such phenomena as *dispersal behaviour, pattern of animal movement, bioenergetics at the landscape scale and ecosystem – landscape sustainability*.

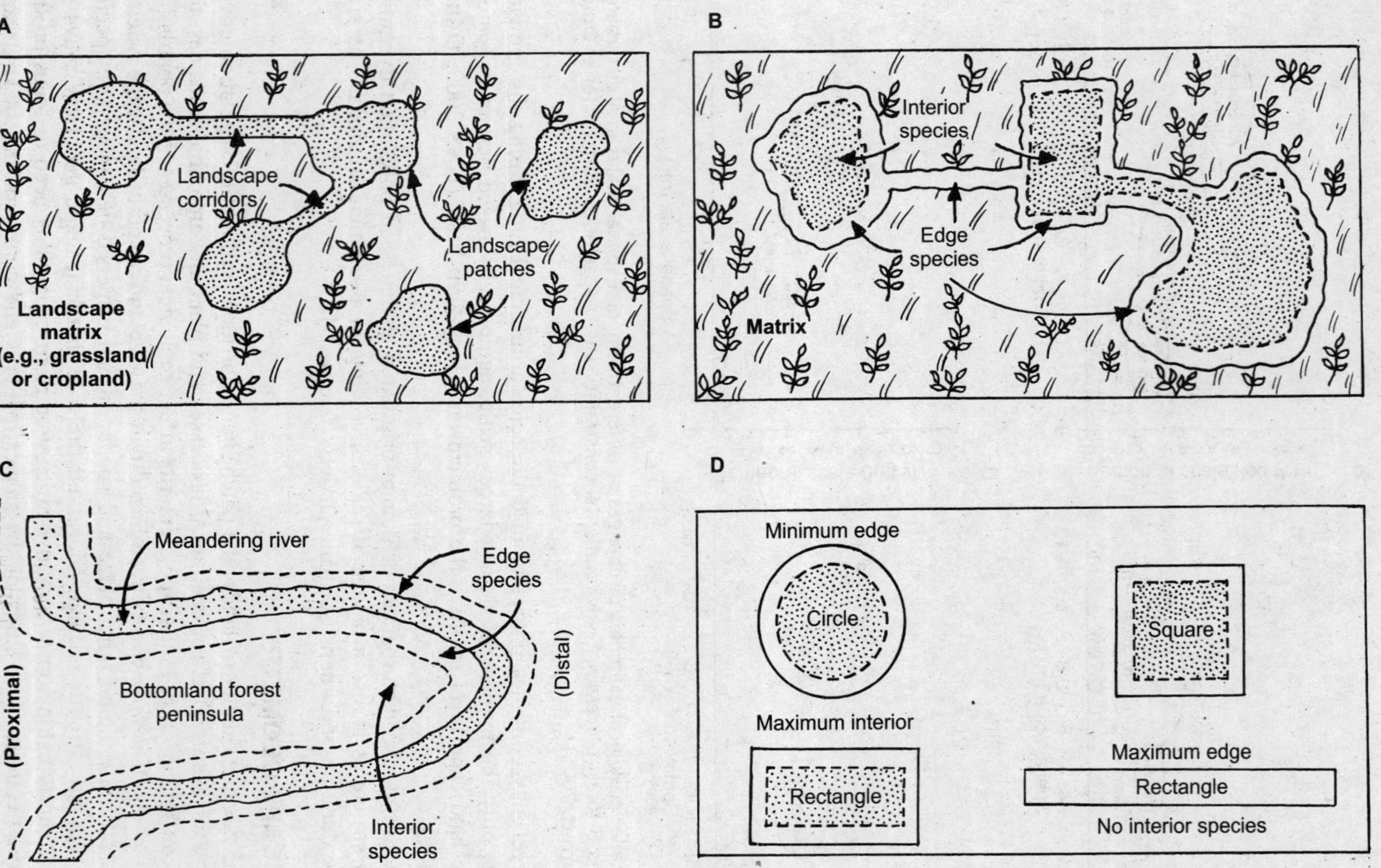

Fig.15.5. Diagram depicting(A) the three major landscape elements (patches, corridors and matrices); B-the relative abundances of edge and interior species in patches and corridors; D- examples of contrasting patch shapes of equal size but different geometry (after Odum and Barrett 2005).

Figure 15.5A summarizes the three major landscape elements (patches, corridors and matrices) found within the landscape mosaic. Figure 15.5B illustrates how the abundances of interior and edge species are related to patch and corridor shape and size. Bird edge species, for example, include the indigo buntings (*Passerina cyanea*), the eastern bluebird (*Sialia sialis*), and the cardinal (*Cardinalis cardinalis*); examples of interior bird species include the wood thrush (*Hylocichla mustelina*), red-eyed vireo (*Vireo olivaceus*) and the downy woodpecker (*Picoides pubescens*). **Kendeigh** (1944) noted that patches of eastern deciduous forest (USA) larger than 26 hectares (65 acres) warranted consideration as valid avifaunal census patches to evaluate forest edge versus interior bird species. Although two of the patches on the left of Fig.15.5B are connected by a narrow corridor, the corridor is too narrow to permit the movement of interior species between patches. Figure 15.5C illustrates a bottom-land forest peninsula, created by a meandering (*meander* is a bend of a river that curves back on itself) stream or river. Peninsula may be large in scope (such as the Baja Peninsula of California) or small in scope (such as the neck of a forest extending into farmland). Figure 15.5D illustrates several patches of equal size but of different shapes. A circular patch maximizes habitat for interior species, whereas a long linear, narrow patch maximizes habitat for edge species. In fact, it is likely that interior habitat is entirely eliminated within these narrow, linear corridors, thus, severely limiting or preventing interior plant and animal species from inhabiting a landscape patch of this configuration.

Patch **connectivity** (that is, corridor presence) was found to be more important than corridor width to the dispersal of meadow voles (*Microtus pennsylvanicus*) between patches (**LaPolla** and **Barrett**, 1993). **Rosenberg** et al. (1997) discussed the geometry, function and efficacy of biological corridors, including the ability of such corridors to mitigate (to make less severe, serious or painful) high local rates of extinction. They also note that linear corridors may function as linear patches (that is, habitat patches) depending on the size and species-specific responses to these configurations.

In the "real world", **circular patches** or ecosystems are abundant where radial-arm irrigation is used (Such as in the western United states, where annual rainfall is limiting), where as **square** or **rectangular patches** are abundant in the farming landscape of the midwestern United States. In some cases, there are roads about every square mile.

Since the amount of **edge habitat** changes with patch shape, it is only natural that landscape ecologists have increasingly investigated the role of edge habitat within the landscape mosaic. Because the *variety* and *diversity* of life is often greatest in and about edges and ecotones, it follows that changes in patch shape, resulting in increasing or decreasing species abundance and biotic diversity.

Thus, **habitat fragmentation** (changes caused by both natural and human-dominated processes) has resulted in a landscape mosaic dominated by change in the size, shape, and frequency of landscape elements (patches, corridors and matrix). These changes resulted from habitat fragmentation have affected plant and animal diversity and abundance, numbers of edge versus interior species and changes in biodiversity also affect ecological processes at several levels of organisation. Long-term evolutionary processes at the organism, population and community levels have resulted in a series of ecosystem, landscape and biome types at the global scale.

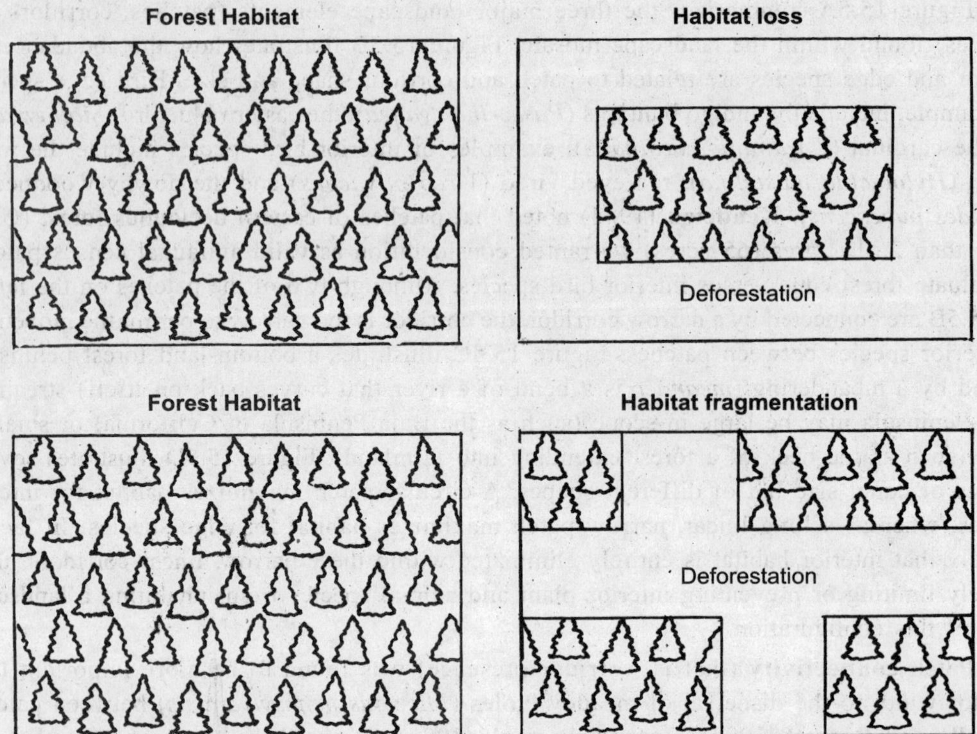

Fig.15.6 Diagram showing the difference between habitat loss (top) and habit fragmentation (bottom). The same total area of deforestation can result in the loss of species requiring large home ranges when habitat is fragmented, for example (after **Odum** and **Barrett** 2005).

15.9 CONCEPT OF LANDSCAPE SUSTAINABILITY

The word sustainability has been variously defined as—"to keep in existence" "to support" "to endorse without failing or yielding", " to maintain" or "to supply with necessities or nourishment to prevent falling below a given threshold of health or vitality". This threshold could be viewed as the **carrying capacity** (*K*) concept discussed in Chapter 5 (Population). Goodland's (1995) definition of sustainability as *"maintaining natural capital"* and *"maintenance of resources"* more clearly defines sustainability at higher levels of ecological organization (*i.e.*, at the level of ecosystem, landscape and global).

Benefits supplied to human societies by natural ecosystem form the **natural capital**. In addition to the production of goods (such as timber, game species, fruits and nuts), ecosystem and landscape services support life through such functions as the purification of air and water, cycling of nutrients, the pollination of crops, the preservation and renewal of soil fertility, the partial stabilisation of climate, the maintenance of biodiversity, the provision of aesthetics, and the control of pests, among others. These services are so fundamental to life that they are, unfortunately, easily taken for granted. Thus, these services which underpin the concept of landscape sustainability, are greatly undervalued and poorly understood by human societies, If awareness and understanding are not increased and current trends continue, humankind will dramatically alter Earth's natural ecosystems and landscapes, to the extent that these services of natural capital will greatly diminish.

15.10 DOMESTICATED LANDSCAPE

Civilization seems to reach its most intense development in what was originally forest and grassland, especially in temperate regions. Consequently, most temperate forests and grasslands have been greatly modified from their primeval condition, but the basic nature of these ecosystems has by no means changed. Humans, in fact, tend to combine features of both grasslands and forests into habitats that might be termed as **forest edge**. Forest edge may be defined as an *ecotone* between forest and grass or shrub communities. When humans settle in grassland regions, we plant trees around our homes, villages and farms, so that small patches of forests become dispersed in what may have been treeless country. Likewise, when humans settle in forests, we replace most of it with grasslands and croplands (as lesser amounts of human food can be obtained from a forest), but leave patches of original forest on farms and around residential areas. Many of the smaller plants and animals originally found in both forest and grassland are able to adapt thrive in close association with humans, as are domestic or cultivated species. The American robin (*Turdus migratorius*), for example, once a bird of the forest, has become so well adapted to the human-made forest edge that it has not only increased in numbers, but has extended its geographic range. Most forest birds (the thrushes in Europe, for example) have switched from the forest to gardens, cities and hedgerows or else they have become extinct, because there are no longer many tracks of unbroken forest. Most native species that persists in regions heavily settled by humans become useful members of the forest-edge landscape, but a few become pests.

If we deem croplands and pastures as modified grasslands of early successional types, then we can say that we depend on grassland for food but like to live and play in the shelter of the forest, from which we also collect useful wood products. Thus, we might say humans, in common with other heterotrophs, seek production and protection from the landscape. In many cases, the *monetary value* of wood, if harvested all at once, is less than the value of the **intact forest** that provides recreation, watershed protection and other life-support services such as home sites and a sustainable harvest of wood.

Agroecosystems and Agrolandscapes

Agroecosystems are domesticated ecosystems that are in many ways intermediate between natural ecosystems, such as grasslands and forests, and fabricated ecosystems such as cities. Like natural ecosystems, agroecosystems are solar-powered, but they differ from natural ecosystems in several ways; processed fossil fuels, along with auxiliary energy sources that boost productivity but also increase pollution; diversity is greatly reduced by human management in order to maximize the yield of specific foods or other products; the dominant plants and animals are under artificial rather than natural selection; and control is external and goal-oriented, rather than internal via subsystem feedback, as in natural ecosystems.

The simplified graphic models in Figure 15.7 depict the stages in the development of agriculture, namely, pre-industrial (similar to a natural ecosystem), industrial and reduced-input conservation tillage (see chapter16, Human Modified Ecosystems). The food web in **pre-industrial agriculture** –as still practiced in less developed countries–is very similar to that in natural ecosystems; with domestic animals replacing wild animals in the grazing food chain. Pre-industrial agriculture is energy efficient and diverse, with a variety of crops grown together, including fish fed on plant residues, and it feeds local villages but does not produce enough surplus for export, or to feed cities. In **industrial agriculture** (Fig. 15.7B), the microbial recycling loop is almost eliminated and replaced by large subsidies (aids) of fertilizer, pesticides and water, which increases yields but produces widespread pollution, erosion and loss of habitat and soil quality. Crop and meat production are decoupled with the development of feed lots, so manure becomes a pollutant rather

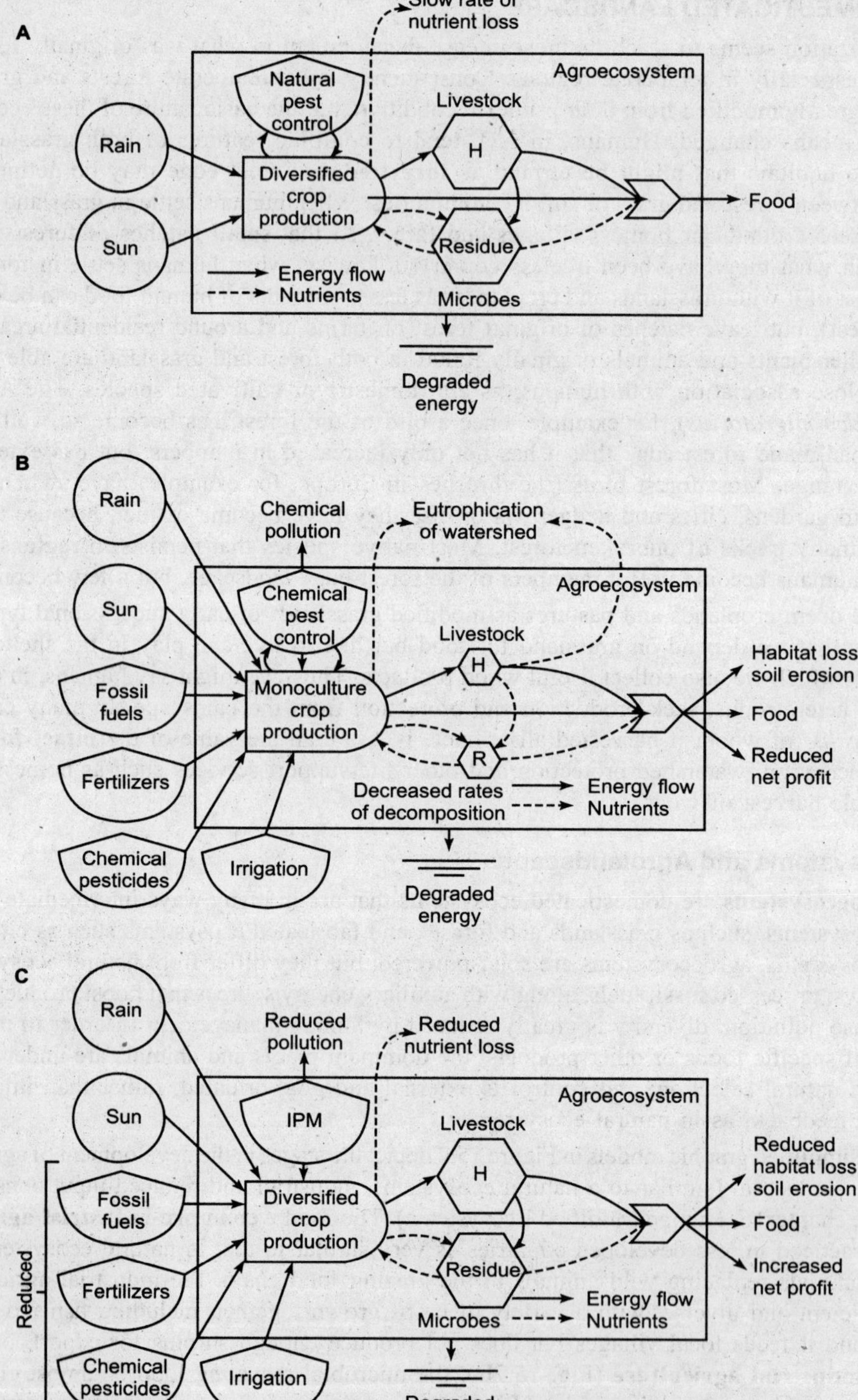

Fig.15.7 Agroecosystem phases: A- Pre-industrial agriculture; B- Industrial agriculture; C-Low- input sustainable agriculture. IPM= integrated pest management (after **Odum** and **Barrett** 2005).

than a fertilizer resource (**Brummer** 1998). Because industrial agriculture is not sustainable, it is being redesigned along the lines of conservation tillage, frequently termed **low-input sustainable agriculture**.

Current unsustainable industrial agriculture can be redesigned to sustain soil quality and high yields by combining the new technologies of residue management, conservation tillage, and polyculture with the soil-building processes of natural ecosystems and pre-industrial agroecosystems. There is increased emphasis on managing agroecosystems as **agrolandscapes**, with greater focus on reconnecting agricultural and urban landscapes (**Barrett** and **Skelton** 2002). There is also increased interest in **agroforestry** – a practice that involves the cultivation of small, fast-growing trees and food crops in alternate rows, in ecologically based pest management and in opportunities to integrate soil, crop and weed management in low-input farming systems.

15.11 TROPHIC CASCADE HYPOTHESIS

The removal of top predator can alter the landscape substantially, in part because herbivores whose population were once controlled by the predator may increase in number and negatively affect vegetation dynamics. In Yellowstone National Park (of Greater Yellowstone Ecosystem or GYE of USA), the growth and reproduction of riparian tree species, such as cottonwoods, aspens, and willows, declined after wolves (*Canis lumus*) were removed (**Ripple** and **Beschta** 2007). A possible reason was that the trees experienced heavy browsing by herbivores such as elk(*Cervus canadensis*), which roamed freely along rivers and streams once the wolves were gone.

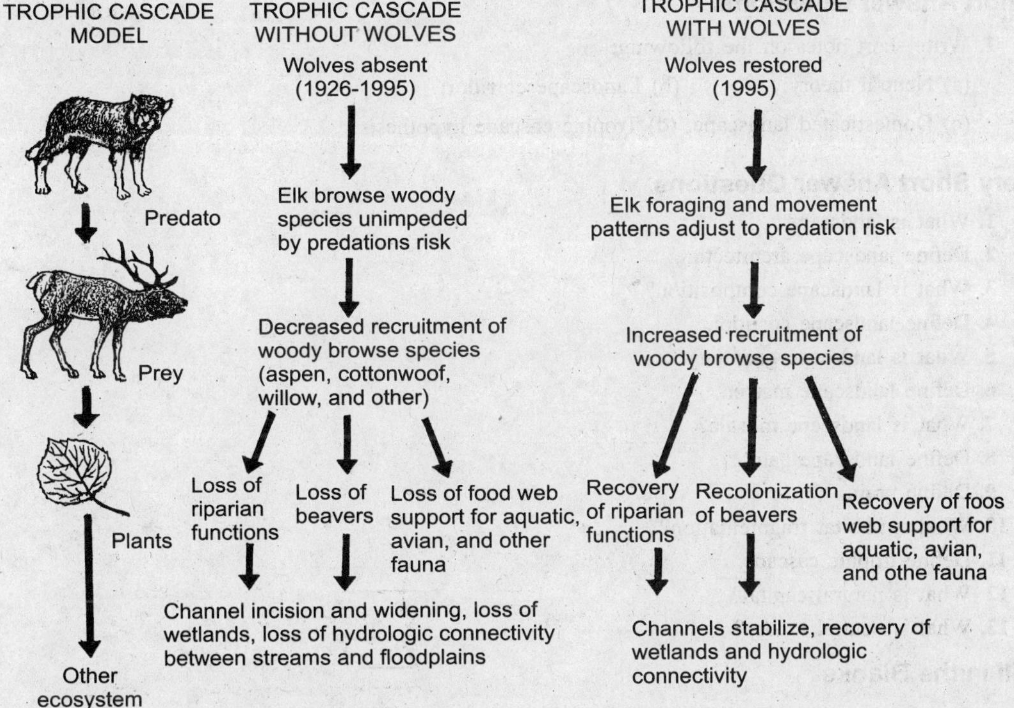

Fig.15.8 The trophic cascade hypothesis of (**Ripple** and **Beschta** 2004). Wolves are top predators and their reintroduction to the Greater Yellowstone Ecosystem (GYE) has the potential to cause cascading trophic effects. According to the hypothesis shown here, elk now avoid those sites where they are most vulnerable to predation and trees and shrubs are now returning to those sites after decades of suppression by elk (Cain *et al.*, 2011).

The reintroduction of wolves began in the winter of 1995-1996, when 31 wolves captured in Canada were released into the park. Their number increased rapidly; by 2004, there were about 250 wolves in the park. Following the reintroduction, population of elk, the wolves' principal prey, have declined by 50%. Elk were initially naive and very vulnerable to wolf predation, but they have since modified their behaviour, showing a preference for foraging in places that provide high visibility. Furthermore, cottonwoods, aspens and willows have begun to recover in some areas. In some cases, the early signs of recovery appeared to be concentrated in areas where elk face a high risk of predation, such as locations where visibility is poor, escape routes are lacking, or ambush sites are common. Thus, elk may be avoiding areas where they are most vulnerable to attack by wolves, allowing trees in those areas to recover- and possibly leading to a series of other, cascading effects (Fig. 15.8). Some researches such as **Kaufman** *et al.* (2010) and **Creel** and **Christianson** (2009) have contradicted trophic cascase hypothesis.

QUESTIONS

Long Answer Questions

1. Write an essay on landscape ecology.

2. Give an account of landscape elements.

3. Describe landscape geometry.

Short Answer Questions

1. Write short notes on the following:

(a) Neutral theory; (b) Landscape corridor;

(c) Domesticated landscape; (d) Trophic cascade hypothesis.

Very Short Answer Questions

1. What is landscape?

2. Define landscape architecture.

3. What is landscape composition?

4. Define landscape corridor.

5. What is landscape geometry.

6. Define landscape matrix.

7. What is landscape mosaic?

8. Define landscape patch.

9. Define grain.

10. What is habitat fragmentation?

11. Define trophic cascade.

12. What is natural capital?

13. What is neutral theory?

Fill in the Blanks

1. Benefits supplied to human societies by natural ecosystem form the ....

2. ....is ecotone between forest and grass communities.

3. .... are intermediate between natural ecosystems and fabricated ecosystems.

4. Grasslands and forests are .... ecosystems.

5. .... are fabricated ecosystems.

ANSWERS

Very Short Answer Questions

1. Landscape is an area that often includes multiple ecosystems and that is spatially heterogeneous in one or more features of the environment, such as the number or arrangement of different habitat types. (2) Landscape is a regional level of organization between the ecosystem and the biome.
2. Landscape architecture is the study of structure and the three-dimensional use of habitat space in the field of landscape ecology.
3. In landscape ecology, the kinds of elements or patches comprised by a landscape and how much of each kind is present.
4. Landscape corridor is the strip of vegetation that differs from the matrix and frequently connects two or more patches of similar habitat.
5. Landscape geometry is the study of shapes, patterns and configurations at the landscape scale.
6. Landscape matrix is large area of similar ecosystem or vegtation types (agricultural or forest, for example) in which landscape patches are embedded.
7. Landscape mosaic are quilt-like patches of different types of vegetative cover across a landscape or 2. Landscape mosaic is cluster of ecosystems of different types.
8. Landscape patch is relatively homegenous area that differs from the surrounding matrix (such as a woodlot embedded in an agricultural matrix).
9. In landscape ecology, the grain is the size of the smallest homogeneous unit of study (such as a pixel in a digital image), which determines the resolution at which the landscape is observed.
10. The breaking up of one continuous habitat into a complex pattern of spatially isolated habitat patches amid a matrix of human dominated landscape.
11. Trophic cascade is a change in the rate of consumption at one trophic level that results in a series of changes in species abundance or composition at lower trophic level.
12. Natural capital is benefits and services supplied to human societies by natural ecosystem.
13. Neutral theory is extension of the theory of island biogeography, proposed by Stephen P. Hubbell in 2001, in which all species are treated as if they possess the same per capita rates of birth, death, dispersal and speciation.

Fill in the Blanks

1. Natural Capital 2. Forest edge 3. Agroecosystems 4. Natural
5. Cities.

16

Human Modified Ecosystems (Agroecosystems, Technoecosystems and Ecological Footprints)

16.1 EXAMPLE OF A HUMAN-MADE ECOSYSTEM: CROPLAND ECOSYSTEM

Cropland ecosystems are artificial or human-engineered ecosystems, where, in order to obtain more food, cloth, pigments, timber, medicines, biofuels and other useful products, human beings become responsible for the replacement of natural systems. To secure maximum production humans make much planned manipulations in the physicochemical environment. These include addition of fertilizers to soil, use of chemicals for disease control, proper irrigation practices, etc. Thus, a cropland ecosystem is an artificial system aimed primarily to grow a single species of one's choice (monoculture). We have ecosystems of dominant crop species such as wheat, maize, jowar, paddy, sugarcane, chicken pea, pea, cotton, jute, vegetables, etc., under most suitable condition of their growth and yield. It is generally argued that in a natural ecosystem, the nature maximizes for gross production, whereas in an artificial ecosystem, human being maximizes for net production. In a cropland ecosystem, there is not necessarily an increase in the total dry matter production of the **whole plants**, but generally most of the production goes into grain and less into leaves, stem and roots. Thus, in agriculture (where all the systems are artificial), there is an objective to achieve **high rates of production** (P) of readily harvestable products with little **standing crop**, *i.e.*, **biomass** (B) left in the field for accumulation, or we may say that there is high P/B effenciency. Nature (natural ecosystem), on the other hand, follows just a reverse efficiency, *i.e.*, a high B/P ratio, where the standing crop is generally accumulated to its maximum. Here we can consider an example of ecology of maize cropland ecosystem.

Maize Cropland Ecosystem

1. Abiotic Components. These include the climatic conditions of the region, where the crop may grow most successfully, and the various minerals particularly C,H,O,N,P,K in soil and atmosphere. In such a field humans generally make additions of a number of chemical fertilizers to soil. Maize generally grows best in slightly alkaline soil with good aeration.

2. Biotic Components. The various living organisms in the food chain of maize cropfield occur as follows:

(i) Producers. The dominant plant species would naturally be maize (*Zea mays*).Besides maize, a number of weeds such as *Cynodon dactylon, Launaea nudiculis, Euphorbia hirta, Cyperus rotundus, Digitaria* spp. and *Alysicarpus* spp. also contribute to primary production of the cropfield.

(ii) Consumers. These are as follows:

(a) Primary consumers. They are herbivores represented by a variety of animals, big as well as small. The smaller animals include chiefly the insects as aphids, thrips, beetles, etc. which feed and lay their eggs on maize leaves. Larger animals are rabbits, rats, birds (parrots) and humans, all feeding on leaves, flower and fruits of maize.

(b) Secondary consumers. They are carnivores such as frogs, toads and some birds that eat insects.

(c) Tertiary consumers. They are carnivores such as snakes and hawks which feed on the secondary consumers, frogs, toads and smaller birds.

(iii) Decomposers. These are microorganisms present in the soil as well as air, that decompose the dead organic matter of plant and animals. These are chiefly bacteria, actinomycetes and fungi, responsible for decay, decomposition and humification, making the minerals available again to the producers.

16.2 ECOLOGY OF AGROECOSYSTEMS

Agroecosystems (short for agriculture ecosystems) differ from natural or seminatural solar-powered ecosystems, such as lakes and forests in three basic ways:

1. The **auxiliary energy** that boost or subsidizes the solar energy input is under the control of humankind and consists of human and animal labour, fertilizers, pesticides, irrigation water, fueled-powered machinery and so on.

2. The **diversity** of organisms and crops is greatly reduced (by human management) to maximize yield of specific food crops or other products.

3. The dominant plants and animals are under **artificial selection** rather than natural selection.

In other words, agroecosystems are designed and managed to channel as much conversion of solar energy and energy subsidies as possible into edible or other marketable products by a twofold process: (i) by **employing additional energy** to do maintenance work that in natural systems would be accomplished by solar energy, thus allowing more solar energy to be converted directly into food, and (ii) By **genetic selection** of food plants and domestic animals to optimize yield in the specialized, energy-subsidized environment. As in all intensive and specialized **land use**, there are costs as well as benefits, including **soil erosion**, **pollution** from pesticide and fertilizer runoff, **high cost** of fuel subsidies, **reduced biodiversity**, and increased vulnerability to **weather changes** and **pests**.

About 10 percent of the world's ice-free land areas is cropland, converted mostly from natural grasslands and forests, but also from deserts and wetlands. Another 20 percent of the land area is pasture, designed for animal rather than plant production. Thus about 30 percent of the terrestrial surface is devoted to agriculture in the broadest sense. According to recent data, all the best land (that is, the land most easily farmed by existing technology) is now in use (**Odum** and **Barrett** 2005).

In USA, recent agriculture practice has switched from Conventional Agriculture practice, with emphasis on crop yield based on increased subsidies, to sustainable agriculture, with emphasis on low-input, sustainable agriculture, decreased soil erosion, and increased biodiversity (National Research Council or NRC, USA 1989).

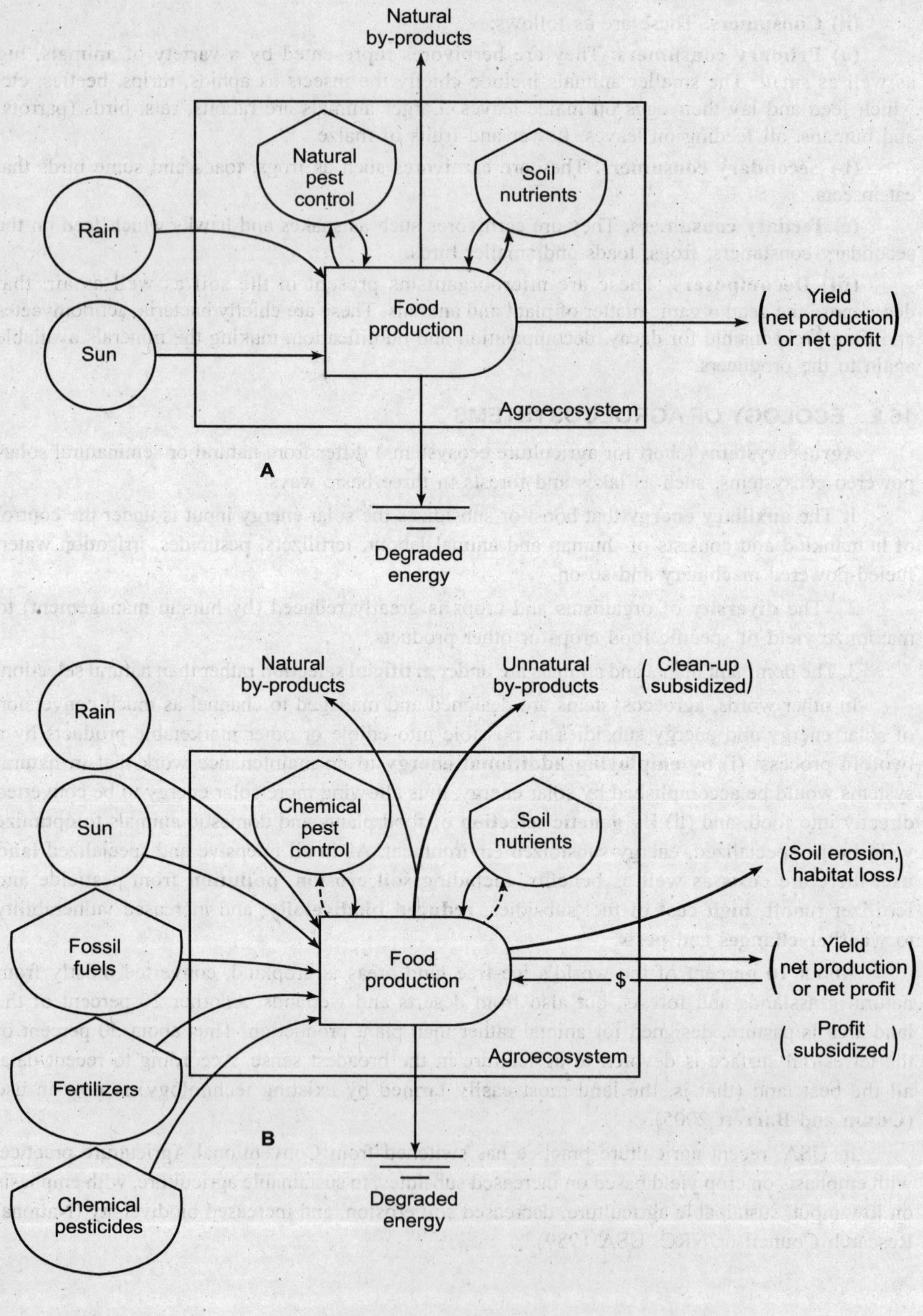

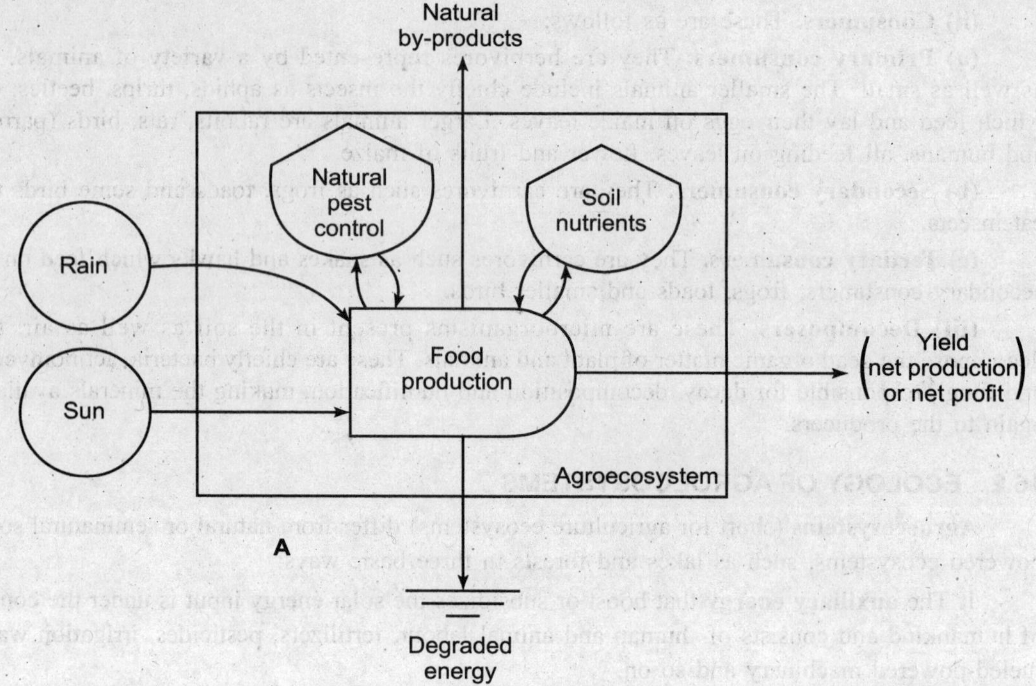

A

Natural
by-products

Unnatural
by-products

(Clean-up
subsidized)

Rain

Sun

Chemical
pest
control

Soil
nutrients

(Soil erosion,
habitat loss)

Fossil
fuels

Food
production

$

Yield
(net production
or net profit)

Fertilizers

(Profit
subsidized)

Agroecosystem

Chemical
pesticides

B

Degraded
energy

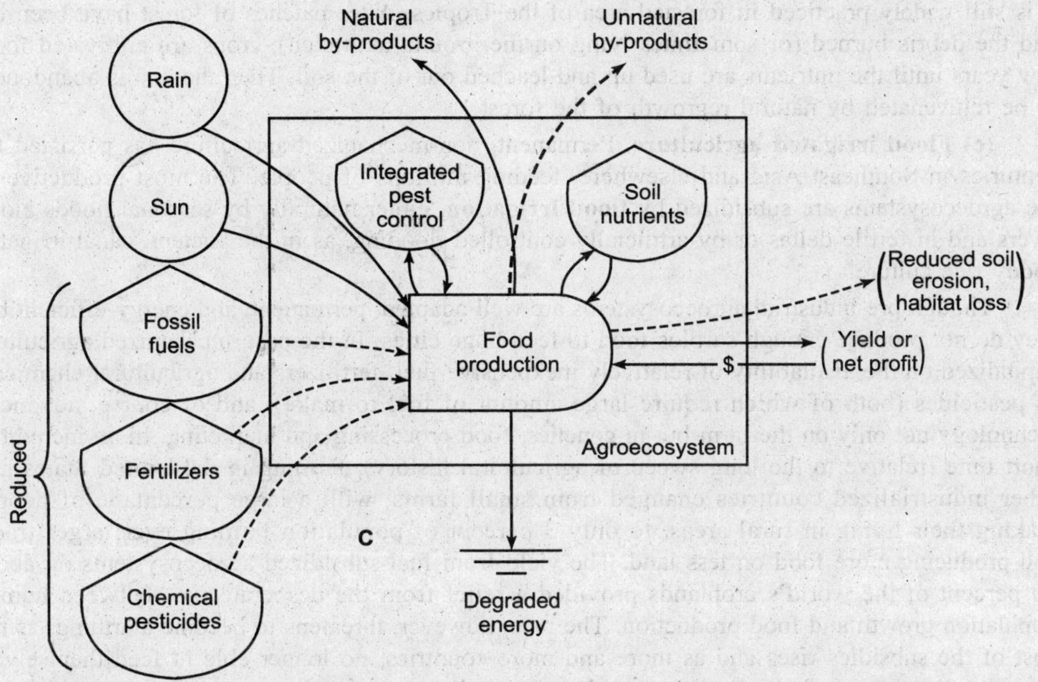

Fig.16.1 A- Diagram depicting a natural, unsubsidized, solar-powered pre-industrial agroecosystem: B-Diagram depicting a human subsidized, solar –powered industrial agroecosystem. C- Example of a low-input sustainable agroecosystem (e.g., soybean cropfield) (after Odum and Barrett 2005).

Agroecosystems are broadly divided into following three types:

1. Pre-industrial agriculture. It is self-sufficient and labour intensive (human and animal provide the energy subsidy). It provides food for farmer and family and for sale or barter in local markets, but does not produce a large surplus for export (Fig. 16.1A).

2. Conventional or Industrial agriculture. It is intensive mechanized, fuel-subsidized agriculture. In it machines and chemicals provide the energy subsidy and produces food exceeding local needs for export and trade, thus making food a commodity and a major market force in the economy rather than providing only life-support goods and services (Fig.16.1B).

3. Alternative agriculture. It is lower-input sustainable agriculture (LISA). It places emphasis on sustaining crop yields and profits while reducing inputs of fossil fuels, pesticides and fertilizer subsidies (Fig. 16.1C).

About 60 percent of the world's croplands are in the pre-industrial category, a large proportion of them in the less developed countries of Asia, Africa and South America that have large human populations. A great variety of types of agroecosystems have been adapted to soil, water and climate conditions, but for the purpose of general discussion, three types predominate: 1. **Pastoral systems**, 2. **Shifting** or **swidden** (slash and burn agriculture), and 3. **Flood-irrigated** and other nonmechanical systems.

(a) Pastoralism. It involves herding cattle or other domestic animals in arid and semiarid regions (especially savanna and grassland regions of Africa), with people subsisting on livestock products such as milk, meat, and hides.

(b) Shifting agriculture. This type of agriculture once was practised throughout the world, it is still widely practiced in forested area of the Tropics. After patches of forest have been cut and the debris burned (or sometimes lying on the ground as mulch), crops are cultivated for a few years until the nutrients are used up and leached out of the soil. Then the site is abandoned, to be rejuvenated by natural regrowth of the forest.

(c) Flood irrigated agriculture. Permanent, non-mechanized agriculture has persisted for centuries in Southeast Asia and elsewhere, feeding millions of people. The most productive of the agroecosystems are subsidized by **flood irrigation**, either naturally by seasonal floods along rivers and in fertile deltas or by artificially controlled flooding, as in the ancient, canal-irrigated paddy rice culture.

Though pre-industrial agroecosystems are well adapted, permanent, and energy-efficient but they do not produce enough surplus food to feed huge cities. In the past, mechanized agriculture capitalized on the availability of relatively inexpensive fuel, fertilizers and agricultural chemicals as pesticides (both of which require large amount of fuel to make), and of course, advanced technology not only on the farm but in genetics, food processing and marketing. In an incredibly short time (relative to the long sweep of agricultural history), farming in the United States and other industrialized countries changed from small farms, with a large percentage of people making their living in rural areas, to only 3 percent of population farming ever larger tracts and producing more food on less land. The yield from fuel-subsidized agroecosystems on about 40 percent of the world's croplands provided a relief from the desperate race between human population growth and food production. The race, however, threatens to become alarming, as the cost of the subsidies rises and as more and more countries, no longer able to feed themselves, are forced to import from countries, such as United States, that have a surplus to export. For example, a field located in Iowa state of USA being cultivated to plow, with large equipment typical of agricultural practice. This times before planting, resulting in extensive loss of soil due to wind and rainfall erosion. On the other hand, in conservation tillage (e.g., soybeans are no-till planted in the mulched residue of last year's corn crop without the need for plowing) which greatly reduces soil erosion, promotes increased soil biotic diversity, and enhances nutrient (detritus) recycling.

Further, where the number of farmers in the developed world has declined dramatically, the number of farm animals has not, and the intensity of production of animal products has increased, paralleling that of crops. Thus, grain-fed beef has replaced grass-fed beef , and chickens are bred and managed as egg – and meat-producing machines, encased in wire cages under artificial light, and supplied with growth – promoting food mixtures and drugs. In the United States, most of the corn crop (and other grains and soybeans) is fed to domestic animals, which in turn feed and perhaps overfeed- the affluent of the world (Science, 7.2.2013).

When considering the population pressure on the environment and resources, one should not forget that not only are there many more domestic animals than people worldwide, but these animals also consume about five times more calories than do people. That is, the biomass of domestic animals (bovinity, equinity, swine, sheep, poultry) is about five times that of humans in **population equivalents**, a quantity based on the number of calories consumed by the average human. This does not include pets, which also consume a large amount of food.

Thus, industrial agriculture has greatly increased the yield of food and fiber per unit of land. This is bright side of the technology, but there are two dark sides: (i) many small farms go out of business worldwide, and these families gravitate towards cities, where they become consumers rather than producers of food; (ii) industrial agriculture has greatly increased nonpoint pollution and soil loss. To counteract the latter, a new technology called **low-input sustainable agriculture (LISA)** is coming into increase use (Fig.16.1C) and soybean cultivation (see **Odum** and **Barrett** 2005).

16.3 TECHNOSYSTEMS AND ECOLOGICAL FOOTPRINTS

Current urban-industrial society not only affects natural life-support ecosystems, but it has created entirely new arrangements, called **technoecosystems** (**Odum** and **Barrett** 2005). Technoecosystems are competitive with and parasitic on natural ecosystems. These new systems involve advanced technology and powerful energy sources. If urban–industrial societies are to survive in a finite world, it is imperative that technoecosystems interface with (meet and interact) natural life-support ecosystems in a more positive or mutualistic manner than is the case at present.

Total Human Ecosystem

Prior to the Industrial Revolution, humans were a part of natural ecosystems. In the ecosystem model of Figure 16.2, humans were top predators and omnivores (the terminal H box in the food web). Early agriculture – as in the case with traditional or preindustrial agricultural system–was compatible with natural systems and often enriched the landscape in addition to providing food. However, with the increasing use of fossil fuels and nuclear fission – energy sources many times more powerful than sunlight – together with the growth of cities and the increase of money –based market economics, the model depicted in Figure 16.2 is no longer adequate. Modern ecologists need to create a model for this new **technosystem** –a term suggested by pioneer landscape ecologist **Zev Naveh** (1982). More recently, Naveh has used the term **total human ecosystem** to describe the relationship of the industrial society (the technosystem) to the total ecosphere (**Naveh** 2000).

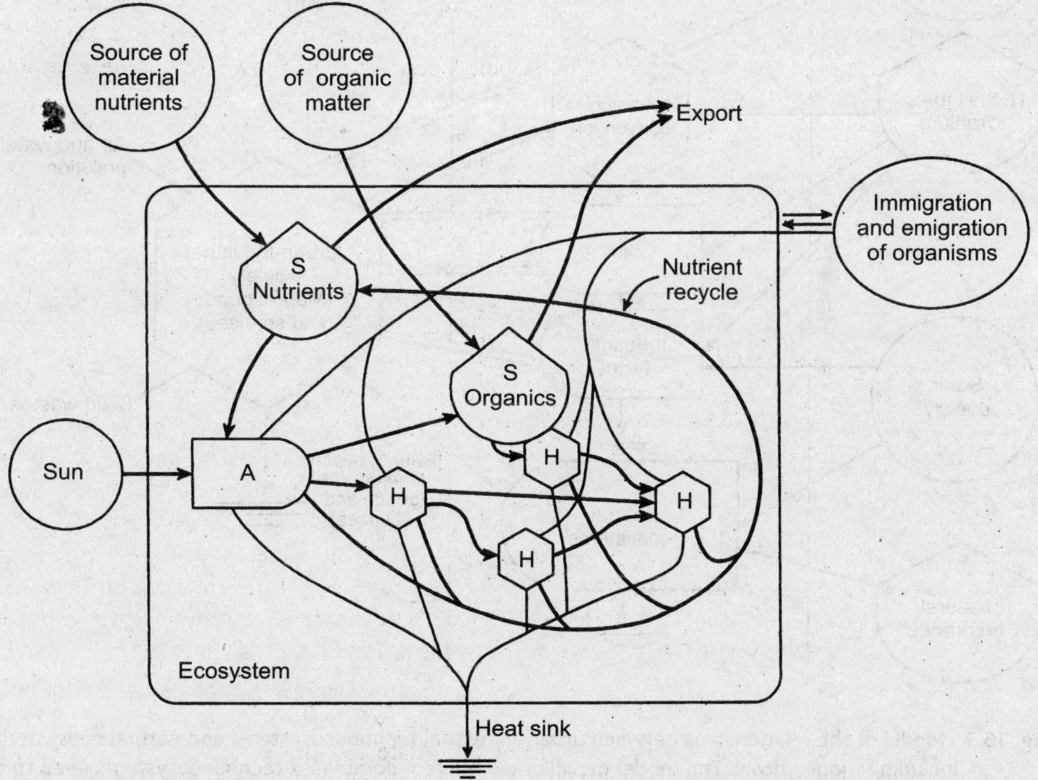

Fig. 16.2 Functional diagram of an ecosystem with emphasis on internal dynamics involving energy flow and material cycles. S= storage, A=autotrophs, H =heterotrophs (after **Odum** and **Barrett** 2005).

Figure 16.3 shows our graphic model for these new (in terms of human history) systems. Shown here the inputs of fossil fuels and uranium energy sources and of natural resources, and the increasing outputs of air, water, and solid waste pollution, which are much larger and more toxic than anything that comes out of natural ecosystem.

In Figure 16.3, the model is completed with natural ecosystems that provide life – supporting goods and services (such as breathing, drinking and eating) and that maintain **homeorhetic global balance** (sustainability) in the atmosphere, soils, fresh water, and oceans (**Homeorhesis**: From Greek " maintaining the flow") is tendency of a system to maintain itself in a pulsing state of equilibrium). Note that money circulates as a reverse flow between society and human-made systems, but not natural systems, there by creating a vast market failure when a society fails to pay for ecosystem services.

Modern City as Human–made Ecosystem

A modern city is the major component of the fabricated technoecosystem. It is an energetic hot spot that requires a large area of low-energy density natural and seminatural countryside to maintain it. Current cities grow little or no food and generate a large waste stream that affects

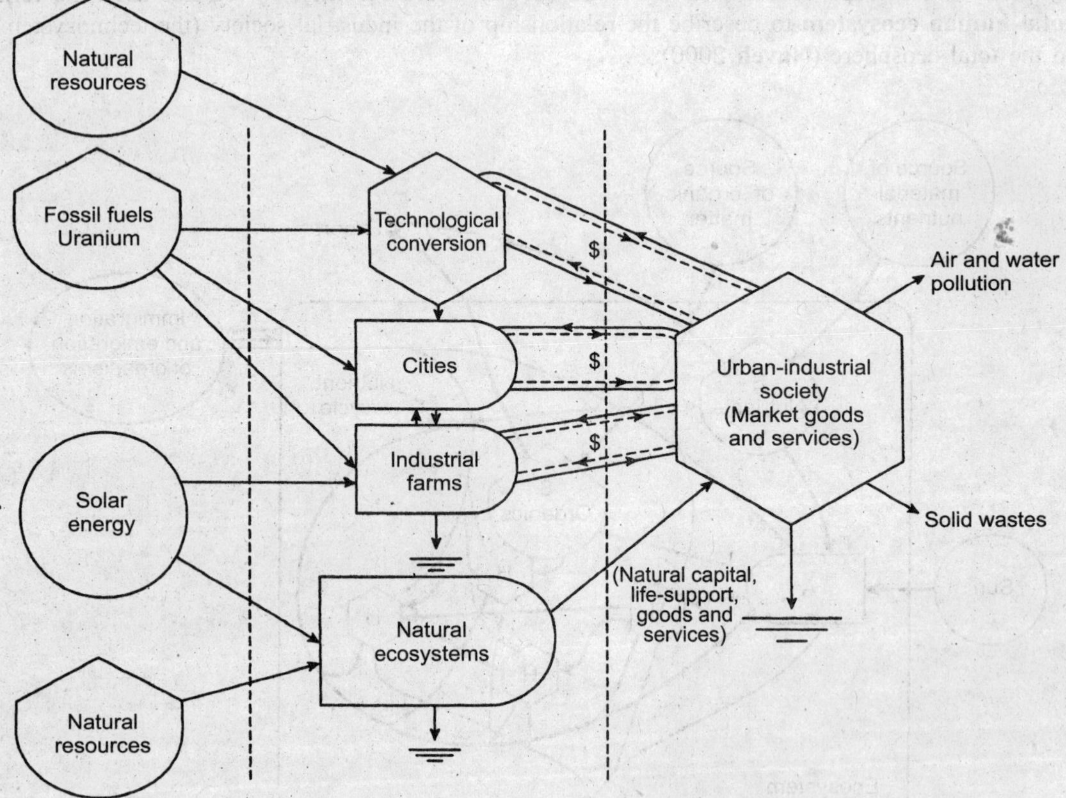

Fig. 16.3 Model of the relationships between urban-industrial technoecosystems and natural ecosystems, including money flows. The model depicts how human–dominated technoecosystems need to be coupled with the goods and services (natural capital) provided by natural ecosystems in order to enhance landscape sustainability (after **Odum** and **Barrett** 2005).

large areas of downstream rural landscapes and oceans. The city does export money that pays for some natural resources, and the city provides many desirable cultural institutions not available in rural areas, such as museums and symphonies. Figure 16.4 compares a human city with a much lower-powered oyster reef, a natural analog to a city. It can be noted that the energy requirements for the urban technosystems are about 70 times greater than for the natural ecosystem.

Odum and **Barrett** (2005) has regarded the cities as parasites on the low-energy countryside. Parasites and hosts in nature tend to coevolve for coexistence; otherwise, if the parasite takes too much from its host, both die. **John Cairns** (1997) has expressed hope that somehow natural ecosystems and technoecosystems will coevolve to prevent such a doomsday. **Wackernagel** and **Rees** (1996) used the term **ecological footprint** to describe the impact of and resources needed by a city in order to provide for its citizenry in a sustainable manner.

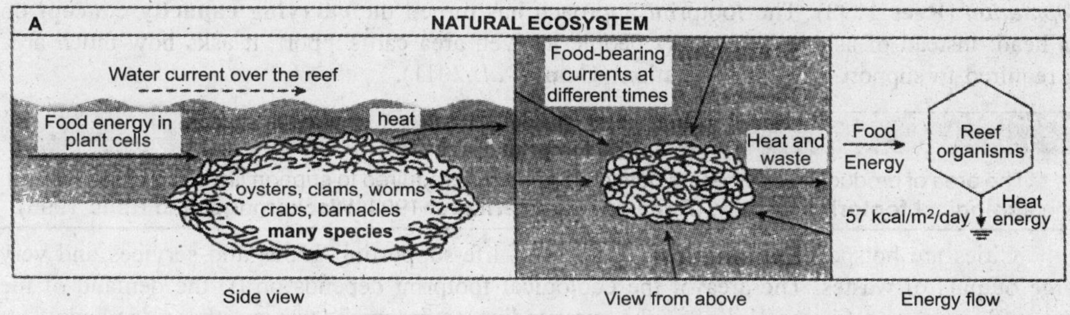

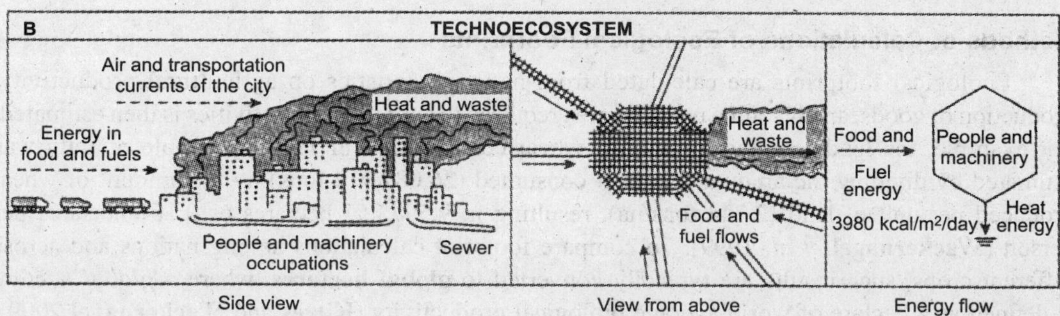

Fig.16.4 Heterotrophic ecosystems. A – One of the "cities" of nature – an oyster reef that is dependent on the inflow of food energy from a large area of the surrounding environment . B – A human - built technoecosystem (industrialized city) maintained by a huge inflow of fuel and food, with a correspondingly large outflow of waste and heat. Its energy requirement, on a per m^2 area basis, is about 70 times that of the reef, or about 4000 kcal / m2/ day, which comes about 1.5 million of kcal per year (after **Odum** and **Barrett** 2005).

Especially threatening to the global life-support systems is the **explosive growth of megacities** in the developing nations, caused atleast in part by the increasing dominance of another technosystem, the **industrialized agriculture**, with its often excessive consumption of water and its use of toxic and eutrophicating chemicals. These systems not only pollute, but essentially put the small farms out of business. Worldwide relocating most of these families into the cities,which are not able to assimilate them at the rate that they are moving into urban areas. This situation reflects what **Paul Gray** (1992) has written *"A paradox of our time is the mixed blessing of almost every technological dark as well as bright sides"*.

Concept of Ecological Footprint

When we turn on a light, purchase an appliance, drive a car, or eat fruit imported from another country, we may not think about the effects our actions have on the natural world (crop and forest land, bodies of water, and undeveloped natural areas). How does driving to the store to get groceries affect forests or coral reef?

To answer this question, we must account for the resources required to support our actions. The grains we eat require farmland; the wood products we use require natural forests or plantations; the fish we eat require productive fishing grounds; the machines and appliances we purchase require raw materials and energy to build, as well as energy for their operation. Ultimately, every aspect of our economy depends on the land and waters of Earth. Recognizing this, **William Rees** proposed that we measure the environmental impact of a population as its **ecological footprint**, which is *the total area of productive ecosystem required to support that population* (**Rees** 1992). The footprint approach has turned the **carrying capacity concept** on its head: instead of asking how many people a given area can support, it asks how much area is required to support a given population (**Cain** *et al.* 2011).

Box 16.1

The area of productive ecosystem outside a city that is required to support life in the city is termed the **ecological footprint** of a city. (**Rees** and **Wackernagel** 1994; **Wackernagel** and **Rees** 1996).

Cities are hotspots with very large inputs of life-supporting goods and services and very large outputs of wastes. The area of the ecological footprint depends on (i) the demand of the city (affluence) and (ii) the ability of the surrounding environment to meet these demands.

Methods of Calculations of Ecological Footprints

Ecological footprints are calculated from national statistics on agricultural productivity, production of goods, and resource use. The area required to support these activities is then estimated. For example, the land required to support wheat consumption in 1993 by people of Italy was estimated by dividing the amount of wheat consumed (26,087,912 tons) by the amount of wheat produced per unit of land (2.744 tons/ha), resulting in 9,507,257 hectares or 0.167 hectares per person (**Wackernagel** *et al.* 1999). To compare footprint calculations among nations and across different crops, such results are typically converted to **global hectares**, where a *global hectare* is defined as a hectare of world average biological productivity (**Kitzes** and **Wackernagel** 2009).

Folke *et al.*, (1997) and **Jansson** *et al.* (1999) estimated that the resource consumption and waste assimilation footprint for the 27 affluent cities that ring the Baltic Sea was 50 to 1000 times the area of the cities themselves. The ecological footprint of Vancouver, Canada, located in a fertile, well-watered region, has been estimated to be 22 times larger than the area of the city itself. The ecological footprints of cities of less developed countries are much smaller (**Odum** and **Barrett** 2005).

Methods of calculating ecological footprints are still being refined, but results to date are sobering. In 2006, there were 11.9 billion global hectares of productive land available, and the ecological footprint of an average person was 2.6 global hectares (**Ewing** *et al.*, 2009). This result suggests that Earth could have supported 4.6 billion people (11.9 billion ha/2.6 ha per person) for a long time. In fact, the human population in 2006 was 6.6 billion, more than a 40% overshoot of its carrying capacity. An overshoot of this magnitude indicates that in 2006, environmental resources were being used more rapidly than they could be regenerated, a pattern of use that cannot be sustained (**Cain** *et al.*, 2011).

Similar calculations of ecological footprints can be made for nations, cities and even companies, schools or individuals. In the United States, for example, the average ecological footprint was 9.0 global hectare per person in 2006. Since there were 1,330 million global hectares of productive land available in the nation, this calculation suggests that the carrying capacity of the United States in 2006 was 148 million people (1,330 million ha 19.0 ha per person); the actual population was 303 million, more than double the carrying capacity.

Factors Affecting Ecological Footprints

Human use of resources changes from year to year, depending on (i) population size, (ii) per capita rate of resource use, and (iii) technology (*i.e.*, efficiency of production). In addition, the total area of productive ecosystem available to support our activities changes overtime due to factors such as (i) gain or loss of farmland, (ii) destruction of natural habitat, (iii) pollution and (iv) extinction of species. As a result, our ecological footprint changes overtime. People have now begun to use our changing footprint as a way to assess whether our current population size and resource use can be sustained. This approach highlights the fact that all of our actions – what we eat, how big a house we buy, how much we drive or fly, the goods we purchase (*e.g.*, clothes, car, cell phones)- depend on and effect the natural world.

Lastly, if highly developed countries would reduce their excessive resource and energy consumption, then international conflicts and terrorists threats would likely be reduced (**Odum** and **Barrett** 2005).

QUESTIONS

Long Answer Questions

1. Describe the maize cropland ecosystem.

2. Give an account of ecology of agroecosystems.

3. Describe the concept of ecological footprint.

Short Answer Questions

1. Give a short account of alternative agriculture.

2. Describe factors which affect ecological footprint.

3. Write short notes on the following:

(a) Megacities and ecology

(b) Total human ecosystem

(c) LISA

Very Short Answer Questions

1. What is agroecosystem?

2. Define sustainability.

3. What is sustainable development?

4. What is ecological footprint?

Fill in the Blanks

1. Canal-irrigated paddy rice culture is .... agriculture.

2. .... means herding cattle or other domestic animals for milk, meat and hides.

3. .... is practiced generally in forested area of tropics.

4. Term total human ecosystem was coined by .... in 2000.

5. .... is tendency of system to maintain itself in a pulsing state of equilibrium.
6. Term ecological footprint was coined by .... and Rees in 1996.
7. .... agriculture involves excessive consumption of water and toxic and eutrophicating chemicals.

Multiple Choice Questions

1. The key nutrients responsible for eutrophication are

   (*a*) nitrogen and phospnorus (*b*) nitrogen and calcium

   (*c*) phosphorus and sulphur (*d*) magnesium and iron

ANSWERS

Very Short Answer Questions

1. Agroecosystem is land used for crops, pasture, and livestock.
2. Sustainability means ensuring that rate of use or consumption of any resource does not exceed the capacity of the natural systems to regenerate it. It also means that development meets the needs of the present without compromising the ability of future generations to meet their own need.
3. Sustainable development is the development and/or growth that meets the need of the generation without compromising the ability of future generations to meet their own needs. It also includes what is mentioned under sustainability.
4. Ecological footprint is the area of productive ecosystems outside a city that is required to support life in the city.

Fill in the Blanks

1. Flood irrigated 2. Pastoralism 3. Shifting agriculture 4. Zev Naveh
5. Homeorhesis 6. Wackernagel 7. Industrialized.

Multiple Choice Questions

1. (*a*).

17

Biodiversity

17.1 DEFINITION

The term *biodiversity* is a concise form of "biological diversity" and was coined by **Walter G. Rosen** in 1986 (**Wilson** 1994). The term *'biological diversity'* was first used by **A.E. Norse** and **R.E. McManus** (1980). Term biodiversity encompasses all forms of life: plants, animals and microorganisms in any region.

The definition of biodiversity, as provided in Article 2 of convention on Biodiversity (CBD; Box 17.1) is as follows " Biological diversity means the variability among living organisms from all sources including, *inter alia* (L., among other things), terrestrial, marine and other ecosystems and the ecological complexes of which they are part, this includes diversity within species, between species and of ecosystems (**Singh** 2014).

Box 17.1 Convention of Biodiversity (CBD)

The global convention of Biodiversity (CBD) was adopted at the UN conference on Environment and Development in 1992. The CBD officially came into force in December 1993 and by January 2000, over 170 nations had ratified the CBD, which is a unique international agreement for conservation and sustainable use of biodiversity.

To illustrate the level of concern for biodiversity issues that has emerged internationally, the Organization for Economic Cooperation recently approved plans for creating the world's largest databank. The Global Biodiversity Information Facility (GBIF) will contain not only a catalogue of all the known species, but will also include scores of database related to molecular and genetic collections, geospatial information and natural history data (**MacCable** 1999).

According to **H.B.Lillywhite** (2002), the term biodiversity refers to 'the variety and variability among living organisms and the ecological complexes in which they occur.'

The interdependence and coexistence of various forms maintain the balance of nature. Biodiversity thus maintains the populations, food chains and nutrient cycles in ecosystem. Humans depend on biodiversity for food, medicines, material or for ecological services. A minimum level of biodiversity is necessary for proper functioning of the ecosystem.

17.2 BIODIVERSITY IN INDIA

India is very rich in biodiversity. It is one of the twelve mega-diversity countries of the world. The other countries are Mexico, Columbia, Brazil, Peru, Ecuador, Zaire, Madagascar, Indonesia, Malaysia, China and Australia. India's biological wealth becomes apparent from the fact that it represents ten biogeographic zones (Table 17.1; Fig. 17.1)

The Indian region is one of the most diverse biogeographic regions of the world, embracing a wide range of topography; from perpetually snow covered high Himalayan ranges to plains to sea level, low lying swamps and mangroves, island systems, tropical evergreen rain forests, fertile alluvial plains, hot deserts and high altitude cold deserts. There are almost rainless areas, as well as the world's highest rainfall areas. The climate ranges from tropical and subtropical in India - Gangetic plains and peninsular regions to temperate and arctic in the Himalayan region. The habitat types vary from the humid tropical Western Ghats to the hot deserts of Rajasthan, from cold deserts of Ladakh to the long, warm coast line stretches of peninsular India. Thus a great variety of climatic and altitudinal variations coupled with varied ecological habitats have contributed immensely to the rich vegetational wealth, and varied flora and fauna generating very unique biodiversity in India.

Table 17.1: Major biogeographic zones of India (ZSI, 1997)

	Biogeographic zone	Biotic province	Total area (Km^2)
1.	Trans-Himalayan	Upper region	186,200
2.	Himalayan	Nort-West Himalaya	69,000
		West Himalaya,	72,000
		Central Himalaya	123,000
		East Himalaya	83,000
3.	Gangetic plan	Upper Gangetic plain	206,400
		Lower Gangetic plain	153,000
4.	Desert	Kutch	45,000
		Thar	180,000
		Ladakha (cold)	–
5.	Semi-Arid	Central India	107,600
		Gujarat-Rajwara	400,400
6.	Deccan Peninsula	Deccan Plateau South	378,000
		Central Plateau	341,000
		Eastern Plateau	198,000
		Chhota Nagpur	217,000
		Central Highlands	287,000
7.	Western Ghats	Malabar Coast	59,700
		Western Ghat Mountains	99,300
8.	North-East India	Brahamputra Valley	65,200
		North-Eastern Hills	106,200
9.	Islands	Andaman Islands	6,397
		Nicobar Islands	1,930
		Lakshadweep Islands	180
10.	Coasts	West coast	6500
		East coast	6500
		East coast	6500

India is also one of the 12 **primary centers of origin** of cultivated plants and domesticated animals. It is considered to be the homeland of 167 important plant species of cereals, millets, fruits, condiments, vegetables, pulses, fiber crops and oil seeds and 114 breeds of domesticated animals. **Endemism** of Indian biodiversity is significant. About 4900 species of flowering plants are endemic genera belonging to 47 families. These are concentrated in the floristically rich areas of North-East India, Western Ghats, North-West Himalayas and Andaman and Nicobar islands. These areas constitute 2 of the 25 hotspots identifies in the world. It is estimated that 62% of the known amphibian species are endemic to India of which a majority is found in Western

Ghats. Thus, endemism is an important characteristics of India's biodiversity. The number and percentage of species of bacteria, fungi, plants and animals are shown in Table 17.2.

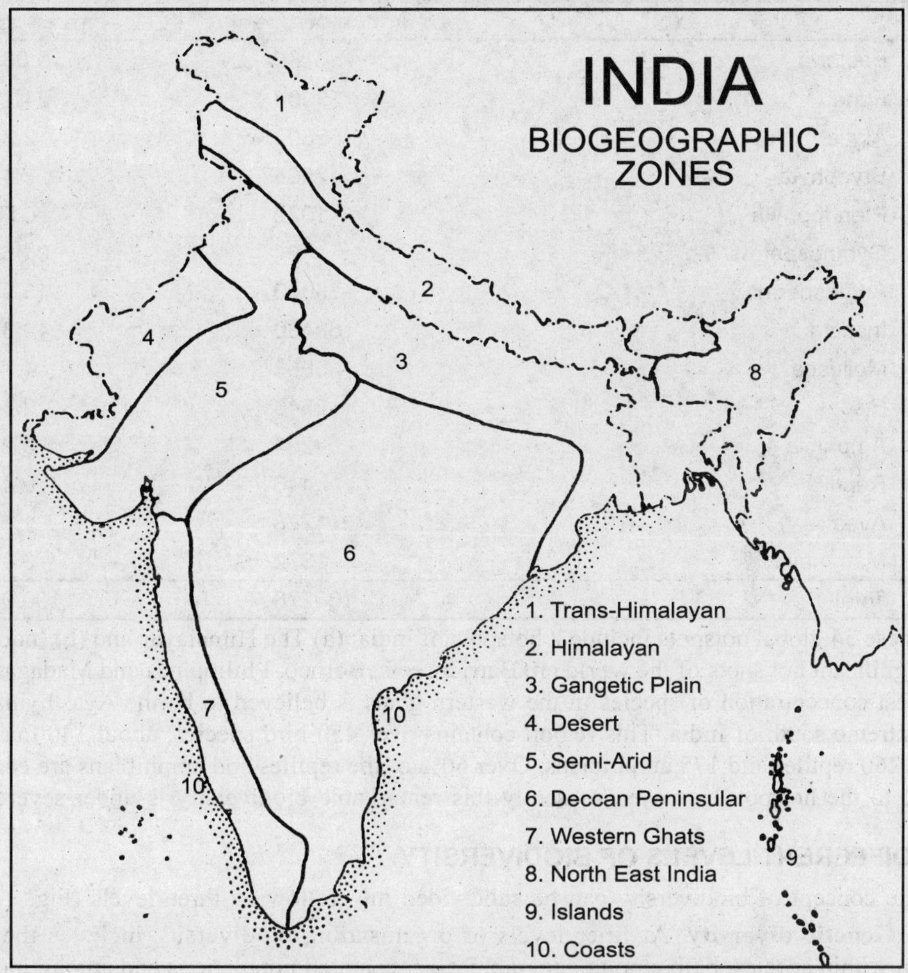

Fig. 17.1 Biogeographic zones of India.

Biodiversity Hotspots

The concept of biodiversity hotspots was originated by **Norman Myers** in 1988. To qualify a biodiversity hotspot a region must meet two vital criteria: (a) it must contain at least 0.5% or 15000 species of vascular plants as endemics and (b) it has to have at least 70% of its primary vegetation. Around the world at least 25 areas qualify under this definition, with nine other possible candidates. These sites support nearly 60% of the world's plant and animal biodiversity with a share of endemic species. The present distribution of hotspots is as follows

1. Africa–8
2. Asia Pacific–13
3. Europe and Central Asia–4
4. North and Central America–4
5. South America–5

Total = 34

Table 17.2. Number and percentage of species of bacteria, fungi, plants and animals in India (source B.S.I and Z.S.I.,1994; Singh 2014).

	Taxon	Number of Species	Percentage
1.	Bacteria	85	0.08
2.	Fungi	23000	21.2
3.	Algae	2500	2.3
4.	Bryophyta	2564	2.4
5.	Pteridophyta	1022	0.9
6.	Gymnosperms	64	0.06
7.	Angiosperms	15000	13.9
8.	Insects	53430	49.3
9.	Mollusca	5050	4.7
10.	Pisces	2546	2.4
11.	Amphibia	204	0.2
12.	Reptilia	446	0.4
13.	Aves	1228	1.1
14.	Mammalia	372	0.3
	Total	108276	100.0

These 34 global hotspots include 2 hotspots of India: (a) The Himalayas and (b) Indo.Burma. Other significant hot spots of the world are Barrier reefs, Borneo, Philippines and Madagascar, etc. The higest concentration of species in the western ghats is believed to be the Agasthymalai hills in the extreme south of India. This region contains over 450 bird species, about 140 mammalian species, 260 reptiles and 175 amphibians. Over 60% of the reptiles and amphibians are completely endemic to the hotspot. However, presently this remarkable biodiversity is under severe threat.

17.3 DIFFERENT LEVELS OF BIODIVERSITY

The concept of biodiversity can be subdivided into following three levels (Fig. 17.2):

1. Genetic diversity. At finer levels of organisation, biodiversity includes the genetic variation within species, both among geographically separated populations and among individuals within single population.

2. Species diversity. Biodiversity at its most basic level includes the full range of species on earth, from microorganisms such as viruses, bacteria and protists through the multicellular kindoms of plants, animals and fungi.

3. Community or Ecosystem diversity. On a wider scale, biodiversity includes variations in the biological communities in which species live, the ecosystem in which communities exist and the interactions among these levels.

17.4 MEASURING OF BIODIVERSITY

At its simplest level, diversity can be defined as the number of species found in a community, a measure known as **species richness.** Diversity is a single statistic in which the number of species and evenness are compounded. Many methods of calculating diversity have been proposed that combine these two types of information. Mathematical indices of biodiversity have also been developed to connate species diversity at different geographical scales as follows:

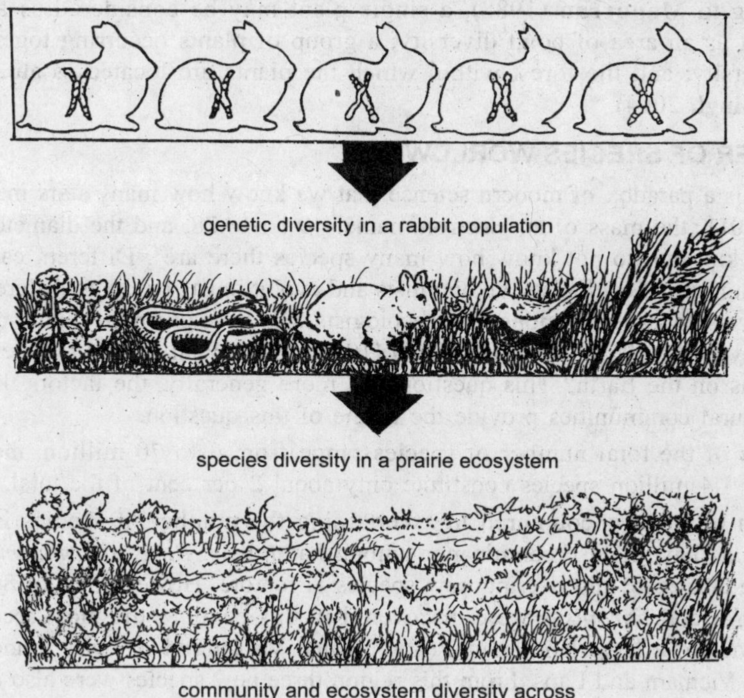

Fig. 17.2 Different levels of biodiversity. *Genetic diversity* refers to genetic variation found within each species; *Species diversity* includes the range of species in any ecosystem; *Community/ Ecosystem diversity* refers to the variety of habitat type and ecosystem processes extending over a given region.

1. Point diversity. This is the diversity on the smallest scale, *i.e.*, the diversity of a micro-habitat or sample taken from within a homogenous habitat. There exists differences in point diversity of two sites (or micro-habitats) of a habitat.

2. Alpha (α) diversity. This refers to number of species in a single community. This diversity comes closest to the popular concept of species richness and can be used to compare the number of species in different ecosystem types.

3. Beta (β) diversity. This pertains to the degree to which species composition changes along an environmental gradient. Beta diversity is high for example, if the species composition of moss communities changes at successively higher elevations on a mountain slope, but is low if the same species occupy the whole mountain side.

Beta diversity is also called **differentiation diversity**. For example, faunal diversity of fish and insects is based on gradient and altitude. A zonation of animals (*e.g.*, fish *Simulium* spp.) within a hill stream river has been reported (see **Singh** 2005).

4. Gamma (γ) diversity. It is defined as the diversity of a larger unit such as an island or landscape. Gamma diversity is the rate at which additional species are encountered as geographical replacements with a habitat type in different localities. Thus, gamma diversity is a species turnover rate with distance between sites of similar habitat or with expanding geographical areas.

5. Epsilon diversity. The epsilon or regional diversity is defined as the total diversity of a group of areas of gamma diversity.

Some ecologists also identify the **delta diversity** which belongs to gamma diversity. **Whittaker** (1977) has recognised four types of inventory species diversity, *i.e.*, alpha diversity, gamma diversity, epsilon diversity and beta diversity.

According to **Magurran** (1988), a single plant may be considered as a unit of alpha diversity, a leaf as an area of point diversity; a group of plants occurring together as an area of gamma diversity; and the forest within which the plants are located as an area of epsilon diversity (see **Singh** 2014).

17.5 NUMBER OF SPECIES WORLDWIDE

Indeed it is a paradox of modern science that we know how many stars in the milky way, namely some 1011, the mass of an electron, namely 9.1 x10.28, and the diameter of the Earth, namely 12,752 km, but do not know how many species there are . Different calculations have been made by number of ecologists, taxonomists and evolutionists based on different assumptions and yielding diverse numbers. Among those ecologists who have given the matter critical attention are **Robert May** (1988,1992) and **E.O.Wilson** (1988,1992). What is total number of living (and dead) organisms on the Earth? This question and more generally, the factors that regulate the diversity of natural communities provide the theme of this question.

Estimates of the total number of species range from 3 to 70 million, meaning that the presently know 1.4 million species constitute only about 2 per cent of the total.

The main problem encountered in ascertaining the number of species is that different groups of animals have held predominance among naturalists. *Birds* and *mammals* are so well studied that the chance of discovering new species is remote. However, 11 of 80 known living species of *whales* and *porpoises* were discovered in this century, the most recent in 1991. In 1992 a new genus of a large land animal (now named **Vu Quand ox**) was found in the rugged region dividing Vietnam and Laos. From this region three new species were also discovered, i.e., two new species of deerlike animals and a new species of fish resembling carp (**Lindeu** 1994).

As well studied as butterflies and beetles have been, only a portion of their actual number, are probably known; terrestrial vascular plants are pretty well catalogued but not so for their marine counterparts- however, plant taxonomists recently discovered three new families of flowering plants in Central America and Southern Mexico (**Raven** and **Wilson** 1992). When we turn to bacteria, fungi and even protozoa the unknown far outnumbers the recorded species. Likewise, some environments are well studied (*e.g.*, temperate forests and grasslands) and he component species are well catalogued; however, organisms of polar regions, the ocean benthos coral reefs and soils are not so well known.

Table 17.3. Numbers of described species of living organisms.

Kingdom and major subdivision	Common name	Number of described species	Totals
1. Viruses			
	Viruses	1,000	**1000**
2. Monera			
(i) Bacteria	Bacteria	3000	
(ii) Myxoplasma	Bacteria	60	
(iii) Cyanophycota	Blue-green algae	1,700	**4,760**
3. Fungi			
(i) Zygomycota	Zygomycete fungi	665	
(ii) Ascomycota (including 18,000 lichen fungi)	Cup fungi	28,650	
(iii) Basidomycota	Basidomycete fungi	16,000	
(iv) Oomycota	Water molds	580	

Kingdom and major subdivision	Common name	Number of described species	Totals
(v) Chytriodomycota	Chytrids	575	
(vi) Acrasiomycota	Cellular slime molds	13	
(vii) Myxomycota	Plasmodial slime mold	500	**46,983**
4. Algae			
(i) Chlorophyta	Green algae	7,000	
(ii) Phaeophyta	Brown algae	1,500	
(iii) Rhodophyta	Red algae	4,000	
(iv) Chrysophyta	Chrysophyte algae	12,500	
(v) Pyrrophyta	Dinoflagellates	1,100	
(vi) Euglenophyta	Euglenoids	800	**26,900**
5. Plantae			
(i) Bryophyta	Mosses, liverworts, hornworts	16,600	
(ii) Psilophyta	Psilopsids	9	
(iii) Lycopodiophyta	Lycophytes	1,275	
(iv) Equisetophyta	Horsetails	15	
(v) Filicophyta	Ferns	10,000	
(vi) Gymnosperma	Gymnosperms	529	
(vii) Dicotyledonae	Dicots	1,70,000	
(viii) Monocotyledonae	Monocots	50,000	**248,428**
6. Protozoa			
	Protozoans	30,800	**30,800**
7. Animalia			
(i) Porifera	Sponges	5,000	
(ii) Cnidaria, Ctenophora	Jelly fish, Corals, Comb jellies	9,000	
(iii) Platyhelminthes	Flatworms	12,200	
(iv) Nematoda	Nematodes (Roundworms)	12,000	
(v) Annelida	Annelids (Earthworms and relatives)	12,000	
(vi) Mollusca	Mollucs	50,000	
(vii) Echinodermata	Echinoderms (starfish and relatives)	6,100	
(viii) Arthropoda	Arthropods		
(1) Insecta	Insects	751,000	
(2) Other arthropods		123,161	
(ix) Minor invertebrate phyla		9,300	**989,761**
8. Chordata			
(i) Tunicata	Tunicates	1,250	
(ii) Cephalochordata	Acorn worms	23	
(iii) Vertebrata	Vertebrates		
1. Agnatha	Lampreys and Hagfishes	63	
2. Chondrichthyes	Sharks and other cartilaginous fishes	843	

Kingdom and major subdivision	Common name	Number of described species	Totals
3. Osteichthyes	Bony fishes	18,150	
4. Amphibia	Amphibians	4,184	
5. Reptilia	Reptiles	6,300	
6. Aves	Birds	9,040	
7. Mammalia	Mammals	4,000	**43,853**
TOTAL: All organisms			1,392,485

17.6 FACTORS DETERMINING BIODIVERSITY WITHIN COMMUNITIES

For an individual community there are many factors which can affect the number of species present. Some of the influences on a community are external ones: the **abiotic effects** of the habitat. Other influences will come from within the community: the individuals present will themselves affect the species diversity (**biotic effects**).

1. Size of habitat. Size of the habitat obviously affects the number of species which can live in it. For a rock pool, which is only a few square meters, only a limited number of species can maintain populations there and all must be small species. A huge habitat is large enough to maintain populations of many different species, *i.e.,* habitat supporting oak woodland.

2. Spatial patchiness. Spatial patchiness in a habitat allows organisms with different requirement to live together. A rocky pool with crevices and a sandy floor with small rocks will provide more microhabitats for species than does a completely smooth walled, rock-bottomed pool.

3. Harshness of habitat. Harshness of habitat may affect the number of species able to survive in a community. Rock pools show so much daily change in their environment that open sea animals cannot survive there and a rock pool which is frequently raked has very few species.

For example, censuses of birds in small areas (usually 5-20 hectares) of relatively uniform habitat reveals about 6 species of breeding birds in grasslands, 14 in scrublands and 24 in floodplain deciduous forests.

In determining species diversity, habitat structure apparently overrides productivity. Marshes are productive but structurally uniform, and they have relatively few species. By contrast, desert vegetation is less productive but its greater variety of structure apparently make room for more kinds of habitats.

4. Predictability of change in the habitat. If the changes occurring in an environment are **cyclic** and therefore **predictable,** then different species will be able to use the habitat at different times in the cycle. Typical cyclic changes include seasonality in temperature or rainfall. In oak woods, for example, the trees lose their leaves each year because of low winter temperatures. Because of this, there is always a spring window in which light-demanding species of the ground flora can grow and flower. As spring progresses, the woodland floor becomes darker as tree's leaves develop and shade-tolerant species take over the space.

5. Disturbance in a habitat. Occasional severe or frequent low-level disturbance affects the population size of species in the community. If population numbers are kept low by disturbance, then there will be more "room" in the habitat for other populations. For example, in the rock pool community a heavy winter storm which removes algae from the rocks will provide space for other species to colonize the bare area. However, if there is too much disturbance, then the habitat becomes too harsh, and species diversity may decline.

6. Isolation of habitat. The absence of similar habitats close by decreases the chance of some species reaching the habitat. The further away an area is from similar habitats, the fewer species are likely to colonize it. This is especially noticeable on islands at a large distance from the mainland.

MacArthur and **Wilson** developed their famous equilibrium theory of island biogeography, which states that the number of species on island balances regional processes governing immigration against local processes of extinction. On islands which are too small to support speciation through geographical isolation of populations, the number of species increases solely by immigration from other islands or from mainlands.

Biotic Factors of Habitat

1. Age of community type. The length of time the community type has existed in evolutionary time may affect the diversity. The various species normally found in the community will be evolving in association together. One theory is that longer the community type has existed, the more species will have had the opportunity to join the community. *Niche volumes will have decreased as more species enter the community structure.*

The way in which community richness can increase with time has been shown by looking at tree species in Britain. **John Birks** (1980) showed that the longer a tree species had been in the British woodland communities, the more insect species were associated with it. The two oak species, which have been in Britain for about 9000 years, have 284 species of insects living on them. The hawthorns which have been in Britain for 7000 years, have 149 species of insects, while the hornbeam (*Carpinus betulus*) and maple (*Acer campestre*), both in Britain about 5000 years have 28 and 26 species of insects respectively. Introduced species have even fewer insects: sycamore (*Acer pseudoplatanus*), introduced 650 years ago, has 15 species of insects and the horse chestnut(*Aesculus hippocastanum*), in Britain for 400 years, has only 4 species of insects.

The overall picture is not as simple as these data might suggest. Some species are better at rejecting grazers and the harmful effects of insects or have a scattered distribution which makes insect colonization difficult. For example, rowen trees (*Sorbus aucuparia*) seems to have been in Britain about 12,000 years, but only has 28 insect species on it.

The **time hypothesis** suggests that more diverse areas are older. Thus, one explanation for the diversity of the tropics is that tropical conditions appeared earlier on the earths surface than polar environments, allowing time for the evolution of a greater variety of plants and animals (**Fischer** 1960; **Margalef** 1963, 1968). Because the tropical zone girdles the earth about its equator–the earth's widest point–the tropics include more area than temperate and arctic regions. The earth's climate has undergone several cycles of warming and cooling, which have been discovered by records, in sediments and fossils, of their influence on vegetation and ocean temperature.

Restriction and fragmentation of the rainforest habitat could have caused the extinction of many species; conversely, the isolation of populations in patches of rainforest could have facilitated the formation of new species (speciation). **Allen** and **Keast** (1961) proposed a scheme of alternate expansion and contraction of humid habitats to account, in part for diversity of the present Australian avifauna. Other animals and plants were probably similarly affected; for example, there are about 500 species in the genus *Eucalyptus* a type of tree or shrub native only to Australia, but now widely planted around the world.

2. Age of particular community. The time any one community has existed at a site affects its species richness. A recently developed community may contain only a few species and the number will increase as more species invade and get established. Rock pool communities have animals which are very slow to establish and disperse to other pools. So it may take many years for a full community to develop in a new pool, or one devastated by storms. The degree of isolation, mentioned earlier, will also alter colonization time. The importance of the age of a single community is different from that of the age of the general community type. The diversity in the general community type often depends on the co-evolution of organisms while the diversity of a particular community is primarily due to the immigration of and colonization by existing species.

3. Primary productivity. Primary productivity of a community may influence species richness. If primary productivity is high, the food for the herbivores will be abundant. Many species of herbivore may be able to utilize the food without competition for a scarce resource. High herbivore numbers and diversity will influence the carnivores and so on along the food chains.

4. Community structure. If community structure is complex, then the community will contain more niches. For instance, the trees, bush and ground flora structure of oak woods provides nest sites for birds, a huge leaf and bark area for insects, support for climbers and ground cover for small mammals. Many of these niches are not available in grasslands.

Most ecologists agree that the high diversity of the tropics result at least in part from there being a greater variety of ecological roles. That is, the total niche space occupied is greater near the equator than it is toward the poles. For example, **Gordon Orians** (1969) suggested that part of the increase in number of species of birds toward the tropics is related to an increase in frugivorous and nectarivorous species, and in insectivorous species that hunt by searching for their prey while quietly sitting on perches–a type of behaviour uncommon in species in temperate regions. Among mammals, the increase in number of species between temperate and tropical areas result primarily from the addition of bats to tropical communities at the equator than they are in the United States and other temperate regions at a similar latitude.

In streams and rivers, the number of species in most taxonomic groups increase from the headwaters to the mouth of the river. We can presume that as the river increases in size, it presents a great variety of ecological opportunities and its physical conditions become more stable and therefore reliable. Local communities reflect these changes. For example, a headwater spring in the Rio Tamesi drainage of east central Mexico supported only one species of fish, a detritus-feeding platyfish (Fig. 17.3). Further downstream, three species occurred: the platyfish,

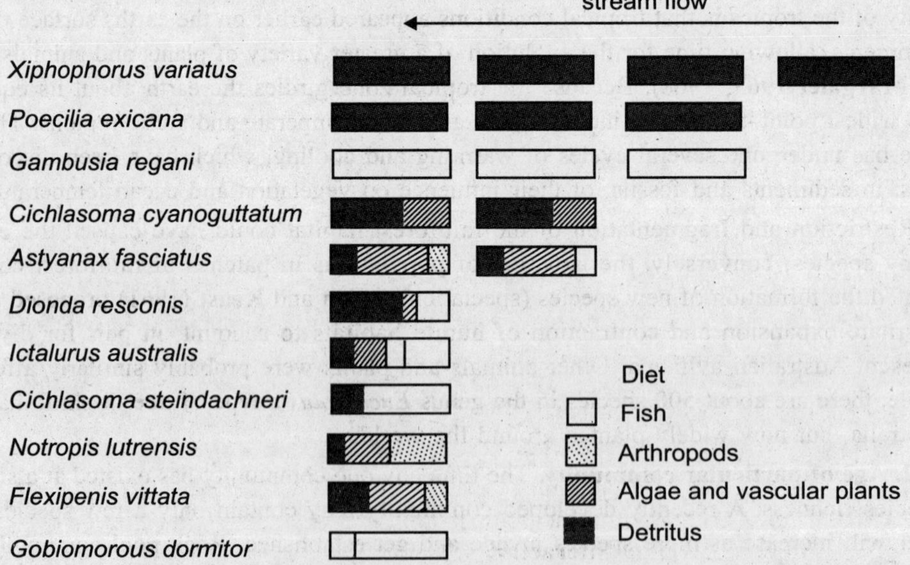

Fig. 17.3. Food habits of fish species in four communities (vertical columns) from a headwater spring with 1 species (*right*) to downstream communities with up to 11 species (*left*). The communities sampled were in the Rio Tamesi drainage of east-central Mexico.

plus a detritus-feeding molly (*Poecilia*) that prefers slightly deeper water than the platyfish, and a mosquito fish (*Gambusia*) that eats mostly insect larvae and small crustacea. Species that appear in the community farther downstream include additional carnivores–among them, fish-eaters — and other fish that feed primarily on filamentous algae and vascular plants. None of the fish species drops out of community downstream from any of the sampling localities. Thus, diversity increases as the stream becomes larger and presents more kinds of habitats and a great variety of food items.

5. Competition. The competition between species in a community may alter species richness. If some species are highly competitive for resources such as space or food, others may be excluded. Sometimes a new species invades a community and becomes very abundant, taking the place of individuals of other species. The population of these species decrease and may disappear.

In fact, competition is related to the degree of disturbance in a community. If disturbance prevents a competitive species from dominating a community, then species that might otherwise disappear may be able to survive. The harshness of the environment in some habitats alters the relative competitive ability of species, giving species which are well adapted to such environments the chance to survive there when they would be excluded from an easier one.

Some or all of these factors affect every community and influence species richness.

Fragility of Biodiversity

Although naturally occurring phenomena such as fire, volcanic action and floods can adversely affect community structure and thereby diversity, it is without question that human activity, especially on fragile environments, has far greater impact on the maintenance of biodiversity, primarily through habitat destruction. Tropical rainforests, presumably the richest source of biodiversity, are also among the most fragile and hence vulnerable habitats for several reasons. **First** , the predominant soils are **oxisols,** typically acidic, nutrient poor and rich in iron oxides. These soils tend to harden, forming lateritic boundaries, from which vital elements such as calcium and potassium are leached and in which phosphorus form insoluble compounds with iron and aluminium, thus, reducing its availability. **Secondly**, the high rates of decomposition and primary production mean most of the carbon and nutrients are bound up in the biomass, thus, precluding rapid regeneration in the case of clear-cutting. **Finally**, because seeds of most rain forest species germinate within a few days or weeks, most sprout and die in the hot, nutrient poor oxisols.

Thus, along with the foregoing factors, the regeneration of a mature rain forest takes centuries, and natural restoration never occur.

Coral reefs are the marine counterpart to tropical rain forests. They are also being subjected to degradation, largely through human activity.

17.7 SPECIES INTERACTIONS IN COMMUNITY: DOMINANT SPECIES, ECOSYSTEM ENGINEERS AND KEYSTONE SPECIES

Species interactions in a community vary greatly in strength and direction (**Cain** *et al.*, 2011). Some species have a strong negative or positive effect on the community, while others probably have little or no effect. **Interaction strength**, the effect of one species on the abundance of another species, can be measured experimentally by removing one species (called **interactor species**) from the community and looking at the effect on the other species (called **target species**).

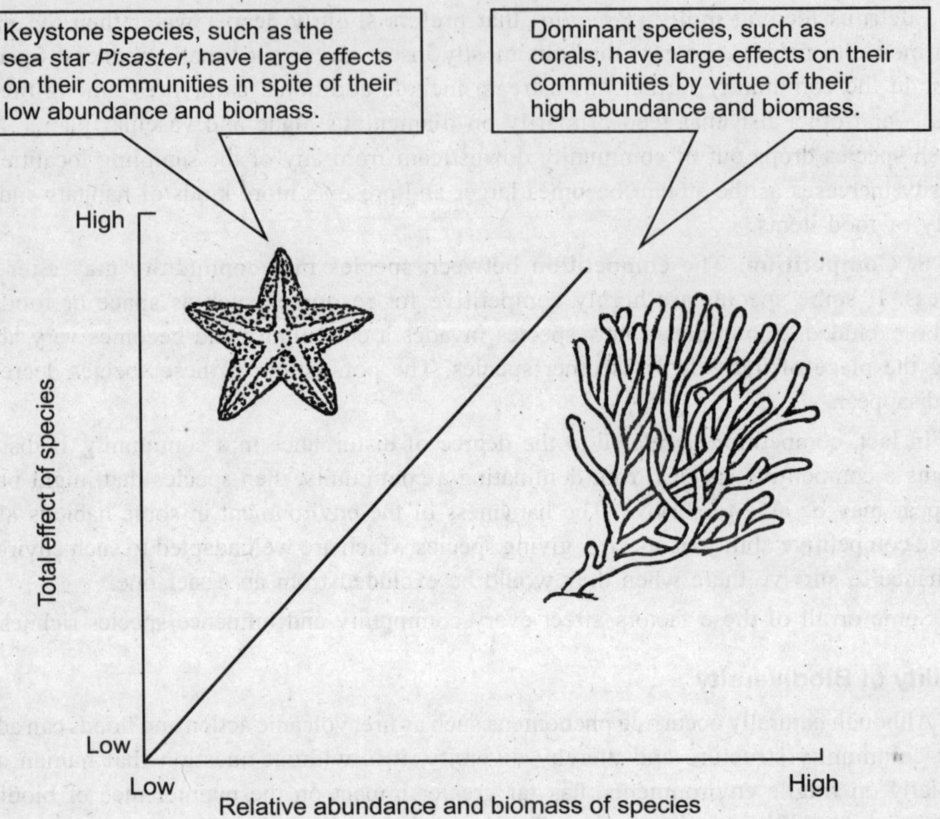

Fig.17.4 Dominant versus Keystone species. Species that have large effects on their communities may or may not do so by virtue of their high abundance or biomass (after **Cain** *et al.*, 2011).

If the removal of the interactor species results in a large decrease in the target species, it indicates that the interaction is strong and positive. However, if the abundance of the target species increases significantly, it means that the interactor species has a strong negative effect on the target species. The interaction strength "dynamic" (*i.e.*, the relative proportion of strong to weak interactions or positive to negative interactions) is not well understood for any community because of the numbers of species involved and the many indirect interactions that emerge.

There are some large or abundant species, such as trees, that are likely to have large community-wide effects by virtue of providing habitat or food for other species. They may also be good competitors for space, nutrients, or light. These so-called **dominant species**, also known as **foundation species** (**Dayton** 1971), have large effects on other species, and thus on the **species diversity** of communities by virtue of their considerable abundance or biomass (Fig. 17.4).

Some dominant species act by "bioengineering" the environment. These species, known as **ecosystem engineers** (**Jones** *et al*., 1994). are able to create, modify, or maintain physical habitat for themselves and other species. For example, trees provide food for other organisms and compete for resources just like any other species. However, trees also engineer their environment refined but important ways (**Jones** *et al.*1997). The trunk, branches, and leaves of a tree provide habitat for a multitude species, from birds to insects to lichens. The physical structure of the tree reduces sunlight, wind and rainfall, influencing temperature and moisture levels in the forest. The roots of tree can increase weathering and creation of the soil and they can anchor and bind rocks that stabilize surrounding substrates. The trees leaves fall to the forest floor, where they add moisture and nutrients to the soil and provide habitat for soil-dwelling invertebrates, seeds and microorganisms. If the tree falls, it can become a "nurse log" providing space, nutrients and moisture for tree seedlings. Thus, trees can have a large physical influence on the structure of a forest community, which obviously changes over time as trees grow, mature and die.

Other strong interactors, so-called **keystone species**, have large effects not because of their abundance, but because of the roles they play in their communities (Box 17.2). They differ

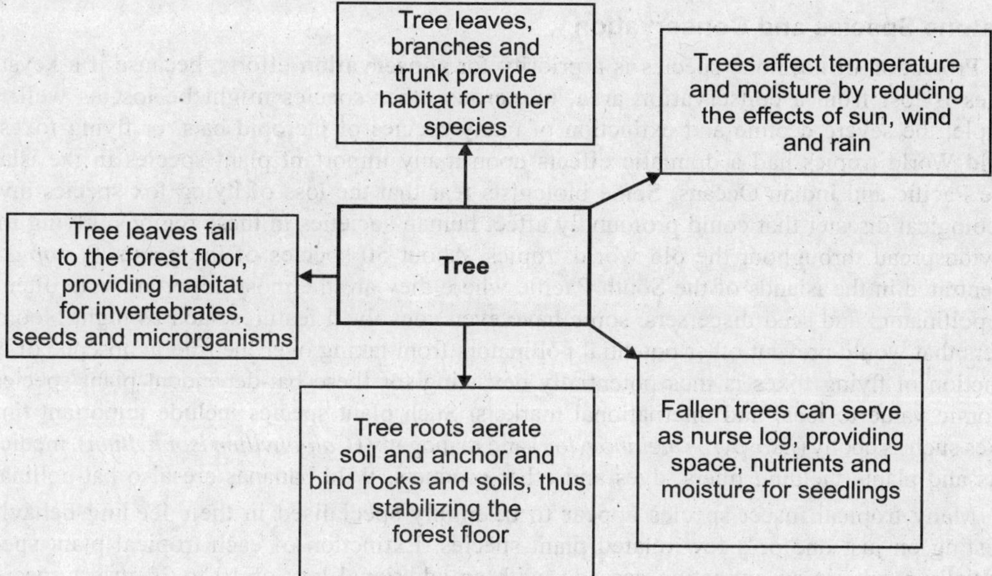

Fig. 17.5. Trees are dominant species and ecosystem engineers. Trees not only provide food for and compete with other species, but also act as ecosystem engineers by creating, modifying, or maintaining physical habitat for themselves and other species.

Box 17.2

Within biological communities, some species may be important in determining the ability of large number of other species to persist in the community. These crucial species have been termed **keystone species** (**Paine 1996; Terborgh 1986; Howe 1984**). Keystone species is a functional group or population without redundancy, a species (such as a predator) having a dominating influence on the structure and functioning of a community or ecosystem (**Odum** and **Barrette** 2005).

It is essential to maintain **redundancy** in an ecosystem – that is, to have more than one species or group of species capable of carrying out major functions or providing links in food web. In assessing the effects of the removal or addition of species, it is important to know if they might be *Keystone species* as a functional type without redundancy. The loss of such a species or group of species will cause major changes in community structure and ecosystem function. An exotic species added to a naturally diverse ecosystem often becomes a keystone species that reduce diversity.

from dominant species in that their effect is large in proportion to their abundance or biomass (Fig. 17.5). Keystone species usually influence community structure indirectly, via trophic means. For example, sea otters are considered keystone species because, by preying on sea urchins, they indirectly control the presence of kelp, which provides important habitat for many species.

There are also keystone species that act as ecosystem engineers. An example is the beaver, a species in which just a few individuals can have dramatic effects on the landscape. Beavers dam streams with cut trees and woody debris. Very quickly, flooding ensues and sediment accumulates as the increasing number of woody obstacles slow the water flow. Eventually, the once swiftly flowing stream is replaced by a wetland; containing plants that can deal with flood conditions; plants that cannot do so, such as trees, lost from the community. At the landscape level, by creating a mosaic of wetlands within a larger forested community, beavers can increase regional species diversity significantly (**Naiman** *et al.*, 1988).

Keystone Species and Conservation

Protection of keystone species is a priority for **conservation** efforts, because if a keystone species is lost from a conservation area, numerous other species might be lost as well. For example, the severe decline and extinction of many species of pteropid bats, or flying foxes, in the old World tropics had a dramatic effects upon many important plant species in the islands of the Pacific and Indian Oceans. Some biologists fear that the loss of flying fox species invites an ecological disaster that could profoundly affect human societies in these regions. Flying foxes are widespread throughout the old world tropics. About 50 species of the genus *Pteropus* are concentrated in the islands of the South Pacific where they are the most important, and often the only pollinators and seed dispersers, some have even coevolved features such as night-blooming flowers that would prevent other potential pollinators from taking over the role in absence of bats. Extinction of flying foxes is thus potentially devasting for these bat-dependent plant species of economic value to local and international markets. Such plant species include important timber species such as ebony (*Diospyros melanoxylon*) and mahogany (*Calophyllum isophyllum*), medicinal plants and plants yielding fibers, dyes and other products. Wild bananas are also bat-pollinated.

Many tropical insect species appear to be highly specialised in their feeding behaviour, subsisting on just one or a few related plant species. Extinction of each tropical plant species potentially results in an extinction cascade, with an additional loss of 10 to 30 insect species.

Among the most obvious keystone species are top predators, since they are often important in controlling herbivore populations. In many localities where grey wolves have been hunted to extinction by human beings, deer populations have exploded. The deer have then severely overgrazed the habitat, eliminating many herbaceous plant species. The loss of these plants was detrimental to the deer and other herbivores, including the insect community that feed on the plants.

The importance of a keystone species may hinge on highly specialised relationships between the keystone species and other organisms. In many tropical forests, fig trees and fig vines (*Ficus spp.*) appear to be keystone species in the functioning of vertebrate communities. Fig flowers are pollinated by small, highly specialised fig wasps, which mature inside the developing fig fruit. Fig also provide a reliable source of food to primates, birds and other fruit-eating vertebrates, even during dry seasons. Thus fig trees are keystone species.

Role of Keystone Species in Conservation

The identification of a Keystone species has several important implications for conservation biology.

1. The elimination of a keystone species from a community may precipitate the loss of many other species (**Sharma** 2009).
2. In order to protect a species of particular interest such as a monkey, it may be necessary to protect the keystone species on which it depends either directly or indirectly such as fig trees and dung beetles.
3. If the few keystone species of a community can be identified, these could be carefully protected or even encouraged if the area is being affected by human activity. For example, during selective logging, figs and other important fruit trees that are keystone species could possibly be reduced in abundance with little loss of biodiversity.

QUESTIONS

Long Answer Questions

1. Explain biodiversity of species in community (*Kerala 2001*)

Short Answer Questions

1. Write short note on biodiversity
2. Why are beavers both keystone species and ecosystem engineers?

Very Short Answer Questions

1. Define keystone species.
2. What is dominant species?
3. What is ecosystem engineer?

Fill in the Blanks

1. Who coined the term biodiversity? ....
2. In hill-stream (or river) the fish *Simulium spp*. show the .... biodiversity.

Multiple Choice questions

1. Those organisms which are dominant in community and affect the structure of microclimate and community, are called (*CCSU Meerut 2014*)
   (*a*) keystone species (*b*) invasion species
   (*c*) recessive species (*d*) dominant species
2. Which ecosystem shows maximum genetic diversity (*CCSU Meerut 2014*)
   (*a*) tropical rain forest (*b*) coniferous forest
   (*c*) temperate forest (*d*) subtropical forest
3. Number of biogeographical regions of India is (*CCSU Meerut 2014*)
   (*a*) 8 (*b*) 10
   (*c*) 16 (*d*) 18
4. Diversity of the habitat over the total landscape or geographical area is called (*CCSU Meerut 2014*)
   (*a*) α-diversity (*b*) β-diversity
   (*c*) γ-diversity (*d*) all the above
5. The term hot spots for high diversity ecological region was coined by (*CCSU Meerut 2014*)
   (*a*) IUCN (*b*) Myer
   (*c*) Odum (*d*) Kormandy
6. The most biodiversity rich zone in India (*CCSU Meerut 2014*)
   (*a*) gangetic planes (*b*) trans-Himalayas
   (*c*) western ghats (*d*) central India
7. Which of the following is called β diversity ? (*CCSU Meerut 2014*)
   (*a*) ecological diversity (*b*) genetic diversity
   (*c*) organismal diversity (*d*) forest diversity
8. Biodiversity of a region includes (*CCSU Meerut 2014*)
   (*a*) gene (*b*) species
   (*c*) ecosystems (*d*) all of the above

9. The term biodiversity crisis was proposed by
(*a*) E.O. Wilson
(*b*) R.H. Whittaker
(*c*) A.E. Magurran
(*d*) D.M. Raup

10. Differentiation diversity is also called
(*a*) beta diversity
(*b*) point diversity
(*c*) alpha diversity
(*d*) delta diversity

ANSWERS

Very Short Answer Questions

1. Keystone species is a strong interactor (*e.g.*, sea star, corals, fig tree, kelp, etc.) that has an effect on energy flow and community structure that is disproportionate to its abundance or biomass.

2. Dominant species is a species that has large, communitywide effects by virtue of its size or abundance, its strong competitive ability, or its provision of habitat or food for other species; It is also called a **founder species**.

3. Ecosystem engineer is a species (*e.g.*, beavers) that influences its community by creating, modifying, or maintaining physical habitat for itself and other species.

Fill in the Blanks

1. W.G. Rosen **2.** Differentiation.

Multiple Choice Questions

1. (*a*) **2.** (*a*) **3.** (*b*) **4.** (*c*)
5. (*b*) **6.** (*c*) **7.** (*b*) **8.** (*d*)
9. (*a*) **10.** (*a*).

18

Wildlife of India and Endangered Species

18.1 WILDLIFE OF INDIA

The term wildlife commonly refers to non-domesticated animals living in a natural habitat. However, in its widest sense, wildlife includes all flora and fauna of the natural habitat.

Indian subcontinent is unique in having great natural beauty in its different biomes and also in possessing a rich and diverse wildlife fauna. It includes about 123 families of terrestrial vertebrates. According to an estimate, there are 400 species of mammals, 1200 species of birds, 350 species of reptiles and more than 29,70,000 species of insects (**Khajuria**, 1957).

Wildlife management requires autecological informations about game animals and some basic facts of game species are the following:

1. Deer

The deer belong to the family 'Cervidae' of the mammalian order 'Artiodactyla' and they are herbivores and form significant component of grazing food-chains of the Indian forests. They control the overgrowth of forest vegetation and themselves become food of carnivores. Due to limited shelter and food supply, deforestation, urbanisation, mass hunting, etc., there has been a great reduction in the number of deer population in India. Sangai of Manipur or **brow-antlered deer** (*Cervus eldi eldi*) is at the verge of extinction. **Mouse deer,** *Tragulus meminna* stands just 30 cm off the forest floor. It has no antlers but male has downward pointed canines that look like tusks. Its small size helps them to evade predators by hiding beneath low bushes. Timid and gentle, these dwarfs lend enchantment to Indian forests and are subject of Indian folk tales.

The **musk deer** (*Moschus moschiferus*) is found at 2500 to 4000 metres altitudes in Himalayas, Himachal Pradesh, Northern Uttar Pradesh, Nepal and Sikkim. It is about 50 cm high, has long ears, short tail and gall-bladder, but no antlers and no pit-glands (suborbital glands) below the eyes. Male or buck has downwardly projecting canines of the upper jaw. It is a shy animal and lives a solitary life. This deer has been almost hunted out mainly for musk which is a valuable product used as perfume fixative and as important ingredient of several Ayurvedic and Homeopathic drugs and is obtained from the musk pouch existing below the skin of abdomen of buck. For the conservation of musk deer a National Park of 1,000 sq. km has been established at Manali in Kulu Valley by Himachal Pradesh Government.

Swamp deer are commonly known as Barasingha because of their large beautiful set of antlers (Fig. 18.1). *They are found only in India* and are declared as highly endangered species. There are two sub-species of barasingha which are found in very small areas of India:

Fig. 18.1. Deer found in India.

1. Swamp dwelling *Cervus duvauceli duvauceli* found in the Terai region of U.P., Kaziranga (Assam) and Sunderbans (Bengal). This subspecies lives in swamps and do not like to come out of water.

2. Hard ground *Cervus duvauceli branderi* is found in the grass covered hard open dry ground of Kanha National Park, M.P.

These animals live in groups of 20-80 individuals. The males (stags) grow impressive pair of antlers which drop after each breeding season. The females (hinds) do not grow antlers. Both sexes bear shining thick fur coat of brownish colour. Swamp deer are active in early mornings and late evenings and breed in winters. They have a strong sense of smell and when alarmed, the heard gives a sharp call and disappears quickly. They are hunted for skin, antlers (for trophies and medicine) and meat.

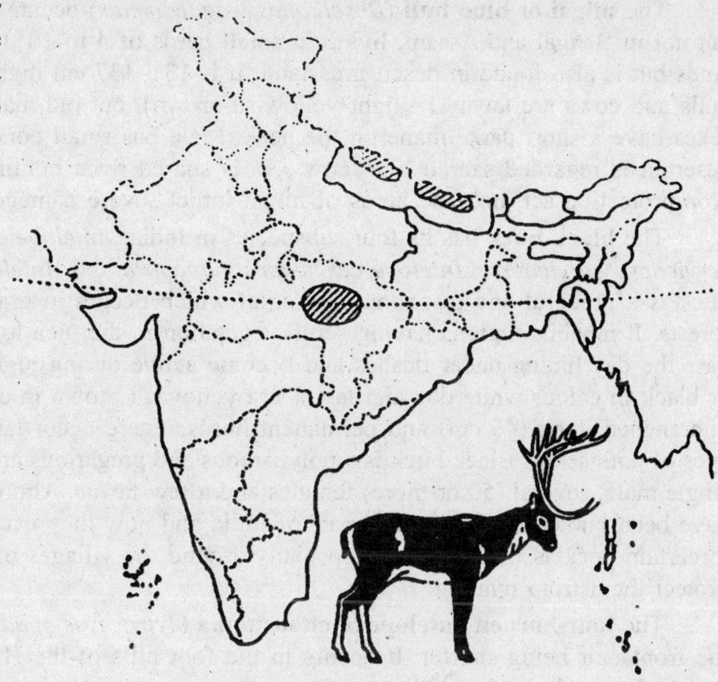

Fig. 18.2. Distribution of swamp deer.

The **barking deer** or **rib-faced deer** (*Cervulus muntijac*) is found at the altitudes of 2000 to 3000 metres and has four subspecies in India. It is 61 cm in height and has suborbital glands below eyes. Buck has antlers and non-protruding canines. It lives in pairs or small families in thickets and being diurnal, grazes in the morning and evening hours. The **dancing deer** or **sambhar** (*Cervus unicolour*) is large-sized (150–160 cm high), browsing and gregarious deer, occurring in the herds of 4 to 12 animals and is found in forests all over the country. It has very long (up to 100 cm) antlers. The **spotted deer** or **cheetal** (*Axis axis*) is 76 to 91 cm high, grazing and gregarious deer, living in the herds of 20 to 30. It is spotted with white on rich brown coat. The buck has long slender antlers, each of which has three branches or tines and is shed annually. It is mostly found in the forests of Indian plains and in hill upto 1333 metres but are not found in Assam, Punjab and Rajasthan. It prefers open type forests with good grazing sites and running streams. The **hog deer** (*Axis porcinus*) is smaller and stouter in built and has hog-like appearance. It lives either solitary or in pair and prefers to live in grass patches bordering on forests. It is diurnal and comes out to feed only during morning and evening.

2. Antelopes and Other Herbivores

(1) Antelopes. In India there are four species of antelopes : the Indian gazelle, nilgai, black buck and chowsingha.

The **Indian gazelle or chinkara** (*Gazella dorcas bennetii*) is mostly found throughout desert region of southern part of Uttar Pradesh and Rajasthan. It lives in herds of 3 to 5 animals and has adaptations to live in extremely arid conditions, kilometres away from drinking water sources (**Prakash**, 1977). It prefers to live in open grasslands with thick bush cover. They feed on a variety of grasses and browse on shrubs.

The **nilgai** or **blue bull** (*Boselaphus tragocamelus*) occurs in plains and Rajasthan deserts but not in Bengal and Assam, living in small herds of 4 to 10. It prefers to stay near cultivated lands but is also found in desert grassland. It is 131–137 cm high, females being smaller. Young bulls and cows are tawny (= light yellowish-brown), but old males have iron gray colour. Both sexes have a short dark mane on the neck. Male has small cone-shaped curved horns. In Thar desert it is regarded sacred like 'cow', so is spared from hunting. However, it is killed under crop-protection act, because herds of nilgai inflict severe damage to crops.

The **black buck** has its four subspecies in India: *Antelope cervicapra rajputanae, Antelope cervicapra cervicapra, Antelope cervicapra rupicapra* and *Antelope cervicapra centralis.* Black buck is a graceful and fast-running animal which occurs in grassy plains of India and avoids forests. It inhabits regions having scrub vegetation in the vicinity of tanks and lakes where they pass the day hiding under bushes and become active during night. The male antelope is brown or black in colour while doe and fawns are yellowish brown in colour. Male has spirally ringed, unbranched, long (46 cm) and permanent horns. There occur large preorbital glands below the eyes of both sexes. Black buck is a polygamous and gregarious animal and its herd often includes single male, several (50 or more) females and a few fawns. Almost all subspecies of black bucks have been hunted out from most part of India and now they occur in appreciable numbers only in certain pockets of Rajasthan, especially around the villages of 'Visnoi caste' of Hindus, who protect them from hunting.

The **four-horned antelope** or **chausingha** (*Tetracerus quadricornis*) has two pairs of horns, the front pair being shorter. It occurs in the foot hills of the Himalayas and in Bundel Khand region. It mostly leads a solitary life.

(2) Wild-buffalo. The wild buffalo (*Bubalus bubalis*) is a large-sized, robust animal having streamlined body. It prefers tall grassy forests close to marshy areas in the vicinity of rivers and lakes. It was once distributed over the grass jungles and river in forests of the Gangetic plain, the terai, Assam and eastern peninsular India down to the river Godavari. Great hunting of wild buffalo has restricted them now only to Assam (Manas and Kaziranga sanctuaries), the Nepal Terai and Bastar district in Madhya Pradesh. According to **Sheshadari** (1969), there are only 1425 wild buffaloes in India.

(3) Bison. Mithun, bull gaur or bison (*Bos gaurus*) is a relative of ox and is found in Andhra Pradesh, Orissa, Madhya Pradesh and Bihar. Due to certain epidemic diseases such as rinderpest and enteric viral disease numerous individuals were wiped out some years ago and now a few bisons are mainly confined to Bandipur in Karnataka.

(4) Wild goat and sheep. In India, wild goat and sheep are restricted to north. **Ibex** is found in the Himalayas. The **Nilgiri tahr** (*Hemitragus hylocrius*) is the only one to be in south, in Nilgiris, Tamil Nadu. It moves about in herds and grazes on grasses. This species is greatly endangered due to habitat disturbance, predation by black panther and wild dogs (dhole) and hunting by poachers and only 450 tahr were present in 1975 (**Davidar**, 1975).

(5) Elephant. The Indian elephant, *Elephas maximus* is largest terrestrial mammal and is confined to the Terai and the foot-hills because of its dependence on succulent grass, bamboo and plenty of water. It is gregarious and its herds move constantly in the search of new feeding grounds. Its population is hunted out mainly for tusk and now Periyar sanctuary (Madras) has some elephants.

(6) Rhinoceros. The Indian rhinoceros (*Rhinoceros unicornis*) is found in the grasslands and jungle area of the foot hills of the Himalayas (central Nepal) and certain pockets of plains of West Bengal and Assam. It prefers swamps and open savannah, covered with the tall elephant grass. A full grown rhinoceros is 1.8 metres in height and 3–9 metres in length. It has a single horn on the nose. The horn is made of keratinized skin and matted hair, it is said to have aphrodisiac and other medicinal values, so that large sums are paid for it, this results in indiscriminate shooting and its near extinction in India. It is a slow, solitary animal and is strictly territorial.

(7) Wild ass. The wild ass (*Equs hemionus khur*) had a fairly wide distribution in the dry regions of North-West India (Jaisalmer and Bikaner), but their distribution now become restricted to the Little Rann in the southern part of the Thar desert, where **Gee** (1963) estimated their number to be 870. In past, the population of wild ass severely reduced due to hunting, catching for breeding mules, surra disease and south African Horse sickness disease. However, people around the Little Rann are orthodox and vegetarian and do not molest the wild ass inspite of the fact they inflict serious damages to crops. Because **it is the only true 'wild ass' of the world,** so it needs protection and conservation (**Prakash,** 1977).

(8) Wild boar. The wild boar, *Sus scrofa cristatus,* has a long mobile snout with terminal nostrils. The canines of both jaws grow continuously and form long triangular and up curved tusks for defence and digging roots, they are better developed in males. The wild boar or pig is omnivorous, it feeds on carrion, snakes, insects, roots, tubers and cultivated crops. It is gregarious living in families in grass or bush of marshy places and is now rare and is mostly found in small numbers along the Aravalli ranges in the eastern most part of Thar desert where they inhabit rocky slopes.

3. Big Cats and Other Carnivores

India has a variety of world-known large-sized cats and other carnivores which are as follows:

The **Asiatic lion,** *Panthera leo persica* was found in the Thar desert, in the arid plains of Sind, and Rajasthan and Punjab. It is on record that the last lions were shot at Anadra and at Jaswantpura of Rajasthan desert in about 1876. In India lions are now preserved only in the Gir sanctuary, in Gujarat, on the South-West of the Thar desert. Its extermination is well known.

The **cheetah,** *Acinonyx jubatus venaticus* was found in Central India, Deccan region, and in Thar desert upto Jaipur in Rajasthan. **Sheshadri** (1969) stated that it disappeared from Rajasthan about the turn of 20th century, from Central India in 1920's and in Deccan it was last reported in 1952. Now the cheetah, one of the most beautiful, agile and fastest animal, is regarded to be extinct from India.

The **tiger** *Panthera tigris tigris* is a gracious carnivore which is distributed in Uttar Pradesh from the Himalayas to the Vindhya forests in the South. In Himalayas it is found up to an elevation of about 2,000 metres. It lives in a variety of habitats, from thorn forests to the dense terai forests.

The **leopard,** *Panthera pardus* is the smaller spotted cousin of tiger which can live in all types of forests. Once it was found all over India but now its number has been drastically reduced. Leopard is given full protection in the Mount Abu sanctuary, however, can be hunted under licence from the Indian Wildlife Act. The **snow leopard** (*Panthera uncia*) is found in the high Himalayas from Kashmir to Sikkim near snow line. It has a creamy grey coat with large black rings. It is greatly hunted for its fur and now is endangered species. Other important cats are **clouded leopard** (*Neofelis nebuleosa*), **leopard cat** (*Felis bengalensis*), **pallas's cat** (*Felis manul*), **caracal** (*Felis caracal*), etc. Caracal once was common in Rajasthan deserts, now it has become extremely rare and is suspected to occur only in the Sirohi-Jalore and Bikaner regions (**Prakash**, 1977).

The **sloth bear**, *Melursuu ursinus*, is a stout carnivore which is found in the hilly forests such as Mount Abu and Erinpura ranges. It lives mostly on fruits, honey and insects, it sucks up white ants from their hills which are broken by its claws with great speed. It is rapidly vanishing due to ruthless hunting and it needs conservation. The black Himalayan bear, has V-shaped white mark on the chest and is carnivorous.

Dolphin. In the Indus river, in the North-West and in the rivers Chenab and Sutlaj in the north of the Thar desert, the freshwater gangetic dolphin, *Platanista gangetica* (Cetacean mammal) was once found in good numbers (**Murray**, 1884), but is now totally extinct (**Prakash**, 1977).

4. Birds

The important wild bird-fauna of India include the following species: pea fowl (*Pavo cristatus*), grey jungle fowl (*Gallus sonnerati*), red spur fowl (*Galloperdix spadicea*), ducks (*Anser indicus, Sarkidiornis melanotos, Anaspoecilorhyncha*), pigeons (*Columba livia*), sand grouses (*Pterocles exustus, P. orientalis, Streptopelia decaocta*), storks (*Leptotilos dubuis*), egrets (*Egretta garetta*), grey partridge (*Francolinus pondieirianus*), black partridge (*Francolinus francolinus*), golden eagle (*Aquilla chrysaetos*), vultures (*Gyps bengalensis, Neophron percnopterus*), spotted billed pelican (*Pelacanus philippensis*), sarus crane (*Grus antigone*), quails (*Coturnix coturnix, C. coromandelica*),and great Indian bustard (*Choriotes nigriceps*). Certain birds such as osprey, great Indian bustard, cranes, mountain quail, horned owl, red start avocet, pinkheaded duck, etc., are on the verge of extinction, if not saved. Great Indian Hornbill with its peculiar nesting behaviour is the pride of India and needs to be saved.

The great Indian bustard is a large Indian game bird having a heavy body, long neck and long bare running legs. It is essentially an inhabitant of wide, open dry scrubby plains and waste, broken undulating lands of the Rajasthan, West Punjab and Gujarat. It is ruthlessly hunted out for its delicious flesh and is almost at the verge of extinction. In 1980, an International Conference was held at Jaipur to assess its present position and to suggest conservation steps to save it from extinction.

5. Crocodiles and Other Reptiles

The following three Indian aquatic reptiles have been found endangered: the estuarine crocodile (*Crocodilus porosus*); subnosed crocodile, freshwater or marsh crocodile or muggar (*Crocodilus palustris*) and gharial (*Gavialis gangeticus*). The estuarine crocodile is found in Bhitarkanika Island and adjacent mangrove areas of Orissa and in the Sunderbans of West Bengal. The marsh crocodile is widely distributed in India and has been greatly reduced in number in the past due to indiscriminate shooting for its valuable hide. *Gavialis* was once common in most large Indian rivers such as Indus, Ganga, Brahmputra and Mahanadi, but now has very sparse distribution in Ganga and larger feeders in the Nepal-Terai, Bihar and Bengal.

Sea turtles migrate to "**Gahirmatha**" in Orissa to carry on nesting popularly known as"**Arribada**". The deep sea fishing is a continuous threat to this reptilian species. When turtles float on surface during copulation, certain tribals which feed on them trap and sell them in the market. Their nesting grounds are also disturbed by humans who collect turtle's eggs for consumption.

A rare limbless lizard is the *Barkudia insularis* which is mistook as the snake and killed indiscriminately. It is subterranean in habit and found at Barkudia inland of Chilka lake and needs protection.

In India, snakes are killed by some and worshipped by others. Generally they are feared. They are believed to be farmer's best friends as a single rat snake can eat hundreds of rats and mice per year. Indiscriminate killing of snakes for skin results in a loss of ten million rat snakes, cobras and pythons every year.

6. Frog

The "State of India's Environment" 1984–85 reports that every year Indians kill an estimated 100 million frogs to satisfy the palates of well fed westerners. In 1981, 4368 tonnes of frog legs were exported from India; these have been finding their way to Canada, Saudi-Arabia, Japan and UAE. Two common species of frogs *viz., Rana tigrina* and *Rana hexadactyla* have been affected by this trade and deserve protection.

18.2 ENDANGERED SPECIES OF INDIA

Red Data Book (RDB) comprises of long directories compiled by IUCN where rare and/or endangered species are listed along with relevant information from all over world. IUCN stands for International Union for Conservation of Nature and Natural Resources (It is now called World Conservation Union or WCU with headquarters at Gland, Switzerland).

The threatened species categories now used in Red Data Book and Red Lists have been in use for almost 34 years. The Red Data Book categories provide an easily and widely understood method for highlighting those species which are under higher extinction risk so as to focus attention on conservation measures to protect them. Based on history of distribution, decline in number of individuals, nature of habitat and potential value of species, IUCN has classified the plant and animal species into following **nine** groups (**Mace** and **Stuart** 1994):

1. Extinct (Ex). A taxon is considered **extinct** when there is no reasonable doubt that its last individual has died. For example, in India, an estimated 13,000 species of mammals and birds were known in 17th century but now more than 130 are already gone for ever; some of the extinct animal species of India are **Mountain quail**, **Pink headed duck**, **Lesser one horned rhino**, **Jerdon's courser** (this bird was rediscovered in 1986 at Cuddapah in Andhra Pradesh) and the streamlined spotted feline **cheetah** (Fig. 18.3). The abundance of cheetah in India could be judged by the fact that Akbar had a stable of over 1,000 cheetahs in his army.

Fig. 18.3. Some extinct animals of India.

More than 20 wild plants are believed to be extinct or facing extinction. These plants have an actual or potential medicinal value, *e.g., Coptis teeta, Dioscorea deltoidea, Drosera* sp., *Paphiopedilum drurui, Rauvolfia serpentina* (*Sarpgandha*) and *Saussurea lappa* (*Brahmakamal*). Similarly *Acorus calamus* which was found in Kashmir, Manipur and Nagaland only a few years ago, is now being imported to meet the medicinal demand in the country.

2. Extinct in the wild (EW). A taxon is considered **extinct in wild** when it is known only to survive in cultivation, in captivity, or as a naturalized population well outside the past range. It is presumed extinct in wild when exhaustive surveys in known and/or expected habitat, at appropriate times (*i.e.,* diurnal, seasonal, annual), throughout its historic range have failed to record an individual. Such surveys should be conducted over a time frame appropriate to the taxon's life cycle and life form.

3. Critically endangered (CR). A species is considered **critically endangered** when it is facing an extremely high risk of extinction in the wild in the immediate future. Its population is estimated to number less than 50 mature individuals. For example, the pink pigeon (*Columba mayri*) of Mauritius is a CR species where it is declined to a tiny population of 20 birds.

4. Endangered (EN). A species is regarded as **endangered** when it is not critically endangered but facing a very high risk of extinction in the wild in the near future. Its population is assessed to number less than 250 mature individuals, *e.g.*, the black-winged Indonesian parrot *Eos cyanogenis,* the tropical hawksbill turtle *Eretmochelys imbricata,* the land snail *Partula rosea* (endemic to the island of Huahine in French polynesia), the Mexican cactus *Aztekium ritteri*, the Malaysian slipper orchid *Paphiopedilum stonel.*

In India, the number of official endangered animal species has risen from 13 in 1952 to 240 today (**Mathur** 1998). Brow antlered deer, hispid hare, lion tailed macaque and pygmy hog are considered very rare and are near to extinction other Indian endangered animals are Indian *Rhinoceros*, asiatic lion, great Indian bustard, crocodile, musk deer and blue whale.

5. Vulnerable (VU). A taxon is considered vulnerable when it contains sufficient number of individuals in its habitat. However, in near future, it may represent the category of endangered species, if unfavourable conditions in the environment continue to operate, *e.g.,* crab eating macaque, golden langur, blood pheasant, chinkara, etc.

6. Conservation dependent (CD). Taxa that do not currently qualify as critically endangered, endangered or vulnerable may be classified as **conservation dependent**. A species under this category must be the focus of a continuing taxon specific or habitat specific conservation programme which directly affects the species (taxon) in question. Discontinuity of this conservation programme would result in the species qualifying for one of the threatened categories above (Box 18.1). For example, the South African white rhinoceros *Ceratotherium simum* is a CD species which has been increased for many years under strict protection and now numbers more than 6000 individuals.

Box 18.1

Previously following two categories of taxons were commonly considered:

1. Rare (R) species. The taxon (species) with small world population that are not presently endangered or valnerable, but are at risk. Such species are usually localised within restricted geographical areas or habitats or are thinly scattered over more extensive range. Examples of rare species include Indian egg eating snake, Olive Ridley turtle, Indian pied hornbill, Crestless Himalayan porcupine, Golden cat.

2. Threatened. Taxa which are in one of the categories of *endangered, vulnerable* and *rare.*

7. Low risk (LR). A taxon is at **low risk** when it has been evaluated and does not qualify for any of the categories such as *critically endangered, endangered, vulnerable, conservation dependent* or *data deficient.*

8. Data deficient (DD). A taxon is **data deficient** when there is inadequate information to make a direct or indirect assessment of its risk of extinction based on its distribution or population status. Listing of taxa in this category indicates that more information is required.

9. Note evaluated (NE). A taxon is under the category **not evaluated** when it is not yet been assessed against the criteria. It is also called **least concern** according to IUCN (2012) Red Data List of Threatened Species (**Sharma** 2014).

Threatened Indian Fauna

Presently wildlife resources are poorer than ever. Some of our major threatened mammals, amphibians, reptiles and birds as enlisted in schedule I of the Wildlife Protection Act 1972 and updated in 1979 are as follows (This list also includes five extinct animals of India):

I. Endangered Mammals of India

1. Andaman wild pig (*Sus scrofa* and *Sus amanensis*)
2. Bharal (*Ovis nahura*)
3. Binturong (*Arctictis binturong*)
4. Bison, Gaur or Mithun (*Bos gaurus*)
5. Black buck (*Antelope cervicapra*)
6. Blue whale (*Balaenoptera musculus*)
7. Brow-antlered deer or Thamin (*Cervus eldi*)
8. Capped langur (*Presbytis pileatus*)
9. Caracal (*Felis caracal*)
10. Cetecean species
11. Cheetah (*Acinonyx jubatus*)
12. Chinese pangolin (*Manis pentadactyla*)
13. Chinkara or Indian gazelle (*Gazella gazella bennetti*)
14. Chital (*Axis axis*)
15. Clouded leopard (*Neofelis nebuleosa*)
16. Crab-eating macaque (*Macaca irus umbrosa*)
17. Desert cat (*Felis libyca*)
18. Dugong (*Dugong dugon*)
19. Fishing cat (*Felis viverrina*)
20. Flying squirrels (all species of genera *Petaurista, Eupetaurus, Belomys, Hylopetes*)
21. Four horned antelope (*Tetraceros quadricornis*)
22. Gangetic dolphin (*Platanista gangetica*)
23. Giant squirrels (*Ratufa macroura, Ratufa indica, Ratufa bicolor*)
24. Golden cat (*Felis temminki*)
25. Golden langur (*Presbytis geei*)
26. Gorals (*Nemorhaedus goral, Nemorhaedus hodgsoni*)
27. Himalayan black bear (*Selenarctos thibetanus*)
28. Himalayan brown bear (*Ursus arctos*)
29. Himalayan crestless porcupine (*Hystrix hodgsoni*)
30. Himalayan ibex (*Capra ibex*)
31. Himalayan tahr (*Hemitragus jemlahicus*)
32. Hispid hare (*Caprolagus hispidus*)
33. Hog badger (*Arctonyx collaris*)
34. Hoolock or gibbon (*Hylobates hoolock*)

35. Hump-backed whale (*Megaptera nevaeangliae*)
36. Hyaena (*Hyaena hyaena*)
37. Indian elephant (*Elephas maximus*)
38. Indian lion (*Panthera leo persica*)
39. Indian wild ass (*Equus hemionus khur*)
40. Indian wolf (*Canis lupus pallipes*)
41. Kashmir stag or hangul (*Cervus elephus hangul*)
42. Leaf monkey (*Presbytis melanophos*)
43. Leopard cat (*Felis bengalensis*)
44. Leopard or panther (*Panthera pardus*)
45. Lesser or red panda (*Wilurus fulgens*)
46. Lion-tailed macaque (*Macaca silenus*)
47. Loris (*Loris tardigradus*)
48. Lynx (*Felix lynx isabellinus*)
49. Malabar civet (*Viverra magaspila*)
50. Malay or sun bear (*Helarctos melayanus*)
51. Marbled cat (*Felix mormorate*)
52. Markhor (*Capra falconeri*)
53. Mouse deer (*Tragulus meminna*)
54. Musk deer (*Moschus moschiferus*)
55. Nilgai (*Boselaphus tragocamelus*)
56. Nilgiri langur (*Presbytis johni*)
57. Nilgiri thar (*Hemitragus hylocrius*)
58. Otter (*Lutra lutra, L.perspicilata, Aonyx cinerea*)
59. Ovis ammon or nyan (*Ovis ammon hodgsoni*)
60. Pallas's cat (*Felis manul*)
61. Pangolin (*Manis crassicaudata*)
62. Pig-tailed macaque (*Macaca memestrina*)
63. Pygmy hog (*Sus salvanius*)
64. Ratel (*Mellivora capensis*)
65. Red fox (*Vulpes vulpes*)
66. Rhinoceros (*Rhinoceros unicornis*)
67. Rusty-spotted cat (*Fellis rubiginosa*)
68. Sambhar (*Cervus unicolor*)
69. Serow (*Capricornis sumatraensis*)
70. Sloth bear (*Melursus ursinus*)
71. Slow loris (*Nycticebus coucang*)
72. Small travancore flying squirrel (*Petinomys fuscocapillus*)
73. Snow leopard (*Panthera uncia*)
74. Spotted linsang (*Priondon pardicolor*)
75. Swamp deer or gond (*Cervus duvauceli*)
76. Takin or mishmi takin (*Budorcas taxicolor*)

77. Tibetan antelope or Chiru (*Panthelopes hodgesoni*)
78. Tibetan fox (*Vulpes ferrilatus*)
79. Tibetan gazelle (*Procapra picticaudata*)
80. Tibetan wild ass (*Equus/Asinus hemionus kiang*)
81. Tiger (*Panthera tigris*)
82. Urial or shapu (*Ovis vignei*)
83. Wild buffalo (*Babalus bubalis*)
84. Wild dog or dhole (*Cuon alpinus*)
85. Wild yak (*Bos grunniens*)

II. Endangered Amphibians of India

86. Himalayan newt or salamander (*Tryletotritor verrucosus*)
87. Viviparous toads (*Nectophyrynoides* sp.)

III. Endangered Repitles of India

88. Agra monitor lizard (*Varanus griseus*)
89. Atlantic ridley turtle or Kemp's ridley turtle (*Lepidochelys kembii*)
90. Barred, oval or yellow monitor lizard (*Varanus flavescens*)
91. Estuarine crocodile (*Crocodylus porosus*)
92. Salt water crocodile (*Crocodylus palustris*)
93. Gharial (*Gavialis gangeticus*)
94. Ganges soft-shelled turtle (*Trionyx gangeticus*)
95. Green sea turtte (*Chelonia mydas*)
96. Hawksbill turtle (*Eretmochelys imbricata inlscata*)
97. Indian egg-eating snake (*Elachistodon westermanni*)
98. Indian soft-shelled turtle (*Lissemys punctata punctata*)
99. Indian tent turtle (*Kachuga tecta tecta*)
100. Large Bengal monitor lizard (*Varanus bengalensis*)
101. Leathery turtle (*Dermochelys coriacea*)
102. Loggerhead turtle (*Caretta caretta*)
103. Olive-back loggerhead turtle (*Lepidochelys olivacea*)
104. Pythons or Azgar (all species of genus *Python*)
105. Three keeled turtle (*Geoemyda tricarinata*)
106. Tortoise (genera of families of Testudinidae, Trionvchidae)
107. Water lizard (*Varanus salvator*)

IV. Endangered Birds of India

108. Andaman teal (*Anas gibberiferon albogularis*)
109. Assam bamboo partridge (*Bambusicola fytechii*)
110. Bazas (*Aviceda jerdoni* and *Aviceda leuphotes*)
111. Bengal florian (*Eupodotis bengalensis*)
112. Black necked crane (*Grus nigricollis*)
113. Blood pheasants (*Ithaginis cruentus tibetanus; Ithaginis cruentus kuseri*)

114. Brown headed gull (*Larus brunnicephalus*)
115. Cheer pheasant (*Catreus wallichii*)
116. Comb duck (*Sarkidiornis melanotos*)
117. Forest spotted owlet (*Athene blewitti*)
118. Great Indian bustard (*Choriotis/Ardeotis nigriceps*)
119. Great Indian hornbill (*Buceros bicornis*)
120. Hooded crane (*Grus monacha*)
121. Hornbills (*Ptilolaemus tickelli austeni, Aceros nipalensis, Rhyticeros undulatus ticehursti*)
122. Houbara bustard (*Chlamydotis undulata*)
123. Hume's bar-backed pheasant (*Syrmaticus humiae*)
124. Indian pied hornbill (*Anthracocerus malabaricus*)
125. Jerdon's courser (*Cursorius bitorquantus*)
126. Lammergeier (*Gypaetus barbatus*)
127. Large falcons (*Falco peregrinus, Falco biarmicus* and *Falco chicquera*)
128. Large whistling teal (*Dendrocygna bicolor*)
129. Monal pheasants (*Lophophorus impejanus*)
130. Mountain quail (*Ophrysia superciliosa*)
131. Narcondom hornbill (*Rhyticeros undulatus narcondami*)
132. Nicobar megapoda (*Megapodius freycinet*)
133. Nicobar pigeon (*Caloenas nicobarica pelewensis*)
134. Peacock pheasant (*Polyplectron bicalcaratum*)
135. Pea fowl (*Pavo cristatus*)
136. Pinkheaded duck (*Rhodonessa caryophyllacea*)
137. Sclater's monal (*Lophophorus sclateri*)
138. Siberian white crane (*Grus leucogeranus*)
139. Spur fowl (*Gallaperdix* spp.)
140. Tibetan snow-cock (*Tetraogallus tibetanus*)
141. Tragopan pheasants (*Tragopan melanocephalus, Tragopan blythii, Tragopan satyra, Tragopan temminckii*)
142. Whitebellied sea eagle (*Haliaeetus leucogaster*)
143. White-eared pheasant (*Crossoptilon crossoptilon*)
144. White spoon bill (*Platalea leucorodia*)
145. White-winged wood duck (*Caririna scutulata*)

18.3 WORLD'S MOST ENDANGERED SPECIES

In global terms, the total number of endangered species is alarming. **Conservation Monitoring Centre (CMC)** of IUCN currently holds records on computers of around 2200 plants and 800 animals. The numbers increase as the CMC data base expands to cover less studied taxonomic groups of relatively unknown geographic areas. With passing time, **vulnerable** species are moving into category of endangered species and species currently regarded as **rare** can become critically threatened with extinction.

Criteria of IUCN

Martyn Murray and **Sara Oldfield** have suggested the following criteria to decide the conservation value of a species:

1. Urgency. Urgency is one of the most significant factor for the conservation of a species; since some species in the endangered category are already closer to extinction than others. The **Plant Red Data Book** (or **Plant RDB**) lists 31 species of endangered plants each with a critically low population in danger of breeding collapse. Six of them are reduced to a single specimen in the wild and rank among the most endangered species in the world. For example, the last remaining specimen of *Ramosmania heterophylla*, an endemic tree of Rodrigues (an island having dependency on Mauritus) in the South Indian Ocean, is now protected by a fence and cuttings will be taken for propagation when conditions are right. In Saint Helena (British Island in South Atlantic Ocean) propagation has already helped to save several critically endangered species including the Saint Helena ebony *Trochetiopsis melanoxylon* which was thought to be extinct until two small specimens were discovered in 1980.

2. Prominence. The prominence of a species in public awareness is a good measure for the conservation of a species. Prominence of a taxon might be due to following factors:

***(i)* Attractiveness.** Because of unusual, striking or appealing looks, behaviour or habits some species are likely to attract conservation interest, *e.g.,* giant panda and all types of bears. A lady's slipper orchid such as *Paphiopedilum druryi*, or an exotic butterfly such as the birdwing *Ornithoptera alexandrae*, both are highly liked by collectors due to some economic value and so need conservation.

***(ii)* Crisis appeal.** Need for conservation of some species are caused by occurrence of some recent environmental disaster, such as Mauritus endemic birds followed by recent hurricane.

***(iii)* Usefulness to mankind.** Having economic value as foods, medicines, or industrial substances, such as the medicinal leech *Hirudo medicinalis.* In the plant RDB, six of the examples selected as sources of economic products are endangered including the nutritious yehebnut *Cordeauxia edulis* and the cypress yehebnut *Cupressus dupreziana*, a valuable Algerian timber species.

***(iv)* Symbolic or figurehead appeal.** Such as the African violet *Saintpaulia inonantha,* is the symbol of the WWF/IUCN Plant Campaign.

3. Biological value. The ecological role (*i.e.,* biological value of an endangered species forms an important criterion for selecting the most endangered species. One aspect of biological value, which is highlighted in the world conservation strategy, is the degree of genetic differentiation of a species. Those species which differ most from their nearest relative, or those which are the sole representative of a higher taxonomic group (genus, family or order) should be given priority just because their genes are so different. For example, kagu *Rhynochetos jubatus,* an egret-like bird which is endemic to the central mountain range of New Caledonia, is the only species in its family and clearly of major taxonomic interest. The aye-aye *Daubentonia madagascariensis* which is the unique Madagascur lemur is also the sole surviving member of the family Daubentoniidae.

Indeed protecting an endangered species *in-situ* by protecting the habitat leads in turn to protection for other species. For example, if the kagu's habitat is conserved, some of the endangered endemic plants of the island will also be protected; protecting the aye-aye's habitat means preserving some of the last remnants of Madagascar's forests. Likewise the protecting the pygmy hog *Sus salvanius* in Asia in the tall grass thatchland vegetation which is its natural habitat, will also help the endangered hispid hare *Caprolagus hispidus.*

Thus, conservation of one species is linked with conservation of other threatened species, or to an area of special interest such as threatened habitat, a site rich in species, or a site with endemic species. Representative species of different geographical areas, such as small oceanic islands, can be selected as *indicators* of general threats. On the Hawaiian island of Oahu all the 19 endemic species of tree snails of the genus *Achatinella* are endangered by the same threats such as loss of forest habitat and pressure from introduced snails. Thus, the whole genus could be selected to show the threats to island ecosystems, or alternatively one species could be chosen to highlight the plight of the genus.

4. Operational considerations. Following operational considerations are also important in species selection for conservation: 1. need for international support; 2. cost of conservation measures; 3. likelihood of success which depends on possibility of using a variety of conservation measures: *central support* (*i.e.,* whether Government is receptive of assistance); *local support* (*i.e.,* whether poeple in vicinity of reserve are sympathetic to or benefitted from the conservation programme); and finally the involvement of IUCN/WWF through species survival commission (SSC).

18.4 REASONS OF DEPLETION OF WILDLIFE

The depletion of wildlife is attributable to the following factors (Box 18.2):

Box 18.2 Causes of Extinction of Wildlife

Many wild animals became extinct due to various human and natural activities: 1. Absence of cover or shelter to wild animals. 2. Due to deforestation for cultivation, road-building, railway routes, dam construction or for urbanisation, occurs reduction in the area for free movement of wild animals which retard reproductive capacity of certain wild animals such as deer, bison, elephant, rhino, tiger, etc. 3. Destruction of wild plants of forests for timber, charcoal and firewood often deprive wild animals their most palatable food and affects their survival. 4. Noise pollution by different transporting media (trucks, buses, rails, aeroplanes, etc.) and polluting river water have adversely affected wild animals. 5. Various natural calamities such as floods, droughts, fires, epidemics, etc., have also caused great destruction of wildlife. 6. Hunting methods of all kinds and for any purpose (*i.e.,* for food, recreation, hide, fur, plumage, musk, tusk, horn, etc.) have caused destruction of wildlife; and 7. Unrestricted grazing, spreading of urbanization and environmental changes, all add to extinction of wildlife.

1. Destruction of habitat. The natural habitat may be destroyed by human beings for his settlement, grazing grounds, agriculture, mining, industries, highway construction, drainage, dam building, etc., with the cutting of forests and eradication of marshes and wetlands, the habitats of wildlife went on shrinking. They were gradually pushed on to comparatively unproductive and unsafe patches of their habitat where they not only faced the scarcity of food and water, but were exposed to poachers. As a result, their numbers have declined considerably; so much so that at present some species are in a rather precarious state and it is feared that further reduction in their numbers would mean their extinction.

2. Hunting for commercial products. One of the main causes of depletion of wildlife in India, as also throughout the world, is unregulated commercial exploitation of wildlife. The wild animals are killed for their products such as hides and skin, tusks, fur, meat, horns, bones, for pharmaceuticals, perfumes (*e.g.,* musk), cosmetics and decoration purposes. For example, in Africa, in recent years 95 per cent of the black rhino populations have exterminated by poachers for their horn. Today, rhino horn fetches up to \$ 15,000 in the pharmaceutical market (**Agarwal,** 1998). In last 10 years alone, over one-third of Africa's elephants have been wiped out for some 3000 tonnes of ivory. In the international market the cost of ivory is \$ 150 per kg, whereas in the Indian market it varies between Rs. 2000-2500 per kg (in 1992). Africa produces about 700 tonnes of raw ivory every year. The major buyers of ivory are Japan, Hong Kong, followed by US, Germany and UK. Carving and sale of African ivory has been banned in India from April 2, 1992. In January, 1987 the central government also banned the trade in Indian ivory.

The scarlet macaw, which is a large tropical American parrot (genus *Ara*), with a long tail, harsh voice and brilliant plumage, once common throughout South America, has been eliminated from most of its range in Central America. Several species of spotted cats such as the ocelot and jaguar have been jeopardised by the demand for their fur. In 1962, nearly 70,000 whales were slaughtered. However, international trade in whale products is banned now.

In India, *rhino* is hunted for its horns and *musk deer* for musk (both have medicinal value), elephant for ivory, *gharial* and *crocodile* for their skin, and jackal for thriving fur trade in Kashmir. Tiger, leopard and snake too are killed for their skin. CITES (Convention on International Trade in Endangered Species of Wild Fauna and Flora) has listed nine Indian species which have been severely depleted due to international trade. These are fin whales (*Balenoptera physalus*), Himalayan musk deer (*Moschus moschiferus*), green turtle (*Chelonia mydas*), hawksbill turtle (*Eretmochelya imbricata*), Olive Ridley turtle (*Dermochelys olivacea*), salt-water crocodile (*Crocodylus porosus*), desert monitor lizard (*Varanus griseus*), yellow monitor lizard (*Varanus flavescens*) and Bengal monitor lizard (*Varanus bengalensis*).

3. Overexploitation. Excessive harvesting of marine organisms such as fishes, molluscs, sea-cows and sea-turtles has resulted in extinction of these animals.

4. Collection for zoo, research or laboratory exercise. Animals and plants are collected throughout the world for zoos, botanical gardens, biological laboratories for study and research in science and medicine. For example, primates such as monkeys and chimpanzees are sacrificed for research as they have anatomical, genetic and physiological similarities to human beings.

5. Introduction of exotic species. Due to introduction of exotic species in an ecosystem, native species are subjected to competition for food and space. For example, introduction of goats and rabbits in Pacific and Indian area has resulted in destruction of habitats of several plants, birds and reptiles.

6. Control of pests and predators. Predator and pest control measures, generally kill predators that are a component of balanced ecosystem and may also indiscrimately poison non-target species.

7. Pollution. Pollution alters the natural habitat. Water pollution is especially injurious to the biotic components of estuary and coastal ecosystems. Toxic wastes, fertilizers and detergents entering the water bodies disturb the foodchain and so to the aquatic ecosystems. Insecticides and pesticides too affect adversely the plant and animal species. Gulf war of 1991 has caused great ecological disaster in the form of oil spill in sea which have killed tens of thousands of birds (*e.g.,* gerbes and cormorants) and threatened sea grasses which form feeding areas of marine organisms as sea-cows and green turtle.

Lastly, there are certain other ecological factors which may also contribute to extinction of plant and animal species:

1. Distribution range. The smaller the range of distribution, the greater the threat of extinction.

2. Degree of specialization. The more specialized an organism is the more vulnerable it is to extinction.

3. Position of organism in the food chain. The higher the organism in food chain, the more susceptible it becomes.

4. Reproductive rate. Large-sized organisms (elephants, whales, etc) tend to produce fewer offsprings at widely spaced intervals.

18.5 WILDLIFE OF RAJASTHAN

Rajasthan does not mean desert alone, it includes hills of Aravalli, slopes of Vindhyas, maze of ravines, Chambal river, many lakes, marsh lands, vast lands and ofcourse above all—

extremely hot and arid Thar desert. The Thar desert (Box 18.3) consists of wind-blown sands, fixed and shifting sand dunes, intersand dunal area and rocky outcrops. Inspite of harsh climatic conditions, Rajasthan supports a variety of flora and fauna.

Box 18.3

Thar desert or Great Indian Desert is the region of sandy desert. It exists in North-West India and South-Pakistan, between Aravalli Range on east and Indus River on west, Sutlaj River is on its north and Arabian sea on its south. Thar desert is 645 km long and 360 km wide (*i.e.,* 259,000 sq./km). It has annual rainfall less than 15 inch (38 cm) and has continuous high temperature. It includes portions of Rajasthan and Gujrat states in India, and Punjab and Sind in Pakistan.

The vegetation in desert regions of Rajasthan is mainly of Rohira (*Tecoma undulata*), Khejri (*Prosopis sinreria*), Thor (*Euphorbia cadvca*), *Acacia leucophloea,* Ber (*Ziziphus mauritiana*), Aak (*Calotropis procera*) and perenial succulents (Cactaceae and Euphorbiaceae). Representatives from most of insect orders are found in Rajasthan. There are velvet mites, grasshopers (some areas of Rajasthan deserts are the permanent breeding grounds of locust *Acridium perigrinum*), variety of beetles, butterflies and moths. A variety of spiders and scorpions too occur in Rajasthan deserts, 43 species of reptiles including snakes, tortoise, varanus, uromastix, skinks and crocodiles are found in the deserts of Rajasthan.

Avifauna of the state is perhaps the richest in the country. The world renowned Bird Sanctuary-**Keoladeo National Park** is located in Bharatpur, Rajasthan. Outside this park, there exists a rich avifauna in certain other regions of Rajasthan. Thus, thousands of Greater Flamingoes and Lesser Flamingoes visit Rann of Kutch, Cranes, Herones, Ducks, King fishers, Woodpeckers, Shrikes, Drongoes, Larks, Babblers, etc., are regular visitors of various ponds and lakes of Rajasthan such as Chhaparwada, Jalmahal, Ramgarh, Jaisamand, Ramsagar and Silisesh. Among other birds, the most important are the **Great Indian Bustards** (*Ardeotis nigriceps*) which are inhabitants of deserts. This bird has been declared as "**the State Bird**" of Rajasthan. Other significant birds of Rajasthan include Sand grouse, Quails, Parakeets, Ring doves, Pigeons, Bee eaters and Peafowls (*Pavo cristatus*), Vultures, etc. (Table 18.4).

Among mammals Rajasthan is famous all over the world for tigers. This desert state contains large populations of Black bucks (an antelope) which enjoys religious protection of *Bishnoi community* of Rajasthan. Various types of deer such as Chinkara, Neelgai, Sambhar, Chital, etc., roam freely in the protected forests of Rajasthan. Small population of highly endangered *wild ass* is found in Rann of Kutch. Gerbils (rodents), Hares, Rabbits, Hedgehogs, Porcupines, Wolves, Jackals, Hyenas, Wild Boars, Desert Camels, etc., are also found in this arid state.

Protected Areas of Rajasthan

Rajasthan state has three national parks, many sanctuaries, reserves and protected forests (Box 18.4).

1. Keoladeo National Park, Bharatpur. Earlier this park is known as **Bharatpur Bird Sanctuary** or *Ghana Bird Sanctuary*. In 1982, it was declared as a *Keoladeo National Park.* In 1985, this National Park was honoured by UNESCO with a title of "**World Heritage**". This worldrenowned park is considered one of the finest and richest **heronries** (Heronry is a place where herons breed). This park spreads around a shallow freshwater marsh surrounded by grasses, herbs, shrubs and mixed deciduous trees. It is full of Babul trees (*Acacia nilotica*). This national park is located at 370 meters above sea level and has total area of about 29 sq. km. The lakes of this park get their water from rains (average rainfall is 26 inches per year) and also through rivers *Gambhir* and *Banganga.*

Fig. 18.4. Some important wild animals of Rajasthan

Fig. 18.5. Some important wild animals of Rajasthan.

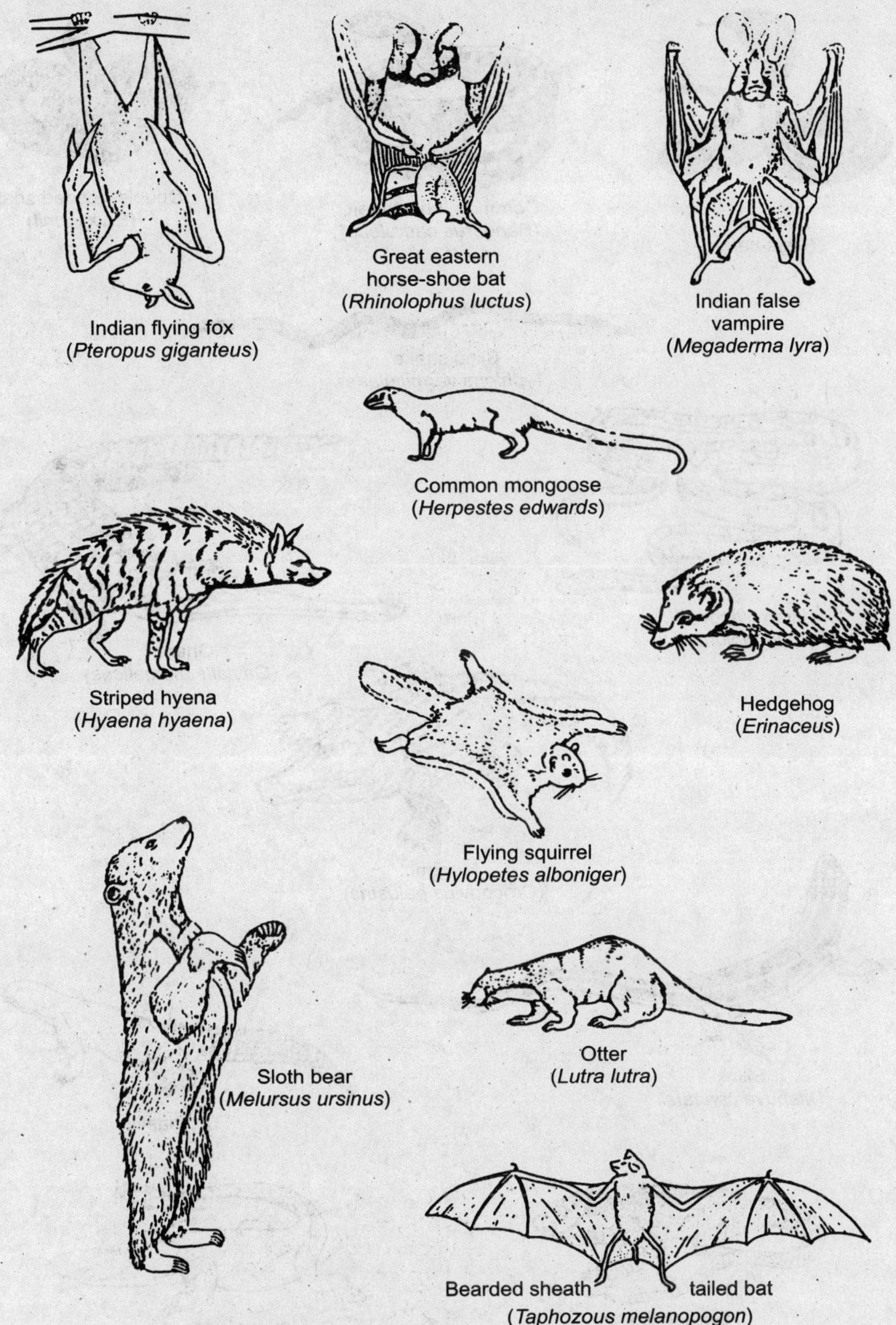

Fig. 18.6. Some important wild animals of Rajasthan.

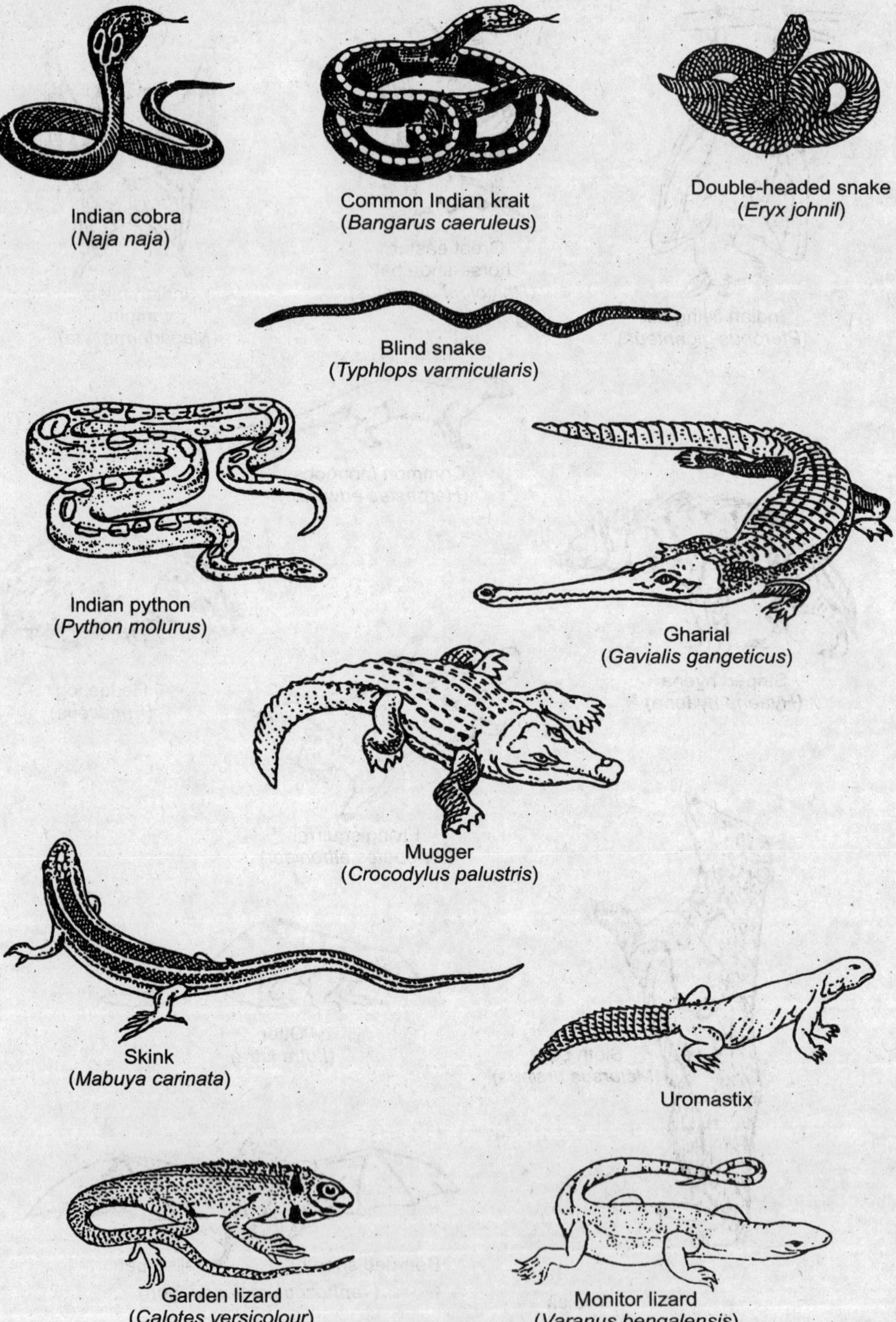

Fig. 18.7. Some important wild animals of Rajasthan.

Table 18.1. Some wild animals of Rajasthan deserts and adjoining areas.

Mammals

1. Tiger (*Panthera tigris*)
2. Hyena (*Hyaena hyaena humacus*)
3. Desert fox (*Vulpes vulpes pusill*)
4. Desert cat (*Felis silvestris ornata*)
5. Desert hare (*Lepas nigriollis*)
6. Desert rat (*Rattus gleadowi*)
7. Caracal (*Caracal caracal caracal*)
8. Flying squirell (*Hylopetes alboniger*)
9. Desert camel (*Camelus dromedarius*)
10. Black buck (*Antilope cervicapra*)
11. Chinkara (*Gazella bennettii*)
12. Cheetal/Chital (*Axis axis*)

Birds

1. Red crested pochard (*Netta rufina*)
2. Shovellar (*Spatula clypeata*)
3. Sand grouse (*Pterocles exustus*)
4. Ring dove (*Streptopelia decaocto*)
5. Pigeon (*Columba livia*)
6. Hubra bustard (*Chamydotis undulata*)
7. Imperial sand grouse (*Pterocles orientalis*)
8. Gray partridge (*Francolinus pondicerianus*)
9. Quail (*Turnix suscitator*)
10. Peafowl (*Pavo cristatus*)
11. Great Indian Bustard (*Choriotis nigriceps*)

Repitles

1. Python (*Python molurus*)
2. Mice-eating snake (*Gongylophis conicus*)
3. Earth snake (*Eryx johnii*)
4. Indian fringe toad (*Acantho dactylus*)
5. Sand lizard (*C. cantoris*)
6. Spiny tailed lizard (*Uromastix hardwickii*)

Some significant birds of Keoladeo National Park include strokes, egrets, spoonbills, cormorants, plovers, sandpipers, snipes, warblers, pipits, vultures, eagles and cranes. Pride of the park is the Siberian cranes.

Indepth studies of BNHS have reported the presence of following animals in this national park: pythons, hyenas, wildboars, black bucks, mongoose, wild cats, toddy cat (common palm civet), jackals, foxes, porcupines, hares, sambhar and neelgai.

2. Ranthambore National Park. This world famous tiger reserve is located admist the Aravali and Vindhya mountain range. This park contains a lake and is divided into a **core area** (285 sq. km) and a **buffer area** (107 sq. km) for better management. Besides the notable population of tigers, this park is rich in other predatory and prey animals: prey populations include sambhar, chital, chinkara, neelgai, Indian hare, wildboar, flying foxes, langurs; predator populations include leopards, jungle cats, caracal, sloth bears, jackals, hyenas, foxes, crocodiles, monitor lizards and snakes. Avian fauna of this national park includes storks, spoonbills, green pigeons, variety of ducks, jacana, fly catcher, quails, sandgrouse, partridges, owls and eagles.

3. Sariska National Park. This national park has an area approximately 800 sq. km. The park spreads into the forests of Aravali hills near Alwar. In 1958, this forest was declared as a **Sanctuary,** then it became a **Tiger Reserve** in 1979. The present status of **National Park** was given to it in the year 1982. The total area of Sariska National Park is also divided into two zones (1) **Core area** (480 sq./km) and (2) **Buffer area** (320 sq./km). Sariska is also famous for its rich tiger population. Other predators of this park include leopards, jungle cats, jackals, hyenas, and occasional caracals.

Among its prey population are the sambhar, chital, chinkara, neelgai and occasional four-horned antelopes. Sariska has a moderate population of wildboars, porcupines, bats, chamelions, skinks and snakes. Among birds, peafowls, partridges and other birds are most common.

4. Desert sanctuary. Large area (3210 sq./km) of Thar desert has varied flora and fauna, but only a small area (360 sq./km) of it was declared as sanctuary in 1980. Desert sanctuary is 30 km away from Jaisalmer.

Desert Sanctuary has moderate numbers of black bucks, chinkara, wolf, desert cat, desert fox, gerbils, long nosed hedgehogs, saw-scaled viper, sand lizards and varanus. Its avian fauna includes desert courser, great Indian bustard, lesser bustard, legger falcon, tawny eagle, short toed eagle, spotted eagle, sand grouse, sparrows and demoiselle cranes. Desert Sanctuary is very famous for its rich source of **fossils** of plants and aquatic animals.

There is a proposal to convert this area to a Biosphere reserve to conserve the biological diversity under the Man and Biosphere Programme of UNESCO.

Box 18.4

Some Protected Areas of Rajasthan and their fauna

Protected area	Fauna
1. **Shergarh sanctuary** (Declared sanctuary in 1982-83)	Sambhar, chital, jackal, foxe, hyena, snake, skink, chameleon, leopard.
2. **Ramgarh sanctuary** (Declared sanctuary in 1987)	Leopard, sloth bear, mongoose, wildcat, wildboar, neelgai, chinkara.
3. **Darrah sanctuary**	Neelgai, sambhar, chital, sloth bear, jackal, hyena, langur, hare, porcupine, hedgehog, crocodiles.
4. **National gavial sanctuary** (This is unique '*flowing sanctuary*' of chambal river)	Gharial, various fish population, water sanke, tortoise, insects.
5. **Jai samand sanctuary**	Crocodile, tortoises, fish, chital, chinkara, wildboar, birds
6. **Kumbhalgarh sanctuary**	Sloth bear, four horned antelope, wolf, leopard, wildboar, porcupine, neelgai, chital, chinkara, hyena, jackal.
7. **Sajjangarh sanctuary**	Sambhar, neelgai, chital, wildboar, langur, rhesus monkey.
8. **Sitamata sanctuary**	Flying squirrel, sloth bear, wild cat, wildboar, fox, jackal, sambhar, neelgai.
9. **Mount Abu sanctuary**	Sloth bear, wildboar, jackal, fox, langur, grey jungle fowl, water birds, crocodile, tortoise.
10. **Talchaper sanctuary**	Black bucks, varanus, uromastix, partridge, snake, large variety of arthropods.

11. Jamva Ramgarh sanctuary	Chital, sambhar, neelgai, hyena, wildboar, desert cat, jungle cat, toddy cat, jackal, fox, mongoose, porcupine,hedgehog, cobra, krait, viper, molluscs, arthropods. Occasionally are found leopard, caracal, wolf, varanus, skink, hare, spur fowl and bats. Its lake is rich in fish such as carps, catfishes, channa, crocodiles and tortoise.
12. Nahargarh sanctuary (Declared sanctuary in 1980)	Leopard, neelgai, jungle cat, wildboar, fox, desert cat, jackal, porcupine, hare, mongoose, varanus, snakes chameleon, skink and a variet of bats (*i.e.,* Indian flying fox, Great horse shoe bat, Indian false vampire, painted bat and Indian pipisterella).

18.6 ECONOMIC IMPORTANCE OF WILDLIFE

Animals and animal products have been of great use to human beings. Association of animals such as elephants, horses, cows, goats, buffaloes, dogs, cats, fish, pigs, poultry with human beings and use of their products such as meat, milk, fur, skin, feathers and bones have been as old as civilization itself. For economic benefit (*e.g.,* amusement or recreation), use of animals or animal products is acceptable to some extent for those animals which breed well and are found in large numbers. Wildlife trade involves the following trades of great economic importance.

1. Ivory trade. Ivory is regarded as **white gold** and has been the biggest trade among wildlife products. Invory is obtained from two upper incisor teeth of elephants (called tuskers). Ivory has been in great demand due to its whiteness and easy to carve the finest possible motifs on it. Ivory is used in jwellery, statues, tables, chairs and chess boards. To acquire an ivory item has become a status symbol. As a result human beings have become biggest threat to elephants. In nieteenth centuary, according to figures compiled by Natural History Museum in Paris, England alone imported annually about 550,000 tons of ivory. By 1900 between 60,000 and 70,000 elephants were slaughtered every year to satisfy the demands of the European market. To meet growing demand of market, elephants were massacred in China, Africa, Far East Asia and India. Indian ivory carving and handicrafts are famous all over the world (There are about 75,000 ivory craftman in India). Capturing and poaching of tuskers for ivory have taken a heavy toll in north-earthern and southern Indian. Once Periyar (Kerala) was famous for its tuskers, today not many of them are left. In 1972, Government of India banned the trade in wildlife products; today trade in ivory is controlled. There are registered wildlife product dealers who procure licence; the buyers of ivory goods have to carry photocopy of that licence (or permission from Government of India to persue this trade) to get clearance at major world airports.

Bones, teeth, horns of many wild animals form lucrative trade items. For example, teeth of whales and tusks of walrus are used as decorative pieces and for carving. Dik-diks are killed to get horns which are sold for use as buttons and as good-luck charms. (Dik-dik is a small North Eastern antelope of genera *Madoqua* and *Rhynchotragus*). Bone jewllery has been common among tribal people of various countries.

2. Decoration. Display of **animal trophies** and **stuffed** animals, specially, tigers, lions, cheetah, bear, leopards, foxes, pythons, elephants, etc., was a royal hobby. Palaces of kings of India and abroad still retain and display them. Leopards are killed, stuffed and displayed as decorative pieces in the drawing rooms of rich and wealthy persons. Hands and feets of bears and apes are used as decorative pieces and as ashtrays. On Isles des Saints in Gaudeloupe there were once many Iguanas (Iguana is an edible, tropical American lizard of Family Iguanidae, especially *Iguana iguana*, which sometimes attains a length of six feet; it is partly arboreal partly herbivorous and unless attacked, is harmless), but the high tourist demand for stuffed Iguanas as souvenirs has pushed these lizards to near extinction. In South America, the young crocodiles are killed, stuffed and are used for sitting. Feathers of peacocks are used to decorate many Indian houses. Polynesians use unusual tail feathers of colourful birds to decorate their hair. Antlers and horns of sambhar, cheetal, swamp deers and wild buffaloes are displayed in houses. People also make handles of cutlery, umbrellas and revolvers using bones and horns of rare animals to make them unique and expensive.

3. Medicinal property. Rhino horn is believed to cure headache, cold and increases sexual power. This is myth that the horn of the rhino is a powerful **aphrodisiac;** this legand is based on the phallic shape of the horn and also on the fact that mating act in rhino lasts for 1 to 2 hours. The rhino horn was also used as a drinking glass by royals because it was believed that it would split if poison was added to drink.

People in South America, far east Asia and South India eat brain and meat of rare breeds of monkeys with an assumption that it cures insane people and give vitality. The tiger bones are also believed to carry aphrodisiac property if eaten in powdered form. Tiger nails worn around the neck are belived to give boundless courage to the bearer (Box 18.5).

Box 18.5

In the world, there is great demand of parts of tigers specially in China. The Chinese system of medicine uses all the dried and powdered parts of body of tiger:

	Organ of tiger	Usage
1.	Nose	For epilepsy and convulsions
2.	Teeth	For rabies, asthma and sore in penis
3.	Whiskers	For toothaches
4.	Blood	For strong body and mind
5.	Gall bladder	For healthy eyes
6.	Stomach	For stomach aches and better digestion
7.	Brain	For laziness
8.	Eyeballs	For epilepsy, nervousness, convulsions
9.	Testes	For tuberculosis
10.	Flesh	For nausea, malaria, vitality
11.	Fat	For vomitting
12.	Tail	For skin diseases
13.	Skin	For mental illness
14.	Hair	For driving away of pests

Rhesus monkey, chimpanzee and other non-human primates are extensively used as test-animals in the field of biomedical research such as vaccine, contraception, cosmetics, space research and in biological warfare.

In south India, peacocks are killed by local people to eat meat and to take out oil for some medicinal value. Tiger fat, Rhino's urine, uromastix oil are used as cure for rheumatoid arthritis, gout, purification of blood and for increasing sexual powers, respectively.

4. Perfume industry. This industry, thrives on scent collected from animals such as musk deer, civet and many other such animals. To get just a few milligrams of scent, glands are extracted from hundreds of animals.

The musk is used as an important ingradient in cologne soaps, hair, tonics, body lotions and other cosmetics. The musk deer is heavily killed for the musk pod.

5. Dress material. Skins and furs of wild animals such as mink, panther, fox, sambhar, jungle cats and reptiles are important items to make coats, shoes, bags, gloves, belts, watch straps and caps from animals. The reptilian skins and furs are the symbols of status among the ladies of high aristocratic societies. Asian crocodiles are killed for skin to make shoes and purses. The reptilian skins remain in demand because of the natural designs and patterns over them which can be tanned in different colours.

Box 18.6 Pathetic Tale of Skinning

Some animals are killed in a particularly horrifying way, *e.g.*, when a leopard is trapped, it is tied, and then pressed by a plate from top or dorsal side so that it spreads on all four limbs called **eagle spread**. In next step, a red hot iron bar is inserted into the leopard's anus and pushed recklessly here and there to destory some vital organs causing death of the animal but leaving its fur uncut intact and shinning for the buyer to get a dress stitched out of it.

Asian crocodiles are skinned alive because skin comes off more easily without stretching and cracking (Mathur 1998).

6. Pet and amusements. India has remained as one of the major bird exporting countries in the world. In India, birds are trapped primarily in the northern states of Bihar and Uttar Pradesh and also from Orissa, Madhya Pradesh, Maharashtra and Assam. The birds are caught and killed for meat and feathers to make fancy sleeping bags, pillows, jackets, greeting cards and decorative pieces. However, the best trade is export of live birds. The total number of birds exported from India between 1970-76 was nearly 13 million, an average of about 1,850,000 birds per year. Chief importer countries include USA, UK, Japan, Italy, France and Belgium. Of 1,160 species of birds, 289 species were exported. Their value ranged from few to hundred dollars. For example, ring necked parakeet (*Psittacula krameri*) and Sarus crane (*Grus antigone*) were sold in U.K. for $ 27.00 and $ 700.00 respectively.

Initially there was no check on export and trade of birds, but in 1976 CITES imposed ban on the export of eight rare and endangered species such as Hill Mynaha (*Gracula religiosa, G. intermedia, G. peninsularia, G. indica*), Green Avadavat (*Amandava formosa*), Finn's Baya Weaver (*Pioceus megarhynchus*). Hill blue flycather (*Cyornis banyumas*) and Malabar parakeet (*Psittacula columboider*).

Export of live and rare animals for the purpose of display in the zoo and circus too are a common man's window to the animal's world. It involves handling and capturing of wild animals and their transportation over long distances, during which there are chances of their starvation, suffocation and death. Keeping a **pet** is common habit of humans. Parrots, parakeets, koalas, snakes, lizards, apes and monkeys are some of the pets owned by many persons. **Hunting** is also a source of amusement for human beings. Rare animals such as deer, partridge, peacock, bear, shark, octopus, turtle and fur bearing animals are killed for eating their meat and are considered a delicacy. Sea turtle shells are used to make jewellery and wall hangings. Whales provide many products: their oil is used for cosmetics, fat is used in soap, skin is used for making leather goods and their meat is eaten.

QUESTIONS

Long Answer Questions

1. Define the term wildlife and write an essay on the present status of wildlife in India.
2. Write an essay on wildlife of India and its management.

3. Describe the Indian endangered fauna from amphibia to mammals. Discuss their protective measures.
4. What is the criteria of IUCN to decide the conservation status of a species? Discuss.
5. How can we estimate that particular wildlife populations are declining? How declining wildlife populations can be conserved.
6. What is an endangered species? How are they conserved?
7. Name the few endangered species? Give a brief account of project tiger.
8. Give an account of endangered species of reptiles and mammals and add a note on the causes of their being endangered.
9. Explain the concept of "Threatened species". Cite four examples of Indian Endangered species. What is endangered species? Cite two examples.
10. Write an essay on the endangered animals of India. What reasons would you assign for their decline.
11. Discuss the causes of extinction of wildlife.
12. Give reasons for the depletion of wildlife in India and different methods for its conservation.
13. Give reasons for wildlife depletion and necessity of its conservation.
14. What is wildlife? What are its values for mankind? What are the various methods for conservation of wildlife.
15. What is wildlife? Describe various factors that necessitates its conservation.
16. Describe the factors influencing carrying capacity of the wildlife habitat.

Short Answer Questions

1. Write short notes on the following:
   (*a*) Endangered species
   (*b*) Economic importance of wildlife
   (*c*) Threatened species
   (*d*) Wildlife resources
   (*e*) Wild life

Very short Answer Questions

1. Give two examples of endangered species of each of mammals, bird and reptiles.
2. Name the Government bodies engaged in the publication of Endangered Flora and Fauna of India.
3. What are WWF and IUCN?
4. What is the most appropriate term used for the following descriptions?
   (*i*) Species in danger of being eliminated
   (*ii*) Species no longer exist.
   (*iii*) Species likely to become endangered species.
5. Explain RDB.

Fill in the Blanks

1. National gavial sanctuary is a .... of Chambal river in India.
2. Nilgiri tahr is found in .... state.
3. Lesser one horned rhinoceros is an .... animal of India.
4. DD taxon stands for ....
5. Stae bird of Rajasthan is ....

Multiple Choice Questions

1. Which of the following animal has become extinct from India? (*CCSU Meerut 2014*)

(*a*) snow leopard (*b*) hippopotamus

(*c*) wolf (*d*) cheetah

2. Source of ivory is? (*CCSU Meerut 2014*)

(*a*) tusk of elephant (*b*) bones of whale

(*c*) antlers of deer (*d*) bones of elephant

3. From the biological point of view wildlife plays an important role in the proper maintenance of (*CCSU Meerut 2014*)

(*a*) biogeochemical cycles (*b*) food chain

(*c*) balance between population (*d*) all of the above

4. Scientic name of Royal Bengal Tiger is (*CCSU Meerut 2014*)

(*a*) *Panthera tigris balica* (*b*) *Panthera tigris tigris*

(*c*) *Panthera tigris virgata* (*d*) *Panthera tigris amoyenis*

5. Major cause of wildlife extinction is (*CCSU Meerut 2014*)

(*a*) introduction of exotic species

(*b*) destruction of natural habitat

(*c*) excessive hunting and killing of wild animals

(*d*) all of the above

6. Those species whose population is depleted seriously and whose security is not assured (*CCSU Meerut 2014*)

(*a*) threatened species (*b*) endangered species

(*c*) vulnerable species (*d*) rare species

7. Wild-life week is observed every year from (*CCSU Meerut 2014*)

(*a*) 1st March to 8th March (*b*) 1st July to 8th July

(*c*) 1st October to 8th October (*d*) none of the above

8. Dodo an extinct flightless bird belonged to (*CCSU Meerut 2014*)

(*a*) Mauritus (*b*) Lakshadweep

(*c*) Canada (*d*) Ireland

9. Which of the following animal has become extinct in India?

(*a*) asiatic lion (*b*) cheetah

(*c*) snow leopard (*d*) rhinoceros

10. Red data book is famous for

(*a*) extinct animals (*b*) endangered animals

(*c*) extinct animals and plants (*d*) important wildlife

11. The great Indian bustard is a

(*a*) rare species (*b*) extinct species

(*c*) endangered species (*d*) vulnerable species

12. Which of the following is endangered species in India?

(*a*) asiatic lion (*b*) spotted deer

(*c*) black buck (*d*) musk deer

13. Red data books are produced by

(*a*) IUCN (*b*) WWF

(*c*) IBWL (*d*) none

14. What is the major cause of diminishing wildlife numbers?
(*a*) cannibalism (*b*) felling of trees
(*c*) habitat destruction (*d*) paucity of drinking water

15. The species become extinct most easily by
(*a*) urbanization (*b*) heavy rain
(*c*) deforestation (*d*) sliding of hills

16. Which of the following has became extinct in India?
(*a*) jerdon's courser (*b*) pink headed duck
(*c*) mountain quail (*d*) all

17. Which of the following include endangered species in India?
(*a*) lion tailed macaque and blue rhino (*b*) crocodile and musk deer
(*c*) musk deer and rhino (*d*) all

18. The species loses in tropical countries is mainly due to
(*a*) pollution (*b*) soil erosion
(*c*) deforestation (*d*) urbanization

19. Who gave the concept that all states of our country should have a state animal and a state plant?
(*a*) M.S. Swaminathan (*b*) Menka Gandhi
(*c*) S.Z. Qasim (*d*) T.N. Khoshoo

ANSWERS

Very short Answer Questions

1. (*i*) Mammals: Tiger and Black buck.
(*ii*) Birds: Great Indian bustard and Siberian crane.
(*iii*) Reptiles, Gharial and Tortoise.

2. The two Government Bodies are:
(*i*) Botanical Survey of India (BSI) which publishes the National Flora of the India.
(*ii*) Zoological Survey of India (ZSI) which publish the fauna of India.

3. **WWF=** It stands for World Wildlife Fund, now called World Wide Fund for Nature (WWFON). It was formed in 1961, with its Headquarters at Glands, Switzerland. It is a NGO.

In India, WWF was found with Board of eight trustees and has its Head Quarters at Bombay .

IUCN= It stands for International Union for Conservation of National Resources. Its function is to provide upto date information about every species of animal in danger of extinction. It is an international body and India is an active participant in it.

4. (*i*) Endangered species; (*ii*) Extinct species; (*iii*) Threatened species.

5. RDB stands for Red Data Book. It is also called IUCN, Red Data List of Threatened Species, 2012.

Fill in the Blanks

1. Flowing sanctuary **2.** Tamil Nadu **3.** Extinct **4.** Data Deficient
5. Great Indian Bustard.

Multiple Choice questions.

1. (*d*)	**2.** (*a*)	**3.** (*d*)	**4.** (*b*)
5. (*d*)	**6.** (*b*)	**7.** (*b*)	**8.** (*a*)
9. (*b*)	**10.** (*c*)	**11.** (*c*)	**12.** (*a*)
13. (*c*)	**14.** (*c*)	**15.** (*c*)	**16.** (*d*)
17. (*d*)	**18.** (*c*)	**19.** (*d*).	

19

Sanctuaries and National Parks of India

Conservation (L.*con.* = together + *server* = guard) means planning and management of resources in a way so as to secure their wider and continuous supply to present generations while maintaining its potential to meet the needs and aspirations of future generations. The conservation of wildlife is required for the following benefits; 1. The wildlife helps us in maintaining the 'balance of nature.' Once this equilibrium is disturbed it leads to many problems. The destruction of carnivores or insectivores often leads to the increase in herbivores which in turn affects the forest vegetation or crops. 2. The wildlife can be used commercially to earn more and more money. It can increase our earning of foreign exchange, if linked with tourism. 3. The preservation of wildlife helps many naturalists and behaviour biologists to study morphology, anatomy, physiology, ecology, behaviour biology of the wild animals under their natural surroundings. 4. The wildlife provides best means of sports and recreation. 5. The wildlife of India is our cultural asset and has deep-rooted effect on Indian art, sculpture, literature and religion.

The term **wildlife** is commonly referred to the non-domesticated animals found in a natural habitat. But in its widest sense it includes all flora and fauna of the natural habitat. For the protection and conservation of wildlife following global efforts have been done:

19.1 IN-SITU CONSERVATION

In-situ conservation deals with conservation of living resources through their maintenance within the natural ecosystems in which they occur. This approach is preferable on account of the fact that not much diversity can be conserved outside the centres of biodiversity. It is true that one may be able to conserve and introduce some measure of diversity, but not the original environment, including pollinators, the peculiar plant-animal-microbe complexes, etc., where the species actually originated and have lived through many generations. The in-situ conservation includes an extensive system of **protected areas** such as the *National Parks, Sanctuaries, Nature Reserves, Natural Monuments, Cultural Landscapes, Biosphere Reserves,* etc. These protected areas vary remarkably in size, design, aim and effectiveness of management but indeed they serve as a repository for much of the world's biological diversity.

Protected areas. In India, the network of protected areas consisting of 19 National Parks and 202 sanctuaries extending over 74,763 km^2 in 1981 and had increased by 1983 to 44 National Parks and 207 sanctuaries covering 88,000 sq. km. Up to 1997, India possessed 441 wildlife sanctuaries and 80 National Parks, with the total protected area notching 4.5 per cent of the total land mass. The protected areas also include 20 tiger reserves and 14 proposed biosphere reserves.

This network of protected areas covers most of the representative habitat types in the country and affords protection both to wild species of flora and fauna. National Parks and sanctuaries have been set up within the legal framework of Wildlife (Protection) Act 1972, thus affording to these areas the greatly required legal support.

National Parks and Wildlife Sanctuaries

A **national park** is an area dedicated to conserve the scenery (or environment) and natural objects and wildlife therein. In national parks, all private rights are non-existant and all forestry operations and other usages such as grazing of domestic animals are prohibited. However, the general public may enter it for the purpose of observation and study. Certain parts of the national parks are developed for tourism in such a way that enjoyment will not disturb or scare the animals (Box 19.1).

Box 19.1 Definition of National Park

IUCN (1975) has adopted the following definition of a national park:

A national park is a relatively large area (*i*) where one or several ecosystems are not materially altered by human exploitation and occupation, where plant and animal species, geomorphological sites and habitats are of special scientific, educative and recreative interest or which contains a natural landscape of great beauty and (*ii*) where the highest competent authority of the country has taken steps to prevent or eliminate as soon as possible exploitation or occupation in the whole area and to enforce effectively the respect of ecological, geomorphological or aesthetic features which have led to its establishment, and (*iii*) where visitors are allowed to enter, under special conditions, for inspirational, cultural and recreative purposes.

A **wildlife sanctuary,** similar to national park, is dedicated to protect the wildlife, but it considers the conservation of species only and also the boundary of it is not limited by state legislation. In the sanctuary, killing, hunting or capturing of any species of birds and animals is prohibited except by or under the control of highest authority in the department responsible for management of the sanctuary. Private ownership may be allowed to continue in a sanctuary, and forestry and other usages permitted to the extent that they do not adversely affect wildlife.

Some of the well-known wildlife sanctuaries and national parks of India are as follows (Table 19.1):

1. Kaziranga wildlife sanctuary. It was established in 1926 in district Sibsagar, subdivision Jorhat (Assam) on the south bank of the Brahmputra river. It consists of 430 sq km of forest grasslands and swamps and supports a fauna of 700 rhinoceros, besides elephant, wild buffalo, bison, tiger, leopard, sloth bear, sambhar, swamp deer, hog deer, barking deer, wild boar, gibbon and birds such as pelican, stork and ring-tailed fishing eagles.

2. Manas wildlife sanctuary. It is located in district Kamrup in Assam, has an area of 540 sq km and is situated at an altitude of 80 metres. River Manas passes through it. It contains the following wild animals: tiger, panther, wild dog, wild boar, rhinoceros, gaur, wild buffalo, sambhar, swamp deer, hispid hare and golden langur.

3. Jaldapara wildlife sanctuary. It is situated in Jalpaiguri district of West Bengal and is a 65 sq km stretch of grassland. Its wildlife fauna includes animals such as rhinoceros, gaur, elephant, tiger, leopard, deer, and a variety of birds and reptiles.

4. Palamau national park. It is situated in Daltongunj district in Bihar and has an area of 345 sq km. Its flora is thick tropical forests. The fauna of this park includes tiger, panther, sloth bear, elephant, chital, gaur, nilgai, chinkara, chowsingha and mouse-deer.

5. Hazaribagh national park. This national park was established in 1954 in Bihar. It has an area of 184 sq km of thick tropical forests. The typical fauna of this park includes wild boar, sambhar, nilgai, tiger, leopard, sloth bear, hyena and gaur.

6. Simlipal national park. It is situated in district Mayur Bhanj in Orissa and has an area of 2750 sq. km. It is covered over by dense sal forests and is chosen for project tiger. Its typical fauna includes tiger, elephant, deer, pea fowl, talking mynas, chital, sambhar, panther, gaur, hyena and sloth bear.

7. Chilka lake Bird Sanctuary. This largest inland lake has an area of 1000 sq km and is about 100 km from Bhubaneshwar (Orissa). Its typical fauna includes water fowls, ducks, cranes, ospreys, golden plovers, sandpipers, stone curlews, flamingoes, etc.

8. Kolameru bird sanctuary. This is a small bird sanctuary near Tadepallegudam in Andhra State. It is a breeding place for pelicans and many marine birds visit this place.

9. Vedanthangal bird sanctuary. This small-sized but very old sanctuary extend over a lake of about 0.30 sq km, 85 km south of Madras. Many migratory birds regularly visit this temporary lake. The regular visitors of this lake are birds such as spoon bill, open billed stork, egrets, ibis, cormorant, darter, grey heron, pelican, snipes and dab chick.

10. Guindy deer park. It is situated near Madras city and has mainly chitals and black bucks. A few albinos of black buck are also found here.

11. Point calimera wildlife sanctuary. This is situated at the southern tip of the Thanjavur district of Tamil Nadu abutting the Palk strait. Its backwater and lagoon is visited by flamingoes and pelicans. Its nearby vedaranyam forests has a fauna of numerous black bucks, chitals and wild boars.

12. Mundanthurai sanctuary. It was established in 1962 in Tirunelveli district of Tamil Nadu and has an area of 520 sq km. It has evergreen forests and Tamaraparani river flows through this sanctuary. Its typical fauna includes panther, tiger, sambhar and chital.

13. Periyar wildlife sanctuary. This is situated in Kerala State and has an area of 777 sq km. It was established in 1940 around the artificial lake which arose behind the dam built across the Periyar river is 1900. This sanctuary of great scenic beauty supports a fauna of wild elephants, gaurs, leopards, sloth bear, sambhars, barking deer, wild dogs, wild boars, black Nilgiri langurs and water birds such as grey hornbills, egrets.

14. Mudumalai wildlife sanctuary. This wildlife sanctuary was established in 1940 in North Western part of Nilgiris in Tamil Nadu. It is known for its rich forests and diversity of fauna that includes wild elephant, gaur, sambhar, chital, barking deer, mouse-deer, four-horned antelope, tiger, panther, bonnet monkey, langur, giant squirrel, flying squirrel, wild dog, jackal, wild cat, sloth bear, porcupine, pangolin, flying-lizard, monitor-lizard, rat snake, python and various birds.

15. Ranganthittoo bird sanctuary. It covers an area of 166 sq km and includes a series of islands in the Cauvery river 15 km off the Banglore-Mysore Road, near Shrirangapattnam. Its avial fauna includes open bill stork, white ibis, egret, spoon bill, wild duck, peafowl, night heron birds.

16. Bandipur wildlife sanctuary. This sanctuary was established in 1941 by the then ruler of Mysore (Karnataka State). It is situated 80 km south of Mysore city enroute to Otacamund. It has an area of 874 sq km and is at an altitude of 1454.4 metres. Its forest is very thick and has plenty of rainfall. Its wildlife fauna includes plenty of gaurs and animals such as elephant, leopard, sloth bear, wild dog, chital, panther, barking deer, porcupine and langur.

17. Cotigao wildlife sanctuary. It is located in South Goa and has an area of 105 sq km. It has wet evergreen forest of Elve and Berla trees and supports rich fauna of the following animals: gaur, sambhar, chital, hog-deer, barking deer, wild boar, hyaena, panther, leopard, jackal, otter, porcupine and birds such as parakeet, lorikeet, woodpecker, kingfisher, bulbul, jungle fowl, egret, etc.

18. Bhagwan Mahadev wildlife sanctuary. It is located in North Goa and has an area of 240 sq km and supports fauna almost similar to Cotigao wildlife sanctuary.

19. Sesar Gir. This famous wildlife sanctuary for the Asiatic lion is situated in Gujarat state, 468 km from Ahmedabad and 43 km from Veraval. It has an area of 1295 sq km of semiarid country with patches of thorn scrub and deciduous trees. Its fauna includes Asiatic lion, spotted deer, blue bull (nilgai), four-horned antelope, chinkara, striped hyena, wild boar, porcupine, langur, python, crocodiles and birds such as green pigeon, partridge, rock-grouse, etc.

20. Kanha national park. This national park was established in 1955 in former Banjar Valley Reserve (Madhya Pradesh). This park has an area of 939.94 sq km and includes hilly terrains and streams. It is 175 km away from Jabalpur and has forests of sal trees. Its typical fauna includes animals such as tiger, chital, panther, sambhar, black buck and bara singha.

21. Tandoba national park. It is located in Chandrapur district (Maharastra) and has an area of 116 sq km. Its typical fauna includes tiger, sambhar, sloth bear, bison, chital, chinkara, barking deer, blue bull, four horned deer, langur, peafowl and few crocodiles.

22. Sariska. This is one of the most beautiful wildlife sanctuary of Rajasthan near Alwar. It has an area of 800 sq km and has dense Dohokra and Solar forests. Its typical fauna includes tiger, leopard, spotted deer, jungle cat, four-horned antelope, langur, porcupine, hedgehog, peafowl, etc.

23. Bharatpur bird sanctuary (Koeldeo Ghana). It is located at Bharatpur (Rajasthan), has an area of 29 sq km and harbour all kinds of indigenous nesting water birds, water side birds and migratory birds. More than 328 varieties of birds including cormorants, spoonbills, white ibis, Indian darters, egrets, painted storks, open billed storks, great black necked storks, etc. Many migratory birds such as ducks, geese, siberian cranes, etc., regularly visit this sanctuary. Drier parts of this marshy sanctuary have animals such as spotted deer, black buck, sambhar, blue bull, wild boar and python.

24. Sultanpur lake bird sanctuary. This small sized (2 sq km area) bird sanctuary is located in Gurgaon district (Haryana), about 30 km from Delhi. Its typical avian fauna includes crane, sarus, spot-bill, rudyshel and drake.

25. Shikari Devi wildlife sanctuary. It is located in Mandi district in Himachal Pradesh. It has an area of 213 sq km and has the following animals : snow leopard, flying fox, black bear, barking deer, musk deer, chakor, partridge, etc.

26. Bir Motibagh sanctuary. It is located near Patiala in Punjab. The typical fauna of this wildlife sanctuary includes black buck, blue bull, hog deer, hare, jackal and birds such as peafowl, partridge, sparrow, babbler, myna, parakeet, pigeon, dove, etc.

27. Dachigam wildlife sanctuary. It was established in 1951 in Kashmir 26 km away from Srinagar. It has an area of 89 sq km and has two levels : upper Dachigam at 3692.3 metres altitudes and Lower Dachigam at 1846.2 metres altitudes. It mainly preserves hangul or Kashmir stag, but also has animals such as musk-deer, leopard, black bear, brown bear and baboon.

28. Corbett national park. It is one of India's most famous wildlife sanctuary and was constituted in 1935 as the *first national park* of India. It is situated between Nainital and Garhwal districts in Uttar pradesh. It has an area of 525 sq km and is located within west-to-south bend of river Ramganga. It supports a rich and diverse fauna of the following animals : tiger, panther, sloth bear, hyaena, elephant, blue bull, swamp deer, barking deer, Indian antelope, porcupine, birds such as bulbul, woodpecker, barbet, babbler and bee eater, and reptiles such as crocodile, python, etc.

29. Shivpuri sanctuary. It is located in Madhya Pradesh and is an asylum for tigers.

30. Anamalai sanctuary. This sanctuary was established in 1972 in the southern part of Coimbatore District in Tamil Nadu. It has a vast area of 958 sq km and supports rich fauna of animals such as elephant, gaur, sambhar, spotted deer, barking deer, Nilgiri tahr, lion-tailed macaque, tiger, panther, sloth bear, langur, porcupine and pangolin.

Table 19.1. List of National Parks/Sanctuaries of India.

	Name of National Park/Sanctuary	District	Wildlife
	1	2	3
		Andhra Pradesh	
1.	Pakhal Wildlife Sanctuary	Warangal	Tiger, panther, sambhar, chital, nilgai
2.	Eturnagaram	Warangal	Tiger, panther, gaur, jungle cat, sambhar, black buck
3.	Pocharam Wildlife Sanctuary	Medak	Panther, chital, chinkara, pea fowl, water birds
4.	Kawal Wildlife Sanctuary	Adilabad	Tiger, panther, sloth bear, wild boar, gaur, sambhar, chital, black buck
5.	Kolleru Pelicanary	Elleru	Pelican, flamingo, heron, painted stork, avocet, teal, tern
6.	Neelayattu Bird Sanctuary	Nellore	Gray pelican, heron, duck, cormorant, teal
7.	Kinnersan Wildlife Sanctuary	Khaman	Tiger, panther, wolf, sloth bear, gaur, chital, sambhar, nilgai
8.	Lanjamadagu (Sivaram) Sanctuary	Adilabad and Karimnagar	Tiger, panther, sloth bear, wild boar, chital, sambhar, nilgai, langur
9.	Manjiri Sanctuary	Medak	Marsh crocodile and water birds
10.	Papikonda Sanctuary	East Godavari, west Godavari and Khamam	Tiger, panther, wild dog, hyaena, jackal, wolf, gaur, fourhorned antelope, chital, sambhar, nilgai
11.	Corianga Sanctuary	East Godavari	Estuarine crocodile, otter, fishing cat, jackal, sea gull, pelican, stork, heron, flamingo
12.	Nagarjun Sagar Srisailam Sanctuary	Guntur, Prakasam, Kurunool, Mahabubnagar and Nalgonda	Tiger, panther, sloth bear, wild boar, chital, sambhar, nilgai, black buck, jackal, fox, wolf, and muger (crocodile)
13.	Pulicat Sanctuary	Nillore District	Flamingo, pelican, duck, teal, stork, crane, bear heron
14.	Pranahita Sanctuary	Adilabad	Black buck, tiger
		Andaman and Nicobar	
15.	Narchondam Island Wildlife Sanctuary	—	—
16.	North Reef Island Wildlife Sanctuary	—	—
17.	Baren Island Wildlife Sanctuary	—	—
18.	South Sentinal Island Wildlife Sanctuary	—	—
		Arunachal Pradesh	
19.	Namidapha Wildlife Sanctuary (Updated as Project Tiger)	Tirap	Tiger, panther, snow leopard, clouded leopard, golden cat, binturong, wild dog, wild buffalo, gaur, swamp deer, hog deer, barking deer, elephant, malabar squirrel, sloth bear, python, king cobra, Indian cobra, monitor lizard, birds of many kinds

	1	2	3
20.	Lali Wildlife Sanctuary	Siang	Wild buffalo, tiger, panther, hog deer, elephant, python, barking deer, sambhar, water birds
21.	Itanagar Wildlife Sanctuary	Subansiri	Elephant, gaur, tiger, panther, sambhar, hog deer, barking deer, sloth bear
22.	Pakkui Wildlife Sanctuary	Nawgong	Elephant, gaur, sambhar, hog deer, barking deer, python
		Assam	
23.	Garampani Wildlife Sanctuary	Diphu	Elephant, wild buffalo, leopard, common langur, hoolock
24.	Kaziranga National Park (Updated as Project Tiger)	Jorhat	Rhinoceros, wild buffalo, elephant, gaur, tiger, leopard, cat, wild boar, civet cat, otter, swamp deer, hog deer, sambhar, tiger, python, pelican, partridge, florican
25.	Lakhawa Wildlife Sanctuary	Nawgong	Rhinoceros, wild buffalo, swamp deer, hog deer, water duck, cormorant
26.	Manas Wildlife Sanctuary (Updated as Project Tiger)	Barpeta	Elephant, tiger, panther, gaur, wild buffalo, rhinoceros, golden langur, civet cat, otter, swamp deer, hog deer, oog deer, sambhar, pigmy hog, water monitor, wild boar, great pied horn-bill, florien, hispid hare
27.	Orang Wildlife Sanctuary	Tezpur	Rhinoceros, elephant, leopard, sambhar, barking deer, swamp deer, tiger, water birds, green pigeon, florican, quail, whistling teal, geese, hawks
28.	Poha Wildlife Sanctuary	North Lakhim-pur	Wild buffalo, barking deer
29.	Sonal-Rupa Wildife Sanctuary	Tezpur	Rhinoceros, elephant, wild buffalo, sloth bear, wild dog, wild boar, sambhar, barking deer, swamp deer, leopard, tiger, hornbill, imperial pigeon
		Bihar	
30.	Hazaribagh National Park	Hazaribagh	Tiger, leopard, sambhar, chital, nilgai, wild boar, wild cat, pea fowl
31.	Bhimbandh Wildlife Sanctuary	Monghyra	Tiger, leopard, sambhar, chital, wild boar, wolf, water birds
32.	Mahudaur Wildlife Sancturary	Daltongunj	Tiger, leopard, wolf, chital, barking deer, wild boar
33.	Dalma Wildlife Sanctuary	Singhbhum	Elephant, leopard, wild boar, mouse deer, bark-ing deer, sloth bear
34.	Palamau (Bihar) Wildlife Sanctuary	Daltongunj	Elephant, panther, leopard, wild boar, wild dog, barking deer, gaur, chital, sambhar, pea fowl
35.	Gautam Budha Wildlife Sanctuary	Gaya	Tiger, leopard, sambhar, chital, barking deer, pea fowl,
36.	Rajgir Wildlife Sanctuary	Patna	-do-
37.	Topchanahi Wildlife Sanctuary	Dhanbad	Variety of ducks
38.	Lawalang Wildlife Sanctuary	Hazaribagh	Tiger, leopard, sloth bear, sambhar, chital
39.	Lanaul, Madanpur and Udaipur Wildlife Sanctuary	Champaran	Tiger, leopard, sambhar, chital, wild boar, nilgai, wolf
40.	Kaimur Wildlife Sanctuary	Rohtas	Tiger, leopard, chinkara, sambhar, nilgai and crocodile

	1	2	3
41.	Bamiaburu Wildlife Sanctuary	Singhbhum	Tiger, panther, sambhar, elephant, wild boar
42.	Tabo Wildlife Sanctuary	Singhbhum	-do-
43.	Kodarama Wildlife Sanctuary	Hazaribagh	Tiger, leopard, sloth bear, sambhar, chital, four-horned antelope
44.	Valmiki Nagar Wildlife Sanctuary	—	—
		Goa, Daman and Diu	
45.	Mollen Wildlife Sanctuary	Goa	Gaur, sambhar, mouse deer, barking deer, panther, flying squirrel, porcupine, civet cats, ant-eater, slender loris, grey-jungle fowl
46.	Bondla Wildlife Sanctuary	Goa	-do-
47.	Cotigoa Wildlife Sanctuary	Goa	-do-
		Gujarat	
48.	Gir National Park	Junagarh	Asiatic lion, panther, striped hyaena, sambhar, nilgai, chital, four-horned antelope, chinkara, wild boar, crocodile
49.	Velavadar National Park	Bhavnagar	Black buck, wolf
50.	Nal Sarovar Bird Sanctuary	Ahmedabad	Water birds
51.	Wild Ass Sanctuary	Surendra	Wild ass, nilgai, wolf, chinkara
52.	Jessore Sloth Bear Sanctuary	Banaskantha	—
53.	Vansada National Park	Valsad	—
54.	Marine Sanctuary	Gulf of Kutch	—
		Haryana	
55.	Sultanpur Lake Birds Sanctuary	Gurgaon	Sarus crane, spot bill, ruddy shell duck
		Jammu – Kashmir	
56.	Dachigam Wildlife Sanctuary	Srinagar	Leopard, black bear, brown bear, serow (large goat antelope), musk deer, hangul
57.	Raj Parian Wildlife Sanctuary	Anantnag	Brown bear, black bear, musk deer, serow, hangul, chakor
58.	Shankarachari Wildlife Sanctuary	Srinagar	Chakor
59.	Chumnai Basin Wildlife Sanctuary	Anantnag	Brown bear, black bear, leopard, musk deer
		Himachal Pradesh	
60.	Sechu-tun-Nallah Sanctuary	Chamba	Ibex, tahr, musk deer, brown bear, black bear, goral, serow, monal, tragopan, snow cock, kalij, koklass, chakor, snow leopard, lynx, marten, fox, panther, marmots, weasels
61.	Gamgul-Siya Behii	Chamba	Tahr, brown bear, musk deer, goral, kashmir stag, koklass, chakor, black bear, monal, tragopan, snow cock, kaliy; panther, weasel, marten, civet, fox, flying fox
62.	Tundah	Chamba	Tahr, brown bear, black bear, musk deer, goral, serow, ibex, monal, tragopan, snow cock, kalij, chakor and chir, snow leopard, panther, weasel, marten, civet, fox, flying fox
63.	Kugti	Chamba	Tahr, brown bear, black bear, musk deer, goral, serow, ibex, monal, tragopan, snow cock, kalij, koklass, chakor, panther, snow leopard, marten, civet, fox, flying fox

	1	2	3
64.	Kelaton and Camo	Chamba	Goral, barking deer, serow, black bear, koklass, kalij, chir, panther, snow leopard, marten, civet, fox, flying fox
65.	Nergu and Winch Mandi Camo	Mandi	Goral, black bear, serow, musk deer, monal, kalij, koklass and chir, panther, snow leopard, marten, civet, fox, flying fox
66.	Shikari Devi	Mandi	Balck bear, goral, barking deer, serow, musk deer, monal, kalij, chir, chakor, partridge, panther, snow leopard, marten, civet, fox, flying fox
67.	Bandli	Mandi	Goral, barking deer, balck bear, kalij, chakor, partridge, panther, snow leopard, marten, civet, fox, flying fox
68.	Gobindsagar	Bilaspur	Duck, teal, goose, crane
69.	Shri Naina	Bilaspur	Sambhar, barking deer, black bear, wild boar, nilgai, kalij, red jungle fowl, partridge, panther, snow leopard, marten, civet, fox, flying fox
70.	Majathal and Harseng	Solan	Goral, barking deer, black bear, kalij, red jungle fowl, partridge, panther, snow leopard, marten, civet, fox, flying fox
71.	Daranghali I and II	Kinnam	Goral, black bear, serow, monal, kalij, koklass, panther, snow leopard, marten, civet, fox, flying fox
72.	Shilli	Solan	Barking deer, chakor, kalij, partridge, panther, snow leopard, marten, civet, fox, flying fox
73.	Water Supply Catchment area Simla	Simla	Goral, barking deer, serow, kalij, koklass, monal (rare), panther, snow leopard, marten, civet, fox, flying fox
74.	Talra	Simla	Goral, black bear, sambhar, musk deer, kalij, koklass, chakor, monal, panther, flying fox,civet, marten
75.	Darlaghat	Solan	Goral, barking deer, chakor, kalij, partridge, panther, snow leopard, marten, civet, fox, flying fox
76.	Raksham and chitkul	Kinnaur	Bharal, goral, black bear, brown bear, monal, koklass, snow cock, chakor, panther, fox, marten, flying fox
77.	Lipa Asrang	Kinnaur	Ibex, baral, goral, monal, koklass, snow cock, chakor, panther, fox, marten, flying fox
78.	Renuka	Sirmpur	Sambhar, goral, barking deer, wild boar, kalij, pea fowl, red jungle fowl, partridge, panther, fox, marten, civet, flying fox
79.	Simbalbara	Sirmpur	Sambhar, goral, barking deer, wild boar, kalij, pea fowl, red jungle fowl, partridge, panther, fox, marten, civet, flying fox, chital and monal
80.	Manali	Kulu	Tahr, serow, goral, black bear, kalij, koklass, chakor, partrige, chir, panther, civet, marten, flying fox
81.	Kias	Kulu	-do-
82.	Khokhan	Kulu	-do-

1		2	3
83.	Kanswar	Kulu	-do-
84.	Thirthan	Kulu	-do-
85.	Chail	—	Goral and barking deer
86.	Taradevi (game reserve)	—	—
87.	Ghana Ki Hatto (game reserve)	Solan	-do-
88.	Basa	Kangra	—
		Karnataka	
89.	Bandipur National Park (Updated as Project Tiger)	Mysore	Elephant, tiger, gaur, sambhar, chital, barking deer, wild dog, wild boar, jackal, sloth bear, panther, four-horned antelope, malabar squirrel, jungle fowl, partridge, green pigeon, bush quail
90.	Bannarghatta National Park	Bangalore	Elephant, sloth bear, chital, barking deer, grey partridge, bush quail, jungle fowl
91.	Nagerhole National Park	Coorg	Elephant, tiger, panther, chital, sambhar, sloth bear, jungle fowl, partridge
92.	Someshwara Wildlife Sanctuary	South Kanara	Chital, gaur, sambhar, panther, tiger, sloth bear, wild dog
93.	Brahmagiri Wildlife Sanctuary	Coorg	Tiger, panther, sambhar, chital, barking deer, sloth bear, cobra, flying squirrel, Malabar squirrel, mouse, deer
94.	Arabithitu Game Reserve	Mysore	Chital, wild boar, porcupine, hare, jackal, partridge, quail, peacock
95.	Malkote Temple Wildlife Sanctuary	Mandya	Panther, wolf, sloth bear, black buck, hare, porcupine, wild boar
96.	Ghataprabha Bird Sanctuary	Belgaum	Egret, cormorant, heron
97.	Nugu Wildlife Sanctuary	Mysore	Elephant, tiger, chital, sambhar, barking deer, gaur, wild boar, sloth bear
98.	Tungabhadra Wildlife Sanctuary	Bellary	Black buck, chinkara, panther, chital, four-horned antelope, fox, grouse, florican, great Indian bustard, sloth bear, sambhar
99.	Ranebennur Black Buck Sanctuary	Dharwar	Black buck, chital, wolf, hare, great Indian bustard
100.	Mukambika Wildlife Sanctuary	South Kanara	Panther, sambhar, elephant, wild boar, porcupine
101.	Sharawathy Valley Wildlife	Shimoga	Elephant, tiger, panther, gaur, sambhar, chital, barking deer, wild dog, wild boar, jackal, sloth bear, four-horned antelope, malabar squirrel, jungle fowl, partridge, green pigeon, bush quail, owl, lion-tailed monkey, tortoise, crocodile, python, cobra, krait, viper, green snake
102.	Biliglri Rangaswami Wildlife Sanctuary	Mysore	Elephant, gaur, chital, barking deer, sambhar, panther, sloth bear
103.	Bhadra Wildlife Sanctuary	Chickmangalore	Gaur, elephant, panther, wild boar, sloth bear, sambhar, chital, barking deer
104.	Shattihalli Wildlife Sanctuary	Shimoga	Gaur, elephant, panther, wild boar, sloth bear, sambhar, chital, barking deer
105.	Ranganthittoo Bird Sanctuary	Mysore	Open-bill stork, white ibis, little egret, cattle egret, dartar, cormorant, pond heron, river tern, spoon bill, crocodile

	1	2	3
106.	Dandeli Wildlife Sanctuary	Dharwar	Tiger, panther, elephant, gaur, sloth bear, sambhar, chital, wild boar
		Kerala	
107.	Eravikulan Rajmellay National Park	Idukki	Elephant, gaur, sambhar, barking deer, tiger, panther, civet, jungle cat, nigai, park langur, wild boar, lion-tailed macaque, malabar squirrel, wild dog, nilgiri thar, imperial pigeon, grey jungle fowl
108.	Periyar Wildlife Sanctuary (Updated as Project Tiger)	Idukki	Elephant, tiger, panther, wild dog, gaur, sloth bear, nilgai, wild boar, sambhar, barking deer, tailed macaque, nilgiri langur, flying squirrel
109.	Parambikulam Wildlife Sanctuary	Pelghat	Elephant, gaur, leopard, tiger, sloth bear, nilgai, chital, sambhar, crocodile
110.	Neyyer Wildlife Sanctuary	Thiruvananthapuram	Elephant, tiger, leopard, wild boar, sambhar, chital, nilgai, tahr
111.	Prechi-vazhani Wildlife Sanctuary	Trichur	Elephant, wild boar, chital
112.	Wynad Wildlife Sanctuary	Cannonore and Kozhikode	Elephant, gaur, sambhar, wild boar, chital, barking deer
113.	Idukki Wildlife Sanctuary	Idukki	Elephant, sambhar, sloth bear, wild bear, wild boar, gaur
		Madhya Pradesh	
114.	Bandhavgarh National Park (Updated as Project Tiger)	Shahdol	Tiger, panther, gaur, chital, sambhar, nilgai, chinkara, chowsingha, barking deer, bear, wild boar, a variety of upland birds (ducks)
115.	Kanha National Park (Updated as Project Tiger)	Mandla and Balaghat	Tiger, panther, gaur, barasingha, chital, sambhar, black buck, chowsingha, barking deer, mouse dear, nilgai, wild dog, boar
116.	Shivpuri National Park	Shivpuri	Tiger, panther, sloth bear, hyaena, sambhar, spotted deer, four-horned antelope, black buck, nilgai, chinkara, wild boar, crocodile
117.	Bori Wildlife Sanctuary	Hoshangbad	Tiger, panther, bison sambhar, cheetal, nilgai, barking deer, chinkara, wild boar, bear
118.	Kutree Wild Buffaloes Game Sanctuary	Baster	Wild buffalo, tiger, panther, sloth bear, nilgai, chital, sambhar, wild boar, barking deer, bison, wild dog, chowsingha, hill myana
119.	Tamor-pigta Sanctuary	Sarguja	Tiger, panther, gaur, cheetal, sambhar, etc.
120.	Semarsot Sanctuary	Sarguja	-do-
121.	Panna National Park	Panna	Tiger, panther, sambhar, chital, wild pig, nilgai, pea fowl
122.	Gomardha Game Sanctuary	Raigarh	Tiger, panther, sambhar, chital, nilgai, gaur, barking deer
123.	Badalkhol Game Sanctuary	Raigarh	Tiger, panther, sambhar, chital, nilgai
124.	Barnawapara Game Sanctuary	Raipur	Tiger, panther, gaur, sambhar, chital, bear
125.	Sirpar Bird Sanctuary	Raipur	Variety of upland birds, wild animals
126.	Sitanadi Sanctuary	Raipur	Tiger, panther, sambhar, chital, bison, wild boar, barking deer, peacock
127.	Noradehi Game Sanctuary	Sagar, Damoh and Narsinghgarh	Tiger, panther, sambhar, chinkara, chital, nilgai, bear
128.	Narsinghgarh Game Sanctuary	Raigarh	Panther, sambhar, chital, wild pig, peacock, nilgai

1		2	3
129.	Dubari Game Sanctuary	Sidhi	Tiger, panther, sambhar, chital, nilgai, barking deer, chinkara
130.	Bagdara Game Sanctuary	Sidhi	Panther, black buck, chinkara, sambhar, nilgai, wild pig
131.	Panchmari Game Sanctuary	Hoshangbad	Tiger, panther, bear, bison, spotted deer, nilgai
132.	Achanakmar Game Sanctuary	Bilaspur	Tiger, bison, sambhar, chital, boar, peacock
133.	Ratapani Game Sanctuary	Risen	Tiger, panther, sambhar, chital, blue bull, chinkara
134.	Pench Game Sanctuary	Chindwara and Seoni	Tiger, panther, gaur, wild dog, chital, sambhar, nilgai, bison, antelope, porcupine, bear
135.	Sindhari Sanctuary	Raisen	-do-
136.	Gandhi Sagar Sanctuary	Mandsaur	Water bird, chital, sambhar, chinkara, barking deer
137.	Khconi Sanctuary	Morena	Gharial
		Maharashtra	
139.	Tadoda National Park	Chandrapur	Tiger, panther, sloth bear, gaur, sambhar, chital, nilgai, chinkara, crocodile
140.	Panch National Park	Nagpur	Tiger, panther, gaur, sloth bear, sambhar, chital, barking deer, four-horned antelope, nilgai, chinkara, jungle fowl, pea fowl
141.	Nawegaon National Park	Bhandara	Tiger, panther, sloth bear, gaur, sambhar, chital, barking deer, nilgai, migratory birds
142.	Borivli National Park	Bombay suburban	Panther, sambhar, four-horned antelope, mouse deer, wild boar, langur
143.	Dhaka-Kolkaz Project Tiger	Amravati	Tiger, panther, sloth bear, sambhar, barking deer, four-horned antelope, wild boar, chital, birds
144.	Yawal Wildlife Sanctuary	Jalgaon	Panther, jungle cat, nilgai, bonnet macaque, common langur, wild dog, sambhar, hyaena, tiger, chital, chinkara, pea fowl, grey jungle fowl, green pigeon, sand grouse, partridge
145.	Rahanagari (Bison) Sanctuary	Kolhapur	Gaur, panther, sambhar, wild boar, lavish birdlife
146.	Karnala (Bird) Sanctuary	Raigad	Ashy minivet, paradise fly catcher, shama, malabar whistling thrush, racket-tailed drongo, woodpecker, panther, four-horned antelope, common langur
147.	Tansa Wildlife Sanctuary	Thane	Panther, four-horned antelope, chital, sambhar, wild boar, diverse bird population
148.	Nagzira Wildlife Sanctuary	Bhandara	Tiger, panther, bison, sloth bear, sambhar, four-horned antelope, nilgai, chital, barking deer, mouse deer, pea fowl, grey jungle fowl, other birds
149.	Bor Wildlife Sanctuary	Wardha	Panther, sloth bear, sambhar, nilgai, chital, barking deer, wild boar, tiger (occasional)
150.	Kinwat Wildlife Sanctuary	Yeotmal and Nanded	Tiger, panther, nilgai, sloth bear, sambhar, chital, chinkara, barking deer, wild boar
151.	Great Indian Bustard Sanctuary	Ahmednagar	Great Indian bustard
152.	Deulgaon Dehekuri (Black buck) Sanctuary	-do-	Black buck
		Manipur	
153.	Keibul Lamjao National Park	Central	Brow antlered deer, wild goat, water birds

1	2	3
	Meghalaya	
154. Sijee Sanctuary	Garo Hills	-do-
155. Balpakram Sanctuary	—	—
	Mizoram	
156. Dampa Wildlife Sanctuary	Aizawal	Elephant, tiger, leopard, sambhar, barking deer, himalayan bear, wild boar, wild dogs, wild cat, gaur, leopard cat, king cobra, python, horn bill pheasant
	Nagaland	
157. Intangki Wildlife Sanctuary	Kohima	Gaur, boar, elephant barking deer, wild boar, clouded-leopard, panther, tiger, pangolin, various kinds of birds and reptiles
158. Pulie Bodze Wildlife Sanctuary	—	-do-
	Orissa	
159. Satkosia Gorge Sanctuary	Dhenkanal Puri, Cuttack and Phulbani	Gharial, muger, tiger, leopard, jungle cat, civet, gaur, ratel, sloth bear, sambhar, chital, nilgai, four-horned antelope, hornbill, jungle fowl, other birds
160. Bhitar Kanika Sanctuary	Cuttack	Salt-water crocodile, leopard, hyaena, jungle cat, leopard cat, wild boar, chital, sambhar, giant squirrel, water monitor, sea-turtle, king cobra, painted stork, adjutant stork, open billed stork, white ibis, black ibis
161. Chilka Lake Bird Sanctuary	Ganjam and Puri	Black buck, chital, abises, cranes, cormorant, egret, flamingo, pelican
162. Kuldiha Wildlife Sanctuary	Balasore	Elephant, gaur, chital, sambhar
163. Karlapat Wildlife Sanctuary	Kalahandi	Tiger, leopard, gaur, chital, sambhar
164. Saptasajya Wildlife Sanctuary	Dhenkanal	Elephant, gaur, leopard, sambhar, chital
165. Kapilash Wildlife Sanctuary	-do-	-do-
166. Ushakothi Wildlife	Sambalpur	Elephant, gaur, tiger, leopard, sambhar, chital, barking deer
167. Simplipal National Park (Updated as Project Tiger)	Mayurbhanj	Tiger, elephant, gaur, sambhar, chital, sloth bear, hyaena, leopard, mouse deer, flying squirrel, muger, peacock
168. Khalasuni Wildlife Sanctuary	Sambalpur	Elephant, gaur, sambhar, chital and sloth bear
169. Balukhand Wildlife Sanctuary	Puri	Black buck, chital
170. Chandaka Wildlife Sanctuary	Puri	Elephant, panther, chital, sambhar, barking deer, civet cat
171. Debrigarh Wildlife Sanctuary	Sambalpur	leopard, sambhar, gaur
172. Hadagarh	Keoenjhar	-do-
173. Nandankanan	Puri	-do-
174. Mahanadi and Baisipally	Nayagarh	-do-
175. Padmalota	Phulbani	-do-
176. Raigoda	Angul	-do-
	Punjab	
177. Bir Moti Bagh Wildlife Sanctuary	Patiala	Nilgai, wild boar, hog deer, hare, pea fowl, partridge, pigeon, dove, sparrow, babbler, woodpecker, myna, parakeet

	1	2	3
178.	Bir Bhunnerheri Wildlife Sanctuary	-do-	-do-
179.	Harike Headwork	Patiala	Duck, poacherd, teal, common coot, water hen, geese, nilgai, wild boar
180.	Abhor Wildlife Sanctuary	Ferozpur	Black buck, hare, partridge, pigeon, dove
		Rajasthan	
181.	Ranthambor National Park (Updated as Project Tiger)	Sawai-Madhopur	Tiger, panther, hyaena, jungle cat, fox, civet, sambhar, chital, nilgai, sloth bear, wild boar, langur, mangoose, partridge, green pigeon, red-spur fowl
182.	Sariska Wildlife Sanctuary (Updated as Project Tiger)	Alwar	Tiger, panther, hyaena, jungle cat, civets, sambhar, chital, chinkara, nilgai, four-horned antelope, langur, porcupine, hedge-hog, partridge, green pigeon, red spur fowl, peacock
183.	Ghana Bird Sanctuary	Bharatapur	Siberian crane, cormorant, stork, spoon bill, quail, coot, heron, teal, tern, sambhar, chital, black wild boar, civet
184.	Ramsagar Van Vihar	Bharatapur	Nilgai, chital, sambhar, panther, tiger, wild boar, hare, partridge
185.	Darrah Sanctuary	Kota	Tiger, panther, sambhar, chital, nilgai, wild boar, hare, partridge, grouse
186.	Jaisamand Sanctuary	Udaipur	Panther, chital, jackal, nilgai, chinkara, hare partridge
187.	Mount Abu Sanctuary	Sirohi	Sambhar, nilgai, hare, jungle-fowl, partridge
188.	Talchappar Sanctuary	Churu	Black buck
189.	Kumbhal garh-Ranakpur Sanctuary	Udaipur and Jodhpur	Wild bear, sambhar, panther, nilgai, wild boar, jungle fowl, red spur fowl
190.	Jawahar Sagar Sanctuary	Kota	Crocodile, chinkara, hare
191.	Doli closed area (cluster of 23 Bishnoi villages)	Jodhpur	Black buck, chinkara
192.	Sitamata Sanctuary	Udaipur	-do-
193.	Desert Sanctuary	Jaisalmer Barmer	Great Indian bustard, Black buck, chinkara
		Sikkim	
194.	Kanchenjunga National Park	Gangtok	Snow leopard, clouded leopard, marbled cat, civet, binturong, himalayan black bear, red panda, tibetan wild ass, blue sheep, serow, goral, takin, musk deer, pheasant, partridge, green pigeon
		Tamil Nadu	
195.	Guindy National Park (includes a snake park)	Madras	Chital, black buck, snakes
196.	Mudumalai Wildlife Sanctuary	Nilgiris	Elephant, gaur, chital, sambhar, tiger, panther, sloth bear, wild dog
197.	Annamalai Wildlife Sanctuary	Coimbatore	-do-
198.	Kalakad-Mundanthurai Project Tiger	Tirunelveli	Tiger, chital, sambhar, wild boar, lion-tailed macaque
199.	Point Calimere Sanctuary	Thanjavur	Black buck, chital, wild boar
200.	Vedanthangal Water Bird Sanctuary	Chingleput	Different water birds
201.	Vettangudi Bird Sanctuary	Ramanatha-puram	Birds

	1	2	3
202.	Corbett National Park (Updated as Project Tiger)	Nainital/ Utterakhand	Elephant, tiger, panther, sloth bear, nilgai, goral, four-horned antelope, rhesus monkey, langur, porcupine, pea fowl, red jungle fowl, partridge, both species of Indian inland crocodiles
		Uttar Pradesh	
203.	Dudhwa National Park (Updated as Project Tiger)	Lakhimpur Kheri	Tiger, panther, sloth bear, sambhar, swamp deer, barking deer, nilgai, peafowl, jungle fowl, partridge
204.	Govind Pashu Vihar	Uttarkashi	Brown and black Himalayan bear, sambhar, swamp deer, chital, hog deer, barking deer, nilgai, peafowl, jungle fowl, partridge
205.	Nanda Devi Sanctuary	Chamoli	-do-
206.	Rajaji National Park	Saharanpur	Elephant, tiger, panther, sloth bear, nilgai, sambhar, chital, kakar, wild boar, porcupine, peafowl, red jungle fowl, partridge
207.	Motichur Sanctuary	Dehradun	-do-
208.	Kishanpur Sanctuary	Lakhimpur Kheri	Tiger, panther, sloth bear, sambhar, swamp deer, barking deer, nilgai, peafowl, jungle fowl, partridge
209.	Kedarnath Sanctuary	Chamoli	Brown and black himalayan bear, snow leopard, bharel, musk deer, tahar, serow, gorel
210.	Chandraprabha Sanctuary	Varanasi	Tiger, panther, sambhar, Indian gazelle, sloth bear, nilgai, peafowl, partridge, sand grouse
211.	Katarniaghat Sanctuary	Baharaich	Tiger, panther, sambhar, bear, chital, black buck
212.	Ranipur Sanctuary	Banda	Tiger, panther, jungle cat, sambhar, hyaena, fox, jackal, chital, chinkara, black buck
213.	Chila Sanctuary	Garhwal	Tiger, panther, bear, elephant, chital, sambhar, goral, nilgai
214.	Mahavir Swami Sanctuary	Lalitpur	Panther, nilgai, bear, chital, chinkara, black buck, bear
215.	Chambal Sanctuary	Lalitpur	Gharial, muger
		West Bengal	
216.	Lothian Island Wildlife Sanctuary	24-Parganas	Wild pigs, chital, otter, estuarine crocodile, gangetic dolphin, water birds
217.	Halliday Island Wildlife Sanctuary	-do-	Royal Bengal tiger, chital, water birds
218.	Sajnakhali Wildlife Sanctuary	24-Parganas	Tiger, wild boar, chital, cormonant, open bill, stork, white ibis, purple heron, grey heron, green bitter pelican, snakes
219.	Gorumara Jalpaiguri Sanctuary	Jalpaiguri	Rhino, elephant, tiger, gaur, sambhar, hog deer, wild boar, a variety of birds
220.	Charramari Wildlife Sanctuary	Jalpaiguri	Elephant, tiger, gaur, sambhar, barking deer, wild boar, a variety of birds
221.	Jaladpara Wildlife Sanctuary	Jalpaiguri	Rhino, elephant, tiger, leopard, wild boar, gaur, barking deer, hog deer, sambhar, a variety of birds
222.	Mahanadi Wildlife Sanctuary	Darjeeling	Tiger, elephant, gaur, sambhar, hog deer, wild boar, gibbons, a variety of birds
223.	Sanchal Wildlife Sanctuary	-do-	Serow, Himalayan black bear, barking deer, goral, a variety of hill birds
224.	Dear Park Ballavpur (Shantiniketan)	Birbhum	Black buck, chital
225.	Sunderbans Project Tiger	-do-	Tiger, leopard, boar, chital, otter, python, crocodile

Table 19.2. Comparison of National Parks, Wildlife sanctuaries and Biosphere reserves.

National Park	Wildlife Sanctuary	Biosphere Reserve
1. Attention is not given on biotic community as a whole, *e.g.*, conservation being connected to habitat for particular wild animal species such as lion, tiger, rhino, etc.	1. Attention is not given on biotic community, *i.e.*, conservation is species oriented *e.g.*, citrus, pitcher plant, great Indian bustard.	1. Attention is focussed on biotic community as a whole, *i.e.*, conservation is ecosystem oriented.
2. The approach is not based on scientific principles. The size ranges from 0.04 to 3162 sq. km; the usual size being between 100 and 500 sq. km. (in about 39%) and between 500 to 1000 sq. km. (in about 16%).	2. The approach is not based on scientific principles. The size ranges from 0.61 to 7818 sq. km.; usual size being between 100 and 500 sq. km. (in about 39%) and between 500 and 1000 sq. km. (in about 24%).	2. The approach is based on sound scientific principles. The size is well over 5670 sq. km.
3. Boundaries are circumscribed by state legislation.	3. Limits are not sanctified.	3. Boundaries are circumscribed by state ligislation.
4. There occurs no biotic interference except in buffer zone.	4. There occurs limited biotic interference.	4. There occurs no biotic interference except in buffer zone.
5. Tourism is though permissible but often is discouraged.	5. Tourism is permissible.	5. Tourism is not permissible.
6. Research and scientific management are lacking.	6. Research and scientific management are lacking.	6. Research and scientific management are carried out.
7. Proper attention is not given to gene pool conservation of economic species, particularly of plants.	7. Proper attention is not given to gene pool conservation of economic species, particularly in plants.	7. Due attention is given to conservation of plants as well as animal species.

QUESTIONS

Long Answer Questions

1. Give total numbers of National parks and sanctuaries located in India and how much land area is covered by them. In what ways both of these differ from each other.

2. Enumerate steps taken in India for the protection of wildlife.

3. Describe objectives and strategies of conservation of wildlife.

4. What do you understand about protected area? Discuss its advantages and efforts being done in India for the protection of wildlife.

5. Write an essay on conservation of wildlife in India.

6. What is a sanctuary? Give detailed account of national parks and their wildlife in Uttar Pradesh.

7. Define national park, Describe the location, area, form and fauna of two national parks of Uttar Pradesh.

8. What is a sanctuary? Describe the location, area, flora and fauna of any two national parks?

9. Give brief accounts of national parks and sanctuaries of Uttar Pradesh.

10. Name ten sanctuaries and two national parks of Himachal Pradesh along with district where they are found.

11. Write an account of any three wildlife sanctuaries in Tamil Nadu.
12. What is difference between sanctuary and national park? Describe the location, area, flora and fauna of any two of the following national parks: 1. Corbett National Park; 2. Hazaribagh National Park; 3. Kanha National Park.
13. Write a brief account on management and protection of wildlife.
14. Write an essay on wildlife management and conservation with special reference to endangered flora and fauna.
15. Mention two important sanctuaries in India and describe the important animals of those areas. Discuss the role of wild animals in ecological balance.
16. Give the importance of wildlife management. Enumerate the efforts by our country in this direction.
17. Give an account of (*a*) Kanha National Park; (*b*) Sunderban tiger reserve.

Short Answer Questions

1. Briefly describe the following:
   (*i*) Dudhwa and Kaziranga National Park;
   (*ii*) Periyar and Bandipur Wildlife Sanctuaries.
2. How do national parks and sanctuaries differ?
3. Give an account of the following;
   (*i*) Corbett National Park
   (*ii*) Palamau Tiger Reserve
   (*iii*) Dachigam Sanctuaries.
4. Write short notes on the following:
   (*i*) Environmental monitoring
   (*ii*) Wildlife management
   (*iii*) Wildlife sanctuaries in India
   (*iv*) Conservation of crocodiles.
   (*v*) Biosphere reserve
   (*vi*) Name the major sanctuaries in India, where they are found
   (*vii*) Conservation of wildlife in India
   (*viii*) Special projects for conservation of endangered animal species initiated in India.
   (*ix*) Location of wildlife of important national parks of Himalayan region.
   (*x*) Wildlife conservation
   (*xi*) W.W.F
   (*xii*) Kevladeo Ghana National Park
   (*xiii*) Red data book
   (*xiv*) Bandipur-Mudumalai Sanctuary
   (*xv*) Gir National Park
   (*xvi*) Tadoba National Park
   (*xvii*) Write the full name of WWF and its role.
5. Differentiate between the following:
   1. Protected forest and National Park
   2. National Park and Game Sanctuary.

Very Short Answer Questions

1. What are the protected areas?
2. What is a national park?

3. What is a wildlife sanctuary?

4. Give one word for the following:

(*i*) An area where special protection is given to wildlife.

(*ii*) An area where killing, hunting or shooting of animals is prohibited but private ownership rights may be allowed.

(*iii*) An area dedicated to conserve the environment, natural and historical objects and where all private rights are non–existent.

Fill in the Blanks

1. Dudhwa National Park is situated in .... in Uttar Pradesh.
2. Grazing of domestic animals is prohibited in ....
3. Corbett National Park is situated in district .... of Utterakhand.

Multiple Choice Questions

1. Brow-antlered deer is one of the rarest mammal found in India. It is found in (*CCSU Meerut 2014*)

(*a*) Nanda Devi-Himachal Pradesh
(*b*) Keibul Lamjao National Park-Manipur
(*c*) Dudhwa National Park- Uttar Pradesh
(*d*) Periyar Wildlife Sanctuary –Kerala

2. Breeding place of flamingo (Hanswar) in india (*CCSU Meerut 2014*)

(*a*) Chilka lake (*b*) Sambar lake
(*c*) Ran of Kutch (*d*) Ghana vihar

3. Ranthambor Reserve is situated in (*CCSU Meerut 2014*)

(*a*) Rajasthan (*b*) Bihar
(*c*) West Bengal (*d*) Orissa

4. First National Park of India was (*CCSU Meerut 2014*)

(*a*) Jim Corbett National Park (*b*) Kaziranga National Park
(*c*) Panna National Park (*d*) Gir National Park

5. Kevala Dev National Park Bharatpur is famous for (CCSU Meerut 2014)

(*a*) migratory birds (*b*) winter visitors
(*c*) cranes (*d*) herons

6. Identify the odd combination of the habitat and the particular animal concerned (*CCSU Meerut 2014*)

(*a*) Dachigam National Park–snow leopard (*b*) Sunderbans–Bengal tiger
(*c*) Periyan–Elephant (*d*) Rann of Kutch–wild ass

7. Manas Sanctuary is famous for (*CCSU Meerut 2014*)

(*a*) hawling monkey (*b*) golden langur
(*c*) panther (*d*) hangul

8. Dachigam Wildlife Sanctuary is famous for protection of? (*CCSU Meerut 2014*)

(*a*) Kashmir stag or hangul (*b*) *Loris*
(*c*) red panda (*d*) sloth bear

9. In a national park protection is provided to

(*a*) flora only (*b*) fauna only
(*c*) both flora and fauna (*d*) entire ecosystem

10. Conservation is
(*a*) proper use of natural resources
(*b*) protection of natural resources
(*c*) management of natural resources
(*d*) all the above

11. Animal species should be preserved mainly because
(*a*) zoologists want to study them
(*b*) they are lovely creatures
(*c*) they are useful to mankind
(*d*) man cannot recreate a species of animals after its destruction

12. Kanha National park is situated in
(*a*) Uttar Pradesh
(*b*) Kerala
(*c*) Karnataka
(*d*) Madhya Pradesh

13. In which state is the Dhudhwa National park
(*a*) Rajasthan
(*b*) Uttar Pradesh
(*c*) Madhya Pradesh
(*d*) Gujrat

14. Nandan-Kanan Zoo is Known for the
(*a*) Nilgiri Tahr
(*b*) Hippopotamus
(*c*) White Tiger
(*d*) Whale

15. Scientific name of our national bird is
(*a*) *Corvus splendens*
(*b*) *Pavo cristatus*
(*c*) *Gallus gallus*
(*d*) *Columba Livia*

16. The sanctuary having maximum number of rare animals is
(*a*) Manas National Park
(*b*) Kaziranga National Park
(*c*) Dudhwa National Park
(*d*) Corbett National Park

17. One of the best tiger reserves in India is in
(*a*) Kashmir
(*b*) Jim Corbett National Park
(*c*) Kanha National Park
(*d*) Bandipur

18. Ranganathittu Sanctuary of Karnataka is famous for
(*a*) tiger reserve
(*b*) birds
(*c*) elephants
(*d*) antelope

19. Which is the state of India having all the three species of crocodiles and which started save Crocodile Project?
(*a*) Uttar Pradesh
(*b*) Karnataka
(*c*) Kerala
(*d*) Maharashtra

20. Bandhavgarh National Park is found in
(*a*) Karnataka
(*b*) Maharashtra
(*c*) Uttar Pradesh
(*d*) Madhya Pradesh

21. One of the following langur is associated with Lord Hanuman and is given special sanction
(*a*) *Presbytis entellus*
(*b*) *Hyalobates hoolock*
(*c*) *Presbytis geei*
(*d*) *Macaca silenus*

22. The famous 'Chipko Movement' is associated with
(*a*) Salim Ali
(*b*) Indira Gandhi
(*c*) Jawahar Lal Nehru
(*d*) S.L. Bahuguna

23. 'Hangul Project' was launched by Government of India with the help of IUCN/WWF in 1970 to save Hangul (*Cervus hangul*). Name the sanctuary in which it was started
(*a*) National Chambal Sanctuary
(*b*) Dachigam Sanctuary
(*c*) Corbett National Park
(*d*) Bandipur National Park

24. Which of the following is related with 'Silent Valley National Park' in Kerala
(*a*) lion tailed macaque (*b*) gharial
(*c*) wild ass (*d*) swamp deer
25. Chilka Lake is largest lake in Orissa. It is famous for its wildlife which is mainly
(*a*) birds (*b*) lion
(*c*) rhino (*d*) tiger
26. Bharatpur National Park is famous for its
(*a*) migratory birds (*b*) winter visitors
(*c*) cranes (*d*) herons
27. The Siberian crane from Russia is a regular visitor to the bird sanctuary in which place in India
(*a*) Lallbagh, Bangalore (*b*) Vedanthgol sanctuary, Tamil Nadu
(*c*) Ranganathittis Sanctuary, Karnataka (*d*) Bharatpur sanctuary, Rajasthan
28. Which of the following is correctly matched
(*a*) Corbett Park-Cats (*b*) Gir Forest- Rhino
(*c*) Kaziranga - Elephant (*d*) Runn of Kutch –Wild Ass.

ANSWERS

Very Short Answer Questions

1. Any area where special protection is offered to wildlife is declared as protected area, such as (*a*) River valley projects and other irrigational works or wetlands that attract water birds and other wild life. (*b*) An area where protection is offered to wildlife in or around large town of sacred places. National parks, wildlife sanctuaries and biosphere reserves are some protected areas.
2. A national park is area dedicated to conserve the environment, national and historical objects and to conserve the wildlife there in and at the same time to provide for enjoyment from them in such a way and by such means, as will leave the unimpaired for the enjoyment of future generations. No private rights exist in a national park. All forestry operations and other usages such as grazing of domestic animals are prohibited.
3. The wildlife sanctuaries are the forest areas where killing, hunting, shooting or capturing of animals is prohibited except by or under the control of the highest authority in the department responsible for management of the sanctuary. Private ownership may be allowed to continue in a sanctuary and forestry and other usages are permitted to the extent that they do not adversely affect wildlife.
4. (*i*) Protected area; (*ii*) wildlife sanctuary: (*iii*) National park.

Fill in the Blanks

1. Lakhimpuri Kheri **2.** National Park **3.** Nainital.

Multiple Choice Questions Answers

1. (*b*)	**2.** (*a*)	**3.** (*a*)	**4.** (*a*)
5. (*a*)	**6.** (*a*)	**7.** (*b*)	**8.** (*a*)
9. (*d*)	**10.** (*c*)	**11.** (*d*)	**12.** (*d*)
13. (*b*)	**14.** (*a*)	**15.** (*b*)	**16.** (*a*)
17. (*c*)	**18.** (*b*)	**19.** (*c*)	**20.** (*d*)
21. (*a*)	**22.** (*d*)	**23.** (*b*)	**24.** (*a*)
25. (*a*)	**26.** (*d*)	**27.** (*d*)	**28.** (*d*).

20

Wildlife Conservation

20.1 BIOSPHERE RESERVES

Biosphere reserves are undisturbed natural areas for scientific study as well as areas where any sort of disturbance is under control. They have been set aside for ecological research and habitat preservation. **Ramde** (1984) described them as the means to protect ecosystems whether natural or modified by human activity, in order to preserve ecological "*evidence*" for the purpose of scientific research. In fact, biosphere reserves are very good methods for implementation of world conservation strategy (WCS), 1980.

Biosphere Reserve Network Programme was launched by UNESCO in 1971 under its **Man and Biosphere (MAB)** programme. The objectives of this programme are to

1. Conserve representative samples of ecosystems (*i.e.*, conservation of biotic diversity for ecological evidence).
2. Provide long-term *in-situ* conservation of genetic diversity.
3. Provide natural areas for basic and applied research in ecology and environmental biology.
4. Impart opportunities for environmental education and training.
5. Promote appropriate sustainable managements of the living resources.
6. Broadcast the experience so as to promote sustainable development else where, and
7. Promote international cooperation.

A protected area which can be declared as a biosphere reserve, should satisfy the following essential characteristics (**Maldague** 1984):

1. Should occur in each of the 193 biogeographical regions (or provinces) of the world distinguished in the classification of **Udvardy** (1975), so as to exhibit a sum total of the biological communities interacting in a single life zone where climate is similar (Biome).
2. A biosphere reserve should contain abundant genetic diversity.
3. Each biosphere reserve should be unique in itself.
4. The biosphere reserve should be size-wise effective to maintain natural populations in numbers which should not lead to genetic drift in future.
5. The proposed biosphere reserves should be natural home of endangered species of plants and animals. These may be endangered on account of a number of reasons such as habitat destruction due to human interference, or biotype depletion due to their relict nature or being quite young and evolving species which did not have enough time to spread their ranges. (Relict means a plant or animal species persisting in a given area as a survival from an earlier period or type).
6. The biosphere reserve should have sufficient resources available for ecological education, training and research to be carried on in respect to conservation of nature.
7. A biosphere reserve should have adequate long term legal protection.

Upto January 1989, 274 biosphere reserves have been established in 68 countries. The country's first biosphere reserve came into being in 1986 in Niligiri, covering 5670 km^2 in Tamil Nadu, Kerala and Karnataka (Table 20.1). It comprises forest areas including the famous **Silent Valley** and **Western Ghats**. It falls under the category of Malabar Rain Forest with 20 different tribal groups living in it. This area is rich biologically with almost all important groups of terrestrial and freshwater animals existing there.

Besides those in the above table, biosphere reserves are envisaged in Little Rann of Kutch, Gujarat; the cold desert area adjoining Himachal Pradesh and Jammu and Kashmir; Abujhmarh, Kanha and Amarkantak in Madhya Pradesh, Valley of flowers (Uttaranchal).

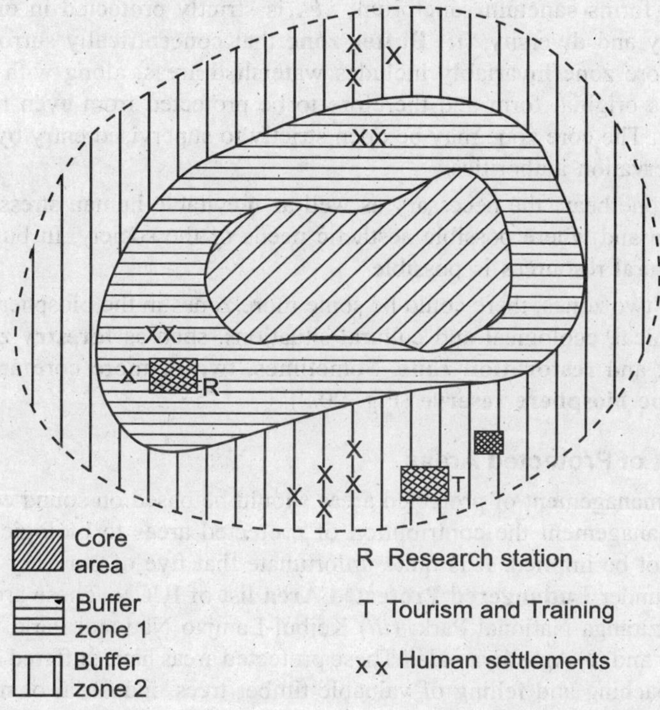

Fig. 20.1. Diagrammatic representation of a typical biosphere reserve.

From biospheric point of view, the Indian subcontinent falls into two realms and 12 biogeographic provinces such as 1. Ladakh; 2. Himalayan highlands; 3. Malabar rainforest; 4. Bengal rain forest; 5. Indus-Ganga-monsoon forest; 6. Assam-Burma monsoon forest; 7. Mahanandian; 8. Coromondal Deccan thorn forest; 9. Thar desert; 10. Lakshadweep Islands; 11. Andaman and Nicobar Islands.

Table 20.1. Sites of biosphere reserves in India.

	Name of the Site	Year of Setting up	Location
1.	Nilgiri	1986	Parts of Wynad, Nagarhole, Bandipur and Madumalai, Nilambur, Silent Valley and Siruvani Hills (Tamil Nadu, Kerala and Karnataka)
2.	Nanda Devi	1988	Parts of Chamoli, Pithoragarh and Almora district.
3.	Nokrek	1988	Parts of Garo Hills (Meghalaya)
4.	Manas	1989	Parts of Kokrajhar, Bongaigaon. Barpeta, Nalbari, Kamrup and Darang districts (Assam)
5.	Sunderbans	1989	Parts of delta of Ganges and Brahmaputra river system (West Bengal)
6.	Gulf of Mannar	1989	Indian part of Gulf of Mannar between India and Sri Lanka (Tamil Nadu)
7.	Great Nicobar	1989	Southern-most islands of Andaman and Nicohbar (A & N) Islands
8.	Similipal	1994	Parts of Mayurbhanj district (Orissa)
9.	Dibru-Saikhowa	1997	Parts of Dibrugarh and Tinsukia district (Assam)
10.	Dehang Debang	1998	Parts of Siang and Debang Valley in Arunachal Pradesh
11.	Pachmarhi	1999	Parts of Betul, Hoshangabad and Chindwara districts of Madhya Pradesh
12.	Kanchenjunga	2000	Parts of Kanchenjunga Hills in Sikkim

The first two provinces belong to the palaearctic realm while the remaining 10 to the Indo-Malayan (**Khoshoo** 1991).

Structure of biosphere reserve. Basically the biosphere reserve consists of two zones—(*i*) **Core zone** which forms sanctum sanctorum, *i.e.,* is strictly protected in order to maintain its ecological integrity and diversity, (*ii*) **Buffer zone** that concentrically surrounds the core zone (Fig. 20.1). The core zone invariably includes watershed areas, along with specific habitats to be conserved in its original form and therefore to be protected from even non-native or exotic plants and animals. The core zone may be open strictly to supervised entry by scientists, research workers and conservation authorities.

The buffer zone bears the necessary as well as inevitable human stress and also meets the research, education and where possible aesthetic needs of the society. In buffer zone controlled exploitation of natural resources is possible.

Beside these two zones, there could be some more zones in the biosphere reserve depending upon the geographical, ecological and cultural situations, such as **forestry zone, tourism zone, agriculture zone**, and **restoration zone.** Sometimes, two or more core areas to be protected form a **cluster type biosphere reserve** (Fig. 20.2).

A. Management of Protected Areas

Essentially, management of protected areas should be based on sound ecological principles. Without proper management the contribution of protected areas to biological conservation and development cannot be implied. It is quite unfortunate that five of our very important Protected Areas have come under **Endangered Protected Area** list of IUCN. These are (*i*) Manas Wildlife Sanctuary; (*ii*) Kaziranga National Park; (*iii*) Keibul-Lamjao National Park; (*iv*) Naryan Sarovar Wildlife Sanctuary and (*v*) Agasthyamalai. These protected areas have suffered due to encroachment of forest areas, poaching and felling of valuable timber trees, industrial or mining activities and indifference of people due to ethenic agitations.

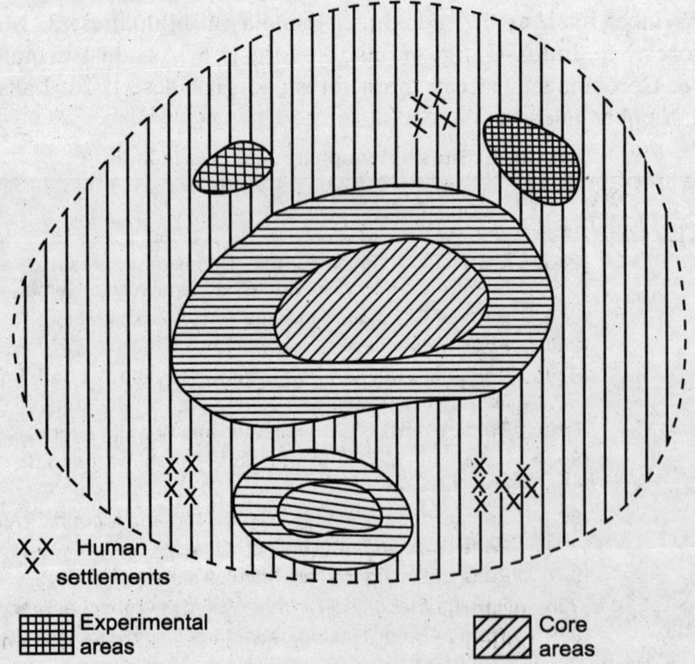

Fig. 20.2. Diagrammatic representation of a cluster type biosphere.

There are many challenges for maintaining our Protected Areas. Perhaps best results can be obtained from economic, social and political reforms and involvement of the people.

B. Ex-Situ Conservation

Ex-situ conservation of biodiversity implies conservation outside the habitats by perpetuating sample populations in genetic resource centres, zoos, botanical gardens, culture collection, etc., or in the form of gene pools and gamete storage for fish. Ex-situ conservation also includes setting of germ plasm banks for seeds, pollen, semen, ova, cells, etc. Ex-situ conservation includes the following methods:

1. Zoos. Zoos or zoological gardens are the places where wild animals are kept for public showing. Some of the zoos contain rare animals. They have recorded success with captive breeding of animals (*e.g.*, birth of a calf to rhino in a Zoo of Bihar in Sept. 2000). The Pere David's Deer and the European Bison, for example, have lost their natural habitats many decades ago, but they have been kept alive solely through zoos.

In India, there are 15 large zoos (total area 2317.11 ha), 17 medium zoos (total area 983.82 ha), 32 small zoos (total area 3148.31 ha) and many mini zoos (total area 2827.88 ha). The Central Zoo Authority (CZA) of India provides financial assistance to zoos and also approves exchanges of donations of animals both domestic and foreign for breeding programmes.

2. Botanical gardens. They play an important role in the conservation of plant species. There are several instances where plants believed to be extinct were found living only in a botanical garden, *e.g., Sophora toroniro*. The Botanical Survey of India (BSI) with its main botanic garden at Sibpur and the regional experimental gardens established at different places to suit the acclimatization of plants occurring in different ecological conditions in the country have in their holding rare, endemic and threatened plants. Records of threatened plants that are in cultivation have been kept in **"Green Books"**. The Indian Green Book prepared by BSI lists 10 such species which are rare, endangered or endemic but all of which are growing in living state in some botanical gardens of BSI in the country. Currently botanical gardens have assumed a special role as centres where active research and action oriented plan for conservation can be developed.

3. Seed banks. The seed bank is an excellent method of conserving diversity. The seeds have to be stored under "*minimal life*" so that they can be kept for several years without accumulations of **mutations** and with a minimum loss of viability. Such sort of conservation is practiced by the help of **cold storage** in seed banks where seeds are stored for long durations.

The preservation of **"recalcitrant seeds"**, *i.e.,* those which cannot be stored at low moisture content, present problems for certain species, especially many fruit and forest species. Efforts to develop improved methods for storing recalcitrant seeds involve a three-fold strategy: **dry storage, cryogenic storage** and **storage of fully hydrated seeds** or near fully hydrated seed at ambient or subambient temperatures.

Quite recently, a seed bank has been set up at the **National Burean of Plant Genetic Resources (NBPGR)**, New Delhi for the preservation of land races and cultivars have replaced recently by modern pedigree strains and wild relatives of agrihorticultural crops. This seed bank is associated with a world network of gene resource centres coordinated by the **International Bureau of Plant Genetic Resources (IBPGR)** of FAO. Seed bank of NBPGR is also involved in the preservation of plants of wild origin of potential economic value, or of scientific, aesthetic or ecological importance.

4. Pollen storage. Pollen storage is of great importance in the conservation of genetic diversity. However, the life of pollen is much shorter than seed and a maximum of 3 years in some species has been reported.

5. Tissue culture. It is best method of preservation of genes through establishment of **tissue banks**. Tissue culture technique can be extended to endangered species or for those which may otherwise require very varied climatic conditions and can be maintained at one place in aseptic cultures. The advantages of this technique are economy of space, relatively low maintenance cost, protection of genotypes from natural calamities and diseases if growing in open and easy international exchange of healthy stock.

Box 20.1 Plant Tissue Culture

1. The growth of cells, tissues and organs, outside the organism in artificial media of salts and nutrients is called **tissue culture**.

2. The technique of plant tissue culture includes growing of plant species in a nutrient medium, *i.e., in vitro*. Unorganised tissue or **callus** is cultured under this technique. These days, the term **plant tissue culture** is used for protoplast, cell, tissue, organ or whole plant culture in a nutrient medium under aseptic conditions. Under this technique the **explants** (*i.e.,* cells, tissues or organs of a plant) are carefully separated and grown aseptically on a nutrient medium under controlled conditions of light and temperature.

3. The technique of **meristem culture** (*i.e.,* shoot apex and root apex) is important in obtaining virus-free materials from many vegetatively propagated crop plants. It is so because meristem of even virus-infected plants are normally virus free.

4. In technique of **micropropagation**, the shoot apices are cultured to promote the growth of a mass of proliferating lateral shoots.

5. The technique of **embryo culture** is used in rescuing such embryos that might otherwise abort.

6. The technique of **anther culture** is used in producing haploid cultures or haploid plantlets. When treated with **colchicine**, such haploid plantlets or cultures may be developed into diploid plants.

7. The technique of **protoplast culture** is used in several experiments of plant genetic manipulation.

Meristem culture technique (Box 20.1) is of greatest significance in vegetatively propagated plants in which it permits the long-term preservation of clones in a virus-free condition. This technique has proved useful in the preservation of some species of orchids, grapes, etc. Many ornamental plants such as *Gerbera, Cordyline, Dracaena* and *Hemerocallis* are now commercially propagated in this manner. There are following three methods by which plants can be **multiplied clonally** through tissue culture: meristem culture, organogenesis and somatic embryogenesis.

6. Genetic engineering. Various modern techniques of genetic engineering or biotechnology such as *recombinant DNA*, *gene splicing, gene synthesis,* etc., are making it possible to transfer genetic material irrespective of the usual taxonomic boundaries and genetic barriers.

Other Conservation Measures

Other important steps need to preserve wildlife include the following measures: (*i*) For the preservation of species wildlife management staff should have a correct idea about the exact habitat which the species under consideration needs. (*ii*) Natural habitat of wild animals should be carefully protected. (*iii*) Habitats of wildlife should be improved by constructing water holes, saltlicks and by raising plantation of better and nourishing fodder grasses and trees. (*iv*) Effective means of census operation should be adopted to measure population sizes of various wild animals. (*v*) The enforcement of Wildlife Protection Acts should be observed strictly. (*vi*) Shooting and hunting of endangered species should be totally banned. (*vii*) To take care of wildlife in case of epidemics, veterinary efforts should be made. (*viii*) To understand the biology and behaviour of wild animals research on wildlife should be encouraged. (*ix*) Public should be educated about the advantages and disadvantages of wildlife.

20.2 CONSERVATION THROUGH NATIONAL/STATE ANIMALS AND PLANTS

An alternative way suggested by ecologists to conserve important indigenous mammals, birds, wild flowers and trees is to allot one each of these items to each state for conservation by declaring these as State Animal, Bird, Flower and Tree (**Khoshoo** 1986). For example, **leopard** (*Panthera pardus*), **sarus crane** (*Grus antigone*), **brahmakamal** (*Saussurea obvallata*) and **sal** (*Shorea robusta*) could be alloted to Uttar Pradesh to be conserved as the 'State animal, bird, flower and tree' respectively. **Musk deer** (*Moschus moschiferus*), **monal pheasant** (*Lophophorus impejanus*), **bish** or **atis** (*Aconitus heterophyllum*) and **deordar** (*Cedrus deodara*) could be alloted to Himachal Pradesh to be conserved as the State animal, bird, flower and tree respectively. **Asiatic lion** (*Panthera leo persica*), **flamingo** (*Phoenicopterus roseus*), **Indian laburnum** (*Cassia fistula*) and **neem** (*Azardirachta indica*) could be alloted to Gujrat for their conservation.

Based on criteria such as nativity/endemicity, general importance, etc., a list of 108 such species (*i.e.,* 27 each of mammals, birds, wild flowers and trees) has been prepared with the help of BSI, ZSI, NBRI and NBPGR alloting four species to each of 27 states of India. However, the states will need to take legislative steps to declare them so. They have to make it obligatory on their part to take practical steps for their conservation and building sizeable populations of these species in the relevant conservation areas within the state.

Government of India has declared **tiger** (*Panthera tigris*) and **peacock** (*Pavo cristatus*) as National animal and bird respectively. However, there is no announcement for National flower and National tree. **Khoshoo** (1986) has suggested that for this purpose the **lotus** (*Nelumbo nucifera*) and **pipal** (*Ficus religiosa*) could be considered.

20.3 CONSERVATION PROJECTS

1. Save the Barasingha Campaign

Government of Madhya Pradesh, supported by the Government of India, in 1969 launched the "*SAVE the Barasingha Campaign*". The Barasingha or swamp deer (*Cervus duvauceli duvauceli*) was once plentiful in central, northern and eastern India. In 1970, only 66 Barasingha were left in and around Kanha. The action plan adopted to save the animal species from extinction, included the following measures: (*i*) regeneration of grasslands; (*ii*) reclamation of the habitat; (*iii*) removal of human population from the area; (*iv*) construction of enclosure; and (*v*) taking-up of anti-poaching measures. All these activities resulted in the increase in Barasingha population from 66 in 1970 to 574 in 1988.

Box 20.2

Presently the total population of three subspecies of Barasingha (*i.e., Cervus duvauceli duvauceli, C.d. branderi* and *C.d. ranjitsinhi*) in India is about 3500-4000 animals (**H.R. Singh** 1997). The species is listed as *endangered species* by the IUCN Cervid Specialist Group.

The PHVA workshop for Barasingha, organised in 1995 by Wildlife Institute of India, Dehradun, has suggested measures for proper management of Barasingha to achieve viable population.

2. Gir Lion Project

Asiatic lions (*Panthera leo persica*) is protected in small area of 1515 sq. km. in Gir Forest of Kathiawar, Gujrat (India) which is natural habitat of this big cat (Fig. 20.3). Following factors have endangered the very existence of Gir lion: (*i*) unabated shooting; (*ii*) reduction of its natural habitat due to rapid development of agriculture; (*iii*) water pollution by locals which result in the death of herbivore animals (prey species of lions); (*iv*) drastic reduction in the wild ungulate population which forms natural prey of lion. Due to scarcity of natural food, lions prey upon livestock, which creates direct conflict between cattle owners and park managers.

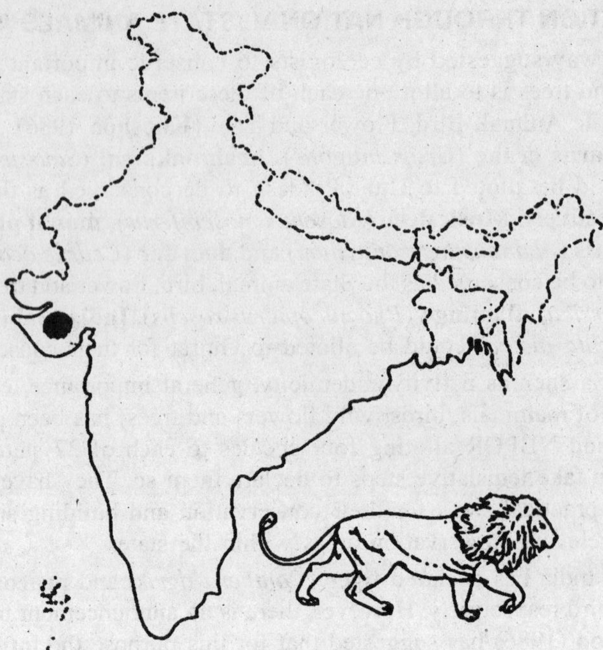

Fig. 20.3. Distribution of lion in India.

A five year plan was proposed in 1972 by the Government of Gujrat to protect Gir lions in the Gir Sanctuary. The lion population has increased from 177 in 1968 to 180 in 1984. Now a core area of 258.7 sq. km. of Gir sanctuary is declared as National Park and the ecological balance of the habitat is properly protected. Due to these recent efforts, now there are about 300 lions.

3. Project Tiger

Tiger (*Panthera tigris tigris*) is an endangered species and is listed in Red Data Book. The population of wild tigers reduced from 40,000 at the turn of the century to 1827 in 1972. Decline of tiger population was mainly due to hunting, habitat destruction by deforestation and taming the rivers for human needs. In response to alarming decrease a long-planning, called **Project Tiger** was proposed to keep the tigers with us in perpetuaty. The scheme of Project Tiger was so named because the tiger is an animal which symbolises the biological diversity of the Indian subcontinent. The project was launched on April 1, 1973 following the recommendations of a special task force of the Indian Board of Wildlife. The project was initiated with 9 Tiger Reserves (total area 13017 sq. km.) located in 9 different states. A *core zone* of minimum of 300 sq. km has been identified in each reserve along with a suitable *area of buffer*. Forest operations and cattle grazing have been completely banned in the core zones and these are being reoriented and also curtailed to suite the wildlife needs in the buffer zones. Villages have been relocated from the core zones and where necessary, also from buffer zones.

The main objectives of the Project Tiger were: (*i*) to ensure the maintenance of viable population of tigers for scientific, economic, aesthetic, cultural and ecological values, and (*ii*) to preserve the habitats of biological importance as a national heritage for the benefits, education and enjoyment of the people. The Project Tiger focuses its activities on protecting good tiger habitats, by creating new ones and extending the existing ones (Box 20.3). Besides tigers, the project helps to preserve and multiply some other animal species too, such as swamp deer, elephant, rhino, wild buffalo, hispid hare, pygmy hog, Bengal floricans and ghariyal. The floral diversity of the tiger reserves has also shown a significant improvement. Thus, programme of Project Tiger has had a direct impact on the conservation of biodiversity. Since tiger is a top-consumer and is at the apex of food-chain, so its growth symbolises the health of natural ecosystem.

The tiger population in the country has gone up from the threatened level of 1827 in 1972 to 4334 by mid 1990 (1989 census), both in tiger reserves and other areas of wildlife. In 1989 the largest number of tigers, 269 was reported from Sunderbans, followed by 97 in Kanha, 97 in Nagarjunsagar, 93 in Simplipal and 92 in Manas. However, quite alarmingly population of tiger has decreased in some tiger reserves. The drop in the tiger population in Kanha has been quite significant from 109 in 1984 to 97 in 1989. Similarly in the case of Manas tiger reserve the number of tigers came down from 123 in 1984 to 92 in 1989. Also, in Sariska National Park, the number of tigers declined from 26 to 19 in the same period. Outside the tiger reserves, the sharpest decline was reported in two north-eastern states of Meghalaya and Arunachal Pradesh.. Thus, decline in tiger population in Maghalaya between 1984 and 1989 was from 125 to 34 and in Arunachal Pradesh the decline was from 219 to 135 in the same period. The official 1993 tiger census estimated a total of 3750 tigers, of these only 1266 tigers were found in the boundaries of 19 tiger reserves. The 1993 census in M.P. estimated a population of 910 tigers, the highest density of tigers anywhere in the world.

Box 20.3

In 1972, there were 268 tigers in 9 tiger reserves and in 1972, 711 tigers were present in 11 tiger reserves. In 1982, it was decided by the Steering Committee of Project Tiger, headed by then Prime Minister Indira Gandhi to *enlarge* 6 tiger reserves and create 5 new tiger reserves. With this the number of reserves became 16 covering a total forest area of about 25000 sq. km. The new tiger reserves were **Namdapha** (Arunachal Pradesh), **Nagarjuna Sagar Shrishailm** (Andhra Pradesh), **Indrawati** (M.P.) and **Buxa** (West Bengal). The tiger reserves whose core areas had been extended were *Corbett, Manas, Simplipal, Palamau* and *Bandipur*. The tiger reserve whose buffer zone were extended were *Palamau, Corbett* and *Ranthambhor*.

To date 20 tiger reserves (now 25; **Rajaram** and **Suri,** 2000) have been established in 13 states covering around 29,976 sq. km forest area, the largest area in the world (Table 20.2).

The **Second Phase** of the Project Tiger is being launched in 1993 to refocus, restructure and reformulate its strategies to save not only the tiger and its habitat but also to conserve the entire biodiversity, rich in flora and fauna. The enhanced programmes introduced in second phase of project tiger include:

1. Establishment of guidelines for tourism in the tiger reserves;
2. Management of buffer areas to ensure availability of adequate firewood and fodder for the people living around the reserves;
3. Integration of local population through ecological development programmes; and
4. Establishment of Nature Interpretation Centres.

Table 20.2. Tiger Reserves in India.

	Name of the Tiger Reserve	Year of declaration	Area (sq. km.)			Wildlife protected
			Core	Buffer	Total	
1.	Bandipur (Karnataka)	1973	523	343	866	Tiger, Elephant, Gaur (or Bison), Leopard, Wild dog, Hyena, Langur, Jackal, Fox
2.	Corbett (Utranchal)	1973	338	183	521	Tiger, Elephant, Leopard, Rhesus monkey, Common langur, Black bear, Sambhar, Chital, Palam civet, Hyena, Gharial

	Name of the Tiger Reserve	Year of declaration	Area (sq. km.)			Wildlife protected
			Core	Buffer	Total	
3.	Kanha (Madhya Pradesh)	1973	940	1005	1945	Tiger, Chital, Sambhar, Black buck, Peacock
4.	Manas (Assam)	1973	470	2370	2840	Tiger, Clouded leopard, Wild buffalo, Golden langur, Black bear, Sambhar, Swamp deer, Gharial
5.	Melghat (Maharashtra)	1973	448	1170	1618	Tiger, Gaur, Hyena, Fox, Wild dog, Sambhar, Nilgai, Boar, Peacock
6.	Palamau (Bihar)	1973	213	715	928	Tiger, Leopard, Wild dog, Elephant, Gaur, Barking deer, Sambhar, Chital, Rhesus monkey, Boar, Peacock
7.	Ranthambore (Rajasthan)	1973	392	433	825	Tiger, Leopard, Sloth bear, Langur, Nilgai, Chital Mangoose, Hyena, Fox
8.	Simlipal (Orissa)	1973	845	1905	2750	Tiger, Elephant, Gaur, Sambhar, Chital, Sloth bear, Hyena, Peacock
9.	Sunderbans (West Bengal)	1973	1330	1255	2585	Tiger, Leopard, Boar, Chital, Otter, Python, Crocodile
10.	Periyar (Kerala)	1982	501	276	777	Tiger, Elephant, Gaur, Lion, Tailed macaque, Nilgiri langur, Sambhar, Flying squirrel
11.	Sariska (Rajasthan)	1982	498	302	800	Tiger, Jungle cat, Leopard, Langur, Porcupine, Hedgehog, Sambhar, Chital, Peacock
12.	Buxa (West Bengal)	1983	315	444	759	Tiger, Leopard, Rhino, Elephant, Chital, Gaur, Barking deer, Civet
13.	Indravati (Madhya Pradesh)	1983	1258	1541	2799	Tiger, Gaur, Leopard, Sambhar, Chinkara, Jackal
14.	Nagarjunasagar Shrishailam (Andhra Pradesh)	1983	1200	2368	3568	Tiger, Leopard, Gaur, Nilgai, Chinkara, Hyena, Sambhar, Peacock
15.	Namdapha (Arunachal Pradesh)	1983	1808	177	1985	Tiger, Leopard, Elephant, Gaur, Swamp deer, Sloth bear, Malabar squirrel
16.	Dudhwa (Uttar Pradesh)	1987	648	163	811	Tiger, Leopard, Sloth bear, Swamp deer, Sambhar, Chital, Hog deer, Braking deer, Nilgai
17.	Kalakad-Mundanthuari (Tamil Nadu)	1988	571	229	800	—
18.	Valmiki (Bihar)	1990	336	504	840	—
19.	Pench (Madhya Pradesh)	1992	300	500	800	Tiger, Leopard, Sloth bear, Sambhar, Wild dog, Porcupine, Chowsingha, Barking deer, Bison, Antelope, Nilgai, Spotted deer
20.	Bandhavgarh (Madhya Pradesh)	1993	700	459	1159	Tiger, Panther, Gaur, Boar, Chital, Bear, Sambhar, Nilgai, Chowsingha, Chinkara, Barking deer, Black buck, Ducks
		Total	13634	16342	29976	

Note: Five new project tiger sites are Tadoba Andheri (Maharastra), Bandhavgarh Panna (MP), Dhampa (Mizoram) and Bhadra (Karnataka)

4. Project Crocodile (Crocodile Breeding Project)

To regulate the alarming decline of Ghariyal population, the Government of India started the Crocodile Breeding and Management Project in 1975 in collaboration with FAO/UNDP under the technical guidance of **Dr. H.R. Bustard**, a FAO expert. Aim of this project was to save all the endangered crocodile species: 1. the salt-water or estuarine crocodile, *Crocodylus porosus,* found in the mangrove areas of the Sunderbans; 2. the fresh water or swamp crocodile, *Crocodylus palustris* which is called the "muger" or "muggar"; and 3. the ghariyal *Gavialis gangeticus.* Their population declined because of uncontrolled and all season hunting for skin, flesh and sport. Loss of habitat due to construction of dams, diversion of rivers and human interference were some other important factors.

Under crocodile project, ghariyal eggs were collected and taken to Orissa for expert incubation. They were hatched at Tickerpada in Dhenkanal District (Orissa) for the first time in June 1975. The Tikerapada hatched ghariyals were successfully reared for subsequent release back into wild. In the same year, a small batch of eggs was also hatched at Kukrail, near Lucknow. Since 1975, ghariyal and crocodile eggs are being collected from Chambal valley (Chambal river) and Katarina Ghat (Girwa river). Thus, crocodile project has not only saved this animal (*i.e.,* Ghariyal) but their number is also increasing.

So far around 3865 captive raised ghariyals have been released in 12 rivers in the country. Presently there are several "protected areas" for ghariyals such as National Chambal Sanctuary (spread over M.P., U.P. and W. Bengal); Corbett National Park, Katarina Ghat Sanctuary (U.P.); Ken Sanctuary (M.P.); Satkosia George Sanctuary (Orissa) and Son Sanctuary (M.P.). The captive Breeding Centres for ghariyal are Nandankanan Biological Park, Orissa; Kukrail Crocodile Rehabilitation Centre, Lucknow and Madras (Chennai). Crocodile Bank, Tamil Nadu, the biggest non-government Crocodile farm is managed by **R. Whitakar** in Chennai.

In 1995, the PHVA (Population and Habitat Viability Analysis) workshop for ghariyal was organised in Gwalior (M.P.) to review the progress of the Crocodile Project and made several recommendations for both in-situ and ex-situ conservation and management of ghariyals.

5. Operation Rhino

The Indian one-horned rhinoceros (*Rhinoceros unicornis*) is an endangered species. It is thriving mainly in Kaziranga National Park of Assam. Once they were widely distributed on the low marshlands on the banks of the Brahmaputra, in a small area in West Bengal and the low tarai lands along the rivers Rapti and Narayani in western-central Nepal. The tall elephant-grasses of Kaziranga serve as excellent shelter for the rhinos. According to a census carried in 1978, there were about 931 rhinos in the Kaziranga National Park (Assam). In 1992, there were less than 1500 Indian rhinos restricted almost entirely to Kaziranga National Park (1080) in Assam and Royal Chitwan National Park (375) in Nepal.

In last 200 years, rhinos were ruthlessly hunted for sport or poached for horn and hide. The Maharaja of Cooch-Behar shot 207 rhinos between 1871 to 1907 in West Bengal and Assam. In Kaziranga National Park out of 603 rhinos lost, during 1983 to 1989, 235 were killed by poachers for horn. The rest died due to flood, disease and old age. During the first two months of 1990, 16 rhinos are lost to poachers. In addition, an equal number of rhinos were lost in the flood in Kaziranga National Park in 1990. Likewise, the entire population of rhinos in the Laokhowa Sanctuary has been wiped out in recent memory. Rhinos were present in Dudhwa (U.P.) and adjacent alluvial plains until the 1817, the last of which shot in 1878 in Pilibhit District. Another reason for the disappearance of rhinos from most of its earlier range and a reduction in their number where they exist presently, was the conversion of its habitat into cultivation and tea plantation.

To protect the Indian species, centrally sponsored rehabilitation programme was undertaken in Dudhwa National Park. As a part of this programme, on April 1, 1984, five rhinos comprising two males and three females were air-lifted from the forest of Assam and released in Dudhwa Park. One year later, four female rhinos were brought from Chitwan National Park (Nepal) and released in Dudhwa. During 1989, four calves were born to four cows. One of the calves, a male died in January 1990. In 1991, two more calves were born and total number of calves rose to 5. The present number of reintroduced rhinos in the park is 10. Further attempts to rehabilitate rhino in Dudhwa are in progress.

6. Project Elephant

The Indian elephant, *Elephas maximus*, is another endangered species that demands for immediate action. The tuskers are mainly poached for ivory. The ivory trade is banned in India from 1992. Presently there are about 25,000 wild elephants in India. Elephant is distributed in forests of Uttar Pradesh, Uttaranchal, Bihar, Orissa, West Bengal, Assam, Arunachal Pradesh, Meghalya, Tripura, Karnataka, Kerala and Tamil Nadu (Fig. 20.4). Elephant in Kerala number about 4200, existing as seven populations, six of which are in ecological continuity with adjacent elephant-habitats of either Tamil Nadu or Karnataka. According to 1999 census of elephants in Uttaranchal, there were only 1911 elephants (*i.e.,* 509 males, 896 females and 506 baby elephants; source: *Denik Jagaran* Feb., 13rd, 2001).

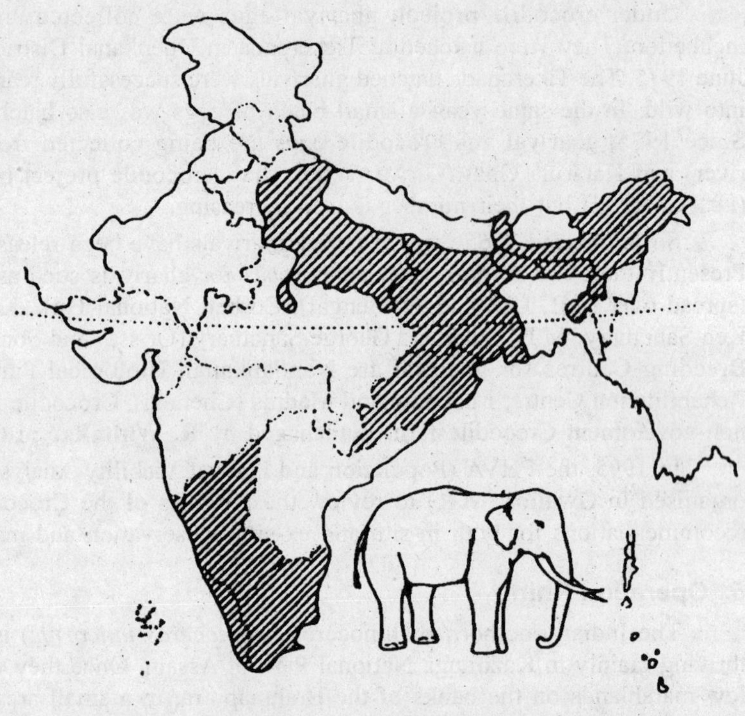

Fig. 20.4. Distribution of elephant in India.

Man-elephant conflict. Livestock grazing, firewood collection, encroachments and poorly planned developmental activities continue to destroy elephant's range/habitat. This is compounded by the fact that elephants driven by hunger, are forced to move out of disturbed natural habitats to raid crops, thus aggravating the already serious **man-elephant conflict.** This conflict becomes worse if a herd of elephant is isolated from the original habitat. An urgency to launch the Project Elephant therefore, was realised *not to increase the elephant population but to protect their habitats* so that elephants do not move out to raid the crop and kill human beings.

Project elephant, a Rs. 20 crore ambitious programme, was launched recently (1991-92) by Ministry of Environment and Forest for restoring the lost and degraded habitat of the elephant. The project would be executed in 10 States of the country, *viz.*, Assam, Andhra Pradesh, Bihar, Karnataka, Kerala, Meghalaya, Nagaland, Orissa, Tamil Nadu and West Bengal. The project is designed on the lines of Project Tiger and will continue through the eight plan period. Unlike the Project Tiger, however, Project Elephant will focus on solving the problems of humans and elephants competing to share the same habitat (Box 20.4).

Box 20.4 Salient Features of Project Elephant

1. Protection of all the viable populations and their habitat.
2. Providing migration corridors linking different parts of the composite elephant range.
3. Improving the quality of life of the people living in the fringe areas through appropriate eco-development programmes so that their dependence on forests becomes minimized.
4. Eradication of weeds such as mikania and lantena and augmenting the number of fodder trees through appropriate plantation programmes.
5. Decreasing the problems being caused by the isolated populations of elephants to the local human population.
6. Compensating people for the loss of life and property caused by elephants.
7. Strengthening the anti-poaching infrastructure.
8. Improved veterinary care and immunisation of domestic live stock against communicable diseases.
9. Restoring the feeling of compassion towards elephants through education campaign and extension programmes.
10. Scientific research on various aspects of management of elephants and their habitats.

7. Project Hangul

The hangul or Kashmir stag *Cervus elephus hanglu* is an endangered species. It is found in Jammu and Kashmir and Himachal Pradesh (Fig. 20.5). Its main concentration is in Dachigam Sanctuary. Hangul has become rare mainly due to human interference in the form of wood cutting, poaching, grazing, etc. From an estimated number of 3000 in 1940, it was reduced to 170 by 1970. According to an estimate only about 30 hanguls have survived outside Dachigam (**Saharia** 1981).

Project Hangul is started in 1970 to save hangul from extinction. The joint efforts of the experts, especially **Dr. Fred Kurt** and Jammu and Kashmir Government have ensured the survival of the hangul. In a census carried out in March 1980, their number was estimated to be 347.

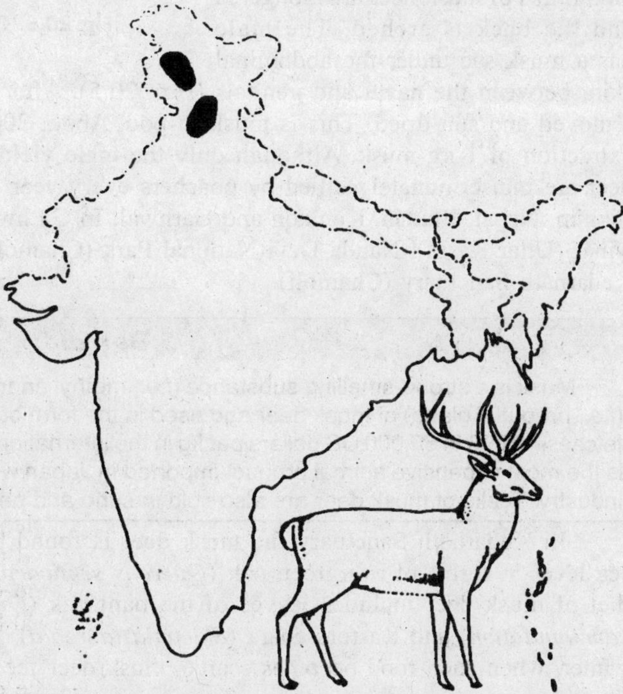

Fig. 20.5. Distribution of hungul of Kashmir stage in India.

8. Project Musk Deer

Globably musk deer contain following three species:

1. *Moschus chrysoga-ster* (in western China, Tibet and Himalayas);

2. *Moschus moschiferus* (in eastern Russia, northern China and Korea);

3. *Moschus berezovskii* (in southern China and North Vietnam).

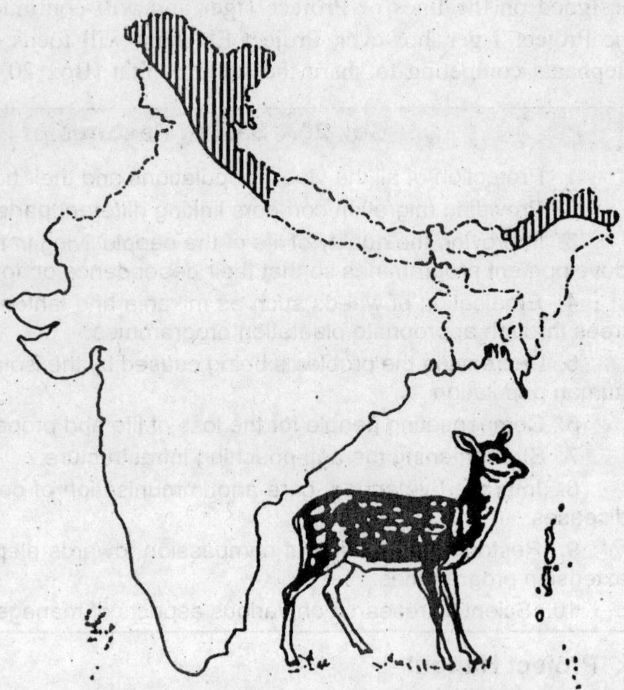

Fig. 20.6. Distribution of musk deer in India

The Himalayan musk deer, *Moschus chrysogaster,* is small-sized deer and is distributed in forests adjoining alpine meadows and in alpine scrubs, *i.e.,* in Afghanistan, Pakistan, Jammu and Kashmir, Himachal Pradesh, Nepal, Sikkim, Bhutan, Arunachal Pradesh and Burma. The male musk deer has no antlers, instead it has well developed canines (tusks) projecting below the lower jaws. The hind limbs of musk deer are elongated and the back is arched. The male has a musk sac under the abdominal skin, between the naval and genitals (Box 20.5). After the animal is dead, the entire gland is removed and sun dried. This is musk in pod. About 200 male musk deer have to be killed for extraction of 1 kg musk. Although only the male yields musk, thousands of male and female deer are indiscriminately killed by poachers every year in Kashmir, Ladakh, Himachal Pradesh, Sikkim, Nepal, Bhutan, Kumaon and Garhwal. In Garhwal, musk deer is found in Govind Pashu Vihar (Uttar Kashi), Nanda Devi National Park (Chamoli), Valley of Flowers National Park and Kedarnath Sanctuary (Chamoli).

Box 20.5

Musk is a strong-smelling substance (*i.e.,* methyl androsterone) obtained from the musk glands (*i.e.,* preputial gland) of musk deer and used in the form of a tincture as a fixative for perfume. Musk fetches 40,000 to 60,000 US dollars per kg in the international market (**Green** and **Singh** 1981). Musk is the most expensive animal product imported in Japan where it is used in cosmetic and perfumery industry. Tusks of musk deer are also sold in curio and perfumery industry.

In Kedarnath Sanctuary the musk deer is found between 2700 and 3700 m above mean sea level in forest of evergreen oak (*Quercus semicarpifolia*) and birch (*Betula utilis*). Natural diet of musk deer includes leaves of the banj oak (*Quercus incana*), semaru (*Rhododendron campanulatum*) and Kastura-chara (*Skimmia laureola*). It also eats tender herbs and grasses. In winter, when such food becomes scarce, musk deer feeds on lichens and mosses too.

Musk deer is a **threatened species** and efforts should be made to save this animal from extinction. The threatened status of the Himalayan musk deer is recognised by the IUCN, which lists the species as *vulnerable* in Red Data Book (IUCN, 1974). A taxa listed in this category are liable to become in danger of extinction if the causal factors such as overexploitation, extensive destruction of the habitat or other environmental disturbances continue operating. **Green** (1986) has suggested the following measures for future conservation of Himalayan musk deer: 1. Protection within National Parks and Sanctuaries; 2. Active involvement of local people in the conservation programme; 3. Effective control on musk trade through CITES both at national and international levels and 4. Sustainable exploitation. It has also been suggested that musk deer should be reared in semi-captivity and the musk could be extracted via the external orifice of the gland.

9. Project Brow-antlered Deer

Brow-antlered deer (*Cervus eldi eldi*) is locally called **sangai**. It is considered one of the rarest mammal found only in India. This deer survived mainly because of the protection provided by the rulers of Manipur State, but later during the British Raj, due to over-hunting by local people and army personnel, sangai was declared extinct in 1950. Later, it was located once again in a small area of Manipur, called **Keibul Lamjao** (Fig. 20.7) which was declared as sanctuary in 1954 and in 1977, it was given a status of a National Park. The survival of sangai in this area is because of the floating swamp through which this deer manages to move far better than humans and cattle. Keibul Lamjao National Park covers an area of 35 sq. km only. From an estimated number of 100 deer in 1960, it was reduced to 18 in 1977 which increased to 23 in 1980's. However, more numbers (49) of brow-antlered deer have been preserved in captivity.

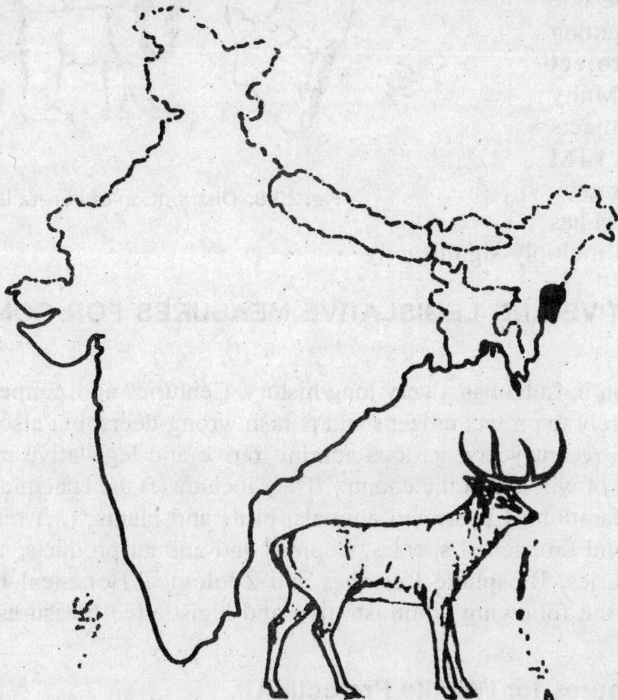

Fig. 20.7. Distribution of Manipur brow-antlered deer in India.

10. Project Lion-tailed Macaque

LTM or lion-tailed macaque (*Macaca silenus*) is a highly endangered species of monkey which is found only in India. It is distributed in region of Western Ghats of Kerala (Fig. 20.8). As the name suggests this monkey has a tail resembling to that of a lion but more than that it has a mane around the head like a lion. The colour of the mane is brownish gray, whereas, colour of body is dark black. LTM are shy animals and live on high canopy of trees of wet evergreen forests. They feed on fruits, insects, frogs and sometimes fungus. The total population of LTM is only 735 to 750 in the entire state of Kerala. In Western Ghat its range is restricted to Nilgiri, Anamalai, Cardamom hills, Peryiar National Park and Silent Valley. In fact, LTM came to limelight when the controversy over the Silent Valley hydro-electric project was raging the country. These monkeys become one of the key animals in getting this controversial project ultimately dropped. Many important research projects are going on to protect LTM in their geographical range. Captive breeding of LTM has caused their number to multiply significantly.

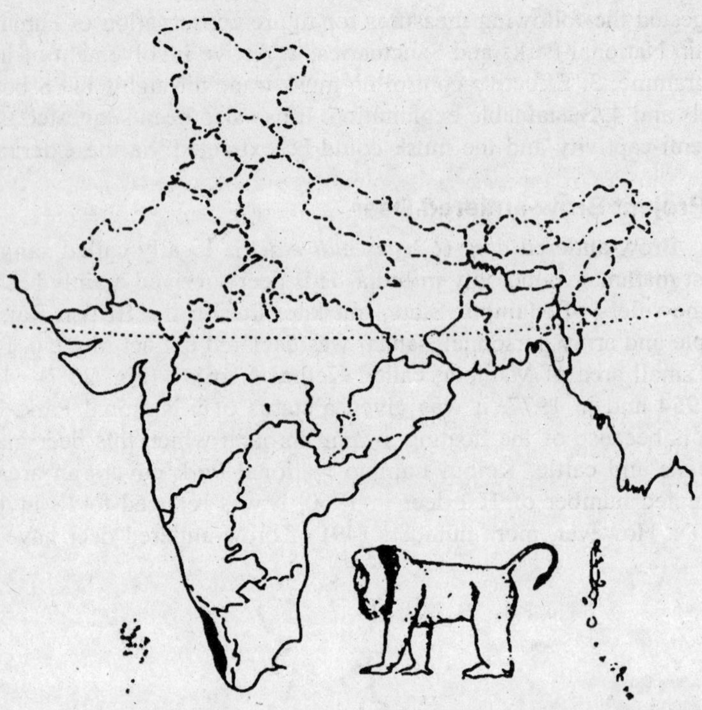

Fig. 20.8. Distribution of lion-tailed macaque.

20.4 ADMINISTRATIVE AND LEGISLATIVE MEASURES FOR CONSERVATION OF WILDLIFE

Wildlife protection in India has a very long history. Centuries ago, Emperor Ashoka defined a king's duty as not merely to protect citizens and punish wrong-doers, but also to preserve animal life and forest trees. In recent years, various administrative and legislative measures have been taken for the protection of wildlife in the country. They include (*i*) the enactment and enforcement of game laws; (*ii*) declaration of protected animals, birds and plants; (*iii*) restriction on trading in and export of rare and threatened species, trophies and animal products; and (*iv*) creation of National Park, Sanctuaries, Biosphere Reserves and Zoological/Botanical Parks/Gardens, etc. Here we can consider the following administrative and legistlative measures for the protection of Indian wildlife.

Administrative Measures for Wildlife Protection

A. International and National Bodies

1. World Conservation Union (WCU)

the WCU was formerly called **International Union for Conservation of Nature and Natural Resources (IUCN).** The IUCN, Morges (Switzerland) was established in 1948. It is independent international agency having a powerful membership, including 60 government and more than 500 nongovernment organisations (NGO's) among its 650 members from 120 countries. The WCU represents those who are *concerned* at the modification of the natural environment by humankind through the rapid growth of urban and industrial development and excessive exploitation of the earth's natural resources, upon which rest the very foundations of his survival.

The **main objective** of WCU is to promote and support action that will ensure the perpetuation of wild nature and natural resources in all possible parts of the world. This is done not only for their intrinsic cultural or scientific values but also for the long term economic and social welfare of humankind. To fulfil this objective the WCU promotes the following steps:

(*i*) Awareness through education, so that as many people as possible may understand the value and importance of renewable natural resources and appreciate the need to use them wisely. This includes the broadcasting of information through the press, radio, films and television and also through the WCU's own publications.

(*ii*) Research to discover the best measures for conservation and to advance the study of ecology upon which all practical conservation depends.

(*iii*) Assistance in providing technical data for the conduct of practical conservation programmes, and

(*iv*) Action on a national and international scale by enlisting the cooperation of governments and international agencies in support of conservation programmes as well as in strengthening legislation and improving its enforcement.

The WCU calls a *General Assembly* every third year in order to act upon issues of current importance, and to serve as forum for discussion of conservation problems. In November 1969, the 10th General Assembly Meeting was held in New Delhi; this was first time in Asia. The WCU is also the Scientific Adviser to World Wildlife Fund (WWF). Recently, WCU has started several conservation campaigns, including a global programme on the conservation of wetlands.

2. World Wildlife Fund (WWF)

The WWF was established in 1961 at Gland, Switzerland and the Giant Panda (*Ailuropoda melanoleuca*) was chosen its symbol. The WWF has the membership of a group of National Organizations representing many countries. India is one of them. The **WWF International** is managed by a board of trustees. The founder trustee is **Mr. Guy Mountfort** who had proposed to save the endangered tiger in India. In September 1986, WWF celebrated its Silver Jubilee at Assisi (Italy). WWF has grown from a basically fund raising organization for protection of individual species such as **Arabian oryx** (one of the first species to be saved) to a much important role "*care for nature*". At the start WWF was a "*fire brigade*" operation, *i.e.,* meeting the immediate needs. These could be through land protection, as at Coto Donana in Spain or by special mediation in defense of directly threatened species, such as Arabian oryx which has now been successfully reintroduced to Oman and Jordan.

The Giant Panda (Fig. 20.9) has proved a befitting symbol of WWF and what it has achieved over the past 25 years. Efforts to safeguard survival of Giant Panda (*Ailuropoda melanoleuca*), which is found only in Kansu and Szechwan provinces in North West China, have evolved along with WWF's. Its approach to conservation from simple protection measures to a project for systematic approach and management based on the realization that *effective conservation will need to consider the relationship between animal and the local people*. This new approach of WWF is reflected in the projects such as "Project Tiger" which

Fig. 20.9. The giant panda.

successfully mobilised funds of Government of India to save the species through its habitat and consequently a whole forest ecosystem.

In fact, the scope and activities of WWF are not restricted to animals only. Plants form an integral part of **WCS** (World Conservation Stratergy) which was launched in March 1988 (Box 20.6). WWF along with WCU has taken up a "*Plants Conservation Campaign*" which includes the following steps: (*i*) spreading the message; (*ii*) building the capacity to conserve; (*iii*) conserving plant genetic resources; (*iv*) conserving wild plants of economic value; (*v*) development of botanical gardens; and (*vi*) promoting plant conservation in selected countries.

Box 20.6

The three main objectives of the WCS are

1. to maintain essential ecological processes and life support system;
2. to preserve the genetic diversity; and
3. to ensure the sustainable utilization of species and ecosystems.

Worldwide Fund for Nature (WFN). After its silver Jubilee Celebration at the town of Assisi in 1986, World Wildlife Fund (WWF) has been renamed as "**Worldwide Fund for Nature**" or **WFN**. Of the more than 4000 projects in 130 countries that WWF supported in these years, the *Project Tiger* provides best example of its approach. In 1972, **Mr. Guy Mountfort**, founder and trustee of WWF, proposed a campaign to save the endangered tiger. In that year wildlife census had revealed that the tiger population had reduced to 1827 from an estimated 40,000 in 1900. Due to efforts of Tiger Project, on the 10th anniversary of the project in 1983, the count of tiger was about 4000. WWF has turned over about 1 million U.S. dollars to India's "Project Tiger" and spent most of the rest on tiger conservation in Bangladesh, Bhutan, Malaysia, Indonesia and Thailand.

3. WWF-India (Worldwide fund for Nature, India)

WWF-India was founded in 1969 with its head-quarters at Bombay. It is registered charitable trust established with the objective of safeguarding and restoring the country's natural environments. Its main aims include **conservation, education** and **research**. The activities of WWF-India involve (*i*) financing specific conservation projects and related research studies, promoting conservation, education among youth, creating public awareness and eliciting mass support for conservation; (*ii*) focusing the Government's attention on special conservation, and (*iii*) raising funds to support these activities. WWF-India has undertaken and supported various important projects, some of which are the following: (*i*) Andman Pig study; (*ii*) Crane Conservation Project; (*iii*) Crocodile Breeding Project; (*iv*) Ecological Survey of Western Ghats; (*v*) Sloth Bear Project, (*vi*) Status Survey of Sea Turtle, etc.

4. Wildlife Preservation Society of India

The Wildlife Preservation Society of India, Dehradun is a non-Government Organization founded in 1958 by a group of wildlife lovers. It is a member of WCU or World Conservation Union. The aims of the society are (*i*) to promote interest in wildlife and impart knowledge in the protection of wildlife; (*ii*) to cooperate with the government in the matter of wildlife protection; (*iii*) to assist in enforcing wildlife protection acts; (*iv*) to promote wildlife tourism; and (*v*) to advise and help the Government and administrators in the formation, maintenance and protection of National Parks and Sanctuaries. Wildlife Preservation Society of India is publishing a journal entitled **Cheetal** which has rendered a great service to arouse public interest in wildlife and to save wildlife.

5. Bombay National History Society (BNHS)

BNHS is founded in 1883 and it is a non-Government body dedicated to the cause of nature conservation. The society has been actively engaged in collecting information and specimens of fauna and flora throughout India, Burma and Shri Lanka. BNHS has been active in focusing attention of public and government officials on the need for proper understanding, conserving and developing the rich heritage of India's wildlife. The prestigious journal of the society is very well serving the cause of conservation of nature through publication of articles and new reports on fauna and flora.

6. Indian Board of Wildlife (IBWL)

With the initiation of Lt. Col. **Burton**, Government of India set up an adhoc committee to consider the steps to be taken for ensuring wildlife protection. On the recommendations of this committee, IBWL was constituted in 1952. Its main objectives include:

1. To sponsor the setting up of National Parks, Sanctuaries and Zoological Gardens for preserving wildlife in their habitats.

2. To device ways and means to protect the wildlife through coordinated legislature and practical measures. They have to declare certain species of animals as protected animals, and check the indiscriminate killing.

3. To promote public interest in wildlife and the need for its preservation in harmony with natural and human environments.

4. To prevent cruelty to beasts and birds caught alive or without injury and

5. To advise the Government on policy in respect of living animals, trophies, skins, furs, feathers, ivory and other wildlife products.

IBWL comprises an Executive Committee, a Bird Wing, a Zoo Wing to deal respectively with the study and preservation of wild birds and wild animals and the development of Zoological gardens and Zoological parks in the country.

State Boards for Wildlife (SBWL) have similar functions to discharge in every State of India. These State Boards have additional responsibility of enforcement of legislation effectively against poaching and illegal killing of game birds and other animals in the protected areas. With the aim to educate the people on the subject of wildlife IBWL celebrates a "Wildlife Week" in the first week of October each year.

7. National Wildlife Action Plan (NWAP)

The NWAP was adopted in 1983 to provide the framework of a strategy involving programmes and projects for wildlife conservation in the future. Under this plan, the protected area proposed to be increased to cover 4 per cent of the geographical area, as against 3 per cent at present. The *Central Directorate of Wildlife Preservation* and the *Wildlife Institute of India,* Dehradun are the nodal agencies for **initiating** and **monitoring** the programmes and projects set out in the Action Plan with the help of States and Union Territories. Both of them are directly responsible for the actual protection and management of wildlife in the country. NWAP adopts the following measures:

1. Survey of all National Parks, Sanctuaries and other areas deserving protection status.
2. Guidelines for the preparation of management plans of wildlife reserves have been drawn up and circulated to States and Union Territories.
3. Review and revision of National Forest Policy has been taken up.
4. Amendments to the Wild-Life (Protection) Act 1975 examined.
5. Captive breeding and rehabilitation programmes have been launched.

8. Red Data Book (RDB)

For wildlife preservation in the 7th and subsequent plans, '*Red Data Book*' (RDB) will be introduced. RDB will provide a complete list of all endangered animals and plants in the country for the first time. About 253 species, subspecies of wild fauna including mammals, birds and reptiles have so far been included in schedule I of Wildlife (Protection) Act which was prepared in 1972 and updated in 1979. About 2000 species of flowering plants were reported to be endangered. Botanical Survey of India has found that there are about 135 species of endangered plants.

9. DNA fingerprinting

The Third International conference on DNA fingerprinting held in Hyderabad on Dec. 13-16, 1994 debated the more use of the techniques for wildlife management. For both in-situ and ex-situ populations it is essential for determining subspecies, level of genetic variability and paternity. DNA fingerprinting in wildlife forensics has been used to identify animal products and in legal arrest of poachers and smugglers.

Legislative Measures for Wildlife Protection

In India, the following legislative measures have been adopted for the protection of wildlife in the country:

1. Directive Principles and State Policy. Founding fathers of our constitution had realised the need of wildlife protection. In the section of Directive Principles of State Policy, the following duties for the state and the citizens are laid down:

"*The State shall endeavour to protect and improve the environment and to safeguard the forests and wildlife of the country.*"

Article 51-A *inter alia* states that:

"*It shall be the duty of every citizen of India to protect and improve the natural environment including forests, lakes, rivers and wildlife and to have passion for living creatures.*"

2. Forests and wildlife in Concurrent List. Forests and wildlife have been included in the Concurrent List in the constitution by the 42th Amendment in 1976. By virture of this provision, not only can Central Government Control protection and preservation of the wildlife, but certain forests and wildlife sanctuaries of national importance can also be acquired by the Central Government under the powers of acquisition of property (Entry 42 of Union List).

3. Acts to protect wildlife. To prevent wildlife depletion, several Acts have been made from time to time, by State as well as Central Government. Some important acts are:

(*i*) Madras Wild Elephant Preservation Act, 1873;
(*ii*) All-India Elephant Preservation Act, 1879;
(*iii*) Wild Birds and Animals Preservation Act, 1912;
(*iv*) Bengal Rhinoceros Preservation Act, 1932;
(*v*) Bombay Wild Animals and Wild Birds Protection Act, 1951;
(*vi*) Assam Rhinoceros Protection Act, 1954;
(*vii*) Indian Wildlife (Protection) Act, 1972 (amended in 1983, 1986 and 1991).
(*viii*) The Government banned the exhibition and training of bears, monkeys, tigers, panthers and lions in 1998 (Indian Express, Oct. 21st, 2000).

Based on recommendations of the IBWL, the Indian Wildlife (Protection) Act was passed by Parliament in 1972. The act is adopted by all states except Jammu and Kashmir, which has its own Act. This Act provides special legal protection to wildlife in general and to the endangered species of fauna in particular. It has a provision for setting up of National Parks and Sanctuaries where our wildlife can receive the fullest protection.

The Wildlife (Protection) Act 1972 has several chapters dealing with the hunting of wild animals, authorities to be appointed under the Act, declaration of Sanctuaries, National Parks, Nature Reserves and closed areas, regulation of trade in wild animals, animal products and trophies, and detection and prevention of offences. Under the Wildlife (Protection) Act, wild animals, birds, amphibians and reptiles are given specific status under 5 different schedules:

1. **Schedule I** includes 41 rare and threatened species of mammals, 2 species of amphibians and 18 species of birds.
2. **Schedule II** includes the animals which enjoy special game status.
3. **Schedule III** includes big games (wild animals).
4. **Schedule IV** includes small games (wild animals).
5. **Schedule V** includes vermin (wild animals).

The animals listed in these schedules are provided rigid protection under the legislation. The hunting of these animals is restricted more than ever before and strict control measures are imposed on it.

The Wildlife (Protection) Act, 1972 has a provision for the regulation and control of trade in wild animals, animal products and trophies. It lays down regulations governing the declaration of stocks, preparation of inventories; issue of certificates of ownership; regulation of transfer of animals, etc., from one state to another and from one owner to another; issue of licences for conducting trade, maintenance and inspection of records and necessary limitations and penal measures.

4. Convention of International Trade in Endangered Species (CITES). The profitable returns and the great demands of the products of wild animals, such as musk pods of musk deer, horns of rhinoceros, skin of crocodile, tusks of elephant, and skin of large wild mammals and snakes, etc., have attracted poachers and smugglers to engage in clandstine trade with other nations for fetching more money in the international market. Keeping in view the regulation of global legal and illegal trade in endangered species, India became a party to the International Convention on Trade in Endangered Species (of Wild Fauna and Flora) in 1976. The convention and legal procurement of certificates to control the trade of all the species listed in the "Appendices" to the convention. The appendix I of CITES, for example, has species threatened with extinction through trade. Appendix II lists species that may become threatended with extinction if trade involving them is not regulated. These lists are likely to increase in future. Under the convention, countries which are party to it cannot import or export the species listed in the CITES without proper licences.

In each country, there is a separate management for implementing the provisions of the convention. **Inspector General of Forests** and the **Director of Wildlife Preservation,** Government of India have been designated as the management authorities. The **Director of Zoological Survey of India,** the **Director of Botanical Survey of India** and the **Director Central Marine Research Institute** have been appointed as the Scientific Authorities for India. All international trade related with wildlife can be operated only from the four major ports of Mumbai, Calcutta, Delhi and Madras. These ports are controlled with the help of four Assistant Directors posted at these places for supervising the trade. The species permitted for the export or import trade, are included in the Export Control Policy from year to year.

A bilateral **Convention on Migratory Birds** has also been negotiated between India and Russia. This convention aims at protecting the birds which migrate annually between India and Russia.

QUESTIONS

Long Answer Questions

1. What do you understand by biosphere reserve? Name few important biosphere reserves of India and give their role also.
2. What do you mean by biosphere reserves? What is their role? Name some biosphere reserves of India.
3. Describe various conservation projects of India for different endangered species.
4. Give an account of Gir Lion Sanctuary Project.
5. Give an account of Crocodile Breeding Project.
6. Describe various legislative and non-legislative measures for conservation of wildlife.
7. What is the role of wildlife legislation for protection of wildlife in india? Give its salient features.

Short Answer Questions

1. Write a note/short note on 'Project Tiger'
2. What is the name of the State Animal (mammal) of H.P.

Very Short Answer Questions

1. What do you understand by MAB?
2. What are Biosphere Reserves?

Fill in the Blanks

1. Hazaribag National Park is located at __________.
2. Dudhwa National Park is situated at __________.
3. Name the Tiger sanctuary situated in U.K. __________.

Multiple Choice Questions

1. Ex-situ conservation of musk deer in Uttar Pradesh is being done at (*CCSU Meerut 2014*)
   (*a*) Nanda Devi (*b*) Valley of flower
   (*c*) Kukrail (*d*) Kanchula Kharak
2. Red Data Book (RDB) gives a complete list of (*CCSU Meerut 2014*)
   (*a*) endangered animals (*b*) endangered plants
   (*c*) both endangered animals and plants (*d*) none of the above
3. The Earth summit at Rio de Janerio in 1992 resulted into (*CCSU Meerut 2014*)
   (*a*) compilation of red list (*b*) establishment of biosphere reserves
   (*c*) convention of biodiversity (*d*) IUCN
4. A high density of a protected animal in a National Park can resulted into (*CCSU Meerut 2014*)
   (*a*) emigration (*b*) predation
   (*c*) mutualism (*d*) intraspecific competition
5. Wild Life Protection Act was enacted by Indian Government in (*CCSU Meerut 2014*)
   (*a*) 1952 (*b*) 1983
   (*c*) 1972 (*d*) 1986
6. Red Panda is state symbol of (*CCSU Meerut 2014*)
   (*a*) Manipur (*b*) Sikkim
   (*c*) Mizoram (*d*) Kashmir
7. Identify the correct match of a tiger reserve and state in which it is located (*CCSU Meerut 2014*)
   (*a*) Corbett—Madhya Pradesh (*b*) Palamau--Orissa
   (*c*) Manas ---Assam (*d*) Bandipur-Tamil Nadu

8. Asiatic lion is protected in (*CCSU Meerut 2014*)
(a) Kaziranga National Park
(*b*) Gir National Park
(*c*) Kanha National Park
(*d*) Desert National Park

9. Which animal is the symbol of the world wildlife Fund? (*CCSU Meerut 2014*)
(*a*) tiger
(*b*) hornbill
(*c*) giant-panda
(*d*) white bear

10. Red Data Book is produced by? (*CCSU Meerut 2014*)
(*a*) IBWL
(*b*) ZSI
(*c*) WWF
(*d*) IUCN

11. Main objectives of Man and Biosphere programme are (*CCSU Meerut 2014*)
(*a*) conservation
(*b*) research and education
(*c*) local involvement
(*d*) all of the above

12. Project tiger in India started in (*CCSU Meerut 2014*)
(*a*) 1970
(*b*) 1973
(*c*) 1982
(*d*) 1985

13. 'MABP' stands for (*CCSU Meerut 2014*)
(*a*) Man and Biology Programme
(*b*) Man and Biosphere Programme
(*c*) Man and Biosphere Project
(*d*) Mammals and Biology Programme

14. Name the sanctuary in which rare animal Hispid hare (*Caprolagus hispidus*) is found
(*a*) Dachigam National Park
(*b*) Kaziranga National Park
(*c*) Manas National Park
(*d*) Kanha National Park

15. 'Project Musk Deer' was launched in 1974 in
(*a*) Dudhwa National Park–U.P.
(*b*) Corbett National Park–U.P.
(*c*) Kedarnath Sanctuary–U.P.
(*d*) Dachigam National Park–J. & K.

ANSWERS

Very Short Answer Questions

1. MAB means Man and Biosphere programme. It was pioneered by United Nations Educational Scientific and Cultural Organization (UNESCO) in 1971. Under MAB programme hundreds of Biosphere Reserves have been designated all over the world.
2. Biosphere Reserves are land, water and/or coastal environments with their living organisms that together constitute a world- wide network of scientific information and include significant examples of natural biomes or an unique representative biological area throughout the world.A total 425 Biosphere Reserves have been created till 2003 in 95 countries. Till the year 2005 fourteen biosphere reserves were set up in the country (*i.e.*, India) (**Sharma** 2009).

Fill in the Blanks

1. Hazaribagh (Bihar); **2.** Lakhimpuri Kheri (U.P); **3.** Corbett National Park.

Multiple Choice Questions

1. (*b*)	**2.** (*c*)	**3.** (*c*)	**4.** (*d*)
5. (*c*)	**6.** (*b*)	**7.** (*c*)	**8.** (*b*)
9. (*c*)	**10.** (*d*)	**11.** (*d*)	**12.** (*b*)
13. (*b*)	**14.** (*c*)	**15.** (*c*).	

21

Pollution: Air Pollution

21.1 DEFINITION

Any undesirable change in the physical, chemical or biological characteristics of air, water and soil, that may create a hazard to the health, safety or welfare of any living species is called **pollution**. *Environmental pollution* is the act of introduction by humans of extraneous substances or energy into the environment that induce unfavourable changes. These changes may affect humans directly or indirectly by endangering his health, harming his living resources and ecosystem or by interfering with legitimate uses of the environment (**Holdgate** 1979; **Kannan** 1997). In a broad sense, pollution means the presence of any substance (solid, liquid or gas) or agent (noise or heat) in the atmosphere in such a concentration that may be injurious to environment affecting living or non-living things (Box 21.1).

Box 21.1

Environmental pollution is a global problem and is common to both developed as well as developing countries. The decline in environmental quality as a result of pollution is evidenced by loss of vegetation cover and biological diversity, excessive concentration of heavy chemicals in the ambient atmosphere and in foodgrains, growing risks of environmental accidents and threats to life support system.

Fertilisers, detergents, biocides, chlorofluorocarbons (CFCs), plastics and pesticides, solvent, fuel, paints, dyes, medicines, food additives, etc., are some examples of the multiplicity of chemical products made and disseminated for the benefit of humans. All these **xenobiotic substances** have the inherent capacity to disturb the life support system. Garbage, sewage, smoke and factory effluents all are threatening the life line of ecosphere.

The problem of pollution have gained serious dimensions in industralized countries and are spreading fast world-wide. The build-up of carbon dioxide concentration in the atmosphere is threatening to change the global climate pattern. Acid rains add to the problem by damaging lakes, rivers and forests. CFCs are destroying the ozone layer. Pollution is seriously damaging the environment of planet Earth at a fast pace, threatening the very survival of life in its entirety.

Causes of Pollution

Population explosion along with urbanisation and industralisiation have greatly increased the intensity of pollution with more people, there has been more sewage, more solid waste, more fuel burning, more use of fertilizers and biocides (pesticides, insectisides, etc.).

21.2 POLLUTANTS

Every human society, be it rural, urban, industrial and most technologically advanced society, dispose of certain kinds of by-products and waste products which when are injected into the biosphere in quantities so great that they affect the normal functioning of ecosystems and have an adverse effect on plants, animals, and humans are collectively called **pollutants**. In **Environment (Protection) Act, 1986 of India** "*environmental pollutant*" means any solid, liquid or gaseous substance present in such concentration as may be, or tend to be injurious to environment; and "*environmental pollution*" means the presence in the environment of any environmental pollutant.

In a way, a pollutant is a resource in the wrong amount, at the wrong place or at the wrong time. For example, nitrogen and phosphorus are essential nutrients for living organisms and are extensively used in agriculture to increase crop yields but they can also cause pollution in lakes and rivers when found in excess by promoting undue algal growth (eutrophication). Thus, it can be said that pollution is resource at wrong place.

Principal Pollutants

1. **Gaseous pollutants.** Oxides of nitrogen (NO, NO_2), sulphur compounds (SO_2, H_2S), carbon compounds (CO_2,CO), Ozone, Halogens (chlorine, bromine, iodine, etc.) and Chlorofluro-carbons (CFCs).
2. **Fluoride compounds**
3. **Metals.** Mercury, lead, iron, zinc, nickel, tin, cadmium, etc.
4. **Agricultural pollutants.** Pesticides, herbicides, fungicides and fertilizers (Box 21.2).
5. **Complex organic pollutants.** Benzene, benzpyrens, acetic acid, ether, etc.
6. **Biotic components.** Pollens.
7. **Deposited matter.** Soot, smoke, tar, dust, grit, etc.
8. **Solid waste**
9. **Radioactive waste**
10. **Noise**
11. **Heat**

Types of Pollutants

Pollutants primarily are grouped into the following two types:

1. **Natural pollutants.** Certain pollutants such as carbon dioxide, carbon monoxide, sulphur dioxide, lead, mercury and other trace elements are the consequence of life processes. These are produced through respiration, faeces, urine and body decomposition. With an increase in human population, the pollutants are increasing with the alarming rate.

Geo-chemical substances (dust, sediment, grit, etc.), biotic component (pollens, products of microbial activity) or physical agents (heat, sound, etc.) are other examples of natural pollutants.

2. **Synthetic, man-made, anthropogenic or xenobiotic pollutants.** A vast array of synthetic pollutants are increasing continuously with urbanization and industrial growth. They include pesticides, detergents, radionuclides, pharmaceuticals, cosmetic products, organic acids, aerosols, and metals, etc. Several of these compounds are extremely stable and persist in the environment for a considerable period posing serious environmental hazards.

From the ecosystem viewpoint, these pollutants can be classified into two basic types: **non-biodegradable pollutants** and **biodegradable pollutants**. The materials and poisons, such as aluminium cans, mercurial salts, long-chain phenolic chemicals and biocides (*e.g.,* DDT) that either do not degrade or degrade only very slowly in the natural environment, are called **non-biodegradable pollutants**. Such non-biodegradable pollutants not only accumulate but are often "biologically magnified" as they move in biogeochemical cycles and along food chains. Also they frequently combine with other compounds in the environment to produce additional toxins. The **biodegradable pollutants** include domestic sewage, heat, etc. The domestic sewage can be rapidly decomposed by natural processes or in engineered systems (such as a municipal sewage treatment plant) that enhance nature's great capacity to decompose and recycle. Problems arise with the biodegradable pollutants when their input into the environment exceeds the decomposition or dispersal capacity.

Box 21.2 Biocides

Biocides are biologically active organic molecules synthesized by chemists. They are introduced into the environment to kill certain plants, insects or rodents. **Pesticides** are substances used to kill or control undesirable organisms (*i.e.*, pests). Pesticides are specifically termed (as shown below) according to the group of pests against which they are employed.

1. **Acaricides** are used to control undesirable **Mites**.
2. **Algicides** are used to control undesirable **Algae**.
3. **Avicides** are used to control undesirable **Birds**.
4. **Bactericides** are used to control undesirable **Bacteria**.
5. **Fungicides** are used to control undesirable **Fungi**.
6. **Herbicides** are used to control undesirable **Weeds**.
7. **Insecticides** are used to control undesirable **Insects**.
8. **Molluscicides** are used to control undesirable **Molluscs**.
9. **Nematicides** are used to control undesirable **Nematodes**.
10. **Piscides** are used to control undesirable **Fishes**.
11. **Rodenticides** are used to control undesirable **Rodents**.

Costs of Pollution

Costs of pollution can be estimated by following ways:

1. Costs of human health. A number of fatal human diseases such as tuberculosis, typhoid, lung cancer, skin cancer, respiratory diseases, hypertension, etc., are cansed by pollution.

2. Loss of efficiency. There is loss of efficiency of workers as a result of unfavourable environment in which they work. This causes reduction in working hours and thereby reduction in output causing economic loss.

3. Costs of pollution control. Money, manpower, etc., are used for disposal of pollutants.

4. Decline in soil fertility. There is decline in fertility of soil that results in reduction in crop production and loss of vegetation cover.

5. Corrosion of metals. Metals such as iron, steel, copper, brass, nickel, zinc, lead, etc., are being corroded.

6. Soiling of buildings, clothes, etc. Costs incurred towards cleaning, painting, resurfacing of buildings.

Kinds of Pollutions

The pollution have been classified and studied in several ways. Two of the most widely used approaches are the following:

A. Classification according to pollutants

1. Sewage pollution
2. Pesticide pollution
3. Smoke pollution
4. Smog pollution
5. Carbon monoxide pollution

6. Chemical pollution
7. Effluent pollution (solid or liquid waste pollution)
8. Radioactive pollution
9. Acid rain pollution

B. Classification according to environment. When different parts of the environment are affected by pollution, the problem is generally discussed under the following headings:

1. Air pollution
2. Water pollution
3. Soil (agricultural land) pollution

When pollution of different habitats are discussed, terms such as **freshwater pollution, marine pollution, river pollution** and **lake pollution**, etc., may be used. Besides these, the **thermal pollution** and **noise pollution** are also alarming ecologists and environmentalists.

21.3 AIR POLLUTION

Air pollution is the presence in the atmosphere, or injection into it, of substances that are not present naturally, or present naturally but is in much smaller concentrations, and that may harm living organisms directly or indirectly (**Allaby,** 1995). Thus, when due to some natural processes or human activities the amount of solid waste or concentration of gases other than O_2 increase in the air which normally has constant percentage of different gases in it, the air is said to be polluted and this phenomenon is referred to as **air** or **atmospheric pollution.** Air pollution is one of the most dangerous and common kind of environmental pollution that is reported in most industrial towns and metropolitans of India and abroad such as Delhi, Bombay, Calcutta, Kanpur, Madras, Hyderabad, Jaipur, Ahmedabad, Nagpur, Firozabad and also in London, New York, Tokyo, Pittsburg, etc.

Air Quality

In our country, data on air quality have been collected by NEERI (1978-79, 1980-81). Although there exist several parameters to determine air quality, but only three, *i.e.,* SO_2, NO_x and SPM (= suspended particulate matter) are used; these give fair idea of load of pollution carried by the air. NEERI has collected data from ten Indian cities/Metropolitans such as Delhi, Kanpur, Calcutta, Nagpur, Hyderabad, Madras, Cochin, Bombay, Ahmedabad and Jaipur to assess the extent and nature of degradation of air quality due to industrialization and urbanization. Each of these ten Indian cities was surveyed for residential, commercial and industrial situations.

Thus, for sulphur dioxide (SO_2), Calcutta is found to be most polluted city, followed in a descending order by Bombay, Delhi, Ahmedabad, Kanpur, Hyderabad, Madras, Nagpur and Jaipur. However, yearly averages did not exceed 80 $\mu g/m^3$. Levels of oxides of nitrogen (NOx) ranged from 4$\mu g/m^3$ in residential and industrial areas of Jaipur to the highest 40 $\mu g/m^3$ in industrial areas of Ahmedabad (1980) and commercial areas in Kanpur (1980). Level of SPM was highest in Delhi and Calcutta and the lowest in Madras and Bombay (coastal cities). In general, SPM levels in all cities were much above the international levels. Residents of Delhi are exposed to some of the highest levels of air pollution in the country and perhaps the world.

Methods of Detection and Measurement of Air Pollution

Air pollution is usually measured by sampling of air by *thermal* and by *electrostatic precipitation, Sonkin impactor* and *electrostatic dust collectors.* The particulate pollution is measured by the instrument called *deposit gauge* or by *Owen's dust counter.* The thickness of the smoke is measured by *Liegean sphere* and by *Ringelmann chart.* The rough estimation of

SO_2 in air can be made by chemical analysis of the dust collected in a deposit gauge or by a *bubbler method.* Fluorides are estimated by colour reactions.

Sources of Air Pollution

1. Air Pollution by Natural Means

Nature adds few natural pollutants such as pollen, hydrocarbons released by vegetation, dusts from deserts, storms, and volcanic activity. Thus, **volcanic eruptions** may eject large amounts of gases and particulate matter. Settling volcanic ash can kill vegetation by coating leaves and preventing photosynthesis and transpiration (*e.g.,* following the 1980 eruption of Mt. St. Helens, USA). Fine particles, mainly of sulphates, may penetrate the stratosphere, spread widely and reflect significant amounts of solar radiation, leading to climatic cooling (*e.g.,* the 1991 eruption of Mt. Pinatubo, Phillippines). Likewise, **dust storms** sometimes carry fine sand for thousands of kilometers and favourable weather conditions stimulaté the release of pollen, affecting people sensitive to it.

2. Air Pollution of Human Activities

(*a*) Industrial chimney wastes. There are a number of industries which are potent sources of air pollution. **Petroleum refineries** are the major sources of gaseous pollutants (*e.g.,* SO_2, NO_x, etc.) Mathura-based petroleum refinery has been accused to aggravate the pollution-related decay of Taj Mahal in Agra and other historical monuments of Fatehpur Sikri Complex. **Industrial processors** such as metallurgical plants and smelters, chemical plants, pulp and paper mills, sugar mills, cotton mills, and synthetic rubber manufacturing plants are responsible for about one-fifth of the air pollution. **Cement factories** emit plenty of dust, which is potential health hazard. Stone crushers and hot mix plants also create a menace. The SPM levels in such stone crushing areas are found to be five time the industrial safety limits. **Chemical manufacturing industries** emit acid vapours in air.

(*b*) Thermal power stations. The coal consumption of thermal power stations of India (*e.g.,* Delhi has three thermal power stations, one at Indraprastha Estate, others at Rajghat and Badarpur) is several million tonnes. The chief pollutants of coal burning are fly ash, SO_2, and other gases (CO, NO_2), aldehydes and hydrocarbons (**Chaudhuri**, 1982).

(*c*) Automobiles. The transportation industry exclusive of automobiles and including railroads, ships, aircrafts, trucks, buses, tractors, etc., contribute the same type of pollutants as cars, scooters and motorbikes. The vehicular exhausts are toxic being a source of considerable air pollution, next only to thermal power plants. The ever increasing vehicular traffic density poses continued threat to the surrounding air quality. At the global level, there are over 300 million cars, trucks and buses and their number is increasing rapidly. India too has millions of vehicles, of which more than 65 per cent are two wheelers operating on petrol.

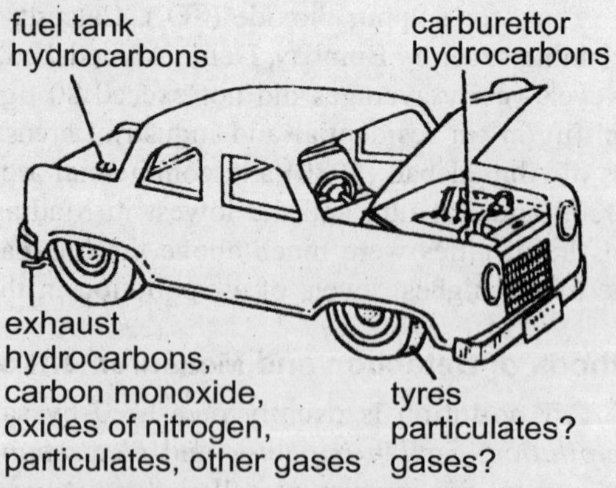

Fig. 21.1. Hydrocarbons and other particulate and gaseous emissions from automobile.

The sources of emission in the automobiles are (*i*) exhaust system; (*ii*) fuel tank and carburettor and (*iii*) crankcase (Fig. 21.1). The exhaust produces many air pollutants including unburnt hydrocarbons, CO, NO_x and lead oxides. There are also traces of aldehydes, esters, ethers, peroxides and ketones; these are chemically active and combine to form smog in presence of light. Due to volatile nature of petrol, evaporation from fuel tank goes on constantly and results in emission of hydrocarbons. The evaporation through carburettor occurs when engine is stopped and heat builds up and as much as 12 to 40 ml of fuel (petrol/diesel) is lost during each long stop causing emission of hydrocarbons. Some gas vapour escapes between walls and the piston, which enters the crankcase and then discharges into the atmosphere. Thus, the total hydrocarbon emission of an engine reaches upto 25 per cent.

Other sources of air pollution are minor in quantities but bear significance due to the harmful substances they release, these are **agriculture,** which is responsible for pesticides and dust from agriculture practices, field burning, and the construction industry.

Types of air pollutants. All the just described sources of air pollution release the following air pollutants: 1. **Carbon compounds** (*e.g.,* CO_2, CO); 2. **Sulphur compounds** (*e.g.,* SO_2, H_2S and H_2SO_4); 3. **Nitrogen oxides** (*e.g.,* NO, NO_2 and HNO_3); 4. **Ozone** (O_3); 5. **Flurocarbons;** 6. **Hydrocarbons** (*e.g.,* benzene, benzypyrene, etc.); 7. **Metals** (e.g., lead, nickel, arsenic, beryllium, tin, vanadium, titanium, cadmium, etc.); 8. **Photochemical products** (*e.g.,* olefins, aldehydes, photochemical smog, PAN, PB_2N, etc.); 9. **Particulate matter** (*e.g.,* fly ash, dust, grit and SPM); and 10. **Toxicants.**

Primary Air Pollutants and Secondary Air Pollutants

Air pollutants are classified into *primary* and *secondary pollutants.*

(*i*) **Primary air pollutants.** Certain air pollutants such as **gases** (CO, CO_2, SO_2, HC (hydrocarbons), NO_2, N_2O_4, H_2S, etc.), **Chemicals** (As, C_6H_6, Cd, Cl_2, F, HCHO, HCL, HF, Hg, Mn, Ni, Pb, SiF_4, etc.) and particulates (grit, fly ash, dust, soot, smoke, fumes, aerosols, etc.) form the example of **primary air pollutants**, since they are emitted directly from sources and pollute the air.

(*ii*) **Secondary air pollutants.** When above mentioned primary air pollutants take part in wide range of photochemical reactions, they form **secondary pollutants.** Such new products often cause greater damage than is caused directly by the primary pollutants.

The main products that result from light striking the hydrocarbons (HC) from exhaust are the **olefins** (also known as *alkenes, ethylenes* or unsaturated hydrocarbons), carbon compound containing double bonds (*e.g.,* ethylene, $CH_2 = CH_2$); the **aldehydes** and the **aromatics** (or benzenes). The aromatics produced photochemically are considered among the most serious of all air pollutants. One of these compounds, **benzpyrene** has been proved to induce cancer. Two other photochemical pollutants that are known to cause serious health problems are **peroxybenzoil nitrate** and **peroxyacetyl nitrate (PAN).**

Other examples of secondary pollutants are **nitrogen monoxide (NO), nitrogen dioxide (NO_2), hydroxyl radical (OH), nitrous acid (HONO), nitric acid (HNO_3).**

Some of the most common air pollutants, their sources and their effects on human health have been tabulated in Table 21.1.

Table 21.1. Common air pollutants, their sources and pathological effects on human.

	Pollutants	Where they come from (source)	Pathological effect on human
1.	Aldehydes	Thermal decomposition of fats, oil, or glycerol.	Irritate nasal and respiratory tracts.
2.	Ammonias	Chemical processes—dye-making; explosives, lacquer; fertilizer.	Inflame upper respiratory passages.
3.	Arsines	Processes involving metals or acids containing arsenic soldering.	Break down red cells in blood, damage kidneys; cause jaundice.
4.	Carbon monoxides	Gasoline motor exhausts; burning of coal.	Reduce oxygen-carrying capacity of blood.
5.	Chlorines	Bleaching cotton and flour; many other chemical processes.	Attack entire respiratory tract and mucous membranes of eyes; cause pulmonary edema.
6.	Hydrogen cyanides	Fumigation; blast furnaces, chemical manufacturing; metal plating.	Interfere with nerve cells; produce dry throat, indistinct vision, headache.
7.	Hydrogen fluorides	Petroleum refining; glass etching; aluminium and fertilizer production.	Irritate and corrode all body passages.
8.	Hydrogen sulphides	Refineries and chemical industries; bituminous fuels.	Smell like rotten eggs; cause nausea; irritate eyes and throat.
9.	Nitrogen oxides	Motor vehicle exhausts; soft coal.	Inhibit cilia action so that soot and dust penetrate far into the lungs.
10.	Phosgenes (carbonyl chloride, $COCl_2$)	Chemical and dye manufacturing.	Induce coughing, irritation, and sometimes fatal pulmonary edema.
11.	Sulphur	Coal and oil combustion.	Cause chest constriction, headache, vomiting, and death from respiratory ailments.
12.	Suspended particles (ash, soot, smoke)	Incinerators; almost any type of manufacturing.	Cause emphysema, eye irritations and possibly cancer.

Ecology of Air Pollution

Once injected into the atmosphere, pollutants enter the biogeochemical cycles by different routes. The air above many cities can assimilate and disperse great quantities of fine particulate and gaseous pollutants as long as air can move and disperse (Fig. 21.2). But if air masses over cities become stagnant, pollutants accumulate quickly and deteriorate air quality which cause many respiratory diseases in human and other animals. Air pollutants also accumulate during **temperature inversions**, when cooler surface layers of air become trapped under warmer upper

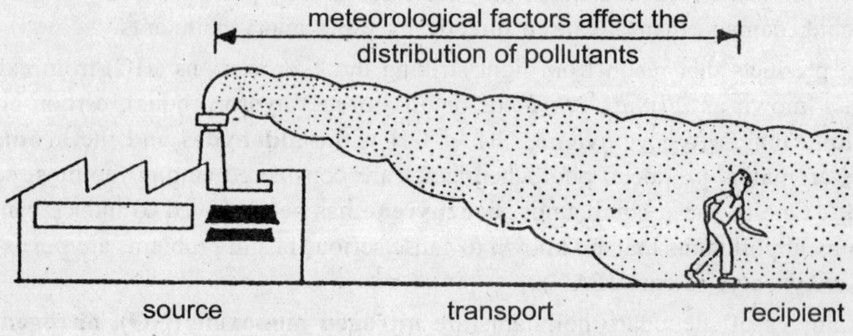

Fig. 21.2. Mode of movement of air pollutant from source to recipient.

layers (Fig. 21.3). In these situations, the upper layers of warm air prevent the vertical rise and dispersal of pollutants which are held near the ground. Temperature inversions commonly occur in cities surrounded by mountains or bordered by mountains on the leeward side.

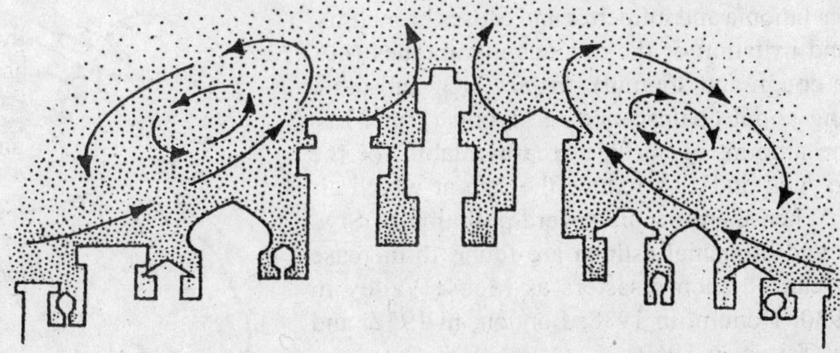

Fig. 21.3. Pattern of circulation of warm and cool air over a city and formation of the heat island.

Further a portion of air pollutants reaches land as dry fallouts; it may then enter various nutrient cycles and food chains through water and soil. Other contaminants of air react chemically or photochemically with each other and produce such **secondary pollutants** as sulphuric acid, ozone, and peroxyacetyl nitrate or PAN. Aerosols and other forms of fine particulate matter act as **condensation nuclei**, to which water vapours present in the air quickly surround to form droplets of fog or rain.

Moreover, different air pollutants adversely affect flora, fauna and climate of a given area variously and some of the common air pollutants and their specific affects on humans, vegetation, climate, etc., have been discussed as follows:

A. Gaseous Pollutants

1. Sulphur oxides and hydrogen sulphide. These gaseous pollutants are naturally released by the biological decomposition and from volcanic eruptions. They are also released artificially due to human activities such as smelting of sulphide-containing ores, combustion of sulphur-containing fuels such as coal and oil, petroleum refining and obtaining of geothermal energy.

According to a report the sulphur dioxide (SO_2) content of the atmosphere in Delhi city has reached the level of 0.233 ppm, whereas in USA and West Germany the permissible limit is only 0.1 and 0.05 ppm. Concentrations of SO_2 as low as 0.3 ppm may damage plants. Lichens are particularly sensitive to SO_2 and in polluted regions one does not find lichens growing on the tree trunks. Thus, lower quantities of sulphur dioxide suppresses the overall vegetative as well as reproductive growth and yield. Its high atmospheric concentrations produce various injuries to leaves such as interveinal and blade damage, necrosis of leaves and cellular collapse (Fig. 21.4). However, moderate SO_2 pollution results in chlorosis of leaves without cellular collapse. Pine trees are more susceptible than broad leaved trees and react by partial defoliation and reduced growth. Plant's exposure to hydrogen sulphide results in leaf lesions, mottling, defoliation and reduced growth.

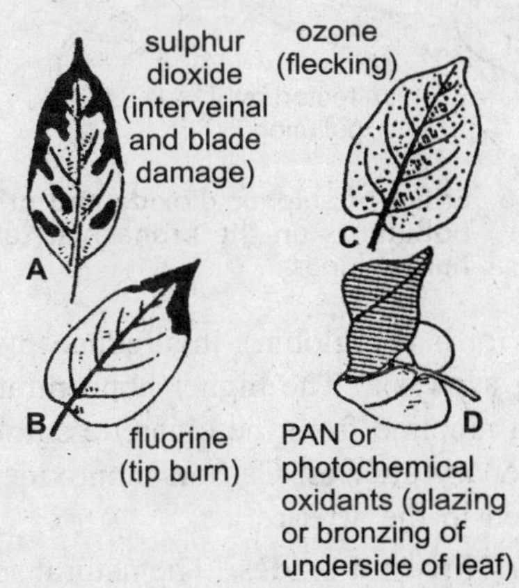

Fig. 21.4. Injuries caused to plant leaves by A–SO_2, B–HF, C– O_3 and D– PAN.

Sulphur dioxide pollution causes in human beings various types of injuries such as eye irritation, chest constriction, headache, vomiting and death from respiratory ailment (Fig. 21.5). It paralyzes or destroys bronchial cilia in air passages of human beings, constricts bronchii, damages lungs, lowers resistance to pneumonia and influenza and causes bronchitis, emphysema and irritation of the mucous membranes (*i.e.*, an increase in cough and sputum). In fact, SO_2 and other pollutants bring about coalescence of alveoli (Fig. 21.6) and reduce the amount of surface area available for the transport of oxygen and also reduce the rate at which air is exchanged. When there occurs severe pollution of SO_2, the death rate and bronchial asthma are found to increase and in past it caused such disasters as Meuse Valley in Belgium in 1930; Donora, in 1938; London, in 1952; and New York and Tokyo in 1960s.

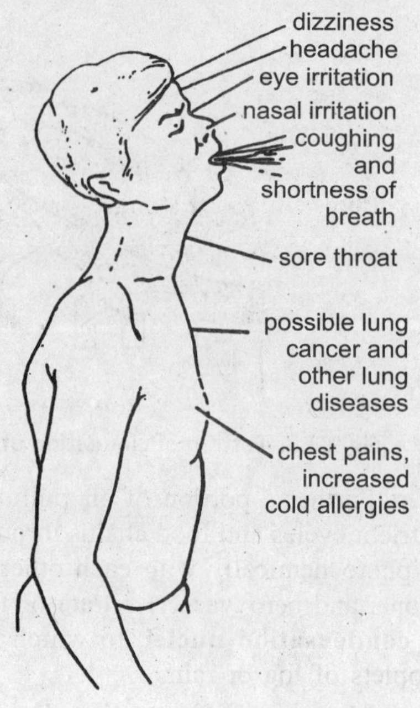

Fig. 21.5. Ailments caused in humans by air pollution.

Further, in the atmosphere, SO_2 does not remain in the gaseous state for long time, but very soon it reacts with moistrue to form sulphuric acid or H_2SO_4. Sulphuric acid causes many respiratory diseases in human and also produces **acid rainfalls** over parts of the earth. In Scandinavia, downwind from the industrial centers of Britain and the Ruhr Valley, the acidity of the rainfall has increased 200 fold since 1966, with pH values as low as 2.8 being recorded (**Oden** and **Ahl**, 1970). This acid rainwater has increased the acidity of Scandinavian streams, interfering with salmon reproduction and destroying salmons runs. It has reduced forest growth and increased the amount of calcium and other nutrients leached from agricultural soil. The acid rainfall is also involved in the erosion of building materials as limestone marbel, the slate used in roofing, mortar and deterioration of statues.

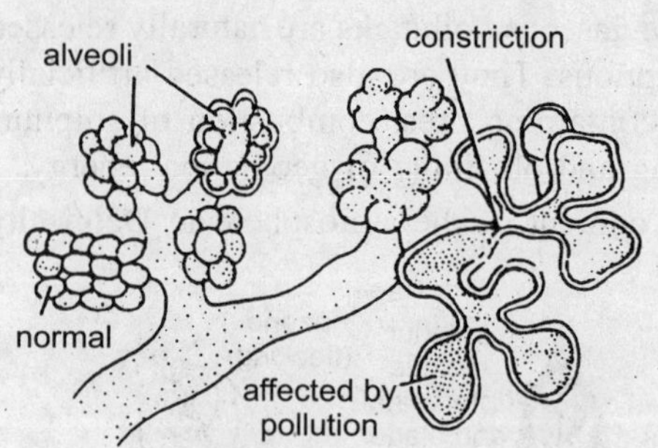

Fig. 21.6. Effect of sulphur dioxide and other air pollutants on the bronachial tubes of human lungs.

2. Carbon monoxide. It is released chiefly from gasoline engines and burning of coal in the defective furnaces. In human, CO produces headache, dizziness, inability to distinguish time intervals, nausea, ringing in the ears, heart palpitation, pressure in chest and difficulty in breathing. This gas combines with haemoglobin of blood to form carboxy-heamoglobin in RBC (In comparison to O_2, CO has 200 times greater affinity for haemoglobin), this reduces its oxygen carrying capacity to all parts of body and, thus, causing asphyxia. The higher concentrations of carbon monoxide (CO) may be even fatal. The CO and nicotine from the cigarette smoke increase the tendency of the blood to clot and block the coronary arteries. Carbon monoxide also increases the rate at which the fatty materials are deposited in the arteries.

3. Nitrogen oxides. The natural sources of this pollutant are anaerobic bacterial breakdown of nitrogenous compounds, forest fires and lightning. Its chief anthropogenic sources are power generators and automobile exhausts (motor vehicles). Burning of organic wastes, and manufacture of explosives and nitrogenous fertilizers further add to nitrogen oxide pollution.

Nitrogen dioxide (NO_2), a pungent gas that produces a brownish haze, causes nose and eye irritations and pulmonary discomfort in humans. It also produces general and pulmonary oedema and hemorrhange. In plants, NO_2 brings about bifacial necrosis leading to collapse of leaves, enhancement of green colour followed by chlorosis and extensive leaf drop. Ultimately there occurs an increase in fruit drop and decrease in fruit crop.

Photochemical smog. In the atmosphere nitrogen dioxide is reduced by ultraviolet light to nitrogen monoxide and atomic oxygen:

$$NO_2 \longrightarrow NO + O$$

Atomic oxygen reacts with oxygen to form ozone:

$$O_2 + O \longrightarrow O_3$$

Ozone reacts with nitrogen monoxide to form nitrogen dioxide and oxygen, thus, closing the cycle:

$$NO + O_3 \longrightarrow NO_2 + O_2$$

Sometimes, in the presence of sunlight, atomic oxygen from the photochemical reduction of NO_2 also reacts with a number of reactive hydrocarbons (such as methane, ethane, toluene, etc., all of which originate from burning of fossil fuels or directly from plants) to form reactive intermediates called **radicals.** These radicals then take part in a series of reactions to form still more radicals that combine with oxygen, hydrocarbons, and NO_2. As a result nitrogen dioxide is regenerated, nitric oxide disappears, ozone accumulates and a number of secondary pollutants are formed such as formaldehyde, aldehydes and peroxyacetly nitrate or PAN ($C_2H_3O_5N$). All of these collectively form **photochemical smog.**

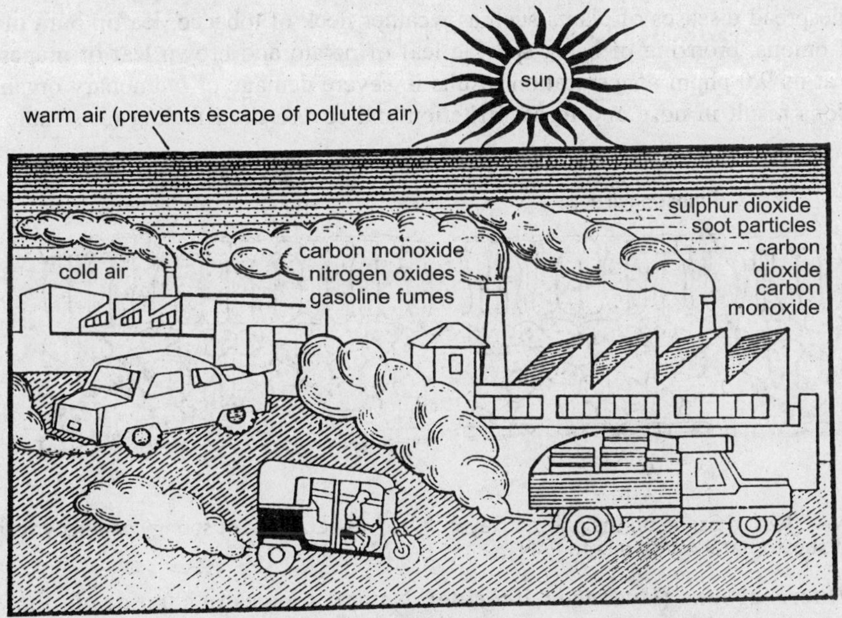

Fig. 21.7. Formation of the smog and its climatic effects.

Ozone, PAN and nitrogen dioxide severely injure many forms of plant life, destroying the cells of leaves, damaging the chloroplasts, and interfering with the plant's metabolic processes.

PAN is known to block "Hill reaction" of photosynthesis. It results in bronzing and glazing of abaxial leaf surface (Fig. 21.4D) which is due to plasmolysis and collapse of mesophyll cells around substomatal chambers (Fig. 21.4B). Epidermal and guard cells are not injured by PAN.

4. Ozone. Levels of ozone (O_3) may rise in atmosphere due to human activities. It is also formed by NO_2 under UV-radiations effect. Minor amounts of ozone are also added to the atmosphere by electric discharges such as lightning flashes, by vertical flux of stratospheric ozone and by tropospheric storms.

Ozone near the earth's surface in the troposphere creates pollution problem. Increase in O_3 concentration near earth's surface is toxic to plants reducing crop yields significantly. It also has adverse effects on human health (see Table 21.2). Thus, while higher levels of O_3 in the atmosphere protects us, it is harmful when it comes in direct contact with us and plant at earth's surface.

Table 21.2. Effects of ozone (O_3) on human health.

Concentration (ppm)	Effects observed
0.2	No ill effects
0.3	Nose and throat irritation
1.0-3.0*	Extreme fatigue after 2 hrs.
9.0	Severe pulmonary oedema

In plants, O_3 enter through stomata and produces visible damage to leaves and results in decrease in yield and quality of plant products. Thus, O_3 results in necrotic flecking of upper surface of leaf (Fig. 21.4C), general chlorosis and bronzing, premature senescence of plants, precocious dropping of older leaves, reduced growth of shoots and roots, suppression of nodulation, reduction in seed set and yield. Ozone causes shrinkage of nuclei and cytoplasm or mesophyll cells which become granular and results in increased inter-cellular space (Fig. 21.8A). At 0.02 ppm, O_3 damages tobacco, tomato, bean, pine and other plants. It is now known to be the cause of several widespread diseases of plants such as weather fleck of tobacco, leaftip burn of carnations, tip burn of onions, bronzing of bean, speckle-leaf of potato and brown leaf of grapes. In human beings, O_3 at its 9.0 phpm concentration results in severe damage of pulmonary organs. Its lower concentrations result in nose and throat irritation and severe tired feeling (see Table 21.2).

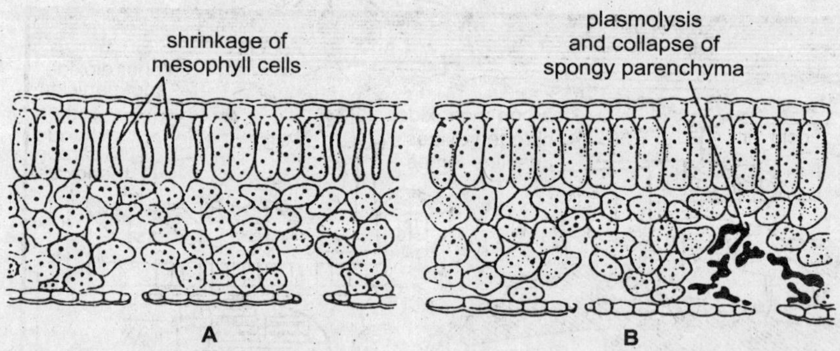

Fig. 21.8. A—Effect of ozone on the palisade layer of leaf; B—Collapse of spongy parenchyma in the region of substomatal cavity due to PAN.

5. Fluorocarbons (Hydrogen fluoride). Natural sources of fluorides in the atmosphere are active volcanoes. Their human-made or artificial sources are petroleum refining, aluminium, steel and electro-chemical reduction plants, blast furnaces, brick-kilns, and tile, glass etching and superphosphate fertilizer industries and combustion of coal.

Fluoride burns the tip of plant leaves (Fig. 21.8B). It's low amounts impair plant growth, result in excessive dropping of bloom and young fruits, development of small, partially or completely seedless fruits and premature formation of soft red flesh and splitting of peach. In human being, it irritates and corrode all body passages.

6. Hydrocarbons. Biological decomposition of organic matter, spill and seepage from natural gas and oilfields and volatile emissions from plants are some major natural causes for the release of hydrocarbons such as methane, terpenes, ethylene and aniline. Incomplete combustion of fuels, automobile exhausts (Fig. 21.1), petroleum-refineries, agricultural burning, motor fuel marketing, manufacture of explosives and cracking of natural gas in petrochemical plants (as a blow-off emissions) constitute the man-made sources that emit hydrocarbons.

The hydrocarbon **ethylene** causes yellowing and occasional necrosis of leaves, chlorosis of floral buds, inhibition of terminal growth, epinasty of leaves, shortening of internodes, thickening of stems, lack of apical dominance, stunted growth, dry sepal disease of orchids and decrease in the amount of chlorophyll and carotenoids. Another hydrocarbon **aniline** results in the appearance of bands on leaves as if they are water-soaked, necrotic spots and abscission of leaves. In humans, hydrocarbons bring about irritation of mucous membrane, bronchial constriction and eye irritation. One of the hydrocarbons released due to incomplete combustion, is **3-4 benzpyrene** which is said to cause lung cancer. **Methyl isocyanate** gas when accidentally leaked out from the storage tanks of the pesticide factory in Bhopal on December 2, 1984, had killed over 3,000 persons and seriously affecting lakhs of residents.

7. Hydrogen chloride. This pollutant is released from combustion of coal, paper, plastics, chlorinated hydrocarbons, accidental spills from the chemical manufacturing plants and ignition of solid-fuel rocket engines. In plants, hydrogen chloride causes plasmolysis and collapse of epidermal cells of leaves and thereby results in abaxial glazing of leaves.

8. Ammonia. The main anthropogenic sources of this gaseous pollutant are refrigerator, precooler systems of cold storages, manufacturing of dyes, explosives, lacquers (varnishes) and anhydrous ammonium fertilizers and nitric acids and domestic incineration. Ammonia causes in plants bleaching of leaves, rusty sports on leaves and flowers, reduction of root and shoot growth, browning and softening of fruits, development of dark, corky lenticels in apples, and reduction in the rate of seed germination. In humans, it inflames upper respiratory passages.

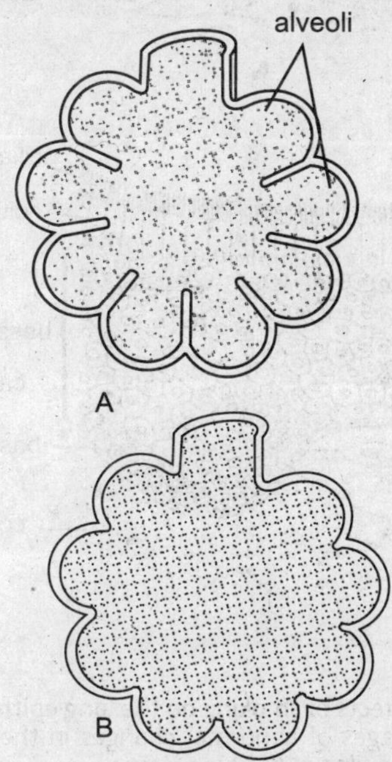

Fig. 21.9. A—The appearance of alveoli of a normal lung; B—The appearance of alveoli in the lung of the person suffering with emphysema.

9. Tobacco smoke. Tabacco smoke contains about 300 chemical compounds including nicotine and carcinogens such as tar ("aromatic hydrocarbon"). It is mainly produced by smoking cigarettes and *beedis*. It is gradually becoming a pollutant especially in closed atmospheres such as buses, trains, auditoria, discotheques and so on. **Nicotine** stimulates some types of synapse of nervous system, increase blood pressure and heart rate by the production of adrenaline, causes vasodilation in the muscles and vaso-constriction in the skin. When a person smokes, tiny particles get caught in the smoke on the lining of the manuwindpipe and bronchial tubes. Extra mucus is produced and the cilia stop beating. The mucus collects in the bronchial tube and this gives rise to a "smoker's cough".

Repeated coughing may cause the delicate walls of the alveoli to break down into larger spaces. This cuts down the surface area over which gaseous exchange can take place, so the person gets very short of breath. Doctors call this condition **emphysema** (Fig. 21.9). Tabacco smoke also brings about thickening of bronchial epithelial layer, loss of ciliated cells and appearance of cells with bizarre nuclei, which are probably the precursors of cancerous cells (Fig. 21.10). Although smoking mainly affects the lungs, it can also cause cancer of other organs such as mouth, throat, oesophagus and bladder. It is also associated with heart disease and stomach ulers and a woman who smokes while she is pregnant is more likely to have spontaneous abortion of stillbirth or to give birth to an under-sized baby.

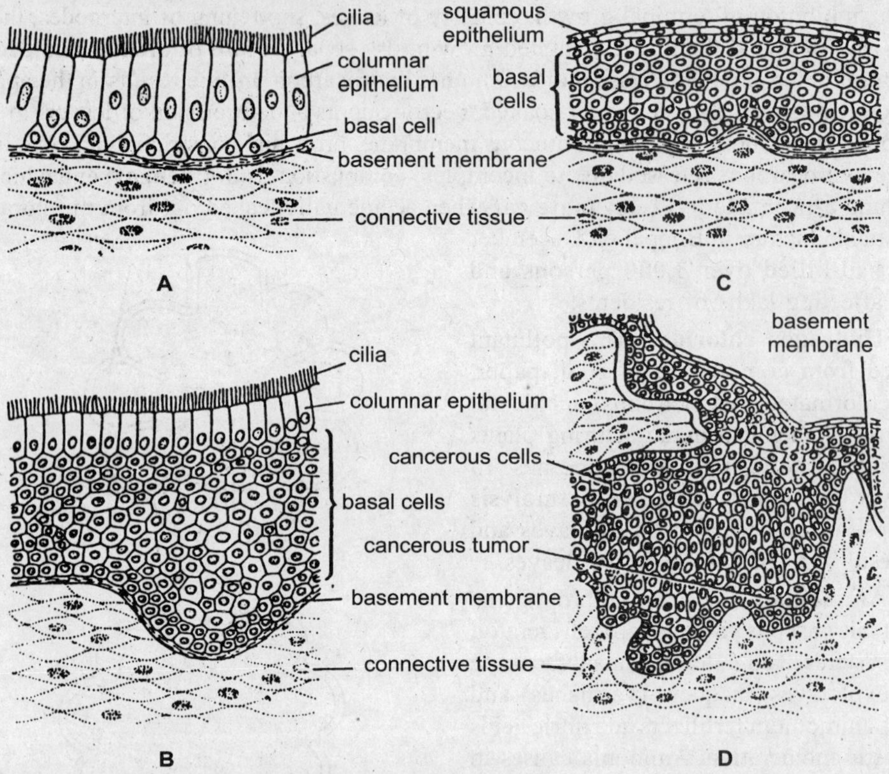

Fig. 21.10. Effects of smoking on the lung epithelium of smokers. A—Normal lung epithelium; B and C—Two stages of abnormal changes in the lung epithelium found in smokers; D—An epithelial tumor growing in the lung tissue.

B. Particulate Pollutants

1. Fluorides. The particulate fluorides originate in the same way as the gaseous fluorides. They settle and accumulate on the grass and other vegetation. They are less toxic to these plants causing occasional leaftip burns. However, ingestion by cattle of various fluorine compounds falling on forage, causes **fluorosis**, a disease characterized by abnormal calcification of bones and teeth, eventually resulting in loss of teeth, body-weight and in lameness.

Fluoride pollution in human and animals is mainly through water. In our country, fluorosis is a public health problem in the states of Gujarat, Rajasthan, Punjab, Haryana, Uttar Pradesh, Andhra Pradesh, Tamil Nadu, Karnataka, and some areas of Delhi. Globally, it is a problem of various other countries such as USA, Italy, Holland, France, Germany, Spain, Switzerland, China, Japan and some African and Latin American countries.

2. Lead. Lead, a heavy metal, is injected in the atmosphere mainly from automobile exhaust. Automobile gasoline contains tetraethyl lead that is used as an antiknock additive. Lead is emitted into the air with the exhaust as volatile **lead halides** (bromides and chlorides). About 75 per cent of lead burnt in gasoline comes out as lead halides through tail pipe in exhaust gases. Of this about 40 per cent settles immediately on the ground and the rest (60 per cent) goes into air. That is why, its concentration is higher in urban areas where automotive and industrial exhausts are more (Fig. 21.11).

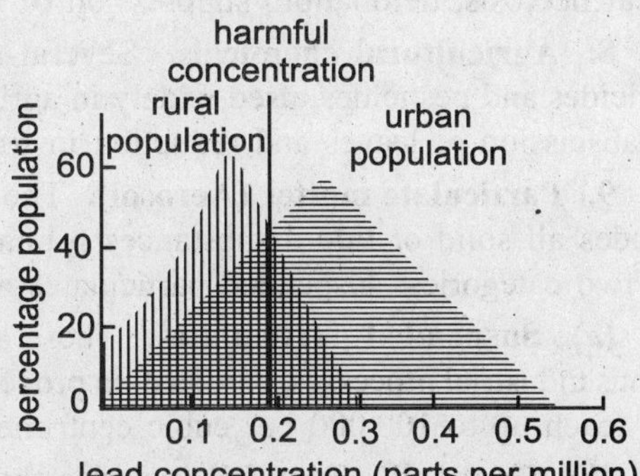

Fig. 21.11. Comparative accumulation of lead in the tissues of rural and urban population.

Further, the use of lead-lined vessels for cooking and the storage of wine resulted in heavy lead burdens in the bodies of Roman citizens. Some have attributed the decline of the Roman empire to chronic lead poisoning. In fact, analysis of bones of Roman citizens have revealed high lead concentrations (**Kimball**, 1975). Lead pollution is also caused by occupational manipulation of putty and paints by painters; habitual or accidental nibbling of peels of old paints; smelting complexes, ceramics, pesticides and solder used for sealing. The lead level of air-quality guide of WHO is 2μg/m^3. This level is already crossed in many Indian cities and in various countries of world. For example, in Kanpur and Ahmedabad lead level vary between 1.05 to 8.3 μg/m^3 and 0.59 to 11.38 μg/m^3, respectively. Roadside plants and meadow mice living along major highways are found to contain high concentrations of lead in tissues, and this has a sublethal effect on the health and longevity of the animal. Traffic policemen and others who are exposed for long periods to heavy traffic have higher than average lead in their blood and body tissues. High lead accumulation in the tissues of human body interferes with development and maturation of erythrocytes (RBC) and causes anaemia. Even 0.2 parts per million (0.2 ppm) concentration of lead in human body creates metabolic disturbances. Chronic exposure to lead leads to stippled red blood cells having impaired capacity for oxygen transportation. Being a cumulative poison, it disrupts the functioning of cells and organs of the muscular, circulatory and nervous systems binding with the cellular enzymes and coagulating proteins, thereby results in nausea, weakness and dizziness. It also damages liver, kidney and gastro-intestine and induces abnormalities in fertility and pregnancy.

3. Mercury. It is a liquid volatile heavy metal which is found in rocks and soil. It is present in air due to human activities such as the use of mercury compounds in production of fungicides, paints, cosmetics, paper pulp, etc. Inhalation of 1 mg/m^3 of mercury in air for three months may lead to human death. Nervous system, liver and eyes are damaged. Infant may be deformed. Other symptoms of mercury toxicity are headache, fatigue, anxiety, lethargy, loss of appetite, etc.

4. Zinc. Zinc in air occurs mostly as white zinc oxide fumes and is toxic to human. It exists in air around zinc smelters and scrap zinc refineries.

5. Cadmium. This metal is emitted to air by human activities and industries (*e.g.*, electroplating and welding of cadmium containing materials; industries producing pesticides and phosphate fertilizers). Cadmium occurs in the air in the form of oxide, sulphate or chloride compounds. It is poisonous at very low levels and is known to accumulate in human liver and kidney. Cadmium causes hypertension, emphysema and kidney damage. It may also act as carcinogen in mammals.

6. Potassium salts. These particulate pollutants are derived mainly from potash mines and cause in plants abnormalities such as branch tip death, chlorosis and necrosis of leaves.

7. Sodium chloride. Certain de-icing salts such as sodium chloride, used to remove ice and snow from roads in winters, are found to cause multiple damage to the roadside trees such as leaf necrosis, defoliation, suppression of flowering, and dieback of terminal shoots in apple.

8. Agricultural chemicals. Several types of chemicals such as insecticides, herbicides, fungicides and pesticides, used widely in agriculture are found to result in foliar lesions, chlorosis and abscission of leaves and reduction in fruit set.

9. Particulate matter (Aerosol). The word particulate has been derived from particle and includes all solid or liquid substances primarily in the air. Particulate matter is usually divided into two categories: suspended particulates and dust fall.

(*a*) Suspended particulates. These are smaller than 10 microns. They are generated by various industrial processes, combustion processes and black soot. In polluted air, their concentration may reach up to 100,000 per cubic centrimetre.

(*b*) Dust fall. Dust fall includes the particles larger than about 10 microns. They are emitted into the air by physical processes such as grinding and abrasion, a soot and fly ash (or PFA = Pulverised fly ash) from fuel combustion. Dust particles settle at the ground quickly and cause nuisance to certain industries requiring aseptic conditions such as drug industries and food-processing plants. The average level of dust and particulate matter in Delhi is about 600 mg/m^3, in Bombay about 200 mg/m^3, in Calcutta about 300 mg/m^3, whereas it is just 150 mg/m^3 in other polluted western cities.

Dust may be inorganic and organic in nature: **1. Inorganic dust** containing silica and trapped heavy metals is the main pollutant in mining, quarrying and stone cutting operations. **2. Organic dust** raised in cotton textile mills, ginning plants, coir retting and processing, jute and hemp processing, saw mills and plywood industries are also potentially toxic with properties of sensitization of persons exposed to them.

Particulates emitted from cement manufacturing units cause in plants premature fall of needles, higher puberty of leaves, formation of more stomata and trichomes, reduction in number and size of cobs and weight of seeds and increase in number of infertile seeds. Dust from stone crushers, lime kilns and slate making units is also hazardous to plants and human beings. Particles of asbestos that are released from factories can wear off brakes and gears of cars. Coal dust and asbestos have been found to cause in plants, necrotic lesions, reduction in fruit sets and silicosis, asbestosis and lung cancer in human beings. The common dust particles sometime become health hazards as they may lead to diseases such as allergic asthma, bronchitis, emphysema and even fibrosis of lungs. **Pulverised fly ash (PFA)** is getting accumulated in such large quantities particularly in areas where thermoelectric generators are installed that it has proved to be one of the major sources of solid wastes. Since it contains boron, its deposition by wind in agricultural tracts often results in deleterious effects.

In fact, living organisms are rarely exposed to a single pollutant in nature. The influence of mixtures of pollutants is usually synergistic, *i.e.,* greater than the additive effects of the different pollutants alone.

C. Effects of Air Pollution on Weather, Climate, and Atmospheric Processes

At gross level, air pollution causes two worldwide problems—contamination of the upper atmosphere and the alteration of weather and climate.

Air pollution also affects weather on a continental or global basis. Many gaseous pollutants and fine aerosols reach the upper atmosphere, where they have basic effects on the penetration and absorption of sunlight. Modern environmental biologists feel that increasing particulate pollution may be reducing the amount of sunlight energy reaching the earth's surface, thereby, lowering solar radiation at the earth's surface and producing a cooling effect on world climates which could ultimately trigger another ice age. In fact, **Thompson** (1975) has reported a decrease in mean annual temperatures in the northern hemispheres and an increase in the north polar ice cap.

Greenhouse effect. Carbon dioxide is a natural constituent of the atmosphere, but, its concentration is increasing in the air with an alarming rate. A byproduct of the burning of fossil fuel, it is not necessarily a pollutant. It produces adverse physiological effects only at very high levels. It is estimated that approximately one-half of the CO_2 input stays in the atmosphere and other half of its is removed by the oceans and by plants. The increased amount of CO_2 in atmosphere is found to increase the temperature of earth.

The spectral properties of CO_2 in the atmosphere are such that it tends to prevent the long wave radiations (*i.e.,* infra-red heat radiation) from earth from escaping into outer space and deflect it back to earth. The latter has an increased temperature at surface. This phenomenon is called **atmospheric effect** or **greenhouse effect**.

The simultaneous cooling and heating effects of air pollution on earth have increased variability in the world-wide weather patterns which may be a serious threat to global food production (**Thompson**, 1975). Recently, certain ecologists have tried to correlate air pollution with serious and prolonged droughts, heavier rains and floods, and more serious hurricanes and tornadoes.

Scientists fear that increased level of CO_2 will increase the greenhouse effect and thereby increase the temperature of the earth. Only a slight temperature rise would cause the polar ice caps to melt and to cause an enough rise in sea level submerging a number of major cities of the world, because many are along sea coasts.

Simultaneously, along with carbon dioxide humans have been adding solid particles and droplets into the air which increase the **albedo** or shineness of the earth. This should act contrary to the greenhouse effect, reduce the sunlight reaching the earth and tend to lower the temperature of earth.

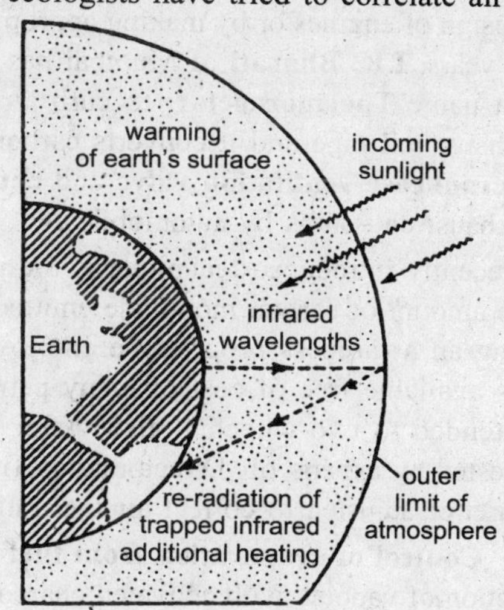

Fig. 21.12. The greenhouse effect of carbon dioxide.

Peeling of ozone umbrella by CFMs. Certain fluorocarbon compounds which are called chlorofluoromethanes or CFMs or "freon" are used as propellants in pressurized aerosol cans. They are inert in normal chemical and physical reactions, but they get accumulated in greater amounts at high altitudes and there in the stratosphere these inert gaseous compounds (*i.e.,* CFMs) release chlorine atoms under the influence of intense short-wave ultraviolet radiation. Each atom of chlorine chain then reacts with more than 1,00,000 molecules of ozone, converting ozone to oxygen. The reduction in stratospheric ozone permits greater penetration of ultraviolet light, which intensifies UV radiation at the earth's surface. This intensified UV radiation has caused a significant increase in skin cancer and eventually has a lethal effects on many organisms, including humans. In plants such enhanced level of UV radiations are found to cause stunted growth, short, thick stems, smaller leaves, plasmolysis of cells, destruction of anthocyanin, chlorophyll and nuclei. They are known to cause bronzing of leaves, injury to various fruits, somatic aberrations, discolouring of staminal hairs, inhibition of pollen germination and pollen tube growth.

The protective ozone layer of the stratosphere is also considered by many ecologists to be endangered by supersonic jets, the SSTs. The jet engines of supersonic aircraft flying at high altitudes release nitrogen oxides (NO_x) which catalytically destroy ozone molecules.

Control of Air Pollution

Most kinds of air pollutions can be controlled by modern technology, but the costs ultimately be borne by the public in the form of higher prices for manufactured goods, higher taxes, reduced profit margins in industry, and more restrictions on individual activities such as burning leaves and trash and use of automobiles. Following steps have to be taken to control pollution at source (prevention) as well as after the release of pollutants in the atmosphere.

1. Prevention and control of vehicular pollution. For controlling the air pollution because of vehicles, the following measures are adopted:

(*a*) Curbing the pollutant emission from vehicular exhaust. This type of control can be attained by (1) using new proportion of gasoline and air; (2) more exact timing of fuel feeding; (3) using gas additives to improve combustion; (4) by injecting air into the exhaust to convert exhaust compounds into less toxic substances, and by (5) correcting the engine design and/or fixing cessation device to improve combustion with the existing design. Complete elimination of three main pollutants, namely CO, NO_x and hydrocarbons can be attained by either updating the present design of engines or by making appropriate changes in devices for improving combustion. In recent years, **I.K. Bharati** of Mumbai has claimed to have devised a simple attachment (by the patent name **Thermoreactor**) to curb air pollution by motor vehicle. The reactor is fitted to the exhaust tail pipe and it converts carbon monoxide into oxygen. Various devices such as **positive crankcase ventilation valve** and **catalytic convertor** have been developed in USA to reduce exhaust emissions by automobiles.

A recently introduced technique is the use of **lead-free petrol** and low-sulphur fuels. To check the amount of carbon monoxide emitted by a vehicle, the Delhi Administration is already conducting an awareness programme employing most modern **Exhaust Gas Analysers.** This facility is available free of cost of many petrol stations in different parts of Delhi and is now being extended to two-wheelers also. After getting the level of CO in a car or two-wheeler exhaust tested by experts of Directorate of Transport, one can go to a service station for getting the carburettor adjusted to correct air-fuel ratio.

(*b*) Control of evaporation from fuel tank and carburetter. This can be performed by (1) collection of vapours with activated charcoal when the engine is turned off; (2) subjecting the gasoline in the tank to slight pressure to prevent the gas from evaporation, and (3) developing low volatile gasoline that does not evaporate easily.

(*c*) Use of filters. Filters can be used to capture and recycle the escaped gases (hydrocarbons) from the engine (*i.e.,* the gas vapours which escape between walls and the piston and reach the crankcase).

2. Prevention and control of industrial pollution. To check the air pollution by industrial and power plant chimney wastes, measures are taken for the removal of particulate matter and gaseous pollutants. In different industries, air pollution can be checked at various steps (Fig. 21.13).

(*a*) Removal of particulate matter. This process involves collection of particulate matter under the influence of different forces, thereby, moving them continuously out of the gas stream. This step involves the use of following equipments:

(*i*) Cyclone collector. In this case the waste gas containing particles is subjected to **contrifugation.** The suspended particles move towards the wall of cyclone body and then to its bottom and finally are discarded out. The cyclone collectors can remove up to 70 per cent of the particles.

***(ii)* Electrostatic precipitators (ESPs).** To remove the suspended particles from gas stream, the **electrical forces** are applied within the chamber in the precipitator. The particles become charged or ionized, and they are attracted to charged electrodes and removed. ESPs can remove 99 per cent of the particulate pollutants from the chimney exhaust. ESPs work very efficiently in power plants, paper mills, cement mills, carbon black plants, etc. To control air pollution recently electrostatic precipitators have been fitted to the chimneys of Indraprastha Power Station in New Delhi. Sometimes, high resistivity dust may make separation in an ESP difficult. To overcome this difficulty, **fibric filters** or **beg filters** are employed. Further, tall chmineys are used for vertical dispersion of air pollutants.

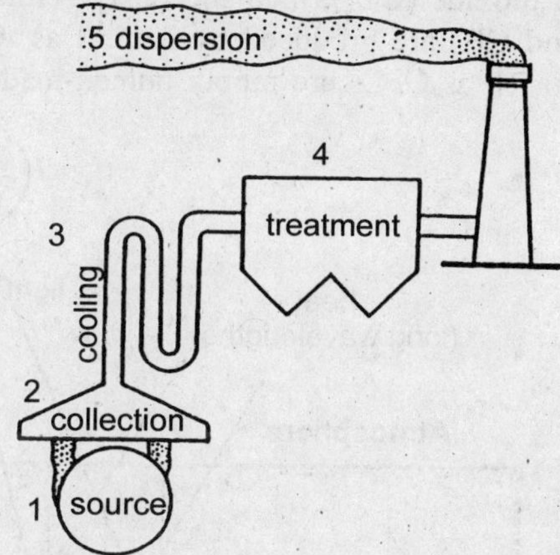

Fig. 21.13. Various possibilities to control the air pollution.

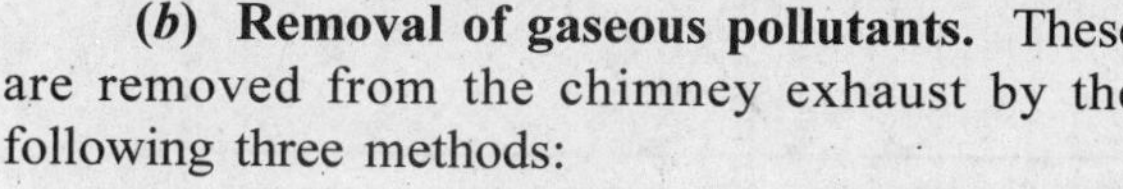

***(b)* Removal of gaseous pollutants.** These are removed from the chimney exhaust by the following three methods:

***(i)* Wet systems.** These are used as **washing towers** in which alkali fluid circulate continuously. This liquid react with SO_2 to produce a precipitate. In **gas scrubber**, a fine spray of water can effectively separate many gases such as ammonia and sulphur dioxide.

***(ii)* Dry systems.** In this case, the gas pollutants are allowed to react with an **absorbent** under a dry phase. Absorbents such as dolomite, lime (CaO) and limestone (CaOH) are placed in the way of the flowing gas (SO_2). The process is cheap and does not involved any spray of water.

***(iii)* Wet dry systems.** In these cases water in the absorbent reacts with the acid components. The absorbent calcium hydroxide slurry is spread into the hot gas stream in the form of small droplets. Calcium reacts with SO_2 and the hot gases cause the water to evaporate simultaneously. The end product is a dry powder containing mostly fly ash and salts.

Control of air pollution through law. In our country there have been several legislative measures both at state and central government levels to prevent and control different types of air pollutions:

1. Bengal Smoke Nuisance Act, 1905;
2. The Motor Vehicle Act, 1938;
3. The Gujarat Smoke Nuisance Act, 1953;
4. The Air (Prevention and Control of Pollution Act, 1981) (It was amended in 1987);
5. The Environment (protection) Act, 1986;
6. The Motor Vehicles Act, 1988 (This Act came into force from 1.7.1989).

21.4 GREENHOUSE EFFECT AND GLOBAL WARMING

The atmosphere is able to keep the planet Earth habitable because some of its constituent gases can absorb heat radiated from the earth's surface preventing it from escaping into space. These gases, called **greenhouse gases (GHGs)**, act as the glass pans of a greenhouse which allows solar radiations to pass through and heat the surface of the earth. However, it does not allow the heat radiated from the ground to pass through, thereby trapping heat in the process (Fig. 21.14). Such a heat trapping phenomenon is known as **greenhouse effect**. It is a natural phenomenon without which the earth would be too cold to support life. Some of the sun's rays that penetrate the layer of CO_2 strike the earth and are converted into heat. The heated earth reradiate this absorbed energy as radiations of longer wavelengths. Much of this energy does not pass through CO_2 and water vapour in the atmosphere and adds to the heating of the earth's atmosphere. There are several greenhouse gases. However, the major and important ones are carbon dioxide (CO_2), methane (CH_4), chlorofluorocarbons (CFCs) and nitrogen oxide (N_2O). CO_2 and CH_4 are produced by natural as well as human-made (anthropogenic) sources while others such as CFCs are purely human-made.

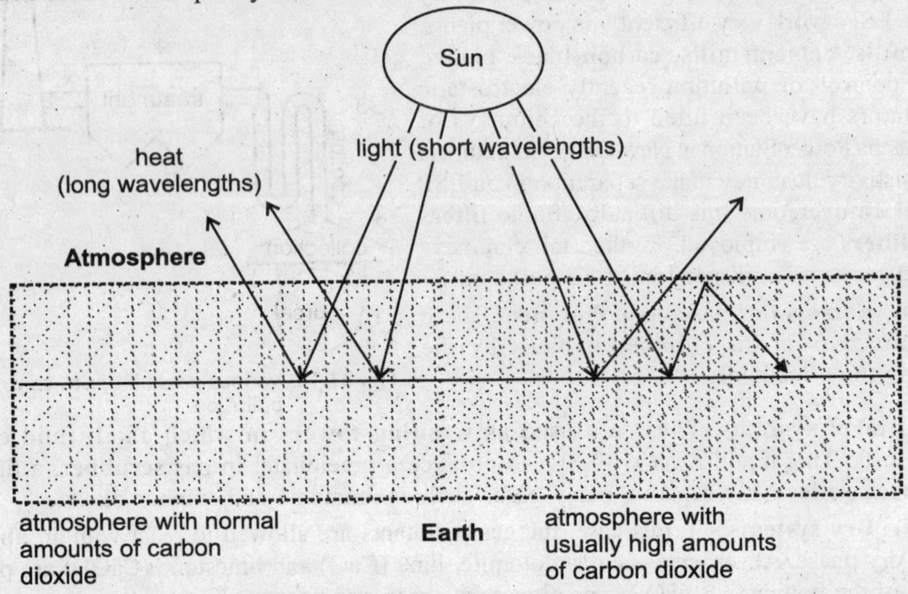

Fig. 21.14. Greenhouse effect of carbon dioxide.

Greenhouse Gases and Their Role in Global Warming

1. Carbon dioxide (CO_2). The single largest contributor to the global greenhouse effect is CO_2 which is produced wherever fossil fuels burn. Up to 1950, two-third of the CO_2 injected into the atmosphere came from biospheric sources and one-third from anthropogenic sources. Since 1950, most CO_2 has come from the burning of fossil fuels, the destruction of the world's forests, and change in land use. An increase of about 12 per cent in the mean annual concentration of CO_2 in 32 years has been confirmed by data collected from worldwide monitoring station (**Smith** 1996).

Only about 40 to 50 per cent of the total CO_2 injected into the atmosphere remains there. What happens to the rest? *This question is the object of considerable research.* Such research is critical because the answer relates directly to global warming. Current assessment of the global carbon budget assumes a balance between earth and atmosphere. The global budget, however, is not balanced (**Edmonds** 1992). A significant fraction, about 22 per cent, of atmospheric carbon cannot be accounted for.

The ocean is the major carbon sink. The uptake of anthropogenic CO_2 by oceans depends heavily upon eddy current circulation between the surface and deep water. Oceans account for about 28 per cent of the total CO_2 deposition. The other potential major sink is the terrestrial biota. Its ability to absorb a significant fraction is controversial. Some ecologists argue that terrestrial vegetation is a net source of CO_2, not a sink.

Anyhow, atmospheric CO_2 has increased about 22 per cent over the past 100 years or so. If atmospheric CO_2 continues to increase at its present rate of 0.39 per cent per year (**Smith** 1996) or 0.5 per cent per year (**Agarwal** 1998), its concentration in the atmosphere will reach

400 ppm by about the year 2030. In view of our increasing dependence on fossil carbon, our need to understand the fate of excess carbon and the nature of carbon sink is pressing (Box 21.3).

Box 21.3 CO_2 Leves in India

On India's part, an estimated 6.4 × 10^{15} gm of carbon has been added to the atmosphere (**Hingane** *et at.*, 1981). In India, the burning of firewood (including agriculture waste) for cooking still remains to be the major source of CO_2 (Table 21.3) and during 1950-80, the average growth rate of CO_2 emissions from fossil fuels in India has been 4.8 per cent (Table 21.4) compared to global average of 4.2 per cent. The sum total of CO_2 emissions by the various fuels consumed in India has contributed 3.78 per cent to the global release during the 23 years period since 1958 when systematic observations begun at Mauna Loa (Hawaii, USA).

Table 21.3. Contribution of various sources to CO_2 levels in India.

Source	Contribution %
1. Firewood	69.2
2. Coal	23.7
3. Petroleum byproducts	6.2
4. Cement industry	0.8
5. Natural gas	0.3

Table 21.4. Relative shares of commercial and non-commercial sources of energy for the period: 1953-54 to 1975-76 (in %)

Form of energy	1953-54	1960-61	1965-66	1970-71	1975-76
1. Commercial	32.4	41.0	47.9	53.4	56.5
2. Non-commercial	67.6	59.0	52.1	46.6	43.5
Total	100.00	100.00	100.00	100.00	100.00

2. Methane (CH_4). Another important greenhouse gas which could affect the climate of the earth is the methane (CH_4), the level of which is increasing alarmingly. It is significant to note that the CH_4 gas is only about 5 per cent of that of CO_2, but is more effective than CO_2 in producing greenhouse effect.

Major natural sources of methane are ruminant animals (*i.e.*, bacterial action in the gut of cattle), microbial decomposition in swamps, marshes and tundra and industrial gases related to the atmosphere. Over time (resident time of 3.2 years) methane oxidizes to H_2O, especially in the stratosphere. Atmospheric methane has approximately doubled over the past 200 years (**Stauffer** *et al.*, 1991). This increase is linked to human population growth, to increased cattle ranching and rice paddy production, garbage dumps and sanitary landfills. To some extent CH_4 is also produced by mining of coal and oil and distribution of natural gas.

3. Chloroflurocarbons (CFCs). These versatile chemicals came on the scene only in the 1960's. Since then they have been indiscriminately used as refrigerants, plastic foaming agents, aerosol propellants and cleaning agents. CFCs first attracted the attention of scientists in 1970's when it was discovered that they could have destructive effects on the stratospheric ozone layer. As a greenhouse gas, the CFCs (*e.g.*, CFC-11 and CFC-12) are thousand times as powerful as CO_2. Moreover they are extremely persistant chemical and can remain in the stratosphere for decades.

Table 21.5. Major greenhouse gases.

Name	Composition	Source of emission	Percentage contribution
1. Carbon dioxide	CO_2	1. Fossil fuel combustion (coal, oil, natural gas) 2. Deforestation and land use changes	61
2. Methane	CH_4	1. Enteric fermentation in cattle and insects 2. Biomass burning and garbage land-fills 3. Coal mines and natural gas leaks 4. Rice paddies 5. Swamps and Tundra	15
3. Chlorofluorocarbons (CFCs)	CFC-11, CFC-12, CFC-113, CFC-114, CFC-115, etc.	1. Aerosols (spray propellants) 2. Refrigeration and air-conditioning 3. Plastic foams 4. Industrial solvents 5. Sterilants for medical supplies	11
4. Nitrous oxide	N_2O	1. Fertilizer use 2. Fossil fuel combustion 3. Biomass burning 4. Changing land use	4
5. Others	CO, NO, NO_2, SO_2, C_2H_4, etc.	—	9

CFCs emissions contribute global warming through two different routes: (*i*) a greenhouse gas like the CO_2 can trap heat resulting into rise in temperature, (*ii*) with the destruction of ozone more solar radiations enter and heat the troposphere.

4. Nitrous oxide (N_2O). It is another greenhouse gas which is produced by burning of coal, oil and biomass, along with CO_2. The N_2O is also given out by nitrogen-based fertilizers in agriculture soil.

5. Water vapour. Water vapour (H_2O) contributes the most to the greenhouse effect (**Cain** *et al.*, 2011).

Extent of Global Warming

The greenhouse effect phenomenon is a reality. From mid- 1700 to 1988, all the major greenhouse gases have increased at a rapid rate: CFCs by 5 per cent a year, methane by 1 per cent, CO_2 by 0.5 per cent and N_2O by 0.3 to 0.4 per cent a year. Due to the heat trapping characteristic of the greenhouse gases, the temperature of planet Earth has been kept at an average temperature of about 15ºC which otherwise could have been at temperature of about —18ºC without CO_2 envelope and other greenhouse gases. Thus, the green house effect for the earth is 33ºC.

Earth's greenhouse effect turn out to be consistent with temperatures on the two neighbouring planets, Mars and Venus. **Mars** has very thin atmosphere and virtually no CO_2 and maintain an average temperature of —53ºC, while **Venus** with its dominated CO_2 atmosphere, about 70,000 times larger than that of the earth, suffers from a greenhouse effect and experiences a rise of temperature from —46ºC to + 477ºC (*i.e.,* the greenhouse effect for Venus is 523ºC).

Environmental Effects of Global Warming

According to Stockholm Environment Institute, natural ecosystems can adopt to a temperature rise of at the most 0.1ºC per decade. However, if rise in temperature is greater than this, it will result in following adverse environmental, consequences:

1. Many species, particularly plants would be unable to migrate to new and suitable habitats, leading to rapid loss of biological diversity and resultant loss of natural resources of greater economic value (Box 21.4).

2. One degree of rise in temperature is expected to be equivalent to a change in latitude of 100 km and this could cause the migration of prairic forest border by this amount each decade.

3. The increasing temperatures will cause more frequent storms to many parts of the world, including regions which have never experienced such storms earlier.

4. As the planet Earth warms, the patterns of rainfall and monsoon would shift dramatically. Some regions of the world could dry out, while the others could receive too much rains flooding rivers and increasing soil erosion.

5. With the rise in temperature the oceans will warm and expand, causing sea levels to rise. As a result, they will flood many of world's highly productive deltic areas, including vast tracts of land in Bangladesh, Egypt, Indonesia, China and India.

6. Melting of glaciers and ice caps, as a consequence of global warming, would also contribute to rise in sea level.

7. The IPCC (Intergovernmental Panel on Climate Change) estimates that the world's oceans could rise about 20 cm by the year 2030, and about 65 cm by the end of 21st century. Due to this, some 300 atolls are likely to disappear and the very existence of many island countries in the pacific and Indian ocean and the Caribbeans could be threatened.

8. Global warming could severely disrupt marine ecosystems. The flooding of many of the world's coastal wetlands as the oceans rise would result loss of nurseries for many fish, shrimp and birds.

9. Rise in sea level will also cause flooding the sewerage and sanitation system of the coastal cities of infectious diseases and epidemics.

10. Groundwater supplies in several parts of the world could become contaminated with sea water as the sea moves in the islands. This could also create drinking water problem.

Box 21.4 Greenhouse Effect on Crop Production

The global warming due to greenhouse effect affects climate. The most obvious effect of climatic change will be on agriculture. A warmer climate is likely to move the areas suitable for growing specific crops such as wheat towards poles. There will also be widespread changes in natural ecosystems with grasslands and desert areas expanding and forests shrinking and moving polewards.

Because CO_2 is a natural fertilizer, most plants would grow larger and faster in a high CO_2 world. Bigger plants mean larger yields but they would demand more fertilizers since the soil could be depleted of them faster. If the CO_2 levels double by the end of 21st century, the yield of many crops and weeds could increase by an average of about 30 to 50 per cent. Such a response depends on how a plant photosynthesizes. Plants that produce intermediary metabolites with three carbon atoms (the C_3 plants) would respond well to an increase in atmospheric levels of CO_2; but those plants that produce metabolic intermediaries with four carbon atoms (the C_4 plants) would respond less. Of the world's 20 major food crops, maize, sorghum, millet and sugarcane are C_4 plants whereas the other 16 crops (*e.g.,* wheat, rice, soybeans) are of C_3 type whose yield would increase in a CO_2-enriched atmosphere. This would change the economic viability of farming, agricultural employment, commodity prices and patterns of world trade.

Measures to reduce the build-up of GHGs

Global warming due to GHGs can be checked by adopting the following measures:

1. Improving the efficiency of energy production and use;

2. Consumption of less carbonaceous fuels, switching up from carbon intensive fuels such as coal to hydrogen-rich fuels, such as natural gas.

3. Speedy development of carbon-free energy sources, such as solar, wind, tidal, hydro-geothermal and nuclear energy.

4. All-out ban on production and use of CFCs; and

5. Curbing deforestation and encouraging afforestation.

21.5 ACID RAIN

Acid rain is a main problem of western countries (Box 18-5) and can be attributed mainly to atmospheric pollutants such as *oxides of sulphur* and *nitrogen*. Oxides of sulphur are released from the smoke-stacks of coal-fired power plants, smelters and other industries (*e.g.,* oil furnaces). The oxides of nitrogen come from combustion of fuels in automobiles (cars, trucks, buses, tempos, three-wheelers and two-wheelers) as well as power plants (*e.g.,* gensets). Once in the air, these oxides may react with moisture to form **sulphuric acid (H_2SO_4)** and **nitric acid (HNO_3)** as shown in the following reactions:

SO_2 (oxidised during its long stay in atmosphere) $\longrightarrow SO_3 + H_2O \longrightarrow H_2SO_4$

NO (oxidised during its long stay in atmosphere) $\longrightarrow NO_2 + H_2O \longrightarrow HNO_3$

Both H_2SO_4 and HNO_3 dissolve in the water in the atmosphere and fall to ground as **acid rain** or **acid snow.**

Box 21.5

The acid rain problem has become a matter of increasing concern in some European countries, USA and Canada since 1950's. The strange aspect of the acid rain is that the culprit is one locality but the victim is another because the pollutants are carried away by prevailing winds to elsewhere where precipitation takes place. The victims of acid rains are perhaps Sweden and Canada. While Canada gets acid rains from petrochemical units in Northern America, heavy winds take acid rains from factories in Britain and France to Sweden reducing the timber growth there by at least 15 per cent.

The acid rain is fast spreading to the developing world whose tropical soils may be more vulnerable than Europe's. In India, industrial areas with the pH value of rain below or close to critical value had been recorded in Delhi, Nagpur, Pune, Bombay and Calcutta. According to a study conducted by BARC (= Bhabha Atomic Research Centre), the average pH value of acid rain at Calcutta is 5.80, Hyderabad 5.73, Madras 5.85, Delhi 6.21 and Bombay 4.80. A rain which has a pH as low as 4 or 4.5 is considered as acid rain.

Effects of acid rain. Acid rain has been observed to cause a number of adverse implications. It increases acidity of soil, threatens human and aquatic life and destroys forests and crops reducing agricultural productivity. Some well-known effects of acid rain can be described as follows:

1. Effects on fish. Due to acid rain many lakes in Germany and other European countries have a pH of less than 5, an acidity level considered lethal for many species of fish. Acidic water can leach *aluminium* from the soil, and ground water and run-off can carry dissolve aluminium to lakes and streams. Because dissolved aluminium clogs the gills of fish and deprives them of oxygen, it is highly toxic to fish and is actually the cause of death of fish in acidic field waters (Box 21.6).

Box 21.6

Acid rain mobilizes heavy metals such as *calcium* and *mercury* in soils, rocks and sediments, which are then leached out by rain and enter surface water. Acidified lakes have high levels of *cadmium, lead, aluminium, manganese, zinc, copper* and *nickel.* All these can kill living organisms if present in sufficient quantities, and it is thought that it is their presence rather than the actual acidity of the water that causes the death of fish and other life.

In fact, acid rain has rendered "dead" more than 200 lakes in the New York State. Nearly 9,000 lakes of Sweden and 1500 lakes of Norway are fishless. These fishless lakes are now known as **"fish graveyards"**.

2. Effect on lake ecosystems. Acid rain falls directly on ponds and lakes. It also accumulates as acid snow; in spring snow melt gives a pulse of acid water to lakes due to rapid melting of accumulated acid snow. **Acid lakes** have low levels of phytoplankton or algae. Snails, clams and other animals with shells of calcium carbonate, which are readily dissolved by acid water, are among the first animals to die in acidified lakes. On the other hand, black flies, mosquitoes and deer flies are quite abundant in those lakes from where fish are eliminated so they appear to thrive in acid conditions. Some insects such as dragonfly larvae and boatman are replacing fish at the top of the food chain in acidified lakes.

Aquatic plants such as broad-leafed pondweeds do not survive acidic conditions. At the bottom of some acidified lakes, thick algal or moss mats become common. In acidified waters the activity of bacteria and other microscopic animals is reduced. As a result, the dead leaves and other accumulated materials on the bottom of lake (*i.e.,* detritus) are not rapidly decomposed and essential nutrients such as nitrogen, phosphorus and sulphur stay locked up in animal and plant remains. Therefore, biomass production is reduced and fish population decline.

3. Effect on terrestrial ecosystem. In plants acid rain causes (*i*) reduced rates of photosynthesis and growth; and (*ii*) increased sensitivity to drought and disease. Acid rain results in retarded growth in a variety of crops such as beans, broccoli, radish and spinach. Root systems are damanged by the uptake of aluminium released from the soil. The activity of nitrogen-fixing bacteria is inhibited, thus reducing soil fertility. Nutrients such as nitrates may be leached from the soil by acid runoff waters.

Trees such as spruce, pine, ashes and birch appear to be highly vulnerable to acid deposition. Acid deposition weakens trees, so they could be easily killed by drought and pathogens.

4. Damage of buildings. Acid rain corrodes buildings, monuments, statues, bridges, fences and railings that costs world 1450 million dollars a year.

5. Effect on health. Acid rain poses a serious threat to human health also, since it contaminates not only the breathing air but also the drinking water and even food.

Control of acid deposition. Short-term control of the problem of acid rain may be achieved by the use of lime. New York has been liming lakes since 1959. Some ecologists have suggested applying lime to entire watershed, including forests rather than just to individual lakes.

21.6 OZONE LAYER

Ozone (O_3) is a triatomic form of oxygen (*i.e.,* it is an isotope of oxygen). It is found largely in the **ozone layer** at a height of 15-60 km in the stratosphere of atmosphere. O_3 levels are low at over equator and high near the poles. Normally O_3 is present in the atmosphere at about 0.5 ppm at sea level. It shows remarkable variation from an average of about 0.02 ppm in winter to about 0.07 in summer.

Ozone has been the most reactive form of molecular oxygen and the fourth most powerful **oxidising agent** known, exceeded by only fluorine (F_2), oxygen difluoride (OF_2) and atomic oxygen. It has distinctive pleasant odour at concentration of about 2 ppm, but it is irritating at higher concentrations. Ozone is used as a disinfectant for air and water. Industrially O_3 is used for bleaching waxes and oils, as well as for organic synthesis.

In nature, O_3 is formed in the stratosphere when UV-light strikes an oxygen molecule. A photon splits the oxygen molecule into two highly reactive oxygen atoms (O^{-}*). These combine quickly with an oxygen molecule to form ozone. The O_3 readily absorb ultraviolet light and dissociates into its constituent component.

$$O_2 \xrightarrow{\text{UV-light}} 2O^{-*} + 2H^+$$

$$O^{-*} + O_2 \xrightleftharpoons{\text{UV-light}} O_3$$

Thus, being a natural constituent of the stratosphere, O_3 is regularly formed and destroyed in a cyclic manner with solar radiations as the driving force.

Ecological role of ozone. As an umbrella, ozone protects the biosphere from the harmful UV-radiations coming from cosmic rays to earth. In spite of being in such a small proportion, it plays a major role in climatology and biology of the earth. It filters out all radiations below 3000A° (UV-B-radiations) which are biologically harmful, and thereby control the 'thermal budget' of the earth. These UV-B-radiations are energetic enough to break apart biological molecules, including DNA leading to an array of health problems. For example, UV causes sunburn, skin cancer, cataracts and can supress the immune system. Evidence indicates that even 1 per cent decline in the O_3 layer could lead to 4-6 per cent increase in the rate of some skin cancers. It could severly damage the oceanic food chains by destruction of phytoplanktons, leading to a reduction in oxygen output. The amount of ozone, therefore, must have intimate connection with the life-sustaining process. Any large scale depletion of ozone content thus would have ill-effects on biotic systems including plants.

Ozone hole. Over last few years it has become clear that the stratospheric ozone layer is thinning out due to the excessive use of CFCs by humans. Once CFCs were considered as miracle compounds being used as coolants for refrigerators and air-conditioners, propellents for airosol sprays, agents for producing foam and cleansers for electronic paints. Now CFCs are considered as obnoxious inert compounds which are emitted mainly by aerosol and escape to stratosphere where atomic chlorine released from them as a result of photochemical dissociations reacting with ozone:

$$CFCl_2 \xrightarrow{\text{UV-light}} Cl^{-*} + CF_2Cl$$

$$Cl^{-*} + O_3 \longrightarrow ClO + O_2$$

$$ClO + O^{-*} \longrightarrow Cl + O_2$$

Chlorine is also liberated from ammonium percholate which is used as an oxidising agent in the rocket of space shuttle. The other areas where human activity can influence the O_3 concentration are direct emission of nitrogen oxides by supersonic transport (SSTs) flying above the tropopause and transport of nitrous oxides as a result of increase use of nitrogen fertilizers. The nuclear tests and atomic explosion too depletes the ozone layer.

The extreme thinning out of ozone-layer, idiomatically known as **ozone hole**, apparently began around 1976. But the problem of peeling of ozone umbrella was identified very late in 1985, when atmospheric scientists of the British Antarctic Survey at cambridge found that the spring time amount of ozone in the atmosphere over the Halley Bay in Antarctica had decreased by more than 40 per cent between 1977 to 1984. US satellite photographs showed that the ozone hole was as large as United States in areas and as deep as Mount Everest (*Tibetan Chomolungma*). According to World Meteorological Organization (WMO), the hole in the ozone layer on the antarctica grew at a record rate in October 1995, reaching 20 million km^2, a twice size of Europe. Recently ozone depletion has also been detected in Arctic or North Pole (**Ernst Augstein** 1987).

If timely measures are not adopted to check ozone depletion, there will be following two sort of problems: (*i*) US Environmental Protection Agency expects that in the next 50 years as a result of ozone depletion nearly 2 lac people may die due to skin cancer, (*ii*) According to **Dr. Robert Watron**, due to ozone depletion in the next few decades, temperature around the world could rise as much as 5.5°C (*i.e.,* it will cause global warming).

QUESTIONS

Long Answer Questions

1. Define the term 'environment pollution'. Discuss the various environmental pollutions and their effects on human beings and other animals.
2. Define the term "Pollution". Illustrate the impact of human action on environment.
3. Define the term 'Environmental Pollution'. Discuss various environmental pollutions and their affects on human beings directly or indirectly.
4. Define pollution? Give a concise account of its source and control measures.
5. Write an essay on environmental pollution.
6. What do you mean by "Pollution is a resource out of place".
7. Define pollution? Write an essay on air pollution.
8. What is air pollution? What are the different sources of air pollution? Describe the effects of air pollution on human life and plants.
9. Give an account of sources of air pollution in our country. Add notes on their impact and control measures.
10. Name the major air pollutants and determine their sources. Discuss the action of pollutants on organisms. Mention the effective measures for the prevention of air pollution.
11. What are the several major processes that contribute to air pollution. Explain the phenomenon of " Green house effect" "Acid rain" and "Ozone hole" What is EL NINO? How is it operating? Write down the effects of air pollution on human health. Suggest some measures to protect indoor air quality.
12. Write an essay on air pollution. Mention its effect on public health.
13. Describe sources of air pollution and control measures.
14. Give an account of air pollution affecting human health.
15. Explain in detail the causes and consequences of air pollution. Suggest remedial measures.
16. Discuss 'Causes and consequences of acid rain.'
17. Give an account of air pollution and add a note on their effects.
18. Explain in detail the relative importance of air pollution in our environment.
19. Automobiles are the major cause of air pollution in India. Comment upon it with reference to animals.
20. Explain the reason of ozone layer depletion. How it may affect human health.
21. What are automobile emmissions? How they affect the human health.
22. Name the 'greenhouse' gas? How does it produce 'greenhouse effect'.
23. Explain 'greenhouse effect'.

Short Answer Questions

1. **Write short notes on the following:**
   (*i*) Air pollution
   (*ii*) Environmental pollution
   (*iii*) Pollutants
   (*iv*) Common air pollutants
   (*v*) Automobile emission
   (*vi*) Greenhouse effect
   (*vii*) Global warming
   (*viii*) Role of methane in greenhouse effect
   (*ix*) Acid rain
   (*x*) Ozone layer

2. What is meant by an acid rain?
3. How ozone layer helps human beings?
4. Comment upon the following:
   (*i*) PAN is a primary air pollutant.
   (*ii*) Acid rain is caused mainly due to the release of combustion products into the atmosphere.
5. What is aerial pollution.
6. PAN is known as secondary pollutant. Why?
7. Differentiate between fog and smog.

Very Short Answer Questions

1. Enlist the green house gases.
2. Which gas was responsible for Bhopal Gas Tragedy? (*CCSU Meerut 2005*)
3. What is fly-ash?
4. What is greenhouse effect?
5. Define global climate change.
6. What is ozone hole?
7. What are chlorofluoromethanes (CFMs) and Chlorofluorocarbons (CFCs)?
8. What is the photochemical smog?

True and False Statements

1. Mark the following statements true or false.
   (*i*) Greenhouse effect is the name given to green revolution..................

Fill in the Blanks

(*i*) Seventy percent of acid precipitation is __________acid, formed from burning____________and 30 per cent is _________acid, formed mainly from burning___________.

(*ii*) A problem with non-biodegradable toxic pollutants is that they undergo biologic__________as they pass through food web.

Multiple Choice Questions

1. The most harmful types of environmental pollutants are
   (*a*) human organic wastes (*b*) wastes from feedlots
   (*c*) non-biodegradable chemicals (*d*) natural nutrients present in excess
2. Acid precipitation kills fish by causing release of
   (*a*) aluminium ions (*b*) carbon monoxide
   (*c*) anticoagulants (*d*) mercury
3. Photochemical smog always contains
   (*a*) aluminium ions (*b*) methane
   (*c*) ozone (*d*) phosphorus
4. Taj Mahal is threatened to the effect of
   (*a*) chlorine (*b*) sulphur dioxide
   (*c*) oxygen (*d*) hydrogen
5. The chemical that contribute to the destruction of ozone layer above the earth's surface is
   (*a*) mercury (*b*) carbon monoxide
   (*c*) sulphur dioxide (*d*) chlorofluorocarbons

6. Carbon monoxide is a pollutant because it
(*a*) inactivates nerves (*b*) combines with oxygen
(*c*) combines with haemoglobin (*d*) inhibits growth

7. Fly ash is environmental pollutant produced by
(*a*) thermal power plant (*b*) oil refinery
(*c*) fertilizer plant (*d*) strip mining

8. Two-wheelers cause this
(*a*) air pollution (*b*) noise pollution
(*c*) air and noise pollution (*d*) water pollution

9. Most hazardous metal pollutant of automobile exhaust is
(*a*) cadmium (*b*) mercury
(*c*) copper (*d*) lead

10. Greenhouse effect is enhanced in the environment by the gas
(*a*) CO_2 (*b*) CO
(*c*) fluorocarbons (*d*) NO_2

11. Acid rain occurs in areas
(*a*) where citrus plants are grown
(*b*) there are large plantations of eucalyptus trees
(*c*) there are large plantations of pine trees
(*d*) there are big industries and the atmosphere is polluted with SO_2

12. In coming years skin related disorders will be more common due to
(*a*) pollutants in air (*b*) use of detergents
(*c*) water pollution (*d*) depletion of ozone layer

13. Formation of ozone hole is maximum over
(*a*) India (*b*) Europe
(*c*) Antarctica (*d*) Africa

14. Air pollution effects are usually found on
(*a*) leaves (*b*) flowers
(*c*) stems (*d*) roots

15. Most harmful types of environmental pollutants
(*a*) human organic waste (*b*) waste from feedlots
(*c*) non-biodegradable chemicals (*d*) natural nutrients present in excess

16. 'Ozone day' is observed on
(*a*) January 30 (*b*) April 21
(*c*) September 16 (*d*) December 25

17. Carbon dioxide content in our atmosphere has increased during the past 150 years approximately from
(*a*) 270 to 340 ppm (*b*) 20 to 35 ppm
(*c*) 0.2 to 0.3 ppm (*d*) 0.027 to 0.034 ppm

18. The basic component of the smog may be
(*a*) O_3 (*b*) PAN
(*c*) O_3 and PAN (*d*) PPN and PBN

19. Acid rains occur when atmosphere is heavily polluted with
(*a*) CO, CO_2 (*b*) smoke particle
(*c*) ozone (*d*) SO_2

20. Carbon monoxide is a major pollutant of
(*a*) water (*b*) air
(*c*) noise (*d*) soil

21. Increasing skin cancer and higher mutation rates are generally the consequence of
(*a*) CO_2 (*b*) ozone depletion
(*c*) ozone biome (*d*) acid rains

22. Acid rain is due to
(*a*) sulphur dioxide pollution (*b*) carbon monoxide pollution
(*c*) pesticide pollution (*d*) dust particles

23. "Catalytic converter" is used in automobiles
(*a*) for removing the poisonous lead
(*b*) for converting CO and harmful hydrocarbons into CO_2 and water vapours
(*c*) for removing water
(*d*) none of the above

24. Which city has most 'Arsenic Pollution'
(*a*) Delhi (*b*) Calcutta
(*c*) Maharastra (*d*) Ahmedabad

25. The toxic effect of carbon monoxide is due to its greater affinity for haemoglobin as compared to oxygen by (approx)
(*a*) 20 times (*b*) 1000 times
(*c*) 2 times (*d*) 200 times

26. Which of the following causes photochemical smog
(*a*) methane (*b*) sulphur dioxide
(*c*) peroxyacetyl nitrates (*d*) ethylene

27. The gaseous air pollutant which combines with blood haemoglobin and reduces the O_2 carrying capacity of the blood is
(*a*) CO_2 (*b*) CO
(*c*) NO_2 (*d*) SO_2

28. Ozone layer is present above
(*a*) stratosphere (*b*) troposphere
(*c*) inosphere (*d*) mesosphere

29. Cyclone collector is used to control
(*a*) air pollution (*b*) noise pollution
(*c*) water pollution (*d*) radioactive pollution

30. The trace gas which is produced in rice paddies and is associated with global warming is
(*a*) methane (*b*) chlorine
(*c*) CO_2 (*d*) hydrogen sulphide

31. Pollutant from motor car exhaust that causes mental disease is
(*a*) Hg (*b*) Pb
(*c*) NO_2 (*d*) SO_2

32. Major aerosol pollutant present in the jet plane emission is
(*a*) sulphur dioxide (*b*) fluorocarbons
(*c*) carbon tetrachloride (*d*) carbon monoxide

33. In upper atmosphere ultra-violet radiation from sunlight causes the reaction that produces
(*a*) CO (*b*) SO_2
(*c*) ozone (O_3) (*d*) fluorides

34. Main air pollutant is
(*a*) CO (*b*) CO_2
(*c*) N_2 (*d*) sulphur

35. Which of the following is normally not an atmospheric classified pollutant
(*a*) CO (*b*) CO_2
(*c*) N_2 (*d*) sulphur

36. The gas which was released in Bhopal Tragedy was
(*a*) ethylene (*b*) methylisocynate
(*c*) mustard gas (*d*) potassium isothiocynate

37. Environmental planning organisation is
(*a*) CSIR (*b*) NEERI
(*c*) ICAR (*d*) CEPHERI

38. NEERI is situated at
(*a*) Bhopal (*b*) New Delhi
(*c*) Nagpur (*d*) Baroda

39. Natural sink of stratospheric ozone layer is
(*a*) SO_3 (*b*) SO_2
(*c*) ferons (*d*) sulphur flux of oceans

40. Pollution control board in India
(*a*) IUCN (*b*) NEERI
(*c*) IUBN (*d*) UNEP

41. Mosses are indicators of
(*a*) air pollution (*b*) water pollution
(*c*) radiation pollution (*d*) soil pollution

42. The main indicators of SO_2 pollution are
(*a*) lichens (*b*) ferns
(*c*) algae (*d*) bryophytes

43. Which of the following serves as indicator of atmosphere pollution?
(*a*) liverworts (*b*) ferns
(*c*) epiphytic lichens (*d*) hornworts

44. Indicator plants which can be used to indicate atomspheric pollution by SO_2 are
(*a*) grass like *deschampsia* (*b*) moss like *sphagnum*
(*c*) lichens like *usnea* (*d*) climbers like *cucurbita*

45. Pollutant of jet plane
(*a*) CO (*b*) NH_3
(*c*) CO_2 (*d*) Fluorocarbons

46. Major source of methane in India is
(*a*) fruit orchards (*b*) sugarcane plantation
(*c*) rice fields (*d*) wheat fields

47. Which pollutant causes leaf curling?
(*a*) SO_2 (*b*) CO
(*c*) H_2S (*d*) O_3

48. Smog is
(*a*) smoke (*b*) other name for dust storm
(*c*) moistened air gases (*d*) smoke and fog

49. Which gas will not cause any atmospheric pollution? (*CCSU Meerut 2014*)
(*a*) hydrogen (*b*) sulphur dioxide
(*c*) carbon dioxide (*d*) carbon monoxide

50. It is said that Taj Mahal be destroyed. It is due to (*CCSU Meerut 2014*)
(*a*) high temperature (*b*) pollution of Yamuna
(*c*) pollutant of Mathura refinery (*d*) none of the above

51. If global warming continues the organism which may face more threat is (*CCSU Meerut 2014*)
(*a*) cow (*b*) snow leopard
(*c*) elephant (*d*) dolphin

52. A gas produced by paddy field and concerned with global warming is (*CCSU Meerut 2014*)
(*a*) methane (*b*) chlorine
(*c*) H_2S (*d*) CO_2

53. Acid rain occur due to atmospheric pollution of (*CCSU Meerut 2014*)
(*a*) SO_2 (*b*) NH_3
(*c*) CO_2 (*d*) N_2O

54. Which of the following causes depletion of ozone layer? (*CCSU Meerut 2014*)
(*a*) trichloethane (*b*) carbon tetrachloride
(*c*) chlorofleurocarbon (*d*) polyvinyl chloride

55. Which of the following group of gases cause photochemical smog? (*CCSU Meerut 2014*)
(*a*) O_3, PAN and NO_2 (*b*) O_3, PAN and CO
(*c*) O_2, PAN and NO_2 (*d*) O_2, PAN and NO_3

56. Maximum green house gas released by which country? (*CCSU Meerut 2014*)
(*a*) India (*b*) USA
(*c*) France (*d*) Pakistan

57. In 1984, Bhopal gas tragedy was caused due to leakage of (*CCSU Meerut 2014*)
(*a*) sodium thiocyanate (*b*) methyl isocyanate
(*c*) sodium monoxide (*d*) potassium isocyanate

58. Which is Ozone day? (*CCSU Meerut 2014*)
(*a*) 30 January (*b*) 08 September
(*c*) 16 September (*d*) 07 April

59. SO_2 pollution affect which part of the plant?
(*a*) lysosome (*b*) mitochondria
(*c*) endoplasmic reticulum (*d*) chloroplast

ANSWER

Very Short Answer Questions

1. Water vapour (H_2O), carbon dioxide (CO_2), methane (CH_4), and nitrous oxide (N_2O).
2. It was due to sudden vaporization and discharge of Methyl Isocyanide (MIC) gas as well as phosgene gas from pesticide plant of Union Carbide (India).
3. Fly-ash is the finely divided particles of ash in fuel gases resulting from combustion of coal (in thermal power units).
4. The warming of Earth by gases (CO_2) in the atmosphere that absorb and reradiate infrared energy emitted by Earth's surface.
5. Modification of the global climate resulting from the increased proportion of greenhouse gases, especially CO_2, emitted as by-products of human activities.

6. An area of the atmosphere with an ozone concentration of less than 220 Dobson units (= 27 × 10^{16} molecules of ozone) per square centimeter; found primarily over the Antarctic region.

7. CFMs and CFCs are compounds of carbon and halogens are named "freons". There are several CFMs and CFCs released into troposphere, where they dissociate to release chlorine that causes depletion of ozone (O_3).

8. Photochemical smog is complex mixture of air pollutants (oxidants) produced in the atmosphere by the reaction of hydrocarbons and nitrogen oxides under the influence of sunlight. Three of the most harmful photochemical oxidants are ozone (O_3), peroxyacetyl nitrate (PAN) and various aldehydes.

Fill in the Blanks

1. Sulphuric acid; coal, nitric, gasoline.   2. Magnification.

Multiple Choice Questions

1. (*c*)	**2.** (*a*)	**3.** (*c*)	**4.** (*b*)
5. (*d*)	**6.** (*c*)	**7.** (*a*)	**8.** (*a*)
9. (*d*)	**10.** (*a*)	**11.** (*d*)	**12.** (*d*)
13. (*c*)	**14.** (*a*)	**15.** (*c*)	**16.** (*c*)
17. (*a*)	**18.** (*c*)	**19.** (*d*)	**20.** (*b*)
21. (*b*)	**22.** (*a*)	**23.** (*a*)	**24.** (*d*)
25. (*d*)	**26.** (*c*)	**27.** (*b*)	**28.** (*b*)
29. (*a*)	**30.** (*a*)	**31.** (*b*)	**32.** (*b*)
33. (*c*)	**34.** (*a*)	**35.** (*b*)	**36.** (*b*)
37. (*b*)	**38.** (*c*)	**39.** (*c*)	**40.** (*b*)
41. (*a*)	**42.** (*a*)	**43.** (*c*)	**44.** (*c*)
45. (*d*)	**46.** (*c*)	**47.** (*a*)	**48.** (*d*)
49. (*a*)	**50.** (*c*)	**51.** (*b*)	**52.** (*a*)
53. (*a*)	**54.** (*c*)	**55.** (*a*)	**56.** (*b*)
57. (*b*)	**58.** (*c*)	**59.** (*d*).	

22

Water Pollution

22.1 DEFINITION

The great solvent power of water makes the creation of absolutely pure water a theoretical rather than a practical goal. Even the highest-quality distilled water contains dissolved gases and to a slight degree, solids. The problem, therefore, is one of determining what quality of water is needed to meet a given purpose and then finding practical means of achieving that quality. The problem is further compounded because every use to which water is put—washing, irrigation, flushing away wastes, cooling, making paper, etc., adds something to the water. In fact for centuries rivers and lakes have been used as dumping grounds for human sewage and industrial wastes of every conceivable kind, many of them are highly toxic. Added to this are the materials leached and transported from land by water percolating through the soil and running off its surface to aquatic ecosystems.

The term **water pollution** is referred to any type of aquatic contamination between following two extremes: (1) a highly enriched, over productive biotic community, such as a river or lake with nutrients from sewage or fertilizer (**cultural eutrophication**), or (2) a body of water poisoned by toxic chemical which eliminate living organisms or even exclude all forms of life (**Southwick,** 1976).

Normally water contains two types of impurities—dissolved and suspended. Dissolved impurities are gases (H_2S, CO_2, NH_3, etc.) and minerals (Ca, Mg, Na, salts). Suspended matter includes clay, silt and sand and even microbes. Polluted water are turbid, unpleasant, foul smelling, unfit for drinking, bathing and washing or other purposes. They are harmful and means of many diseases as cholera, dysentery, typhoid, hepatitis, etc.

Types of Water Pollution

Water pollution may be classified by the *medium* in which they occur, such as **surface water pollution, ground water pollution, soil water pollution,** etc.; the *habitat* in which they occur, such as **river pollution, lake pollution, estuarine pollution, coastal water pollution, open ocean pollution,** etc.; and *source* or *type* of contamination, such as **nutrient pollution, bacterial pollution, viral pollution, metallic pollution, petrochemical pollution, pesticide pollution, thermal pollution, radioactive pollution,** etc.

Kinds and Sources of Water Pollutants

Pollutants entering water sources are classified broadly into following categories: domestic sewage and oxygen-demanding wastes; infectious agents; plant nutrients; chemicals such as insecticides, herbicides, and detergents; other minerals and chemicals; sediment from land erosions; radioactive substances; and heat from power and industrial plants.

These aquatic pollutants come from many sources. Excessive nutrients, such as nitrates and phosphates, commonly originate in domestic sewage, run-off from agricultural fertilizer, waste materials from animal feed lots, packing plants, etc. Toxic chemicals as agents of water pollution originate in industrial operations, acid mine drainage, surface erosion from strip mines, washing of herbicides and insecticides, radioactive fall out from atomic explosion, and commercial accidents such as oil spills or the rupture of chemical tanks. Besides the pollutants which come from **point sources** such as sewage, factory or industry, there are many pollutants which come from watershed run-off. Urban and sub-urban run-off, for example, contains many pollutants such as oil, pesticides, radioactive dust, salt, fertilizers, miscellaneous chemicals, and nematodes, pathogenic protozoans, bacteria and viruses. However, **Kimbal** (1975) recognized only three major sources of pollution: *domestic, industrial* and *agricultural.*

22.2 ECOLOGY OF WATER POLLUTION

Each type of water pollution affects the abiotic and biotic factors of different aquatic systems in different degrees and its ultimate effect on humans remains quite drastic in medical, aesthetic, and economical sense. Some of the well known ecological effects of water pollution are following:

1. Sewage Pollution

Contamination of freshwaters and shallow offshore seas by sewage is a common occurrence. Domestic sewage and waste-water is about 99.9 per cent water and 0.02-0.04 per cent solids of which proteins and carbohydrates each comprise 40–50 per cent and fats 5–10 per cent (**Simmons**, 1974). In other words, sewage includes mostly **biodegradable pollutants** such as human faecal matter, animal wastes, and certain dissolved organic compounds (*e.g.,* carbohydrates, urea, etc.) and inorganic salts such as nitrates and phosphates of detergents and sodium, potassium, calcium and chloride ions. Under natural processes most of the biodegradable pollutants of sewage are rapidly decomposed, but, when they accumulate in large quantities, they create problem, *i.e.*, when their input into environment exceeds the decomposition or dispersal capacity of the latter. Most cities of well developed countries such as USA, Britain, etc., and some cities of developing countries such as India have evolved certain engineering systems, such as, **septic tanks, oxidation ponds, filter beds, waste water treatment plants** and **municipal sewage treatment plants** for the removal of many harmful becteria and other microbes, organic wastes and other pollutants from the sewage, before it is tipped into river or sea.

Sewage treatment is usually performed in following three stages: (1) **Primary treatment,** which removes large objects and suspended undissolved solids of raw sewage and converts them into a biologically inactive and aesthetically inoffensive state, the **sludge**, a valuable fertilizer. (2) **Secondary treatment**, which supplies aeration and bacteriological action to decompose organic compounds into harmless substances such as CO_2, sulphate and water. During later stages of secondary treatment, whole waste water is chlorinated (*i.e.,* treated with chlorine) to reduce its content of bacteria, (3) **Tertiary treatment**, which removes nitrates and phospates and releases pure water. These three stages of sewage treatment have become increasingly expensive and only in most advanced countries all the three treatments of sewage are done. However, most Indian cities either lack any sewage or waste-water treatment plant or have inadequate sewage treatment facilities. Consequently, normally and especially during heavy downpour and floods, raw sewage or incompletely treated sewage is dumped into rivers which cause severe water pollution problems in following ways:

(i) **Bacterial and viral contamination.** Sewage wastes may contain pathogenic bacteria and viruses which are a threat to human health. Waterborne diseases such as typhoid, bacillary dysentery, amoebic dysentery, botulism, poliomyelitis, and hepatitis all represent potential health hazards in sewage-contaminated waters. Due to such kinds of sewage pollution waters of many ponds, lakes, rivers, sea beaches in India and abroad have been prohibited for human use, whether for drinking, bathing, swimming or other sort of water recreation.

BOD test. **BOD** or **biological oxygen demand** is the amount of oxygen required for biological oxidation by microbes in any unit volume of water. The test is done at 20ºC for at least five days. BOD value generally approximates the amount of oxidisable organic matter (such as sewage and other organic wastes, animals and human excreta, all of which are called **oxygen demanding wastes**) and is, therefore, used as a measure of degree of water pollution and waste level. Thus, due to addition of sewage and waste, oxygen levels are depleted which are reflected in terms of BOD values of water. The number of microbes as *Escherichia coli* (bacterium) also increases tremendously and these also consume most of the oxygen. The number of bacteria as *E.coli* in unit volume of water is also taken (called ***E.coli*** **index**) as a parameter of water pollution. The quantity of oxygen in water (called **dissolved oxygen** or **DO**) along with BOD is indicated by the kind of organism present in water. For example, fish become rare at DO value of 4 to 5 ppm of water. Further decrease in DO value may lead to increase in anaerobic bacteria. Typical BOD value for raw sewage run from 200 to 400 mg of oxygen per litre of water (therefore, 200–400 ppm). Water for drinking should have a BOD less than 1.

Table 22.1. Comparison of BOD and COD.

	Biological oxygen Demand (BOD)		Chemical Oxygen Demand (COD)
1.	It is the amount of oxygen used for biochemical oxidation by microorganisms in a unit volume of water. This test has been developed for five days at 30ºC.	1.	It is the amount of oxygen required by organic matter in a sample of water for its oxidation by a strong chemical oxidant, and is expressed as ppm of oxygen taken from a solution of potassium dichromate in two hours.
2.	BOD value approximates the amount of oxdizable organic matter and therefore, used as a measure of degree of water pollution and waste strength.	2.	This value is a poor measure of strength of organic matter as oxygen is also consumed in the oxidation of inorganic matter such as nitrates sulphates, reduced metal ions, and also that some organic molecules such as benzene, pyridine and few other cyclic organic compounds are not oxidized by this test.
3.	BOD values are useful in process design and loading calculations, measure of treatment efficiency and operation, stream pollution control and in evaluating self purification capacity of a stream.	3.	It is very important parameter in management and design of the treatment plants because of its rapidity in determination value are taken as basis for calculation of efficiency of treatment plants and also figure in the standards for discharging industrial/domestic effluents in various kinds of waters.
4.	BOD test is influenced by following factors: types of microorganisms, pH, presence of toxins, some reduced mineral matter and nitrification of microorganisms process.	4.	Presence of toxins and other such unfavourable conditions for the growth of microorganisms do not affect COD values.

(*ii*) Eutrophication. According to **Hutchinson** (1969), the eutrophication is a natural process which literally means "well nourished or enriched." It is a natural state in many lakes and ponds which have a rich supply of nutrients, and it also occurs as part of the aging process in lakes, as nutrients accumulate through natural succession. Eutrophication becomes excessive, however, when abnormally high amounts of nutrients from sewage, fertilizer, animal wastes and detergents, enter streams and lakes, causing excessive growth or 'bloom' of microorganisms and aquatic vegetation.

Most secondary sewage treatment plants, through, precipitate solids and inactivate most bacteria in domestic sewage, yet they do not remove the basic nutrients such as ammonia, nitrogen, nitrates, nitrites and phosphates. These nutrients stimulate algal growth and lead to **plankton blooms**. Some plankton blooms, particularly those of blue-green algae produce obnoxious odours and tastes in waters. Others, such as the dinoflagellate blooms or "**the red tide**" of southern coastal regions, produce toxic metabolic products which can result in major fish kills.

Plankton blooms of green algae do not always produce undesirable ordours or toxic products, but still create problems of oxygen supply in the water. While these blooms exist under abundant sunlight, they contribute oxygen to the water through photosynthesis, but under conditions of prolonged cloudiness, they begin to decay and consume more oxygen and with heavy load the oxygen contents of the water may diminish below the point where most fish can not survive. As the conditions in the water become anaerobic due to increased oxygen depletion by bacterial decomposition or planktonic blooms, the breakdown products become reduced rather than oxidized molecules, many of which (*e.g.,* hydrogen sulphide) produce offensive odours and tastes.

Excessive nutrient levels in aquatic systems can also cause two other kinds of ecological problems. Primarily, they may lead to extensive growth of aquatic weeds such as Eurasion milfoil, water hyacinth, water chestnut, etc. Excessive growth of these weeds can impair fishing, bathing, fish spawning, shell fish production, and even navigation. Secondarily, nitrates can be converted in the human digestive tract by certain bacteria to nitrites, and the same transition may occur in opened cans of food even if they are subsequently refrigerated. Nitrites react with haemoglobin, forming **methemoglobin** which will not take up oxygen. Laboured breathing and occasional suffocation result most severely in human infants. Nitrites may also react with creathinine (present in the vertebrate muscles) to form nitrosarcosine which can be carcinogenic.

Effects of organic pollution on aquatic animal life. Organic pollution tends to bring about changes in faunal composition in a freshwater ecosystem. Nymphs of stone flies and may-flies are the first to disappear from water which has high organic pollution. As pollution increases, caddishfly larvae and many fish which require high levels of environmental oxygen move into less polluted area of the stream. Shrimps, water fleas, leeches, snails and most of the fish vanish as the pollution becomes severe. At such levels of pollution there is very little of dissolved oxygen and the animals present are chironomid larvae (blood worms) and the oligochaete worms *Tubifex* (Fig. 22.1). Some decomposing plants are known to produce toxins as **strychnine** which kills animals including cattle.

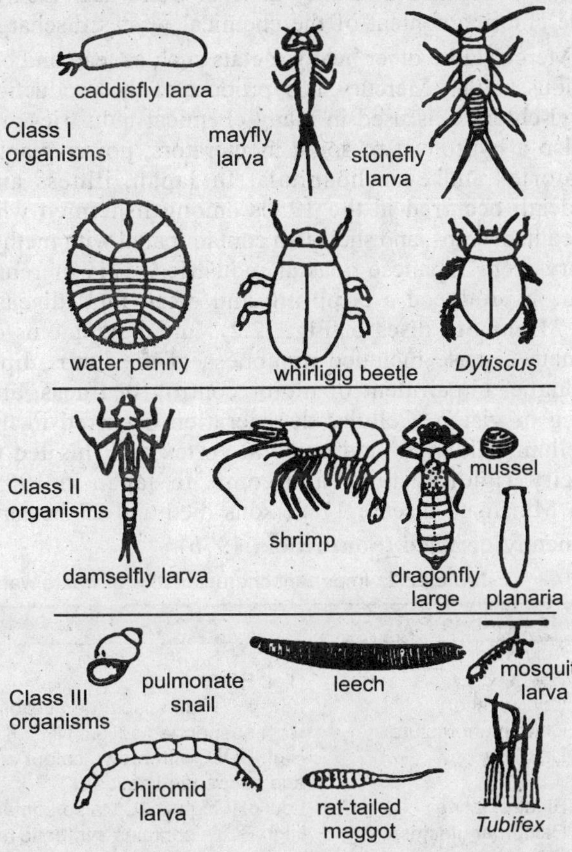

Fig. 22.1. Classes of stream fauna used as pollution indicators.

2. Industrial Pollution

Most of the Indian rivers and freshwater streams are seriously polluted by industrial waste of effluents (see Table 22.2). **Effluents** are waste products in a liquid form resulting from industrial processes and domestic activities. They are released by different industries such as petro-chemical complexes; fertilizer factories; oil refineries; pulp, paper, textile, sugar and steel mills, tanneries, distilleries, coal washeries, synthetic material plants for drugs, fibres, rubber, plastics, etc. The industrial wastes of these industries and mills include metals (copper, zinc, lead, mercury, etc.), detergents, petroleum, acids, alkalis, phenols, carbamates, alcohols, cyanide, arsenic, chlorine and many other inorganic and organic toxicants (Table 22.2 and 22.3). All of these chemicals of industrial waste are toxic to animals and many cause death or sublethal pathology of the liver, kidneys, reproductive systems, respiratory systems, or nervous systems in both invertebrate and vertebrate aquatic animals. Chlorine which is added to water to control growth of algae and bacteria in the cooling system of power station, may persist in streams to cause mortality of plankton and fish. Heavy fish mortality in river **Sone** near Dehri-on-sone in Bihar is reported to cause by free chlorine content of the chemical wastes discharged by factories near Mirzapur in U.P.

Mercury like other heavy metals such as lead and cadmium has cropped up as a toxic agent of serious nature. Mercury, a byproduct of the production of vinyl-chloride, is used in many chemical industries and it is also a byproduct of some incinerators, power plants, laboratories and even hospitals. In Japan, illness and even death occurred in the 1950s among fishermen who ingested fish, crabs, and shell-fish contaminated with methyl mercury from Japanese coastal industries. This mercury poisoning produced a crippling and often fatal disease called **Minamata disease** (Fig. 22.2). Initial symptoms of minamata disease included numbness of the limbs, lips, and tongue, impairment of motor control, deafness, and blurring of vision. Cellular degeneration occurred in the cerebellum, midbrain, and cerebral cortex and this led to spasticity, rigidity, stupor and coma. In Japan in 1953, due to Minamata disease 17 persons died and 23 became permanently disabled (**Southwick**, 1976).

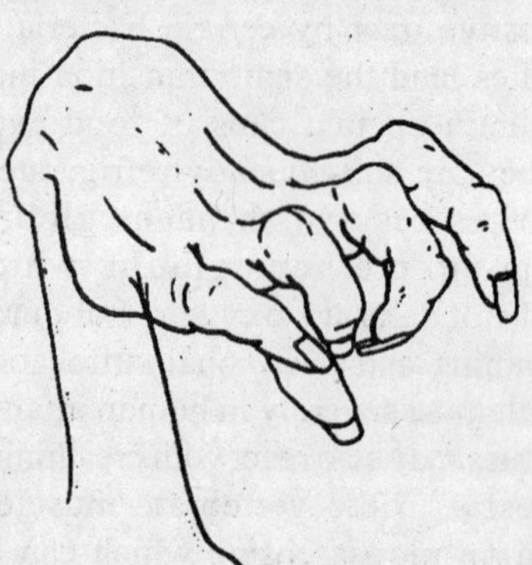

Fig. 22.2. Minamata disease caused by mercury pollution.

Table 22.2. Important characteristics of waste water from some major industries.

	Industries	Characteristics
1.	Acid manufacture	Low pH
2.	Beet sugar	High BOD
3.	Coal washery	Low pH, High suspended solids, H_2SO_4 and $FeSO_4$ present
4.	Coke manufacture	High suspended solids, NH_3, H_2S, phenol and oils
5.	Distillery	High BOD, with brown colour and disagreeable odour, high dissolved and suspended solids
6.	Electroplating	Low pH, high COD*, also contains heavy metals and toxic substances
7.	Paint manufacture	High BOD; contains synthetic resins, solvents, pigments, and heavy metal such as Al, Cr, Pb, etc.
8.	Petroleum refinery	High COD/BOD ratio, also contains hydrocarbons, alcohols, aldehydes, phenols, oils, metals etc.
9.	Plastic manufacture	Acids, formaldehyde and phenols
10.	Pulp, paper and board industry	Intense brown coloured with characteristic odour, alkaline contains highly dissolved and suspended solids
11.	Steel industry	Low pH; phenols, suspended solids and metals present
12.	Tannery	High BOD and COD; contains highly dissolved and suspended solids and oil and grease, and heavy metals such as Cr
13.	Textile processing	High BOD; suspended and dissolved solids, phenolic substances, acids, chlorine dyes and Cr present.

*COD is a test used to assess the strength of sewage and trade waste.

Table 22.3. Some Indian rivers and their major sources of pollution.

	Name of the river	Sources of pollution
1.	Kali at Meerut (U.P.)	Sugar mills; distilleries; paint, soap, rayon, silk, yarn, tin and glycerine industries.
2.	Jamuna near Delhi	D.D.T. factory, sewage, Indraprastha Power Station, Delhi.
3.	Ganga at Kanpur	Jute, chemical, metal, and surgical industries; tanneries, textile mills and great bulk of domestic sewage of highly organic nature.
4.	Gomati near Lucknow (U.P.)	Paper and pulp mills; sewage.
5.	Dajora in Bareilly (U.P.)	Synthetic rubber factories.
6.	Damodar between Bokaro and Panchet	Fertilizers, fly ash from steel mills, suspended coal particles from washeries, and thermal power station.
7.	Hooghly near Calcutta	Power Stations; paper pulp, jute, textile, chemical mills, paint, varnishes, metal, steel, hydrogenated vegetable oils, rayon, and soap, match, shellace, and polythene industries and sewage.
8.	Sone at Dalmianagar (Bihar)	Cement, pulp and paper mills.
9.	Bhadra (Karnataka)	Pulp, paper and steel industries.
10.	Cooum, Adyar and Buckinghum canal (Madras)	Domestic sewage, automobile workshops.
11.	Cauvery (Tamil Nadu)	Sewage, tanneries, distilleries, paper and rayon mills.
12.	Godavari	Paper mills.
13.	Siwan (Bihar)	Paper, sulphur, cement, sugar mills.
14.	Kulu (between Bombay and Kalyan)	Chemical factories, rayon mills and tanneries.
15.	Suwao (in Balrampur)	Sugar industries.

Quite recently, **Dash** and coworkers (1988) studied mercury pollution due to effluents of a paper mill in the river Ib of the Mahanadi water system in Orissa. The paper mill uses mercury as a cathode in chloroalkali electrostatic cells and mercury leakage can be as high as 200 mg/tonne of chlorine produced (**Fergusson**, 1982). An analysis of the mercury in some fish showed 500 times biomagnification (or bioamplification) over the downstream surface water. Generally benthic fish and those higher in the food chain accumulated more mercury. People of the villages of that region, who used river water for drinking, suffered from diseases of eye, skin rashes and anemia.

3. Thermal Pollution

Various industrial processes may utilize water for cooling, and resultant warmed water has often been discharged into streams or lakes. Coal-or oil-fired generators and atomic energy plants cause into large amount of waste heat which is carried away as hot water and cause **thermal pollution** or **calefaction** (warming). Thermal pollution produces distinct changes in aquatic biota. A body of water at 30–35ºC is essentially a biological desert and many game fish require temperature of 10ºC for successful reproduction, although they will survive above that temperature. A temperature rise of 10ºC will double the rate of many chemical reactions and so the decay of the organic matter, the rusting of iron and the solution rate of salts are also accelerated by calefaction. Since the rate of exchange of salt in organisms increase, any toxin is liable to exert greater effects and temperature fluctuations are likely to affect organisms. Some plants and animals are killed outright by the very hot water. Other adverse effects of aquatic pollution on aquatic life include (*i*) early hatching of fish eggs, (*ii*) failure of trout eggs to hatch; (*iii*) failure of salmon to spawn, (*iv*) increase in BOD, *i.e.*, solubility of oxygen is reduced causing deoxygenation; (*v*) change in diurnal and seasonal behaviour and metabolic responses of organisms; (*vi*) significant shift in algal forms and other organisms towards more heat tolerant forms (this leads to decrease in species diversity); (*vii*) affect changes in macrophytes and (*viii*) migration of some aquatic forms.

4. Silt Pollution

A result of intensive agriculture, earth moving for construction projects, poor conservation practices and downpour with resultant floods, is the increased production of silt in streams and lakes. This load of particulate matter cuts down primary productivity by decreasing the depth of light penetration. Silt may also interrupt or prevent the reproduction of fish, by smothering eggs laid on the bottom.

5. Water Pollution by Agrochemicals

Water that flows on the surface of crop fields, where agrochemicals such as fertilizers, pesticides and herbicides are used, contributes to heavy water and soil pollution.

Pollution by agriculture waste. Depletion of dissolved oxygen caused by **phosphate-induced** algal growth leads to death of fish and other aquatic biota. In the presence of phosphates, **nitrates** too bring about an exaggrerate growth of vegetation. Excessive and indiscriminate use of chemical fertilizers and synthetic feed for livestock often leads to accumulation of nitrates in water. When such waters are drunk by cattle or humans, these are taken into body and converted into toxic **nitrites** by intestinal bacteria. Nitrites in turn combine with the haemoglobin to form **methanohaemoglobin**, which interferes with oxygen-carrying capacity of the blood, producing a serious disease called **methanohaemoglobinaemia**. The various ailments that result from this disease include damage of respiratory and vascular system, blue colouration of the skin and even cancer.

Nitrate poisoning has been frequently noted in various desert districts of Rajasthan as a result of hardness and salinity of water and alleged to have caused infant death in the state. Some other adverse effects of fertilizers use are:

1. Changes in mobility status of nutrients in soils;
2. Deterioration of water resources as a result of eutrophication.
3. Stimulation of weed growth in crop fields, and
4. Disturbance in the ionic balance and equilibrium in soils, often leading to high acidity, nutritional imbalance, shortage of certain trace elements and molybdenum and selenium toxicity.

Pesticides and weedicides are used by human beings to control crop diseases by the pests or to kill the weeds and, hence, to increase the productivity. The use of these toxic chemicals has created health hazards not only for livestock and wild life but also for fish, other, aquatic organisms, birds and mammals including humans.

Apart from killing the living organisms present on the surface of the soil, they reach even the deeper layers through tilling and irrigation of the land, killing still more living forms which might be involved in soil formation or humus formation, (*e.g.,* earthworms, centipede, millipede, bacteria, fungi, etc.). With their continuous use the soil microorganisms lose their ability of nitrogen fixation. Moreover, when these chemicals find their way into water supplies, they contaminate and disrupt the aquatic ecosystem as well.

Any substance or mixture of substances which prevents, repels, destroys or mitigates any pest (*e.g.,* bacteria, viruses, fungi, insects, nematodes, rodents, etc.) is called a **pesticide**. Pesticides have proved tremendously beneficial to human populations, in reducing or eliminating the target organisms such as insects, snails, rats, and other animals which transmit disease, destroy agricultural crops, damage homes and stored products, and directly or indirectly affect human health and welfare. Hence, the significance of chemical pesticides controlling mosquitoes, termites, houseflies, cockroaches, house crickets, weevils, locusts and grasshoppers, borers, snails, rats, rabbits, and a multitudes of other animals has been great, and it is difficult to imagine modern disease control and agricultural programmes without some forms of chemical control. Like pesticides, **herbicides** are specifically designed chemicals for the control of weed pests and unwanted plant growth.

Ecologically pesticides and herbicides have created two major serious problems which were not previously anticipated. In the first place many of them have persisted and accumulated in the environment and have harmed or contaminated numerous animals or plants not intended to be targets. Secondly, many of them have directly or indirectly affected human health.

Biomagnification or bioamplification. Many of pesticides, such as DDT, aldrin and dieldrin, have a long life time in the environment. They are fat-soluble and generally not biodegradable. They get incorporated into the food chain and ultimately deposited in the fatty tissues of animals and humans. In the food chain, because of their build up, they get magnified in the higher trophic levels (called **biological magnification** (**Carson** 1962) or **biological amplification**). The pesticides have been in use during last 50 years. Their targets are insect pests, fungi, nematodes and rodents which damage crops. But these pesticides have created great problems for non-target organisms consisting largely beneficial species such as earthworms, honeybees, fish, amphibia, some reptiles, birds, mammals and humans.

The phenomenon of biological magnification is also reported for certain other pollutants such as heavy metals such as lead, mercury and copper and radioactive substances (or radionuclids) as strontium-90.

Toxicity of pesticides. The toxicity of organo-chlorine pesticides (*i.e.*, DDT or Dichlorodiphenyl trichloroethane, hexachlorocyclohexane, chlordane, aldrin, dieldrin, etc.) lies in their inhibiting Na+, K+ and Mg+ adenosine triphosphatase activity in the nerve endings of animals particularly insects. It affects the sensory, motor nerve fibers and motor cortex (**Matsuma** and **Patil**, 1961). In the giant axons of cockroach, DDT is known to influence the efflux of potassium ions from the axon.

DDT and other organochlorine pesticides are absorbed from the intestinal tract, from the alveoli of lungs and also through the skin, if the pesticides are in solution. A high concentration of DDT causes brain damage, centrolobular necrosis of the liver, and liver enlargement in small mammals. Concentration as low as 5–10 ppm in diet cause liver damage. In some birds, DDT concentration of 1–3 ppm destroys the female sex hormone and the egg shell becomes so thin that the eggs break when the parents sit on them for hatching.

Organophosphorus insecticides are absorbed by the gastrointestinal and respiratory tracts and the skin. These insecticides inhibit the **acetylcholinesterase (Ach E)** enzyme. Thus, an abnormal accumulation of endogenous acetylocholine occurs. This causes excessive activity of the para-sympathetic system, in the form of sweating, abdominal cramp, chest discomfort, vomiting, over-activity of smooth muscles, and so on. It also causes headache, nervousness, and over-activity of voluntary muscles. The nervous system is also adversely affected.

The carbamate insecticides are absorbed by all portals, including the skin. These pesticides are reversible inhibitors of Ach E.

6. Marine Pollution

The marine ecosystems (*i.e.,* seas, oceans and estuaries) are so grand and vast but interacting in various ways with human life. How mighty and majestic ocean may appear to us, it was once challenged and tamed by Lord Rama during his march towards Ravana's Lanka. The oceans are used for navigation, fisheries, aquaculture, mining, acquisition of water, naval and military exercises as well as for the discharge and dumping of a variety of wastes. The very vastness of the ocean has led to the assumption that all wastes dumped into them can be harmlessly absorbed. Such an assumption now belongs to the past. The ever increasing range and volume of polluting activities tend to seriously affect marine production. **Marine pollution is most evident in coastal waters and estuarine areas.** The oceans have in fact, become the final settling basin for millions of tons of waste products from human activities. For example, in the late 1960s

West Germany was dumping 375 tons of sulphuric acid, 750 tons of iron sulphate, 20 tons of chlorinated hydrocarbons and 16,000 tons of gypsum wastes into the North sea and North Atlantic Ocean every day. Table 22.4 shows a few of the industrial and agricultural pollutants reaching the ocean annually in the mid-1970s.

Thus, petroleum and its distillates, halogenated hydrocarbons including pesticides, organic compounds including detergents, domestic raw sewage, agricultural run-offs, inorganic chemicals including metals such as mercury, lead and copper, dumped containers, radioactive wastes, cooling waters from industries and hydel projects and coastal constructions are the major pollutants entering the sea. Often the noxious pollutants are not dispersed and diluted but tend to remain concentrated and localised, or when diluted may be reconcentrated by living organisms.

This tremendous burden of pollutants is evidently affecting the health and integrity of the world's oceans. Due to oceanic pollution the marine biota has been seriously affected. **Cousteau** (1974) reported a 30 to 40 per cent decline in the over all productivity of pelagic organisms from shrimps to whales in the past 25 years. He also observed a serious shrinkage of coral reefs in many tropical areas of world, and a displacement of these rich and varied communities with turbid and relatively barren waters. There has occurred a decline in populations of many fishes due to oceanic pollution. Oil spills have killed water birds, mammals, fish and vegetation.

Oil is the most apparent pollutant of the ocean. It is discharged into the sea either deliberately when oil tankers are washed out prior to reloading or in accidental spills when tankers are wrecked (during war or otherwise). In coastal areas when oil spreads on the water surface, it clogs the feathers of diving birds making their flight impossible. While preening themselves in an attempt to clear the plumage, they swallow enough oil to poison themselves. Apart from these, oil interferes with the insulation provided by the feathers and the birds die of cold or become susceptible to pneumonia. When oil covers the rocks and sea weeds, molluscs and crustaceans growing on them die. Oil-spills on the coasts of sea-side resorts, drive away holiday makers, affecting the economy of the place.

Table 22.4. Examples of industrial and agricultural pollutants discharged annually into the world's oceans (Source: **Southwick**, 1976).

	Pollutant	Estimated annual discharge, 1970-75 (in metric tons)	Source
1.	Petroleum and industrial hydrocarbons	3,405,000	Offshore wells, oil tankers, industrial wastes.
2.	Hydrocarbons (airborne)	15,000,000	Vehicles, industries, power plants.
3.	Airborne lead	350,000	Vehicles.
4.	Mercury	100,000	Industrial operations.
5.	Aldrin-Toxaphene (converted to dieldrin)	25,000	Agricultural and public health operations.
6.	Benzene hexachloride (BHC)	50,000	Agricultural and public health operations.
7.	DDT	25,000	Agricultural and public health operations.
8.	Polychlorinated biphenols (PCBs)	25,000	Plastic industries.

Control of Water Pollution

Most cities of world have evolved certain engineering systems such as **septic tanks, oxidation ponds, filter beds, waste water treatment plants** and **municipal sewage treatment plants** for the removal of various pollutants from the sewage before it is tipped into river or sea. It is essential to have modern sewage treatment plants for every town and city of India so that the biodegradable as well as non-biodegradable pollutants can be removed from it and pure water obtained for recirculation.

The urban sewerage condition of India is quite distressing. At present only 15 class-I cities out of 142 are fully sewered, whereas, only 7 of the 190 class-II towns claim to have this facility. Still the Indian cities continue to pour their waste water (286 thousand million cubic metre) into natural water sources.

In India, the enactment of 'Prevention and Control of Water Pollution Act' in 1974 and Environmental (Protection Act, 1986) have helped to a certain extent to prevent water pollution. However, adequate legislative measures have to be adopted by every State to ensure: (*a*) proper disposal of sewage and industrial wastes; (*b*) prevention of abuse of water resources; (*c*) recycling of waste waters through proper methods of purifications; and (*d*) punishment of erring industries which do not install effluent devices in their factories. A large number of international and inter-governmental agencies are now busy in controlling marine pollution. A number of highly hazardous pollutants are **black listed** and their discharge in the sea is totally prohibited. These are halogenated organic compounds, tin, mercury and cadmium compounds, radioactive isotopes and carcinogenic material. Somewhat less toxic substances are **gray listed** such as compounds of arsenic, zinc, and antimony, cyanides, organosilicon compounds, crude oils, foam forming detergents and surfactants.

Realizing the devastation caused by the population of cities, towns and villages of 8 states and the union territory located along the coasts of river Ganga, through its run of 2525 km from Gangotri (Himalayas) to Ganga Sagar (Bay of Bengal), the Government of India has launched **Ganga Action Plan** to restore its water quality. The first phase of this gigantic cleaning operation had been initiated in year 1985 and may take several years in its completion.

Further, since **xenobiotics**, *i.e.,* synthetic herbicides and pesticides and fertilizers pose a challenge to the natural degradative process. Therefore, we should encourage a judicious, efficient and optimum use of organic manures. Recently, certain microbes have been used for the breakdown of these synthetic compounds. For instance, specific, microorganisms are being evolved by molecular breeding process to convert the chief ingredients of herbicide agents, 2, 4, 5-T, into carbon dioxide and chloride.

The waste water polluted with cyanides and heavy metals can be purified by certain bacteria which can not only withstand but are also able to grow at high level of cyanides. Millions of these bacteria are introduced in each of the several rotating discs of the plant's main processing units (Fig. 22.3). The stickly body surfaces of these bacteria pick up zinc, iron and other metals from the industrial waste water as it passes over the plates. These bacteria also absorb cyanide which is toxic to the fish and other aquatic life. Subsequently this water is passed through clarifier, and filtered and then released into the streams.

In the past, attempts have been made to get rid of oil slicks by spraying them with **detergents** which break up the oil into drops. The trouble is that detergents are even more deadly than the oil and they kill many marine organisms that might have escaped the oil. Nowadays less destructive methods are used. Recently, a product called "**bregoil**", resembling sawdust from paper industry waste has been employed. It rapidly absorbs oil. The resulting residue can be conveniently handled and even burnt as a fuel. Bacteria which thrive on oil eventually break it down and this brings about a regeneration of fauna and flora even within a year. Recently a strain of the bacterium *Pseudomonas putida* has been created by genetic engineering which breaks down octane, xylene and camphor (**Green** *et al.,* 1990).

Currently, water hyacinth (*Eichhornia crassipes*), an otherwise pernicious aquatic weed, has come into prominence for purifying domestic and industrial waste water. The plant regenerates rapidly and has a tremendous capacity to accumulate heavy and even radioactive metals. It is efficient in absorbing nitrogen, phosphorus and similar chemical pollutants. It reduces BOD, COD and organic carbon. The polluted water fed into reservoirs or lagoons with water hyacinth, becomes markedly clean and free from 75–90 per cent of its pollutants. This plant has also been used as a new source of food, fertilizer and biogas (energy).

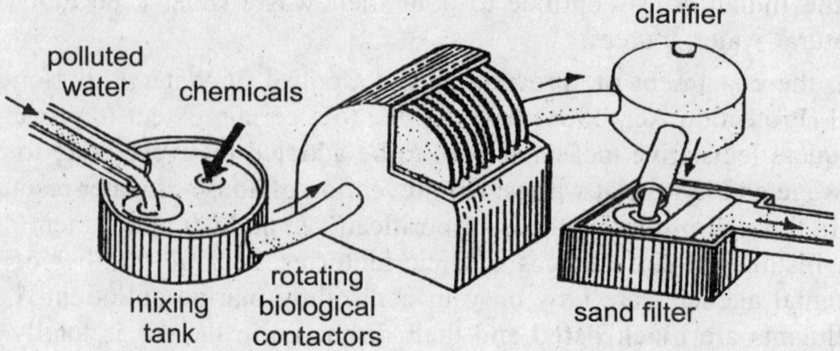

Fig. 22.3. The method of purification of water polluted with cyanide and heavy metals through bacteria.

Reutilization and Recycling of waste

Various kinds of wastes such as paper pulp, municipal and industrial effluents, sewage and thermal pollutants can be recycled to advantage. For example, urban waste could be recycled to generate cheaper fuel gas and electricity. Scientists at the NEERI, Nagpur, have evolved cheap know-how for the management of radioactive wastes and chemical effluents of atomic power plants. They have reclaimed waste water and have supplied cheaper piped gas and generated electricity by recycling urban waste. One large size urban waste recycling plant is already existing at Okhla near New Delhi. The total gas generation from the plant is about 0.6 million cubic feet per day having a heat value of 700-800 BTU per cubic foot. Waste water reclamation, through aquaculture, optimum utilization of domestic and industrial waste water in agriculture, detoxification of phenol and cyanide bearing waste by biologicel methods have been the major areas of investigation at NEERI. CSIR have devised techniques for the control of water pollution by **reverse-osmosis method, adsorption, electrodialysis, ion exchange**, etc. By using these methods various chemicals such as *ammonia, mercury, phenolics, sodium salts, proteins, lactose,* etc., have been removed from effluents of various industries.

22.3 SEWAGE TREATMENT

Sewage is waste matter from industrial and domestic sources disposed of through sewers. More aptly, sewage is the waterborne waste derived from home (domestic waste) and animal or food processing plants. It includes human excreta, paper, cloth, soap, detergents, etc. About two-thirds of the total solids in sewage are **organic** in origin and contain 250-400 ppm organic carbon and 80-120 ppm of total nitrogen. Traces of Zn, Cu, Cr, Mn, Ni and Pb have also been found in sewage. In order to reduce the pollution load or the extent of pollution, the sewage/waste water has to be cleaned/treated before being drained into a stream or river (Box 22.1).

Box 22.1 Misra's Waste Stabilisation Pond Technology

This sewage treatment technology has been suggested quite recently by **Prof. Veer Bhadra Misra** for cleaning the Ganga river. Misra's stellar efforts to clean the Ganga ensured **Time** magzine chose him as one of the 7 heros for the planet in 1998. Misra is head priest of the Sankat Mochan Mandir at Varanasi and formerly headed Civil Engineering Department of Banaras Hindu University. He is still professor of Hydraulic Engineering there. Central Pollution Control Board (CPCB) and **Prof. G.D. Agarwal** (Ex-Professor, IIT, Kanpur) have found Misra's proposal quite attractive since it has lower capital and recurring costs and treat effluent by this method has much lower faecal matter and bacterial load coliforms (**Jain,** in *Indian Express*, 10th March 2001).

Sewage treatment. The treatment of municipal sewage includes the following three steps:

A. Primary treatment. This involves the removal of floating and suspended solids, paper, rags, etc., by physical and chemical means, *i.e.,* screening, shredding, flocculating and sedimentation. In all developed countries about 30 per cent of domestic sewage receives primary treatment. Suspended matter as well as some oxygen demanding wastes are removed in the process through the following steps:

1. Screening. A series of screens removes large objects from the inflowing raw sewage.

2. Grit chamber. Any fairly large objects that may have passed through the screens are ground up to facilitate their processing.

3. Settling or sedimentation tank. From grit chamber sewage is piped into settling tank where it is detained for several hours so that sand and suspended organic material can settle to the bottom. The fluid that is left is removed from the tank. Chlorinated to kill disease organisms and ultimately discharged into a stream or lake.

4. Sludge digester. The sedimented solids are then piped to a sludge digester, where the organic matter is decomposed by microorganisms in the absence of oxygen. One decomposition product, **methane gas**, is frequently used to serve up to 90 per cent power needs of the primary treatment plant.

The dried sludge may be used as soil conditioner in lawns and flower gardens.

Primary treatment removes about 60 per cent of settlement solids as sludge and about 35 per cent of the oxygen-demanding waste. Thus, it results in 35 per cent reduction of BOD value (Fig. 22.4).

B. Secondary treatment. This is biological in nature and the organic material of sewage (*i.e.,* suspended organic materials that remain in the effluent after the primary treatment) through metabolic action of organisms which is accelerated with abundant of oxygen. In all developed countries about 60 to 70 per cent of domestic sewage receives the secondary treatment. Secondary treatment includes the following two steps (Fig. 22.5):

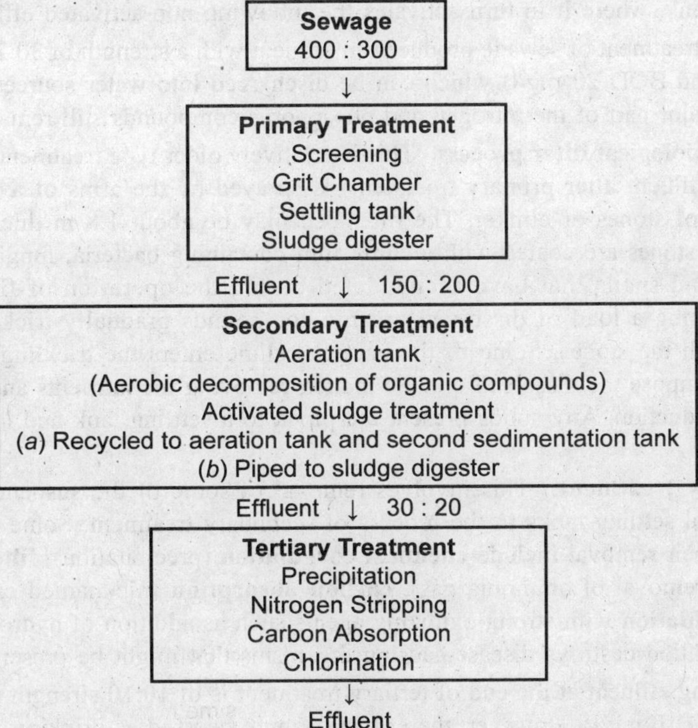

Fig. 22.4. Domestic sewage treatment methods. Sewage with suspended solids 400 mg/l and BOD 300 mg/l (*i.e.,* 400/300) gives an effluent of 10:10 strength, as a result of treatment.

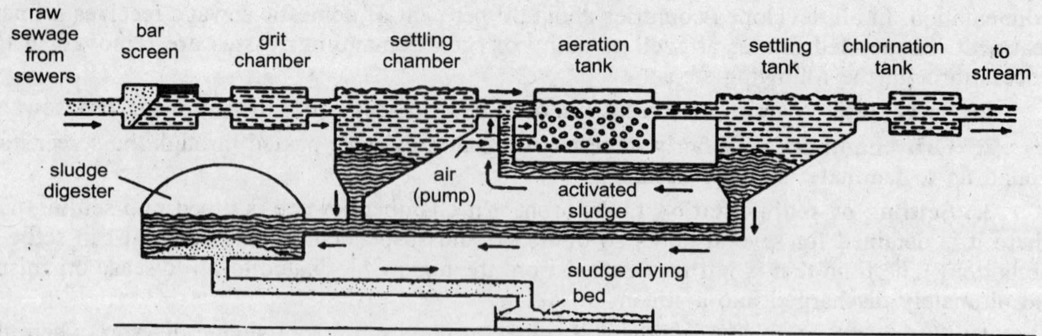

Fig. 22.5. Various steps of secondary sewage treatment.

1. Aeration. The effluent from the first settling tank is piped to an aeration tank in which air is bubbled to provide maximum supply of oxygen. This enables the aerobic bacteria to decompose the organic compounds. The liquid that accumulates at the top of second sedimentation tank is then chlorinated and discharged.

2. Activated sludge treatment. The sludge that settles at the bottom of the second settling tank is called **activated sludge** because it bears a high number of bacteria that have become conditioned to the unique environment of system. Algae in the sludge generate oxygen to be required for the growth and multiplication of the bacteria. A part of this activated sludge is piped to the sludge digester. The remaining part is recycled back to the aeration tank and the second settling tank, where it in turn activates the inflowing non-activated effluent.

Secondary treatment of sewage produces an effluent with a strength of 30:20 (i.e., suspended solids 30 mg/l and BOD 20 mg/l) which can be discharged into water sources. However, toxic chemicals and major part of the nitrogen and phosphorus compounds still remain in the effluent.

Trickling/biological filter process. This is relatively older type treatment method in which sewage (or the effluent after primary treatment) is sprayed by the arms of a rotating sprinkler into a filter bed of stones or clinker. The filter bed may be about 1.8 m thick and upto 60 m in diameter. The stones are coated with a slimy film containing bacteria, fungi, protozoa, insect larvae, worms and snails that have accumulated during the operation of filter. The sewage (effluent) containing a load of dissolved organic compounds gradually trickles downward or percolates through the stones. Some of the bacterial slime enters the trickling waste water and the bacteria decompose the dissolved organic matter, removing the nutrients and energy for their growth and reproduction. Any solids present are piped to a settling tank and later transferred to a sludge digester.

3. Tertiary treatment. This involves removal of some of the suspended matter which escapes from final settling tanks in the process of secondary treatment. Some of the techniques which assist in their removal include **chemical coagulation** (precipitation), **filteration, nitrogen stripping** (*e.g.,* removal of ammonia gas), **carbon absorption** (nicknamed **carbon polishing**) and **chemical oxidation** with strong oxidizing agents such as addition of hydrogen peroxide and chlorine. Chlorination destroys disease-causing organisms that might be present in the effluent.

The resulting effluent at the end of tertiary treatment is of 10:10 strength (suspended solids 10 mg/l and BOD 10 mg/l). However, the effluent cannot be used as drinking water without the normal chemical treatment given to natural waters.

QUESTIONS

Long Answer Questions

1. Define pollution? Explain the causes and consequences of water pollution. State some remedial measures for the abatement of pollution.
2. Describe the water pollution in detail and give measures to control it.
3. Define pollution? What are the effects of water pollutants? How water pollution is controlled?
4. Write an essay on water pollution.
5. Discuss the sources and impact of water pollution.
6. Give an account of aquatic pollution and its effect on human life. Mention the important control measures also.
7. What are the sources of water pollution? Briefly describe the effects of polluted water on human health and aquatic animals.
8. Give an account of the effects of the heavy metals, dyes and detergents on aquatic biota.
9. Give an account of the aquatic pollution by considering agricultural and industrial effluents?
10. Explain one method of determining industrial pollution.
11. Write notes on the following keeping in mind the aspects of water pollution.
    (i) Sewage and domestic wastage;
    (ii) Detergent and fertilizer;
    (iii) Thermal pollution.
12. What is pollution? Describe the types of pollution and discuss the effects of water pollution on human life.
13. Write an essay on microbial degradation of hydrocarbons.
14. Give three examples of organochlorine, organophosphorus and carbamate insecticides? How do they differ in their effect on aquatic biota?

(Hint:

1. Organochlorine. DDT, Endrin, Dieldrin, Chlordane, Aldrin, Endosulphan, Heptochlor, etc.
2. Organophosphate. Parathion, Malathion, Ethion, Dursban, Phosdrin, Fonofos. etc.
3. Carbamates. Carbaryl, Propoxur, Aminocrab, Fenuron, Zectran, Aldicarb, etc.)

15. Explain the term pollution. Classify the pollutants encountered in air as well as in water. How degradability of the pollutants can affect the fresh water environment. Estimate the pollutant level in drinking water. Comment on the significance of Ganga Action Plan.
16. What do you mean by point and non-point source of pollution? Classify three types of pollution considering the sources in nature. Elucidate the pollution caused due to the following: 1. Eutrophication 2. Heavy metal toxicity 3. Chlorofluorocarbons (CFCs) 4. Acid rain and 5. Pesticides.
17. Suggest any two methods to indicate the degradation of water quality. How the extent of pollution of river can be reduced.

Short Answer Questions

1. What is organic pollution?
2. What BOD stands for? Also mention its significance.
3. Comment upon the following
    (i) B.O.D. acts as indicator of water pollution.

(ii) Minamata disease was caused by insecticide pollution.
(iii) Eutrophication is caused by insecticide pollution.
(iv) Sewage is non-degradable pollutant.

4. Write short notes on the following
(i) Water pollution
(ii) Aquatic pollution
(iii) Water borne disease
(iv) Minamata disease
(v) Effect of water pollution
(vi) Biomagnification
(vii) Biological magnification
(viii) Thermal pollution
(ix) B.O.D
(x) Detergents
(xi) Fertilizer
(xii) Sewage culture
(xiii) Pesticides
(xiv) Effects of pesticides on non-target organisms
(xv) Recycling of wastes
(xvi) Sewage pollution
(xvii) Waste disposal in rural and urban settlement

5. Differentiate between the following.
(i) B.O.D and dissolved oxygen (DO).
(ii) Detergent and fertilizer
(iii) Rodenticide and Pesticide
(iv) BOD and COD.

Very Short Answer Questions

1. What is COD?

2. What is DO?

3. What do you mean by BOD?

4. What causes Itai-Itai disease? *(CCSU Meerut 2005)*

5. What caused Minamata disease?

Fill in the Blanks

1. The term eutrophication was coined by .... in 1969.

2. Some decomposing aquatic plants produce toxins such as .... which kills cattle and other animals.

3. Thermal pollution is also known as ....

4. .... is the increase in the concentration of a persistent chemical (*e.g.*, DDT) by the organisms at successive levels in a food chain.

Multiple Choice Questions

1. When huge amount of sewage is dumped into a river the BOD will *(CCSU Meerut 2014)*
(*a*) increased
(*b*) decreased
(*c*) slightly decreased
(*d*) remain unchanged

2. Eutrophication is caused by (*CCSU Meerut 2014*)
(*a*) acid rain
(*b*) sulphates and carbonates
(*c*) CO_2 and CO
(*d*) nitrates and phosphates

3. The high amount of *E.coli* in water is the indicator of (*CCSU Meerut 2014*)
(*a*) sewage pollution
(*b*) hardness of water
(*c*) industrial pollution
(*d*) presence of chlorine in water

4. Which of the following is the main factor of water pollution (*CCSU Meerut 2014*)
(*a*) ammonia
(*b*) detergent
(*c*) smoke
(*d*) industrial waste

5. Minamata disease was caused due to the consumption of (*CCSU Meerut 2014*)
(*a*) food containing cadmium
(*b*) fish contaminated with mercury
(*c*) oysters contaminated with pesticide
(*d*) fish contaminated with selanium

6. Polluted water can be purified by using (*CCSU Meerut 2014*)
(*a*) microorganisms
(*b*) fishes
(*c*) pesticide
(*d*) fungus

7. Biological Oxygen Demand is a measure of (*CCSU Meerut 2014*)
(i) industrial wastes poured into river
(ii) extent to which river is polluted with organic compounds
(iii) amount of O_2 in haemoglobin
(iv) amount of O_2 needed by green plants
(*a*) (i) statement is correct
(*b*) (ii) statement is correct
(*c*) (iii) statement is correct
(*d*) (iv) statement is correct

8. Which of the following is the main factor of water pollution:
(*a*) pesticides
(*b*) ammonia
(*c*) industrial wastes
(*d*) detergents

9. Water pollution is due to
(*a*) industrial discharges
(*b*) oxygen
(*c*) carbon dioxide
(*d*) sulphur dioxide

10. Decomposition of domestic wastes under natural process is known as
(*a*) industrial pollution
(*b*) thermal pollution
(*c*) non biodegradable pollution
(*d*) biodegradable pollution

11. Sewage water can be purified for recycling with action of
(*a*) microorganisms
(*b*) penicillin
(*c*) fishes
(*d*) aquatic plants

12. If water pollution continue at its present rate, it will eventually
(*a*) stop the water cycle
(*b*) make oxygen molecules unavailable to water plants
(*c*) make nitrate molecules unavailable to plants.
(*d*) prevent precipitation

13. The discharge of raw sewage into water ways causes the collapse of ecosystems, primarily because of
(*a*) lack of dissolved oxygen
(*b*) lack of nutrients
(*c*) increased level of carbon dioxide
(*d*) growth of algae

14. Often in water bodies subjected to sewage pollution, fishes die because of the
(*a*) pathogens related by sewage
(*b*) reduction of dissolved oxygen caused by microbial activity
(*c*) clogging of their gills by solid substances.
(*d*) foul smell

15. Sewage infection of water will be demonstrated by cysts of
(*a*) *Escherichia coli*
(*b*) *Leishmania*
(*c*) *Entamoeba*
(*d*) *Pseudomonas*

16. Phosphate pollution is caused by
(*a*) phosphate rock only (*b*) agricultural fertilizers only
(*c*) sewage and phosphate rocks (*d*) sewage and agricultural fertilizers

17. Biodegradable pollutant is
(*a*) plastic (*b*) sewage
(*c*) asbestos (*d*) all of these

18. If large quantities of domestic sewage are continuously emptied into a small stream, it leads to
(*a*) algal bloom (*b*) increase in temperature
(*c*) eutrophication (*d*) depletion of oxygen content in stream water

19. Which of the following does not cause pollution?
(*a*) automobiles (*b*) hydroelectric power plant
(*c*) thermal power plant (*d*) nuclear power plant

20. Spraying of DDT on crop plants produce pollution of
(*a*) air only (*b*) air and soil only
(*c*) air and water only (*d*) air, soil and water

21. The excessive discharge of fertilizers into water bodies results in
(*a*) silt (*b*) death of hydrophytes
(*c*) eutrophication (*d*) growth of fish

22. Shallow lakes with rich organic products are called
(*a*) heterotrophic (*b*) eutrophic
(*c*) mesotrophic (*d*) oligotrophic

23. Eutrophic lakes are
(*a*) rich in nutrients (*b*) poor in nutrients
(*c*) rich in light (*d*) rich in oxygen

24. Human activities speed up the eutrophication process of a
(*a*) stream (*b*) sea
(*c*) lake (*d*) river

25. Eutrophication causes reduction in
(*a*) dessolved oxygen (*b*) dissolved hydrogen
(*c*) dessolved salts (*d*) all of the above

26. Complete eutrophication of a lake renders it
(*a*) nutrient rich and productive (*b*) nutrient poor and unproductive
(*c*) nutrient rich but unproductive (*d*) nutrient poor but productive

27. Population explosion of small organisms due to nutrient enrichment of a lake causes a visual aspect called
(*a*) primary production (*b*) bloom
(*c*) climax state (*d*) dominance

28. In a polluted lake, the index of pollution is
(*a*) frog (*b*) *Artemia*
(*c*) *Daphnia* (*d*) none

29. BOD stands for
(*a*) biotic oxidation demand (*b*) biochemical organic decay
(*c*) biological organisms death (*d*) biological oxygen demand

30. BOD of a river water is found very high. This means water
(*a*) is clean (*b*) is highly polluted
(*c*) contain algae (*d*) contain many dissolved minerals

31. By what method can the quantity of organic pollutants in water be determined?
(*a*) pH measurement (*b*) measuring change of colour
(*c*) measuring BOD (*d*) transparency measurements

32. When huge amount of sewage is dumped into a river the BOD will
(*a*) increase (*b*) remain unchanged
(*c*) slightly decrease (*d*) decrease

33. BOD is a measure of polluting organic matter present in sample of
(*a*) water (*b*) air
(*c*) soil (*d*) bottom mud from a pond

34. BOD is connected with
(*a*) microbe only (*b*) microbes and organic matter
(*c*) organic matter only (*d*) none of the above

35. Which of these is biondegradable pollutant?
(*a*) mud (*b*) leaves
(*c*) plastic things (*d*) fruits and vegetables

36. Indiscriminate use of DDT is undesirable because
(*a*) it is harmful (*b*) it is degradable
(*c*) it is accumulated in food chain (*d*) it causes mutation

37. Of the various pesticide residues which of the following shows peristence in the soil and is known to persist even upto four years? Identify the same
(*a*) DDT (*b*) parathion
(*c*) malathion (*d*) 2, 4-*D*

38. Which of the following causes water pollution?
(*a*) 2, 4-D and pesticides (*b*) smoke
(*c*) automobile exhaust (*d*) aeroplanes

39. Which of the following pollutants affect more to organisms of the higher trophic level of a food chain due to biological amplification?
(*a*) poisonous cyanides
(*b*) detergents
(*c*) sewage and plant fertilizers
(*d*) heavy metals, mercury salts and non-biodegradable phenolic chemicals

40. If the concentration of insecticide DDT in a pond water is 0.04 ppm, which of the following sequences of DDT concentration would be correct for algae-fishes-humans food chain?
(*a*) algae > fishes < humans (*b*) algae > fishes > humans
(*c*) algae < fishes < humans (*d*) algae < fishes > humans

41. Increasing concentration of DDT in organisms of a food chain in higher trophic levels is known as
(*a*) biological chain (*b*) biological value
(*c*) biotic potential (*d*) biological amplification

42. 'Minamata disease' is produced due to the pollution of water with
(*a*) silver nitrate (*b*) mercuric chloride
(*c*) methyl isocyanate (*d*) crude oil

43. Minamata disease is a pollution related disease which results from
(*a*) oil spills into sea
(*b*) release of human organic waste into drinking water
(*c*) accumulation of arsenic into atmosphere
(*d*) release of industrial waste of mercury into fishing water

44. Among the following which one is likely to have highest concentration of DDT in the body?
(*a*) phytoplankton (*b*) zooplankton
(*c*) sea gull (*d*) eel

45. The major drawback of DDT as a pesticide is that
(*a*) it is not degraded in nature
(*b*) its cost of production is high
(*c*) it is significantly less effective than other pesticides
(*d*) organisms at once develop resistance to it

46. 'Itai-itai' ('Ouch-Ouch') disease is caused by the poisoning of
(*a*) mercury (*b*) lead
(*c*) chromium (*d*) cadmium

47. Which of the following systems is affected by 'Minamata disease'?
(*a*) circulatory system (*b*) skeletal system
(*c*) nervous system (*d*) respiratory system

48. 'Heat islands' are produced due to
(*a*) air pollution (*b*) water pollution
(*c*) land pollution (*d*) all the above

49. Thermal pollution of water bodies is due to
(*a*) discharge of waste from mining
(*b*) dishcarge of agricultural run-off
(*c*) discharge of chemicals from industries
(*d*) discharge of heat (hot water) from power plants

50. The cause of thermal pollution of lakes, rivers and estuaries is
(*a*) nuclear power plants (*b*) nuclear submarines
(*c*) offshore oil rigs (*d*) moving of steamers

51. Pollution can be controlled by
(*a*) sewage treatment (*b*) by checking atomic blasts
(*c*) manufacturing electrically operated vehicles (*d*) all of the above

52. BOD of drinking water is
(*a*) more than dirty water (*b*) less than dirty water
(*c*) equal to dirty water (*d*) all

53. Water pollution causes
(*a*) increased photosynthesis (*b*) increased oxygenation
(*c*) decreased turbidity (*d*) increased deoxygenation and turbidity

54. The most common indicator organisms that represent polluted water is

(*a*) *E. coli* (*b*) *P. typhi*
(*c*) *C. vibrio* (*d*) *Entamoeba*

55. Which of the following type of pollution causes outbreak of jaundice

(*a*) air (*b*) thermal
(*c*) water (*d*) land

56. Sewage drained into water bodies kill fishes because

(*a*) excessive CO_2 is added in water
(*b*) it gives off a bad smell
(*c*) it removes the competition with fishes to dissolve oxygen
(*d*) it increases competition with fishes to dissolve oxygen

57. The first biocide used on a large scale was

(*a*) DDT (*b*) 2, 4-D
(*c*) mercury (*d*) melathion

ANSWERS

Very Short Answer Questions

1. The amount of oxygen required for oxidation of chemicals present in unit volume of water of lakes, ponds and rivers, etc., is called chemical oxygen demand (COD).
2. The amount of oxygen in mg per liter water which sustains the aquatic life in a water body, is called dissolved oxygen (DO). The DO level requirement of most of the aquatic organisms is around 5 mg/liter or ppm.
3. The amount of oxygen in mg needed by bacteria (microorganisms) to oxidise one liter of organic waste is called biochemical oxygen demand (BOD). It is an important index which gives a measure of sewage strength.
4. Itai-Itai disease is caused due to cadmium (Cd) toxicity (pollution). It was first reported from Japan.
5. Minamata disease was caused due to consumption of heavily contaminated fish by mercury in Japan.

Fill in the Blanks

1. Hutchinson **2.** Strychnine **3.** Calefaction
4. Biomagnification (or Bioconcentration/ Bioaccumulation).

Multiple Choice Questions

1. (*a*)	**2.** (*d*)	**3.** (*a*)	**4.** (*d*)
5. (*b*)	**6.** (*a*)	**7.** (*b*)	**8.** (*c*)
9. (*a*)	**10.** (*d*)	**11.** (*a*)	**12.** (*b*)
13. (*a*)	**14.** (*b*)	**15.** (*c*)	**16.** (*d*)
17. (*b*)	**18.** (*d*)	**19.** (*b*)	**20.** (*d*)
21. (*c*)	**22.** (*b*)	**23.** (*a*)	**24.** (*c*)
25. (*a*)	**26.** (*c*)	**27.** (*b*)	**28.** (*c*)
29. (*d*)	**30.** (*b*)	**31.** (*c*)	**32.** (*a*)
33. (*a*)	**34.** (*b*)	**35.** (*c*)	**36.** (*c*)
37. (*a*)	**38.** (*a*)	**39.** (*d*)	**40.** (*c*)
41. (*d*)	**42.** (*c*)	**43.** (*d*)	**44.** (*c*)
45. (*a*)	**46.** (*d*)	**47.** (*c*)	**48.** (*b*)
49. (*d*)	**50.** (*a*)	**51.** (*d*)	**52.** (*b*)
53. (*d*)	**54.** (*a*)	**55.** (*c*)	**56.** (*d*)
57. (*a*).			

23

Land Pollution and Noise Pollution

23.1 SOIL (LAND) POLLUTION

The land pollution is caused by solid wastes and chemicals. There are many examples of land that has been stripped of vegetation by industrial development and disposal of waste. Some common soil contaminants, their sources and wide range impacts on the biota have been summarized as follows:

The slag heaps from mines bear witness to the destructive effects which this can have on our environment. Areas around smelting and mining complexes are usually soiled by metals such as cadmium, zinc, lead, copper, arsenic and nickel. These are not only phytotoxic even in small amounts but also render plants unsafe for human and animal consumption. Zinc, often with cadmium, is released into the environment during the use or breakdown of lubricating oils, vehicle tyres, galvanized metals and fertilizers.

The major sources of land pollution are the industries such as pulp and paper mills, sugar mills, oil refineries, power and heating plants, chemicals and fertilizer manufacturing units, iron and steel plants, plastic and rubber producing complexes and so on. Huge amounts of solid wastes are either dumped, burnt, or emptied into rivers and seas. Most industrial furnaces produce a grey, powdery residue of unburnt material known as fly ash. The fly ash, cinders, solid wastes and litter all are thrown away by industries and form huge mounds which spoil the landscape.

Our households too contribute a large amount of solid rubbish in the form of domestic wastes. Some common examples are groceries, food scraps, vegetable remains, packing materials, cans, cardboard cartons, rags, paper, cinders, ash, broken gadgets, wood, worn-out furniture, metals, bones of dead animals, plastics, polythene bags, ceramics, glass, aluminium, rubber, leather, construction rubbish, brick, sand and other junk. Some man-made materials can be used again such as paper, scrap metal, glass, polythene, plastic, etc. But the majority of these cannot be reused and must be got rid of somehow. All these go to constitute heaps of municipal refuse. If not properly disposed, this rubbish can prove perilous, filthy and unhygienic. Such places often become a home for rats, flies, mosquitoes, bacteria and many other vectors, which may spread numerous human diseases.

Such an abundance of solid waste examplifies enormous problem of solid waste production and the problem of its disposal, especially in affluent countries where labour charges of waste collection is high. In our country, however, all old junks are purchased by professional hawkers from door to door to be resold on marginal profit.

Disposal of Solid Waste

Some of the junk is readily degraded by the activity of microorganisms in nature and the materials are recycled but some others as metals, plastics, nylon and polythenes are not degradable. Solid wastes can be minimized by adopting the following methods:

1. Sanitary landfill. About 90 per cent of the solid waste (garbage) in the city is dumped into a natural or constructed pit or depression and left to disintegrate slowly. It is also dumped in natural gulleys, canyons or ravines, in artificial long and shallow trenches, or in swampy and marshy areas. Each day the trash is covered with dust from the surrounding areas and compacted. In this way odours and ugliness can be considerably reduced. However, with the increased urbanization problem of ever strinking dumping sites pose serious problems.

2. Compostings. Composting is a common practice of solid waste disposal in rural areas for the production of manure and biogas. Large scale composting operations are expensive as they need first the separation of metals, glass, rubber and plastics. The remaining trash has to be shredded, subjected to aerobic periods to increase bacterial action and hasten decomposition and then allowed to incubate for several months. The final product of the process is used as **manure** to increase the fertility of soil.

3. Incineration. This is most convenient and quick method of solid waste disposal in thickly populated areas of the town (*e.g.,* Timarpur in Delhi where electricity is generated from garbage). Burning of waste and utilization of heat to warm residential units is a common practice at several places.

4. Recycling. As a result of waste management, efforts are now being made to separate materials from the garbage for recycling. Materials like glass cannot be burnt but recyled by melting and preparing new articles. Old paper cans, news papers and other waste paper can be recycled to produce fresh paper or other useful materials. Since cattle can easily degrade cellulose, researches are preparing paper trash in a manner appealing to the cattle. Waste could also be reduced into sturdy blocks that can be used in construction work. Valuable components of discarded and unserviceable vehicle especially metals (such as steel, cast iron, zinc, aluminium, copper, lead) if could be processed by inexpensive means, then its disposal would also be facilitated. German industry is now prepared to recycle old cars. Recycling methods to retrieve valuable oils, chemicals, gas and the tar from old tyres are also being perfected.

23.2 NOISE POLLUTION

The word **noise** is derived from the Latin word '*nausea*' which means feeling of sickness at the stomach with an urge to vomit. Noise is now considered as a potential hazard to health and communication. Noise is usually defined as 'unwanted sound or sound without value that causes discomfort to listner'. **Noise pollution** refers to the unwanted sound dumped into the atmosphere leading to health hazards. A particular sound may be music to one but noise to another; pleasant when soft but noise when loud.

Loudness of Sound

Sound is often described in terms of "loudness" which cannot be measured like the sound frequency, sound pressure and sound intensity. Loudness is lisner's auditory impression of the strength of a sound and it is therefore a subjective judgement. Loudness of sound generated by several sources of noise has been given in Table 23.1.

Generally 80 dB sound may be considered the critical level for human ear damage. The ear is able to analyse sounds into frequency components and the range of a young person with perfect hearing is from about 20 to 20,000 Hz (= Hertz). Normally hearing is most acute in the frequency range 2000 to 550 Hz, but it is reduced rapidly below 200 Hz and above 10,000 Hz. The ear is potentially liable to damage if it receives high intensity noise.

Table 23.1. Loudness of sound generated by some sources.

	Source	Sound level in dB (= decibel) (Approximate)	Sound level as understood by common man
1.	Soft whisper	15 to 25	Very quiet
2.	Ordinary conversation	35 to 60	Moderately loud
3.	Vaccum cleaner	70	Moderately loud
4.	Food blender	85	Very loud
5.	Motor cycle (about 25 feet away)	90	Very loud
6.	Jet plane (about 100 feet away)	105	Uncomfortably loud
7.	Train whistle at 50 feet distance	110	Uncomfortably loud
8.	Thunder clap	120	Uncomfortably loud
9.	Jet-plane take off	150	Painful
10.	Rocket engine	180	Painful

Box 23.1

India is a noise loving country; on every occasion, from birth to death loud noise, is the part of our culture. Playing of loudspeaker at full pitch is the most common practice at all the religious, cultural and social functions. With the industrial revolution, population explosion and increase in autovehicles, the situation has become alarming.

Sources of Noise Pollution

Noise may be either **natural** as thunder or human-made. Human-made sources of noise may broadly be classified into two types (*i*) Industrial, and (*ii*) Non-industrial.

A. Industrial source of noise. This is also called **occupational noise** and is largely produced by industrial machines and processes. In an industry, workers are subjected to noise for about eight hours per day. Thus, industries of various kinds such as textile, iron and steel, utensils, automobile, fertilizer, paper, ceramics, and thermal power stations are major contributor of noise pollution. Noise orginates there from various machines in connection with crushing of different materials, grinding, drilling, bunching, weaving, boiler making, forging, pressing and blasting operations. Noise level in some Indian industries have been tabulated in Table 20-2. The close examination of this data indicates that the noise levels in most of the Indian industries are quite high.

B. Non-industrial source of noise. The chief non-industrial noise sources have been classified as follows:

(*i*) Domestic noise. Entertainment devices, *e.g.,* television, radio, transistor, stereo, tape recorders, CD player, telephone, musical instruments (*e.g.,* tabla, harmonium), fans, air conditioner, cooler, washing machine, exhaust fan, lawn movers, vaccum cleaner, kitchen appliances (*e.g.,* food mixture, pressure cooker, metal utensils), etc., are common sources of domestic noise. Effects of sound produced by T.V. and radio (called electromagnetic interference (EMI) are very slow but long lasting. They slowly cause deafness and other bodily changes in those who hear them with a very high volume. On the other hand, shouting, children playing rock music, etc., create noise that may not only affect adversely the users but neighbours too.

Table 23.2. Noise level in some Indian industries.

	Industry	Noise levels [dB (A)]		Industry	Noise levels [dB(A)]
1.	Naval ships (Control room)	100-200	13.	Vanaspati (Filling staton)	76-87
2.	Naval ships (engine room)	93-94	14.	Auto vehicles (engine testing)	80-90
3.	Aircraft repair shop	100-110	15.	Hindustan Petroleum Corporation	74-98
4.	Heavy vehicle factory	110-115	16.	Synthetic power industry	90-116
5.	Glass blowing shop	70-108	17.	Power plant	90-100
6.	Ammunition factory	85-101	18.	Sugar industry	81-104
7.	Saw mills	90-112	19.	Light engineering unit	78-82
8.	Brass band units	90-112	20.	Plastic industry	87-94
9.	Silver foil manufacture	90-112	21.	Leather industry	75-80
10.	Foundry	90-100	22.	Fabrication unit	75-80
11.	Ceramics	90-100	23.	Heavy engineering unit	84-93
12.	Paper	88-96			

***(ii)* Loud-speaker.** This is one of the most common sources of human-made noise pollution. For every occasion, religious or non-religous, public or private, the use of loud-speaker is a common practice. Screaming loud speakers during festivals, *melas* and election meetings produce sound ranging between 55 dB (A) to 90 dB(A). Sometimes, one is liking or not, he/she is bound to listen different religious discources due to use of loud-speakers.

***(iii)* Construction work.** Noise from construction sites (*i.e.,* construction of roads, bridges, and huge residential and commercial complexes) has been generally far worse than noise originating from industries or factories.

***(iv)* Transport noise.** Transport or traffic noise is created by a continuous running of autovehicles and is generated by engines in jets, motor in trucks and contact of tyres on roads and wheels on rails. The steam engines used commonly by Indian railways produce a lot of noise. This affects not only those who lives adjacent to roads/ railway lines but also occupants of transportation means as well. Traffic noise has been increasing over the last few decades due to increased density of vehicles and their higher speeds. The noise is at maximum (often unbearable) during take off and landing of an aircraft (Jet). Often no regulation is observed in blowing horns and use of defective silencers. Sometimes, people even use their vehicles without silencers which generate annoying noise.

Survey conducted by AIIMS reported that the average noise levels in Delhi, Bombay and Calcutta were around 95 dB.

***(v)* Crowded markets.** In metropolitan cities such as Bombay, Calcutta, Delhi and Madras, the crowded markets have become great sources of noise pollution. The chattering and bargaining of shoppers, foot falls on pavements, loud-speakers urging for buyers, stopping and starting of vehicle engines, etc., contribute a lot in creating noise pollution.

***(vi)* Theaters, circus and carnivals.** Theaters, dance halls, discotheques, circus and carnivals all cause noise pollution to nearby residents, especially in the evening.

***(vii)* Religious rituals and festivals.** Various religious rituals such as Yagna, Bhagwati Jagaran, Kirtan, Akhandpath, Navaratra, Gurubani, Azan and religious festivals such as Deepawali,

Durga Puja, Ganeshotsav, Urs fairs, all cause noise pollution. For example, the din and noise of crackers during Deepawali are so loud and unbearable that it is necessary to think seriously about the health hazard due to uncontrollable increase in the deafening noise associated with this festival. The survey of noise levels at various religious and cultural functions were conducted by **Lal** (1984) in Lucknow (Table 23.3).

(viii) **Projection of satellite in space.** Satellites are thrown into space with the aid of high explosive rockets. During this, deafening noise is produced at the time of "lifting off" a satellite. Tonnes of TNT and other explosives used in these operations produce noise pollution.

(ix) **Atomic explosions.** Huge noise is emitted from the site of experimental atomic explosion.

Table 23.3. Noise levels in Lucknow at various religious and cultural functions.

	Functions	Noise levels (in dB)		Functions	Noise levels (in dB)
1.	Deepawali	73-114	9.	Fairs	80-90
2.	Ram Leela	78-85	10.	Fireworks	92-99
3.	Ravan Vadh	90-100	11.	Visarjans	89-102
4.	Durga Puja	85-100	12.	Bhagwati Jagaran	84-94
5.	Akhand Ramayan	85-96	13.	Kirtans	90-102
6.	Urs	93-101	14.	Lottery stalls	82-90
7.	Religious discourses	82-88	15.	Music functions	86-96
8.	Group dances	77-93			

Measurement of Noise

Sound consists of wave motion in an elastic medium. These waves travel through the medium from the source of sound to the listner. Sound waves consist of vibrations in pressure, or oscillations of the medium in which they travel. The pitch of the sound is determined by frequency and intensity of sound. The apparent noise that is perceived by human ear depends on both frequency and intensity of the sound. The **intensity** of the noise (or sound) is measured in **decibels** (dB) (*deci*=10 and *bel* = after the name of scientist **Alfred Graham Bell**). The dB measures sound intensity (or loudness). It is tenth of the largest unit, the **bel**. A decibel is the amount of sound pressure that equals to 0.0002 microbars (= dynes cm^{-2} or energy of about 10^{-16} watts).

Human ear is found sensitive to an extremely wide intensity from 0 to 180 dB. Here 0 dB is the threshold of hearing while 140 dB is the threshold of pain. Threshold of hearing means the lowest intensity at which stimulus gets perceptible. Some people feel discomfort even with sound of 85 dB, while most do not feel discomfort even with sound of 115 dB. Pain is usually felt at 140 dB.

Another scale, which has been adopted as dB (A)—'A' indicating the weightage used, gives more weight to high frequency. The dB (A) scale is the most commonly used scale for measurements of general noise level. The sound intensity is measured by instruments such as **Larn barometer or sonometer**.

The effect of sound on human beings depends upon its frequency (or pitch). The frequency can be defined as the number of vibrations $second^{-1}$. It is denoted by **Hertz (hz).** One hz is equal to 1 vibration $sound^{-1}$. People can hear sound from 16 to 20,000 hz, but this range is reduced with age and other subjective factors. The frequencies 500, 1000 and 2000 hz are of greatest importance because human speech mostly occurs at this range. The range of vibrations below 16 hz are **infra-audible** and those about 20,000 hz **ultrasonic**. Some persons can hear frequencies that others are not able to detect.

Effects of Noise Pollution

The loudest sound that a person can stand without much discomfort is about 80 dB. The optimum levels as prescribed by WHO are 45 dB by day and 35 dB by night and noise level about 85 dB is hazardous because it harms the hearing faculty. Ambient Air Quality Standards as prescribed in **Environment (Protection) Third Amendment Rules, 1989** has been given in Table 23.4.

Table 23.4. Ambient air quality standards.

	Area	Limits in dB (A)	
		Day time	Night time
1.	Industrial area	75	70
2.	Commercial area	65	55
3.	Residential area	55	45
4.	Silent zone	50	40

In many industries around the world, however, workers have to bear sound levels as high as 110 to 115 dB. Generally, beyond 100 dB it is painful. A survey of the AIIMS, New Delhi (1986) has revealed that noise level of Indian metropolitan cities such as Bombay, Madras, Delhi and Calcutta, is about 90 dB, more than twice the level as prescribed by the WHO.

A. Effects of Noise Pollution on Human Beings

The one or more following factors are responsible for the adverse effects of noise pollution: (*i*) sound intensity; (*ii*) frequency; (*iii*) exposure duration, and (*iv*) intermittance or continuation of sound. The various ill-effects of noise pollution on human beings may be of following types (Table 23.5):

Table 23.5. High intensity noise and human health hazards.

Noise intensity (dB)	Health hazards
80	Annoying
90	Hearing damage (8 hr)
95	Very annoying
110	Stimulation of reception in skin
120	Pain threshold
130–135	Nausea, vomiting, dizziness, interference with touch
140	Pain in ear, prolonged exposure causing insanity
150	Burning of skin, significant change in pulse rate
160	Other minor permanent damage if prolonged
180	Eardrum ruptures, major permanent damage
194	Lung damage

1. Auditory Effects

1. Auditory fatigue. At the noise level of 90 dB symptoms of auditory fatigue appear such as whishtling and buzzing in the ears. In Indian metropolitan cities such as Calcutta, Mumbai, Madras, and Delhi more than 40 per cent among cobblers, fruitsellers, shopkeepers and drivers complain of tinnitus in the ears.

2. Deafness. Continuous noise may lead to gradual decline in hearing ability and eventually temporary or permanent deafness (or noise-induced hearing loss or NIHL). Long-term exposure to rock music too cause the following disorders—quivering hands, increased excitability and irregular pulse.

II. Non-auditory Effects

1. Communication interference. Frequencies in the range of 300-500 hz (or 50 to 60 dB) commonly produced by road traffic interferes with speech, sometimes making it unintelligible.

2. Sleep interference. Noise reduces the depth and quality of sleep and thereby adversely affects the overall mental and physical health. The noise may not disturb the sleep of young and middle-aged people, but it may wake older people and prevent them from returning to sleep. The WHO taskforce on environmental health criteria for noise, therefore, recommended sound level at 35 dB for sound sleep.

3. Loss of working efficiency. Noise has adverse effect on the efficiency of workers in factory or industry. In noisy atmosphere for longer duration they become more fatigued, make more errors and lead to lowered efficiency that may result into reduced output, increased absentecism and higher rates of accidents and injuries.

4. Annoyance. Neurotic people (about 80 per cent population) are more sensitive to noise than balanced people, which express great annoyance with even low levels of noise as crowd, highway traffic, radio, etc. This results in ill-temper, brickering and even enemity.

5. Physiological disorders. A number of physiological disorders in human body are caused due to over-exposure to noise. These are **neurosis, anxiety, insomnia, hypertension, increase sweating, giddiness, nausea** and **fatigue**. Cardiovascular system is especially vulnerable to high levels of noise. Workers exposed to high levels of noise were found to have higher incidence of circulatory problems, cardiac disease, hypertension, peptic ulcers and neurosensory and motor impairment. Chronic noise pollution may lead to abortions and congenital defects in urban children.

B. Effects of Noise Pollution on Wildlife

Noise has various adverse effects on wildlife too. There is decline in migration of birds to a habitat if it becomes noisy. Zoo animals such as deers, lions, rhinoceros, etc., are reported to be worst affected from the traffic noise. They become dull and inactive and their health deteriorate.

C. Effects on Non-living Things

The high intensity of noise, such as vibrations eminating from heavy machinery cause shattering of window glasses, loosening of plaster of walls, cracks in walls and house-hold crockery and breaking down of the wallhangings in the house.

Control of Noise Pollution

Noise pollution can be regulated by adopting the following measures:

1. Control of noise at source. This can be achieved by (*i*) proper lubrication of the wheels and designing and fabricating silencing devices and their use in air-craft engines, motor cycles, cars, trucks, industrial machines and home appliances; (*ii*) segregating the noisy equipments; (*iii*) introducing the less noisy machines in place of noisy ones; (*iv*) conducting noisy operations in an open place far from any residential colony; (*v*) setting of industries at least 5 km away from residential areas; and (*vi*) switching over to diesel engines by railways.

2. Control of transmission. This is achieved by covering the room walls with sound absorbing materials such as installation of acoustical tiles and baffles and construction of enclosures around the noisy machinery.

3. Protection of exposed persons. Under noisy situations, one may hold his hands over his ears or may run away from the source, or may simply stuff a bit of cotton or ear plugs in the ears to reduce much hazards of noise pollution. Workers exposed to noise louder than 85 dB as in factories, jet planes, etc., should wear devices such as ear plugs and ear muffs.

4. Creation of vegetation buffer zones. For abatement of noise pollution in cities and towns, noise absorbing plants such as Ashok, Banyan, Neem, Casuarina, Tamarind, Coconut, etc., can be planted along roadsides and in front of buildings, such as colleges, hospitals, libraries, etc.

5. Education and awareness. People must be made aware of health hazards of noise pollution through newspapers, articles, T.V., radio, news reels in cinema halls, workshops and popular lectures.

6. Legislation. Creation of silent zones (noise abatement areas up to 100 metres around schools, nursing homes, colleges and hospitals); prohibition of the entry of heavy vehicles in some central areas and of indiscriminate use of loudspeakers in religious and other public places, regulation of use of pressure horns by truck/bus drivers, etc., could be sought to be controlled through legislation. There should be a complete ban on the use of loudspeakers during 10.00 p.m. and 5.00 a.m., as is prescribed under the **Rajasthan Noise Control Act,** 1963. Hawkers at odd hours cause intolerable noise. This poses a serious challenge to the comfort of residents of the area. Their movements through residential areas too should be regulated.

QUESTIONS

Long Answer Questions

1. Describe solid waste management. Write short notes on the following:
   (i) Solid waste pollution
   (ii) Soil pollution

2. Describe the main source of noise pollution and its control measures in brief. (*Kurukshetra 1999*)

3. Define pollution. Describe measures to control noise pollution.

4. What is pollution? Discuss the impact of noise pollution.

Short Answer Questions

1. Write short notes on the following:
   (i) Noise pollution
   (ii) Noise abatement

Very Short Answer Questions

Answer briefly the following questions:

1. Decibel unit (dB) measures what?

Multiple Choice Questions

1. The noise pollution is measured in

(*a*) Fathoms (*b*) Decibels

(*c*) Tons (*d*) Kilogram

2. Most important cause of soil pollution is

(*a*) Iron junks (*b*) Glass junks

(*c*) Detergents (*d*) Plastics

ANSWERS

Very Short Answer Questions

1. The dB unit measures sound intensity (or loudness).

Multiple Choice Questions

1. (*b*) **2.** (*d*).

24

Radioactive Pollution

Radioactive substances are among the most toxic substances known. The magnitude of their injurious effects is terrifically high as compared to ordinary organic poison, *e.g.,* radium is 25,000 times more lethal than arsenic (Box 24.1). Nuclear war materials, test explosions, increase use of power reactors and radioactive materials in medical industry and research are the principal sources of radioactive exposures that threaten environmental health.

24.1 RADIONUCLIDES

The smallest unit of an element (such as hydrogen, carbon, oxygen) that can exist while retaining the characteristic of that element is called the **atom.** Each atom has elementary particles such as **proton** (+), **neutron** (uncharged) and **electron** (–). Atom of each element has characteristic numbers of protons, neutrons and electrons. Most elements, however, in nature contain atoms that are not exactly like the predominant form. These atoms have different number of neutrons. These different forms of the same element are called **isotopes.** Some isotopes of common elements are stable under ordinary conditions. Others have various degree of instability and some of them disintegrate with the emission of radiations of one kind or another. The atoms that undergo disintegration are called **radioactive**. Radioactivity is shown by about 18 types of naturally occurring elements, but all the elements may become radioactive when processed in a **nuclear reactor**.

Box 24.1 Some Classical cases of Radiation Toxicity

During the early 1900s thousands of people were exposed to **radium**, either as participants in faddish medical therapies or as radium dial painters who "sharpened" the tips of small paint brushes by twirling the bristles on their tongue. Epidemiological studies of these two populations show an increased incidence of **bone cancer** (osteogenic carcinoma).

Epidemiological studies of persons exposed to radioactive Ra^{224} as a treatment for *tuberculosis*, X-rays as a treatment for *ringworms* (both early to mid 1900s) and atomic bomb surviviors (1945) also reveal these forms of ionizing radiation to be powerful genotoxic carcinogens. Dramatic increases in childhood **thyroid cancer** rates in countries most contaminated by the 1986 nuclear accident at Chernobyl have been linked to radiation exposure-most likely ra-dioactive iodine released during the catastrophe.

***Epidemiology.** The study of the prevalence and spread of disease and death in population.

In **radioactive decay**, a very small amount of mass is converted into energy in the proportion required by Einstein's equation and is released as **heat**. Radiactive decay is a spontaneous process that cannot be influenced by external factors. However, in nuclear reactor humans are able to create new radioactive isotopes. Under neutron bombardment inside a reactor the nuclei of a wide variety of elements can absorb neutrons to form new isotopes. Most isotopes formed in this way are unstable. They undergo decay releasing dangerous radiations as well as heat and energy.

Spontaneous radioactive disintegration begins with a **parent isotope**, such as U^{238} and leads to formation of another unstable isotope known as **daughter products**. The daughter product of U^{238} is Th^{234} which is produced by emission of an alpha particle. In this series the next daughter product is Pa^{234} (Protactinium-234) produced by emission of a beta particle. Thus, one daughter product leads to another in a series that may produce a number of elements in succession; and ultimately chain ends in a stable isotope, which in this case is Pb^{206} (lead-206). During radioactive decay the three forms of emission (alpha particles, beta particles and gamma rays) are absorbed in the surrounding matter after a short-distance travel and converted to heat.

Radiation is energy which travels in waves or particles. One form of radiation energy, which causes injury to the living tissue, is called the **high-energy radiation**. Radiations that are able to remove electrons from atoms and attach them to other atoms, thereby creating positive and negative ion pairs, are known as **ionizing radiations**. Ionizing radiation produces ions as it passes through the protoplasm. It is chief cause of injury to the living matter and the damage is to the number of ion pairs produced in the absorbing material. Isotopes of elements that emit ionizing radiation are called **radioisotopes** or **radionuclides**.

Half-life

The half-life of a radionuclide refers to its period of radioactivity and has a bearing on the length of time that radioactive substance might remain active as environmental pollutant. For example, if one starts with one atom of C^{14} (Carbon-14) then within 5,568 years it will shoot out its beta particle and in the process change from one atom of C^{14} to an atom of N^{14} (Nitrogen-14). This means that if one starts with 100 atoms of C^{14}, it will take an average of 5,568 years for 50 of them to expel their beta particles. It will take another 5,568 years for 25 of the remaining 50 to emit their beta particles. Thus, 5,568 years is said to be the half life of C^{14}. Some important radio-isotopes and their half lives have been tabulated in Table 24.1.

Table 24.1. Some important radionuclides (= radio-isotopes) and their half-lives.

	Radio-Isotope	Half lives
1.	Tritium	15 hours
2.	Carbon-14	5568 years
3.	Sodium-22	2.6 Years
4.	Sodium-24	16 hours
5.	Manganese-54	300 days
6.	Iron-55	2.9 years
7.	Cobalt-58	71 days
8.	Cobalt-60	5.2 years
9.	Strontium-90	27 years
10.	Zirconium-95	65 days
11.	Iodine-129	15,800,000 years
12.	Iodine-131	8 days
13.	Copper-64	12.8 hours
14.	Caesium-135	2,300,000 years
15.	Caesium-137	30 years
16.	Hafnium-181	40 days
17.	Tantalum-182	111 days

	Radio-Isotope	Half lives
18.	Polonium-210	138.4 days
19.	Radon-222	3.83 days
20.	Radium-226	1622 years
21.	Thorium-230	77,000 years
22.	Thorium-232	14,000,000,000 years
23.	Uranium-233	1,62,000 years
24.	Uranium-235	710,000 years
25.	Uranium-238	4,500,000 years
26.	Uranium-239	23.5 minutes
27.	Neptunium-237	2,100,000 years
28.	Neptunium-239	2.35 days
29.	Plutonium-239	24,400 years
30.	Plutonium-240	6,540 years
31.	Plutonium-244	3,80,000 years

The half-life of a radioactive isotope is considered important from the viewpoint of environmental health. The half -life is not affected by any environmental factors. If the half-life is very long, then the isotope will persist for a very long time, posing a constant hazard to human life. Isotopes having long half-lives decay very slowly releasing radiation at a very low level.

Ecological Classification of Radionuclides

From the ecological point of view there are three groups of radionuclides. The first group includes naturally occurring radionuclides which contribute to background radiation, *e.g.,* Uranium-235, Radium-226, Potassium-40. The second group includes nuclides of metabolically important elements which are used as "tracers", *e.g.,* Cobalt-60, Iodine-131, Phosphorus-32 and many more. The third group of radionuclides are those produced by the fission of uranium and certain other elements, *e.g.,* Strontium-90, Cesium-137, Plutonium-239 and Iodine-131. They enter the environment through fall out and waste disposal.

24.2 UNITS OF RADIOACTIVITY

1. Curie (Ci). It was named after **Marie Curie** (1867-1934). Each radionuclide decay with the emission of radiation at a fixed and measurable rate. Curie is the unit used to express the intensity of radioactivity or rate of nuclear disintegration of a particular sample of material. One curie is equal to 37 billion nuclear integration per second (dps).

The term curie is also used to express the quantity of a nuclide having one curie of radioactivity. For example, one gram of radium has a radioactivity of one curie.

2. Roentgen (R). It was named after **Wilhelm Conard Roentgen** (1845-1923; German physicist, recipient of the first Nobel Prize for Physics for his discovery of X-rays). It is the unit for radiation by X-rays and gamma rays. One roentgen is the amount of radiation that ionizes one atom in about 10 billion atom of air or the energy that will produce 2,083 billion pairs in one cubic centimetre of air under standard conditions.

3. Rad (= Radiation absorbed dose). It is used for doses of alpha and beta particles, other fast neutrons and kinds of radiations. It is the radiation dose. It is the amount of radiation that gives an energy (100 ergs) per gram of substance (*i.e.,* living tissue). With X-rays, roentgen amounts to an energy absorption of 84 ergs per gram. Thus, practically, the rad is nearly equivalent of the roentgen.

4. Rem (= Roentgen equivalent man). It is an estimate of the amount of radiation of any type which produces the same biological injury in human being as that resulting from the absorption of a given amount of X-radiation or gamma-radiation. Rem gives an indication of biological damage.

International Systems of Units has changed the radiation units, roentgen, rad and rem by new SI units of **Coulomb Kg^{-1}, Gray (Gy) Kg^{-1}** and **Joule Kg^{-1}** respectively (**Sharma** 1996).

24.3 KINDS OF RADIATION

Following two main types of radiations are released by nuclear disintegration or man-made devices:

1. Electromagnetic radiations. These include a broad-spectrum of energy. These are as follows: (*i*) ultraviolet rays; (*ii*) X-rays; (*iii*) gamma rays; (*iv*) infra-red rays; (*v*) radio waves; (*vi*) visible light rays. The **near UV** extends from visible violet, about 390 nm to about 170 nm, whereas **far UV** is roughly 100 nm. **X-rays** include a broad spectrum of wave lengths, from about 100 nm to less than 0.00001 nm. An X-ray has a wavelength of about 0.1 nm. **Gamma rays** are similar to X-rays. They, however, have high energy and are more penetrating than X-rays.

2. Particulate radiations. They consist of the particles ejected from atoms at high speed and often with tremendous energy. These have electrons, proton or neutron. This type of radiation include:

(*i*) Beta particles. These are high-speed electrons, emitted by many radionuclides. They ionize any substance with which they collide in their path, *e.g.*, C^{14}, P^{32}, H^{3}, etc.

(*ii*) Alpha particles. These are fast-moving particles, containing 2 protons and 2 neutrons. They are positively charged and less penetrating than X-rays, gamma rays and beta particles. They can be stopped by a sheet of paper.

(*iii*) Proton particles. Their effect is similar to alpha particles.

(*iv*) Energetic neutrons

(*v*) Cosmic rays. These are energetic particles from the sun and outer space. They strike the earth at high velocity, some penetrating several thousand feet of solid rock. Many of the cosmic ray particles are charged atomic nuclei called **primaries**. Additional types of cosmic rays are called **secondaries**. At sea level, the cosmic radiation is about 40 millirem per year. As per rough estimate, this value increases at the rate of 1 with increase in allitude of every 100 feet. For different Indian cities, cosmic radiation values have been tabulated in Table 24.2.

Table 24.2. Cosmic radiation values of various Indian cities.

	Name of City	Value of Cosmic Radiation
1.	Delhi (at 210 metre above sea level)	31.0 m rem/yr.
2.	Calcutta (sea level)	28.0 m rem/yr.
3.	Bombay (sea level)	28.0 m rem/yr.
4.	Madras (sea level)	28.0 m rem/yr.
5.	Bangalore (at 920 metre above sea level)	44.0 m rem/yr.

This means that people living at higher elevations are likely to be exposed to greater cosmic radiations. Their intensity in the biosphere is very low, 35 m rads/year, yet they pose a great hazard during space travel. At about 20 km height cosmic radiation becomes very intense. As a result a commercial pilot receives about 300 m rad/yr of cosmic radiations.

24.4 SOURCES OF IONIZING RADIATIONS

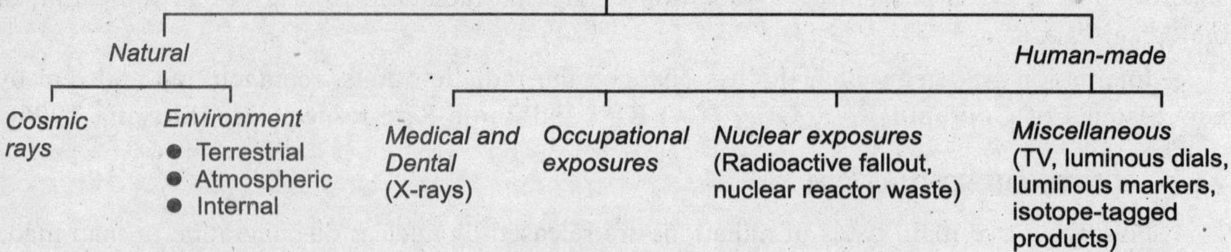

1. Atmospheric radiation. It emanates from both natural and human-made sources. For instance, radioactive gases such as **thoron** and **radon** are present in air. However, they have low values of roughly 2m rad/yr.

2. Background radiation. The so-called background radiation comes from following three natural sources:

(*i*) **Cosmic rays** from the outer space enter the atmosphere.

(*ii*) **Nuclides from earth crust** such as Radium-223 and Uranium-238 emit radiation.

(*iii*) **Nuclides present in food** grown in earth (soil) and drinking water percolated through soil such as Carbon-14, Radon-222 and Potassium-40. Marine sediments generally have higher concentrations of radionuclides.

Humans are also exposed to **internal radiations** from radioactive substances present in the body tissues. For example, uranium, thorium and isotopes of potassium, strontium and carbon exist in small amounts in the body. Internal radiation values vary from 25 to 75 m rads/yr.

3. Man-made radiations. These include (*i*) X-ray machines (diagnostic and radiotherapeutic); (*ii*) radioactive fall-out (nuclear tests), (*iii*) nuclear reactor wastes; (*iv*) industrial, medical and research uses of radioactive materials and (*v*) miscellaneous (*e.g.,* T.V., luminious watch dials, etc.). Of the abovesaid sources, we shall elaborate here only two, *i.e.,* radioactive fall-out and nuclear power.

(A) Radioactive fall-out. In an nuclear explosion about 50 per cent of the energy goes to the blast, 33 per cent as heat (having 10^8C temperature) and the rest 17 per cent to radioactivity. The radioactive dust that falls to earth after atomic explosions is called **radioactive fallout.** Explosive tests of nuclear weapons conducted in air, on the ground, underground and under the sea have produced significant quantities of Carbon-14, Strontium-90, Iodine-131 and Caesium-137. The radioisotopes are dispersed in the atmosphere and eventually are deposited upon the earth's surface as radioactive fall-out. Caesium-137 is found to contaminate the food chain consisting of the lichens, reindeer and humans in Arctic Tundra.

(B) Emmissions from nuclear reactors. The largest contribution of radiation comes from the operation of nuclear reactors and the complimentary nuclear fuel processing (Box 21-2). The production of nuclear power is based on three types of operations. Some reactors use Uranium-238 ore concentrate as the basic raw material. This has to be processed by purification, chemical conversions and enrichment with Uranium-235, and then manufactured into fuel elements. Such sort of fabrication of the elements produces solid, liquid and gaseous wastes (*i.e.,* three kinds of radioactive pollutants) having a low total radioactivity.

Box 24.2

Conventional steam-electric power plants (Thermal Power Stations) use a fossil fuel such as coal, oil or natural gas. The fuel is burnt in a boiler that produces steam which, in turn, drives a steam turbine generator, called a **turbogenerator.** In a nuclear power plant, the heat is produced in a device, called a **nuclear reactor** instead of a boiler.

The fuel of a nuclear electric generator is atomic pellets of uranium metal. On weight for weight basis, uranium contains three million times as much potential energy as coal. One gram of fissionable material releases 23,000 K watt hours of heat. A ton of uranium would provide as much as energy as 3 million tons of coal or 12 million barrels of oil. A **reactor** is one in which fission of atomic fuel occurs.

The second operation takes place in the reactor itself, where fission of the nuclear fuel releases heat energy which is used to drive a steam turbine for electric generator. Radioactive emissions to the air vary with the type of reactor. They include *nuclear fission products, neutron generated nuclides induced in the coolant medium* and the *products of erosion and corrosion processes* in the reactor circuits. The gaseous wastes consist of several nuclides such as Argon-41, Carbon-14, Tritium-3, Iodine-131, Krypton-85 and Xenon-133. Tritium-3 containing liquid waste and radioactive isotopes of iron, cobalt and other corrosion products are also produced.

In the third operation the spent fuel elements are reprocessed. This involves chemical treatment to separate the reusable components of Uranium-235, Uranium-238 and Plutonium-239 from waste fossil products. In this process gaseous waste products such as Tritium-3, Iodine-131 and Krypton-85 and large quantities of liquid waste containing long half-life nuclides and some high activity solid wastes are also produced.

24.5 EFFECT OF IONIZING RADIATION

Most of the radiations have a high penetrating power. Thick sheets of steel, cement concrete walls, etc., can not contain them. They can easily penetrate to deep seated organs of organisms (including humans).

Regardless of the form–alpha, beta or gamma the radiation usually affects living tissues by causing ionizations that, in turn, cause cellular damage. When a water molecule in the cell is irradiated, an electron is knocked out of orbit. The ejected electron may then attach to the normal water molecule, making it unstable. Such unstable molecules split into hydrogen ions (H^+), hydroxide ions (OH^-) and the free radicals OH and H. These free radicals react with various molecules in the cell, which can no longer function normally and ultimately die.

Biological Effects of Ionizing Radiation

A biological system is unable to distinguish between a radioactive and a normal isotope of an element as their physical and chemical properties are similar. Radioactive isotopes are therefore absorbed and incorporated within the bodies of living organisms as normal isotopes are. This lodges a radioactive source within the bodies of living organisms itself. Like any other element radioactive isotopes are also absorbed, accumulated and biomagnified thousands of times.

Biological effects of radiation can be classified as somatic or genetic effects:

1. Somatic or non-genetic effects. The direct results of action of radiation on the body cells and tissues are called **somatic effects**. Somatic effects occur within an individual's life time and appear as skin damage, eye cataracts, damage to liver, spleen and thyroid and reduced fertility.

Experiments on laboratory animals such as mice, rats and rabbits have indicated that radiation can induce cancer in a variety of organs such as lymph glands, mammary glands, lungs, ovaries and skin, etc. Other effects are blood cancer or **leukemia**, where the red bone marrow produces uncontrolled numbers of white blood cells and various forms of **carcinoma**. Somatic effects of radiation in human beings have been tabulated in Table 24.3.

Table 24.3. Somatic effects of radiation in human beings.

Dose (in Rems)	Somatic effect
100	Temporary to permanent sterility; reduced life expectancy from cancer and leukemia
200	Death within months for 10 per cent of those exposed; 90 per cent survive
700	Death within months for 90 per cent of those exposed; 10 per cent survive
1000	Death within few days
10,000	Death within few hours
100,000	Death within few minutes

Immediate or delayed effects. A dose of 400 to 500 R on whole human body is fatal in 50 per cent of cases, and 600 to 700 R in practically every case. A dose of 25-50 R to the whole body resulted in mild weariness and softening of muscles. A lethal dose causes an acute form of **radiation sickness**. The symptoms of radiation sickness include reddened and ulcerated skin, loss of hair, nausea and anaemia. Within a few hours the patient feels nausea and fatigue. There is gradual decrease in the number of red and white blood corpuscles eventually causing anaemia. Victim feels declined in vigour and ultimately he dies from anaemia and haemorrhage. In delayed effects, the patient may survive and the radiation effects take time (months or years) to develop. Delayed effects include leukemia, cardiovascular disorders, eye cataracts, premature aging and reduced longevity.

2. Genetic effects. These are various and result from damage to chromosomes and genes. Nucleic acids (DNA and RNA) effectively absorb these radiations. Even low level radiations which do not cause any visible damage are completely absorbed by nuclear materials which causes carcinogenic, mutagenic and teratogenic effects.

Genetic damage is not apparent in one generation, but can appear in offspring and in subsequent generations. Radiation affects the reproductive organs and so the gametes produced contain deleterious gene mutations which may be inherited. Within the gametes, the genetic mechanism of the chromosomes becomes damaged by ionization and breakage. Sometimes radiation may bring about changes within growing cells which deprive them of the ability to divide. The stimulus to divide is inhibited or destroyed and so the cells continue to grow until they become giant cells and eventually die because they become inefficient due to large size.

Human females who have received abnormal doses of X-rays for diagnostic purposes during their pregnancy tend to produce an above average number of mongol children. It has been stated that X-rays probably cause between 2500 and 100,000 genetically defective births, and 200 to 300 leukemia deaths annually at the global scale.

Variation in sensitivity to radiation. Embryos of animals and humans are more sensitive to radiation damage because they contain rapidly growing and dividing cells in all parts of body. The degree of sensitivity varies with the embryonic stage, the younger embryos being the most sensitive because their cells grow and divide most rapidly. A great variation is found in the radiosensitivity of different species of organisms. For example, a dose of about 600 R to the whole body, when administered within a period of a day or so would be sufficient to kill a human being; a dose of about 900 R will be required to kill adult rats; however, it requires more than 25,000 R to kill an onion plant and some protozoa can survive a dose more than this. Generally, the less complex the organism the greater its resistance to radiation damage. Mammals rate as the most sensitive and microbes as the most resistant of organisms. Since complexity is correlated with the amount of DNA within the cells, this factor might explain the difference in resistance, at least to some extent. Thus, cells with more DNA content and greater chromosome area, would have a greater chance of damage of DNA and of chromosome as a whole.

Effects of Radiation on Plants

Plants are also affected by radiation. An intense radiation kills plants but differentially. Variations in radiosensitivity among trees and shrubs are due, in part, to differences in their chromosome number and size. It has been suggested that higher plants with a small number of large chromosomes are most likely to be damaged by radiation than those with a large number of small chromosomes, simply because the former provide a larger "target".

Effects of Radiation on Ecosystem

According to **Dr. Bill** (1986) radiation induced genetic malformations could also occur in plants, animals and birds destroying the whole biosphere. Radiation stress is found to lead to reduction in species diversity. It also alter predator-prey equilibria or bring on pest irruption. Radioactive pollution of the soils, oceans and other environments where humans do not actually live, nevertheless, affects their vital life-support system. Any radioactive substance, with a long half-life, introduced in the environment anywhere in the biosphere is likely to find its way into the body of animals (including humans).

24.6 CONTROL OF RADIOACTIVE POLLUTION

Protection of workers, improvement of radiation techniques and reduction in the amount of dose given will reduce the health hazards (WHO 1965). Effective measures include regular use of lead shields and lead rubber aprons by all workers associated with X-ray procedures; lead aprons (0.5 mm of lead) will reduce the intensity of scattered X-rays over 90 per cent (**Little** 1966). Workers must wear a film badge or dosimeter.

Disposal of Radioactive Wastes

Only the process of natural decay which requires hundreds of thousands or even millions of years reduces the radioactivity of nuclear wastes. There is no physical, chemical or biological process which can reduce this span of time. All the human beings can do is to store the radioactive material at some remote place away from the biosphere for a million years — a task which is too difficult for the mankind. Further, high level irradiated waste produce a lot of heat and hence have to cooled to avoid explosion of the cask or the containers in which they are kept.

The present disposal problem of radioactive wastes can be tackled by three basic approaches. These are (*i*) *dilute and disposal*, used for the high volume, low activity wastes; (*ii*) *delay and decay*, used for medium activity wastes where slow movements through soil allows time for radioactive decay; and (*iii*) *Concentrate and contain*, used for high activity wastes. As they are too dangerous to be released anywhere in the biosphere, they are usually stored under-ground in tanks (Fig. 24.1; Box 24.3).

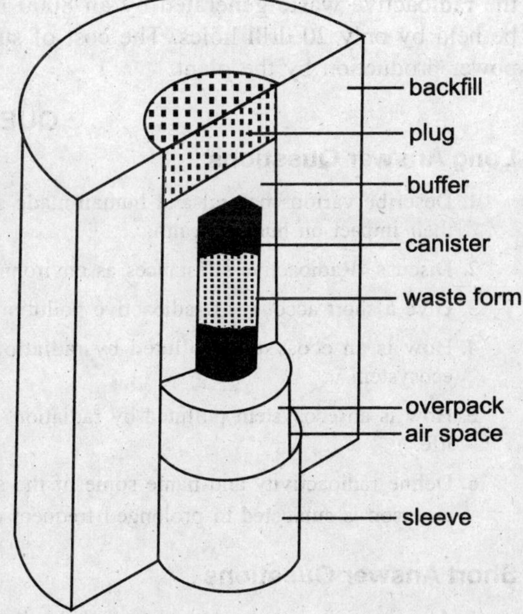

Fig. 24.1. A model of package of radioactive wastes prepared for long-term underground storage.

Box 24.3

Todate following disposal methods have been considered for isolation of radioactive wastes from the biosphere:

1. **Burial under Antarctica Ice caps.** A leakage could be catastrophic. With temperature change ice could melt disturbing the burial. Abandoned.
2. **Burial under sea bed.** The radioactivity may leak out in the surrounding waters and from there to other places. Abandoned.
3. **Disposal in space outside earth's gravitational field.** A single launch failure could introduce radioactivity in the atmosphere from where it can contaminate the entire planet. Abandoned.
4. **Disposal deep underground surface.** This involves difficulties arising due to ground water flows (chance of acquifer contamination) and uncertain geology which may cause tectonic disturbances, volcanic activity, etc. Moreover, human intrusion over such long periods of time may not be ruled out. However, this proposal has been followed till date.
5. **Burial in specially constructed buildings under human care.** The proposal involves too much dependence on human care which may not be possible for such long periods of time as required.
6. **Reprocessing of nuclear wastes.** Chemical separation of uranium and plutonium from nuclear wastes reduces the radioactivity by about 3 per cent but increases the volume of wastes almost 150 times. Hence abandoned.
7. **Transmutation to isotopes with shorter half-lives.** Bombardment by neutrons may convert isotopes with long half lives into isotopes with shorter half-lives. This method is very expensive, yet is under active study.

Recently Australian scientists have developed a technique for radioactive waste disposal that might be the safest devised to date. It involves "locking" radioactive waste in the crystalline structure of a special type of rock that the scientists have synthesized. The rock is relatively insoluble in water, is instable and is capable of imprisoning the waste for millions of years. The waste holding rocks should then be enclosed in containers made of a nickel-iron alloy and inserted into holes drilled into solid granite to a depth of more than two kilometres. By this method all the radioactive waste generated by an 8000 MW power plant over a period of 40-years could be held by only 20 drill holes. The cost of such burial would be only 1 per cent of the cost of power production by the plant.

QUESTIONS

Long Answer Questions

1. Describe various natural and human-made sources of radioactive substances giving an account of their impact on human health.
2. Discuss "Radioactive substances as environment toxicants".
3. Give a short account of radioactive pollution.
4. How is an ecosystem polluted by radiation? Trace the movement of radioactive element in an ecosystem.
5. How is an ecosystem polluted by radiation? Describe the effect of radioactive pollutants in animal life.
6. Define radioactivity and name some of the sources of radioactive pollution. What would happen if a person is subjected to prolonged frequent exposures to radioactive substances.

Short Answer Questions

1. Atomic explosion should be stopped. Justify.
2. How UV rays are harmful to us?

3. Write short notes in the following:
   (*i*) Radioactive pollution
   (*ii*) Nuclear pollution
   (*iii*) Radioactive substances
   (*iv*) Radioactive contamination
   (*v*) Hazards of radiation
   (*vi*) Biological effects of ionizing radiation
4. Explain difference between Thermal pollution and Radioactive pollution.

Very Short Answer Questions

1. What are radionuclides?
2. Define radiation ecology?
3. What is rad?

Fill in the Blanks

1. Strontium–90 is the .... which causes .... cancer in human beings.
2. The half-life of a radionuclide refers to its period of ....
3. Uranium–235 is the example of .... occurring radionuclides.
4. Rad stands for .... dose.
5. Particulate radiations include .... , ... and neutrons.
6. Beta particles are high–speed ....

Multiple Choice Questions

1. The most outstanding danger at present for survival of living beings on earth is
   (*a*) glaciation — (*b*) deforestation
   (*c*) radiation hazards — (*d*) desertification
2. Chernobyl disaster was caused by a
   (*a*) nuclear weapon accident — (*b*) nuclear waste disposal leak
   (*c*) nuclear reactor accident — (*d*) nuclear test
3. Which part of human body is the first to be affected by nuclear radiations?
   (*a*) brain — (*b*) lungs
   (*c*) liver — (*d*) bone marrow
4. The term 'Nuclear winter' is associated with
   (*a*) nuclear war — (*b*) nuclear disarmament
   (*c*) nuclear weapon testing — (*d*) aftermath of nuclear holocaust
5. The animals in the deep sea water are severely affected by
   (*a*) sewage — (*b*) nuclear wastes
   (*c*) oil spills — (*d*) heavy metals
6. The affect of today's radioactive fall-out will probably be more harmful to children of future generations than to children now living because
   (*a*) mutated genes are frequently recessive
   (*b*) susceptibility to radiation increases with age
   (*c*) contamination of milk supply is not cumulative
   (*d*) infants are more susceptible to radiation

7. One of the most dangerous radioactive pollutant is
(*a*) phosphorus-32 (*b*) strontium-90
(*c*) calcium-40 (*d*) sulphur-32

8. Radioactive strontium as result of
(*a*) Sr-80 (*b*) Sr-85
(*c*) Sr-90 (*d*) Sr-95

9. Which of the following chemical causes bone cancer and degeneration of tissue?
(*a*) iodine-131 (*b*) calcium-40
(*c*) iodine-127 (*d*) strontium-90

10. Which of the following radioactive isotopes is used in the detection of thyroid cancer?
(*a*) iodine-131 (*b*) carbon-14
(*c*) uranium -238 (*d*) phosphorus-32

11. Chernobyl tragedy occurred in (*CCSU Meerut 2014*)
(*a*) 26th April, 1980 (*b*) 26th April, 1986
(*c*) 26th May, 1966 (*d*) 26th June, 1966

ANSWERS

Very Short Answer Questions

1. Radionuclides are isotopes of elements that emit ionizing radiations.
2. Radiation ecology is the branch of ecology concerned with the effects of radioactive materials on living systems and with the pathways by which these materials are dispersed within ecosystems.
3. Rad is the unit of radiation defined as an absorbed dose of 100 ergs of energy per gram of tissue.

Fill in the Blanks

1. Bone **2.** Radioactivity **3.** Naturally
4. Radiation Absorbed **5.** Electrons, Protons **6.** Electrons.

Multiple Choice Questions

1. (*c*) **2.** (*c*) **3.** (*d*) **4.** (*d*)
5. (*b*) **6.** (*a*) **7.** (*b*) **8.** (*c*)
9. (*d*) **10.** (*a*) **11.** (*b*).

INDEX

A

B

C

D

E

F

G

I

M

N

O

Q

R

S

T

U

V

W

X

Y

Z